Prealgebra

Annotated Instructor's Edition

Robert H. Prior

Riverside Community College

Addison-Wesley

Boston San Francisco New York
London Toronto Sydney Tokyo Singapore Madrid
Mexico City Munich Paris Cape Town Hong Kong Montreal

Editorial Director	Christine Hoag
Editor in Chief	Maureen O'Connor
Sponsoring Editor	Jennifer Crum
Executive Project Manager	Kari Heen
Project Editor	Katie Nopper DePasquale
Editorial Assistant	Jonathan Wooding
Developmental Editor	Peg Monahan-Pashall
Production Manager	Ron Hampton
Cover and Interior Design	Leslie Haimes
Production Services	Pre-Press PMG
Media Producer	Audra Walsh
Software Development	TestGen: Mary Durnwald MathXL: Tanya Farber
Marketing Manager	Marlana Voerster
Marketing Coordinator	Nathaniel Koven
Prepress Services Buyer	Caroline Fell
Manufacturing Manager	Evelyn Beaton
Senior Media Buyer	Ginny Michaud
Cover Photo	Frog on plant, isolated black. © Sascha Burkard

NOTICE: This work is protected by U.S. copyright laws and is provided solely for the use of college instructors in reviewing course materials for classroom use. Dissemination or sale of this work, or any part (including on the World Wide Web), will destroy the integrity of the work and is not permitted. The work and materials from it should never be made available to students except by instructors using the accompanying text in their classes. All recipients of this work are expected to abide by these restrictions and to honor the intended pedagogical purposes and the needs of other instructors who rely on these materials.

Copyright © 2010 Pearson Education, Inc., publishing as Addison-Wesley, 75 Arlington Street, Boston, MA 02116. All rights reserved. Manufactured in the United States of America. This publication is protected by copyright and permission should be obtained from the publisher prior to any prohibited reproduction, storage in a retrieval system, or transmission in any form or by any means, electronic, mechanical, photocopying, recording, or likewise. To obtain permission(s) to use material from this work, please submit a written request in Pearson Education, Inc., Permissions Department, 501 Boylston Street, 9th floor, Boston, MA 02116.

Many of the designations used by manufacturers and sellers to distinguish their products are claimed as trademarks. Where those designations appear in this book, and the publisher was aware of a trademark claim, the designations have been printed in initial caps or all caps.

Library of Congress Cataloging-in-Publication Data
Prior, Robert H.
Prealgebra / Robert H. Prior — 1st ed.
p. cm. Annotated Instructor's Edition: ISBN-13: 978-0-321-59408-2 ISBN-10: 0-321-59408-8
Includes index.
Student Edition: ISBN-13: 978-0-321-21378-5 ISBN-10: 0-321-21378-5
1. Mathematics—Textbooks. 2. Mathematics—Problems, exercises.
I. Title.
QA39.3.P756 2010
513'.12—dc22 2008023682

1 2 3 4 5 6 7 8—WB—11 10 09 08

Addison-Wesley
is an imprint of

www.pearsonhighered.com

ISBN-10: 0-321-21378-5
ISBN-13: 978-0-321-21378-5

Contents

Preface vi

Basic Skill Pretest xviii

Chapter 1 Number Sense 1

1.1 Whole Numbers and Their Properties 2
1.2 Adding and Subtracting Whole Numbers 17
1.3 Multiplying Whole Numbers 30
1.4 Dividing Whole Numbers 41
1.5 Exponents, Square Roots, and the Order of Operations 51
1.6 Factors 64
1.7 Equations 77
1.8 Solving Applications Using Equations 84
Chapter 1 Review 95
Chapter 1 Review Exercises 104
Chapter 1 Test 110

Chapter 2 Integers and Algebraic Expressions 111

2.1 Introduction to Algebra 112
2.2 Adding Signed Numbers 122
2.3 Subtracting Signed Numbers 133
2.4 Multiplying and Dividing Signed Numbers 142
2.5 The Order of Operations for Signed Numbers 151
2.6 Formulas 158
2.7 Algebraic Expressions: Combining Like Terms 165
2.8 Multiplying Algebraic Expressions 173
Chapter 2 Review 179
Chapter 2 Review Exercises 185
Chapter 2 Test 189
Chapters 1–2 Cumulative Review 191

Chapter 3 Equations 195

3.1 Solving Equations Involving One Operation 196
3.2 Solving Equations Involving Two Operations 203
3.3 Solving Equations Involving More Than Two Operations 206
3.4 Problem Solving 210
3.5 Solving Applications Involving One Unknown Value 217
3.6 Solving Applications Involving Two Unknown Values 226
Chapter 3 Review 233
Chapter 3 Review Exercises 237
Chapter 3 Test 240

Chapter 4 Fractions 241
4.1 Common Factors 242
4.2 Introduction to Fractions 249
4.3 Equivalent Fractions 259
4.4 Multiplying and Dividing Fractions 269
4.5 Adding and Subtracting Like Fractions 281
4.6 Common Denominators 291
4.7 Adding and Subtracting Unlike Fractions 298
4.8 Equations and Applications Involving Fractions 307
Chapter 4 Review 316
Chapter 4 Review Exercises 325
Chapter 4 Test 329
Chapters 1–4 Cumulative Review 331

Chapter 5 Decimals 333
5.1 Introduction to Decimals 334
5.2 Rounding Decimals 344
5.3 Adding and Subtracting Decimals 350
5.4 Multiplying Decimals 361
5.5 Dividing Decimals 368
5.6 Applications Involving Decimals 378
Chapter 5 Review 388
Chapter 5 Review Exercises 392
Chapter 5 Test 396

Chapter 6 Ratios, Proportions, and Percent 397
6.1 Ratios 398
6.2 Rates 408
6.3 Proportions 415
6.4 Percents 425
6.5 Solving Percent Problems Using Proportions 434
6.6 Solving the Percent Equation 441
6.7 Applications Involving Percents 450
Chapter 6 Review 459
Chapter 6 Review Exercises 465
Chapter 6 Test 471
Chapters 1–6 Cumulative Review 473

Chapter 7 Polynomials 477
7.1 Rules of Exponents 478
7.2 Introduction to Polynomials 485
7.3 Adding and Subtracting Polynomials 497
7.4 Multiplying Polynomials 502
7.5 Dividing Polynomials 512
7.6 Factoring Polynomials 518
Chapter 7 Review 526
Chapter 7 Review Exercises 530
Chapter 7 Test 533

Chapter 8 Units of Measure and Geometry 535

8.1 U.S. Measures 536
8.2 Metric Measures 547
8.3 Converting U.S. and Metric Measures 554
8.4 Lines and Angles 561
8.5 Geometric Shapes 574
8.6 Perimeter 586
8.7 Area 596
8.8 Volume 610
Chapter 8 Review 618
Chapter 8 Review Exercises 626
Chapter 8 Test 633
Chapters 1–8 Cumulative Review 635

Chapter 9 Graphing Lines, Statistics, and Probability 639

9.1 The Rectangular Coordinate System 640
9.2 Graphing Lines 655
9.3 Graphing Data: Line Graphs and Bar Graphs 669
9.4 Graphing Data: Circle Graphs and Histograms 686
9.5 Mean, Median, and Mode 701
9.6 Probability 711
Chapter 9 Review 718
Chapter 9 Review Exercises 727
Chapter 9 Test 737
Chapters 1–9 Cumulative Review 743

Appendixes 749

A.1 Scientific Notation: Large Numbers 749
A.2 Integer Exponents 754
A.3 Scientific Notation: Small Numbers 760
A.4 Scientific Notation: Multiplication 764
A.5 Factoring Trinomials of the form $x^2 + bx + c$ 768

Answers AN-1

Glossary G-1

Photograph Credits G-6

Index of Applications IA-1

Index I-1

Preface

Dear Students and Instructors,

I'm so glad you have chosen to take the time to read through the preface of my text. I hope that you find this book not only easy to read but easy to *use*. Things are changing in mathematics education today. The class formats, teaching styles, and student learning styles are all driving that change. Because the courses, students, and teachers are evolving, textbooks and their support resources must evolve as well. In the spirit of this change, I've written *Prealgebra.*

You can see right away that this book is different! The spiral binding is intended to make the book easier to write in and to use whether at a desk or next to a computer—but the differences don't end with the book's binding. I first wrote this text for my students in a self-paced course. I had to write it in a way that those students could learn on their own, mostly by reading the book and doing the exercises. Through class testing the material, I found that students in a traditional classroom, as well as online students, also benefited from the thorough explanations and detailed development of the content.

I've written this text as if I were speaking directly to the student, keeping in mind that the goal is for these students to continue on through algebra. To that end, I prepare students by integrating algebraic concepts throughout the text. For all students, I want them to see the *why* behind much of what they are learning, so the rules are gradually developed and rarely just given outright.

In addition, I include many opportunities for students to practice what they are learning through the *You Try It* exercises that follow the examples and the *Think About It* questions embedded in the text. This interactive format provides students with a chance to self-assess their understanding and gain confidence in the content they are learning. This has proven to be quite successful through usability tests and class testing of my book.

Though there are many other aspects of my book that are different, another obvious one comes with the way the book and supplements are distributed. In an effort to be more environmentally friendly, we have changed all printed paper supplements to a digital format, with a DVD resource for students and online access for instructors. Providing the supplements in a digital format not only saves trees and energy, but makes the instructor and student resources very convenient to use.

Thank you again for taking the time to read through the preface of my text and for making it a part of your learning or teaching experience. I encourage you to take the time to read through the rest of the preface, paying close attention to the Anatomy of a Textbook and Teaching and Learning Package sections. Please feel free to contact me if you have any questions or comments about the book.

Take care, and have a great semester!

Bob Prior
Bob@bobprior.com

Anatomy of a Textbook

Pedagogy: How to Use This Textbook

Most people don't think about dissecting the parts of a textbook when they are about to use it, but why not? It is important to be aware of the structure and organization of a textbook so that you can make the most of all the guidance the author has provided for you. Keep in mind that the organization and layout of this textbook—every element, every color choice, and every sentence—were carefully developed and are provided as a way to make the mathematics more clear and the learning easier. Take a moment to familiarize yourself with the different pedagogical elements and the general layout of this book, as presented on the following pages.

(page 195)

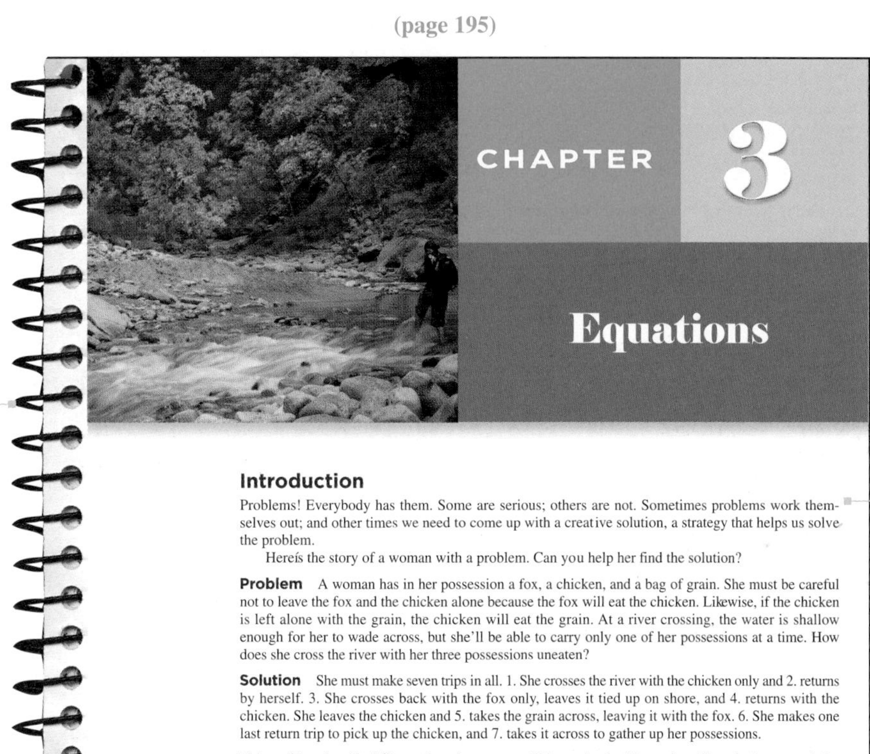

CHAPTER 3

Equations

Introduction

Problems! Everybody has them. Some are serious; others are not. Sometimes problems work themselves out; and other times we need to come up with a creative solution, a strategy that helps us solve the problem.

Here's the story of a woman with a problem. Can you help her find the solution?

Problem A woman has in her possession a fox, a chicken, and a bag of grain. She must be careful not to leave the fox and the chicken alone because the fox will eat the chicken. Likewise, if the chicken is left alone with the grain, the chicken will eat the grain. At a river crossing, the water is shallow enough for her to wade across, but she'll be able to carry only one of her possessions at a time. How does she cross the river with her three possessions uneaten?

Solution She must make seven trips in all. 1. She crosses the river with the chicken only and 2. returns by herself. 3. She crosses back with the fox only, leaves it tied up on shore, and 4. returns with the chicken. She leaves the chicken and 5. takes the grain across, leaving it with the fox. 6. She makes one last return trip to pick up the chicken, and 7. takes it across to gather up her possessions.

This problem is a bit different than the ones you'll be seeing in this section. There isn't any math involved in this problem, but there is a strategy.

In Chapter 3, you'll be using many of the ideas you learned in Chapters 1 and 2 to solve equations and applications. For each type of problem that you encounter, a strategy is given so that you will know how to approach a problem with confidence. For some of the problems, the strategy will be to remove something from one side and put it on the other side, just like the woman with the fox, the chicken, and the bag of grain.

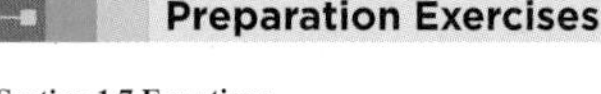

Preparation Exercises

Section 1.7 Equations
Solve.

1. $p + 19 = 32$ $p = 13$
2. $45 = 3 \cdot y$ $y = 15$
3. $153 = 86 + x$ $x = 67$
4. $7 \cdot n = 105$ $n = 15$

Section 2.2 Adding Signed Numbers
Evaluate.

5. $-10 + 8$ -2
6. $-3 + (-12)$ -15

Section 2.3 Subtracting Signed Numbers
Evaluate.

7. $-7 - 9$ -16
8. $-6 - (-18)$ 12

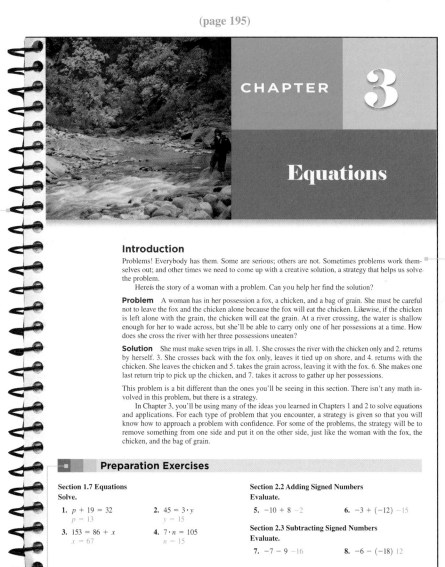

The **spiral binding** of this textbook makes it easier for you to use the book in class, at the computer, or while doing homework on your own. Take advantage of this flexibility! Take notes right on the pages during class, keep the book next to you as you watch the chapter test solutions or lectures on video, or have it open while working through extra practice exercises on the computer.

The **Introduction** of the chapter is your chance to preview the overarching goals and topics to be covered throughout the chapter, along with some related applications.

The **Preparation Exercises** are provided to give you a sense of the skills you should already know in order to be successful in this chapter. If you struggle with any of these skills, don't worry. The section numbers listed next to each problem reference the section where that topic was originally covered so that you can refresh your skills prior to starting this new chapter.

Each section of the book has a set of **objectives**. These are essentially the goals of what you are going to learn within the section. All of the examples, text, and exercises in the chapter are there to help you reach these goals.

Each section opens with an **Introduction** that provides you with an overview of the topic or a relevant application. Math is used often in our everyday lives to solve problems or to get results more quickly, and these openers highlight when you might use a particular skill or how it pertains to other skills you have already learned.

Many students feel that they understand math best when they learn by **example**. Each example in this text is designed to support and supplement the written explanation that it follows. Pay close attention to the example **procedure**, which essentially lays out the steps you would take to solve the problem, as well as the **answer**, which often has further explanation. Keep in mind that the procedures are the steps you will want to follow when you solve your own problems in the You Try Its and exercise sets.

(pages 210–211)

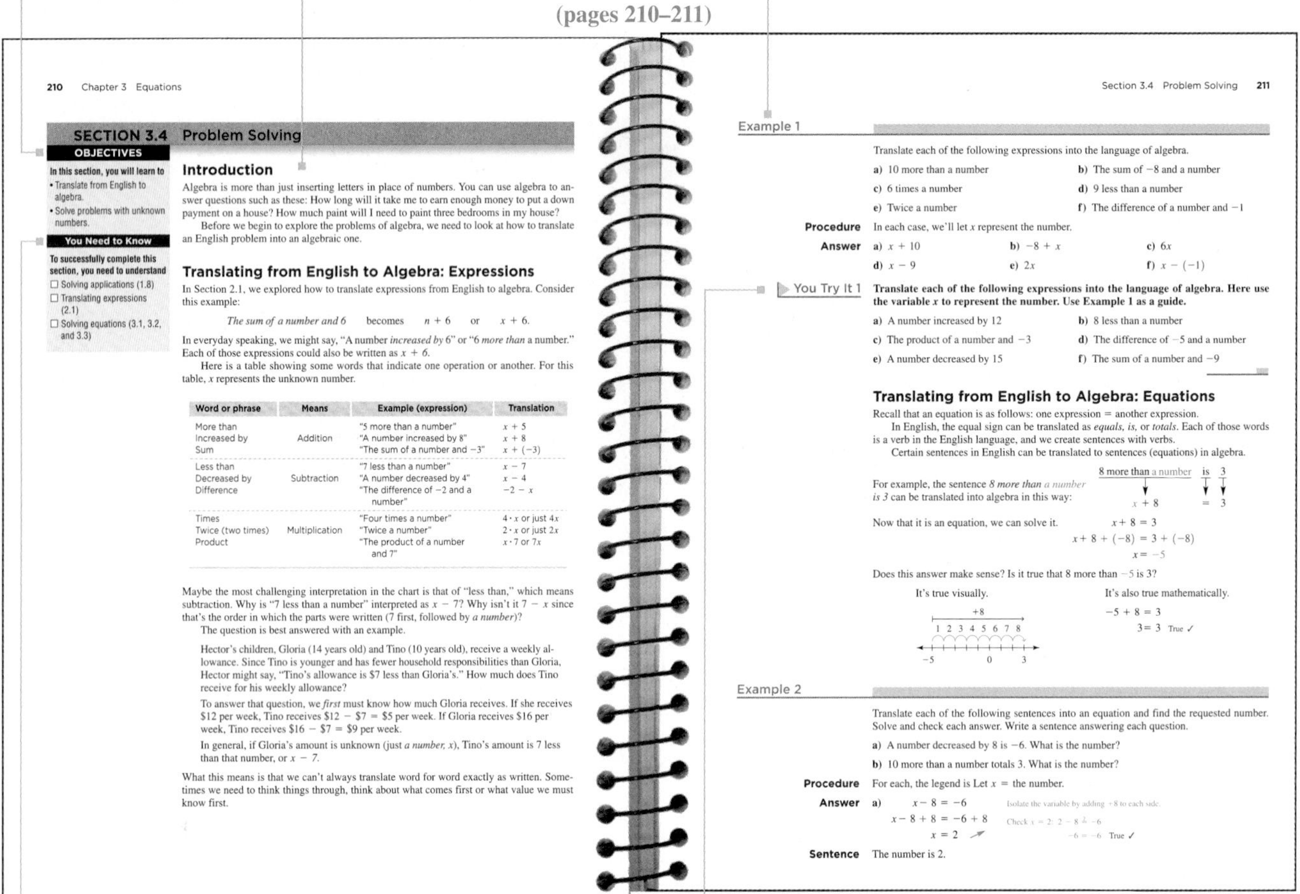

210 Chapter 3 Equations

SECTION 3.4 Problem Solving

OBJECTIVES

In this section, you will learn to
- Translate from English to algebra.
- Solve problems with unknown numbers.

You Need to Know

To successfully complete this section, you need to understand
- ☐ Solving applications (1.8)
- ☐ Translating expressions (2.1)
- ☐ Solving equations (3.1, 3.2, and 3.3)

Introduction

Algebra is more than just inserting letters in place of numbers. You can use algebra to answer questions such as these: How long will it take me to earn enough money to put a down payment on a house? How much paint will I need to paint three bedrooms in my house?

Before we begin to explore the problems of algebra, we need to look at how to translate an English problem into an algebraic one.

Translating from English to Algebra: Expressions

In Section 2.1, we explored how to translate expressions from English to algebra. Consider this example:

The sum of a number and 6 becomes $n + 6$ or $x + 6$.

In everyday speaking, we might say, "A number *increased by* 6" or "6 *more than* a number." Each of those expressions could also be written as $x + 6$.

Here is a table showing some words that indicate one operation or another. For this table, x represents the unknown number.

Word or phrase	Means	Example (expression)	Translation
More than Increased by Sum	Addition	"5 more than a number" "A number increased by 8" "The sum of a number and −3"	$x + 5$ $x + 8$ $x + (-3)$
Less than Decreased by Difference	Subtraction	"7 less than a number" "A number decreased by 4" "The difference of −2 and a number"	$x - 7$ $x - 4$ $-2 - x$
Times Twice (two times) Product	Multiplication	"Four times a number" "Twice a number" "The product of a number and 7"	$4 \cdot x$ or just $4x$ $2 \cdot x$ or just $2x$ $x \cdot 7$ or $7x$

Maybe the most challenging interpretation in the chart is that of "less than," which means subtraction. Why is "7 less than a number" interpreted as $x - 7$? Why isn't it $7 - x$ since that's the order in which the parts were written (7 first, followed by *a number*)?

The question is best answered with an example.

Hector's children, Gloria (14 years old) and Tino (10 years old), receive a weekly allowance. Since Tino is younger and has fewer household responsibilities than Gloria, Hector might say, "Tino's allowance is $7 less than Gloria's." How much does Tino receive for his weekly allowance?

To answer that question, we *first* must know how much Gloria receives. If she receives $12 per week, Tino receives $12 − $7 = $5 per week. If Gloria receives $16 per week, Tino receives $16 − $7 = $9 per week.

In general, if Gloria's amount is unknown (just *a number*, x), Tino's amount is 7 less than that number, or $x - 7$.

What this means is that we can't always translate word for word exactly as written. Sometimes we need to think things through, think about what comes first or what value we must know first.

Section 3.4 Problem Solving 211

Example 1

Translate each of the following expressions into the language of algebra.

a) 10 more than a number b) The sum of −8 and a number
c) 6 times a number d) 9 less than a number
e) Twice a number f) The difference of a number and −1

Procedure In each case, we'll let x represent the number.

Answer a) $x + 10$ b) $-8 + x$ c) $6x$
d) $x - 9$ e) $2x$ f) $x - (-1)$

You Try It 1 **Translate each of the following expressions into the language of algebra. Here use the variable x to represent the number. Use Example 1 as a guide.**

a) A number increased by 12 b) 8 less than a number
c) The product of a number and −3 d) The difference of −5 and a number
e) A number decreased by 15 f) The sum of a number and −9

Translating from English to Algebra: Equations

Recall that an equation is as follows: one expression = another expression.

In English, the equal sign can be translated as *equals*, *is*, or *totals*. Each of those words is a verb in the English language, and we create sentences with verbs.

Certain sentences in English can be translated to sentences (equations) in algebra.

For example, the sentence *8 more than a number is 3* can be translated into algebra in this way:

8 more than a number is 3
$x + 8 = 3$

Now that it is an equation, we can solve it.

$$x + 8 = 3$$
$$x + 8 + (-8) = 3 + (-8)$$
$$x = -5$$

Does this answer make sense? Is it true that 8 more than −5 is 3?

It's true visually. It's also true mathematically.

$-5 + 8 = 3$
$3 = 3$ True ✓

Example 2

Translate each of the following sentences into an equation and find the requested number. Solve and check each answer. Write a sentence answering each question.

a) A number decreased by 8 is −6. What is the number?
b) 10 more than a number totals 3. What is the number?

Procedure For each, the legend is Let x = the number.

Answer a)
$x - 8 = -6$ Isolate the variable by adding +8 to each side.
$x - 8 + 8 = -6 + 8$ Check $x = 2$: $2 - 8 \stackrel{?}{=} -6$
$x = 2$ $-6 = -6$ True ✓

Sentence The number is 2.

Following each list of objectives is a **You Need to Know** list of topics that highlights the skills in which you should already be proficient to start this section. If there are any skills on this list that you are not confident you know well, refer back to the section that is noted next to the topic (in parentheses) for a refresher. Being proficient with these skills will make learning the content of this section far easier.

The most valuable way to assess whether you really understand a concept or have mastered a skill is to try a problem yourself. The **You Try It** exercises following each example give you just this opportunity. Each You Try It exercise has several practice problems that are similar to the preceding example for you to try immediately after you have read the example. If you are not successful in solving the You Try It, then reread the explanation given, reevaluate the example provided, or try watching the video that accompanies that particular section of the book.

Think About It exercises pose thought-provoking questions that encourage discussion of the concepts and topics, making learning more interactive and memorable. Some of the concepts behind these exercises are revisited in the *Think Again* exercises that begin each exercise set.

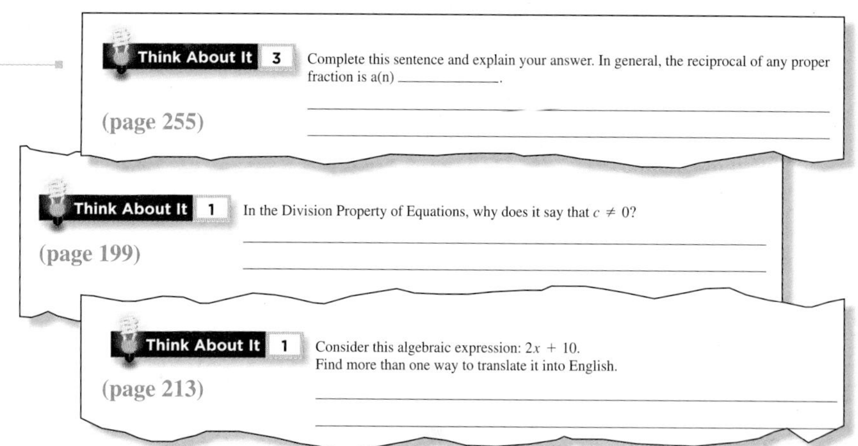

Think About It 3 Complete this sentence and explain your answer. In general, the reciprocal of any proper fraction is a(n) ________.

(page 255)

Think About It 1 In the Division Property of Equations, why does it say that $c \neq 0$?

(page 199)

Think About It 1 Consider this algebraic expression: $2x + 10$. Find more than one way to translate it into English.

(page 213)

(page 214)

Answers: You Try It and Think About It

You Try It: **1. a)** $x + 12$ **b)** $x - 8$ **c)** $x \cdot (-3)$ or $-3x$ **d)** $-5 - x$ **e)** $x - 15$ **f)** $x + (-9)$ **2.** Legend: Let x = the number. **a)** $x + 12 = -6$; the number is -18. **b)** $2x = -24$; the number is -12. **c)** $x - 8 = -1$; the number is 7. **d)** $-6x = 96$; the number is -16. **3. a)** $4x + 9$ **b)** $5x + 10$ **4.** Legend: Let x = the number. **a)** $3x + 8 = 2$; the number is -2. **b)** $5x + 2 = -18$; the number is -4. **c)** $4x - 3 = 21$; the number is 6. **d)** $2(x + 3) = -10$; the number is -8.

Think About It: **1.** Answers may vary. Here are two possible answers: The sum of twice a number and 10; and 10 more than twice a number.

Answers to the You Try It and **Think About It exercises** are provided so that you can immediately assess your understanding. This instant feedback allows you to know whether you need to practice these skills more or whether you can move onto the next topic with confidence.

(page 451)

Occasionally, a **You Finish It** box will appear within an example. This is a chance for you to get involved in the example and apply your arithmetic skills while solving a more complex problem.

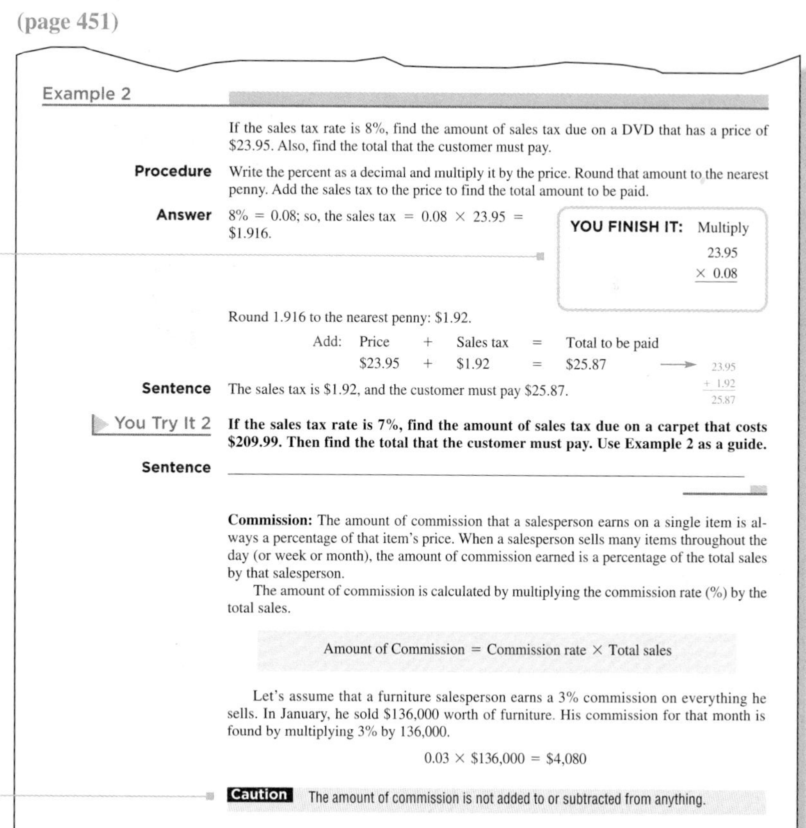

Example 2

If the sales tax rate is 8%, find the amount of sales tax due on a DVD that has a price of \$23.95. Also, find the total that the customer must pay.

Procedure Write the percent as a decimal and multiply it by the price. Round that amount to the nearest penny. Add the sales tax to the price to find the total amount to be paid.

Answer 8% = 0.08; so, the sales tax = 0.08×23.95 = \$1.916.

YOU FINISH IT: Multiply 23.95 × 0.08

Round 1.916 to the nearest penny: \$1.92.

Add: Price + Sales tax = Total to be paid

\$23.95 + \$1.92 = \$25.87 → 23.95 + 1.92 = 25.87

Sentence The sales tax is \$1.92, and the customer must pay \$25.87.

You Try It 2 **If the sales tax rate is 7%, find the amount of sales tax due on a carpet that costs \$209.99. Then find the total that the customer must pay. Use Example 2 as a guide.**

Sentence ________

Commission: The amount of commission that a salesperson earns on a single item is always a percentage of that item's price. When a salesperson sells many items throughout the day (or week or month), the amount of commission earned is a percentage of the total sales by that salesperson.

The amount of commission is calculated by multiplying the commission rate (%) by the total sales.

Amount of Commission = Commission rate × Total sales

Let's assume that a furniture salesperson earns a 3% commission on everything he sells. In January, he sold \$136,000 worth of furniture. His commission for that month is found by multiplying 3% by 136,000.

$0.03 \times \$136{,}000 = \$4{,}080$

Caution The amount of commission is not added to or subtracted from anything.

Just as your instructor does in class, these **Caution** boxes point out missteps or misconceptions to avoid while learning new concepts.

(page 224)

224 Chapter 3 Equations

Section 3.5 Exercises

FOR EXTRA HELP: Student Resources on DVD-ROM — Includes ▸ Student's Solutions Manual ▸ Video Lectures ▸ Chapter Test Prep Video | MyMathLab | MathXL

Think Again

1. What purpose does the legend serve when we are setting up an application problem?
   The legend helps define the number that the variable represents.

2. When we are setting up an application problem, why is a diagram helpful?
   Answers will vary. Some possibilities: The diagram helps us visualize the problem. It helps us place numbers and letters on the diagram. It helps us recognize a formula and develop the equation.

3. What information should be included in the concluding sentence of an application problem?
   Answers will vary. One possibility: The sentence should include many of the same words in the question and the answer along with the unit of measure.

4. After reading an application problem through once, what should a student do if he or she does not know the answer to the question?
   Answers will vary. One possibility: Read the problem again and look for important information.

Focus Exercises

Solve each application and answer with a complete sentence. (*Hint:* Set up a legend, draw a diagram, identify the formula, and create and solve the equation.)

5. Two of the angles in a triangle measure 108° and 37°. What is the measure of the third angle?
   The measure of the third angle is 35°.

6. In a right triangle, the measure of the smallest angle is 26°. What is the measure of the middle angle?
   The measure of the middle angle is 64°.

7. The perimeter of a rectangle is 178 inches. If the length is 58 inches, what is the width?
   The width of the rectangle is 31 inches.

8. The perimeter of a square is 152 inches. What is the length of each side of the square?
   The length of each side is 38 inches.

9. Lisa wants to plant a garden in her backyard. Her plan is to make the garden rectangular so that the length is 16 feet. She's planning to use 58 feet of chicken wire to surround the garden (as a perimeter). What should be the width of the garden so that she can use all 58 feet of fencing?
   The width of the garden should be 13 feet.

10. Kay wants to bend a 108-inch wire into a rectangle so that the width is 24 inches. What is the length of this rectangle?
    The length of the rectangle is 30 inches.

11. Adrienne did not score well on her first test. She studied much harder for her second test and got a score of 95. Her teacher indicated that her two test scores totaled 168 points. What was Adrienne's score on the first test?
    Adrienne's score on the first test was 73.

12. Anthony received scores of 87 points and 85 points on his first two tests. By the end of the third test, his goal was to have a total score of 270 points. (This would give him an average score of 90.) How many points will Anthony need to score on his third test to reach his goal?
    Anthony will need to score 98 points on his third test.

13. This Saturday, Margo plans to ride her bike 42 miles from her home to the beach. If she is able to average 14 miles per hour, how long will it take Margo to get to the beach?
    It will take Margo 3 hours to get to the beach.

14. Every summer, Andy hikes a mountain trail in southwestern Colorado. The route he takes is 72 miles long. If he can average 18 miles per day, how many days will it take Andy to hike this trail?
    It will take Andy 4 days to hike this trail.

15. Silvio wants to bend a 48-inch wire into a triangle so that the longest side is 21 inches and the shortest side is 12 inches. What is the length of the third side?
    The length of the third side is 15 inches.

16. Reggie owns a triangular plot of land. One side is 185 yards long, and another side is 212 yards long. If the total perimeter is 626 yards, what is the length of the third side?
    The length of the third side is 229 yards.

17. Kacey has 126 square feet of ceramic tile with which to cover a rectangular floor. She wants to build herself an office so that she can use this tile. If the floor's width is to be 9 feet, what will be the length of the floor?
    The length of the floor will be 14 feet.

At the beginning of each exercise set, a **For Extra Help** box appears, reminding you of all of the resources available if you get stuck on something. See the Teaching and Learning Package section of this Preface for more details.

As a way to ensure that you have not only mastered the skills but that you fully understand the concepts behind those skills, each exercise set begins with a few **Think Again** exercises. These are modeled after the *Think About It* exercises and are designed to make you think again about a particular concept.

Focus Exercises follow the *Think Again* exercises as a way to practice and apply the skills you have just learned throughout the section. Be sure to read the specific directions just prior to each problem, and feel free to check your answers to the odd-numbered exercises in the back of the book.

(page 225)

Section 3.5 Exercises 225

18. A developer has built a house on a rectangular piece of land with an area of 1,200 square yards. If the width of this land is 25 yards, what is its length?
    The length of this land is 48 yards.

19. Jenna is an artist who makes her own canvases. For her next project, she wants a rectangular canvas with an area of 720 square inches and a width of 15 inches. What will be the length of the canvas?
    The length of the canvas will be 48 inches.

20. Derek wants to build a computer desk that has a rectangular top. The top is to have an area of 1,080 square inches and a width of 24 inches. What will be the length of the top?
    The length of the top will be 45 inches.

21. Hannah and Therese were driving from their home in Phoenix, Arizona, to their college in San Antonio, Texas. The total trip was 983 miles. If Therese drove 498 miles, how many miles did Hannah drive?
    Hannah drove 485 miles.

22. Jimi and Stephyn took a train and bus trip through Canada from Vancouver, British Columbia, to Toronto, Ontario. The total trip was 2,762 miles. If they traveled 747 miles on a bus, how many miles did they ride on the train?
    They rode 2,015 miles on the train.

23. Roberto and Luisa Barron-Lopez each work full-time jobs; and Luisa's mother, Carmena, collects a monthly pension of $628. Between the three of them, their total household income is $4,238 each month. If Luisa earns $1,772, how much does Roberto earn each month?
    Roberto earns $1,838 each month.

24. Three candidates were vying for Commissioner of Clay County, West Virginia: Katie DePasquale, Jenny Crum, and Ron Hampton. Of the 2,543 votes cast, Katie received 684 votes and Ron received 802 votes. How many votes did Jenny receive?
    Jenny received 1,057 votes.

Think Outside the Box

Solve each application and answer with a complete sentence. (*Hint:* Set up the legend, draw a diagram, identify the formula, and create and solve the equation.)

25. The width of a rectangle is 2 feet, and the perimeter is 134 inches. What is the length of the rectangle?
    The length of the rectangle is 43 inches.

26. Together, Mia and Damon earned a total average of $3,819 each month in 2006. If Mia earned $24,936 that year, how much did Damon earn?
    Damon earned $20,892.

No math book would be complete without some problems that ask you to **Think Outside the Box**. These exercises ask you to take a step or two beyond the basics and either apply what you have learned to a more challenging situation or simply think about a problem in a slightly different way.

Studying for a chapter test can often be the time when concepts gel and understanding reaches a higher level. The **Chapter Review** pages are provided at the end of each chapter to help you organize the topics that you have studied and provide definitions and examples of the key topics. Make this detailed review of the chapter your own by adding your own notes or by creating study cards out of it!

Each key topic from the chapter is listed here by section. If you struggle with a concept or skill, refer back to that specific section.

Each key topic in the chapter is illustrated with a definition or description in the **concept** section as well as with a specific **example**.

(page 233)

Chapter 3 Review 233

Chapter 3 Review

Section 3.1 Solving Equations Involving One Operation

Concept	Example
An **equation** is a mathematical sentence in which one expression equals another expression. To solve an equation, we must *isolate the variable* on one side of the equal sign. The **solution** of an equation makes the equation true, and we can check the solution by replacing the variable with that number.	Is 3 the solution to $5x - 4 = 2x + 5$? Check $x = 3$: $5(3) - 4 \stackrel{?}{=} 2(3) + 5$; $15 - 4 \stackrel{?}{=} 6 + 5$; $11 = 11$ True ✓
The Addition Property of Equations: We may add any number, c, to *each side* of an equation. We can use this property to *clear a constant* by adding its opposite to each side.	If $a = b$ then $a + c = b + c$; $x - 5 = -9$, $x - 5 + 5 = -9 + 5$
The Division Property of Equations: We may divide *each side* of an equation by any number, c (as long as c isn't 0). We can use this property to *clear a coefficient* by dividing each side by the coefficient.	If $a = b$ then $\frac{a}{c} = \frac{b}{c}$; $-3x = 12$, $\frac{-3x}{-3} = \frac{12}{-3}$

Section 3.2 Solving Equations Involving Two Operations

Concept	Example
When an equation has two operations, first isolate the variable term by clearing the constant.	$-2x + 9 = 15$; $-2x + 9 + (-9) = 15 + (-9)$; $-2x + 0 = 6$; $-2x = 6$; $\frac{-2x}{-2} = \frac{6}{-2}$; $x = -3$

Practice all the skills that you have learned throughout the chapter with the **Chapter Review Exercises**. These exercises are broken out by section so that you can refer back to any particular section of the text if you need some guidance. In addition, the answers to all Chapter Review Exercises are provided in the back of the text so you can evaluate your mastery of these skills.

(page 237)

Chapter 3 Review Exercises

True or false.

1. In an equation, to isolate the variable and clear a constant, we add the opposite of the constant to each side of the equation. ____True____ **(3.1)**
2. In an equation, to isolate the variable and clear a coefficient, we divide each side of the equation by the opposite of the coefficient. ____False____ **(3.1)**

Fill in each blank with the word that correctly completes the sentence.

3. When solving an application problem, we write a(n) ____legend____ to define or describe what the variable represents. **(3.5)**
4. When solving an application problem, we develop the equation from a(n) ____formula____. **(3.5)**

Word List
legend
formula

Section 3.1

Determine if the replacement value given is a solution to the equation.

5. $3x - 8 = -2; x = -2$ No, −2 is not a solution.
6. $c + 5 = 2c - 3; c = 8$ Yes, 8 is a solution.

Solve. Make sure you check each answer.

7. $w + 9 = 17$ $w = 8$
8. $d + 12 = 31$ $d = 19$
9. $5 = n + 18$ $n = -13$
10. $2 = 6 + h$ $h = -4$
11. $13 + x = -4$ $x = -17$
12. $-10 = 5 + y$ $y = -15$
13. $m - 8 = 15$ $m = 23$
14. $7 = c - 13$ $c = 20$
15. $-12 + x = -2$ $x = 10$
16. $-4 = -5 + w$ $w =$
17. $m - 7 = -10$ $m = -3$
18. $y - 5$ ...
19. $-8 = k - 11$ $k = 3$
20. $-5 =$ ...
21. $-2 = -1 + y$ $y = -1$
22. $-4 =$ ...
23. $2n = -22$ $n = -11$
24. $9x =$ ...
25. $-3y = -87$ $y = 29$
26. $-40 =$ ...
27. $72 = -8y$ $y = -9$
28. $60 =$ ...
29. $28 = 7p$ $p = 4$
30. $108 =$ ...

Section 3.2

Solve. Make sure you check each answer.

31. $6x - 12 = 42$ $x = 9$
32. $4w - 16 = -48$ $w = -8$
33. $-6 = 12p + 30$ $p = -3$
34. $3 = 8k + 19$ $k = -2$
37. $-5y - 6 = -11$ $y = 1$
38. $16 =$ ...
39. $8 = -5k - 7$ $k = -3$
40. $-7y$ ...
41. $12 = -p + 20$
42. $1 =$ ...

Each chapter ends with a comprehensive **Chapter Test**, giving you the opportunity to assess your mastery of the skills taught in the chapter. These chapter tests are also a great opportunity for you to practice test-taking skills.

(page 240)

Chapter 3 Test

Solve and simplify. Check each answer.

1. $y - 16 = -9$ $y = 7$
2. $-9x = 36$ $x = -4$
3. $-7 = -8 + k$ $k = 1$
4. $-4x = -32$ $x = 8$
5. $6x - 12 = 30$ $x = 7$
6. $24 = -18 - 3x$ $x = -14$
7. $15 - 5w = -75 - 10w$ $w = -18$
8. $8x - 5 = 25 - 2x$ $x = 3$
9. $2x + 5 - 9x = 3x - 15$ $x = 2$
10. $30 - y = 4(6 - y)$ $y = -2$

For each of the following, the legend is *Let x = the number*. Translate the sentence into an equation and solve. Write a sentence that answers the question.

11. 14 more than a number is −11. What is the number? $x + 14 = -11$ The number is −25.
12. 9 less than twice a number is 13. What is the number?
15. A 240-centimeter board is to be cut into two pieces so that the longer piece is twice as long as the shorter piece. Find the length of each piece. Let x = the length of the shorter piece. $2x$ = the length of the longer piece. $x + 2x = 240$ The shorter piece is 80 cm. The longer piece is 160 cm.
16. The perimeter of a triangle is 42 inches. One side is 18 inches long. The second side is 2 inches longer than the third side. What are the lengths of the second and third sides? Let x = the length of the third side. $x + 2$ = the length of the second side. $18 + (x + 2) + x = 42$ The length of the second side is 13 inches. The length of the third side is 11 inches.
17. To get to her cousin's wedding, Greta drove a total of 1,050 miles. On the first day, she drove 180 miles more than on the second day. How many miles did she drive each day? Let x = the number of miles on the second day. $x + 180$ = the number of miles on the first day. $(x + 180) + x = 1,050$ Greta drove 435 miles on the second day and 615 miles on the first day.

Teaching and Learning Package

Things are changing in mathematics education. The class formats, teaching styles, student learning styles are all affecting those changes. Because *Prealgebra* has been written for these new formats and styles, so too should the resources that support the book. To that end, we have created an extremely comprehensive support package without unnecessarily wasting paper, trees, or energy.

Designed for the Student!

Student Resources on DVD-ROM for Prealgebra,
ISBN-13: 978-0-321-58895-1; ISBN-10: 0-321-58895-9

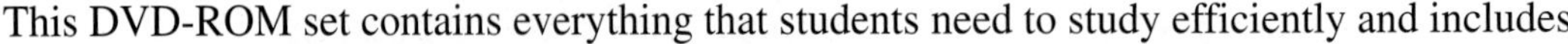

This DVD-ROM set contains everything that students need to study efficiently and includes

- **Chapter Test Prep Videos**, which provide step-by-step solutions to every problem in the Chapter Tests. These videos provide guidance and support when students need the most help — the week, or night, before an exam.
- **Video Lectures**, which provide an instructional lecture for each section of the textbook from the author. These lectures cover important definitions, procedures, and concepts from the section by working through examples and exercises from the textbook. Videos have optional closed captioning.
- **Student's Solutions Manual**, includes PDFs of the detailed, worked-out solutions to all You Try It exercises, to all odd-numbered exercises, and to all Chapter Review and Chapter Test exercises in the text.

MathXL Tutorials on CD for Prealgebra,
ISBN-13: 978-0-321-59955-1; ISBN-10: 0-321-59955-1
This interactive tutorial CD-ROM provides algorithmically generated practice exercises that are correlated at the objective level to the exercises in the textbook. Every practice exercise is accompanied by an example and a guided solution designed to involve students in the solution process. Selected exercises may also include a video clip to help students visualize concepts. The software provides helpful feedback for incorrect answers and can generate printed summaries of your progress.

Designed for the Instructor!

Annotated Instructor's Edition, ISBN-13: 978-0-321-59408-2; ISBN-10: 0-321-59408-8
This version of the text includes answers directly next to or below all exercises presented in the book, as well as helpful teaching tips called Instructor Insight. These **Instructor Insight** boxes, located *only* in the Annotated Instructor's Edition, provide some quick tips for instructors, highlighting areas where students may need to spend more time or areas that often cause confusion for students.

The following materials are available to the instructor free of charge via the Instructor Resource Center at PearsonHigherEd.com. If you need access to this site, please contact your local Pearson representative.

Printable Test Bank, ISBN-13: 978-0-321-59956-8; ISBN-10: 0-321-59956-X
This supplement contains PDFs of 1 short quiz per **section**, 3 free-response, and 2 multiple-choice tests per **chapter**, and 4 free-response and 4 multiple-choice **final exams**.

Instructor's Solutions Manual, ISBN-13: 978-0-321-59954-4; ISBN-10: 0-321-59957-8
This manual includes PDFs of the detailed, worked-out solutions to all You Try It exercises, to all even exercises, and to all Chapter Review Exercises, and Chapter Test exercises in the text.

Instructor's Resource and Support Manual,
ISBN-13: 978-0-321-59957-5; ISBN-10: 0-321-59957-8
This manual presents resources designed to help both new and experienced instructors with course preparation and classroom management. This includes chapter-by-chapter teaching tips and support for media supplements, among other resources.

TestGen® (www.pearsonhighered.com/testgen) enables instructors to build, edit, print, and administer tests using a computerized bank of questions developed to cover all the objectives of the text. TestGen is algorithmically based, allowing instructors to create multiple, but equivalent, versions of the same question or test with the click of a button. Instructors can also modify test bank questions or add new questions. Tests can be printed or administered online. The software and testbank are available for download from Pearson Education's online catalog.

Designed for Students and Instructors!

MyMathLab® is a series of text-specific, easily customizable online courses for Pearson Education's textbooks in mathematics and statistics. Powered by CourseCompass™ (our online teaching and learning environment) and MathXL® (our online homework, tutorial, and assessment system), MyMathLab gives you the tools you need to deliver all or a portion of your course online, whether your students are in a lab setting or working from home. MyMathLab provides a rich and flexible set of course materials featuring free-response exercises that are algorithmically generated for unlimited practice and mastery. Students can also use online tools, such as video lectures, animations, and a multimedia textbook, to independently improve their understanding and performance. Instructors can use MyMathLab's homework and test managers to select and assign online exercises correlated directly to the textbook, and they can also create and assign their own online exercises and import TestGen tests for added flexibility. MyMathLab's online gradebook—designed specifically for mathematics and statistics—automatically tracks students' homework and test results and gives the instructor control over how to calculate final grades. Instructors can also add offline (paper-and-pencil) grades to the gradebook. MyMathLab also includes access to the **Pearson Tutor Center** (www.pearsontutorservices.com). The Tutor Center is staffed by qualified mathematics instructors who provide textbook-specific tutoring for students via toll-free phone, fax, email, and interactive Web sessions. MyMathLab is available to qualified adopters. For more information, visit our website at www.mymathlab.com or contact your sales representative.

MathXL® is a powerful online homework, tutorial, and assessment system that accompanies Pearson Education's textbooks in mathematics or statistics. With MathXL, instructors can create, edit, and assign online homework and tests using algorithmically generated exercises correlated at the objective level to the textbook. They can also create and assign their own online exercises and import TestGen tests for added flexibility. All student work is tracked in MathXL's online gradebook. Students can take chapter tests in MathXL and receive personalized study plans based on their test results. The study plan diagnoses weaknesses and links students directly to tutorial exercises for the objectives they need to study and retest. Students can also access supplemental animations and video clips directly from selected exercises. MathXL is available to qualified adopters. For more information, visit our website at www.mathxl.com or contact your sales representative.

The Development of This Text

Insight, Research, and Determination

The creation of this textbook and support package is the result of the author's determination to create a better learning resource for students; data received from a usability study with students on how they use textbooks; and the willingness of over 50 instructors to provide insight through reviewing, focus groups, and class testing into how student and instructor materials can be improved for mathematics education.

Along with all of the great insight provided from reviewers and focus group participants, this book has been through an extensive accuracy checking process. Likely, only those who have been involved in the production of a mathematics textbook can fully appreciate all that goes into ensuring its accuracy. However, in an effort to illustrate all the steps that we've taken to reach the level of accuracy that you would expect from Pearson Education, we've outlined the accuracy checking process on the opposite page.

About the Author

Bob Prior is an associate professor of mathematics at Riverside Community College's Norco Campus in southern California. He received his bachelor's degree from the University of California at San Diego and his master's degree from California State University at Fullerton. Along with his teaching at the community college level, he has presented intensive summer institutes to high school algebra teachers and to upper elementary school educators. He has also presented at AMATYC, ICTCM, and at CMC3 (California), both North and South.

Bob has authored developmental math texts for use in three online classes for Riverside Community College: Basic Math, Prealgebra, and Elementary Algebra. He is restructuring each of those texts, along with Intermediate Algebra, for Pearson. Bob says his main goal with his writing style is that he wants students to see the 'why' behind the math they are learning, so the rules are developed and rarely just given outright.

Bob lives with his wife, Laura, in the semirural town of Norco, in western Riverside County, California. His daughter, Katie, is a graduate of UC Davis, and his son, Nate, is completing his senior year at Sterling College, in the town of Sterling, Kansas. He is active in the community as a member of the Norco Kiwanis Club and in his church where he sings and plays guitar, neither of which will make him a millionaire. Bob appeared (many years ago) on two different TV game shows where he won \$1,500 and "lovely parting gifts," and more recently, Bob was a featured extra on the TV show *NUMB3RS*.

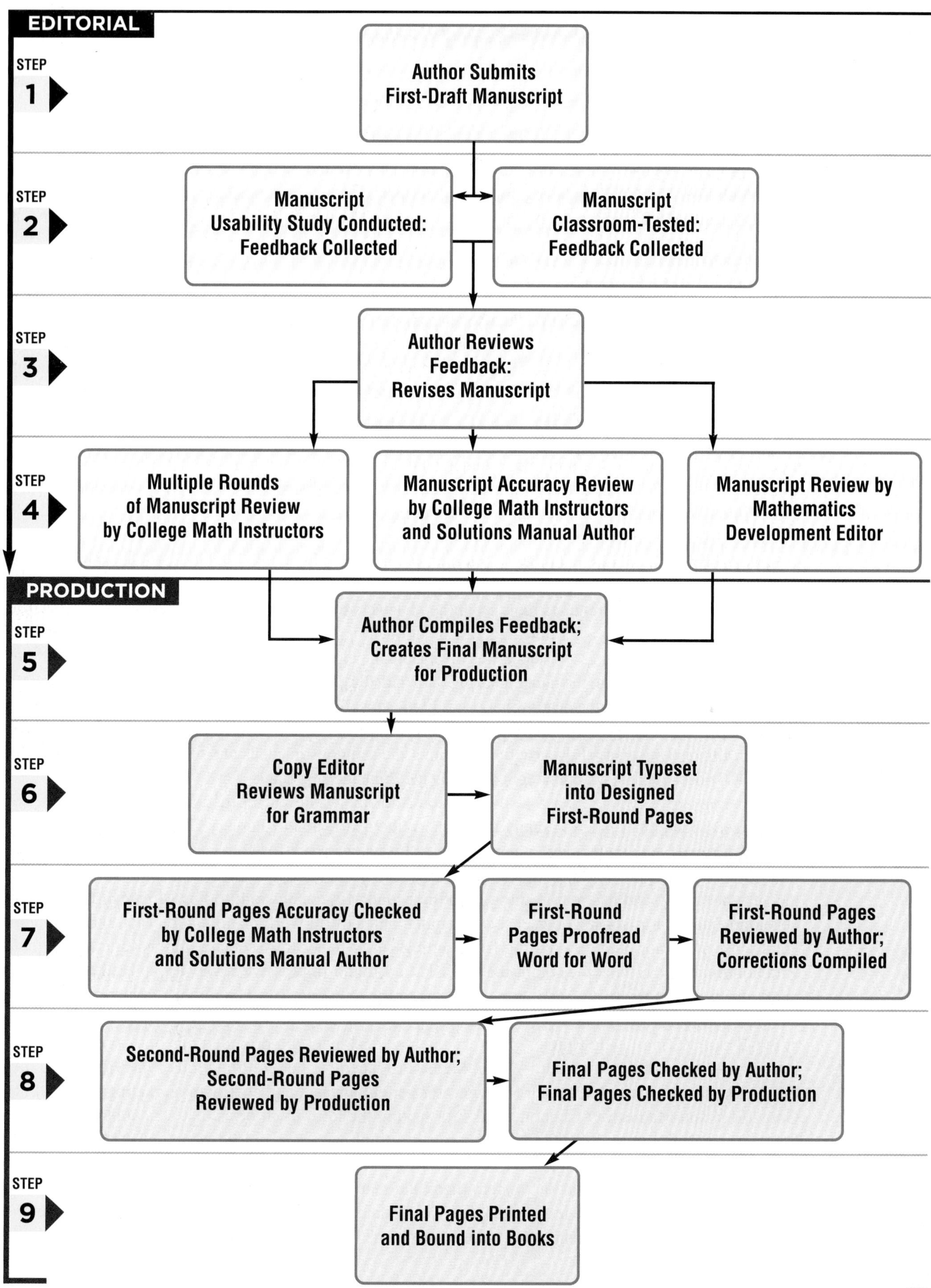
EDITORIAL
STEP 1
Author Submits First-Draft Manuscript
STEP 2
Manuscript Usability Study Conducted: Feedback Collected
Manuscript Classroom-Tested: Feedback Collected
STEP 3
Author Reviews Feedback: Revises Manuscript
STEP 4
Multiple Rounds of Manuscript Review by College Math Instructors
Manuscript Accuracy Review by College Math Instructors and Solutions Manual Author
Manuscript Review by Mathematics Development Editor
PRODUCTION
STEP 5
Author Compiles Feedback; Creates Final Manuscript for Production
STEP 6
Copy Editor Reviews Manuscript for Grammar
Manuscript Typeset into Designed First-Round Pages
STEP 7
First-Round Pages Accuracy Checked by College Math Instructors and Solutions Manual Author
First-Round Pages Proofread Word for Word
First-Round Pages Reviewed by Author; Corrections Compiled
STEP 8
Second-Round Pages Reviewed by Author; Second-Round Pages Reviewed by Production
Final Pages Checked by Author; Final Pages Checked by Production
STEP 9
Final Pages Printed and Bound into Books

Acknowledgments

Dedication

This book is dedicated to my wife, Laura, and to my children, Katie and Nate. Your patience and support for this project have been immeasurable. (My picture can be found on page xiv, in case you have forgotten what I look like. That's right, I have a little less hair than you may remember.)

I also dedicate this book to the memory of my mother, Marilyn Prior, my role model as a parent and as a teacher.

Acknowledgments

This project was based on the belief that a better math textbook could be written, and I want to thank those who encouraged me, supported me, and believed that I could actually write one. So to my colleagues at Riverside Community College, who have class-tested my textbooks, I give a hearty thanks: Elisa Chung, Deborah Smith, Brian Johnson, Janet Frewing, Joseph DeGuzman, Jason Rey, Richard Ries, Kathy Saxon, and John "Biff" Pietro.

To my students, who told me time and again that they liked math—and understood it—for the first time ever, I say thanks for helping me to understand the way a student thinks so that I could then answer your questions in the book, before you even asked them. Also, thanks for catching my occasional errors, even without the benefit of extra credit points for doing so.

To my friends who took the time to share their ideas, their experience, and their wisdom, I say thanks to Harley Hahn, an author of a number of Internet and programming books, and to H. H. Hanson, someone who everyone wishes they knew.

A project like this also requires input from the reviewers throughout the country who took time to read through one or two chapters at a time and pass along their experience and thoughts about the writing style and the flow of the text. Their assistance was invaluable in helping me to decide on the proper way to introduce some topics. It seems to me rather amazing that, with the collective talents of all involved, errors and typos still find their way into the nearly final pages of the text. To combat these errors, the service of accuracy checkers, Samantha Hodges and Betty Vix Weinberger, proved invaluable. Thanks also to Julia Simms at Southern Illinois University–Edwardville for writing the solutions manuals and to Laura Hoye and Harriet Merkel at Trident Technical College for compiling the Printable Test Bank.

I must also give a huge thanks to the Pearson Addison-Wesley team that supported me throughout the project. I remember meeting many members of the team back in January 2002. Publisher Greg Tobin said, "You have to understand, you don't work for us, we work for you," and that has certainly rung true. From Edgar Espina who "discovered" me; to my editor, Jenny Crum, who shared my vision from the very beginning and has seen it through to this wonderful conclusion; to Peg Monahan-Pashall, my developmental editor whose good sense of grammar and skillful word manipulation (not to mention her good sense of humor) kept my writing from getting out of hand; to Associate Editor Katie DePasquale who coordinated the schedules, kept me on task, and found impressive reviewers to read through the many writings and rewritings of the manuscript; to the amazing Ron Hampton, Production Manager, who has a keen eye for detail and really knows how to manage a project; to Sam Blake and Tim Rodes at Pre-PressPMG; to Leslie Haimes, the very talented designer who provided this book with its unique interior and cover design; to Audra Walsh, Media Producer, for her coordination of the videos and media; to Marlana Voerster, Marketing Manager, for her enthusiasm and energy in marketing this book; to Greg Tobin and Maureen O'Connor, thanks for having the faith in all of us that we could make a textbook unique in design and distinct in its approach. I think we have, indeed, accomplished what we set out to do.

The author and Pearson Education wish to thank the many instructors and students who participated in making this text such a unique learning experience. Certainly this book would not be what it is without the countless conversations and insightful feedback that came from these wonderful instructors.

Reviewers

Froozan Pourboghnat Afiat	Community College of Southern Nevada
Nkechi Agwu	Borough of Manhattan Community College
Darryl Allen	Solano Community College
Pam Baenziger	Kirkwood Community College
Lynn Beckett-Lemus	El Camino College
Monika Bender	Central Texas College
Caroline Best	Pellissippi State Technical Community College
Becky Blackwell	Pellissippi State Technical Community College–Magnolia Avenue Campus
Robert Blythe	Eastern Kentucky University
Wayne Brown	Oklahoma State University–Oklahoma City
Gary Chumley	St. Philip's College
John F. Close	Salt Lake Community College
John Coburn	St. Louis Community College–Florrisant Valley Campus
William W. Coe	Montgomery College of Rockville
Ann Davis	Northeastern Technical College
Babette Dickelman	Jackson Community College
Gigi Drent	Kaua'i Community College
David Ellsworth	Chandler-Gilbert Community College
Jean Freeman	Kent State University–Ashtabula
Milton Gatewood	Austin Community College
John Hake	St. Louis Community College–Florrisant Valley Campus
Doris Holland	Tarrant County College
Diane Hollister	Reading Area Community College
Marilyn Jacobi	Gateway Community College
Karen Jensen	Southeastern Community College
Nancy R. Johnson	Manatee Community College
Fred Katiraie	Montgomery College
Patrick Kent	North Central State College
Kathy Kopelousos	Lewis and Clark Community College
Carla Kulinsky	Salt Lake Community College
Jan LaTurno	Rio Hondo College
Femar Lee	Honolulu Community College
Paul Wayne Lee	St. Philip's College
Mikal McDowell	Cedar Valley College
Roberta Pardo	Chandler-Gilbert Community College
Faith Peters	Miami Dade Community College
Melinda Rudibaugh	Chandler-Gilbert Community College
Glenn R. Sandifer	San Jacinto College
Becki Saylor	Sanford Brown College
Jeanette Shea Shotwell	Central Texas College
Julia Simms	Southern Illinois University–Edwardsville
Claire Suddeth	Pellissippi State Technical Community College
Cory Takemoto	Honolulu Community College
Shae Thompson	Montana State University
Linda Tucker	Rose State College
Barbara Van Allen-Satterwhite	Essex County College
James Vogel	Sanford-Brown College
Edward Wagner	Central Texas College
Mary Jo Westlake	Itasca Community College
Carol L. Williams	Des Moines Area Community College
Flo Wilson	Central Texas College
Chock Wong	Chaminade University
Michelle Wyatt	Community College of Southern Nevada
Richard Zucker	Irvine Valley College

Accuracy Checkers

Samantha Hodges, BettyVix Weinberger

Accuracy Coordinator

Deborah J. Smith

Basic Skills Pretest

Add.

1. 416 + 511
927

2. 358 + 64
422

3. 6,173 + 827
7,000

Subtract.

4. 4,976 − 2,735
2,241

5. 1,164 − 239
925

6. 3,000,000 − 1,069,502
1,930,498

Multiply.

7. 8 × 5
40

8. 6 × 6
36

9. 6 × 9
54

10. 7 × 80
560

11. 57 × 6
342

12. 183 × 15
2,745

Divide.

13. 49 ÷ 7
7

14. 36 ÷ 9
4

15. 72 ÷ 8
9

16. 84 ÷ 6
14

17. 192 ÷ 4
48

18. 243 ÷ 9
27

Use long division to divide. Be sure to write any remainder next to the quotient.

19. 90 ÷ 7
12 r 6

20. 512 ÷ 6
85 r 2

21. 1,053 ÷ 13
81

22. 1,878 ÷ 39
48 r 6

Round each number to the place shown.

23. 935 *ten*
940

24. 862 *ten*
860

25. 809 *hundred*
800

26. 3,953 *hundred*
4,000

27. 40,256 *thousand*
40,000

28. 539,701 *thousand*
540,000

29. 275,608 *ten thousand*
280,000

30. 139,420 *hundred thousand*
100,000

Work each application and answer with a complete sentence.

31. Tamari works two part-time jobs. Last month, she earned $749 at her morning job and $876 at her evening job. How much did Tamari earn last month?
Tamari earned $1,625 last month.

32. Markos rented 12 tables for a banquet. The total cost of the table rental was $216. What was the rental price for each table?
The rental price for each table was $18.

33. Gilberto is sharing an apartment with two other people, and his share of the monthly rent is $425. How much rent will Gilberto pay for one year?
Gilberto will pay $5,100 rent for one year.

34. Shayenne is shopping at a large furniture store and is deciding whether to buy a couch or a recliner. The price of the couch is $1,029, and the price of the recliner is $783. How much more is the price of the couch than the price of the recliner?
The price of the couch is $246 more than the price of the recliner.

Place the correct letter of the property in column B next to the property being demonstrated in column A.

Column A	Column B
35. 46 × 1 = 46 __B__	**A.** Identity of Addition
36. 3 + (4 + 8) = (3 + 4) + 8 __C__	**B.** Identity of Multiplication
37. 83 + 12 = 12 + 83 __E__	**C.** Associative Property of Addition
38. 0 + 59 = 59 __A__	**D.** Associative Property of Multiplication
39. 4 × (7 × 2) = (4 × 7) × 2 __D__	**E.** Commutative Property of Addition
40. 0 × 6 = 0 __G__	**F.** Commutative Property of Multiplication
41. 3 × (5 + 8) = 3 × 5 + 3 × 8 __H__	**G.** Zero Product Property
42. 12 × 3 = 3 × 12 __F__	**H.** Distributive Property

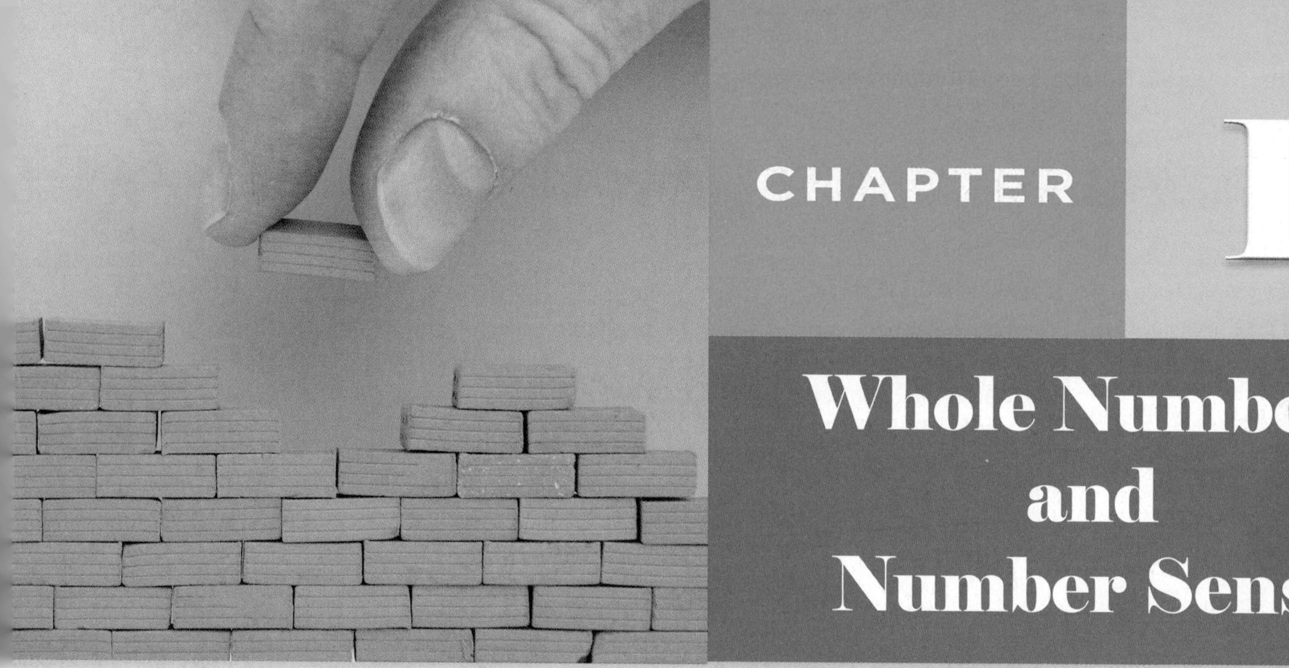

CHAPTER 1

Whole Numbers and Number Sense

Introduction

The topics in math can be compared to bricks in a wall. Some bricks lie next to each other, and some bricks are on top of two or more lower bricks.

The bricks represented in this wall are presented in Chapter 1. Together they form the foundation for the rest of your preparation for algebra, including fractions, decimals, percents, equations, and problem solving.

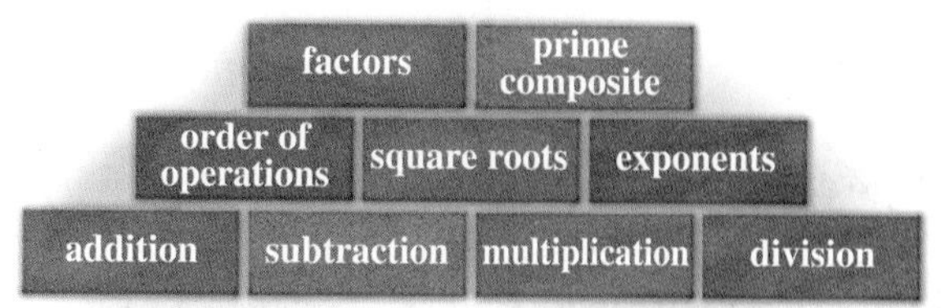

Some of the material in this chapter may be familiar to you. If so, that's great! Use that knowledge to build on ideas presented later in the book. If some of the material is new to you, take the time to understand *why* it works as well as *how* it works.

Above all, make a commitment to yourself, committing to your education and your goals. Set up a daily (or weekly) study routine, find a study partner, and do not be afraid to ask questions when something is unclear to you. Use this book to its fullest, making it your own so that you can use it as a reference when you take that next step to algebra and beyond.

Preparation Exercises

Every chapter opens with a set of Preparation Exercises. For Chapters 2–9, the exercises are from previous chapters in the text. They are designed to help prepare you for the material in the upcoming chapter.

In preparation for Chapter 1, a Basic Skills Pretest is included to help you and your teacher get an idea of what basic math skills you already have. When you complete this test, your teacher can analyze the results and integrate the material in Chapter 1 based on the skills of the whole class.

In preparation for the course, you may want to spend some time thinking about and discussing the following questions:

1. Why are you taking this course?
   a) What plans do you have for your education?
   b) What career plans do you have in mind?
   c) Where will you be in five years?
2. How do you plan to be successful in this course?
   a) Where is your favorite place to study?
   b) Do you like to study by yourself or in a group?
   c) How much time should be spent on math each week, including time in class?
   d) How important is it to do the assigned homework?
   e) How important is it to read the textbook?
   f) What is the best advice you can give to someone to help him or her succeed in this class?

SECTION 1.1 Whole Numbers and Their Properties

OBJECTIVES

In this section, you will learn to

- Identify the base-ten numbers.
- Find the place value of a digit in a whole number.
- Write a whole number in expanded form and in words.
- Identify and apply the four basic operations.
- Evaluate a mathematical expression.
- Identify and apply the Commutative Properties.
- Identify and apply the Associative Properties.
- Recognize the identities for addition and multiplication.
- Round whole numbers.

Introduction

Numbers have been around since the beginning of language. People first used numbers to count things, especially when items were being traded, such as five sheep for three pigs. As time progressed, numbers were used to measure things, such as the length of an ark, the distance between two cities, and the number of daylight hours in a month.

Eventually, people started doing calculations with numbers, and it was called "arithmetic." They started drawing circles and triangles, and it was called "geometry." They started using letters to represent numbers, and it was called "algebra." They gave all of it a name—"mathematics."

Mathematics is at the heart of every computer, every television, and every cell phone. In short, mathematics is a hidden part of your everyday life. As complex as the world has become, even the simplest mathematics plays a role.

As students of prealgebra, most of you will continue your studies through algebra. Some of you may even go on to study higher levels of math, such as statistics or calculus. Whatever your goal, learning prealgebra is an important step.

Numbers and Numerals

In the world of language, a **number** is an adjective. It describes *how many* items there might be, as in *Foofi has* six *puppies.*

A **numeral** is a symbol that represents a number. For example, the number *six* can be expressed by any of these symbols (and there are others):

In this text, we often use the word *number* in place of the word *numeral* to make reading easier.

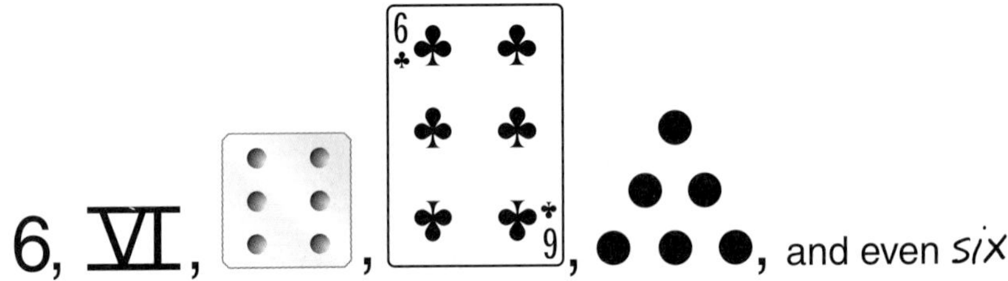

Our way of counting is built on a **base-ten numbering system** that uses ten **digits,** or numerals:

0, 1, 2, 3, 4, 5, 6, 7, 8, and 9

These digits are the first of the **whole numbers.** When we get to 9, we have exhausted all of the single-digit whole numbers. So, to represent the next whole number, ten, we must use two digits, 10.

The whole numbers: 0, 1, 2, 3, 4, 5, 6, 7, 8, 9, 10, 11, 12, 13, 14, and so on.

Even though 0 is written as the first whole number, it is not the first *counting number.*

The **counting numbers,** or **natural numbers:** 1, 2, 3, 4, 5, . . . , 10, 11, 12, and so on.

We can represent whole numbers visually along a **number line.**

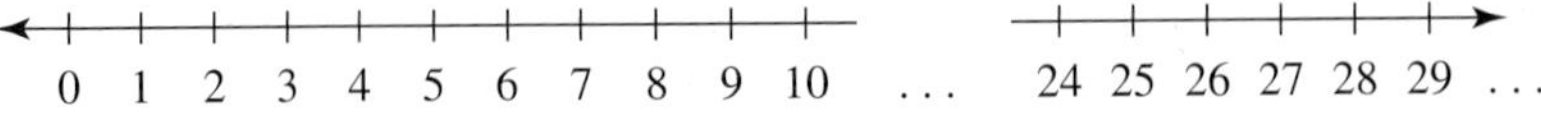

The farther to the right along the number line, the higher the value of each number.

Place Value

Some numbers have meaning but no value. For example, a telephone number such as 313-555-2406 is not considered a higher number than 313-555-2405. It's just a different number.

Consider, though, numbers that do have value, such as the mid-2005 population of New York State: 19254630. In this number, each digit has a value, called a **place value,** based on its position, or *place,* in the number. For example, in this number, 2 represents *two hundred thousand;* but that can be difficult to determine without seeing the number written in a more organized way. That organization is found in the form of a place value chart.

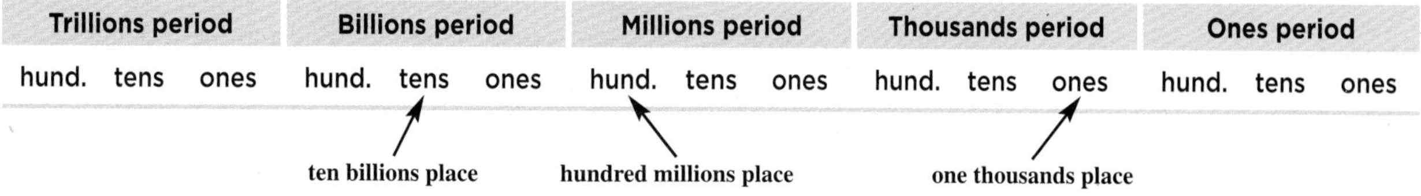

Notice the groups of threes, called **periods,** and the pattern of repetition (hundreds, tens, ones) within each period. When writing the number in an organized way, we use commas—starting from the right—to separate the periods from each other.

You will probably agree that the number 19254630 can be more easily read using commas: 19,254,630. Let's place the number, one digit for each place, in the following place-value chart.

Trillions period			Billions period			Millions period			Thousands period			Ones period		
hund.	tens	ones	hund.	tens	ones	hund.	tens	ones	hund.	tens	ones	hund.	tens	ones
							1	9	2	5	4	6	3	0

Example 1

Identify in which period and place you find the digit 2. Then write the value the 2 represents.

a) 62,439 **b)** 524,093,186

Procedure Let's use a place-value chart to help us read the numbers more easily.

	Millions period			Thousands period			Ones period		
	hund.	tens	ones	hund.	tens	ones	hund.	tens	ones
a)					6	2	4	3	9
b)	5	2	4	0	9	3	1	8	6

Answer **a)** In 62,439, the 2 is in the thousands period and in the *ones place*. So, the 2 represents two thousand (2,000).

b) In 524,093,186, the 2 is in the millions period and in the *tens place*. So, the 2 represents 20,000,000 (twenty million).

You Try It 1 **Identify in which period and place you find the digit 5. Then write the value the 5 represents. Use Example 1 as a guide.**

	Period	Place	Value
a) 70,951	________	________	________
b) 539,418	________	________	________
c) 65,423,189	________	________	________

The **You Try It** exercises are presented within the section because you are expected to do these exercises when you encounter them, while the material is fresh in your mind. These exercises sometimes lead to other ideas that rely on your having worked them. The answers to each You Try It are found at the end of the section.

Only the ones place determines whether a whole number is *odd* or *even*. A whole number is **even** when it contains a 0, 2, 4, 6, or 8 in the ones place. A whole number is **odd** when it contains a 1, 3, 5, 7, or 9 in the ones place.

Writing a Number in Expanded Form

We can write a number in *expanded form* when we consider (1) the value of each digit and (2) the place of each digit. For example, consider the digit 6 in the number 26,358. It is in the thousands place; so, it represents 6,000 (six thousand).

Likewise, the digit 3 is in the hundreds place and represents 300 (three hundred). In fact, we can break down the entire number into its individual digits, each according to its place, in the following manner:

Millions period			Thousands period			Ones period		
hund.	tens	ones	hund.	tens	ones	hund.	tens	ones
				2	6	3	5	8
				↓ 20,000	↓ 6,000	↓ 300	↓ 50	↓ 8

= 20,000 + 6,000 + 300 + 50 + 8
This is the expanded form.

You Try It 2 **Write each number in expanded form.**

a) 3,075 = ____________ b) 503,142 = ____________

Writing Whole Numbers in Words

Sometimes it is appropriate to write a number in word form, such as on a check or in an essay. When writing a number in word form, we should follow these guidelines:

1. If the number is one thousand or more, a comma is used to separate each period.
2. If the number is one thousand or more, the numerical period is written before each comma.
3. When spelled out, two-digit numbers, such as 28, are hyphenated: *twenty-eight.*

Recall the mid-2005 population of New York State, 19,254,630. Written in words, this is

nineteen *million,* two hundred fifty-four *thousand,* six hundred thirty.

A simpler example is 307. We write this as three hundred seven.

Caution You might be tempted to include the word *and* in three hundred seven, but this is not appropriate when writing whole numbers. We reserve the word *and* for mixed numbers that include whole numbers along with fractions or decimals. In this sense, the word *and* means "plus."

For example, compare two hundred six million, which is 206,000,000

to two hundred *and* six million, which is 200 + 6,000,000 = 6,000,200.

To eliminate this confusion, we do not use the word *and* when writing out whole numbers.

Example 2

Write each number in words.

a) 8,104 **b)** 403,005 **c)** 49,023,000

Procedure Write the number, reading from left to right, using a comma to indicate the conclusion of one period and the beginning of a new period.

Answer **a)** 8,104 is eight thousand, one hundred four.

Once we write the word *thousand,* we place a comma *after* it and finish writing the number.

b) 403,005 is four hundred three thousand, five.

After the thousands, the ones period contains only 5; it is the comma that shows us the separation.

c) 49,023,000 is forty-nine million, twenty-three thousand.

Because there are no nonzero digits in the ones period, we stop at the thousands.

You Try It 3 **Write each number in words. Use Example 2 as a guide.**

a) 863 ______

b) 62,009 ______

c) 5,003,102 ______

Example 3

Write the whole number in numeral form.

a) Seven thousand, five hundred nine

b) Six million, eighteen

Procedure Look for the numerical period and the comma.

Answer **a)** 7,509

There are no tens between the five hundred and the nine. No tens is represented by 0 in the tens place.

b) 6,000,018

Because we never see the word *thousand,* there must be no thousands at all. We can't skip them, but represent that fact with three 0's in the thousands place.

You Try It 4 **Write the whole number in numeral form. Use Example 3 as a guide.**

a) Two thousand, forty-eight ______

b) Seventy-five thousand, four ______

c) Six million, one hundred four thousand, thirty ______

Basic Math Terminology

There are four basic **operations** in mathematics: addition (+), subtraction (−), multiplication (×), and division (÷). The *written form* of an operation, such as **5 + 4,** is called an **expression.**

We also have words for the *results* of the operations, as shown in the chart below. To get the result of an operation we must *apply* the operation.

Operation	Name	As an expression	Result	Meaning
Addition (plus)	Sum	$3 + 5$	$= 8$	The sum of 3 and 5 is 8.
Subtraction (minus)	Difference	$9 - 5$	$= 4$	The difference between 9 and 5 is 4.
Multiplication (times)	Product	2×3	$= 6$	The product of 2 and 3 is 6.
Division (divided by)	Quotient	$6 \div 3$	$= 2$	The quotient of 6 and 3 is 2.

For multiplication, the numbers in a product are called **factors.** For example, 2 and 3 are factors of 6 because $2 \times 3 = 6$. 1 and 6 are also factors of 6 because $1 \times 6 = 6$.

For division, the quotient can be expressed three ways:

In standard division form	In long division form	In fractional form
Dividend ÷ Divisor = Quotient	$\text{Divisor}\overline{)\text{Dividend}}$ with Quotient above	$\dfrac{\text{Dividend}}{\text{Divisor}} = \text{Quotient}$
$6 \div 3 = 2$	$3\overline{)6}$ with 2 above	$\dfrac{6}{3} = 2$

Example 4

Write both the expression and the result of each of the following.

Answer

	As an expression		As a result
a) the sum of 5 and 1	$5 + 1$	=	6
b) the difference between 10 and 6	$10 - 6$	=	4
c) the product of 4 and 7	4×7	=	28
d) the quotient of 20 and 4	$20 \div 4$	=	5

You Try It 5 **Write both the expression and the result of each of the following. Use Example 4 as a guide.**

	As an expression		As a result
a) the sum of 9 and 3	________	=	________
b) the difference between 12 and 4	________	=	________
c) the product of 5 and 8	________	=	________
d) the quotient of 30 and 5	________	=	________

Evaluating an Expression

To **evaluate** means to "find the value of." When we find the sum of the expression $3 + 5$, we are simply finding a different way to express the *value* of $3 + 5$, which is 8.

You Try It 6 **Evaluate each expression.**

a) $2 + 6 =$ ________ **b)** $9 - 2 =$ ________

c) $4 \times 5 =$ ________ **d)** $12 \div 3 =$ ________

Parentheses, ( ), are considered **grouping symbols.** Parentheses group different values together so that they can be treated as one *quantity*.

A **quantity** is an expression that is considered to be *one value*. For example, $(7 + 3)$ is a quantity; it has just one value: 10.

Generally, in any expression that has parentheses, the quantity within those parentheses should be evaluated *first*.

Example 5

Evaluate each expression.

a) $(2 \times 4) - 3$ **b)** $4 + (9 \div 3)$ **c)** $2 \times (5 + 4)$

Procedure Evaluate the quantity first. Then evaluate the new expression.

Answer

a) $(2 \times 4) - 3$
$= 8 - 3$
$= 5$

b) $4 + (9 \div 3)$
$= 4 + 3$
$= 7$

c) $2 \times (5 + 4)$
$= 2 \times 9$
$= 18$

You Try It 7 **Evaluate each expression. Use Example 5 as a guide.**

a) $(4 + 6) \div 2$ **b)** $9 - (1 \times 4)$ **c)** $8 + (6 \div 3)$

The Commutative Properties

Evaluate $2 + 4 =$ ______ and $4 + 2 =$ ______. For each of these expressions, the resulting sum is 6. These two sums illustrate a simple yet important property of mathematics, the *Commutative Property of Addition*.

The **Commutative Property of Addition** states the following: When two numbers are added, it doesn't matter which number is written first. The resulting sum will be the same.

The Commutative Property is true for multiplication as well. For example, $3 \times 5 = 15$ and $5 \times 3 = 15$.

The Commutative Properties are more formally written using letters, a and b, to represent any two numbers:

The Commutative Properties of Addition and Multiplication

If a and b are any numbers, then

Addition

$$a + b = b + a$$

The order in which we add two numbers doesn't affect the resulting sum.

Multiplication

$$a \times b = b \times a$$

The order in which we multiply two numbers doesn't affect the resulting product.

Caution Division is *not* commutative. If we switch the order of the numbers when dividing, we don't get the same result. For example, $10 \div 5$ is *not* the same as $5 \div 10$. Here's an illustration:

At a youth car wash, when 5 youths wash a truck for \$10, they get \$2 each ($\$10 \div 5 = \$2$). However, if 10 youths wash a car for \$5, they get only \$0.50 each ($\$5.00 \div 10 = \$0.50$).

Similarly, subtraction is *not* commutative.

Example 6

If possible, use a commutative property to rewrite each expression.

a) $3 + 4$ **b)** 6×5 **c)** $9 - 5$ **d)** $15 \div 3$

Procedure Switch the order of the numbers to arrive at an equivalent expression.

Answer
a) $3 + 4 = \underline{4 + 3}$

b) $6 \times 5 = \underline{5 \times 6}$

c) $9 - 5$ cannot be rewritten; subtraction is not commutative.

d) $15 \div 3$ cannot be rewritten; division is not commutative.

You Try It 8 **If possible, use a commutative property to rewrite each expression. Use Example 6 as a guide.**

a) $2 + 6$ **b)** $9 - 2$ **c)** $12 \div 3$ **d)** 4×5

What could you tell a classmate to help him or her remember why the two previous properties are called *commutative* properties?

____

____

The Associative Properties

Another important property of mathematics is the *Associative Property*. There is an Associative Property for addition and an Associative Property for multiplication.

Whenever we need to add three numbers together, such as $3 + 2 + 4$, we must choose to add two of them first. We can use parentheses to group two of them to create a quantity, then evaluate the quantity first.

For example, with the sum $3 + 2 + 4$, we can group the first two numbers or the last two numbers.

$(3 + 2) + 4$ (Add 3 and 2 first.) $= 5 + 4 = 9$ or $3 + (2 + 4)$ (Add 2 and 4 first.) $= 3 + 6 = 9$

Notice that the resulting sum, 9, is the same no matter which grouping we choose.

The same is true of multiplying three numbers together, such as $3 \times 2 \times 4$.

$(3 \times 2) \times 4$ (Multiply 3 and 2 first.) $= 6 \times 4 = 24$ or $3 \times (2 \times 4)$ (Multiply 2 and 4 first.) $= 3 \times 8 = 24$

Notice that the resulting product, 24, is the same no matter which grouping we choose.

The Associative Property uses letters *a*, *b*, and *c* to represent any numbers we want to put in their place.

The Associative Properties of Addition and Multiplication

If ***a***, ***b***, and ***c*** are any three numbers, then

Addition	**Multiplication**
$(a + b) + c = a + (b + c)$	$(a \times b) \times c = a \times (b \times c)$
If the only operation is addition, we can change the grouping of the numbers without affecting the resulting sum.	If the only operation is multiplication, we can change the grouping of the numbers without affecting the resulting product.

Example 7

Write each expression two different ways using parentheses. Then evaluate it.

a) $1 + 6 + 3$ **b)** $5 \times 2 \times 3$

Procedure Use the Associative Property. Notice that the order in which the numbers are written doesn't change.

Answer

		$7 \quad + 3$		$1 + \quad 9$		
a) $1 + 6 + 3$	$=$	$(1 + 6) + 3$	$=$	$1 + (6 + 3)$	$=$	**10**
		$10 \quad \times 3$		$5 \times \quad 6$		
b) $5 \times 2 \times 3$	$=$	$(5 \times 2) \times 3$	$=$	$5 \times (2 \times 3)$	$=$	**30**

You Try It 9 **Write each expression two different ways using parentheses. Then evaluate it. Use Example 7 as a guide.**

a) $2 + 8 + 6$ = ________ = ________ = ________

b) $4 + 6 + 9$ = ________ = ________ = ________

c) $3 \times 2 \times 4$ = ________ = ________ = ________

d) $6 \times 5 \times 2$ = ________ = ________ = ________

Think About It 2 What could you tell a classmate to help him or her remember why the two previous properties are called *associative* properties?

____

____

The Associative Properties will be used throughout this text in a variety of areas. One example of how the Associative Property of Addition is used is in adding two numbers with a sum greater that 10.

For example, to find the sum $6 + 7$, we can think of 7 as $4 + 3$, making the sum $6 + (4 + 3)$.

The Associative Property allows us to regroup this as $(6 + 4) + 3$.

This type of grouping allows for the quantity to equal 10: $10 + 3$.

And this sum is easy to evaluate: 13.

The Associative Property of Addition is used throughout Section 1.2. The Associative Property of Multiplication is used in a similar way in Section 1.3.

The Identities

The notion of *identity* is another important property of addition and multiplication. An **identity** is a number that, when applied, doesn't change the value of another number or quantity. In other words, the identity value maintains the value of the other number.

For addition, the identity is 0 (zero) because

$6 + 0 = 6$ $0 + 3 = 3$ 0 is called the **additive identity**.

Notice that adding 0 (zero) to a number doesn't change the value of the number.

For multiplication, the identity is 1 (one) because

$5 \times 1 = 5$ $1 \times 8 = 8$ 1 is called the **multiplicative identity**.

Notice that multiplying a number by 1 doesn't change the value of the number.

The Identities for Addition and Multiplication

If ***a*** is any number, then

Addition	Multiplication
$a + 0 = 0 + a = a$	$a \times 1 = 1 \times a = a$
Adding 0 to a number doesn't change the number's value.	Multiplying a number by 1 doesn't change the number's value.

Notice that each identity is expressed two ways using the Commutative Property.

You Try It 10 **Apply the idea of identity by filling in the blank.**

a) $4 + 0 =$ ______ **b)** $9 +$ ______ $= 9$ **c)** $0 +$ ______ $= 12$

d) $7 \times 1 =$ ______ **e)** $23 \times$ ______ $= 23$ **f)** $1 \times$ ______ $= 15$

Think About It 3 Is 1 an identity for division? Explain your answer.

____

____

Because $5 is the middle number between two $10 values, such as $280 and $290, it might seem that we can round up or down. For the sake of consistency, we always round *up* when the rounding digit is 5. The main reason for this is that there are five digits (0, 1, 2, 3, and 4) that cause a number to be rounded down; so, there should also be five digits that cause a number to be rounded up (5, 6, 7, 8, and 9).

Rounding Values

Sometimes when working with a large number, we find it easier to think of the number in *rounded* terms. For example, if the price of a lawn mower is $283, we could round that number to the nearest $10 value and think of it as $280 or we could round it to the nearest $100 value and think of it as $300. These rounded numbers, $280 and $300, are called *approximations* or *estimates*.

In rounding $283 to the nearest $10, we round *down* to $280.

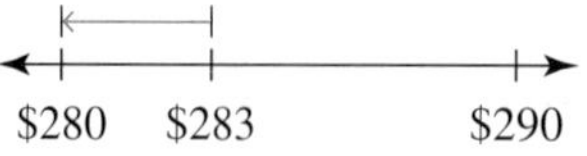

In rounding $283 to the nearest $100, we round *up* to $300.

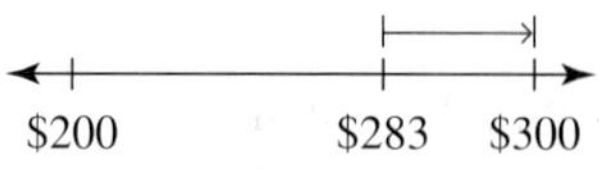

When rounding to the nearest ten, we use the digit in the ones place to determine whether we should round the number up or down. This ones place digit is called the **rounding digit.** When we are rounding to the nearest hundred, the rounding digit is in the tens place.

- When the rounding digit is 0, 1, 2, 3, or 4, we round *down*.
- When the rounding digit is 5, 6, 7, 8, or 9, we round *up*.

Following are the rules for rounding. We use two examples—one that rounds up and one that rounds down—to show the application of the various steps.

Rules for Rounding

1. Identify the place digit that is to be rounded.

a) *Round 28,714 to the nearest thousand* means that the 8 in the thousands place is the place digit.

28,714

b) *Round 28,714 to the nearest hundred* means that the 7 in the hundreds place is the place digit.

28,714

2. Identify the number to its *immediate right*, called the rounding digit.

a) Rounding 28,714 to the nearest thousand means that the rounding digit is the 7.

28,⑦14

b) Rounding 28,714 to the nearest hundred means that the rounding digit is the 1.

28,7①4

3. • If the rounding digit is 5 or higher, round *up* (add 1 to the place digit).
• If the rounding digit is 4 or lower, round *down* (add 0 to the place digit).

4. All digits to the right of the place digit are written as zeros. Rewrite the number showing the appropriate approximation.

a) In rounding 28,714 to the nearest thousand, the rounding digit, 7, is *more than* 5. So, we must add 1 to the place digit 8, making a 9 in the thousands place; and all of the digits following it are zeros.

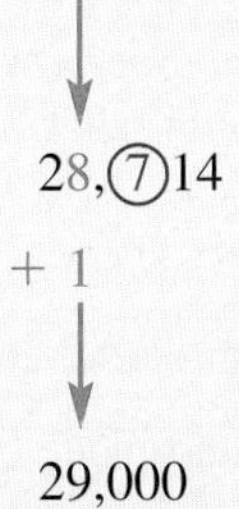

b) In rounding 28,714 to the nearest hundred, the rounding digit, 1, is *less than* 5. So we must add 0 to the place digit, meaning the hundreds place will remain a 7; and all of the digits f ol-lowing it are zeros.

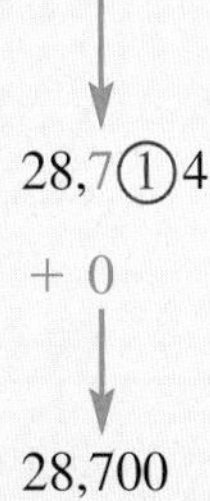

So, 28,714 can be approximated as 29,000 when rounded to the nearest *thousand* and as 28,700 when rounded to the nearest *hundred*.

Notice that the digits to the *left* of the place digit remain unchanged. This is true for most rounding. As you will see, when the place digit is 9, rounding up will change more than just the place digit.

Example 8

For each of the following numbers, round to the indicated place.

a) 726 *nearest ten* **b)** 726 *nearest hundred*

c) 4,082 *nearest thousand* **d)** 235,471 *nearest ten-thousand*

Procedure Identify the place digit and the rounding digit, decide whether the number will round up or down, and round the number.

Answer a) 730 **b)** 700 **c)** 4,000 **d)** 240,000

You Try It 11 **For each of the following numbers, round to the indicated place. Use Example 8 as a guide.**

a) 528 *nearest ten* **b)** 4,609 *nearest hundred* **c)** 4,609 *nearest thousand*

d) 75,406 *nearest ten-thousand* **e)** 1,925,046 *nearest hundred-thousand*

Rounding When the Place Digit Is 9

What happens when the place digit is 9? If the number needs to be rounded up, it rounds up to 10. This means that more than just the place digit is affected. The digit to the left also increases by 1. (If the number is to be rounded down, the 9 will stay a 9.)

Rounding 928 to the nearest hundred means rounding *down* to 900.

Rounding to the nearest hundred:

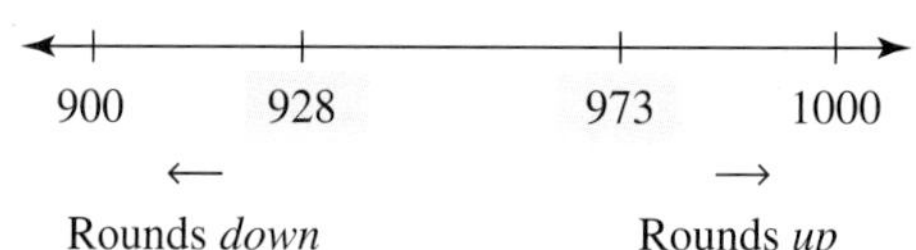

Rounding 973 to the nearest hundred means rounding *up* to "one more than 9" hundred. This means rounding it to ten hundred (1000), or 1,000.

Likewise, rounding 2,962 to the nearest hundred means rounding *up* to "one more than 29" hundred. This means rounding it to *thirty* hundred (3000), or 3,000.

Rounding to the nearest hundred:

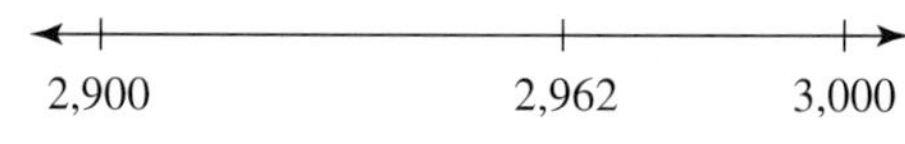

Twenty-nine hundred Thirty hundred

So, when the place digit is 9, rounding up affects more than just the place digit. It affects the digit to its left as well.

Example 9

For each of the following numbers, round to the indicated place.

a) 97 *nearest ten* **b)** 928 *nearest hundred* **c)** 4,951 *nearest hundred*

d) 29,950 *nearest hundred* **e)** 9,850 *nearest thousand*

Procedure For each of these numbers, the place digit is 9. Think about them carefully.

Answer **a)** 100 **b)** 900 **c)** 5,000 **d)** 30,000 **e)** 10,000

You Try It 12 **For each of the following numbers, round to the indicated place. Use Example 9 as a guide.**

a) 596 *nearest ten* b) 953 *nearest hundred* c) 3,947 *nearest hundred*

d) 689,810 *nearest thousand* e) 396,085 *nearest ten-thousand*

Applications

Sometimes it's helpful when working with large numbers to round them first. What follows are some statistical data involving large numbers. Your job is to round them and write a sentence that shows the approximation.

Example 10

The U.S. census estimates that in the year 2015, the population of the state of Illinois will be 12,808,000. Round this number to the nearest million and write a sentence indicating the approximation. *Source:* **www.census.gov**

Procedure First, round the number to the nearest million. It rounds to 13,000,000. Then write a sentence for the approximation. Use the word *approximately* in the sentence and write the number in words.

Answer In 2015, the population of Illinois will be approximately thirteen million.

You Try It 13 **The average distance from the Earth to the Moon is 239,062 miles. Round this number to the nearest *ten thousand* and write a sentence indicating the approximation.** *Source:* **www.nasa.gov**

Sentence ______________________________

You Try It 14 **The average distance from the Earth to the Sun is 149,597,871 kilometers. Round this number to the nearest *million* and write a sentence indicating the approximation.** *Source:* **www.nasa.gov**

Sentence ______________________________

Do you round numbers in any of your activities at work, at the store, for your hobbies, or during other daily routines? Write a few examples.

____

____

Answers: You Try It and Think About It

You Try It:

1.	**Period**	**Place**	**Value**
a)	ones	ten	50 (fifty)
b)	thousands	hundred	500,000 (five hundred thousand)
c)	millions	one	5,000,000 (five million)

2. a) 3,000 + 70 + 5 **b)** 500,000 + 3,000 + 100 + 40 + 2 **3. a)** eight hundred sixty-three **b)** sixty-two thousand, nine **c)** five million, three thousand, one hundred two **4. a)** 2,048 **b)** 75,004 **c)** 6,104,030 **5. a)** $9 + 3 = 12$ **b)** $12 - 4 = 8$ **c)** $5 \times 8 = 40$ **d)** $30 \div 5 = 6$ **6. a)** 8 **b)** 7 **c)** 20 **d)** 4 **7. a)** 5 **b)** 5 **c)** 10 **8. a)** $6 + 2$ **b)** Cannot be rewritten **c)** Cannot be rewritten **d)** 5×4 **9. a)** $(2 + 8) + 6 = 2 + (8 + 6) = 16$ **b)** $(4 + 6) + 9 = 4 + (6 + 9) = 19$ **c)** $(3 \times 2) \times 4 = 3 \times (2 \times 4) = 24$ **d)** $(6 \times 5) \times 2 = 6 \times (5 \times 2) = 60$ **10. a)** 4 **b)** 0 **c)** 12 **d)** 7 **e)** 1 **f)** 15 **11. a)** 530 **b)** 4,600 **c)** 5,000 **d)** 80,000 **e)** 1,900,000 **12. a)** 600 **b)** 1,000

c) 3,900 **d)** 690,000 **e)** 400,000 **13.** The average distance from the Earth to the Moon is approximately two hundred forty thousand miles. **14.** The average distance from the Earth to the Sun is approximately one hundred fifty million kilometers.

Think About It: **1.** Answers may vary. **2.** Answers may vary. **3.** No. An identity must be expressed two ways using the Commutative Property, and division is not commutative. In other words, although it is true that $a \div 1 = a$, $1 \div a$ does not equal a. **4.** Answers may vary.

Section 1.1 Exercises

FOR EXTRA HELP
Student Resources on DVD-ROM
Includes
➤Student's Solutions Manual
➤Video Lectures
➤Chapter Test Prep Video
MyMathLab
MathXL

Think Again

1. How can you use the words *commute* and *associate* to help you remember the commutative and associative properties? (*Refer to Think About It 1 and 2*)
Answers will vary.

2. Under what circumstances might it be necessary to write a number in word form?
Answers will vary. One possibility: When writing a check.

3. If an expression includes parentheses, why is it appropriate to evaluate within the parentheses first?
The parentheses indicate a single value, and that one value should be known before the whole expression is evaluated.

4. When rounding a number, why is it appropriate to round up when the rounding digit is 5?
Even though 5 is the halfway point between 0 and 10, there are already five digits that round down (0, 1, 2, 3, and 4); so, there should also be five digits that round up (5, 6, 7, 8, and 9).

Focus Exercises

Identify in which period and place you find the digit 7. Then write the value that the 7 represents.

	Period	Place	Value
5. 53,278	ones	ten	70 (seventy)
6. 352,716	ones	hundred	700 (seven hundred)
7. 6,703,214	thousands	hundred	700,000 (seven hundred thousand)
8. 17,480,300	millions	one	7,000,000 (seven million)

Write each number in expanded form.

9. 486
400 + 80 + 6

10. 4,065
4,000 + 60 + 5

11. 203,058
200,000 + 3,000 + 50 + 8

12. 1,500,043
1,000,000 + 500,000 + 40 + 3

Write each number in words.

13. 498
four hundred ninety-eight

14. 6,204
six thousand, two hundred four

15. 507,093
five hundred seven thousand, ninety-three

16. 1,013,000
one million, thirteen thousand

Write the whole number.

17. Five hundred eighteen
518

18. Two thousand, three hundred six
2,306

19. Two hundred eighty thousand, thirty-four
280,034

20. Nine hundred five thousand, eight
905,008

21. One million, four hundred twenty-six
1,000,426

22. Three million, two thousand
3,002,000

Write both the expression and the result of each of the following.

	As an expression		As a result
23. the quotient of 8 and 4	$8 \div 4$	=	2
24. the difference between 10 and 3	$10 - 3$	=	7
25. the sum of 2 and 5	$2 + 5$	=	7
26. the product of 2 and 10	2×10	=	20

Determine whether the expression shown is a sum, difference, product, or quotient.

27. $41 - 13$
difference

28. $57 \div 3$
quotient

29. 19×42
product

30. $126 + 379$
sum

Evaluate.

31. $(9 - 4) + 3$
8

32. $9 - (4 + 3)$
2

33. $4 + (5 - 3)$
6

34. $(4 + 5) - 3$
6

35. $(2 + 6) \div 2$
4

36. $2 + (6 \div 2)$
5

37. $9 - (6 \div 3)$
7

38. $(9 - 6) \div 3$
1

39. $7 - (3 \times 2)$
1

40. $(7 - 3) \times 2$
8

41. $8 \times (4 \times 2)$
64

42. $(8 \times 4) \times 2$
64

Which property does each expression represent?

43. $5 + (8 + 2) = (5 + 8) + 2$
Associative Property of Addition

44. $12 \times 5 = 5 \times 12$
Commutative Property of Multiplication

45. $0 + 11 = 11$
Identity of Addition

46. $8 \times (5 \times 4) = (8 \times 5) \times 4$
Associative Property of Multiplication

47. $25 + 19 = 19 + 25$
Commutative Property of Addition

48. $13 \times 1 = 13$
Identity of Multiplication

Round each of these numbers to the nearest ten.

49. 67
70

50. 92
90

51. 683
680

52. 345
350

53. 1,941
1,940

54. 2,318
2,320

55. 199
200

56. 795
800

Round each of these numbers to the nearest hundred.

57. 638
600

58. 2,049
2,000

59. 1,708
1,700

60. 2,350
2,400

61. 14,386
14,400

62. 22,708
22,700

63. 6,951
7,000

64. 9,962
10,000

Round each of these numbers to the nearest thousand.

65. 2,906
3,000

66. 8,061
8,000

67. 36,407
36,000

68. 49,513
50,000

69. 159,370
159,000

70. 509,423
509,000

71. 289,516
290,000

72. 699,850
700,000

Round each of these numbers to the nearest ten thousand.

73. 25,408
30,000

74. 47,583
50,000

75. 99,361
100,000

76. 192,805
190,000

77. 150,764
150,000

78. 324,612
320,000

79. 898,053
900,000

80. 3,996,416
4,000,000

Round each of these numbers to the nearest hundred thousand.

81. 624,058
600,000

82. 709,655
700,000

83. 952,407
1,000,000

84. 990,521
1,000,000

85. 784,672
800,000

86. 850,043
900,000

87. 4,966,381
5,000,000

88. 9,980,376
10,000,000

Round each of these numbers to the nearest million.

89. 3,580,416
4,000,000

90. 5,929,300
6,000,000

91. 4,277,095
4,000,000

92. 9,308,416
9,000,000

93. 9,301,448
9,000,000

94. 46,283,000
46,000,000

95. 29,508,416
30,000,000

96. 199,540,926
200,000,000

Rewrite the underlined words with the requested approximation.

97. In 2005, the Rialto Unified School District's budget revenues were **$192,863,877**. Round this number to the nearest million. ***Source:*** **www.rialto.k12.ca.us**
The Rialto Unified School District's budget revenues were approximately $193,000,000.

98. In 2005, the population of Florida was **17,789,864**. Round this number to the nearest hundred thousand.
Source: **http://factfinder.census.gov**
The population of Florida was approximately 17,800,000.

99. In 2002, gas consumption statistics for all international flights showed that the total number of gallons of gas consumed was **4,990,797,640**. Round this number to the nearest hundred million.
Source: **www.bts.gov**
The total number of gallons of gas consumed was approximately 5,000,000,000.

100. It is estimated that the world's rain forests are being destroyed at a rate of **77,893,900** acres per year. Round this yearly acreage to the nearest million.
Source: **www.rain-tree.com**
The world's rain forests are being destroyed at a rate of approximately 78,000,000 acres per year.

101. In the 2001–2002 school year, **6,248,610** students were enrolled in grades K–12 in California. Round this number to the nearest hundred thousand.
Source: **http://nces.ed.gov**
Approximately 6,200,000 students were enrolled in grades K–12 in California.

102. Statistics compiled by the Centers for Disease Control through December 2004 indicate that the total number of reported AIDS cases was **944,305**. Round this number to the nearest ten thousand.
Source: **www.cdc.gov**
The total number of reported AIDS cases was approximately 940,000.

Think Outside the Box

103. If each number between 1 and 400 is rounded to the nearest ten, how many different (unique) rounded numbers will there be?
There will be 41 different rounded numbers.

104. If each number between 1 and 400 is rounded to the nearest ten, how many of the different (unique) rounded numbers will contain at least one digit of 3?
There will be 13 different rounded numbers that contain at least one digit of 3.

SECTION 1.2 Adding and Subtracting Whole Numbers

OBJECTIVES

In this section, you will learn to
- Add whole numbers.
- Find the perimeter of a geometric figure.
- Subtract whole numbers.
- Read bar graphs.

You Need to Know

To successfully complete this section, you need to understand
- ☐ Place value (1.1)
- ☐ The Associative Property (1.1)
- ☐ The Commutative Property (1.1)
- ☐ Adding by grouping to 10 (1.1)

Introduction

The purpose of this section is to reintroduce you to adding and subtracting whole numbers. Another purpose of this section is to explain some of the steps, such as *carrying* in addition and *regrouping* in subtraction.

When adding two whole numbers, you may need to rely on the technique of using the Associative Property to add numbers whose sum is more than 10, such as 7 + 8.

$$\begin{aligned} & 7 + 8 \\ &= 7 + (3 + 5) \\ &= (7 + 3) + 5 \\ &= 10 + 5 \\ &= 15 \end{aligned}$$

Adding Whole Numbers

When adding numbers with two or more digits, you must align them, one above the other, so that the ones place is aligned, the tens place is aligned, and so on.

For example, 23 + 64 should be written as

$$\begin{array}{r} 23 \\ +\ 64 \\ \hline \end{array}$$

and 1,325 + 2,503 should be written as

$$\begin{array}{r} 1{,}325 \\ +\,2{,}503 \\ \hline \end{array}$$

Once the numbers in a sum are aligned, you may add directly, starting with the ones column (shown at the right).

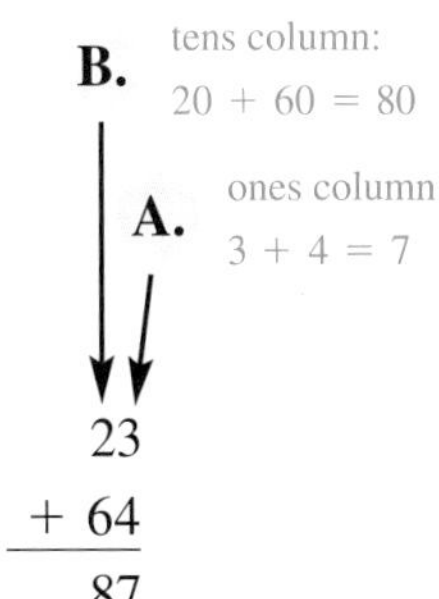

We could also expand each number, allowing us to see the place values.

$$\begin{array}{rcrcr} 23 & \longrightarrow & 20 + 3 & \text{or} & 2 \text{ tens} + 3 \text{ ones} \\ +\ 64 & \longrightarrow & +\ 60 + 4 & & +\ 6 \text{ tens} + 4 \text{ ones} \\ \hline & & 80 + 7 = 87 & & 8 \text{ tens} + 7 \text{ ones} = 87 \end{array}$$

If one number has more digits than the other, it is appropriate to place 0's in front of the lesser number so that the two numbers have the same number of digits. For example, before adding, 1,325 + 64 can be written as

$$\begin{array}{r} 1{,}325 \\ +\,0{,}064 \\ \hline 1{,}389 \end{array}$$

You Try It 1 **Add each pair of numbers. Align the numbers vertically so that one number is directly above the other.**

a) 157 + 230 **b)** 2,754 + 1,121 **c)** 133 + 4,426 **d)** 135,604 + 2,261

Adding by Carrying Numbers Over

You probably noticed that none of the individual column sums in You Try It 1 was higher than 9. The problems were carefully chosen so that you could add directly without having to worry about *carrying over*.

Let's consider what happens when two of the digits add to 10 or more, such as 6 + 7 = 13. Written vertically, it looks like this:

$$\begin{array}{r} 6 \\ +\ 7 \\ \hline 13 \end{array}$$ **13 is 1 ten and 3 ones**

What's especially important to recognize is that although both 6 and 7 are in the ones place, we get a sum that extends itself into the tens place.

If we add, for example, 26 + 57, the extra ten that we get from 6 + 7 must be figured into the sum. Here's the long way to look at it:

26 → 2 tens and 6 ones
+ 57 → + 5 tens and 7 ones
7 tens and 13 ones **13 ones can be rewritten as 1 ten and 3 ones.**
= 7 tens and 1 ten and 3 ones = 70 + 13
= 7 tens and 1 ten and 3 ones = 70 + 10 + 3
= 8 tens and 3 ones = 80 + 3 = **83**

Shown this way, the addition process is a bit of work. A shorter way to find this sum is to show that extra 10 (from the 13) in the tens column.

In other words, the 13 we get from 6 and 7 must be thought of as 1 ten and 3 ones. Then the 3 can stay in the ones column and the 1 ten can *carry over* into the tens column. Example 1 demonstrates how it is typically written.

Example 1

Add 26 and 57. Follow the steps outlined here.

A. Start with the ones by column, 6 + 7 = 13.

B. Carry the 1 ten into the tens column.

C. Complete the addition by adding the tens column.

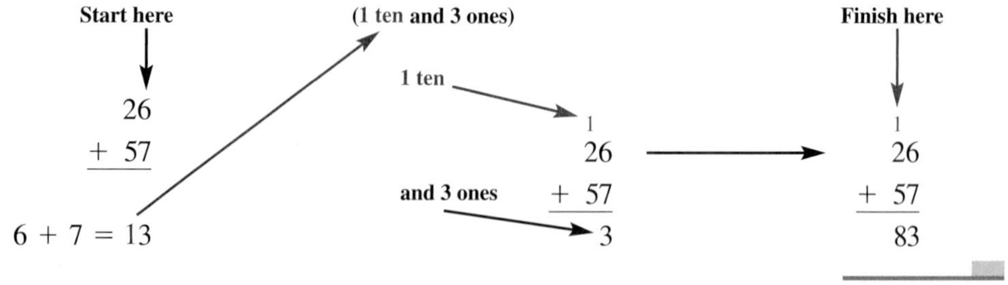

Start here

$$\begin{array}{r} 26 \\ +\ 57 \\ \hline \end{array}$$

6 + 7 = 13

(1 ten and 3 ones)

1 ten

and 3 ones

$$\begin{array}{r} {}^{1} \\ 26 \\ +\ 57 \\ \hline 3 \end{array}$$

Finish here

$$\begin{array}{r} {}^{1} \\ 26 \\ +\ 57 \\ \hline 83 \end{array}$$

The idea of carrying over can extend to other place values as well.

Example 2

Add 358 and 294.

Procedure Follow the steps outlined here.

A. Starting with the ones column,

B. Carry the 1 ten into the tens column.

C. Adding in the tens column, $1 + 5 + 9 = 15$ tens (1 hundred and 5 tens).

D. Complete the addition by adding in the hundreds column.

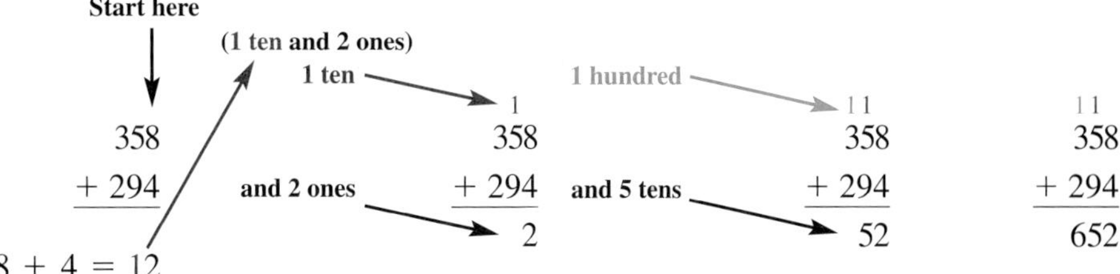

Answer $358 + 294 = 652$

You Try It 2 **Add. Use Examples 1 and 2 as guides.**

a) $17 + 58$ **b)** $525 + 387$ **c)** $375 + 6{,}925$ **d)** $3{,}708 + 6{,}533$

Adding More Than Two Whole Numbers

There are occasions when we must add a list of numbers, whether they be the prices of items in a grocery cart or scores on tests or the weights of boxes to be shipped to customers.

Adding more than two numbers is similar to what we've seen already.

1. We align the numbers so that the ones place is above the ones place, the tens place is above the tens place, and so on.
2. We carry the tens (or hundreds) value to the next column for any sum over 9.

When four numbers are added, the sum in any one column could be more than 20 or 30, as shown in Example 3. In that case, we must carry the 2 or 3 to the next column.

Example 3

Add 186, 395, 478, and 294. Follow the steps outlined here.

Procedure Add one column at a time starting with the ones column.

Answer

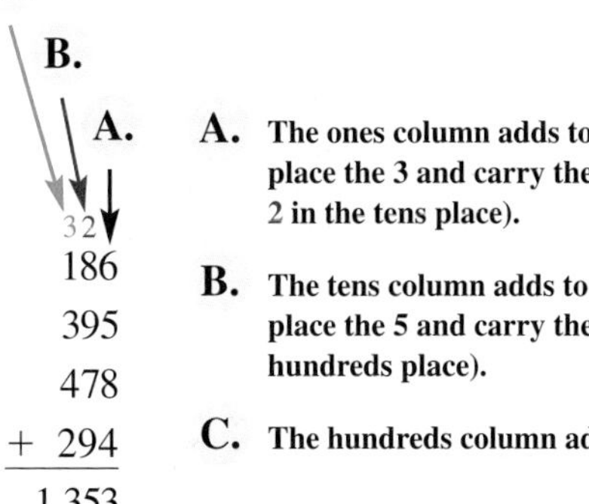

A. The ones column adds to 23; so, place the 3 and carry the 20 (as a 2 in the tens place).

B. The tens column adds to 35; so, place the 5 and carry the 3 (in the hundreds place).

C. The hundreds column adds to 13.

You Try It 3 **Add. Use Example 3 as a guide. (You may want to place zeros in front of numbers that have fewer digits.)**

a)

$$\begin{array}{r} 7 \\ 36 \\ +\ 98 \\ \hline \end{array}$$

b)

$$\begin{array}{r} 8 \\ 346 \\ 95 \\ +\ 673 \\ \hline \end{array}$$

c)

$$\begin{array}{r} 1{,}634 \\ 2{,}976 \\ 9{,}597 \\ +\ 8{,}967 \\ \hline \end{array}$$

d)

$$\begin{array}{r} 36{,}525 \\ 7{,}489 \\ 20{,}946 \\ 877 \\ +\ 48{,}103 \\ \hline \end{array}$$

Applications Involving Addition

In a story problem (or word problem), the last sentence is usually in the form of a question asking you to find an amount of something. In the question, you should look for a word or phrase that indicates which operation is to be used.

In an addition problem, you will likely see words such as *total*, *combined*, *sum*, and *in all*. There are many situations to which addition can apply.

Example 4

The South Orange County Community College District has two colleges, Irvine Valley College and Saddleback College. Last summer Irvine Valley had an enrollment of 5,087, Saddleback had an enrollment of 9,953, and no one attended both colleges. How many total students were enrolled in the entire district last summer?

Procedure The key word in the last sentence is *total*. This indicates that we should add the student enrollments at each college.

Answer

Irvine Valley	5,087
+ Saddleback	+ 9,953
Total	15,040

Sentence Last summer a total of 15,040 students were enrolled in the district.

You Try It 4 **Juan's monthly salary is $2,038, and his wife Angelica's monthly salary is $1,979. What is their combined monthly salary?**

Sentence ______________________

You Try It 5 **Allison often travels out of state for business. On her last trip, she flew from Seattle to Houston (1,848 miles), from Houston to Detroit (1,083 miles), and from Detroit to Seattle (1,927 miles). How many total miles did she travel on that trip?**

Sentence ______________________

Addition in Geometry: Perimeter

The **perimeter** of a geometric figure—such as a triangle or rectangle—is the total measure around the figure. The perimeter is found by adding the lengths of all of the sides.

For the triangle at the right,

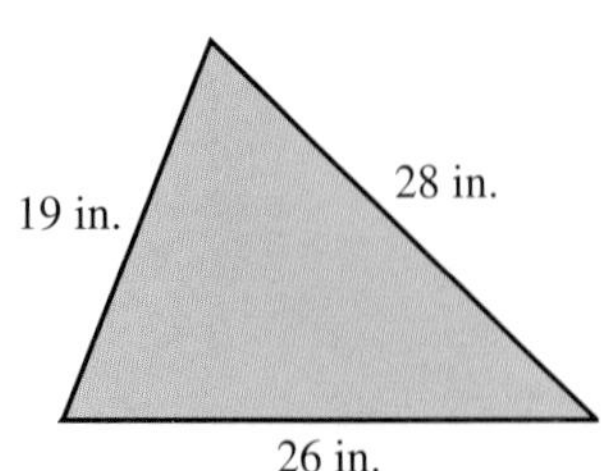

Perimeter = 19 inches + 28 inches + 26 inches

Perimeter = 73 inches

$$\begin{array}{r} 19 \\ 28 \\ +\ 26 \\ \hline 73 \end{array}$$

In a rectangle, the top and bottom sides have the same length and the left and right sides have the same length. So, if we know, for example, the top length and the right side length, we can fill in the other two side measures and then find the perimeter of the rectangle.

In the rectangle at the right, we can find the perimeter by adding together all four side lengths.

10 in. / 7 in. → 10 in., 7 in., 7 in., 10 in.

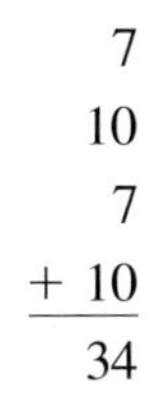

The perimeter is 34 inches.

You Try It 6 **Find the perimeter of each shape.**

a)

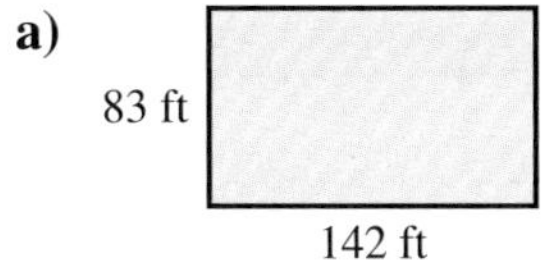

b)

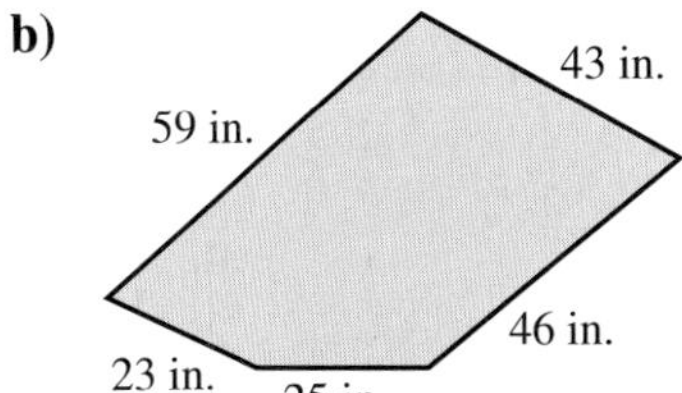

Subtracting Whole Numbers

Addition and subtraction are *inverse* operations. This means, among other things, that we can check our subtraction answers using addition and vice versa. Let's see how this is so by exploring a simple example, $8 - 5 = 3$.

Because of the inverse nature of addition and subtraction, understanding addition is key to understanding subtraction. Following is an example of how we can use addition to check whether our subtraction result (the difference) is correct.

$$\begin{array}{r} 8 \\ -\ 5 \\ \hline 3 \end{array} \quad \text{because} \quad \begin{array}{r} 3 \\ +\ 5 \\ \hline 8 \end{array}$$

You Try It 7 **Subtract. Check each answer (on paper or mentally) by addition.**

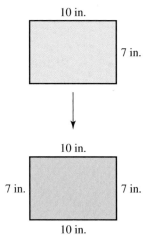

a) $\begin{array}{r} 7 \\ -\ 4 \\ \hline \end{array}$ **b)** $\begin{array}{r} 8 \\ -\ 6 \\ \hline \end{array}$ **c)** $\begin{array}{r} 8 \\ -\ 1 \\ \hline \end{array}$ **d)** $\begin{array}{r} 4 \\ -\ 4 \\ \hline \end{array}$

Subtraction of larger numbers follows the same process as single-digit subtraction. As in addition, we must align the place values.

Example 5

Subtract 57 − 26.

Procedure Align the numbers, the first above the second, and follow the steps outlined here.

A. Start with the ones column. 7 − 6 = 1

B. Then subtract in the tens column: 5 − 2 = 3.

C. Check by addition.

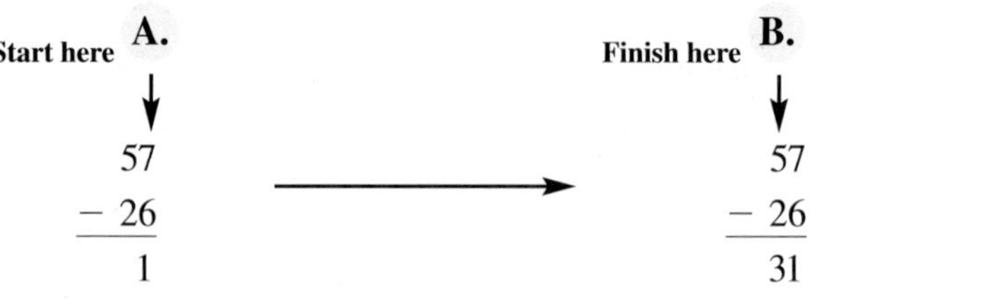

Answer 57 − 26 = 31

Example 6

Subtract.

a) 683 − 31 **b)** 759 − 725

Procedure **a)** We can rewrite 31 as 031 so that every place is represented by a digit.

b) Notice in the answer that the 0 in the hundreds place is unnecessary.

$$\begin{array}{r} 683 \\ -\ 31 \\ \hline \end{array} \rightarrow \begin{array}{r} 683 \\ -\ 031 \\ \hline 652 \end{array} \qquad \text{Check } \begin{array}{r} 652 \\ +\ 31 \\ \hline \end{array}$$

$$\begin{array}{r} 759 \\ -\ 725 \\ \hline 034 \end{array} \rightarrow \begin{array}{r} 759 \\ -\ 725 \\ \hline 34 \end{array} \qquad \text{Check } \begin{array}{r} 34 \\ +\ 725 \\ \hline \end{array}$$

Answer →

You supply the answer to the check.

You Try It 8 **Subtract. Check each answer by addition. Use Examples 5 and 6 as guides.**

a) 78 − 41

b) 694 − 321

c) 2,156 − 41

d) 36,815 − 36,514

Subtracting by Regrouping

When we subtract, as when we add, we do so one place at a time. In subtraction, we always start in the ones column, then go the tens column, and so on.

If we come to a place in which we can't subtract—because the first number is less than the second number—we must use regrouping to adjust the first number, as shown in the next problem.

Consider $\begin{array}{r} 45 \\ -\ 16 \\ \hline \end{array}$

We cannot subtract directly in the ones column. So, we need to rewrite 45 in a way that creates a larger ones column value in the first number.

To do this, think of 45 as 4 tens and five ones.

45 = 4 tens and 5 ones

We can use the Associative Property to separate 1 ten from the 40 and give it to the 5.

= 3 tens + 1 ten + 5 ones
= 3 tens + 1 ten + 5 ones
= 3 tens + 15 ones

We can then write 45 as 3^15, as shown here.

B. 30 − 10 = 20

A. 15 − 6 = 9

			Check by addition.
45	⟶	3^15	29
− 16		− 1 6	+ 16
		2 9	45 ✓

This checks out.

Regrouping in subtraction extends to other places as well. For example, in 926 − 372, we can subtract directly in the ones place but not in the tens place. So, we must take 1 hundred from the 900 and give it to the 2 tens.

926
− 372
?4

C. 800 − 300 = 500

B. 120 − 70 = 50

A. 6 − 2 = 4

			Check by addition.
926	⟶	8^126	554
− 372		− 372	+ 372
		554	926 ✓

This checks out.

Example 7

Subtract.

a) 81 − 27 **b)** 458 − 376

Procedure In part a, regroup by taking 1 ten from the tens place in 81 and giving it to the ones place. In part b, regroup by taking 1 hundred from the hundreds place in 458 and giving it to the tens place.

Answer

a)

81	⟶	7^11	**You show the check:**
− 27		− 2 7	
		5 4	

b)

458	⟶	3^158	**You show the check:**
− 376		− 3 76	
		82	

You Try It 9 **Subtract. Check each answer by addition. Use Example 7 as a guide.**

a) 96 − 68 **b)** 70 − 36 **c)** 519 − 327 **d)** 805 − 745

Sometimes we must regroup twice (or more) in the same subtraction. For example, in 9,845 − 1,758 we cannot subtract directly in either the ones column or the tens column.

$$\begin{array}{r} 9{,}845 \\ -\ 1{,}758 \\ \hline ?? \end{array}$$

We start by making sure we have enough in the ones place by taking 1 ten from the 4 tens in the first number. We then determine whether we can subtract directly in the tens column. If we cannot, we must regroup again.

Take 1 ten from the 4 and give it to the 5. **Take 1 hundred from the 8 and give it to the 3.**

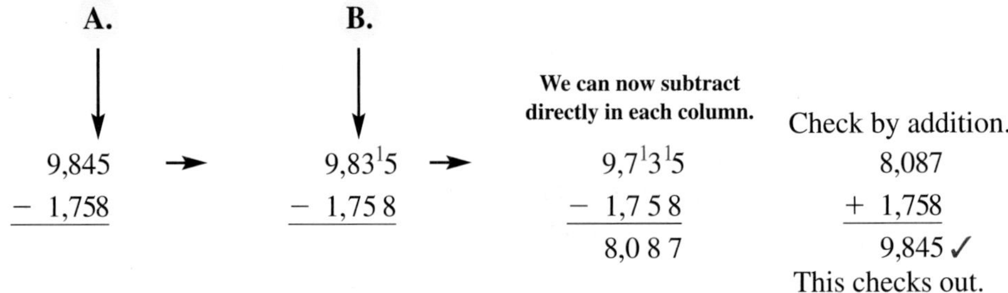

A.
B.
9,845
− 1,758
9,83¹5
− 1,75 8
We can now subtract directly in each column.
9,7¹3¹5
− 1,7 5 8
8,0 8 7
Check by addition.
8,087
+ 1,758
9,845 ✓
This checks out.

Typically, it's common to see fewer steps than what was just shown. The value that is being taken from may be crossed out and the new value written above it, as seen here in 45 − 18.

$$\begin{array}{r} 45 \\ -\ 18 \\ \hline \end{array}$$

A. **When we take 1 ten from the 40 (represented by the 4 tens), we are left with 3 in the tens place.**

B. **Adding the 10 to the 5, we get 15 in the ones place.**

$$\begin{array}{r} {}^{3\ 15} \\ \not{4}\ \not{5} \\ -\ 1\ 8 \\ \hline 2\ 7 \end{array}$$

Sometimes in the regrouping process, we come across a 0 in the tens place and there are no tens to give to the ones place. When that happens, we must first take 1 hundred from the hundreds place and give it to the tens place digit. Then we have enough in the tens place and can take 1 ten from it to give to the ones place.

We can't take 1 ten from the 0; so, we take 1 hundred from the 6. **Now there are 10 tens, and we can take 1 ten from the 10 and give it to the 2.**

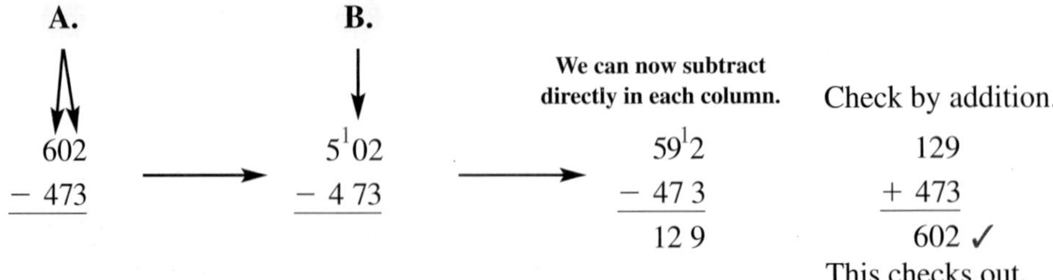

A.
B.
602
− 473
5¹02
− 4 73
We can now subtract directly in each column.
59¹2
− 47 3
12 9
Check by addition.
129
+ 473
602 ✓
This checks out.

This next example shows how it might look when the values are crossed out and written above.

Example 8

Subtract 306 − 149. Check using addition.

Procedure For each problem, align the numbers vertically and regroup as necessary. Check by addition.

Answer

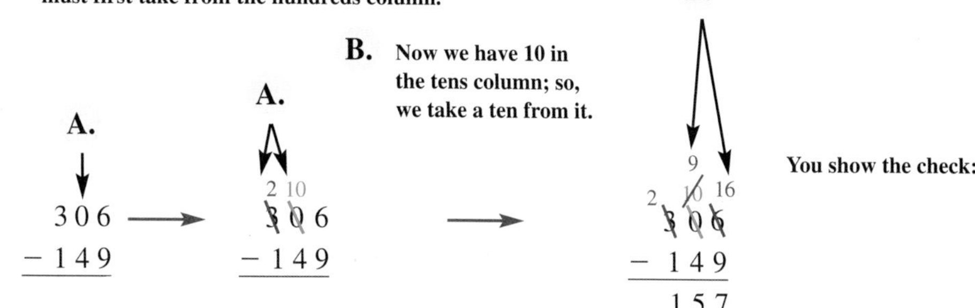
A. Here we need to take from the tens column; but only 0 is there, meaning no tens at all. So, we must first take from the hundreds column.
B. Now we have 10 in the tens column; so, we take a ten from it.
A.
A.
B.
306
− 149
2 10
306
− 149
9
2 10 16
306
− 149
157
You show the check:

Caution When regrouping, it is very important to be neat in your work and to keep the columns properly aligned.

You Try It 10 **Subtract. Some of these require regrouping more than once. Be careful. Use Example 8 as a guide.**

a) 906 − 348 **b)** 1,302 − 835 **c)** 2,006 − 1,619

Applications Involving Subtraction

Here are some applications involving subtraction. Generally, subtraction is used when we are asked to compare two numbers. *How much more than* . . . or *What is the difference between* . . . or *What was the change in* . . . usually indicate subtraction.

For each application, check the answer with addition and write a sentence answering the question.

You Try It 11 **Marcus traveled 1,385 miles on business last month while Ruben traveled 859 miles. How many more miles did Marcus travel last month than Ruben?**

Sentence ______________________________

You Try It 12 **Last year, Arash's annual salary was $25,457. This year he got a raise and is now making $27,350. What was the amount of Arash's raise?**

Sentence ______________________________

You Try It 13 **Two years ago, Sam's Neighborhood Market had annual receipts totaling $1,887,518. Last year sales were better and the market had annual receipts totaling $1,960,710. How much higher were the market receipts last year than the year before?**

Sentence ______________________________

Reading a Bar Graph

Many addition and subtraction applications can be found on a **bar graph,** a visual way to quickly compare numbers of things in a particular category.

For example, if we ask the 23 children in Mrs. Dennison's first-grade class what their favorite flavor of ice cream is (the category), their answers may be represented by the bar graph shown here.

From this bar graph, we can quickly see that 9 of the children prefer chocolate and 5 of the children prefer vanilla.

We can see, by simple subtraction, that 4 more children prefer chocolate than vanilla: $9 - 5 = 4$. We can also see the total number of children that were asked: $5 + 9 + 6 + 3 = 23$.

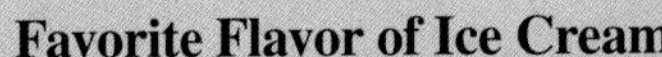

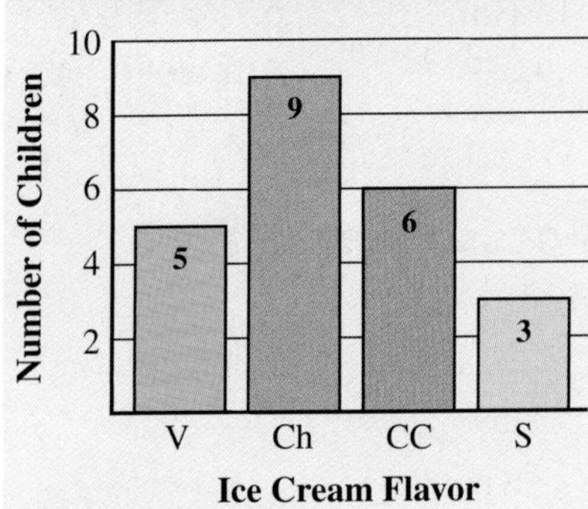

V = Vanilla CC = Chocolate Chip
Ch = Chocolate S = Strawberry

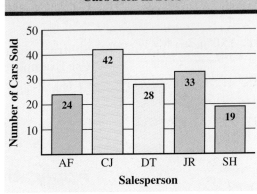

You Try It 14 **The bar graph at the right represents the number of cars sold in 2008 by five different salespersons at Cramer Honda.**

Answer the following questions based on this bar graph. Write each answer as a sentence.

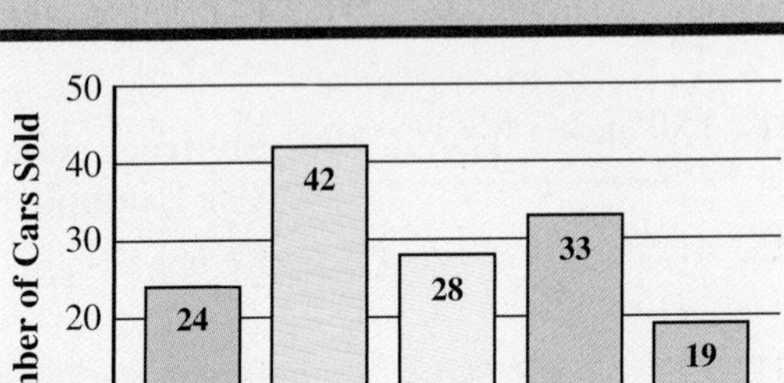

AF = Alan Franks CJ = Corine Jackson
DT = Danny Torres JR = Janice Rennock
SH = Steve Houshmand

a) How many more cars did Corine sell than Steve?

____

b) Together, how many cars did Alan and Danny sell?

____

c) How many cars did all five salespersons sell in 2006?

____

Answers: You Try It and Think About It

You Try It:

1. a) 387 **b)** 3,875 **c)** 4,559 **d)** 137,865 **2. a)** 75 **b)** 912 **c)** 7,300 **d)** 10,241 **3. a)** 141 **b)** 1,122 **c)** 23,174 **d)** 113,940 **4.** Their combined monthly salary is \$4,017. **5.** Allison traveled a total of 4,858 miles. **6. a)** Perimeter = 450 feet **b)** Perimeter = 196 inches **7. a)** 3 **b)** 2 **c)** 7 **d)** 0 **8. a)** 37 **b)** 373 **c)** 2,115 **d)** 301 **9. a)** 28 **b)** 34 **c)** 192 **d)** 60 **10. a)** 558 **b)** 467 **c)** 387 **11.** Marcus traveled 526 miles more than Ruben. **12.** Arash's raise was \$1,893. **13.** Last year's receipts were \$73,192 higher than receipts from the year before. **14. a)** Corine sold 23 more cars than Steve. **b)** Together, Alan and Danny sold 52 cars. **c)** All five salespersons sold 146 cars in 2006.

Think About It: No Think About Its in this section.

Section 1.2 Exercises

FOR EXTRA HELP

Student Resources on DVD-ROM — Includes ➤ Student's Solutions Manual ➤ Video Lectures ➤ Chapter Test Prep Video | MyMathLab | MathXL

Think Again

1. How can we find the perimeter of a rectangle if we add only the length to the width?
   If we add the length and width together, we can double that value to find the perimeter.

2. Why can we use addition to check the results of a subtraction problem?
   Because addition and subtraction are inverse operations of each other.

Focus Exercises

Add.

3. 17 + 42
59

4. 41 + 56
97

5. 425 + 132
557

6. 416 + 23
439

7. 52 + 835
887

8. 5,461 + 36
5,497

9. 72 + 6,803
6,875

10. 1,581 + 3,709
5,290

11. 4,706 + 58,219
62,925

12. 24,360 + 74,654
99,014

13. 15,086 + 742,511
757,597

14. 230,495 + 604,201
834,696

15. 41,580 + 63,219
104,799

16. 34,962 + 81,050
116,012

17. 875,213 + 124,787
1,000,000

18. 263,819 + 612,086
875,905

19.
$$\begin{array}{r} 28 \\ 35 \\ +\ 41 \\ \hline 104 \end{array}$$

20.
$$\begin{array}{r} 36 \\ 94 \\ 28 \\ +\ 50 \\ \hline 208 \end{array}$$

21.
$$\begin{array}{r} 129 \\ 214 \\ 78 \\ +\ 396 \\ \hline 817 \end{array}$$

22.
$$\begin{array}{r} 512 \\ 418 \\ 91 \\ +\ 229 \\ \hline 1{,}250 \end{array}$$

23.
$$\begin{array}{r} 4{,}426 \\ 9{,}508 \\ +\ 3{,}077 \\ \hline 17{,}011 \end{array}$$

24.
$$\begin{array}{r} 11{,}581 \\ 6{,}215 \\ +\ 23{,}024 \\ \hline 40{,}820 \end{array}$$

25.
$$\begin{array}{r} 52{,}681 \\ 17{,}938 \\ 33{,}075 \\ +\ 40{,}206 \\ \hline 143{,}900 \end{array}$$

26.
$$\begin{array}{r} 146{,}819 \\ 253{,}022 \\ 346{,}795 \\ +\ 361{,}364 \\ \hline 1{,}108{,}000 \end{array}$$

Subtract. Check each answer (on paper or mentally) by addition.

27. 285 − 33
252

28. 389 − 269
120

29. 147 − 45
102

30. 964 − 920
44

31. 3,452 − 140
3,312

32. 48,839 − 614
48,225

33. 42 − 17
25

34. 63 − 25
38

35. 53 − 48
5

36. 80 − 29
51

37. 100 − 46
54

38. 100 − 73
27

39. 156 − 99
57

40. 274 − 93
181

41. 621 − 528
93

42. 512 − 133
379

43. 230 − 157
73

44. 406 − 392
14

45. 316 − 23
293

46. 800 − 352
448

47. 5,461 − 36
5,425

48. 2,754 − 1,121
1,633

49. 40,216 − 15,381
24,835

50. 604,201 − 230,495
373,706

51. 6,803 − 72
6,731

52. 2,374 − 515
1,859

53. 4,426 − 133
4,293

54. 25,053 − 624
24,429

55. 135,604 − 2,261
133,343

56. 742,511 − 5,086
737,425

57. 300,000 − 106,578
193,422

58. 1,000,000 − 361,047
638,953

Work each application and answer with a complete sentence.

59. Throughout 2004, Dionne gave monetary donations to two charities: $1,258 to the Salvation Army and $875 to the United Way. What was Dionne's total contribution to these two charities?
Dionne contributed $2,133 to the two charities.

60. Ron's electric bill in July was $212 but only $87 in June. How much more was Ron's electric bill in July than June?
Ron's bill was $125 more in July than June.

61. Debbie has an adjustable rate mortgage that can change from year to year. In 2007, her monthly mortgage payment was $1,426. In 2008, her monthly payment was $1,347. How much less was Debbie's monthly payment in 2008 than in 2007?
Debbie's monthly payment was $79 less in 2008 than in 2007.

62. Throughout their NBA careers, Kareem Abdul-Jabbar scored a total of 38,387 points and Michael Jordan scored a total of 29,277 points. How many more points did Kareem Abdul-Jabbar score than Michael Jordan? ***Source:*** **http://sportsillustrated.cnn.com**
Kareem Abdul-Jabbar scored 9,110 points more than Michael Jordan.

63. The Detroit office of Globe Realty employs four agents. Mike, the owner, has 21 years of experience in real estate; Ann has 23 years of experience; Uta has 19 years; and Francisco has 17 years of experience. How many combined years of experience in real estate do the four agents have?
The four agents have 80 combined years of experience.

64. Eugenia ordered new furniture for her Portland, Oregon, office. The order included an executive desk, $327; a computer desk, $133; a swivel chair, $148; and a client chair, $92. Along with shipping charges of $116, what was the total amount of the order? (By the way, there is no sales tax in Oregon.)
The total amount of the order was $816.

65. Enerio is shopping for a new car for his young family. The Hyundai Elantra has caught his attention. The car has a sticker price of $14,695. To add a sunroof and an antilock braking system is an extra $1,148. Add to that sales tax of $1,228 and dealer charges of $862. What is the total cost of the car?
The total cost of the car is $17,933.

66. South Orange County Community College District has two colleges, Irvine Valley College and Saddleback College. Last summer Irvine Valley had an enrollment of 5,087, Saddleback had an enrollment of 9,953, and no one attended both colleges. How many more students attended Saddleback than Irvine Valley?
4,866 more students attended Saddleback than Irvine Valley.

67. Baseball attendance figures for a weekend series between the Houston Astros and the Florida Marlins were 35,403 on Friday, 41,292 on Saturday, and 28,515 on Sunday.

a) What was the total attendance for the three-game series?
The total attendance was 105,210 for the three-game series.

b) How many more fans were in attendance on Saturday than on Sunday?
There were 12,777 more fans in attendance on Saturday than on Sunday.

68. The elevation of Mount Everest, in Nepal, is 29,035 feet (8,850 meters). The elevation of Mount Kilimanjaro, in Tanzania, is 19,340 feet (5,895 meters). How much higher is Mount Everest than Mount Kilimanjaro

a) in meters?
Mount Everest is 2,955 meters higher than Mount Kilimanjaro.

b) in feet?
Mount Everest is 9,695 feet higher than Mount Kilimanjaro.

69. Find the perimeter of each figure.

a)

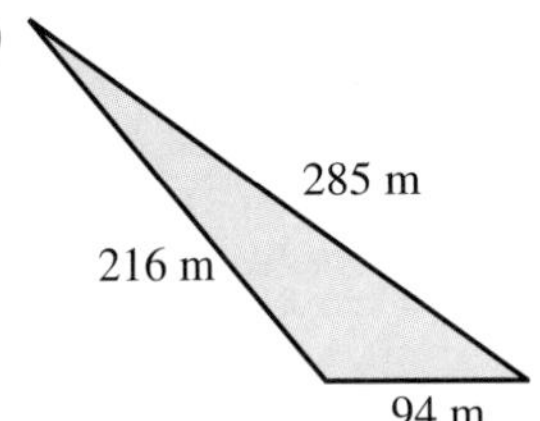

The perimeter is 595 meters.

b)

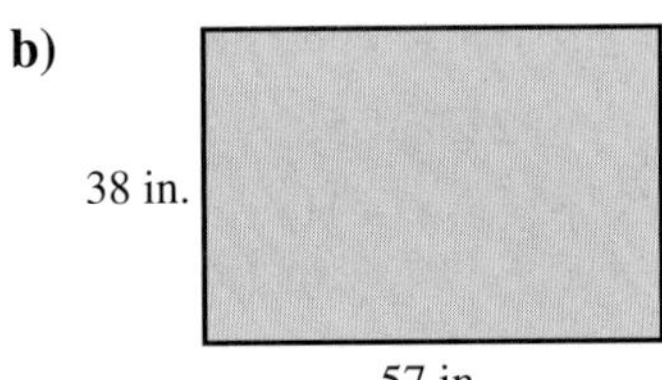

The perimeter is 190 inches.

c)

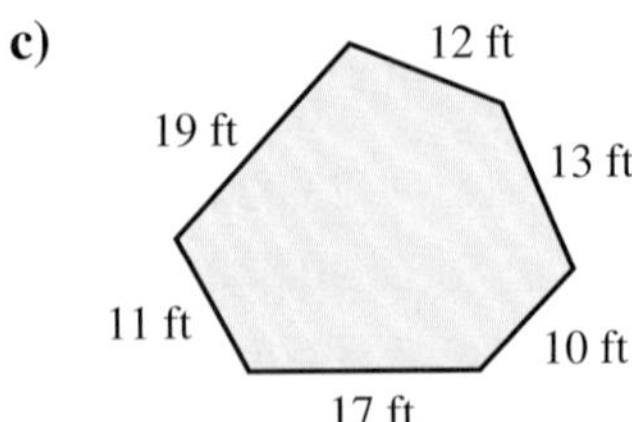

The perimeter is 82 feet.

70. The bar graph at the right represents the size of the four Galilean moons of Jupiter based on each one's radius at the equator. (The Galilean moons were discovered by the famous astronomer Galileo in 1610.)

Answer the following questions based on this bar graph. Write each answer as a sentence.

a) How much longer is the radius of Callisto than the radius of Io?
The radius of Callisto is 366 miles longer than the radius of Io.

b) How much longer is the radius of the largest moon than the radius of the smallest moon?
The radius of the largest moon is 663 miles longer than the radius of the smallest moon.

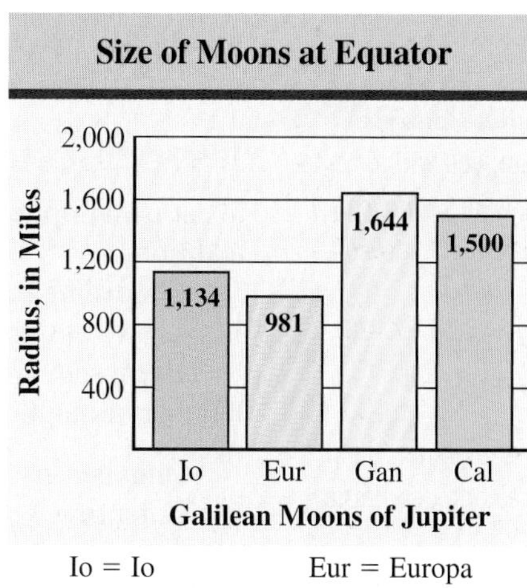

Io = Io Eur = Europa
Gan = Ganymede Cal = Callisto

Think Outside the Box

71. Consider this list of four numbers: 1,208, 12,476, 317, and 29,638.

a) Find the sum of the four numbers and then round the result to the nearest thousand.
The sum is 43,639. Rounded to the nearest thousand, it is 44,000.

b) Round each number to the nearest thousand and then find the sum of the rounded numbers.
The rounded numbers are 1,000, 12,000, 0, and 30,000. The sum of the rounded numbers is 43,000.

c) Compare the results in parts a and b. Which of the two results appears to be the best approximation for the sum? Do you prefer one approximation method over the other? Explain your answer.
Answers may vary.

72. Can the Associative Property apply to subtraction? Show an example to justify your answer.
No.
Answers may vary. One possibility:
$10 - (6 - 1) = 10 - 5 = 5$
$(10 - 6) - 1 = 4 - 1 = 3$

SECTION 1.3 Multiplying Whole Numbers

OBJECTIVES

In this section, you will learn to

- Define the term *multiplication*.
- Write numbers in factored form.
- Multiply by 10 and by 100.
- Identify and apply the Distributive Property.
- Multiply two-digit numbers.
- Find the area of a geometric figure.

You Need to Know

To successfully complete this section, you need to understand

- ☐ The Commutative Property (1.1)
- ☐ The Associative Property (1.1)
- ☐ Adding whole numbers (1.2)

Introduction

What is multiplication? For whole numbers, **multiplication** is an abbreviation for repeated addition.

For example, in the lobby of an apartment building is a rectangular mailbox center that has 4 rows of 6 mailboxes each. How many mailboxes are there altogether?

Because there are four rows of 6 mailboxes, we can answer the question by adding the number of mailboxes in each row.

number in 1st row	+	number in 2nd row	+	number in 3rd row	+	number in 4th row	
6	+	6	+	6	+	6	= 24 mailboxes

We can also think of this as *four* 6's.

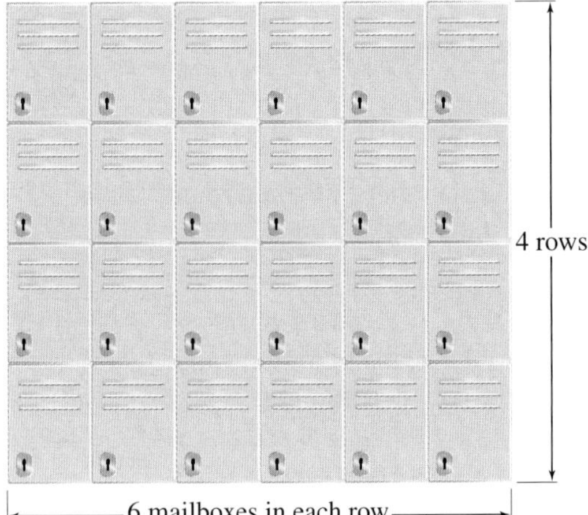

This sum of four 6's is an example of *repeated* addition, and it can be abbreviated using multiplication.

$$\textit{four } 6\text{'s} = 4 \times 6 = \mathbf{24}$$

Furthermore, multiplication can be represented in a variety of ways. For example, *4 times 6* can be written as follows:

4×6 ← typical arithmetic

$4 \cdot 6$, $(4)6$, $4(6)$, and $(4)(6)$ ← typical algebra

In this text, we often (although not always) use the raised dot for multiplication.

Caution The raised dot is *not* a decimal point. When we get to multiplying decimals (in Section 5.4), we'll use the arithmetic multiplication sign, ×, to avoid confusion.

Following is a 12 by 12 multiplication table. It will be useful in a variety of ways throughout this section. You may refer to it as needed, but it is best if you memorize the multiplication facts in this table.

×	1	2	3	4	5	6	7	8	9	10	11	12
1	1	2	3	4	5	6	7	8	9	10	11	12
2	2	4	6	8	10	12	14	16	18	20	22	24
3	3	6	9	12	15	18	21	24	27	30	33	36
4	4	8	12	16	20	24	28	32	36	40	44	48
5	5	10	15	20	25	30	35	40	45	50	55	60
6	6	12	18	24	30	36	42	48	54	60	66	72
7	7	14	21	28	35	42	49	56	63	70	77	84
8	8	16	24	32	40	48	56	64	72	80	88	96
9	9	18	27	36	45	54	63	72	81	90	99	108
10	10	20	30	40	50	60	70	80	90	100	110	120
11	11	22	33	44	55	66	77	88	99	110	121	132
12	12	24	36	48	60	72	84	96	108	120	132	144

Numbers in Factored Form

The numbers in a product are called *factors*.

Factor × Factor = Product

From the multiplication table, we see that $4 \times 6 = 24$. We can say that 4 and 6 are factors of 24.

Also, a number is in **factored form** when it is written as a product of factors. So, we can say that 4×6 is a factored form of 24.

There are other factored forms of 24. In fact, in the multiplication table, we can locate 24 as a product a total of six times: 2×12, 3×8, 4×6, and—by the Commutative Property—12×2, 8×3, and 6×4.

Also, because 1 is the identity for multiplication, $24 = 1 \cdot 24$ and $24 = 24 \cdot 1$.

So, in factored form, $24 = 1 \cdot 24$, $24 = 2 \cdot 12$, $24 = 3 \cdot 8$, $24 = 4 \cdot 6$, and so on.

Example 1

Using the multiplication table and the identity for multiplication, find six ways to write 18 in factored form.

Answer $18 = 2 \cdot 9$ $18 = 9 \cdot 2$ $18 = 3 \cdot 6$ $18 = 6 \cdot 3$ $18 = 1 \cdot 18$ $18 = 18 \cdot 1$

You Try It 1 **Using the multiplication table, find different ways to write 20 in factored form. Use Example 1 as a guide.**

20 = ________ ________ ________ ________ ________ ________

Multiples of a Number

The **multiples** of any number, ***a***, are all of the products involving ***a*** and some other whole number.

For example, the first five multiples of 3 from the multiplication table are as follows:

3, 6, 9, 12, and 15

Each of these numbers is a product of 3 and some other whole number.

$$\underbrace{3}_{1 \cdot 3} \quad \underbrace{6}_{2 \cdot 3} \quad \underbrace{9}_{3 \cdot 3} \quad \underbrace{12}_{4 \cdot 3} \quad \underbrace{15}_{5 \cdot 3}$$

You Try It 2 **Starting with 4, list the first eight multiples of 4.**

First eight multiples of 4: 4, ______________________________

The Multiplication Property of 0

Consider this: $4 \cdot 0 =$ four 0's $= 0 + 0 + 0 + 0 = 0$.

In other words, $4 \cdot 0 = 0$. This is true of any multiple of 0. Also, the Commutative Property allows us to say that $0 \cdot 4 = 0$.

The Multiplication Property of 0

$$\boldsymbol{a \cdot 0 = 0} \quad \text{and} \quad \boldsymbol{0 \cdot a = 0}$$

The product of 0 and any number is always 0.

Think About It 1 Is 0 a multiple of 3? Explain your answer.

____

____

Multiplication by 10 and by 100

Recall that 1 is the multiplicative identity. So, multiplying by 1 is very easy: Any number times 1 is itself.

$6 \cdot 1 = 6$ and $1 \cdot 19 = 19$ In general, $a \cdot 1 = a$ and $1 \cdot a = a$.

Multiplying by 10 and multiplying by 100 is just as easy. The 12×12 multiplication table shows the first twelve multiples of 10.

You Try It 3 **Write the first twelve multiples of 10. Below each multiple, write the number in factored form. (The first few are given to get you started.)**

10	20	___	___	___	___	___	___	___	___	___	___
$1 \cdot 10$	$2 \cdot 10$	$3 \cdot 10$	___	___	___	___	___	___	___	___	___

Multiplying by 10

When one factor is 10, the product is the *other* factor with a 0 placed on the end (in the ones place).

For example, $3 \cdot 10$ is *three* 10's: $10 + 10 + 10 = 30$. So, $3 \cdot 10 = 30$.

The Commutative Property also allows us to say that $10 \cdot 3 = 30$.

Multiplying by 100 is similar.

Multiplying by 100

When one factor is 100, the product is the *other* factor with *two* 0's placed on the end.

Because $100 = 10 \cdot 10$, multiplying by 100 is the same as multiplying by 10 twice. That's why we place *two* 0's after the other factor.

Example 2

Evaluate.

a) $16 \cdot 10$ **b)** $10 \cdot 30$ **c)** $9 \cdot 100$ **d)** $100 \cdot 58$

Procedure Multiply by placing one or two zeros at the end of the other factor, whichever is appropriate.

Answer **a)** $16 \cdot 10 = 160$ (16 with one 0 placed on the end is 160.)

b) $10 \cdot 30 = 300$ (30 with one 0 placed on the end is 300.)

c) $9 \cdot 100 = 900$ (9 with two 0's placed on the end is 900.)

d) $100 \cdot 58 = 5{,}800$ (58 with two 0's placed on the end is 5800 or 5,800.)

You Try It 4 **Evaluate. Use Example 2 as a guide.**

a) $8 \cdot 10 =$ ______ **b)** $10 \cdot 32 =$ ______ **c)** $604 \cdot 10 =$ ______

d) $100 \cdot 7 =$ ______ **e)** $100 \cdot 50 =$ ______ **f)** $298 \cdot 100 =$ ______

Multiplying by Multiples of 10

Any number that ends in 0 (has 0 in the ones place) is a multiple of 10. For example, 120 is $12 \cdot 10$; so, 120 is a multiple of 10. Likewise, any number that ends in two zeros, such as 800, is a multiple of 100.

Multiplying by Multiples of 10

When one or more factors in a product are a multiple of 10 (or 100), we can temporarily ignore any ending zeros and include the total number of ending zeros together in the final product.

For example, the product $30 \cdot 6$ includes one multiple of 10. We can temporarily ignore the ending zero and multiply $3 \cdot 6$, which is 18, and place the 0 on the end of that product: 180. So, $30 \cdot 6 = 180$.

We can use the Associative Property of Multiplication to understand why this is so.

	$30 \cdot 6$
Because $30 = 10 \cdot 3$, we can replace 30 with $(10 \cdot 3)$	$= (10 \cdot 3) \cdot 6$
We can apply the Associative Property to regroup	$= 10 \cdot (3 \cdot 6)$
$3 \cdot 6 = 18$; so, we get	$= 10 \cdot 18$
	$= 180$

What is important to see here is that the end product is just $3 \cdot 6$ multiplied by 10: $18 \cdot 10 = 180$.

This rule can be expanded to include two multiples of 10 or 100, as demonstrated in the next example.

Example 3

Evaluate each expression.

a) $3 \cdot 20$ **b)** $40 \cdot 7$ **c)** $80 \cdot 60$ **d)** $300 \cdot 90$ **e)** $500 \cdot 400$

Procedure Ignore the ending zeros of each number and multiply. Place the total number of zeros after the result.

Answer

a) $3 \cdot 20 = 60$ Ignore the 0 after the 2, treat the expression as $3 \cdot 2$ to get 6, and place the 0 at the end: 60.

b) $40 \cdot 7 = 280$ Ignore the 0 after the 4, treat the expression as $4 \cdot 7$ to get 28, and place the 0 at the end: 280.

c) $80 \cdot 60 = 4{,}800$ Ignore the 0 after the 8 and the 6, treat the expression as $8 \cdot 6$ to get 48, and place the two 0's at the end: 4800 = 4,800.

d) $300 \cdot 90 = 27{,}000$ Ignore all of the 0's—there are three of them—and treat the expression as $3 \cdot 9$ to get 27; then place all three 0's at the end: 27000 = 27,000.

e) $500 \cdot 400 = 200{,}000$ Ignore all of the 0's—there are four of them—and treat the expression as $5 \cdot 4$ to get 20; then place all four 0's at the end. Notice that the 0 in 20 is not counted as one of the four 0's—they were all placed *after* the 20. This gives the number a total of five 0's. This will sometimes happen when 5's (as in 500) are involved.

You Try It 5 **Evaluate each expression. Use Example 3 as a guide.**

a) $3 \cdot 50 =$ ______ **b)** $90 \cdot 7 =$ ______ **c)** $50 \cdot 8 =$ ______

d) $300 \cdot 30 =$ ______ **e)** $800 \cdot 200 =$ ______ **f)** $6{,}000 \cdot 500 =$ ______

The Distributive Property of Multiplication over Addition

As you learned in Section 1.1, parentheses group a quantity and make it one value. For example, in the expression $3 \cdot (10 + 2)$, the parentheses suggest that $(10 + 2)$ should be treated as one value, 12.

The expression can become $3 \cdot (12)$, which is just three 12's: $12 + 12 + 12 = 36$. This means that $3 \cdot 12 = 36$.

However, treating $3 \cdot (12)$ as $3 \cdot (10 + 2)$ suggests that we have three 10's and three 2's.

$$
\begin{aligned}
3 \cdot (10 + 2) &= (10 + 2) + (10 + 2) + (10 + 2)\\
3 \cdot (10 + 2) &= (10 + 10 + 10) + (2 + 2 + 2)\\
3 \cdot (10 + 2) &= (\text{three 10's}) + (\text{three 2's})\\
3 \cdot (10 + 2) &= 3 \cdot 10 + 3 \cdot 2\\
3 \cdot (10 + 2) &= 30 + 6\\
3 \cdot (10 + 2) &= 36
\end{aligned}
$$

The Commutative and Associative Properties allow us to rearrange the numbers into groups of three 10's and three 2's.

We can say that the multiplier 3 is being distributed to both the 10 and the 2.

This diagram is a shortcut to the preceding work.

$$
\begin{aligned}
3 \cdot (10 + 2) &= 3 \cdot 10 + 3 \cdot 2\\
&= 30 + 6\\
&= 36
\end{aligned}
$$

This idea helps to introduce an important property of mathematics, the Distributive Property, formally known as the **Distributive Property of Multiplication over Addition.**

Usually, when evaluating an expression, we evaluate within the parentheses first. However, when specifically applying the Distributive Property, we multiply first.

The Distributive Property of Multiplication over Addition

We can distribute a multiplier, a, over a sum $(b + c)$ so that it multiplies both numbers in the sum.

$$a \cdot (b + c) = a \cdot b + a \cdot c$$

In other words, a is a multiplier of both b and c.

The notion of the Distributive Property is quite useful in multiplying by numbers with two or more digits. We'll use the fact that every two-digit number can be separated into the sum of its tens and ones.

For example, $14 = 10 + 4$ and $59 = 50 + 9$. We can use this information and the Distributive Property when multiplying horizontally, as in Example 4.

Example 4

Multiply. **a)** $6 \cdot 14$ **b)** $7 \cdot 59$

Procedure Treat 14 as $(10 + 4)$ and 59 as $(50 + 9)$ and use the Distributive Property.

Answer

a) $6 \cdot 14$
$= 6 \cdot (10 + 4)$

b) $7 \cdot 59$
$= 7 \cdot (50 + 9)$

Apply the Distributive Property as follows:

$$6 \cdot (10 + 4) = 6 \cdot 10 + 6 \cdot 4 = 60 + 24 = 84 \qquad \begin{array}{r} 60 \\ +\ 24 \\ \hline 84 \end{array}$$

$$7 \cdot (50 + 9) = 7 \cdot 50 + 7 \cdot 9 = 350 + 63 = 413 \qquad \begin{array}{r} 350 \\ +\ 63 \\ \hline 413 \end{array}$$

You Try It 6 **Multiply using the technique demonstrated in Example 4.**

a) $4 \cdot 18 = 4 \cdot (10 + 8)$

b) $5 \cdot 46 = 5 \cdot (40 + 6)$

c) $7 \cdot 19 = 7 \cdot (\qquad)$

d) $8 \cdot 73 = 8 \cdot (\qquad)$

Multiplication by a Single-Digit Number

Multiplication of a two-digit number by a single-digit number often requires us to write a product, such as $26 \cdot 3$, vertically with the columns aligned, as in addition and subtraction.

We use the Distributive Property to multiply 6×3 and then 20×3:

$$\begin{array}{r} 26 \\ \times\ 3 \\ \hline \end{array} \longrightarrow \begin{array}{r} 20 + 6 \\ \times \qquad 3 \\ \hline 60 + 18 = 78 \end{array}$$

To simplify the work, we don't show the actual distribution process. Instead, we directly multiply 6×3 and then 2×3, as demonstrated here:

The reason the 7 in part B belongs in the tens place:

1. The 1 we carried is really 10 (from $18 = 10 + 8$).
2. The 2 we multiplied by the 3 is really **20**.

So, the actual multiplication is $(\mathbf{20} \cdot 3) + 10 = 60 + 10 = \mathbf{70}$.

A. **Multiplying the ones, we get $6 \times 3 = 18$. As in addition, we can put the 8 in the ones place (in the answer) and carry the 1 to the next column.**

$$\begin{array}{r} {}^{+1} \\ 26 \\ \times\ 3 \\ \hline 8 \end{array}$$

B. **We now multiply 2×3 and add the 1 that we carried.**

$$(2 \times 3) + 1 = 6 + 1 = 7$$

Then we put the 7 in the tens place in the answer.

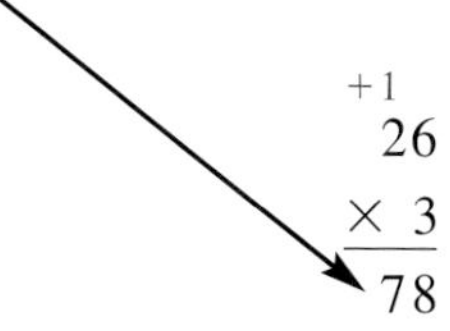

The procedure for multiplying a three-digit number by a one-digit number is the same; it just continues one more place.

You Try It 7 **Multiply using the preceding technique.**

a) $\begin{array}{r} 23 \\ \times\ 4 \\ \hline \end{array}$ **b)** $\begin{array}{r} 19 \\ \times\ 5 \\ \hline \end{array}$ **c)** $\begin{array}{r} 286 \\ \times\ \ 3 \\ \hline \end{array}$ **d)** $\begin{array}{r} 429 \\ \times\ \ 5 \\ \hline \end{array}$

Two-Digit by Two-Digit Multiplication

The multiplication process we just learned can be extended to two-digit by two-digit multiplication. For example, if we are to multiply $57 \cdot 63$, also written as

$$\begin{array}{r} 57 \\ \times\ 63 \\ \hline \end{array}$$

we can multiply the tens place digit, 6 (which is really 60), in the same manner we multiply the ones place digit, 3.

This is a three-step process, giving us two *partial products* that we'll add to get the final product

1. Multiply 57×3. (This gives us the first partial product.)
2. Multiply 57×60. (This gives us the second partial product.)
3. Add the two partial products to get the final product.

Here is the step-by-step outline for 57×63:

A. **First partial product: Multiply to the ones place digit.**

$$\begin{array}{r} {\scriptstyle +2} \\ 57 \\ \times\ 3 \\ \hline 171 \end{array}$$

B. **Second partial product: Multiply to the tens place digit.**

$$\begin{array}{r} {\scriptstyle +4} \\ 57 \\ \times\ 60 \\ \hline 3{,}420 \end{array}$$

This is what your work might look like when you are done.

$$\begin{array}{r} {\scriptstyle +4\ \not{+2}} \\ 57 \\ \times\ 63 \\ \hline 171 \\ +\ 3{,}420 \\ \hline 3{,}591 \end{array}$$

C. **Add the partial products: First partial product** ⟶ 171
Second partial product ⟶ + 3,420
Final product ⟶ 3,591

Think About It 2 $57 \cdot 6 = 342$; but in the outline above, it is written as 3,420. Why is the zero in the ones place?

____

____

The process of multiplying two-digit numbers can be extended to numbers with more than two digits.

You Try It 8 **Multiply.**

a) $\begin{array}{r} 417 \\ \times\ 25 \\ \hline \end{array}$ **b)** $\begin{array}{r} 941 \\ \times\ 33 \\ \hline \end{array}$ **c)** $\begin{array}{r} 534 \\ \times\ 146 \\ \hline \end{array}$

Applications Involving Multiplication

There are many situations to which multiplication can apply. Since multiplication is an abbreviation for repeated addition, we use it whenever the same number is added over and over a certain number of times.

For each of these word problems

- Read the problem carefully (maybe two or three times).
- Think about the situation (imagine yourself in the situation).
- Remember that in multiplication, there are two numbers:
  - ➤ One number that will be repeated
  - ➤ Another number that indicates the number of times it is repeated
- Multiply appropriately.
- Write a sentence answering the question.

Example 5

Bindee just completed her associate's degree and is looking for a job. Searching the Internet, she found a job as an entry-level secretary that pays $1,783 per month. If she is hired for the job, how much will she earn in 12 months?

Procedure The monthly wage is the same each month—and Bindee would earn that wage 12 *times*—so, multiplication is the operation to use.

Multiply the monthly wage ($1,783) by the number of months (12).

Answer

Bindee's monthly wage: ⟶	1,783
Times 12 months: ⟶	× 12
	3,566
	17,830
	21,396

Sentence If Bindee gets the job, she will earn $21,396 in 12 months.

You Try It 9 **George sells sports cards. In each Upper Deck *Stars* box there are 24 packs of cards. If George has 5 of these boxes, how many packs of Upper Deck *Stars* does he have?**

Sentence ______________________________

You Try It 10 **At the restaurant where Leilani is a waitress, there is a buffet special for $8 per person. A group of 15 patrons came in one afternoon, and they all ordered the buffet special. Before tax and tip, what was the total amount of their bill?**

Sentence ______________________________

You Try It 11 **Sam, a truck driver, is asked to pick up 93 boxes, each weighing 38 pounds. What is the total weight of this load?**

Sentence ______________________________

Multiplication in Geometry: Area

In geometry, **area** is the amount of surface in an enclosed region. Area is always measured in square units, such as square feet (*sq ft*) and square centimeters (*sq cm*).

We can use the area of a rectangle to illustrate the idea that multiplication is repeated addition. This illustration will also develop the formula for the area of a rectangle.

First, consider the single unit square at the right. (It has no specific measure, such as an inch; so, we say that each side length is one *unit*.)

1 unit square

We can put 12 of these small squares together to form a *3 by 4* rectangle. The following illustration also shows the area formula for a rectangle.

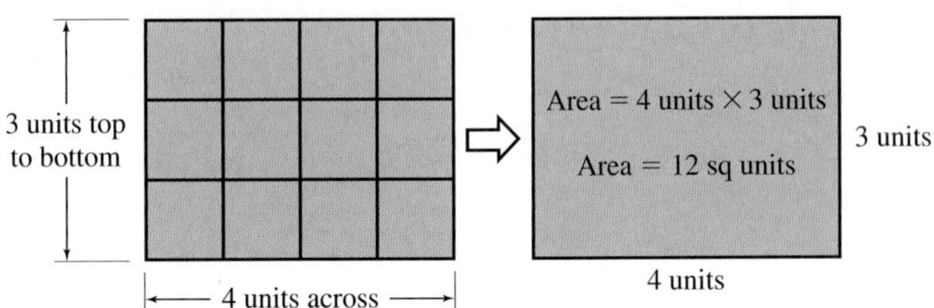

In general,

Area of a rectangle = Length × Width.

W (width) Area = L × W

L (length)

Example 6

Find the area of the rectangle.

Answer Area = Length × Width

Area = 8 feet × 5 feet

Area = 40 sq ft

5 ft

8 ft

You Try It 12 **Find the area of each rectangle. Use Example 6 as a guide.**

a) 7 in. 16 in.

b) 95 m 45 m

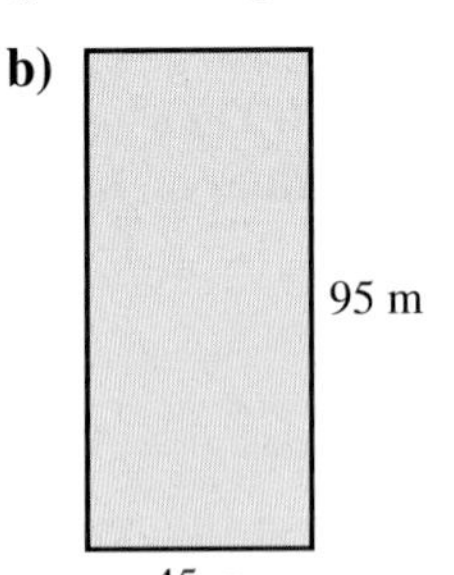

Answers: You Try It and Think About It

You Try It: 1. 20 = 2 · 10, 20 = 10 · 2, 20 = 4 · 5, 20 = 5 · 4, 20 = 1 · 20, 20 = 20 · 1

2. 4, 8, 12, 16, 20, 24, 28, and 32

3.

10	20	30	40	50	60	70	80	90	100	110	120
1 · 10	2 · 10	3 · 10	4 · 10	5 · 10	6 · 10	7 · 10	8 · 10	9 · 10	10 · 10	11 · 10	12 · 10

4. a) 80 **b)** 320 **c)** 6,040 **d)** 700 **e)** 5,000 **f)** 29,800 **5. a)** 150 **b)** 630 **c)** 400 **d)** 9,000 **e)** 160,000 **f)** 3,000,000 **6. a)** $40 + 32 = 72$ **b)** $200 + 30 = 230$ **c)** $7 \cdot (10 + 9) = 70 + 63 = 133$ **d)** $8 \cdot (70 + 3) = 560 + 24 = 584$ **7. a)** 92 **b)** 95 **c)** 858 **d)** 2,145 **8. a)** 10,425 **b)** 31,053 **c)** 77,964 **9.** George has 120 packs of Upper Deck *Stars.* **10.** The total amount of the bill, before tax and tip, was $120. **11.** The total weight of the load is 3,534 pounds. **12. a)** 112 square inches **b)** 4,275 square meters

Think About It: **1.** Yes, because 0 is a whole number and $0 \cdot 3 = 0$. **2.** A 0 is in the ones place because we are really multiplying 57 by 60, a multiple of 10: $57 \cdot 60 = 3{,}420$.

Section 1.3 Exercises

FOR EXTRA HELP
Student Resources on DVD-ROM
Includes
➤Student's Solutions Manual
➤Video Lectures
➤Chapter Test Prep Video
MyMathLab
MathXL

Think Again

1. Is 0 a multiple of 3? Explain your answer. *(Refer to Think About It 1)*
Yes. The multiples of 3 are all of the products involving 3 and some other whole number, and 0 is a whole number. So, $0 \cdot 3 = 0$ is a multiple of 3.

2. Are there any odd number multiples of 2? Explain your answer or show an example that supports your answer.
No. Any whole number times 2 is an even number.

3. Consider a square with side length of 4 inches. Find the perimeter and the area of the square. Are the perimeter and area the same? Explain your answer or show an example that supports your answer.
No. The perimeter is 16 inches, but the area is 16 *square* inches. Those are not the same.

4 in.
4 in.

4. Can the product of two odd numbers ever be an even number? Show examples that support your answer.
No. Examples will vary.

Focus Exercises

Multiply.

5. $5 \cdot 4$ 20	**6.** $6 \cdot 3$ 18	**7.** $4 \cdot 7$ 28	**8.** $8 \cdot 5$ 40
9. $9 \cdot 4$ 36	**10.** $3 \cdot 8$ 24	**11.** $9 \cdot 6$ 54	**12.** $8 \cdot 7$ 56
13. $9 \cdot 3$ 27	**14.** $8 \cdot 4$ 32	**15.** $7 \cdot 5$ 35	**16.** $6 \cdot 8$ 48
17. $8 \cdot 8$ 64	**18.** $9 \cdot 8$ 72	**19.** $7 \cdot 7$ 49	**20.** $9 \cdot 9$ 81

Multiply.

21. $3 \cdot 80$ 240	**22.** $4 \cdot 90$ 360	**23.** $60 \cdot 7$ 420	**24.** $50 \cdot 9$ 450
25. $6 \cdot 50$ 300	**26.** $80 \cdot 5$ 400	**27.** $6 \cdot 400$ 2,400	**28.** $8 \cdot 200$ 1,600
29. $900 \cdot 7$ 6,300	**30.** $500 \cdot 5$ 2,500	**31.** $30 \cdot 90$ 2,700	**32.** $70 \cdot 40$ 2,800
33. $600 \cdot 60$ 36,000	**34.** $400 \cdot 50$ 20,000	**35.** $600 \cdot 800$ 480,000	**36.** $200 \cdot 900$ 180,000

Multiply.

37. 18×19 342	**38.** 26×15 390	**39.** 43×43 1,849	**40.** 62×62 3,844
41. 218×45 9,810	**42.** 307×92 28,244	**43.** 425×209 88,825	**44.** 528×303 159,984
45. 153×112 17,136	**46.** 216×144 31,104	**47.** 256×128 32,768	**48.** $1{,}153 \times 224$ 258,272

Which property is being demonstrated?

49. $6 \cdot 9 = 9 \cdot 6$
Commutative Property of Multiplication

50. $3 \cdot (5 + 8) = 3 \cdot 5 + 3 \cdot 8$
Distributive Property of Multiplication over Addition

51. $4 \cdot (7 \cdot 2) = (4 \cdot 7) \cdot 2$
Associative Property of Multiplication

52. $18 \cdot 0 = 0$
Multiplication Property of 0

Work each application and answer the question with a complete sentence.

53. The parking lot at the Hillside Craft Fair has 14 rows of parking with 12 parking spaces in each row. How many cars can fit in the parking lot?
168 cars can fit in the parking lot.

54. Monica has a one-year lease on her apartment and pays a monthly rent of $835. How much rent will Monica pay for the entire year (12 months)?
Monica will pay $10,020 in rent for the entire year.

55. Ignacio's truck gets 13 miles per gallon of gas. He just filled the gas tank with a total of 28 gallons. How many miles will Ignacio's truck go before it runs out of gas?
Ignacio's truck will go 364 miles before it runs out of gas.

56. Sandy is counting inventory at The Office Station. In the pen and pencil section, there are 33 boxes of Script Magic gel pens. If each box holds 18 pens, how many Script Magic gel pens does The Office Station have in its inventory?
The Office Station has 594 Script Magic gel pens in its inventory.

57. A jumbo jet is flying from San Francisco to Australia and is averaging 492 miles per hour. How many miles will the jet travel in 16 hours?
The jet will travel 7,872 miles in 16 hours.

58. Rico spends $25 every week on lottery tickets. How much does Rico spend in a year (52 weeks) on lottery tickets?
Rico spends $1,300 in a year on lottery tickets.

59. Diane is a massage therapist. During her first week working at Dr. Kent's chiropractic office, she saw 33 patients. If she earns $18 for every patient, how much did Diane earn that first week?
Diane earned $594 that first week.

60. Toby is preparing a 23-pound turkey for his family's Thanksgiving gathering. The recipe he's using says to roast the turkey for 12 minutes per pound. For how many minutes should Toby roast the turkey?
Toby should roast the turkey for 276 minutes.

61. There are 306 families in the Magnolia Elementary PTA. At the beginning of the year, each family must contribute $18 to support the PTA's field trip program. What is the total income for the PTA's field trip program this year?
The total income for the PTA's field trip program is $5,508.

62. Soo works for the purchasing department of a mid-sized company. Last week she bought 12 computers for the new West Valley office, which opens in two months. Each computer is priced at $1,289. What is the total price for the 12 computers?
The total price for the 12 computers is $15,468.

63. A youth soccer field is in the shape of a rectangle. The length is 60 yards, and the width is 28 yards. What is the area of the soccer field?
The area of the soccer field is 1,680 square yards.

64. An official college basketball court is 94 feet long and 50 feet wide. What is the area of the basketball court?
The area of the basketball court is 4,700 square feet.

65. Wyoming is in the shape of a rectangle. It is 357 miles long and 274 miles wide. Round each of these dimensions to the nearest ten. Use the rounded numbers to approximate the area of Wyoming.
The approximate area of Wyoming is 97,200 square miles.

66. Colorado is in the shape of a rectangle. It is 376 miles long and 282 miles wide. Round each of these dimensions to the nearest ten. Use the rounded numbers to approximate the area of Colorado.
The approximate area of Colorado is 106,400 square miles.

Think Outside the Box

67. A soup company stores its cans of soup in boxes and crates before delivering it to stores around the state. 24 cans of soup are placed in a box, and 80 boxes are placed in a crate. The crates are then stacked in a warehouse as shown in the diagram at the right. According to the diagram, how many cans of soup are stored in the warehouse?
There are 92,160 cans of soup stored in the warehouse.

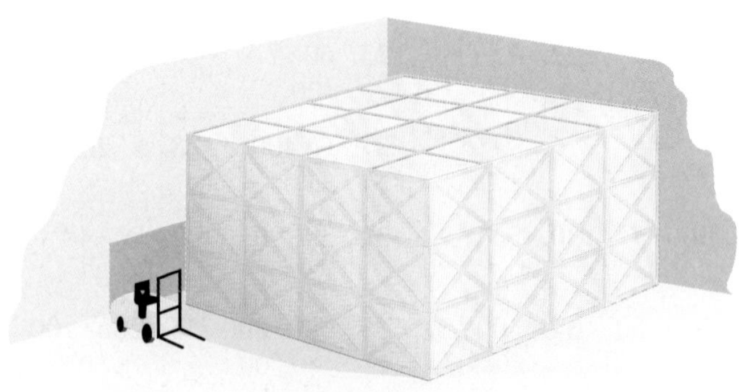

68. Can the Distributive Property be applied to subtraction? Show an example to justify your answer.
Yes. Answers may vary. One possibility:
$10 \cdot (6 - 1) = 10 \cdot (5) = 50$
$10 \cdot 6 - 10 \cdot 1 = 60 - 10 = 50$

SECTION 1.4 Dividing Whole Numbers

OBJECTIVES

In this section, you will learn to
- Define the term *division.*
- Perform short division.
- Perform long division.

You Need to Know

To successfully complete this section, you need to understand
- ☐ Rounding whole numbers (1.1)
- ☐ Adding whole numbers (1.2)
- ☐ Subtracting whole numbers (1.2)
- ☐ Multiplying whole numbers (1.3)

Introduction

Consider the following situations

1. Ruben took 8 children from the Boys and Girls Club to the county fair. He purchased 120 ride tickets for the children. How many ride tickets will each child receive?
2. Three neighbors decide to have a common garage sale and split the proceeds evenly. If they earn a total of $1,728, how much will each neighbor get?
3. Guy used his credit card too much and now owes $4,536. To avoid more interest charges, he agreed to cut up his card and pay back all $4,536 in 24 months. How much must Guy pay each month?
4. Taiyana is purchasing wooden boxes with premade mail slots for the employees in her office. Each box contains mail slots for 12 employees. If the office has 163 employees, how many boxes must she purchase?

Each of these problems can be answered using division. As you'll recall from Section 1.1, the result of division is called the quotient. Here are the parts of a division problem, along with an example of how to read each one.

Standard form: Dividend ÷ Divisor = Quotient $15 \div 5$ is read "15 divided *by* 5."

Long division form: $\text{Divisor}\overline{)\text{Dividend}}$ with Quotient above $5\overline{)15}$ is read "5 divided *into* 15."

Fraction form: $\dfrac{\text{Dividend}}{\text{Divisor}} = \text{Quotient}$ $\dfrac{15}{5}$ is read "15 divided *by* 5."

In the fraction form, the line separating the dividend and the divisor is called the **division bar.**

$$\frac{\text{Dividend}}{\text{Divisor}} \leftarrow \text{division bar}$$

quotient
divisor)dividend

Because we use these words—dividend, divisor, and quotient—throughout this section, it's important that you become familiar with them. You'll often see this little diagram as a continual reminder of their proper placement.

What Is Division?

Division is *the inverse operation of multiplication.* Just as multiplication is repeated addition, division can be thought of as repeated subtraction.

When we think of division as repeated subtraction, we subtract the divisor until we end at 0 or at another number less than the divisor, called the **remainder.** When the repeated subtraction ends in 0, we say that the remainder is 0 or that there is no remainder.

For example, to find out how many times 3 will divide into 12, we can subtract 3 repeatedly until we get a number less than 3.

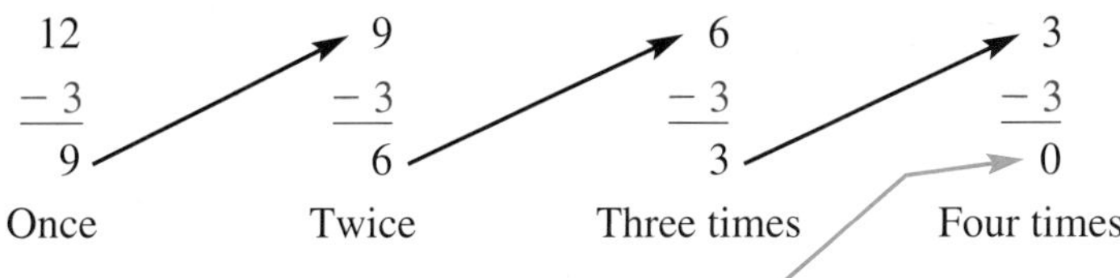

When there is no remainder, as in $12 \div 3 = 4$, we say that 3 *divides evenly* into 12 four times. Another term that means "divides evenly" is *divisible.* We can say that 12 is divisible by 3.

Likewise, to find out how many times 3 will divide into 13, we can subtract 3 repeatedly until we get a number less than 3.

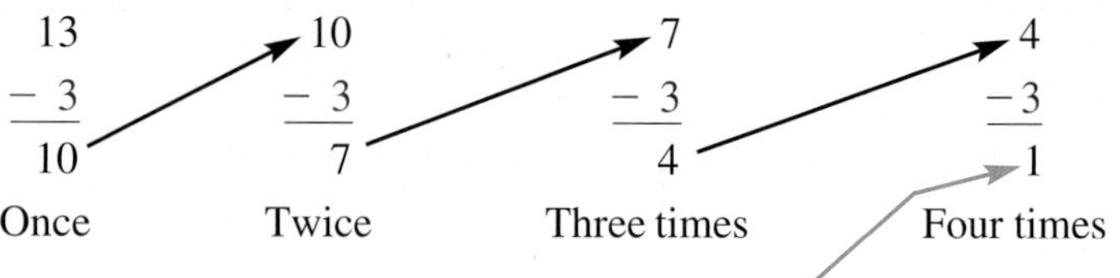

We can't subtract any further. There is a remainder of 1.

When there is a remainder (besides 0), as in $13 \div 3$, we say that 3 does *not* divide evenly into 13. 13 is *not* divisible by 3.

The remainder must be smaller than the divisor. For example, in repeatedly subtracting 3 from 13, we can't stop when we get to 10, 7, or 4 because it is still possible to subtract 3 at least one more time.

For exact division, we can use a circular argument of inverses to see how multiplication and division work together.

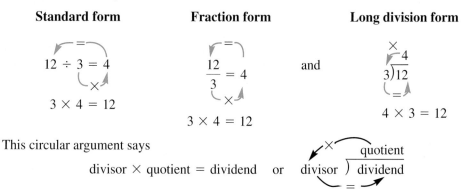

Standard form	**Fraction form**		**Long division form**
$12 \div 3 = 4$	$\frac{12}{3} = 4$	and	$3\overline{)12}$ with quotient 4
$3 \times 4 = 12$	$3 \times 4 = 12$		$4 \times 3 = 12$

This circular argument says

divisor × quotient = dividend or $\text{divisor}\overline{)\text{dividend}}$ with quotient above

Because of the circular nature of division and multiplication, we can use multiplication to verify the accuracy of our division result, the quotient.

Short Division

When we are able to divide in one step, as demonstrated in Examples 1 and 2, we call it **short division.** When the division is not obvious (it cannot be done in just one step), we call it **long division.**

In demonstrating short division, we can use the standard form or the fraction form (as shown in Example 2) or the long division form (as shown in Example 3).

Example 1

Reminder:
$\text{divisor}\overline{)\text{dividend}}$ with quotient above

Use short division to evaluate. Verify each answer by multiplying the divisor and the quotient.

a) $35 \div 7$ **b)** $\frac{18}{6}$ **c)** $5 \div 5$ **d)** $\frac{0}{4}$ **e)** $8 \div 1$

Procedure Think about what number will multiply by the divisor to get the quotient. In other words, use the circular argument for division, divisor × quotient = dividend, to find the quotient.

Answer

a) $35 \div 7 = \underline{5}$ because $7 \times \underline{5} = 35$

b) $\frac{18}{6} = \underline{3}$ because $6 \times \underline{3} = 18$

c) $5 \div 5 = \underline{1}$ because $5 \times \underline{1} = 5$ Rule: A natural number divided by itself is 1.

d) $\frac{0}{4} = \underline{0}$ because $4 \times \underline{0} = 0$ Rule: 0 divided by a natural number is 0.

e) $8 \div 1 = \underline{8}$ because $1 \times \underline{8} = 8$ Rule: A number divided by 1 is itself.

Parts c, d, and e of Example 2 demonstrate three basic principles of division, as follows

1. A natural number divided by itself is 1: $c \div c = 1; \frac{c}{c} = 1$.
2. 0 divided by a natural number is 0: $0 \div c = 0; \frac{0}{c} = 0$ (c cannot be 0).
3. A number divided by 1 is itself: $c \div 1 = c; \frac{c}{1} = c$.

You Try It 1 **Use short division to evaluate. Verify each answer by multiplying the divisor by the quotient, as in Example 1.**

a) $50 \div 5 =$ ______ **b)** $18 \div 3 =$ ______ **c)** $\frac{4}{4} =$ ______

d) $\frac{40}{5} =$ ______ **e)** $0 \div 2 =$ ______ **f)** $\frac{7}{1} =$ ______

In standard form, to verify that $28 \div 4 = 7$, we might write because $4 \times 7 = 28$.

To verify the same division using the long division form, we can show the product of 4 and 7 *below* the dividend. Because they are the same, the result is 0 when we subtract.

$$\begin{array}{r} 7 \\ 4\overline{)\,28} \\ -\,28 \\ \hline 0 \end{array}$$

Example 2

Use short division to evaluate. Verify each answer by multiplying the divisor and the quotient. Write that product below the dividend and subtract.

a) $2\overline{)18}$ **b)** $10\overline{)60}$

Procedure To find the quotient, think about what number you will multiply by the divisor to get the dividend. In other words, use the circular argument for division,

quotient × divisor = dividend, to find the quotient.

Answer

a) $$\begin{array}{r} 9 \\ 2\overline{)\,18} \\ -\,18 \\ \hline 0 \end{array}$$ ← This 0 indicates that 2 divides evenly into 18.

b) $$\begin{array}{r} 6 \\ 10\overline{)\,60} \\ -\,60 \\ \hline 0 \end{array}$$

You Try It 2 **Use short division to evaluate. Verify each answer by multiplying the divisor by the quotient, as in Example 2.**

a) $7\overline{)49}$ **b)** $6\overline{)54}$ **c)** $8\overline{)8}$ **d)** $1\overline{)6}$

Can the Divisor Ever Be 0?

We can *never* divide by 0 (zero). We have seen that 0 can be the dividend, but it can never be the divisor. The circular argument shows us why.

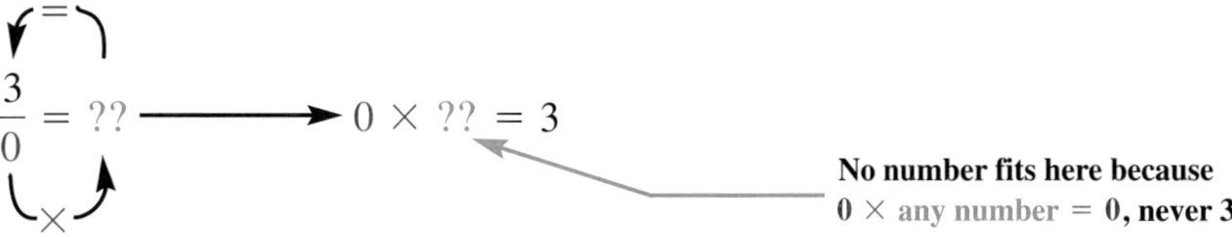

In general, we say that $a \div 0$, or $\frac{a}{0}$, **is undefined.**

As an example, you might think of division by 0 in terms of distributing equal lottery winnings among a group of people, as follows:

Lottery winnings	÷	Number of people	=	Each person gets	
\$0	÷	3	=	\$0	$0 \div 3 = 0$
\$35	÷	0	=	??	$35 \div 0 = ??$ (impossible)

You can't distribute \$35 among 0 people. It isn't possible.

The Remainder

Earlier in this section we saw that 3 does not divide evenly into 13 and that when we subtracted 3 repeatedly, we ended with a remainder of 1. Here is a simple example of how the remainder is found using the division process.

Consider $17 \div 5$.

5 does not divide into 17 evenly; so, there will be a remainder. Using the long division form, $5\overline{)17}$, we can start the quotient by choosing a number that when multiplied by 5, will give a product close to 17.

The number we choose to start the quotient can't be too large—it can't give us a product larger than the dividend (because then we can't subtract to find the remainder). It can't be too small either—the remainder we get must be less than the divisor.

Let's try 4:

$$\begin{array}{r} 4 \\ 5\overline{)\,17} \\ -\,20 \\ \hline ?? \end{array}$$

4 is too large because 20 can't be subtracted from 17.

Let's try 2:

$$\begin{array}{r} 2 \\ 5\overline{)\,17} \\ -\,10 \\ \hline 7 \end{array}$$

2 is too small because we get a remainder that is larger than the divisor.

Let's try 3:

$$\begin{array}{r} 3 \\ 5\overline{)\,17} \\ -\,15 \\ \hline 2 \end{array}$$

Just right!

When we divide correctly and get a remainder *less* than the divisor, the remainder is shown next to the quotient—a whole number—and is abbreviated with the letter **r.** When a remainder exists, we will see the long division answer as shown here.

$$\begin{array}{r} \text{quotient r remainder} \\ \text{divisor}\overline{)\text{dividend}} \end{array}$$

For $17 \div 5$, we show the final result as

$$\begin{array}{r} 3 \text{ r } 2 \\ 5\overline{)\,17} \\ -\,15 \\ \hline 2 \end{array}$$

The remainder is shown to the right of the quotient.

Example 3

Divide. If after multiplying the quotient and the divisor there is a remainder, show it next to the quotient.

a) $6\overline{)27}$ **b)** $7\overline{)42}$

Procedure If the divisor doesn't divide evenly into the dividend, think about what number it will divide into evenly.

Answer a)
```
   4 r 3
6)27
 - 24
    3
```
$6 \times 4 = 24$ is less than 27, and $6 \times \mathbf{5} = 30$ is too large.

The quotient is 4, and the remainder is 3.

(Because this remainder is less than the divisor, 6, we have divided properly.)

b)
```
    6
7)42
 - 42
    0
```
$7 \times 6 = 42$; so, 42 is divisible by 7 and the remainder is 0.

You Try It 3 **Divide. Make sure you show any remainder (other than 0) next to the quotient. (Do each division in pencil. Have an eraser handy in case your first try gives a quotient that is too large or too small.) Use Example 3 as a guide.**

a) 7)60 **b)** 8)15 **c)** 4)28 **d)** 11)70

The Long Division Algorithm

Reminder:
quotient r remainder
divisor)dividend

As you might imagine, not every division problem can be done so quickly. For example, it is true that 4 divides evenly into 948 (as you'll soon see), but how many times?

To discover the answer, you'll need to learn a process called the **Long Division Algorithm.** (An *algorithm* is a set of repeated rules that leads to a desired result.)

The Long Division Algorithm	Examples	
	978 ÷ 6	259 ÷ 7
1. If possible, divide the divisor into the first (leftmost) digit in the dividend whether or not it divides into the digit evenly. If the first digit of the dividend is too small, place a zero in the quotient (above the first digit in the dividend) and divide the divisor into the first two or more (leftmost) digits. Place the quotient over the last digit used.	1 6)(9)78	03 7)(25)9
2. Multiply the quotient and the divisor. Place this product directly under the digits used in this division.	×1 6)978 = 6	×03 7)259 = 21
3. Subtract.	1 6)978 − 6 3	03 7)259 − 21 4
4. "Bring down" the first unused digit in the dividend. Repeat this process (starting at step 1) until you can divide no further. Write the remainder next to the quotient. The division process stops when the quotient "covers" the last digit in the dividend. The **remainder** is the amount left over after the last digit in the dividend is covered by the quotient.	1 6)975 − 6↓ 37	03 7)259 − 21↓ 49

The remainder—which must be less than the divisor—is considered only after the last digit in the dividend is covered by the quotient.

The example that follows shows the steps one at a time. The explanation requires a great deal of space, but your actual work won't be as long.

Example 4

Use the long division algorithm to divide 948 by 4.

Answer

> *Reminder:*
> $\text{divisor}\overline{)\text{dividend}}$ with quotient r remainder above

Steps **1.** and **2.**	**Recognize that 4 will divide into 9 *two times*.** **Place the 2 above the 9 and multiply: 2 × 4 = 8.**	2 4)948
Step **3.**	**Subtract this product from 9.**	− 8 1
Step **4.**	**Bring down the first unused digit, 4. Start the division process over with a new dividend, 14.**	2 4)948 − 8↓ 14
Repeat steps **1.** and **2.**	**Recognize that 4 will divide into 14 *three times*.** **Place the 3 above the 4 (the last digit used) and multiply: 3 × 4 = 12.**	23 4)948 − 8 14
Step **3.**	**Subtract this product from 14.**	− 12 2
Step **4.**	**Bring down the unused digit, 8. Start the division process over with a new dividend, 28.**	23 4)948 − 8 14 − 12↓ 28
Repeat steps **1.** and **2.**	**4 divides evenly into 28 *seven times*.** **Place the 7 above the 8 (the last digit used) and multiply: 7 × 4 = 28.**	237 4)948 − 8 14 − 12 28
Step **3.**	**Subtract this product from 28 and get a remainder of 0.**	− 28 0

Notice that the last digit of the dividend is "covered" by the 7 in the quotient; so, we know that we're finished dividing.

This next example shows all of the steps combined into one division problem.

Example 5

Use the long division algorithm to divide 863 ÷ 5. (You may want to cover up part of the problem so that you can see the progress step-by-step.)

Answer

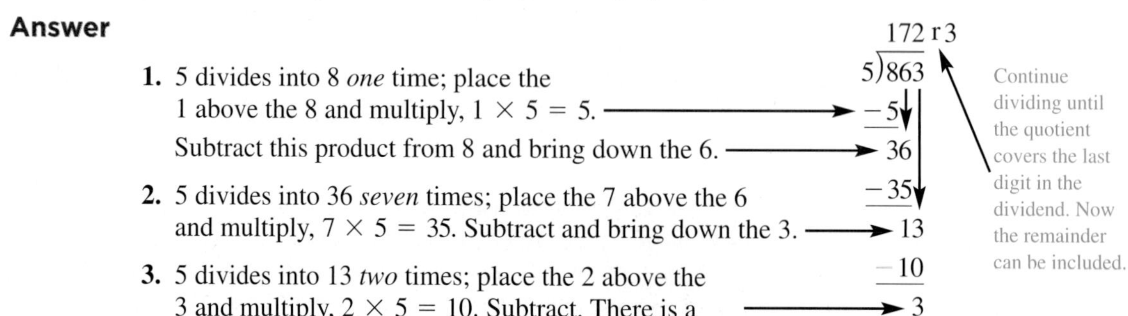

1. 5 divides into 8 *one* time; place the 1 above the 8 and multiply, 1 × 5 = 5.
   Subtract this product from 8 and bring down the 6.
2. 5 divides into 36 *seven* times; place the 7 above the 6 and multiply, 7 × 5 = 35. Subtract and bring down the 3.
3. 5 divides into 13 *two* times; place the 2 above the 3 and multiply, 2 × 5 = 10. Subtract. There is a remainder of 3.

We know that we're finished dividing because the last digit in the dividend is covered by the quotient.

The next example shows what to do when a 0 appears in the quotient.

Example 6

Use the long division algorithm to divide 2,461 ÷ 8.

Answer

1. 8 won't divide into 2; so, place a 0 above the 2. 8 does divide **evenly** into 24 three times; place the 3 above the 4 and multiply, 3 × 8 = 24. Subtracting gives 0; bring down the 6.

```
   03
8)2461
 - 24↓
    06
```

2. At this point, the new dividend is 6 (same as 06); but it is less than 8, and 8 can't divide into 6. So, we say that 8 divides into 6 zero times; place the 0 above the 6.
3. We can multiply the 0 and the 8, but we'll get 0 and the remainder will still be 6. We then bring down the 1.
4. 8 divides into 61 seven times; place the 7 above the 1 and multiply, 7 × 8 = 56. Subtract. The remainder is 5.

```
   0307 r 5
8)2461
 - 24↓
    06
   - 0↓
     61
   - 56
      5
```

Showing the multiplication by 0 is not necessary. Instead, once we get a 0 in the quotient, we can bring down the next digit in the dividend.

Now that we are finished, we don't need the starting 0; so, 2,461 ÷ 8 = 307 r 5.

You Try It 4 **Evaluate each expression using long division. Use Examples 4, 5, and 6 as guides.**

a) $\frac{372}{4}$ **b)** 1,628 ÷ 7 **c)** 7,835 ÷ 6 **d)** $\frac{40{,}016}{8}$

When the Divisor Is a Two-Digit Number

So far in the long division process, all of the divisors have been single-digit numbers. When the divisor contains more than one digit, we need to do some estimation and, at times, some trial and error.

For example, when dividing 1,167 ÷ 38, we'll first set it up as 38)1167 and prepare to use long division. We know that 38 won't divide into 1 and 38 won't divide into 11. Here we'll need to use the first *three* digits in the dividend before we can start to divide. In other words, we'll try to divide 38 into 116.

This can prove to be a challenge. But if we round each number (the divisor 38 and the three digits 116) to the nearest ten, we can make an educated guess as to what the first digit of the quotient will be.

We can estimate that 38 rounds up to 40 and that 116 rounds up to 120. So, we might think of this as 40)120. When we ignore the 0's, it is like dividing 4)12: 12 ÷ 4 = 3.

This suggests that our choice for the first digit of the quotient should be 3.

Caution Keep an eraser handy because often you'll need to make a second educated guess. Also, the rounded numbers, 40 and 120, are used only for the purpose of making an educated guess. We do not use them in any other part of the division process (although we might need to round 38 again later in the problem).

Example 7

Use the long division algorithm to divide 1,167 by 38.

Answer

1. 38 won't divide into 1 or 11, but it does divide into 116. After rounding to the nearest ten, we'll make an educated guess that 38 divides in three times; place the 3 above the 6 and multiply, $3 \times 38 = 114$. Subtracting gives 2; bring down the 7.

$$\begin{array}{r} {}^{+2}\\ 38 \\ \times\ 3 \\ \hline 114 \end{array} \qquad \begin{array}{r} 0030 \text{ r } 27 \\ 38\overline{)1167} \\ -\,114\downarrow \\ \hline 27 \end{array}$$

2. At this point, the new dividend is 27, but it is less than 38. So, 38 divides into 27 zero times; place the 0 above the 7.

Caution The 0 above the 7 is necessary because without it, the quotient wouldn't cover the last digit in the dividend.

Sometimes the estimation will get us close to the correct answer, but it won't be quite right. If we had chosen to start this quotient

with 4, we'd get the following:

$$\begin{array}{r} {}^{+3}\\ 38 \\ \times\ 4 \\ \hline 152 \end{array} \qquad \begin{array}{r} 004 \\ 38\overline{)1167} \\ -152 \\ \hline \end{array}$$

152 is larger than 116; so, we should start with a quotient value smaller than 4.

and with 2, we'd get the following:

$$\begin{array}{r} {}^{+1}\\ 38 \\ \times\ 2 \\ \hline 76 \end{array} \qquad \begin{array}{r} 002 \\ 38\overline{)1167} \\ -76 \\ \hline 40 \end{array}$$

Here the remainder, 40, is larger than the divisor, 38. This means that 38 will divide into 116 at least one more time; so, we should start with a quotient value larger than 2.

You Try It 5 **Evaluate each expression using long division. Use Example 7 as a guide.**

a) $\frac{936}{18}$ **b)** $8{,}912 \div 32$ **c)** $12{,}728 \div 43$

Applications Involving Division

Applications that involve division often require that a number of items be divided equally among individuals of a group. A key word meaning division is *each*. It is often found in the last sentence.

Example 8

Kayla is making arrangements for 93 basketball fans to attend an out-of-town high school basketball game. She needs to know how many vans to rent so that everyone has a ride. If each van seats 6, how many vans are needed so that everyone has a seat?

Procedure Here we want to divide up the 93 people into groups of 6 so that each group can fit in a van. Notice that the last sentence includes the word *each*. We will divide to find the answer.

Answer

$$\begin{array}{r} 15 \text{ r } 3 \\ 6\overline{)93} \\ -\ 6 \\ \hline 33 \\ -\ 30 \\ \hline 3 \end{array}$$

Notice that 6 did not divide evenly into 93. What should be done with the remainder of 3?

The answer suggests that each of the 15 vans will be full with 6 people and that one more van will be needed to carry the remaining 3 people. So, 16 vans are needed in all.

Sentence 16 vans are needed so that everyone has a seat.

You Try It 6 **Ruben took 8 children from the Boys and Girls Club to the county fair. He purchased 120 ride tickets for the children. How many ride tickets will each child receive?**

Sentence ______________________

You Try It 7 **Three neighbors decide to have a common garage sale and split the proceeds evenly. If they earn a total of $1,729, how much will each neighbor get?**

Sentence ______________________

You Try It 8 **Guy used his credit card too much and now owes $4,536. To avoid more interest charges, he agreed to cut up his card and pay back all $4,536 in 24 months. How much must Guy pay each month?**

Sentence ______________________

You Try It 9 **Taiyana is purchasing wooden boxes with premade mail slots for the employees in her office. Each box contains mail slots for 12 employees. If the office has 163 employees, how many boxes must she purchase?**

Sentence ______________________

Answers: You Try It and Think About It

You Try It: **1. a)** 10 **b)** 6 **c)** 1 **d)** 8 **e)** 0 **f)** 7 **2. a)** 7 **b)** 9 **c)** 1 **d)** 6 **3. a)** 8 r 4 **b)** 1 r 7 **c)** 7 **d)** 6 r 4 **4. a)** 93 **b)** 232 r 4 **c)** 1,305 r 5 **d)** 5,002 **5. a)** 52 **b)** 278 r 16 **c)** 296 **6.** Each child will receive 15 ride tickets. **7.** They will each get $576, but $1 will be left over (remainder). Since they're friends of mine, they'll give the $1 to me. **8.** Guy must pay $189 each month. **9.** Taiyana must purchase 14 boxes. (13 of the boxes will be full, and 1 box will hold mail for the 7 remaining employees.)

Think About It: No Think About Its in this section.

Section 1.4 Exercises

FOR EXTRA HELP
Student Resources on DVD-ROM
Includes
➤ Student's Solutions Manual
➤ Video Lectures
➤ Chapter Test Prep Video
MyMathLab
MathXL

Think Again

1. When a number is subtracted from itself, the result is 0. When a number is divided by itself, the result is 1. What do these two results have in common?
   The results are both identities. 0 is the identity for addition, and 1 is the identity for multiplication.
2. In division, what should we do when the remainder is larger than the divisor?
   When the remainder is larger than the divisor, we must try a larger number in the quotient.
3. We know that a natural number divided by itself is 1. Is that also true for 0? Explain your answer or show an example that supports your answer.
   No. 0 divided by itself, 0 ÷ 0, is undefined.
4. In a complete division, can the remainder ever be larger than the quotient? Show an example to support your answer.
   Yes. One example is 23 ÷ 8 is 2 r 7.

Focus Exercises

Divide. Check by multiplying the divisor and the quotient.

5. 20 ÷ 4 — 5
6. 28 ÷ 7 — 4
7. 0 ÷ 6 — 0
8. 0 ÷ 5 — 0

9. $\frac{36}{9}$ — 4
10. $\frac{24}{8}$ — 3
11. $\frac{54}{9}$ — 6
12. $\frac{56}{7}$ — 8

13. 27 ÷ 3 — 9
14. 32 ÷ 4 — 8
15. 35 ÷ 5 — 7
16. 18 ÷ 6 — 3

17. $\frac{80}{8}$ — 10
18. $\frac{90}{9}$ — 10
19. $\frac{40}{10}$ — 4
20. $\frac{50}{10}$ — 5

Divide.

21. $90 \div 6$
15

22. $85 \div 5$
17

23. $87 \div 3$
29

24. $76 \div 4$
19

25. $\frac{74}{2}$
37

26. $\frac{96}{8}$
12

27. $\frac{91}{7}$
13

28. $\frac{68}{4}$
17

29. $115 \div 9$
12 r 7

30. $137 \div 6$
22 r 5

31. $166 \div 5$
33 r 1

32. $183 \div 4$
45 r 3

33. $951 \div 8$
118 r 7

34. $966 \div 7$
138

35. $1{,}218 \div 4$
304 r 2

36. $2{,}516 \div 5$
503 r 1

37. $18{,}029 \div 3$
6,009 r 2

38. $16{,}509 \div 8$
2,063 r 5

39. $27{,}036 \div 9$
3,004

40. $35{,}042 \div 7$
5,006

41. $\frac{78{,}300}{6}$
13,050

42. $\frac{90{,}080}{4}$
22,520

43. $\frac{345}{15}$
23

44. $\frac{594}{18}$
33

45. $876 \div 12$
73

46. $2{,}756 \div 13$
212

47. $24{,}054 \div 24$
1,002 r 6

48. $40{,}016 \div 32$
1,250 r 16

49. $\frac{1{,}386}{33}$
42

50. $\frac{1{,}431}{27}$
53

51. $\frac{6{,}300}{75}$
84

52. $\frac{8{,}645}{91}$
95

53. $258{,}387 \div 129$
2,003

54. $375{,}250 \div 125$
3,002

55. $209{,}100 \div 204$
1,025

56. $625{,}770 \div 306$
2,045

Work each application and answer it with a complete sentence.

57. The Lazy *W* summer camp has 238 campers and 17 counselors. The campers are divided evenly into family groups. If each counselor is in charge of one family group, how many campers are in each group?
There are 14 campers in each group.

58. Jorge is in charge of scheduling trash pickup in Timonium, Maryland. He has 18 trash trucks to cover 5,310 homes. If he divides the homes equally among his drivers, how many homes will each driver be assigned?
Each driver will be assigned 295 homes.

59. 96 people attended a concert in Memorial Park. Each person paid the same amount for a ticket. If the total of the receipts was $2,208, how much did each person pay to attend the concert?
Each person paid $23 to attend the concert.

60. Alicia ordered 54 new computers for the teachers at her school. If the total bill (before tax and shipping) came to $53,082, what was the price of each computer?
The price of each computer was $983.

61. A Harley-Davidson distributor needs to ship (by train) 365 new motorcycles. If each boxcar can hold 16 motorcycles, how many boxcars are needed to ship the motorcycles?
23 boxcars are needed to ship the motorcycles.

62. In planning for the next semester, Mr. Tom anticipates 900 students will want to take Elementary Algebra. If each section can contain up to 42 students, how many sections of Elementary Algebra should he schedule?
22 sections of Elementary Algebra should be scheduled.

63. Priority Express printed 1,350 booklets for a large company. Carrie's job is to box them all. If she can fit 24 booklets into each box, how many boxes will she need for all of the booklets?
She will need 57 boxes for all of the booklets.

64. A running club is planning to raise money to help preserve a historic part of Yosemite National Forest. The 27 runners in the club will take turns—relay style—running the entire length of California, from Oregon to Mexico, 783 miles in all. If the distance is divided equally among all of the runners, how many miles will each person run?
Each person will run 29 miles.

Think Outside the Box

Divide.

65. $3{,}992{,}934 \div 3{,}594$
1,111

66. $2{,}107{,}672 \div 1{,}708$
1,234

67. $4{,}640{,}754 \div 1{,}074$
4,321

68. Show an example to justify your answer. Can the Associative Property apply to division?
No. Answers may vary. One possibility:
$24 \div (6 \div 2) = 24 \div 3 = 8$
$(24 \div 6) \div 2 = 4 \div 2 = 2$

SECTION 1.5 Exponents, Square Roots, and the Order of Operations

OBJECTIVES

In this section, you will learn to

- Identify perfect squares.
- Use exponents to abbreviate repeated multiplication.
- Evaluate powers of 10.
- Find square roots of perfect squares.
- Evaluate expressions by applying the order of operations.

You Need to Know

To successfully complete this section, you need to understand

- ☐ Evaluating quantities (1.1)
- ☐ Adding and subtracting whole numbers (1.2)
- ☐ The multiplication table (1.3)
- ☐ Multiplying by 10 and 100 (1.3)
- ☐ The definition of *factors* (1.3)
- ☐ Multiplying whole numbers (1.3)
- ☐ Dividing whole numbers (1.4)

Introduction

Exponents and square roots are found in many formulas in math, chemistry, physics, economics, and statistics. The ability to perform calculations with exponents and square roots, for instance, helps astronomers provide a safe flight for the space shuttle and for the many satellites orbiting the earth.

Research biologists often use exponents to explain the increase or decrease in the number of cells they are observing. When cells divide (split in two), the researcher can count twice as many cells as before.

For example, Sheila is observing the growth of a bacteria used in making yogurt, and she is taking notes on what she sees. Sheila notices that the cells double in number every minute. Here is a summary of her notes:

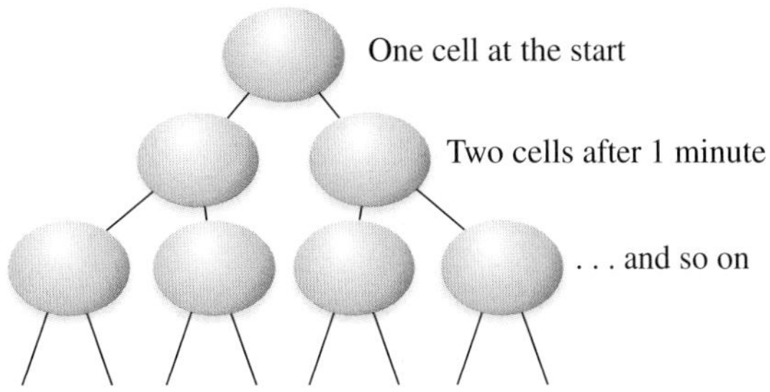

Number of cells at start	At 1 minute	At 2 minutes	At 3 minutes	At 4 minutes	At 5 minutes
1	$1 \cdot 2$	$2 \cdot 2$	$4 \cdot 2$	$8 \cdot 2$	$16 \cdot 2$
1 cell	= 2 cells	= 4 cells	= 8 cells	= 16 cells	= 32 cells

As part of her research, Sheila must make a prediction about the number of cells present at 8 minutes. Assuming the pattern will continue, Sheila notices that the number of cells can be predicted each minute by multiplying by another factor of 2.

Number of cells at start	At 1 minute	At 2 minutes	At 3 minutes	At 4 minutes	At 5 minutes
1	2	$2 \cdot 2$	$2 \cdot 2 \cdot 2$	$2 \cdot 2 \cdot 2 \cdot 2$	$2 \cdot 2 \cdot 2 \cdot 2 \cdot 2$

She realizes that at 3 minutes, there are three factors of 2; at 4 minutes, there are four factors of 2; and so on. With this in mind, Sheila predicts that at 8 minutes, there will be eight factors of 2 bacteria.

$$2 \cdot 2 \cdot 2 \cdot 2 \cdot 2 \cdot 2 \cdot 2 \cdot 2 = 256$$

As you might imagine, representing the number of bacteria by this repeated multiplication can become a bit tedious. Instead, there is a way to abbreviate repeated multiplication—with an exponent.

For values raised to the second power, such as 6^2, it's most common to say that the base is "squared," such as "6 squared." The expression *"6 squared"* comes from the area of a square in which each side has a length of 6 units, as you will see later in this section.

Exponents

Eight factors of 2 can be abbreviated as 2^8. The repeated factor, 2, is called the **base;** and the small, raised ^{8} is called the **exponent** or **power.** 2^8 is read as "2 to the eighth power."

2^8 is called the exponential form, and $2 \cdot 2 \cdot 2 \cdot 2 \cdot 2 \cdot 2 \cdot 2 \cdot 2$ is called the expanded form.

An exponent indicates the number of factors of the base.

Example 1

Write each in exponential form.

a) $7 \cdot 7 \cdot 7$ **b)** $10 \cdot 10 \cdot 10 \cdot 10 \cdot 10 \cdot 10$ **c)** $5 \cdot 5$

Procedure To identify the exponent, count the number of factors of the base.

Answer **a)** 7^3 There are *three* factors of 7.

b) 10^6 There are *six* factors of 10.

c) 5^2 There are *two* factors of 5.

You Try It 1 **Write each in exponential form. Use Example 1 as a guide.**

a) $4 \cdot 4 \cdot 4 \cdot 4 \cdot 4 \cdot 4 \cdot 4$ **b)** $3 \cdot 3 \cdot 3 \cdot 3$ **c)** $1 \cdot 1 \cdot 1 \cdot 1 \cdot 1 \cdot 1 \cdot 1 \cdot 1$

Example 2

Expand each notation and find its value.

a) 2^3 **b)** 3^4 **c)** 7^2

Answer **a)** $2^3 = 2 \cdot 2 \cdot 2 = 8$ 2^3 means three factors of 2.

b) $3^4 = 3 \cdot 3 \cdot 3 \cdot 3$
$= (3 \cdot 3) \cdot (3 \cdot 3)$
$= 9 \cdot 9 = 81$

If you use the Associative Property to group two factors at a time and then multiply them, you'll get $9 \cdot 9$, which is 81.

c) $7^2 = 7 \cdot 7 = 49$ 7^2 means two factors of 7.

You Try It 2 **Expand each notation and find its value. Use Example 2 as a guide.**

a) 6^2 **b)** 2^4 **c)** 9^3

How should we interpret 5^1? 5^1 means "one factor of 5." In this case, there are no repeated factors of 5; there is only 5 itself. In other words, $5^1 = 5$.

This principle is true for any base:

If b represents any base, then $b^1 = b$.

You Try It 3 **Rewrite each notation without an exponent.**

a) 2^1 **b)** 9^1 **c)** 17^1 **d)** 1^1

The Powers of 10

In Section 1.3, we discussed multiples of 10 (such as 20, 30, and 40) and multiples of 100 (such as 200, 300, and 400). We also learned that when multiplying by 10 or 100, we need only place the appropriate number of zeros on the "end" of the other factor.

For example, $7 \cdot 10 = 70$ and $4 \cdot 100 = 400$.

Numbers such as 100, 1,000, and 10,000 are called **powers of 10** because

$$10^2 = 10 \cdot 10 = 100.$$
$$10^3 = 10 \cdot 10 \cdot 10 = 1{,}000.$$
$$10^4 = 10 \cdot 10 \cdot 10 \cdot 10 = 10{,}000.$$

Based on the previous results, notice the following:

The exponent of 10 indicates the number of zeros that follow the 1.

So, 10^6 is a 1 followed by six zeros: 1,000,000; in other words, 1,000,000 is the 6th power of 10. Likewise, 10,000,000 is a 1 followed by seven zeros and is abbreviated as 10^7.

Example 3

For each power of 10, write its value or abbreviate it with an exponent, whichever is missing.

a) 1,000 **b)** 100,000 **c)** 10^4 **d)** 10^8

Procedure For parts a and b, count the number of zeros; for parts c and d, the exponent indicates the number of zeros following a 1.

Answer
a) $1{,}000 = 10^3$ **b)** $100{,}000 = 10^5$

c) $10^4 = 10{,}000$ **d)** $10^8 = 100{,}000{,}000$

You Try It 4 **For each power of 10, write its value or abbreviate it with an exponent, whichever is missing. Use Example 3 as a guide.**

a) 10^5 **b)** 10^7 **c)** 10^1

d) 100 **e)** 10,000 **f)** 1,000,000

Three hundred can be written as 300 and as $3 \cdot 100$. This number can be further abbreviated as $3 \cdot 10^2$. In other words, $300 = 3 \cdot 10^2$.

When a number ends in one or more zeros, it can be written as a product of a whole number and a power of 10.

For example, 45,000,000 has six zeros following the 45. So, the *sixth* power of 10, or 10^6, is a factor: $45{,}000{,}000 = 45 \cdot 10^6$.

Example 4

Identify the power of 10 and rewrite the number as a product of a whole number and a power of 10.

a) 6,000 **b)** 290,000 **c)** 506,000,000 **d)** 80

Procedure Count the number of zeros following the nonzero digit(s); that is the power of 10.

Answer
a) $6{,}000 = 6 \cdot 10^3$ 6 followed by three zeros

b) $290{,}000 = 29 \cdot 10^4$ 29 followed by four zeros

c) $506{,}000{,}000 = 506 \cdot 10^6$ 506 followed by six zeros

d) $80 = 8 \cdot 10^1$ 8 followed by one zero

You Try It 5 **As outlined in Example 5, rewrite the number as a product of a whole number and a power of 10.**

a) 240 = ____________ **b)** 5,600 = ____________

c) 308,000,000 = ____________ **d)** 7,260,000 = ____________

Perfect Squares

This is a diagram of a *unit* square, just as we saw in Section 1.3.

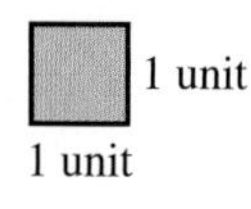

In the following squares, notice that a number is associated with each square—the number of unit squares within. Each number is called a **perfect square** because the unit squares form a square.

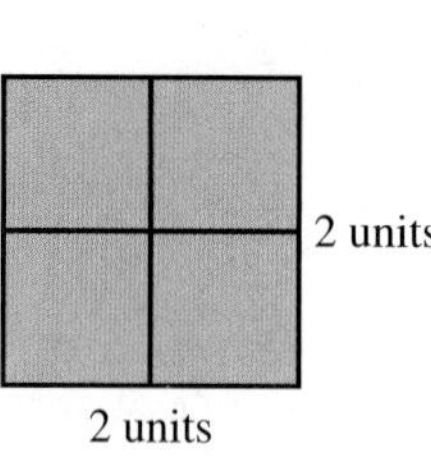

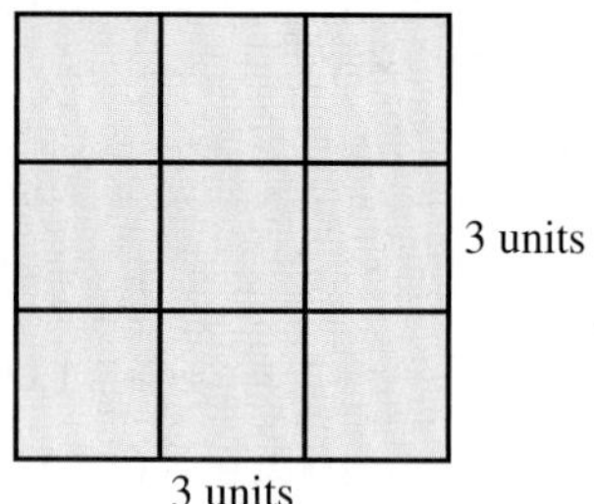

2 units $\times$ 2 units = 4 square units
4 is a perfect square.

3 units $\times$ 3 units = 9 square units
9 is a perfect square.

Whenever we create a square with a number, n, as the length of one side, that same number, n, must appear as the length of the other side. The product of these two numbers, $n \cdot n$, is called a perfect square number or a perfect square.

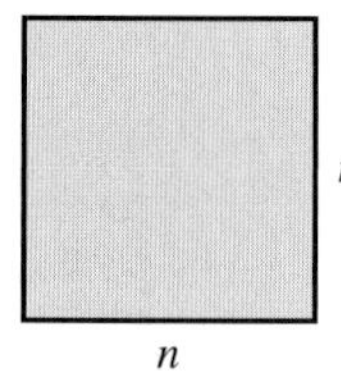

In other words, because

$1 \cdot 1 = 1^2 = 1,$ 1 is a perfect square.

$2 \cdot 2 = 2^2 = 4,$ 4 is a perfect square.

$3 \cdot 3 = 3^2 = 9,$ 9 is a perfect square.

$4 \cdot 4 = 4^2 = 16,$ 16 is a perfect square.

$5 \cdot 5 = 5^2 = 25,$ 25 is a perfect square.

The list of perfect squares goes on and on. Anytime you multiply a whole number by itself, you get a result that is (automatically) a perfect square.

Notice, also, that many numbers are *not* perfect squares: 2, 3, 5, 6, 7, and the list goes on and on. 2 is not a perfect square because no whole number multiplied by itself equals 2. The same is true for all of the numbers in this list.

Example 5

Is it possible to draw a square that has

a) 81 unit squares within it? **b)** 24 unit squares within it?

Answer **a)** Yes, a square with 9 units on each side will have 81 square units within it because $9 \cdot 9 = 81$.

b) No, there is no whole number that when multiplied by itself will give 24.

You Try It 6 **Use Example 5 as a guide to answer the following questions. Is it possible to draw a square that has**

a) 36 unit squares within it? ______

b) 12 unit squares within it? ______

c) 49 unit squares within it? ______

Example 6

Find the perfect square number associated with each product.

a) 7^2 **b)** 11^2 **c)** 13^2 **d)** 20^2

Procedure Expand each product and evaluate.

Answer **a)** $7 \cdot 7 = \mathbf{49}$ **b)** $11 \cdot 11 = \mathbf{121}$ **c)** $13 \cdot 13 = \mathbf{169}$ **d)** $20 \cdot 20 = \mathbf{400}$

You Try It 7 **Find the perfect square number associated with each product. Use Example 6 as a guide.**

a) 6^2 **b)** 10^2 **c)** 18^2

Square Roots

Consider this perfect square of 16. We know from our discussion of perfect squares that the number 4 makes the perfect square 16. For this reason, we call 4 a **square root** of 16.

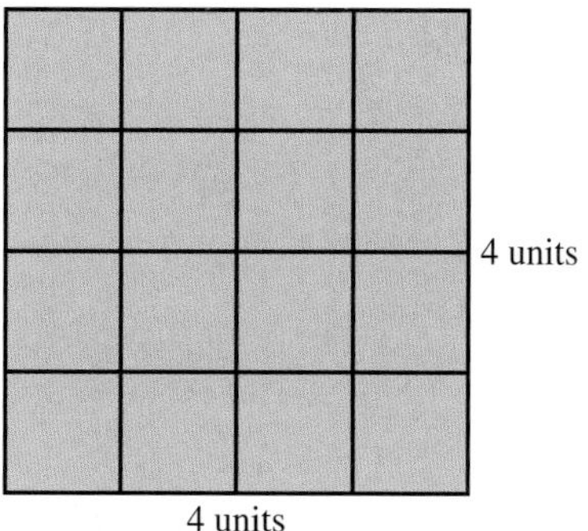

The Square Root Property

If $p = r^2$ (p is a perfect square using r as one side), then r is a square root of p.

Example 7

What is a square root of the following numbers?

a) 49 **b)** 36 **c)** 1 **d)** 25

Procedure What number multiplied by itself gives each of those numbers?

Answer **a)** A square root of 49 is **7**. **b)** A square root of 36 is **6**.

c) A square root of 1 is **1**. **d)** A square root of 25 is **5**.

You Try It 8 **What is a square root of the following numbers? Use Example 7 as a guide.**

a) A square root of 4 is __________. **b)** A square root of 9 is __________.

c) A square root of 64 is __________. **d)** A square root of 100 is __________.

The Square Root Symbol

The square root symbol, $\sqrt{}$, called a **radical,** makes it easy to write square roots.

So, instead of writing "A square root of 16 is 4," we can simply write $\sqrt{16} = 4$. The number within the radical (in this case 16) is called the **radicand.**

When we use the radical to find a square root of a number such as 16, the radical becomes an operation. Just like the plus sign tells us to apply addition to two numbers, the radical tells us to take the square root of a number.

Example 8

Evaluate each square root.

a) $\sqrt{1}$ **b)** $\sqrt{25}$ **c)** $\sqrt{144}$

Procedure Because each of these radicands is a perfect square, applying the radical leaves a number with no radical symbol.

Answer **a)** $\sqrt{1} = 1$ **b)** $\sqrt{25} = 5$ **c)** $\sqrt{144} = 12$

You Try It 9 **Evaluate each square root. Use Example 8 as a guide.**

a) $\sqrt{4} =$ **b)** $\sqrt{9} =$ **c)** $\sqrt{121} =$

d) $\sqrt{36} =$ **e)** $\sqrt{100} =$ **f)** $\sqrt{81} =$

The Order of Operations

Let's review the operations we've worked with so far.

Operation	Example	Name
Multiplication	$3 \cdot 5 = 15$	product
Division	$14 \div 2 = 7$	quotient
Subtraction	$12 - 9 = 3$	difference
Addition	$2 + 8 = 10$	sum
Exponent	$2^3 = 8$	power
Radical	$\sqrt{25} = 5$	square root

As we learned earlier, to *evaluate a mathematical expression* means to find the value of the expression. Consider this expression: $3 + 4 \cdot 5$. How could it be evaluated?

If we apply the operation of multiplication first and then addition, we get

$$3 + 4 \cdot 5$$
$$= 3 + 20 \quad \text{We shall soon see that this is } \textit{correct.}$$
$$= \mathbf{23}$$

If, however, we evaluate using addition first and then multiplication, we get

$$3 + 4 \cdot 5$$
$$= 7 \cdot 5 \quad \text{We shall soon see that this is } \textit{incorrect.}$$
$$= \mathbf{35}$$

Notice that depending on which operation is applied first, we get two different results. Math, however, is an exact science and doesn't allow for two different values of the same expression.

Because of this, a set of guidelines, called the **order of operations,** was established. The order of operations tells us which operation should be applied first in an expression. The order of operations was developed with these thoughts in mind

1. Any quantity within grouping symbols should be applied first. Grouping symbols include parentheses ( ); brackets [ ]; braces { }; and the radical, or the square root symbol, $\sqrt{}$.

There are basically two types of grouping symbols

- Those that form a quantity, such as ( ), [ ], and { }
- Those that are actual operations, such as $\sqrt{}$

The radical is *both* a grouping symbol *and* an operation.

2. Because an exponent is an abbreviation for repeated multiplication, exponents should rank higher than multiplication.
3. Because multiplication and division are inverse operations they should be ranked together. Also, because multiplication is an abbreviation for repeated addition (and division is an abbreviation for repeated subtraction), multiplication (and division) should rank higher than addition (and subtraction).
4. Because addition and subtraction are inverse operations they should be ranked together.

In summary,

The Order of Operations

1. Evaluate within all grouping symbols (one at a time) if there are any.
2. Apply any exponents.
3. Apply multiplication and division *reading from left to right.*
4. Apply addition and subtraction *reading from left to right.*

We sometimes refer to the order of the operations by their rank. For example, we might say that an exponent has a *higher rank* than multiplication.

Because multiplication and division have the same rank, we must apply them (carefully) from *left to right*. You'll see how we do this in the following examples.

When evaluating an expression, is it ever possible to apply addition before multiplication? Explain.

The best way to understand these guidelines is to put them to work. We'll find that there is only one way to evaluate an expression using the rules, but we'll also find that some steps can be combined in certain situations. For now, though, let's evaluate each expression one step at a time.

Example 9

Evaluate each expression according to the order of operations.

a) $14 - 6 \div 2$ **b)** $(14 - 6) \div 2$ **c)** $7 + 3^2$

d) $(7 + 3)^2$ **e)** $24 \div 4 \cdot 2$ **f)** $24 \div (4 \cdot 2)$

Answer Each expression has two operations; some of the expressions have grouping symbols that will affect the order in which the operations are applied.

a) $14 - 6 \div 2$ Two operations, subtraction and division: divide first.

$= 14 - 3$ Notice that the minus sign appears in the second step.

$= 11$ That's because it hasn't been applied yet.

b) $(14 - 6) \div 2$ Here are the same two operations as in part a, this time with grouping symbols. Evaluate the expression inside the parentheses first.

$= 8 \div 2$ Because we've already evaluated within the grouping symbols, we don't need the parentheses anymore.

$= 4$

c) $7 + 3^2$ Two operations, addition and exponent: apply the exponent first, then add.
$= 7 + 9$
$= 16$

d) $(7 + 3)^2$ Here are the same two operations as in part c; work within the grouping symbols first.
$= (10)^2$ $(10)^2$ is also 10^2; at this point, the parentheses are no longer necessary.
$= 100$

e) $24 \div 4 \cdot 2$ Two operations: division and multiplication. Because they have the same rank and there are no grouping symbols to tell us which to apply first, we apply them in order from *left to right*. That means that division is applied first.
$= 6 \cdot 2$
$= 12$

f) $24 \div (4 \cdot 2)$ This time we do have grouping symbols; so, we begin by evaluating the expression within the parentheses.
$= 24 \div 8$
$= 3$

You Try It 10 **Evaluate each expression according to the order of operations. First, identify the two operations; then identify which is to be applied first. Show all steps. Use Example 9 as a guide.**

a) $24 \div 6 + 2$ **b)** $24 \div (6 + 2)$ **c)** $10 - 3 \cdot 2$

d) $(10 - 3) \cdot 2$ **e)** $12 \div 2^2$ **f)** $(12 \div 2)^2$

Some expressions contain more than two operations. In those situations, we need to be even more careful when we apply the order of operations.

Example 10

Evaluate each expression according to the order of operations.

a) $36 \div 3 \cdot 6 - 2$ **b)** $36 \div (3 \cdot 6) - 2$ **c)** $36 \div [3 \cdot (6 - 2)]$

Answer Each expression has three operations. In part c, there is a smaller quantity $(6 - 2)$ within the larger quantity of the brackets [ ].

a) $36 \div 3 \cdot 6 - 2$ Because multiplication and division have the same rank, we apply them from left to right. We divide first.
$= 12 \cdot 6 - 2$ Notice that we are applying only one operation at a time and rewriting everything else.
$= 72 - 2$
$= 70$

b) $36 \div (3 \cdot 6) - 2$ Evaluate the expression within the grouping symbols first.
$= 36 \div 18 - 2$ Divide.
$= 2 - 2$ Subtract.
$= 0$

c) $36 \div [3 \cdot (6 - 2)]$ Start with what is inside the large brackets. Inside those grouping symbols is another quantity, and we must evaluate it first: $6 - 2 = 4$.
$= 36 \div [3 \cdot 4]$ Evaluate within the brackets: $3 \cdot 4 = 12$.
$= 36 \div 12$ Divide.
$= 3$

Example 10c illustrates that when one quantity is within another one, the inner quantity is to be evaluated first.

Sometimes an expression has two sets of grouping symbols that are unrelated to each other, meaning that evaluation of one does not affect evaluation of the other. In other words, some quantities can be evaluated at the same time.

For example, in the expression $(8 - 3) \cdot (12 \div 4)$, we can evaluate within each grouping symbol regardless of the operation.

$(8 - 3) \cdot (12 \div 4)$ There are three operations: subtraction, multiplication, and division.

$= (5) \cdot (3)$ Subtraction and division can be applied at the same time because the order of operations tells us to begin by evaluating what is inside the grouping symbols.

$= 15$

You Try It 11 **Evaluate each expression according to the order of operations. Use Example 10 as a guide.**

a) $36 \div 3 + 3 \cdot 2$ **b)** $36 \div (3 + 3) \cdot 2$ **c)** $36 \div (3 + 3 \cdot 2)$

d) $11 + 4 \cdot 6 - 1$ **e)** $(11 + 4) \cdot (6 - 1)$ **f)** $11 + [4 \cdot (6 - 1)]$

g) $2 \cdot 3^2 \div (6 + 3)$ **h)** $(2 \cdot 3)^2 \div (6 + 3)$

Now let's look at some examples that contain a radical.

Example 11

Evaluate each expression completely. Remember, the radical is both a grouping symbol and an operation.

a) $\sqrt{5 + 11}$ **b)** $\sqrt{3^2 + 4^2}$ **c)** $13 - 2 \cdot \sqrt{9}$

Procedure Because the radical is a grouping symbol, we must evaluate within it first.

Answer

a) $\sqrt{5 + 11}$ First, apply addition.

$= \sqrt{16}$ Now apply the square root.

$= 4$

b) $\sqrt{3^2 + 4^2}$ Apply both exponents within the same step.

$= \sqrt{9 + 16}$ Next, apply addition.

$= \sqrt{25}$ Apply the square root.

$= 5$

c) $13 - 2 \cdot \sqrt{9}$ The radical is a grouping symbol; so, we should apply it—the square root—first. Next, apply multiplication and then subtraction.

$= 13 - 2 \cdot 3$

$= 13 - 6$

$= 7$

You Try It 12 **Evaluate each expression according to the order of operations. Use Example 11 as a guide.**

a) $\sqrt{4 \cdot 9}$ **b)** $\sqrt{25} - \sqrt{9}$ **c)** $\sqrt{1 + (12 \cdot 4)}$ **d)** $\sqrt{(6 - 2) \cdot 5^2}$

Calculator Tip **Scientific Calculator**

Is your calculator a nonscientific calculator or a scientific calculator? Input the expression $3 + 4 \times 5$ to find out.

A nonscientific calculator is programmed to evaluate an expression in the order the operations are entered. On this calculator, each time you push an operation key (or the = key) the expression is evaluated, whether you are finished or not.

For example, on a nonscientific calculator, the expression $3 + 4 \times 5$ is evaluated this way:

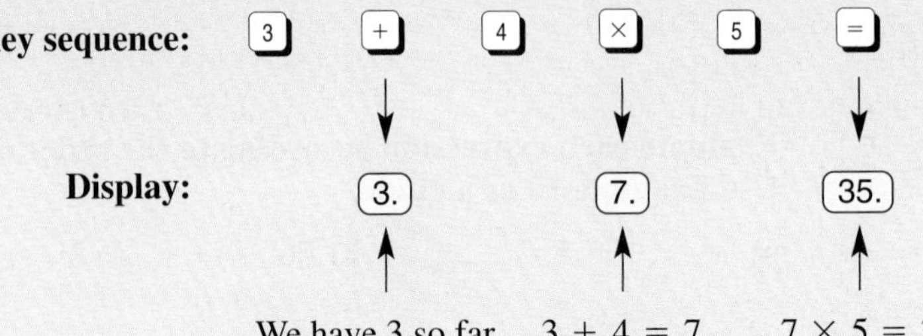

A scientific calculator, though, is programmed to use the order of operations correctly. In doing so, it waits to see what operation button is pushed next before evaluating.

For example, on a scientific calculator, the expression $3 + 4 \times 5$ would be evaluated this way:

To determine whether your calculator is a scientific calculator, input the expression $3 + 4 \times 5$ and see which answer you get.

Answers: You Try It and Think About It

You Try It: **1. a)** 4^7 **b)** 3^4 **c)** 1^8 **2. a)** $6 \cdot 6 = 36$ **b)** $2 \cdot 2 \cdot 2 \cdot 2 = 16$ **c)** $9 \cdot 9 \cdot 9 = 729$ **3. a)** 2 **b)** 9 **c)** 17 **d)** 1 **4. a)** 100,000 **b)** 10,000,000 **c)** 10 **d)** 10^2 **e)** 10^4 **f)** 10^6 **5. a)** $24 \cdot 10^1$ **b)** $56 \cdot 10^2$ **c)** $308 \cdot 10^6$ **d)** $726 \cdot 10^4$ **6. a)** Yes, a square with 6 units on each side will have 36 square units. **b)** No, there is no whole number that when multiplied by itself equals 12. **c)** Yes, a square with 7 units on each side will have 49 square units. **7. a)** 36 **b)** 100 **c)** 324 **8. a)** 2 **b)** 3 **c)** 8 **d)** 10 **9. a)** 2 **b)** 3 **c)** 11 **d)** 6 **e)** 10 **f)** 9 **10. a)** 6 **b)** 3 **c)** 4 **d)** 14 **e)** 3 **f)** 36 **11. a)** 18 **b)** 12 **c)** 4 **d)** 34 **e)** 75 **f)** 31 **g)** 2 **h)** 4 **12. a)** 6 **b)** 2 **c)** 7 **d)** 10

Think About It: **1.** Yes. If the operation within grouping symbols is addition and multiplication is outside the grouping symbols, addition is to be applied before multiplication.

Section 1.5 Exercises

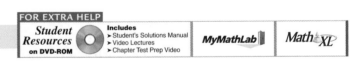

FOR EXTRA HELP
Student Resources on DVD-ROM
Includes
➤ Student's Solutions Manual
➤ Video Lectures
➤ Chapter Test Prep Video
MyMathLab
MathXL

Think Again

1. When we are evaluating an expression, is it ever possible to apply addition before multiplication? Explain your answer. *(Refer to Think About It 1)*
   Yes, we can apply addition before multiplication when the addition is within a pair of grouping symbols.

2. How many factors of 10 are in $10^3 \cdot 10^5$? Explain your answer or show an example that supports your answer.
   There are eight factors of 10. When each number is expanded, there are three factors of 10 and five more factors of 10, all multiplied together.

3. 20 is not a perfect square; so, it does not have a whole number square root. Between which two consecutive whole numbers is the value of $\sqrt{20}$? Explain your answer or show an example that supports your answer.
$\sqrt{20}$ is between 4 and 5 because $\sqrt{20}$ is between $\sqrt{16}$ and $\sqrt{25}$.

4. If an expression with three whole numbers has subtraction as the only operation, does it matter which subtraction we apply first? Show an example to support your answer.
Yes, it does matter which subtraction we apply first. For example, in the expression $9 - 6 - 1$, we must subtract in order from left to right, giving us an answer of 2. However, if we subtract 1 from 6 first, we get 4 as an answer, which is incorrect.

Focus Exercises

Expand each notation and find its value.

5. $6^3 =$
$6 \cdot 6 \cdot 6 = 216$

6. $2^5 =$
$2 \cdot 2 \cdot 2 \cdot 2 \cdot 2 = 32$

7. $15^2 =$
$15 \cdot 15 = 225$

8. $3^3 =$
$3 \cdot 3 \cdot 3 = 27$

9. $12^1 =$
12

10. $4^3 =$
$4 \cdot 4 \cdot 4 = 64$

11. $5^4 =$
$5 \cdot 5 \cdot 5 \cdot 5 = 625$

12. $8^1 =$
8

13. $10^3 =$
$10 \cdot 10 \cdot 10 = 1{,}000$

14. $10^5 =$
$10 \cdot 10 \cdot 10 \cdot 10 \cdot 10 = 100{,}000$

Express each number as a power of 10.

15. 100,000 =
10^5

16. 1,000,000,000 =
10^9

17. 10,000 =
10^4

18. 1,000,000 =
10^6

Rewrite each number as a product of a whole number and a power of 10.

19. 300 =
$3 \cdot 10^2$

20. 5,000 =
$5 \cdot 10^3$

21. 48,000 =
$48 \cdot 10^3$

22. 710,000 =
$71 \cdot 10^4$

23. 9,500,000 =
$95 \cdot 10^5$

24. 200,000,000 =
$2 \cdot 10^8$

Evaluate the following square roots.

25. $\sqrt{49}$
7

26. $\sqrt{25}$
5

27. $\sqrt{36}$
6

28. $\sqrt{1}$
1

29. $\sqrt{81}$
9

30. $\sqrt{16}$
4

31. $\sqrt{144}$
12

32. $\sqrt{64}$
8

Evaluate each expression according to the order of operations.

33. $30 \div 5 + 1$
7

34. $30 \div (5 + 1)$
5

35. $(8 + 5) \cdot 2$
26

36. $8 + 5 \cdot 2$
18

37. $5 \cdot 6 \div 3$
10

38. $5 \cdot (6 \div 3)$
10

39. $5 \cdot 3^2$
45

40. $2^3 \cdot 2^2$
32

41. $2^3 \cdot 3^2$
72

42. $28 \div (7 \cdot 2)$
2

43. $28 \div 7 \cdot 2$
8

44. $16 \div 4 - 2$
2

45. $30 \div 2 \cdot 3$
45

46. $30 \div (2 \cdot 3)$
5

47. $7^2 + 5 - 3$
51

48. $5 \cdot 2^2 - 7$
13

49. $(5 \cdot 2)^2 - 7$
93

50. $4^2 \div 2 + 2$
10

51. $(5 + 3) \cdot 9$
72

52. $(6 + 12) \div (2 \cdot 3)$
3

53. $6 + [12 \div (2 \cdot 3)]$
8

54. $12 + [28 \div (7 - 3)]$
19

55. $24 \div (6 - 2) \cdot 3$
18

56. $6 + 12 \div 2 \cdot 3$
24

57. $(12 + 28) \div (7 - 3)$
10

58. $[(6 - 2) \cdot 3]^2$
144

59. $(6 - 2) \cdot 3^2$
36

60. $3 + \sqrt{16}$
7

61. $9 \cdot \sqrt{25}$
45

62. $11 - \sqrt{49}$
4

63. $6^2 - \sqrt{25}$
31

64. $\sqrt{64} - \sqrt{25}$
3

65. $9 + \sqrt{2 \cdot 2}$
11

66. $\sqrt{4 \cdot 9}$
6

67. $3 \cdot \sqrt{9} + 7$
12

68. $\sqrt{8 \cdot 6 + 1}$
7

69. $24 - 8 \div 2 \cdot 4$
8

70. $22 - 2^2 \cdot 5$
2

71. $8 + 2 \cdot 3^2$
26

72. $6 + 4 \cdot 5^2$
106

73. $(24 - 8) \div 2 \cdot 4$
32

74. $(22 - 2)^2 \cdot 5$
2,000

75. $(8 + 2) \cdot 3^2$
90

76. $6 + (4 \cdot 5)^2$
406

77. $[9 - (3 + 1)]^2$
25

78. $[9 + (3 - 1)]^2$
121

79. $5 \cdot 2^2 - 4 \cdot (5 - 2)$
8

80. $44 - 4 \cdot (5 - 2)^2$
8

81. A square has a side length of 4 yards.
a) Draw and label all four sides of the square.

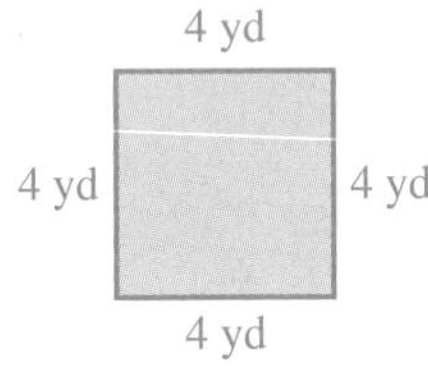

b) Find the perimeter of the square.
The perimeter is 16 yards.
c) Find the area of the square.
The area is 16 square yards.

82. A square has a side length of 9 inches.
a) Draw and label all four sides of the square.

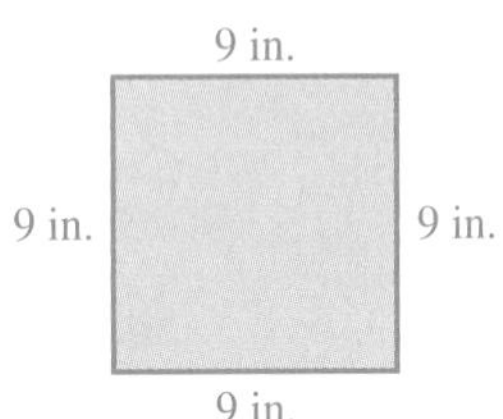

b) Find the perimeter of the square.
The perimeter is 36 inches.
c) Find the area of the square.
The area is 81 square inches.

83. The floor of a cottage is in the shape of a square. Each side is 20 feet long.
a) Draw and label all four sides of the square.

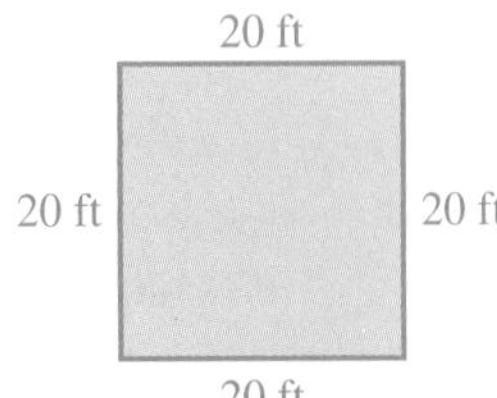

b) Find the perimeter of the square.
The perimeter is 80 feet.

c) Find the area of the square.
The area is 400 square feet.

84. A city park is in the shape of a square. Each side is 100 feet long.
a) Draw and label all four sides of the square.

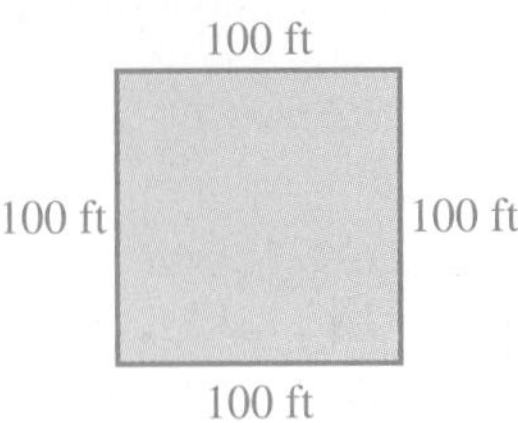

b) Find the perimeter of the square.
The perimeter is 400 feet.
c) Find the area of the square.
The area is 10,000 square feet.

85. A square has an area of 36 square feet.
a) What is the length of each side of the square?
The length of each side is 6 feet.
b) What is the perimeter of the square?
The perimeter is 24 feet.

86. A square has an area of 100 square inches.
a) What is the length of each side of the square?
The length of each side is 10 inches.
b) What is the perimeter of the square?
The perimeter is 40 inches.

87. A tabletop is in the shape of a square and has an area of 9 square feet.
a) What is the length of each side of the tabletop?
The length of each side is 3 feet.
b) What is the perimeter of the tabletop?
The perimeter is 12 feet.

88. A sandbox is in the shape of a square and has an area of 49 square yards.
a) What is the length of each side of the sandbox?
The length of each side is 7 yards.
b) What is the perimeter of the sandbox?
The perimeter is 28 yards.

Think Outside the Box

Predict whether the value of the expression will be odd or even. Explain your answer.

89. $(3)^{14}$
Odd. The product of any two or more odd numbers is another odd number.

90. $(6)^{17}$
Even. The product of any two or more even numbers is another even number.

Given the numbers 1, 2, 3, and 4, using all four numbers exactly once, grouping symbols (if needed), and one or more of the operation symbols (+, −, ×, and ÷), make an expression equal to the underlined number. The numbers may be rearranged in any order, and one or more of the numbers may be used as an exponent.

Example A: If the answer is <u>18</u>, we can create the expression $4^2 + 3 - 1 = 18$.

Example B: If the answer is <u>19</u>, we can create the expression $(4 + 2) \times 3 + 1 = 19$.

91. <u>11</u> **92.** <u>15</u> **93.** <u>21</u> **94.** <u>45</u>

Answers may vary. Some possibilities:
$4 \times 3 - 2 + 1$ $(4 - 1)(3 + 2)$ $4 \times (3 + 2) + 1$ $3^2 \times (4 + 1)$

SECTION 1.6 Factors

OBJECTIVES

In this section, you will learn to

- Identify factors and factor pairs.
- Identify prime numbers and composite numbers.
- Use the divisibility tests to determine if one number is a factor of another number.
- Find the prime factorization of composite numbers.

You Need to Know

To successfully complete this section, you need to understand

- ☐ The multiplication table (1.3)
- ☐ Multiples of a number (1.3)
- ☐ Dividing whole numbers (1.4)
- ☐ Square roots (1.5)

Introduction

Chemists want to know which atoms make up a particular molecule. They know, for example, that

- Water is two parts hydrogen (H) and one part oxygen (O), written H_2O.
- Ammonia is three parts hydrogen (H) and one part nitrogen (N), written NH_3.
- Methane is four parts hydrogen (H) and one part carbon (C), written CH_4.

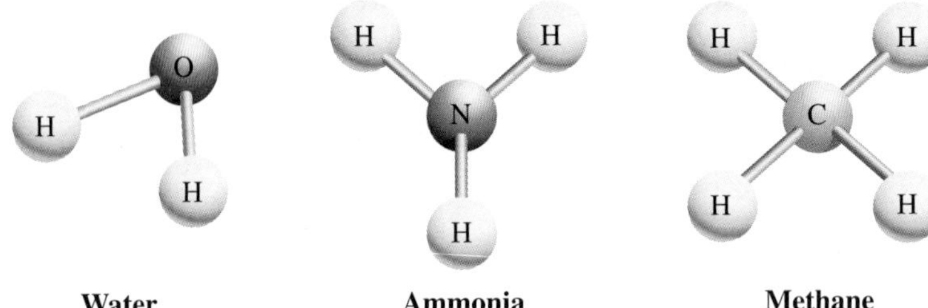

In math, we want to understand the factors that make up a number. For example, $15 = 3 \cdot 5$. Two of the factors of 15 are 3 and 5. But what about a number such as 105? Does it have a factor of 3? Does it have a factor of 5? What other factors does 105 have? Before the end of this section, you will be able to answer each of those questions.

To prepare us to answer those questions, let's recall three definitions about multiplication and division.

From Section 1.3, recall the following

1. The product of any two numbers is the result when those two numbers are multiplied together, and the numbers in a product are called *factors*.

   For example, the product of 3 and 5 is 15, $3 \times 5 = 15$. This means that 3 and 5 are factors of 15.

2. A *multiple* of any number, $\boldsymbol{a}$, is a product involving $\boldsymbol{a}$ and some other whole number.

   For example, 60 is a multiple of 3 because $60 = 20 \cdot 3$.

From Section 1.4, recall the following

3. If a number, $\boldsymbol{a}$, divides evenly into a number, $\boldsymbol{b}$, we say that $\boldsymbol{b}$ is *divisible* by $\boldsymbol{a}$.

   For example, 7 divides evenly into 35; so, 35 is *divisible* by 7.

Combining all of those definitions, if two whole numbers, m and n, multiply to get a product, p, that is, $m \cdot n = p$, then

1. p is a *multiple* of both m and n.
2. p is *divisible by* both m and n.
3. Both m and n *divide evenly into* p.
4. Both m and n are *factors* of p.

Example 1

Because $6 \cdot 7 = 42$

- 42 is a multiple of both 6 and 7.
- 42 is divisible by both 6 and 7.
- both 6 and 7 divide evenly into 42.
- both 6 and 7 are factors of 42.

You Try It 1 **Use Example 1 as a guide to complete each of the following.**

a) Because $4 \cdot 5 = 20$:

- ______________________
- ______________________
- ______________________
- ______________________

b) Because $9 \cdot 8 = 72$:

- ______________________
- ______________________
- ______________________
- ______________________

Factors

Any whole number can be written as the product of two numbers, called a **factor pair.** For example,

- 12 can be written as $2 \cdot 6$; so, $2 \cdot 6$ is a factor pair of 12.
- 9 can be written as $3 \cdot 3$; so $3 \cdot 3$ is a factor pair of 9.
- 7 can be written as $1 \cdot 7$; so, $1 \cdot 7$ is a factor pair of 7.

Many numbers have more than one factor pair, and we can use a *factor pair table* to help us find all of the factors of a particular number.

Consider this factor pair table of 24. Notice how it is organized to find all of the factor pairs of 24.

24	
1	24
2	12
3	8
4	6

Start with 1 on the left and write the other factor of the pair, 24, on the right. Do the same for 2, 3, and so on. Do not include 5 because 5 isn't a factor of 24.

Now that the factor pair table is complete, we can list all of the factors of 24.

The factors of 24 are 1, 2, 3, 4, 6, 8, 12, and 24.

24	
1	24
2	12
3	8
4	6
6	4
8	3
12	2
24	1

Consider the complete list of the factor pairs of 24 at the right. Notice that every factor pair is written twice; this is not necessary. In other words, if we include the factor pair 3 and 8, we don't need to include the factor pair 8 and 3.

When creating a factor pair table, you should write the left side in numerical order starting with 1 so that you don't accidentally skip any factor pairs. But how do you know when to stop searching for factor pairs? Is there a number that tells you when to stop searching for factors?

Yes. This number is *the square root of the first perfect square larger than the given number.*

For example, if we're trying to find all of the factor pairs of 24, the first perfect square after 24 is 25. Because $\sqrt{25} = 5$, we don't need to go beyond 5 in our search for factor pairs.

There are factors of 24 beyond 5, such as 8; but this factor is paired with a number less than 5, namely 3.

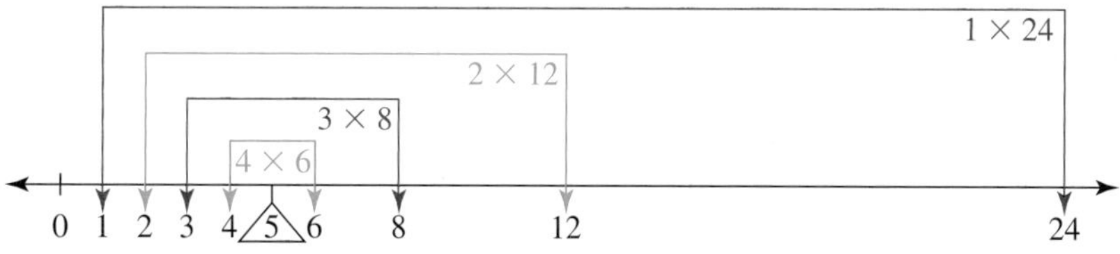

Example 2

Use a factor pair table to find all of the factor pairs of 30. From the table, write a list of the factors of 30.

Procedure Think of 30 as a product of two numbers. Start with 1 on the left side, then 2, then 3, and so on, and decide whether those numbers are factors of 30.

Answer

30		
1	30	$1 \cdot 30 = 30$
2	15	$2 \cdot 15 = 30$
3	10	$3 \cdot 10 = 30$
5	6	$5 \cdot 6 = 30$

Notice that 4 is not on this list because 4 doesn't divide evenly into 30.

So, the factors of 30 are 1, 2, 3, 5, 6, 10, 15, and 30.

You Try It 2 **Use a factor pair table to find all of the factor pairs of each number. From the factor pair table, write the list of factors of that number. Use Example 2 as a guide.**

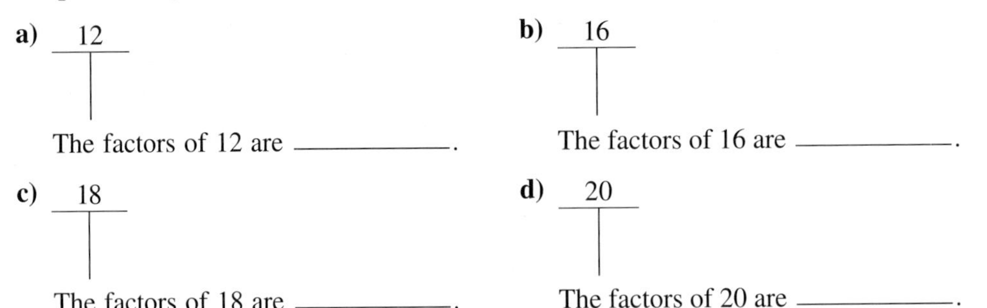

a) 12

The factors of 12 are ____________.

b) 16

The factors of 16 are ____________.

c) 18

The factors of 18 are ____________.

d) 20

The factors of 20 are ____________.

Prime and Composite Numbers

Whole numbers greater than 1 fall into one of two categories: They are either *prime* or *composite*.

A whole number is a **prime number** when it has exactly two distinct whole number factors: 1 and itself.

A whole number that has *more* than two distinct factors is a **composite number.** A composite number is any whole number (greater than 1) that is not prime.

Caution *1 is not a prime number* because it has only *one* factor.

Example 3

For each number, determine whether it is prime or composite.

a) 7 **b)** 12 **c)** 9

Answer

a) 7 is a prime number because the only factors of 7 are 1 and 7. Or we might say that 7 is prime because the only whole numbers that divide evenly into 7 are 1 and 7.

b) 12 is a composite number because it has more than two factors. The factors of 12 are 1, 2, 3, 4, 6, and 12.

c) 9 is a composite number because it has more than two factors. The factors of 9 are 1, 3, and 9.

You Try It 3 **For each number, determine whether it is prime or composite. Use Example 3 as a guide.**

a) 15 **b)** 13 **c)** 1 **d)** 4

You Try It 4 **Here is the list of prime numbers less than 100.**

2, 3, 5, 7, 11, 13, 17, 19, 23, 29, 31, 37, 41, 43, 47, 53, 59, 61, 67, 71, 73, 79, 83, 89, 97

a) What do you notice about the first prime number? ____________

b) What do you notice about all of the other prime numbers? ____________

c) Are all odd numbers prime numbers? Support your answer with some examples.

____

Refer to the list of prime numbers less than 100 in You Try It 4. Notice that the only even prime number is 2 and that no *two-digit* prime number ends in 5. Also notice that there is no pattern to the list of primes; in other words, you can't predict what the next prime number is going to be just by looking at the list.

Divisibility Tests: 2, 3, 5, 9, and 10

Sometimes it is not easy to know what factors a composite number has, especially when the number has three or more digits. To assist in identifying some factors, there are some tests for divisibility. What follows are divisibility tests for 2, 3, 5, 9, and 10. We can use these tests to find some factors of some composite numbers.

Divisibility Test for 2

2 is a factor of a whole number if and only if* the number is even (has 0, 2, 4, 6, or 8 in the ones place).

*This means that if the number is odd, *not* even, then 2 is *not* a factor. (An odd number has a 1, 3, 5, 7, or 9 in the ones place.)

You Try It 5 **Without trying to find any other factors or factor pairs, determine whether 2 is a factor of each number. Explain your answer.**

a) 52 **b)** 61 **c)** 70

Divisibility Test for 5

5 is a factor of a whole number *if and only if* the number has either 5 or 0 in the *ones* place.

You Try It 6 **Without trying to find any other factors or factor pairs, determine whether 5 is a factor of each number. Explain your answer.**

a) 90 **b)** 175 **c)** 608

Divisibility Test for 10

10 is a factor of a whole number *if and only if* the number has 0 in the *ones* place.

10 is a factor of 30, 160, 1,420, and 700 because each number has 0 in the ones place.

Divisibility Test for 3

3 is a factor of a whole number *if and only if* the number's digits *add to a multiple of 3.*

Example 4

Determine whether 3 is a factor of the number. Verify each answer by dividing the number by 3.

a) 285 **b)** 473

Procedure

a) Add the digits: $2 + 8 + 5 = 15$; 3 is a factor of 15.

b) Add the digits: $4 + 7 + 3 = 14$; 3 is not a factor of 14.

Answer

Yes, 3 *is* a factor of 285.

No, 3 is *not* a factor of 473.

Check

a)
```
  95
3)285
- 27
  15
- 15
   0
```
So, 285 is divisible by 3, and 3 is a **factor** of 285.

b)
```
 157 r 2
3)473
- 3
  17
- 15
   23
 - 21
    2
```
Because the remainder is not 0, 473 is not divisible by 3. Therefore, 3 is *not* a factor of 473.

You Try It 7 **Determine whether 3 is a factor of the number. Verify each answer by dividing the number by 3. Use Example 4 as a guide.**

a) 87: ____________ **b)** 671: ____________

c) 8,395: ____________ **d)** 25,074: ____________

Divisibility Test for 9

9 is a factor of a whole number if and only if the number's digits add to a multiple of 9.

Example 5

Determine whether 9 is a factor of the number. Verify each answer by dividing the number by 9.

a) 675 **b)** 1,983

Procedure

a) Add the digits: $6 + 7 + 5 = 18$; 18 is a multiple of 9.

b) Add the digits: $1 + 9 + 8 + 3 = 21$; 21 is *not* a multiple of 9.

Answer

Yes, 9 is a factor of 675.

No, 9 is *not* a factor of 1,983.

Check

$$\begin{array}{r} 75 \\ 9\overline{)675} \\ -63 \\ \hline 45 \\ -45 \\ \hline 0 \end{array}$$

So, 675 is divisible by 9, and 9 is a **factor** of 675.

$$\begin{array}{r} 220 \text{ r } 3 \\ 9\overline{)1983} \\ -18 \\ \hline 18 \\ -18 \\ \hline 03 \\ -00 \\ \hline 3 \end{array}$$

Because the remainder is not 0, 1,983 is not divisible by 9, and 9 is ***not*** a factor of 1,983.

You Try It 8 **Determine whether 9 is a factor of the number. Verify each answer by dividing the number by 9. Use Example 5 as a guide.**

a) 548: ______________ **b)** 3,582: ______________

c) 8,511: ______________ **d)** 20,142: ______________

Think About It 1 If we add the digits of 25, we get $2 + 5 = 7$. Does this mean that 7 is a factor of 25?

____

____

Let's put the divisibility tests for 2, 3, and 5 together in the following example.

Example 6

Which of the first three prime numbers—2, 3, and 5—are factors of the following numbers?

a) 42
- 42 is even (2)
- $4 + 2 = 6$ (3)
- doesn't end in 0 or 5

b) 135
- 135 is not even
- $1 + 3 + 5 = 9$ (3)
- ends in 5 (5)

c) 570
- 570 is even (2)
- $5 + 7 + 0 = 12$ (3)
- ends in 0 (5)

d) 91
- 91 is not even
- $9 + 1 = 10$
- doesn't end in 0 or 5

Answer **a)** 2 and 3 **b)** 3 and 5 **c)** 2 and 3 and 5 **d)** none of these

You Try It 9 **Which of the first three prime numbers—2, 3, and 5—are factors of the following numbers? Use Example 6 as a guide.**

a) 213 ______________ **b)** 390 ______________ **c)** 419 ______________ **d)** 2,835 ______________

Think About It 2 If a number such as 169 doesn't have 2, 3, or 5 as one or more of its factors, is it a prime number? Explain your answer.

____

____

On page 64 of this section, the following questions were asked about 105. Can you answer them now?

Does 105 have a factor of 3? ______________________

Does 105 have a factor of 5? ______________________

What other factors does 105 have? ______________________

Prime Factorization

Just as a chemist breaks up an element into its most basic atoms, we will break up a composite number into its prime factors. When we write a composite number as a product of its prime factors, we call it **prime factorization.**

To help us understand prime factorization, we look at this analogy about primes and composites:

In paints, we have three primary colors: red, blue, and yellow. For the purposes of this analogy, we can think of these colors as representing the prime numbers.

We can mix any two of the primary colors to get other colors, called secondary colors. In particular,

mixing equal amounts of red and yellow makes orange.

mixing equal amounts of red and blue makes purple.

mixing equal amounts of blue and yellow makes green.

These secondary colors are like composite numbers. You can't get green without blue and yellow, just as a whole number can't have 6 as a factor without also having 2 and 3 as factors.

We can even mix all three primary colors together to form brown; so, we can say that brown is composed of red, blue, and yellow. Brown is like 30; 30 is a composite number. It is composed of the prime factors 2, 3, and 5: $30 = 2 \cdot 3 \cdot 5$. (This is the prime factorization of 30.)

Also, we can mix yellow and blue to get green; then we can mix in more yellow for lime green. Lime green is a composite color with two amounts of yellow and one amount of blue. Similarly, 12 is a composite number with two factors of 2 and one factor of 3: $12 = 2 \cdot 2 \cdot 3$. (This is the prime factorization of 12.)

To complete this analogy, consider how a paint scientist might break down the color puce. She might discover that puce is a mixture of brown and purple, then recognize that brown is composed of red and green, and so on.

Here is a diagram of what this breakdown into primary colors might look like. The primary colors are circled to indicate that they can't be broken down further.

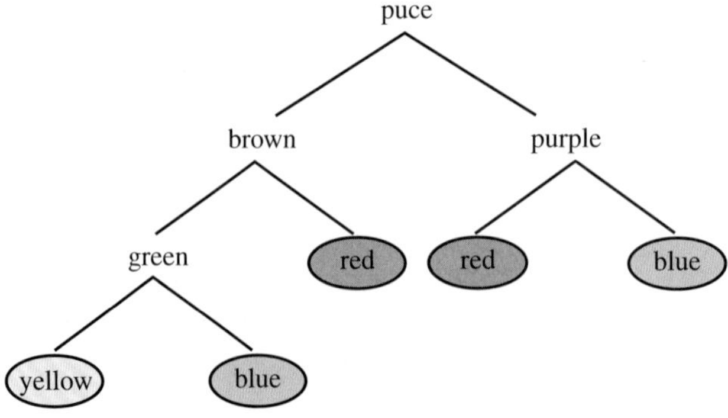

We see that puce is composed of one amount of yellow, two amounts of red, and two amounts of blue.

Example 7

Find the prime factorization of each composite number.

a) 6 **b)** 14 **c)** 15

Procedure Think of the factor pairs that make up the composite number. Because 1 is not a prime number, it should not be included.

Answer **a)** $6 = 2 \cdot 3$ **b)** $14 = 2 \cdot 7$ **c)** $15 = 3 \cdot 5$

You Try It 10 **Find the prime factorization of each composite number. Use Example 7 as a guide.**

a) 21 **b)** 22 **c)** 35 **d)** 77

Think About It 4 Although it's true that $7 = 7 \cdot 1$, we can't use this as the prime factorization. Why?

___

___

Factor Trees

A **factor tree** is a visual method used to look at the factors of a number. First, lets look at the variety of factors of 24 using a factor tree.

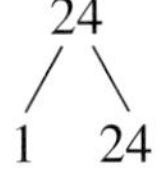

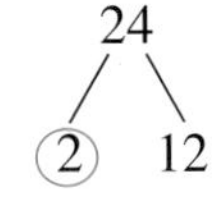

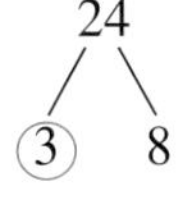

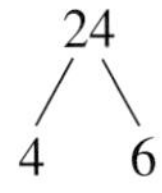

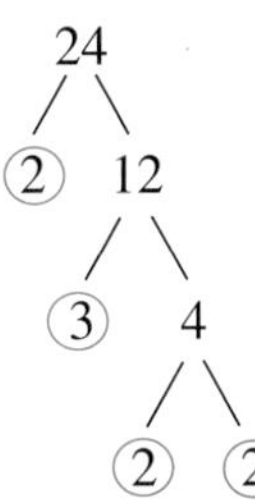

Think of the number—24 in this case—as a flower or plant. The lines leading to the factors are like *branches*. The circled numbers are prime numbers, like the fruit of the tree, which is the purpose of a factor tree: to identify prime factors of the original number.

Each branch that leads to a prime number bears fruit, and each branch that leads to a composite number must branch again until it bears the fruit of a prime number. At the right is the completed factor tree for 24. This factor tree indicates that $24 = 2 \cdot 2 \cdot 2 \cdot 3$.

Think About It 5 Is this factor tree a good start for the prime factorization for 36? Why or why not?

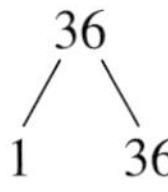

___

___

Example 8

Find the prime factorization of 30.

Procedure We'll use a factor tree to find any two factors of 30. We'll circle any primes that appear to indicate that the branch can't be factored further.

If we arrive at any composite numbers, we must factor them further to continue our search for prime factors.

Here are three different paths to finding the prime factors of 30:

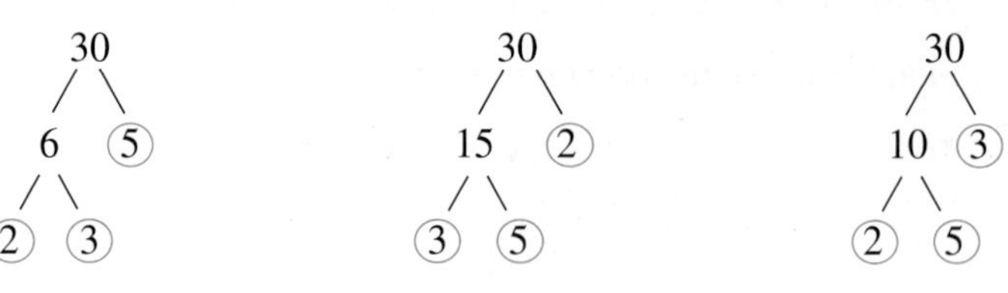

Answer $30 = 2 \cdot 3 \cdot 5$ $30 = 3 \cdot 5 \cdot 2$ $30 = 2 \cdot 5 \cdot 3$

Notice that the result (the prime factorization) is the same no matter which factor path we choose. Generally, though, we write the prime factorization in numerical order starting with the lowest number. So, we'd write $30 = 2 \cdot 3 \cdot 5$.

You Try It 11 **Find the prime factorization of the following numbers. Create a factor tree and put a circle around any prime factor. Use Example 8 as a guide.**

a) 42 **b)** 54 **c)** 70

42 = 54 = 70 =

Example 9

Find the prime factorization of 24. Write the answer two ways, with and without exponents.

Procedure This time you are given one of three paths. It is up to you to find two more. Any correct path you choose will give the same prime factorization.

1. Our path: **2. Your first path:** **3. Your second path:**

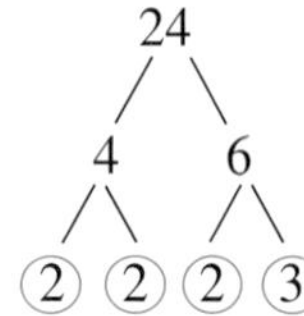

Notice this time that we have repeated prime factors, and we must list them all when writing the prime factorization. We can write the answer as follows:

Answer $24 = 2 \cdot 2 \cdot 2 \cdot 3$, or $2^3 \cdot 3$.

You Try It 12 **Find the prime factorization of the following numbers using a factor tree. Write the answer two ways: with and without exponents, as shown in Example 9. Make sure you show the factor tree and circle the prime factors as they appear.**

a) 12 **b)** 50

12 = __________ or __________ 50 = __________ or __________
(*without* exponents) or (*with* exponents)

c) 36 **d)** 27

36 = __________ or __________ 27 = __________ or __________

Example 10

Find the prime factorization of 280.

Procedure When a number is rather large, don't be intimidated by it. Identify at least one number (prime *or* composite) that is a factor and begin the process. The other factors will quickly become smaller and easier to work with.

1. Is 2 a factor of 280?

2. Is 3 a factor of 280?

3. Is 5 a factor of 280?

4. Is 9 a factor of 280?

5. Is 10 a factor of 280?

10 is a good factor to start with.

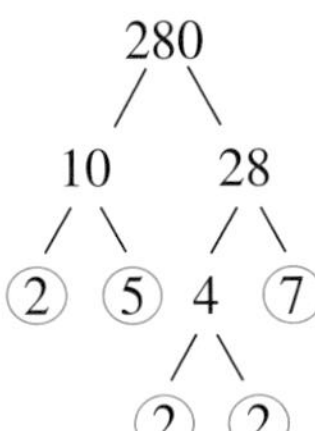

Answer $280 = 2 \cdot 2 \cdot 2 \cdot 5 \cdot 7$, or $2^3 \cdot 5 \cdot 7$

You Try It 13 **Find the prime factorization of the following numbers. Write it with and without exponents. Use Example 10 as a guide.**

a) 100 **b)** 260 **c)** 1,540

100 = ______________ (without exponents) 260 = ______________ (without exponents) 1,540 = ______________ (without exponents)

100 = ______________ (with exponents) 260 = ______________ (with exponents) 1,540 = ______________ (with exponents)

The Division Method

An alternative to using a factor tree is to divide repeatedly by the lowest possible prime number, generating a new quotient each time. In this division method, the primes we divide by are called **prime divisors.**

To find the *lowest possible prime number,* you must divide by 2 when the number is even and continue to divide by 2 as long as the resulting quotients are even. Once the quotients stop being even, see if you can divide by 3; if not, try 5, then 7, and so on, until the final quotient is a prime number. The prime factorization is the product of all prime divisors and the final prime quotient.

Example 11

Find the prime factorization of each number using the division method of dividing by the lowest possible prime.

a) 90 **b)** 189 **c)** 240

Procedure If the number is even, we'll divide by 2. We'll then look at the quotient and decide whether we can divide by 2 again. If not, we'll see if 3 is a factor using the divisibility test for 3, we'll see if 5 is a factor using the divisibility test for 5 and so on.

Each new quotient appears *below* the dividend.

Answer **a)** 90 is even; so, the first prime divisor is 2: $90 \div 2 = 45$.
45 is not even; but because $4 + 5 = 9$, we know that 3 is a factor: $45 \div 3 = 15$.
3 is a factor of 15, giving a quotient of 5.
Because 5 is a prime number, we won't be able to divide and discover any new primes.

$$\begin{array}{r|l} 2 & 90 \\ \hline 3 & 45 \\ \hline 3 & 15 \\ \hline & 5 \end{array}$$

The prime factorization of 90 is the product of all of the prime divisors and the last prime, 5.
So, the prime factorization of 90 is $2 \cdot 3 \cdot 3 \cdot 5$, or $2 \cdot 3^2 \cdot 5$.

b) Because 189 is not even, let's see if 3 works: $1 + 8 + 9 = 18$.
Yes, $189 \div 3 = 63$.
Try 3 again: $63 \div 3 = 21$.
3 is also a factor of 21, giving a quotient of 7, a prime number.
The prime factorization of 189 is the product of all of the prime divisors and the last prime, 7.
So, the prime factorization of 189 is $3 \cdot 3 \cdot 3 \cdot 7$, or $3^3 \cdot 7$.

$$\begin{array}{r|l} 3 & 189 \\ 3 & 63 \\ 3 & 21 \\ & 7 \end{array}$$

c) 240 is even; so, the first prime divisor is 2: $240 \div 2 = 120$.
120 is also even; so, we divide by 2 again: $120 \div 2 = 60$.
Because 60 is even, we divide by 2 again: $60 \div 2 = 30$.
30 is also even; so, we divide by 2 again: $30 \div 2 = 15$.
Finally, there are no more even quotients; we know that 3 is a factor of 15, giving a quotient of 5, which is prime.
The prime factorization of 240 is the product of all of the prime divisors and the last prime, 5.
So, the prime factorization of 240 is $2 \cdot 2 \cdot 2 \cdot 2 \cdot 3 \cdot 5$, or $2^4 \cdot 3 \cdot 5$.

$$\begin{array}{r|l} 2 & 240 \\ 2 & 120 \\ 2 & 60 \\ 2 & 30 \\ 3 & 15 \\ & 5 \end{array}$$

You Try It 14 **Find the prime factorization of the following numbers. Use the division method outlined in Example 11.**

a) 24 **b)** 60 **c)** 175

24 = ________ (without exponents) 60 = ________ (without exponents) 175 = ________ (without exponents)

24 = ________ (with exponents) 60 = ________ (with exponents) 175 = ________ (with exponents)

d) 252 **e)** 405 **f)** 660

252 = ________ (without exponents) 405 = ________ (without exponents) 660 = ________ (without exponents)

252 = ________ (with exponents) 405 = ________ (with exponents) 660 = ________ (with exponents)

For additional practice with the division method, go back to You Try It 3 and You Try It 4 and use the division method. You will come up with the same prime factorizations if everything is done correctly.

Answers: You Try It and Think About It

You Try It: **1. a)** • 20 is a multiple of both 4 and 5. • 20 is divisible by both 4 and 5. • Both 4 and 5 divide evenly into 20. • Both 4 and 5 are factors of 20. **b)** • 72 is a multiple of both 9 and 8. • 72 is divisible by both 9 and 8. • Both 9 and 8 divide evenly into 72. • Both 9 and 8 are factors of 72.

2. a) 12

1	12
2	6
3	4

Factors of 12: 1, 2, 3, 4, 6, and 12

b) 16

1	16
2	8
4	4

Factors of 16: 1, 2, 4, 8, and 16

c) 18

1	18
2	9
3	6

Factors of 18: 1, 2, 3, 6, 9, and 18

d) 20

1	20
2	10
4	5

Factors of 20: 1, 2, 4, 5, 10, and 20

3. a) Composite **b)** Prime **c)** Neither prime nor composite **d)** Composite **4. a)** The first prime number is even. (It is the only even prime number.) **b)** All other prime numbers are odd numbers. **c)** No, many odd numbers are not prime. The number 9 is an example of an odd number that is not prime. Other examples of odd numbers that are not prime are 15, 21, 25, 27, and 33. **5. a)** 2 is a factor of 52 because 52 is an even number. **b)** 2 is not a factor of 61 because 61 is an odd number. **c)** 2 is a factor of 70 because 70 is an even number. **6. a)** 5 is a factor of 90 because 90 has a 0 in the ones place. **b)** 5 is a factor of 175 because 175 has a 5 in the ones place. **c)** 5 is not a factor of 608 because 608 does not have a 0 or 5 in the ones place. **7. a)** Because $8 + 7 = 15$ and because 3 is a factor of 15, 3 is a factor of 87. **b)** Because $6 + 7 + 1 = 14$ and because 3 is not a factor of 14, 3 is not a factor of 671. **c)** Because $8 + 3 + 9 + 5 = 25$ and because 3 is not a factor of 25, 3 is not a factor of 8,395. **d)** Because $2 + 5 + 0 + 7 + 4 = 18$ and because 3 is a factor of 18, 3 is a factor of 25,074. **8. a)** Because $5 + 4 + 8 = 17$ and because 9 is not a factor of 17, 9 is not a factor of 548. **b)** Because $3 + 5 + 8 + 2 = 18$ and because 9 is a factor of 18, 9 is a factor of 3,582. **c)** Because $8 + 5 + 1 + 1 = 15$ and because 9 is not a factor of 15, 9 is not a factor of 8,511. **d)** Because $2 + 0 + 1 + 4 + 2 = 9$ and because 9 is a factor of 9, 9 is a factor of 20,142. **9. a)** 3 **b)** 2, 3, and 5 **c)** none **d)** 3 and 5 **10. a)** $3 \cdot 7$ **b)** $2 \cdot 11$ **c)** $5 \cdot 7$ **d)** $7 \cdot 11$ **11. a)** $2 \cdot 3 \cdot 7$ **b)** $2 \cdot 3 \cdot 3 \cdot 3$ **c)** $2 \cdot 5 \cdot 7$ **12. a)** $2 \cdot 2 \cdot 3$ or $12 = 2^2 \cdot 3$ **b)** $2 \cdot 5 \cdot 5$ or $50 = 2 \cdot 5^2$ **c)** $2 \cdot 2 \cdot 3 \cdot 3$ or $36 = 2^2 \cdot 3^2$ **d)** $3 \cdot 3 \cdot 3$ or $27 = 3^3$

13. a) $2 \cdot 2 \cdot 5 \cdot 5$, $2^2 \cdot 5^2$ **b)** $2 \cdot 2 \cdot 5 \cdot 13$, $2^2 \cdot 5 \cdot 13$ **c)** $2 \cdot 2 \cdot 5 \cdot 7 \cdot 11$, $2^2 \cdot 5 \cdot 7 \cdot 11$

14. a) $2 \cdot 2 \cdot 2 \cdot 3$, $2^3 \cdot 3$ **b)** $2 \cdot 2 \cdot 3 \cdot 5$, $2^2 \cdot 3 \cdot 5$ **c)** $5 \cdot 5 \cdot 7$, $5^2 \cdot 7$ **d)** $2 \cdot 2 \cdot 3 \cdot 3 \cdot 7$, $2^2 \cdot 3^2 \cdot 7$ **e)** $3 \cdot 3 \cdot 3 \cdot 3 \cdot 5$, $3^4 \cdot 5$ **f)** $2 \cdot 2 \cdot 3 \cdot 5 \cdot 11$, $2^2 \cdot 3 \cdot 5 \cdot 11$

Think About It: **1.** No, the divisibility rule of adding digits together is a rule for 3 and 9 only. **2.** Not necessarily. The number may have a prime factor besides 2, 3, and 5. (In fact, $169 = 13 \cdot 13$.) **3.** Yes, 105 has a factor of 3; yes, 105 has a factor of 5; the other factors of 105 are 1, 7, 15, 21, 35, and 105. **4.** Because 1 is not a prime number, it cannot be used in any prime factorization. **5.** No. Reasons may vary. One possibility: A factor tree should not include 1 because 1 is neither prime nor composite.

FOR EXTRA HELP
Student Resources on DVD-ROM
Includes
➤Student's Solutions Manual
➤Video Lectures
➤Chapter Test Prep Video
MyMathLab
MathXL

Section 1.6 Exercises

Think Again

1. If a number such as 169 doesn't have 2, 3, or 5 as one or more of its factors, is it a prime number? Explain your answer. *(Refer to Think About It 2)*
   Not necessarily. The number might be prime, but it might have larger prime factors. For example, $169 = 13 \times 13$; so, 169 is not prime.

2. What type of number has one factor pair that contains two of the same number? Show an example to support your answer.
   A perfect square number. For example, a factor pair of 36 is $6 \cdot 6$.

3. If 1 was considered a prime number, what would be the prime factorization of 7? (Is there more than one option for the prime factorization of 7?) Explain your answer or show an example that supports your answer.
   In that case, the prime factorization of 7 could be $7 \cdot 1$, $7 \cdot 1 \cdot 1 \cdot 1$, and so on. In other words, we can have as many factors of 1 as we want. (This is why 1 is not a prime number.)

4. What could you write as a Divisibility Rule for 6? Show some examples that support this rule.
   6 is a factor of a number if both 2 and 3 are factors of that number. As for examples, answers will vary.

Focus Exercises

List the first eight multiples of the given number.

5. 4
   4, 8, 12, 16, 20, 24, 28, 32
6. 5
   5, 10, 15, 20, 25, 30, 35, 40
7. 7
   7, 14, 21, 28, 35, 42, 49, 56
8. 9
   9, 18, 27, 36, 45, 54, 63, 72

Use a factor pair table to find all of the factor pairs of each number.

9. 32

32	
1	32
2	16
4	8

10. 40

40	
1	40
2	20
4	10
5	8

11. 28

28	
1	28
2	14
4	7

12. 42

42	
1	42
2	21
3	14
6	7

Of the first three prime numbers—2, 3, and 5—which are factors of the following?

13. 32
2

14. 80
2 and 5

15. 127
None

16. 414
2 and 3

17. 76
2

18. 57
3

19. 125
5

20. 390
2, 3, and 5

21. 315
3 and 5

22. 860
2 and 5

23. 156
2 and 3

24. 4,231
None

25. 7,287
3

26. 41,592
2 and 3

27. 322,980
2, 3, and 5

28. 994,515
3 and 5

Determine whether 9 is a factor of the number.

29. 372
No

30. 4,797
Yes

31. 7,506
Yes

32. 20,601
Yes

Of the following, determine which are prime, which are composite, and which are neither.

33. 0, 7, 9, 23, 8, 40, 15, 33, 32, 12, 41, 51, 50
Prime: 7, 23, 41
Composite: 9, 8, 40, 15, 33, 32, 12, 51, 50
Neither: 0

34. 6, 17, 2, 27, 31, 38, 1, 29, 41, 49, 55, 57, 61, 71
Prime: 17, 2, 31, 29, 41, 61, 71
Composite: 6, 27, 38, 49, 55, 57
Neither: 1

Find the prime factorization of the following numbers using a factor tree. Write the answers two ways: with and without exponents.

35. 18
$2 \cdot 3 \cdot 3$ or $2 \cdot 3^2$

36. 48
$2 \cdot 2 \cdot 2 \cdot 2 \cdot 3$ or $2^4 \cdot 3$

37. 63
$3 \cdot 3 \cdot 7$ or $3^2 \cdot 7$

38. 75
$3 \cdot 5 \cdot 5$ or $3 \cdot 5^2$

39. 105
$3 \cdot 5 \cdot 7$

40. 256
$2 \cdot 2 \cdot 2 \cdot 2 \cdot 2 \cdot 2 \cdot 2 \cdot 2$ or 2^8

41. 496
$2 \cdot 2 \cdot 2 \cdot 2 \cdot 31$ or $2^4 \cdot 31$

42. 588
$2 \cdot 2 \cdot 3 \cdot 7 \cdot 7$ or $2^2 \cdot 3 \cdot 7^2$

43. 720
$2 \cdot 2 \cdot 2 \cdot 2 \cdot 3 \cdot 3 \cdot 5$ or $2^4 \cdot 3^2 \cdot 5$

44. 735
$3 \cdot 5 \cdot 7 \cdot 7$ or $3 \cdot 5 \cdot 7^2$

45. 945
$3 \cdot 3 \cdot 3 \cdot 5 \cdot 7$ or $3^3 \cdot 5 \cdot 7$

46. 1,050
$2 \cdot 3 \cdot 5 \cdot 5 \cdot 7$ or $2 \cdot 3 \cdot 5^2 \cdot 7$

Find the prime factorization of the following numbers using the division method. Write the answers two ways: with and without exponents.

47. 20
$2 \cdot 2 \cdot 5$ or $2^2 \cdot 5$

48. 45
$3 \cdot 3 \cdot 5$ or $3^2 \cdot 5$

49. 52
$2 \cdot 2 \cdot 13$ or $2^2 \cdot 13$

50. 72
$2 \cdot 2 \cdot 2 \cdot 3 \cdot 3$ or $2^3 \cdot 3^2$

51. 76
$2 \cdot 2 \cdot 19$ or $2^2 \cdot 19$

52. 88
$2 \cdot 2 \cdot 2 \cdot 11$ or $2^3 \cdot 11$

53. 98
$2 \cdot 7 \cdot 7$ or $2 \cdot 7^2$

54. 111
$3 \cdot 37$

55. 124
$2 \cdot 2 \cdot 31$ or $2^2 \cdot 31$

56. 135
$3 \cdot 3 \cdot 3 \cdot 5$ or $3^3 \cdot 5$

57. 200
$2 \cdot 2 \cdot 2 \cdot 5 \cdot 5$ or $2^3 \cdot 5^2$

58. 224
$2 \cdot 2 \cdot 2 \cdot 2 \cdot 2 \cdot 7$ or $2^5 \cdot 7$

Think Outside the Box

59. Identify a factor pair of 24 for which the sum of the two factors is 10.
6 and 4

60. Identify a factor pair of 36 for which the sum of the two factors is 15.
3 and 12

61. Identify a factor pair of 12 for which the difference of the two factors is 1.
4 and 3

62. Identify a factor pair of 30 for which the difference of the two factors is 13.
15 and 2

Here are the first ten prime numbers:

2, 3, 5, 7, 11, 13, 17, 19, 23, and 29.

Using division by prime numbers, determine which of the following numbers are prime and which are composite.

63. 91 Composite: $91 = 7 \times 13$

64. 119 Composite: $119 = 7 \times 17$

65. 197 Prime

66. 247 Composite: $247 = 13 \times 19$

SECTION 1.7 Equations

OBJECTIVES

In this section, you will learn to

- Define the term *equation.*
- Apply the Subtraction Property of Equality.
- Apply the Division Property of Equality.
- Solve simple equations.

You Need to Know

To successfully complete this section, you need to understand

- ☐ The identities (1.1)
- ☐ The Commutative Properties (1.1)
- ☐ Adding whole numbers (1.2)
- ☐ Subtracting whole numbers (1.2)
- ☐ Multiplying whole numbers (1.3)
- ☐ Dividing whole numbers (1.4)

Introduction

To this point, we have looked at some application situations that involve only one operation: addition, subtraction, multiplication, or division. Here is a situation that involves two operations: addition and subtraction.

> Wendy's monthly budget includes \$529 for rent, \$175 for food, \$192 for utilities and transportation, and \$125 for entertainment. If she takes home (after taxes and other deductions) \$1,238 per month, how much does she have left at the end of the month for savings?

Such situations—those that involve more than one operation—are more common than you might think. This section and the next will introduce you to algebra and some fairly consistent techniques for solving applications. First, though, you need to become familiar with the notion of *equations*.

The Vocabulary of Equations

An **equation** is a mathematical sentence in which one expression equals another expression.

We use an equation when we are asked to find a number that is not yet known. In this way, we say that we are seeking an **unknown value.**

The unknown value that we seek is a number that must be represented in the equation. Since we don't know what the number is, we use a letter to represent it. This letter is called a **variable.** We might use the letter n, for *number*, or v, for *value*, or some other letter of our choosing.

A known value such as 8 is called a **constant.**

Think About It 1

The word *variable* can be thought of as *vary-able*. How would you explain to a classmate the difference between the words *variable* and *constant*?

____

____

Here are some examples of equations, each with its own variable:

$$n + 3 = 8$$
$$18 = 6 \cdot v$$
$$6 + 5 = x$$
$$28 + 38 + 65 + y = 178$$

Notice that each equation contains an equal sign (=). This is used to indicate that the two "sides" of the equation—the left side and the right side—are **equivalent;** that is, they have equal value. Notice further that the variable can be on either side of the equal sign.

The purpose of writing an equation is to find the value of the variable, the unknown value. Finding the value of the variable means finding a *solution* to the equation. A **solution** to an equation is a number that makes the equation true.

Example 1

For each equation, is 6 the solution?

a) $n + 5 = 11$ **b)** $12 = 8 + x$ **c)** $w \cdot 5 = 35$ **d)** $24 = 4 \cdot y$

Procedure For each equation, replace the variable with 6. Evaluate one side and see if it equals the other side. If the two sides are equal, 6 makes the equation true and 6 is the solution. If the two sides are not equal, 6 does *not* make the equation true and 6 is *not* the solution.

Answer

Equation	Does replacing the variable with 6 make the equation true?
a) $n + 5 = 11$	
$6 + 5 = 11$	
$11 = 11$	Yes, 6 is the solution.
b) $12 = 8 + x$	
$12 = 8 + 6$	
$12 = 14$	No, 6 is not the solution.
c) $w \cdot 5 = 35$	
$6 \cdot 5 = 35$	
$30 = 35$	No, 6 is not the solution.
d) $24 = 4 \cdot y$	
$24 = 4 \cdot 6$	
$24 = 24$	Yes, 6 is the solution.

You Try It 1 **For each equation, is 5 the solution? Use Example 1 as a guide.**

a) $18 = p + 11$ **b)** $15 + m = 20$ **c)** $50 = d \cdot 10$ **d)** $8 \cdot x = 45$

Balancing Equations

To **solve** an equation means to find the solution, to find the number that makes the equation true.

To do so, we must get the variable, say x, by itself on one side of the equation and all constants on the other side. This is called **isolating the variable.** The best way to isolate the variable is to remove the constants that are on the same side of the equal sign as x.

For example, in the equation $x + 3 = 8$, we want to isolate the variable, x, by removing 3 from the left side. This will leave x alone (isolated) on that side.

In the process of isolating the variable, we'll write a new equation based on the original equation. It is important to know that the solution to the new equation must be the same as the solution to the original equation.

Before learning how to isolate the variable, let's explore the idea of a **balanced equation.** All equations must be balanced.

This scale will help you visualize a balanced equation.

The equation $x + 3 = 8$ has a solution of 5. In other words, the equation is true when x is replaced with 5, but it isn't true for any other number.

What do you think would happen if we subtracted some "weight" from just one side of the scale?

Caution Subtracting weight from just one side of a balanced scale means *that* side becomes lighter and goes up. We should *not* write a new equation based on what we see because the sides are not in balance.

So how do we keep an equation in balance?

> Whatever value is subtracted from one side must also be subtracted from the other side at the same time. This will keep the equation in balance.

So, we may remove 3 by subtracting it from *each* side to isolate the variable. Notice that in subtracting 3 from *each* side, we get a new equation.

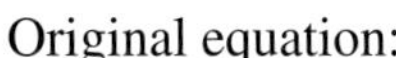

Original equation:	$x + 3 = 8$
New equation (subtract 3 from each side):	$x + 3 - 3 = 8 - 3$
This simplifies to:	$x + 0 = 5$
And this simplifies to just:	$x = 5$

The solution is 5.

Although the solution is 5, it's often appropriate to state it as $x = 5$.

When an equation involves addition, such as $x + 3 = 8$, we can *isolate the variable* using the Subtraction Property of Equality. After solving, we can check the answer to show that it is the solution by replacing the variable in the original equation with the answer to see if it makes the equation true.

The Subtraction Property of Equality

If $\boldsymbol{x + a = b}$, we can isolate the variable, x, by subtracting the same value, $\boldsymbol{a}$, from each side—whether the variable is written first or second—and still have a balanced equation.

	Example A	Example B
$x + a = b$	$x + 8 = 20$	$18 = 5 + x$
$x + a - a = b - a$	$x + 8 - 8 = 20 - 8$	$18 - 5 = 5 - 5 + x$
↑ $a - a$ is 0, the identity for addition. Adding 0 helps isolate the variable.	$x + 0 = 12$	$13 = 0 + x$
	$x = 12$	$13 = x$
		or $\boldsymbol{x = 13}$*
	Check: $12 + 8 = 20$ True!	Check: $18 = 5 + 13$ True!

* Notice in Example B that the variable is isolated on the right side. In this text, all final results will be written with the variable on the left side and the number on the right.

Example 2

Solve the equation $y + 6 = 9$ by using the Subtraction Property of Equality.

Answer

$y + 6 = 9$	**To isolate the variable we need to subtract 6 from each side.**
$y + 6 - 6 = 9 - 6$	**Apply subtraction to each side.**
$y + 0 = 3$	**Because 0 is the identity for addition, $y + 0$ is y.**
$y = 3$	**We now know the value of the variable; it is no longer unknown.**

Check We can show that 3 is the solution by replacing y with 3 in the original equation.

$y + 6 = 9$	**becomes**
$3 + 6 = 9$	**which is true. So, 3 is the solution.**

Whether the constant is written before or after the variable, we can place the subtraction right after the constant.

Example 3

Solve the equation $23 = 15 + w$ by using the Subtraction Property of Equality. Check the answer to show that it is the solution.

Procedure Isolate the variable by subtracting 15 from each side.

Answer

	$23 = 15 + w$	**To isolate the variable, we need to subtract 15 from each side.**
	$23 - 15 = 15 - 15 + w$	**Apply subtraction to each side.**
	$8 = 0 + w$	**Because 0 is the identity for addition, $0 + w$ is w.**
	$8 = w$	Check **Replace w with 8 in the original equation:**
Write w on the left:	$w = 8$	**$15 + 8 = 23$ which is true. So, the solution is 8.**

You Try It 2 **Solve each equation by using the Subtraction Property of Equality. Show all steps. Use Examples 2 and 3 as guides. Check each answer to show that it is the solution.**

a) $p + 8 = 12$ **b)** $20 = 7 + v$ **c)** $45 + x = 73$ **d)** $51 = y + 39$

In some equations, more than one constant is on the same side as the variable. Before subtracting a constant from each side, we must add those constants.

Example 4

Solve the equation $15 + 6 + x = 30$.

Procedure First, simplify the left side by adding $15 + 6$.

Answer

$15 + 6 + x = 30$	**$15 + 6 = 21$. Rewrite the equation.**
$21 + x = 30$	**Now we can subtract 21 from each side.**
$21 - 21 + x = 30 - 21$	**Apply subtraction to each side.**
$0 + x = 9$	**$0 + x$ is x.**
$x = 9$	

Check
$$\begin{array}{r} 15 \\ 6 \\ +\ 9 \\ \hline 30\ ✓ \end{array}$$
True!

Check Replace x with 9: $15 + 6 + 9 = 30$ The solution is 9.

You Try It 3 **Solve the following equations by isolating the variable. Use Example 4 as a guide.**

a) $29 + 61 + c = 96$ **b)** $113 = n + 46 + 28$

Solving Equations Involving Multiplication

When a variable, say x, is multiplied by a number, we call that number a **coefficient,** pronounced "co-ee-fish′-unt."

Example 5

In each equation, identify the coefficient.

a) $x \cdot 6 = 8$ **b)** $8 \cdot x = 40$ **c)** $12 = 4 \cdot y$

Answer **a)** Coefficient: 6 **b)** Coefficient: 8 **c)** Coefficient: 4

You Try It 4 **In each equation, identify the coefficient, the number multiplied by the variable. Use Example 5 as a guide.**

a) $m \cdot 5 = 30$ Coefficient: ______

b) $7 \cdot r = 28$ Coefficient: ______

c) $24 = 3 \cdot k$ Coefficient: ______

d) $18 = w \cdot 9$ Coefficient: ______

When an equation involves multiplication instead of addition, we can isolate the variable by *dividing* each side by the same number, the coefficient. This rule is the Division Property of Equality.

The Division Property of Equality

If $x \cdot a = b$, we can isolate the variable by dividing each side by the same value, a (the coefficient)—whether the variable is written first or second—and still have a balanced equation.

	Example C	Example D
$x \cdot a = b$	$x \cdot 8 = 40$	$36 = 4 \cdot x$
$x \cdot \underbrace{a \div a} = b \div a$	$x \cdot 8 \div 8 = 40 \div 8$	$36 \div 4 = 4 \div 4 \cdot x$
$a \div a$ is 1, the identity for multiplication. Multiplying by 1 helps isolate the variable.	$x \cdot 1 = 5$	$9 = 1 \cdot x$
	$x = 5$	$9 = x$
		Write x on the left: $x = 9$

This time, instead of subtracting to get 0 (the identity for addition), we want to divide to get 1 (the identity for multiplication).

Example 6

Solve the equation $n \cdot 6 = 18$ using the Division Property of Equality.

Procedure Isolate the variable by dividing each side by the coefficient, 6.

Answer

$n \cdot 6 = 18$ **To isolate the variable, we need to divide each side by the coefficient 6.**

$n \cdot 6 \div 6 = 18 \div 6$ **Apply division to each side.**

$n \cdot 1 = 3$ **Because 1 is the identity for multiplication, $n \cdot 1$ becomes n.**

$n = 3$ **We now know the value of the variable; it is no longer unknown.**

Check **Replace n with 3 in the original equation:**

$3 \cdot 6 = 18$ ✓ which is true. So, 3 is the solution.

Whether the coefficient is written before or after the variable, we can place the division right after the coefficient.

Example 7

Solve the equation $4 \cdot x = 156$ using the Division Property of Equality.

Answer

$4 \cdot x = 156$ Divide each side by the coefficient.

$4 \div 4 \cdot x = 156 \div 4$ Divide each side by 4.

$1 \cdot x = 39$ $156 \div 4 = 39$

$x = 39$

So, the solution is 39.

$$\begin{array}{r} 39 \\ 4\overline{)156} \\ -12 \\ \hline 36 \\ -36 \\ \hline 0 \end{array}$$

Check Does $4 \cdot 39 = 156$?

$$\begin{array}{r} 39 \\ \times 4 \\ \hline 156 \end{array}$$ ✓

You Try It 5 **Solve each equation by using the Division Property of Equality. Use Examples 6 and 7 as guides. Show each step.**

a) $p \cdot 5 = 30$ **b)** $21 = 3 \cdot y$ **c)** $12 \cdot w = 3{,}768$ **d)** $675 = 25 \cdot n$

In some equations, the variable is already isolated, but one side needs to be evaluated.

Example 8

Solve each equation.

a) $17 + 26 + 58 = p$ **b)** $36 \cdot 5 = y$

Procedure Evaluate the left side. Here the only check necessary is our addition work in part a and our multiplication work in part b.

Answer

a) $17 + 26 + 58 = p$ — Because p is already isolated (on the right side), we simply need to evaluate the left side.

$101 = p$

Write p on the left: $p = 101$ — So, the solution is 101.

$$\begin{array}{r} 17 \\ 26 \\ +\ 58 \\ \hline 101 \end{array}$$

b) $36 \cdot 5 = y$ — Because y is already isolated (on the right side), we simply evaluate the left side; apply multiplication directly.

$180 = y$

Write y on the left: $y = 180$ — So, the solution is 180.

$$\begin{array}{r} 36 \\ \times\ 5 \\ \hline 180 \end{array}$$

You Try It 6 **Solve each equation. Use Example 8 as a guide.**

a) $49 + 57 + 38 = v$ **b)** $x = 559 + 467$ **c)** $v = 25 \cdot 32$ **d)** $40 \cdot 15 = x$

Answers: You Try It and Think About It

You Try It: **1. a)** $18 = 16$ No **b)** $20 = 20$ Yes **c)** $50 = 50$ Yes **d)** $40 = 45$ No **2. a)** $p = 4$ **b)** $v = 13$ **c)** $x = 28$ **d)** $y = 12$ **3. a)** $c = 6$ **b)** $n = 39$ **4. a)** 5 **b)** 7 **c)** 3 **d)** 9 **5. a)** $p = 6$ **b)** $y = 7$ **c)** $w = 314$ **d)** $n = 27$ **6. a)** $v = 144$ **b)** $x = 1{,}026$ **c)** $v = 800$ **d)** $x = 600$

Think About It: **1.** Answers may vary. One possibility: A variable is something that can vary, can change in value; but something that is constant doesn't change its value, such as a dollar is always a dollar.

Section 1.7 Exercises

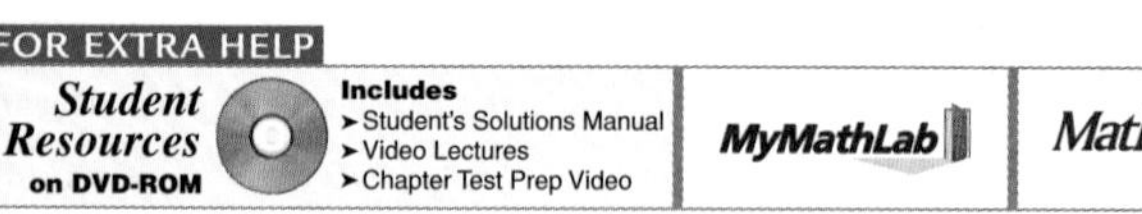

Think Again

1. The word *variable* can be thought of as *vary-able.* How would you explain to a classmate the difference between the words *variable* and *constant*? *(Refer to Think About It 1)*
Something that is *constant* is never changing.

2. Is an equation the same as an expression? Explain your answer or show an example that supports your answer.
No. Answers will vary. One possibility: An equation has an equal sign between two expressions, and an expression has no equal sign. Another possibility: Equations are solved but expressions are simplified.

3. What number do we always attain when we apply the Subtraction Property of Equality to isolate the variable? Explain your answer or show an example that supports your answer.
When we apply the Subtraction Property of Equality, we always attain the number 0.

4. What number do we always attain when we apply the Division Property of Equality to isolate the variable? Explain your answer or show an example that supports your answer.
When we apply the Division Property of Equality, we always attain the number 1.

Focus Exercises

For each equation, replace the variable with 16 and decide whether 16 is the solution.

5. $29 + m = 55$
   $45 = 55$ No
6. $114 = k + 98$
   $114 = 114$ Yes
7. $5 \cdot d = 80$
   $80 = 80$ Yes
8. $210 = x \cdot 15$
   $210 = 240$ No

Solve each equation. Check each answer to show that it is the solution.

9. $8 + n = 21$
   $n = 13$
10. $16 + x = 49$
    $x = 33$
11. $63 = p + 28$
    $p = 35$
12. $129 = m + 45$
    $m = 84$
13. $q = 38 + 84 + 76$
    $q = 198$
14. $925 + 110 + 640 = w$
    $w = 1{,}675$
15. $191 + 186 + v = 500$
    $v = 123$
16. $1{,}000 = 306 + 471 + c$
    $c = 223$
17. $2{,}817 + 3{,}199 = x$
    $x = 6{,}016$
18. $4{,}608 + 3{,}392 = h$
    $h = 8{,}000$
19. $x = 29 + 56 + 34$
    $x = 119$
20. $y = 18 + 12 + 33 + 17$
    $y = 80$
21. $5{,}208 + 3{,}691 + y = 10{,}000$
    $y = 1{,}101$
22. $8{,}156 + 7{,}519 + n = 20{,}000$
    $n = 4{,}325$
23. $c \cdot 3 = 57$
    $c = 19$
24. $52 = q \cdot 4$
    $q = 13$
25. $98 = 7 \cdot n$
    $n = 14$
26. $8 \cdot x = 96$
    $x = 12$
27. $a \cdot 12 = 156$
    $a = 13$
28. $405 = 15 \cdot d$
    $d = 27$
29. $w = 35 \cdot 4$
    $w = 140$
30. $42 \cdot 5 = k$
    $k = 210$
31. $37 \cdot 40 = y$
    $y = 1{,}480$
32. $r = 23 \cdot 30$
    $r = 690$
33. $356 = 2 \cdot w$
    $w = 178$
34. $740 = 5 \cdot v$
    $v = 148$
35. $6 \cdot m = 216$
    $m = 36$
36. $9 \cdot p = 486$
    $p = 54$
37. $7 \cdot x = 399$
    $x = 57$
38. $1{,}895 = 11 \cdot y$
    $y = 172$ r3
39. $18 \cdot x = 252$
    $x = 14$
40. $32 \cdot y = 1{,}344$
    $y = 42$
41. $25 \cdot x = 800$
    $x = 32$
42. $700 = 35 \cdot p$
    $p = 20$
43. $b \cdot 53 = 2{,}491$
    $b = 47$
44. $k \cdot 48 = 3{,}168$
    $k = 66$

Think Outside the Box

For each equation, determine whether 7 is the solution.

45. $5 \cdot x + 17 = 52$
    Yes, 7 is the solution.
46. $63 = 9 + 8 \cdot m$
    No, 7 is not the solution.

Each equation contains addition and multiplication. Solve each equation. (*Hint:* Determine whether it is best to subtract and then divide or divide and then subtract.)

47. $7 \cdot w + 19 = 96$
    $w = 11$
48. $98 = 23 + 5 \cdot y$
    $y = 15$

SECTION 1.8 Solving Applications Using Equations

OBJECTIVES

In this section, you will learn to

- Solve applications involving addition and subtraction.
- Solve applications involving multiplication and division.
- Solve applications involving averages.
- Solve applications involving perimeter and area.

You Need to Know

To successfully complete this section, you need to understand

- ☐ Rounding whole numbers (1.1)
- ☐ Adding whole numbers (1.2)
- ☐ Subtracting whole numbers (1.2)
- ☐ Multiplying whole numbers (1.3)
- ☐ Dividing whole numbers (1.4)
- ☐ Solving equations (1.7)

Introduction

Now we are ready to apply our understanding of solving equations to applications. Here is a situation we might find in the workplace:

> Tony needs to make sure his truck isn't carrying too much weight for the road conditions ahead. His truck weighs 3,125 pounds, and it is carrying a 586-pound load. What is the total weight of the truck and the load?

An application is an equation written in sentence form. Approach the problem as if it were an assignment given to you by a boss or supervisor.

One formula that we use very often is

The sum of all of the parts equals the whole.

As you might imagine, the "whole" is always larger than any individual "part."

Some words that might help you identify the whole are *total, combined* (as in *combined* weight), and *altogether*. Mostly, though, you should rely less on focusing on words and more on *thinking* about the situation to identify the whole and the parts.

If an application has only two parts (say Part 1 and Part 2) and the unknown value is the whole, the rule could be written as follows:

Part 1 + Part 2 = Unknown whole (w)

If, instead, the unknown value is one of the parts, the formula could be written as follows:

Part 1 + Unknown part (p) = Whole

Key Steps for Solving Application Problems

1. Think about the application
   - **a)** by putting yourself in the situation.
   - **b)** by drawing a diagram.
   - **c)** by making estimates about the answer.
   - **d)** by using smaller numbers and making a model of the situation.
2. Decide what is known and what is unknown.
   - **a)** Each number is a part or the whole. Usually, the last sentence in the problem indicates what is unknown; this is the variable.
   - **b)** Write a **legend** identifying what the variable represents.
3. Write an equation based on a formula.
4. Solve the equation by isolating the variable.
5. Write a complete sentence to answer the question in the application problem. Usually, it's possible to reword the question and put it in the form of an answer.

The **legend**, in this regard, is like that found on a map. On a map, a legend lists symbols and explains what each symbol represents.

In the problem-solving process, it's a good idea to read the problem once to get an idea of the situation. Then read it a second time and look for the important information, thinking about what might be the whole and what might be the parts.

As you read the problem a second time, underline or put a box around important information. Also, in the last sentence, underline the unknown value and write a variable (maybe p, w, or x) under it. You will use this for your legend.

> Tony needs to make sure his truck isn't carrying too much weight for the road conditions ahead. His truck weighs [3,125 pounds], and it is carrying a [586-pound] load. What is the total weight of the truck and the load?
> w

When selecting a variable to represent the unknown, we might choose w because the unknown value is the total *weight*. However, in general situations, it's quite common to use x as the variable.

Applications Involving Addition and Subtraction

Let's use the key steps for solving applications to answer the questions in Examples 1 and 2.

Example 1

Tony needs to make sure his truck isn't carrying too much weight for the road conditions ahead. His truck weighs 3,125 pounds, and it is carrying a 586-pound load. What is the total weight of the truck and the load?

Procedure First, reread the problem and put a [box] around important information. Put yourself in Tony's spot. Picture the truck and picture the load going into the truck.

In this case, the parts are as follows:

1. The weight of the truck when there is no load (3,125 pounds)
2. The weight of the load that Tony's truck is carrying (586 pounds)

From the last sentence, we know that the unknown value, the total weight, is the whole.

Answer We start with the legend: Let w = the total weight.

	Part 1 + Part 2 = Unknown whole
Write the equation.	$3{,}125 + 586 = w$
Apply addition.	$3{,}711 = w$

$$\begin{array}{r} 3{,}125 \\ +\ \ 586 \\ \hline 3{,}711 \end{array}$$

Because the legend says that w is the total weight, we can write a sentence based on the question, What is the total weight of the truck and the load?

Sentence The total weight is 3,711 pounds.

Example 2

Connie is buying a new car and is considering adding some of the optional features. One option is a CD stereo system for $589. Another option includes the stereo system along with a navigation system for a total of $1,074. What is the cost of the navigation system alone?

Procedure First, reread the problem and put a [box] around important information. Use the following outline. These steps aren't necessary, but they are a useful aid in thinking about the situation.

a) Write down any known information.
   - The total costs is $1,074.
   - The cost of the stereo system is $589.

b) Also write what we're trying to find.
   - Find the cost of the navigation system.

c) Write the legend: Let x = the cost of the navigation system. We know the whole; so, the unknown value is one of the parts.

Answer **d)** Write the formula. Part 1 + Unknown part = Whole

Stereo + Navigation = Total cost

Write the equation. $589 + x = 1{,}074$

Subtract 589 from each side. $589 - 589 + x = 1{,}074 - 589$

$x = 485$

$$\begin{array}{r} 1{,}074 \\ -\ 589 \\ \hline 485 \end{array}$$

Sentence **e)** The cost of the navigation system alone is $485.

Caution Although it's true that the whole will be larger than any of the individual parts, just because one number is larger than another doesn't make it the whole.

The You Try It exercises in this section show how you might organize the written information given in a problem. Use these exercises as guides to help you solve the Focus Exercises at the end of the section.

You Try It 1 **Mark owns a courier service and uses his small plane to make deliveries. His plane can carry cargo that has a combined weight of 1,280 pounds. One customer has asked him to deliver several large packages that total 891 pounds. How much more cargo weight can Mark's plane carry? (*Hint:* His plane is not full; so, space is available.)**

a) Write down any known information.

Whole: ____________ **Parts:** ____________

b) **Legend:** ____________

c) Write the equation and solve by isolating the variable.

d) **Sentence:** ____________

You Try It 2 **Barbara, a florist, has received many orders for valentines roses. To fill all of the orders, she needs a total of 312 roses. She already has 138 roses at the store. How many more roses does Barbara need?**

a) Write down any known information. Also write what we're trying to find.

Whole: ____________ **Parts:** ____________

b) **Legend:** ____________

c) Write the equation and solve by isolating the variable.

d) **Sentence:** ____________

Applications Involving More Than Two Parts

Many applications require adding more than two numbers together. Also, some applications involve both addition and subtraction.

Example 3

Wendy's monthly budget includes $529 for rent, $175 for food, $192 for utilities and transportation, and $125 for entertainment. Wendy takes home (after taxes and other deductions) $1,238 each month. At the end of the month, how much does Wendy have left for savings?

Procedure First, reread the problem and put a box around important information. This problem has five parts: rent, food, utilities and transportation, entertainment, and savings. The whole is the total amount she brings home each month, $1,238.

Answer **Legend:** Let x = how much Wendy has for savings each month.

Equation:

$$\text{Part 1} + \text{Part 2} + \text{Part 3} + \text{Part 4} + \text{Part 5} = \text{Whole}$$
$$529 + 175 + 192 + 125 + x = 1{,}238$$
$$1{,}021 + x = 1{,}238$$
$$x + 1{,}021 - 1{,}021 = 1{,}238 - 1{,}021$$
$$x = 217$$

529
175
192
+ 125
1,021

1,238
− 1,021
217

Sentence Wendy has $217 left for savings.

You Try It 3 **At her bookstore this month, Gena sold 167 books the first week, 228 books the second week, and 174 books the third week. How many books must Gena sell during the fourth week to reach her monthly goal of 700?**

a) Write down any known information.

Whole: ______________ **Parts:** ______________

b) Write what we're trying to find in the legend: ______________

c) Write the equation and solve by isolating the variable.

d) Sentence: ______________

Applications Involving Multiplication

In this section, we will solve applications that involve multiplication. As in situations that involve addition, some values are known and some are unknown. As with addition, we use an equation to solve for the unknown value.

The formula for addition is as follows: The sum of all of the parts equals the whole.

For multiplication, we use the same formula; the difference is that the parts are *repeated* parts. The formula can be shown as follows:

Part + Same part + Same part + · · · + Same part = Whole

This formula can also be rewritten as a product:

(Number of same parts) × Part = Whole

For example, Kahlil earns $87 per day working as a plumber's assistant. How much does he earn in a 5-day week?

His earnings can be looked at as

$$\frac{\$87}{\text{Monday}} + \frac{\$87}{\text{Tuesday}} + \frac{\$87}{\text{Wednesday}} + \frac{\$87}{\text{Thursday}} + \frac{\$87}{\text{Friday}}$$

This is the same part added repeatedly, and there are 5 of them. So, the equation is

$$5 \times \$87 = \text{whole}.$$
$$5 \times \$87 = \$435$$

In the application where a number is added repeatedly, the *unknown value* is one of the following

1. The part
2. The number of parts (number of repeats)
3. The whole

Identifying the unknown value is relatively simple because it is usually mentioned in the question. However, determining whether the unknown is the whole can be a little more challenging.

Guidelines for Recognizing the Whole, the Part, and the Number of Parts

1. The whole and the part always have the same unit measure. They might be inches or cookies or dollars or miles or any other unit of measure, but they must be the same.
2. The whole is always larger than the part.
3. The number of parts is usually a different unit of measure, such as the number of days, the number of boxes, or the number of months.

Yet with all of the helpful information you find on these pages, it's still best to approach the problem by *thinking*, by putting yourself in the situation and looking for key words.

The examples presented in this section will give you some insight into the best approach. However, it is up to you to think about the situation beyond simply reading the words.

Example 4

Union dues are \$432 a year, and they are split into 12 equal monthly payments. How much must Gloria pay each month for union dues?

Procedure First, reread the problem and put a box around important information. Use the previous guidelines to think about the problem and to set up the equation.

The unknown value is d = how much dues (in dollars) Gloria pays each month.

The whole is the total dues (also dollars) she pays for the year: \$432.

The part is how much she pays each month. The part occurs 12 times.

Notice that the whole and the part are measured in dollars. The other measure is number of months.

Answer **Legend:** Let d = the amount Gloria pays each month for union dues.

The equation is this: $12 \cdot d = 432$

Solve it. $12 \div 12 \cdot d = 432 \div 12$

$d = 36$

$$\begin{array}{r} 36 \\ 12\overline{)432} \\ -36 \\ \hline 72 \\ -72 \\ \hline 0 \end{array}$$

Sentence Gloria must pay \$36 each month for union dues.

You Try It 4 **Keith works in a factory that produces 600 nails per hour, and 40 nails fit into each box. How many boxes does Keith need for the 600 nails?**

a) What is the unknown value? Legend: ________________

b) What value (known or unknown) represents the whole? ________________

c) What value (known or unknown) represents the part? ________________

d) What value (known or unknown) represents the number of times the part occurs? ________________

e) Write the equation and solve it.

f) Sentence: ________________

Example 5

It cost Adrian \$29 per day to rent a car. He rented the car for 7 days. How much total rent did Adrian pay for the car?

Procedure First, reread the problem and put a box around important information. The unknown value is the total paid (in dollars), the whole.

The part is \$29 per day. (Notice that it's also in dollars, same as the whole.)

The part is repeated 7 times (7 days).

Answer: **Legend:** p = the total rent Adrian paid for the car

$$7 \cdot 29 = p$$
$$203 = p$$
$$p = 203$$

$$\begin{array}{r} 29 \\ \times 7 \\ \hline 203 \end{array}$$

Sentence Adrian paid \$203 to rent the car.

You Try It 5 **Mary contributes \$15 from her paycheck each month to the United Way. How much does she contribute in a year (12 months)?**

a) What is the unknown value? Legend: ______

b) What value (known or unknown) represents the whole? ______

c) What value (known or unknown) represents the part? ______

d) What value (known or unknown) represents the number of times the part occurs? ______

e) Write the equation and solve it.

f) Sentence: ______

Applications Involving Averages

For this next situation, you will see the key words *on average*. An average suggests that the part was divided *evenly* among the whole. We also usually find the word *each* with the words *on average*; this, again, indicates a part.

Example 6

Sally, a salesperson at an electronics store, worked 20 days in February. She sold a total of 140 televisions during that month. How many televisions did she sell, on average, each day?

Procedure First, reread the problem and put a box around important information. The key words appearing in the question are *on average*. This means that the unknown value is the part.

The measure that appears twice in the problem is televisions. The unknown value is the average number of televisions sold (the part).

The whole is 140 televisions.

The part is repeated 20 times (20 days).

Answer **Legend:** a = the average number of televisions sold per day

$$20 \cdot a = 140$$
$$20 \div 20 \cdot a = 140 \div 20$$
$$a = 7$$

Sentence Sally sold, on average, 7 televisions each day.

Does this mean that Sally actually sold 7 televisions each day? Probably not.

Most likely Sally had some good selling days and some poor selling days. One day she may have sold as many as 15 televisions, and another day she may have sold as few as 2. But over the course of one full month, she was able to sell an average of 7 televisions per day.

You Try It 6 **At her bookstore, Gena sold 165 books in one 5-day period. How many books did Gena sell, on average, each day?**

a) What is the unknown value? Legend: ____________

b) What value (known or unknown) represents the whole? ____________

c) What value (known or unknown) represents the part? ____________

d) What value (known or unknown) represents the number of times the part occurs?

____

e) Write the equation and solve it.

f) Sentence: ____________

Applications Involving Geometry

Recall from Section 1.2 that the perimeter of a figure is the sum of the lengths of its sides. When we know the perimeter (P) of a triangle but only two of the side measures, we can use an equation to find the length of the third side.

We can let x be the length of the unknown side and solve for it using *the sum of the parts equals the whole.*

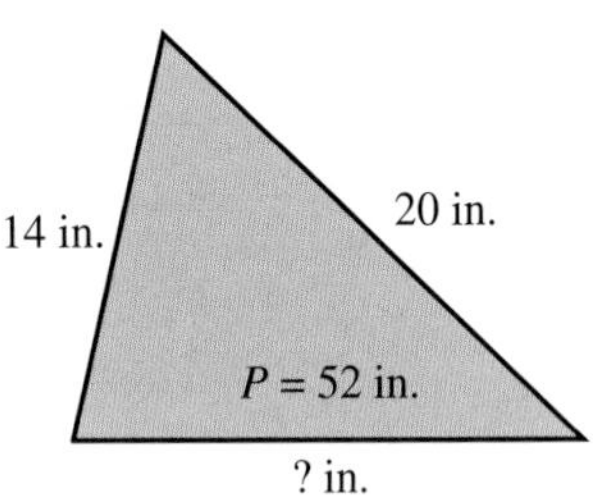

Example 7

The perimeter of a triangle is 52 inches. The longest side measures 20 inches, and the shortest side measures 14 inches. What is the length of the third side?

Procedure First, reread the problem and put a box around important information. If a triangle is not provided, you can draw one of your own and label the sides.

Answer **Legend:** x = the measure of the third side

$$20 + 14 + x = 52$$ Simplify the left side by combining the numbers.

$$34 + x = 52$$ Subtract 34 from each side.

$$34 - 34 + x = 52 - 34$$ Simplify.

$$x = 18$$

Sentence The third side is 18 inches long.

You Try It 7 **A park is in the shape of a triangle. The perimeter of the sidewalk around the park is 168 yards. The shortest side is 42 yards, and the longest side is 70 yards. What is the length of the third side? Draw a triangle and place the known measures around it.**

Legend ____________

Sentence ____________

Recall from Section 1.3 that the area of a rectangle is the product of its length and width:

Area of a rectangle = Length × Width

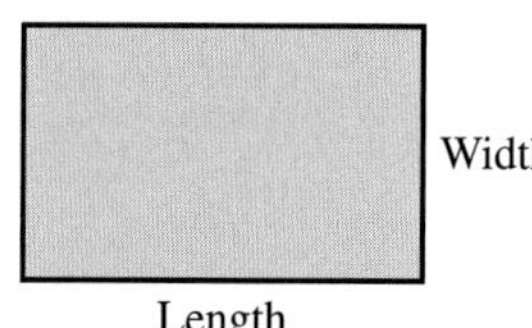

When the area and the measure of one side are known, we can find the measure of the other side using an equation.

Example 8

The area of a rectangle is 78 square inches. The width is 6 inches. What is the length?

$A = 78$ in.^{2} 6 in. L

Procedure We can draw a rectangle and label the width as 6 and the length as L. We can even use L instead of x in the equation: $A = L \cdot W$.

Answer **Legend:** Let L = the length of the rectangle.

$$78 = L \cdot 6 \qquad \text{Divide each side by 6.}$$
$$78 \div 6 = L \cdot 6 \div 6 \qquad 78 \div 6 = 13$$
$$13 = L$$

$$\begin{array}{r} 13 \\ 6\overline{)78} \\ -6 \\ \hline 18 \\ -18 \\ \hline 0 \end{array}$$

Sentence The length of the rectangle is 13 inches.

You Try It 8 **A rectangular dance floor has an area of 255 square feet. The length is 17 feet. What is the width? Draw a rectangle and place the known measures around it.**

Legend ______________________________

Sentence ______________________________

Answers: You Try It and Think About It

You Try It: **1. a)** Whole is 1,280 pounds; part is 891 pounds. **b)** Let x = how much more weight can be carried. **c)** $891 + x = 1{,}280$ **d)** Mark's plane can carry 389 more pounds. **2. a)** Whole is 312 roses; part is 138 roses. **b)** Let x = the number of roses needed. **c)** $138 + x = 312$ **d)** Barbara needs to order 174 more roses. **3. a)** Whole is 700 books; parts are 167 books, 228 books, and 174 books. **b)** Let b = the number of books Gena must sell during the fourth week. **c)** $167 + 228 + 174 + b = 700$ **d)** Gena must sell 131 books during the fourth week. **4. a)** Let b = the number of boxes needed. **b)** Whole is 600 nails. **c)** Part is 40 nails. **d)** The number of times the part occurs is b. **e)** $40 \cdot b = 600$ **f)** Keith needs 15 boxes for the 600 nails. **5. a)** Let x = the amount (in dollars) she will contribute in one year. **b)** Whole is x. **c)** Part is \$15. **d)** The number of times the part occurs is 12 months. **e)** $12 \cdot 15 = x$ **f)** In one year, Mary contributes \$180 to the United Way. **6. a)** Let a = the average number of books sold each day. **b)** Whole is 165 books. **c)** Part is a. **d)** The number of times the part occurs is 5 days. **e)** $5 \cdot a = 165$ **f)** Gena sold, on average, 33 books each day.

7. Let x = the length of the third side.
$42 + 70 + x = 168$
The length of the third side is 56 yards.

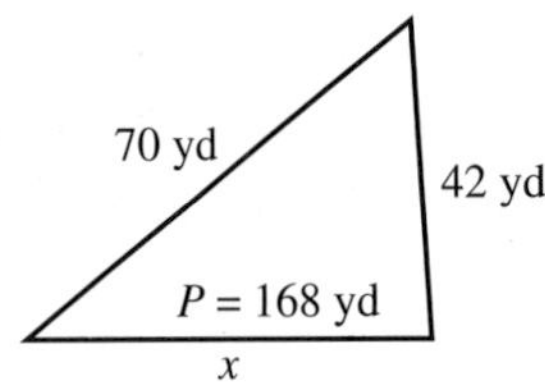

8. Let W = the width of the rectangle.
$17 \cdot W = 255$
The width is 15 feet.

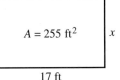

$A = 255 \text{ ft}^2$
x
17 ft

Think About It: No Think About Its in this section.

Section 1.8 Exercises

FOR EXTRA HELP
Student Resources on DVD-ROM
Includes
➤ Student's Solutions Manual
➤ Video Lectures
➤ Chapter Test Prep Video
MyMathLab
MathXL

Think Again

1. In an application problem, what is the purpose of the legend?
The legend helps define what number the variable represents.

2. In an application problem, where does the equation come from?
The equation comes from a formula that pertains to the problem.

3. In an application problem, why is it important to write a sentence at the end?
Answers will vary.

4. In an application problem involving multiplication, what do the part and the whole always have in common?
They always have the same unit of measure.

Focus Exercises

Work each application. Write a legend and an equation. Solve the equation and answer the question in a complete sentence.

5. Nate is saving his money to buy an Xbox. So far he has saved \$73. The model he wants sells for \$162 including tax. How much more money does Nate need to pay for the Xbox?
Let m = how much more money Nate needs.
$73 + m = 162$
Nate needs \$89 more dollars.

6. Shika needs 132 yards of material to make enough doll dresses to sell at the craft fair. She already has 58 yards of material. How many more yards of material does Shika need?
Let x = how much more material Shika needs.
$58 + x = 132$
Shika needs 74 more yards of material.

7. Adam has saved \$900 for his vacation. He has figured that he'll spend \$380 on the hotel room, \$140 on meals, and \$115 on transportation. How much will Adam have left to spend on other things during his vacation?
Let m = how much money Adam will have left over.
$380 + 140 + 115 + m = 900$
Adam will have \$265 left to spend on vacation.

8. A cab company spends \$18,652 per year on insurance for its drivers. Insuring a new driver costs \$1,208. How much total insurance will the company pay if it adds one more driver?
Let x = total insurance costs.
$18{,}652 + 1{,}208 = x$
The company will pay a total of \$19,860 for insurance.

9. Sandra works for the city accounting office. She must prepare a report for the city manager listing a variety of costs that must be paid. For example, it cost the city \$21,000 to operate and maintain the public pool. So far the city has collected \$19,629 from user fees. How much more must the city collect to equal the costs of operating the pool?
Let x = how much more money the city must collect.
$19{,}629 + x = 21{,}000$
The city must collect \$1,371 more.

10. Through a regular deduction from her paycheck, Alexis paid a total of \$2,867 last year for federal taxes. In filing her income tax return, she should have paid \$3,052. How much does Alexis still owe on her federal taxes?
Let a = how much Alexis still owes.
$2{,}867 + a = 3{,}052$
Alexis still owes \$185 on her taxes.

11. Kinko's is putting together 5,000 brochures for a company that are due tomorrow at noon. Kelley has assigned the job to three different workers, each using a different photocopier. Heather's copier has produced 1,758 copies, Omar's copier has produced 1,365 copies, and Carla's copier has produced 1,259 copies. How many copies are left to produce?
Let c = the number of copies left to produce.
$1{,}758 + 1{,}365 + 1{,}259 + c = 5{,}000$
There are 618 copies left to produce.

12. The billing department in a large company had \$15,000 in its annual supplies budget. So far this year they have purchased a computer system for \$3,193, new furniture for \$2,607, and various office supplies for \$612. Rika's job is to update the department's budget. How much is left in the supplies budget after these purchases?
Let m = how much money is left in the supplies budget.
$3{,}193 + 2{,}607 + 612 + m = 15{,}000$
There is \$8,588 left in the supplies budget.

13. Ajay belongs to the local carpenter's union. He pays \$252 per year in union dues. If the payments are spread equally over 12 months, how much does Ajay pay each month in union dues?
Let a = the amount Ajay pays each month.
$12 \cdot a = 252$
Ajay pays \$21 each month for dues.

14. Kami owns an embroidery business, and she just received a rush order. She has to embroider 168 caps in 7 days. How many caps, on average, must Kami embroider each day?
Let c = the number of caps embroidered each day.
$7 \cdot c = 168$
Kami must embroider, on average, 24 caps each day.

15. Carnell's job is to box up jars of jam and prepare them for shipping.

a) It takes Carnell 72 minutes to prepare 18 boxes for shipping. How many minutes, on average, does Carnell take to prepare 1 box for shipping?
Let m = the number of minutes to prepare 1 box for shipping.
$18 \cdot m = 72$
Carnell takes, on average, 4 minutes to prepare 1 box.

b) If each box holds 24 jars of jam, how many jars of jam are in the 18 boxes?
Let j = the number of jars of jam in all of the boxes.
$24 \cdot 18 = j$
There are 432 jars of jam in the 18 boxes.

16. Beth is a nutritionist. She has many overweight clients. Most of her clients are referred to her by doctors who work at a nearby hospital for eating disorders.

a) One of her heaviest clients has been put on a strict diet and exercise program by his doctor. His total weight loss over a 6-month period was 138 pounds. How many pounds did he lose, on average, each month?
Let w = the amount of weight lost, on average, each month.
$6 \cdot w = 138$
He lost, on average, 23 pounds each month.

b) Beth has another client, Lona, who is on a plan to lose 15 pounds a month for 12 months. Assuming that Lona stays on the plan, what will her total weight loss be?
Let w = total weight loss.
$15 \cdot 12 = w$
Lona's total weight loss will be 180 pounds.

17. Mansour owns a courier service and uses his small plane to make deliveries.

a) One customer has asked him to deliver 25 boxes, each weighing 42 pounds. How much cargo will Mansour's plane be carrying?
Let x = the total weight of the cargo.
$25 \cdot 42 = x$
Mansour's plane will be carrying 1,050 pounds.

b) A customer has 24 boxes weighing a total of 864 pounds. What is the average weight of each box?
Let x = the average weight of each box.
$24 \cdot x = 864$
The average weight of each box is 36 pounds.

18. Aimee pays \$375 in rent each month. How much does Aimee pay in rent for a full year?
Let r = the amount of rent paid in a year.
$375 \cdot 12 = r$
Aimee pays \$4,500 in rent for the year.

19. Allison works at a printer. The business often receives large printing orders for pamphlets and flyers. One customer wants 3,200 flyers printed and needs them right away. Allison decides to use 5 printing machines to do the job. On average, how many flyers will each machine print?
Let f = the number of flyers printed on each machine.
$5 \cdot f = 3{,}200$
Each machine will print, on average, 640 flyers.

20. Marcus is planning to drive across the country from San Diego to Philadelphia, a total of 2,778 miles. He plans to make this trip in 6 days. How many miles, on average, will Marcus drive each day.
Let m = the number of miles driven each day.
$6 \cdot m = 2{,}778$
Marcus will drive, on average, 463 miles each day.

21. Three steel bridge support girders form a triangle with a perimeter of 90 feet. The two longer sides measure 42 feet and 31 feet. What is the measure of the shortest side?
Let x = the measure of the shortest side.
$42 + 31 + x = 90$
The shortest side measures 17 feet.

22. A triangular window has a perimeter of 121 inches. The two shorter sides measure 27 inches and 38 inches. What is the measure of the longest side?
Let x = the measure of the longest side.
$27 + 38 + x = 121$
The longest side measures 56 inches.

23. The home plate in a youth sports league is shown at the right. The perimeter of the plate is 48 inches. What is the length of the top side?

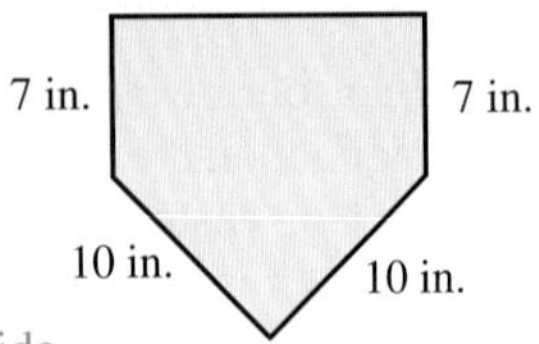

Let x = the length of the top side.
$7 + 10 + 10 + 7 + x = 48$
The length of the top side is 14 inches.

24. The Bermuda Triangle, known for the unexplained disappearance of ships and planes that have crossed through it, forms a triangle between three Atlantic locations: Miami, Florida; Bermuda; and San Juan, Puerto Rico. The perimeter of this triangle is 2,992 miles. The distance from Bermuda to Miami is 1,059 miles. The distance from Bermuda to San Juan is 955 miles. What is the distance from San Juan to Miami?
Let d = the distance from San Juan to Miami.
$1{,}059 + 955 + d = 2{,}992$
The distance from San Juan to Miami is 978 miles.

25. Carleen needs to rent a sectional dance floor that can be placed in the park. The dance floor is 30 feet long and 24 feet wide. What is the area of the dance floor?
Let a = the area of the dance floor.
$30 \cdot 24 = a$
The area of the dance floor is 720 square feet.

26. A rectangular parking lot is 23 yards wide and 59 yards long. What is the area of the parking lot?
Let a = the area of the parking lot.
$23 \cdot 59 = a$
The area of the parking lot is 1,357 square yards.

27. A rectangular plot of land is a full acre: 4,840 square yards. If the width is 40 yards, what is the length of the rectangle?
Let x = the length of the rectangle.
$40 \cdot x = 4{,}840$
The length of the rectangle is 121 yards.

28. A hotel pool is in the shape of a rectangle. The rectangle is 15 yards wide and has an area of 540 square yards. What is the length of the pool?
Let x = the length of the pool.
$15 \cdot x = 540$
The length of the pool is 36 yards.

29. A candy maker uses 7 ounces of peppermint extract to make 945 candy canes. How many candy canes can he make using just 1 ounce of peppermint extract?
Let c = the number of candy canes.
$7 \cdot c = 945$
He can make 135 candy canes.

30. At a banquet, the caterer served 6 coconut shrimp to each attendee. If she served 438 total coconut shrimp, how many people were served at the banquet?
Let p = the number of people served.
$6 \cdot p = 438$
There were 73 people served at the banquet.

Think Outside the Box

31. The area of a rectangle is 273 square inches, and the width is 13 inches. What is the perimeter of the rectangle?
The perimeter of the rectangle is 68 inches.

32. Liz started with a stack of 500 flyers announcing her Kiwanis Club's annual silent auction. She distributed 12 flyers to each of several businesses around town. When she was done for the day, she still had 44 flyers left. To how many businesses did Liz distribute the flyers?
Liz distributed the flyers to 38 businesses.

Chapter 1 Review

Section 1.1 Whole Numbers and Their Properties

Concept	Example
Whole numbers	0, 1, 2, 3, 4, 5, 6, 7, 8, 9, 10, 11, 12, 13, . . .
Counting numbers, or **natural numbers**	1, 2, 3, 4, 5, . . . , 10, 11, 12, . . .
Every digit in a whole number has a **place value** based on the position it holds in the number. In the place value chart at the right, the value of 8 is 8 hundred: 800. The value of 9 is 9 ten thousands: 90,000. The value of 3 is millions: 3,000,000.	See place value chart below.
To write a whole number in words 1. Hyphenate most two-digit numbers. 2. Never use the word *and*. 3. Separate the periods with commas.	1. 34 is written *thirty-four.* 2. 306 is written *three hundred six.* 3. 5,073,109 is written *five million, seventy-three thousand, one hundred nine.*
The four basic **operations** are addition (+), subtraction (−), multiplication (×), and division (÷).	$5 + 2 = 7$ $7 - 2 = 5$ $3 \times 2 = 6$ $6 \div 2 = 3$
A **sum** is the written addition and the result of adding.	Addend + Addend = Sum
A **difference** is the written subtraction and the result of subtracting.	Minuend − Subtrahend = Difference
A **product** is the written multiplication and the result of multiplying.	Factor × Factor = Product
A **quotient** is the written division and the result of dividing.	Dividend ÷ Divisor = Quotient $\text{Divisor}\overline{)\text{Dividend}}$ with Quotient above
The written form of an operation is called an **expression.**	$5 + 4$
To **evaluate** means to "find the value of."	Evaluate 5 + 4: $5 + 4 = 9$
Parentheses group different values together so that they can be treated as one **quantity.**	$(5 + 2) \cdot 4 = 7 \cdot 4$

Place value chart

	Millions Period			Thousands Period			Ones Period		
Place	hund.	tens	ones	hund.	tens	ones	hund.	tens	ones
		2	5	3	1	9	4	6	7

continued

Section 1.1 Whole Numbers and Their Properties

Concept	Example
The **Commutative Property** allows us to change the order of a sum or product.	
The Commutative Property of Addition: $a + b = b + a$	$6 + 2 = 2 + 6$
The Commutative Property of Multiplication: $a \times b = b \times a$	$6 \cdot 2 = 2 \cdot 6$
The **Associative Property** allows us to change the grouping in a sum or product.	
The Associative Property of Addition: $(a + b) + c = a + (b + c)$	$(6 + 4) + 3 = 6 + (4 + 3)$
The Associative Property of Multiplication: $(a \cdot b) \cdot c = a \cdot (b \cdot c)$	$(5 \cdot 2) \cdot 4 = 5 \cdot (2 \cdot 4)$
An **identity** is a number that when applied, won't change the value of another number or quantity.	
The identity for addition, or **additive identity,** is 0. $a + 0 = a$ and $0 + a = a$	$12 + 0 = 12$; also $0 + 12 = 12$
The identity for multiplication, or **multiplicative identity,** is 1. $a \cdot 1 = a$ and $1 \cdot a = a$	$15 \cdot 1 = 15$; also $15 \cdot 1 = 15$
To round a whole number **1.** Identify the place digit. **2.** Identify the rounding digit. **3. a.** If the rounding digit is 5 or higher, round *up*. **b.** If the rounding digit is 4 or lower, round *down*. **4.** Write all digits after the place digit as zeros.	Round 63,514 to the nearest *thousand*. 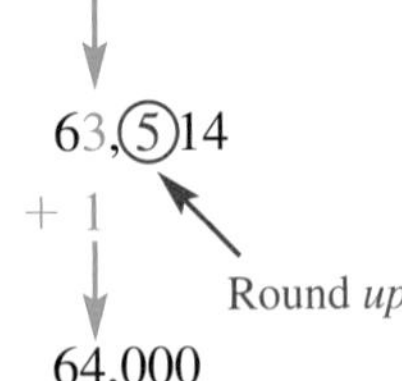

Section 1.2 Adding and Subtracting Whole Numbers

Concept	Example
To add whole numbers **1.** Align the numbers one above the other. If they don't have the same number of digits, place zeros in front of the number with the fewest digits. **2.** Add each column starting with the ones column. **3.** If any column adds to more than 9, carry over the sum's tens digit to the next column.	Add 68 + 274: $\begin{array}{r} {}^{1\,1}\\ 068 \\ +\ 274 \\ \hline 342 \end{array}$
The **perimeter** of a geometric figure is the total measure around the figure. The perimeter is found by adding the lengths of all of the sides.	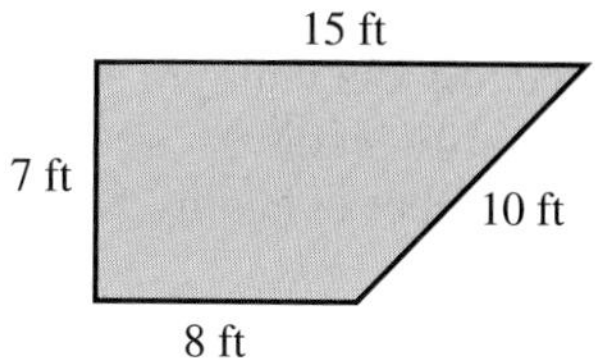 The perimeter is: 7 feet + 8 feet + 10 feet + 15 feet = 40 feet.
To subtract whole numbers **1.** Align the numbers one above the other. If they don't have the same number of digits, place zeros in front of one of them. **2.** Subtract each column starting with the ones column. **3.** If subtraction cannot be performed in a certain column, regroup the number on top. **4.** Check subtraction by adding the result (the difference) to the bottom number.	Subtract 624 − 79: $\begin{array}{r} {}^{5\ \overset{11}{\not 1}\ 14} \\ \not 6\,\not 2\,\not 4 \\ -\ 0\,7\,9 \\ \hline 5\,4\,5 \end{array}$ **Check** $\begin{array}{r} {}^{+1\ +1} \\ 5\ 4\ 5 \\ +\ 0\ 7\ 9 \\ \hline 6\ 2\ 4 \end{array}$

Section 1.3 Multiplying Whole Numbers

Concept	Example
Multiplication is an abbreviation for repeated addition. Numbers in a product are called **factors.**	$5 \cdot 8 = 8 + 8 + 8 + 8 + 8 = 40$; $5 \cdot 8 = 40$ 5 and 8 are factors of 40.
A number is in **factored form** when it is written as a product of factors.	Factored forms of 40: $1 \cdot 40$, $2 \cdot 20$, $4 \cdot 10$, and $5 \cdot 8$
If a and b are two whole numbers, their product, $a \cdot b$, is a **multiple** of a and a multiple of b.	5, 10, 15, 20, and 25 are multiples of 5 because each has a factored form that includes 5. 5 10 15 20 25 $1 \cdot 5$ $2 \cdot 5$ $3 \cdot 5$ $4 \cdot 5$ $5 \cdot 5$
The Multiplication Property of 0: $a \cdot 0 = 0$ and $0 \cdot a = 0$	$5 \cdot 0 = 0$ and $0 \cdot 5 = 0$

continued

Section 1.3 Multiplying Whole Numbers

Concept	Example
Multiplying by 10 When one factor is 10, the product is the *other* factor with a 0 placed on the end.	$15 \cdot 10 = 150$
Multiplying by 100 When one factor is 100, the product is the *other* factor with two 0's placed on the end.	$9 \cdot 100 = 900$
Multiplying by multiples of 10 If one or more factors is a multiple of 10, such as 30 or 60 or 700, temporarily ignore such (ending) zeros and include the total number of ending zeros together in the final product.	$30 \cdot 400 = 12{,}000$
The Distributive Property of Multiplication over Addition We can *distribute* a multiplier, a, to a sum $(b + c)$ so that it multiplies both numbers in the sum: $a \cdot (b + c) = a \cdot b + a \cdot c$	$3 \cdot (6 + 4) = 3 \cdot 6 + 3 \cdot 4$ $= 18 + 12$ $= 30$
To multiply a two-digit number by a single-digit number, write the product vertically with the columns aligned. 1. Multiply the ones place digit times the single digit. Carry any tens place value in this product and place it above the tens place digit. 2. Multiply the tens place digit times the single digit and add any carried value. This process can extend to three or more digits.	Multiply 86 × 9: $\begin{array}{r} {}^{5} \\ 86 \\ \times\ 9 \\ \hline 774 \end{array}$
To multiply a two-digit number by a two-digit number, write the product vertically with the columns aligned. We will get two partial products that will be added at the end of the process. 1. The first partial product comes from multiplying the top number by the ones place digit. 2. The second partial product comes from multiplying the top number by the tens place digit. Because the tens place value is a multiple of 10, this second partial product will have a zero in the ones place. This process can extend to three or more digits.	Multiply 86 × 39: $\begin{array}{r} 86 \\ \times\ 39 \\ \hline 774 \\ +\ 2{,}580 \\ \hline 3{,}354 \end{array}$
Area is the amount of surface in an enclosed region and is always measured in square units. The area of a rectangle is the product of its width and length: $Area = Length \times Width$	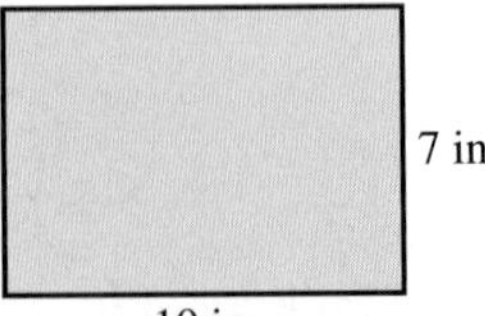 The area is 7 inches · 10 inches = 70 square inches.

Section 1.4 Dividing Whole Numbers

Concept	Example
Division is the inverse operation of multiplication. It can be thought of as repeated subtraction. When the repeated subtraction results in 0, the divisor *divides evenly into* the dividend; we also say that the dividend is *divisible* by the divisor.	$15 \div 5$: $15 - 5 = 10$ Once $10 - 5 = 5$ Twice $5 - 5 = 0$ Three times $15 \div 5 = 3$ 5 divides evenly into 15, and 15 is divisible by 5.
When the repeated subtraction doesn't result in 0, there is a **remainder.** When the repeated subtraction ends in 0, we can say that the remainder is 0 or that there is no remainder.	$17 \div 5$: $17 - 5 = 12$ Once $12 - 5 = 7$ Twice $7 - 5 = 2$ Three times, remainder of 2 $17 \div 5 = 3$ r 2
Division by 0 is not allowed: $a \div 0$, or $\frac{a}{0}$, is undefined.	$12 \div 0$ is undefined.
The Long Division Algorithm **1.** If possible, divide the divisor into the first (leftmost) digit in the dividend whether or not it divides into the digit evenly. If the first digit of the dividend is too small, place a zero in the quotient (above the first digit in the dividend) and divide the divisor into the first two or more (leftmost) digits. Place the quotient over the last digit used. **2.** Multiply the quotient and the divisor. Place this product directly under the digits used in this division. **3.** Subtract. **4.** "Bring down" the first unused digit in the dividend. Repeat this process (starting at step 1) until you can divide no further. Write the remainder next to the quotient. The division process stops when the quotient "covers" the last digit in the dividend. The **remainder** is the amount left over after the last digit in the dividend is covered by the quotient.	$72 \div 5$: $\begin{array}{r} 14 \text{ r } 2 \\ 5\overline{)72} \\ -5 \\ \hline 22 \\ -20 \\ \hline 2 \end{array}$

Section 1.5 Exponents, Square Roots, and the Order of Operations

Concept	Example
Exponents give us a way to abbreviate repeated multiplication using the powers of a whole number. In the notation 2^3, 2 is the **base** and ^{3} is the **exponent** or **power.** The whole number 2 is raised to the power of 3.	2^3 represents the repeated multiplication $2 \cdot 2 \cdot 2$ and is read "2 to the *third power*."

continued

Section 1.5 Exponents, Square Roots, and the Order of Operations

Concept	Example
The Power of 1 If b represents any base, then $b^1 = b$.	$5^1 = 5$
The Powers of 10 The exponent of 10 indicates the number of zeros that follow the 1.	10^5 is a 1 followed by five zeros: 100,000. 10,000 can be abbreviated as 10^4, the number of zeros indicating the power of 10.
Numbers that end in one or more zeros can be abbreviated using powers of 10.	$700 = 7 \cdot 10^2$ $970{,}000 = 97 \cdot 10^4$
A perfect square is a rectangle with equal side measures and a number. The number is equivalent to the area of the square. The **square root** of a perfect square is the length of one side of the square. If $r^2 = p$, then r is a **square root** of p. Also, if r is a square root of p, then $r^2 = p$.	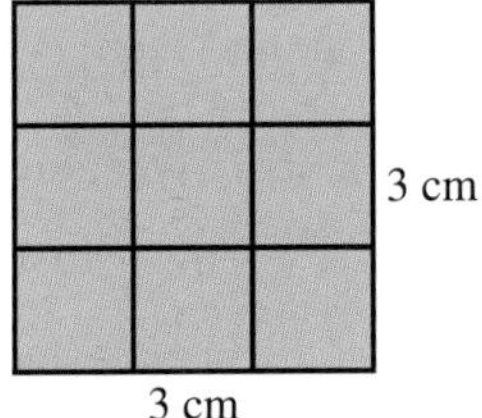 Area $= 9 \text{ cm}^2$ A square root of 9 is 3. Because $3^2 = 9$, 3 is a square root of 9.
The square root symbol, a **radical** $\sqrt{\ }$, is used to represent a square root of a number.	$\sqrt{9} = 3$ 9 is the **radicand.**
Some grouping symbols form quantities.	( ), [ ], and { }
The radical is both a grouping symbol and an operation.	$\sqrt{25} + 10 = 5 + 10 = 15$
The Order of Operations 1. Evaluate within all grouping symbols (one at a time) if there are any. 2. Apply any exponents. 3. Apply multiplication and division *reading from left to right.* 4. Apply addition and subtraction *reading from left to right.*	$(3 + 9) \div 2^2 \cdot 3 - 2$ Evaluate within the parentheses. $= 12 \div 2^2 \cdot 3 - 2$ Apply the exponent. $= 12 \div 4 \cdot 3 - 2$ Apply division. $= 3 \cdot 3 - 2$ Apply multiplication. $= 9 - 2$ Apply subtraction. $= 7$

Section 1.6 Factors

Concept	Example
A **factor pair** of a number is two factors whose product is the number.	One factor pair of 18 is 2 and 9 because $2 \cdot 9 = 18$. Other factor pairs of 18 are 1 and 18 and 3 and 6.

Section 1.6 Factors

Concept	Example
A whole number is considered to be a **prime** number when it has exactly two distinct whole number factors: 1 and itself. 1 is not a prime number because it has only one factor.	The first ten prime numbers are 2, 3, 5, 7, 11, 13, 17, 19, 23, and 29.
A whole number that has more than two distinct factors is a **composite** number. A composite number is a whole number (greater than 1) that is not prime. The number 1 is neither prime nor composite.	The first ten composite numbers are 4, 6, 8, 9, 10, 12, 14, 15, 16, and 18.
Divisibility Test for 2 2 is a factor of a whole number if and only if the number is even (has 0, 2, 4, 6, or 8 in the ones place).	2 is a factor of each of these numbers: 28, 46, 174, 382, and 590
Divisibility Test for 5 5 is a factor of a whole number if and only if the number has either 5 or 0 in the ones place.	5 is a factor of each of these numbers: 35, 70, 105, 230, and 775
Divisibility Test for 10 10 is a factor of a whole number if and only if the number has 0 in the ones place.	10 is a factor of each of these numbers: 70, 190, 230, and 900
Divisibility Test for 3 3 is a factor of a whole number if and only if the number's digits add to a multiple of 3.	3 is a factor of each of these numbers: 105 because $1 + 0 + 5 = 6$, a multiple of 3 264 because $2 + 6 + 4 = 12$, a multiple of 3
Divisibility Test for 9 9 is a factor of a whole number if and only if the number's digits add to a multiple of 9.	9 is a factor of each of these numbers: 198 because $1 + 9 + 8 = 18$, a multiple of 9 513 because $5 + 1 + 3 = 9$, a multiple of 9
A **factor tree** shows composite and prime factors of a number. The **prime factorization** of a composite number is the product of the prime factors (including repetitions) of the number.	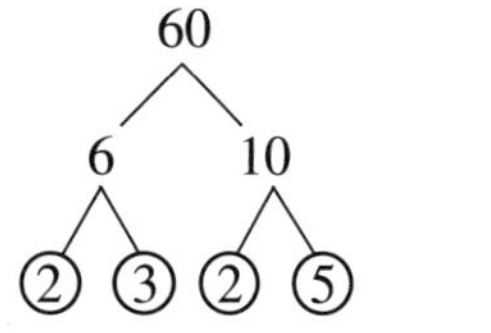 The prime factorization of 60: $60 = 2 \cdot 2 \cdot 3 \cdot 5$ $60 = 2^2 \cdot 3 \cdot 5$
The Division Method Repeatedly divide by *the lowest possible prime number*, generating a new quotient each time. The divisibility tests help determine which prime to use as a divisor. In the division method, the primes we divide by are called **prime divisors.** This process is complete when the new quotient is a prime number.	Find the prime factorization of 60. $60 \div 2 = 30$ $2\overline{)60}$ $30 \div 2 = 15$ $2\overline{)30}$ ← **New quotient** $15 \div 3 = 5$ $3\overline{)15}$ ← **New quotient** 5 ← **New quotient** The prime factorization of 60 is the product of all of the prime divisors and the last prime (5). So, the prime factorization of 60 is $2 \cdot 2 \cdot 3 \cdot 5$, or $2^2 \cdot 3 \cdot 5$.

Section 1.7 Equations

Concept	Example
An **equation** is a mathematical sentence in which one expression equals another expression. A letter such as x is called a **variable;** and it represents a number, an unknown value. It is the unknown value that is to be solved for in an equation. A known value such as 8 is called a **constant.** When a variable is multiplied by a number, we call that number a **coefficient.**	$x + 8 = 15$ and $24 = 6 \cdot x$ are equations. The variable is x. 8, 15, and 24 are constants. 6 is a coefficient.
The Subtraction Property of Equality If $x + a = b$, we can subtract the same value, a, from each side: $x + a = b$ $x + a - a = b - a$	$x + 8 = 15$ $x + 8 - 8 = 15 - 8$ $x + 0 = 7$ $x = 7$
The Division Property of Equality If $x \cdot a = b$, we can divide each side by the same value, a: $x \cdot a = b$ $x \cdot a \div a = b \div a$	$24 = 6 \cdot x$ $24 \div 6 = 6 \div 6 \cdot x$ $4 = 1 \cdot x$ $x = 4$

Section 1.8 Solving Applications Using Equations

Concept	Example
To solve an application problem **1.** Read the problem and think about the situation. **2.** Decide what the unknown value is (usually from the last sentence in the problem) and represent it with a variable by writing a **legend.** **3.** Write an equation based on a formula. **4.** Solve the equation. **5.** Write a complete sentence to answer the question.	A textbook salesperson drove 212 miles on Monday, 87 miles on Tuesday, and 146 miles on Wednesday. What is the total number of miles she drove in those three days? **Legend:** Let x = the total number of miles. **Formula:** Miles + Miles + Miles = Total miles $212 + 87 + 146 = x$ $445 = x$ $x = 445$ **Sentence:** She drove a total of 455 miles.

Section 1.8 Solving Applications Using Equations

Concept	Example
The basic formula for addition: **The sum of all of the parts equals the whole.**	Manuel works for FedEx in the accounts receivable department. One client owes FedEx $4,328 for the month of April, but the client paid only $2,550. How much money does Manuel still need to collect from the client? **Legend:** Let x = the amount still to be collected from the client. **Formula:** $\$ \text{ collected} + \$ \text{ still to be collected} = \text{Total due}$ $2{,}550 + x = 4{,}328$ $2{,}550 - 2{,}550 + x = 4{,}328 - 2{,}550$ $x = 1{,}778$ **Sentence:** Manuel still needs to collect $1,778 from the client.
The basic formula for multiplication: **(Number of same parts) × Part = Whole**	Karea works for a pool supply company. She places an order for 6 identical pool sweepers. The total price for the 6 sweepers is $1,134. How much does each pool sweeper cost? **Legend:** Let x = the cost of each pool sweeper **Formula:** $(\text{Number of same parts}) \times \text{Part} = \text{Whole}$ $6 \cdot x = 1{,}134$ $6 \div 6 \cdot x = 1{,}134 \div 6$ $x = 189$ **Sentence:** Each pool sweeper costs $189.

Chapter 1 Review Exercises

Chapter 1 Vocabulary

Fill in each of the following blanks with the correct word from the word list. Each word in the word list will be used only once.

Word List
addition
approximation
area
coefficient
composite
constant
digits
evaluate
factors
identity
legend
multiple
number
numeral
operations
perimeter
quotient
remainder
solution
variable

1. A ___number___ describes how many of something there is.
2. A ___numeral___ is a symbol that represents a number.
3. There are ten ___digits___ in the base-ten numbering system.
4. Addition and division are two ___operations___.
5. To ___evaluate___ means "to find the value of."
6. For addition, the ___identity___ is 0.
7. A rounded number is called an ___approximation___.
8. The ___perimeter___ of a geometric figure is the sum of the lengths of the sides.
9. Multiplication is an abbreviation for repeated ___addition___.
10. The numbers in a product are called ___factors___.
11. If a and b are two whole numbers, then their product, $a \cdot b$, is a ___multiple___ of a.
12. ___Area___ is the amount of surface in an enclosed region.
13. The answer to an exact division is called the ___quotient___.
14. The amount left over after dividing is the ___remainder___.
15. A ___composite___ number has more than two factors.
16. A ___variable___ is a letter that represents a number.
17. In the expression $y + 7$, the number 7 is called the ___constant___.
18. The ___solution___ of an equation makes the equation true.
19. In the expression $9 \cdot x$, the number 9 is called the ___coefficient___.
20. The ___legend___ describes the unknown value in an application problem.

Section 1.1

Write each number in expanded form.

21. 724
700 + 20 + 4

22. 6,807
6,000 + 800 + 7

Write each number in words.

23. 408
four hundred eight

24. 9,051
nine thousand, fifty-one

25. 206,005
two hundred six thousand, five

26. 5,470,000
five million, four hundred seventy thousand

Write the whole number as a numeral.

27. One hundred seven
107

28. Two thousand, five
2,005

29. Five hundred eight thousand, forty-one
508,041

30. One million, six hundred fifty-two
1,000,652

Which property is being demonstrated?

31. $39 \cdot 1 = 39$
Identity of Multiplication

32. $4 + (3 + 7) = (4 + 3) + 7$
Associative Property of Addition

33. $74 + 15 = 15 + 74$
Commutative Property of Addition

34. $0 + 26 = 26$
Identity of Addition

Round each number to the nearest ten.

35. 642
640

36. 295
300

37. 1,450
1,450

38. 2,996
3,000

Round each number to the nearest hundred.

39. 642
600

40. 30,295
30,300

41. 126,450
126,500

42. 4,949
4,900

Round each number to the nearest thousand.

43. 30,529
31,000

44. 54,067
54,000

45. 249,801
250,000

46. 812
1,000

Rewrite the underlined words using the requested approximation.

47. In 2003, <u>the total number of full-time airline employees was 507,091</u>. Round this number to the nearest *ten thousand*. ***Source:*** **www.bts.gov**
The total number of full-time airline employees was approximately 510,000.

48. <u>In 2004, the U.S. population, was 294,490,706</u>. Round this number to the nearest *hundred thousand*. ***Source:*** **www.census.gov**
In 2004, the U.S. population was approximately 294,500,000.

Section 1.2

Align the numbers, then add.

49. $319 + 211$
530

50. $457 + 93$
550

51. $1{,}934 + 98$
2,032

52. $9{,}184 + 828$
10,012

53. $36 + 51 + 14 + 9$
110

54. $435 + 943 + 25 + 1{,}462$
2,865

Work each application and answer with a complete sentence.

55. On Tuesday morning, Brian rode an exercise bike for one-half hour and burned 387 calories. After a break, he rode another half hour and burned 295 calories. How many total calories did Brian burn on the exercise bike?
Brian burned 682 calories on the exercise bike.

56. Kaira added up her test scores for the first three math tests. Her scores were 89, 75, and 92. What is the total number of points Kaira received on the three tests?
Kaira received a total of 256 points on the three tests.

Find the perimeter of each figure.

57.

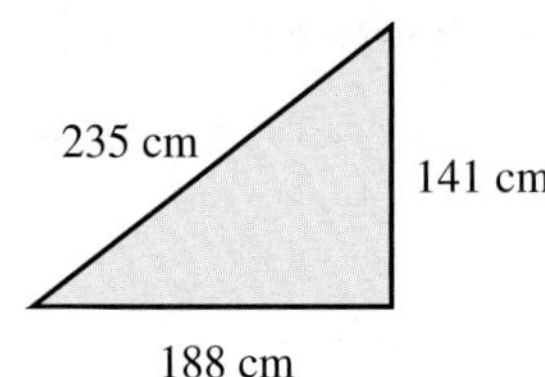

564 cm

58.

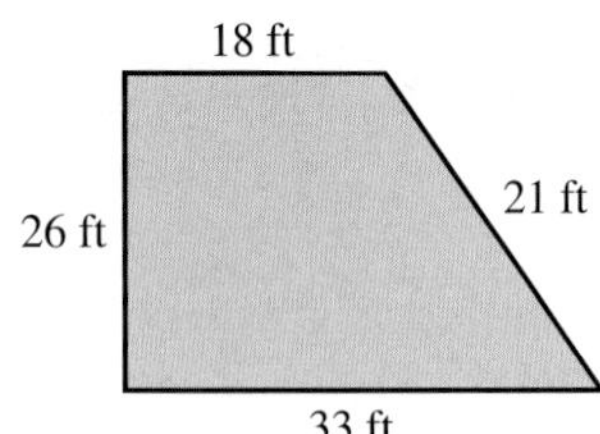

98 feet

Align the numbers, then subtract. Check each answer (on paper or mentally) by addition.

59. 549 − 408
141

60. 132 − 58
74

61. 1,193 − 546
647

62. 16,425 − 841
15,584

63. 6,050 − 5,872
178

64. 4,000,000 − 3,096,205
903,795

Work each application and answer with a complete sentence.

65. In 2007, the average attendance for the New Orleans Saints home games was 70,004. The average attendance for the Oakland Raiders home games was 59,109. On average, how many more fans were in attendance at New Orleans home games than Oakland home games? ***Source:*** **www.sports.espn.com**
There were, on average, 10,895 more fans in attendance at New Orleans home games than Oakland home games.

66. The total area of Nevada is 110,561 square miles, and the total area of Michigan is 96,716 square miles. How much larger (in square miles) is Nevada than Michigan? ***Source:*** **www.infoplease.com**
Nevada is 13,845 square miles larger than Michigan.

The bar graph below shows the monthly snowfall at Mt. Shasta, California, during the 2005–2006 snow season.

67. How many more inches of snow did Mt. Shasta receive in March than in November?
Mt. Shasta received 48 more inches of snow in March than in November.

68. How many total inches of snow did Mt. Shasta receive during the 2005–2006 snow season?

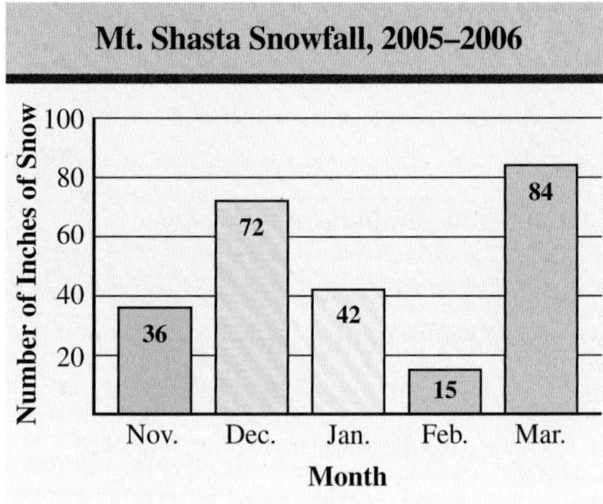

Mt. Shasta received a total of 249 inches of snow.

Section 1.3

Multiply.

69. $7 \cdot 5$
35

70. $7 \cdot 4$
28

71. $5 \cdot 8$
40

72. $9 \cdot 6$
54

73. $8 \cdot 7$
56

74. $9 \cdot 9$
81

75. $7 \cdot 30$
210

76. $80 \cdot 30$
2,400

77. $9 \cdot 800$
7,200

78. $200 \cdot 8$
1,600

79. $90 \cdot 700$
63,000

80. $20 \cdot 500$
10,000

Align the numbers, then multiply.

81. $47 \cdot 8$
376

82. $28 \cdot 43$
1,204

83. $174 \cdot 12$
2,088

84. $192 \cdot 306$
58,752

Which property is being demonstrated?

85. $3 \cdot (8 \cdot 5) = (3 \cdot 8) \cdot 5$
Associative Property of Multiplication

86. $0 \cdot 9 = 0$
Zero Product Property

87. $4 \cdot (2 + 7) = 4 \cdot 2 + 4 \cdot 7$
Distributive Property of Multiplication over Addition

88. $15 \cdot 4 = 4 \cdot 15$
Commutative Property of Multiplication

Work each application and answer the question with a complete sentence.

89. Marley drives 38 miles round trip to work and home each workday. In March, she worked 23 days. How many total miles did Marley drive to and from work in March?
Marley drove 874 miles to and from work in March.

90. Colin's basement floor is in the shape of a rectangle. The width is 19 feet, and the length is 34 feet. What is the area of Colin's basement floor?
The area of Colin's basement floor is 646 square feet.

Section 1.4

Divide. Check the division by mentally multiplying the divisor and the quotient.

91. $36 \div 4$
9

92. $49 \div 7$
7

93. $0 \div 8$
0

94. $40 \div 5$
8

95. $72 \div 9$
8

96. $56 \div 8$
7

97. $24 \div 3$
8

98. $54 \div 6$
9

Use long division to divide. Make sure you write any remainder next to the quotient.

99. $60 \div 9$
6 r 6

100. $43 \div 7$
6 r 1

101. $115 \div 5$
23

102. $168 \div 4$
42

103. $1{,}369 \div 8$
171 r 1

104. $2{,}832 \div 9$
314 r 6

105. $364 \div 14$
26

106. $1{,}024 \div 16$
64

107. $9{,}100 \div 65$
140

108. $1{,}921 \div 26$
73 r 23

109. $4{,}250 \div 59$
72 r 2

110. $7{,}929 \div 73$
108 r 45

Work each application and answer it with a complete sentence.

111. The Jacksonville Rotary Club purchased a trailer it used to sell food at county fairs. Each of the 24 members had to contribute the same amount to pay for the $6,360 trailer. How much was each member's contribution?
Each member's contribution was $265.

112. 405 sixth-grade students are visiting a college. The director of the event wants to divide the students up into different classrooms for the variety of programs planned. If each classroom can hold 32 students, how many classrooms are needed?
13 classrooms are needed.

Section 1.5

Expand each notation and find its value.

113. 1^6
$1 \cdot 1 \cdot 1 \cdot 1 \cdot 1 \cdot 1$
1

114. 2^4
$2 \cdot 2 \cdot 2 \cdot 2$
16

115. 3^5
$3 \cdot 3 \cdot 3 \cdot 3 \cdot 3$
243

116. 4^3
$4 \cdot 4 \cdot 4$
64

117. 16^1
16

118. 17^2
$17 \cdot 17$
289

119. 20^3
$20 \cdot 20 \cdot 20$
8,000

120. 10^7
$10 \cdot 10 \cdot 10 \cdot 10 \cdot 10 \cdot 10 \cdot 10$
10,000,000

Express each number as a power of 10.

121. 1,000
10^3

122. 10,000,000
10^7

123. 100,000
10^5

124. 10
10^1

Rewrite the number as a product of a whole number and a power of 10.

125. 70
$7 \cdot 10^1$

126. 8,400
$84 \cdot 10^2$

127. 300,000
$3 \cdot 10^5$

128. 1,200,000
$12 \cdot 10^5$

Evaluate the following square roots.

129. $\sqrt{36}$
6

130. $\sqrt{4}$
2

131. $\sqrt{9}$
3

132. $\sqrt{100}$
10

Evaluate each problem according to the order of operations.

133. $(18 - 4) \div 2$
7

134. $18 - 4 \div 2$
16

135. $54 \div 3^2$
6

136. $6^2 - 5 \cdot 3$
21

137. $8 \div 2^2 + 7$
9

138. $4^2 \cdot 2 - 2$
30

139. $24 \div 3 \cdot 4 - 2$
30

140. $12 - 30 \div (6 + 4)$
9

141. $(6 - 2) \cdot (12 \div 3)$
16

142. $24 \div 3 \cdot (4 - 2)$
16

143. $23 - \sqrt{16}$
19

144. $\sqrt{4^2 + 9}$
5

Section 1.6

List the first five multiples of each of the following numbers.

145. 3
3, 6, 9, 12, 15

146. 6
6, 12, 18, 24, 30

147. 11
11, 22, 33, 44, 55

148. 12
12, 24, 36, 48, 60

Use a factor pair table to find all of the factor pairs of the following numbers.

149.

18	
1	18
2	9
3	6

150.

36	
1	36
2	18
3	12
4	9
6	6

151.

45	
1	45
3	15
5	9

152.

60	
1	60
2	30
3	20
4	15
5	12

Of the following numbers, determine which are prime, which are composite, and which are neither.

153. 15, 17, 29, 0, 81, 45, 11
Prime: 17, 29, 11
Composite: 15, 81, 45
Neither: 0

154. 2, 61, 70, 43, 62, 1, 31, 57
Prime: 2, 61, 43, 31
Composite: 70, 62, 57
Neither: 1

Of the first three prime numbers—2, 3, and 5—which are factors of the following numbers? Use the divisibility tests for 2, 3, and 5.

155. 147
3

156. 230
2 and 5

157. 625
5

158. 1,782
2 and 3

Determine whether 9 is a factor of each number. Verify your answer by dividing each number by 9.

159. 171
Yes

160. 5,292
Yes

161. 6,708
No

162. 17,451
Yes

Find the prime factorization of the following numbers using any method. Write the answers two ways: with and without exponents.

163. 108
$2 \cdot 2 \cdot 3 \cdot 3 \cdot 3$
$2^2 \cdot 3^3$

164. 162
$2 \cdot 3 \cdot 3 \cdot 3 \cdot 3$
$2 \cdot 3^4$

165. 210
$2 \cdot 3 \cdot 5 \cdot 7$

166. 215
$5 \cdot 43$

Section 1.7

For each equation, replace the variable with 12 and decide whether 12 is the solution.

167. $41 = n + 19$
12 is not the solution.

168. $18 + x = 30$
12 is the solution.

169. $6 \cdot y = 72$
12 is the solution.

170. $350 = w \cdot 25$
12 is not the solution.

Solve the following equations. Check each answer to show that it is the solution.

171. $c + 7 = 15$
$c = 8$

172. $72 = x + 39$
$x = 33$

173. $2{,}094 + 3{,}516 + m = 10{,}000$
$m = 4{,}390$

174. $p \cdot 7 = 98$
$p = 14$

175. $490 = 5 \cdot y$
$y = 98$

176. $35 \cdot w = 1{,}260$
$w = 36$

Section 1.8

Work each application and answer it in a complete sentence.

177. Carlotta sells bedroom furniture. Her company gives her bonus pay if she has sales of $20,000 or more during the Labor Day weekend (Saturday through Monday). On Saturday, she sold $10,560 of merchandise. On Sunday, she had sales of $6,280. What do Carlotta's sales need to be on Monday to reach the $20,000 goal?
Carlotta's sales need to be $3,160 on Monday.

178. Rhani purchased a used car from her parents for $4,500. She has agreed to pay them back over the next three years with 36 equal monthly payments. How much will Rhani pay her parents each month?
Rhani will pay her parents $125 each month.

179. Antonio is a waiter at a pricey restaurant. Last Saturday he waited on 14 tables and earned $266 in tips. On average, how much in tips did Antonio earn from each table?
On average, Antonio earned $19 in tips from each table.

180. The carpeted children's reading room at the library is in the shape of a rectangle. The carpet is 9 yards wide and has an area of 243 square yards. What is the length of the carpet?
The length of the carpet is 27 yards.

Chapter 1 Test

Round each number to the place shown.

1. 749 hundred 700

2. 9,524 thousand 10,000

3. 582,907 ten thousand 580,000

Evaluate each expression.

4. 135×64
8,640

5. $3{,}547 + 958$
4,505

6. $1{,}017 - 459$
558

7. $980 \div 35$
28

Identify the property shown.

8. $16 \cdot 1 = 16$
Identity of Multiplication

9. $4 \cdot (3 + 5) = 4 \cdot 3 + 4 \cdot 5$ Distributive Property of Multiplication over Addition

10. $(14 + 6) + 18 = 14 + (6 + 18)$
Associative Property of Addition

11. $35 + 15 = 15 + 35$
Commutative Property of Addition

Expand each notation and find its value.

12. 5^3 $5 \cdot 5 \cdot 5$
125

13. 20^2 $20 \cdot 20$
400

Rewrite the number as a product using a power of 10 as a factor.

14. 740,000 $74 \cdot 10^4$

15. 900 $9 \cdot 10^2$

Evaluate the following square roots.

16. $\sqrt{16}$
4

17. $\sqrt{81}$
9

Evaluate each expression according to the order of operations. Show all work.

18. $36 \div 4 \cdot 3$
27

19. $2 \cdot 3^2 - 1$
17

20. $2 \cdot (4 + 1)^2$
50

Of the following numbers, determine which are prime, which are composite, and which are neither.

21. 41, 77, 38, 19, 2, 1
Prime: 41, 19, 2
Composite: 77, 38
Neither: 1

Of the first three prime numbers—2, 3, and 5—which are factors of the following numbers?

22. 135 3 and 5

23. 84 2 and 3

24. 149 None

25. 172 2

Find the prime factorization of the following numbers. Write the answer two ways: with and without exponents.

26. 84 $2 \cdot 2 \cdot 3 \cdot 7$
$2^2 \cdot 3 \cdot 7$

27. 80 $2 \cdot 2 \cdot 2 \cdot 2 \cdot 5$
$2^4 \cdot 5$

28. 540
$2 \cdot 2 \cdot 3 \cdot 3 \cdot 3 \cdot 5$
$2^2 \cdot 3^3 \cdot 5$

Solve.

29. $24 = x + 9$
$x = 15$

30. $7 \cdot y = 343$
$y = 49$

31. $25 + 17 + w = 61$
$w = 19$

Solve each application.

32. Alfre needs 500 local voters' signatures to put a no-growth measure on the city's November ballot. In June, Alfre collected 126 signatures; in July, 248. How many more signatures does Alfre need to collect to reach 500?
Alfre needs to collect 126 more signatures.

33. A school's kindergarten playground has a rectangular sandbox with an area of 450 square yards. If the width of the sandbox is 18 yards, what is its length?
The length of the sandbox is 25 yards.

34. Jerry treated the 12 members of his youth basketball team to a Harlem Globetrotters game. Children's tickets were \$14 each. How much did Jerry spend on his team's tickets?
Jerry spent \$168 on his team's tickets.

35. The 17 members of the Lincoln High School Madrigal Choir raised a total of \$3,655 for their annual trip to Chicago. What was the average amount of money that each member raised?
The average amount of money that each member raised was \$215.

CHAPTER 2

Integers and Algebraic Expressions

Introduction

This chapter introduces you to positive and negative numbers and the beginning elements of algebra. As you begin your study of algebra, you may find that it is arithmetic in a different form and that the rules of arithmetic still apply. The order of operations is still the same, and we'll see how it applies to positive and negative numbers.

Negative numbers appear in a variety of situations. A negative temperature is a temperature that is *below zero,* and it's very cold. A negative score means "below par" in golf, and that is very good. In physics, upward is a positive direction and downward is a negative direction. The ocean floor is below sea level and is considered a negative distance from sea level. In accounting, a loss is negative and a gain is positive.

Many students want to know how algebra will apply to their everyday lives. It's difficult to make everything you learn in algebra have a practical, everyday purpose. Instead, look at algebra as the gateway to other areas of study. Learning algebra opens doors to many opportunities that might otherwise be closed, such as science, geography, economics, sports conditioning, nursing, fire technology, and police investigation. And if your interest is not in any way related to math, know that learning algebra makes you smarter because it causes you to think and exercise your brain.

Preparation Exercises

Section 1.2 Addition and Subtraction

Add.

1. $24 + 87$
111

2. $653 + 2{,}397$
3,050

Subtract.

3. $42 - 27$
15

4. $305 - 76$
229

Section 1.3 Multiplying Whole Numbers

Multiply.

5. 13×52
676

6. $27 \cdot 35$
945

Section 1.4 Dividing Whole Numbers

Divide.

7. $\frac{351}{9}$
39

8. $7{,}392 \div 28$
264

Section 1.5 Exponents, Square Roots, and the Order of Operations
Evaluate each expression.

9. 2^3
8

10. 5^2
25

11. $\sqrt{49}$
7

12. $\sqrt{81}$
9

13. $11 - (7 - 4)$
8

14. $24 \div 6 \cdot 2$
8

15. $3 + (8 - 2)^2$
39

16. $5 + 2 \cdot \sqrt{16}$
13

SECTION 2.1 Introduction to Algebra

OBJECTIVES

In this section, you will learn to

- Express ideas through algebraic symbols and abbreviations.
- Translate between English and algebra.
- Identify integers and locate them on the number line.
- Find the absolute value of a number.
- Define *number.*

You Need to Know

To successfully complete this section, you need to understand

- ☐ Exponents and square roots (1.5)
- ☐ The order of operations (1.5)

The Greek mathematician Diophantus was one of the first people to use symbols to *abbreviate* mathematical expressions and thought. Although his symbols were different from what we use today, he is credited with taking the discussion of mathematics and problem solving from sentence form to symbolic form. Could you imagine having to *discuss* everyday mathematics without the use of an addition sign or a multiplication sign?

Introduction

Algebra is all about symbols and abbreviations, using them and manipulating them. In algebra, we use the same symbols for the five basic operations you've already encountered:

Sum: addition (+)
Difference: subtraction (−)
Product: multiplication (×), (·), and (at times) parentheses
Quotient: division (÷ and /) and the fraction bar (—), as in $\frac{8}{2}$
Power: the exponent

Letting symbols represent mathematical ideas allows us to *abbreviate* those ideas. A mathematical sentence written in words can be more easily understood when it is written symbolically as long as you know what the symbols mean.

For example, the simple English expression *the sum of ten and six* can be easily abbreviated as the numerical expression $10 + 6$.

We also use symbols to represent abbreviations and unknown numbers. Some of these abbreviations are

1. **Multiplication:** An abbreviation for repeated addition, such as $3 \cdot 5 = 5 + 5 + 5 = 15$.
2. The **exponent:** An abbreviation for repeated factors, such as $2^5 = 2 \cdot 2 \cdot 2 \cdot 2 \cdot 2 = 32$.
3. The **radical:** An abbreviation for the square root, such as $\sqrt{49} = 7$.

Algebra Vocabulary

Recall from Chapter 1 that a variable is a letter that represents a number. A variable, like an operation sign and the radical, is also an algebraic symbol. This leads us to the definition of an *algebraic expression.*

An **algebraic expression** is a combination of numbers and letters connected by operations and grouping symbols. At its simplest, an algebraic expression can be a single variable, such as *y*, or a single number, such as 3.

Because variables represent numbers, an algebraic expression *expresses* a numerical value.

Here are some other examples of algebraic expressions:

$$w + 8 \qquad 5 \cdot x + 6 \cdot y^2 \qquad \sqrt{3y - x^2}$$

Using variables in algebraic expressions gives us a slightly simpler way to represent multiplication. When there is no other operation between a number and a variable or between two variables, the operation is automatically assumed to be multiplication.

For example, $5 \cdot x$ can be written as $5x$.

Likewise, xy means $x \cdot y$

and $6 \cdot x \cdot y^2$ can be abbreviated as $6xy^2$.

As you know, the Commutative Property of Multiplication allows us to write a product in a different order. For example, $5 \cdot y$ can also be written as $y \cdot 5$. Although we might write $5 \cdot y$ as **5y**, it's *unusual* to write $y \cdot 5$ as **y5**. In other words, when writing the product of a number (coefficient) and a variable, we almost always write the coefficient first (so that we don't mistake the 5 for an exponent).

Variables

Sometimes we know what number a variable represents and sometimes we don't. When we are given the numerical value of a variable, that number is called a **replacement value.** We can also say that we *substitute* the value for the variable.

If an expression contains more than one variable, such as $2W + 2L$, each variable will have its own replacement value, as demonstrated in Example 1.

Example 1

Evaluate each expression using the given replacement values.

a) $w + y$ Replace **w** with 8 and **y** with 9.

b) $2xy$ Replace **x** with 5 and **y** with 6.

c) $a^2 - c$ Replace **a** with 3 and **c** with 7.

d) $2W + 2L$ Replace **W** with 5 and **L** with 8.

Procedure For each application, apply the order of operations.

Answer **a)** $8 + 9 = 17$

b) $2 \cdot 5 \cdot 6 = 60$ Remember, $2xy$ means $2 \cdot x \cdot y$

c) $3^2 - 7 = 9 - 7 = 2$

d) $2 \cdot 5 + 2 \cdot 8 = 10 + 16 = 26$

You Try It 1 **Evaluate each expression using the given replacement values. Use Example 1 as a guide.**

a) $2wc$ Replace **w** with 5 and **c** with 19.

b) $h \cdot b \div 2$ Replace **h** with 12 and **b** with 9.

c) $(r + n)^2$ Replace **r** with 2 and **n** with 6.

d) $(y + 4) \div (3 \cdot k)$ Replace **y** with 26 and **k** with 2.

Sometimes a variable (say w) will appear in more than one part of an expression. When this happens, the replacement value for w is the same for each occurrence of w.

In the expression $3 \cdot w + 20 \div w$, we see two w's. If the replacement value for w is 4, each w has the value of 4. It is not possible for one w to be 4 and the other w to be 7. If we want the variables to have two different values at the same time, the expression must contain two different variables.

Example 2

Evaluate each expression using the given replacement value(s).

a) $3w + 20 \div w$ Replace w with 4.

b) $x^2 + 3x - 7$ Replace x with 5.

c) $4x + xy$ Replace x with 3 and y with 2.

Answer **a)** $3 \cdot 4 + 20 \div 4 = 12 + 5 = 17$

b) $5^2 + 3 \cdot 5 - 7 = 25 + 15 - 7 = 33$

c) $4 \cdot 3 + 3 \cdot 2 = 12 + 6 = 18$

You Try It 2 **Evaluate each expression using the given replacement value(s). Use Example 2 as a guide.**

a) $y^2 + 5y$ Replace y with 6.

b) $3x + 7x$ Replace x with 5.

c) $w \cdot v + 5w$ Replace w with 4 and v with 9.

d) $a + ar$ Replace a with 5 and r with 8.

Translating between English and Algebra

Variables are particularly helpful when they represent numbers whose values we do not yet know. If we need to represent the sum of *a number* and 25, we can write this as $x + 25$. We recognize $x + 25$ as a sum, and we have used the x to represent *a number*.

Why would we need to consider writing expressions such as $x + 25$? Consider this simple example.

Scott's daughter, Jennifer, was born on Scott's 25th birthday. As Jennifer grew older, she began to understand how to calculate her dad's age. She realized that her dad's age was the sum of her age and 25.

To think about how old her dad might be at various stages in her life, Jennifer represented her own age as J and wrote an expression for her dad's age as $J + 25$. Then she thought,

"When I'm 18 and graduate from high school, Dad will be $18 + 25 = 43$;
when I'm 22 and graduate from college, Dad will be $22 + 25 = 47$;
when I'm 30 and start a family of my own, Dad will be $30 + 25 = 55$,
a good age to be a grandpa."

This example also demonstrates that an algebraic expression represents a number; the number represented by $J + 25$ is the age of Jennifer's dad.

In general, when we need to express an unknown number, we can use any variable we choose. A commonly used variable is x.

Example 3

Translate each English expression into an algebraic expression. Use any variable of your liking to represent the unknown number. (Here, x is used.)

a) The sum of a number and 18.
b) The difference of a number and 3.
c) The difference of 17 and a number.
d) The product of 6 and a number.
e) The quotient of a number and 5.
f) The quotient of 20 and a number.
g) The square of a number.
h) The square root of a number.

Answer

a) $x + 18$
b) $x - 3$ (but *not* $3 - x$)
c) $17 - x$ (but *not* $x - 17$)
d) $6 \cdot x$ or $6x$
e) $x \div 5$ or $\frac{x}{5}$ (but *not* $5 \div x$ or $\frac{5}{x}$)
f) $20 \div x$ or $\frac{20}{x}$ (but *not* $x \div 20$ or $\frac{x}{20}$)
g) x^2
h) $\sqrt{x}$

You Try It 3 **Translate each English expression into an algebraic expression. Use any variable of your liking to represent the unknown number. Use Example 3 as a guide.**

a) The product of 5 and a number
b) The quotient of a number and 4
c) The difference of 6 and a number
d) The sum of 11 and a number
e) The square root of a number
f) A number squared.

Parentheses as Separators

As you know, parentheses are considered grouping symbols; they group different values together so that they can be treated as one quantity. Parentheses are also used as separators to keep values apart from each other.

Later in this chapter, you may see an expression such as $(6) + (3)$. In this expression, the parentheses are unnecessary because they don't group anything. However, the parentheses do separate the numbers (from each other and from the plus sign); so, they are *separators*. In Section 2.3, it will be more evident why we use parentheses this way.

Another situation in which parentheses can be used as separators is multiplication. For example, $3 \cdot 5$ can be expressed as $(3) \cdot (5)$ or as $(3)(5)$. Again, what we see are two values with no operation between them; therefore, the operation is *assumed* to be multiplication.

Using parentheses as separators, we can write the product of 3 and 5 in many similar ways:

$$(3)(5) = 15 \qquad 3(5) = 15 \qquad (3)5 = 15$$

Number Lines

We can represent numbers visually along a horizontal number line.

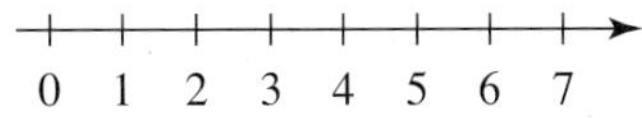

The number zero (0) has a special name on a number line. It is called the **origin,** which means "the beginning." The arrowhead on the number line indicates that the line continues in that direction indefinitely.

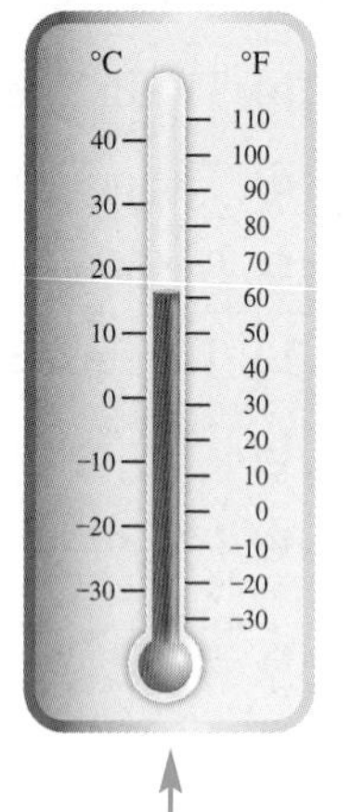

This thermometer has two temperature scales on it: **C** for Celsius and **F** for Fahrenheit.

There are also vertical number lines; one example is a thermometer.

An outdoor thermometer (at the left) includes numbers less than zero to indicate temperatures below zero.

The numbers less than 0 are called **negative numbers.** Numbers greater than 0 are called **positive numbers.**

On a vertical number line (at the right), 0 is still the origin, the positive numbers are above 0, and the negative numbers are below 0.

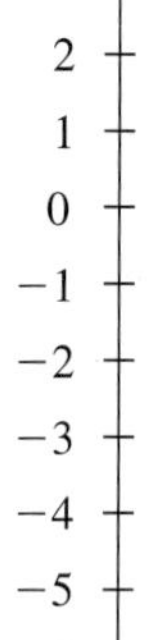

On a horizontal number line, positive numbers are to the right of 0 and negative numbers are to the left of 0. Even though 0 is in the middle here, it is still referred to as the **origin.**

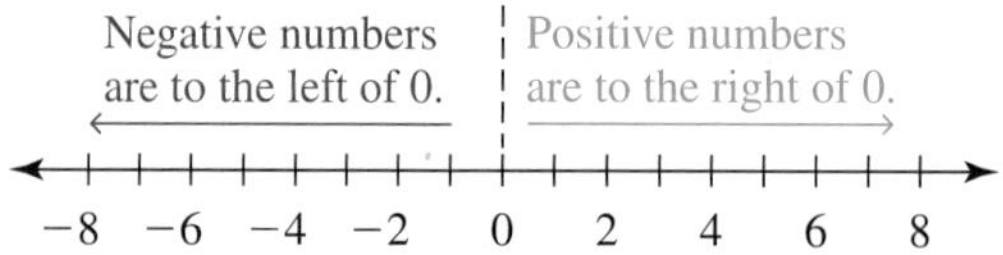

We sometimes write a positive number, such as 6, with a plus sign (+) before the number. The + is not necessary but gives emphasis to the fact that the number is positive. So, +6 is the same as 6. In this case, the + indicates positive, not addition.

Together, positive and negative numbers are called **signed numbers.**

Two signed numbers, one positive and one negative, with the same numeral are called **opposites.** They are on opposite sides of 0 and are the same distance from 0 on the number line.

For example, +2 and −2 are opposites. Each of those numbers is the same distance from 0 on the number line: +2 is 2 units to the right of 0, and −2 is two units to the left of 0.

Example 4

Fill in the blanks.

a) The opposite of +15 is _______.

b) The opposite of −16 is _______.

c) _______ is the opposite of 20.

d) _______ is the opposite of −18.

e) 23 is the _______ of −23.

f) −35 is the _______ of 35.

Answer **a)** −15 **b)** 16 **c)** −20 **d)** 18 **e)** opposite **f)** opposite

You Try It 4 **Fill in the blanks. Use Example 4 as a guide.**

a) The opposite of +6 is _______.

b) The opposite of −9 is _______.

c) _______ is the opposite of 13.

d) _______ is the opposite of −10.

e) −2 is the _______ of 2.

f) +4 is the _______ of −4.

Think About It 1 What is the value of −0? Explain your answer.

____

____

The whole numbers and their opposites are called **integers.**

Integers

Each whole number and its opposite is an integer.

The list of integers can be written as . . . −3, −2, −1, 0, 1, 2, 3, . . .

0 (zero) is an integer but is neither positive nor negative.

You Try It 5 **On the blank number line, write in all of the missing integers.**

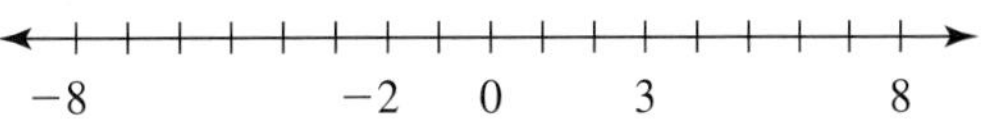

Comparing Signed Numbers

We know that 8 is greater than 3. We can represent this as 8 > 3. The symbol between the 8 and 3, >, is called the **greater than** symbol. (This symbol has the appearance of an arrowhead.)

8 > 3

This means *is greater than.*

In each of these, the arrowhead part is pointing toward the lesser number.

Similarly, 3 is less than 8. We can represent this as 3 < 8. The symbol between the 3 and 8, <, is called the **less than** symbol.

3 < 8

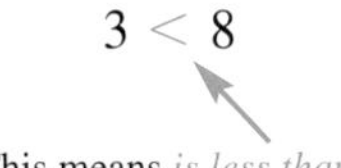

This means *is less than.*

8 > 3
8 is greater than 3

0 3 8

8 is to the right of 3

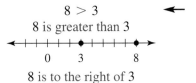

On a typical horizontal number line, *greater than* means "to the right of" and *less than* means "to the left of." So, 8 is greater than 3 because 8 is to the right of 3 on the number line.

Likewise, −5 is less than 2 because −5 is *to the left of* 2 on the number line.

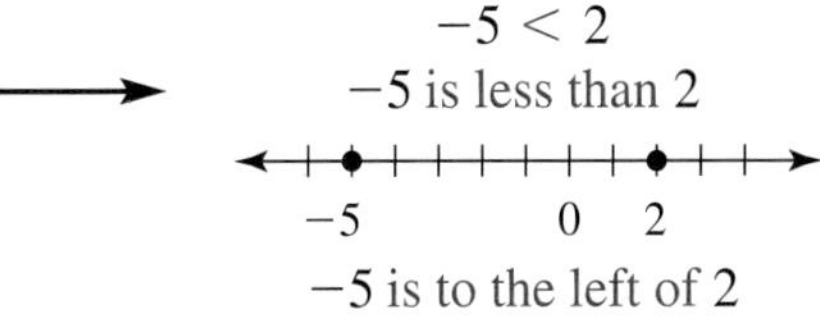

Example 5

Insert the correct symbol between each pair of numbers, either < (less than) or > (greater than).

a) 1 4 **b)** 3 −4 **c)** −8 −5 **d)** −1 −7

Procedure Locate each number on the number line and decide whether the first number is to the left of or to the right of the second number.

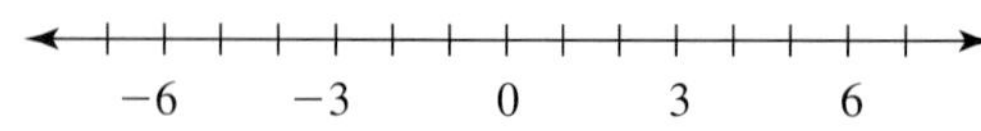

Answer **a)** $1 < 4$ **b)** $3 > -4$ **c)** $-8 < -5$ **d)** $-1 > -7$

You Try It 6 **Insert the correct symbol between each pair of numbers, either < (less than) or > (greater than). Use Example 5 as a guide.**

a) 9 4 **b)** −6 3 **c)** −2 −6 **d)** 4 −8

Absolute Value

The **absolute value** of a number is its distance from 0 on the number line. Whether the number is on the left side of 0 or the right side of 0 doesn't matter; its ***absolute value is positive***.

The symbols we use to represent the absolute value of a number are called *absolute value bars* and look like this: | and |.

For example, both +3 and −3 are three units away from 0.

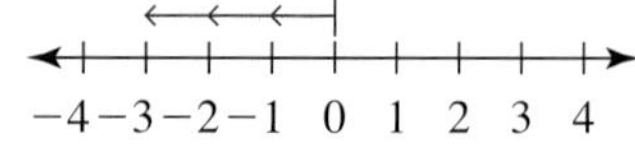

The *absolute value* of 3 is **3**: $|3| = 3$

The *absolute value* of −3 is also **3**: $|-3| = 3$

Caution Notice that the absolute value maintains positivity for positive numbers but it becomes the opposite for negative numbers. **Do not confuse *absolute value* with *the opposite*.** The absolute value of a positive number is *not* its opposite.

Example 6

Evaluate each absolute value expression

a) $|9|$ **b)** $|-7|$ **c)** $|0|$

Answer

a) $|9| = 9$ 9 is a distance of nine units away from the origin.

b) $|-7| = 7$ −7 is a distance of seven units away from the origin.

c) $|0| = 0$ 0 is *no* distance away from itself.

You Try It 7 **Find the absolute value of each number. Use Example 6 as a guide.**

a) $|12|$ **b)** $|-5|$ **c)** $|+24|$ **d)** $|-1|$ **e)** $|-0|$

Number Defined

The following definition of the term *number* is extremely helpful in understanding how positive and negative numbers work together.

> Every nonzero number has a *numerical value* and a *direction*.
>
> - The numerical value is the distance along the number line from 0. The numerical value is another name for the absolute value.
> - The direction is left or right depending on the location on the number line when compared to 0 (zero).
> - Negative numbers are to the left of 0, and positive numbers are to the right of 0.
> - Zero has value but no direction.

Example 7

Identify the numerical value and direction of each number.

a) 6 The value is 6, and the direction is to the right.

b) −8 The value is 8, and the direction is to the left.

c) 0 The value is 0, and it has no direction.

d) +9 The value is 9, and the direction is to the right.

You Try It 8 **Identify the numerical value and direction of each number. Fill in the blanks. Use Example 7 as a guide.**

a) 3 The value is ________, and the direction is ____________________.

b) −5 The ________ is ________, and the direction is ____________________.

c) +7 The ________ is ________, and the direction is ____________________.

d) −4 The ________ is ________, and the ________________ is ______________.

e) +2 The ____.

Answers: You Try It and Think About It

You Try It: **1. a)** 190 **b)** 54 **c)** 64 **d)** 5 **2. a)** 66 **b)** 50 **c)** 56 **d)** 45 **3. a)** $5 \cdot x$ or $5x$ **b)** $x \div 4$ or $\frac{x}{4}$ **c)** $6 - x$ **d)** $11 + x$ **e)** $\sqrt{x}$ **f)** x^2 **4. a)** −6 **b)** 9 **c)** −13 **d)** 10 **e)** opposite **f)** opposite

5.

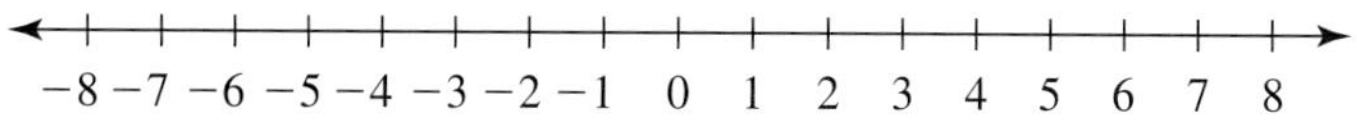

6. a) $9 > 4$ **b)** $-6 < 3$ **c)** $-2 > -6$ **d)** $4 > -8$ **7. a)** 12 **b)** 5 **c)** 24 **d)** 1 **e)** 0 **8. a)** 3, to the right **b)** value, 5, to the left **c)** value, 7, to the right **d)** value, 4, direction, to the left **e)** value is 2, and the direction is to the right

Think About It: **1.** −0 is the same as 0. The negative sign indicates *opposite,* and the opposite of 0 is zero.

Section 2.1 Exercises

FOR EXTRA HELP
Student Resources on DVD-ROM
Includes
➤ Student's Solutions Manual
➤ Video Lectures
➤ Chapter Test Prep Video
MyMathLab
MathXL

Think Again

1. What is the value of -0? Explain your answer. *(Refer to Think About It 1)*
   -0 is the same as 0. It is in the same position as 0 on the number line.

2. Is the opposite of a nonzero absolute value a positive or negative number? Explain your answer or show an example that supports your answer.
   It is a negative number. Because an absolute value is always positive (unless it is 0), its opposite is negative.

3. Is the absolute value of a number the same as the opposite of the number?
   No, the absolute value of a positive number is not the opposite of the number.

4. Why is -8 less than -2?
   -8 is less than -2 because -8 is to the left of -2 on the number line.

Focus Exercises

Evaluate the following according to the replacement value given.

5. $3x + 4$; replace $\boldsymbol{x}$ with 2.
   10

6. $5x - 4$; replace $\boldsymbol{x}$ with 3.
   11

7. $3v \div 9$; replace $\boldsymbol{v}$ with 6.
   2

8. $\sqrt{w}$; replace $\boldsymbol{w}$ with 36.
   6

9. $19 - 5x$; replace $\boldsymbol{x}$ with 1.
   14

10. $20 - 3m$, replace $\boldsymbol{m}$ with 4.
    8

11. $bc \div 2$; replace $\boldsymbol{b}$ with 11 and $\boldsymbol{c}$ with 4.
    22

12. $h \div c$; replace $\boldsymbol{h}$ with 36 and $\boldsymbol{c}$ with 12.
    3

13. $y^2 + 3y$; replace $\boldsymbol{y}$ with 4.
    28

14. $xy + y^2$; replace $\boldsymbol{x}$ with 8 and $\boldsymbol{y}$ with 5.
    65

15. $(P + 10) \div P$; replace $\boldsymbol{P}$ with 5.
    3

16. $(24 - d) \div 10$; replace $\boldsymbol{d}$ with 4.
    2

17. $5u - v$; replace $\boldsymbol{u}$ with 9 and $\boldsymbol{v}$ with 20
    25

18. $3x - 2y$; replace $\boldsymbol{x}$ with 5 and $\boldsymbol{y}$ with 2.
    11

19. $(p + r) \div (p - r)$; replace $\boldsymbol{p}$ with 11 and $\boldsymbol{r}$ with 9.
    10

20. $(x + 6y) \div (4x - y)$; replace $\boldsymbol{x}$ with 2 and $\boldsymbol{y}$ with 3.
    4

Translate each English expression into an algebraic expression. Use any variable to represent the unknown number.

21. The difference of a number and 15
    $x - 15$

22. The difference of 12 and a number
    $12 - x$

23. The sum of 20 and a number
    $20 + x$

24. The sum of a number and five
    $x + 5$

25. The quotient of a number and 18
    $x \div 18$

26. The quotient of 36 and a number
    $36 \div x$

27. The square of a number
    x^2

28. The square root of a number
    $\sqrt{x}$

29. The product of a number and 9
    $x \cdot 9$ or $9x$

30. The product of 7 and a number
    $7 \cdot x$ or $7x$

31. The absolute value of a number
    $|x|$

32. The opposite of a number
    $-x$

Identify the opposite of each number.

33. 8
-8

34. -16
16

35. -18
18

36. $+23$
-23

Insert the correct symbol between each pair of numbers, either < (less than) or > (greater than).

37. $0 < 5$

38. $3 > -9$

39. $-5 < 4$

40. $-10 < -2$

41. $-1 < 7$

42. $-6 > -13$

43. $-4 < 0$

44. $0 > -8$

Find the absolute value.

45. $|-15|$
15

46. $|7|$
7

47. $|+21|$
21

48. $|-1|$
1

49. $|0|$
0

50. $|44|$
44

51. $|-65|$
65

52. $|-0|$
0

Identify the numerical value and direction.

53. 6
6, to the right

54. $+18$
18, to the right

55. -12
12, to the left

56. -2
2, to the left

57. 0
0, it has no direction

58. -0
0, it has no direction

59. $+10$
10, to the right

60. 4
4, to the right

Think Outside the Box

Evaluate the following according to the replacement value given.

61. Is it possible to replace x with 2 in the expression $(6 - x) \div (6 - 3x)$? Explain your answer.
No. Replacing x with 2 would make the expression $4 \div 0$, and that is undefined.

62. Is it possible to replace y with 4 in the expression $(8 - 2x) \div (8 - x)$? Explain your answer.
Yes. Replacing x with 4 would make the expression $0 \div 4$, and that equals 0.

Translate each English expression into an algebraic expression. Use any variable to represent the unknown number.

63. The product of 8 and the quantity of the difference of a number and 9
$8(x - 9)$

64. The square of the quantity of the sum of 15 and a number
$(x + 15)^2$

SECTION 2.2 Adding Signed Numbers

OBJECTIVES

In this section, you will learn to

- Represent signed numbers as vectors.
- Add signed numbers using either the number line or rules of addition.

You Need to Know

To successfully complete this section, you need to understand

- ☐ Adding and subtracting whole numbers (1.2)
- ☐ The integer number line (2.1)
- ☐ Absolute value (2.1)

Introduction

At the end of Section 2.1, you were introduced to the definition of *number*.

> Every nonzero number has both value and direction. Zero has value but no direction.

We can represent this notion of number using a linear length—representing the numerical value of the number—and an arrow—representing the direction of the number.

Vectors

Here is a linear length of 5 units. We can identify its length by counting the number marks from left to right or right to left.

We can give the linear length direction (left or right) by placing an arrowhead at one end or the other. With an arrow, the linear length is called a **vector.**

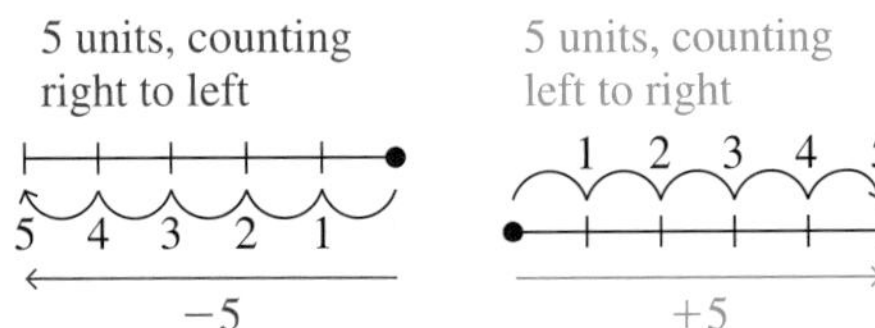

The direction of the vector indicates the sign of the number. A vector pointing left indicates a negative number, and a vector pointing right indicates a positive number.

Example 1

To represent the number +5 as a vector, we start at 0 and draw a vector of length 5 units to the right.

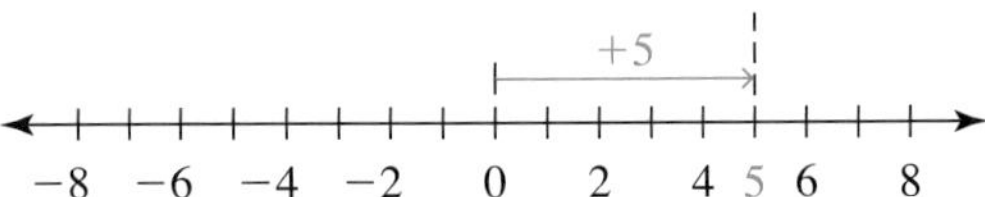

To represent the number −6 as a vector, we start at 0 and draw a vector of length 6 units to the left.

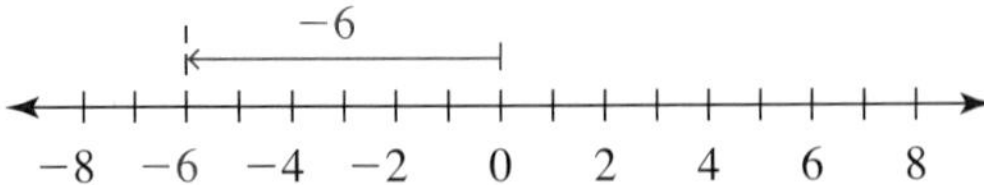

Note As a vector, zero has no length and no direction. In other words, 0 is neither positive nor negative.

You Try It 1

Use the number line to represent the given number as a vector with a linear length and a direction. For each of these number lines, start the vector at the origin, 0. Use Example 1 as a guide.

a) −3

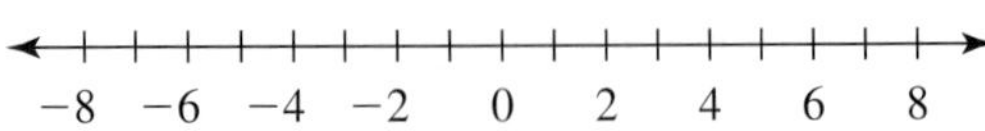

b) +7

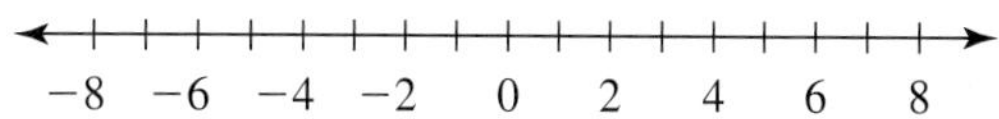

c) −8

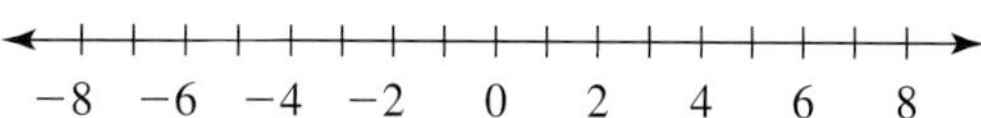

A vector that represents a certain number (say, +5) must always have the same length and direction. +5 will always have a length of 5 (its numerical value) and point to the right (because it is positive).

However, vectors can have other starting points. If the vector representing +5 starts at +3 on the number line, it will still have a length of 5 and point to the right. It will look like this:

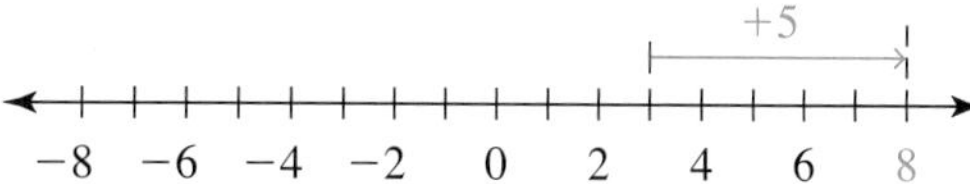

Notice that the vector starts at +3 and extends 5 units to the right. Also notice that it ends at 8. You might guess the reason it ends at 8 is because $3 + 5 = 8$, and you'd be right!

Example 2

Represent each number as a vector starting at the given point. State where the vector stops.

a) Represent +4 starting at −1. **b)** Represent −6 starting at +8.

c) Represent +3 starting at −7. **d)** Represent −3 starting at −2.

Procedure Represent each number as a vector having length and direction.

Answer **a)** The vector for +4 has a length of 4 and points to the right. Here it is starting at −1.

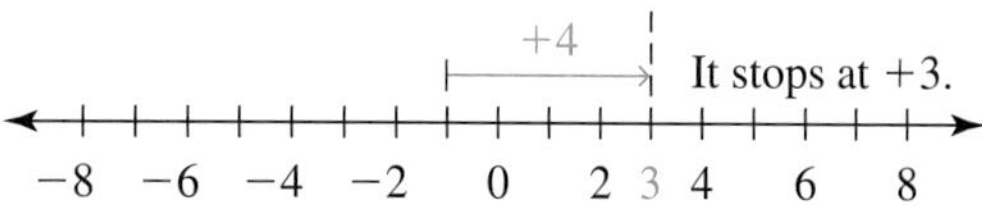

b) The vector for −6 has a length of 6 and points to the left. Here it is starting at +8.

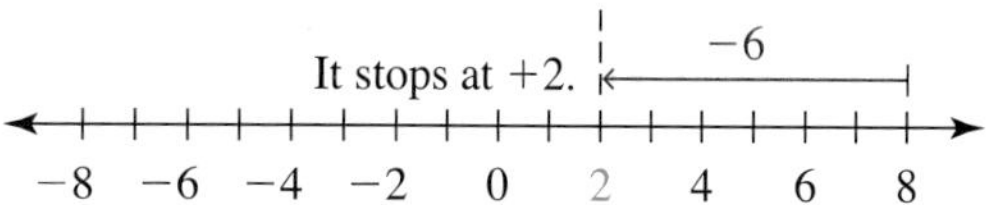

c) The vector for +3 has a length of 3 and points to the right. Here it is starting at −7.

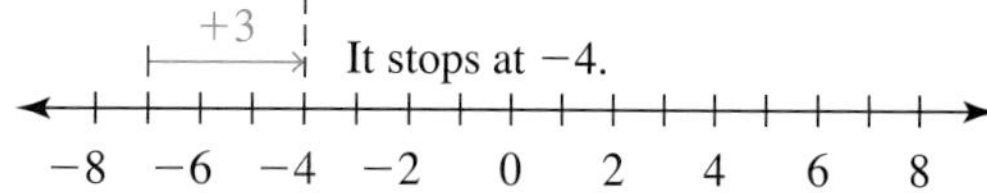

d) The vector for −3 has a length of 3 and points to the left. Here it is starting at −2.

It stops at −5. −3

−8 −6 −4 −2 0 2 4 6 8
−5

You Try It 2 **Use the number line to represent the given number as a vector with a linear length and a direction. Start at the given value and state where it stops. Use Example 2 as a guide.**

a) -3 starting at $+5$

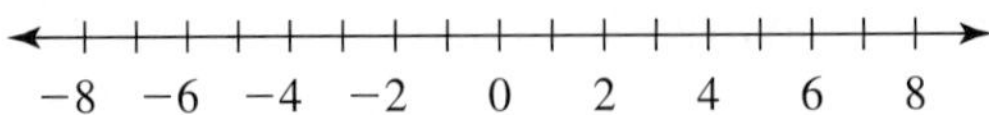

b) $+7$ starting at -4

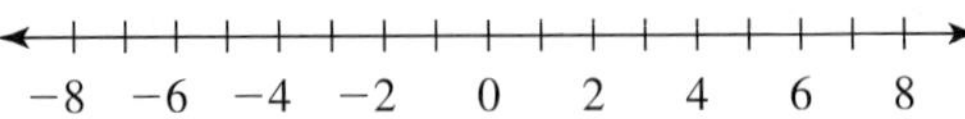

c) -6 starting at -2

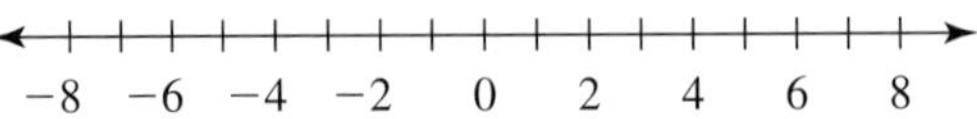

Adding Signed Numbers on the Number Line

We can use the number line to add numbers. For example, to add $3 + 5$ we construct

1. A vector for $+3$ that starts at the origin, 0.

2. A vector for $+5$ that *starts at 3* (where the first vector stopped).

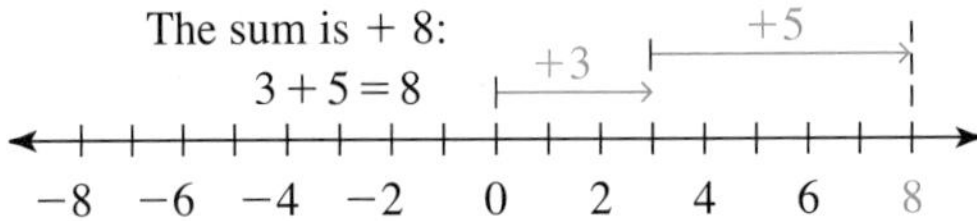

It doesn't matter if the numbers we are adding are positive or negative. Each number can be represented as a vector having a length and a direction. To add them, simply follow these guidelines.

Adding Two Numbers on the Number Line

1. Draw a vector for the first number, starting from the origin, 0.

2. Draw a vector for the second number that starts where the first vector stopped.

3. The number at which the second vector stops is the sum.

We can add $+3$ and -5 by showing the vector for $+3$ (which will start at 0 and stop at $+3$). Then from $+3$, we can draw the vector for -5 (which points to the left). Where the second vector stops is the sum.

The sum is -2:
$3 + (-5) = -2$
-5
$+3$
−8 −6 −4 −2 0 2 4 6 8

The sum of $+3$ and -5 is written $3 + (-5)$. Notice the parentheses around the -5. In this case, the parentheses act as separators, separating the plus sign $(+)$ from the negative sign in -5.

Recall from Section 1.1 the additive identity: $a + 0 = a$ and $0 + a = a$.

Using a number line and vectors, explain why the sum of any number and zero is always that number.

____

____

Example 3

Represent the sum by drawing two vectors. Use the guidelines for adding two numbers on the number line.

a) $4 + (-1)$ **b)** $-6 + 8$ **c)** $3 + (-7)$

Procedure Represent each number as a vector having length and direction.

Answer **a)** As a vector, $+4$ has a length of 4 and points to the right. As a vector, -1 has a length of 1 and points to the left.

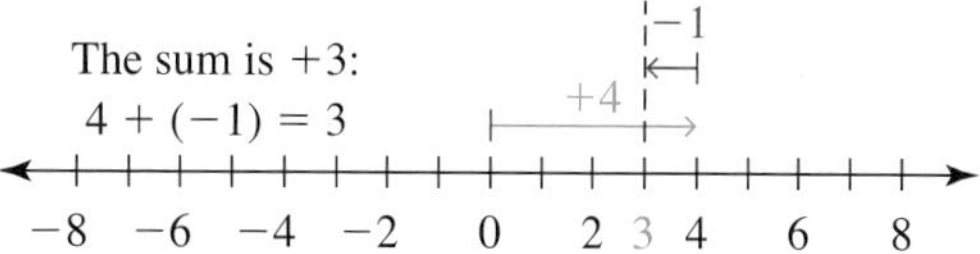

b) As a vector, -6 has a length of 6 and points to the left and $+8$ has a length of 8 and points to the right.

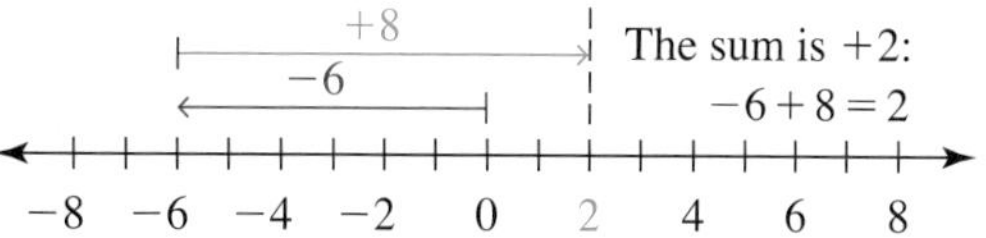

c) As vectors, $+3$ is 3 to the right and -7 is 7 to the left.

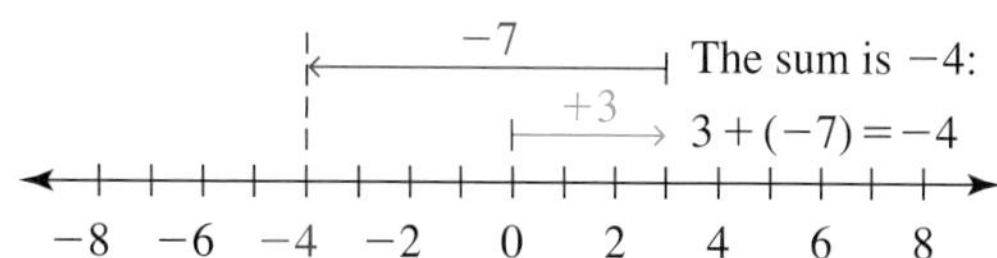

You Try It 3 **Add. You may use the number line above each addition pair to help you evaluate the sum. Use Example 3 as a guide.**

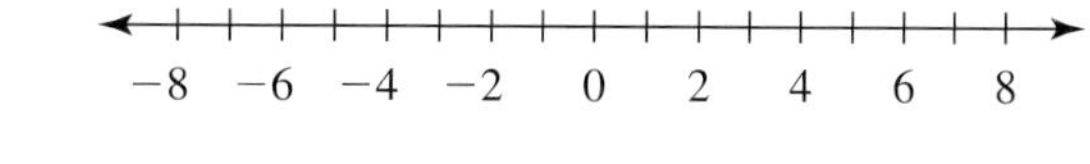

a) $3 + 4 =$ ______ **b)** $-6 + (-2) =$ ______

−8 −6 −4 −2 0 2 4 6 8

c) $8 + (-3) =$ ______ **d)** $2 + (-7) =$ ______

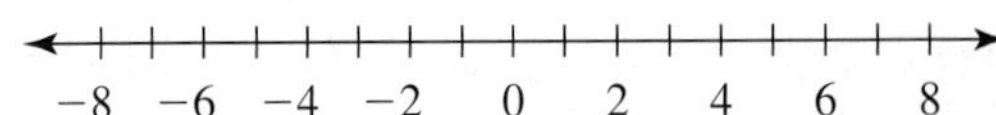

e) $-3 + 6 =$ ______ f) $-8 + 7 =$ ______

−8 −6 −4 −2 0 2 4 6 8

g) $-4 + 4 =$ ______ h) $6 + (-6) =$ ______

As you have seen, the sum of two numbers can be found using the number line and drawing vectors. This time find these sums mentally by drawing the vector in your mind, not with your pencil. Try to visualize the number line and the direction of the vectors.

You Try It 4 **Try to do each of these sums without the aid of a number line. If you need a number line, you may draw one.**

a) $-10 + (-6)$ b) $13 + (-4)$ c) $2 + (-10)$

d) $-4 + 11$ e) $-12 + 7$ f) $10 + (-10)$

Adding Signed Numbers Using the Commutative Property

The Commutative Property of Addition works for integers as well as for whole numbers. You will recall that the Commutative Property of Addition allows the sum $3 + 5$ to be rewritten as $5 + 3$. The same is true for signed numbers.

For example, $2 + (-7)$ can be written as $-7 + 2$. They are both equal to -5.

$2 + (-7)$:

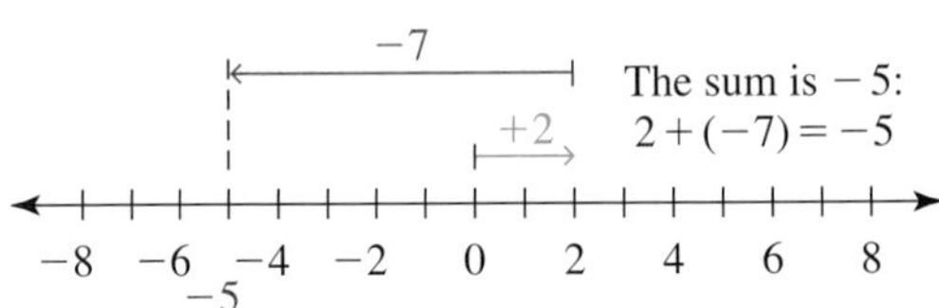

$-7 + 2$:

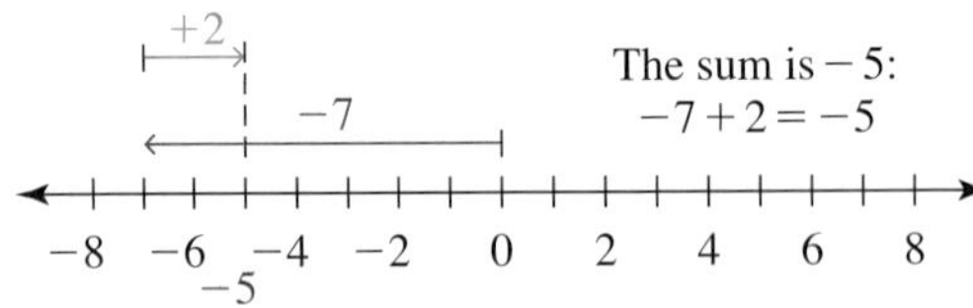

Example 4

Rewrite each sum using the Commutative Property. Then evaluate the original sum and the new sum.

a) $4 + (-1)$ b) $-6 + 0$ c) $3 + (-7)$ d) $-5 + 5$

Answer

a) $4 + (-1) = -1 + 4$
They both $= +3$.

b) $-6 + 0 = 0 + (-6)$
They both $= -6$.

c) $3 + (-7) = -7 + 3$
They both $= -4$.

d) $-5 + 5 = 5 + (-5)$
They both $= 0$.

You Try It 5 **Rewrite each sum using the Commutative Property. Then evaluate the original sum and the new sum. Use Example 4 as a guide.**

a) $-10 + 3 =$ __________
They both = __________.

b) $-12 + (-4) =$ __________
They both = __________.

c) $11 + (-8) =$ __________
They both = __________.

d) $0 + (-9) =$ __________
They both = __________.

Adding Numbers: When the Signs Are the Same

By now, you should have a sense of how to add integers on the number line. However, some integers are so large in value that placing them on a number line can be difficult. For example, to evaluate the sum $-23 + 48$ on the number line would be a challenge.

We need to develop some rules for adding signed numbers that have large values. These rules, though, must be consistent with our understanding of how to add small-valued integers. We'll develop the rules using smaller numbers and the number line.

The sum of two numbers with the same sign—*same direction*—is a number that has the same direction (same sign) and is either *more positive* or *more negative*.

For example, two positive numbers such as 8 and 5 add up to a number that is *more positive*.

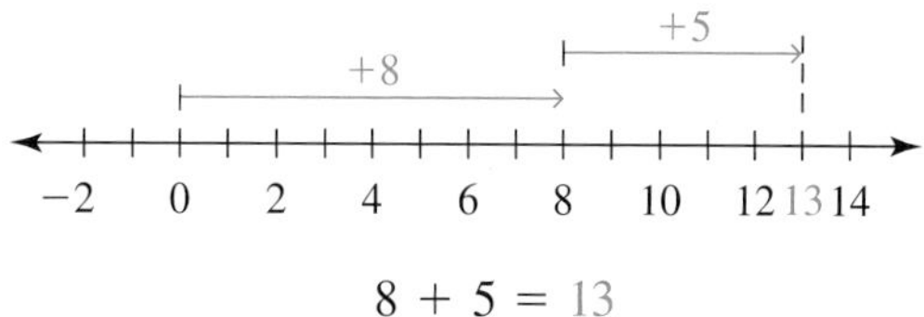

$$8 + 5 = 13$$

Now consider the sum of two negative numbers. Again, because the two numbers have the same direction—to the left—their sum will also be in the same direction. It will be farther to the left and will be more negative.

For example, add -7 and -8. The sum of the two left-pointing vectors is even farther to the left, farther away from 0 in the negative direction. The result is a number that is *more negative*.

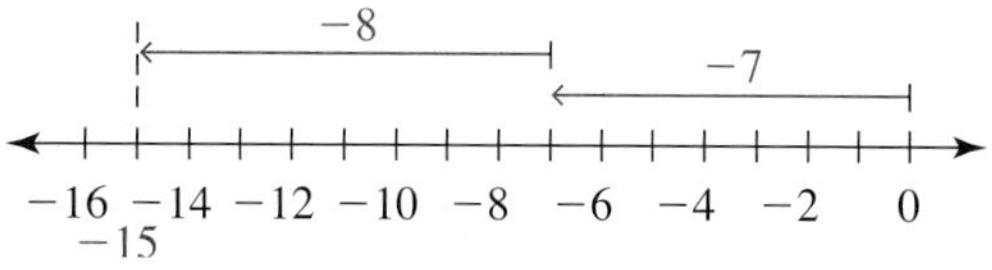

$$-7 + (-8) = -15$$

In each sum, we see the effect of adding numbers with the same sign (the same direction): The vectors go farther in that direction. In fact, before evaluating the sum, it is helpful to determine whether the sum will be positive or negative.

The Sum of Two Numbers with the Same Sign

When finding the sum of two numbers of the same sign,

1. Add the numerical values.

2. Remember that the sign of the sum is the same as the sign of each number

Example 5

Add.

a) $+7 + 4$ **b)** $-7 + (-3)$ **c)** $+12 + 26$

d) $-5 + (-8)$ **e)** $+6 + 9$ **f)** $-35 + (-52)$

Procedure Notice that the numbers have the same sign—both positive or both negative.

Answer **a)** $+7 + 4 = +11$ **b)** $-7 + (-3) = -10$

c) $+12 + 26 = +38$ **d)** $-5 + (-8) = -13$

e) $+6 + 9 = +15$ **f)** $-35 + (-52) = -87$

You Try It 6 **Add. Use Example 5 as a guide.**

a) $+2 + 17$ **b)** $-15 + (-20)$ **c)** $+43 + 86$

d) $-4 + (-12)$ **e)** $+15 + 29$ **f)** $-92 + (-88)$

Adding Numbers: When the Signs Are Different

When adding two numbers of different signs (one positive and the other negative), we can use the Commutative Property and write the sum so that the larger-valued number is first. This is not a requirement, but it will help us develop the rule.

We could write the sum $-5 + 8$ as $8 + (-5)$. Likewise, the sum $3 + (-7)$ could be written as $-7 + 3$.

When written this way, the first vector will go in the direction of the larger-valued number. The second vector will go in the opposite direction, toward 0.

For example, to find the sum of $8 + (-5)$, we can draw two vectors, one stretching away from 0 (zero) to the right and the other going toward zero (heading left) but not all the way to 0. In other words, the result stays on the right side of 0, which is a positive number.

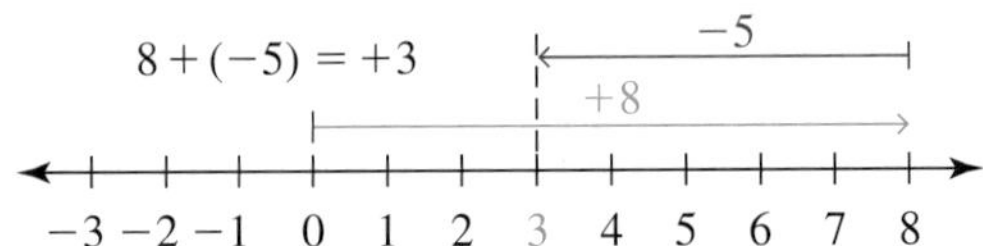

To find the sum of $-7 + 3$, we can draw two vectors, one stretching away from 0 (zero) to the left and the other going toward zero (heading right) but not all the way to 0. In other words, the result stays on the left side of 0, making it a negative number.

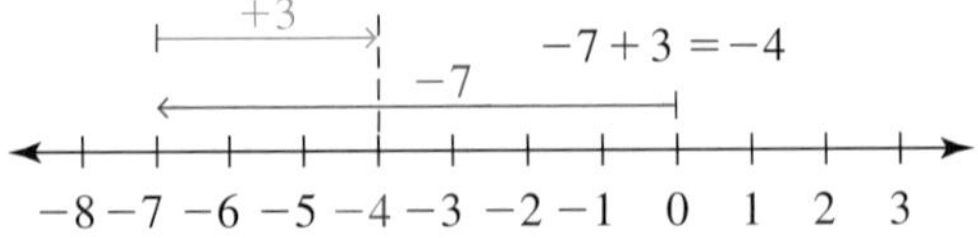

By writing the sum with the larger-valued number first, we get an immediate clue as to whether the result will be positive or negative. The result will have the same sign as the larger-valued number. Before evaluating the sum, it is helpful to determine whether the sum will be positive or negative.

Here are the guidelines for finding the sum of two numbers with different signs.

The Sum of Two Numbers with Different Signs

When finding the sum of two numbers with different signs

1. Use the Commutative Property, if necessary, to write the larger-valued number first.
2. Subtract the two numerical values: larger value $-$ smaller value.
3. Remember that the sign of the sum is the same as the sign of the larger-valued number.

Example 6

Add.

a) $7 + (-5)$ **b)** $29 + (-35)$ **c)** $-37 + 49$

Procedure Notice that the numbers have different signs. Follow the preceding guidelines to find the sum. The answers are shown with the larger-valued number written first.

Answer **a)** $7 + (-5) = +2$

The larger-valued number is positive; so, the result will be positive.
The difference: $7 - 5 = 2$; so, the result is $+2$.

b) First, write the sum as $-35 + 29$.

$-35 + 29 = -6$

The larger-valued number is negative; so, the result will be negative.
The difference: $35 - 29 = 6$; so, the result is -6.

c) $49 + (-37) = +12$

The larger-valued number is positive; so, the result will be positive.
The difference: $49 - 37 = 12$; so, the result is $+12$.

You Try It 7 **Find each sum. Use Example 6 as a guide.**

a) $27 + (-42)$ **b)** $-38 + 52$ **c)** $45 + (-29)$ **d)** $-86 + 42$

Think About It 2 If two numbers have different signs but have the same numerical value, such as $-3 + 3$, which sign should be given to the result? Explain.

____

____

A number and its opposite have vectors that are the same length but in opposite directions. So, when they are added together, the sum is 0.

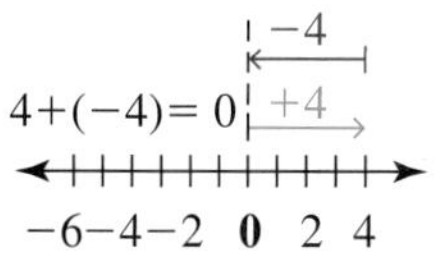

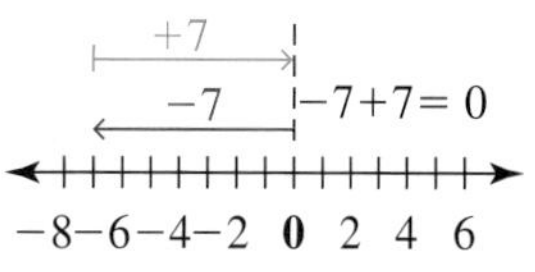

$4+(-4)= 0$ (vectors $+4$ and -4 on a number line from -6 to 4)

$-7+7= 0$ (vectors -7 and $+7$ on a number line from -8 to 6)

The sum of a number and its opposite is 0: $a + (-a) = 0$.

Example 7

Find each sum.

a) $7 + (-7)$ **b)** $-6 + 6$

Answer **a)** $7 + (-7) = 0$ **b)** $-6 + 6 = 0$

You Try It 8 **Find each sum. Use Example 7 as a guide.**

a) $12 + (-12)$ **b)** $-9 + 9$ **c)** $23 + (-23)$

A sum may contain more than two numbers. If there is more than one positive number or more than one negative number, we can use the Commutative Property and add the positive numbers separately from the negative numbers. We can then add the remaining two numbers.

Example 8

Find the sum. $2 + (-3) + (-5) + 4$

Procedure This sum contains two positive numbers (2 and 4) and two negative numbers (-3 and -5). Use the Commutative Property to write the sum with the two positive numbers first.

Answer

$2 + (-3) + (-5) + 4$ — Commute.

$= 2 + 4 + (-3) + (-5)$ — Add the positive numbers separately from the negative numbers.

$= 6 + (-8)$ — This can be written as $-8 + 6$

$= -2$

You Try It 9 **Find each sum. Use Example 8 as a guide.**

a) $6 + (-4) + (-9) + 1$ **b)** $-5 + (-6) + 3 + (-7)$

Answers: You Try It and Think About It

You Try It: *(Note: Positive answers may also be written with a plus sign in front.)*

1. a)

−3

−6 −3 0 3

b)

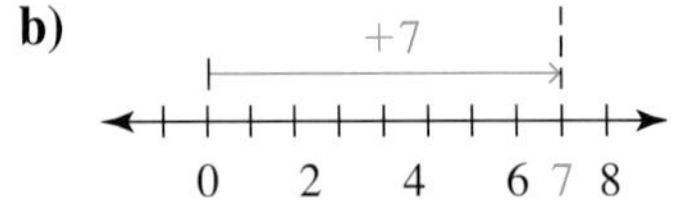

c)

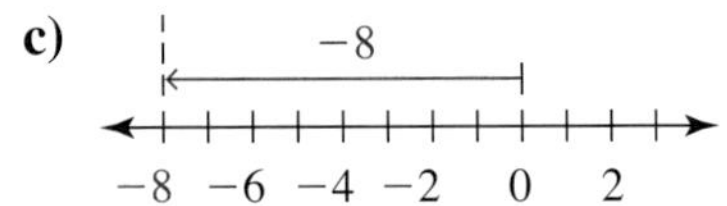

2. a)

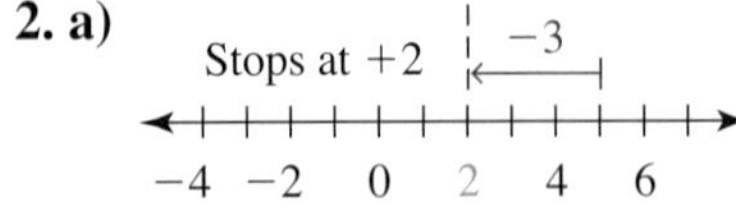

b)

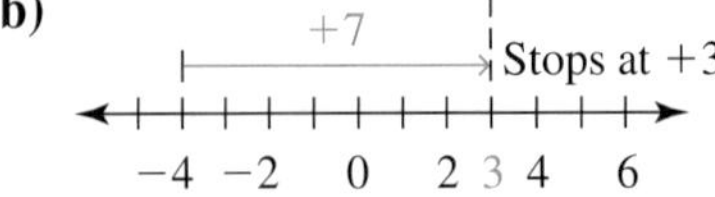

c)

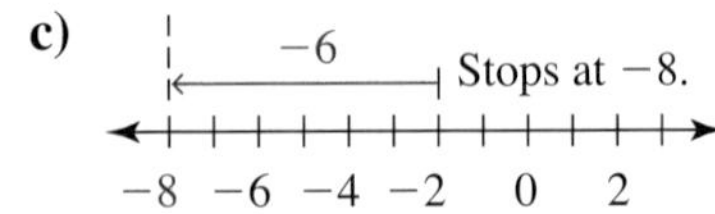

3. a) 7 **b)** -8 **c)** 5 **d)** -5 **e)** 3 **f)** -1 **g)** 0 **h)** 0 **4. a)** -16 **b)** 9 **c)** -8 **d)** 7 **e)** -5 **f)** 0 **5. a)** $3 + (-10)$; both $= -7$ **b)** $-4 + (-12)$; both $= -16$ **c)** $-8 + 11$; both $= 3$ **d)** $-9 + 0$; both $= -9$ **6. a)** 19 **b)** -35 **c)** 129 **d)** -16 **e)** 44 **f)** -180 **7. a)** -15 **b)** 14 **c)** 16 **d)** -44 **8. a)** 0 **b)** 0 **c)** 0 **9. a)** -6 **b)** -15

Think About It: **1.** Answers may vary. One possibility: The zero vector adds nothing to the length of the first vector; so, the sum of the two vectors is the same as the first vector. **2.** The sum is 0; so, the sign does not matter.

Section 2.2 Exercises

FOR EXTRA HELP
Student Resources on DVD-ROM
Includes
➤ Student's Solutions Manual
➤ Video Lectures
➤ Chapter Test Prep Video
MyMathLab
MathXL

Think Again

1. If two numbers have different signs but have the same numerical value, such as $-3 + 3$, which sign should be given to the result? Explain your answer. *(Refer to Think About It 2)*
   Because the result is 0, neither sign should be given to the result.

2. Can the sum of two negative numbers ever be positive? Explain your answer or show an example that supports your answer.
   No. Every negative number has a direction that is to the left. The sum of two negative numbers is two vectors to the left of 0, and the result will always be negative.

3. Is it possible for $a + b$ to be neither positive nor negative? Explain your answer or show an example that supports your answer.
   Yes, but only if a and b are opposites, such as 2 and -2.

4. If a is a positive number, under what circumstances will $-a + b$ be positive?
   $-a + b$ will be positive only when b is greater than a.

Focus Exercises

Find the sum.

5. $6 + (-4)$
   2
6. $5 + (-1)$
   4
7. $11 + (-3)$
   8
8. $10 + (-5)$
   5
9. $3 + (-6)$
   -3
10. $6 + (-8)$
   -2
11. $1 + (-9)$
   -8
12. $2 + (-10)$
   -8
13. $-8 + (-5)$
   -13
14. $-7 + (-6)$
   -13
15. $-6 + (-3)$
   -9
16. $-1 + (-4)$
   -5
17. $5 + 7$
   12
18. $9 + 6$
   15
19. $1 + 18$
   19
20. $3 + 12$
   15
21. $-3 + 8$
   5
22. $-5 + 6$
   1
23. $-1 + 6$
   5
24. $-9 + 13$
   4
25. $-10 + 7$
   -3
26. $-5 + 4$
   -1
27. $-6 + 1$
   -5
28. $-12 + 3$
   -9
29. $-7 + 7$
   0
30. $-3 + 3$
   0
31. $-10 + 10$
   0
32. $-15 + 15$
   0
33. $6 + (-6)$
   0
34. $4 + (-4)$
   0
35. $1 + (-1)$
   0
36. $11 + (-11)$
   0
37. $0 + 1$
   1
38. $0 + 3$
   3
39. $7 + 0$
   7
40. $9 + 0$
   9
41. $0 + (-1)$
   -1
42. $0 + (-8)$
   -8
43. $-5 + 0$
   -5
44. $-12 + 0$
   -12
45. $-1 + 8$
   7
46. $-5 + 2$
   -3
47. $2 + (-7)$
   -5
48. $-5 + 4$
   -1
49. $1 + (-3)$
   -2
50. $10 + (-10)$
   0
51. $5 + (-5)$
   0
52. $9 + (-8)$
   1
53. $+9 + (+15)$
   24
54. $-6 + (-11)$
   -17
55. $-13 + (-5)$
   -18
56. $-26 + 14$
   -12
57. $+14 + (-5)$
   9
58. $+8 + (-12)$
   -4
59. $-1 + (+5)$
   4
60. $-10 + (+3)$
   -7

61. $-7 + 7$
0

62. $-9 + 9$
0

63. $-15 + (-16)$
-31

64. $-11 + (-5)$
-16

65. $-3 + 2$
-1

66. $+20 + 9$
29

67. $13 + 3$
16

68. $-10 + (-3)$
-13

69. $22 + (-15)$
7

70. $17 + (-28)$
-11

71. $-22 + (-18)$
-40

72. $-12 + 43$
31

73. $-31 + 24$
-7

74. $-30 + 14$
-16

75. $-53 + (-33)$
-86

76. $-16 + 42$
26

77. $28 + (-51)$
-23

78. $-20 + (-32)$
-52

79. $-19 + 19$
0

80. $25 + (-25)$
0

Find the sum.

81. $2 + (-6) + 9$
5

82. $8 + (-5) + 7$
10

83. $6 + (-14) + 4$
-4

84. $3 + (-8) + 2$
-3

85. $-3 + (-2) + 8$
3

86. $-1 + (-8) + 10$
1

87. $-4 + 6 + (-7)$
-5

88. $-5 + 11 + (-9)$
-3

89. $-2 + (-5) + (-1)$
-8

90. $-3 + (-8) + (-6)$
-17

91. $-6 + (-4) + 10$
0

92. $3 + 8 + (-11)$
0

93. $5 + (-1) + 3 + (-12)$
-5

94. $2 + (-6) + 7 + (-4)$
-1

95. $8 + (-5) + (-4) + 6$
5

96. $12 + (-9) + (-1) + 5$
7

97. $-6 + 4 + (-2) + 4$
0

98. $-1 + 6 + (-12) + 7$
0

99. $-9 + 6 + 0 + 5 + (-3) + (-5)$
-6

100. $8 + (-10) + 0 + 13 + (-10) + (-4)$
-3

Think Outside the Box

101. Find the perimeter of this figure.

The perimeter is 80 inches.

102. What is the area of this rectangle?

The area is 600 square feet.

SECTION 2.3 Subtracting Signed Numbers

OBJECTIVES

In this section, you will learn to

- Identify the opposite of a number.
- Evaluate a double negative.
- Write subtraction as addition.
- Subtract signed numbers.
- Solve applications involving signed numbers.

You Need to Know

To successfully complete this section, you need to understand

☐ The Commutative Property of Addition (1.1)
☐ Adding signed numbers (2.2)

Introduction

You have been doing subtraction for a long time. In the world of math, we call subtraction *finding the difference*. What you might not know is that subtraction is another form of addition. This section introduces subtraction in a new way, especially as it relates to positive and negative numbers.

The First Three Meanings of the Dash

You have recognized the dash, –, for many years as the subtraction sign or minus sign. More recently, you have seen it used as the negative sign. There is a third meaning that will prove to be very valuable to your understanding of algebra. The dash also means "the opposite of."

1. Minus, as in $9 - 4$ is "9 minus 4"
2. Negative, as in -6 is "negative 6"
3. The opposite of, as in -4 is "the opposite of 4"

Example 1

Rewrite the outer negative sign as "the opposite of" the number within the parentheses.

Answer

a) $-(+7)$ means the opposite of $+7$, which is -7.

b) $-(-5)$ means the opposite of -5, which is $+5$ (or 5).

c) $-(8)$ means the opposite of 8, which is -8.

You Try It 1

Rewrite the outer negative sign as "the opposite of" the number within the parentheses. Also state its value as you might normally say it. Use Example 1 as a guide.

a) $-(+3)$ means ____________________, which is __________.

b) $-(-1)$ means ____________________, which is __________.

c) $-(-12)$ means ____________________, which is __________.

d) $-(16)$ means ____________________, which is __________.

The Double Negative

Refer back to Example 1 and You Try It 1. What was the result of finding the opposite of a negative number? We may say, "The opposite of a negative number is a positive number." Does that seem accurate? When written algebraically, the opposite of -5 becomes $-(-5)$, which has the value of $+5$.

The notation $-(-5)$ is an example of a **double negative.** When two negative signs—two minus signs, two hyphens—are next to each other (possibly separated by a parenthesis), the number can be rewritten as a positive number. In other words, the two negative signs can be replaced by a single positive sign.

In these cases, parentheses are usually quite helpful. For example, it's a little awkward to read $--5$ as anything meaningful; so, we separate the negative signs with parentheses: $-(-5)$.

Example 2

Rewrite each double negative as a positive. Make sure you include the + in front of the positive number.

a) $-(-9)$ **b)** $-(-4)$

Answer **a)** $-(-9) = +9$ **b)** $-(-4) = +4$

It's also appropriate to write a double negative with two plus signs, as in $-(-4) = +(+4)$.

You Try It 2 **Rewrite each double negative as a positive. Make sure you include the + in front of the positive number. Use Example 2 as a guide.**

a) $-(-15)$ **b)** $-(-59)$ **c)** $-(-1)$

We'll return to the double negative in a little while. First, though, let's look at why and how subtraction can be thought of as addition.

Writing Subtraction as Addition

Consider the difference between 7 and 4: $7 - 4$ (which you know is 3). We can see this difference on the number line.

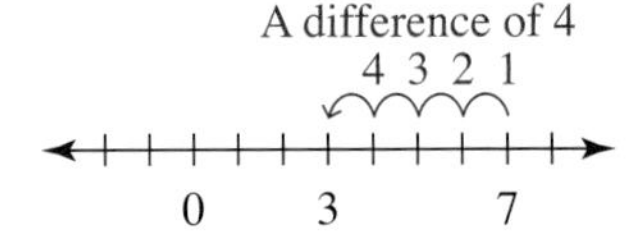

This is the same as $7 + (-4)$ and suggests that $7 - 4 = 7 + (-4)$, or 7 + the *opposite* of 4.

In other words, subtraction can be written as *adding the opposite.*

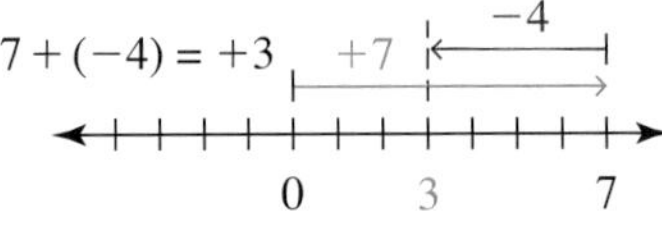

Similarly, consider the difference between 3 and −2: $3 - (-2)$. We can see this difference on the number line.

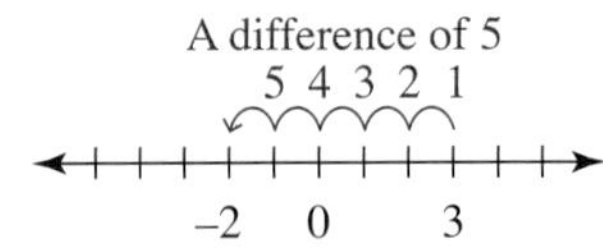

The advantage of being able to write subtraction this way is that we do not need to learn new rules. We can use the rules of addition.

Subtraction can be rewritten as addition, as *adding the opposite.* Symbolically, this is written as $a - b = a + (-b)$.

Let's look at writing subtraction as addition. This means to change two things: change the operation to addition and change the second number to its opposite.

Example 3

Rewrite each subtraction as addition and evaluate.

a) $6 - 2$ **b)** $-1 - 7$ **c)** $5 - (-6)$ **d)** $-9 - (-2)$

Procedure For each subtraction, the operation will become addition and the second number will become its opposite.

Answer **a)** $6 - 2 = \underline{6 + (-2)} = \underline{+4}$ **b)** $-1 - 7 = \underline{-1 + (-7)} = \underline{-8}$

c) $5 - (-6) = \underline{5 + (+6)} = \underline{+11}$ **d)** $-9 - (-2) = \underline{-9 + (+2)} = \underline{-7}$

You Try It 3 **Rewrite each subtraction as addition. Then evaluate the sum. Use Example 3 as a guide.**

a) $12 - 5 =$ ________ $=$ ____ **b)** $6 - (-4) =$ ________ $=$ ____

c) $4 - 10 =$ ________ $=$ ____ **d)** $-3 - (-9) =$ ________ $=$ ____

e) $-5 - 3 =$ ________ $=$ ____ **f)** $-12 - (-2) =$ ________ $=$ ____

g) $-6 - (-6) =$ ________ $=$ ____ **h)** $0 - 7 =$ ________ $=$ ____

If an expression contains more than two integers, we should keep in mind the order of operations and add two integers at a time, working from left to right.

Example 4

Evaluate the expression. $2 - 3 + 4 - (-5)$

Procedure First, write each subtraction as addition. Then use the Commutative Property to get the positive number(s) together and the negative number(s) together.

Answer

$2 - 3 + 4 - (-5)$	Change all subtraction to adding the opposite.
$= 2 + (-3) + 4 + (+5)$	Group the three positive numbers together and add.
$= 2 + 4 + 5 + (-3)$	$2 + 4 + 5 = 11$
$= 11 + (-3)$	Add 11 and −3.
$= 8$	

You Try It 4 **Evaluate each expression. Use Example 4 as a guide.**

a) $6 + (-4) - (-9) - 1$ **b)** $-5 - (-6) - 7 + 3$

Guidelines for Adding and Subtracting Signed Numbers: A Summary

Our work with signed numbers thus far has led us to these guidelines on how to add and subtract two signed numbers.

Change subtraction to addition: Add the opposite.	$3 - 5 \rightarrow 3 + (-5)$ $-4 - (-7) \rightarrow -4 + (+7)$	
If the two numbers have the *same* sign **a)** Add the numerical values of each number. **b)** Remember that the resulting sum will have the same sign.	$+2 + (+8)$ Both positive ↓ $+10$	$-1 + (-6)$ Both negative ↓ -7
If the two numbers have *different* signs **a)** Write the larger-valued number first. **b)** Subtract the two numerical values: larger value – smaller value. **c)** Remember that the sign of the sum is the same as the sign of the larger-valued number.	$3 + (-8)$ $-8 + 3$ $8 - 3 = 5$ Larger-valued number is negative. $= -5$	$-2 + 11$ $11 + (-2)$ $11 - 2 = 9$ Larger-valued number is positive. $= +9$

Example 5

Evaluate.

a) $29 - 42$ **b)** $37 + (-25)$ **c)** $-47 + (-19)$ **d)** $12 - (-38)$

Procedure Change any subtraction to addition. Apply the guidelines in the summary to find the sum.

Answer

a) $29 - 42$ — **First, rewrite as addition: 29 + (−42).**

$= 29 + (-42)$ — **The signs are different; write the larger-valued number first.**

$= -42 + 29$ — **Find the difference: 42 − 29 = 13.**

$= -13$ — **The result is negative: −13.**

b) $37 + (-25)$ — **The signs are different and the larger-valued number is written first. Find the difference: 37 − 25 = 12.**

$= +12$ (or 12) — **The result is positive: +12.**

c) $-47 + (-19)$ — **The signs are the same; they are both negative.**

$= -66$ — **Add the numerical values: 47 + 19 = 66; the result is negative.**

d) $12 - (-38)$ — **First, rewrite subtraction as addition: 12 + (+38).**

$= 12 + (+38)$ — **The signs are the same; they are both positive.**

$= +50$ (or 50) — **Add the numerical values: 12 + 38 = 50; the result is positive.**

You Try It 5 **Evaluate. Use Example 5 as a guide.**

a) $33 + (-95)$ **b)** $-21 + 68$ **c)** $65 - (-15)$ **d)** $-76 - 48$

Applications with Integers

Here are a few situations in which signed numbers are common:

Finances: Borrowing money represents a debt, a negative amount. Credits and payments are positive amounts; debits (and purchases) are negative amounts. (The word *debit* is related to *debt*.)

Temperature: Even though an outside temperature of 5° (5 degrees) is cold, −5° is colder by 10°.

Altitude: Below sea level, the altitude is negative; above sea level, the altitude is positive.

The next few examples and exercises make use of adding and subtracting integers. For each application, set up a numerical expression (using addition or subtraction) and answer the question with a sentence.

Finances

Example 6

On her Home Depot card, Julia has a debit of \$25 and makes a payment of \$14. What is the new balance of her account? Is this new balance a debit or a credit?

Procedure A debit is a negative number (−\$25), and a payment on an account is a positive number (+ \$14) added to the account. Numerically, it looks like this:

Answer *Numerical Expression:* $-25 + 14 = -11$

Sentence The new balance is − \$11; this is a *debit*.

You Try It 6 **Solve each application. Write a numerical expression and write the answer in the form of a sentence. Use Example 6 as a guide.**

a) Bonnie has a debit of \$32 on her Sears card. She likes to stay ahead by making larger-than-needed payments. Her most recent payment was for \$50. What is the new balance of her account? Is it a debit or a credit?

Numerical Expression

Sentence ______________________________

b) Art has a credit balance of \$36 on his Visa card. He makes purchases on this card worth \$100. What is the new balance of his account? Is it a debit or a credit?

Numerical Expression

Sentence ______________________________

Temperature

In the winter, the temperature in some northern states falls below 0°. Such temperatures can be represented by negative numbers.

Example 7

If the temperature was 8° at 4 PM and fell 15° by midnight, what was the temperature at midnight?

Procedure Temperature *falling* means *subtracting* the number of degrees it fell from the starting temperature.

Answer *Numerical expression:* $8 - 15 = -7$

Sentence The temperature at midnight was $-7°$.

You Try It 7 **Solve each application. Write a numerical expression and write the answer in the form of a sentence. Use Example 7 as a guide.**

a) At 2 AM the outside temperature was $-13°$. By noon, the temperature had risen 27°. What was the temperature at noon?

Numerical Expression

Sentence ______________________________

b) At noon the outside temperature was $-11°$. By midnight, the temperature had fallen 18°. What was the temperature at midnight?

Numerical Expression

Sentence ______________________________

Altitude

Altitude means how high or how low something is compared to sea level.

If a hilltop is 50 feet above sea level, it has an altitude of 50 feet, or +50 feet.

If the ocean floor is 40 feet deep, it is 40 feet below sea level and has an altitude of -40 feet.

Sea level has an altitude of 0 feet.

We use subtraction to find the difference in altitudes:

First altitude – Second altitude.

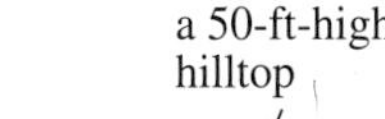

+50

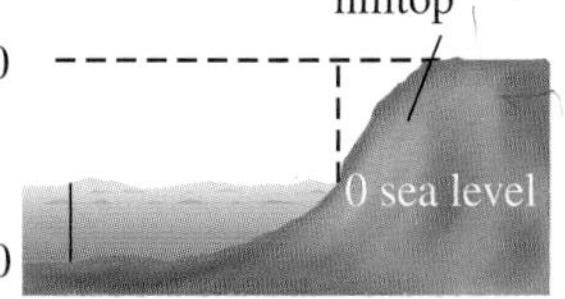

−40

Example 8

Find the difference in altitude between a hill 50 feet above sea level and an ocean floor 40 feet below sea level.

Procedure Subtract *first altitude* − *second altitude*. In this case, that's hill − ocean floor.

Answer *Numerical expression:* $+50 - (-40) = 50 + (+40) = +90$

Sentence The difference in altitude is 90 feet.

You Try It 8

Solve each application. Write a numerical expression and write the answer in the form of a sentence. Use Example 8 as a guide.

a) Find the difference in altitude between an airplane 1,280 feet above sea level and a mountain peak 1,150 feet above sea level.

Numerical Expression

Sentence ______________________________

b) Find the difference in altitude between a cliff 45 feet above sea level and an ocean floor 35 feet below sea level.

Numerical Expression

Sentence ______________________________

c) Use the bar graph to find the difference in altitude between the seaside cliff and the sandbar.

Numerical Expression

Sentence ______________________________

d) Use the bar graph to find the difference in altitude between the sandbar and the ocean floor.

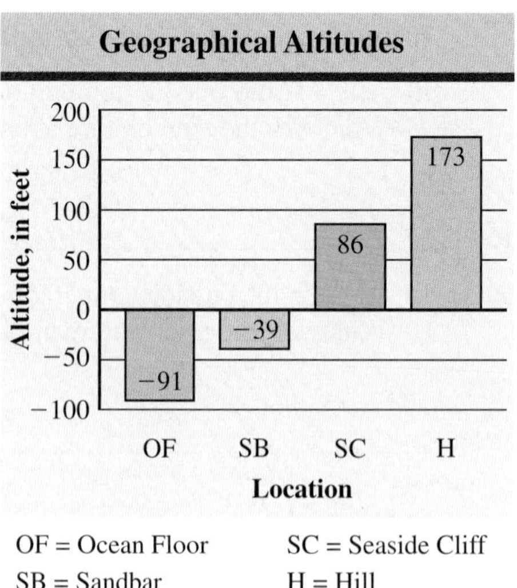

Numerical Expression

Sentence ______________________________

Answers: You Try It and Think About It

You Try It: *(Note: Positive answers may also be written with a plus sign.)*
1. a) the opposite of $+3$, -3 **b)** the opposite of -1, 1 **c)** the opposite of -12, 12 **d)** the opposite of 16, -16 **2. a)** $+15$ **b)** $+59$ **c)** $+1$ **3. a)** $12 + (-5), 7$ **b)** $6 + (+4), 10$ **c)** $4 + (-10), -6$ **d)** $-3 + (+9), 6$ **e)** $-5 + (-3), -8$ **f)** $-12 + (+2), -10$ **g)** $-6 + (+6), 0$ **h)** $0 + (-7), -7$ **4. a)** 10 **b)** -3 **5. a)** -62 **b)** 47 **c)** 80 **d)** -124 **6. a)** $-32 + 50 = 18$ (a credit); Bonnie's new balance is a credit of \$18. **b)** $+36 - 100 = -64$ (a debit); Art's new balance is a debit of \$64. **7. a)** $-13 + 27 = 14$; The noon temperature was 14°. **b)** $-11 - 18 = -29$; The midnight temperature was $-29°$. **8. a)** $1{,}280 - 1{,}150 = 130$; The difference in altitude is 130 feet. **b)** $45 - (-35) = 45 + (+35) = 80$; The difference in altitude is 80 feet. **c)** $86 - (-39) = 86 + (+39) = 125$; The difference in altitude is 125 feet. **d)** $-39 - (-91) = -39 + (+91) = 52$; The difference in altitude is 52 feet.

Think About It: No Think About Its in this section.

Section 2.3 Exercises

FOR EXTRA HELP
Student Resources on DVD-ROM
Includes
➤ Student's Solutions Manual
➤ Video Lectures
➤ Chapter Test Prep Video
MyMathLab
MathXL

Think Again

1. Can the difference of two negative numbers ever be positive? Explain your answer or show an example that supports your answer.
Yes, when the first negative number is to the right of the second negative number.
For example, $-2 - (-8) = -2 + 8 = +6$

2. Subtraction can be written as "adding the opposite." Can addition be written as "subtracting the opposite"? Show some examples that support your answer.
Yes. Examples may vary. Two examples are $5 + 7 = 5 - (-7)$ and $3 + (-4) = 3 - (+4)$.

3. If a is a negative number, what is $a - a$? Show some examples that support your answer.
0. Examples may vary. One example is $-3 - (-3)$.

4. If b is a negative number, then what is $-(-b)$? Show an example that supports your answer.
b. Examples may vary. One example is: When b is -3, $-(-b)$ is $-(-(-3)) = -(+3) = -3$, the same value as b.

Focus Exercises

Evaluate each expression.

5. $8 - 2$
6

6. $3 - 9$
-6

7. $5 - 11$
-6

8. $9 - 6$
3

9. $3 - 11$
-8

10. $7 - 7$
0

11. $-4 - 4$
-8

12. $-6 - 4$
-10

13. $10 - 6$
4

14. $-7 - 6$
-13

15. $1 - 8$
-7

16. $-9 - 5$
-14

17. $5 - (-2)$
7

18. $1 - (-8)$
9

19. $7 - (-1)$
8

20. $3 - (-6)$
9

21. $8 - (-2)$
10

22. $9 - (-3)$
12

23. $-3 - (-5)$
2

24. $-2 - (-10)$
8

25. $-4 - (-6)$
2

26. $-11 - (-5)$
-6

27. $-12 - (-2)$
-10

28. $-8 - (-7)$
-1

29. $-70 - 20$
-90

30. $-26 - 17$
-43

31. $-19 - 35$
-54

32. $31 - 35$
-4

33. $46 - 63$
-17

34. $-21 - 32$
-53

35. $-10 - (-10)$
0

36. $-34 - (-5)$
-29

37. $26 - (-44)$
70

38. $-28 - (-25)$
-3

39. $63 - (-28)$
91

40. $-31 - (-72)$
41

41. $6 + (-2)$
4

42. $6 - (-2)$
8

43. $2 - 3$
-1

44. $9 + (-5)$
4

45. $-8 + (-9)$
-17

46. $-1 - 8$
-9

47. $-1 - 3$
-4

48. $-28 + (-3)$
-31

49. $5 + (-1)$
4

50. $9 - 20$
-11

51. $-28 - 7$
-35

52. $-2 + (-1)$
-3

53. $4 - (-21)$
25

54. $-9 + 5$
-4

55. $3 + (-1)$
2

56. $3 - 7$
-4

57. $1 - 9$
-8

58. $6 + (-8)$
-2

59. $-2 + (-6)$
-8

60. $-5 - 2$
-7

61. $-5 - 9$
-14

62. $-12 + (-9)$
-21

63. $8 + (-5)$
3

64. $6 - 10$
-4

65. $2 - (-8) + 5$
15

66. $7 - (-3) + 9$
19

67. $-5 - 6 - (-11)$
0

68. $7 - (-2) - (-8)$
17

69. $6 - (-3) - 5$
4

70. $-16 - (-2) + 4$
-10

71. $-5 + 3 - 8 + 4$
-6

72. $6 + (-4) - (-3) + 5$
10

73. $-1 + 4 - (-10)$
13

74. $9 - 3 - 2 - 6$
-2

75. $-15 + 9 - (-6) - 8$
-8

76. $-12 - 5 - (-13)$
-4

For each application, write a numerical expression and evaluate. Also write a sentence answering the question.

Finances

77. The ATM showed Mike's checking account balance as $-\$375$ at the end of the July. But on the first day of August, his paycheck of \$2,825 was automatically deposited. What was Mike's new account balance?
$-375 + 2{,}825 = 2{,}450$
Mike's new account balance was \$2,450.

78. Allison had \$25 in her checking account when she wrote a check for \$38. What is Allison's new account balance?
$25 - 38 = -13$
Allison's new account balance is $-\$13$.

79. Joni started the week with a checkbook balance of \$58. On Monday, she wrote a check for \$72. On Tuesday, she made a deposit of \$33. On Wednesday, she wrote a check for \$39. On Thursday, she wrote a check for \$29; and on Friday, she made a deposit of \$115. What was Joni's account balance on Saturday?
$58 - 72 + 33 - 39 - 29 + 115 = 66$
Joni's account balance on Saturday was \$66.

80. Carter started the week with a checkbook balance of \$22. On Monday, he wrote a check for \$40. On Tuesday, he made a deposit of \$18. On Wednesday, he wrote a check for \$26. On Thursday, he wrote a check for \$42; and on Friday, he made a deposit of \$53. What was Carter's account balance on Saturday?
$22 - 40 + 18 - 26 - 42 + 53 = -15$
Carter's account balance on Saturday was $-\$15$.

Temperature

81. At 8 AM, the outside temperature was $-16°$. By 10 AM, the temperature had warmed up by 7°. What was the temperature at 10 AM?
$-16 + 7 = -9$
The temperature was $-9°$ at 10 AM.

82. This morning the temperature was $-10°$ F. By the afternoon, the temperature had risen 18° F. What was the new temperature?
$-10 + 18 = 8$
The new temperature was 8° F.

83. Last night the temperature was $-2°$ F. By morning, the temperature had fallen 16° F. What was the new temperature?
$-2 - 16 = -18$
The new temperature was $-18°$ F.

84. At 9 PM the outside temperature was 8°. By 3 AM the next morning, the temperature had fallen 11°. What was the temperature at 3 AM?
$8 - 11 = -3$
The temperature at 3 AM was $-3°$.

Altitude

85. Find the difference in altitude between a mountain 5,280 feet above sea level and a mountain 1,416 feet above sea level.
$5{,}280 - 1{,}416 = 3{,}864$
The difference in altitude between the two mountains is 3,864 feet.

86. Find the difference in altitude between a hill 152 feet above sea level and an ocean floor 87 feet below sea level.
$152 - (-87) = 152 + (+87) = 239$
The difference in altitude between the hill and the ocean floor is 239 feet.

87. Find the difference in altitude between a mountain 4,638 feet above sea level and an ocean floor 784 feet below sea level.
$4{,}638 - (-784) = 4{,}638 + (+784) = 5{,}422$
The difference in altitude between the mountain and the ocean floor is 5,422 feet.

88. Find the difference in altitude between an ocean floor 192 feet below sea level and an ocean canyon 397 feet below sea level.
$-192 - (-397) = -192 + (+397) = 205$
The difference in altitude between the ocean floor and the ocean canyon is 205 feet.

Think Outside the Box

89. If $5 \cdot 3$ is the sum of five 3's $(3 + 3 + 3 + 3 + 3)$, what is the value of $5 \cdot (-3)$? Explain your answer.
-15; answers may vary.

90. If -6 is the opposite of 6, what is the value of $-6 \cdot 4$? Explain your answer.
-24; answers may vary.

91. Answer each question and explain your answer.

a) What is the difference of 5 and -2?
7

b) What is the distance between 5 and -2 on the number line?
7 units

c) What is the difference of -2 and 5?
-7

d) What is the distance between -2 and 5 on the number line?
7 units

e) Compare your answers to the previous four questions and write a conclusion about how the words *difference* and *distance* are similar and different in their meaning in mathematics.
Answers may vary.

SECTION 2.4 Multiplying and Dividing Signed Numbers

OBJECTIVES

In this section, you will learn to

- Multiply signed numbers.
- Divide signed numbers.

You Need to Know

To successfully complete this section, you need to understand

- ☐ Commutative Property of Multiplication (1.1)
- ☐ Associative Property of Multiplication (1.1)
- ☐ Multiplying whole numbers (1.3)
- ☐ Dividing whole numbers (1.4)
- ☐ Adding signed numbers (2.2)

Introduction

We now turn our attention to multiplication and division of signed numbers. Just as we developed rules for the addition of signed numbers, we will develop rules for their multiplication and division.

The Product of a Positive Number and a Negative Number

As you know, multiplication is an abbreviation for repeated addition. For example, $3 \cdot 4 = 12$ because $3 \cdot 4$ means the sum of *three* 4's: $4 + 4 + 4 = 12$.

With this idea in mind, let's consider what happens when we try to add *two* -5's, or $(-5) + (-5)$. We know that the answer is -10; but when we abbreviate this addition problem and turn it into multiplication, it looks like $2 \cdot (-5)$. This introduces us to the idea of multiplying signed numbers.

Positive × Negative

First: $2 \cdot (-5) = (-5) + (-5) = -10$

Furthermore, $3 \cdot (-5) = (-5) + (-5) + (-5) = -15$

We could add another (-5)—making it $4 \cdot (-5)$—and get -20, a number that is *more* negative, more to the left on the number line.

(−5)+(−5)+(−5)+(−5)

−20 −15 −10 −5 0 5

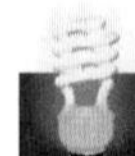

Think About It 1

Look carefully at this series of products. What pattern(s) do you see? Complete the final two products.

$3 \cdot 3 = 9$
$3 \cdot 2 = 6$
$3 \cdot 1 = 3$
$3 \cdot 0 = 0$
$3 \cdot (-1) =$ ____
$3 \cdot (-2) =$ ____

____

____

____

The rules for addition and the rules for multiplication are different, just like rules for chess and checkers are different and just like rules for soccer and football are different. These games may be played on the same checkerboard or on the same playing field, but you can't use the rules of one game to play the other.

Likewise, the rules for addition and the rules for multiplication are different. The numbers look the same, but you can't use the rules of addition to evaluate a problem of multiplication.

The diagram and the Think About It illustrate the first rule for multiplying signed numbers:

> The product of a positive number and a negative number is a negative number.

Example 1

Evaluate each product.

a) $6 \cdot (-3)$ **b)** $12 \cdot (-2)$ **c)** $7 \cdot (-8)$ **d)** $4 \cdot (-1)$ **e)** $0 \cdot (-5)$

Procedure The product of a positive number and a negative number is negative.

Answer
a) $6 \cdot (-3) = -18$ **b)** $12 \cdot (-2) = -24$

c) $7 \cdot (-8) = -56$ **d)** $4 \cdot (-1) = -4$

e) $0 \cdot (-5) = 0$ 0 times any number is 0: $0 \times a = 0$.

You Try It 1 **Evaluate each product. Use Example 1 as a guide.**

a) $9 \cdot (-4)$ **b)** $2 \cdot (-11)$ **c)** $0 \cdot (-8)$ **d)** $3 \cdot (-10)$ **e)** $8 \cdot (-1)$

Negative × Positive

Think About It 2

Look carefully at this series of products. What pattern(s) do you see? Complete the final two products.

____

____

____

$3 \cdot 3 = 9$
$2 \cdot 3 = 6$
$1 \cdot 3 = 3$
$0 \cdot 3 = 0$
$-1 \cdot 3 =$ ____
$-2 \cdot 3 =$ ____

Recall the Commutative Property of Multiplication, which allows us to write $3 \cdot 4 = 4 \cdot 3$.
The Commutative Property is also true for signed numbers so that we may write

$3 \cdot (-5)$ as $(-5) \cdot 3$. The answer, either way we look at it, is -15:
$3 \cdot (-5) = -15$ and $(-5) \cdot 3 = -15$.

The Commutative Property and the Think About It illustrate the second rule for multiplying signed numbers:

> The product of a negative number and a positive number is a negative number.

Example 2

Evaluate each product.

a) $-6 \cdot 4$ **b)** $-2 \cdot 15$ **c)** $-5 \cdot 8$ **d)** $-1 \cdot 4$ **e)** $-9 \cdot 0$

Procedure The product of a negative number and a positive number is negative.

Answer
a) $-6 \cdot 4 = -24$ **b)** $-2 \cdot 15 = -30$

c) $-5 \cdot 8 = -40$ **d)** $-1 \cdot 4 = -4$

e) $-9 \cdot 0 = 0$ Any number times 0 is 0: $a \times 0 = 0$.

You Try It 2 **Evaluate each product. Use Example 2 as a guide.**

a) $-7 \cdot 3$ **b)** $-6 \cdot 0$ **c)** $-2 \cdot 2$ **d)** $-1 \cdot 9$ **e)** $-10 \cdot 5$

The Fourth Meaning of the Dash

Next, consider the special number -1. Notice that -4 is the opposite of 4 and that $(-1)\cdot 4 = -4$.

Therefore, we can think of "multiplication by -1" as "the opposite of" so that

$$(-1)\cdot 3 = -3 \text{ (the opposite of 3) and}$$
$$(-1)\cdot 7 = -7 \text{ (the opposite of 7).}$$

By that same reasoning, if we multiply -1 by a negative number, we'll get the opposite of that negative number (a positive number) as a result:

$$(-1)\cdot(-4) = +4 \text{ (the opposite of } -4)$$

We now have a fourth interpretation of the dash

1. Minus $(6 - 2 = 4)$
2. Negative (-5 is "negative 5")
3. "the opposite of" (-8 is "the opposite of" 8)
4. "-1 times" (-4 is $-1\cdot 4$)

Example 3

Evaluate each product.

a) $-1\cdot 9$ **b)** $-1\cdot(-3)$ **c)** $-1\cdot 5$ **d)** $-1\cdot(-6)$

Procedure Multiplying a number by -1 results in the opposite of the number.

Answer
a) $-1\cdot 9 = -9$ **b)** $-1\cdot(-3) = +3$, or 3
c) $-1\cdot 5 = -5$ **d)** $-1\cdot(-6) = +6$, or 6

You Try It 3 **Evaluate each product. Use Example 3 as a guide.**

a) $-1\cdot 3$ **b)** $-1\cdot 7$ **c)** $-1\cdot(-8)$ **d)** $-1\cdot(-2)$

Likewise, we could think of any negative number as a product of -1 and a positive number, as demonstrated in Example 4.

Example 4

Write each negative number two ways as the product of -1 and a positive number.

a) $-9 = \underline{-1\cdot 9}$ and $\underline{9\cdot(-1)}$ **b)** $-3 = \underline{-1\cdot 3}$ and $\underline{3\cdot(-1)}$

You Try It 4 **Write each negative number two ways as the product of -1 and a positive number. Use Example 4 as a guide.**

a) -4 **b)** -7 **c)** -15

The Product of Two Negative Numbers

Recall the Associative Property of Multiplication. It says that in a product, we can change the grouping of the factors without affecting the resulting product: $(a\cdot b)\cdot c = a\cdot(b\cdot c)$.

Let's use the Associative Property with the fourth meaning of the dash.

Consider, for example, $(-5)\cdot(-4)$. We can think of this as $5\cdot(-1)\cdot(-4)$.

First, let's group the last two factors:

$5 \cdot (-1) \cdot (-4)$ ← Multiply: $(-1) \cdot (-4) = +4$.
$= 5 \cdot (+4)$ ← Now $5 \cdot (+4) = +20$.
$= +20$ ← These products will both be $+20$. →

Second, we'll group the first two factors:

$5 \cdot (-1) \cdot (-4)$
$= (-5) \cdot (-4)$
$= +20$

This shows that $(-5) \cdot (-4) = +20$ and suggests that the product of two negative numbers is positive.

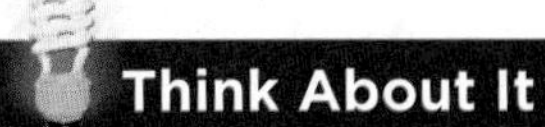
Think About It 3

Look carefully at this series of products. What pattern(s) do you see? Complete the final two products.

____

____

____

$3 \cdot (-3) = -9$
$2 \cdot (-3) = -6$
$1 \cdot (-3) = -3$
$0 \cdot (-3) = 0$
$-1 \cdot (-3) =$ _____
$-2 \cdot (-3) =$ _____

> The product of two negative numbers is a positive number.

Example 5

Evaluate each product.

a) $-10 \cdot (-6)$ **b)** $-5 \cdot (-5)$ **c)** $(-3) \cdot (-12)$ **d)** $(-4) \cdot (-7)$

Procedure The product of two negative numbers is a positive number.

Answer **a)** $-10 \cdot (-6) = +60$, or 60 **b)** $-5 \cdot (-5) = +25$, or 25

c) $(-3) \cdot (-12) = +36$, or 36 **d)** $(-4) \cdot (-7) = +28$, or 28

You Try It 5 **Evaluate each product. Use Example 5 as a guide.**

a) $-2 \cdot (-8)$ **b)** $(-5) \cdot (-4)$ **c)** $(-7) \cdot (-7)$ **d)** $-6 \cdot (-9)$

The next rule is a rule you've been using most of your life; you've just probably never had to think of it this way:

> The product of two positive numbers is a positive number.

Example 6

Evaluate the product.

a) $6 \cdot 4$ **b)** $2 \cdot 15$ **c)** $5 \cdot 8$

Answer **a)** $6 \cdot 4 = +24$, or 24 **b)** $2 \cdot 15 = +30$, or 30 **c)** $5 \cdot 8 = +40$, or 40

The bottom line is that the four previous rules can be broken down into just two rules:

Multiplying Two Signed Numbers

1. If the signs of the factors are different, the product will be negative.
2. If the signs of the factors are the same, the product will be positive.

Also, if zero is a factor, the product will be 0.

Example 7

Find the product of the two given signed numbers.

a) $-8 \cdot 6$ **b)** $3 \cdot 9$ **c)** $-4 \cdot (-2)$ **d)** $5 \cdot (-7)$ **e)** $-1 \cdot 0$

Answer

	Rule
a) $-8 \cdot 6 = -48$	The signs are different; the product is negative.
b) $3 \cdot 9 = +27$, or 27	The signs are the same; the product is positive.
c) $-4 \cdot (-2) = +8$, or 8	The signs are the same; the product is positive.
d) $5 \cdot (-7) = -35$	The signs are different; the product is negative.
e) $-1 \cdot 0 = 0$	If zero is a factor, the product is 0.

You Try It 6 **Find the product. Use Examples 6 and 7 as guides.**

a) $-7 \cdot (-6)$ **b)** $-7 \cdot 8$ **c)** $2 \cdot (-6)$ **d)** $0 \cdot (-9)$ **e)** $5 \cdot 4$

The Product of More Than Two Signed Numbers

What if we were to multiply more than two signed numbers? No problem. The Associative Property and the Commutative Property allow us to multiply in any order we choose.

Example 8

Find each product.

a) $2 \cdot 3 \cdot 4$ (no negatives) **b)** $-2 \cdot (-3) \cdot 4$ (two negatives)

c) $-2 \cdot 3 \cdot 4$ (one negative) **d)** $-2 \cdot (-3) \cdot (-4)$ (three negatives)

Procedure Notice the number of negatives in the expression. Use the Associative Property to group and multiply two numbers at a time.

Answer

a)
$$\begin{aligned} &2 \cdot 3 \cdot 4 \\ &= (2 \cdot 3) \cdot 4 \\ &= 6 \cdot 4 \\ &= 24 \end{aligned}$$

b)
$$\begin{aligned} &-2 \cdot (-3) \cdot 4 \\ &= [-2 \cdot (-3)] \cdot 4 \\ &= +6 \cdot 4 \\ &= 24 \end{aligned}$$

With two negative factors, the end product is positive.

c)
$$\begin{aligned} &-2 \cdot 3 \cdot 4 \\ &= (-2 \cdot 3) \cdot 4 \\ &= -6 \cdot 4 \\ &= -24 \end{aligned}$$

With one negative factor, the end product is negative.

d)
$$\begin{aligned} &-2 \cdot (-3) \cdot (-4) \\ &= [-2 \cdot (-3)] \cdot (-4) \\ &= +6 \cdot (-4) \\ &= -24 \end{aligned}$$

The first two factors make a positive product, +6.

The third negative factor makes the final product negative.

So, three negative factors gives a negative product. What do you think will happen if there are four negative factors? Let's find out.

$$\begin{aligned} &-1 \cdot (-2) \cdot (-3) \cdot (-4) \\ &= [-1 \cdot (-2)] \cdot [(-3) \cdot (-4)] \\ &= (+2) \cdot (+12) \\ &= 24 \end{aligned}$$

There are two pairs of products, each with two negative factors. Each pair results in a positive product.

Did you expect the answer to be positive? Notice that every *pair* of negative factors produces a positive number. We can actually make rules out of this.

Multiplying Two or More Signed Numbers

When there is no zero factor

1. If there is an odd number of negative factors, the end product will be negative.

2. If there is an even number of negative factors, the end product will be positive.

Example 9

Multiply. First, decide whether the product is positive or negative. Then multiply the numerical values of the factors.

a) $(6)(-3)(-2)$ **b)** $(-5)(4)(10)$

c) $(-2)(-8)(-10)$ **d)** $(-4)(-3)(-5)(-1)$

Answer

Product of Factors	Number of Negatives	Result is	Product
a) $(6)(-3)(-2)$	Two (even)	Positive	$(6)(-3)(-2) = +36$
b) $(-5)(4)(10)$	One (odd)	Negative	$(-5)(4)(10) = -200$
c) $(-2)(-8)(-10)$	Three (odd)	Negative	$(-2)(-8)(-10) = -160$
d) $(-4)(-3)(-5)(-1)$	Four (even)	Positive	$(-4)(-3)(-5)(-1) = +60$

You Try It 7 **Multiply. First, decide whether the product will be positive or negative. Then multiply numerical values. Use Examples 8 and 9 as guides. (You may state the answer directly without showing all of your work.)**

a) $(2)(-4)(-5)$ **b)** $(-4)(5)(3)(-2)$ **c)** $(-3)(-1)(-6)$

d) $(-10)(-2)(0)(-3)$ **e)** $(-1)(-4)(-2)(-3)$

Dividing Signed Numbers

We have seen the circular connection between multiplication and division (Section 1.4). In a fraction such as $\frac{15}{3}$ the denominator (divisor) multiplies the quotient to result in the numerator (dividend).

Dividend → $\frac{15}{3} = 5$ ← Quotient
Divisor →

Divisor × Quotient = Dividend
3 × 5 = 15

This circular relationship is also seen in standard division: $15 \div 3 = 5$ because $3 \times 5 = 15$.

How would this be different if some negative numbers were included? It would work the same, but we'd have to put more thought into it.

For example, $15 \div (-3) = -5$ or $\frac{15}{-3} = -5$ because $-3 \times (-5) = +15$.

In other words, positive ÷ negative = negative, or $\frac{\text{positive}}{\text{negative}}$ = negative because

negative × negative = positive.

You Try It 8 **Use your understanding of the rules for multiplying signed numbers—and the circular relationship between multiplication and division—to fill in the missing quotient.**

a) $\dfrac{18}{-3} = \_\_\_$ b) $\dfrac{-30}{6} = \_\_\_$ c) $\dfrac{-28}{-7} = \_\_\_$

The rules of dividing signed numbers are the same as the rules for multiplication.

Dividing Signed Numbers

In any division, either by fraction or standard form,

1. If there is an odd number of negative factors, the quotient will be negative.

2. If there is an even number of negative factors, the quotient will be positive.

Also recall from Section 1.4 that we cannot divide by 0. Zero can be a dividend but not the divisor. Let's work through some examples.

Example 10

Divide.

a) $\dfrac{-21}{3}$ **b)** $\dfrac{36}{-9}$ **c)** $-35 \div (-7)$ **d)** $32 \div 4$ **e)** $\dfrac{0}{-2}$ **f)** $\dfrac{-4}{0}$

Procedure First, decide whether the quotient is positive or negative (by the number of negatives). Then divide the numerical values.

Answer

Quotient	Number of Negatives	Result is	Quotient
a) $\dfrac{-21}{3}$	One (odd)	Negative	$= -7$
b) $\dfrac{36}{-9}$	One (odd)	Negative	$= -4$
c) $-35 \div (-7)$	Two (even)	Positive	$= 5$, or $+5$
d) $32 \div 4$	Zero (even)	Positive	$= 8$, or $+8$
e) $\dfrac{0}{-2}$	One, but 0 is in the numerator	0	$= 0$
f) $\dfrac{-4}{0}$	One, but 0 is in the denominator	Undefined	Zero is in the denominator.

You Try It 9 **First, decide whether the quotient will be positive or negative, and then divide the numerical values. Use Example 10 as a guide.**

a) $45 \div (-5)$ **b)** $\dfrac{-36}{-3}$ **c)** $-42 \div 0$

d) $-28 \div (-4)$ **e)** $\dfrac{0}{-8}$ **f)** $\dfrac{-54}{9}$

Calculator Tip **Negative Numbers**

If you use a scientific calculator to evaluate expressions with negative numbers, you need to be familiar with the *negative* key on the calculator: (−)

This key is not the same as the subtraction key, −; and the subtraction key is not used to represent *negative*.

Here are the calculator keystrokes for different expressions involving negative numbers.

1. $-6 + 9$ (−) 6 + 9 =
2. $-3 - (-4)$ (−) 3 − (−) 4 =
3. $8 \cdot (-5)$ 8 × (−) 5 =
4. $-4 \div (-2)$ (−) 4 ÷ (−) 2 =

Answers: You Try It and Think About It

You Try It: *(Note: Positive answers may also be written with a plus sign.)*
1. a) -36 **b)** -22 **c)** 0 **d)** -30 **e)** -8 **2. a)** -21 **b)** 0 **c)** -4 **d)** -9 **e)** -50 **3. a)** -3 **b)** -7 **c)** 8 **d)** 2 **4. a)** $-1 \cdot 4$ and $4 \cdot (-1)$ **b)** $-1 \cdot 7$ and $7 \cdot (-1)$ **c)** $-1 \cdot 15$ and $15 \cdot (-1)$ **5. a)** 16 **b)** 20 **c)** 49 **d)** 54 **6. a)** 42 **b)** -56 **c)** -12 **d)** 0 **e)** 20 **7. a)** 40 **b)** 120 **c)** -18 **d)** 0 **e)** 24 **8. a)** -6 **b)** -5 **c)** 4 **9. a)** -9 **b)** 12 **c)** Undefined **d)** 7 **e)** 0 **f)** -6

Think About It: **1.** Answers may vary. One possibility: As the second number in the product decreases by 1, the resulting product decreases by 3. The last two numbers are -3 and -6. **2.** Answers may vary. One possibility: As the first number in the product decreases by 1, the resulting product decreases by 3. The last two numbers are -3 and -6. **3.** Answer may vary. One possibility: As the first number in the product decreases by 1, the resulting product increases by 3. The last two numbers are 3 and 6.

Section 2.4 Exercises

FOR EXTRA HELP
Student Resources on DVD-ROM
Includes
➤ Student's Solutions Manual
➤ Video Lectures
➤ Chapter Test Prep Video
MyMathLab
MathXL

Think Again

1. Predict whether the value of $(-6)^{17}$ will be positive or negative. Explain you answer.
It will be negative because there is an odd number of negative factors.

2. If a product has three negative factors, four positive factors, and two factors of zero, is the end result positive or negative? Explain your answer or show an example that supports your answer.
The end result is 0 because of the two factors of 0.

3. In the division of two integers, if the quotient is positive, what do you know about the dividend and the divisor?
The dividend and the divisor must have the same sign, both positive or both negative.

4. In the division of two integers, if the quotient is -1, what do you know about the dividend and the divisor?
The dividend and the divisor must be opposites.

Focus Exercises

Evaluate.

5. $(-5)(-4)$
20

6. $-11 \cdot (-3)$
33

7. $9 \cdot (-7)$
-63

8. $(11)(-9)$
-99

9. $6 \cdot 10$
60

10. $(-2)(0)$
0

11. $(-4)(-4)$
16

12. $8 \cdot 8$
64

13. $0 \cdot (-6)$
0

14. $(-8)(-8)$
64

15. $7(-4)$
-28

16. $(-1)(3)$
-3

17. $-20 \cdot 5$
-100

18. $-6 \cdot (-32)$
192

19. $-3 \cdot 12$
-36

20. $3 \cdot (-5)$
-15

21. $-4 \cdot (-6)$
24

22. $6 \cdot (-6)$
-36

23. $-8 \cdot 3$
-24

24. $-2 \cdot (-9)$
18

25. $9 \cdot (-1)$
-9

26. $-4 \cdot 7$
-28

27. $-12 \cdot 5$
-60

28. $-10 \cdot (-3)$
30

29. $10 \cdot (-17)$
-170

30. $-20 \cdot 5$
-100

31. $14 \cdot 8$
112

32. $19 \cdot 25$
475

33. $-30 \cdot (-4)$
120

34. $-3 \cdot (-25)$
75

35. $-4 \cdot 5 \cdot (-3)$
60

36. $2 \cdot (-7) \cdot (-1)$
14

37. $6 \cdot (-2) \cdot 8$
-96

38. $3 \cdot 4 \cdot (-6)$
-72

39. $-1 \cdot (-9) \cdot (-6)$
-54

40. $-4 \cdot (-2) \cdot (-7)$
-56

41. $-4 \cdot 5 \cdot (-3) \cdot 2$
120

42. $-1 \cdot (-2) \cdot 0 \cdot 11$
0

43. $-2 \cdot (-1) \cdot 0 \cdot (-4)$
0

44. $-5 \cdot 6 \cdot (-2) \cdot (-2)$
-120

45. $-4 \cdot (-1) \cdot (-3) \cdot (-5)$
60

46. $-10 \cdot (-2) \cdot (-2) \cdot (-3)$
120

47. $-25 \div 5$
-5

48. $-42 \div 7$
-6

49. $-21 \div (-3)$
7

50. $-28 \div (-4)$
7

51. $18 \div (-6)$
-3

52. $40 \div (-5)$
-8

53. $-30 \div (-15)$
2

54. $-42 \div (-6)$
7

55. $-20 \div 0$
Undefined

56. $\frac{-36}{-9}$
4

57. $\frac{-60}{3}$
-20

58. $\frac{-72}{9}$
-8

59. $\frac{0}{-8}$
0

60. $\frac{16}{-2}$
-8

61. $\frac{-21}{3}$
-7

62. $\frac{-27}{-9}$
3

63. $\frac{-45}{-5}$
9

64. $\frac{28}{0}$
Undefined

65. $\frac{36}{-4}$
-9

66. $\frac{35}{-7}$
-5

67. $\frac{-30}{5}$
-6

68. $\frac{-65}{5}$
-13

69. $\frac{-54}{6}$
-9

70. $\frac{-60}{10}$
-6

71. $\frac{-8}{-4}$
2

72. $\frac{-16}{-2}$
8

73. $\frac{40}{-8}$
-5

74. $\frac{34}{-2}$
-17

75. $\frac{-42}{-3}$
14

76. $\frac{-45}{-15}$
3

Think Outside the Box

Evaluate.

77. $(-1)^{23}$
-1

78. $(-1)^{32}$
$+1$

79. $[-8 - (-12)] \cdot (-10 + 5)$
-20

80. $\frac{-20 - 12}{-6 + 2}$
$+8$

SECTION 2.5 The Order of Operations for Signed Numbers

OBJECTIVES

In this section, you will learn to

- Apply powers to negative numbers.
- Find negative square roots.
- Apply the order of operations to signed numbers.

You Need to Know

To successfully complete this section, you need to understand

- ☐ Exponents and square roots (1.5)
- ☐ The order of operations (1.5)
- ☐ Absolute value (2.1)
- ☐ Replacement values (2.1)
- ☐ Operations with signed numbers (2.2, 2.3, 2.4)

Introduction

You are already quite familiar with the order of operations—so why are we looking at it again? We understand negative numbers, but we want to see how they work within this system. We do so in preparation for Section 2.6, where you will be introduced to a variety of mathematical formulas that use both positive and negative numbers.

The Order of Operations

Recall from Chapter 1 the order of operations as follows:

The Order of Operations

1. Evaluate within all grouping symbols if there are any.
2. Apply any exponents.
3. Apply multiplication and division reading from left to right.
4. Apply addition and subtraction reading from left to right.

Also remember that we refer to an operation's rank when speaking of which operation to perform first. Because multiplication is applied before addition, multiplication has a higher rank than addition.

Let's practice using the order of operations with a few exercises.

Example 1

Evaluate each expression according to the order of operations.

a) $-6 + 12 \div 3$ **b)** $(4 - 16) \div 2^2$

Answer

a) $-6 + 12 \div 3$ Apply division first, then addition.

$= -6 + 4$

$= -2$

b) $(4 - 16) \div 2^2$ The parentheses have the highest rank; so, apply subtraction first. $4 - 16 = 4 + (-16) = -16 + 4 = -12$.

$= -12 \div 2^2$ The exponent has the higher rank; so, apply it before division.

$= -12 \div 4$

$= -3$

Example 2

Evaluate $\sqrt{24 + 4 \cdot (-2)}$.

Procedure The radical is both a grouping symbol and an operation. As a grouping symbol, we must evaluate within before applying the square root.

Answer

$\sqrt{24 + 4 \cdot (-2)}$ Multiply first: $4 \cdot (-2) = -8$.

$= \sqrt{24 + (-8)}$

$= \sqrt{16}$

$= 4$

Caution The key to applying the order of operations successfully is to do one step at a time. It's very important to show your work every step of the way. This will lead to accurate answers and enable others to read and learn from your work.

You Try It 1 **Evaluate each expression according to the order of operations. Use Examples 1 and 2 as guides.**

a) $-24 \div 6 + 2$ **b)** $(5 - 7) \cdot 2$

c) $-9 \cdot \sqrt{25}$ **d)** $(6 - 12) \div (-2 \cdot 3)$

Exponents and Negative Numbers

Recall from Section 1.5 that an exponent is an abbreviation for repeated multiplication. For example, 5^3 means three factors of 5: $5 \cdot 5 \cdot 5 = 125$.

Let's extend this idea to powers (exponents) of negative numbers. For example, we know that $3^2 = 9$. It's also true that $(-3)^2 = 9$. It's easy to demonstrate this using the definition of exponents: $(-3)^2 = (-3)(-3) = +9 = 9$.

As we know from Section 2.4, an even number of negative factors has a product that is positive. So, when a base is negative and its exponent is an *even* number, such as $(-2)^4$, $(-5)^2$, and $(-10)^6$, there will be an even number of negative factors and the end result will be a positive number.

$$(-2)^4 = +16, \ (-5)^2 = +25, \text{ and } (-10)^6 = +1{,}000{,}000$$

Likewise, an odd number of negative factors has a product that is negative. So, when a base is negative and its exponent is an *odd* number, such as $(-2)^5$, $(-5)^3$, and $(-10)^7$, there will be an odd number of negative factors and the end result will be a negative number.

$$(-2)^5 = -32, \ (-5)^3 = -125, \text{ and } (-10)^7 = -10{,}000{,}000$$

In general:

$$(\text{Negative base})^{\text{even exponent}} = \text{positive number}$$
$$(\text{Negative base})^{\text{odd exponent}} = \text{negative number}$$

Example 3

Evaluate each expression.

a) $(-9)^2$ **b)** $(-3)^3$ **c)** $(-2)^4$ **d)** $(3 - 5)^3$ **e)** $(-8 + 3)^2$

Procedure First, decide if the result is going to be positive or negative based on the exponent.

Answer

a) $(-9)^2 = +81 = 81$ The exponent is 2, an even number; so, the result is positive: $9 \cdot 9 = 81$.

b) $(-3)^3 = -27$ The exponent is 3 (odd); so, the result is negative: $3 \cdot 3 \cdot 3 = 27$.

c) $(-2)^4 = +16 = 16$ The exponent is 4 (even); so, the result is positive: $2 \cdot 2 \cdot 2 \cdot 2 = 16$.

For parts d and e, evaluate inside the grouping symbols and then decide if the result will be positive or negative.

d) $(3 - 5)^3 = (-2)^3 = -8$ The exponent is 3 (odd); so, the result is negative: $2 \cdot 2 \cdot 2 = 8$.

e) $(-8 + 3)^2 = (-5)^2 = +25 = 25$ The exponent is 2 (even); so, the result is positive: $5 \cdot 5 = 25$.

You Try It 2 **Evaluate each expression. Use Example 3 as a guide.**

a) $(-7)^2$ **b)** $(-4)^3$ **c)** $(-10)^4$ **d)** $(-1)^5$

e) $(1 - 9)^2$ **f)** $(-10 + 7)^3$ **g)** $(-9 - 1)^3$

Negative Square Roots

We have seen that $3^2 = 9$ and that $(-3)^2 = 9$. This example demonstrates that there are two numbers for which 9 is a perfect square. It's appropriate, therefore, to say that 9 has *two* square roots:

1. A positive square root $(+3)$
2. A negative square root (-3)

The Square Root

If $r^2 = P$, then r is a square root of P.

Does this change what we've learned about the square root radical? No, $\sqrt{9}$ still means the square root of 9; but now it refers only to the positive square root: $\sqrt{9} = +3$, not -3. We call the positive square root of 9 the *principal* square root.

If we want to use the radical to represent the negative square root of 9, we must place a negative sign outside the radical: $-\sqrt{9} = -3$.

Think About It 1 What is $\sqrt{0}$? Explain your answer.

____

____

Example 4

Evaluate each radical expression.

a) $\sqrt{36}$ **b)** $-\sqrt{36}$ **c)** $\sqrt{25}$ **d)** $-\sqrt{25}$

Answer Apply the square roots before negating.

a) $\sqrt{36} = 6$, or $+6$ **b)** $-\sqrt{36} = -6$

c) $\sqrt{25} = 5$, or $+5$ **d)** $-\sqrt{25} = -5$

You Try It 3 **Evaluate each radical expression. Use Example 4 as a guide.**

a) $-\sqrt{16}$ **b)** $\sqrt{49}$ **c)** $-\sqrt{81}$ **d)** $-\sqrt{1}$

Double Quantities

Sometimes an expression contains two sets of grouping symbols that are unrelated to each other; in other words, evaluating within one set does not affect the evaluation within the other. This means that some quantities can be evaluated at the same time.

For example, in the expression $(3 - 8) \cdot (12 \div 4)$, we can evaluate within each grouping symbol regardless of what operation each expression contains.

$(3 - 8) \cdot (12 \div 4)$ There are three operations: subtraction, multiplication, and division. Because the parentheses have the highest rank, we can apply subtraction and division separately, yet at the same time.

$= (-5) \cdot (3)$

$= -15$

Example 5

Evaluate each expression according to the *order of operations*.

a) $(5 \cdot 6) \div (4 - 7)$ **b)** $(-8 + 3) \cdot (-14 \div 2)$

c) $(24 \div 6) - \sqrt{5 + 11}$ **d)** $\sqrt{4^2} - \sqrt{50 \cdot 2}$

Procedure Single operations within different grouping symbols have equal rank. They can be applied at the same time.

Answer

a) $(5 \cdot 6) \div (4 - 7)$ We can apply the multiplication and the subtraction in the same step.

$= (30) \div (-3)$ A positive divided by a negative results in a negative.

$= -10$

b) $(-8 + 3) \cdot (-14 \div 2)$ We can apply the addition and the division in the same step.

$= (-5) \cdot (-7)$ A negative multiplied by a negative results in a positive.

$= +35 = 35$

c) $(24 \div 6) - \sqrt{5 + 11}$ We can apply the division and the addition in the same step.

$= 4 - \sqrt{16}$ Now apply the radical.

$= 4 - 4$ Subtract.

$= 0$

d) $\sqrt{4^2} - \sqrt{50 \cdot 2}$ We can apply the exponent and the multiplication in the same step.

$= \sqrt{16} - \sqrt{100}$ Now we can apply both radicals at the same time because they have the same rank.

$= 4 - 10$ $\mathbf{4 - 10 = 4 + (-10) = -6}$

$= -6$

You Try It 4 **Evaluate each expression according to the order of operations. Use Example 5 as a guide.**

a) $(2 \cdot 3) + (-42 \div 6)$ **b)** $(6 - 13) \cdot (-5 + 2)$

c) $(6 + 3)^2 - \sqrt{-4 + 5}$ **d)** $\sqrt{12 \cdot 3} - \sqrt{8 + 41}$

Other Grouping Symbols: The Division Bar and Absolute Value Bars

The division bar is a grouping symbol. It groups the dividend (top) separately from the divisor (bottom). When evaluating an expression that involves a division bar, we treat the dividend and the divisor as if they were grouped separately and evaluate within them separately.

$$\frac{\text{Dividend}}{\text{Divisor}}$$

The parentheses shown in $\frac{(3 \cdot 8)}{(5+1)}$ are not necessary; they are there to emphasize the grouping nature of the division bar.

For example, in the expression $\frac{3 \cdot 8}{5 + 1}$, the operations multiplication (in the dividend) and addition (in the divisor) have equal rank because of the grouping provided by the division bar: $\frac{(3 \cdot 8)}{(5 + 1)}$.

So, we can apply multiplication and addition at the same time to get $\frac{24}{6}$. We can then divide: $24 \div 6 = 4$.

Example 6

Evaluate $\dfrac{56 \div 7}{3 - 1}$.

Procedure The dividend and the divisor are separate quantities and must be evaluated separately.

Answer

$\dfrac{56 \div 7}{3 - 1}$ First, apply the division and subtraction individually but at the same time.

$= \dfrac{8}{2}$ Think of this as division: $8 \div 2 = 4$.

$= 4$

Example 7

Evaluate $\dfrac{\sqrt{16} + 5}{2^2 - 7}$.

Procedure Remember that the radical is both a grouping symbol and an operation.

Answer

$\dfrac{\sqrt{16} + 5}{2^2 - 7}$ First, apply the radical in the dividend and square the 2 in the divisor.

$= \dfrac{4 + 5}{4 - 7}$ Next, apply subtraction and addition.

$= \dfrac{9}{-3}$ Think of this as division: $9 \div (-3) = -3$.

$= -3$

You Try It 5 **Evaluate each expression according to the order of operations. Use Examples 6 and 7 as guides.**

a) $\dfrac{7 + 3}{9 - 4}$ **b)** $\dfrac{2 \cdot 3}{-42 \div 7}$ **c)** $\dfrac{-4 - 14}{4^2 - 10}$ **d)** $\dfrac{3 - 15}{-4 - (-2)}$

Absolute value bars are grouping symbols, and we must evaluate within them first. However, the absolute value is also an operation; so once the expression within is simplified, we can apply the absolute value to that number.

Example 8

Evaluate each expression completely.

a) $|5| - |-9|$ **b)** $|4 - 6| + |-3|$

Procedure Remember, the absolute value bars are both a grouping symbol and an operation.

Answer

a) $|5| - |-9|$ Each expression within the absolute value bars is already simplified. Treat this, though, as a double quantity and evaluate each separately.

$= 5 - 9$ The minus sign is unaffected by the absolute values.

$= 5 + (-9)$ Change subtraction to adding the opposite.

$= -4$

b) $|4 - 6| + |-3|$ Evaluate within the first absolute value only: $|4 - 6| = |-2|$.

$= |-2| + |-3|$ Now apply the absolute value, not as a grouping symbol, but as an operation.

$= 2 + 3$ Last, apply addition: $2 + 3 = 5$.

$= 5$

You Try It 6 **Evaluate each expression according to the order of operations. Use Example 8 as a guide.**

a) $|-9 + 3|$ **b)** $|-12| - |-5|$

c) $(6 - 8) \cdot |5 - 10|$ **d)** $|7 - 4| - |-10|$

Answers: You Try It and Think About It

You Try It: *(Note: Positive answers may also be written with a plus sign)*
1. a) -2 **b)** -4 **c)** -45 **d)** 1 **2. a)** 49 **b)** -64 **c)** 10,000 **d)** -1 **e)** 64 **f)** -27 **g)** $-1{,}000$ **3. a)** -4 **b)** 7 **c)** -9 **d)** -1 **4. a)** -1 **b)** 21 **c)** 80 **d)** -1 **5. a)** 2 **b)** -1 **c)** -3 **d)** 6 **6. a)** 6 **b)** 7 **c)** -10 **d)** -7

Think About It: **1.** $\sqrt{0} = 0$. $\sqrt{0}$ does not have a principal square root because a principal square root is positive and 0 is not positive.

Section 2.5 Exercises

FOR EXTRA HELP
Student Resources on DVD-ROM
Includes
➤ Student's Solutions Manual
➤ Video Lectures
➤ Chapter Test Prep Video
MyMathLab
MathXL

Think Again

1. What is $\sqrt{0}$? Explain your answer. *(Refer to Think About It 1)*
   $\sqrt{0} = 0$ because $0 \cdot 0 = 0$.

2. Does $|a + b| = |a| + |b|$? Sometimes, but not always. Under what circumstances is $|a + b|$ equal to $|a| + |b|$?
   When a and b have the same sign, both positive or both negative.

3. Does $|a + b| = |a| + |b|$? Sometimes, but not always. Under what circumstances is $|a + b|$ *not* equal to $|a| + |b|$?
   When one number is positive and the other number is negative.

4. Because 30 is not a perfect square, it does not have a whole number square root. Between which two consecutive integers is the value of $-\sqrt{30}$? Explain your answer or show an example that supports your answer.
   $-\sqrt{30}$ is between -5 and -6 because $-\sqrt{30}$ is between $-\sqrt{25}$ and $-\sqrt{36}$.

Evaluate.

5. $\sqrt{64}$ — 8
6. $-\sqrt{100}$ — -10
7. $\sqrt{81}$ — 9
8. $-\sqrt{1}$ — -1
9. $\sqrt{25}$ — 5
10. $-\sqrt{16}$ — -4
11. $\sqrt{49}$ — 7
12. $-\sqrt{36}$ — -6
13. $(-1)^2$ — 1
14. $(-1)^3$ — -1
15. $(-4)^1$ — -4
16. $(-4)^2$ — 16
17. $(-4)^3$ — -64
18. $(-8)^2$ — 64
19. $(-10)^3$ — $-1{,}000$
20. $(-7)^3$ — -343
21. $(-6)^2$ — 36
22. $(-5)^3$ — -125
23. $(-3)^4$ — 81
24. $(-2)^2$ — 4
25. $(-2)^3$ — -8
26. $(-2)^4$ — 16
27. $(-2)^5$ — -32
28. $(-2)^6$ — 64

Evaluate each expression using the order of operations.

29. $-5 + 6 \cdot 2$ — 7
30. $8 + 5 \cdot (-3)$ — -7
31. $6 \cdot 2 - 4 \cdot 8$ — -20
32. $-9 + (2 - 5) + 4^2$ — 4
33. $6 - 2 \cdot (-2) + 2^3$ — 18
34. $-3(2 - 8)$ — 18
35. $2[-6 + 3(7 - 2)]$ — 18
36. $(-4 + 10) \div (-3)$ — -2
37. $30 \div 5 \cdot 2 - 5$ — 7
38. $6 - 14 \div 2$ — -1
39. $-24 \div 6 \cdot (-2)$ — 8
40. $7 - 4^2$ — -9
41. $(7 - 9)^3 - 6 \cdot 3$ — -26
42. $5^2 + |-6|$ — 31
43. $[-12 \div (-3)]^2$ — 16
44. $(-3)^2 - (4 - 6)$ — 11
45. $(5 - 9)^2$ — 16
46. $-24 \div (6 + 2)$ — -3
47. $-24 \div (3 - 5)$ — 12
48. $3 - 4 \cdot 5$ — -17
49. $3 - 4 \cdot (-5)$ — 23

50. $(4 - 10)^2$
36

51. $-2 \cdot 3 - 8 \div (-4)$
-4

52. $(4 - 6)^3$
-8

53. $20 - (-8)^2$
-44

54. $(3 - 10)^2$
49

55. $-5 \cdot (3 - 4)^2$
-5

56. $-30 + (-7)^2$
19

57. $(7 + 3) \cdot (-9 + 4)$
-50

58. $(-2 \cdot 12) \div (-4 - 2)$
4

59. $(7 + 3)^2 \div (9 - 4)^2$
4

60. $(2 - 3)^2 - (-12 \div 6)^3$
9

61. $(5 - 6)^3 - 4 \cdot (-2)$
7

62. $-\sqrt{81} + 24 \div (-3)$
-17

63. $35 \div (-5) + 2 \cdot (-3)$
-13

64. $6 - [12 \div (-2 \cdot 3)]$
8

65. $|6| - 8$
-2

66. $|-7| + 3$
10

67. $9 - |-8|$
1

68. $-6 + |-5|$
-1

69. $|9 - 5|$
4

70. $|1 - 4|$
3

71. $|3 - 9| + 8$
14

72. $|2 - 8| - 6$
0

73. $|-3| + |8|$
11

74. $|-6| - |4|$
2

75. $|11| - |-15|$
-4

76. $|-12| - |-18|$
-6

77. $|3 + 7| + |4 - 2|$
12

78. $|-4 + 6| + |-5 - 1|$
8

79. $|5 - 9| - |2 - 8|$
-2

80. $|-7 + 2| - |-10 - 1|$
-6

81. $-6 + \sqrt{16}$
-2

82. $-10 \cdot \sqrt{64}$
-80

83. $3 - \sqrt{81}$
-6

84. $-6 - 3\sqrt{4}$
-12

85. $\sqrt{9} + \sqrt{16}$
7

86. $\sqrt{80 \div 5} - 9$
-5

87. $18 - 2 \cdot \sqrt{100}$
-2

88. $-\sqrt{-4 \cdot (-9)}$
-6

89. $\sqrt{-1 + 2 \cdot 5}$
3

90. $\sqrt{(6 - 2) \cdot 5^2}$
10

91. $-\sqrt{5^2 - 3^2}$
-4

92. $\sqrt{3^2 - 4(2)}$
1

Think Outside the Box

Evaluate.

93. $\left(\sqrt{-5 + 9}\right)^4$
16

94. $|-7 + 4|^{|-7+4|}$
27

95. $(-6 + 10 \div 2)^{(-7+10)}$
-1

96. $(-8 + 6)^{(-2)^2}$
16

SECTION 2.6 Formulas

OBJECTIVE

In this section, you will learn to

- Evaluate formulas by using the order of operations.

You Need to Know

To successfully complete this section, you need to understand

- ☐ The order of operations (1.5 and 2.5)

Introduction

If you've ever baked a cake, you probably followed a recipe telling you how much of each ingredient to use, what the oven temperature should be, and how long to bake the cake. You mixed the ingredients together and poured them into the cake pan. If you followed the recipe correctly, the result was a delicious cake for dessert.

Using a formula in math is like following a recipe. We

1. Put certain numbers into the formula.
2. Use the order of operations to evaluate the expression.

If we follow the order of operations correctly, we will have a new value that can be applied to whatever problem we're trying to solve.

Formulas are the best application of the order of operations. Many formulas—in a variety of fields—require the order of operations for their proper evaluation.

For example, you may see Celsius temperatures at bank buildings. The sign out front might say that the temperature is 10°C (Celsius), but is that T-shirt or jacket weather? To convert the temperature reading from Celsius degrees to Fahrenheit degrees, we use this formula:

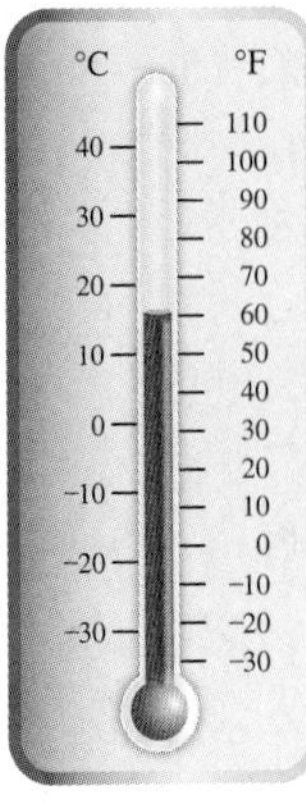

$$F = 9 \cdot \frac{C}{5} + 32, \qquad \text{where} \qquad F = \text{Fahrenheit degrees}$$
$$C = \text{Celsius degrees}$$

Notice that the formula contains two variables, F and C. To evaluate the formula, we need to know the value of *only one* of the variables. We then use that value to find the value of the other variable.

If we substitute the known value of C (10°) into the formula, we can evaluate the expression on the right side using—that's right—the order of operations.

Replacement Values

To **substitute** a value into a formula means to replace the variable with a number. This number is called the **replacement value.** Whatever operation affected the variable now affects—and is applied to—the replacement value.

Example 1

What is the Fahrenheit temperature equivalent of 10° C?

Procedure Use the formula $F = 9 \cdot \frac{C}{5} + 32$ and substitute 10 for C.

Answer

$F = 9 \cdot \frac{10}{5} + 32$ Apply the order of operations to complete the evaluation. Because the division bar is a grouping symbol, divide first.

$F = 9 \cdot 2 + 32$

$F = 18 + 32$

$F = 50$

This means that 10° C is equivalent to 50° F.

If we know the Fahrenheit temperature instead and we want to know the equivalent Celsius value, we use a different form of the same formula:

$$C = 5 \cdot \frac{F - 32}{9}$$

Notice that the numbers and variables are the same as in the Celsius-to-Fahrenheit formula, but they are in a different order.

Example 2

What is the Celsius temperature equivalent of 77° F?

Procedure Use the formula $C = 5 \cdot \frac{F - 32}{9}$ and substitute 77 for F.

Answer $C = 5 \cdot \frac{77 - 32}{9}$

$C = 5 \cdot \frac{45}{9}$ Apply the order of operations to complete the evaluation.

$C = 5 \cdot 5$

$C = 25$

This means that 77° F is equivalent to 25° C, a warm outdoor temperature.

You Try It 1 **Find the equivalent temperature in either Celsius or Fahrenheit. Use Examples 1 and 2 as guides. (Make sure you use the correct formula.)**

a) The temperature is 40° C. **b)** The temperature is 15° C.

c) The temperature is 86° F. **d)** The temperature is 5° F.

Many formulas contain more than one variable. In this case, we substitute for all variables—using their corresponding replacement values—at the same time.

Example 3

Evaluate the numerical value of each formula with the given replacement values.

a) $d = r \cdot t$ $r = 50$, $t = 2$

b) $z = \frac{x - m}{s}$ $x = 63$, $m = 77$, $s = 7$

c) $E = \frac{9 \cdot R}{I}$ $R = 10$, $I = 45$

d) $c = \sqrt{a^2 + b^2}$ $a = 3$, $b = 4$

Answer

a) $d = r \cdot t$

$d = (50) \cdot (2)$

$d = 100$

b) $z = \frac{x - m}{s}$

$z = \frac{63 - 77}{7}$

$z = \frac{-14}{7} = -2$

c) $E = \frac{9 \cdot R}{I}$

$E = \frac{9 \cdot 10}{45}$

$E = \frac{90}{45}$

$E = 2$

d) $c = \sqrt{a^2 + b^2}$

$c = \sqrt{3^2 + 4^2}$

$c = \sqrt{9 + 16}$

$c = \sqrt{25} = 5$

You Try It 2 **Evaluate the numerical value of each formula with the given replacement values. Use Example 3 as a guide.**

a) $P = 2 \cdot L + 2 \cdot W$ $L = 15$, $W = 8$

b) $c = \sqrt{a^2 + b^2}$ $a = 8$, $b = 6$

c) $A = \dfrac{a + b + c}{3}$ $a = 21$, $b = 18$, $c = 27$

d) $m = \dfrac{y - q}{x - p}$ $y = 9$, $q = 3$, $x = 2$, $p = 5$

Where do we get all of these formulas, anyway? In Example 3, we saw four formulas from a variety of disciplines and interests.

a) $d = r \cdot t$ is from physics: Distance = Rate · Time.

b) $z = \frac{x - m}{s}$ is from statistics: a conversion formula.

c) $E = \frac{9 \cdot R}{I}$ is from baseball: a pitcher's earned run average.
R = the number of runs allowed; I = the number of innings pitched.

d) $c = \sqrt{a^2 + b^2}$ is from geometry: the Pythagorean theorem. c is the length of the longest side of a right triangle, and a and b are the lengths of the other two sides.

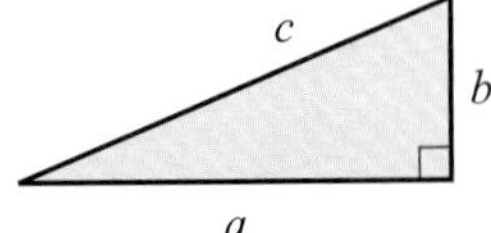

The Distance, Rate, and Time Formula

The formula $d = r \cdot t$, read as "**d**istance equals **r**ate times **t**ime," is used in any problem featuring motion—by car, by plane, by train, by bicycle, or by foot. In each case, a certain *distance* is traveled at a certain *rate* of speed for a certain period of *time*.

A common use of this formula is related to driving a car where the unit of measure for distance is *miles*, for time is *hours*, and for rate is *miles per hour*. The formula is not restricted to those measures, though.

For example, we might want to measure the rate an ant travels in terms of *centimeters per second* or in *feet per minute*. What must be present is a measure of length (miles, feet, centimeters) and of time (seconds, minutes, hours).

Imagine, for example, that you are driving a car at a rate of 60 miles per hour (mph). If you travel for 1 hour, you will have gone 60 miles. Traveling for 2 hours will take you 120 miles.

Using the formula $\boldsymbol{d = r \cdot t}$, we get

a) $r = 60$ mph, $t = 1$ hour
$d = 60 \cdot 1 = 60$ miles

b) $r = 60$ mph, $t = 2$ hours
$d = 60 \cdot 2 = 120$ miles

We don't need to represent the units of measure (mph and hours) throughout the process, but we must represent a unit of measure (miles) at the end.

Example 4

Use the distance formula to determine the distance Janelle traveled at the rate and time shown. Write a sentence indicating the distance.

a) rate = 18 mph, time = 4 hours

b) rate = 45 mph, time = 3 hours

Procedure Use $d = r \cdot t$

Answer **a)** $d = 18 \cdot 4 = 72$ miles

Janelle traveled 72 miles.

b) $d = 45 \cdot 3 = 135$ miles

Janelle traveled 135 miles.

We can use the distance formula to find the distance when the rate and time are known (as in Example 4). However, if we know how far Janelle traveled (say, 310 miles) and for how long she traveled (say, 5 hours), we can use a slight variation of the distance formula to determine her average rate of speed for the trip.

$$r = \frac{d}{t}$$

$$r = \frac{310 \text{ miles}}{5 \text{ hours}}$$

$$r = 62 \text{ miles per hour}$$

$$\text{or} \quad r = 62 \text{ mph}$$

$$\begin{array}{r} 62 \\ 5\overline{)310} \\ -30 \\ \hline 10 \\ -10 \\ \hline 0 \end{array}$$

The formula for rate makes sense in this way: **per** means "divided by." So, "miles *per* hour" suggests "miles *divided by* hours." This is what we create when we use distance (miles) ÷ time (hours).

If, instead, we know how far it is to get from Des Moines, Iowa, to Detroit, Michigan (585 miles), and we know how fast we will go (65 miles per hour), how much *time* will it take to get there? The distance formula now looks like this:

$$t = \frac{d}{r}$$

$$t = \frac{585 \text{ miles}}{65 \text{ miles per hour}}$$

$$t = 9 \text{ hours}$$

$$\begin{array}{r} 9 \\ 65\overline{)585} \\ -585 \\ \hline 0 \end{array}$$

Notice that the unit is *hours*, a measure of time.

Example 5

Use one of the distance formulas to determine the missing value, which will be rate, distance, or time.

a) Rate = 65 mph, time = 4 hr

b) Distance = 48 cm, time = 12 sec

c) Distance = 80 ft, rate = 5 ft per minute

d) Distance = 738 mi, time = 6 hr

Procedure Use the formula for the measure that is not known: $d = r \cdot t$, $r = \frac{d}{t}$, or $t = \frac{d}{r}$.

Answer

a) We know rate and time. We don't know distance; so, we'll use $d = r \cdot t$.

$$d = r \cdot t = 65 \cdot 4 = 260 \text{ miles}$$

b) We know distance and time. We don't know the rate; so, we'll use $r = \frac{d}{t}$. Notice that the units are in centimeters and seconds; so, the rate will be centimeters per second.

$$r = \frac{d}{t} = \frac{48}{12} = 4 \text{ centimeters per second}$$

c) We know distance and rate. We don't know the time; so, we'll use $t = \frac{d}{r}$.

$$t = \frac{d}{r} = \frac{80}{5} = 16 \text{ minutes}$$

d) We know distance and time. We don't know rate; so, we'll use $r = \frac{d}{t}$.

$$r = \frac{d}{t} = \frac{738}{6} = 123 \text{ mph}$$

We say that r represents an *average* rate of speed because we don't consider the speed every second of every minute of the trip. We don't see the slowing down due to heavy traffic, the stopping at a traffic signal, or the accelerating when traffic is lighter. Instead, the formula $r = \frac{d}{t}$ asks you to consider the rate of speed *over the entire trip* (or over a certain portion of the trip).

You Try It 3 **Ben drove from San Diego to Los Angeles, a distance of 147 miles. The traffic was so heavy that it took him 3 hours to get there. What was his average rate (speed) for the trip?**

Sentence ______________________________

You Try It 4 **Veronica was piloting a plane from San Francisco to New Orleans. She averaged 240 miles per hour while traveling the 1,920 miles. How many hours did the flight take?**

Sentence ______________________________

You Try It 5 **Banjo the beagle loves to chase tennis balls. While Banjo and his owner were playing on an empty football field one day, Banjo's owner timed Banjo while he was chasing balls. On one chase, Banjo ran 162 feet in 9 seconds. What was Banjo's rate (in feet per second) on that run?**

Sentence ______________________________

You Try It 6 **A jet is averaging 385 miles per hour. How far will it travel in 6 hours?**

Sentence ______________________________

Answers: You Try It and Think About It

You Try It: **1. a)** 104° F **b)** 59° F **c)** 30° C **d)** −15° C **2. a)** $P = 46$ **b)** $c = 10$ **c)** $A = 22$ **d)** $m = -2$ **3.** Ben's average rate for the trip was 49 miles per hour. **4.** The flight took 8 hours. **5.** Banjo ran 18 feet per second on that run. **6.** The jet will travel 2,310 miles.

Think About It: No Think About Its in this section.

Section 2.6 Exercises

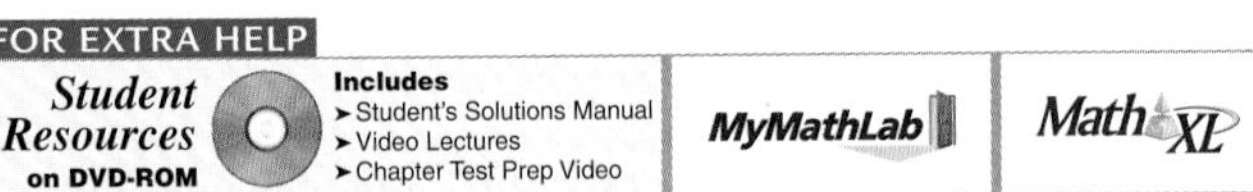

Think Again

1. As presented in this section, the formula for Celsius is $C = 5 \cdot \dfrac{F - 32}{9}$. Which of these two formulas is the same as that one: $C = \frac{5}{9} \cdot (F - 32)$ or $C = 5 \cdot (F - 32) \div 9$? Explain your answer or show an example that supports your answer.
   They are both variations of the same formula. The order of operations groups the quantity in the numerator $(F - 32)$, 5 multiplies this quantity, and it is divided by 9.

2. The following information about a trip is known: The average rate of speed was 55 miles per hour, and the time it took to complete the trip was 120 minutes. How do we calculate the number of miles of the trip?
   First, we must change 120 minutes to 2 hours so that the units of measure are consistent. Second, we can use the formula $d = r \cdot t$ to calculate the distance.

Focus Exercises

Find the equivalent temperature in either Celsius or Fahrenheit.

$$F = 9 \cdot \frac{C}{5} + 32 \quad C = 5 \cdot \frac{F - 32}{9}$$

3. The temperature is 5° C.
   41° F
4. The temperature is 30° C.
   86° F
5. The temperature is 50° F.
   10° C
6. The temperature is 95° F.
   35° C

Evaluate the numerical value of each formula with the given replacement values.

7. $z = \frac{x - m}{s}$ $x = 25$, $m = 16$, $s = 3$
   $z = 3$
8. $z = \frac{x - m}{s}$ $x = 27$, $m = 42$, $s = 5$
   $z = -3$
9. $A = \frac{a + b + c}{3}$ $a = 13$, $b = 41$, $c = 33$
   $A = 29$
10. $A = \frac{a + b + c}{3}$ $a = -8$, $b = 2$, $c = -6$
    $A = -4$
11. $a = \sqrt{c^2 - b^2}$ $c = 10$, $b = 6$
    $a = 8$
12. $a = \sqrt{c^2 - b^2}$ $c = 13$, $b = 12$
    $a = 5$
13. $A = \frac{b \cdot h}{2}$ $b = 5$, $h = 8$
    $A = 20$
14. $A = \frac{b \cdot h}{2}$ $b = 14$, $h = 7$
    $A = 49$
15. $P = 2 \cdot L + 2 \cdot W$ $L = 13$, $W = 8$
    $P = 42$
16. $P = 2 \cdot L + 2 \cdot W$ $L = 41$, $W = 26$
    $P = 134$
17. $r = \frac{d}{t}$ $d = 108$, $t = 9$
    $r = 12$
18. $r = \frac{d}{t}$ $d = 325$, $t = 5$
    $r = 65$
19. $A = \frac{h \cdot (b + c)}{2}$ $h = 3$, $b = 5$, $c = 7$
    $A = 18$
20. $A = \frac{h \cdot (b + c)}{2}$ $h = 4$, $b = 9$, $c = 6$
    $A = 30$
21. $W = \frac{A}{L}$ $A = 120$, $L = 15$
    $W = 8$
22. $W = \frac{A}{L}$ $A = 252$, $L = 9$
    $W = 28$
23. $C = \frac{2 \cdot W}{E^2}$ $W = 18$, $E = 3$
    $C = 4$
24. $C = \frac{2 \cdot W}{E^2}$ $W = 26$, $E = 2$
    $C = 13$
25. $P = 2 \cdot (L + W)$ $L = 9$, $W = 6$
    $P = 30$
26. $P = 2 \cdot (L + W)$ $L = 15$, $W = 22$
    $P = 74$
27. $I = \frac{A}{T - B}$ $A = 36$, $T = 11$, $B = 8$
    $I = 12$
28. $I = \frac{A}{T - B}$ $A = 60$, $T = 10$, $B = 6$
    $I = 15$
29. $m = \frac{y - q}{x - p}$ $y = 9$, $q = 3$, $x = 2$, $p = 5$
    $m = -2$
30. $m = \frac{y - q}{x - p}$ $y = -2$, $q = 10$, $x = 1$, $p = 5$
    $m = 3$

31. $m = \dfrac{y - q}{x - p}$ $y = 5$, $q = -7$, $x = -1$, $p = 2$

$m = -4$

32. $m = \dfrac{y - q}{x - p}$ $y = -6$, $q = 4$, $x = 3$, $p = -2$

$m = -2$

33. $R = \dfrac{E \cdot I}{9}$ $E = 3$, $I = 24$

$R = 8$

34. $R = \dfrac{E \cdot I}{9}$ $E = 4$, $I = 27$

$R = 12$

Use one of the distance formulas to answer the following questions.

$$d = r \cdot t, \quad r = \frac{d}{t}, \quad \text{or } t = \frac{d}{r}$$

35. Mai traveled from Chicago to Cleveland by car. She was able to make the 335 mile trip in 5 hours. What was Mai's average rate of speed?
Mai's average rate of speed was 67 miles per hour.

36. If Reggie averages 65 miles per hour on his trip, how far will he travel in 8 hours?
Reggie will travel 520 miles.

37. If it took Bertie 5 hours to drive her car 295 miles, what was her average rate of speed for the trip?
Bertie's average rate of speed was 59 miles per hour.

38. Jorge works for a cross-town courier service in Atlanta delivering packages and mail throughout the city. One day the freeways were particularly crowded, and it took Jorge 3 hours to drive 78 miles to make a delivery. What was Jorge's average rate of speed?
Jorge's average rate of speed was 26 miles per hour.

39. Hank, a truck driver, is trying to figure out how long it will take him to get from Baltimore, Maryland, to Worcester, Massachusetts. The whole trip is about 495 miles, and the load he is carrying allows him to average about 55 miles per hour. How many hours should it take Hank to complete the trip?
It should take Hank 9 hours to complete the trip.

40. Luisa is planning to fly her small plane from Kansas City, Missouri, to Roanoke, Virginia. The trip is 925 miles, and her plane averages 185 miles per hour. How many hours should it take Luisa to complete the flight?
It should take Luisa 5 hours to complete the flight.

41. Padam plans to drive from San Diego to San Francisco, 512 miles in all. What will Padam's average rate of speed need to be if he wants to complete the trip in 8 hours (not counting rest stops)?
Padam's average rate of speed will need to be 64 miles per hour.

42. Michael drives a bus for a touring company. This weekend he is taking the Riverside Youth Group on a tour to Phoenix. The distance between the cities is about 390 miles. Michael's bus will average 65 miles per hour. How many hours will it take the group to get to Phoenix?
It will take 6 hours for the group to get to Phoenix.

Think Outside the Box

43. Last summer, Tim moved from Anchorage, Alaska, to Corona, California. One day in March, the high temperature in Corona was 86°F and the high temperature in Anchorage was −30° C. On that day, what was the difference between the two cities' high temperatures

a) in Fahrenheit degrees?
108° F

b) in Celsius degrees?
60° C

44. Jackleen is a driver for FedEx in rural Kansas. One day she made deliveries to people in four small towns. From her home base in Hutchinson, she drove 12 miles to Nickerson in 15 minutes. It took her 16 minutes to drive the 11 miles to Sterling and another 14 minutes to drive the 9 miles to Lyons. From Lyons, she drove the 30 miles to McPherson in 39 minutes; and the 28 miles back to Hutchinson took 36 minutes. When her total trip distance and time are considered, what was Jackleen's average rate of speed?
Jackleen's average rate of speed was 45 miles per hour.

SECTION 2.7 Algebraic Expressions: Combining Like Terms

OBJECTIVES

In this section, you will learn to
- Define *numerical coefficient* and *term.*
- Combine like terms.

You Need to know

To successfully complete this section, you need to understand
- ☐ Adding signed numbers (2.2)
- ☐ Subtracting signed numbers (2.3)

Introduction

In algebra, when a number is multiplied by a variable, such as $5 \cdot y$, the number, 5, is called a **numerical coefficient** and the product of a coefficient and a variable creates a **term.** In the term $5y$, the numerical coefficient is 5.

In a term, the variable may have an exponent. For example, $8x^3$ is also a term. Actually, every variable has both a numerical coefficient and an exponent even if we don't see them.

Consider, for example, the simple term x. It *appears* that this term has no coefficient and no exponent. Actually, it has both. In fact,

- x is the same as $1x$ or $1 \cdot x$.
- It's also the same as x^1 (read "x to the first power").
- It can even be thought of as $1x^1$.

This may look as if we're getting carried away, but the 1's that you see here are very important numbers. In both cases, they are referred to as "the invisible 1." Although it may sound a little like the twilight zone, the "invisible 1" will rush to our aid—and make itself visible—in many situations.

Also, a term can be just a number with no variable, such as 6 and -5. In this case, as we saw in Chapter 1, the number is called a constant.

Example 1

Given the term, identify the coefficient, the variable, and the exponent of the variable.

Term	Coefficient	Variable	Exponent
$8x^2$	8	x	2
$-3a$	-3	a	1
m^4	1	m	4
$-y$	-1	y	1
7	7*	None	None

****7 is a term that is a constant term. It is not a coefficient as defined previously.***

You Try It 1

Given the term, identify the coefficient, the variable, and the exponent of the variable. Use Example 1 as a guide.

	Term	Coefficient	Variable	Exponent
a)	$3y^4$			
b)	-4			
c)	$-2c$			
d)	$-x^2$			
e)	d^5			
f)	w			

Algebraic Expressions

The terms in an algebraic expression are often—although not always—separated by a plus sign and/or a minus sign, which allows us to distinguish one term from another.

The expression $-2x^3 + 9x + 5$ has three terms, each separated by a plus sign: $-2x^3$, $9x$, and 5.

Also, many algebraic expressions include subtraction as one of the operations, such as $3x^2 - 8x - 2$. Recall from Section 2.3 that subtraction can be written as *adding the opposite*; so, this expression can be rewritten as $3x^2 + (-8x) + (-2)$. Now that it is written in addition only, the individual terms become clearer. The terms are: $3x^2$, $-8x$, and -2.

What we see is that the coefficient of the subtracted term takes on the minus sign as a negative. This means that the sign in front of a term—the plus sign or minus sign—belongs to the term and makes it positive or negative.

> The sign in front of a term belongs to that term.

Here are two expressions with their terms separated:

Expression:	$7x^2 - 2x - 5$	Expression:	$-4x^2 + x - 6$
Terms:	$+7x^2$ $-2x$ -5	Terms:	$-4x^2$ $+1x$ -6

Example 2

Identify the terms in each algebraic expression.

Expression	Terms (separated by commas)
a) $7x^2 + 6x + 1$	Three terms: $7x^2, 6x, 1$
b) $3x^2 - 5x - 2$	Three terms: $3x^2, -5x, -2$
c) $-7w + 8$	Two terms: $-7w, 8$
d) $3a^5 - 2a^3 - 4a + 9$	Four terms: $3a^5, -2a^3, -4a, 9$
e) $-15y$	One term: $-15y$

You Try It 2 **Identify the terms in each algebraic expression. Use Example 2 as a guide.**

Expression	Terms	Expression	Terms
a) $5x^2 - x + 3$	________	**b)** $c^3 - 6c^2 - 8c$	________
c) $-2x - 6y$	________	**d)** $-5x^3 + 6x^2 + x - 4$	________

Like Terms

Two things are *like* each other if they share the same characteristics. In algebra, two or more terms are **like terms** if they have exactly the same variable and exponent, even if the coefficients are different.

For example, $3x^2$ and $9x^2$ are like terms because they have *exactly* the same variable and exponent, x^2.

However, $4c$ and $-7p$ are *not* like terms because the variables are different. Similarly, $3x^2$ and $8x^4$ are *not* like terms because the variables have different exponents.

Furthermore, constants, which have no variable at all, are "like" all other constants. In the following example, carefully examine which terms are "like" and which are *not* "like." See if you can decide why some terms are not like others within the same expression.

Example 3

In each expression, identify the like terms.

	Expression		Like terms
a)	$7x - 6x + 1$	One pair of terms is **like**	$7x$ and $-6x$
b)	$5x - 2 + 9$	One pair of terms is **like**	-2 and 9
c)	$3a^2 + 2b^3 + 4a^2 + 9b^3$	Two pairs of terms are **like**	$3a^2$ and $4a^2$, $2b^3$ and $9b^3$
d)	$-15y + 6y - 3y$	All three terms are **like**	$-15y$ and $6y$ and $-3y$
e)	$4x + 6y - 7$	No terms are **like**	None
f)	$-2x^3 + 6x^2 + 9x$	No terms are **like**	None

Caution A different exponent on the same variable indicates a different, *unlike*, term.

You Try It 3 **In each expression, identify the like terms. Use Example 3 as a guide.**

	Expression	Terms		Expression	Terms
a)	$5x - x + 3$	________	**b)**	$x^2 - 6x - 8x^2$	________
c)	$4 - y - y$	________	**d)**	$9x^5 + 2x^5 - 6x^5$	________

Combining Like Terms

In algebra, you will be asked to simplify a great deal. Basically, an expression is simplified when it is written in the shortest and least complex form. When working with like terms, an expression is said to be simplified when the like terms are combined. To combine like terms means to find their sum or difference, depending on the operation between the terms.

We can combine like terms because the variable and exponent are the same: $2x + 3x = 5x$.

Notice that the result is "like" the original two terms. In this way, the only change is the coefficient. To get the coefficient of the result, 5, we add the coefficients of the like terms, 3 and 2.

Why does combining like terms work this way? Good question. It goes back to the abbreviation called multiplication. Remember that multiplication is an abbreviation for repeated addition. So, finding the sum of $4h$ and $3h$ is like this:

$4h$ means $h + h + h + h$. Likewise, $3h$ means $h + h + h$.

Therefore,	$4h$	+	$3h$
is the same as	$(h + h + h + h)$	+	$(h + h + h)$
or just	$h + h + h + h$	+	$h + h + h$,
which is abbreviated as	$7h$.		

This is also why we *can't* combine $2y + 5x$. It becomes

$$(y + y) + (x + x + x + x + x).$$

In this case, we can't say that we have 7 of anything that is the *same.*

The combining of like terms is not restricted to addition; we can also subtract like terms. In this way, it is like eliminating terms. Consider cutting 2 feet off a 5-foot board.

5 feet − 2 feet = 3 feet

Or five x's reduced by two x's leaves three x's: $5x - 2x = 3x$

Likewise, we can combine like terms with negative coefficients, as demonstrated in these problems.

Dollars: Chuck's account has $5 in it, but he writes a check for $8. What is the balance of Chuck's account?

5 dollars − 8 dollars = −3 dollars

Just as $5y - 8y = -3y$.

Temperature: At midnight, it was −12° outside. By 10 AM, the temperature had risen by 5°. What was the temperature at 10 AM?

−12 degrees + 5 degrees = −7 degrees

Just as $-12w + 5w = -7w$

> To **combine** like terms, we need only add or subtract their numerical coefficients. The resulting term is *like* the original terms.

Example 4

Simplify each expression by combining like terms.

a) $7x + 6x$ **b)** $2h + 3h + 6h$ **c)** $-2 + 9$

d) $-3a^2 + 4a^2$ **e)** $4y + 1y$ **f)** $4p + p$

g) $6w + (-8w)$ **h)** $-2k + (-3k)$ **i)** $9x^3 + 9x^2$

Procedure If the terms are like terms, find the sum of the coefficients.

Answer

a) $7x + 6x = 13x$ Simply add the coefficients together; the result is another like term.

b) $2h + 3h + 6h = 11h$ Yes, we can combine three terms if they are all like terms.

c) $-2 + 9 = 7$ All constants are *like* each other and can be combined.

d) $-3a^2 + 4a^2 = 1a^2$ This can be simplified to a^2.

e) $4y + 1y = 5y$ We can see that the coefficient of the second term, y, is 1.

f) $4p + p = 5p$ We can *write* the coefficient of the second term, p, as 1: $4p + 1p$.

g) $6w + (-8w) = -2w$ Simply add the coefficients together: $6 + (-8) = -2$. (Don't forget the w.)

h) $-2k + (-3k) = -5k$ Simply add the coefficients together: $-2 + (-3) = -5$. (Don't forget the k.)

i) $9x^3 + 9x^2$ These terms cannot combine because they are not like terms.

Think About It 1 In Example 4i, why aren't $9x^3$ and $9x^2$ like terms?

____

____

You Try It 4 **Simplify each expression by combining like terms. If the terms are not like terms, state so. Use Example 4 as a guide.**

a) $3y + 9y$ **b)** $-6c^2 + 7c^2$ **c)** $3p + (-4p)$ **d)** $5p^2 + 6p$

When subtracting terms, we can either subtract the coefficients directly or rewrite the subtraction as *adding the opposite*.

Example 5

Simplify each expression by combining like terms.

a) $7w - 2w$ **b)** $3x^2 - 2x^2$ **c)** $3c - 7c$ **d)** $5x - 6x$

Procedure You may subtract directly or change subtraction to *adding the opposite*.

Answer
a) $7w - 2w = 5w$
b) $3x^2 - 2x^2 = 1x^2$ or just x^2 — The coefficient 1 doesn't need to be written.
c) $3c - 7c = 3c + (-7c) = -4c$ — We can rewrite the subtraction as *adding the opposite*.
d) $5x - 6x = -1x$ — This could also be written as $-x$ instead of $-1x$.

You Try It 5 **Simplify each expression by combining like terms. If the terms are not like terms, state so. Use Example 5 as a guide.**

a) $11y - 9y$ b) $5w^3 - 8w^3$ c) $-2x - (-6x)$

Remember, if a term has no visible coefficient, the coefficient is 1 or -1.

Example 6

Simplify each expression by combining like terms.

a) $-9y + y$ b) $-c + 7c$ c) $-m - (-6m)$ d) $x - x$

Procedure You may subtract directly or change subtraction to *adding the opposite*.

Answer
a) $-9y + y = -9y + 1y = -8y$ — We can write the second coefficient as 1.
b) $-c + 7c = -1c + 7c = 6c$ — We can write the first coefficient as -1.
c) $-m - (-6m) = -1m + 6m = 5m$
d) $x - x = 1x + (-1x) = 0x = (0)$ — $0x$ means $0 \cdot x = 0$.

Think About It 2 In Example 6d, why does the answer have no variable?

____

____

You Try It 6 **Simplify each expression by combining like terms. If the terms are not like terms, state so. Use Example 6 as a guide.**

a) $-c - 8c$ b) $-4x - (-x)$ c) $-6m + 6m$

Expressions with Two Pairs of Like Terms

Because the sign in front of a term belongs to the term, we can move terms within an expression—using the Commutative Property—as long as the sign moves with each term.

For example, we can write the terms of the expression $5x + 9 - 2x$ in a different order so that the x-terms are together. Similarly, we can rearrange the terms of the expression $-4y + 8x - 3y$ so that the y-terms are together.

Expression:	$5x + 9 - 2x$	Expression:	$-4y + 8x - 3y$
Recognize the terms:	$+5x$ $+9$ $-2x$	Recognize the terms:	$-4y$ $+8x$ $-3y$
Rewrite the expression:	$5x - 2x + 9$	Rewrite the expression:	$8x - 4y - 3y$

If an expression has more than one pair of like terms, we can gather each pair separately and combine only those terms that are like each other.

In the expression $3x^2 + 1 + 5x^2 - 6$, we can see that $3x^2$ and $5x^2$ form a pair of like terms. Likewise, 1 and -6 form a pair of like terms. We can rearrange the expression, gathering the like terms together:

$3x^2 + 1 + 5x^2 - 6$ — Rearrange the terms so that like terms are together.
$= 3x^2 + 5x^2 + 1 - 6$ — Combine the like terms individually.
$= 8x^2 + (-5)$ — This can be written as $8x^2 - 5$. We cannot combine any further.
$= 8x^2 - 5$

Example 7

Combine like terms.

a) $5x - 2x + 2 + 8$ **b)** $2w^2 + 3w + 7w^2 - 5w$ **c)** $4q - 5 - 6q + 2$

Procedure First, rearrange the terms by gathering like terms together. Second, combine any like terms.

Answer **a)** $5x - 2x + 2 + 8$ — The like terms are already gathered together. Let's underline them to see the like terms.
$= \underline{5x - 2x} + \underline{2 + 8}$ — Combine like terms.
$= 3x + 10$ — We can't combine further.

b) $2w^2 + 3w + 7w^2 - 5w$ — Rearrange the terms to gather the like terms in pairs.
$= 2w^2 + 7w^2 + 3w - 5w$ — Combine the like terms. (Underlining is not necessary.)
$= 9w^2 + (-2w)$ — The final expression has two terms: $9w^2$ and $-2w$, which we can write as subtraction.
$= 9w^2 - 2w$

c) $4q - 5 - 6q + 2$ — Rearrange the terms to gather the like terms in pairs.
$= 4q - 6q - 5 + 2$ — We can write subtraction as addition.
$= 4q + (-6q) + (-5) + 2$
$= -2q + (-3)$
$= -2q - 3$

You Try It 7 **Simplify each expression by combining like terms. Use Example 7 as a guide.**

a) $y + 4y + 6y^2 - 2y^2$ **b)** $3x^2 - 4x^2 + 2x - 8x$

c) $2c^2 + 3 + c^2 - 8$ **d)** $3m - 1 + m - 6$

e) $5w - 2 + 1 + 2w$

Answers: You Try It and Think About It

You Try It:

1.

	Term	Coefficient	Variable	Exponent
a)	$3y^4$	3	y	4
b)	-4	-4	None	None
c)	$-2c$	-2	c	1
d)	$-x^2$	-1	x	2
e)	d^5	1	d	5
f)	w	1	w	1

2. a) $5x^2, -x, 3$ **b)** $c^3, -6c^2, -8c$ **c)** $-2x, -6y$ **d)** $-5x^3, 6x^2, x, -4$ **3. a)** $5x$ and $-x$ **b)** x^2 and $-8x^2$ **c)** $-y$ and $-y$ **d)** $9x^5$, $2x^5$, and $-6x^5$ **4. a)** $12y$ **b)** c^2 **c)** $-1p$ or $-p$ **d)** Not like terms **5. a)** $2y$ **b)** $-3w^3$ **c)** $4x$ **6. a)** $-9c$ **b)** $-3x$ **c)** 0 **7. a)** $5y + 4y^2$ **b)** $-1x^2 + (-6x)$, *or* $-x^2 - 6x$ **c)** $3c^2 + (-5)$, *or* $3c^2 - 5$ **d)** $4m + (-7)$, *or* $4m - 7$ **e)** $7w + (-1)$, *or* $7w - 1$

Think About It: **1.** $9x^3$ and $9x^2$ are not like terms because there are different exponents on the variable, x. **2.** The resulting like term is $0x$. But 0 times any number, even a variable, is 0; so, no variable is required.

Section 2.7 Exercises

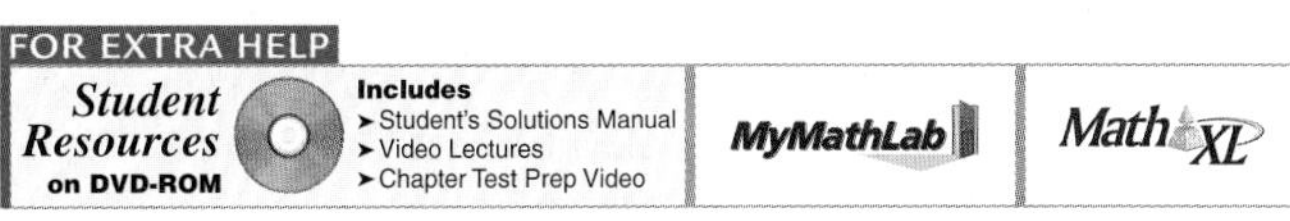

Think Again

1. Why aren't $9x^3$ and $9x^2$ like terms? *(Refer to Think About It 1)*
   $9x^3$ and $9x^2$ aren't like terms because the exponents on the variables are different.

2. What does *the sign in front of a term belongs to that term* mean?
   It means that a plus sign in front of the term makes the term positive and that a minus sign in front of the term makes the term negative.

3. Can $5x$ represent a negative number? Explain your answer or show an example that supports your answer.
   Yes. If x is negative, then $5x$ is negative.

4. Can $-6x$ represent a positive number? Explain your answer or show an example that supports your answer.
   Yes. If x is negative, then $-6x$ is positive.

Focus Exercises

Given the term, identify the coefficient, the variable, and the exponent of the variable.

	Term	Coefficient	Variable	Exponent
5.	$4m^2$	4	m	2
6.	$-7x^3$	-7	x	3
7.	$-3y$	-3	y	1
8.	$6c$	6	c	1
9.	9	9	None	None
10.	-12	-12	None	None
11.	y^3	1	y	3
12.	x^2	1	x	2
13.	$-d^5$	-1	d	5
14.	$-p^4$	-1	p	4
15.	$-n$	-1	n	1
16.	k	1	k	1

17. *Answer each question based on the expression* $3x^4 - 5x^3 + x - 9$.
   a) What is the first term?
   $3x^4$
   b) What is the second term?
   $-5x^3$
   c) What is the coefficient of the third term?
   1
   d) What is the constant term?
   -9
   e) What is the exponent of the first term?
   4
   f) What is the exponent of the third term?
   1

18. *Answer each question based on the expression* $7w^2 - 2 + 4w^3 - w$.
   a) What is the first term?
   $7w^2$
   b) What is the third term?
   $4w^3$
   c) What is the coefficient of the fourth term?
   -1
   d) What is the constant term?
   -2
   e) What is the exponent of the first term?
   2
   f) What is the exponent of the fourth term?
   1

19. *Answer each question based on the expression* $-8y + y^2 - 3y^4 + 1$.
   a) What is the first term?
   $-8y$
   b) What is the second term?
   y^2
   c) What is the coefficient of the third term?
   -3
   d) What is the constant term?
   1
   e) What is the exponent of the first term?
   1
   f) What is the exponent of the third term?
   4

20. *Answer each question based on the expression* $-y^8 + 5y^4 - y - 7$.
   a) What is the first term?
   $-y^8$
   b) What is the second term?
   $5y^4$
   c) What is the coefficient of the third term?
   -1
   d) What is the constant term?
   -7
   e) What is the exponent of the first term?
   8
   f) What is the exponent of the third term?
   1

Simplify. Combine like terms if possible.

21. $2x + 5x$
$7x$

22. $6y - 9y$
$-3y$

23. $4x^2 + 9x^2$
$13x^2$

24. $7y^3 - 6y^3$
y^3

25. $4a - 5a$
$-a$

26. $-6w^3 + w^3$
$-5w^3$

27. $-12x^2 - 10x$
Not like terms

28. $b^3 + 6b^3$
$7b^3$

29. $-9y^2 - (-11y^2)$
$2y^2$

30. $-5x - (-6x)$
x

31. $3m - 6m$
$-3m$

32. $-3y + (-8y)$
$-11y$

33. $-8r + 8r$
0

34. $4h^3 - 4h^3$
0

35. $-p^2 + 9p^2$
$8p^2$

36. $-d - 3d$
$-4d$

37. $k + (-7k)$
$-6k$

38. $2x^2 - 9x$
Not like terms

39. $v^2 - 2v^2$
$-v^2$

40. $-7a - (-a)$
$-6a$

41. $2s - (-4m)$
Not like terms

42. $3n - 4k$
Not like terms

43. $-3v + 3v$
0

44. $-v - v$
$-2v$

45. $5y + 7y - 2y$
$10y$

46. $4w + 3w - 2w$
$5w$

47. $5x - 3x - 7x$
$-5x$

48. $3p^2 - 9p^2 - p^2$
$-7p^2$

49. $-8c + 4c + 10c$
$6c$

50. $-6m + 10m - m$
$3m$

51. $-13x + 8x - 2x$
$-7x$

52. $-5w^3 + 3w^3 - w^3$
$-3w^3$

53. $-5y^2 - 6y^2 + 3y^2$
$-8y^2$

54. $-10c - 5c + c$
$-14c$

55. $-7p - 6p - 15p$
$-28p$

56. $-12b - 8b - b$
$-21b$

Simplify by combining pairs of like terms.

57. $4x + 7x + 8 - 5$
$11x + 3$

58. $6y + 5y + 7 - 6$
$11y + 1$

59. $3x - 4x + 4 - 6$
$-x - 2$

60. $5p - 6p + 3 - 9$
$-p - 6$

61. $-3a + 7a + 4 - 9$
$4a - 5$

62. $-5a + 6a + 2 - 7$
$a - 5$

63. $3w + 6w^2 + 2w - w^2$
$5w^2 + 5w$

64. $4m^2 + 5m + 3m^2 - m$
$7m^2 + 4m$

65. $-6 - 2x + 5 - x$
$-3x - 1$

66. $-x^2 - 6x + 4x^2 - x$
$3x^2 - 7x$

67. $-5c - 3b + c - b$
$-4b - 4c$

68. $-3x - 7y + x - y$
$-2x - 8y$

Think Outside the Box

Find the algebraic expression in terms of x that represents the perimeter of the given rectangle.

69.

$(10x + 7)$ yd

$(20 - 2x)$ yd

$16x + 54$ yards represents the perimeter.

70.

$(4 - 6x)$ m

$(3x + 7)$ m

$-6x + 22$ meters represents the perimeter.

71. Referring to the rectangle in Exercise 69, does the diagram make sense if $x = 1$? Explain your answer.
If $x = 1$, the length is 17 yards and the width is 18 yards. This means that the width is longer than the length.

72. Referring to the rectangle in Exercise 70, does the diagram make sense for any integer value of x? Explain your answer.
If $x = -1$, the length is 10 meters and the width is 4 meters.
If $x = -2$, the length is 16 meters and the width is 1 meter.

SECTION 2.8 Multiplying Algebraic Expressions

OBJECTIVES

In this section, you will learn to

- Multiply terms.
- Evaluate expressions using the Distributive Property.

You Need to Know

To successfully complete this section, you need to understand

- ☐ The Distributive Property (1.3)
- ☐ Multiplying signed numbers (2.4)
- ☐ The definition of *term* (2.7)

Introduction

This section pulls together ideas that are now familiar to you, such as

- Multiplication is an abbreviation for repeated addition.
- The product of a negative number and a positive number is negative.
- The sign in front of a term belongs to that term.
- Subtraction can be rewritten as addition: *adding the opposite*.
- The product of two negative numbers is positive.

Furthermore, algebra is just arithmetic using variables in place of numbers. So, the properties of arithmetic apply to algebra as well. Recall from Chapter 1 the following properties of numbers

1. The Associative Property of Multiplication (Section 1.1)

If the only operation is multiplication, we can change the grouping of the numbers without affecting the resulting product.

2. The Commutative Property of Multiplication (Section 1.1)

The order in which we multiply two numbers doesn't affect the resulting product.

3. The Distributive Property of Multiplication over Addition (Section 1.3)

We can distribute a multiplier (for example, 3) to a sum $(10 + 2)$ so that it multiplies both numbers in the sum.

$$
\begin{aligned}
3 \cdot (10 + 2) &= 3 \cdot 10 + 3 \cdot 2 \\
&= 30 + 6 \\
&= 36
\end{aligned}
$$

3 is a multiplier of both 10 and 2.
3 is being distributed to both the 10 and the 2.

We will explore the role that the Distributive Property plays in algebra a little later in this section. First, we'll apply the Associative Property to find the product of two terms.

Multiplying Terms

Recall from Section 2.7 that a term can be just a number (a constant) or it can be the product of a number and a variable (with an exponent).

Example 1 shows the product of a term that is a constant and a term with a variable.

Example 1

Find the product.

a) $3(5x)$ **b)** $-2(6y^3)$ **c)** $8 \cdot (-4p^2)$ **d)** $-5 \cdot (-9m)$

Procedure Using the Associative Property, we can regroup the multiplication, allowing us to multiply the numbers separately from the variable.

Answer

a) $3(5x)$
$= (3 \cdot 5) \cdot x$
$= 15 \cdot x$
Better written as $15x$

b) $-2(6y^3)$
$= (-2 \cdot 6) \cdot y^3$
$= -12 \cdot y^3$
Better written as $-12y^3$

c) $8 \cdot (-4p^2)$
$= 8 \cdot (-4) \cdot p^2$
$= -32 \cdot p^2$
Better written as $-32p^2$

d) $-5 \cdot (-9m)$
$= (-5) \cdot (-9) \cdot m$
$= +45 \cdot m$
Better written as $45m$

You Try It 1 **Find the product. Use Example 1 as a guide.**

a) $(2)(9c)$ **b)** $-7 \cdot (3x^2)$ **c)** $(-5)(-2a^5)$ **d)** $4 \cdot (-3k)$

When multiplying terms, we multiply the coefficients and include the variable factor in the end result. In other words, we can multiply in just one step.

a) $(-10)(6x^4) = -60x^4$ **b)** $-5 \cdot (-4y) = 20y$

You Try It 2 **Find the product. Try to do these in just one step.**

a) $7 \cdot (3c)$ **b)** $-9 \cdot (-4p)$ **c)** $6(-4m^5)$ **d)** $-1(8y)$

Multiplying Algebraic Expressions Using the Distributive Property

In algebra, the Distributive Property works the same as in arithmetic. The main difference is that we have terms with variables instead of just numbers.

For example, $3(x + y)$ and $3(-2x + 5y)$

$$3 \cdot (x + y) = 3 \cdot x + 3 \cdot y = 3x + 3y$$

$$3 \cdot (-2x + 5y) = 3 \cdot (-2x) + 3 \cdot (5y) = -6x + 15y$$

Here is the general form of the Distributive Property of Multiplication over Addition.

The Distributive Property of Multiplication over Addition

$$b \cdot (c + d) = b \cdot c + b \cdot d$$

Here's why the Distributive Property works. Consider the expression $3 \cdot (x + y)$.

Because this is multiplication and multiplication is an abbreviation for repeated addition, this means three $(x + y)$'s.

This can be written as $= (x + y) + (x + y) + (x + y)$.

The Associative Property allows us to write that expression without parentheses. $= x + y + x + y + x + y$

We can then use the Commutative Property to reorganize the terms. $= x + x + x + y + y + y$

Let's group the x's separately from the y's. $= (x + x + x) + (y + y + y)$

Each grouping has *three* of something. $=$ three x's $+$ three y's

$= 3 \cdot x + 3 \cdot y$

So, from the beginning: $3 \cdot (x + y) = 3 \cdot x + 3 \cdot y$.

Example 2

Use the Distributive Property to rewrite $4 \cdot (5y + 6)$.

Answer $4 \cdot (5y + 6)$ — 4 is the multiplier. Distribute the "4 times" to each term.

$= 4 \cdot 5y + 4 \cdot 6$ ← Writing this step isn't necessary, but it shows the multiplication.

$= 20y + 24$

You Try It 3 **Use the Distributive Property to rewrite each expression. Use Example 2 as a guide.**

a) $3 \cdot (8x + 2)$ **b)** $6 \cdot (5p + 1)$ **c)** $4 \cdot (9k + 2)$

If the second term of a quantity is negative, the term is usually written as subtraction. For example, a quantity with terms $5x$ and -3 is written $(5x - 3)$.

If we want to distribute the multiplier 4 to the quantity $(5x - 3)$, the product, at first, looks like this: $4 \cdot (5x - 3)$. We can, however, think of the quantity $5x - 3$ as $5x + (-3)$, effectively changing the product to $4 \cdot (5x + (-3))$.

Let's look at it one step at a time:

$4 \cdot (5x - 3)$ — Rewrite subtraction as a sum: *adding the opposite.*

$= 4 \cdot (5x + (-3))$ — We can distribute the multiplier to the sum.

$= 4 \cdot 5x + 4 \cdot (-3)$ — $4 \cdot (-3) = -12$

$= 20x + (-12)$ — This can be rewritten as subtraction.

$= 20x - 12$

We don't need to go through all of that work if we keep in mind that *the sign in front of a number belongs to that number.*

$20x$, -12

$4 \cdot (5x - 3)$

$4 \cdot (5x - 3)$ Treat $5x$ as $+5x$ and *minus 3* as -3.

$= 20x - 12$ $4 \cdot (+5x) = +20x$; $4 \cdot (-3) = -12$, or *minus 12.*

This leads to a new form of the Distributive Property.

The Distributive Property of Multiplication over Subtraction

$$b \cdot (c - d) = b \cdot c - b \cdot d$$

Example 3

Use the Distributive Property to rewrite each expression.

a) $5 \cdot (7y - 3)$ **b)** $2 \cdot (-4w - 6)$

Answer **a)** $5 \cdot (7y - 3)$ — 5 is the multiplier. Distribute the "5 times" to both terms.

$= 5 \cdot 7y - 5 \cdot 3$ — This step isn't necessary, but it shows the multiplication.

$= 35y - 15$

b) $2 \cdot (-4w - 6)$ — The multiplier is 2.

$= 2 \cdot (-4w) - 2 \cdot 6$ — This step can be done in your head.

$= -8w - 12$ — You can distribute and multiply directly without writing down the multiplication step.

You Try It 4 **Use the Distributive Property to rewrite each expression. Use Example 3 as a guide.**

a) $6 \cdot (4y - 3)$ **b)** $7 \cdot (9m - 4)$ **c)** $8 \cdot (-2x - 6)$

Think About It 1 Will the distribution property work the same way if the quantity has three terms? You may use $2(3x^2 + 7x - 6)$ or your own example in your discussion.

____

____

The Distributive Property with a Negative Multiplier

We can use the Distributive Property even when the multiplier is negative. We must be more careful when multiplying each term by a negative number, and we need to remember that the sign in front of a number belongs to that number.

Consider, for example, the product of the multiplier -2 and the difference $(4x - 3)$. As a product, this looks like $-2 \cdot (4x - 3)$.

$-8x$ $+6$

$-2 \cdot (4x - 3)$

$-2 \cdot (4x - 3)$	Treat $4x$ as $+4x$ and -3 as -3.
$= -8x + 6$	$-2 \cdot (+4x) = -8x$; $-2 \cdot (-3) = +6$, or *plus 6*.

Example 4

Use the Distributive Property to rewrite each expression.

a) $-5 \cdot (7k + 3)$ **b)** $-2 \cdot (4m - 6)$ **c)** $-8 \cdot (-5x + 2)$ **d)** $-3 \cdot (-6w - 1)$

Answer

a) $-5 \cdot (7k + 3)$ $\quad -5 \cdot (+7k) = -35k$
$= -35k - 15$ $\quad -5 \cdot (+3) = -15$

b) $-2 \cdot (4m - 6)$ $\quad -2 \cdot (+4m) = -8m$
$= -8m + 12$ $\quad -2 \cdot (-6) = +12$

c) $-8 \cdot (-5x + 2)$ $\quad -8 \cdot (-5x) = +40x$
$= 40x - 16$ $\quad -8 \cdot (+2) = -16$

d) $-3 \cdot (-6w - 1)$ $\quad -3 \cdot (-6w) = +18w$
$= 18w + 3$ $\quad -3 \cdot (-1) = +3$

You Try It 5 **Use the Distributive Property to rewrite each expression. Use Example 4 as a guide.**

a) $-5 \cdot (6d + 3)$ **b)** $-2 \cdot (8h - 1)$ **c)** $-6 \cdot (-10v + 4)$ **d)** $-4 \cdot (-11k - 5)$

Answers: You Try It and Think About It

You Try It: **1. a)** $18c$ **b)** $-21x^2$ **c)** $10a^5$ **d)** $-12k$ **2. a)** $21c$ **b)** $36p$ **c)** $-24m^5$ **d)** $-8y$ **3. a)** $24x + 6$ **b)** $30p + 6$ **c)** $36k + 8$ **4. a)** $24y - 18$ **b)** $63m - 28$ **c)** $-16x - 48$ **5. a)** $-30d - 15$ **b)** $-16h + 2$ **c)** $60v - 24$ **d)** $44k + 20$

Think About It: **1.** Yes. Answers may vary.

Section 2.8 Exercises

FOR EXTRA HELP
Student Resources on DVD-ROM
Includes
➤ Student's Solutions Manual
➤ Video Lectures
➤ Chapter Test Prep Video
MyMathLab
MathXL

Think Again

1. Will the distribution property work the same way if the quantity has three terms? You may use $2(3x^2 + 7x - 6)$ or your own example in your discussion. *(Refer to Think About It 1)*
Yes. Answers will vary.

2. Explain to a classmate how to distribute. Write out the steps involved in multiplying $5(8y - 3)$.
Answers will vary.

3. Explain to a classmate why he or she should be careful in distributing when the multiplier is a negative number. You may use $-3(4x + 6)$ or your own example in your discussion.
Answers will vary.

4. What is the opposite of the expression $5x - 2$? Explain how you got your answer.
$-5x + 2$. Answers will vary. One possibility: To find the opposite of an expression, write the opposite of each term.

Focus Exercises

Find the product.

5. $4(5x)$
$20x$

6. $6(3y)$
$18y$

7. $-2 \cdot (-2y)$
$4y$

8. $-4 \cdot (-5w)$
$20w$

9. $7 \cdot (-4c^2)$
$-28c^2$

10. $9 \cdot (-2h^3)$
$-18h^3$

11. $-8(6a^4)$
$-48a^4$

12. $-5(9k^5)$
$-45k^5$

Rewrite each expression using the Distributive Property and simplify the result.

13. $2 \cdot (4w + 5)$
$8w + 10$

14. $3 \cdot (4y + 6)$
$12y + 18$

15. $4 \cdot (5y + 8)$
$20y + 32$

16. $5 \cdot (2x + 9)$
$10x + 45$

17. $7 \cdot (x^2 + 2)$
$7x^2 + 14$

18. $10 \cdot (w^2 + 6)$
$10w^2 + 60$

19. $2 \cdot (x - 8)$
$2x - 16$

20. $5 \cdot (y - 11)$
$5y - 55$

21. $7 \cdot (2p - 3)$
$14p - 21$

22. $6 \cdot (10k - 5)$
$60k - 30$

23. $8 \cdot (4x^2 - 1)$
$32x^2 - 8$

24. $12 \cdot (3c^2 - 1)$
$36c^2 - 12$

25. $2 \cdot (-4c + 3)$
$-8c + 6$

26. $6 \cdot (-2m + 5)$
$-12m + 30$

27. $5 \cdot (-3x^2 - 1)$
$-15x^2 - 5$

28. $9 \cdot (-2w^2 - 1)$
$-18w^2 - 9$

29. $-5 \cdot (2y + 7)$
$-10y - 35$

30. $-6 \cdot (3x + 2)$
$-18x - 12$

31. $-4 \cdot (9x - 3)$
$-36x + 12$

32. $-3 \cdot (2y - 5)$
$-6y + 15$

33. $-4 \cdot (1 - 2b^2)$
$-4 + 8b^2$

34. $-5 \cdot (1 - 3p^4)$
$-5 + 15p^4$

35. $-6 \cdot (-2x + 5)$
$12x - 30$

36. $-2 \cdot (-2x + 9)$
$4x - 18$

37. $-2 \cdot (2y + 2)$
$-4y - 4$

38. $-4 \cdot (-4x - 4)$
$16x + 16$

39. $-1 \cdot (x + 3)$
$-x - 3$

40. $-7 \cdot (-5x^2 - 1)$
$35x^2 + 7$

41. $-8 \cdot (-6w^2 - 1)$
$48w^2 + 8$

42. $-1 \cdot (8 - 4y^3)$
$-8 + 4y^3$

43. $-1 \cdot (m + 8)$
$-m - 8$

44. $-1 \cdot (3g - 4)$
$-3g + 4$

45. $-1 \cdot (7p - 7)$
$-7p + 7$

46. $-1 \cdot (-3m - 2)$
$3m + 2$

47. $-1 \cdot (-9w - 12)$
$9w + 12$

48. $-1 \cdot (5 - 3x^2)$
$-5 + 3x^2$

49. $1 \cdot (c + 6)$
$c + 6$

50. $1 \cdot (8 + x^2)$
$8 + x^2$

51. $1 \cdot (w^2 - 12)$
$w^2 - 12$

52. $1 \cdot (q - 3)$
$q - 3$

53. $1 \cdot (-h + 10)$
$-h + 10$

54. $1 \cdot (-n + 13)$
$-n + 13$

55. $1 \cdot (-y - 4)$
$-y - 4$

56. $1 \cdot (-9 - a^2)$
$-9 - a^2$

57. $5 \cdot (k + 9)$
$5k + 45$

58. $6 \cdot (x^2 + 10)$
$6x^2 + 60$

59. $2 \cdot (p^2 - 1)$
$2p^2 - 2$

60. $4 \cdot (q - 8)$
$4q - 32$

61. $4 \cdot (-m + 8)$
$-4m + 32$

62. $7 \cdot (-c^2 + 9)$
$-7c^2 + 63$

63. $7 \cdot (-y - 2)$
$-7y - 14$

64. $9 \cdot (-p^2 - 1)$
$-9p^2 - 9$

65. $-10 \cdot (q^2 + 2)$
$-10q^2 - 20$

66. $-6 \cdot (x + 11)$
$-6x - 66$

67. $-2 \cdot (n - 5)$
$-2n + 10$

68. $-5 \cdot (y^2 - 4)$
$-5y^2 + 20$

69. $-8 \cdot (-v^2 + 1)$
$8v^2 - 8$

70. $-8 \cdot (-w + 7)$
$8w - 56$

71. $-11 \cdot (-x - 3)$
$11x + 33$

72. $-3 \cdot (-p - 5)$
$3p + 15$

Think Outside the Box

It is true that $x \cdot x = x^2$ and that $x \cdot x \cdot x = x^3$. Use that information to find each product.

73. $x^2 \cdot x^3$
x^5

74. $-2x \cdot (7x^2)$
$-14x^3$

75. $(-8x)^2$
$64x^2$

76. $-5x \cdot (-4x) \cdot (2x) \cdot (-x)$
$-40x^4$

Rewrite each expression using the Distributive Property and simplify the result.

77. $3x \cdot (5x + 4)$
$15x^2 + 12x$

78. $-7x \cdot (3x^2 - 2x)$
$-21x^3 + 14x^2$

79. $2x^2 \cdot (8x + 6)$
$16x^3 + 12x^2$

80. $-4x \cdot (x^2 + 2x - 9)$
$-4x^3 - 8x^2 + 36x$

81. Is this rectangle possible with the given side measures? Explain your answer.

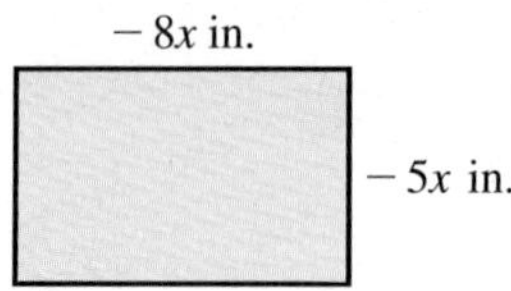

Yes, as long as the value of x is negative.

Chapter 2 Review

Section 2.1 Introduction to Algebra

Concept	Example
An **algebraic expression** is a combination of numbers and letters connected by the operations and grouping symbols. At its simplest, an algebraic expression can be a single variable, such as *y*, or a single number, such as *3*. Because variables represent numbers, an algebraic expression *expresses* a numerical value.	$w + 8 \qquad 5x + 6y^2 \qquad \dfrac{2x^2y + 4\sqrt{w}}{1 - 5z}$
When we are given the numerical value of a variable, that number is called a **replacement value.** We *replace,* or *substitute,* the variable with the number (the value).	Evaluate the expression $x + x \cdot y$ replacing $\boldsymbol{x}$ with 2 and $\boldsymbol{y}$ with 5. $2 + 2 \cdot 5 = 2 + 10 = 12$
We can represent numbers visually along a **number line.** Numbers less than 0 are **negative** numbers. Numbers greater than 0 are **positive** numbers. On a horizontal number line, positive numbers are to the right of 0 and negative numbers are to the left of 0. The number zero (0) is called the **origin,** which means "the beginning."	Negative numbers are to the left of 0. | Positive numbers are to the right of 0. 0
Numbers such as +2 and −2 are called **opposites,** numbers that are the same distance from 0 but on opposite sides of 0. The **integers** include each whole number and its opposite. Zero (0) is an integer but is neither positive nor negative.	Integers . . . , −3, −2, −1, 0, 1, 2, 3, . . .
Two numbers located at different points on the number line are compared to each other based on their location. The number on the left **is less than, <,** the number on the right; and the number on the right **is greater than, >,** the number on the left.	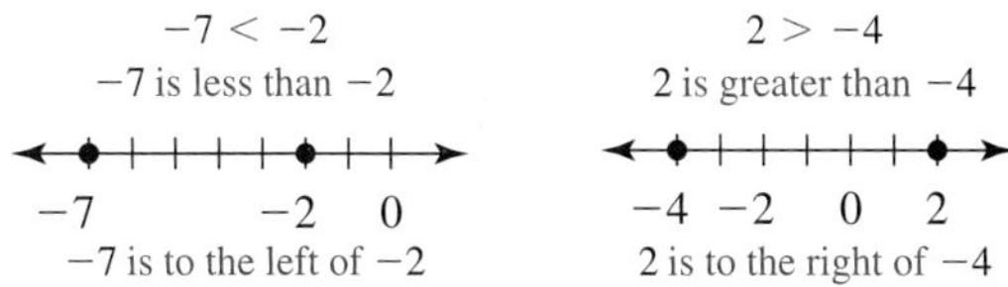
The **absolute value** of a number is its distance from 0 on the number line. For any nonzero number, whether it is on the left side of 0 or the right side of 0 doesn't matter; its absolute value is positive. We use *absolute value bars,* | and |, to express the absolute value of a number.	The *absolute value* of 3 is **3** (+3). The *absolute value* of −3 is also **3** (+3). $\lvert 3 \rvert = 3$ and $\lvert -3 \rvert = 3$ $\lvert 0 \rvert = 0$ 0 is *no* distance away from itself.
Nonzero numbers have a **numerical value** and a **direction.** The **numerical value** is the distance along the number line from 0. The numerical value is another name for the absolute value. The **direction** is left or right depending on the location on the number line when compared to 0 (zero) **1.** Negative numbers are to the left of 0, and positive numbers are to the right of 0. **2.** Zero has value but no direction.	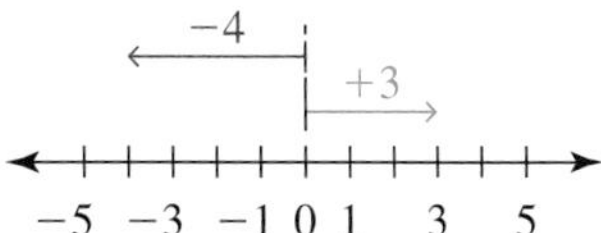 +3; the value is 3, and the direction is to the right. −4; the value is 4, and the direction is to the left.

Section 2.2 Adding Signed Numbers

Concept	Example
Each number can be represented on a number line by a vector. A **vector** has a linear *length* indicating its numerical value and an arrow indicating its direction. As a vector, 0 (zero) has *no length* and *no direction*.	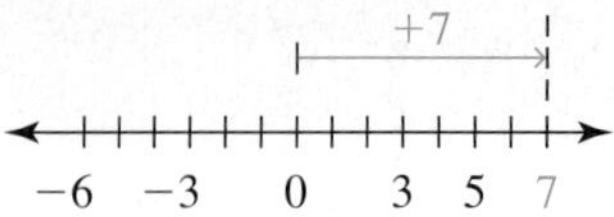
To add two numbers using vectors along the number line **1.** Draw a vector for the first number, starting from the origin, 0. **2.** Draw a vector for the second number that starts where the first vector stopped. The number at which the second vector stops is the sum.	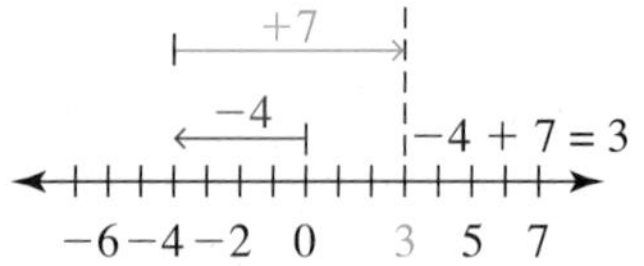
To add two numbers with the same sign **1.** Add their numerical values. **2.** Remember that the sign of the sum is the same as the sign of each number.	$-18 + (-25)$ **1.** The numerical values are 18 and 25: $18 + 25 = 43$ **2.** The sign for each number is *negative;* so, $-18 + (-25) = -43.$
To add two numbers with different signs **1.** Write the larger-valued number first if necessary. **2.** Subtract the numerical values: Larger value − Smaller value **3.** Remember that the sign of the sum is the same as the sign of the larger-valued number.	$18 + (-25)$ **1.** The sum can be written as $-25 + 18$. **2.** The numerical values are 25 and 18: $25 - 18 = 7$ **3.** The larger-valued number is *negative;* so, $-25 + 18 = -7.$
The sum of a number and its opposite is 0 $a + (-a) = 0$	$5 + (-5) = 0$ $-14 + 14 = 0$
If a sum contains more than two integers, use the Commutative and Associative Properties to rewrite the expression to add the positive numbers separately from the negative numbers. Then find the sum of those two results.	$8 + (-5) + (-9) + 4$ $= 8 + 4 + (-5) + (-9)$ $= +12 \quad + (-14)$ $= -2$

Section 2.3 Subtracting Signed Numbers

Concept	Example
The first three meanings of the dash **1.** Minus **2.** Negative **3.** "The opposite of"	**1.** $9 - 5 = 4$ **2.** -7 is "negative 7" **3.** -3 is "the opposite of" 3
$-(-5)$ is an example of a **double negative.**	$-(-5)$ means *the opposite of* -5, which is $+5$. $-(-5)$ can be rewritten as $+(+5)$, $+5$, or just 5.
Write subtraction as addition, as *adding the opposite;* then follow the rules for addition of signed numbers.	**a.** $10 - 17$ $= 10 + (-17)$ $= -7$ **b.** $-3 - (-8)$ $= -3 + (+8)$ $= +5$

Section 2.4 Multiplying and Dividing Signed Numbers

Concept	Example
The fourth meaning of the dash is "-1 times."	-6 is $-1 \cdot 6$.
Multiplication by -1 is the same as "the opposite of."	$(-1) \cdot 7 = -7$ (the opposite of 7). $(-1) \cdot (-4) = +4$ (the opposite of -4).
To multiply two signed numbers **1.** If the signs of the factors are **different,** the product will be **negative.** **2.** If the signs of the factors are the **same,** the product will be **positive.** **3.** If one factor is zero, the product will be 0.	**1.** $(6)(-3) = -18$ **2.** $(-5)(-4) = +20 = 20$ **3.** $(-7)(0) = 0$
When a product has more than two nonzero factors **1.** If there is an *odd* number of negative factors, the end product will be *negative.* **2.** If there is an *even* number of negative factors, the end product will be *positive.*	**1.** $(-6)(-3)(-2) = -36$ **2.** $(-5)(-4)(-3)(-2) = +120$ or 120
The rules for dividing two signed numbers are the same as the rules for multiplication **1.** If the signs of the numbers are **different,** the quotient will be **negative.** **2.** If the signs of the numbers are the **same,** the quotient will be **positive.** **3.** The divisor (denominator) can never be 0.	**1.** $-18 \div 6 = -3$ **2.** $-20 \div (-5) = +4 = 4$ **3.** $\frac{-5}{0}$ is undefined.

Section 2.5 The Order of Operations for Signed Numbers

Concept	Example
The order of operations **1.** Evaluate within all grouping symbols if there are any. **2.** Apply any exponents. **3.** Apply multiplication and division reading from left to right. **4.** Apply addition and subtraction reading from left to right.	Evaluate $(4 - 16) \div 2^2$ $= (-12) \div 2^2$ $= -12 \div 4$ $= -3$
If a negative number is the base of an exponent **1.** An odd exponent results in a negative number. $(-\text{base})^{\text{odd exponent}} = \text{negative number}$ **2.** An even exponent results in a positive number. $(-\text{base})^{\text{even exponent}} = \text{positive number}$	**1.** $(-2)^3 = -8$ **2.** $(-7)^2 = +49 = 49$
Because $3^2 = 9$ and $(-3)^2 = 9$, 9 has two square roots: one positive $(+3)$ and one negative (-3). When using the radical symbol, $\sqrt{9} = +3$ only. This positive square root is called the **principal square root.** To represent the negative square of 9, we must put a negative in front of (outside) the radical symbol.	Negative square root: $-\sqrt{9} = -3$
When an expression contains two sets of grouping symbols that are unrelated to each other, each quantity can be evaluated at the same time.	$(2 - 6) \cdot (-10 \div 2)$ $= (-4) \cdot (-5)$ $= +20 = 20$
The division bar is a grouping symbol. It groups the dividend (top) separately from the divisor (bottom), and each may be evaluated at the same time.	$\dfrac{-2 - 8}{6 \div 3}$ $= \dfrac{-10}{+2}$ $= -5$
Absolute value bars are grouping symbols, and we must evaluate within them first. However, the absolute value is also an operation; so, once the expression within is simplified, we can apply the absolute value to that number.	$\lvert -6 + 2 \rvert - \lvert -7 \rvert$ $= \lvert -4 \rvert - \lvert -7 \rvert$ $= 4 - 7$ $= -3$

Section 2.6 Formulas

Concept	Example
A **formula** is one variable written in terms of one or more other variables. To evaluate a formula, we replace variables with numbers—called **replacement values**—and evaluate according to the order of operations. **Temperature** conversion formulas $F = 9 \cdot \frac{C}{5} + 32$ $C = 5 \cdot \frac{(F - 32)}{9}$ $\begin{cases} F = \text{Fahrenheit degrees} \\ C = \text{Celcius degrees} \end{cases}$	Find the temperature in degrees Fahrenheit for 15° C. Find F when $C = 15$. Use the formula. $F = 9 \cdot \frac{C}{5} + 32$ $F = 9 \cdot \frac{15}{5} + 32$ Divide. $F = 9 \cdot (3) + 32$ Multiply. $F = 27 + 32$ Add. $F = 59$ 15° C is equivalent to 59° F.
Distance, rate, and **time** formulas **1.** $d = r \cdot t$ (Distance = Rate · Time) **2.** $r = \frac{d}{t}$ $\left(\text{Rate} = \frac{\text{Distance}}{\text{Time}}\right)$ **3.** $t = \frac{d}{r}$ $\left(\text{Time} = \frac{\text{Distance}}{\text{Rate}}\right)$	**1.** Find the distance traveled when the rate is 18 miles per hour and the time is 3 hours. Distance = 18 miles per hour · 3 hours = 54 miles **2.** Find the rate when the distance is 531 miles and the time is 9 hours. Rate = 531 miles ÷ 9 hours = 59 miles per hour **3.** Find the time when the distance is 385 miles and the rate is 55 miles per hours. Time = 385 miles ÷ 55 miles per hours = 7 hours

Section 2.7 Algebraic Expressions: Combining Like Terms

Concept	Example
In algebra, a **term** can be a constant or it can be the product of a coefficient and a variable. Also, the variable may have an exponent.	$7, -y, 6x^2$, and $-9w^4$
The **numerical coefficient,** or *coefficient*, of a term is the numerical factor of the term.	For $6x^2$, the numerical coefficient is 6. For $-9w^4$, the numerical coefficient is -9.
If a variable has no visible coefficient, the coefficient is 1 or -1. If the variable has no visible exponent, the exponent is 1.	The coefficient of x^2 is 1: $1x^2$. The coefficient of $-y^3$ is -1: $-1y^3$. The exponent of x in the term $4x$ is 1: $4x^1$.
In algebra, two or more terms are **like terms** if they have exactly the same variable and exponent, even if the coefficients are different. *Caution:* A different exponent on the same variable indicates a different term.	$3x^2$ and $9x^2$ are like terms. $5y$ and $5y^2$ are *not* like terms.

continued

Section 2.7 Algebraic Expressions: Combining Like Terms

Concept	Example
To combine like terms, add or subtract their numerical coefficients.	$-7y + 9y$ $= +2y$ $-4c + c$ $= -4c + 1c$ $= -3c$
The sign in front of a term belongs to that term. Because of the Associative and Commutative Properties, this means that we can move terms within an expression as long the sign moves with the term.	Rewrite the expression so that the x-terms are together; then combine like terms. $4x + 10 - 7x$ $= 4x - 7x + 10$ $= -3x + 10$

Section 2.8 Multiplying Algebraic Expressions

Concept	Example
When a constant is multiplied by a term with a variable, we multiply the constant and the coefficient. The variable and its exponent are otherwise unaffected.	$-2(4x) = -8x$ $3(-7y^2) = -21y^2$ $-9\cdot(-5m) = +45m = 45m$
The Distributive Property of Multiplication over Addition $b\cdot(c + d) = b\cdot c + b\cdot d$ b is called the multiplier.	$3\cdot(-2y + 1)$ $= -6y + 3$
The Distributive Property of Multiplication over Subtraction $b\cdot(c - d) = b\cdot c - b\cdot d$	$2\cdot(5x - 9)$ $= 10x - 18$
When the multiplier is negative, we need to remember that *the sign in front of a number belongs to that number.* This is especially true for terms within parentheses.	$-5\cdot(3p - 4)$ $= -15p + 20$

Chapter 2 Review Exercises

Chapter 2 Vocabulary

Fill in each of the following blank with the word that correctly completes the sentence.

1. A(n) algebraic expression is a combination of numbers and letters connected by operations and grouping symbols. (2.1)

2. When we are given the numerical value of a variable, that number is called a(n) replacement value. (2.1)

3. The numbers less than zero are called negative numbers. (2.1)

4. Whole numbers and their opposites are called integers. (2.1)

5. The sum of two negative numbers is always negative. (2.2)

6. In a term, the numerical factor is called the coefficient. (2.7)

True or false.

7. Zero has no value. False (2.1)

8. The sum of a positive and negative number is always negative. False (2.2)

9. The sum of a number and its opposite is always zero. True (2.2)

10. The product of a positive and negative number is always negative. True (2.4)

Section 2.1

Evaluate the following problems according to the replacement value given.

11. $5x + 9$; replace $\boldsymbol{x}$ with 3.
24

12. $p^2 - 3p$; replace $\boldsymbol{p}$ with 7.
28

13. $\frac{h + k}{2}$; replace $\boldsymbol{h}$ with 21 and $\boldsymbol{k}$ with 39.
30

14. $x^2 + xy$; replace $\boldsymbol{x}$ with 3 and $\boldsymbol{y}$ with 8.
33

Translate each English expression into an algebraic expression. Use any variable to represent the unknown number.

15. The difference of 8 and a number
$8 - x$

16. The sum of a number and 14
$x + 14$

Insert the correct symbol between each pair of numbers, either < (less than) or > (greater than).

17. 9 3
$>$

18. −6 0
$<$

19. −8 −3
$<$

20. 4 −5
$>$

Find the absolute value.

21. $|+9|$
9

22. $|-11|$
11

23. $|-0|$
0

24. $|21|$
21

Section 2.2

Find the sum.

25. $2 + (-8)$
-6

26. $1 + (-5)$
-4

27. $-4 + (-9)$
-13

28. $-3 + (-2)$
-5

29. $-3 + 3$
0

30. $-11 + (-9)$
-20

31. $-13 + 29$
16

32. $-19 + (-12)$
-31

33. $-16 + 87$
71

34. $-70 + 18$
-52

35. $-12 + 86$
74

36. $64 + (-91)$
-27

37. $7 + 4 + (-11) + (-16)$
-16

38. $6 + (-10) + 2 + 17 + (-10) + (-8)$
-3

Section 2.3

Evaluate each expression.

39. $4 - 6$
-2

40. $7 - 11$
-4

41. $0 - 7$
-7

42. $-7 - 5$
-12

43. $6 - (-3)$
9

44. $-6 - (-10)$
4

45. $-30 + 60$
30

46. $-26 + 13$
-13

47. $71 - 79$
-8

48. $82 - 27$
55

49. $-10 - (-10)$
0

50. $27 - (-64)$
91

51. $-9 + 7 - 4 + 8$
2

52. $-19 + 5 - (-2) - 4$
-16

For each application, write a numerical expression; then evaluate. Also write a sentence answering the question.

53. Find the difference in altitude between a seaside cliff 97 feet above sea level and an ocean floor 64 feet below sea level.
$97 - (-64) = 97 + (+64) = 161$
The difference in altitude between the seaside cliff and the ocean floor is 161 feet.

54. At 3 AM, the outside temperature was $-13°$. By 6 AM, the temperature had risen by $8°$. What was the temperature at 6 AM?
$-13 + 8 = -5$
The temperature at 6 AM was $-5°$.

55. Adele has only \$10 in her checking account, and she writes a check for \$17. What is Adele's new checkbook balance?
$10 - 17 = -7$
Adele's new checkbook balance is −\$7.

56. Adele is being foolish by writing checks when she doesn't have enough money. Her check register has a balance of −\$26 when she writes a check for \$15. What is Adele's checkbook balance now?
$-26 - 15 = -41$
Adele's checkbook balance is now −\$41.

Section 2.4

Evaluate.

57. $(-3)(-2)$
6

58. $-11\cdot(-9)$
99

59. $(11)(-7)$
-77

60. $4\cdot 10$
40

61. $(-2)(-2)$
4

62. $0\cdot(-4)$
0

63. $-2\cdot 3\cdot(-5)$
30

64. $8\cdot(-5)\cdot(-1)$
40

65. $6\cdot 2\cdot(-4)$
-48

66. $-1\cdot(-7)\cdot(-4)$
-28

67. $33\div(-3)$
-11

68. $-48\div(-8)$
6

69. $\dfrac{-42}{-7}$
6

70. $\dfrac{-45}{9}$
-5

Section 2.5

Evaluate.

71. $\sqrt{9}$
3

72. $-\sqrt{4}$
-2

73. $(-12)^1$
-12

74. $(-9)^2$
81

Evaluate each expression using the order of operations. Show all steps.

75. $(8-14)\div 3$
-2

76. $-36\div(-9)\cdot(-2)$
-8

77. $(7-9)^3$
-8

78. $-20\div(2-6)$
5

79. $(-18\div 3)\cdot(2-7)$
30

80. $-\sqrt{-3\cdot(-12)}$
-6

81. $|-10|-3$
7

82. $|1-7|+|9-(-2)|$
17

83. $\dfrac{-7-1}{2-4}$
4

84. $\dfrac{-12+2}{-9-(-4)}$
2

Section 2.6

Find the equivalent temperature in either Celsius or Fahrenheit. Use the formulas below as needed.

$$F = 9\cdot\frac{C}{5} + 32 \qquad C = 5\cdot\frac{F-32}{9}$$

85. The temperature is 100° C.
212° F

86. The temperature is 15° C.
59° F

87. The temperature is 122° F.
50° C

88. The temperature is 59° F.
15° C

Evaluate the numerical value of each formula with the given replacement values.

89. $A = \dfrac{a+b}{2}$ $\quad a = 77$, $b = 91$
$A = 84$

90. $W = \dfrac{A}{L}$ $\quad A = 192$, $L = 12$
$W = 16$

91. $A = \dfrac{h\cdot(b+c)}{2}$ $\quad h = 5$, $b = 6$, $c = 4$
$A = 25$

92. $z = \dfrac{x-m}{s}$ $\quad x = 53$, $m = 45$, $s = 4$
$z = 2$

93. $a = \sqrt{c^2 - b^2}$ $\quad c = 13$, $b = 12$
$a = 5$

94. $C = \dfrac{2\cdot W}{E^2}$ $\quad W = 12$, $E = 2$
$C = 6$

Use one of the distance formulas to answer the following questions.

$$\text{Rate} = \frac{\text{Distance}}{\text{Time}} \qquad \text{Time} = \frac{\text{Distance}}{\text{Rate}} \qquad \text{Distance} = \text{Rate} \cdot \text{Time}$$

95. It took Tracey 4 hours to ride her bike 52 miles to the beach. What was her average rate of speed?
Tracey's average rate of speed was 13 miles per hour.

96. When Timara pilots her plane, she usually averages 145 miles per hour. How far can she fly in 6 hours?
Timara can fly 870 miles.

97. Charles is planning to drive his car 495 miles from Charleston, West Virginia, to Charleston, South Carolina, to visit his sisters, Virginia and Caroline. If he is able to average 55 miles per hour, how many hours will it take him to get there?
It will take him 9 hours to get there.

98. Pepito entered his pet snail, Peetey, in a race. The straight racecourse was 87 cm long, and Peetey finished in 3 minutes. What was Peetey's average rate of speed?
Peetey's average rate of speed was 29 cm per minute.

Section 2.7

Answer each question based on the expression $2w - 8w^2 - w^3 + 4$.

99. What is the first term?
$2w$

100. What is the second term?
$-8w^2$

101. What is the coefficient of the third term?
-1

102. What is the constant term?
4

103. What is the exponent of the first term?
1

104. What is the exponent of the third term?
3

Simplify. Combine like terms if possible.

105. $9x + 3x$
$12x$

106. $2y - 8y$
$-6y$

107. $-5w^2 + w^2$
$-4w^2$

108. $b^2 + 5b^2$
$6b^2$

109. $-4x - (-5x)$
x

110. $-7r + 7r$
0

111. $9p - (-3k)$
Not like terms

112. $-12x + 7x - 9x$
$-14x$

113. $4x - 2x - 6x$
$-4x$

114. $-9y + 7y + 4 - 5$
$-2y - 1$

115. $-4c - 8b - c + b$
$-7b - 5c$

116. $-5 - 4x - x + 9$
$-5x + 4$

Section 2.8

Find the product.

117. $3(-4x^2)$
$-12x^2$

118. $-9(-9y)$
$81y$

119. $7(6p)$
$42p$

120. $-5(3a^3)$
$-15a^3$

Rewrite each expression using the Distributive Property.

121. $9 \cdot (y + 2)$
$9y + 18$

122. $6 \cdot (-2y + 3)$
$-12y + 18$

123. $5 \cdot (3h - 1)$
$15h - 5$

124. $-2 \cdot (y + 1)$
$-2y - 2$

125. $-3 \cdot (8y + 7)$
$24y - 21$

126. $-4 \cdot (2p - 7)$
$-8p + 28$

Chapter 2 Test

Translate each English expression into an algebraic expression.

1. The difference of 15 and a number
$15 - x$

2. The square root of a number
$\sqrt{x}$

Insert the correct symbol between each pair of numbers, either < (less than) or > (greater than).

3. $-2 \quad 6$
$<$

4. $0 \quad -5$
$>$

Evaluate the following expression according to the replacement value given.

5. $\dfrac{w + v}{v - 2w}$; replace w with 6, v with 10.
-8

Evaluate each expression.

6. $-5 + 19$
14

7. $18 + (-20)$
-2

8. $-97 + (-3)$
-100

9. $-3 - 7$
-10

10. $-8 - (-2)$
-6

11. $-61 - (-76)$
15

12. $12(-3)$
-36

13. $4 \cdot (-4)$
-16

14. $-25 \div 5$
-5

15. $28 \div (-4)$
-7

16. $-12 \div 0$
Undefined

17. $\dfrac{-30}{-6}$
5

18. $-\sqrt{49}$
-7

19. $(-1)^5$
-1

20. $(-4)^2$
16

21. $(-3)^3$
-27

22. $-5 + 7 - 6 - (-2)$
-2

23. $-3 \cdot (-4) \cdot (-5)(-2)$
120

For each question, write a numerical expression; then evaluate. Also write a sentence answering the question.

24. What is the difference in altitude between an undersea hilltop 162 feet below sea level and the ocean floor 458 feet below sea level?
$-162 - (-458) = -162 + (+458) = 296$
The difference in altitude between the undersea hilltop and the ocean floor is 296 feet.

25. In 2005, the highest temperature in Duluth, Minnesota, was 95° and the lowest temperature was −8°. What was the difference between the highest and lowest temperatures?
$95 - (-8) = 95 + (+8) = 103$
The difference between the highest and lowest temperatures was 103°.

Evaluate each expression using the order of operations. Show all steps.

26. $(-2)^2 - (3 + 5)$
-4

27. $|-6 + 3| - |-2|$
1

Find the equivalent temperature in Fahrenheit using this formula:

$$F = 9 \cdot \frac{C}{5} + 32$$

28. The temperature is 95° C.
203° F

Find the equivalent temperature in Celsius using this formula:

$$C = 5 \cdot \frac{F - 32}{9}$$

29. The temperature is 95° F.
35° C

Evaluate the numerical value of the formula with the given replacement values.

30. $A = \dfrac{h \cdot (b + c)}{2}$ $\quad h = 5, \; b = 6, \; c = 4$
$A = 25$

Use one of the distance formulas to answer the following question.

$$r = \frac{d}{t} \qquad t = \frac{d}{r} \qquad d = r \cdot t$$

31. If Rogelio averages 65 miles per hour riding his motorcycle the 455 miles from Albuquerque to Denver, how many hours will it take him to get there?
It will take him 7 hours to get there.

Simplify by combining like terms if possible.

32. $-p^2 + 8p^2$
$7p^2$

33. $-6a - (-a)$
$-5a$

34. $-9x - 6y + y - x$
$-10x - 5y$

Find the product.

35. $-4(-8k^4)$
$32k^4$

36. $-5(2y)$
$-10y$

Rewrite each expression using the Distributive Property.

37. $3(2w - 5)$
$6w - 15$

38. $-7(2c - 3)$
$-14c + 21$

Chapters 1 and 2 Cumulative Review

1. Write the value that 7 represents in 876,153.
70,000 (seventy thousand)

2. Write 500,026 in words.
Five hundred thousand, twenty-six

Round each number to the nearest hundred.

3. 548
500

4. 7,952
8,000

Round each number to the nearest thousand.

5. 8,489
8,000

6. 209,607
210,000

Which property is being demonstrated?

7. $43 \cdot 6 = 6 \cdot 43$
Commutative Property of Multiplication

8. $14 + 0 = 14$
Identity of Addition

9. $7 \cdot (5 + 8) = 7 \cdot 5 + 7 \cdot 8$
Distributive Property of Multiplication over Addition

10. $(8 + 5) + 9 = 8 + (5 + 9)$
Associative Property of Addition

Align each, then add.

11. 2,549 + 487
3,036

12. 1,908 + 93
2,001

Align each, then subtract.

13. 1,548 − 673
875

14. 10,000 − 571
9,429

15. Find the perimeter of this figure.

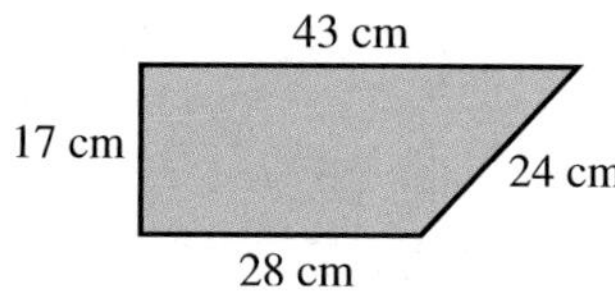

112 cm

16. Find the perimeter of this rectangle.

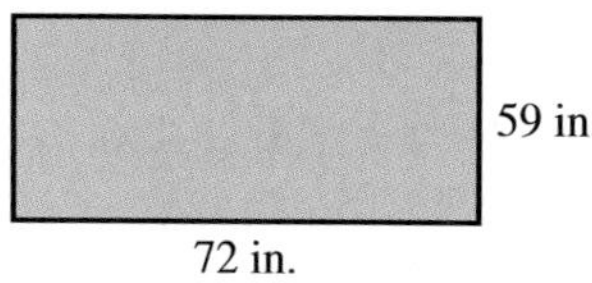

262 in

17. Find the area of this rectangle.

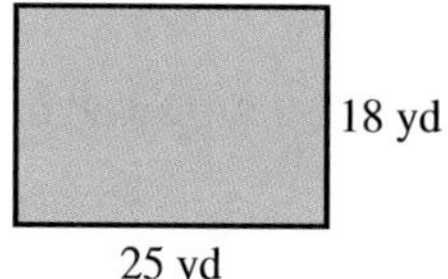

450 square yards

18. In 2000, during the presidential elections, Connecticut voters chose Al Gore—with 816,015 votes—over George W. Bush—with 561,094 votes. How many more votes did Gore receive than Bush? ***Source:* www.infoplease.com**
Gore received 254,921 more votes than Bush.

A group of people were surveyed about their favorite fast-food burger restaurant. The bar graph below shows the results of the survey.

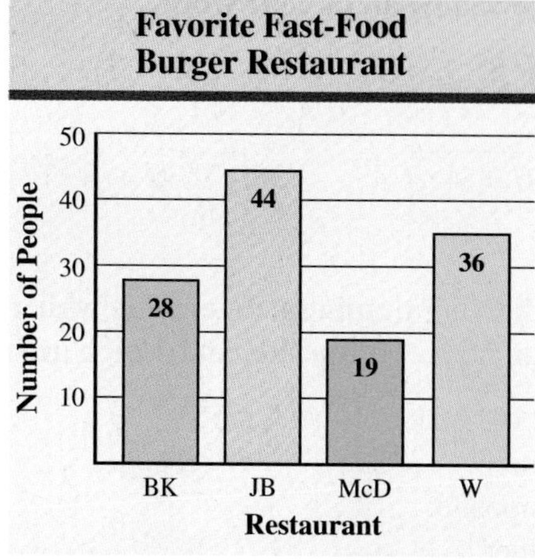

BK = Burger King JB = Jack in the Box
McD = McDonald's W = Wendy's

19. How many more people prefer Jack in the Box than McDonald's?
25 more people prefer Jack in the Box than McDonald's.

20. How many total people were surveyed?
127 total people were surveyed.

Multiply.

21. $80 \cdot 700$
56,000

22. $165 \cdot 28$
4,620

Divide.

23. $379 \div 8$
47 r 3

24. $1{,}456 \div 28$
52

Expand each notation and find its value.

25 9^3
$9 \cdot 9 \cdot 9 = 729$

26. 5^4
$5 \cdot 5 \cdot 5 \cdot 5 = 625$

Express each number as a power of 10.

27. 10,000
10^4

28. 1,000,000,000
10^9

Rewrite the number as a product of a whole number and a power of 10.

29. 60,000
$6 \cdot 10^4$

30. 5,200,000
$52 \cdot 10^5$

Evaluate the following square roots.

31. $\sqrt{64}$
8

32. $\sqrt{121}$
11

Evaluate each problem according to the order of operations. Show all of your work.

33. $36 \div 3^2 + 3$
7

34. $(36 \div 3)^2 + 3$
147

35. $\sqrt{36 + 4 \cdot 7}$
8

36. $\sqrt{36} + 4 \cdot 7$
34

Of the following numbers, determine which are prime, which are composite, and which are neither.

37. 5, 18, 31, 1, 43, 55
Prime: 5, 31, 43
Composite: 18, 55
Neither: 1

Of the first three prime numbers—2, 3, and 5—which are factors of the following numbers? Use the divisibility tests for 2, 3, and 5.

38. 150
2 and 3 and 5

39. 282
2 and 3

40. 397
None

41. 765
3 and 5

42. 861
3

Determine whether 9 is a factor of each number. Verify the answer by dividing the number by 9.

43. 171
Yes

44. 5,292
Yes

Find the prime factorization of the following numbers. Write the answer two ways: with and without exponents.

45. 168
$2 \cdot 2 \cdot 2 \cdot 3 \cdot 7$
$2^3 \cdot 3 \cdot 7$

46. 300
$2 \cdot 2 \cdot 3 \cdot 5 \cdot 5$
$2^2 \cdot 3 \cdot 5^2$

Solve the following expressions by isolating the variable. Check each answer to show that it is the solution.

47. $x + 45 = 72$
$x = 27$

48. $455 = 13 \cdot y$
$y = 35$

Work each application and answer it with a complete sentence.

49. Lydia photocopied a packet containing 34 pages of information for each of the 15 members attending the annual board meeting. How many total pages were photocopied?
A total of 510 pages were photocopied.

50. Citrus Hills High School purchased a total of 540 whiteboard markers to be used during the semester by 45 teachers. If the markers were evenly distributed to the teachers, how many markers did each teacher receive?
Each teacher received 12 markers.

51. Ben and Adrian decide to have a garage sale. Their goal is to raise \$500. On Friday, they raise \$138. On Saturday, they raise \$249. How much do Ben and Adrian need to raise on Sunday to meet their goal?
Ben and Adrian need to raise \$113.

52. At Amaya's slumber party, a bucket of Red Vines (red licorice) was divided evenly among the 7 girls. If the bucket had 245 Red Vines in it, how many did each girl receive?
Each girl received 35 Red Vines.

Evaluate the following expressions according to the replacement value given.

53. $2y \div c$ Replace ***y*** with 12 and ***c*** with 8.
3

54. $\dfrac{20 + x}{x}$ Replace ***x*** with 4.
6

Translate each English expression into an algebraic expression. Use any variable to represent the unknown number.

55. The sum of a number and 9
$x + 9$

56. The difference of six and a number
$6 - x$

57. The quotient of 24 and a number
$24 \div x$

58. The product of 8 and a number
$8 \cdot x$

59. The square root of a number
$\sqrt{x}$

60. The opposite of a number
$-x$

Insert the correct symbol between each pair of numbers, either $<$ (less than) or $>$ (greater than).

61. 4 13
$<$

62. 1 -7
$>$

63. -14 -10
$<$

64. -3 2
$<$

Find the absolute value as indicated.

65. $|7|$
7

66. $|-8|$
8

67. $|0|$
0

68. $|+5|$
5

Evaluate each expression.

69. $2 + (-2)$
0

70. $9 - 11$
-2

71. $-8 - 8$
-16

72. $-10 + 7$
-3

73. $9 - (-6)$
15

74. $-22 + 48$
26

75. $-15 - 79$
-94

76. $60 + (-76)$
-16

77. $-71 - (-36)$
-35

78. $7 \cdot (-5)$
-35

79. $(-6)(-6)$
36

80. $7(-4)$
-28

81. $-5(10)$
-50

82. $-72 \div (-9)$
8

83. $\dfrac{-49}{7}$
-7

84. $\dfrac{30}{-5}$
-6

85. $-1 \cdot (-6)$
6

86. $7 \div (-1)$
-7

87. $5 \cdot (-4) \cdot 6$
-120

88. $-3 \cdot (-7) \cdot (-2)$
-42

89. $2 + (-8) + (-7) + 9$
-4

90. $-6 - 5 - (-12) + 3$
4

For each application, write a numerical expression; then evaluate. Also write a sentence answering the question.

91. At 6 PM, the outside temperature was 6°. By 2 AM, the temperature had dropped 10°. What was the temperature at 2 AM?
$6 - 10 = -4$
The temperature at 2 AM was $-4°$.

92. Yesterday Ajay's checkbook had a balance of $-\$123$. Fortunately, today he received a check for the $75 that his brother owed him, and Ajay deposited it in the bank to cover his account. What is the new balance?
$-123 + 75 = -48$
The new balance is $-\$48$.

Evaluate.

93. $-\sqrt{25}$
-5

94. $(-3)^4$
81

95. $6 - 5^2$
-19

96. $-3 - \sqrt{25}$
-8

97. $-4 - |-5|$
-9

98. $|3 - 9|$
6

99. $\dfrac{8 \cdot 3}{-\sqrt{36}}$
-4

100. $(-3 \cdot 15) \div (-6 - 3)$
5

Find the equivalent temperature in either Celsius or Fahrenheit.

$$F = 9 \cdot \frac{C}{5} = 32 \qquad C = 5 \cdot \frac{F - 32}{9}$$

101. The temperature is 60° C.
140° F

102. The temperature is 113° F.
45° C

Evaluate the numerical value of each formula with the given replacement values.

103. $a = \sqrt{c^2 - b^2}$ $c = 10$, $b = 8$

$a = 6$

104. $c = \dfrac{2 \cdot A}{h} - b$ $A = 15$, $h = 5$, $b = 2$

$c = 4$

Use one of the distance formulas to answer the following questions. Remember, $d = r \cdot t$, $r = \frac{d}{t}$, and $t = \frac{d}{r}$.

105. James is an airline pilot. Recently, it took him 6 hours to fly from Los Angeles to Boston, a 2,610-mile flight. What was the jet's average speed for that flight?
The jet's average speed was 435 miles per hour.

106. Jasper races stock cars. One race in Santa Fe, New Mexico, is a timed race where the racers drive for 3 hours. If Jasper can average 95 miles per hour, how far will he go in that race?
Jasper will go 285 miles.

Simplify by combining like terms if possible.

107. $2x^3 + 8x^3$
$10x^3$

108. $4y^2 - 5y^2$
$-y^2$

109. $-2y + (-7y)$
$-9y$

110. $9x^4 - 10x^3$
Not like terms

111. $3h^2 - 3h^2$
0

112. $3w + 2w - 9w$
$-4w$

113. $-2a + 3a + 3 - 6$
$a - 3$

114. $-x^4 + 5x - 3x^4 - x$
$-4x^4 + 4x$

Find the product.

115. $7(3y)$
$21y$

116. $6(-x^2)$
$-6x^2$

117. $-2(9y^2)$
$-18y^2$

118. $-8(-5y^3)$
$40y^3$

Rewrite each expression using the Distributive Property.

119. $2 \cdot (-3x + 2)$
$-6x + 4$

120. $4 \cdot (-x - 3)$
$-4x - 12$

121. $-6 \cdot (3a + 5)$
$-18a - 30$

122. $-5 \cdot (9w - 2)$
$-45w + 10$

CHAPTER 3

Equations

Introduction

Problems! Everybody has them. Some are serious; others are not. Sometimes problems work themselves out; and other times we need to come up with a creative solution, a strategy that helps us solve the problem.

Here's the story of a woman with a problem. Can you help her find the solution?

Problem A woman has in her possession a fox, a chicken, and a bag of grain. She must be careful not to leave the fox and the chicken alone because the fox will eat the chicken. Likewise, if the chicken is left alone with the grain, the chicken will eat the grain. At a river crossing, the water is shallow enough for her to wade across, but she'll be able to carry only one of her possessions at a time. How does she cross the river with her three possessions uneaten?

Solution She must make seven trips in all. 1. She crosses the river with the chicken only and 2. returns by herself. 3. She crosses back with the fox only, leaves it tied up on shore, and 4. returns with the chicken. She leaves the chicken and 5. takes the grain across, leaving it with the fox. 6. She makes one last return trip to pick up the chicken, and 7. takes it across to gather up her possessions.

This problem is a bit different than the ones you'll be seeing in this section. There isn't any math involved in this problem, but there is a strategy.

In Chapter 3, you'll be using many of the ideas you learned in Chapters 1 and 2 to solve equations and applications. For each type of problem that you encounter, a strategy is given so that you will know how to approach a problem with confidence. For some of the problems, the strategy will be to remove something from one side and put it on the other side, just like the woman with the fox, the chicken, and the bag of grain.

Preparation Exercises

Section 1.7 Equations
Solve.

1. $p + 19 = 32$ $p = 13$

2. $45 = 3 \cdot y$ $y = 15$

3. $153 = 86 + x$ $x = 67$

4. $7 \cdot n = 105$ $n = 15$

Section 2.2 Adding Signed Numbers
Evaluate.

5. $-10 + 8$ -2

6. $-3 + (-12)$ -15

Section 2.3 Subtracting Signed Numbers
Evaluate.

7. $-7 - 9$ -16

8. $-6 - (-18)$ 12

Section 2.4 Multiplying and Dividing Signed Numbers

Evaluate.

9. $8 \cdot (-9)$ -72

10. $-5 \cdot (-10)$ 50

11. $45 \div (-9)$ -5

12. $\dfrac{-28}{-7}$ 4

Section 2.7 Algebraic Expressions: Combining Like Terms

Combine like terms.

13. $-4x + 9x$ $5x$

14. $-5x - 11x$ $-16x$

Section 2.8 Multiplying Algebraic Expressions

Distribute.

15. $4(5x - 7)$
$20x - 28$

16. $-5(3x - 2)$
$-15x + 10$

SECTION 3.1 Solving Equations Involving One Operation

OBJECTIVES

In this section, you will learn to

- Apply the Addition Property of Equations.
- Apply the Division Property of Equations.
- Isolate variables.
- Solve equations involving one operation.

You Need to Know

To successfully complete this section, you need to understand

- ☐ Additive identity (1.1)
- ☐ Multiplicative identity (1.1)
- ☐ Solving equations (1.7)
- ☐ Signed numbers and terms (Chapter 2)

Introduction

We first began solving equations in Section 1.7. We learned the following:

- To solve an equation, we must ***isolate the variable*** on one side of the equal sign.

Example:

$$14 = x + 8$$
$$14 - 8 = x + 8 - 8$$
$$6 = x$$
$$x = 6$$

- The **solution** of an equation makes the equation true, and we can check the solution by replacing the variable with that number.

Check $x = 6$:

$$14 \stackrel{?}{=} x + 8$$
$$14 \stackrel{?}{=} 6 + 8$$
$$14 = 14 \quad \text{True! ✓}$$

(The equal sign with a question mark over it, $\stackrel{?}{=}$, indicates that we are questioning whether the answer is true.)

In Chapter 2, we learned about algebraic expressions such as $-8x$, $6y - 5$, and $7 - 4w$; and we'll see those types of expressions in the equations in Chapter 3. In this section, we will solve simple equations, those that have only one operation. In Section 3.2, we'll solve equations that have more than one operation.

The key to solving any equation is to isolate the variable. Once the variable has been isolated correctly, we have the solution to the equation. Isolating the variable is all about having the variable (say, x) on one side of the equal sign and a single number (constant) on the other side, as in $x = 5$. This means manipulating the equation so that we get either

$x + 0 = 5$ or $1 \cdot x = 5$

$x = 5$ ← Each simplifies to $x = 5$. → $x = 5$

What's special about *plus zero* and *one times*?

You may recall from Section 1.1 that

1. 0 is the additive identity, meaning $a + 0 = a$, as in $8 + 0 = 8$.
2. 1 is the multiplicative identity, meaning $1 \cdot a = a$, as in $1 \cdot 7 = 7$.

To solve an equation means that we must achieve $x + 0$ or $1 \cdot x$ on one side of the equation. We will use the following three properties to help us isolate the variable.

1. $a + (-a) = 0$ The sum of a number and its opposite is 0 (Section 2.2). For example, $3 + (-3) = 0$ and $-9 + 9 = 0$.

2. $\frac{a}{a} = 1$ Any number (except 0) divided by itself is 1 (Section 1.4). For example, $\frac{6}{6} = 1$ and $\frac{-4}{-4} = 1$.

3. Whatever we do to modify one side of an equation—by adding, subtracting, multiplying, or dividing a number or term—we must do to the other side. We must keep the equation balanced.

Isolating the Variable: The Addition Property of Equations

We first look at simple equations in which the only operation is addition or subtraction, such as $x + 8 = 5$ or $w - 9 = -2$. Both of those equations have a constant on each side of the equal sign. To isolate the variable, we need to "clear" the constant on the same side as the variable. This is done using the Addition Property of Equations.

The Addition Property of Equations

We may add any number, c, to *each side* of an equation.

If $a = b$,
then $a + c = b + c$.

This property means that we can add the same positive or negative number to each side of the equation. Typically, we'll add the opposite of the constant term that we want to clear.

Adding the opposite of the constant creates 0. This is called *clearing the constant.*

Example 1

Solve each of these equations by clearing the constant. Check each answer to show that it is the solution.

a) $x - 6 = 3$ **b)** $-9 = y + 4$

Procedure To clear the constant, *add its opposite* to each side of the equation.

Answer **a)**

$x - 6 = 3$ Clear the constant, -6, by adding its opposite, $+6$, to each side.

$x - 6 + 6 = 3 + 6$

$x + 0 = 9$ $-6 + 6 = 0$, the additive identity.

$x = 9$ → Check the answer: $9 - 6 = 3$ True ✓

b)

$-9 = y + 4$ Clear the constant, 4, by adding its opposite, -4, to each side.

$-9 + (-4) = y + 4 + (-4)$

$-13 = y + 0$ $4 + (-4) = 0$, the additive identity.

$-13 = y$

Write the final equation with the variable on the left side. → $y = -13$ → Check the answer: $-9 = -13 + 4$ True ✓

Look back at Example 1 and notice three things

1. Each step in the solving process is directly below the preceding step.
2. The equal signs are lined up one below the other.
3. We achieved the additive identity, 0, to isolate the variable.

You Try It 1 **Solve each of these equations by clearing the constant. Show all steps! Also make sure you check each answer. Use Example 1 as a guide.**

a) $a - 4 = 11$ **b)** $5 = y + 9$ **c)** $-12 = x - 3$

Caution To clear the constant, we always add its opposite, no matter on which side of the equation it is written.

Example 2

Solve each of these equations by clearing the constant. Check each answer to show that it is the solution.

a) $40 + x = -50$ **b)** $-22 = -38 + y$

Procedure To clear the constant, *add its opposite* to each side of the equation.

Answer **a)**

$$40 + x = -50$$ Clear the constant, 40, by adding its opposite, -40, to each side.

$$40 + (-40) + x = -50 + (-40)$$

$$0 + x = -90$$ $40 + (-40) = 0$, the additive identity.

$$x = -90$$ → Check the answer: $40 + (-90) = -50$ True ✓

b)

$$-22 = -38 + y$$ Clear the constant, -38, by adding its opposite, $+38$, to each side.

$$-22 + 38 = -38 + 38 + y$$

$$+16 = 0 + y$$ $-38 + 38 = 0$, the additive identity.

$$16 = y$$

Write the final equation with the variable on the left side. →

$$y = 16$$ → Check the answer: $-22 = -38 + 16$ True ✓

You Try It 2 **Solve each of these equations by clearing the constant. Show all steps! Also make sure you check each answer. Use Example 2 as a guide.**

a) $30 + w = -120$ **b)** $25 = -35 + p$

Isolating the Variable: The Division Property of Equations

Recall from Section 1.7 that we can isolate the variable when it has a coefficient of 1, such as $1x$ and $1y$. To achieve this, we need to clear the coefficient by dividing each side of the equation by that coefficient.

For example, if the equation is $8x = -32$, we divide each side by 8 because that will create $1x$. Likewise, if the equation is $-7y = 14$, we divide each side by -7 because that will create $1y$.

This idea of creating a coefficient of 1 (the identity for multiplication) is shown in the Division Property of Equations.

We use the equal sign with a slash through it, $\neq$, to represent *is not equal to.*

The Division Property of Equations

We may divide each side of an equation by any number, c, $c \neq 0$.

$$\text{If} \quad a = b,$$
$$\text{then} \quad \frac{a}{c} = \frac{b}{c}.$$

Think About It 1 In the Division Property of Equations, why does it say that $c \neq 0$?

____

____

Example 3

Solve each of these equations by clearing the coefficient.

a) $5x = -20$ **b)** $12 = -2w$

Procedure Divide each side by the coefficient. Remember to keep the equation balanced. Mentally check each answer by using it as a replacement value for the variable in the original equation.

Answer

a)

$5x = -20$ The coefficient is 5. Clear it by dividing each side by 5.

$\frac{5x}{5} = \frac{-20}{5}$ Dividing by 5 makes the coefficient equal 1.

$1x = -4$ We get the new coefficient 1 because $\frac{5}{5} = 1$.

$x = -4$ → Check the answer: $5(-4) = -20$ True ✓

b)

$12 = -2w$ The coefficient is -2. Clear it by dividing each side by -2.

$\frac{12}{-2} = \frac{-2w}{-2}$ Dividing by -2 makes the coefficient equal 1.

$-6 = 1w$ We get the new coefficient 1 because $\frac{-2}{-2} = 1$.

$-6 = w$

$w = -6$ → Check the answer: $12 = -2 \cdot (-6)$ True ✓

Look back at Example 3 and notice three things

1. Each step in the solving process is directly below the preceding step.
2. The equal signs are lined up one below the other.
3. We achieved the multiplicative identity, 1, to isolate the variable.

You Try It 3 **Solve each of these equations by clearing the coefficient. Check each answer. Use Example 3 as a guide.**

a) $3y = -24$ **b)** $-5x = -20$ **c)** $18 = -2p$

In a simple equation, when the coefficient is -1, we can clear it by dividing or multiplying each side by -1.

Example 4

Solve $-x = 7$ by

a) dividing by -1 **b)** multiplying by -1

Procedure The coefficient is -1. Remember to modify each side of the equation the same way.

Answer

a) $-x = 7$ ← Divide each side by -1.

$$\frac{-1x}{-1} = \frac{7}{-1}$$

$x = -7$ ← We get the same solution. →

b) $-x = 7$ ← Multiply each side by -1.

$$-1 \cdot (-1x) = -1 \cdot 7$$

$x = -7$

You Try It 4 **Solve each equation by clearing the coefficient. Check each answer. Use Example 4 as a guide.**

a) $-w = -14$ **b)** $12 = -y$

Guidelines to Solving Simple Equations

To isolate a variable in a simple equation, we must clear any numerical value—a constant or coefficient—by applying the appropriate operation to each side of the equation.

1. Clear a constant term by *adding its opposite*.
2. Clear an integer coefficient by *dividing by the coefficient.*
3. Clear a coefficient of -1 by *dividing or multiplying by* -1.

Caution Make sure you keep the equation balanced by modifying each side in the same manner.

Answers: You Try It and Think About It

You Try It: **1. a)** $a = 15$ **b)** $y = -4$ **c)** $x = -9$ **2. a)** $w = -150$ **b)** $p = 60$ **3. a)** $y = -8$ **b)** $x = 4$ **c)** $p = -9$ **4. a)** $w = 14$ **b)** $y = -12$

Think About It: **1.** We are never allowed to divide by 0.

Section 3.1 Exercises

FOR EXTRA HELP
Student Resources on DVD-ROM
Includes
➤ Student's Solutions Manual
➤ Video Lectures
➤ Chapter Test Prep Video
MyMathLab
MathXL

Think Again

1. In the Division Property of Equations, why does it say that $c \neq 0$? *(Refer to Think About It 1)*
   We can never divide by 0; so, we need to make sure that the Division Property of Equations reflects that.

2. When first learning to solve equations in Section 1.7, we used the Subtraction Property of Equality. In this section, we use the Addition Property of Equality. Why didn't we use the Addition Property of Equality when we solved the equations in Section 1.7?
   The Addition Property of Equality requires knowledge of how to add signed numbers (Section 2.2).

3. To clear a constant, we add its opposite to each side. To clear a coefficient, why don't we divide by its opposite?
   To clear a coefficient, we need to attain a coefficient of 1. That can be achieved only by dividing by the exact coefficient, not its opposite.

4. What extra step(s) are required to solve this equation: $-x + 5 = -2$?
   After we add -5 to each side, we multiply each side by -1 to isolate the variable.

Focus Exercises

Solve each equation by clearing the constant. Make sure you check each answer.

5. $x - 12 = 6$
   $x = 18$
6. $p - 2 = 8$
   $p = 10$
7. $w - 4 = -7$
   $w = -3$
8. $m - 3 = -10$
   $m = -7$
9. $y - 9 = -8$
   $y = 1$
10. $x - 12 = -6$
    $x = 6$
11. $x + 5 = 8$
    $x = 3$
12. $h + 4 = 11$
    $h = 7$
13. $p + 8 = 2$
    $p = -6$
14. $a + 9 = 6$
    $a = -3$
15. $m + 6 = -4$
    $m = -10$
16. $x + 9 = -8$
    $x = -17$
17. $12 = x + 6$
    $x = 6$
18. $9 = w + 8$
    $w = 1$
19. $-4 = c - 7$
    $c = 3$
20. $-3 = k - 10$
    $k = 7$
21. $12 = p + 20$
    $p = -8$
22. $1 = h + 12$
    $h = -11$
23. $-9 = x + 5$
    $x = -14$
24. $-6 = y + 1$
    $y = -7$
25. $-8 = w - 4$
    $w = -4$
26. $-11 = p - 6$
    $p = -5$
27. $m - 6 = -6$
    $m = 0$
28. $x - 8 = -8$
    $x = 0$
29. $m - 30 = -120$
    $m = -90$
30. $y - 290 = -350$
    $y = -60$
31. $x + 150 = -60$
    $x = -210$
32. $x + 40 = 185$
    $x = 145$
33. $20 = h + 360$
    $h = -340$
34. $100 = p + 420$
    $p = -320$
35. $y - 42 = 57$
    $y = 99$
36. $c - 92 = 43$
    $c = 135$
37. $k + 36 = 84$
    $k = 48$
38. $w + 65 = 28$
    $w = -37$
39. $m + 98 = -72$
    $m = -170$
40. $y - 61 = -43$
    $y = 18$

Solve each equation by clearing the coefficient. Make sure you check each answer.

41. $6n = 42$
    $n = 7$
42. $5x = 20$
    $x = 4$
43. $7y = -42$
    $y = -6$
44. $4m = -28$
    $m = -7$
45. $-3k = -36$
    $k = 12$
46. $-2x = -18$
    $x = 9$
47. $18 = 9p$
    $p = 2$
48. $35 = 7w$
    $w = 5$
49. $-16 = 2v$
    $v = -8$
50. $-27 = 3m$
    $m = -9$
51. $-12 = -4y$
    $y = 3$
52. $-36 = -9x$
    $x = 4$
53. $9y = 72$
    $y = 8$
54. $4p = 28$
    $p = 7$
55. $-45 = 15h$
    $h = -3$
56. $-18 = 6w$
    $w = -3$
57. $-8c = -64$
    $c = 8$
58. $-11y = -55$
    $y = 5$

59. $10w = -90$
$w = -9$

60. $-30y = 210$
$y = -7$

61. $-x = -6$
$x = 6$

62. $-m = -13$
$m = 13$

63. $-k = 22$
$k = -22$

64. $-a = 15$
$a = -15$

Solve each equation. Make sure you check each answer.

65. $b - 1 = -5$
$b = -4$

66. $h + 8 = -4$
$h = -12$

67. $-20 = -5m$
$m = 4$

68. $-30 = -10x$
$x = 3$

69. $-4 + x = -4$
$x = 0$

70. $7 + y = 7$
$y = 0$

71. $8 + y = -14$
$y = -22$

72. $15 + c = -36$
$c = -51$

73. $23 = -42 + d$
$d = 65$

74. $40 = -38 + n$
$n = 78$

75. $q - 52 = -76$
$q = -24$

76. $x - 90 = -63$
$x = 27$

77. $-4x = -60$
$x = 15$

78. $-42 = -14y$
$y = 3$

79. $184 = -8w$
$w = -23$

80. $225 = -9p$
$p = -25$

81. $-231 = 7h$
$h = -33$

82. $-297 = 11k$
$k = -27$

83. $12x = 300$
$x = 25$

84. $14v = 238$
$v = 17$

85. $-64 + v = 141$
$v = 205$

86. $-129 + c = -97$
$c = 32$

87. $x + 103 = -71$
$x = -174$

88. $m + 119 = 55$
$m = -64$

Think Outside the Box

89. The perimeter of this triangle is 261 feet. Find the length of each side.

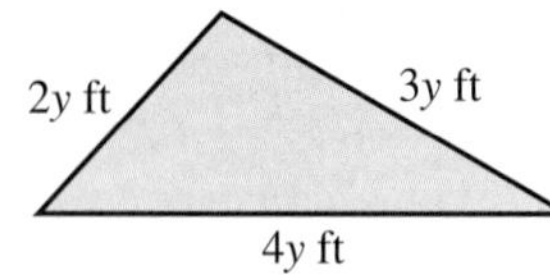

The length of the sides are 58 feet, 87 feet, and 116 feet.

90. The perimeter of this triangle is 374 inches. Find the length of each side (if possible).

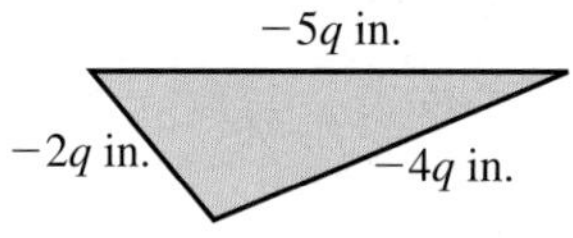

The length of the sides are 68 inches, 136 inches, and 170 inches.

SECTION 3.2 Solving Equations Involving Two Operations

OBJECTIVES

In this section, you will learn to

- Solve equations involving two operations.

You Need to Know

To successfully complete this section, you need to understand

- ☐ Signed numbers (Chapter 2)
- ☐ Solving equations involving one operation (3.1)

Introduction

The simple equations we solved in Section 3.1 had only one operation between the variable and a number—addition, subtraction, or multiplication. To isolate the variable in those equations, we cleared the constant or the coefficient. In this section, we solve equations that have two operations, multiplication and either addition or subtraction.

Solving Linear Equations Involving Two Operations

Some equations have more than one operation to clear. When this occurs, we must have a consistent method of isolating the variable.

For example, in an equation such as $3x + 5 = -7$, we might ask, "To isolate x, which should we clear first, the *constant* term, 5, or the *coefficient*, 3?"

Because the goal is to isolate the variable, we must prepare this equation—and others like it—by isolating the variable *term*. This means that we must clear the constant first. Once the constant has been cleared, we have a simple equation that can be solved by clearing the coefficient.

Example 1

Solve $3x + 5 = -7$. Check the answer to show that it is the solution.

Procedure To isolate the variable term, clear the constant first.

Answer

$3x + 5 = -7$ — Prepare the equation for solving by clearing the constant term, +5: Add −5 to each side.

$3x + 5 + (-5) = -7 + (-5)$

$3x + 0 = -12$ — Adding $5 + (-5)$ gives the additive identity, 0.

$3x = -12$ — This is now a simple equation. Clear the coefficient by dividing each side by 3.

$\frac{3x}{3} = \frac{-12}{3}$ — Because $\frac{3}{3} = 1$, the left side is $1x$, or just x.

$x = -4$ → Check the solution, −4: $3(-4) + 5 \stackrel{?}{=} -7$

$-12 + 5 = -7$ True ✓

Some equations contain the variable term on the right side, as shown in Example 2.

Example 2

Solve $-15 = 6 - 7y$. Check the answer to show that it is the solution.

Procedure To isolate the variable term, clear the constant first.

Answer

$-15 = 6 - 7y$ — Prepare the equation for solving by clearing the constant term, +6: Add −6 to each side.

$-15 + (-6) = 6 + (-6) - 7y$

$-21 = 0 - 7y$ — Adding $-6 + 6$ gives the additive identity, 0. Notice that $0 - 7y$ becomes $-7y$.

$-21 = -7y$ — This is now a simple equation. Clear the coefficient by dividing each side by −7.

$\frac{-21}{-7} = \frac{-7y}{-7}$ — Left side: $-21 \div (-7) = 3$; right side becomes $1y$, or just y.

$3 = y$

$y = 3$ → Check the answer: $-15 \stackrel{?}{=} 6 - 7 \cdot 3$

$-15 = 6 - 21$ True ✓

You Try It 1 **Solve each equation. Check the answer to show that it is the solution. Use Examples 1 and 2 as guides.**

a) $3x - 5 = 19$ **b)** $2y + 15 = 3$

c) $1 = -9 - 2w$ **d)** $-2 = 18 - 5p$

Answers: You Try It and Think About It

You Try It: **1. a)** $x = 8$ **b)** $y = -6$ **c)** $w = -5$ **d)** $p = 4$

Think About It: No Think About Its in this section.

FOR EXTRA HELP
Student Resources on DVD-ROM
Includes
➤Student's Solutions Manual
➤Video Lectures
➤Chapter Test Prep Video
MyMathLab
MathXL

Section 3.2 Exercises

Think Again

1. What extra step(s) would be required to solve this equation: $4p + 9 - 6 = -17$?
The like terms need to be combined on the left side.

2. What extra step(s) would be required to solve this equation: $-1 = 3y - 5y + 9$?
The like terms need to be combined on the right side.

Focus Exercises

Solve each equation. Check each answer to show that it is the solution.

3. $3x + 8 = 20$
$x = 4$

4. $24 = 7w + 3$
$w = 3$

5. $5x + 2 = 2$
$x = 0$

6. $3p - 7 = -7$
$p = 0$

7. $6 - 5w = -34$
$w = 8$

8. $13 = -5 - 9c$
$c = -2$

9. $43 = 4x + 7$
$x = 9$

10. $-23 = 8w + 9$
$w = -4$

11. $-7v - 4 = 17$
$v = -3$

12. $-2y - 3 = -11$
$y = 4$

13. $-y + 5 = 3$
$y = 2$

14. $-k - 8 = -1$
$k = -7$

15. $-4 = -p + 16$
$p = 20$

16. $8 = -m - 3$
$m = -11$

17. $-4 = 3x - 4$
$x = 0$

18. $6 = 6 - 5m$
$m = 0$

19. $2y + 2 = 8$
$y = 3$

20. $-2 - 5v = 8$
$v = -2$

21. $-7 = 4x + 5$
$x = -3$

22. $21 = -3 + 6v$
$v = 4$

23. $4 - 3k = -5$
$k = 3$

24. $6 - y = 1$
$y = 5$

25. $19 = -11 - 5w$
$w = -6$

26. $7 = -9 + 4c$
$c = 4$

27. $4m + 18 = -30$
$m = -12$

28. $6v - 28 = 50$
$v = 13$

29. $103 = -5p + 18$
$p = -17$

30. $-72 = 8x - 48$
$x = -3$

31. $61 + 12k = 1$
$k = -5$

32. $88 - 15y = -2$
$y = 6$

33. $110 = -40 - 5y$
$y = -30$

34. $7 = -91 - 7c$
$c = -14$

35. $-57 - 3x = -123$
$x = 22$

36. $-60 + 9p = 66$
$p = 14$

37. $-41 = 94 - 15m$
$m = 9$

38. $-70 = 95 + 11c$
$c = -15$

Think Outside the Box

39. This rectangle has a perimeter of 132 centimeters. What is the width of the rectangle?

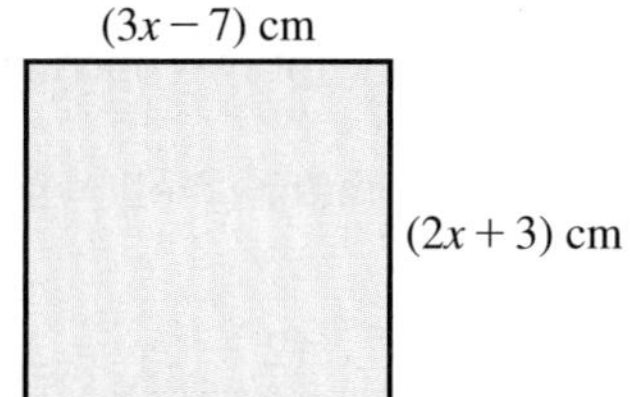

The width of the rectangle is 31 centimeters.

40. The following table shows the number of students in each sixth grade class at Lincoln Elementary School that participated in the annual Arbor Day tree planting at River Road Park. In this chart, x represents the number of students from Mr. Ukpo's class that participated. Based on the information in the chart, how many students participated from each class?

Teacher	Number of students participating	
Miss Garcia	$x + 6$	17
Mrs. Crum	$3x - 8$	25
Mr. Farad	$x + 1$	12
Miss Trinh	$2x - 7$	15
Mr. Ukpo	x	11
Total	**80**	

SECTION 3.3 Solving Equations Involving More than Two Operations

OBJECTIVES

In this section, you will learn to

- Solve equations involving more than one variable term.
- Simplify expressions before solving.

You Need to Know

To successfully complete this section, you need to understand

- ☐ Signed numbers (Chapter 2)
- ☐ Combining like terms (2.7)
- ☐ The Distributive Property (2.8)
- ☐ Solving equations involving two operations (3.2)

Introduction

In Section 3.2, we became familiar with solving equations with two operations and with a variable term and a constant term on the same side.

Sometimes, though, an equation has more than two operations. Following are examples of such equations:

$$4x + 17 = 6x + 5 \qquad 2x + 8 + 4x = -4 + 3 \qquad 12 - 2y = 3(y - 1)$$

We can solve these equations after we put each one in a more familiar form, an equation with only two operations like those in Section 3.2. To do this, we need to *prepare* the equation for solving by simplifying one side or the other, maybe by distributing or by combining like terms.

Solving Linear Equations Containing More than One Variable Term

Some equations have more than one variable term, one on each side.

$$\textcircled{4x} + 17 = \textcircled{6x} + 5$$

Should we isolate this variable term? or this variable term?

For example, $4x + 17 = 6x + 5$ has a variable term on each side. You might ask, "To solve this equation, which variable should we isolate?"

To answer that question, we must remember that any solution we find becomes a replacement value. This replacement value replaces the variable in both variable terms. In the preceding example, the x's are the same variable and must be combined as one term before we can isolate the variable.

However, because the variables are in different expressions—on different sides of the equal sign—the only way we can combine them is to clear one of the variable terms by adding its opposite to each side.

Does it really matter which variable term we choose to clear? No. Because the equation has only one solution, whichever variable term we choose to clear will lead us to the solution.

Example 1 and You Try It 1 solve the same equation two different ways.

Example 1

Solve $4x + 17 = 6x + 5$.

Procedure First, combine the variable terms by adding the opposite of one of them to each side.

Answer Let's choose to clear the $4x$ term by adding $-4x$ to each side.

$4x + 17 = 6x + 5$	Prepare the equation for solving by clearing $+4x$: Add $-4x$ to each side of the equation.
$4x + (-4x) + 17 = 6x + (-4x) + 5$	
$0 + 17 = 2x + 5$	$4x + (-4x) = 0x$, or just 0.
$17 = 2x + 5$	Now isolate the variable term by clearing the constant term, $+5$: Add -5 to each side.
$17 + (-5) = 2x + 5 + (-5)$	
$12 = 2x + 0$	Now clear the coefficient by dividing each side by 2.
$\frac{12}{2} = \frac{2x}{2}$	Check the answer, $x = 6$: $4(6) + 17 \stackrel{?}{=} 6(6) + 5$
$6 = x$	$24 + 17 \stackrel{?}{=} 36 + 5$
$x = 6$	$41 = 41$ True ✓

You Try It 1 **Solve this equation by clearing the $+6x$ term from the right side. The first step is shown. Complete the solving process. Use Example 1 as a guide.**

$$4x + 17 = 6x + 5$$
$$4x + (-6x) + 17 = 6x + (-6x) + 5$$

Check the answer:

Think About It 1 Comparing the work shown in Example 1 and your own work in You Try It 1, which variable term, $4x$ or $6x$, would you clear if you had your choice? Why? Share your answer with a classmate.

You Try It 2 **Solve each equation. Check each answer to show that it is the solution. Use Example 1 as a guide.**

a) $9x - 5 = 10 + 6x$ **b)** $2p + 5 = -4p + 23$

c) $4w - 5 = 16 + 7w$ **d)** $x + 10 = 2 - 3x$

Simplifying Expressions before Solving

Sometimes, in preparing an equation for solving, we need to simplify one (or both) of the sides; after all, each side is an expression and it's common for us to simplify expressions.

Consider this example

1. We may need to distribute a number to a quantity first, as in the expression $3(y - 1)$.
2. We may need to combine like terms, as in the expression $2x + 8 + 4x$.

Example 2

Solve $12 - 2y = 3(y - 1)$ by simplifying each side first.

Answer

$$12 - 2y = 3(y - 1)$$
First, distribute on the right side. Do not try to clear any terms just yet.

$$12 - 2y = 3y - 3$$
Now we can start the process of isolating the variable by getting the variable terms together on the same side. Let's clear $-2y$ by adding its opposite, $+2y$, to each side.

$$12 - 2y + 2y = 3y + 2y - 3$$
$$12 + 0 = 5y - 3$$
$$12 = 5y - 3$$
Isolate the variable term by clearing the constant. Add the opposite of -3 to each side: $-3 + 3 = 0$.

$$12 + 3 = 5y - 3 + 3$$
$$15 = 5y + 0$$
$$15 = 5y$$
Clear the coefficient by dividing each side by 5.

$$\frac{15}{5} = \frac{5y}{5}$$
$$3 = y$$
$$y = 3$$

Check the answer by substituting $y = 3$ into the *original* equation:
$$12 - 2(3) \stackrel{?}{=} 3(3 - 1)$$
$$12 - 6 \stackrel{?}{=} 3(2)$$
$$6 = 6 \quad \text{True ✓}$$

Example 3

Solve $2x + 8 + 5x = -4 + 3x$ by simplifying each side first.

Answer

$$2x + 8 + 5x = -4 + 3x$$

First, combine like terms on the left side: $2x + 5x = 7x$.

$$7x + 8 = -4 + 3x$$

Now we can get the variable terms together on the same side. Let's clear $+3x$ by adding its opposite, $-3x$, to each side.

$$7x + (-3x) + 8 = -4 + 3x + (-3x)$$

$$4x + 8 = -4 + 0$$

$$4x + 8 = -4$$

Isolate the variable term by clearing the constant: Add -8 to each side.

$$4x + 8 + (-8) = -4 + (-8)$$

$$4x + 0 = -12$$

$$4x = -12$$

Clear the coefficient by dividing each side by 4.

$$\frac{4x}{4} = \frac{-12}{4}$$

$$x = -3$$

Check the answer by substituting $x = -3$ into the *original* equation: $2(-3) + 8 + 5(-3) \stackrel{?}{=} -4 + 3(-3)$

$-6 + 8 + (-15) \stackrel{?}{=} -4 + (-9)$

$-13 = -13$ True ✓

You Try It 3 **Solve each equation by simplifying each side first. Use Examples 2 and 3 as guides.**

a) $4(y + 1) = 6y - 6$ **b)** $-6w + 15 = -3(3w - 4)$

c) $x - 10 = 3x + 6 + 2x$ **d)** $1 - 2x + 10 = -4x - 9$

Answers: You Try It and Think About It

You Try It: **1.** $x = 6$ **2. a)** $x = 5$ **b)** $p = 3$ **c)** $w = -7$ **d)** $x = -2$ **3. a)** $y = 5$ **b)** $w = -1$ **c)** $x = -4$ **d)** $x = -10$

Think About It: **1.** Answers may vary.

Section 3.3 Exercises

FOR EXTRA HELP
Student Resources on DVD-ROM
Includes
➤ Student's Solutions Manual
➤ Video Lectures
➤ Chapter Test Prep Video
MyMathLab
MathXL

Think Again

1. Why is it important to simplify each side of an equation before applying any of the properties of equality?
Answers will vary.

2. What would you write to a classmate to explain the importance of doing a check after solving an equation?
Answers will vary.

Focus Exercises

Solve each equation. Check each answer to show that it is the solution.

3. $4x + 6 = 18 - 2x$
$x = 2$

4. $3m + 4 = 7m - 12$
$m = 4$

5. $3 - 2y = 9 - 8y$
$y = 1$

6. $5 + 2x = -7 + 5x$
$x = 4$

7. $2p - 5 = -4p - 23$
$p = -3$

8. $3y - 9 = -5y - 41$
$y = -4$

9. $3x - 7 = 11 + 5x$
$x = -9$

10. $5w + 4 = 9w + 28$
$w = -6$

11. $18 + 8y = 7 - 3y$
$y = -1$

12. $3 - 2n = 21 + 7n$
$n = -2$

13. $14 - 6x = -6 + 4x$
$x = 2$

14. $2x - 5 = 12x + 45$
$x = -5$

15. $3w + 3 = 7w - 1$
$w = 1$

16. $7v - 5 = 5v - 3$
$v = 1$

17. $h + 11 = 5 - h$
$h = -3$

18. $-4c - 3 = 9 - c$
$c = -4$

Solve each equation. Check each answer to show that it is the solution.

19. $3 + m + 7 = 19 - 2m$
$m = 3$

20. $12 - 8a - 6 = -13a + 21$
$a = 3$

21. $-y + 18 + 3y = 6 + 5y$
$y = 4$

22. $4x - 2 - 3x = -x + 16$
$x = 9$

23. $2(5 - x) = 3x + 6 - x$
$x = 1$

24. $4d + 1 - 6d = 3(5 - 3d)$
$d = 2$

25. $3(y - 6) = 4y - 8$
$y = -10$

26. $4(3c + 2) = 2c - 2$
$c = -1$

27. $4x + 15 = 5(x + 4)$
$x = -5$

28. $2x + 2 = 6(x + 3)$
$x = -4$

29. $-c + 6 + 3c = -2c + 6$
$c = 0$

30. $5p + 7 - 3p = 17 - 9p + 12$
$p = 2$

31. $2y + 5 - 4y = 6y - 11$
$y = 2$

32. $-10 - 3w + 7 = 9w + 3 - 6w$
$w = -1$

33. $-10x + 19 + 20x = 4 + 5x$
$x = -3$

34. $15 + 2y - 2 = 6y - 11 - 10y$
$y = -4$

35. $k - 12 + 5k = 3(k + 2)$
$k = 6$

36. $2v + 3 - 4v = -1(v + 9)$
$v = 12$

37. $6(r - 1) = r - 10 + 3r$
$r = -2$

38. $-7(m + 2) = -m - 2 - 5m$
$m = -12$

39. $5(x + 4) = 3x + 5 + 7x$
$x = 3$

40. $-3(2q - 5) = 4q + 6 - q$
$q = 1$

41. $8c - 2 + 5c = -11 + 2(c - 1)$
$c = -1$

42. $14 + 8m - 13 = 3(m - 4) + 3$
$m = -2$

43. $8x - 2 + 10x = 6 - 7x - 8$
$x = 0$

44. $7 + 6p = 10p + 7 - 30p$
$p = 0$

45. $-1(w - 10) = 3w - 2 - 7w$
$w = -4$

46. $2(3 - v) = v - 9 - 6v$
$v = -5$

47. $y + 2 - 6y = -4(y - 3)$
$y = -10$

48. $n + 10 - 2n = -5(n + 2)$
$n = -5$

Think Outside the Box

Solve each equation.

49. $x^2 + 5x - 4 = x^2 - 3x - 52$
$x = -6$

50. $2(4x - 3x^2) - 9 = -3(x + 2x^2) + 46$
$x = 5$

51. The perimeter of the rectangle is the same as the perimeter of the triangle. What is the length of the rectangle?

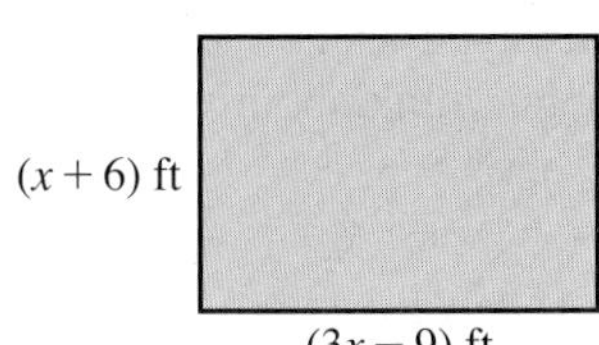

(2x – 3) ft (3x – 7) ft

(4x – 8) ft

The length of the rectangle is 27 feet.

SECTION 3.4 Problem Solving

OBJECTIVES

In this section, you will learn to
- Translate from English to algebra.
- Solve problems with unknown numbers.

You Need to Know

To successfully complete this section, you need to understand
- ☐ Solving applications (1.8)
- ☐ Translating expressions (2.1)
- ☐ Solving equations (3.1, 3.2, and 3.3)

Introduction

Algebra is more than just inserting letters in place of numbers. You can use algebra to answer questions such as these: How long will it take me to earn enough money to put a down payment on a house? How much paint will I need to paint three bedrooms in my house?

Before we begin to explore the problems of algebra, we need to look at how to translate an English problem into an algebraic one.

Translating from English to Algebra: Expressions

In Section 2.1, we explored how to translate expressions from English to algebra. Consider this example:

The sum of a number and 6 becomes $n + 6$ or $x + 6$.

In everyday speaking, we might say, "A number *increased by* 6" or "6 *more than* a number." Each of those expressions could also be written as $x + 6$.

Here is a table showing some words that indicate one operation or another. For this table, x represents the unknown number.

Word or phrase	Means	Example (expression)	Translation
More than Increased by Sum	Addition	"5 more than a number" "A number increased by 8" "The sum of a number and −3"	$x + 5$ $x + 8$ $x + (-3)$
Less than Decreased by Difference	Subtraction	"7 less than a number" "A number decreased by 4" "The difference of −2 and a number"	$x - 7$ $x - 4$ $-2 - x$
Times Twice (two times) Product	Multiplication	"Four times a number" "Twice a number" "The product of a number and 7"	$4 \cdot x$ or just $4x$ $2 \cdot x$ or just $2x$ $x \cdot 7$ or $7x$

Maybe the most challenging interpretation in the chart is that of "less than," which means subtraction. Why is "7 less than a number" interpreted as $x - 7$? Why isn't it $7 - x$ since that's the order in which the parts were written (7 first, followed by *a number*)?

The question is best answered with an example.

Hector's children, Gloria (14 years old) and Tino (10 years old), receive a weekly allowance. Since Tino is younger and has fewer household responsibilities than Gloria, Hector might say, "Tino's allowance is \$7 less than Gloria's." How much does Tino receive for his weekly allowance?

To answer that question, we *first* must know how much Gloria receives. If she receives \$12 per week, Tino receives $\$12 - \$7 = \$5$ per week. If Gloria receives \$16 per week, Tino receives $\$16 - \$7 = \$9$ per week.

In general, if Gloria's amount is unknown (just *a number, x*), Tino's amount is 7 less than that number, or $x - 7$.

What this means is that we can't always translate word for word exactly as written. Sometimes we need to think things through, think about what comes first or what value we must know first.

Example 1

Translate each of the following expressions into the language of algebra.

a) 10 more than a number

b) The sum of -8 and a number

c) 6 times a number

d) 9 less than a number

e) Twice a number

f) The difference of a number and -1

Procedure In each case, we'll let x represent the number.

Answer

a) $x + 10$ **b)** $-8 + x$ **c)** $6x$

d) $x - 9$ **e)** $2x$ **f)** $x - (-1)$

You Try It 1 **Translate each of the following expressions into the language of algebra. Here use the variable x to represent the number. Use Example 1 as a guide.**

a) A number increased by 12

b) 8 less than a number

c) The product of a number and -3

d) The difference of -5 and a number

e) A number decreased by 15

f) The sum of a number and -9

Translating from English to Algebra: Equations

Recall that an equation is as follows: one expression = another expression.

In English, the equal sign can be translated as *equals, is,* or *totals*. Each of those words is a verb in the English language, and we create sentences with verbs.

Certain sentences in English can be translated to sentences (equations) in algebra.

For example, the sentence *8 more than a number is 3* can be translated into algebra in this way:

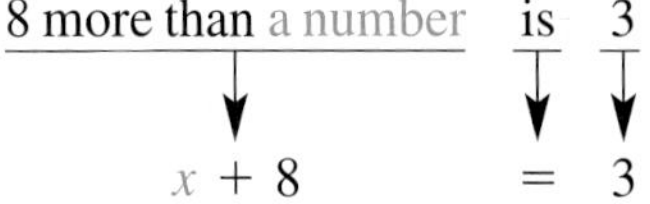

Now that it is an equation, we can solve it.

$$x + 8 = 3$$
$$x + 8 + (-8) = 3 + (-8)$$
$$x = -5$$

Does this answer make sense? Is it true that 8 more than -5 is 3?

It's true visually.

+8

1 2 3 4 5 6 7 8

-5 0 3

It's also true mathematically.

$$-5 + 8 = 3$$
$$3 = 3 \quad \text{True ✓}$$

Example 2

Translate each of the following sentences into an equation and find the requested number. Solve and check each answer. Write a sentence answering each question.

a) A number decreased by 8 is -6. What is the number?

b) 10 more than a number totals 3. What is the number?

Procedure For each, the legend is Let x = the number.

Answer a)

$$x - 8 = -6$$
$$x - 8 + 8 = -6 + 8$$
$$x = 2$$

Isolate the variable by adding +8 to each side.

Check $x = 2$: $2 - 8 \stackrel{?}{=} -6$

$-6 = -6$ True ✓

Sentence The number is 2.

Answer **b)**

$$x + 10 = 3$$
$$x + 10 + (-10) = 3 + (-10)$$
$$x = -7$$

Isolate the variable by adding -10 to each side.

Check $x = -7$: $-7 + 10 = 3$
$3 = 3$ True ✓

Sentence The number is -7.

Example 3

Translate the following sentences into an equation and find the requested number. Solve and check the answer. Write a sentence answering the question.

Four times a number is -48. What is the number?

Procedure The legend is Let x = the number.

Answer

$$4x = -48$$
$$\frac{4x}{4} = \frac{-48}{4}$$
$$x = -12$$

Isolate the variable by dividing each side by 4.

Check $x = -12$: $4 \cdot (-12) \stackrel{?}{=} -48$
$-48 = -48$ True ✓

Sentence The number is -12.

You Try It 2 **Translate each of the following sentences into an equation and find the requested number. Solve and check each answer. Then write a sentence. Use Examples 2 and 3 as guides.**

For each, the legend is ____________________.

a) 12 more than a number is -6. What is the number?

Sentence ____________________

b) Twice a number is -24. What is the number?

Sentence ____________________

c) 8 less than a number is -1. What is the number?

Sentence ____________________

d) The product of -6 and a number is 96. What is the number?

Sentence ____________________

Translating Expressions Containing Two Operations

Some expressions contain more than one operation, as in this expression:

the **sum** of 5 and *twice* a number.

This expression suggests that we will be adding two terms together, 5 and $2x$: $5 + 2x$.

An English expression that means the same thing is "5 **more than** *twice* a number," which is written as $2x + 5$.

Example 4

Translate each of the following expressions into the language of algebra.

a) 7 more than three times a number **b)** 8 less than twice a number

Procedure Let x = the number.

Answer **a)** 7 more than three times a number **b)** 8 less than twice a number

$3x + 7$ $2x - 8$

You Try It 3 **Translate each of the following expressions into the language of algebra. Here use the variable x to represent the unknown number. Use Example 4 as a guide.**

a) The sum of four times a number and 9 **b)** 10 more than five times a number

Think About It 1 Consider this algebraic expression: $2x + 10$.
Find more than one way to translate it into English.

____

____

Translating Equations Containing Two Operations

The expressions containing two operations can also be a part of an equation. Remember that the word *is* means "equals."

Example 5

Translate each of the following sentences into an equation and find the requested number. Solve and check the answer. Write a sentence answering the question.

a) The sum of 3 and twice a number is −9. What is the number?

b) Twice the sum of a number and 5 is 8. What is the number?

Procedure For each, the legend is Let x = the number.

Answer **a)**

$$3 + 2x = -9$$ Isolate the variable by adding −3 to each side.

$$3 + (-3) + 2x = -9 + (-3)$$ The left side becomes $0 + 2x$.

$$2x = -12$$ Divide each side by 2.

$$\frac{2x}{2} = \frac{-12}{2}$$

$$x = -6$$ Check $x = -6$: $3 + 2(-6) \stackrel{?}{=} -9$

$3 + (-12) = -9$ True ✓

Sentence The number is −6.

Answer **b)** *Twice the sum . . .* means "two times a quantity of addition": 2(+). In this case, it means $2(x + 5)$.

$$2(x + 5) = 8$$ On the left side of the equal sign, distribute 2.

$$2x + 10 = 8$$ Isolate the variable by adding −10 to each side.

$$2x + 10 + (-10) = 8 + (-10)$$ The left side becomes $2x + 0$, or $2x$.

$$2x = -2$$ Divide each side by 2.

$$\frac{2x}{2} = \frac{-2}{2}$$ The left side becomes $1x$, or x.

$$x = -1$$ Check $x = -1$: $2(-1 + 5) \stackrel{?}{=} 8$

$2(4) = 8$ True ✓

Sentence The number is −1.

You Try It 4 **Translate each of the following sentences into an equation and find the requested number. Solve and check the answer. Then write a sentence answering the question. Use Example 5 as a guide.**

For each, the legend is ______________________________.

a) The sum of three times a number and 8 is 2. What is the number?

b) 2 more than five times a number is -18. What is the number?

c) 3 less than four times a number is 21. What is the number?

d) Twice the sum of a number and 3 is -10. What is the number?

Answers: You Try It and Think About It

You Try It: **1. a)** $x + 12$ **b)** $x - 8$ **c)** $x \cdot (-3)$ or $-3x$ **d)** $-5 - x$ **e)** $x - 15$ **f)** $x + (-9)$ **2.** Legend: Let $x =$ the number. **a)** $x + 12 = -6$; the number is -18. **b)** $2x = -24$; the number is -12. **c)** $x - 8 = -1$; the number is 7. **d)** $-6x = 96$; the number is -16. **3. a)** $4x + 9$ **b)** $5x + 10$ **4.** Legend: Let $x =$ the number. **a)** $3x + 8 = 2$; the number is -2. **b)** $5x + 2 = -18$; the number is -4. **c)** $4x - 3 = 21$; the number is 6. **d)** $2(x + 3) = -10$; the number is -8.

Think About It: **1.** Answers may vary. Here are two possible answers: The sum of twice a number and 10; and 10 more than twice a number.

Section 3.4 Exercises

FOR EXTRA HELP
Student Resources on DVD-ROM
Includes
➤Student's Solutions Manual
➤Video Lectures
➤Chapter Test Prep Video
MyMathLab
MathXL

Think Again

1. Find more than one way to translate $2x + 10$ into English. *(Refer to Think About It 1)*
Answers will vary. One possibility: The sum of twice a number and 10; 10 more than twice a number. Also, twice a number increased by 10.

2. *Less than* suggests subtraction. Why is the phrase *5 less than a number* translated as $x - 5$ instead of $5 - x$?
Answers will vary. One possibility: To find 5 less than a number, you first need to know what the number is, then take (subtract) 5 from it: $x - 5$.

Focus Exercises

Translate each of the following expressions into the language of algebra. Here use the variable x to represent the number.

3. 9 more than a number
$x + 9$

4. The sum of a number and 4
$x + 4$

5. 5 less than a number
$x - 5$

6. A number decreased by 7
$x - 7$

7. Twice a number
$2x$

8. Five times a number
$5x$

9. 18 decreased by a number
$18 - x$

10. 12 increased by a number
$12 + x$

11. The sum of -9 and a number
$-9 + x$

12. The sum of a number and -20
$x + (-20)$

13. The difference of 3 and a number
$3 - x$

14. The difference of a number and -7
$x - (-7)$

15. The product of 8 and a number
$8x$

16. The product of negative two and a number
$(-2)x$

17. The sum of twice a number and 4
$2x + 4$

18. The sum of 10 and twice a number
$10 + 2x$

19. The sum of six times a number and 7
$6x + 7$

20. The sum of 15 and twice a number
$15 + 2x$

21. 1 more than three times a number
$3x + 1$

22. 3 more than twice a number
$2x + 3$

23. 8 less than twice a number
$2x - 8$

24. 11 less than five times a number
$5x - 11$

For each of the following, find the requested number, and write a sentence that answers the question. (*Hint:* Use the techniques in this section: write a legend for the unknown value, translate the sentence into an equation, solve and check the answer.)

25. 9 more than a number is 6. What is the number?
The number is -3.

26. A number increased by 11 is 5. What is the number?
The number is -6.

27. A number increased by 9 is -5. What is the number?
The number is -14.

28. 6 more than a number is -3. What is the number?
The number is -9.

29. 13 less than a number is -8. What is the number?
The number is 5.

30. A number decreased by 10 is -1. What is the number?
The number is 9.

31. A number decreased by 6 is -12. What is the number?
The number is -6.

32. 5 less than a number is -9. What is the number?
The number is -4.

33. The difference of a number and -2 is 8. What is the number?
The number is 6.

34. The difference of -4 and a number is -9. What is the number?
The number is 5.

35. The sum of a number and -6 is -3. What is the number?
The number is 3.

36. The sum of -10 and a number is -7. What is the number?
The number is 3.

37. Twice a number is -14. What is the number?
The number is -7.

38. The product of 6 and a number is -18. What is the number?
The number is -3.

39. The sum of twice a number and 4 is 18. What is the number?
The number is 7.

40. The sum of 10 and twice a number is 34. What is the number?
The number is 12.

41. The sum of 7 and four times a number is 3. What is the number?
The number is -1.

42. The sum of 2 and nine times a number is 11. What is the number?
The number is 1.

43. The sum of twice a number and 3 is -7. What is the number?
The number is -5.

44. The sum of six times a number and 9 is -9. What is the number?
The number is -3.

45. 1 more than three times a number is 16. What is the number?
The number is 5.

46. 3 more than twice a number is 25. What is the number?
The number is 11.

47. 8 more than five times a number is -7. What is the number?
The number is -3.

48. 9 more than four times a number is 1. What is the number?
The number is -2.

49. 7 less than twice a number is 9. What is the number?
The number is 8.

50. 11 less than twice a number is -3. What is the number?
The number is 4.

51. 3 less than eight times a number is -43. What is the number?
The number is -5.

52. 10 less than seven times a number is -3. What is the number?
The number is 1.

53. Twice the sum of a number and 7 is 26. What is the number?
The number is 6.

54. Twice the sum of a number and 11 is -4. What is the number?
The number is -13.

55. Twice the sum of 9 and a number is 8. What is the number?
The number is -5.

56. Twice the sum of -6 and a number is 20. What is the number?
The number is 16.

57. Three times the sum of a number and 5 is 12. What is the number?
The number is -1.

58. Three times the sum of a number and -2 is -9. What is the number?
The number is -1.

59. Four times the sum of a number and -6 is 20. What is the number?
The number is 11.

60. Five times the sum of a number and 8 is -15. What is the number?
The number is -11.

Think Outside the Box

For each of the following, find the requested number, and write a sentence that answers the question.

61. 5 more than three times a number is 9 less than the number. What is the number?
The number is -7.

62. Twice the sum of a number and -12 is 21 more than five times the number. What is the number?
The number is -15.

63. The difference of -25 and a number is three times the difference of 7 and the number. What is the number?
The number is 23.

64. 16 less than six times a number is twice the sum of the number and -18. What is the number?
The number is -5.

SECTION 3.5 Solving Applications Involving One Unknown Value

OBJECTIVES

In this section, you will learn to

- Set up a legend for one or two unknown values.
- Identify a formula for a given situation.
- Solve an application using algebra.

You Need to Know

To successfully complete this section, you need to understand

- ☐ Solving applications (1.8)
- ☐ Translating expressions (2.1)
- ☐ Solving equations (3.1, 3.2, and 3.3)
- ☐ Solving problems (3.4)

Introduction

Here is a typical application situation:

> A rectangle has a perimeter of 44 inches. The length is 15 inches.
> What is the width of the rectangle?

If you don't know the answer immediately, that's okay. This section will teach you a step-by-step approach to solving application situations, also known as word problems and story problems.

The following guidelines will help you stay organized because they offer a step-by-step approach.

Guidelines for Mastering an Application Problem

1. **Overview** Read the application once to get a general idea of what it involves. (Don't be overwhelmed.) Then read the application more carefully and start identifying key information such as known and unknown values.
2. **Legend** Decide how many unknown values the application is suggesting and use a variable (sometimes x) to represent each unknown value. Write a legend to define or describe what the variable represents.
3. **Diagram** If possible, draw a diagram. Label the diagram appropriately with known and unknown values.
4. **Formula** Identify a formula that fits the information. From that formula, develop an equation. Solve the equation. Check the answer to see if it is a solution to the equation and if it makes sense given the application.
5. **Conclusion** Write the solution in a complete sentence. If possible, use words that indicate the correct unit of measure, such as *feet, miles, hours,* and *pounds.*

Caution It's important to follow all of these guidelines, but it's not necessary to do them in this exact order. After the overview, it might be appropriate to draw a diagram, write the formula, and then set up the legend.

Let's take a look at each guideline one at a time.

Guideline 1: Read the Application Situation

Solving applications is not an instantaneous process. If you tackle solving applications with a step-by-step approach, you can break a large problem into smaller pieces that eventually fit together.

Let's consider this typical application:

> Kylee attends college in Lansing, Michigan; and some weekends she drives 260 miles to get to her home in Columbus, Ohio. She is able to average 65 miles per hour. How much time does it take Kylee to drive home?

Do you know the answer immediately? Most of us do not because we are not able to think quickly enough to know the answer after reading the problem just once. That's why we have these guidelines.

Guideline 2: Determine the Legend for One Unknown Value

Every application provides a great deal of information as well as a question to be answered. In the preceding situation, there is

1. The information about the distance that Kylee drives (260 miles).
2. The fact that she averages 65 miles per hour.
3. A question asking us to find the amount of time it takes her to get home from college.

The question at the end of the problem usually indicates the unknown value. The answer to the question is a number. Since we're going to use algebra to find the answer, we will use a variable—typically x—to represent that number.

In the situation presented, there is only one unknown value: the amount of time it takes Kylee to drive home. We set up a legend to help us define the variable representation.

Legend: Let x = the time (in hours) it takes Kylee to drive home.

Instead of x, we may use h for *hours*.

Caution You may be tempted to use t for time; but as a variable, t isn't recommended because it might be confused with a plus sign in the equation.

Notice this about the legend:

It defines what the variable x (or h) is going to represent.

It says that x (or h) represents a *number*, an amount of *time*.

The legend is important because it helps you organize your thoughts; and when you finish the problem—by solving an equation—you'll know what the solution means.

For example, if you solve the equation and the solution is $x = 4$, you know that is the amount of time and can answer the question: It takes Kylee 4 hours to drive home.

Example 1

Identify the unknown value in the last sentence of each of these applications. Set up a legend to define the variable representation of the unknown value.

a) Two of the angles in a triangle measure 30° and 50°. What is the measure of the third angle?

b) Monica drove 150 miles to get to the Human Resources Trainers Convention. It took her 3 hours to get there. What was Monica's average rate of speed while driving?

c) A rectangle has a perimeter of 54 inches and a length of 20 inches. What is the width of the rectangle?

Procedure Each of these applications has only one unknown value. It can be identified from the last sentence.

Answer **a)** Legend: Let x = the measure of the third angle.

b) Legend: Let r = Monica's average rate of speed.

c) Legend: Let w = the width of the rectangle.

You Try It 1 **Identify the unknown value in the last sentence of each of these applications. Set up a legend to define the variable representation of the unknown value. Use Example 1 as a guide.**

a) Marjorie took a trip up the East Coast. The total distance she drove was 540 miles, and the total amount of time she spent driving was 12 hours. What was Marjorie's average rate of speed for the duration of the trip?

Legend ____________________

b) A triangle has one angle that measures 47° and another angle that measures 79°. What is the measure of the third angle?

Legend ______________________________

c) Bette wants to use 60 yards of fencing to enclose a rectangular corral. She wants the width to be 12 yards. What will be the length of the corral?

Legend ______________________________

Guideline 3: Draw a Diagram

Some diagrams are easy to draw, such as rectangles and triangles:

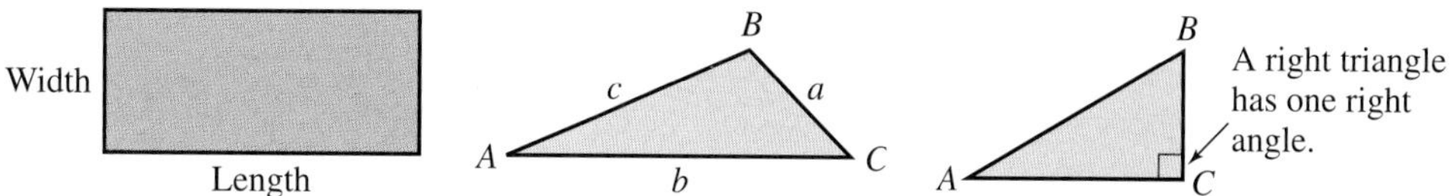

Sometimes you can be creative and show a diagram, or picture, with other situations as well. Drawing a diagram, as in Examples 2 and 3, can help you organize the information.

Example 2

Set up the legend and draw a diagram for the following applications.

a) A rectangle has a perimeter of 54 inches and a length of 20 inches. What is the width of the rectangle?

b) Monica drove 150 miles to get to the Human Resources Trainers Convention. It took her 3 hours to get there. What was Monica's average rate of speed while driving?

Procedure Draw a diagram and label the known and unknown values.

Answer a) Legend: Let w = The width of the rectangle.

b) Legend: Let r = Monica's average rate of speed.

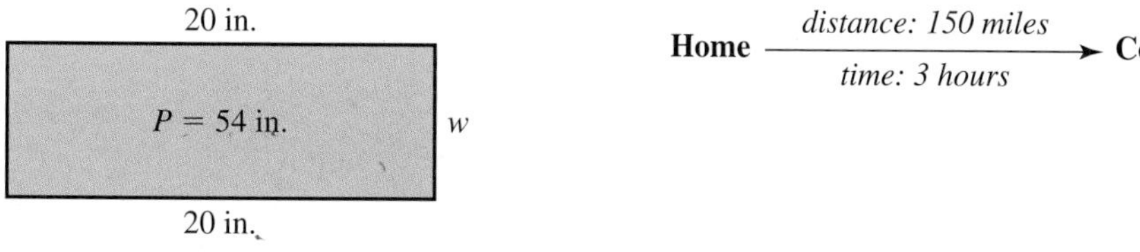

You Try It 2

Set up the legend and draw a diagram for the following applications. Use Example 2 as a guide.

a) In a right triangle, the measure of the smallest angle is 25°. What is the measure of the third angle?

b) A 72-inch board is cut into two pieces so that the longer piece is 48 inches. What is the length of the shorter piece?

c) Bette wants to use 60 yards of fencing to enclose a rectangular corral. She wants the width to be 12 yards. What will be the length of the corral?

d) A new photocopy machine and an older model were used to create 540 manuals. If the newer copier made 321 copies, how many copies did the older machine make?

Guideline 4: Identify a Formula and Develop the Equation

To solve an application, we must approach the problem in a methodical manner. We need to recognize known values and unknown values. We need to identify an equation—a formula—that will allow us to use our algebraic techniques to answer the question.

The generic formula we'll use is this:

The sum of all of the parts equals the whole.

This generic formula is the basis of some of the formulas in geometry. In particular, the generic formula generates the perimeter formulas for both the rectangle and triangle.

Rectangle: Perimeter $= 2 \cdot \text{Length} + 2 \cdot \text{Width}$

Triangle: Perimeter $= a + b + c$

as well as the formula for the sum of the angles in a triangle:

$$A + B + C = 180°.$$

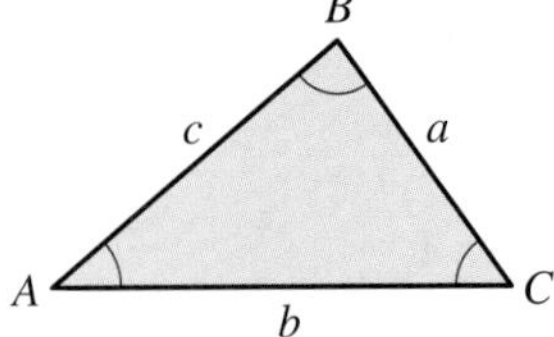

Side lengths are a, b, and c. Angle measures are A, B, and C.

In a right triangle, the right angle is known to be 90°.

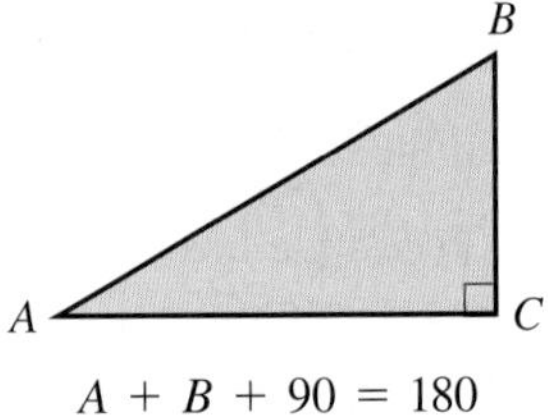

$A + B + 90 = 180$

The generic formula—**the sum of all of the parts equals the whole**—is not restricted to just geometric formulas. It relates to other situations, many of which might not have a ready-made formula.

Each of the following situations uses the generic formula.

Example 3

For each situation, create an equation using the generic formula. Use words to describe the parts and the whole in the equation.

a) A board is cut into two pieces.

b) An inheritance is split between two people.

c) A photocopying job is divided between two machines.

Answer

a) Length of shorter piece + Length of longer piece = Total length of the board

b) First person's share + Second person's share = Total inheritance

c) Number of copies from first + Number of copies from second = Total number of copies

In this section, you will use two formulas that involve multiplication. These two formulas do not rely on the generic formula. One is the motion formula $\boldsymbol{d = r \cdot t}$, distance = rate × time. The other is the formula for the area of a rectangle $\boldsymbol{A = L \cdot W}$, area = length × width. (Recall from Section 1.3 that area is always measured in *square* units.)

Example 4

Identify the formula that pertains to each application.

a) Two of the angles in a triangle are 30° and 50°. Find the measure of the third angle.

b) Monica drove 150 miles to get to the Human Resources Trainers Convention. It took her 3 hours to get there. What was Monica's average rate of speed while driving?

c) A rectangle has a perimeter of 54 inches and a length of 20 inches. What is the width of the rectangle?

d) A rectangular conference room has an area of 120 square feet. The length is 15 feet. What is the width of the room?

e) In a small town's election for mayor, a total of 540 votes were cast. If the winner received 312 votes, how many votes did the losing candidate receive?

Answer

a) The sum of the angles in a triangle is 180°: $\mathbf{A + B + C = 180}$

b) Distance = Rate · Time: $\mathbf{d = r \cdot t}$

c) Perimeter = 2 · Length + 2 · Width: $\mathbf{P = 2L + 2W}$

d) Area = Length · Width: $\mathbf{A = L \cdot W}$

e) Generic: Number of winner's votes + Number of loser's votes = Total votes: $\mathbf{W + L = Total}$

You Try It 3 **Identify the formula that pertains to each application. Use Examples 3 and 4 as guides.**

a) In a right triangle, the measure of the smallest angle is 25°. What is the measure of the middle angle?

b) A 72-inch board is cut into two pieces so that the longer piece is 48 inches. What is the length of the shorter piece?

c) Bette wants to use 60 yards of fencing to enclose a rectangular corral. She wants the width to be 12 yards. What will be the length of the corral?

d) A new photocopy machine and an older model were used to create 540 manuals. If the newer copier made 321 copies, how many copies did the older machine make?

Guideline 5: Write the Solution in Sentence Form

Everything discussed so far leads to an algebraic equation that we must solve to answer the question.

Once the equation has been solved, it's important to determine whether the solution makes sense and then write a complete sentence answering the question. In the sentence, include the correct unit of measure, such as *feet, dollars, pounds,* or *degrees.*

Solving Application Situations

Let's use the guidelines outlined in this section to solve a couple of application situations. Remember, don't try to solve the problem all at once. Also, be neat and organized in your work.

Example 5

Tim was disgusted with his performance on his first test and crumpled up the test without paying much attention to the grade. Determined to do better on the next test, he studied hard and received a score of 88 points. On the test, Tim's instructor wrote Tim's total score for both tests: Total = 146 points. In light of this, how many points did Tim receive on his first test?

Procedure Use the following guidelines to help set up the problem.

Answer **a)** How many unknowns are there? There is one unknown value.

b) Set up the legend: Let x = Tim's score on the first test.

c) Draw and label a diagram.

Teacher's Gradebook

Student	Test 1	Test 2
Tim	x	88

d) Identify a formula.

$$\underbrace{\text{Score on first test} + \text{Score on second test}}_{\text{The sum of the parts}} = \underbrace{\text{Total score}}_{\text{The whole}}$$

$$x + 88 = 146$$

$$x + 88 + (-88) = 146 + (-88)$$

$$x = 58$$

Isolate the variable by adding -88 to each side.

Our legend tells us that x = Tim's score on the first test; so, his first test score was 58 points.

e) Answer the question with a sentence.
Tim received 58 points on the first test.

Does this answer make sense? We can check by adding the two test scores together; they should add to 146.

$$\begin{array}{r} 58 \\ +\ 88 \\ \hline 146 \end{array} \quad \text{Yes!}$$

Example 6

In a right triangle, the middle angle is 58°. What is the measure of the smallest angle?

Procedure Use the following guidelines to help you set up the problem.
Remember, in a right triangle, the largest angle is 90°.

Answer **a)** How many unknowns are there? There is one unknown value.

b) Set up the legend. Let x = the measure of the smallest angle.

c) Draw and label a diagram.

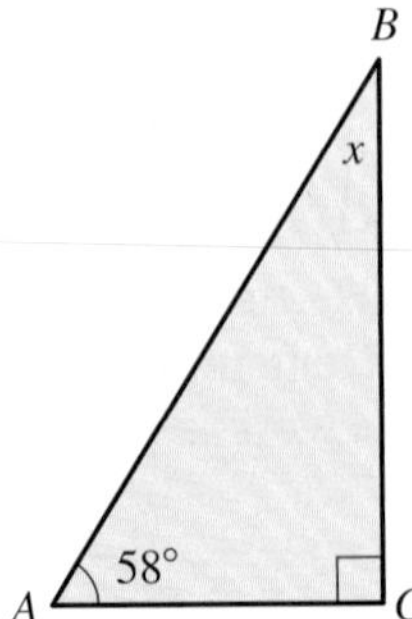

d) Identify a formula. The sum of the angles in a triangle is 180°.

$$A + B + C = 180°$$ Place the angle measures into the formula.

$$58 + x + 90 = 180$$ Combine like terms.

$$x + 148 = 180$$ Isolate the variable by adding −148 to each side.

$$x + 148 + (-148) = 180 + (-148)$$

$$x = 32$$

Our legend tells us that x = the measure of the smallest angle; so, the smallest angle is 32°.

e) Answer the question with a sentence.
The smallest angle is 32°.

Does this answer make sense? We can check by adding the three angle measures together; they should add to 180°.

$$\begin{array}{r} 32 \\ 58 \\ +\ 90 \\ \hline 180 \end{array}$$ Yes!

You Try It 4 **Solve each application completely. Write a legend, draw a diagram, identify a formula, solve the equation, and write a sentence answering the question. Use Examples 5 and 6 as guides.**

a) Marjorie took a trip up the East Coast. The total distance she drove was 540 miles, and the total amount of time she spent driving was 12 hours. What was Marjorie's average rate of speed for the duration of the trip?

b) Bette wants to use 60 yards of fencing to enclose a rectangular corral. She wants the width to be 12 yards. What will be the length of the corral?

c) In a right triangle, the measure of the smallest angle is 25°. What is the measure of the middle angle?

d) A new photocopy machine and an older model were used to create 540 manuals. If the newer copier made 321 copies, how many copies did the older machine make?

Answers: You Try It and Think About It

You Try It: *Variables other than the ones shown here may be used in the legends.*
1. a) Legend: Let x = Marjorie's average rate of speed. **b)** Legend: Let x = the measure of the third angle. **c)** Legend: Let x = the length of the rectangle.

2. a)

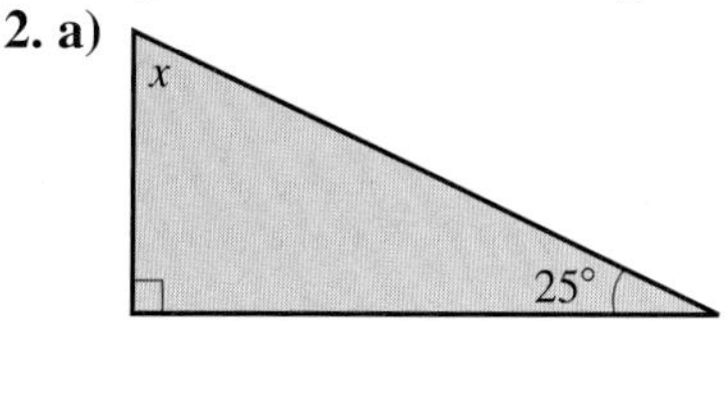

b)

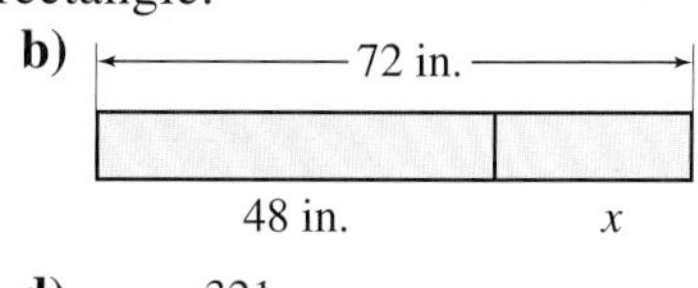

c)

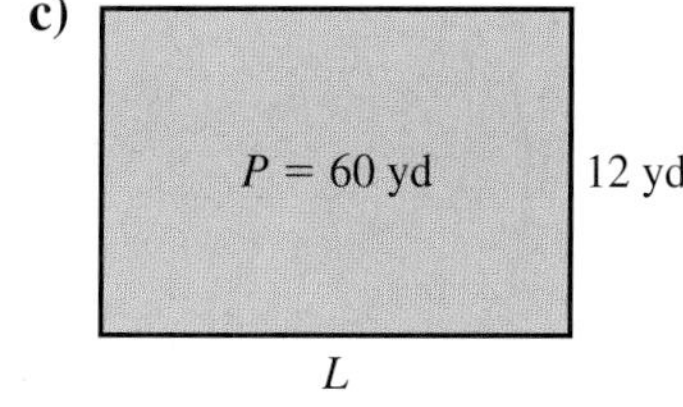

d)

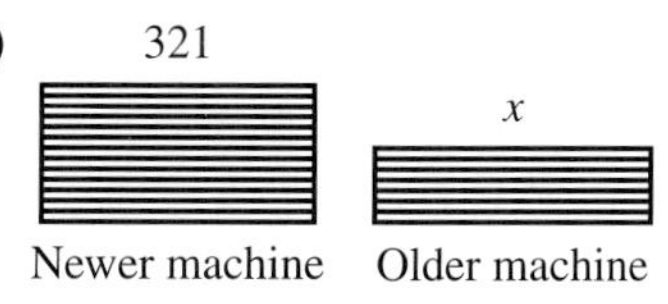

3. a) The sum of the angles in a triangle is 180°: $A + B + C = 180°$. **b)** Sum of the parts = Whole: Length of shorter piece + Length of longer piece = Total **c)** Perimeter = 2 · Length + 2 · Width; $P = 2L + 2W$ **d)** Sum of the parts = Whole: Number of copies from older machine + Number of copies from newer machine = Total number of copies
4. a) Marjorie's average rate of speed was 45 miles per hour. **b)** The length of the corral will be 18 yards. **c)** The middle angle is 65°. **d)** The older machine printed 219 copies.

Think About It: No Think About Its in this section.

Section 3.5 Exercises

FOR EXTRA HELP
Student Resources on DVD-ROM — Includes ➤ Student's Solutions Manual ➤ Video Lectures ➤ Chapter Test Prep Video | MyMathLab | MathXL

Think Again

1. What purpose does the legend serve when we are setting up an application problem?
   The legend helps define the number that the variable represents.

2. When we are setting up an application problem, why is a diagram helpful?
   Answers will vary. Some possibilities: The diagram helps us visualize the problem. It helps us place numbers and letters on the diagram. It helps us recognize a formula and develop the equation.

3. What information should be included in the concluding sentence of an application problem?
   Answers will vary. One possibility: The sentence should include many of the same words in the question and the answer along with the unit of measure.

4. After reading an application problem through once, what should a student do if he or she does not know the answer to the question?
   Answers will vary. One possibility: Read the problem again and look for important information.

Focus Exercises

Solve each application and answer with a complete sentence. (*Hint:* Set up a legend, draw a diagram, identify the formula, and create and solve the equation.)

5. Two of the angles in a triangle measure 108° and 37°. What is the measure of the third angle?
   The measure of the third angle is 35°.

6. In a right triangle, the measure of the smallest angle is 26°. What is the measure of the middle angle?
   The measure of the middle angle is 64°.

7. The perimeter of a rectangle is 178 inches. If the length is 58 inches, what is the width?
   The width of the rectangle is 31 inches.

8. The perimeter of a square is 152 inches. What is the length of each side of the square?
   The length of each side is 38 inches.

9. Lisa wants to plant a garden in her backyard. Her plan is to make the garden rectangular so that the length is 16 feet. She's planning to use 58 feet of chicken wire to surround the garden (as a perimeter). What should be the width of the garden so that she can use all 58 feet of fencing?
   The width of the garden should be 13 feet.

10. Kay wants to bend a 108-inch wire into a rectangle so that the width is 24 inches. What is the length of this rectangle?
    The length of the rectangle is 30 inches.

11. Adrienne did not score well on her first test. She studied much harder for her second test and got a score of 95. Her teacher indicated that her two test scores totaled 168 points. What was Adrienne's score on the first test?
    Adrienne's score on the first test was 73.

12. Anthony received scores of 87 points and 85 points on his first two tests. By the end of the third test, his goal was to have a total score of 270 points. (This would give him an average score of 90.) How many points will Anthony need to score on his third test to reach his goal?
    Anthony will need to score 98 points on his third test.

13. This Saturday, Margo plans to ride her bike 42 miles from her home to the beach. If she is able to average 14 miles per hour, how long will it take Margo to get to the beach?
    It will take Margo 3 hours to get to the beach.

14. Every summer, Andy hikes a mountain trail in southwestern Colorado. The route he takes is 72 miles long. If he can average 18 miles per day, how many days will it take Andy to hike this trail?
    It will take Andy 4 days to hike this trail.

15. Silvio wants to bend a 48-inch wire into a triangle so that the longest side is 21 inches and the shortest side is 12 inches. What is the length of the third side?
    The length of the third side is 15 inches.

16. Reggie owns a triangular plot of land. One side is 185 yards long, and another side is 212 yards long. If the total perimeter is 626 yards, what is the length of the third side?
    The length of the third side is 229 yards.

17. Kacey has 126 square feet of ceramic tile with which to cover a rectangular floor. She wants to build herself an office so that she can use this tile. If the floor's width is to be 9 feet, what will be the length of the floor?
    The length of the floor will be 14 feet.

18. A developer has built a house on a rectangular piece of land with an area of 1,200 square yards. If the width of this land is 25 yards, what is its length?
The length of this land is 48 yards.

19. Jenna is an artist who makes her own canvases. For her next project, she wants a rectangular canvas with an area of 720 square inches and a width of 15 inches. What will be the length of the canvas?
The length of the canvas will be 48 inches.

20. Derek wants to build a computer desk that has a rectangular top. The top is to have an area of 1,080 square inches and a width of 24 inches. What will be the length of the top?
The length of the top will be 45 inches.

21. Hannah and Therese were driving from their home in Phoenix, Arizona, to their college in San Antonio, Texas. The total trip was 983 miles. If Therese drove 498 miles, how many miles did Hannah drive?
Hannah drove 485 miles.

22. Jimi and Stephyn took a train and bus trip through Canada from Vancouver, British Columbia, to Toronto, Ontario. The total trip was 2,762 miles. If they traveled 747 miles on a bus, how many miles did they ride on the train?
They rode 2,015 miles on the train.

23. Roberto and Luisa Barron-Lopez each work full-time jobs; and Luisa's mother, Carmena, collects a monthly pension of \$628. Between the three of them, their total household income is \$4,238 each month. If Luisa earns \$1,772, how much does Roberto earn each month?
Roberto earns \$1,838 each month.

24. Three candidates were vying for Commissioner of Clay County, West Virginia: Katie DePasquale, Jenny Crum, and Ron Hampton. Of the 2,543 votes cast, Katie received 684 votes and Ron received 802 votes. How many votes did Jenny receive?
Jenny received 1,057 votes.

Think Outside the Box

Solve each application and answer with a complete sentence. (*Hint:* Set up the legend, draw a diagram, identify the formula, and create and solve the equation.)

25. The width of a rectangle is 2 feet, and the perimeter is 134 inches. What is the length of the rectangle?
The length of the rectangle is 43 inches.

26. Together, Mia and Damon earned a total average of \$3,819 each month in 2006. If Mia earned \$24,936 that year, how much did Damon earn?
Damon earned \$20,892.

SECTION 3.6 Solving Applications Involving Two Unknown Values

OBJECTIVES

In this section, you will learn to

- Set up a legend for two unknown values.
- Identify a formula for a given situation.
- Solve an application using algebra.

You Need to Know

To successfully complete this section, you need to understand

- ☐ Translating expressions (2.1)
- ☐ Solving equations (3.1, 3.2, and 3.3)
- ☐ Solving problems (3.4)
- ☐ Solving applications (3.5)

Introduction

Here is a typical application involving two unknown values:

> A rectangle has a perimeter of 44 inches. The length is 1 inch longer than twice the width. What are the dimensions of the rectangle?

In general, we solve applications with two unknown values the same way we did in the previous section. We may draw a diagram, we identify a formula and set up and solve an equation, and we still write a sentence answering the question.

The legend, however, will now contain two unknown values. In this section, you will learn how to decide which unknown value will be x (or w or h) and how to write the other unknown value in terms of x (or w or h).

Determining the Legend for Two Unknown Values

Sometimes a situation has more than one unknown value. In this case, there will always be a comparison between the two unknown values. The comparison is found in one of the sentences describing the situation.

Here are some sentences that compare two unknown values. The comparison phrase is underlined, and the unknown values are in bold.

1. The **length** is 3 feet longer than the **width.**
2. Cindy's **score on the second test** was 8 points lower than her **score on the first test.**
3. It took **Trahn** 1 hour longer than **Pam** to complete the bicycle course. (Here, the unknown values are **Trahn's *time*** and **Pam's *time*.**)
4. **Maria** took twice as long as **Mark** to finish the quiz. (Here, the unknown values are **Maria's *time*** and **Mark's *time*.**)

Each of those sentences fits this pattern:

The first unknown value compared to **a second unknown value.**

You might ask, "Which unknown do I make x, the first or the second? Does it matter?"

Yes, it matters. The individual variable, x, is always the *second* unknown mentioned. The first unknown is being compared to x.

Example 1

Identify the first unknown value, the comparison, and the last unknown value in each of these sentences. Also decide which one should be represented by *just* x.

a) The length is 3 feet longer than the width.

b) Cindy's score on the second test was 8 points lower than her score on the first test.

c) It took Trahn 1 hour longer than Pam to complete the bicycle course.

d) Maria took twice as long as Mark to finish the quiz.

e) The height is 4 feet less than twice the length.

Answer

	First Unknown	Comparison	Second Unknown	*Part* of the Legend
a)	Length	3 feet longer than	The width	w = the width
b)	Second test score	8 points lower than	First test score	x = the first test score
c)	Trahn's riding time	1 hour longer than	Pam's riding time	p = Pam's riding time
d)	Maria's time	Twice	Mark's time	m = Mark's time
e)	Height	4 feet less than twice	The length	x = the length

You Try It 1 **Underline the first unknown value, the comparison, and the last unknown value in each of these sentences. Decide which one should be represented by *just x* and set up a legend. Use Example 1 as a guide.**

a) Sarah wants to build a pigpen using 36 feet of fencing. Her design is a simple rectangle in which the width is 4 feet less than the length. What will be the dimensions (length and width) of the pen?

Legend ____________________

b) In a right triangle (in which one of the angles is known to be 90°), the measure of the smallest angle is 10° less than the measure of the middle angle. What are the measures of the smallest and middle angles?

Legend ____________________

c) Tom needs to cut a 6-foot (72 inches) board into two pieces. The bigger piece is to be 12 inches longer than the smaller piece. What will be the lengths of the two pieces?

Legend ____________________

d) Harry just passed away and left $87,000 for his two grandchildren's college funds. Harry anticipated that the younger grandchild, Nate, would pay more for college (when he is old enough to attend) than would his sister, Katie, since she is already a high school senior. For this reason, Harry's will states that Nate's college fund is to receive $15,000 more than Katie's. How much will each child's college fund receive?

Legend ____________________

Interpreting the First Unknown in a Comparison

We have just learned how to write the legend for the *second* unknown in a comparison. Now let's look at the legend for the *first* unknown.

If the second unknown is represented by just x, the first unknown will be written in terms of x based on the statement of comparison.

Example 2

Create a legend for both unknown values based on the comparison given.

a) The length is 3 feet longer than the width.

b) Cindy's score on the second test was 8 points lower than her score on the first test.

c) It took Trahn 1 hour longer than Pam to complete the bicycle course.

d) Maria took twice as long as Mark to finish the quiz.

e) The height is 4 feet less than twice the length.

Procedure Let x represent the second unknown value. Then use the comparison to decide the first unknown value.

Answer

	First Unknown	Comparison	Second Unknown	Legend
a)	Length	3 feet longer than	Width	w = width $w + 3$ = length
b)	Second test score	8 points lower than	First test score	x = first test score $x - 8$ = second test score
c)	Trahn's riding time	1 hour longer than	Pam's riding time	p = Pam's time $p + 1$ = Trahn's time
d)	Maria's time	Twice	Mark's time	m = Mark's time $2m$ = Maria's time
e)	Height	4 feet less than twice	Length	x = length $2x - 4$ = height

You Try It 2 **Underline the first unknown value, the comparison, and the last unknown value in each of these sentences. Create a legend for the two unknown values. Use Example 2 as a guide.**

a) Sarah wants to build a pigpen using 36 feet of fencing. Her design is a simple rectangle in which the length is three times the width. What will be the dimensions (length and width) of the pen?

Legend ______________________________

____

b) In a right triangle (in which one of the angles is known to be 90°), the measure of the first angle is 10° less than the measure of the second angle. What are the measures of the first and second angles?

Legend ______________________________

____

c) Tom needs to cut a 6-foot (72 inches) board into two pieces. The bigger piece is to be 12 inches longer than the smaller piece. What will be the lengths of the two pieces?

Legend ______________________________

____

d) Harry just passed away and left $87,000 for his two grandchildren's college funds. Harry anticipated that the younger grandchild, Nate, would pay more for college (when he is old enough to attend) than would his sister, Katie, since she is already a high school senior. For this reason, Harry's will states that Nate's college fund is to receive $15,000 more than Katie's. How much will each child's college fund receive?

Legend ______________________________

____

Solving Applications Involving Two Unknowns

Let's use the Guidelines for Mastering an Application Problem outlined at the beginning of Section 3.5, on page 217, to solve a couple of application situations. Remember, don't try to solve a problem all at once. Also be neat and organized in your work.

Example 3

In geography, Rami's second test score was 18 points higher than her first test score. If the total of her two scores is 166 points, how many points did she receive on each test?

Procedure Use the following guidelines to help set up the problem.

Answer **a)** How many unknowns are there? **There are two unknown values.**

b) Set up the legend. Because the *first test score* is written second in the comparison,

Let x = Rami's score on the first test.

$x + 18$ = Rami's score on the second test.

c) Draw and label a diagram.

Teacher's Gradebook

Student	Test 1	Test 2
Rami	x	$x + 18$

d) Identify a formula.

The sum of the parts = The whole

Score on first test + Score on second test = Total score

$x + (x + 18) = 166$ — Remove the parentheses from the left side of the equation.

$x + x + 18 = 166$ — Combine like terms.

$2x + 18 = 166$ — Add -18 to each side.

$2x + 18 + (-18) = 166 + (-18)$ — Simplify.

$2x = 148$ — Divide each side by 2.

$\frac{2x}{2} = \frac{148}{2}$ — Simplify.

$x = 74$ — Divide each side by 2.

Our legend tells us that x = Rami's score on the first test; so, her *first* test score was 74 points.

Our legend also tells us that $x + 18$ = her second test score; so, her *second* test score was 92 points.

e) Answer the question with a sentence.

Rami received 74 points on the first test and 92 points on the second test.

Does this answer make sense? We can check by adding the two test scores together. They should add to 166.

$$\begin{array}{r} 74 \\ +\ 92 \\ \hline 166 \end{array}$$ Yes!

Example 4

In a right triangle, the middle angle is 12 more than twice the smallest angle. What are the measures of the smallest and middle angles?

Procedure Use the following guidelines to help set up the problem. Remember, in a right triangle, the largest angle is 90°.

Answer **a)** How many unknowns are there? There are two unknown values.

b) Set up the legend. Let x = the measure of the smallest angle.

$2x + 12$ = the measure of the middle angle.

c) Draw and label a diagram.

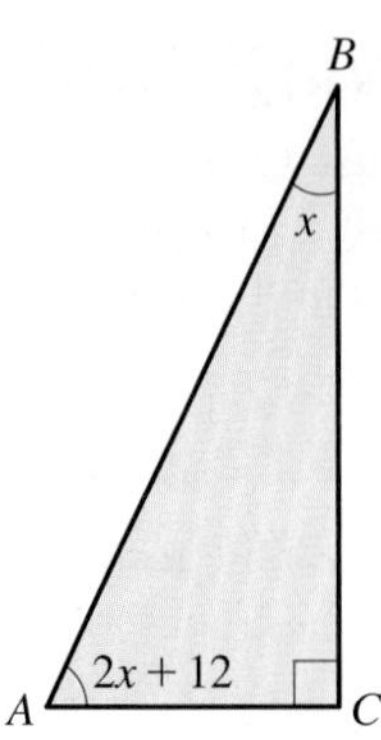

d) Identify a formula. The sum of the angles in a triangle is 180°.

$$A + B + C = 180°$$ Place the angle measures into the formula.

$$(2x + 12) + x + 90 = 180$$ Remove the parentheses.

$$2x + 12 + x + 90 = 180$$ Combine like terms.

$$3x + 102 = 180$$ Isolate the variable term by first adding −102 to each side.

$$3x + 102 + (-102) = 180 + (-102)$$

$$3x = 78$$ Divide each side by 3.

$$\frac{3x}{3} = \frac{78}{3}$$ Simplify.

$$x = 26$$

Our legend tells us that $x =$ the measure of the smallest angle; so, the smallest angle is 26°.

The legend also tells us that the middle angle measures $2x + 12$; so, the middle angle is $2(26) + 12 = 64°$

e) Answer the question with a sentence.

The smallest angle is 26°, and the middle angle is 64°.

Do these answers make sense? We can check by adding the three angle measures together. They should add to 180°.

$$\begin{array}{r} 26 \\ 64 \\ +\ 90 \\ \hline 180 \end{array}$$ Yes!

Caution In a situation with two unknown values, don't stop after finding the value of *x*; *x* applies to one unknown value only. Use the legend to find the value of the other unknown.

You Try It 3 **Solve each application completely and write a sentence answering the question. Use the guidelines outlined in Examples 3 and 4: Write a legend, draw a diagram, identify a formula, and solve the equation.**

a) A new photocopy machine can print twice as many copies as an older machine. If together the machines printed 540 copies, how many copies did each machine print?

b) Tomaso and Hilo were in a hot dog eating contest. Together they ate 33 hot dogs. If Tomaso ate 5 more hot dogs than Hilo, how many hot dogs did each boy eat?

c) In a right triangle, the measure of the middle angle is 18° more than the measure of the smallest angle. What are the measures of the smallest and middle angles?

d) A rectangle has a perimeter of 44 inches. The length is 1 inch longer than twice the width. What are the dimensions of the rectangle?

Answers: You Try It and Think About It

You Try It: **1. a)** Legend: Let x = the length of the rectangle. **b)** Legend: Let x = the measure of the middle angle. **c)** Legend: Let x = the measure of the smaller piece. **d)** Legend: Let x = the amount that Katie's fund receives. **2. a)** Legend: Let w = the width of the rectangle. $3w$ = the length of the rectangle. **b)** Legend: Let x = the measure of the second angle. $x - 10$ = the measure of the first angle. **c)** Legend: Let x = the length of the smaller piece. $x + 12$ = the length of the bigger piece. **d)** Legend: Let k = the amount that Katie's fund receives. $k + 15{,}000$ = the amount that Nate's fund receives. **3. a)** The older machine printed 180 copies; the newer machine printed 360 copies. **b)** Hilo ate 14 hot dogs, and Tomaso ate 19 hot dogs. **c)** The smallest angle is 36°, and the middle angle is 54°. **d)** The width is 7 inches, and the length is 15 inches.

Think About It: No Think About Its in this section.

Section 3.6 Exercises

FOR EXTRA HELP — *Student Resources* on DVD-ROM. Includes ➤Student's Solutions Manual ➤Video Lectures ➤Chapter Test Prep Video | MyMathLab | MathXL

Think Again

1. When an application problem has two unknowns, should they both be represented in the legend? Explain your answer or show an example that supports your answer.
   Yes. Answers will vary.

2. When an application problem has two unknowns, why is it better to express both in terms of one variable, x, rather than one in terms of x and the other in terms of y?
   It is better to express the two unknowns in terms of one variable because the equation is in terms of only one variable.

Focus Exercises

Solve each application and answer with a complete sentence. (*Hint:* Set up the legend, draw a diagram, identify the formula, and create and solve the equation.)

3. In a recent school PTA election, two candidates were running for president. Of the 120 votes cast, Mrs. Jenkins received 16 fewer votes than Mr. Daniels. How many votes did each candidate receive?
   Mr. Daniels received 68 votes, and Mrs. Jenkins received 52 votes.

4. Jorge and Darush participated in the Boys & Girls Clubs Walk for Hunger. Together the boys walked a total of 58 kilometers. If Darush walked 4 kilometers fewer than Jorge, how many kilometers did each boy walk?
   Jorge walked 31 miles, and Darush walked 27 miles.

5. Roberta wants to cut a 24-foot wire into two pieces so that the longer piece is 3 times the length of the shorter piece. What is the length of each piece?
   The shorter piece is 6 feet, and the longer piece is 18 feet.

6. Suki wants to saw a 96-inch board into two pieces so that the smaller piece is 18 inches shorter than the bigger piece. What is the length of each piece?
   The bigger piece is 57 inches, and the shorter piece is 39 inches.

7. Chandra and Eli are tile layers. Eli is an apprentice and is learning the trade from Chandra. Together they are to be paid $2,200 to tile a bathroom. If Chandra is paid $500 more than Eli, how much is each to be paid for this job?
   Eli is to be paid $850, and Chandra is to be paid $1,350.

8. Uncle Jackson has two nieces, Kiesha and Tanya. Tanya is in her first year of college. In his will, he decides to give Tanya $15,000 more than Kiesha so that Tanya can pay for her education. If he leaves them $107,000 in his will, how much will each niece receive?
   Kiesha will receive $46,000, and Tanya will receive $61,000.

9. In a triangle, one of the angles is 100°. The measure of the middle angle is 24° more than the measure of the smallest angle. What are the measures of these two angles?
   The smallest angle is 28°, and the middle angle is 52°.

10. In a right triangle, the measure of the middle angle is 18° more than the measure of the smallest angle. What are the measures of these two angles?
    The smallest angle is 36°, and the middle angle is 54°.

11. After two tests, Carol has a total score of 170. The score on her first test was 8 points higher than the score on her second test. What was Carol's score on each test?
    Carol's score on the second test was 81, and her score on the first test was 89.

12. Omar's score on his third math test was 89. His score on the first test was 5 points higher than his score on the second test. His total for all three tests was 264. What were Omar's scores on the first two tests?
Omar's score on the second test was 85, and his score on the first test was 90.

13. George wants to bend a 30-inch wire into a triangle so that the longest side is 12 inches and the other two sides are the same length. What is the length of each shorter side?
The length of each shorter side is 9 inches.

14. A triangle has a perimeter of 38 inches. One side is 17 inches. The second side is 3 inches less than the third side. What are the lengths of the second and third sides?
The length of the second side is 9 inches, and the length of the third side is 12 inches.

15. A rectangular playground has a perimeter of 180 feet. If the length is 12 feet longer than the width, what are the dimensions of the playground?
The width of the playground is 39 feet, and the length is 51 feet.

16. A rectangular building has a perimeter of 520 feet. If the width is 82 feet shorter than the length, what are the dimensions of the building?
The length of the building is 171 feet, and the width is 89 feet.

17. Two watermelons weigh a total of 34 pounds. The weight of the larger melon is 5 pounds less than twice the weight of the smaller melon. What is the weight of each watermelon?
The smaller watermelon weighs 13 pounds, and the larger watermelon weighs 21 pounds.

18. Moira and Sheila created a film documentary last summer. Together they spent a total of 121 hours on the project. If Sheila's time was 7 hours more than twice Moira's time, how much time did each girl spend on the project?
Moira spent 38 hours, and Sheila spent 83 hours on the project.

19. Together, Shufen and her dad, Zhe, drove 1,472 miles to take Shufen home from college. Zhe drove 155 miles more than twice the number of miles that Shufen drove. How many miles did each of them drive?
Shufen drove 439 miles, and her dad, Zhe, drove 1,033 miles.

20. A rectangular field has a perimeter of 772 feet. If the length is 8 feet longer than twice the width, what are the dimensions of the field?
The width of the field is 126 feet, and the length is 260 feet.

21. Marisa and Shandrell ran for student body president. Of the 1,308 votes cast, Shandrell received 16 less than three times as many votes as Marisa. How many votes did each candidate receive?
Marisa received 331 votes, and Shandrell received 977 votes.

22. One angle of a triangle measures 65°. The largest angle is 9 degrees less than three times the measure of the smallest angle. What are the measures of the largest and smallest angles?
The measure of the smallest angle is 31°, and the measure of the largest angle is 84°.

Think Outside the Box

Solve each application and answer with a complete sentence. (*Hint:* Set up the legend, draw a diagram, identify the formula, and create and solve the equation. Caution: Each application has more than two unknown values.)

23. Geri is an apprentice for a construction firm. She is asked to bend a 93-inch metal bar into the shape of a triangle so that the middle side is 5 inches longer than the shortest side and the longest side is twice as long as the shortest side. What are the lengths of the three sides of this triangle?
The sides of the triangle are 22 inches, 27 inches, and 44 inches.

24. A triangle has a perimeter of 88 inches. The first side is 1 foot longer than the second side, and the third side is twice as long as the second side. What are the lengths of the three sides?
The sides of the triangle are 19 inches, 31 inches, and 38 inches.

25. The three angles in a triangle are such that the measure of the largest angle is three times the measure of the smallest angle. The middle-sized angle is 9° less than the measure of the largest angle. What are the measures of the three angles?
The angles of the triangle are 27°, 72°, and 81°.

26. Uncle Harry has three nieces. In his will, he decides to give Janet $10,000 more than he gives Nancy and to give Cathy twice as much as he gives Janet. If Harry's will leaves $92,000 to the three nieces, how much will each girl receive?
Nancy will receive $15,500, Janet will receive $25,500, and Cathy will receive $51,000.

Chapter 3 Review

Section 3.1 Solving Equations Involving One Operation

Concept	Example
An **equation** is a mathematical sentence in which one expression equals another expression. To solve an equation, we must ***isolate the variable*** on one side of the equal sign. The **solution** of an equation makes the equation true, and we can check the solution by replacing the variable with that number.	Is 3 the solution to $5x - 4 = 2x + 5$? Check $x = 3$ $\quad 5(3) - 4 \stackrel{?}{=} 2(3) + 5$ $15 - 4 \stackrel{?}{=} 6 + 5$ $11 = 11$ True ✓
The Addition Property of Equations: We may add any number, c, to *each side* of an equation. We can use this property to *clear a constant* by adding its opposite to each side.	If $\quad a = b \qquad x - 5 = -9$ then $\quad a + c = b + c \qquad x - 5 + 5 = -9 + 5$
The Division Property of Equations: We may divide *each side* of an equation by any number, c (as long as c isn't 0). We can use this property to *clear a coefficient* by dividing each side by the coefficient.	If $\quad a = b \qquad -3x = 12$ then $\quad \frac{a}{c} = \frac{b}{c} \qquad \frac{-3x}{-3} = \frac{12}{-3}$

Section 3.2 Solving Equations Involving Two Operations

Concept	Example
When an equation has two operations, first isolate the variable term by clearing the constant.	$-2x + 9 = 15$ $-2x + 9 + (-9) = 15 + (-9)$ $-2x + 0 = 6$ $-2x = 6$ $\frac{-2x}{-2} = \frac{6}{-2}$ $x = -3$

Section 3.3 Solving Equations Involving More Than Two Operations

Concept	Example
If an equation has a variable term on each side of the equal sign, add the opposite of one of the terms to each side of the equation. This will clear the variable term from one side and place all variable terms on the other side. The equation then has only two operations, and we can continue to solve by clearing the constant.	$5x + 2 = 8x + 11$ $5x + (-5x) + 2 = 8x + (-5x) + 11$ $0 + 2 = 3x + 11$ $2 = 3x + 11$ $2 + (-11) = 3x + 11 + (-11)$ $-9 = 3x$ $\frac{-9}{3} = \frac{3x}{3}$ $-3 = x$ $x = -3$
Sometimes in preparing an equation for solving, we need to simplify one (or both) of the sides. We might need to distribute a number to a quantity, and we might need to combine like terms.	$2x - 10 + 5x = 2(3x - 4)$ $7x - 10 = 6x - 8$ $7x + (-6x) - 10 = 6x + (-6x) - 8$ $x - 10 = 0 - 8$ $x - 10 = -8$ $x - 10 + 10 = -8 + 10$ $x = 2$

Section 3.4 Problem Solving

Concept	Example
Some expressions contain more than one operation.	"The **sum** of 5 and *twice* a number" suggests adding two terms together, 5 and $2x$: $5 + 2x$.
In English, the equal sign can be translated as *equals, is,* or *totals*.	The sum of 5 and twice a number is 17. $5 + 2x = 17$

Section 3.5 Solving Applications Involving One Unknown Value

Concept	Example
When reading through a problem, we find known and unknown values. We use a **legend** to identify the unknown value(s) in terms of a single variable such as x. When a problem has one unknown value, the legend is Let x = the unknown value.	The workers at a factory cast 300 ballots in an election for a new union president. Running for the position were Henrietta Howard and Benson Fox. Henrietta received 145 votes. How many votes did Benson receive? **Legend:** Let x = the number of votes Benson received.

Concept	Example
Guidelines for mastering an application problem	
1. Overview: Read the application once to get a general idea of what it involves. (Don't be overwhelmed.) Then read the application more carefully and start identifying key information such as known and unknown values.	**1.** The workers at a factory cast 300 ballots in an election for a new union president. Running for the position were Henrietta Howard and Benson Fox. Henrietta received 145 votes. How many votes did Benson receive?
2. Legend: Decide how many unknown values the application is suggesting and use a variable to represent each unknown value. Write a legend to easily identify the representations.	**2. Legend:** Let x = the number of votes Benson received.
3. Diagram: If possible, draw a diagram. Label the diagram appropriately with known and unknown values.	**3.** 
4. Formula: Identify a formula that fits the information. From that formula, develop an equation. Solve the equation. Check the answer to see if it is a solution to the equation and if it makes sense given the application.	**4.** Henrietta's votes + Benson's votes = Total votes $145 + x = 300$ $145 + (-145) + x = 300 + (-145)$ $x = 155$
5. Conclusion: Write the solution in a complete sentence. If possible, use words that indicate correct units of measure, such as *feet, miles, hours,* and *pounds*.	**5.** Benson received 155 votes.

Section 3.6 Solving Applications Involving Two Unknown Values

Concept	Example
When a problem has two unknown values, it will contain a statement of comparison: the first unknown value *compared to* a second unknown value. Use this statement of comparison to help set up the legend. In a comparison, x always represents the second unknown mentioned. Write the other unknown in terms of x based on the comparison.	The workers at a factory cast 300 ballots in their election for a new union vice president. Running for the position were Minnie Drover and Abigail Reyes. Minnie received 70 more votes than Abigail received. How many votes did each candidate receive? **Legend:** Let x = the number of votes Abigail received (the second unknown). $x + 70$ = the number of votes Minnie received (the first unknown).

continued

Sections 3.6 Solving Applications Involving Two Unknown Values

Concept	Example
Guidelines for mastering an application problem	
1. Overview: Read the application once to get a general idea of what it involves. (Don't be overwhelmed.) Then read the application more carefully and start identifying key information such as known and unknown values.	**1.** The workers at a factory cast 300 ballots in their election for a new union vice president. Running for the position were Minnie Drover and Abigail Reyes. Minnie received 70 more votes than Abigail received. How many votes did each candidate receive?
2. Legend: Decide how many unknown values the application is suggesting and use a variable to represent each unknown value. Write a legend to easily identify the representations.	**2. Legend:** Let $x =$ the number of votes Abigail received (the second unknown). $x + 70 =$ the number of votes Minnie received (the first unknown).
3. Diagram: If possible, draw a diagram. Label the diagram appropriately with known and unknown values.	**3.** 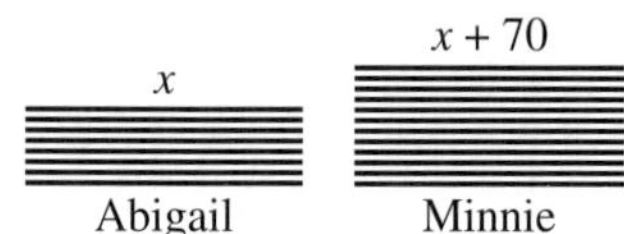
4. Formula: Identify a formula that fits the information. From that formula, develop an equation. Solve the equation. Check the answer to see if it is a solution to the equation and if it makes sense given the application.	**4.** Abigail's votes + Minnie's votes = Total votes $x + (x + 70) = 300$ $x + x + 70 = 300$ $2x + 70 = 300$ $2x + 70 + (-70) = 300 + (-70)$ $2x = 230$ $\frac{2x}{2} = \frac{230}{2}$ $x = 115$
5. Conclusion: Write the solution in a complete sentence. If possible, use words that indicate correct units of measure, such as *feet, miles, hours,* and *pounds*.	**5.** Abigail received 115 votes, and Minnie received 185 votes.

Chapter 3 Review Exercises

True or false.

1. In an equation, to isolate the variable and clear a constant, we add the opposite of the constant to each side of the equation. True **(3.1)**

2. In an equation, to isolate the variable and clear a coefficient, we divide each side of the equation by the opposite of the coefficient. False **(3.1)**

Fill in each blank with the word that correctly completes the sentence.

Word List
legend
formula

3. When solving an application problem, we write a(n) legend to define or describe what the variable represents. **(3.5)**

4. When solving an application problem, we develop the equation from a(n) formula. **(3.5)**

Section 3.1

Determine if the replacement value given is a solution to the equation.

5. $3x - 8 = -2; x = -2$
No, -2 is not a solution.

6. $c + 5 = 2c - 3; c = 8$
Yes, 8 is a solution.

Solve. Make sure you check each answer.

7. $w + 9 = 17$
$w = 8$

8. $d + 12 = 31$
$d = 19$

9. $5 = n + 18$
$n = -13$

10. $2 = 6 + h$
$h = -4$

11. $13 + x = -4$
$x = -17$

12. $-10 = 5 + y$
$y = -15$

13. $m - 8 = 15$
$m = 23$

14. $7 = c - 13$
$c = 20$

15. $-12 + x = -2$
$x = 10$

16. $-4 = -5 + w$
$w = 1$

17. $m - 7 = -10$
$m = -3$

18. $y - 5 = -9$
$y = -4$

19. $-8 = k - 11$
$k = 3$

20. $-5 = 9 + x$
$x = -14$

21. $-2 = -1 + y$
$y = -1$

22. $-4 = -8 + w$
$w = 4$

23. $2n = -22$
$n = -11$

24. $9x = -45$
$x = -5$

25. $-3y = -87$
$y = 29$

26. $-40 = -5m$
$m = 8$

27. $72 = -8y$
$y = -9$

28. $60 = -12x$
$x = -5$

29. $28 = 7p$
$p = 4$

30. $108 = 9y$
$y = 12$

Section 3.2

Solve. Make sure you check each answer.

31. $6x - 12 = 42$
$x = 9$

32. $4w - 16 = -48$
$w = -8$

33. $-6 = 12p + 30$
$p = -3$

34. $3 = 8k + 19$
$k = -2$

35. $-10x + 15 = -35$
$x = 5$

36. $17 = -2x + 23$
$x = 3$

37. $-5y - 6 = -11$
$y = 1$

38. $16 = -8 - 4c$
$c = -6$

39. $8 = -5k - 7$
$k = -3$

40. $-7y - 21 = -35$
$y = 2$

41. $12 = -p + 20$
$p = 8$

42. $1 = -h + 12$
$h = 11$

Section 3.3

Solve. Make sure you check each answer.

43. $6y - 2 = 3y + 13$
$y = 5$

44. $5x - 25 = 15 + 15x$
$x = -4$

45. $6 - 5y = 4 - 3y$
$y = 1$

46. $5p + 8 = -7p + 56$
$p = 4$

47. $-9x + 17 = -7x - 5$
$x = 11$

48. $8 - 5x = -8x + 2$
$x = -2$

49. $-y - 12 + 6y = 9 + 8y$
$y = -7$

50. $7x - 5 - 6x = -x + 19$
$x = 12$

51. $6(y - 5) = 7y - 3$
$y = -27$

52. $5x - 7 = 9(x - 3)$
$x = 5$

53. $7x + 18 = 8(x + 7)$
$x = -38$

54. $8p + 3 - 6p = 12 - 4p + 15$
$p = 4$

Section 3.4

Translate each of the following expressions into the language of algebra. Use the variable *x* to represent the number.

55. Four times a number
$4x$

56. 12 less than a number
$x - 12$

57. 23 decreased by a number
$23 - x$

58. 9 increased by a number
$9 + x$

59. 8 fewer than a number
$x - 8$

60. The sum of 5 and twice a number
$5 + 2x$

61. 6 less than twice a number
$2x - 6$

62. 15 more than six times a number
$6x + 15$

For each of the following, find the requested number and write a sentence that answers the question. (*Hint:* Write a legend for the unknown value, translate the sentence into an equation, solve and check the answer.)

63. 7 more than a number is −3. What is the number?
The number is −10.

64. A number decreased by 12 is −4. What is the number?
The number is 8.

65. Five times a number is 165. What is the number?
The number is 33.

66. Twice a number is −18. What is the number?
The number is −9.

67. The sum of twice a number and 9 is 3. What is the number?
The number is −3.

68. 7 more than twice a number is −15. What is the number?
The number is −11.

69. 5 more than four times a number is 41. What is the number?
The number is 9.

70. 9 less than three times a number is −9. What is the number?
The number is 0.

71. The sum of twice a number and 7 is 33. What is the number?
The number is 13.

72. Twice the sum of a number and 15 is −22. What is the number?
The number is −26.

Section 3.5

Solve each application and answer with a complete sentence. (*Hint:* Set up the legend, draw a diagram, identify the formula, and create and solve the equation.)

73. Enzio drives 455 miles to visit his girlfriend in college. If Enzio averages 65 miles per hour, how long will it take him to get there?
It will take him 7 hours to get there.

74. Berenda has received a total of $1,250 in college scholarships for her freshman year. The first semester she used $783 of the scholarship money. How much scholarship money is available for Berenda's second semester?
$467 is available for Berenda's second semester.

75. In a right triangle, the smallest angle is 19°. What is the measure of the middle angle?
The measure of the middle angle is 71°.

76. The area of Latrice's rectangular painting canvas is 184 square inches. If the width is 8 inches, what is the length of the canvas?
The length of the canvas is 23 inches.

Section 3.6

Solve each application and answer with a complete sentence. (*Hint:* Set up the legend, draw a diagram, identify the formula, and create and solve the equation.)

77. Karl and Kristen are participating in a Relay for Life bicycle event to raise money for cancer research. They agreed before the event that Karl would ride 5 miles more than Kristen. If they rode a total of 63 miles, how many miles did each of them ride?
Kristen rode 29 miles, and Karl rode 34 miles.

78. Luigi's triangular garden has a perimeter of 53 yards. One side is 24 yards. The second side is 7 yards less than the third side. What are the lengths of the second and third sides?
The length of the third side is 18 yards, and the length of the second side is 11 yards.

79. In a triangle, the measure of the largest angle is 100°. The measure of the middle angle is three times the measure of the smallest angle. What are the measures of the middle and smallest angles?
The smallest angle is 20°, and the middle angle is 60°.

80. A rectangular computer memory card has a perimeter of 36 centimeters (cm). The length is 3 cm more than twice the width. What are the dimensions of the computer memory card?
The width of the rectangle is 5 cm, and the length is 13 cm.

Chapter 3 Test

Solve and simplify. Check each answer.

1. $y - 16 = -9$
   $y = 7$
2. $-9x = 36$
   $x = -4$
3. $-7 = -8 + k$
   $k = 1$
4. $-4x = -32$
   $x = 8$
5. $6x - 12 = 30$
   $x = 7$
6. $24 = -18 - 3x$
   $x = -14$
7. $15 - 5w = -75 - 10w$
   $w = -18$
8. $8x - 5 = 25 - 2x$
   $x = 3$
9. $2x + 5 - 9x = 3x - 15$
   $x = 2$
10. $30 - y = 4(6 - y)$
    $y = -2$

For each of the following, the legend is *Let x = the number.* Translate the sentence into an equation and solve. Write a sentence that answers the question.

11. 14 more than a number is -11. What is the number?
    $x + 14 = -11$ The number is -25.
12. 9 less than twice a number is 13. What is the number?
    $2x - 9 = 13$ The number is 11.

Solve each application and answer with a complete sentence.

13. Josh is to receive a total of $450 for trimming the Elsons' trees, a job that will take Josh two days to complete. After the first day, the Elsons paid him $185. How much will the Elsons pay Josh after the second day?
    Let x = the money after the second day.
    $185 + x = 450$
    The Elsons will pay Josh $265 after the second day.
14. After taking his second test, Timon's total score for his first two tests was 185 points. If he scored 89 points on the first test, how many points did Timon score on the second test?
    Let x = the number of points on the second test.
    $89 + x = 185$
    Timon scored 96 points on the second test.
15. A 240-centimeter board is to be cut into two pieces so that the longer piece is twice as long as the shorter piece. Find the length of each piece.
    Let x = the length of the shorter piece.
    $2x$ = the length of the longer piece.
    $x + 2x = 240$ The shorter piece is 80 cm.
    The longer piece is 160 cm.
16. The perimeter of a triangle is 42 inches. One side is 18 inches long. The second side is 2 inches longer than the third side. What are the lengths of the second and third sides?
    Let x = the length of the third side.
    $x + 2$ = the length of the second side.
    $18 + (x + 2) + x = 42$ The length of the second side is 13 inches. The length of the third side is 11 inches.
17. To get to her cousin's wedding, Greta drove a total of 1,050 miles. On the first day, she drove 180 miles more than on the second day. How many miles did she drive each day?
    Let x = the number of miles on the second day.
    $x + 180$ = the number of miles on the first day.
    $(x + 180) + x = 1{,}050$ Greta drove 435 miles on the second day and 615 miles on the first day.
18. Tunde and Marta are married and pool their income. Tunde earns $1,500 per year less than Marta. If their total annual income is $47,500, how much does each earn?
    Let x = the money Marta earns.
    $x - 1{,}500$ = the money Tunde earns.
    $x + (x - 1{,}500) = 47{,}500$
    Marta earns $24,500. Tunde earns $23,000.
19. In a right triangle, the first angle is 12° smaller than twice the measure of the second angle. What are the measures of the first and second angles?
    Let x = the measure of the second angle.
    $2x - 12$ = the measure of the first angle.
    $(2x - 12) + x + 90 = 180$
    The measure of the second angle is 34°.
    The measure of the first angle is 56°.
20. A rectangular field has a total perimeter of 128 feet. The width is 24 feet less than the length. What are the dimensions of the field?
    Let x = the length of the field.
    $x - 24$ = the width of the field.
    $128 = 2x + 2(x - 24)$
    The length of the field is 44 feet.
    The width of the field is 20 feet.

CHAPTER

Fractions

Introduction

We use fractions in a variety of occupations and hobbies, some of which might be obvious and some of which you might find surprising. People use fractions in construction, in the printing industry, in the fashion industry, and in machine repair, to name just a few. People also use fractions at home while cooking, sewing, working with wood, and doing home repair—the lists go on and on. Understanding how to work with fractions quickly and effortlessly makes each of the jobs and tasks easier to perform. Understanding fractions can make an employee more efficient, more productive, and more valuable to an employer.

This chapter builds on our understanding of signed numbers and solving equations. Later in the text, we'll see the connections between fractions, decimals, and percents.

Preparation Exercises

Section 1.1 Whole Numbers and Their Properties
Fill in the blank.

1. $15 \times 1 =$ _______ 15

2. $6 \times$ _______ $= 6$ 1

3. _______ $\times 1 = 12$ 12

Section 1.3 Multiplying Whole Numbers
Multiply.

4. $6 \cdot 4$ 24 **5.** $8 \cdot 3$ 24 **6.** $4 \cdot 9$ 36 **7.** $5 \cdot 12$ 60

Section 1.4 Dividing Whole Numbers
Divide.

8. $72 \div 4$ 18 **9.** $90 \div 5$ 18 **10.** $42 \div 3$ 14 **11.** $38 \div 2$ 19

Section 1.5 Exponents, Square Roots, and the Order of Operations
Evaluate.

12. $7 + 5 \cdot 3$ 22

13. $40 \div 4 \cdot 2$ 20

14. $-24 \div (8 - 6)$ -12

15. $\dfrac{19 - 7}{1 - 5}$ -3

Section 1.6 Factors
Find the prime factorization of each number.

16. 45 $3^2 \cdot 5$ or $3 \cdot 3 \cdot 5$

17. 60 $2^2 \cdot 3 \cdot 5$ or $2 \cdot 2 \cdot 3 \cdot 5$

18. 72 $2^3 \cdot 3^2$ or $2 \cdot 2 \cdot 2 \cdot 3 \cdot 3$

19. 80 $2^4 \cdot 5$ or $2 \cdot 2 \cdot 2 \cdot 2 \cdot 5$

Section 2.2 Adding Signed Numbers
Evaluate.

20. $-3 + (-8)$
-11

21. $2 + (-12)$
-10

Section 2.3 Subtracting Signed Numbers
Evaluate.

22. $4 - 9$
-5

23. $-3 - (-7)$
4

Section 2.4 Multiplying and Dividing Signed Numbers
Evaluate.

24. $-3 \cdot (-8)$
24

25. $2 \cdot (-12)$
-24

26. $54 \div (-9)$
-6

27. $\dfrac{-7}{-7}$
1

Section 3.1 Solving Equations Involving One Operation
Solve by isolating the variable.

28. $x \cdot 15 = 90$
$x = 6$

29. $8 \cdot y = 96$
$y = 12$

30. $54 = 3 \cdot w$
$w = 18$

SECTION 4.1 Common Factors

OBJECTIVES

In this section, you will learn to
- Find the greatest common factor of two numbers.
- Identify relatively prime numbers.

You Need to Know

To successfully complete this section, you need to understand
- ☐ Dividing whole numbers (1.4)
- ☐ Factors of numbers (1.6)
- ☐ Divisibility tests (1.6)
- ☐ Prime factorization (1.6)

Introduction

In Section 1.6, we looked at individual factors of a number as well as the prime factorization of a number. We will now consider *common factors* of two numbers, one of the building blocks of fractions.

The **common factors** of two numbers are all of the numbers that are factors of both numbers. To illustrate what this means, we'll start with an example.

Example 1

Find the common factors of 24 and 36.

Procedure Consider all of the factors of 24 and 36. Then make a list of the factors that are common to both.

Factors of 24: 1, 2, 3, 4, 6, 8, 12, 24

Factors of 36: 1, 2, 3, 4, 6, 9, 12, 18, 36

Answer Factors that are common to 24 and 36: 1, 2, 3, 4, 6, 12

You Try It 1 **Find the common factors of 12 and 30. Use Example 1 as a guide.**

Factors of 12: ____________________

Factors of 30: ____________________

Factors that are common to 12 and 30: ____________________

Finding the Greatest Common Factor Using Prime Factorization

Refer back to Example 1. The highest, or *greatest,* of the common factors of 24 and 36 is 12. So, we say that 12 is the **greatest common factor (GCF)** of 24 and 36.

It's difficult to make a list of common factors for every pair of numbers that may come along. Is there an easier way to find the GCF of two numbers?

There are many methods of finding the GCF of two numbers. Three methods are demonstrated in this section. The first, demonstrated in Example 1, is to list all of the factors of each number. The second method involves recognizing the prime factors of each number. For example, if you were to find the common prime factors of 20 and 30, you could start by using the divisibility tests from Section 1.6. Ask yourself these questions:

1. Is 2 a common factor of 20 and 30? Yes, because 20 and 30 are even numbers.
2. Is 5 a common factor of 20 and 30? Yes, because 20 and 30 have 0 in the ones place.
3. Is 3 a common factor of 20 and 30? No, it's a factor of 30 but not of 20.
4. Both 2 and 5 are common prime factors of 20 and 30. Is there a composite number that is a factor of both? Yes, 10 is a common factor because 20 and 30 end in 0.

Because 1 is a factor of every number, it is also a common factor of 20 and 30. So,

The common factors of 20 and 30 are 1, 2, 5, and 10.

As you can see, the greatest common factor is 10.

Notice that the GCF 10 can be built up from the prime factors 2 and 5, $2 \cdot 5 = 10$. This indicates that we can find the GCF if we identify the common *prime* factors of two numbers. The method we used in Section 1.6 for finding prime factors, *prime factorization*, is useful in recognizing common prime factors of two numbers.

Let's find the GCF of 20 and 30 using prime factorization.

Example 2

Find the GCF of 20 and 30.

Procedure Find the prime factorization of each number. Then look for common prime factors. Their product will be the GCF.

Answer

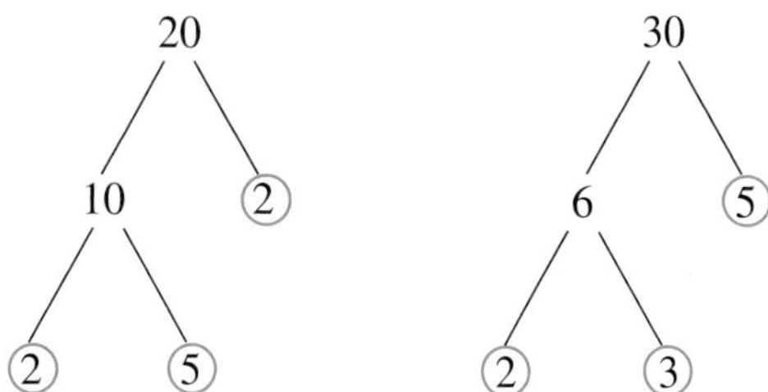

$20 = 2 \cdot 2 \cdot 5$

$30 = 2 \cdot 3 \cdot 5$

Now match up the common prime factors, 2 and 5.

$\text{GCF} = 2 \cdot 5 = 10$ The GCF is the product of all of the common prime factors.

Notice in the previous example that in the prime factorization of 20, the *second* factor of 2 does not match up with any prime factors of 30. The factor of 2 (in 30) is already matched up with the first 2 (in 20) and is no longer available for matching.

Let's use prime factorization to find the GCF of 24 and 36. (We already found it to be 12 in Example 1.)

Example 3

Find the greatest common factor of 36 and 24.

Procedure Find the prime factorization of each number.

Answer

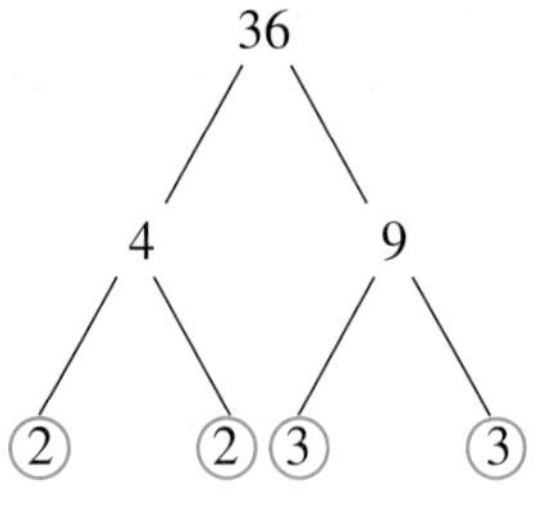

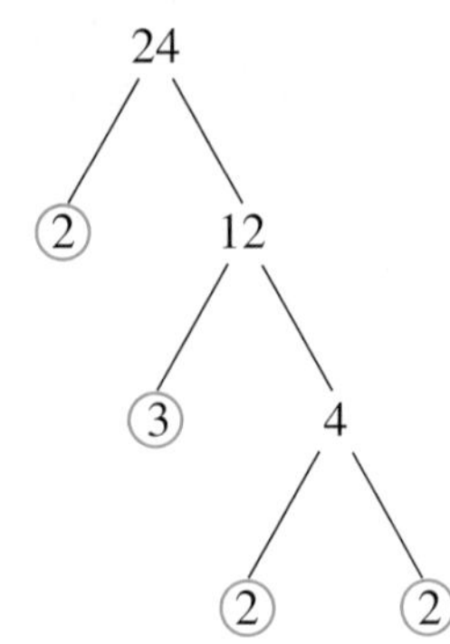

$36 = 2 \cdot 2 \cdot 3 \cdot 3$

$24 = 2 \cdot 2 \cdot 2 \cdot 3$

$\text{GCF} = 2 \cdot 2 \cdot 3 = 12$

There are many matches. Notice that when writing all of the matching prime factors for the GCF, we write only one prime number from each match. We don't write both numbers in the same match.

You Try It 2 **Use prime factorization, as in Examples 2 and 3, to find the GCF of each pair of numbers.**

a) 18 and 24 **b)** 40 and 60

Recognizing Relatively Prime Numbers

What if the two numbers have no prime factors in common? They still have a common factor that isn't prime: namely, 1. Consider the following:

Because 1 is a factor of every number, every pair of numbers will have at least one factor in common, 1. So, if two numbers have no *primes* in common, their greatest common factor will be 1.

When two numbers have no common *prime* factors, we say that the two numbers are **relatively prime.** In this case, the GCF is 1.

Although you might not expect it, two *composite* numbers may be relatively prime.

Even though two numbers, such as 14 and 15, are composite numbers, when they have no common factor other than 1, we say that they are prime compared to (or relative to) each other. In other words, they are *relatively prime*.

Example 4

List the common factors of 12 and 35. What is the GCF of 12 and 35?

Answer The factors of 12 are 1, 2, 3, 4, 6, and 12.
The factors of 35 are 1, 5, 7, and 35.

Therefore, 12 and 35 have only 1 as a common factor. So, 1 is the GCF. They are relatively prime.

Example 5 is the same as Example 4, but this time the prime factorization is shown.

Example 5

Find the GCF of 12 and 35.

Procedure Find the prime factorization of each number.

Answer

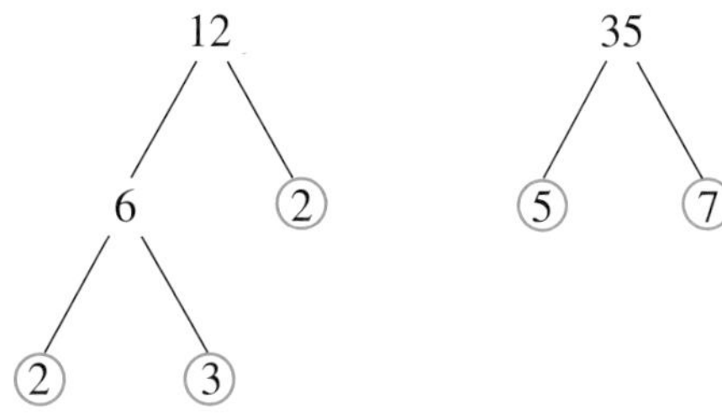

$12 = 2 \cdot 2 \cdot 3$

$35 = 5 \cdot 7$ There are no matches.

Because there are no prime factor matches, the GCF = 1. 12 and 35 are relatively prime.

You Try It 3 **For each pair of numbers, identify the GCF. See if you can do these in your head using the divisibility tests for 2, 3, and 5. If the numbers have no common prime factors, the GCF is 1. Write *relatively prime* underneath the numbers.**

a) 6 and 14 The GCF is ________.

b) 9 and 16 The GCF is ________.

c) 12 and 25 The GCF is ________.

Will two different prime numbers always be relatively prime to each other? Explain your answer or show an example that supports your answer.

____

____

Finding the Greatest Common Factor Using the Division Method

The next example illustrates a different method of finding the GCF based on the following guidelines. The *division method* tackles the common factors one at a time without matching them up.

Guidelines for Finding the Greatest Common Factor Using the Division Method

1. Use the divisibility tests to find a common *prime* factor.
2. Divide both numbers by that common prime factor to get two quotients.
3. See if the quotients have a common prime factor. If they do, divide the quotients by the common prime factor to get new quotients. Repeat this process until the new quotients are relatively prime.
4. Make a list of all of the common prime factors found. The GCF is the product of all of the common prime factors.

This method is very similar to the division method we used to find the prime factorization of a single number (Section 1.6). In this case, we're looking for common prime factors.

Example 6

Find the GCF of 60 and 140.

Procedure Follow the preceding steps. As in Section 1.6, we'll use an upside-down division symbol to assist us in the process.

Answer Because both numbers are even, 2 is a common prime factor.

Divide 60 and 140 by 2. Then see if the quotients have a common prime factor.

$$\begin{array}{r|rr} 2 & 60 & 140 \\ \hline & 30 & 70 \end{array}$$

← **Original numbers**
← **Quotients (new numbers)**

Now find a common prime factor of 30 and 70. Again, because they are even, 2 is a common prime factor. Repeat the process.

$$\begin{array}{r|rr} 2 & 30 & 70 \\ \hline & 15 & 35 \end{array}$$

← **Divide again.**
← **Second pair of quotients**

This time 5 is a common factor.

$$\begin{array}{r|rr} 5 & 15 & 35 \\ \hline & 3 & 7 \end{array}$$

← **Divide again.**
← **Third pair of quotients**

Stop! 3 and 7 are relatively prime.

Because the third pair of quotients, 3 and 7, has no common prime factors, this part is done. Next, we pick all of the common prime factors (the numbers on the left side of the division symbols) and multiply them to find the GCF.

The common prime factors are 2, 2, and 5.

The GCF of 60 and 140 is $2 \cdot 2 \cdot 5 = 20$.

Example 7 shows the steps as you might write them.

Example 7

Find the GCF of 18 and 54.

Procedure Here is the entire process for finding common factors.

Answer

$$\begin{array}{r|rr} 2 & 18 & 54 \\ \hline 3 & 9 & 27 \\ \hline 3 & 3 & 9 \\ \hline & 1 & 3 \end{array}$$

Prime divisor → 2 ← **Divide 18 and 54 by 2.**
Prime divisor → 3 ← **Divide 9 and 27 by 3.**
Prime divisor → 3 ← **Divide 3 and 9 by 3.**
← **Stop! 1 and 3 are relatively prime.**

The GCF of 18 and 54 is $2 \cdot 3 \cdot 3 = 18$.

In Example 7, note that the GCF, 18, is one of the original two numbers. This will happen whenever the smaller number is a factor of the larger number.

You Try It 4 **Use the division method to find the GCF of the following pairs of numbers. Use Example 7 as a guide.**

a) 20 and 28 **b)** 15 and 60 **c)** 36 and 54

When finding the GCF, we don't need to find only common *prime* factors. We can find *any* common factor of the two numbers and start there.

Example 8

Find the GCF of 150 and 240.

Procedure We'll start by finding any common factor between the numbers and dividing each by that factor, continuing until the new quotients are relatively prime.

Because 150 and 240 end in 0, 10 is a factor. So, we'll divide by 10 first.

Answer

10	150	240	← **Divide 150 and 240 by 10 (a composite factor).**
3	15	24	← **Divide 15 and 24 by 3.**
	5	8	← **Stop! 5 and 8 are relatively prime.**

The greatest common factor is still found by multiplying the two common factors (10 and 3): GCF $= 10 \cdot 3 = 30$.

You Try It 5 **Find the GCF of the following pairs of numbers. Use Example 8 as a guide. Remember, you can divide by any obvious common factor, prime or composite.**

a) 80 and 120 **b)** 36 and 42 **c)** 16 and 80

Answers: You Try It and Think About It

You Try It: **1.** Factors of 12: 1, 2, 3, 4, 6, and 12 Factors of 30: 1, 2, 3, 5, 6, 10, 15, and 30 Factors that are common to 12 and 30: 1, 2, 3, and 6 **2. a)** 6 **b)** 20 **3. a)** 2 (Both numbers are even.) **b)** 1 (The numbers are relatively prime.) **c)** 1 (The numbers are relatively prime.) **4. a)** 4 **b)** 15 **c)** 18 **5. a)** 40 **b)** 6 **c)** 16.

Think About It: **1.** Yes. Two different prime numbers have only 1 as a common factor, so the numbers must be relatively prime to each other.

Section 4.1 Exercises

FOR EXTRA HELP
Student Resources on DVD-ROM
Includes
➤Student's Solutions Manual
➤Video Lectures
➤Chapter Test Prep Video
MyMathLab
MathXL

Think Again

1. Will two different prime numbers always be relatively prime to each other? Explain your answer or show an example that supports your answer. *(Refer to Think About It 1)*

Yes. Two prime numbers have only 1 as a common factor, making them relatively prime.

2. Consider two numbers, a and b. If 2 is a common prime factor of a and b and 3 is a common prime factor of a and b, is it certain that $2 \cdot 3 = 6$ is a common composite factor of a and b? Explain your answer or show an example that supports your answer.

Yes. In writing out the GCF, we'd see $2 \cdot 3$, possibly among some other prime factors. This product indicates that $2 \cdot 3 = 6$ is a factor of the GCF.

3. Consider two numbers, a and b. If 6 is a common composite factor of a and b and 9 is a common composite factor of a and b, is it certain that $6 \cdot 9 = 54$ is a common composite factor of a and b? Explain your answer or show an example that supports your answer.

No. In writing out the GCF, we'd definitely see $2 \cdot 3$ (because 6 is a factor) and we'd definitely see $3 \cdot 3$ (because 9 is a factor); but this does not guarantee that we would see $2 \cdot 3 \cdot 3 \cdot 3$ (which is 54). For example, 18 and 36 have 6 and 9 as factors, but 54 is not a factor of either one.

4. Is it possible for the smaller of two numbers to be the greatest common factor of the pair of numbers? Explain your answer or show an example that supports your answer.

Yes. If the smaller number is a factor of the larger number, the smaller number is the GCF. For example, because 12 is a factor of 24, the GCF of 12 and 24 is 12.

Focus Exercises

Use prime factorization to find the GCF of each pair of numbers. If the GCF is 1, write *relatively prime*.

5. 15 and 25
5

6. 18 and 21
3

7. 15 and 24
3

8. 16 and 24
8

9. 15 and 30
15

10. 12 and 24
12

11. 16 and 21
relatively prime

12. 49 and 110
relatively prime

13. 30 and 45
15

14. 30 and 42
6

15. 24 and 40
8

16. 35 and 90
5

17. 36 and 48
12

18. 50 and 125
25

19. 48 and 120
24

20. 60 and 105
15

Use the division method to find the GCF of the following pairs of numbers.

21. 34 and 36
2

22. 16 and 50
2

23. 20 and 45
5

24. 28 and 49
7

25. 30 and 42
6

26. 50 and 70
10

27. 12 and 36
12

28. 8 and 40
8

29. 20 and 50
10

30. 55 and 99
11

31. 25 and 49
relatively prime

32. 35 and 48
relatively prime

33. 54 and 72
18

34. 28 and 98
14

35. 45 and 75
15

36. 42 and 54
6

37. 40 and 72
8

38. 42 and 105
21

39. 80 and 96
16

40. 48 and 64
16

41. 45 and 60
15

42. 60 and 96
12

43. 24 and 72
24

44. 56 and 70
14

45. 35 and 54
relatively prime

46. 64 and 75
relatively prime

47. 21 and 32
relatively prime

48. 42 and 55
relatively prime

49. 60 and 150
30

50. 90 and 120
30

51. 100 and 150
50

52. 120 and 180
60

53. 270 and 720
90

54. 120 and 840
120

55. 54 and 90
18

56. 72 and 108
36

57. 70 and 175
35

58. 105 and 135
15

59. 40 and 96
8

60. 42 and 154
14

Think Outside the Box

Find the greatest common factor of the following sets of three numbers.

61. 8, 12, and 20
4

62. 6, 18, and 30
6

63. 30, 60, and 75
15

64. 80, 140, and 200
20

65. 48, 72, and 120
24

66. 54, 90, and 108
18

SECTION 4.2 Introduction to Fractions

OBJECTIVES

In this section, you will learn to

- Place positive and negative fractions on a number line.
- Rewrite fractions as division.
- Identify proper fractions, improper fractions, and mixed numbers.
- Write mixed numbers as improper fractions and improper fractions as mixed numbers.
- Use the principles of fractions to evaluate fractions and expressions.
- Apply the product rule of fractions.
- Identify the reciprocal of a fraction.
- Use the multiplicative inverse.

You Need to Know

To successfully complete this section, you need to understand

- ☐ The multiplication table (1.3)
- ☐ Long division (1.4)
- ☐ Dividing signed numbers (2.4)
- ☐ Multiplying signed numbers (2.4)

To remember which is the denominator, remember that *D* is the first letter of both ***Down*** and ***Denominator***.

Once you know which component is the denominator, the other must be the numerator.

Introduction

Understanding fractions is important in the study and practice of the sciences, business, and finance. In these areas of study, the fractions of arithmetic and algebra are important. In this chapter, we will explore the fractions of arithmetic while seeing the connections to algebra. Later in this chapter we will see how fractions are helpful in solving equations and applications.

A **fraction** is the comparison of a part of something to its whole, where all of the parts are the same size.

For example, a whole week has 7 days, and each day is one-seventh—one equal part—of the week. The weekend includes two of those days; so, the weekend is *two-sevenths* of the week. It is common to express a fraction such as two-sevenths as $\frac{2}{7}$.

A fraction consists of three components: a **fraction bar,** a number above the fraction bar, and a number below the fraction bar. The number *above* the fraction bar is called the *numerator,* and the number *below* the fraction bar is called the *denominator.*

The **numerator** represents the number of equal-sized parts being considered, and the **denominator** represents the number of equal-sized parts contained in the whole.

Let's take a look at visual representations of some fractions. Each fraction represented here, except for part d, is based on $\frac{\textbf{part}}{\textbf{whole}}$, where part of the whole has been shaded. On the small ruler in part d, the whole is represented by 1 inch and the fractions shown are parts of an inch.

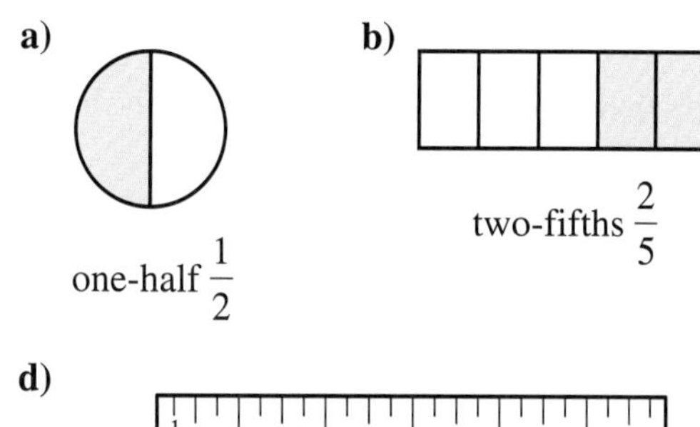

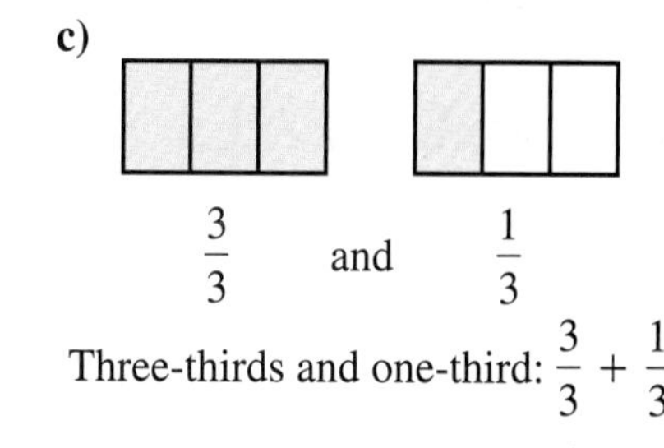

or four-thirds: $\frac{4}{3}$

or one whole and one-third: $1\frac{1}{3}$

d)

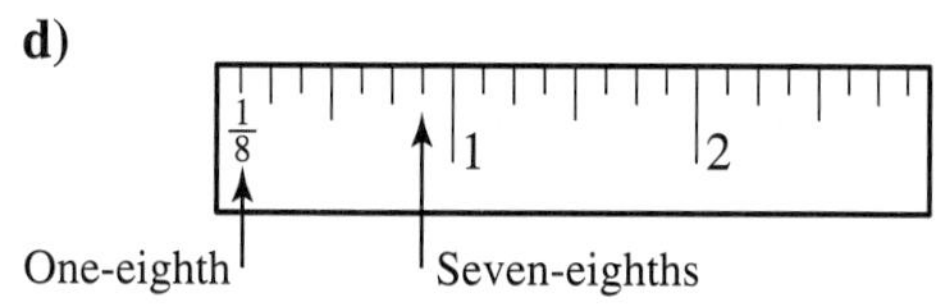

Even though one of two parts is shaded, does this diagram represent $\frac{1}{2}$? Explain your answer.

Fractions on the Number Line

On a number line, an integer can be divided into parts just like an inch can be divided into fourths and eighths. This allows us to represent both positive and negative fractions on the number line.

In the following diagram, on the positive side (to the right of zero), the distance between each small mark and the one next to it is one-eighth of a whole. On the negative side (to the left of zero), the distance between each small mark and the one next to it is one-sixth of a whole.

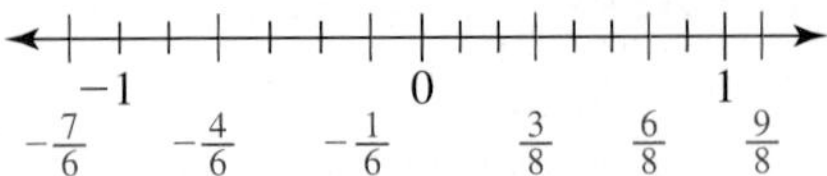

You Try It 1 **Place each of these numbers on the number line. (On the positive side, the distance between each small mark and the one next to it is one-seventh of a whole. On the negative side, the distance between each small mark and the one next to it is one-fifth of a whole.)**

a) $\frac{3}{7}$ **b)** $-\frac{1}{5}$ **c)** $-\frac{4}{5}$ **d)** $\frac{5}{7}$ **e)** $\frac{9}{7}$ **f)** $-\frac{6}{5}$

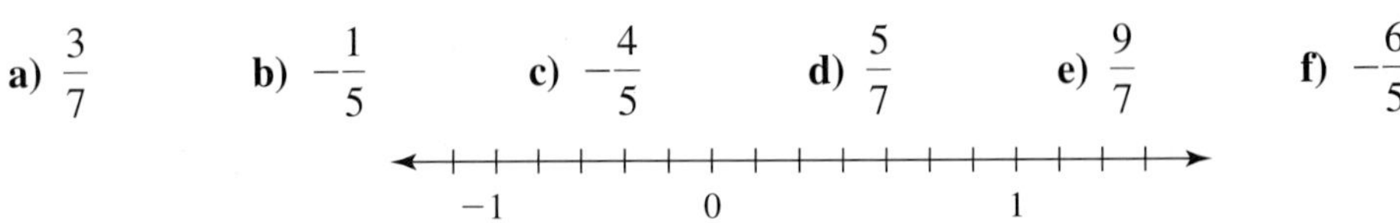

Fractions as Division

Fractional notation is another way to indicate the operation of division. For example, $\frac{8}{2}$ can also be written as $8 \div 2$. Sometimes the fraction bar is referred to as the division bar.

Example 1

Rewrite each fraction as a numerical expression with division, and evaluate the expression.

a) $\frac{12}{4} = \underline{12 \div 4} = \underline{3}$ **b)** $\frac{10}{-5} = \underline{10 \div (-5)} = \underline{-2}$

You Try It 2 **Rewrite each fraction as a numerical expression with division, and evaluate the expression. Use Example 1 as a guide.**

a) $\frac{20}{4} =$ ________ = ______ **b)** $\frac{-18}{2} =$ ________ = ______

Recall from Section 1.4 the circular (inverse) nature of division and multiplication. In fractional form, we can see it as follows:

$$\frac{\text{Dividend}}{\text{Divisor}} = \text{Quotient} \qquad \frac{\text{Numerator}}{\text{Denominator}} = \text{Quotient} \qquad \frac{12}{3} = 4$$

Divisor × Quotient = Dividend   Denominator × Quotient = Numerator $3 \times 4 = 12$

Example 2

Rewrite each division problem as a multiplication problem.

a) $\frac{16}{8} = 2$ $\underline{8 \cdot 2 = 16}$ **b)** $\frac{-15}{3} = -5$ $\underline{3 \cdot (-5) = -15}$

You Try It 3 **Rewrite each division problem as a multiplication problem. Use Example 2 as a guide.**

a) $\frac{20}{4} = 5$ ________ **b)** $\frac{18}{-6} = -3$ ________ **c)** $\frac{-35}{-5} = 7$ ________

Recall from Section 1.4 that we may never divide by 0. That means that the denominator of a fraction may never be 0.

Recall that we use the equal sign with a slash through it, $\neq$, to represent "is not equal to." We can express a fraction as $\frac{a}{b}$, $b \neq 0$.

Division by Zero

If a stands for *any number*, then $\boldsymbol{a \div 0}$, or $\frac{a}{0}$, is ***undefined.***

Another way to express this is

The denominator of a fraction can never be zero.

The circular argument does not support the denominator being 0.

$$\frac{4}{0} = ? \longrightarrow 0 \times ? = 4$$

Instructor Insight

Ask the students to use the circular argument to show that $\frac{0}{a} = 0$. They can start with $\frac{0}{a} = ?$.

If you were to use a calculator to evaluate $\frac{4}{0}$ (entering it as $4 \div 0$), you'd get an error message in the display. This means that there is an ***error*** in the division. There is no such value. As a fraction, we say that $\frac{4}{0}$ is **undefined.**

However, the numerator may be 0.

Zero as a Numerator

If $\boldsymbol{a}$ is any number *(except 0)*, then $\frac{0}{a} = 0$.

Think About It 2

Why does it say "except 0" in the preceding box?

____

____

Types of Fractions

Let's classify the types of fractions you will work with in this chapter.

A **proper fraction** has a numerator that is *less* than the denominator.

$\frac{1}{7}, \frac{2}{3}$, and $\frac{6}{11}$ are examples of proper fractions. { 1 is less than 7, 2 is less than 3, and 6 is less than 11.

An **improper fraction** has a numerator that is *equal to* or *greater than* the denominator.

$\frac{7}{7}, \frac{5}{3}$, and $\frac{22}{11}$ are examples of improper fractions. { 7 is equal to 7, 5 is greater than 3, and 22 is greater than 11.

A **mixed number** is the *sum* of a whole number and a fraction.

$1\frac{3}{8}$ and $4\frac{1}{5}$ are examples of mixed numbers: $1\frac{3}{8} = 1 + \frac{3}{8}$ and $4\frac{1}{5} = 4 + \frac{1}{5}$.

We can say that a mixed number has a whole number part and a fractional part.

You Try It 4 **Identify each fraction as a proper fraction, an improper fraction, or a mixed number.**

a) $\frac{13}{4}$ ________________ **b)** $\frac{5}{8}$ ________________

c) $5\frac{3}{4}$ ________________ **d)** $\frac{3}{3}$ ________________

Writing Mixed Numbers as Improper Fractions

Consider the mixed number $2\frac{3}{5}$. It can be written as $1 + 1 + \frac{3}{5}$. And because $\frac{5}{5} = 1$ $(5 \div 5 = 1)$, we can write this mixed number as $\frac{5}{5} + \frac{5}{5} + \frac{3}{5}$, as shown in the diagram. This makes for a total of $\frac{13}{5}$, an improper fraction.

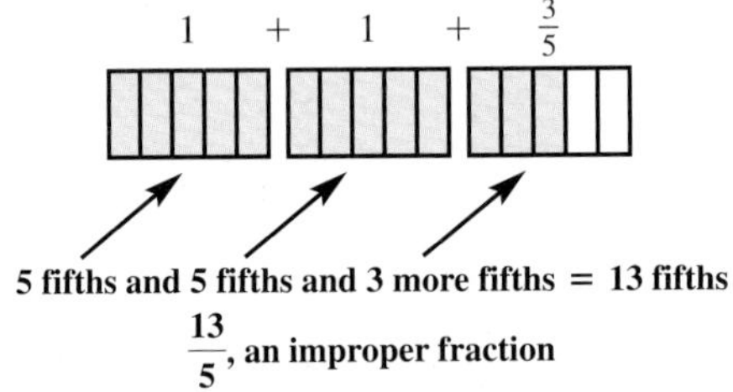

In the diagram, the denominator, 5, indicates how many equal parts each *bar* will have and the numerators indicate how many of the parts in the bar are to be shaded. In the mixed number $2\frac{3}{5}$, the whole number, 2, accounts for $2 \times 5 = 10$ shaded parts. The additional 3 shaded parts of the fraction give us 13 shaded parts in all.

This idea leads us to a formula that converts mixed numbers into improper fractions.

$$\text{Whole} + \frac{\text{Numerator}}{\text{Denominator}} = \frac{\text{Whole} \times \text{Denominator} + \text{Numerator}}{\text{Denominator}}$$

$$2 + \frac{3}{5} = \frac{2 \times 5 + 3}{5}$$

Example 3

Write each mixed number as an improper fraction using the conversion formula.

a) $3\frac{2}{7}$ **b)** $6\frac{3}{4}$

Procedure Use the conversion formula.

Answer

a) $3\frac{2}{7} = \frac{3 \times 7 + 2}{7} = \frac{21 + 2}{7} = \frac{23}{7}$

b) $6\frac{3}{4} = \frac{6 \times 4 + 3}{4} = \frac{24 + 3}{4} = \frac{27}{4}$

If a mixed number is negative, its equivalent improper fraction is also negative. To maintain the negative, we can place a negative sign in front of the fraction formula while making the conversion to an improper fraction.

Example 4

Write each mixed number as an improper fraction using the conversion formula.

a) $-4\frac{7}{9}$ **b)** $-12\frac{1}{5}$

Procedure Place a negative sign in front of the fraction formula before converting.

Answer

a) $-4\frac{7}{9} = -\frac{4 \times 9 + 7}{9} = -\frac{36 + 7}{9} = -\frac{43}{9}$

b) $-12\frac{1}{5} = -\frac{12 \times 5 + 1}{5} = -\frac{60 + 1}{5} = -\frac{61}{5}$

You Try It 5 **Write each mixed number as an improper fraction. Use Examples 3 and 4 as guides.**

a) $2\frac{3}{8}$ **b)** $9\frac{5}{6}$ **c)** $-5\frac{1}{4}$ **d)** $-11\frac{2}{7}$

Writing Improper Fractions as Mixed Numbers

Let's look again at fractions and division. We have seen, for example, that the improper fraction $\frac{45}{5}$ means $45 \div 5$. As you know, 5 divides evenly into 45 and we get $45 \div 5 = 9$.

There are other improper fractions where the denominator does not divide evenly into the numerator, such as $\frac{45}{7}$. In this case, when we divide, we get a remainder.

$$\begin{array}{r} 6 \text{ r } 3 \\ 7\overline{)45} \\ -42 \\ \hline 3 \end{array}$$

This shows us that 7 divides into 45 six whole times with a remainder of 3.

The remainder is a *part* of the whole, not an entire whole. In other words, the remainder is a fractional part of the whole and we can write the answer as a mixed number, $6\frac{3}{7}$.

Here's a conversion diagram from improper fraction (as long division) to mixed number.

$$\begin{array}{r} \text{quotient r remainder} \\ \text{denominator}\overline{)\text{numerator}} \end{array} \longrightarrow \text{quotient} + \frac{\text{remainder}}{\text{denominator}}$$

When an improper fraction is negative, its equivalent mixed number also is negative. When converting a negative improper fraction, we can temporarily ignore the negative sign during the division process.

Example 5

Write each improper fraction as a mixed number using long division.

a) $\frac{21}{4}$ **b)** $\frac{38}{5}$ **c)** $-\frac{74}{3}$

Procedure Use long division. For part c, ignore the negative sign when dividing.

a)
$$\begin{array}{r} 5 \text{ r } 1 \\ 4\overline{)21} \\ -20 \\ \hline 1 \end{array}$$

b)
$$\begin{array}{r} 7 \text{ r } 3 \\ 5\overline{)38} \\ -35 \\ \hline 3 \end{array}$$

c)
$$\begin{array}{r} 24 \text{ r } 2 \\ 3\overline{)74} \\ -6 \\ \hline 14 \\ -12 \\ \hline 2 \end{array}$$

Answer $\frac{21}{4} = 5\frac{1}{4}$ $\frac{38}{5} = 7\frac{3}{5}$ $-\frac{74}{3} = -24\frac{2}{3}$

You Try It 6 **Write each improper fraction as a mixed number. Use Example 5 as a guide.**

a) $\frac{43}{8}$ **b)** $-\frac{25}{6}$ **c)** $\frac{95}{7}$

The Principles of Fractions

There are certain principles of numbers—first introduced in Chapter 1—that we can apply to fractions.

1. Recall from Section 1.1 that the *multiplicative identity* is 1 because $a \cdot 1 = a$ no matter the value of a.

   For example, $7 \cdot 1 = 7$ and $\frac{3}{10} \cdot 1 = \frac{3}{10}$. Multiplying by 1 doesn't change the value.

2. Recall from Section 1.4 that a natural number divided by itself is 1: $a \div a = 1$.

   For example, $5 \div 5 = 1$. In fraction form, $\frac{5}{5} = 1$. $\frac{a}{a} = 1$

3. Also recall from Section 1.4 that a number divided by 1 is itself. $a \div 1 = a$.

   For example, $9 \div 1 = 9$. In fraction form, $\frac{9}{1} = 9$. $\frac{a}{1} = a$

You Try It 7 **Evaluate the following expressions.**

a) $\frac{9}{9} =$ ______ **b)** $\frac{6}{1} =$ ______ **c)** $\frac{3}{5} \cdot 1 =$ ______

A principle new to this section is the **product rule for fractions.**

The Product Rule for Fractions

$$\frac{a}{b} \cdot \frac{c}{d} = \frac{a \cdot c}{b \cdot d} = \frac{\text{The product of the numerators}}{\text{The product of the denominators}}$$

Remember the multiplication rules for integers. If one or more of the fractions is negative, determine whether the end product will be positive or negative; then complete the multiplication.

Example 6

Multiply the following fractions using the product rule. If the answer is an improper fraction, write it as a mixed number.

a) $\frac{3}{7} \cdot \frac{2}{3}$ **b)** $-\frac{3}{2} \cdot \frac{5}{4}$ **c)** $-\frac{2}{5} \cdot \left(-\frac{4}{3}\right)$ **d)** $\frac{3}{7} \cdot \frac{8}{5}$

Procedure Multiply the numerators and multiply the denominators to result in one new fraction.

Answer a) $\frac{3}{7} \cdot \frac{2}{3} = \frac{5 \cdot 2}{7 \cdot 3} = \frac{10}{21}$

b) $-\frac{3}{2} \cdot \frac{5}{4} = -\frac{3 \cdot 5}{2 \cdot 4} = -\frac{15}{8} = -1\frac{7}{8}$

c) $-\frac{2}{5} \cdot \left(-\frac{4}{3}\right) = +\frac{2 \cdot 4}{5 \cdot 3} = \frac{8}{15}$

d) $\frac{3}{7} \cdot \frac{8}{5} = \frac{3 \cdot 8}{7 \cdot 5} = \frac{24}{35}$

You Try It 8 **Multiply the following fractions using the product rule. If the answer is an improper fraction, write it as a mixed number. Use Example 6 as a guide.**

a) $\frac{7}{6} \cdot \frac{5}{3}$ b) $\frac{8}{3} \cdot \left(-\frac{2}{7}\right)$ c) $-\frac{1}{4} \cdot \left(-\frac{3}{10}\right)$ d) $\frac{1}{3} \cdot \frac{2}{1}$

The Reciprocals of Fractions

Consider this:

> The **reciprocal** of a fraction, $\frac{a}{b}$, is the fraction, $\frac{b}{a}$, as long as a and b are not 0. That is, the reciprocal of a fraction results from interchanging the numerator and the denominator.

The reciprocal of a negative number is also negative. Let's use Example 7 to apply the definition of reciprocal to three numbers.

Example 7

Identify the reciprocal of each of these numbers: $\frac{2}{3}$, $-\frac{1}{4}$, and 7.

Procedure First, write any whole number as a fraction before writing its reciprocal. Then interchange the numerator and the denominator.

Answer We'll use the table at the right.

Fraction	Reciprocal
$\frac{2}{3}$	$\frac{3}{2}$
$-\frac{1}{4}$	$-\frac{4}{1} = -4$
$7 = \frac{7}{1}$	$\frac{1}{7}$

You Try It 9 **Identify the reciprocal. Use Example 7 as a guide.**

a) The reciprocal of $-\frac{6}{7}$ is ________.

b) The reciprocal of $\frac{9}{4}$ is ________.

c) The reciprocal of $\frac{1}{8}$ is ________.

d) The reciprocal of -5 is ________.

Complete this sentence and explain your answer. In general, the reciprocal of any proper fraction is a(n) ______________.

____

____

The reciprocal of a fraction is also called the multiplicative *inverse* of the fraction. When finding the reciprocal of a fraction, we say that the fraction has been inverted. As an **inverse,** the reciprocal has this special property:

Multiplicative Inverse

The product of a fraction and its reciprocal is 1 (the multiplicative identity).

$$\frac{a}{b} \cdot \frac{b}{a} = 1, \text{ provided } a \neq 0 \text{ and } b \neq 0.$$

Example 8

First, identify the reciprocal of the fraction.
Second, find the product of the fraction and its reciprocal.

a) $\frac{4}{5}$ **b)** $-\frac{9}{2}$ **c)** $\frac{1}{7}$ **d)** 8

Procedure For part b, recall that the product of two negative numbers is positive.

Answer

	Fraction	Reciprocal	Product
a)	$\frac{4}{5}$	$\frac{5}{4}$	$\frac{4}{5} \cdot \frac{5}{4} = \frac{20}{20} = 1$
b)	$-\frac{9}{2}$	$-\frac{2}{9}$	$-\frac{9}{2} \cdot \left(-\frac{2}{9}\right) = +\frac{18}{18} = 1$
c)	$\frac{1}{7}$	$\frac{7}{1}$	$\frac{1}{7} \cdot \frac{7}{1} = \frac{7}{7} = 1$
d)	$8 = \frac{8}{1}$	$\frac{1}{8}$	$\frac{8}{1} \cdot \frac{1}{8} = \frac{8}{8} = 1$

You Try It 10 **First, identify the reciprocal of each fraction. Second, find the product of each fraction and its reciprocal. Use Example 8 as a guide.**

	Fraction		Reciprocal		Product		Fraction		Reciprocal		Product
a)	$\frac{3}{10}$	×	______	=	______	b)	$-\frac{11}{7}$	×	______	=	______
c)	$\frac{1}{5}$	×	______	=	______	d)	9	×	______	=	______

Answers: You Try It and Think About It

You Try It: **1.**

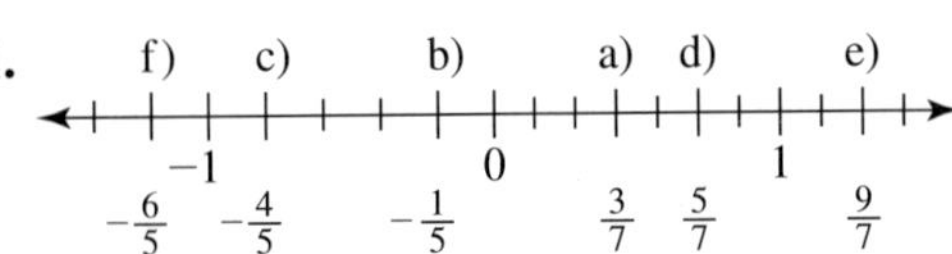

2. a) $20 \div 4 = 5$ **b)** $-18 \div 2 = -9$ **3. a)** $4 \cdot 5 = 20$ **b)** $-6 \cdot (-3) = 18$ **c)** $-5 \cdot 7 = -35$ **4. a)** Improper fraction **b)** Proper fraction **c)** Mixed number **d)** Improper fraction **5. a)** $\frac{19}{8}$ **b)** $\frac{59}{6}$ **c)** $-\frac{21}{4}$ **d)** $-\frac{79}{7}$ **6. a)** $5\frac{3}{8}$ **b)** $-4\frac{1}{6}$ **c)** $13\frac{4}{7}$ **7. a)** 1 **b)** 6 **c)** $\frac{3}{5}$ **8. a)** $\frac{35}{18} = 1\frac{17}{18}$ **b)** $-\frac{16}{21}$ **c)** $\frac{3}{40}$ **d)** $\frac{2}{3}$ **9. a)** $-\frac{7}{6}$ **b)** $\frac{4}{9}$ **c)** $\frac{8}{1}$ or 8 **d)** $-\frac{1}{5}$ **10. a)** $\frac{3}{10} \times \frac{10}{3} = \frac{30}{30} = 1$ **b)** $-\frac{11}{7} \times -\frac{7}{11} = +\frac{77}{77} = 1$ **c)** $\frac{1}{5} \times \frac{5}{1} = \frac{5}{5} = 1$ **d)** $\frac{9}{1} \times \frac{1}{9} = \frac{9}{9} = 1$

Think About It: **1.** No. Because the two parts are not of equal size, the diagram does not represent $\frac{1}{2}$. **2.** Because 0 can never be in the denominator, a cannot be 0; $a \neq 0$. **3.** *An improper fraction.* In a proper fraction, the denominator is greater than the numerator. In its reciprocal, the numerator is greater than the denominator, making it an improper fraction.

Section 4.2 Exercises

FOR EXTRA HELP
Student Resources on DVD-ROM
Includes
➤ Student's Solutions Manual
➤ Video Lectures
➤ Chapter Test Prep Video
MyMathLab
MathXL

Think Again

1. In this section, we see that 0 can be in the numerator: If a is any number *(except 0)*, then $\frac{0}{a} = 0$. Why does it say "except 0"? *(Refer to Think About It 2)*
We may never have 0 in the denominator, and that is true for the fraction $\frac{0}{0}$.

2. What positive number is equal to its own reciprocal?
1

3. Complete this sentence and explain your answer or show an example that supports your answer: The reciprocal of any proper fraction is a(n) ___________. *(Refer to Think About It 3)*
An improper fraction. In a proper fraction, the denominator is greater than the numerator. In its reciprocal, the numerator is greater than the denominator, making it an improper fraction.

4. Can a fraction that has 0 in the numerator, such as $\frac{0}{3}$, have a reciprocal? Explain your answer or show an example that supports your answer.
No. The reciprocal would place a 0 in the denominator, and that is not allowed.

Focus Exercises

Identify each fraction as a proper fraction, an improper fraction, or a mixed number.

5. $\frac{7}{3}$ Improper fraction

6. $\frac{13}{10}$ Improper fraction

7. $\frac{7}{15}$ Proper fraction

8. $\frac{1}{9}$ Proper fraction

9. $7\frac{3}{4}$ Mixed number

10. $1\frac{1}{2}$ Mixed number

11. $\frac{4}{4}$ Improper fraction

12. $\frac{6}{1}$ Improper fraction

Write each mixed number as an improper fraction.

13. $1\frac{4}{9}$ $\frac{13}{9}$

14. $1\frac{2}{7}$ $\frac{9}{7}$

15. $-2\frac{3}{8}$ $-\frac{19}{8}$

16. $-2\frac{2}{3}$ $-\frac{8}{3}$

17. $3\frac{1}{3}$ $\frac{10}{3}$

18. $3\frac{5}{6}$ $\frac{23}{6}$

19. $4\frac{4}{7}$ $\frac{32}{7}$

20. $4\frac{5}{9}$ $\frac{41}{9}$

21. $-5\frac{1}{4}$ $-\frac{21}{4}$

22. $-5\frac{2}{7}$ $-\frac{37}{7}$

23. $6\frac{1}{10}$ $\frac{61}{10}$

24. $6\frac{4}{11}$ $\frac{70}{11}$

25. $7\frac{1}{3}$ $\frac{22}{3}$

26. $8\frac{5}{6}$ $\frac{53}{6}$

27. $-9\frac{3}{4}$ $-\frac{39}{4}$

28. $-10\frac{2}{3}$ $-\frac{32}{3}$

29. $11\frac{7}{8}$ $\frac{95}{8}$

30. $12\frac{1}{5}$ $\frac{61}{5}$

31. $15\frac{3}{8}$ $\frac{123}{8}$

32. $20\frac{3}{4}$ $\frac{83}{4}$

Write each improper fraction as a mixed number.

33. $\frac{11}{5}$ $2\frac{1}{5}$

34. $\frac{13}{8}$ $1\frac{5}{8}$

35. $-\frac{13}{4}$ $-3\frac{1}{4}$

36. $-\frac{9}{2}$ $-4\frac{1}{2}$

37. $\frac{24}{7}$ $3\frac{3}{7}$

38. $\frac{17}{6}$ $2\frac{5}{6}$

39. $\frac{10}{3}$ $3\frac{1}{3}$

40. $\frac{14}{9}$ $1\frac{5}{9}$

41. $-\frac{19}{8}$ $-2\frac{3}{8}$

42. $-\frac{15}{4}$ $-3\frac{3}{4}$

43. $\frac{32}{5}$ $6\frac{2}{5}$

44. $\frac{40}{9}$ $4\frac{4}{9}$

45. $\frac{37}{5}$ $7\frac{2}{5}$

46. $\frac{75}{8}$ $9\frac{3}{8}$

47. $\frac{82}{7}$ $11\frac{5}{7}$

48. $\frac{97}{4}$ $24\frac{1}{4}$

Multiply. If the result is an improper fraction, write it as a mixed number.

49. $\frac{8}{9} \cdot \frac{1}{5}$
$\frac{8}{45}$

50. $\frac{3}{7} \cdot \frac{9}{2}$
$1\frac{13}{14}$

51. $-\frac{3}{5} \cdot \frac{3}{5}$
$-\frac{9}{25}$

52. $-\frac{7}{6} \cdot \frac{1}{4}$
$-\frac{7}{24}$

53. $\frac{2}{5} \cdot \frac{4}{3}$
$\frac{8}{15}$

54. $\frac{5}{7} \cdot \frac{3}{2}$
$1\frac{1}{14}$

55. $\frac{7}{8} \cdot \left(-\frac{5}{6}\right)$
$-\frac{35}{48}$

56. $\frac{2}{3} \cdot \left(-\frac{11}{5}\right)$
$-1\frac{7}{15}$

57. $\frac{7}{3} \cdot \frac{1}{2}$
$1\frac{1}{6}$

58. $\frac{21}{10} \cdot \frac{3}{2}$
$3\frac{3}{20}$

59. $-\frac{9}{4} \cdot \left(-\frac{5}{2}\right)$
$5\frac{5}{8}$

60. $-\frac{7}{2} \cdot \left(-\frac{11}{6}\right)$
$6\frac{5}{12}$

First, identify the reciprocal if possible. Second, find the product of each fraction and its reciprocal.

61. $\frac{3}{5}$
$\frac{5}{3}$, product is 1

62. $\frac{2}{9}$
$\frac{9}{2}$, product is 1

63. $\frac{7}{6}$
$\frac{6}{7}$, product is 1

64. $\frac{9}{4}$
$\frac{4}{9}$, product is 1

65. $-\frac{7}{10}$
$-\frac{10}{7}$, product is 1

66. $-\frac{10}{3}$
$-\frac{3}{10}$, product is 1

67. $\frac{1}{6}$
6, product is 1

68. $\frac{1}{4}$
4, product is 1

69. 8
$\frac{1}{8}$, product is 1

70. $\frac{3}{1}$
$\frac{1}{3}$, product is 1

71. 0
No reciprocal

72. $\frac{0}{5}$
No reciprocal

Think Outside the Box

Multiply. If the result is an improper fraction, write it as a mixed number. (*Hint:* First, write each mixed number as an improper fraction.)

73. $4\frac{1}{2} \cdot 1\frac{1}{4}$
$5\frac{5}{8}$

74. $2\frac{1}{3} \cdot 3\frac{1}{5}$
$7\frac{7}{15}$

First, identify the reciprocal if possible. Second, find the product of each fraction and its reciprocal.

75. $2\frac{3}{4}$
The reciprocal is $\frac{4}{11}$; the product is 1.

76. $3\frac{1}{3}$
The reciprocal is $\frac{3}{10}$; the product is 1.

SECTION 4.3 Equivalent Fractions

OBJECTIVES

In this section, you will learn to

- Simplify fractions.
- Build up fractions: find equivalent fractions by multiplying by a fractional form of 1.

You Need to Know

To successfully complete this section, you need to understand

- ☐ The Multiplicative Identity (1.1)
- ☐ Division (1.4)
- ☐ The greatest common factor (4.1)
- ☐ Relatively prime numbers (4.1)
- ☐ Negative fractions (4.2)
- ☐ The principles of fractions (4.2)

Introduction

Some fractions look different from one another but have the same value—they are *equivalent* fractions. For example, $\frac{20}{64}$ and $\frac{2}{5}$ are equivalent fractions (as we'll see in Example 1), and $\frac{2}{5}$ is a *simplified* form of $\frac{24}{60}$. In this section, we learn how to simplify fractions.

Here is one of many situations in which the knowledge of simplifying a fraction is important.

> Olivia has written her first sales report for her new company. In her report, she mentioned that 30 out of 48, or $\frac{30}{48}$, of her sales last week were phone orders. Her manager has asked her to edit the report to include a simplified fraction. What fraction should Olivia use?

We will help Olivia answer that question. First, though, let's look at what it means for two or more fractions to be equivalent.

Fractions and Lowest Terms

Recall from Section 4.2 that a fraction is a comparison of a part of something to its whole, where all of the parts are the same size.

Consider this chocolate bar. It has 6 rows of 3 squares each, 18 squares in all. We can treat the whole as having 6 equal parts (each row) or 18 equal parts (each square).

1 whole bar of chocolate
or
6 rows of chocolate
or
18 squares of chocolate

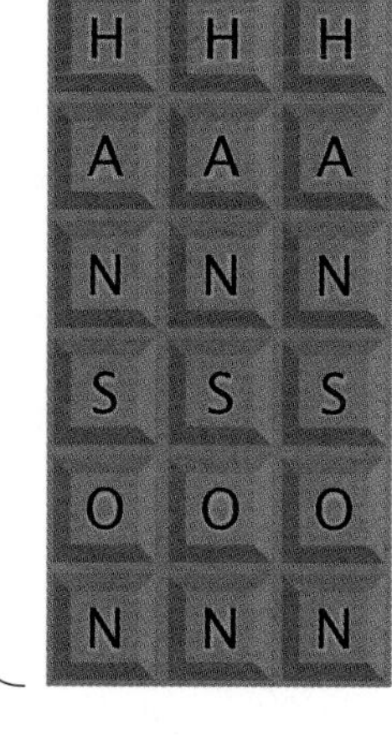

one-half of the whole bar
or
3 rows of chocolate
or
9 squares of chocolate

If we break off half the bar, we can say that we have $\frac{1}{2}$ of the bar. In terms of rows, we have 3 of the 6 rows, or $\frac{3}{6}$ of the original rows in the bar. In terms of squares, we have 9 of the 18 squares, or $\frac{9}{18}$ of the original squares in the bar.

Each of these fractions indicates the same amount of chocolate: $\frac{1}{2}$, $\frac{3}{6}$, and $\frac{9}{18}$ of the bar. The fractions are equivalent. Notice that in each fraction, the numerator is half the denominator.

Of the three fractions $\frac{1}{2}$, $\frac{3}{6}$, and $\frac{9}{18}$, the one that is in *lowest terms* is $\frac{1}{2}$. **Lowest terms** means that the numerator and the denominator are relatively prime.

To find an equivalent fraction in lowest terms, we *simplify* the fraction.

> To **simplify** a fraction means to divide out any factors that are common to the numerator and the denominator.

In other words, if the numerator and the denominator have a common factor (other than 1), the fraction can be simplified.

Simplifying Fractions

Recall from Section 4.2 the product rule. We can look at it two ways.

1. Multiplying two fractions together:

$$\frac{a}{b} \cdot \frac{c}{d} = \frac{a \cdot c}{b \cdot d} \qquad \frac{5}{7} \cdot \frac{2}{3} = \frac{5 \cdot 2}{7 \cdot 3}$$

2. Separating one fraction into two fractions:

$$\frac{a \cdot c}{b \cdot d} = \frac{a}{b} \cdot \frac{c}{d} \qquad \frac{5 \cdot 2}{7 \cdot 3} = \frac{5}{7} \cdot \frac{2}{3}$$

We'll use the idea of separating one fraction into two fractions to help us understand how to simplify fractions.

In the problem posed at the beginning of this section, Olivia must simplify the fraction $\frac{30}{48}$. A behind-the-scenes look at this simplifying will include some of the principles of numbers. Let's see!

1. First, consider common factors of the numerator and the denominator. The common factors of 30 and 48 are 2, 3, and 6. $\qquad \frac{30}{48} \qquad$ **GCF = 6**

2. Rewrite the numerator and the denominator as a product with the GCF, 6, as a common factor. $\qquad = \frac{5 \cdot 6}{8 \cdot 6}$

3. Separate this fraction into the product of two fractions. $\qquad = \frac{5}{8} \cdot \frac{6}{6}$

4. Replace $\frac{6}{6}$ with 1. $\qquad = \frac{5}{8} \cdot 1$

5. Multiply $\frac{5}{8}$ by 1. Notice that the value doesn't change. $\qquad = \frac{5}{8}$

We can say that $\frac{30}{48}$ has been "simplified by a factor of 6" to $\frac{5}{8}$ and that $\frac{30}{48}$ and $\frac{5}{8}$ are equivalent fractions.

In simplifying the fraction $\frac{30}{48}$, we have extracted a common factor of 6 from the numerator and the denominator. In other words, we divided the numerator and the denominator by 6 to get the equivalent fraction $\frac{5}{8}$.

$$\frac{30}{48} = \frac{30 \div 6}{48 \div 6} = \frac{5}{8}$$

We could have simplified the fraction by a different common factor, such as 2, but the fraction would not have been completely simplified.

$$\frac{30}{48} = \frac{30 \div 2}{48 \div 2} = \frac{15}{24}$$

To get the answer to lowest terms, we must simplify $\frac{15}{24}$ by a factor of 3.

$$\frac{15}{24} = \frac{15 \div 3}{24 \div 3} = \frac{5}{8}$$

Simplifying a fraction by its GCF requires the fewest steps, but simplifying by one common factor at a time is okay too.

Caution To simplify a fraction means to simplify it *completely*—to lowest terms. If you simplify a fraction by one factor, you should check to see if the new (equivalent) fraction can also be simplified.

Example 1

Simplify each of the following fractions.

a) $\frac{7}{21}$ **b)** $\frac{24}{60}$

Procedure Identify a common factor of the numerator and the denominator and divide each by that number. Continue this process until the fraction is simplified to its lowest terms.

Answer **a)** $\frac{7}{21}$ The GCF of 7 and 21 is 7.

$$\frac{7}{21} = \frac{7 \div 7}{21 \div 7} = \frac{1}{3}$$ $\frac{7}{21}$ has been simplified by a factor of 7.

b) $\frac{24}{60}$ A common factor of 24 and 60 is 3.

$$\frac{24}{60} = \frac{24 \div 3}{60 \div 3} = \frac{8}{20}$$

Because 8 and 20 have a common factor of 4, we can simplify further.

$$\frac{8}{20} = \frac{8 \div 4}{20 \div 4} = \frac{2}{5}$$

We can say that $\frac{24}{60}$ simplified to lowest terms is $\frac{2}{5}$.

In Example 1, the fraction $\frac{24}{60}$ was simplified by a factor of 3 and then by a factor of 4. Could this fraction have been simplified by a different factor in just one step? If so, by what factor? Explain.

____

____

You Try It 1 **Simplify each of the following fractions. Use Example 1 as a guide.**

a) $\frac{6}{14}$ **b)** $\frac{35}{45}$

c) $\frac{3}{24}$ **d)** $\frac{30}{72}$

A quicker method of simplifying fractions requires mental division, as we are still *dividing out* a common factor. In this method, we recognize the common factor and cross out the numerator and the denominator, writing the simplified values to the side. It's important that we simplify the numerator and the denominator by the same factor.

Example 2

Simplify $\frac{6}{15}$.

Procedure The numerator and the denominator have a common factor of 3; so, we may divide each by 3. The division is done mentally. We *think* $\frac{\div 3}{\div 3}$, but we don't actually write it.

Answer $\frac{\overset{2}{\cancel{6}}}{\underset{5}{\cancel{15}}} = \frac{2}{5}$

When we simplify a fraction by one factor, if the fraction is not completely simplified, we must perform the step again.

Example 3

Simplify $\frac{32}{48}$.

Procedure Sometimes this process must be performed more than once because the resulting fraction isn't completely simplified. Below is the simplifying process for $\frac{32}{48}$.

Answer **1.** First in the division format:

$$\frac{32}{48} = \frac{32 \div 8}{48 \div 8} = \frac{4}{6} = \frac{4 \div 2}{6 \div 2} = \frac{2}{3}$$

This fraction ($\frac{4}{6}$) **can be simplified by a factor of 2.**

2. Now in the mental-dividing-out format:

$$\frac{\overset{4}{\cancel{32}}}{\underset{6}{\cancel{48}}} = \frac{\overset{2}{\cancel{4}}}{\underset{3}{\cancel{6}}} = \frac{2}{3}$$

Simplify by a factor of 8. Simplify by a factor of 2.

Either way you choose to simplify is fine, as long as you do it correctly. Also, make sure you have simplified the fraction to its lowest terms.

You Try It 2 **Simplify each fraction. You may use any method. Use Examples 1, 2, and 3 as guides.**

a) $\frac{12}{52}$ **b)** $\frac{60}{84}$ **c)** $\frac{30}{96}$ **d)** $\frac{60}{150}$

In Section 4.1, we learned the division method of finding the GCF of two numbers. The division method has an interesting connection to simplifying fractions. Consider simplifying $\frac{24}{60}$ one prime factor at a time.

Simplifying		Division method
$\frac{24}{60} = \frac{24 \div 2}{60 \div 2} = \frac{12}{30}$	← **Divide 24 and 60 by 2.** →	$2 \,\lfloor\, 24 \;\; 60$ → $12 \;\; 30$
$\frac{12}{30} = \frac{12 \div 2}{30 \div 2} = \frac{6}{15}$	← **Divide 12 and 30 by 2.** →	$2 \,\lfloor\, 12 \;\; 30$ → $6 \;\; 15$
$\frac{6}{15} = \frac{6 \div 3}{15 \div 3} = \frac{2}{5}$	← **Divide 6 and 15 by 3.** →	$3 \,\lfloor\, 6 \;\; 15$ → $2 \;\; 5$

(In the division method, the left column is the Numerator and the right column is the Denominator; the final 2 is the Numerator and 5 is the Denominator.)

Notice that each time we extract a factor using the division method, we get the same two numbers as the numerator and the denominator of the simplified fraction.

This is not a coincidence. Dividing the numerator and the denominator of a fraction by 2 is the same as extracting 2 using the Division Method.

In other words, the end result of the Division Method is the numerator and the denominator in simplified form.

Example 4

Simplify each fraction.

a) $\dfrac{18}{30}$ **b)** $\dfrac{56}{84}$

Procedure Apply the division method to the numerator and the denominator.

Answer

a)
$$\begin{array}{r|rr} 2 & 18 & 30 \\ \hline 3 & 9 & 15 \\ \hline & 3 & 5 \end{array}$$

$\dfrac{18}{30} = \dfrac{3}{5}$

b)
$$\begin{array}{r|rr} 2 & 56 & 84 \\ \hline 2 & 28 & 42 \\ \hline 7 & 14 & 21 \\ \hline & 2 & 3 \end{array}$$

$\dfrac{56}{84} = \dfrac{2}{3}$

You Try It 3 **Simplify each fraction. You may use any method you wish. Use Examples 1, 2, 3, and 4 as guides.**

a) $\dfrac{56}{70}$ **b)** $\dfrac{54}{72}$

c) $\dfrac{75}{90}$ **d)** $\dfrac{32}{104}$

Fractions Containing Negative Signs

In a fraction, we can put a negative sign and a positive sign in one of three places: in front of the fraction, in the numerator, or in the denominator. In fact, each of those places does have a sign although it's not always visible.

For example, $-\frac{3}{8}$ could be represented as $-\frac{+3}{+8}$. If a fraction has one negative sign, that sign can appear anywhere in the fraction.

$$-\frac{+3}{+8} = +\frac{-3}{+8} = +\frac{+3}{-8}$$ These three fractions are equivalent to each other.

Without the positive signs, the equivalent fractions can be written as $-\frac{3}{8} = \frac{-3}{8} = \frac{3}{-8}$. Although those three fractions are equivalent, this text will treat the simplified form to be the one with the negative sign in front of the entire fraction, $-\frac{3}{8}$.

Because a fraction is another form of division, the rules of negatives in division (from Section 2.4) apply to fractions as well.

1. When there is an odd number of negative signs in the fraction, the fraction is negative.
2. When there is an even number of negative signs in the fraction, the fraction is positive.

When simplifying a fraction with one or more negative signs, it is best to decide whether the fraction is positive or negative before simplifying.

Example 5

Simplify each fraction.

a) $-\dfrac{-6}{10}$ **b)** $\dfrac{3}{-9}$ **c)** $-\dfrac{-14}{-21}$

Procedure First, decide whether the fraction is positive or negative.

Answer a) $-\dfrac{-6}{10} = +\dfrac{6}{10} = \dfrac{3}{5}$ $\quad \dfrac{6}{10}$ **can be simplified by a factor of 2.**

Two negative signs; the fraction is *positive*.

b) $\dfrac{3}{-9} = -\dfrac{3}{9} = -\dfrac{1}{3}$ $\quad \dfrac{3}{9}$ **can be simplified by a factor of 3.**

One negative sign; the fraction is *negative*.

c) $-\dfrac{-14}{-21} = -\dfrac{14}{21} = -\dfrac{2}{3}$ $\quad \dfrac{14}{21}$ **can be simplified by a factor of 7.**

Three negative signs; the fraction is *negative*.

You Try It 4 **Simplify each fraction. Use Example 5 as a guide.**

a) $\dfrac{24}{-64}$ b) $\dfrac{-35}{-42}$ c) $-\dfrac{-30}{-45}$

Fractions Containing Variables

If a fraction has the same variable factor in both the numerator and the denominator, that variable factor can be divided out.

For example, $\dfrac{5x}{7x} = \dfrac{5 \cdot x \div x}{7 \cdot x \div x} = \dfrac{5}{7}$. We can also show dividing out x mentally:

$$\frac{5 \cdot \overset{1}{\cancel{x}}}{7 \cdot \underset{1}{\cancel{x}}} = \frac{5}{7}$$

If a variable factor is raised to a power, such as x^2, we should write it in expanded form, $x \cdot x$, before we simplify the fraction.

For example, $\dfrac{5x^2}{7x}$ should be written as $\dfrac{5 \cdot x \cdot x}{7 \cdot x}$. Then we can simplify:

$$\frac{5x^2}{7x} = \frac{5 \cdot \overset{1}{\cancel{x}} \cdot x}{7 \cdot \underset{1}{\cancel{x}}} = \frac{5x}{7}$$

Example 6

Simplify $\dfrac{2wy}{10y^2}$.

Procedure First, expand y^2 to be $y \cdot y$. Then divide out any common factors.

Answer
$$\frac{2wy}{10y^2} = \frac{\overset{1}{\cancel{2}} \cdot w \cdot \overset{1}{\cancel{y}}}{\underset{5}{\cancel{10}} \cdot y \cdot \underset{1}{\cancel{y}}} = \frac{1 \cdot 1 \cdot w}{5 \cdot 1 \cdot y} = \frac{w}{5y}$$

You Try It 5 **Simplify each fraction. Use Example 6 as a guide.**

a) $\dfrac{8a}{12a}$ b) $\dfrac{-18wx}{45w}$ c) $\dfrac{-12p^2}{-60mp}$

Simplifying Improper Fractions

Because of the possibility of variables in the numerator or the denominator, it is common in algebra to leave improper fractions as improper and not change them to mixed numbers, even if the improper fraction has no variable within it.

Example 7

Simplify each fraction.

a) $-\dfrac{14}{6}$ **b)** $\dfrac{15h}{10h^2}$

Procedure In part b, expand h^2 to $h \cdot h$; then divide out any common factors. Do not write an improper fraction as a mixed number.

Answer a) $-\dfrac{\overset{7}{\cancel{14}}}{\underset{3}{\cancel{6}}} = -\dfrac{7}{3}$ b) $\dfrac{15h}{10h^2} = \dfrac{\overset{3}{\cancel{15}} \cdot \overset{1}{\cancel{h}}}{\underset{2}{\cancel{10}} \cdot h \cdot \underset{1}{\cancel{h}}} = \dfrac{3 \cdot 1}{2 \cdot h \cdot 1} = \dfrac{3}{2h}$

You Try It 6 **Simplify each fraction. Use Example 7 as a guide.**

a) $-\dfrac{24}{18}$ **b)** $\dfrac{42bc}{24c}$ **c)** $\dfrac{-35x^2}{-14wx}$

Building Up Fractions

We can simplify $\frac{4}{6}$ to an equivalent fraction by dividing the numerator and the denominator by 2.

$$\frac{4 \div 2}{6 \div 2} = \frac{2}{3}$$

In a similar way, we can *build up* $\frac{2}{3}$ to an equivalent fraction by multiplying the numerator and the denominator by 2.

$$\frac{2 \cdot 2}{3 \cdot 2} = \frac{4}{6}$$

Multiplying the numerator and the denominator by 2 is the same as multiplying the fraction $\frac{2}{3}$ by $\frac{2}{2}$, which is a form of 1.

$$\frac{2}{3} \cdot \frac{2}{2} = \frac{4}{6}$$

In fact, we can build up any fraction by multiplying by a form of 1, such as $\frac{3}{3}$, $\frac{4}{4}$, and $\frac{5}{5}$.

> Any fraction $\frac{a}{b}$, where $b \neq 0$, can be built up to an equivalent fraction by multiplying by a fractional form of 1.
>
> $$\frac{a}{b} = \frac{a}{b} \cdot \frac{c}{c} = \frac{a \cdot c}{b \cdot c}, c \neq 0$$

Caution When we build up a fraction, we should not simplify the result. Simplifying would undo the building-up process.

Example 8

Use the product rule to multiply, but do not simplify. (Notice that the second fraction in each case is equivalent to 1.)

a) $\frac{3}{4} \cdot \frac{3}{3} = \underline{\frac{9}{12}}$ **b)** $-\frac{5y}{7} \cdot \frac{4}{4} = \underline{-\frac{20y}{28}}$

You Try It 7 **Use the product rule to multiply, but do not simplify. Use Example 8 as a guide.**

a) $\frac{5}{6} \cdot \frac{3}{3} =$ ________ **b)** $-\frac{7x}{4} \cdot \frac{5}{5} =$ ________ **c)** $\frac{1}{2p} \cdot \frac{6}{6} =$ ________

Next, we can build up a fraction so that it has a specific denominator. For example, if we want $\frac{2}{3}$ to be built up to a fraction that has a denominator of 12, we need to find the right fractional form of 1 to multiply times $\frac{2}{3}$.

Example 9

Build up $\frac{2}{3}$ to an equivalent fraction that

a) has a denominator of 12.

b) has a denominator of 21.

Procedure We need to multiply $\frac{2}{3}$ by a fraction equivalent to 1. Look at each denominator to decide by what number to multiply the numerator and the denominator of $\frac{2}{3}$.

Answer

a)

Start with	Multiply by 1	Result
$\frac{2}{3}$ ×	$\frac{?}{?}$ =	$\frac{?}{12}$

Because 3 × **4** = 12, use $\frac{4}{4}$ for 1.

$$\frac{2}{3} \times \frac{4}{4} = \frac{8}{12}$$

b)

Start with	Multiply by 1	Result
$\frac{2}{3}$ ×	$\frac{?}{?}$ =	$\frac{?}{21}$

Because 3 × **7** = 21, use $\frac{7}{7}$ for 1.

$$\frac{2}{3} \times \frac{7}{7} = \frac{14}{21}$$

You Try It 8 **Multiply by the appropriate value of 1 to build up each fraction to one with the denominator shown. Use Example 9 as a guide.**

	Start with	Multiply by 1	Result
a)	$\frac{3}{4}$ ×	____ =	$\frac{\quad}{24}$
c)	$\frac{5m}{8}$ ×	____ =	$\frac{\quad}{24}$
c)	$-\frac{9c}{7}$ ×	____ =	$-\frac{\quad}{28}$
d)	$\frac{3}{1}$ ×	____ =	$\frac{\quad}{8}$

Answers: You Try It and Think About It

You Try It: **1. a)** $\frac{3}{7}$ **b)** $\frac{7}{9}$ **c)** $\frac{1}{8}$ **d)** $\frac{5}{12}$ **2. a)** $\frac{3}{13}$ **b)** $\frac{5}{7}$ **c)** $\frac{5}{16}$ **d)** $\frac{2}{5}$ **3. a)** $\frac{4}{5}$ **b)** $\frac{3}{4}$ **c)** $\frac{5}{6}$ **d)** $\frac{4}{13}$ **4. a)** $-\frac{3}{8}$ **b)** $\frac{5}{6}$ **c)** $-\frac{2}{3}$ **5. a)** $\frac{2}{3}$ **b)** $-\frac{2x}{5}$ **c)** $\frac{p}{5m}$ **6. a)** $-\frac{4}{3}$ **b)** $\frac{7b}{4}$ **c)** $\frac{5x}{2w}$ **7. a)** $\frac{15}{18}$ **b)** $-\frac{35x}{20}$ **c)** $\frac{6}{12p}$ **8. a)** $\frac{3}{4} \times \frac{6}{6} = \frac{18}{24}$ **b)** $\frac{5m}{8} \times \frac{3}{3} = \frac{15m}{24}$ **c)** $-\frac{9c}{7} \times \frac{4}{4} = -\frac{36c}{28}$ **d)** $\frac{3}{1} \times \frac{8}{8} = \frac{24}{8}$

Think About It: **1.** Yes. $\frac{24}{60}$ can be simplified in one step by a factor of 12. This is because $3 \cdot 4 = 12$.

FOR EXTRA HELP
Student Resources on DVD-ROM
Includes
➤ Student's Solutions Manual
➤ Video Lectures
➤ Chapter Test Prep Video
MyMathLab
MathXL

Section 4.3 Exercises

Think Again

1. Is it allowable to divide out the 2's in this fraction: $\frac{2+1}{2+3}$? Explain your answer.
No. We may divide out only common factors, and 2 is not a factor of the numerator or the denominator because it is not being multiplied.

2. Are there certain circumstances when we can build up a fraction by multiplying it by $\frac{2}{1}$? Explain your answer or show an example that supports your answer.
No. We can build up a fraction (to an equivalent fraction) only by multiplying it by 1, not by any other value, such as $\frac{2}{1}$.

Focus Exercises

Simplify completely.

3. $\frac{9}{15}$ $\frac{3}{5}$
4. $\frac{3}{12}$ $\frac{1}{4}$
5. $\frac{8}{18}$ $\frac{4}{9}$
6. $\frac{6}{21}$ $\frac{2}{7}$
7. $\frac{21}{28}$ $\frac{3}{4}$
8. $\frac{20}{60}$ $\frac{1}{3}$
9. $-\frac{9}{24}$ $-\frac{3}{8}$
10. $-\frac{15}{35}$ $-\frac{3}{7}$
11. $\frac{8}{-32}$ $-\frac{1}{4}$
12. $\frac{14}{-35}$ $-\frac{2}{5}$
13. $\frac{10}{25}$ $\frac{2}{5}$
14. $\frac{22}{77}$ $\frac{2}{7}$
15. $-\frac{18}{27}$ $-\frac{2}{3}$
16. $-\frac{36}{42}$ $-\frac{6}{7}$
17. $\frac{-20}{-45}$ $\frac{4}{9}$
18. $\frac{-60}{-100}$ $\frac{3}{5}$
19. $\frac{12}{30}$ $\frac{2}{5}$
20. $\frac{16}{24}$ $\frac{2}{3}$
21. $\frac{10}{-45}$ $-\frac{2}{9}$
22. $\frac{8}{-36}$ $-\frac{2}{9}$
23. $\frac{3}{15}$ $\frac{1}{5}$
24. $\frac{2}{18}$ $\frac{1}{9}$
25. $-\frac{-5}{35}$ $\frac{1}{7}$
26. $-\frac{7}{-42}$ $\frac{1}{6}$
27. $-\frac{-5}{-25}$ $-\frac{1}{5}$
28. $-\frac{-7}{-14}$ $-\frac{1}{2}$
29. $\frac{44}{16}$ $\frac{11}{4}$ or $2\frac{3}{4}$
30. $\frac{60}{18}$ $\frac{10}{3}$ or $3\frac{1}{3}$
31. $\frac{18}{54}$ $\frac{1}{3}$
32. $\frac{32}{56}$ $\frac{4}{7}$
33. $\frac{28}{42}$ $\frac{2}{3}$
34. $\frac{36}{60}$ $\frac{3}{5}$
35. $\frac{20}{72}$ $\frac{5}{18}$
36. $\frac{14}{42}$ $\frac{1}{3}$
37. $\frac{24}{72}$ $\frac{1}{3}$
38. $\frac{24}{96}$ $\frac{1}{4}$
39. $\frac{90}{75}$ $\frac{6}{5}$ or $1\frac{1}{5}$
40. $\frac{120}{96}$ $\frac{5}{4}$ or $1\frac{1}{4}$
41. $\frac{30}{96}$ $\frac{5}{16}$
42. $\frac{48}{80}$ $\frac{3}{5}$
43. $\frac{8w}{20w}$ $\frac{2}{5}$
44. $\frac{12n}{42n}$ $\frac{2}{7}$
45. $\frac{6a}{24a}$ $\frac{1}{4}$
46. $\frac{9c}{45c}$ $\frac{1}{5}$
47. $\frac{44xy}{20y}$ $\frac{11x}{5}$
48. $\frac{35cd}{21c}$ $\frac{5d}{3}$
49. $\frac{18ab}{63bc}$ $\frac{2a}{7c}$
50. $\frac{40xy}{64wx}$ $\frac{5y}{8w}$
51. $\frac{15x}{60x^2}$ $\frac{1}{4x}$
52. $\frac{14k}{42k^2}$ $\frac{1}{3k}$
53. $\frac{12w^2}{48wv}$ $\frac{w}{4v}$
54. $\frac{18pq}{90q^2}$ $\frac{p}{5q}$

Multiply by a fraction equivalent to 1 to build up each fraction to one with the denominator shown.

	Start with		Multiply by 1		Result
55.	$\frac{7}{10}$	×	$\frac{3}{3}$	=	$\frac{21}{30}$
56.	$\frac{5}{12}$	×	$\frac{2}{2}$	=	$\frac{10}{24}$
57.	$-\frac{4}{9}$	×	$\frac{2}{2}$	=	$\frac{-8}{18}$
58.	$-\frac{3}{8}$	×	$\frac{5}{5}$	=	$\frac{-15}{40}$
59.	$\frac{9}{7}$	×	$\frac{6}{6}$	=	$\frac{54}{42}$
60.	$\frac{1}{9}$	×	$\frac{8}{8}$	=	$\frac{8}{72}$
61.	$\frac{4}{15}$	×	$\frac{3}{3}$	=	$\frac{12}{45}$
62.	$\frac{6}{1}$	×	$\frac{9}{9}$	=	$\frac{54}{9}$
63.	$-\frac{3}{20}$	×	$\frac{5}{5}$	=	$\frac{-15}{100}$

	Start with		Multiply by 1		Result
64.	$-\frac{14}{15}$	$\times$	$\frac{4}{4}$	$=$	$\frac{-56}{60}$
65.	$\frac{3p}{8}$	$\times$	$\frac{10}{10}$	$=$	$\frac{30p}{80}$
66.	$\frac{4y}{1}$	$\times$	$\frac{15}{15}$	$=$	$\frac{60y}{15}$
67.	$-\frac{7w}{12}$	$\times$	$\frac{5}{5}$	$=$	$\frac{-35w}{60}$

	Start with		Multiply by 1		Result
68.	$-\frac{5x}{6}$	$\times$	$\frac{11}{11}$	$=$	$\frac{-55x}{66}$
69.	$\frac{8m}{25}$	$\times$	$\frac{4}{4}$	$=$	$\frac{32m}{100}$
70.	$\frac{17c}{30}$	$\times$	$\frac{4}{4}$	$=$	$\frac{68c}{120}$

Think Outside the Box

Multiply by a fraction equivalent to 1 to build up each fraction to one with the denominator shown.

	Start with		Multiply by 1		Result
71.	$\frac{5}{8}$	$\times$	$\frac{4x}{4x}$	$=$	$\frac{20x}{32x}$
72.	$\frac{7}{9}$	$\times$	$\frac{3y^2}{3y^2}$	$=$	$\frac{21y^2}{27y^2}$

	Start with		Multiply by 1		Result
73.	$\frac{k}{6mp}$	$\times$	$\frac{7mp}{7mp}$	$=$	$\frac{7kmp}{42m^2p^2}$
74.	$\frac{8v}{15q}$	$\times$	$\frac{4rq}{4rq}$	$=$	$\frac{32rqv}{60rq^2}$

SECTION 4.4 Multiplying and Dividing Fractions

OBJECTIVES

In this section, you will learn to
- Multiply fractions.
- Divide fractions.
- Multiply and divide mixed numbers.
- Solve applications involving multiplying fractions.

You Need to Know

To successfully complete this section, you need to understand
- ☐ Multiplying whole numbers (1.3)
- ☐ Writing mixed numbers as improper fractions (4.2)
- ☐ The product rule (4.2)
- ☐ Simplifying fractions (4.3)

Introduction

Recall from Section 1.3 that multiplication of whole numbers is an abbreviation for repeated addition. For example, if a recipe makes 12 cookies, we can make twice as many cookies, 24, by doubling the recipe, multiplying all of the ingredients by 2.

However, cutting the recipe in half (that is, multiplying by $\frac{1}{2}$) results in only 6 cookies. This is not repeated addition. In fact, there are fewer cookies than in the original recipe.

When multiplying with fractions, we have learned a different rule, the product rule.

Multiplying Fractions

The product rule of fractions, introduced in Section 4.2, can be extended to include more than two fractions. The fractions may also include variables.

The Extended Product Rule of Fractions

$$\frac{a}{b}\cdot\frac{c}{d}\cdot\frac{e}{f}=\frac{a\cdot c\cdot e}{b\cdot d\cdot f}=\frac{\text{The product of the numerators}}{\text{The product of the denominators}},\ b,\ d,\text{ and } f\neq 0$$

If one or more fractions are negative, decide whether the product will be positive or negative before you apply the product rule. The product rule can be extended to multiplying three fractions, as demonstrated in Example 1.

If one or more fractions in the product has a variable, those variables may be placed at the end of the multiplication in the numerator and the denominator, as in Example 1b.

Example 1

Multiply.

a) $3\cdot\frac{7}{5}\cdot\frac{1}{8}$ **b)** $-\frac{5x}{9}\cdot\left(-\frac{1}{3y}\right)\cdot\left(-\frac{7}{2}\right)$

Procedure The product rule says to multiply the numerators together and then multiply the denominators together. In part a, first write 3 as $\frac{3}{1}$. In part b, the product of three negatives is negative.

Answer

a) $\frac{3}{1}\cdot\frac{7}{5}\cdot\frac{1}{8}=\frac{3\cdot 7\cdot 1}{1\cdot 5\cdot 8}=\frac{21}{40}$

b) $-\frac{5x}{9}\cdot\left(-\frac{1}{3y}\right)\cdot\left(-\frac{7}{2}\right)=-\frac{5\cdot 1\cdot 7\cdot x}{9\cdot 3\cdot 2\cdot y}=-\frac{35x}{54y}$

Notice that x is placed as the last factor of the numerator and that y is placed as the last factor of the denominator.

You Try It 1 **Multiply. Use Example 1 as a guide.**

a) $\frac{1}{3}\cdot\frac{4}{7}\cdot\frac{5}{3}$ **b)** $-\frac{7a}{11}\cdot\left(-\frac{1}{2}\right)\cdot\frac{5}{3b}$

Multiplying and Simplifying Fractions

In Section 4.3, we simplified fractions by dividing out factors that were common to the numerator and the denominator. In this section, we combine simplifying with multiplying to solve applications involving fractions.

Example 2

Find the product. Simplify the answer.

a) $-\frac{5}{6}\cdot\frac{4}{15}$ b) $\frac{3}{2}\cdot\frac{11}{4}$ c) $\frac{9w}{6}\cdot\frac{1}{2w}\cdot\frac{2x}{3}$

Procedure Apply the product rule to each of these. If the result can be simplified, do so. If the result cannot be simplified, leave it as it is. The answer to part a is negative.

Answer You might be able to do these in fewer steps. All steps are shown here.

a) $-\frac{5}{6}\cdot\frac{4}{15} = -\frac{5\cdot 4}{6\cdot 15} = -\frac{20}{90} = -\frac{20 \div 10}{90 \div 10} = -\frac{2}{9}$

This fraction ↑ can be simplified by a factor of 10.

b) $\frac{3}{2}\cdot\frac{11}{4} = \frac{3\cdot 11}{2\cdot 4} = \frac{33}{8}$ This fraction cannot be simplified because 33 and 8 are relatively prime.

c) $\frac{9w}{6}\cdot\frac{1}{2w}\cdot\frac{2x}{3} = \frac{18wx}{36w} = \frac{\overset{3}{\cancel{18}}\cdot\overset{1}{\cancel{w}}\cdot x}{\underset{6}{\cancel{36}}\cdot\underset{1}{\cancel{w}}} = \frac{\overset{1}{\cancel{3}}\cdot x}{\underset{2}{\cancel{6}}\cdot 1} = \frac{x}{2}$

↑ Simplify 18 and 36 by a factor of 6. ↑ Also divide out w. ↑ Simplify 3 and 6 by a factor of 3.

You Try It 2 **Find the product. Simplify your answer. Use Example 2 as a guide.**

a) $\frac{6}{7}\cdot\frac{2x}{3}$ b) $-\frac{8c}{11}\cdot\left(-\frac{3}{4c}\right)$

c) $\frac{3}{9}\cdot\frac{10}{5}\cdot\frac{1}{2}$ d) $-\frac{7}{4m}\cdot\frac{2}{5}\cdot\frac{3m}{7}$

Simplifying Fractions before Multiplying

Two fractions that cannot be simplified are $\frac{5}{7}$ and $\frac{9}{10}$. Their product, though, can be simplified.

$\frac{5}{7}\cdot\frac{9}{10} = \frac{5\cdot 9}{7\cdot 10} = \frac{45}{70}$, which can be simplified by a factor of 5: $\frac{45 \div 5}{70 \div 5} = \frac{9}{14}$

That method finds the full product before simplifying. It is possible, though, to simplify the fraction before multiplying by looking at the prime factorization of each number in the numerator and the denominator.

$$\frac{5}{7}\cdot\frac{9}{10} = \frac{5\cdot 9}{7\cdot 10} = \frac{5\cdot 3\cdot 3}{7\cdot 2\cdot 5}$$

↑ **The product rule** ↑ **Prime factorizations of each numerator and denominator**

At this point, we may divide out the common factor of 5 directly.

$$\frac{\overset{1}{\cancel{5}}\cdot 3\cdot 3}{7\cdot 2\cdot\underset{1}{\cancel{5}}} = \frac{9}{14}$$

Example 3

Find the product. Simplify before multiplying.

a) $-\frac{5}{6}\cdot\left(-\frac{4}{15}\right)$ b) $\frac{3}{5}\cdot\frac{7}{6}\cdot\frac{10}{14}$ c) $\frac{6h}{8}\cdot\frac{14k}{21h}$

Procedure Apply the product rule to each expression and write it as one fraction. Find the prime factorization of each number and divide out any common factors.

Answer

a) $-\frac{5}{6}\cdot\left(-\frac{4}{15}\right) = +\frac{5\cdot 4}{6\cdot 15} = \frac{\overset{1}{\cancel{5}}\cdot\overset{1}{\cancel{2}}\cdot 2}{\underset{1}{\cancel{2}}\cdot 3\cdot\underset{1}{\cancel{5}}\cdot 3} = \frac{2}{9}$

b) $\frac{3}{5}\cdot\frac{7}{6}\cdot\frac{10}{14} = \frac{3\cdot 7\cdot 10}{5\cdot 6\cdot 14} = \frac{\overset{1}{\cancel{3}}\cdot\overset{1}{\cancel{7}}\cdot\overset{1}{\cancel{2}}\cdot\overset{1}{\cancel{5}}}{\underset{1}{\cancel{5}}\cdot\underset{1}{\cancel{3}}\cdot\underset{1}{\cancel{2}}\cdot\underset{1}{\cancel{7}}\cdot 2} = \frac{1}{2}$

c) $\frac{6h}{8}\cdot\frac{14k}{21h} = \frac{6\cdot 14\cdot h\cdot k}{8\cdot 21\cdot h} = \frac{\overset{1}{\cancel{2}}\cdot\overset{1}{\cancel{3}}\cdot\overset{1}{\cancel{2}}\cdot\overset{1}{\cancel{7}}\cdot\overset{1}{\cancel{h}}\cdot k}{\underset{1}{\cancel{2}}\cdot 2\cdot\underset{1}{\cancel{2}}\cdot\underset{1}{\cancel{3}}\cdot\underset{1}{\cancel{7}}\cdot\underset{1}{\cancel{h}}} = \frac{k}{2}$

Caution When dividing out prime factors directly, it's important to write a 1 above and below the common factor being divided out.

In Example 4b, if we had *not* written the 1's, it might appear as though the entire numerator was gone and we might think that we no longer have a fraction.

A common student error: $\frac{\cancel{3}\cdot\cancel{7}\cdot\cancel{2}\cdot\cancel{5}}{\cancel{5}\cdot\cancel{3}\cdot\cancel{2}\cdot\cancel{7}\cdot 2} = 2$ Incorrect!

You Try It 3 **Find the product. Simplify before multiplying. Use Example 3 as a guide.**

a) $\frac{6}{7}\cdot\frac{4}{9}$ b) $-\frac{12}{15}\cdot\left(-\frac{5}{7}\right)\cdot\left(-\frac{3}{4}\right)$ c) $\frac{2p}{9v}\cdot\frac{15}{10p}$

Another option is to divide out common factors directly instead of first writing a factored form.

Example 4

Multiply $\frac{30}{42}\cdot\frac{14}{20}$. Simplify before multiplying.

Procedure Apply the product rule. Identify some common factors of the numerator and the denominator and divide them out. Make sure you simplify.

Answer

30 and 20 have a common factor of 10.

$\frac{30}{42}\cdot\frac{14}{20} = \frac{\overset{3}{\cancel{30}}\cdot\overset{2}{\cancel{14}}}{\underset{6}{\cancel{42}}\cdot\underset{2}{\cancel{20}}} = \frac{6}{12} = \frac{1}{2}$

42 and 14 have a common factor of 7.

And this fraction can be simplified by a factor of 6.

You Try It 4 **Find the product. Simplify before multiplying. Use Example 4 as a guide.**

a) $-\frac{9}{16}\cdot\frac{8}{15}$ b) $-\frac{14}{21}\cdot\left(-\frac{20}{15}\right)$ c) $\frac{4}{15}\cdot\frac{10}{21}\cdot\frac{7}{8}$

Rewriting Division as Multiplication

Because division is the inverse operation of multiplication (Section 1.4) and because the reciprocal is the multiplicative inverse (Section 4.2), we can rewrite a division problem as multiplication. With fractions, we don't just replace the division sign (÷) with the multiplication sign (× or ·); we also must rewrite the second number as its reciprocal.

This process is called **invert and multiply.**

Writing Division as Multiplication $\frac{a}{b}\div\frac{c}{d}=\frac{a}{b}\cdot\frac{d}{c}$

1. Keep the first number (or fraction) the same.
2. Write the multiplication sign.
3. Invert (write the reciprocal of) the second number (or fraction).

This means, for example, that $8 \div 2$ can be rewritten as $8\cdot\frac{1}{2}$.

In general, if both numbers are already fractions, we can apply the Quotient Rule of Fractions.

Quotient Rule of Fractions

$$\frac{a}{b}\div\frac{d}{c}=\frac{a}{b}\cdot\frac{c}{d},\ b, c, \text{and } d \neq 0$$

Example: $\frac{5}{7}\div\frac{2}{3}=\frac{5}{7}\cdot\frac{3}{2}$

We have changed the problem from one of division to one of multiplication.

Caution Remember to write the problem as a product before attempting to simplify.

Example 5

Divide. Simplify if possible.

a) $\frac{4}{7}\div\frac{3}{5}$ b) $3 \div 4$ c) $-\frac{4}{9}\div\frac{5}{4}$ d) $\frac{9x}{10}\div\frac{6x}{5}$

Procedure Use the quotient rule. Invert the second fraction and multiply. Do not try to divide out common factors before writing the problem as multiplication.

Answer a) $\frac{4}{7}\div\frac{3}{5}$ Invert and multiply.

$=\frac{4}{7}\cdot\frac{5}{3}=\frac{4\cdot5}{7\cdot3}=\frac{20}{21}$

b) $3 \div 4$ Treat 3 as $\frac{3}{1}$.

$=\frac{3}{1}\cdot\frac{1}{4}=\frac{3}{4}$

c) $-\frac{4}{9} \div \frac{5}{4}$ The 4's cannot divide out.

$= -\frac{4}{9} \cdot \frac{4}{5} = -\frac{4 \cdot 4}{9 \cdot 5} = -\frac{16}{45}$

d) $\frac{9x}{10} \div \frac{6x}{5}$ Invert and multiply.

$= \frac{9x}{10} \cdot \frac{5}{6x} = \frac{\overset{3}{\cancel{9}} \cdot \overset{1}{\cancel{5}} \cdot \overset{1}{\cancel{x}}}{\underset{2}{\cancel{10}} \cdot \underset{2}{\cancel{6}} \cdot \underset{1}{\cancel{x}}} = \frac{3}{4}$

You Try It 5 **Divide using the quotient rule. Simplify if possible. Use Example 5 as a guide.**

a) $\frac{3}{4} \div \frac{4}{5}$ **b)** $\frac{5m}{6} \div \left(-\frac{7m}{12p}\right)$ **c)** $\frac{8}{x} \div \frac{4y}{3x}$

d) $-12 \div (-9)$ **e)** $-6 \div \frac{4}{5}$ **f)** $\frac{10}{9} \div (-5)$

Multiplying and Dividing Mixed Numbers

Section 4.3 states that because of variables in fractions, it is common not to change improper fractions to mixed numbers even if the fraction doesn't have a variable. However, if we are given mixed numbers to add, subtract, multiply, or divide, it's appropriate to write the answer as a mixed number as well.

In Example 2b earlier in this section, we saw the product of two improper fractions resulting in a third improper fraction: $\frac{3}{2} \cdot \frac{11}{4} = \frac{33}{8}$. If each of these fractions was written as a mixed number, the product would look like this: $1\frac{1}{2} \times 2\frac{3}{4} = 4\frac{1}{8}$.

Question Is there any way to multiply the mixed numbers $1\frac{1}{2} \times 2\frac{3}{4}$ as they are to get $4\frac{1}{8}$? Look at the whole numbers and look at the fractions.

Answer The whole numbers 1 and 2, by themselves, don't multiply to get 4. Also, the fractions $\frac{1}{2}$ and $\frac{3}{4}$, by themselves, don't multiply to get $\frac{1}{8}$.

The point is this: It is not practical to multiply (or divide) mixed numbers while they are still mixed numbers. Instead, we must rewrite them as improper fractions and then multiply (or divide).

To Multiply and Divide Mixed Numbers

1. Rewrite each mixed number as an improper fraction.
2. Multiply (or divide) these fractions as you would any others and simplify the result.
3. Rewrite the answer, if appropriate, as a mixed number.

Example 6

Evaluate and simplify. Write the answer as a mixed number if possible.

a) $2\frac{1}{4} \times 3\frac{1}{3}$ **b)** $3\frac{2}{3} \div 1\frac{1}{4}$

Procedure First, rewrite each fraction as an improper fraction.

Answer

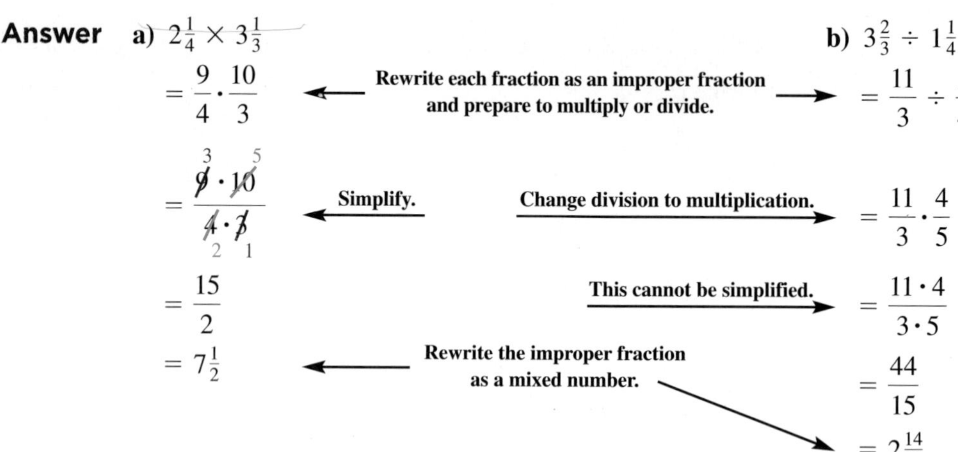

a) $2\frac{1}{4} \times 3\frac{1}{3}$
$= \frac{9}{4} \cdot \frac{10}{3}$
$= \frac{\overset{3}{\cancel{9}} \cdot \overset{5}{\cancel{10}}}{\underset{2}{\cancel{4}} \cdot \underset{1}{\cancel{3}}}$
$= \frac{15}{2}$
$= 7\frac{1}{2}$
Rewrite each fraction as an improper fraction and prepare to multiply or divide.
Simplify.
Change division to multiplication.
This cannot be simplified.
Rewrite the improper fraction as a mixed number.
b) $3\frac{2}{3} \div 1\frac{1}{4}$
$= \frac{11}{3} \div \frac{5}{4}$
$= \frac{11}{3} \cdot \frac{4}{5}$
$= \frac{11 \cdot 4}{3 \cdot 5}$
$= \frac{44}{15}$
$= 2\frac{14}{15}$

You Try It 6 **Evaluate and simplify. Write the answer as a mixed number if possible. Use Example 6 as a guide.**

a) $1\frac{2}{5} \times 3\frac{1}{3}$ **b)** $6\frac{2}{3} \times 1\frac{7}{8}$ **c)** $4\frac{2}{3} \div 3\frac{1}{2}$ **d)** $8 \div 9\frac{1}{3}$

When *of* Means Multiply

You may have heard this expression: In mathematics, *of* means "multiply." More appropriately,

> A *fraction of* means "multiply."

For example, if a class has 48 students and three-fourths are women, we could say,

"Three-fourths *of* the students are women."

Mathematically, we can interpret this as

$\frac{3}{4} \times$ the number of students $=$ the number of women.

In other words, $\frac{3}{4} \times 48 =$ the number of women.
We can evaluate this to find that the number of women in the class is 36.

$$\frac{3}{4} \cdot \frac{48}{1} = \frac{3 \cdot 48}{4 \cdot 1} = \frac{3 \cdot \overset{12}{\cancel{48}}}{\underset{1}{\cancel{4}} \cdot 1} = \frac{36}{1} = 36$$

> ***Of*** means **"multiply"** when we are referring to
>
> - A *fraction of* a number.
> - A *decimal of* a number.
> - A *percent of* a number.

In this section, we'll work with fractions only.

Example 7

Evaluate each.

a) $\frac{3}{7}$ of $\frac{5}{4}$ **b)** half of $\frac{8}{5}$ **c)** two-thirds of 15 **d)** three-fourths of $1\frac{7}{9}$

Procedure Rewrite each as multiplication, evaluate, and simplify if possible.

Answer **a)** $\frac{3}{7}$ of $\frac{5}{4}$ means $\frac{3}{7}\cdot\frac{5}{4} = \frac{15}{28}$ which cannot be simplified.

b) one-half of $\frac{8}{5}$ means $\frac{1}{2}\cdot\frac{8}{5} = \frac{1\cdot 8}{2\cdot 5} = \frac{1\cdot \cancel{8}^{4}}{\cancel{2}_{1}\cdot 5} = \frac{4}{5}$

c) two-thirds of 15 means $\frac{2}{3}\cdot 15 = \frac{2}{3}\cdot\frac{15}{1} = \frac{2\cdot 15}{3\cdot 1} = \frac{2\cdot \cancel{15}^{5}}{\cancel{3}_{1}\cdot 1} = \frac{10}{1} = 10$

d) three-fourths of $1\frac{7}{9}$ means $\frac{3}{4}\times 1\frac{7}{9}$ Before we multiply, we must write the mixed number as an improper fraction.

$$= \frac{3}{4}\cdot\frac{16}{9} = \frac{\cancel{3}^{1}\cdot\cancel{16}^{4}}{\cancel{4}_{1}\cdot\cancel{9}_{3}} = \frac{4}{3} = 1\frac{1}{3}$$

You Try It 7 **Evaluate each by rewriting it as a product of fractions. Use Example 7 as a guide.**

a) $\frac{2}{3}$ of $\frac{4}{5}$ **b)** $\frac{5}{6}$ of 9 **c)** three-fourths of $\frac{16}{21}$ **d)** five-sixths of $4\frac{4}{5}$

Applications Involving Multiplication of Fractions

In applications involving multiplication of fractions, we look for *a fraction of* something. Sometimes these situations are straightforward and don't require a variable or an equation. However, a sentence that answers the question is still required.

Example 8

A cupcake recipe calls for $5\frac{1}{4}$ cups of flour (among other ingredients). The recipe makes 24 cupcakes. Aisha wants to make two-thirds of this recipe.

a) How many cupcakes will the recipe make? **b)** How much flour is needed?

Procedure We see the *fraction of* as two-thirds of. This means that we must multiply (a) the number of cupcakes by $\frac{2}{3}$ and (b) the amount of flour by $\frac{2}{3}$.

Answer

a) $\frac{2}{3}\times 24 = \frac{2}{3}\cdot\frac{24}{1} = \frac{2\cdot\cancel{24}^{8}}{\cancel{3}_{1}\cdot 1} = \frac{16}{1} = 16$

b) $\frac{2}{3}\times 5\frac{1}{4} = \frac{2}{3}\cdot\frac{21}{4} = \frac{\cancel{2}^{1}\cdot\cancel{21}^{7}}{\cancel{3}_{1}\cdot\cancel{4}_{2}} = \frac{7}{2} = 3\frac{1}{2}$

Sentence Two-thirds of the recipe makes 16 cupcakes.

Two-thirds of the recipe requires $3\frac{1}{2}$ cups of flour.

You Try It 8 **A recipe for chili makes 36 servings. The recipe calls for $5\frac{1}{3}$ pounds of ground beef. Willie wants to make three-fourths of the recipe.**

a) How many servings will it make?

b) How many pounds of ground beef are necessary?

Sentence ______________________ ______________________

Example 9

Andres must find the middle of a piece of wood trim that is $3\frac{1}{4}$ inches wide. How far from the edge must he measure to find the middle of the wood trim?

Procedure Finding the *middle* of any length or width means finding *half* of that measure. Make sure you write the mixed number as an improper fraction.

Answer $\frac{1}{2} \times 3\frac{1}{4} = \frac{1}{2} \cdot \frac{13}{4} = \frac{1 \cdot 13}{2 \cdot 4} = \frac{13}{8} = 1\frac{5}{8}$

Sentence Andres must measure $1\frac{5}{8}$ inches from the edge to find the middle of the wood trim.

You Try It 9 **Carmen must find the middle of the logo she is stitching. It is $2\frac{1}{2}$ inches in height and $3\frac{1}{8}$ inches long. How far from the bottom (height) and how far from the side (length) will she find the middle of the logo?**

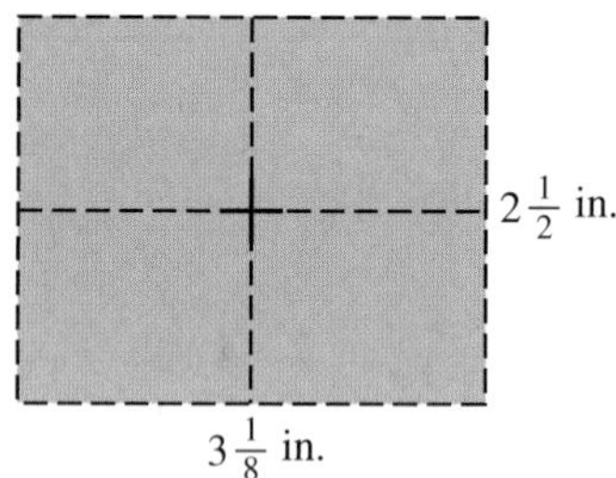

Applications Involving Geometry

Fractions are often used in situations involving geometry. Recall from Section 1.3 that the formula for the area of a rectangle is area = length × width.

Example 10

Find the area of the rectangle that is $3\frac{3}{4}$ inches long and $1\frac{2}{3}$ inches wide, as shown.

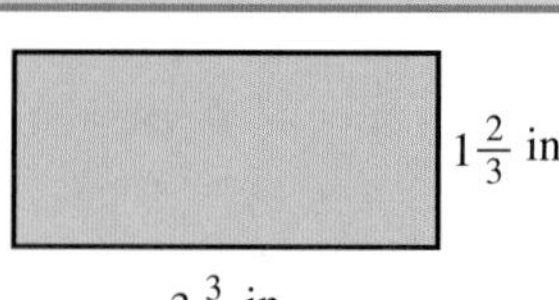

Procedure Use the formula for the area of a rectangle: area = length × width.

Answer Area $= 3\frac{3}{4} \times 1\frac{2}{3} = \frac{15}{4} \cdot \frac{5}{3} = \frac{\overset{5}{\cancel{15}} \cdot 5}{4 \cdot \underset{1}{\cancel{3}}} = \frac{25}{4} = 6\frac{1}{4}$

Sentence The area is $6\frac{1}{4}$ in^2.

You Try It 10 **Find the area of each rectangle. Use Example 10 as a guide.**

a) 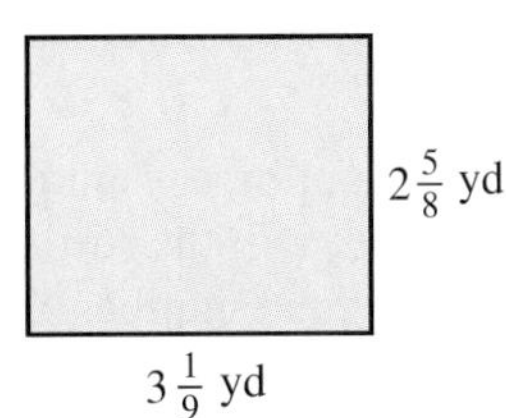

b) 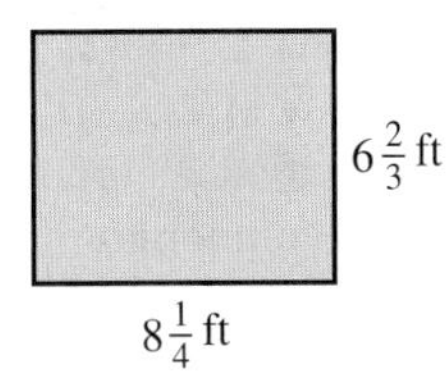

The area formula for a triangle is **area** $= \frac{1}{2} \times$ **base** $\times$ **height.**

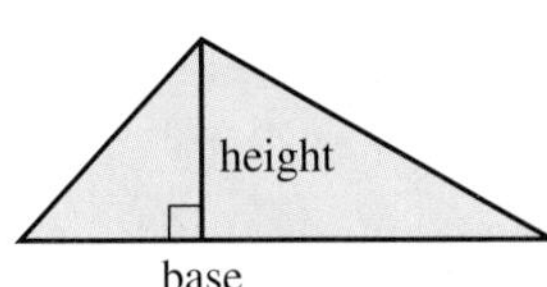

Example 11

Find the area of the triangle with base $3\frac{3}{4}$ feet and height $1\frac{2}{3}$ feet, as shown.

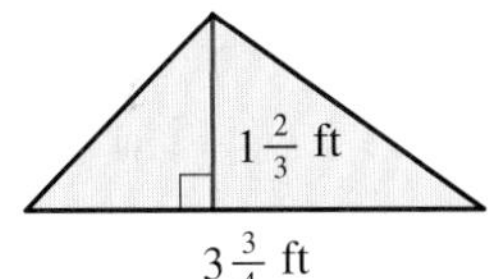

Procedure Use the formula for the area of a triangle:

area $= \frac{1}{2} \times$ base $\times$ height.

Answer Area $= \frac{1}{2} \cdot 3\frac{3}{4} \cdot 1\frac{2}{3} = \frac{1}{2} \cdot \frac{15}{4} \cdot \frac{5}{3} = \frac{1 \cdot \cancel{15}^{5} \cdot 5}{2 \cdot 4 \cdot \cancel{3}_{1}} = \frac{25}{8} = 3\frac{1}{8}$

Sentence The area is $3\frac{1}{8}$ ft^2.

You Try It 11 **Find the area of each triangle. Use Example 11 as a guide.**

a) 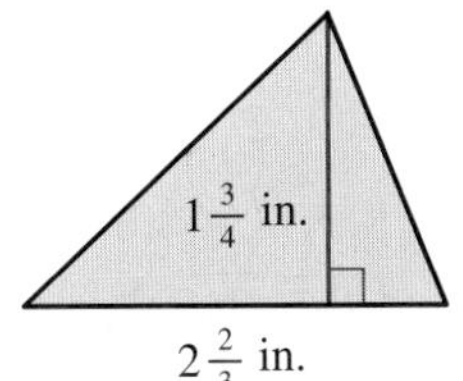

b) 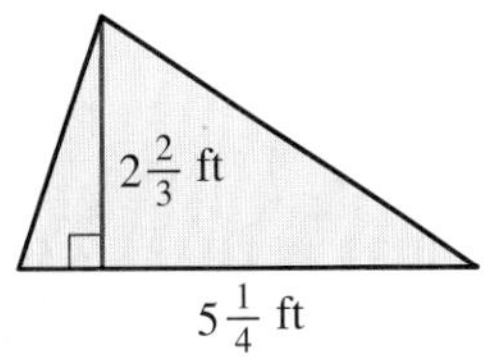

Answers: You Try It and Think About It

You Try It: **1. a)** $\frac{20}{63}$ **b)** $\frac{35a}{66b}$ **2. a)** $\frac{4x}{7}$ **b)** $\frac{6}{11}$ **c)** $\frac{1}{3}$ **d)** $-\frac{3}{10}$ **3. a)** $\frac{8}{21}$ **b)** $-\frac{3}{7}$ **c)** $\frac{1}{3v}$ **4. a)** $-\frac{3}{10}$ **b)** $\frac{8}{9}$ **c)** $\frac{1}{9}$ **5. a)** $\frac{15}{16}$ **b)** $-\frac{10p}{7}$ **c)** $\frac{6}{y}$ **d)** $\frac{4}{3}$ **e)** $-\frac{15}{2}$ **f)** $-\frac{2}{9}$ **6. a)** $4\frac{2}{3}$ **b)** $12\frac{1}{2}$ **c)** $1\frac{1}{3}$ **d)** $\frac{6}{7}$ **7. a)** $\frac{8}{15}$ **b)** $\frac{15}{2}$ or $7\frac{1}{2}$ **c)** $\frac{4}{7}$ **d)** 4 **8. a)** Three-fourths of the recipe will make 27 servings. **b)** Three-fourths of the recipe will require 4 pounds of ground beef. **9.** The middle of the logo is $1\frac{1}{4}$ inches from the bottom and $1\frac{9}{16}$ inches from the side. **10. a)** $8\frac{1}{6}$ yd^2 **b)** 55 ft^2 **11. a)** $2\frac{1}{3}$ in.^{2} **b)** 7 ft^2

Think About It: No Think About Its in this section.

Section 4.4 Exercises

FOR EXTRA HELP
Student Resources on DVD-ROM
Includes
➤ Student's Solutions Manual
➤ Video Lectures
➤ Chapter Test Prep Video
MyMathLab
MathXL

Think Again

1. How can we simplify a fraction that has the same variable with an exponent in the numerator and the denominator? Write out the steps and simplify each of the fractions.

a) $\frac{m^7}{m^3}$ b) $\frac{x^8}{x^2}$ c) $\frac{p^5}{p^5}$ d) $\frac{y^4}{y^9}$ e) $\frac{w^1}{w^4}$

We can simplify the fraction by expanding the numerator and the denominator and dividing out common factors.

a) m^4 b) x^6 c) 1 d) $\frac{1}{y^5}$ e) $\frac{1}{w^3}$

2. In your own words, describe to a classmate how to divide fractions. You may use $\frac{8}{15} \div \frac{2}{3}$ or your own example in your description.
Answers will vary.

3. If a number is multiplied by a positive proper fraction, will the result be a number that is less than the original number, greater than the original number, or the same as the original number? Explain your answer or show an example that supports your answer.
The result will be less than the original number. Multiplying by 1 keeps the value the same, and multiplying by a number less than (a proper fraction) gives a result that is less than the original number.

4. If a number is multiplied by a positive mixed number, will the result be a number that is less than the original number, greater than the original number, or the same as the original number? Explain your answer or show an example that supports your answer.
The result will be greater than the original number. Multiplying by 1 keeps the value the same, and multiplying by a number less than 1 (a proper fraction) gives a result that is less than the original number.

Focus Exercises

Multiply. Simplify each answer completely.

5. $\frac{4}{9} \cdot \frac{1}{5}$ $\frac{4}{45}$

6. $\frac{7}{12} \cdot \frac{1}{3}$ $\frac{7}{36}$

7. $-\frac{5}{6} \cdot \frac{5}{7}$ $-\frac{25}{42}$

8. $-\frac{3}{11} \cdot \frac{4}{5}$ $-\frac{12}{55}$

9. $\frac{5a}{8} \cdot \frac{4}{3b}$ $\frac{5a}{6b}$

10. $\frac{5}{7w} \cdot \frac{2y}{15}$ $\frac{2y}{21w}$

11. $-\frac{5}{21} \cdot \left(-\frac{3}{4}\right)$ $\frac{5}{28}$

12. $-\frac{8}{9} \cdot \left(-\frac{7}{12}\right)$ $\frac{14}{27}$

13. $\frac{14}{21} \cdot \left(-\frac{10}{55}\right)$ $-\frac{4}{33}$

14. $\frac{25}{35} \cdot \left(-\frac{18}{24}\right)$ $-\frac{15}{28}$

15. $\frac{9c}{15} \cdot \frac{28}{40c}$ $\frac{21}{50}$

16. $\frac{8}{32x} \cdot \frac{25x}{45}$ $\frac{5}{36}$

17. $\frac{2}{15} \cdot 3$ $\frac{2}{5}$

18. $\frac{7}{18} \cdot 9$ $\frac{7}{2}$ or $3\frac{1}{2}$

19. $-18 \cdot \frac{2}{3}$ -12

20. $-10 \cdot \frac{3}{5}$ -6

21. $\frac{7}{10} \cdot \frac{5}{14} \cdot \frac{2}{3}$ $\frac{1}{6}$

22. $\frac{5}{8} \cdot \frac{4}{9} \cdot \frac{3}{5}$ $\frac{1}{6}$

23. $\frac{10q^2}{9} \cdot \frac{3}{4q} \cdot \frac{2}{5}$ $\frac{q}{3}$

24. $\frac{2}{3b^2} \cdot \frac{9}{15} \cdot \frac{5b}{3}$ $\frac{2}{3b}$

25. $-\frac{8}{5} \cdot \frac{3}{21} \cdot \left(-\frac{3}{4}\right)$ $\frac{6}{35}$

26. $-\frac{1}{9} \cdot 24 \cdot \frac{5}{4}$ $-\frac{10}{3}$ or $-3\frac{1}{3}$

27. $8 \cdot \frac{9y}{12x} \cdot \left(-\frac{2}{15y}\right)$ $-\frac{4}{5x}$

28. $\frac{3k}{4} \cdot \frac{12h}{45} \cdot \frac{5}{4k}$ $\frac{h}{4}$

Divide. Simplify if possible.

29. $\frac{3m}{5} \div \frac{2}{7}$ $\frac{21m}{10}$

30. $\frac{2}{9} \div \frac{3r}{4}$ $\frac{8}{27r}$

31. $\frac{8}{21} \div \frac{4}{7}$ $\frac{2}{3}$

32. $\frac{6}{5} \div \frac{9}{25}$ $\frac{10}{3}$ or $3\frac{1}{3}$

33. $\frac{8d}{15} \div \frac{4d}{5}$ $\frac{2}{3}$

34. $\frac{2}{21x} \div \frac{6}{9x}$ $\frac{1}{7}$

35. $\frac{9}{10} \div \frac{12}{20}$ $\frac{3}{2}$ or $1\frac{1}{2}$

36. $\frac{14}{15} \div \frac{7}{10}$ $\frac{4}{3}$ or $1\frac{1}{3}$

37. $\frac{3}{4} \div \frac{15}{16}$ $\frac{4}{5}$

38. $\frac{12}{25} \div \frac{8}{15}$ $\frac{9}{10}$

39. $\frac{8p^2}{35} \div \frac{2p}{7}$
$\frac{4p}{5}$

40. $\frac{4y}{15} \div \frac{16y^2}{20}$
$\frac{1}{3y}$

41. $-\frac{20}{21} \div \frac{15}{28}$
$-\frac{16}{9}$ or $-1\frac{7}{9}$

42. $-\frac{3}{8} \div \frac{9}{16}$
$-\frac{2}{3}$

43. $\frac{5}{14} \div \left(-\frac{10}{21}\right)$
$-\frac{3}{4}$

44. $\frac{9}{16} \div \left(-\frac{12}{32}\right)$
$-\frac{3}{2}$ or $-1\frac{1}{2}$

45. $-\frac{2a}{15b} \div \left(-\frac{8a}{9}\right)$
$\frac{3}{20b}$

46. $-\frac{27m}{28} \div \left(-\frac{18m}{35q}\right)$
$\frac{15q}{8}$

Evaluate and simplify completely. Write each answer as a mixed number whenever possible.

47. $3\frac{3}{4} \times 2\frac{2}{3}$
10

48. $\frac{3}{4} \times 2\frac{2}{9}$
$1\frac{2}{3}$

49. $1\frac{7}{8} \times 3\frac{1}{5}$
6

50. $2\frac{1}{10} \times \frac{5}{6}$
$1\frac{3}{4}$

51. $2\frac{3}{4} \times \frac{4}{11}$
1

52. $11\frac{1}{4} \times 1\frac{7}{9}$
20

53. $1\frac{1}{14} \times 4\frac{1}{5}$
$4\frac{1}{2}$

54. $4\frac{2}{3} \times 2\frac{1}{10}$
$9\frac{4}{5}$

55. $2\frac{1}{7} \times 21$
45

56. $15 \times 1\frac{3}{10}$
$19\frac{1}{2}$

57. $9 \times 1\frac{1}{15}$
$9\frac{3}{5}$

58. $2\frac{3}{4} \times 6$
$16\frac{1}{2}$

59. $2\frac{2}{3} \div 2\frac{1}{2}$
$1\frac{1}{15}$

60. $1\frac{4}{5} \div \frac{3}{8}$
$4\frac{4}{5}$

61. $3\frac{3}{5} \div \frac{9}{10}$
4

62. $\frac{4}{15} \div 2\frac{2}{5}$
$\frac{1}{9}$

63. $\frac{8}{15} \div 1\frac{1}{9}$
$\frac{12}{25}$

64. $\frac{7}{9} \div 4\frac{2}{3}$
$\frac{1}{6}$

65. $2\frac{5}{8} \div 1\frac{3}{4}$
$1\frac{1}{2}$

66. $\frac{4}{15} \div 3\frac{5}{9}$
$\frac{3}{40}$

67. $6 \div 4\frac{1}{5}$
$1\frac{3}{7}$

68. $20 \div 1\frac{1}{3}$
15

69. $4\frac{2}{7} \div 12$
$\frac{5}{14}$

70. $2\frac{7}{10} \div 9$
$\frac{3}{10}$

Evaluate each by rewriting it as a product of fractions. Write each answer as a mixed number whenever possible.

71. One-fourth of $\frac{8}{15}$
$\frac{1}{4} \cdot \frac{8}{15} = \frac{2}{15}$

72. One-third of $\frac{9}{10}$
$\frac{1}{3} \cdot \frac{9}{10} = \frac{3}{10}$

73. One-fifth of $\frac{15}{16}$
$\frac{1}{5} \cdot \frac{15}{16} = \frac{3}{16}$

74. Three-fourths of $\frac{8}{9}$
$\frac{3}{4} \cdot \frac{8}{9} = \frac{2}{3}$

75. Two-thirds of $\frac{9}{16}$
$\frac{2}{3} \cdot \frac{9}{16} = \frac{3}{8}$

76. Five-sixths of $\frac{12}{25}$
$\frac{5}{6} \cdot \frac{12}{25} = \frac{2}{5}$

77. One-third of 18
$\frac{1}{3} \cdot \frac{18}{1} = 6$

78. Two-fifths of 30
$\frac{2}{5} \cdot \frac{30}{1} = 12$

79. Three-eighths of 48
$\frac{3}{8} \cdot \frac{48}{1} = 18$

80. Four-ninths of $\frac{3}{8}$
$\frac{4}{9} \cdot \frac{3}{8} = \frac{1}{6}$

81. One-tenth of $\frac{1}{100}$
$\frac{1}{10} \cdot \frac{1}{100} = \frac{1}{1{,}000}$

82. One-eighth of $\frac{1}{25}$
$\frac{1}{8} \cdot \frac{1}{25} = \frac{1}{200}$

83. Two-thirds of $2\frac{5}{8}$
$\frac{2}{3} \cdot 2\frac{5}{8} = 1\frac{3}{4}$

84. Three-fifths of $3\frac{8}{9}$
$\frac{3}{5} \cdot 3\frac{8}{9} = 2\frac{1}{3}$

85. One half of $3\frac{1}{5}$
$\frac{1}{2} \cdot 3\frac{1}{5} = 1\frac{3}{5}$

86. Seven-tenths of $3\frac{1}{3}$
$\frac{7}{10} \cdot 3\frac{1}{3} = 2\frac{1}{3}$

87. Three-sevenths of $4\frac{1}{5}$
$\frac{3}{7} \cdot 4\frac{1}{5} = 1\frac{4}{5}$

88. One half of $1\frac{3}{8}$
$\frac{1}{2} \cdot 1\frac{3}{8} = \frac{11}{16}$

For each application, write a sentence answering the question.

89. $\frac{2}{3}$ of a cup of laundry detergent washes a full load of laundry. How much detergent will wash $\frac{1}{4}$ of a load?
$\frac{1}{6}$ of a cup will wash $\frac{1}{4}$ of a load.

90. Every day, Janet walks her beagle once around the neighborhood. One day she measured the route in her car and found it to be $1\frac{3}{10}$ miles. How many miles does Janet walk her dog in 30 days?
Janet walks her dog 39 miles in 30 days.

91. To pass his history class, Pedro needs to get three-fifths of the problems correct on the final exam. If 60 questions are on the final, how many must Pedro get correct to pass the class?
Pedro must get 36 correct to pass the class.

92. Connie owns a $6\frac{1}{9}$-acre plot of land in the country. She decides to split it up so that each smaller plot is one-fifth the original size. How many acres will each new plot be?
Each new plot will be $1\frac{2}{9}$ acres.

Marcus has a photo that he has scanned into his computer. The photo measures $3\frac{3}{4}$ inches high and $4\frac{3}{8}$ inches wide.

93. How far from the side and how far from the bottom is the middle of the photo in its original size?
The middle of the photo is $2\frac{3}{16}$ inches from the side and $1\frac{7}{8}$ inches from the bottom.

94. To use the photo on an advertising flyer, Marcus wants to make the photo 2 times its original dimensions. What will be the dimensions of the large photo?
The dimensions of the large photo will be $7\frac{1}{2}$ inches high and $8\frac{3}{4}$ inches wide.

Find the area of each geometric figure.

95.

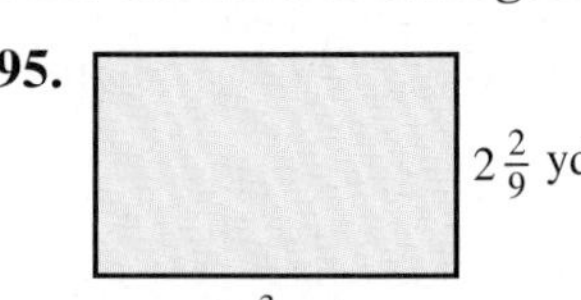

$8\frac{1}{3}$ square yards

96.

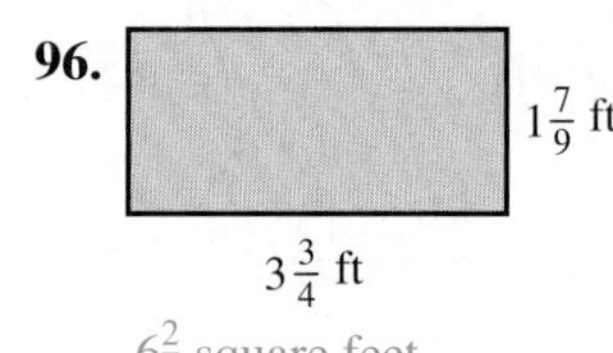

$6\frac{2}{3}$ square feet

97.

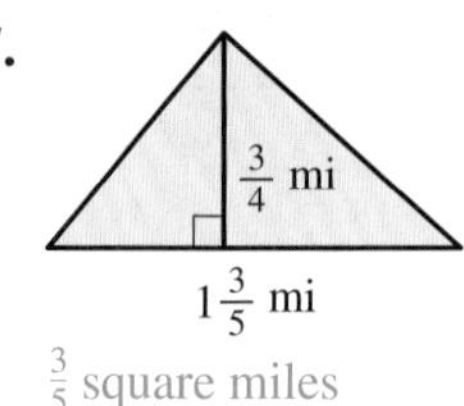

$\frac{3}{5}$ square miles

98.

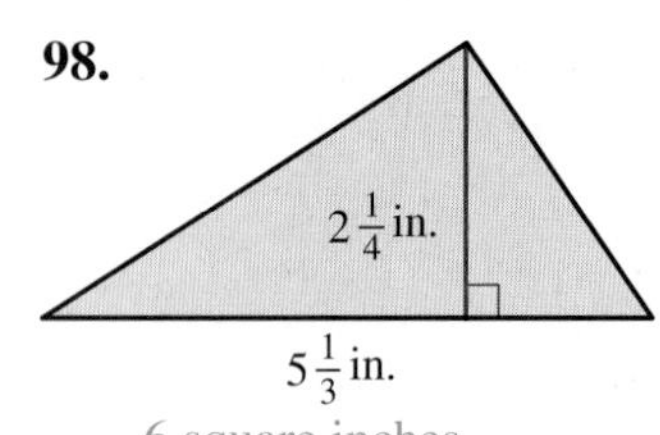

6 square inches

Think Outside the Box

Evaluate and simplify. Write each answer as a mixed number if possible.

99. $-\frac{4}{21} \div \frac{5}{9} \div \left(-\frac{6}{7}\right)$
$\frac{2}{5}$

100. $\frac{7}{9} \div \left(-\frac{4}{25} \div \frac{6}{35}\right)$
$-\frac{5}{6}$

101. $1\frac{7}{8} \div 3\frac{1}{3} \div 2\frac{1}{4}$
$\frac{1}{4}$

102. $2\frac{1}{10} \div (8\frac{1}{6} \div 8\frac{1}{3})$
$2\frac{1}{7}$

SECTION 4.5 Adding and Subtracting Like Fractions

OBJECTIVES

In this section, you will learn to

- Apply rules for adding and subtracting like fractions.
- Add and subtract positive and negative like fractions.
- Add and subtract fractions containing variables.
- Add and subtract mixed numbers containing like fractions.

You Need to Know

To successfully complete this section, you need to understand

- ☐ Adding signed numbers (2.2)
- ☐ Subtracting signed numbers (2.3)
- ☐ Combining like terms (2.7)
- ☐ Writing mixed numbers (4.2)
- ☐ Simplifying fractions (4.3)

Introduction

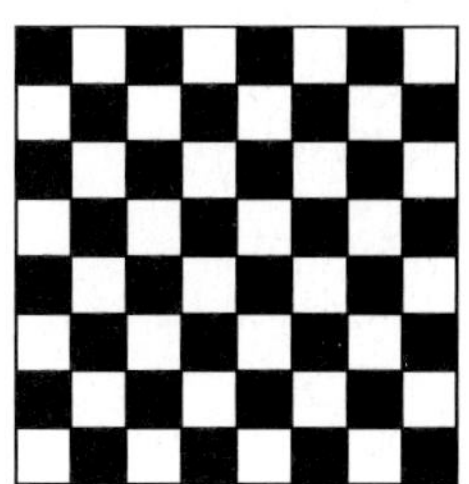

Checkers and chess are played on the same board. Even though the games use the same board, the rules for checkers and the rules for chess are different.

Likewise, addition and multiplication use the same numbers, including fractions, but the rules for addition and the rules for multiplication are different.

We will prepare for the discussion of adding and subtracting fractions by reviewing how to combine like units.

Combining Like Units of Measure

Recall from Section 2.7 that two things are "like" each other when they share the same characteristics. When units of measure are involved, we can add or subtract only if the units are "like" each other. If units are not like, we must convert them into like units before we can combine them.

Just as units must be like before they are combined (added or subtracted), so must fractions. In other words, two fractions can be added or subtracted as long as they are **like fractions**, having the same denominator.

To emphasize how we can add like fractions, consider this:

Just as	2 feet	+	3 feet	=	5 feet,
so it is that	2 sevenths	+	3 sevenths	=	5 sevenths.
This means that	$\frac{2}{7}$	+	$\frac{3}{7}$	=	$\frac{5}{7}$.

However, what we really need to know is that $\frac{2}{7} + \frac{3}{7} = \frac{\mathbf{2+3}}{7} = \frac{5}{7}$.

Notice that just like adding feet, the result is feet; likewise, just like adding sevenths, the result is sevenths.

Combining Like Fractions

To combine like fractions means to add them together or to subtract one from the other.

Adding and Subtracting Like Fractions

Adding Fractions	Subtracting Fractions
$\frac{a}{b} + \frac{c}{b} = \frac{a+c}{b}, b \neq 0$	$\frac{a}{d} - \frac{c}{d} = \frac{a-c}{d}, d \neq 0$

To add (or subtract) two like fractions, add (or subtract) the numerators. The resulting fraction is *like* the original fractions and has the same denominator.

$$\frac{2}{9} + \frac{5}{9} = \frac{2+5}{9} = \frac{7}{9} \qquad \frac{6}{7} - \frac{4}{7} = \frac{6-4}{7} = \frac{2}{7}$$

Caution Notice that the denominators are the same throughout each problem. We don't perform any operation on them.

If the result of adding or subtracting two fractions can be simplified, simplify at the very end. If the result is an improper fraction, it can be written as a mixed number.

Example 1

Combine each pair of like fractions (add or subtract as indicated).

a) $\dfrac{8}{10} + \dfrac{1}{10}$ b) $\dfrac{7}{9} + \dfrac{4}{9}$ c) $\dfrac{9}{11} - \dfrac{3}{11}$ d) $\dfrac{7}{8} - \dfrac{5}{8}$

Answer a) $\dfrac{8}{10} + \dfrac{1}{10} = \dfrac{8+1}{10} = \dfrac{9}{10}$ b) $\dfrac{7}{9} + \dfrac{4}{9} = \dfrac{7+4}{9} = \dfrac{11}{9}$ or $1\frac{2}{9}$ Written as a mixed number

c) $\dfrac{9}{11} - \dfrac{3}{11} = \dfrac{9-3}{11} = \dfrac{6}{11}$ d) $\dfrac{7}{8} - \dfrac{5}{8} = \dfrac{7-5}{8} = \dfrac{2}{8} = \dfrac{1}{4}$ Simplified by a factor of 2

You Try It 1 **Combine each pair of like fractions (add or subtract as indicated). Simplify if possible. Use Example 1 as a guide.**

a) $\dfrac{7}{15} + \dfrac{1}{15}$ b) $\dfrac{11}{12} + \dfrac{5}{12}$ c) $\dfrac{10}{13} - \dfrac{5}{13}$ d) $\dfrac{7}{10} - \dfrac{1}{10}$

Think About It 1 Take a look again at Example 1a: $\frac{8}{10} + \frac{1}{10}$. Notice that $\frac{8}{10}$ can be simplified. Would it be a good idea to simplify $\frac{8}{10}$ in the problem? Why or why not?

____

____

Caution In addition and subtraction, only the *final* answer should be simplified.

Example 2

Combine each pair of like fractions as indicated.

a) $\dfrac{3}{15} + \dfrac{8}{15}$ b) $\dfrac{10}{20} - \dfrac{5}{20}$

Procedure To add or subtract fractions, the fractions must have the same denominator. Even though you can simplify at least one of the original fractions before combining them, don't do it.

Answer a) $\dfrac{3}{15} + \dfrac{8}{15} = \dfrac{3+8}{15} = \dfrac{11}{15}$ This result can't be simplified.

b) $\dfrac{10}{20} - \dfrac{5}{20} = \dfrac{10-5}{20} = \dfrac{5}{20} = \dfrac{1}{4}$ $\frac{5}{20}$ can be simplified by a factor of 5.

You Try It 2 **Combine each pair of like fractions as indicated. Simplify the result if possible. Use Example 2 as a guide.**

a) $\dfrac{8}{20} + \dfrac{2}{20}$ b) $\dfrac{9}{18} - \dfrac{5}{18}$ c) $\dfrac{18}{25} - \dfrac{3}{25}$

Combining Negative Fractions

Recall from Section 2.3 that subtraction can also be written as "adding the opposite." So, $\frac{6}{7} - \frac{4}{7}$ can be written as $\dfrac{6}{7} + \dfrac{-4}{7} = \dfrac{6+(-4)}{7} = \dfrac{2}{7}$. When we are subtracting a negative fraction such as $\frac{4}{11} - \frac{-2}{11}$, we can write it as $\dfrac{4}{11} + \dfrac{+2}{11} = \dfrac{4+2}{11} = \dfrac{6}{11}$.

If the first fraction is negative, such as $-\frac{3}{5}$, the negative sign can be placed in the numerator, $\frac{-3}{5}$, before the fraction is combined with another fraction.

Example 3

Combine these fractions. Simplify the result if possible.

a) $\frac{3}{20} - \frac{9}{20}$ b) $-\frac{4}{15} + \frac{7}{15}$ c) $-\frac{13}{30} - \frac{-4}{30}$

Procedure Be aware of the negative signs and make sure you simplify carefully.

Answer

a) $\frac{3}{20} - \frac{9}{20}$

$= \frac{3}{20} + \frac{-9}{20}$

$= \frac{3 + (-9)}{20}$

$= \frac{-6}{20}$ This can be simplified by a factor of 2.

$= -\frac{3}{10}$

b) $-\frac{4}{15} + \frac{7}{15}$

$= \frac{-4}{15} + \frac{7}{15}$

$= \frac{-4 + 7}{15}$

$= \frac{3}{15}$ This can be simplified by a factor of 3.

$= \frac{1}{5}$

c) $-\frac{13}{30} - \frac{-4}{30}$

$= \frac{-13}{30} + \frac{+4}{30}$

$= \frac{-13 + 4}{30}$

$= \frac{-9}{30}$ This can be simplified by a factor of 3.

$= -\frac{3}{10}$

You Try It 3 **Combine these fractions. Simplify the result if possible. Use Example 3 as a guide.**

a) $-\frac{14}{25} + \frac{4}{25}$ b) $\frac{11}{24} - \frac{-5}{24}$ c) $-\frac{17}{40} - \frac{13}{40}$ d) $-\frac{23}{36} - \frac{-3}{36}$

Combining Fractions Containing Variables

Recall from Section 2.7 that two or more terms are considered to be *like terms* when they have the same variable and exponent, even if the coefficients are different. This is true even when the coefficient is a fraction.

Also recall that to combine like terms, we add their coefficients. For example, $2x$ and $3x$ are like terms and $2x + 3x = 5x$. Similarly, $\frac{2}{7}x$ and $\frac{3}{7}x$ are like terms and $\frac{2}{7}x + \frac{3}{7}x = \frac{5}{7}x$.

If the fractional coefficients of like terms add to an improper fraction, it is appropriate to leave the fraction as improper and not change it to a mixed number.

Example 4

Combine these fractions. Simplify the result if possible.

a) $\frac{7}{12}y + \frac{11}{12}y$ b) $\frac{7}{18}c^2 - \frac{5}{18}c^2$ c) $-\frac{11}{16}p - \frac{-3}{16}p$

Procedure To combine, we must have like fractions (same denominator) and like terms (same variable).

Answer

a) $\frac{7}{12}y + \frac{11}{12}y$

$= \frac{7 + 11}{12}y$

$= \frac{18}{12}y$ This can be simplified by a factor of 6.

$= \frac{3}{2}y$

b) $\frac{7}{18}c^2 - \frac{5}{18}c^2$

$= \frac{7}{18}c^2 + \frac{-5}{18}c^2$

$= \frac{7 + (-5)}{18}c^2$

$= \frac{2}{18}c^2$ This can be simplified by a factor of 2.

$= \frac{1}{9}c^2$

c) $-\frac{11}{16}p - \frac{-3}{16}p$

$= \frac{-11}{16}p + \frac{+3}{16}p$

$= \frac{-11 + 3}{16}p$

$= \frac{-8}{16}p$ This can be simplified by a factor of 8.

$= -\frac{1}{2}p$

You Try It 4 **Combine these fractions. Simplify the result if possible. Use Example 4 as a guide.**

a) $\frac{4}{9}x + \frac{8}{9}x$ **b)** $\frac{1}{14}w - \frac{-9}{14}w$ **c)** $-\frac{17}{30}b^2 + \frac{11}{30}b^2$ **d)** $-\frac{7}{20}q - \frac{13}{20}q$

Adding Mixed Numbers with Like Denominators

When adding mixed numbers, we add the whole numbers separately from the fractions. We are able to do this because of the Associative and Commutative Properties.

Consider $1\frac{2}{9} + 4\frac{5}{9}$. We could write each mixed number as a sum: → $\left(1 + \frac{2}{9}\right) + \left(4 + \frac{5}{9}\right)$

The Associative Property allows us to remove the parentheses. $= 1 + \frac{2}{9} + 4 + \frac{5}{9}$

The Commutative Property allows us to change the order of the numbers. $= 1 + 4 + \frac{2}{9} + \frac{5}{9}$

The Associative Property allows us to regroup. $= (1 + 4) + \left(\frac{2}{9} + \frac{5}{9}\right)$

And the order of operations allows us to add the whole numbers separately from the fractions. $= 5 + \frac{7}{9}$

Giving us a mixed number as an answer. $= 5\frac{7}{9}$

In practice, we commonly write the two numbers vertically, one above the other, before adding.

$$\begin{array}{r} 1\frac{2}{9} \\ +\ 4\frac{5}{9} \\ \hline 5\frac{7}{9} \end{array}$$

You Try It 5 **Evaluate. Simplify the answer if possible.**

a) $1\frac{3}{7} + 4\frac{2}{7}$ **b)** $3\frac{5}{8} + 1\frac{1}{8}$ **c)** $15\frac{4}{9} + 23\frac{2}{9}$

Sometimes the fractional parts of mixed numbers add to an improper fraction. In that case, we must *adjust* the mixed number by rewriting the improper fraction as a mixed number. We may then combine the whole number parts.

Example 5

Each mixed number contains an improper fraction. Adjust the mixed number.

a) $1\frac{8}{5}$ **b)** $7\frac{6}{6}$

Procedure Rewrite the improper fraction as a mixed number and combine the whole numbers. In part b, the improper fraction simplifies to a whole number with no fractional part.

Answer **a)** $1\frac{8}{5} = 1 + 1\frac{3}{5} = 2\frac{3}{5}$ **b)** $7\frac{6}{6} = 7 + 1 = 8$

You Try It 6 **Each mixed number contains an improper fraction. Adjust the mixed number. Use Example 5 as a guide.**

a) $4\frac{8}{7}$ **b)** $16\frac{7}{4}$ **c)** $44\frac{3}{3}$ **d)** $12\frac{9}{5}$

When adding mixed numbers, if the fractional part becomes improper, we must adjust the mixed number.

Example 6

Add each pair of fractions. Simplify and adjust the mixed number if the fraction is improper.

a) $4\frac{5}{6} + 1\frac{3}{6}$ **b)** $5\frac{3}{8} + 7\frac{5}{8}$

Answer

a)

$$\begin{array}{r} 4\frac{5}{6} \\ +\ 1\frac{3}{6} \\ \hline 5\frac{8}{6} \end{array}$$

Adjust the mixed number.

$= 5 + 1\frac{2}{6}$

$= 6\frac{2}{6}$ ← **Simplify the fraction by a factor of 2.**

$= 6\frac{1}{3}$

b)

$$\begin{array}{r} 5\frac{3}{8} \\ +\ 7\frac{5}{8} \\ \hline 12\frac{8}{8} \end{array}$$

$12 + 1$

$= 13$

You Try It 7 **Add each pair of mixed numbers. Simplify and adjust the mixed number if the fraction is improper. Use Example 6 as a guide.**

a) $1\frac{3}{4} + 3\frac{2}{4}$ **b)** $5\frac{5}{8} + 14\frac{7}{8}$ **c)** $39\frac{2}{7} + 16\frac{5}{7}$

Subtracting Mixed Numbers with Like Denominators

As in addition, when subtracting mixed numbers, we subtract whole number from whole number and fraction from fraction. Sometimes we can subtract directly, as in $8\frac{7}{9} - 2\frac{5}{9}$.

$$\begin{array}{r} 8\frac{7}{9} \\ -\ 2\frac{5}{9} \\ \hline 6\frac{2}{9} \end{array}$$

We can subtract directly because the numerator in the first fraction is greater than the numerator in the second fraction.

Example 7

Evaluate. Simplify the answer if possible.

a) $6\frac{5}{9} - 1\frac{2}{9}$ **b)** $5\frac{5}{8} - \frac{1}{8}$

Procedure Rewrite this subtraction **vertically** so that the first mixed number is above the other mixed number. In each of these problems, we will be able to subtract directly without needing to regroup.

Answer

a)

$$\begin{array}{r} 6\frac{5}{9} \\ -\ 1\frac{2}{9} \\ \hline 5\frac{3}{9} \end{array} = 5\frac{1}{3}$$

$\frac{3}{9}$ can be simplified by a factor of 3.

b)

$$\begin{array}{r} 5\frac{5}{8} \\ -\ \frac{1}{8} \\ \hline 5\frac{4}{8} \end{array} = 5\frac{1}{2}$$

$\frac{4}{8}$ can be simplified by a factor of 4.

You Try It 8 **Evaluate. Simplify the answer if possible. Use Example 7 as a guide.**

a) $\begin{array}{r} 4\frac{3}{7} \\ -\ 1\frac{2}{7} \\ \hline \end{array}$ **b)** $\begin{array}{r} 7\frac{3}{8} \\ -\ 1\frac{1}{8} \\ \hline \end{array}$ **c)** $\begin{array}{r} 17\frac{5}{12} \\ -\ 13 \\ \hline \end{array}$ **d)** $\begin{array}{r} 20\frac{7}{9} \\ -\ \frac{5}{9} \\ \hline \end{array}$

Regrouping within Mixed Numbers

We know that 8 is greater than 2, and it follows that $8\frac{4}{9}$ is greater than $2\frac{5}{9}$. Because of that, the difference $8\frac{4}{9} - 2\frac{5}{9}$ is a positive mixed number.

Yet because $\frac{4}{9}$ is less than $\frac{5}{9}$, when we subtract the fractional parts, $\frac{4}{9} - \frac{5}{9}$, we get $-\frac{1}{9}$. However, it becomes too challenging to work with a negative fraction within a positive mixed number. For that reason, we cannot directly subtract $8\frac{4}{9} - 2\frac{5}{9}$ with the fractions in their current form.

$$\begin{array}{r} 8\frac{4}{9} \\ -\ 2\frac{5}{9} \\ \hline \end{array}$$

We cannot directly subtract the fractions.

We must *regroup* the first mixed number so that its fractional part is greater than $\frac{5}{9}$. Here is the regrouping process:

1. Take 1 from the whole number: $8\frac{4}{9} = 7 + 1 + \frac{4}{9}$
2. Make the 1 into an equivalent fraction, such as $\frac{9}{9}$, so that it can be combined with the fraction:

$$= 7 + \frac{9}{9} + \frac{4}{9}$$

3. Add the fractions to create a new, equivalent mixed number:

$$= 7 + \frac{13}{9} = 7\frac{13}{9}$$

Now we can subtract $7\frac{13}{9} - 2\frac{5}{9}$:

$$\begin{array}{r} 7\frac{13}{9} \\ -\ 2\frac{5}{9} \\ \hline 5\frac{8}{9} \end{array}$$

To prepare for subtracting mixed numbers, let's practice the regrouping process.

Example 8

Regroup the mixed number $5\frac{3}{8}$.

Procedure Take 1 from the whole number and write it as a fraction. Add the fractions to create a regrouped mixed number.

	Mixed number		Take 1 from the whole number		Write 1 as a fraction		Regroup the mixed number
Answer	$5\frac{3}{8}$	=	$4 + 1 + \frac{3}{8}$	=	$4 + \frac{8}{8} + \frac{3}{8}$	=	$4\frac{11}{8}$

You Try It 9

Regroup the mixed number. Use Example 8 as a guide.

Mixed number	Take 1 from the whole number	Write 1 as a fraction	Regroup the mixed number
a) $4\frac{5}{9} =$	______	______	______
b) $10\frac{1}{6} =$	______	______	______
c) $2\frac{5}{8} =$	______	______	______

Subtracting Mixed Numbers Using Regrouping

When subtracting mixed numbers with like fractions, if the first fraction's numerator is less than the second fraction's numerator, we must regroup the first mixed number.

Example 9

Subtract $7\frac{1}{4} - 5\frac{3}{4}$.

Procedure First, write the problem vertically.

Second, notice that the first numerator is less than the second numerator. This means that we must regroup the first mixed number.

Third, after the problem has been rewritten, subtract as in Example 7.

Answer We must take 1 from the whole number. We will rewrite 1 as $\frac{4}{4}$.

$$\begin{array}{r} 7\frac{1}{4} \\ -\ 5\frac{3}{4} \\ \hline \end{array} \longrightarrow 7\frac{1}{4} = 6 + \frac{4}{4} + \frac{1}{4} \longrightarrow \begin{array}{r} 6\frac{5}{4} \\ -\ 5\frac{3}{4} \\ \hline 1\frac{2}{4} \end{array}$$

This allows us to subtract directly.

This answer's fraction can be simplified by a factor of 2: $1\frac{2}{4} = 1\frac{1}{2}$.

You Try It 10 **Evaluate. Simplify whenever possible. Use Example 9 as a guide.**

a) $\begin{array}{r} 4\frac{3}{7} \\ -\ 1\frac{5}{7} \\ \hline \end{array}$ **b)** $\begin{array}{r} 7\frac{3}{8} \\ -\ 1\frac{7}{8} \\ \hline \end{array}$ **c)** $\begin{array}{r} 10\frac{1}{12} \\ -\ 3\frac{5}{12} \\ \hline \end{array}$

Applications

Here are some situations that involve adding or subtracting fractions and mixed numbers. Recall from Section 1.2 that words such as *total, combined, sum,* and *in all* mean addition. Also recall that expressions such as *How much more than . . .* or *What is the difference between . . .* or *What was the change in . . .* usually indicate subtraction.

Example 10

A cookie recipe calls for $2\frac{7}{8}$ cups of flour, $2\frac{3}{8}$ cups of brown sugar, and $1\frac{1}{8}$ cups of white sugar. Before any other ingredients are included, how many combined cups of mixture are there?

Procedure The question asks for the *combined* number of cups. This means that we must add the three amounts. Add the whole numbers separately from the fractions. If necessary, adjust the resulting mixed number.

Answer $2\frac{7}{8} + 2\frac{3}{8} + 1\frac{1}{8} = 5\frac{11}{8} = 5 + 1\frac{3}{8} = 6\frac{3}{8}$

Sentence There are $6\frac{3}{8}$ cups of combined mixture.

You Try It 11 **Shira is almost finished installing the Guzmans' new hardwood floor. All that remains is for her to install baseboard along three short walls. The measures of the three spaces are $5\frac{5}{16}$ inches, $7\frac{9}{16}$ inches, and $4\frac{7}{16}$ inches. What is the total length of these three measures?**

Sentence ______________________________

You Try It 12 **Natalie made two custom bookcases. One bookcase is $38\frac{1}{4}$ inches high, and the other is $32\frac{3}{4}$ inches high. What is the difference in the heights of the two bookcases?**

Sentence ______________________________

Answers: You Try It and Think About It

You Try It: **1. a)** $\frac{8}{15}$ **b)** $\frac{4}{3}$ or $1\frac{1}{3}$ **c)** $\frac{5}{13}$ **d)** $\frac{3}{5}$ **2. a)** $\frac{1}{2}$ **b)** $\frac{2}{9}$ **c)** $\frac{3}{5}$ **3. a)** $-\frac{2}{5}$ **b)** $\frac{2}{3}$ **c)** $-\frac{3}{4}$ **d)** $-\frac{5}{9}$ **4. a)** $\frac{4}{3}x$ **b)** $\frac{5}{7}w$ **c)** $-\frac{1}{5}b^2$ **d)** $-1q$ or $-q$ **5. a)** $5\frac{5}{7}$ **b)** $4\frac{3}{4}$ **c)** $38\frac{2}{3}$ **6. a)** $5\frac{1}{7}$ **b)** $17\frac{3}{4}$ **c)** 45 **d)** $13\frac{4}{5}$ **7. a)** $5\frac{1}{4}$ **b)** $20\frac{1}{2}$ **c)** 56 **8. a)** $3\frac{1}{7}$ **b)** $6\frac{1}{4}$ **c)** $4\frac{5}{12}$ **d)** $20\frac{2}{9}$ **9. a)** $4\frac{5}{9} = 3 + 1 + \frac{5}{9} = 3 + \frac{9}{9} + \frac{5}{9} = 3\frac{14}{9}$ **b)** $10\frac{1}{6} = 9 + 1 + \frac{1}{6} = 9 + \frac{6}{6} + \frac{1}{6} = 9\frac{7}{6}$ **c)** $2\frac{5}{8} = 1 + 1 + \frac{5}{8} = 1 + \frac{8}{8} + \frac{5}{8} = 1\frac{13}{8}$ **10. a)** $2\frac{5}{7}$ **b)** $5\frac{1}{2}$ **c)** $6\frac{2}{3}$ **11.** The total length of these three measures is $17\frac{5}{16}$ inches. **12.** The difference in the heights of the two bookcases is $5\frac{1}{2}$ inches.

Think About It: **1.** Answers may vary. One possibility is: No, simplifying $\frac{8}{10}$ would make the denominator different from the fraction $\frac{1}{10}$, and then the fractions could not be combined.

FOR EXTRA HELP
Student Resources on DVD-ROM — Includes: ➤ Student's Solutions Manual ➤ Video Lectures ➤ Chapter Test Prep Video | MyMathLab | MathXL

Section 4.5 Exercises

Think Again

1. Are $\frac{2}{3}x$ and $\frac{1}{5}x$ like terms? Explain your answer or show an example that supports your answer.
Yes, like terms have the same variable with the same exponent regardless of the coefficient.

2. Chris, a student in your class, added $\frac{7}{8} + \frac{1}{8}$ this way: $\frac{7}{8} + \frac{1}{8} = \frac{7+1}{8+8} = \frac{8}{16} = \frac{1}{2}$. What would you tell Chris to help correct this inaccurate work?
Answers will vary.

Focus Exercises

Add. Simplify the result whenever possible.

3. $\frac{1}{2} + \frac{1}{2}$ 1

4. $\frac{2}{3} + \frac{1}{3}$ 1

5. $\frac{2}{3} + \frac{2}{3}$ $\frac{4}{3}$ or $1\frac{1}{3}$

6. $\frac{5}{6} + \frac{4}{6}$ $\frac{3}{2}$ or $1\frac{1}{2}$

7. $\frac{3}{6} + \frac{5}{6}$ $\frac{4}{3}$ or $1\frac{1}{3}$

8. $\frac{4}{9} + \frac{8}{9}$ $\frac{4}{3}$ or $1\frac{1}{3}$

9. $\frac{3}{7} + \frac{1}{7}$ $\frac{4}{7}$

10. $\frac{4}{7} + \frac{5}{7}$ $\frac{9}{7}$ or $1\frac{2}{7}$

11. $\frac{4}{7} + \frac{6}{7}$ $\frac{10}{7}$ or $1\frac{3}{7}$

12. $\frac{3}{4} + \frac{3}{4}$ $\frac{3}{2}$ or $1\frac{1}{2}$

13. $\frac{4}{5} + \frac{3}{5}$ $\frac{7}{5}$ or $1\frac{2}{5}$

14. $\frac{3}{10} + \frac{2}{10}$ $\frac{1}{2}$

15. $\frac{1}{8} + \frac{3}{8}$ $\frac{1}{2}$

16. $\frac{7}{10} + \frac{1}{10}$ $\frac{4}{5}$

17. $\frac{6}{11} + \frac{5}{11}$ 1

18. $\frac{2}{9} + \frac{7}{9}$ 1

19. $\frac{1}{9} + \frac{5}{9}$ $\frac{2}{3}$

20. $\frac{7}{10} + \frac{8}{10}$ $\frac{3}{2}$ or $1\frac{1}{2}$

21. $\frac{3}{10} + \frac{1}{10}$ $\frac{2}{5}$

22. $\frac{3}{12} + \frac{1}{12}$ $\frac{1}{3}$

23. $\frac{9}{20}k + \frac{7}{20}k$ $\frac{4}{5}k$

24. $\frac{11}{24}x + \frac{7}{24}x$ $\frac{3}{4}x$

25. $\frac{3}{14}y + \frac{4}{14}y$ $\frac{1}{2}y$

26. $\frac{11}{12}c + \frac{5}{12}c$ $\frac{4}{3}c$

27. $\frac{3}{10}v + \frac{7}{10}v$ v

28. $\frac{8}{13}p + \frac{5}{13}p$ p

29. $\frac{17}{20}n + \frac{13}{20}n$ $\frac{3}{2}n$

30. $\frac{13}{14}w + \frac{9}{14}w$ $\frac{11}{7}w$

31. $\frac{4}{15}hk + \frac{2}{15}hk$ $\frac{2}{5}hk$

32. $\frac{9}{16}mp + \frac{3}{16}mp$
$\frac{3}{4}mp$

33. $\frac{7}{15}xy + \frac{2}{15}xy$
$\frac{3}{5}xy$

34. $\frac{11}{15}bc + \frac{7}{15}bc$
$\frac{6}{5}bc$

35. $\frac{17}{21}q^2 + \frac{11}{21}q^2$
$\frac{4}{3}q^2$

36. $\frac{19}{20}d^2 + \frac{13}{20}d^2$
$\frac{8}{5}d^2$

37. $\frac{26}{30}x^2 + \frac{9}{30}x^2$
$\frac{7}{6}x^2$

38. $\frac{19}{24}y^2 + \frac{17}{24}y^2$
$\frac{3}{2}y^2$

Subtract. Simplify the result whenever possible.

39. $\frac{5}{8} - \frac{3}{8}$
$\frac{1}{4}$

40. $\frac{7}{9} - \frac{1}{9}$
$\frac{2}{3}$

41. $\frac{1}{2} - \frac{1}{2}$
0

42. $\frac{7}{9} - \frac{7}{9}$
0

43. $\frac{11}{12} - \frac{7}{12}$
$\frac{1}{3}$

44. $\frac{13}{18} - \frac{5}{18}$
$\frac{4}{9}$

45. $\frac{17}{20} - \frac{9}{20}$
$\frac{2}{5}$

46. $\frac{31}{36} - \frac{25}{36}$
$\frac{1}{6}$

47. $\frac{8}{9}h - \frac{4}{9}h$
$\frac{4}{9}h$

48. $\frac{9}{11}q - \frac{3}{11}q$
$\frac{6}{11}q$

49. $\frac{11}{15}x - \frac{8}{15}x$
$\frac{1}{5}x$

50. $\frac{9}{10}y - \frac{7}{10}y$
$\frac{1}{5}y$

51. $\frac{11}{12}ab - \frac{2}{12}ab$
$\frac{3}{4}ab$

52. $\frac{13}{16}mp - \frac{5}{16}mp$
$\frac{1}{2}mp$

53. $\frac{19}{20}d^2 - \frac{7}{20}d^2$
$\frac{3}{5}d^2$

54. $\frac{17}{24}x^2 - \frac{11}{24}x^2$
$\frac{1}{4}x^2$

55. $\frac{23}{30}m^2 - \frac{11}{30}m^2$
$\frac{2}{5}m^2$

56. $\frac{31}{40}p^2 - \frac{7}{40}p^2$
$\frac{3}{5}p^2$

Evaluate. Simplify if possible.

57. $\frac{5}{12} - \frac{11}{12}$
$-\frac{1}{2}$

58. $\frac{7}{20} - \frac{13}{20}$
$-\frac{3}{10}$

59. $-\frac{5}{6} - \frac{1}{6}$
-1

60. $-\frac{1}{8} - \frac{7}{8}$
-1

61. $-\frac{7}{18} + \frac{13}{18}$
$\frac{1}{3}$

62. $-\frac{4}{15} + \frac{13}{15}$
$\frac{3}{5}$

63. $-\frac{11}{16} + \frac{3}{16}$
$-\frac{1}{2}$

64. $-\frac{16}{21} + \frac{7}{21}$
$-\frac{3}{7}$

65. $\frac{7}{18} - \left(-\frac{5}{18}\right)$
$\frac{2}{3}$

66. $\frac{5}{36} - \left(-\frac{13}{36}\right)$
$\frac{1}{2}$

67. $-\frac{11}{24}b + \left(-\frac{7}{24}b\right)$
$-\frac{3}{4}b$

68. $-\frac{5}{32}p + \left(-\frac{23}{32}p\right)$
$-\frac{7}{8}p$

69. $-\frac{11}{24}v - \left(-\frac{5}{24}v\right)$
$-\frac{1}{4}v$

70. $-\frac{13}{30}m - \left(-\frac{1}{30}m\right)$
$-\frac{2}{5}m$

71. $-\frac{19}{40}ab - \left(-\frac{3}{40}ab\right)$
$-\frac{2}{5}ab$

72. $-\frac{31}{45}qr - \left(-\frac{16}{45}qr\right)$
$-\frac{1}{3}qr$

73. $\frac{17}{40}y^2 - \frac{25}{40}y^2$
$-\frac{1}{5}y^2$

74. $-\frac{31}{60}x^2 - \frac{17}{60}x^2$
$-\frac{4}{5}x^2$

Evaluate. Simplify and adjust the mixed number if the answer contains an improper fraction.

75. $3\frac{7}{10} + 6\frac{1}{10}$
$9\frac{4}{5}$

76. $1\frac{5}{12} + 4\frac{1}{12}$
$5\frac{1}{2}$

77. $10\frac{2}{9} + 2\frac{5}{9}$
$12\frac{7}{9}$

78. $6\frac{1}{5} + 3\frac{2}{5}$
$9\frac{3}{5}$

79. $2\frac{3}{8} + 9\frac{5}{8}$
12

80. $6\frac{3}{7} + 2\frac{4}{7}$
9

81. $5\frac{1}{10} + 8\frac{9}{10}$
14

82. $1\frac{2}{5} + 3\frac{3}{5}$
5

83. $2\frac{3}{4} + 34\frac{3}{4}$
$37\frac{1}{2}$

84. $9\frac{5}{6} + 1\frac{3}{6}$
$11\frac{1}{3}$

85. $4\frac{5}{8} + \frac{7}{8}$
$5\frac{1}{2}$

86. $28\frac{9}{10} + 7\frac{7}{10}$
$36\frac{3}{5}$

Subtract. Simplify whenever possible.

87. $5\frac{7}{9} - 3\frac{2}{9}$
$2\frac{5}{9}$

88. $8\frac{7}{15} - 2\frac{3}{15}$
$6\frac{4}{15}$

89. $12\frac{5}{7} - 3\frac{1}{7}$
$9\frac{4}{7}$

90. $14\frac{7}{11} - 13\frac{5}{11}$
$1\frac{2}{11}$

91. $4\frac{8}{15} - 3\frac{2}{15}$
$1\frac{2}{5}$

92. $12\frac{13}{16} - 3\frac{9}{16}$
$9\frac{1}{4}$

93. $7\frac{1}{9} - 2\frac{5}{9}$
$4\frac{5}{9}$

94. $1\frac{5}{11} - \frac{9}{11}$
$\frac{7}{11}$

95. $12\frac{4}{15} - 4\frac{8}{15}$
$7\frac{11}{15}$

96. $12\frac{1}{5} - 4\frac{3}{5}$
$7\frac{3}{5}$

97. $10\frac{1}{6} - 5\frac{5}{6}$
$4\frac{1}{3}$

98. $16\frac{1}{10} - 6\frac{7}{10}$
$9\frac{2}{5}$

Solve each application using addition or subtraction and write a sentence answering the question.

99. Marius is a tile layer, and he is completing the job of laying tile for a new kitchen counter. On top of the cabinets, he is putting down plywood that is $\frac{7}{8}$ inch thick. He is also putting down a $\frac{5}{8}$-inch-thick piece of construction board. How much total thickness is Marius adding to the top of the cabinets?
Marius is adding $1\frac{1}{2}$ inches to the top of the cabinets.

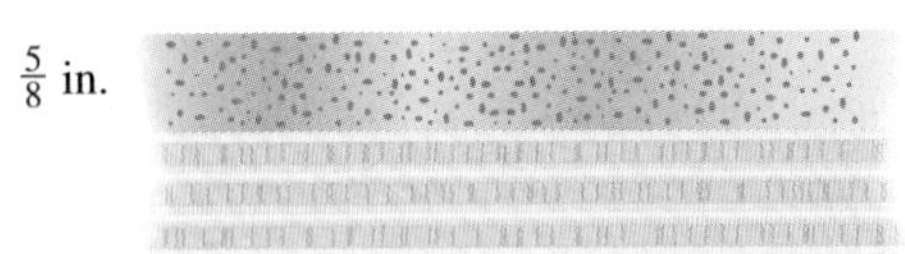

100. Terry built three new steps leading to her cabin's porch. The first step is $5\frac{7}{8}$ inches high (from the ground), the second step is $10\frac{3}{8}$ inches high, and the third step is $17\frac{1}{8}$ inches high.

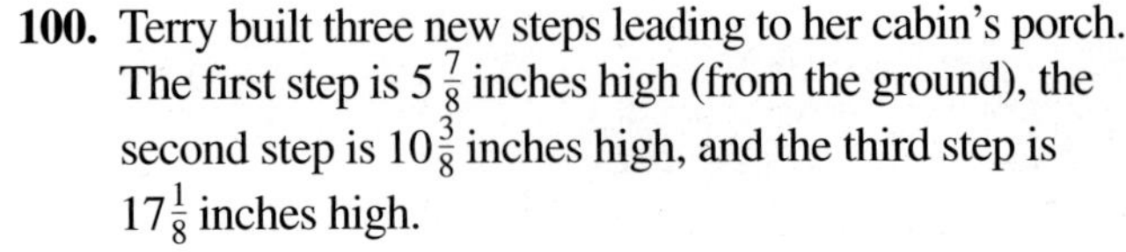

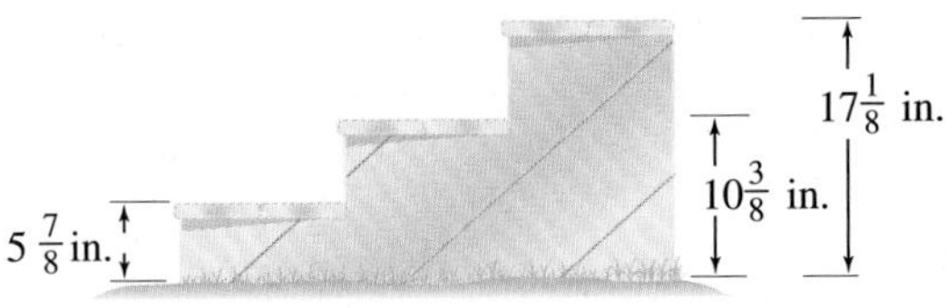

a) What is the difference in height between the second and first step?
The difference in height is $4\frac{1}{2}$ inches.

b) What is the difference in height between the third and second step?
The difference in height is $6\frac{3}{4}$ inches.

101. Juanita and Carlita are twins. On their second birthday, Juanita was $24\frac{3}{8}$ inches tall and Carlita was $26\frac{1}{8}$ inches tall. On their second birthday, how much taller was Carlita than Juanita?
Carlita was $1\frac{3}{4}$ inches taller than Juanita.

102. Find the perimeter of this triangle.
The perimeter is $10\frac{2}{5}$ meters.

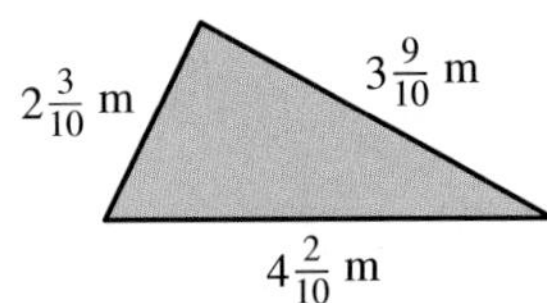

Think Outside the Box

Evaluate.

103. $\frac{7}{10} - \left(\frac{3}{10} - \frac{9}{10}\right)$
$\frac{13}{10}$ or $1\frac{3}{10}$

104. $\frac{8}{15} \cdot \left(-\frac{17}{24} - \left(-\frac{5}{24}\right)\right)$
$-\frac{4}{15}$

105. $\left(\frac{11}{30} - \left(-\frac{7}{30}\right)\right) \div \frac{9}{20}$
$\frac{4}{3}$ or $1\frac{1}{3}$

106. $-\frac{7}{10} \div \left(-\frac{11}{15} + \frac{4}{15}\right)$
$\frac{3}{2}$ or $1\frac{1}{2}$

SECTION 4.6 Common Denominators

OBJECTIVES

In this section, you will learn to

- Identify the least common denominator.
- Build up unlike fractions to have a common denominator.
- Find the least common denominator by three different methods.

You Need to Know

To successfully complete this section, you need to understand

- ☐ Multiples of numbers (1.3)
- ☐ Prime factorization (1.6)
- ☐ The greatest common factor (4.1)
- ☐ Relatively prime numbers (4.1)
- ☐ Building up fractions (4.3)

Introduction

Are $\frac{1}{4}$ and $\frac{1}{6}$ like fractions? Recall that *like* fractions have the same denominator. So, $\frac{1}{4}$ and $\frac{1}{6}$ are *not* like fractions. Because they are not like fractions, we cannot add them together, at least not in their current form.

In Section 4.3, we learned that we can build up a fraction by multiplying by a form of 1 to an equivalent fraction with a new denominator. This means that we can find a *common denominator* to which both $\frac{1}{4}$ and $\frac{1}{6}$ can be built up. To identify such a common denominator, we need to find a number into which 4 and 6 can divide evenly.

For example, 4 and 6 divide evenly into 24; so, we can multiply each fraction by a form of 1 that yields a new denominator of **24**.

$$\frac{1}{4} \cdot \frac{6}{6} = \frac{6}{\mathbf{24}} \qquad \frac{1}{6} \cdot \frac{4}{4} = \frac{4}{\mathbf{24}}$$

There are other numbers into which 4 and 6 divide evenly, such as 12, 36, and 48. Each of those possible common denominators is a multiple of 4 and 6.

$\mathit{12} = \mathbf{6} \cdot 2$ and $\mathit{12} = \mathbf{4} \cdot 3$; $\mathit{24} = \mathbf{6} \cdot 4$ and $\mathit{24} = \mathbf{4} \cdot 6$; $\mathit{36} = \mathbf{6} \cdot 6$ and $\mathit{36} = \mathbf{4} \cdot 9$; and so on.

In building up two denominators to a single common denominator, it is usually most efficient to build them up to the *least common multiple* of the two denominators. This is referred to as the **least common denominator (LCD).** In the preceding possibilities, the LCD for $\frac{1}{4}$ and $\frac{1}{6}$ is 12.

$$\frac{1}{4} \cdot \frac{3}{3} = \frac{3}{\mathbf{12}} \qquad \frac{1}{6} \cdot \frac{2}{2} = \frac{2}{\mathbf{12}}$$

Finding the Least Common Denominator by Listing Multiples

There are many techniques for identifying the LCD for two fractions. One is to simply write out some of the multiples of each denominator until a match is found. For example, we could write out the multiples of 4 and 6, and it would not take us very long to find a match.

Write the first four multiples of 4: 4, 8, 12, 16, . . .

Write the first four multiples of 6: 6, 12, 18, 24, . . .

We could write more if needed, but we can see a match at 12. So, 12 is the LCD to which we should build up each fraction.

Example 1

Find the LCD for $\frac{5}{6}$ and $\frac{4}{15}$. Then build up each fraction to have that same denominator.

Procedure Finding the LCD means finding the least common multiple of the two denominators. List the first few multiples of 6, then 15. If necessary, go back to 6, then to 15 again.

Answer The first few multiples of 6: 6, 12, 18, 24, **30**
The first few multiples of 15: 15, **30**, 45, 60
} The LCD for $\frac{5}{6}$ and $\frac{4}{15}$ is 30.

Now let's build up each fraction to have a denominator of 30.

$$\frac{5}{6} \cdot \frac{5}{5} = \frac{\mathbf{25}}{\mathbf{30}} \quad \text{and} \quad \frac{4}{15} \cdot \frac{2}{2} = \frac{\mathbf{8}}{\mathbf{30}}$$

You Try It 1 **Find the LCD of the two given fractions. Then build up each fraction to have that same denominator. Use Example 1 as a guide. (Use the space provided to write the multiples of each number.)**

a) $\frac{1}{6}$ and $\frac{7}{8}$

Multiples of 6: ________________

Multiples of 8: ________________

LCD: ________ Build up each fraction:

b) $\frac{9}{10}$ and $\frac{7}{12}$

Multiples of 10: ________________

Multiples of 12: ________________

LCD: ________ Build up each fraction:

This technique is more challenging when the numbers are larger. It can be more difficult to quickly identify the multiples of each, or the first matching multiple might be far into each list.

For example, the least common multiple of 24 and 20 is 120, but we won't find it among the first four multiples of each number.

The first four multiples of 24: 24, 48, 72, 96, . . .?
The first four multiples of 20: 20, 40, 60, 80, . . .?

Is it possible to find the *greatest* common denominator for $\frac{5}{6}$ and $\frac{4}{15}$? Explain your answer. ________________________

Finding the Least Common Denominator Using Prime Factorization

Just as we found the greatest common factor (GCF) using prime factorization (Section 1.6), we can do the same for finding the least common multiple of two denominators.

A number is the product of all of its prime factors. We might say that a number is composed of its prime factors. For example, 15 isn't 15 without factors 3 and 5 ($3 \cdot 5 = 15$), and 6 isn't 6 without factors 2 and 3 ($2 \cdot 3 = 6$).

This means that every common multiple of 6 and 15 must have prime factors 2 and 3 (for the 6) and prime factors 3 and 5 (for the 15).

We saw in Example 1 that the least common multiple of 6 and 15 is 30. The prime factorization of $30 = 2 \cdot 3 \cdot 5$.

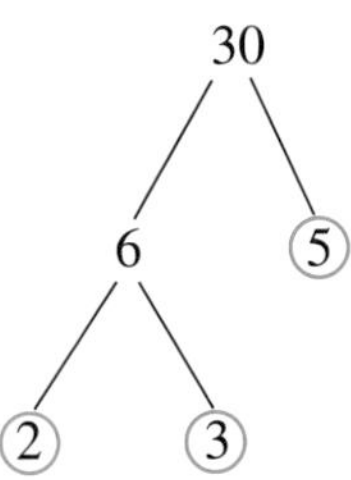

$$30 = \overbrace{2 \cdot 3}^{6} \cdot 5 \quad\text{with}\quad \underbrace{3 \cdot 5}_{15}$$

Within 30, we can see the prime factors of 6 (2 and 3) and the prime factors of 15 (3 and 5). The prime factorization of 30 shows us that the 3 is being shared between the 6 and 15. That's okay; sharing is good.

If in the common multiple we include all of the factors of 6 along with all of the factors of 15, we get $(2 \cdot 3) \cdot (3 \cdot 5) = 90$. It is true that 90 is a common multiple of 6 and 15, but it is not the *least* common multiple. In other words, 90 has too many prime factors to be the *least* common multiple of 6 and 15.

$$\overbrace{2 \cdot 3}^{6} \cdot \underbrace{3 \cdot 5}_{15} = 90$$

Therefore, when finding the *least* common multiple, we need to list as few common prime factors as possible.

Consider another example: 8 isn't 8 without *three* factors of 2 $(2 \cdot 2 \cdot 2 = 8)$, and 12 isn't 12 without *two* factors of 2 and *one* factor of 3 $(2 \cdot 2 \cdot 3 = 12)$. This means that every common multiple of 8 and 12 must have prime factors $2 \cdot 2 \cdot 2$ (for the 8) and prime factors $2 \cdot 2 \cdot 3$ (for the 12).

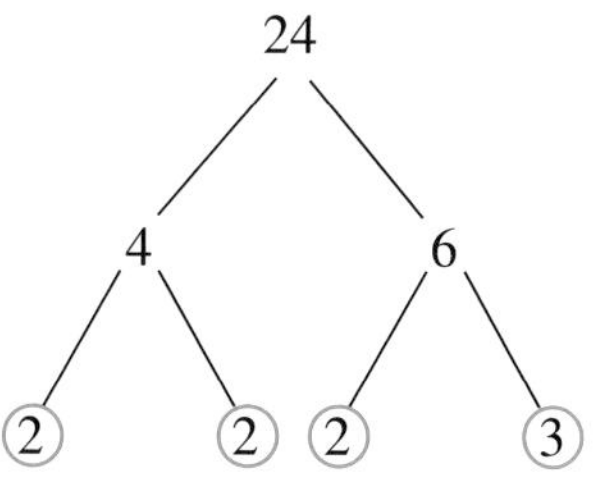

The least common multiple of 8 and 12 is 24, and the prime factorization of $24 = 2 \cdot 2 \cdot 2 \cdot 3$.

Within 24, we can see the prime factors of 8 (2 and 2 and 2) and the prime factors of 12 (2 and 2 and 3). The prime factorization of 24 shows us that *two* factors of 2 are being shared by 8 and by 12.

$$24 = \overbrace{2 \cdot 2 \cdot 2}^{8} \cdot 3 \quad (12 = 2 \cdot 2 \cdot 3)$$

Here are the guidelines for finding the least common denominator using prime factorization.

Finding the Least Common Denominator Using Prime Factorization

1. Find the prime factorization of each denominator. (Use a factor tree if needed.)
2. Write "LCD = ________________ = ________."
3. On the line, write all of the prime factors (with multiplication signs) of the first number.
4. Include any *prime* factors of the second number that have not yet been written down.
5. Multiply the prime factors and write that product after the second equal sign.

Example 2

Use the previous guidelines for finding the LCD for $\frac{3}{8}$ and $\frac{5}{12}$.

Procedure

1. Find the prime factorization of each denominator: $8 = 2 \cdot 2 \cdot 2$; $12 = 2 \cdot 2 \cdot 3$
2. Write: LCD = ________ =
3. Write all primes of the first number, 8: LCD = $\mathbf{2 \cdot 2 \cdot 2}$ = — We start with all of the factors of 8.
4. Include all primes of the second number, 12, not already included: LCD = $2 \cdot 2 \cdot 2 \cdot \mathbf{3}$ = — 12 requires *two* factors of 2 and one factor of 3. Because both 2's are already on the line, we need to include only the 3.
5. Multiply: LCD = $2 \cdot 2 \cdot 2 \cdot 3 = \mathbf{24}$ — Find the LCD by multiplying.

Answer The LCD for $\frac{3}{8}$ and $\frac{5}{12}$ is **24**.

In practice, lines 2 through 5 of Example 2 are actually written on just one line.

Find the LCD of $\frac{3}{8}$ and $\frac{5}{12}$.

$8 = 2 \cdot 2 \cdot 2$

$12 = 2 \cdot 2 \cdot 3$ LCD $= \underline{2 \cdot 2 \cdot 2 \cdot 3 = 24}$

Example 3

Find the LCD for the following pairs of fractions. Then build up each fraction to have that common denominator.

a) $\dfrac{9}{20}$ and $\dfrac{11}{30}$

b) $\dfrac{1}{4}$ and $\dfrac{8}{15}$

Answer

a) The LCD for 20 and 30:

$20 = 2 \cdot 2 \cdot 5$

$30 = 2 \cdot 3 \cdot 5$

LCD $= \underline{2 \cdot 2 \cdot 5 \cdot 3 = 60}$

$$\frac{9}{20} \cdot \frac{3}{3} = \mathbf{\frac{27}{60}}$$

$$\frac{11}{30} \cdot \frac{2}{2} = \mathbf{\frac{22}{60}}$$

b) The LCD for 4 and 15:

$4 = 2 \cdot 2$

$15 = 3 \cdot 5$

4 and 15 are relatively prime and share no factors. So, the LCD is just their product.

LCD $= \underline{2 \cdot 2 \cdot 3 \cdot 5 = 60}$

$$\frac{1}{4} \cdot \frac{15}{15} = \mathbf{\frac{15}{60}}$$

$$\frac{8}{15} \cdot \frac{4}{4} = \mathbf{\frac{32}{60}}$$

You Try It 2 **Find the LCD for the following pairs of fractions. Then build up each fraction to have that common denominator. Use Examples 2 and 3 as guides.**

a) $\dfrac{3}{10}$ and $\dfrac{4}{15}$

$10 = 2 \cdot 5$

$15 = 3 \cdot 5$

LCD = ________ = ____

Build up the fractions:

b) $\dfrac{1}{12}$ and $\dfrac{3}{20}$

LCD = ________ = ____

Build up the fractions:

c) $\dfrac{7}{9}$ and $\dfrac{3}{10}$

LCD = ________ = ____

Build up the fractions:

Finding the Least Common Denominator Using the Division Method

There is a third method of finding the LCD of two numbers, and you already have experience using it: the division method. In Section 4.1, you were introduced to the division method for finding the GCF of two numbers. We'll use that same method to find the least common multiple as well.

Here's a quick reminder of how to find the GCF of 12 and 30 using the division method. Recall that the division method stops when the bottom quotients are relatively prime.

- 2 is a common factor of 12 and 30; so, divide by 2. ⟶ $2 \,|\, 12 \quad 30$
- 3 is a common factor of 6 and 15; so, divide by 3. ⟶ $3 \,|\, 6 \quad 15$
- 2 and 5 are relatively prime; so, there are no more common factors. $2 \quad 5$

Multiply the common factors. The greatest common factor of 12 and 30 is $\mathbf{2 \cdot 3 = 6}$.

The division method for finding the least common multiple of the denominators works the same way. The LCD, though, is found by multiplying the common factors on the side along with the two relatively prime numbers at the bottom (the last pair of quotients).

To find the LCD using the division method:

1. Divide out any common factor of the two denominators, getting a new pair of numbers called *quotients*. The common factors may be prime or composite numbers.
2. Continue dividing until the new quotients are relatively prime.
3. Multiply each of the common factors and each number in the relatively prime quotients.

Let's use the division method to find the least common multiple of two denominators. We'll start with some of the same examples already demonstrated in this section.

Consider $\frac{3}{8}$ and $\frac{5}{12}$ (from Example 2) and $\frac{9}{20}$ and $\frac{11}{30}$ (from Example 3a)

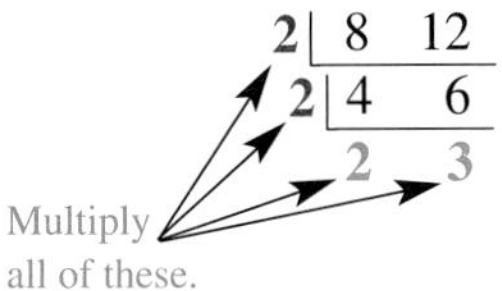

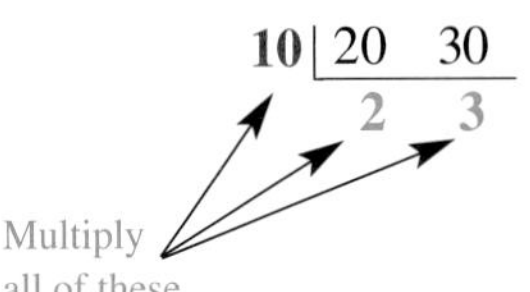

LCD for $\frac{3}{8}$ and $\frac{5}{12}$ is $\mathbf{2} \cdot \mathbf{2} \cdot 2 \cdot 3 = 24$.

LCD for $\frac{9}{20}$ and $\frac{11}{30}$ is $\mathbf{10} \cdot 2 \cdot 3 = 60$.

Each of these results is consistent with what we already know from Examples 2 and 3a.

You might find it helpful to circle all of the factors (as shown), making a large L, indicating the factors of the LCD.

2 | 8 12
2 | 4 6
2 3

LCD = (Side factors) × (Bottom factors)

Example 4

Find the LCD for each pair of fractions. Then build up each fraction to have that common denominator.

a) $\dfrac{5}{12}$ and $\dfrac{11}{30}$ **b)** $\dfrac{1}{18}$ and $\dfrac{49}{90}$

Procedure Use the division method. Draw the large L around the side primes and bottom quotients.

Answer **a)** For 12 and 30:

2	12	30	← Divide by **2**.
3	6	15	← Divide by **3**.
	2	5	← The last pair of quotients is relatively prime.

Multiply the factors along the side with those along the bottom (the last pair of quotients).

$$\mathbf{LCD} = (2 \cdot 3) \times (2 \cdot 5) = \mathbf{60}$$

- Build up the fractions: $\dfrac{5}{12} \cdot \dfrac{5}{5} = \dfrac{\mathbf{25}}{\mathbf{60}}$ $\dfrac{11}{30} \cdot \dfrac{2}{2} = \dfrac{\mathbf{22}}{\mathbf{60}}$

b) For 18 and 90:

2	18	90	← Divide by **2**.
3	9	45	← Divide by **3**.
3	3	15	← Divide by **3** again.
	1	5	← The last pair of quotients is relatively prime.

$$\mathbf{LCD} = (2 \cdot 3 \cdot 3) \cdot (1 \cdot 5) = \mathbf{90}$$

- Build up the fractions: $\dfrac{1}{18} \cdot \dfrac{5}{5} = \dfrac{\mathbf{5}}{\mathbf{90}}$ $\dfrac{49}{90} \cdot \dfrac{1}{1} = \dfrac{\mathbf{49}}{\mathbf{90}}$

Notice in Example 4b that the LCD, 90, is one of the denominators. When this happens, we don't need to multiply by 1 to build up that fraction.

Caution This division method will work for finding the *least* common multiple of two denominators only if we complete the division process. In other words, the last pair of quotients must be *relatively prime*.

In Example 4a, we found the LCD of 12 and 30. If we stopped the division process early and then multiplied, we would have gotten *a* common denominator but not the *least* common denominator:

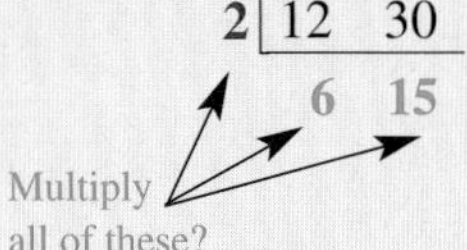

Does the LCD $= 2 \cdot 6 \cdot 15 = 180$?

No. It is 3 *times larger* than the real LCD of 60, and 3 is the value of the common factor that we didn't yet divide out.

You Try It 3 **Find the LCD of the following pairs of fractions. Then build up each fraction to have that common denominator. Use Example 4 as a guide.**

a) $\frac{3}{4}$ and $\frac{5}{14}$ **b)** $\frac{1}{12}$ and $\frac{8}{9}$ **c)** $\frac{13}{30}$ and $\frac{8}{45}$ **d)** $\frac{11}{18}$ and $\frac{7}{30}$

Answers: You Try It and Think About It

You Try It: **1. a)** LCD = 24; $\frac{1}{6} \cdot \frac{4}{4} = \frac{4}{24}$ and $\frac{7}{8} \cdot \frac{3}{3} = \frac{21}{24}$ **b)** LCD = 60; $\frac{9}{10} \cdot \frac{6}{6} = \frac{54}{60}$ and $\frac{7}{12} \cdot \frac{5}{5} = \frac{35}{60}$ **2. a)** LCD $= 2 \cdot 5 \cdot 3 = 30$; $\frac{3}{10} \cdot \frac{3}{3} = \frac{9}{30}$ and $\frac{4}{15} \cdot \frac{2}{2} = \frac{8}{30}$ **b)** LCD $= 2 \cdot 2 \cdot 3 \cdot 5 = 60$; $\frac{1}{12} \cdot \frac{5}{5} = \frac{5}{60}$ and $\frac{3}{20} \cdot \frac{3}{3} = \frac{9}{60}$ **c)** LCD $= 3 \cdot 3 \cdot 2 \cdot 5 = 90$; $\frac{7}{9} \cdot \frac{10}{10} = \frac{70}{90}$ and $\frac{3}{10} \cdot \frac{9}{9} = \frac{27}{90}$ **3. a)** LCD = 28; $\frac{3}{4} \cdot \frac{7}{7} = \frac{21}{28}$ and $\frac{5}{14} \cdot \frac{2}{2} = \frac{10}{28}$ **b)** LCD = 36; $\frac{1}{12} \cdot \frac{3}{3} = \frac{3}{36}$ and $\frac{8}{9} \cdot \frac{4}{4} = \frac{32}{36}$ **c)** LCD = 90; $\frac{13}{30} \cdot \frac{3}{3} = \frac{39}{90}$ and $\frac{8}{45} \cdot \frac{2}{2} = \frac{16}{90}$ **d)** LCD = 90; $\frac{11}{18} \cdot \frac{5}{5} = \frac{55}{90}$ and $\frac{7}{30} \cdot \frac{3}{3} = \frac{21}{90}$

Think About It: **1.** No. There are infinitely many common denominators of $\frac{5}{6}$ and $\frac{4}{15}$.

Section 4.6 Exercises

FOR EXTRA HELP
Student Resources on DVD-ROM
Includes
➤ Student's Solutions Manual
➤ Video Lectures
➤ Chapter Test Prep Video
MyMathLab
MathXL

Think Again

1. Do you think it's possible to find the *greatest* common denominator for $\frac{5}{6}$ and $\frac{4}{15}$? Explain your answer. *(Refer to Think About It 1)*
There are many common denominators for $\frac{5}{6}$ and $\frac{4}{15}$, including 60, 90, 120, 600, and 900. There is no *greatest* common denominator.

2. In your own words, how do you find the least common denominator using prime factorization?
Answers will vary.

3. If two denominators are relatively prime, what is the quickest way to identify their least common denominator? You may use $\frac{1}{8}$ and $\frac{4}{15}$ or your own example in your explanation.
The LCD is the product of the two denominators.

4. The LCD for $\frac{1}{6}$, $\frac{3}{8}$, and $\frac{2}{9}$ is 72. Can we use the division method as described in this section to find the LCD? Explain your answer or show an example that supports your answer.
No. As described, the division method will only show that the greatest common factor of all three numbers is 1. It won't lead us to the least common denominator.

Find the LCD of each pair of fractions. Then build up each fraction to have that common denominator.

5. $\frac{1}{4}$ and $\frac{1}{6}$ LCD is 12
$\frac{3}{12}$ and $\frac{2}{12}$

6. $\frac{3}{4}$ and $\frac{7}{10}$ LCD is 20
$\frac{15}{20}$ and $\frac{14}{20}$

7. $\frac{3}{4}$ and $\frac{9}{14}$ LCD is 28
$\frac{21}{28}$ and $\frac{18}{28}$

8. $\frac{1}{4}$ and $\frac{11}{18}$ LCD is 36
$\frac{9}{36}$ and $\frac{22}{36}$

9. $\frac{5}{6}$ and $\frac{9}{10}$ LCD is 30
$\frac{25}{30}$ and $\frac{27}{30}$

10. $\frac{1}{6}$ and $\frac{8}{15}$ LCD is 30
$\frac{5}{30}$ and $\frac{16}{30}$

11. $\frac{1}{6}$ and $\frac{13}{20}$ LCD is 60
$\frac{10}{60}$ and $\frac{39}{60}$

12. $\frac{5}{6}$ and $\frac{16}{21}$ LCD is 42
$\frac{35}{42}$ and $\frac{32}{42}$

13. $\frac{3}{8}$ and $\frac{1}{10}$ LCD is 40
$\frac{15}{40}$ and $\frac{4}{40}$

14. $\frac{1}{8}$ and $\frac{7}{18}$ LCD is 72
$\frac{9}{72}$ and $\frac{28}{72}$

15. $\frac{7}{8}$ and $\frac{11}{20}$ LCD is 40
$\frac{35}{40}$ and $\frac{22}{40}$

16. $\frac{5}{8}$ and $\frac{9}{28}$ LCD is 56
$\frac{35}{56}$ and $\frac{18}{56}$

17. $\frac{3}{4}$ and $\frac{7}{15}$ LCD is 60
$\frac{45}{60}$ and $\frac{28}{60}$

18. $\frac{2}{5}$ and $\frac{11}{12}$ LCD is 60
$\frac{24}{60}$ and $\frac{55}{60}$

19. $\frac{7}{8}$ and $\frac{3}{10}$ LCD is 40
$\frac{35}{40}$ and $\frac{12}{40}$

20. $\frac{3}{8}$ and $\frac{4}{5}$ LCD is 40
$\frac{15}{40}$ and $\frac{32}{40}$

21. $\frac{3}{4}$ and $\frac{7}{12}$ LCD is 12
$\frac{9}{12}$ and $\frac{7}{12}$

22. $\frac{3}{5}$ and $\frac{22}{35}$ LCD is 35
$\frac{21}{35}$ and $\frac{22}{35}$

23. $\frac{13}{15}$ and $\frac{23}{30}$ LCD is 30
$\frac{26}{30}$ and $\frac{23}{30}$

24. $\frac{5}{12}$ and $\frac{17}{36}$ LCD is 36
$\frac{15}{36}$ and $\frac{17}{36}$

25. $\frac{9}{14}$ and $\frac{17}{42}$ LCD is 42
$\frac{27}{42}$ and $\frac{17}{42}$

26. $\frac{5}{18}$ and $\frac{23}{54}$ LCD is 54
$\frac{15}{54}$ and $\frac{23}{54}$

27. $\frac{3}{5}$ and $\frac{1}{4}$ LCD is 20
$\frac{12}{20}$ and $\frac{5}{20}$

28. $\frac{1}{2}$ and $\frac{7}{9}$ LCD is 18
$\frac{9}{18}$ and $\frac{14}{18}$

29. $\frac{3}{4}$ and $\frac{5}{9}$ LCD is 36
$\frac{27}{36}$ and $\frac{20}{36}$

30. $\frac{7}{8}$ and $\frac{4}{9}$ LCD is 72
$\frac{63}{72}$ and $\frac{32}{72}$

31. $\frac{1}{4}$ and $\frac{21}{25}$ LCD is 100
$\frac{25}{100}$ and $\frac{84}{100}$

32. $\frac{1}{8}$ and $\frac{4}{15}$ LCD is 120
$\frac{15}{120}$ and $\frac{32}{120}$

33. $\frac{5}{6}$ and $\frac{2}{5}$ LCD is 30
$\frac{25}{30}$ and $\frac{12}{30}$

34. $\frac{8}{11}$ and $\frac{6}{7}$ LCD is 77
$\frac{56}{77}$ and $\frac{66}{77}$

35. $\frac{37}{60}$ and $\frac{11}{15}$ LCD is 60
$\frac{37}{60}$ and $\frac{44}{60}$

36. $\frac{41}{90}$ and $\frac{13}{18}$ LCD is 90
$\frac{41}{90}$ and $\frac{65}{90}$

37. $\frac{5}{16}$ and $\frac{13}{24}$ LCD is 48
$\frac{15}{48}$ and $\frac{26}{48}$

38. $\frac{9}{16}$ and $\frac{7}{20}$ LCD is 80
$\frac{45}{80}$ and $\frac{28}{80}$

39. $\frac{17}{36}$ and $\frac{7}{24}$ LCD is 72
$\frac{34}{72}$ and $\frac{21}{72}$

40. $\frac{1}{6}$ and $\frac{9}{25}$ LCD is 150
$\frac{25}{150}$ and $\frac{54}{150}$

Think Outside the Box

Find the least common denominator of each set of fractions. Then build up each fraction in the set to have that common denominator.

41. $\frac{2}{3}$, $\frac{3}{4}$, and $\frac{1}{6}$ LCD = 12
$\frac{8}{12}$, $\frac{9}{12}$, and $\frac{2}{12}$

42. $\frac{1}{4}$, $\frac{5}{6}$, and $\frac{7}{8}$ LCD = 24
$\frac{6}{24}$, $\frac{20}{24}$, and $\frac{21}{24}$

43. $\frac{1}{6}$, $\frac{11}{12}$, and $\frac{13}{16}$ LCD = 48
$\frac{8}{48}$, $\frac{44}{48}$, and $\frac{39}{48}$

44. $\frac{3}{4}$, $\frac{5}{12}$, and $\frac{7}{20}$ LCD = 60
$\frac{45}{60}$, $\frac{25}{60}$, and $\frac{21}{60}$

SECTION 4.7 Adding and Subtracting Unlike Fractions

OBJECTIVE

In this section, you will learn to

- Add and subtract fractions with different denominators.

You Need to Know

To successfully complete this section, you need to understand

- ☐ Simplifying fractions (4.3)
- ☐ Building up fractions (4.3)
- ☐ Adding and subtracting like fractions (4.5)
- ☐ Adding and subtracting mixed numbers (4.5)
- ☐ The least common denominator (4.6)

Introduction

In Section 4.5, we learned that like fractions can be added or subtracted. In Section 4.6, we learned about building up two fractions to have a common denominator, making them *like* fractions. In this section, we combine those two ideas and learn to add and subtract *unlike* fractions, fractions with *different* denominators.

Using the LCD to Add and Subtract Unlike Fractions

Before we can add $\frac{1}{6}$ and $\frac{2}{15}$, we need to identify the least common denominator (LCD) and then build up each fraction to that denominator. The fractions are then like fractions and can be added or subtracted. If the end result is an improper fraction, it may be written as a mixed number.

Example 1

Evaluate each expression by finding the LCD. Simplify the result.

a) $\frac{1}{6} + \frac{2}{15}$ **b)** $\frac{3}{4} + \frac{7}{10}$

Procedure First, find the LCD. Then build up each fraction to have that denominator. Part a shows the division method, and part b shows the prime factorization method.

Answer

a)
$$\begin{array}{r|ll} 3 & 6 & 15 \\ \hline & 2 & 5 \end{array}$$

$\text{LCD} = 3 \cdot 2 \cdot 5 = 30$

$$\frac{1}{6} \cdot \frac{5}{5} + \frac{2}{15} \cdot \frac{2}{2}$$
$$= \frac{5}{30} + \frac{4}{30}$$
$$= \frac{9}{30}$$ ← **This simplifies by a factor of 3.**
$$= \frac{3}{10}$$

b)
$4 = 2 \cdot 2$
$10 = 2 \cdot 5$
$\text{LCD} = 2 \cdot 2 \cdot 5 = 20$

$$\frac{3}{4} \cdot \frac{5}{5} + \frac{7}{10} \cdot \frac{2}{2}$$
$$= \frac{15}{20} + \frac{14}{20}$$
$$= \frac{29}{20}$$ ← **This is an improper fraction, and it can be written as a mixed number.**
$$= 1\frac{9}{20}$$

Subtraction works the same way.

Example 2

Evaluate $\frac{4}{5} - \frac{7}{15}$.

Procedure First, find the LCD:
$$\begin{array}{r|ll} 5 & 5 & 15 \\ \hline & 1 & 3 \end{array}$$

$\text{LCD} = 5 \cdot 1 \cdot 3 = 15$

The second fraction already has this denominator and does not need to be built up.

Answer

$$\frac{4}{5}\cdot\frac{3}{3}-\frac{7}{15}$$

$$=\frac{12}{15}-\frac{7}{15}$$

$$=\frac{5}{15}$$ ← **This can be simplified by a factor of 5.**

$$=\frac{1}{3}$$

You Try It 1 **Evaluate each sum or difference by finding the LCD and building up the fractions appropriately. Simplify each answer. Use Examples 1 and 2 as guides.**

a) $\frac{7}{10}+\frac{1}{8}$ **b)** $\frac{4}{15}+\frac{9}{10}$ **c)** $\frac{3}{4}-\frac{2}{5}$

Think About It 1 Write a set of steps for adding unlike fractions.

Adding and Subtracting Negative Fractions

When subtracting a fraction, whether positive or negative, we can rewrite it as *adding the opposite.*

Example 3

Combine these like terms.

a) $\frac{1}{4}-\frac{9}{20}$ **b)** $-\frac{1}{6}-\frac{-4}{15}$

Procedure For each expression we can rewrite the subtraction as adding the opposite, then find the LCD. Simplify the result.

Answer

a) $\frac{1}{4}+\frac{-9}{20}$

$$\begin{array}{r|rr} 2 & 4 & 20 \\ \hline 2 & 2 & 10 \\ \hline & 1 & 5 \end{array}$$

$$\text{LCD} = 2\cdot 2\cdot 1\cdot 5 = 20$$

$$\frac{1}{4}\cdot\frac{5}{5}+\frac{-9}{20}$$

$$=\frac{5}{20}+\frac{-9}{20}$$

$$=\frac{5+(-9)}{20}$$

$$=\frac{-4}{20}$$ ← **This simplifies by a factor of 4.**

$$=-\frac{1}{5}$$

b) $\frac{-1}{6}+\frac{4}{15}$

$$6 = 2\cdot 3$$

$$15 = 3\cdot 5$$

$$\text{LCD} = 2\cdot 3\cdot 5 = 30$$

$$-\frac{1}{6}\cdot\frac{5}{5}+\frac{4}{15}\cdot\frac{2}{2}$$

$$=\frac{-5}{30}+\frac{8}{30}$$

$$=\frac{-5+8}{30}$$

$$=\frac{3}{30}$$ ← **This simplifies by a factor of 3.**

$$=\frac{1}{10}$$

You Try It 2 **Combine these like terms. Use Example 3 as a guide.**

a) $\frac{7}{12} - \frac{-5}{24}$ **b)** $\frac{9}{25} + \frac{-9}{10}$ **c)** $-\frac{3}{4} - \frac{1}{10}$

Combining Unlike Fractions with Variables

When the coefficients of like terms are fractions, we can combine the terms only if the fractions have like denominators. In finding the LCD and building up the fractions, the variables are generally unaffected. Also, when the resulting coefficient is an improper fraction, it is appropriate to leave it improper.

Example 4

Combine these like terms.

$$\frac{1}{12}x + \frac{7}{15}x$$

Procedure First, find the LCD of the fractions. Simplify the result.

Answer

$\frac{1}{12}x + \frac{7}{15}x$ $\quad$ $3\,\underline{|\ 12\quad 15}$ $\quad 4 \quad 5$ $\quad$ LCD $= 3 \cdot 4 \cdot 5 = 60$

$$= \frac{1}{12} \cdot \frac{5}{5}x + \frac{7}{15} \cdot \frac{4}{4}x$$

$$= \frac{5}{60}x + \frac{28}{60}x$$

$$= \frac{5 + 28}{60}x$$

$$= \frac{33}{60}x \quad \longleftarrow \textbf{This simplifies by a factor of 3.}$$

$$= \frac{11}{20}x$$

You Try It 3 **Combine these like terms. Use Example 4 as a guide.**

a) $\frac{4}{9}b + \frac{7}{18}b$ **b)** $-\frac{11}{8}w^2 + \frac{1}{4}w^2$ **c)** $-\frac{7}{8}y - \frac{3}{10}y$

Adding Mixed Numbers with Unlike Denominators

To add or subtract mixed numbers with unlike denominators, we must identify a common denominator and build up each fraction before combining.

Example 5

Evaluate the sum. Simplify if possible.

a) $4\frac{3}{4} + 5\frac{1}{6}$ **b)** $3\frac{7}{8} + 1\frac{13}{16}$

Procedure First, identify a common denominator.
Second, build up each fraction to have that common denominator.
Third, combine the mixed numbers as in the examples presented earlier in this section.

Answer a) For $\frac{3}{4}$ and $\frac{1}{6}$, the LCD is 12. $\begin{array}{r|ll} 2 & 4 & 6 \\ \hline & 2 & 3 \end{array}$ LCD $= 2 \cdot 2 \cdot 3 = 12$

$4\frac{3}{4}$ **Build up the fraction.** ⟶ $4\frac{3}{4} \cdot \frac{3}{3} =$ $4\frac{9}{12}$

$+\ 5\frac{1}{6}$ **Build up the fraction.** ⟶ $5\frac{1}{6} \cdot \frac{2}{2} =$ $+\ 5\frac{2}{12}$

$9\frac{11}{12}$

b) For $\frac{7}{8}$ and $\frac{13}{16}$, the LCD is 16 because 16 is a multiple of 8.

$3\frac{7}{8}$ **Build up the fraction.** ⟶ $3\frac{7}{8} \cdot \frac{2}{2}$ ⟶ $3\frac{14}{16}$

$+\ 1\frac{13}{16}$ **Keep this fraction as is.** ⟶ $1\frac{13}{16}$ ⟶ $+\ 1\frac{13}{16}$

$4\frac{27}{16}$

We must take this one step further by rewriting the improper fraction and adjusting the mixed number.

$$4\frac{27}{16} = 4 + 1\frac{11}{16} = 5\frac{11}{16}$$

You Try It 4 **Add these mixed numbers. Find a common denominator, simplify, and adjust the mixed number if the fraction is improper. Use Example 5 as a guide.**

a) $1\frac{3}{8} + 3\frac{1}{4}$ b) $12\frac{2}{5} + 6\frac{3}{4}$

Subtracting Mixed Numbers with Unlike Denominators

When subtracting two mixed numbers with unlike denominators, we must find common denominators before combining.

Example 6

Evaluate. Simplify the answer if possible. $8\frac{3}{4} - 5\frac{1}{6}$

Procedure First, identify the LCD. $\begin{array}{r|ll} 2 & 4 & 6 \\ \hline & 2 & 3 \end{array}$ LCD $= 2 \times 2 \times 3 = 12$

Second, build up each fraction to have a denominator of 12.

Third, combine the mixed numbers as in the examples presented earlier in this section.

Let's rewrite this difference, $8\frac{3}{4} - 5\frac{1}{6}$, vertically. It is easier to work with the fractions that way.

Answer

$8\frac{3}{4}$ **Build up the fraction.** ⟶ $8\frac{3}{4} \times \frac{3}{3} =$ $8\frac{9}{12}$ We can subtract directly without regrouping

$-\ 5\frac{1}{6}$ **Build up the fraction.** ⟶ $5\frac{1}{6} \times \frac{2}{2} =$ $-\ 5\frac{2}{12}$

$3\frac{7}{12}$

You Try It 5 **Evaluate. Simplify the answer if possible. Use Example 6 as a guide.**

a) $10\frac{4}{9}$ $-\ 6\frac{5}{18}$

b) $12\frac{5}{6}$ $-\ 4\frac{3}{4}$

Example 7

Evaluate the difference: $3\frac{5}{12} - 1\frac{2}{3}$. Simplify the answer if possible.

Procedure At first glance, it may look as though we won't need to regroup, but that can't be decided until after we find the common denominator.

First, identify the LCD: 12 is a multiple of 3; so, LCD = 12.
Second, build up each fraction to have a denominator of 12.
Third, combine the mixed numbers as in the examples presented earlier in this section.

Answer Write the problem vertically. It is easier to work with the fractions that way.

$3\frac{5}{12}$ **Keep this fraction.** ⟶ $3\frac{5}{12} = 3\frac{5}{12}$

$-1\frac{2}{3}$ **Build up this fraction.** ⟶ $1\frac{2}{3} \times \frac{4}{4} = -1\frac{8}{12}$

Now we must regroup before we can subtract.

$3\frac{5}{12}$ ⟶ $3\frac{5}{12} = 2 + \frac{12}{12} + \frac{5}{12}$ ⟶ $2\frac{17}{12}$

$-1\frac{8}{12}$ ⟵ **Keep this mixed number the same.** ⟶ $-1\frac{8}{12}$

$1\frac{9}{12}$

Finally, simplify the fraction by a factor of 3: $1\frac{9}{12} = 1\frac{3}{4}$

You Try It 6 **Evaluate. Find a common denominator and regroup if needed. Simplify the answer if possible. Use Example 7 as a guide.**

a) $11\frac{5}{8} - 6\frac{3}{4}$

b) $8\frac{1}{3} - 4\frac{1}{2}$

Example 8

Evaluate the difference. Simplify the answer if possible. $7 - 1\frac{2}{3}$

Procedure Because the first number has no fraction, we need to take 1 from the 7 and make a fraction. It is a simple matter of choosing 1 to be $\frac{3}{3}$ so we can subtract like fractions.

We must regroup in order to subtract.

Answer

7 ⟶ $7 = 6 + \frac{3}{3}$ ⟶ $6\frac{3}{3}$

$-1\frac{2}{3}$ This allows us to subtract directly. $-1\frac{2}{3}$

$5\frac{1}{3}$

You Try It 7 **Evaluate the difference. Take 1 from the whole number and choose an appropriate denominator. Simplify the answer if possible. Use Example 8 as a guide.**

a) $12 - 9\frac{7}{12}$

b) $15 - 10\frac{3}{7}$

Applications

Recall from Section 3.5 the generic formula *The sum of all of the parts equals the whole*. In some situations, we are given two parts and are asked to find the whole. In such instances, we can apply addition directly without having to create an equation.

When working with applications, when an answer is an improper fraction, it is common to write it as a mixed number.

Example 9

Joan is a carpenter. She needs one piece of molding that is $28\frac{1}{2}$ inches long and another that is $31\frac{3}{4}$ inches long. How many total inches of molding does Joan need?

Procedure Finding the *total* indicates that we need to add the amounts.

Answer

$$28\frac{1}{2} + 31\frac{3}{4} \longrightarrow 28\frac{1}{2} \longrightarrow 28\frac{1}{2}\cdot\frac{2}{2} \longrightarrow 28\frac{2}{4}$$

$$\text{Write it vertically: } +31\frac{3}{4} \longrightarrow \longrightarrow \longrightarrow +31\frac{3}{4}$$

$$59\frac{5}{4}$$

Adjust the mixed number: $59\frac{5}{4} = 59 + 1\frac{1}{4} = 60\frac{1}{4}$

Sentence Joan needs a total of $60\frac{1}{4}$ inches of molding.

You Try It 8 **To get a pale blue shade of paint, Connie mixed $\frac{2}{3}$ gallon of white paint with $\frac{3}{4}$ gallon of blue paint. What is the total, in gallons, of the mixture?**

Sentence ______________________________

You Try It 9 **During the weekend, Sharla went hiking in Yellowstone National Park. On Saturday, she hiked $7\frac{1}{10}$ miles. On Sunday, she hiked $4\frac{3}{4}$ miles. How many more miles did Sharla hike on Saturday than on Sunday?**

Sentence ______________________________

Answers: You Try It and Think About It

You Try It: **1. a)** $\frac{33}{40}$ **b)** $\frac{7}{6}$ or $1\frac{1}{6}$ **c)** $\frac{7}{20}$ **2. a)** $\frac{19}{24}$ **b)** $-\frac{27}{50}$ **c)** $-\frac{17}{20}$ **3. a)** $\frac{5}{6}b$ **b)** $-\frac{9}{8}w^2$ **c)** $-\frac{47}{40}y$ **4. a)** $4\frac{5}{8}$ **b)** $19\frac{3}{20}$ **5. a)** $4\frac{1}{6}$ **b)** $8\frac{1}{12}$ **6. a)** $4\frac{7}{8}$ **b)** $3\frac{5}{6}$ **7. a)** $2\frac{5}{12}$ **b)** $4\frac{4}{7}$ **8.** The mixture totals $1\frac{5}{12}$ gallons of paint. **9.** Sharla hiked $2\frac{7}{20}$ more miles on Saturday than on Sunday.

Think About It: **1.** Answers may vary. One possibility: (1) determine the least common denominator (LCD); (2) build up each fraction to have the LCD; (3) add the two like fractions; (4) simplify the result, if possible.

FOR EXTRA HELP
Student Resources on DVD-ROM
Includes
➤ Student's Solutions Manual
➤ Video Lectures
➤ Chapter Test Prep Video
MyMathLab
MathXL

Section 4.7 Exercises

Think Again

1. Write a set of steps for adding unlike fractions.
Answers will vary.

2. The least common denominator for the difference $\frac{5}{6} - \frac{2}{9}$ is 18. If a student chose a common denominator of 54 $(6 \cdot 9 = 54)$ instead and built up each fraction to have a denominator of 54, would this be incorrect? Explain your answer.
No. Answers will vary.

3. Is there any benefit to simplifying each fraction before adding $\frac{15}{40} + \frac{10}{16}$? Explain your answer.
Yes. When simplified, each fraction will have a denominator of 8, thereby making the denominators common denominators.

4. What is the coefficient of the term $\frac{x}{8}$? Explain your answer.
$\frac{1}{8}$. The term $\frac{x}{8}$ can be written as $\frac{1 \cdot x}{8}$, which is the same as $\frac{1}{8}x$. This means that the coefficient of x is $\frac{1}{8}$.

Focus Exercises

Add or subtract as indicated. Simplify the results if possible. You may write any answer that contains an improper fraction as a mixed number.

5. $\frac{3}{5} + \frac{3}{10}$ $\frac{9}{10}$

6. $\frac{1}{6} + \frac{1}{3}$ $\frac{1}{2}$

7. $\frac{5}{9} + \frac{1}{3}$ $\frac{8}{9}$

8. $\frac{9}{16} + \frac{3}{8}$ $\frac{15}{16}$

9. $\frac{5}{8} - \frac{1}{3}$ $\frac{7}{24}$

10. $\frac{7}{3} - \frac{6}{5}$ $1\frac{2}{15}$

11. $\frac{10}{11} - \frac{2}{3}$ $\frac{8}{33}$

12. $\frac{10}{9} - \frac{4}{5}$ $\frac{14}{45}$

13. $\frac{5}{6} + \frac{1}{3}$ $1\frac{1}{6}$

14. $\frac{3}{4} + \frac{5}{8}$ $1\frac{3}{8}$

15. $\frac{11}{12} + \frac{1}{3}$ $1\frac{1}{4}$

16. $\frac{1}{2} + \frac{7}{8}$ $1\frac{3}{8}$

17. $\frac{7}{12} - \frac{1}{4}$ $\frac{1}{3}$

18. $\frac{9}{8} - \frac{1}{2}$ $\frac{5}{8}$

19. $\frac{2}{3} - \frac{4}{9}$ $\frac{2}{9}$

20. $\frac{3}{4} - \frac{7}{16}$ $\frac{5}{16}$

21. $\frac{3}{7} + \frac{1}{4}$ $\frac{19}{28}$

22. $\frac{2}{3} + \frac{4}{5}$ $1\frac{7}{15}$

23. $\frac{4}{5} + \frac{3}{4}$ $1\frac{11}{20}$

24. $\frac{4}{9} + \frac{3}{4}$ $1\frac{7}{36}$

25. $\frac{5}{6} - \frac{4}{9}$ $\frac{7}{18}$

26. $\frac{13}{15} - \frac{1}{10}$ $\frac{23}{30}$

27. $\frac{11}{12} - \frac{4}{9}$ $\frac{17}{36}$

28. $\frac{5}{6} - \frac{7}{10}$ $\frac{2}{15}$

29. $\frac{9}{10} + \frac{1}{6}$ $1\frac{1}{15}$

30. $\frac{1}{12} + \frac{5}{8}$ $\frac{17}{24}$

31. $\frac{7}{30} + \frac{3}{10}$ $\frac{8}{15}$

32. $\frac{13}{18} + \frac{4}{9}$ $1\frac{1}{6}$

33. $\frac{4}{9}c + \frac{1}{6}c$ $\frac{11}{18}c$

34. $\frac{1}{12}k + \frac{5}{9}k$ $\frac{23}{36}k$

35. $\frac{7}{12}y + \frac{5}{18}y$ $\frac{31}{36}y$

36. $\frac{3}{14}x + \frac{5}{21}x$ $\frac{19}{42}x$

37. $\frac{7}{10}q - \frac{1}{6}q$ $\frac{8}{15}q$

38. $\frac{7}{8}y - \frac{3}{10}y$ $\frac{23}{40}y$

39. $\frac{8}{9}x - \frac{7}{15}x$ $\frac{19}{45}x$

40. $\frac{11}{12}p - \frac{2}{9}p$ $\frac{25}{36}p$

41. $\frac{3}{10}ab + \frac{4}{15}ab$ $\frac{17}{30}ab$

42. $\frac{3}{20}xy + \frac{7}{30}xy$ $\frac{23}{60}xy$

43. $\frac{8}{25}p^2 + \frac{1}{10}p^2$ $\frac{21}{50}p^2$

44. $\frac{4}{15}m^2 + \frac{11}{20}m^2$ $\frac{49}{60}m^2$

Evaluate. Simplify, if possible.

45. $\frac{3}{4} + \left(-\frac{5}{8}\right)$ $\frac{1}{8}$

46. $\frac{1}{3} + \left(-\frac{5}{6}\right)$ $-\frac{1}{2}$

47. $\frac{2}{3} + \left(-\frac{2}{9}\right)$ $\frac{4}{9}$

48. $\frac{9}{8} + \left(-\frac{3}{10}\right)$ $\frac{33}{40}$

49. $-\frac{9}{10}m + \left(-\frac{3}{4}m\right)$ $-\frac{33}{20}m$

50. $-\frac{7}{10}d + \frac{8}{15}d$ $-\frac{1}{6}d$

51. $-\frac{1}{5} - \frac{3}{4}$ $-\frac{19}{20}$

52. $-\frac{1}{6} - \frac{5}{9}$ $-\frac{13}{18}$

53. $\frac{1}{6} - \left(-\frac{3}{8}\right)$ $\frac{13}{24}$

54. $\frac{1}{9} - \left(-\frac{5}{12}\right)$ $\frac{19}{36}$

55. $-\frac{4}{15}x - \left(-\frac{1}{6}x\right)$ $-\frac{1}{10}x$

56. $-\frac{3}{14}y - \left(-\frac{11}{21}y\right)$ $\frac{13}{42}y$

Evaluate. Simplify completely and adjust the mixed number if the answer contains an improper fraction.

57. $20\frac{1}{4} + 5\frac{3}{8}$ $25\frac{5}{8}$

58. $7\frac{1}{6} + 11\frac{3}{4}$ $18\frac{11}{12}$

59. $12\frac{1}{6} + 13\frac{2}{9}$ $25\frac{7}{18}$

60. $1\frac{7}{10} + 2\frac{1}{5}$ $3\frac{9}{10}$

61. $4\frac{7}{12} + 5\frac{5}{6}$ $10\frac{5}{12}$

62. $1\frac{13}{18} + 1\frac{4}{9}$ $3\frac{1}{6}$

63. $7\frac{7}{9} + 13\frac{2}{3}$ $21\frac{4}{9}$

64. $32\frac{5}{6} + \frac{7}{9}$ $33\frac{11}{18}$

65. $17\frac{3}{8} + \frac{11}{12}$ $18\frac{7}{24}$

66. $5\frac{2}{3} + 4\frac{1}{2}$ $10\frac{1}{6}$

67. $21\frac{2}{3} + 44\frac{3}{4}$ $66\frac{5}{12}$

68. $38\frac{5}{8} + 13\frac{5}{6}$ $52\frac{11}{24}$

69. $3\frac{7}{10} - 1\frac{1}{4}$ $2\frac{9}{20}$

70. $5\frac{2}{3} - 3\frac{1}{6}$ $2\frac{1}{2}$

71. $1\frac{11}{12} - \frac{5}{9}$ $1\frac{13}{36}$

72. $6\frac{5}{6} - 5\frac{3}{8}$ $1\frac{11}{24}$

73. $9 - 6\frac{3}{5}$ $2\frac{2}{5}$

74. $6 - 3\frac{5}{8}$ $2\frac{3}{8}$

75. $8 - 2\frac{4}{9}$ $5\frac{5}{9}$

76. $4 - 1\frac{5}{12}$ $2\frac{7}{12}$

77. $5\frac{1}{10} - 3\frac{7}{20}$ $1\frac{3}{4}$

78. $11\frac{4}{15} - 9\frac{7}{10}$ $1\frac{17}{30}$

79. $25\frac{5}{6} - 13\frac{11}{12}$ $11\frac{11}{12}$

80. $38\frac{5}{9} - 13\frac{5}{6}$ $24\frac{13}{18}$

Solve each application using addition or subtraction and write a sentence answering the question.

81. A meatball recipe calls for $\frac{5}{8}$ pound of beef and $\frac{3}{4}$ pound of pork. How much meat is in this recipe?
There are $1\frac{3}{8}$ pounds of meat in this recipe.

82. Katie needs $3\frac{1}{4}$ yards of material for a dress she is making for her niece. Katie also needs $2\frac{3}{8}$ yards of the same fabric to make her daughter a skirt. How many total yards of material does Katie need for the two projects?
Katie needs a total of $5\frac{5}{8}$ yards of material.

83. Last winter, Rohin started with $1\frac{1}{4}$ cord (a measure used for a big stack of cut-up logs) of wood to burn in his fireplace. By the end of winter, he had only $\frac{1}{3}$ cord of wood left. How much wood did Rohin burn during the winter?
Rohin burned $\frac{11}{12}$ of a cord during the winter.

84. Tom is raising a platform by $1\frac{3}{4}$ inches by stacking two boards together. The bottom board is $\frac{5}{8}$ inch thick. How thick is the top board?
The top board is $1\frac{1}{8}$ inches thick.

85. Allen purchased two watermelons for a family picnic. One melon weighed $15\frac{7}{8}$ pounds, and the other weighed $17\frac{3}{4}$ pounds. What is the total weight of the two watermelons?
The total weight of the two watermelons is $33\frac{5}{8}$ pounds.

86. Giselle needs to keep track of the number of hours she works on various projects. One day she spent $4\frac{3}{4}$ hours counting inventory and $2\frac{1}{2}$ hours doing data entry on the computer. How many total hours did Giselle work on those projects that day?
Giselle worked a total of $7\frac{1}{4}$ hours.

87. Carin needs $1\frac{1}{3}$ cups of milk for a pancake recipe. From one carton, she was able to pour $\frac{3}{4}$ cup of milk before the carton was empty. To have enough for the recipe, how much milk must Carin pour from a new carton?
Carin must pour $\frac{7}{12}$ cup of milk from a new carton.

88. Carla owned a $1\frac{1}{4}$-acre plot of land before she sold $\frac{3}{5}$ of an acre to a neighbor. How much land (in acres) did Carla own after the sale?
Carla owned $\frac{13}{20}$ of an acre after the sale.

89. Henry bought a door for his hallway closet that measures $33\frac{3}{4}$ inches wide. As it turns out, the door is not wide enough; in fact, it needs to be $1\frac{7}{16}$ inches wider. How wide should the door be that Henry buys?
The door should be $35\frac{3}{16}$ inches wide.

90. At birth, Jason measured $19\frac{3}{8}$ inches long. During his first year, he grew $6\frac{3}{4}$ inches. How tall was Jason on his first birthday?
Jason was $26\frac{1}{8}$ inches tall on his first birthday.

91. Sanjeer took his aluminum cans to be recycled. After placing the cans in the wire-framed bin, he had the bin weighed. The total weight came to $21\frac{1}{4}$ pounds. The empty bin weighs $9\frac{7}{8}$ pounds. How many pounds of aluminum cans did Sanjeer take to be recycled?
Sanjeer took $11\frac{3}{8}$ pounds of aluminum cans to be recycled.

92. To ride on the new Angel's Flight roller coaster, riders must be at least $63\frac{1}{2}$ inches tall. Kaitlin is only $61\frac{3}{4}$ inches tall. To meet the minimum height requirement for the ride, how many inches taller does Kaitlin need to be?
Kaitlin needs to be $1\frac{3}{4}$ inches taller.

93. Bob bought a set of new golf clubs, but they were too short for him. The driver is $34\frac{11}{16}$ inches long, but the golf pro at the shop determines that Bob's ideal driver length is $36\frac{1}{8}$ inches. To meet Bob's ideal driver length, how many inches should be added to the length of the driver?
$1\frac{7}{16}$ inches should be added to the length of the driver.

94. Marci is making her own rectangular planter box for a window garden. For the top of the planter, the width is $3\frac{3}{4}$ feet and the length is $6\frac{2}{3}$ feet. What is the perimeter of the top of the planter?

$6\frac{2}{3}$ ft

$3\frac{3}{4}$ ft

The perimeter of the top of the planter is $20\frac{5}{6}$ feet.

Think Outside the Box

Evaluate. Simplify if possible.

95. $\frac{2}{3} - \frac{3}{4} + \frac{1}{6}$
$\frac{1}{12}$

96. $\frac{1}{4} + \frac{3}{7} - \frac{11}{14}$
$-\frac{3}{28}$

97. $1\frac{1}{6} - 2\frac{7}{10} - 3\frac{8}{15}$
$-5\frac{1}{15}$

98. $\frac{5}{6}x - \frac{4}{15}x - \frac{17}{20}x$
$-\frac{17}{60}x$

SECTION 4.8 Equations and Applications Involving Fractions

OBJECTIVE

In this section, you will learn to

- Solve equations and applications involving multiplying and dividing fractions.

You Need to Know

To successfully complete this section, you need to understand

- ☐ Solving equations (Chapter 3)
- ☐ Simplifying fractions (4.3)
- ☐ Multiplying and dividing fractions (4.4)
- ☐ Adding and subtracting fractions (4.5, 4.7)

Introduction

Recall the Addition Property of Equations.

The Addition Property of Equations

We may add any number, c, to *each side* of an equation.

If $a = b$,

then $a + c = b + c$.

Notice that this property allows c to be any number, including fractions.

Example 1

Solve each equation by clearing the constant. Check the answer to show that it is the solution.

a) $w - \frac{2}{5} = \frac{8}{5}$ **b)** $\frac{5}{6} = x + \frac{1}{4}$

Procedure Clear the constant by adding its opposite to each side.

Answer **a)**

$w - \frac{2}{5} = \frac{8}{5}$ Clear the constant $-\frac{2}{5}$ by adding its opposite to each side.

$w - \frac{2}{5} + \frac{2}{5} = \frac{8}{5} + \frac{2}{5}$ The opposite of $-\frac{2}{5}$ is $+\frac{2}{5}$.

$w + 0 = \frac{10}{5}$ $-\frac{2}{5} + \frac{2}{5} = 0$; this is what we want.

$w = 2$ ⟶ Check the answer: $\frac{10}{5} - \frac{2}{5} \stackrel{?}{=} \frac{8}{5}$ True ✓

b)

$\frac{5}{6} = x + \frac{1}{4}$ Add the opposite of $\frac{1}{4}$ to each side.

$\frac{5}{6} + \frac{-1}{4} = x + \frac{1}{4} + \frac{-1}{4}$ On the left side, the LCD is 12.

$\frac{2}{2}\cdot\frac{5}{6} + \frac{3}{3}\cdot\frac{-1}{4} = x + 0$

$\frac{10}{12} + \frac{-3}{12} = x$

$x = \frac{7}{12}$

Check the answer: $\frac{5}{6} \stackrel{?}{=} \frac{7}{12} + \frac{1}{4}$

$\frac{5}{6} \stackrel{?}{=} \frac{7}{12} + \frac{3}{12}$

$\frac{5}{6} = \frac{10}{12}$ True ✓

You Try It 1 **Solve each of these equations by clearing the constant. Show all steps. Also check each answer. Use Example 1 as a guide.**

a) $p - \frac{1}{2} = \frac{5}{2}$ **b)** $-\frac{8}{15} = x - \frac{11}{15}$ **c)** $y + \frac{3}{4} = -\frac{3}{2}$

Equations with Fractional Solutions

Recall the Division Property of Equations.

The Division Property of Equations

We may divide each side of an equation by any number, c, $c \neq 0$.

$$\text{If} \quad a = b,$$
$$\text{then} \quad \frac{a}{c} = \frac{b}{c}.$$

In Section 3.1, we learned that to solve an equation involving multiplication, such as $6x = 18$, we divide each side by the coefficient of x to isolate the variable. This causes the coefficient of x to be just 1, our main goal in isolating the variable.

$6x = 18$ Divide each side by 6.

$\frac{6x}{6} = \frac{18}{6}$ This gives x a coefficient of 1.

$1x = 3$

$x = 3$ The solution is 3.

The solution for each equation in Section 3.1 was a whole number. This section introduces us to equations in which the solution is a fraction or a mixed number.

Example 2

Solve the equation $12 = 9x$.

Procedure Divide each side by the coefficient so that it isolates the variable.

Answer

$12 = 9x$ Divide each side by the coefficient, 9.

$\frac{12}{9} = \frac{9x}{9}$ $\frac{12}{9}$ can be simplified by a factor of 3 to become $\frac{4}{3}$.

$\frac{4}{3} = x$ $\frac{4}{3}$ can be written as a mixed number, $1\frac{1}{3}$.

$x = 1\frac{1}{3}$ Use the improper fraction $\frac{4}{3}$ when doing the check.

Check

$12 \stackrel{?}{=} 9 \cdot \frac{4}{3}$

$12 \stackrel{?}{=} \frac{36}{3}$

$12 = 12$ ✓

You Try It 2 **Solve each equation by dividing by the coefficient so that it isolates the variable. If the solution is an improper fraction, write it as a mixed number. Use Example 2 as a guide.**

a) $14y = 21$ **b)** $-28 = 42w$

When the Coefficient Is a Fraction

What is the coefficient in the expression $\frac{2x}{3}$? You might think that it is just 2 because that is the number we see being multiplied directly to x. However, the actual coefficient is the fraction $\frac{2}{3}$, as demonstrated here.

$$\frac{2x}{3} = \frac{2 \cdot x}{3 \cdot 1} = \frac{2}{3} \cdot \frac{x}{1} = \frac{2}{3} \cdot x \quad \text{or} \quad \frac{2}{3}x$$

Similarly, the coefficient of $\frac{x}{5}$ is $\frac{1}{5}$: $\frac{x}{5} = \frac{1 \cdot x}{5 \cdot 1} = \frac{1}{5} \cdot \frac{x}{1} = \frac{1}{5} \cdot x$ or $\frac{1}{5}x$

You Try It 3 **For each fraction, identify the coefficient.**

a) $\frac{6w}{7}$ **b)** $\frac{4x}{9}$ **c)** $\frac{y}{5}$ **d)** $-\frac{x}{8}$

In an equation, if the coefficient is a fraction, such as $\frac{2y}{3} = 8$, the coefficient should be shown as being multiplied directly to the variable. In other words, we should rewrite the equation as $\frac{2}{3}y = 8$.

The Multiplication Property of Equations

As you might imagine, there is also a Multiplication Property of Equations.

The Multiplication Property of Equations

We may multiply each side of an equation by any number, c, $c \neq 0$.

$$\text{If} \quad a = b,$$
$$\text{then} \quad c \cdot a = c \cdot b.$$

We typically use the Multiplication Property of Equations when the coefficient is a fraction, such as in $\frac{2}{3}y = 8$. In this case and others like it, we can multiply each side by the reciprocal of the coefficient, as demonstrated in Example 3.

Example 3

Solve this equation by clearing the coefficient: $\frac{2y}{3} = 8$.

Procedure First, write the equation with the coefficient $\frac{2}{3}$. Then multiply each side by the reciprocal of the coefficient. Check the answer by using it as a replacement value for the variable in the original equation.

Answer

$\frac{2}{3}y = 8$ — The coefficient is $\frac{2}{3}$. Clear it by multiplying each side by $\frac{3}{2}$.

$\frac{3}{2} \cdot \frac{2}{3}y = \frac{8}{1} \cdot \frac{3}{2}$ — Multiplying $\frac{3}{2} \cdot \frac{2}{3}$ makes the coefficient $= 1$.

$1y = \frac{24}{2}$ — The right side simplifies to 12.

$y = 12$

Check $\frac{2}{3} \cdot \frac{12}{1} \stackrel{?}{=} 8$

$\frac{24}{3} = 8$ True ✓

If the coefficient is a mixed number, write it as an improper fraction.

Example 4

Solve $1\frac{3}{5}y = 2\frac{2}{5}$.

Procedure First, write each mixed number as an improper fraction.

Answer

$\frac{8}{5}y = \frac{12}{5}$ — The coefficient is $\frac{8}{5}$. Clear it by multiplying each side by $\frac{5}{8}$.

$\frac{5}{8} \cdot \frac{8}{5}y = \frac{12}{5} \cdot \frac{5}{8}$ — Multiplying $\frac{5}{8} \cdot \frac{8}{5}$ makes the coefficient $= 1$.

$1y = \frac{60}{40}$ — The right side simplifies by a factor of 20.

$y = \frac{3}{2}$

$y = 1\frac{1}{2}$

Check $\frac{8}{5} \cdot \frac{3}{2} \stackrel{?}{=} \frac{12}{5}$

$\frac{24}{10} \stackrel{?}{=} \frac{12}{5}$ True ✓

You Try It 4 **Solve each equation by clearing the coefficient. Check each answer. Use Examples 3 and 4 as guides.**

a) $\dfrac{7w}{8} = -28$ **b)** $12 = -\dfrac{x}{3}$ **c)** $1\frac{1}{9}y = \dfrac{2}{3}$

If the coefficient is an integer, we can clear the coefficient by multiplying each side of the equation by the reciprocal of the coefficient. This is especially helpful when the constant on the other side of the equal sign is a fraction.

Example 5

Solve $6x = \dfrac{3}{4}$.

Procedure Clear the coefficient by multiplying each side by its reciprocal, $\frac{1}{6}$.

Answer

$6x = \dfrac{3}{4}$ Write 6 as a fraction: $\frac{6}{1}$. Clear $\frac{6}{1}$ by multiplying each side by $\frac{1}{6}$.

$\dfrac{1}{6}\cdot\dfrac{6}{1}x = \dfrac{3}{4}\cdot\dfrac{1}{6}$ Multiplying $\frac{1}{6}\cdot\frac{6}{1}$ makes the coefficient $= 1$.

$x = \dfrac{3}{24}$ The right side simplifies by a factor of 3.

$y = \dfrac{1}{8}$

Check $\dfrac{6}{1}\cdot\dfrac{1}{8} \stackrel{?}{=} \dfrac{3}{4}$

$\dfrac{6}{8} \stackrel{?}{=} \dfrac{3}{4}$ True ✓

You Try It 5 **Solve each equation by clearing the coefficient. Check each answer. Use Example 5 as a guide.**

a) $8v = \dfrac{12}{5}$ **b)** $\dfrac{15}{4} = -5x$

Applications Involving Multiplying and Dividing Fractions

In Section 1.8, you were introduced to some situations that require using multiplication. At that time, we explored the formula

(Number of same parts) × Part = Whole

Recall that the whole and the part are always the same unit of measure.

Although this formula involves multiplication, some of the situations you will see involve dividing the whole into equal parts. Still, this formula can be used for both division and multiplication problems.

In each problem, one of the "pieces" is unknown. It may be the whole, the part, or the number of repetitions of the part. The key to setting up the formula is recognizing which pieces have the common measure. They will be the whole (larger number) and the part (smaller number).

For example, a dozen (12) cookies can be divided easily among 4 children. The question might be, "How many cookies will each child get?" Let's use this simple situation to illustrate the proper way to approach this problem and others like it.

Example 6

If a dozen cookies are to be divided evenly among four children, how many cookies will each child receive?

Procedure First, determine the common measure, the whole, and the part.

Common Measure Cookies. So, the part and the whole are the cookies and the part is repeated by the 4 children.

The **whole** is the total number of cookies.

The **part**, the unknown, is how many cookies each child will receive.

Since the part is divided evenly among 4 children, the number of times the part is repeated is 4.

Answer Legend: Let x = the number of cookies each child receives.

$$4 \cdot x = 12 \quad \text{Divide each side by 4.}$$

$$\frac{4x}{4} = \frac{12}{4}$$

$$x = 3$$

Sentence Each child will receive 3 cookies.

This example may seem a bit too easy, especially since it doesn't involve dividing fractions. Consider this next example, though. It is similar, and we can use what we know about fractions to find the answer.

Example 7

Brian has a strip of wood that is $2\frac{1}{4}$ inches wide, and it needs to be divided into 3 smaller strips of equal width. How wide will each smaller strip be?

Procedure First, we can write $2\frac{1}{4}$ as an improper fraction: $\frac{9}{4}$.

Next, picture yourself in the situation. You've got a strip of wood that is $\frac{9}{4}$ inches wide. You are going to divide it evenly into smaller pieces.

Common Measure We're given a width $\left(\frac{9}{4}\text{ inches}\right)$, and we're asked to find a second width. This must mean that the widths (in inches) are the part and the whole.

The larger (original) piece must be the **whole** (total width of $\frac{9}{4}$ inches).

Each smaller strip is a **part**, the width is unknown, and the part is repeated 3 times. (There are to be 3 smaller strips.)

Answer Legend: Let x = the width of each smaller strip.

$$3 \cdot x = \frac{9}{4} \quad \text{Divide each side by 3.}$$

$$\frac{1}{3} \cdot \frac{3}{1} \cdot x = \frac{9}{4} \cdot \frac{1}{3}$$

$$x = \frac{9}{12} \quad \text{Simplify } \tfrac{9}{12} \text{ by a factor of 3.}$$

$$x = \frac{3}{4}$$

Sentence Each smaller strip will be $\frac{3}{4}$ inches wide.

You Try It 6 **Gail has a stack of flyers she wants to fold, and she asks Barbara and Jaynelle to help her. They decide to measure the height (thickness) of the stack. It turns out to be $4\frac{1}{8}$ inches high. If they divide the full stack evenly 3 ways, how high will each person's pile be?**

Sentence ______________________________

Sometimes the number of repetitions is unknown. We can determine this when we identify the common measure. When the known values have the same common measure, they must be the part and the whole and the unknown value is the number of repetitions.

Example 8

Dimitri needs small strips of wood that are only $\frac{3}{8}$ inches wide. From a board that is $5\frac{1}{4}$ inches wide, how many smaller strips of wood can Dimitri get?

Procedure First, let's look at the common measure to help us identify the part and the whole. Also write $5\frac{1}{4}$ as $\frac{21}{4}$.

Common Measure Dimitri has a board that is $\frac{21}{4}$ inches wide, and the smaller strips are $\frac{3}{8}$ inches wide. The width of the board is the **whole** $\left(\frac{21}{4}\text{ inches}\right)$, and the width of each strip $\left(\frac{3}{8}\text{ inches}\right)$ is the **part**. Both of these measurements are in inches and are known values. So, the unknown value is the number of parts, which is mentioned in the question.

Answer Legend: Let x = the number of strips. The equation is $x \cdot \frac{3}{8} = \frac{21}{4}$, or $\frac{3}{8} \cdot x = \frac{21}{4}$.

$$\frac{3}{8} \cdot x = \frac{21}{4}$$ Multiply each side by the reciprocal of the coefficient.

$$\frac{8}{3} \cdot \frac{3}{8} x = \frac{21}{4} \cdot \frac{8}{3}$$ Simplify the right side.

$$x = \frac{21 \cdot 8}{4 \cdot 3}$$ 21 and 3 have a common factor of 3, and 8 and 4 have a common factor of 4.

$$x = \frac{7 \cdot 2}{1 \cdot 1} = 14$$

Sentence Dimitri can get 14 smaller strips.

You Try It 7 **Marie works in a candy shop that advertises homemade candy. One specialty is a brick of fudge. Marie must cut a 6-inch piece of fudge into slices, each having a width of $\frac{3}{4}$ inches. From the 6-inch brick, how many slices of fudge can Marie cut?**

Sentence ______________________________

There are some situations in which the whole is unknown. When that occurs, the variable, n, represents the whole.

Example 9

At a bookstore, Devra needs to place 32 copies of the new best-selling paperback book side by side on a shelf. If each book is $1\frac{1}{4}$ inches thick, how many inches of the shelf will the 32 books take up?

Common Measure $1\frac{1}{4} = \frac{5}{4}$; so, each book is $\frac{5}{4}$ inches thick. The question asks, "How many inches . . ."; so, inches of books is the common measure. The **part** is the thickness, in inches, of each book; the whole (the number of inches of books) is unknown; and the number of parts is 32 books.

Answer Legend: Let x = the total number of inches of books.

$$32 \cdot \frac{5}{4} = x$$

$$\frac{32}{1} \cdot \frac{5}{4} = x$$

$$\frac{32 \cdot 5}{1 \cdot 4} = x \quad \text{Simplify 32 and 4 by a factor of 4.}$$

$$\frac{8 \cdot 5}{1 \cdot 1} = x$$

$$40 = x$$

Sentence The 32 books will take up 40 inches of the shelf.

You Try It 8 **Katie uses $\frac{5}{6}$ yards of fabric for every doll dress she makes. To make 12 doll dresses, how many yards of fabric does Katie need?**

Sentence ______________________________

Answers: You Try It and Think About It

You Try It: **1. a)** $p = 3$ **b)** $x = \frac{1}{5}$ **c)** $y = -\frac{9}{4} = -2\frac{1}{4}$ **2. a)** $y = \frac{3}{2} = 1\frac{1}{2}$ **b)** $w = -\frac{2}{3}$ **3. a)** $\frac{6}{7}$ **b)** $\frac{4}{9}$ **c)** $\frac{1}{5}$ **d)** $-\frac{1}{8}$ **4. a)** $w = -32$ **b)** $x = -36$ **c)** $y = \frac{3}{5}$ **5. a)** $v = \frac{3}{10}$ **b)** $x = -\frac{3}{4}$ **6.** Each person's pile will be $1\frac{3}{8}$ inches high. **7.** Marie can cut 8 slices of fudge. **8.** Katie needs 10 yards of fabric.

Think About It: No Think About Its in this section.

Section 4.8 Exercises

Think Again

1. In an equation, if the coefficient is a mixed number, why must we first rewrite it as an improper fraction?
   We must first rewrite the mixed number as an improper fraction so that we can identify the reciprocal of the coefficient.

2. In Example 5 in this section, in the equation $6x = \frac{3}{4}$, the variable was isolated by first multiplying each side by $\frac{1}{6}$. How would isolating the variable be different if we had divided each side by 6 instead?
   On the right side, we would divide $\frac{3}{4}$ by 6, which is the same as multiplying by $\frac{1}{6}$.

Focus Exercises

Solve.

3. $y - \frac{3}{8} = \frac{5}{8}$
   $y = 1$

4. $c - \frac{2}{3} = \frac{4}{3}$
   $c = 2$

5. $k + \frac{3}{12} = \frac{7}{12}$
   $k = \frac{1}{3}$

6. $w + \frac{1}{12} = \frac{7}{12}$
   $w = \frac{1}{2}$

7. $\frac{2}{15} = m + \frac{8}{15}$
   $m = -\frac{2}{5}$

8. $\frac{1}{10} = y + \frac{9}{10}$
   $y = -\frac{4}{5}$

9. $-\frac{10}{9} = x - \frac{4}{9}$
   $x = -\frac{2}{3}$

10. $-\frac{16}{21} = c - \frac{2}{21}$
    $c = -\frac{2}{3}$

11. $p + \frac{11}{15} = -\frac{3}{5}$
    $p = -\frac{4}{3} = -1\frac{1}{3}$

12. $h + \frac{7}{12} = -\frac{3}{4}$
    $h = -\frac{4}{3} = -1\frac{1}{3}$

13. $x + \frac{13}{8} = -\frac{5}{24}$
    $x = -\frac{11}{6} = -1\frac{5}{6}$

14. $y + \frac{3}{10} = -\frac{1}{30}$
    $y = -\frac{1}{3}$

15. $9y = 12$
    $y = \frac{4}{3} = 1\frac{1}{3}$

16. $4p = 18$
    $p = \frac{9}{2} = 4\frac{1}{2}$

17. $-21 = 14h$
    $h = -\frac{3}{2} = -1\frac{1}{2}$

18. $-8 = 6w$
    $w = -\frac{4}{3} = -1\frac{1}{3}$

19. $-15c = -24$
    $c = \frac{8}{5} = 1\frac{3}{5}$

20. $-12y = -30$
    $y = \frac{5}{2} = 2\frac{1}{2}$

21. $12m = 8$
$m = \frac{2}{3}$

22. $4w = 14$
$w = \frac{7}{2} = 3\frac{1}{2}$

23. $6n = 9$
$n = \frac{3}{2} = 1\frac{1}{2}$

24. $9k = 24$
$k = \frac{8}{3} = 2\frac{2}{3}$

25. $35p = 21$
$p = \frac{3}{5}$

26. $45y = 20$
$y = \frac{4}{9}$

27. $12 = 32x$
$x = \frac{3}{8}$

28. $16 = 40n$
$n = \frac{2}{5}$

29. $15 = 20w$
$w = \frac{3}{4}$

30. $12 = 27y$
$y = \frac{4}{9}$

31. $18 = 24v$
$v = \frac{3}{4}$

32. $18 = 30x$
$x = \frac{3}{5}$

33. $\frac{3}{5}y = -21$
$y = -35$

34. $\frac{2}{3}w = -16$
$w = -24$

35. $-\frac{5}{6}x = -10$
$x = 12$

36. $-\frac{1}{4}m = -6$
$m = 24$

37. $\frac{9}{10}k = \frac{18}{5}$
$k = 4$

38. $\frac{3}{8}a = -\frac{9}{4}$
$a = -6$

39. $\frac{11}{5}x = 22$
$x = 10$

40. $\frac{4}{9}y = 24$
$y = 54$

41. $\frac{w}{4} = -6$
$w = -24$

42. $\frac{m}{3} = -8$
$m = -24$

43. $\frac{1}{6} = \frac{v}{2}$
$v = \frac{1}{3}$

44. $\frac{6}{10} = \frac{x}{5}$
$x = 3$

45. $\frac{8}{21} = \frac{4n}{7}$
$n = \frac{2}{3}$

46. $\frac{25}{42} = \frac{10w}{7}$
$w = \frac{5}{12}$

47. $\frac{4}{7} = 8x$
$x = \frac{1}{14}$

48. $\frac{14}{15} = 7y$
$y = \frac{2}{15}$

49. $-6y = \frac{3}{7}$
$y = -\frac{1}{14}$

50. $-9y = \frac{15}{8}$
$y = -\frac{5}{24}$

Solve each equation. Simplify.

51. $1\frac{2}{3}x = 10$
$x = 6$

52. $2\frac{2}{3}y = 8$
$y = 3$

53. $3\frac{1}{5}v = 32$
$v = 10$

54. $4\frac{1}{2}p = 18$
$p = 4$

55. $2\frac{2}{5}m = 6$
$m = \frac{5}{2} = 2\frac{1}{2}$

56. $3\frac{1}{8}w = 5$
$w = \frac{8}{5} = 1\frac{3}{5}$

57. $2 = 3\frac{1}{5}x$
$x = \frac{5}{8}$

58. $16 = 1\frac{1}{15}k$
$k = 15$

59. $1\frac{1}{9} = 1\frac{2}{3}y$
$y = \frac{2}{3}$

60. $1\frac{1}{15} = 1\frac{1}{3}w$
$w = \frac{4}{5}$

61. $10\frac{1}{2} = 1\frac{1}{2}n$
$n = 7$

62. $1\frac{1}{4} = 1\frac{7}{8}x$
$x = \frac{2}{3}$

Solve each application and write a sentence answering the question. (*Hint:* Identify the common measure, the whole, the part, and the number of parts; set up the legend, and set up and solve the equation.)

63. Marnay gives guitar lessons $7\frac{1}{2}$ hours each day. Each lesson lasts $\frac{3}{4}$ hour. How many lessons does Marnay give each day?
Marnay gives 10 lessons each day.

64. At an amusement park, employees rotate from location to location so that they have a variety of duties to do each day. One location, a hot dog cart, is open for 12 hours. If 8 different employees evenly share the time they work during 12 hours the cart is open, how many hours does each employee work at the hot dog cart?
Each employee works $1\frac{1}{2}$ hours at the hot dog cart.

65. A mother has a sick child and needs to divide liquid cough medicine into 12 equal doses. If the mother has $1\frac{1}{2}$ cups of cough syrup, how much cough syrup will each dose be?
Each dose will be $\frac{1}{8}$ cup of syrup.

66. Sandra works for the city's Parks and Recreation Department. Each spring her job is to mark six soccer playing fields on one large field. The large field is $\frac{3}{8}$ of a mile long, and each of the smaller soccer fields must be of equal length. How long (in miles) will each of the six soccer fields be?
Each of the six soccer fields will be $\frac{1}{16}$ of a mile long.

67. At a gift wrapping booth, each package is topped with a bow made of $2\frac{2}{3}$ feet of ribbon. If a spool holds 24 feet of ribbon, how many bows can be made from one spool?
9 bows can be made from one spool.

68. Joachin, a cross-country skier, has fully recovered from a broken leg. His coach has put him on a strict training schedule in which he gradually increases the distance he skis each week. In the third week of this schedule, Joachin must ski a total of 45 miles while skiing the same $7\frac{1}{2}$-mile course each day. To complete the 45 miles, how many days does Joachin need to ski this course?
Joachin needs to ski this course 6 days.

69. Yuan must complete 45 hours of counseling high school students before she can become a certified educational adviser. Her assignment at Lincoln High School allows her to counsel students for $2\frac{1}{4}$ hours in the afternoon. How many of these afternoon sessions will Yuan need to complete to receive her certificate?
Yuan will need 20 afternoon sessions.

70. Nate's dog Banjo, a golden retriever, eats $3\frac{1}{2}$ cups of dog food each day. How many cups of dog food does Banjo eat each month (30 days)?
Banjo eats 105 cups of dog food each month.

71. Tim is a janitor for an elementary school. It takes him $\frac{1}{4}$ hour to clean each classroom at the end of the school day. How many hours does Tim take to clean all 25 classrooms?
Tim takes $6\frac{1}{4}$ hours to clean all 25 classrooms.

72. Ramon has written a book that contains colorful graphs and charts. His printer can print the entire book in $1\frac{1}{3}$ hours. He needs to print 6 copies for his team of editors. How long will it take to print 6 copies of the book?
It will take 8 hours to print 6 copies of the book.

Think Outside the Box

Solve.

73. $\frac{5}{6}x + 5 = -5$
$x = -12$

74. $\frac{3}{4}y - \frac{1}{2} = -1$
$y = -\frac{2}{3}$

75. $\frac{1}{6}w + 10 = -\frac{2}{3}w$
$w = -12$

76. $\frac{5}{12}p + \frac{10}{9} = -\frac{1}{4}p$
$p = -\frac{5}{3}$ or $p = -1\frac{2}{3}$

Chapter 4 Review

Section 4.1 Common Factors

Concept	Example
The **common factors** of two numbers are all of the numbers that are factors of both numbers. The **greatest common factor (GCF)** of two numbers is the highest, or greatest, of their common factors.	Common factors of 20 and 30: 1, 2, 5, and 10 The GCF of 20 and 30: 10
If the only common factor is 1, the numbers are **relatively prime** and the GCF is 1.	Factors of 15: 1, 3, 5, and 15 Factors of 28: 1, 2, 4, 7, 14, and 28 15 and 28 are relatively prime.
Three methods of finding the GCF of two numbers **1.** List all of the factors of the two numbers, identify which factors are common to both, and identify the greatest of the common factors.	Find the GCF of 54 and 72. **1.** Factors of 54: 1, 2, 3, 6, 9, 18, 27, 54 Factors of 72: 1, 2, 3, 4, 6, 8, 9, 12, 18, 24, 36, 72 GCF = 18
2. Find the prime factorization of each number and match up all common prime factors.	**2.** $72 = 2 \cdot 2 \cdot 2 \cdot 3 \cdot 3$ $54 = 2 \cdot 3 \cdot 3 \cdot 3$ $\text{GCF} = 2 \cdot 3 \cdot 3 = 18$
3. Use the division method and divide by common factors (prime or composite) until the quotients are relatively prime.	**3.** 2⌊ 54 72 ← Divide by 2. 3⌊ 27 36 ← Divide by 3. 3⌊ 9 12 ← Divide by 3 again. 3 4 ← Relatively prime $\text{GCF} = 2 \cdot 3 \cdot 3 = 18$

Section 4.2 Introduction to Fractions

Concept	Example
A **fraction** is the comparison of a part of something to its whole, where all of the parts are the same size.	$\frac{\textbf{part}}{\textbf{whole}}$
A fraction consists of three components: a **fraction bar,** a number above the fraction bar, and a number below the fraction bar. The number above the fraction bar is called the **numerator,** and the number below the fraction bar is called the **denominator.**	$\frac{\text{numerator}}{\text{denominator}}$ ← the fraction bar

Section 4.2 Introduction to Fractions

Concept	Example
The **numerator** represents the number of equal-sized parts being considered, and the **denominator** represents the number of equal-sized parts contained in the whole.	There are 8 equally sized rectangles, and 3 of them are shaded. $\frac{3}{8}$ of the rectangles are shaded. 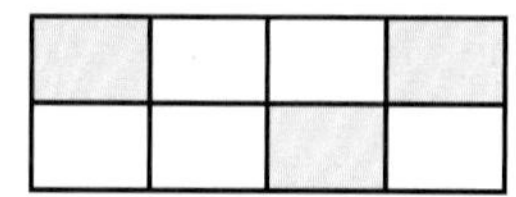
On a number line, we can represent both positive and negative fractions.	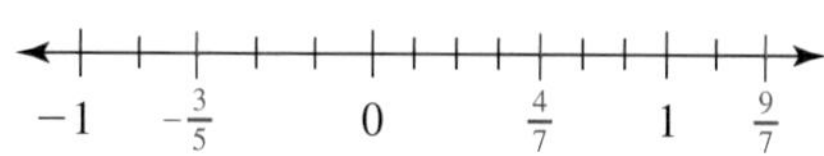
Fractional notation is another way to indicate the operation of division. Sometimes the fraction bar is referred to as the division bar.	$\frac{12}{3}$ can also be written as $12 \div 3$.
If b is any number (*except 0*), then $\frac{0}{b} = 0$.	$\frac{0}{9} = 0$
The denominator of a fraction can never be zero.	$\frac{a}{0}$ is *undefined.*
A **proper fraction** has a numerator that is *less* than the denominator.	$\frac{1}{18}, \frac{3}{5}, \frac{7}{19},$ and $\frac{34}{35}$
An **improper fraction** has a numerator that is *equal to* or *greater than* the denominator.	$\frac{4}{4}, \frac{7}{5}, \frac{26}{13},$ and $\frac{42}{42}$
A **mixed number** is the sum of a whole number and a fraction.	$1\frac{4}{5} = 1 + \frac{4}{5}$
Every mixed number can be written as an improper fraction. $\text{Whole} + \frac{\text{Numerator}}{\text{Denominator}} = \frac{\text{Whole} \times \text{Denominator} + \text{Numerator}}{\text{Denominator}}$	$2\frac{3}{8} = \frac{2 \times 8 + 3}{8} = \frac{16 + 3}{8} = \frac{19}{8}$
Every improper fraction can be written as a mixed number by using long or short division.	$\frac{36}{5} = 36 \div 5 = 7\frac{1}{5}$ $\begin{array}{r} 7 \text{ r } 1 \\ 5\overline{)36} \\ -35 \\ \hline 1 \end{array}$

continued

Section 4.2 Introduction to Fractions

Concept	Example
A natural number divided by itself is 1: $\frac{a}{a} = 1$.	$\frac{5}{5} = 1$
A number divided by 1 is itself. $\frac{a}{1} = a$.	$\frac{5}{1} = 5$
The Product Rule for Fractions $\frac{a}{b} \cdot \frac{c}{d} = \frac{a \cdot c}{b \cdot d} = \frac{\text{The product of the numerators}}{\text{The product of the denominators}}$, b and $d \neq 0$	$\frac{5}{3} \cdot \frac{2}{7} = \frac{5 \cdot 2}{3 \cdot 7} = \frac{10}{21}$
The **reciprocal** of a fraction, $\frac{a}{b}$, is the fraction $\frac{b}{a}$, as long as $a \neq 0$ and $b \neq 0$.	The reciprocal of $\frac{5}{3}$ is $\frac{3}{5}$.
The reciprocal of a fraction is also called the **multiplicative inverse** of the fraction. When finding the reciprocal of a fraction, we say that the fraction has been *inverted*. As an inverse, the reciprocal has this special property: The product of a fraction and its reciprocal is 1. $\frac{a}{b} \cdot \frac{b}{a} = 1$, provided that $a \neq 0$ and $b \neq 0$.	The product of reciprocals $\frac{5}{3}$ and $\frac{3}{5}$: $\frac{5}{3} \cdot \frac{3}{5} = \frac{15}{15} = 1$

Section 4.3 Equivalent Fractions

Concept	Example
Two fractions are **equivalent** when they have the same value.	$\frac{1}{2}, \frac{2}{4}, \frac{3}{6}$, and $\frac{5}{10}$—each equals $\frac{1}{2}$.
A fraction is in **lowest terms** when the numerator and the denominator have no common factors other than 1.	$\frac{2}{3}$ is in lowest terms because 2 and 3 have no common factor (other than 1).
To **simplify** a fraction means to divide out any factors greater than 1 that are common to the numerator and the denominator. A fraction is simplified when it is in lowest terms.	8 and 12 have a common factor of 4. So, $\frac{8}{12}$ can be simplified by a factor of 4: $\frac{8 \div 4}{12 \div 4} = \frac{2}{3}$.
A fraction can be simplified by any common factor of the numerator and the denominator. To simplify a fraction to its lowest terms requires that we divide out the greatest common factor (GCF).	The GCF of 24 and 40 is 8. Simplify $\frac{24}{40}$ by dividing the numerator and the denominator by 8: $\frac{24 \div 8}{40 \div 8} = \frac{3}{5}$.

Section 4.3 Equivalent Fractions

Concept	Example
In a fraction, a negative sign and a positive sign can be placed in one of three places: in front of the fraction, in the numerator, or in the denominator. Each of those places has a sign although it's not always visible.	$-\frac{+3}{+8} = +\frac{-3}{+8} = +\frac{+3}{-8}$: all represent $-\frac{3}{8}$. $-\frac{3}{8} = \frac{-3}{8} = \frac{3}{-8}$: all represent $-\frac{3}{8}$.
The rules of negatives in division apply to fractions **1.** When there is an odd number of negative signs in the fraction, the fraction is negative. **2.** When there is an even number of negative signs in the fraction, the fraction is positive.	**1.** $-\frac{-5}{-6} = -\frac{5}{6}$ an odd number of negative signs **2.** $-\frac{-5}{6} = +\frac{5}{6}$ an even number of negative signs
If a fraction has the same variable factor in the numerator and the denominator, that variable factor can be divided out.	$\frac{4m}{9m} = \frac{4 \cdot m \div m}{9 \cdot m \div m} = \frac{4}{9}$
Multiplying a fraction by a fractional form of 1, such as $\frac{3}{3}$, is called **building up** a fraction. This means that the numerator and the denominator become larger—are built up—but the result is an equivalent fraction.	To build up $\frac{3}{5}$ to an equivalent fraction with a denominator of 20, multiply $\frac{3}{5}$ by a fractional form of 1, $\frac{4}{4}$: $\frac{3}{5} \cdot \frac{4}{4} = \frac{12}{20}$. $\frac{12}{20}$ is equivalent to $\frac{3}{5}$.

Section 4.4 Multiplying and Dividing Fractions

Concept	Example
The extended product rule of fractions $$\frac{a}{b} \cdot \frac{c}{d} \cdot \frac{e}{f} = \frac{a \cdot c \cdot e}{b \cdot d \cdot f} = \frac{\text{The product of the numerators}}{\text{The product of the denominators}}, b, d, \text{and } f \neq 0$$	$\frac{2}{3} \cdot \frac{4}{5} \cdot \frac{1}{7} = \frac{2 \cdot 4 \cdot 1}{3 \cdot 5 \cdot 7} = \frac{8}{105}$
Write a whole number as a fraction, $\frac{\text{whole number}}{1}$, before multiplying.	$4 \cdot \frac{2}{7} = \frac{4}{1} \cdot \frac{2}{7} = \frac{4 \cdot 2}{1 \cdot 7} = \frac{8}{7}$
Quotient Rule of Fractions $$\frac{a}{b} \div \frac{d}{c} = \frac{a}{b} \cdot \frac{c}{d} = \frac{a \cdot c}{b \cdot d}, b, c, \text{and } d \neq 0$$ **To divide by a fraction, $\frac{a}{b} \div \frac{d}{c}$, invert and multiply** **1.** Keep the first number (or fraction) the same. **2.** Write the multiplication sign. **3.** *Invert* (write the reciprocal of) the *second* number (or fraction).	$\frac{4}{9} \div \frac{3}{5} = \frac{4}{9} \cdot \frac{5}{3} = \frac{20}{27}$

continued

Section 4.4 Multiplying and Dividing Fractions

Concept	Example
To multiply and divide mixed numbers **1.** Rewrite each mixed number as an improper fraction. **2.** Multiply (or divide) these fractions as you would any others. Simplify the result. **3.** Write the answer, if appropriate, as a mixed number.	$4\frac{1}{6}\cdot 2\frac{7}{10} = \frac{25}{6}\cdot\frac{27}{10} = \frac{\overset{5}{\cancel{25}}\cdot\overset{9}{\cancel{27}}}{\underset{2}{\cancel{6}}\cdot\underset{2}{\cancel{10}}} = \frac{45}{4} = 11\frac{1}{4}$ $2\frac{7}{9} \div 3\frac{1}{3} = \frac{25}{9} \div \frac{10}{3} = \frac{25}{9}\cdot\frac{3}{10} = \frac{\overset{5}{\cancel{25}}\cdot\overset{1}{\cancel{3}}}{\underset{3}{\cancel{9}}\cdot\underset{2}{\cancel{10}}} = \frac{5}{6}$
A *fraction of* means "multiply."	*Two-thirds of* $\frac{3}{4}$ is $\frac{2}{3}\cdot\frac{3}{4} = \frac{2\cdot 3}{3\cdot 4} = \frac{6}{12} = \frac{1}{2}$.
The area of a triangle is $\frac{1}{2}\times$ base $\times$ height: $A = \frac{1}{2}\cdot b\cdot h$	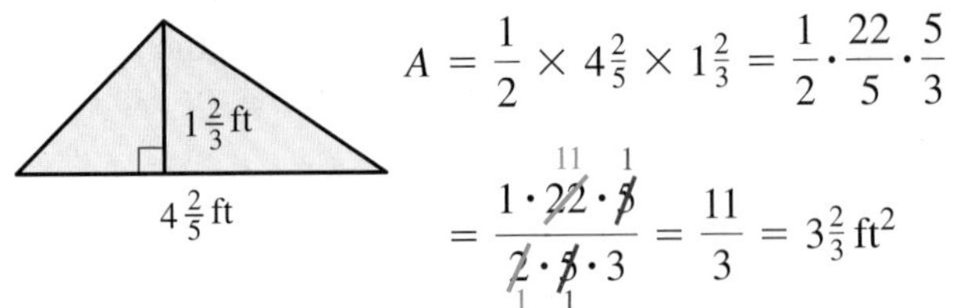 $A = \frac{1}{2}\times 4\frac{2}{5}\times 1\frac{2}{3} = \frac{1}{2}\cdot\frac{22}{5}\cdot\frac{5}{3}$ $= \frac{1\cdot\overset{11}{\cancel{22}}\cdot\overset{1}{\cancel{5}}}{\underset{1}{\cancel{2}}\cdot\underset{1}{\cancel{5}}\cdot 3} = \frac{11}{3} = 3\frac{2}{3}\text{ ft}^2$

Section 4.5 Adding and Subtracting Like Fractions

Concept	Example
To add any two *like* fractions, add the numerators: $\frac{a}{b} + \frac{c}{b} = \frac{a+c}{b},\ b \neq 0$	$\frac{2}{9} + \frac{5}{9} = \frac{2+5}{9} = \frac{7}{9}$
To subtract any two *like* fractions, subtract the numerators: $\frac{a}{b} - \frac{c}{b} = \frac{a-c}{b},\ b \neq 0$	$\frac{8}{15} - \frac{5}{15} = \frac{8-5}{15} = \frac{3}{15}$ $\frac{3}{15}$ can be simplified to $\frac{1}{5}$.
Subtraction of fractions can be written as "adding the opposite." When the first fraction is negative, the negative sign can be placed in the numerator before combining the fraction with another fraction.	$\frac{1}{15} - \frac{3}{15} = \frac{1}{15} + \frac{-3}{15} = \frac{1+(-3)}{15} = \frac{-2}{15} = -\frac{2}{15}$ $-\frac{8}{21} - \frac{-10}{21} = \frac{-8}{21} + \frac{+10}{21} = \frac{-8+10}{21} = \frac{2}{21}$

Section 4.5 Adding and Subtracting Like Fractions

Concept	Example
When two or more terms are *like terms* with fractional coefficients, combine the terms by combining their coefficients. If the coefficient of the resulting terms is an improper fraction, it is appropriate to leave the fraction as improper and not change it to a mixed number.	$\frac{7}{9}x + \frac{5}{9}x = \frac{7+5}{9}x = \frac{12}{9}x = \frac{4}{3}x$
When adding mixed numbers, add the whole numbers separately from the fractions. The sum will be another mixed number.	$4\frac{3}{8} + 2\frac{1}{8} = 6\frac{4}{8} = 6\frac{1}{2}$
If the fractions add to an improper fraction, adjust the mixed number by rewriting the improper fraction and adding the whole numbers.	$4\frac{3}{8} + 2\frac{7}{8} = 6\frac{10}{8} = 6 + 1\frac{2}{8} = 7\frac{2}{8} = 7\frac{1}{4}$
When subtracting mixed numbers, subtract the fractions first, then the whole numbers. The fractions must have a common denominator, and the first fraction cannot be less than the second fraction.	$7\frac{3}{8} - 3\frac{1}{8} = 4\frac{2}{8} = 4\frac{1}{4}$
If two fractions cannot subtract directly, we must regroup the first number, making its fraction an improper fraction.	$8\frac{1}{4} = 7 + \frac{4+1}{4} = 7\frac{5}{4}$ (Adjust the top fraction.) $8\frac{1}{4} - 6\frac{3}{4} \rightarrow 7\frac{5}{4} - 6\frac{3}{4} = 1\frac{2}{4} = 1\frac{1}{2}$

Section 4.6 Common Denominators

Concept	Example
The **least common denominator (LCD)** is the *least common multiple* of two denominators. We can find the LCD of two denominators by writing out some of the multiples of each denominator until the first match is found.	Find the LCD for $\frac{1}{4}$ and $\frac{5}{6}$: Multiples of 4: 4, 8, **12**, 16, 20, 24, 28, . . . Multiples of 6: 6, **12**, 18, 24, 30, . . . The LCD is **12**.
To find the LCD using prime factorization **1.** Find the prime factorization of each denominator. **2.** Write "LCD = ______________ = ___." **3.** On the line, write all of the prime factors (with multiplication signs) of the first number. **4.** Include any *prime* factors of the second number that have not yet been written down. **5.** Multiply the prime factors, and write that product after the second equal sign.	Find the LCD of $\frac{5}{12}$ and $\frac{7}{40}$. $12 = 2 \cdot 2 \cdot 3$ $40 = 2 \cdot 2 \cdot 2 \cdot 5$ LCD = $2 \cdot 2 \cdot 3 \cdot 2 \cdot 5$ = **120**
To find the LCD using the division method **1.** Divide out any common factor of the two denominators, getting a new pair of numbers called *quotients*. The common factors may be prime or composite numbers. **2.** Continue dividing until the new quotients are relatively prime. **3.** Multiply each factor that was divided out and each number in the relatively prime quotients.	Find the LCD of $\frac{5}{12}$ and $\frac{7}{40}$: 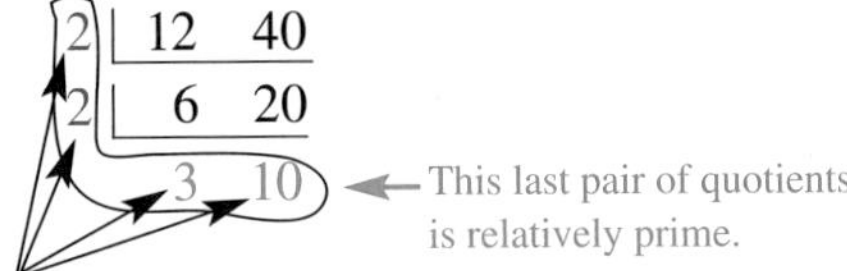 Multiply all of these: LCD = $2 \cdot 2 \cdot 3 \cdot 10 = 120$.
Once the LCD is known, each fraction can be built up to have the LCD as its denominator.	The LCD for $\frac{1}{4}$ and $\frac{5}{6}$ is 12. $\frac{1}{4} \cdot \frac{3}{3} = \frac{3}{12}$ and $\frac{5}{6} \cdot \frac{2}{2} = \frac{10}{12}$

Section 4.7 Adding and Subtracting Unlike Fractions

Concept	Example
We can build up two unlike fractions to have the same denominator, called a **common denominator.** Once two unlike fractions are built up to have the same denominator, they can be combined either by addition or subtraction. When two mixed numbers contain unlike fractions, the fractions must be built up to a common denominator before the mixed numbers can be added or subtracted.	Add: $\frac{1}{4} + \frac{1}{6}$. The LCD 12. $= \frac{1}{4}\cdot\frac{3}{3} + \frac{1}{6}\cdot\frac{2}{2}$ $= \frac{3}{12} + \frac{2}{12} = \frac{3+2}{12} = \frac{5}{12}$ Subtract: $\frac{2}{3} - \frac{1}{6}$. The LCD 6. $= \frac{2}{3}\cdot\frac{2}{2} - \frac{1}{6}$ $= \frac{4}{6} - \frac{1}{6} = \frac{4-1}{6} = \frac{3}{6} = \frac{1}{2}$ Combine like terms: $-\frac{3}{8}y - \frac{1}{6}y$. The LCD 24. $= \frac{-3}{8}\cdot\frac{3}{3}y + \frac{-1}{6}\cdot\frac{4}{4}y$ $= \frac{-9}{24}y + \frac{-4}{24}y = \frac{-9+(-4)}{24}y = \frac{-13}{24}y = -\frac{13}{24}y$

Section 4.8 Equations and Applications Involving Fractions

Concept	Example
If a variable, x, is in the numerator of a fraction, it can be separated from the fraction so that we can identify the coefficient of x.	$\frac{3x}{8} = \frac{3}{8}x$ and $\frac{x}{4} = \frac{1}{4}x$
The Multiplication Property of Equations We may multiply each side of an equation by any number, c, $c \neq 0$. If $a = b$, then $c \cdot a = c \cdot b$. If the coefficient of x is a fraction or a mixed number, we can isolate the variable by multiplying each side by the reciprocal of the coefficient.	$\frac{5x}{2} = 20$ $\frac{5}{2}x = 20$ $\frac{2}{5}\cdot\frac{5}{2}x = \frac{2}{5}\cdot 20$ $x = 8$

continued

Section 4.8 Equations and Applications Involving Fractions

Concept	Example
A basic formula involving multiplication and division of fractions is (Number of *same parts*) × Part = Whole The whole and the part are always the same unit of measure.	For a pizza party, Patti has figured that each guest will eat $\frac{3}{8}$ of a pizza. If Patti plans to serve 6 pizzas, how many guests can she invite? **Whole** = Total **Number** of pizzas **Part** = $\frac{3}{8}$ of a pizza Let x = the number of guests Patti can invite. $\frac{3}{8} \cdot x = 6$ $\frac{8}{3} \cdot \frac{3}{8} \cdot x = \frac{8}{3} \cdot 6$ $x = 16$ Patti can invite 16 guests.

Chapter  Review Exercises

Section 4.1

Find the greatest common factor (GCF) of each pair of numbers. If the GCF is 1, write *relatively prime*. You may use any method.

1. 40 and 48
8

2. 14 and 56
14

3. 36 and 90
18

4. 35 and 48
Relatively prime

Section 4.2

Identify each fraction as a proper fraction, an improper fraction, a mixed number, or a complex fraction.

5. $2\frac{5}{6}$
Mixed number

6. $\frac{19}{8}$
Improper fraction

7. $\frac{4}{6}$
Proper fraction

8. $\frac{7}{7}$
Improper fraction

First, identify the reciprocal if possible. Second, find the product of each fraction and its reciprocal.

13. $\frac{9}{4}$
Reciprocal is $\frac{4}{9}$; Product is 1

14. $\frac{1}{8}$
Reciprocal is 8; Product is 1

15. -4
Reciprocal is $-\frac{1}{4}$; Product is 1

16. $\frac{0}{5}$
No reciprocal

Write each mixed number as an improper fraction and each improper fraction as a mixed number.

9. $3\frac{4}{9}$
$\frac{31}{9}$

10. $-5\frac{1}{6}$
$-\frac{31}{6}$

11. $\frac{14}{3}$
$4\frac{2}{3}$

12. $-\frac{70}{9}$
$-7\frac{7}{9}$

Section 4.3

Simplify completely. If the fraction is improper, write it as a mixed number.

17. $\frac{18}{24}$
$\frac{3}{4}$

18. $\frac{8}{-36}$
$-\frac{2}{9}$

19. $-\frac{-14}{35}$
$\frac{2}{5}$

20. $\frac{75xy}{60y}$
$\frac{5x}{4}$

Multiply by a fraction equivalent to 1 to build up each fraction to one with a denominator as shown.

	Start with		Multiply by 1		Result with
21.	$\frac{3}{5}$	$\times$	$\frac{4}{4}$	$=$	$\frac{12}{20}$
23.	$-\frac{5}{12}$	$\times$	$\frac{4}{4}$	$=$	$\frac{-20}{48}$

	Start by 1		Multiply		Result
22.	$-\frac{1}{10}$	$\times$	$\frac{6}{6}$	$=$	$\frac{-6}{60}$
24.	$\frac{4}{9}$	$\times$	$\frac{7}{7}$	$=$	$\frac{28}{63}$

Section 4.4

Multiply. Simplify the answer completely. Write all improper fraction answers as mixed numbers.

25. $\frac{7}{5} \cdot \frac{10}{21}$
$\frac{2}{3}$

26. $\frac{1}{12m} \cdot \frac{15m}{7} \cdot \frac{6}{5}$
$\frac{3}{14}$

27. $-\frac{5}{18y} \cdot 3y$
$-\frac{5}{6}$

28. $-8 \cdot \frac{9}{24} \cdot \left(-\frac{5}{3}\right)$
5

Divide. Simplify if possible.

29. $\frac{4}{18} \div \frac{2}{9}$
1

30. $\frac{15k}{8} \div \frac{5k}{16}$
6

31. $\frac{6}{10} \div \left(-\frac{12}{70}\right)$
$-3\frac{1}{2}$

32. $-\frac{25}{18v^2} \div \left(-\frac{5}{12v}\right)$
$\frac{10}{3v}$

Evaluate and simplify completely. Write each answer as a mixed number whenever possible.

33. $2\frac{1}{2} \times 3\frac{3}{5}$
9

34. $3\frac{3}{5} \times 2\frac{1}{9}$
$7\frac{3}{5}$

35. $1\frac{7}{8} \div 3\frac{1}{3}$
$\frac{9}{16}$

36. $8\frac{3}{4} \div 2\frac{5}{8}$
$3\frac{1}{3}$

Evaluate each by rewriting it as a product of fractions. Write each answer as a mixed number whenever possible.

37. Two-thirds of $\frac{15}{14}$
$\frac{2}{3} \cdot \frac{15}{14} = \frac{5}{7}$

38. Three-eighths of $\frac{4}{15}$
$\frac{3}{8} \cdot \frac{4}{15} = \frac{1}{10}$

39. Three-fourths of $2\frac{2}{9}$
$\frac{3}{4} \cdot 2\frac{2}{9} = \frac{3}{4} \cdot \frac{20}{9} = \frac{5}{3} = 1\frac{2}{3}$

40. Five-sixths of $4\frac{1}{5}$
$\frac{5}{6} \cdot 4\frac{1}{5} = \frac{5}{6} \cdot \frac{21}{5} = \frac{7}{2} = 3\frac{1}{2}$

Solve each application and write a sentence answering the question.

41. Two-thirds of the students in Ms. Grecu's algebra class are women. If she has 42 students in her class, how many women are in Ms. Grecu's algebra class?
There are 28 women in Ms. Grecu's algebra class.

42. Jasper's Meats prepares hamburger patties that weigh $\frac{2}{3}$ pounds each. How many pounds of meat are there in 12 patties?
There are 8 pounds of meat in 12 patties.

Find the area of each geometric figure.

43.

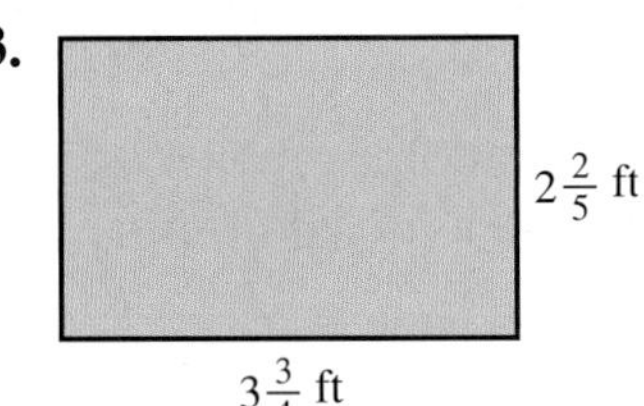

9 square feet

44.

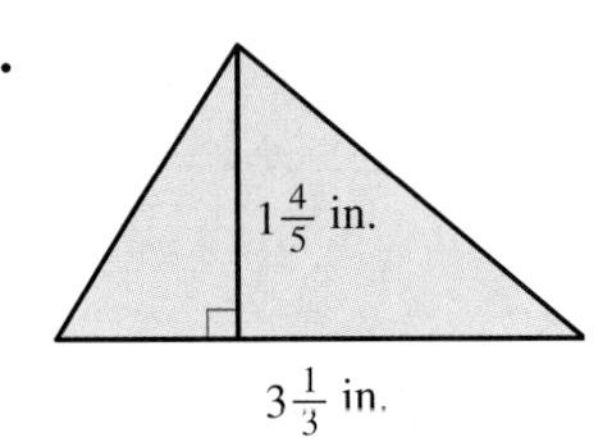

3 square inches

Section 4.5

Add. Simplify the result whenever possible.

45. $\frac{10}{28} + \frac{11}{28}$
$\frac{3}{4}$

46. $\frac{19}{45} + \frac{17}{45}$
$\frac{4}{5}$

47. $\frac{7}{10}k + \frac{9}{10}k$
$\frac{8}{5}k$

48. $\frac{13}{15}ab + \frac{17}{15}ab$
$2ab$

Subtract. Simplify the result whenever possible.

49. $\frac{7}{9} - \frac{1}{9}$

$\frac{2}{3}$

50. $\frac{17}{24} - \frac{5}{24}$

$\frac{1}{2}$

51. $\frac{31}{25}q - \frac{16}{25}q$

$\frac{3}{5}q$

52. $\frac{17}{30}d^2 - \frac{12}{30}d^2$

$\frac{1}{6}d^2$

Evaluate. Simplify completely.

53. $\frac{2}{27} - \frac{11}{27}$

$-\frac{1}{3}$

54. $-\frac{2}{3}n - \frac{1}{3}n$

$-n$

55. $6\frac{5}{8} + 4\frac{7}{8}$

$11\frac{1}{2}$

56. $6\frac{1}{12} - 5\frac{5}{12}$

$\frac{2}{3}$

Solve each application using addition or subtraction and write a sentence answering the question.

57. In Garberville, California, the one-day record rainfall showed $2\frac{3}{10}$ inches before noon and $1\frac{9}{10}$ inches after noon. How many total inches of rain fell in Garberville that day?

A total of $4\frac{1}{5}$ inches of rain fell in Garberville that day.

58. Marika weighed $6\frac{7}{8}$ pounds when she was born. By her first birthday, she weighed $23\frac{1}{8}$ pounds. How much weight did Marika gain during her first year?

Marika gained $16\frac{1}{4}$ pounds during her first year.

Section 4.6

Find the LCD of each pair of fractions. Then build up each fraction to have that common denominator.

59. $\frac{5}{6}$ and $\frac{4}{9}$

LCD is 18 $\frac{15}{18}$ and $\frac{8}{18}$

60. $\frac{5}{14}$ and $\frac{10}{21}$

LCD is 42 $\frac{15}{42}$ and $\frac{20}{42}$

61. $\frac{11}{18}$ and $\frac{29}{54}$

LCD is 54 $\frac{33}{54}$ and $\frac{29}{54}$

62. $\frac{3}{20}$ and $\frac{7}{15}$

LCD is 60 $\frac{9}{60}$ and $\frac{28}{60}$

Section 4.7

Add. Simplify the results if possible. Write any answers that contain improper fractions as mixed numbers.

63. $\frac{1}{2} + \frac{1}{6}$

$\frac{2}{3}$

64. $\frac{9}{20} + \frac{7}{15}$

$\frac{11}{12}$

65. $\frac{8}{15}x + \frac{4}{9}x$

$\frac{44}{45}x$

66. $\frac{7}{9}h^2 + \frac{5}{12}h^2$

$\frac{43}{36}h^2$

Subtract. Simplify the results if possible.

67. $\frac{11}{18} - \frac{4}{9}$

$\frac{1}{6}$

68. $\frac{13}{15} - \frac{7}{10}$

$\frac{1}{6}$

69. $\frac{17}{20}ab - \frac{1}{4}ab$

$\frac{3}{5}ab$

70. $\frac{13}{20}q - \frac{5}{12}q$

$\frac{7}{30}q$

Evaluate. Find the least common denominator and build up each fraction as necessary. Simplify if possible.

71. $-\frac{9}{10} + \left(-\frac{3}{4}\right)$

$-1\frac{13}{20}$

72. $-\frac{7}{10}m + \frac{8}{15}m$

$-\frac{1}{6}m$

73. $-\frac{1}{5}d^2 - \frac{3}{4}d^2$

$-\frac{19}{20}d^2$

74. $-\frac{4}{15} - \left(-\frac{1}{6}\right)$

$-\frac{1}{10}$

Evaluate. Simplify completely and adjust the mixed number if the answer contains an improper fraction.

75. $1\frac{4}{5} + 1\frac{9}{20}$

$3\frac{1}{4}$

76. $10\frac{5}{6} - 2\frac{8}{15}$

$8\frac{3}{10}$

Solve each application and write a sentence answering the question.

77. Last winter, Mary started with $\frac{5}{6}$ of a cord of wood. At the end of the winter, she had $\frac{2}{15}$ of a cord of wood. How much wood did Mary burn during the winter?

Beginning of winter

End of winter

Mary burned $\frac{7}{10}$ of a cord of wood during the winter.

78. A bolt of fabric has $8\frac{1}{4}$ yards on it. A customer buys $5\frac{5}{12}$ yards from that bolt. How many yards of fabric are left on the bolt?

$2\frac{5}{6}$ yards of fabric are left on the bolt.

Section 4.8

Solve.

79. $\frac{3}{6} = y - \frac{1}{6}$

$y = \frac{2}{3}$

80. $m + \frac{11}{20} = \frac{3}{20}$

$m = -\frac{2}{5}$

81. $x - \frac{13}{18} = -\frac{7}{18}$

$x = \frac{1}{3}$

82. $p + \frac{7}{12} = \frac{2}{3}$

$p = \frac{1}{12}$

83. $-10w = 15$

$w = -1\frac{1}{2}$

84. $-21 = -30m$

$m = \frac{7}{10}$

85. $\frac{8}{15} = \frac{4x}{9}$

$x = 1\frac{1}{5}$

86. $2\frac{7}{10}w = 3\frac{3}{5}$

$w = 1\frac{1}{3}$

Solve each application and write a sentence answering the question.

87. As part of his punishment for painting graffiti on the school wall, Keith has to perform 80 hours of community service. Each Saturday, Keith spends $3\frac{1}{3}$ hours at a local soup kitchen. How many Saturdays must Keith work to fulfill his 80-hour obligation?

Keith must work 24 Saturdays.

88. Armand was asked to create a presidential wall of portraits of U.S. presidents. Each row on the wall can display 7 portraits side by side. If each portrait is $1\frac{1}{3}$ feet wide, how long is each row in the display?

Each row in the dispay is $9\frac{1}{3}$ feet long.

Chapter 4 Test

Find the greatest common factor of each pair of numbers. You may use any method.

1. 16 and 24
8

2. 27 and 45
9

3. 70 and 112
14

Write each fraction as directed.

4. Write $7\frac{2}{5}$ as an improper fraction.
$\frac{37}{5}$

5. Write $\frac{52}{7}$ as a mixed number.
$7\frac{3}{7}$

Evaluate each expression. Simplify completely.

6. $\frac{42}{60}$
$\frac{7}{10}$

7. $-\frac{36}{-45}$
$\frac{4}{5}$

8. $\frac{24x^2}{16xy}$
$\frac{3x}{2y}$

Multiply. Simplify each answer completely.

9. $-\frac{5}{6} \cdot \left(-\frac{9}{10}\right)$
$\frac{3}{4}$

10. $\frac{8c}{9} \cdot \frac{15d}{16c} \cdot \frac{3}{5}$
$\frac{d}{2}$

Evaluate each by rewriting it as a product of fractions.

11. Five-eighths of 12
$7\frac{1}{2}$

12. Two-ninths of $6\frac{3}{4}$
$1\frac{1}{2}$

Find the area of the rectangle.

13.
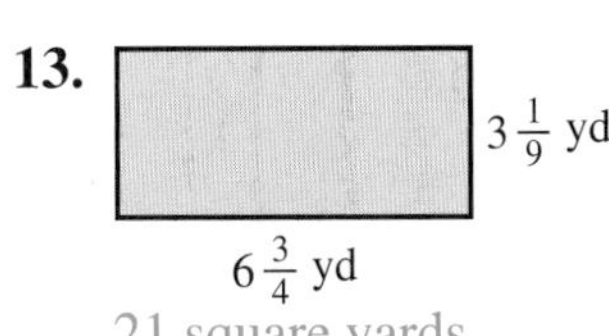
$3\frac{1}{9}$ yd

$6\frac{3}{4}$ yd

21 square yards

Solve the application and write a sentence answering the question.

14. Kareem went to Kinko's to have a photo enlarged into a poster that is 36 inches high. The original photo is $\frac{2}{9}$ the size of the poster. How high is the original picture?
The original picture is 8 inches high.

Divide. Simplify. If possible, write each answer as a mixed number.

15. $\frac{14}{15} \div \left(-\frac{7}{5}\right)$
$-\frac{2}{3}$

16. $5\frac{1}{3} \div 2\frac{2}{9}$
$2\frac{2}{5}$

Find the least common denominator of each pair of fractions. Then build up each fraction to have that common denominator.

17. $\frac{4}{15}$ and $\frac{5}{9}$
LCD is 45 $\frac{12}{45}$ and $\frac{25}{45}$

18. $\frac{7}{12}$ and $\frac{11}{20}$
LCD is 60 $\frac{35}{60}$ and $\frac{33}{60}$

Evaluate. Simplify the result whenever possible. Write any answers that contain improper fractions as mixed numbers.

19. $\frac{8}{15} + \frac{2}{15}$
$\frac{2}{3}$

20. $\frac{19}{16}y - \frac{7}{16}y$
$\frac{3}{4}y$

21. $3\frac{4}{9} + 1\frac{2}{9}$
$4\frac{2}{3}$

22. $6\frac{3}{10} - 4\frac{9}{10}$
$1\frac{2}{5}$

Evaluate. Simplify the result whenever possible. Write any answers that contain improper fractions as mixed numbers.

23. $\frac{7}{10} + \frac{1}{6}$
$\frac{13}{15}$

24. $\frac{11}{18} - \frac{5}{12}$
$\frac{7}{36}$

25. $3\frac{9}{10} + 6\frac{4}{15}$
$10\frac{1}{6}$

26. $9\frac{3}{8} - 6\frac{1}{12}$
$3\frac{7}{24}$

27. $-\frac{1}{4} - \left(-\frac{3}{4}\right)$
$\frac{1}{2}$

28. $\frac{1}{6}p - \frac{7}{15}p$
$-\frac{3}{10}p$

Solve each application and write a sentence answering the question.

29. For her morning workout, Timina walked $2\frac{3}{5}$ miles. For her afternoon workout, she walked $3\frac{2}{3}$ miles. How many total miles did Timina walk that day?
Timina walked $6\frac{4}{15}$ total miles that day.

30. In June of last year, Haines, Oregon, recorded its total one-day record rainfall of $3\frac{1}{4}$ inches. By noon that day, it had rained $1\frac{7}{10}$ inches. How many more inches did it rain that day?
It rained $1\frac{11}{20}$ more inches that day.

Solve each equation. Simplify any fraction answer.

31. $y - \frac{5}{12} = -\frac{7}{12}$
$y = -\frac{1}{6}$

32. $27p = -36$
$p = -1\frac{1}{3}$

33. $\frac{5}{7}y = 15$
$y = 21$

34. $1\frac{1}{2}x = 10\frac{1}{2}$
$x = 7$

Solve each application and write a sentence answering the question.

35. A special gift box contains 18 chocolate bars. If the chocolate weighs $22\frac{1}{2}$ pounds in total, how much does each chocolate bar weigh?
Let $x =$ the weight of each chocolate bar.
$18 \cdot x = 22\frac{1}{2}$
Each chocolate bar weighs $1\frac{1}{4}$ pounds.

Chapters 1–4 Cumulative Review

1. Total 2005 home game attendance for the Cleveland Indians was 1,973,185. Round this to the nearest
   a) ten thousand 1,970,000
   b) hundred thousand 2,000,000

2. Which property is being demonstrated?
   a) $5 = 5 + 0$ Identity for Addition
   b) $9 \cdot 8 = 8 \cdot 9$ Commutative Property of Multiplication

Based on the bar graph, answer each question with a sentence.

During her campaign for mayor, Alexandra Bohler raised both small and large contributions.

3. How much more did Ms. Bohler receive from business groups than from police/fire groups?

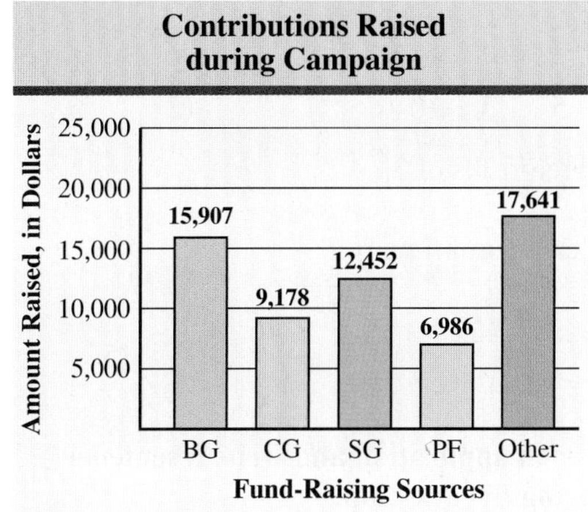

BG = Business Groups CG = Civic Groups
SG = School Groups PF = Police/Fire Groups
Other = All Other

Ms. Bohler received $8,921 more from business groups than from police/fire groups.

4. How much in total contributions did Ms. Bohler receive?
   Ms. Bohler received $62,164 in total contributions.

5. Find the area of this rectangle.

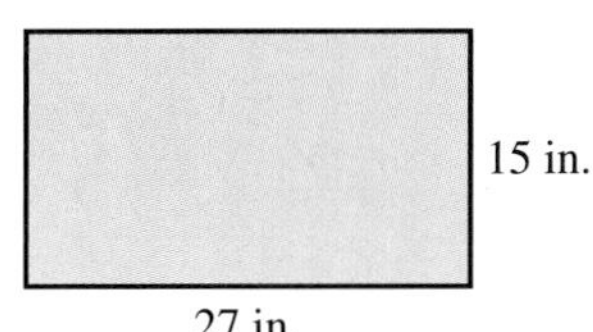

405 square inches

6. Expand 3^4 and find its value. $3 \cdot 3 \cdot 3 \cdot 3 = 81$

7. The 48 employees of Miller, Inc., pooled their money and bought lottery tickets. They didn't win the big prize, but they won a number of smaller prizes totaling $7,296. If the winnings are divided evenly, how much will each employee receive?
   Each employee will receive $152.

Of the following numbers, determine which are prime, which are composite, and which are neither.

8. 2, 9, 42, 43, 1, 115 Prime 2, 43 Composite 9, 42, 115 Neither 1

Of the first three prime numbers—2, 3, and 5—which are factors of the following numbers? Use the divisibility tests for 2, 3, and 5.

9. 170 2 and 5

10. 528 2 and 3

Determine whether 9 is a factor of the following number. Verify the answer by dividing the number by 9.

11. 618 9 is not a factor.

Find the prime factorization of the following number. Write the answer two ways: with and without exponents.

12. 108 $2 \cdot 2 \cdot 3 \cdot 3 \cdot 3$ $2^2 \cdot 3^3$

Evaluate by replacing x with 6 and y with 2.

13. $24 \div x - y$ 2

Evaluate each expression.

14. $-9 - 15 + 17 - (-4)$ -3

15. $5 \cdot (-2) \cdot (-7) \cdot (-3)$ -210

16. $-49 \div (-7)$ 7

17. $-32 \div 4$ -8

Evaluate each expression using the order of operations. Show all steps.

18. $\sqrt{16} - 9$ -5

19. $|3 - 5| + |-8|$ 10

Find the equivalent temperature in either Celsius or Fahrenheit.

$$F = 9 \cdot \frac{C}{5} + 32 \quad C = 5 \cdot \frac{F - 32}{9}$$

20. The temperature is $-10°C$. 14°F

21. The temperature is 5°F. $-15°$ C

Simplify by combining like terms wherever possible.

22. $-10y - 5y + 9y$ $-6y$

23. $-5p^3 + 8p + p^3 - 4p$ $-4p^3 + 4p$

Find the product.

24. $-9(-4k)$ $36k$

Rewrite the expression using the Distributive Property.

25. $-2(8m - 15)$ $-16m + 30$

Solve. Make sure you check each answer.

26. $7y - 35 = 28$
$y = 9$

27. $3y - 6 = 7y + 14$
$y = -5$

Translate the sentence into an equation and solve.

28. 7 more than twice a number is -15. What is the number?
$7 + 2x = -15$
$x = -11$

Solve each application and write a sentence answering the question. (*Hint:* Set up the legend, draw a diagram, identify the formula, and write and solve the equation.)

29. Two photocopiers were used to create 350 copies of a brochure. If the older machine created 80 fewer copies than the newer machine, how many brochures did each copier create?
There were 215 copies from the newer machine and 135 copies from the older machine.

30. The perimeter of a triangle is 53 inches. The first side is 18 inches, and the second side is 7 inches less than twice the third side. What are the measures of the second and third sides?
The second side is 21 inches, and the third side is 14 inches.

31. Find the greatest common factor of 56 and 42. 14

Simplify completely.

32. $\dfrac{18hk}{54h^2}$ $\dfrac{k}{3h}$

Write the improper fraction as a mixed number and the mixed number as an improper fraction.

33. $\dfrac{19}{4}$ $4\frac{3}{4}$

34. $6\frac{5}{8}$ $\frac{53}{8}$

Find the LCD of this pair of fractions. Then build up each fraction to have that common denominator.

35. $\dfrac{7}{30}$ and $\dfrac{8}{45}$ LCD is 90 $\frac{21}{90}$ and $\frac{16}{90}$

Evaluate. Simplify completely. Write improper fractions as mixed numbers.

36. $\dfrac{3}{2} \cdot \dfrac{5}{6} \cdot \dfrac{8}{15}$ $\frac{2}{3}$

37. $\dfrac{10}{21} \div \dfrac{15}{14}$ $\frac{4}{9}$

38. $4\frac{1}{6} \div 2\frac{2}{9}$ $1\frac{7}{8}$

39. $\dfrac{13}{24}m + \dfrac{5}{24}m$ $\frac{3}{4}m$

40. $\dfrac{29}{35} - \dfrac{2}{5}$ $\frac{3}{7}$

41. $2\frac{7}{10} + 3\frac{7}{15}$ $6\frac{1}{6}$

42. $-\dfrac{3x}{5} \div \left(-\dfrac{9x}{10}\right)$ $\frac{2}{3}$

43. $-\dfrac{5}{12}r^2 + \dfrac{1}{6}r^2$ $-\frac{1}{4}r^2$

44. Three-fifths of $3\frac{8}{9}$ $2\frac{1}{3}$

Solve each equation.

45. $-28 = 42x$
$x = -\frac{2}{3}$

46. $\dfrac{3m}{10} = \dfrac{2}{15}$
$m = \frac{4}{9}$

Solve each application and write a sentence answering the question.

47. The last event in the Policeman's Olympics is the relay race in which six team members must run an equal portion of a $10\frac{1}{2}$-mile course. How many miles must each team member run in the relay?
Each team member must run $1\frac{3}{4}$ miles.

48. Before yesterday, Kahlil's best long jump was 7 feet $4\frac{5}{8}$ inches. Yesterday he set a new personal best by jumping 7 feet $6\frac{1}{16}$ inches. By how many inches did Kahlil surpass his previous best mark?
Kahlil surpassed his previous mark by $1\frac{7}{16}$ inches.

CHAPTER 

Decimals

Introduction

We encounter decimals all the time. We see them when dealing with money, such as \$4.549 for a gallon of gas. Decimals are in sports statistics. For instance, Hank Aaron's lifetime batting average is .305. We also find them in the nutrition guide on the side of a cereal box, saying, for example, that each serving contains 1.5 grams of fat.

Adding with decimals is the same as adding with whole numbers; and this is also true for subtracting, multiplying and dividing with decimals. The only difference between evaluating with decimals and evaluating with whole numbers is that in Chapter 5, we have a decimal point to keep track of.

Fear not. Every procedure is explained step-by-step, and you'll be taught *how* to evaluate with decimals as well as *why* the procedures work the way they do.

Preparation Exercises

Section 1.1 Whole Numbers

1. Round 71,486 to the nearest thousand.
   71,000
2. Round 6,952 to the nearest hundred.
   7,000

Section 1.2 Adding and Subtracting Whole Numbers
Evaluate.

3. $75 + 542$
   617
4. $28 + 9 + 672$
   709
5. $926 - 349$
   577
6. $2{,}000 - 153$
   1,847

Section 1.3 Multiplying Whole Numbers
Evaluate.

7. 127×8
   1,016
8. 24×65
   1,560

Section 1.4 Dividing Whole Numbers
Evaluate.

9. $7\overline{)329}$
   47
10. $23\overline{)805}$
   35

Section 1.7 Equations
Solve for x.

11. $326 + x = 509$
    $x = 183$
12. $25x = 475$
    $x = 19$

Section 4.3 Equivalent Fractions
Completely simplify each fraction.

13. $\dfrac{35}{100}$
    $\frac{7}{20}$
14. $\dfrac{44}{60}$
    $\frac{11}{15}$

Section 4.4, Multiplying and Dividing Fractions
Multiply and simplify.

15. $\frac{8}{9} \cdot \frac{15}{4}$ $3\frac{1}{3}$

16. $\frac{7}{10} \cdot \frac{9}{10}$ $\frac{63}{100}$

Section 4.4 Multiplying and Dividing Fractions
Divide and simplify.

17. $\frac{27}{100} \div \frac{9}{10}$ $\frac{3}{10}$

18. $\frac{7}{10} \div \frac{35}{100}$ 2

Section 4.5 Adding and Subtracting Like Fractions
Add or subtract. Simplify.

19. $\frac{11}{18} - \frac{5}{18}$ $\frac{1}{3}$

20. $\frac{43}{100} + \frac{22}{100}$ $\frac{13}{20}$

Section 4.7 Adding and Subtracting Unlike Fractions
Add or subtract. Simplify.

21. $\frac{5}{6} - \frac{11}{24}$ $\frac{3}{8}$

22. $\frac{27}{100} + \frac{3}{10}$ $\frac{57}{100}$

SECTION 5.1 An Introduction to Decimals

OBJECTIVES

In this section, you will learn to
- Identify and define the parts of a decimal.
- Name decimal numbers.
- Write decimal numbers as decimal fractions.
- Write decimal fractions as decimal numbers.
- Abbreviate repeating decimal numbers.
- Locate decimal numbers on the number line.

You Need to Know

To successfully complete this section, you need to understand
- ☐ Writing whole numbers (1.1)
- ☐ Powers of 10 (1.5)
- ☐ Integers on the number line (2.1)
- ☐ Mixed numbers (4.2)
- ☐ Simplifying fractions (4.3)
- ☐ Building up fractions (4.3)

Introduction

You've seen decimals used in amounts of money. *One dollar and fifty-three cents* is represented as $1.53. The dot that separates the dollars from the cents is the **decimal point,** a separator between the whole number and the fractional part of a decimal number.

In this text, we refer to numbers such as 0.5, 0.28, and 3.406 as *decimal numbers*, or just *decimals*.

The whole number is to the *left* of the decimal point, and the fractional part is to the *right* of the decimal point.

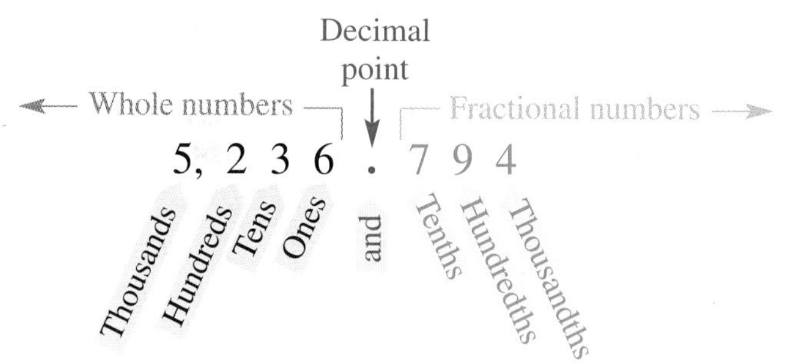

Notice that every place in the fractional part ends in **-ths**.

Example 1

In 5,236.794, name the place and the value represented by

a) 5 **b)** 2 **c)** 7 **d)** 9 **e)** 4

Answer **a)** 5 is in the **thousands** place and represents **five thousand.**

b) 2 is in the **hundreds** place and represents **two hundred.**

c) 7 is in the **tenths** place and represents **seven-tenths.**

d) 9 is in the **hundredths** place and represents **nine-hundredths.**

e) 4 is in the **thousandths** place and represents **four-thousandths.**

You Try It 1 **In 914.673, name the place and the value represented by the given digit. Use Example 1 as a guide.**

a) 6 **b)** 9 **c)** 3

When there is no visible whole number, as in .47, it's common to write such a decimal with a whole number of 0, as in 0.47. Although either way is acceptable, in this text, such decimals usually include the 0 whole number.

Naming Decimals

When reading a decimal, it is most common to use the last (rightmost) decimal place in its name. For example, 0.85 contains 8 tenths and 5 hundredths. The last decimal place is the hundredths, and 0.85 should be read as eighty-five *hundredths*. In other words, when reading 0.85, we usually do not mention "tenths."

Example 2

Name each decimal by writing it in word form.

a) 0.3 **b)** 0.15 **c)** 0.70 **d)** 0.028

Procedure Identify the last decimal place shown and use that place in its name.

Answer **a)** Three-tenths **b)** Fifteen-hundredths

c) Seventy-hundredths **d)** Twenty-eight thousandths

You Try It 2 **Name each decimal by writing it in word form. Use Example 2 as a guide.**

a) 0.04 **b)** 0.009 **c)** 0.031 **d)** 0.20

Recall from Section 1.1 that we did not use the word *and* in writing whole numbers in words. The word *and* is reserved for numbers with both whole number and decimal values.

Example 3

Name each number by writing it in word form.

a) 5.7 **b)** 17.04 **c)** 51.309

Procedure Separate the whole number from the decimal number by using the word *and*.

Answer **a)** Five and seven-tenths

b) Seventeen and four-hundredths

c) Fifty-one and three hundred nine-thousandths

You Try It 3 **Name each decimal by writing it in word form. Use Example 3 as a guide.**

a) 6.8 **b)** 14.19 **c)** 35.002

Think About It 1 What is the difference between one hundred twelve-thousandths and one hundred and twelve-thousandths?

____

____

Decimal Fractions

As mentioned, the part of a decimal to the right of the decimal point is called the fractional part. To understand why, consider *seven-tenths*.

Seven-tenths can be represented as **0.7** or as $\mathbf{\frac{7}{10}}$. This means that **0.7** is equivalent to the fraction $\mathbf{\frac{7}{10}}$.

Also, *three-hundredths* can be represented as **0.03** or as $\mathbf{\frac{3}{100}}$, and *six-thousandths* can be represented as **0.006** or as $\mathbf{\frac{6}{1{,}000}}$.

> Recall from Section 1.5:
> - 100 is the second power of 10: $100 = 10 \times 10 = 10^2$.
> - 1,000 is the third power of 10: $1{,}000 = 10 \times 10 \times 10 = 10^3$.

Because of their direct relationship to decimals, fractions such as $\frac{7}{10}$, $\frac{3}{100}$, and $\frac{6}{1{,}000}$ are called *decimal fractions*. A **decimal fraction** is any fraction with a denominator that is a power of 10 (such as 10, 100, and 1,000).

When a decimal number is named (written in words), the place value (such as hundredths) indicates the denominator of the decimal fraction. For example, 0.36, which is thirty-six-*hundredths*, can be represented by a decimal fraction, $\frac{36}{100}$.

Notice the connection between the number of decimal places and the number of zeros in the denominator. 0.36 has two decimal places, and there are two zeros in the denominator because 100 is 10^2.

Example 4

Write each decimal number as a decimal fraction.

a) 0.239 **b)** 0.8 **c)** 0.71 **d)** 0.006

Procedure Recognize the rightmost place in the decimal. That place indicates the denominator of the equivalent decimal fraction.

Answer

a) $0.239 = \frac{239}{1{,}000}$ — three decimal places. The denominator is 1,000.

b) $0.8 = \frac{8}{10}$ — one decimal place. The denominator is 10.

c) $0.71 = \frac{71}{100}$ — two decimal places. The denominator is 100.

d) $0.006 = \frac{6}{1{,}000}$ — three decimal places. The denominator is 1,000.

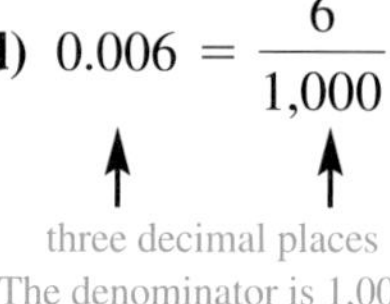

You Try It 4 **Write each decimal number as a decimal fraction. Use Example 4 as a guide.**

a) 0.3 **b)** 0.197 **c)** 0.07 **d)** 0.0086

Writing Mixed Numbers as Decimal Fractions

Consider *one and four-tenths*. As a mixed number, it can be written two different ways.

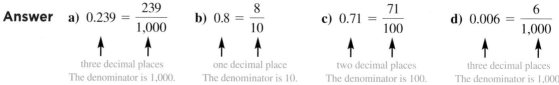

1. As a fraction: $1\frac{4}{10}$ or $1 + \frac{4}{10}$

2. As a decimal: 1.4 or 1 + .4

So, $1.4 = 1\frac{4}{10}$.

As an improper fraction, $1\frac{4}{10} = \frac{14}{10}$; so, it makes sense that $1.4 = \frac{14}{10}$ too.

The point is this: A mixed decimal number can be written as an improper decimal fraction.

For example,

Mixed decimal ⟶ $9.8 = 9\frac{8}{10} = \frac{98}{10}$ ⟵ **Improper decimal fraction**

Mixed decimal ⟶ $6.93 = 6\frac{93}{100} = \frac{693}{100}$ ⟵ **Improper decimal fraction**

Let's look more closely at $6.93 = \frac{693}{100}$. Even though there are three digits in all, only two of them are decimal digits (*after* the decimal point). Those two decimal places indicate the power of 10 in the denominator, $10^2 = 100$.

In other words, no matter how many digits are in the whole number part, only the number of digits in the decimal part indicates the power of 10 in the denominator.

Example 5

Write each mixed decimal number as an improper decimal fraction. Do not simplify.

a) 4.7 **b)** 8.41 **c)** 17.2

Procedure First, keep in mind that the number of decimal places determines the power of 10 in the denominator.

Second, the numerator of the fraction will be a whole number that comes from *all* digits in the decimal number.

Answer **a)** $4.7 = \frac{47}{10}$ There is one decimal place; so, the denominator is 10. The numerator is the whole number 47.

b) $8.41 = \frac{841}{100}$ There are two decimal places; so, the denominator is 100. The numerator is 841.

c) $17.2 = \frac{172}{10}$ Again, only one decimal place means a denominator of 10 no matter how many digits are in the whole number.

You Try It 5 **Write each mixed decimal number as an improper decimal fraction. Do not simplify. Use Example 5 as a guide.**

a) 9.8 **b)** 3.51 **c)** 89.3 **d)** 7.012

Writing Decimal Fractions as Decimals

In converting a decimal fraction to a decimal form, the number of zeros in the denominator determines the number of decimal places needed to the right of the decimal point. For example, a decimal fraction with three zeros in the denominator, such as $\frac{967}{1{,}000}$, indicates that there will be three places after the decimal point, 0.967.

In a decimal fraction, the number of zeros in the denominator's power of 10 indicates the number of decimal places the decimal number will contain.

Still, we must make sure that there are at least as many digits in the numerator as there are zeros in the denominator. One of three situations can happen based on the number of zeros in the denominator.

1. The numerator has the same number of digits as the number of zeros in the denominator.

   For example, $\frac{14}{100}$ will have two decimal places because there are two zeros in the denominator. The numerator has two digits to fit into those two decimal places: $\frac{14}{100} = 0.14$.

2. The numerator has fewer digits than the number of zeros in the denominator.

A fraction such as $\frac{8}{100}$ doesn't have enough digits in the numerator to cover the two required decimal places. To compensate, we can rewrite $\frac{8}{100}$ as $\frac{08}{100}$ without changing the value of the fraction. This also allows us to write $\frac{08}{100}$ as 0.08.

3. The numerator has more digits than the number of zeros in the denominator.

Consider $\frac{34}{10}$. In this case, the 4 fits into the one required decimal place and the 3 is the whole number in the ones place: $\frac{34}{10} = 3.4$.

Example 6

Write each decimal fraction as a decimal number.

a) $\dfrac{5}{100}$ **b)** $\dfrac{81}{10}$ **c)** $\dfrac{3}{1{,}000}$ **d)** $\dfrac{618}{100}$ **e)** $\dfrac{276}{10}$

Procedure Write out the numerator completely. Then decide how many decimal places it will have by looking at the denominator.

The number of zeros in the denominator indicates the number of decimal places.

Answer

a) $\dfrac{5}{100} = 0.05$ The denominator of 100 indicates two decimal places. Think of 5 as 05, making the fraction $\frac{05}{100}$, so that we have two digits to place after the decimal point.

b) $\dfrac{81}{10} = 8.1$ There is one zero in the denominator; so, there is one decimal place.

c) $\dfrac{3}{1{,}000} = 0.003$ There are three zeros in the denominator; so, there are three decimal places. This time think of 3 as 003, giving us $\frac{003}{1{,}000}$.

d) $\dfrac{618}{100} = 6.18$ There are two zeros in the denominator; so, there are two decimal places.

e) $\dfrac{276}{10} = 27.6$ One zero in the denominator allows the number to have only one decimal place.

You Try It 6

Write each decimal fraction as a decimal number. Use Example 6 as a guide.

a) $\dfrac{87}{10}$ **b)** $\dfrac{619}{100}$ **c)** $\dfrac{504}{10}$

d) $\dfrac{6}{100}$ **e)** $\dfrac{98}{1{,}000}$ **f)** $\dfrac{5}{1{,}000}$

Place Value and Powers of 10

Consider one-half. This can be represented in many ways. In fact, we can build it up to be written with any denominator we choose.

For now, though, let's choose one-half to be written as a decimal fraction.

Decimal fractions that are equivalent to one-half

$$\frac{1}{2} \times \frac{5}{5} = \mathbf{\frac{5}{10}} = .5$$

$$\frac{5}{10} \times \frac{10}{10} = \mathbf{\frac{50}{100}} = .50, \quad \text{which is .5 followed by one zero.}$$

$$\frac{50}{100} \times \frac{10}{10} = \mathbf{\frac{500}{1{,}000}} = .500, \quad \text{which is .5 followed by two zeros.}$$

$$\frac{500}{1{,}000} \times \frac{10}{10} = \mathbf{\frac{5{,}000}{10{,}000}} = .5000, \quad \text{which is .5 followed by three zeros.}$$

Notice that multiplying a decimal fraction by $\frac{10}{10}$ simply places one more zero at the end of the decimal number.

We can place as many zeros as we choose at the *end* of a decimal number without changing its value.

Caution We may not place zeros *between* the decimal point and other digits. For example, 1.5 = 1.50, but 1.5 ≠ 1.05.

Example 7

Write each decimal by including enough zeros at the end so that it has a total of *three* decimal places.

a) 0.8 **b)** 7.54 **c)** 3

Answer **a)** 0.8 = 0.800 **b)** 7.54 = 7.540 **c)** 3 = 3.000

You Try It 7 **Write each decimal by including enough zeros at the end so that it has a total of *four* decimal places. Use Example 7 as a guide.**

a) 6.43 **b)** 9.108 **c)** 0.1 **d)** 5

If a decimal number has at least one ending zero, it can be simplified by eliminating that zero (and any other zeros that follow it). To understand *why* we can eliminate those ending zeros, consider 0.80.

As it is, $0.80 = \dfrac{80}{100}$, which can be simplified by a factor of 10:

$$\frac{80 \div 10}{100 \div 10} = \frac{8}{10} = 0.8 \text{ (eight-tenths).}$$

This simplifying process is quickened by dropping any ending zeros. For example, 8.9100 = 8.91 and 16.70000 = 16.7.

Caution Only ending zeros may be eliminated, not zeros *within* the decimal number.

1.30600 = 1.306 and 0.0040 = 0.004

Example 8

Simplify each decimal by eliminating any ending zeros.

a) 0.2000 **b)** 8.630 **c)** 15.20700

Answer **a)** 0.2000 = 0.2 **b)** 8.630 = 8.63 **c)** 15.20700 = 15.207

Notice that in part c, the 0 that remains in 15.207 is not an ending zero; so, it cannot be eliminated.

You Try It 8 **Simplify each decimal by eliminating any ending zeros. Use Example 8 as a guide.**

a) 6.5000 **b)** 9.4600 **c)** 0.102000 **d)** 0.000800

Terminating and Repeating Decimals

Many decimals have a definite end. We call these **terminating decimals.**

For example, 98.5 terminates after *one* decimal place, 9.632 terminates after *three* decimal places, and 0.19 terminates after *two* decimal places.

Some decimals never terminate; they just go on and on. Such decimal numbers are called **nonterminating decimals.** However, many of these nonterminating decimals repeat a pattern of the same digits, such as 8.56565656. Here the pattern of digits *56* repeats continuously. Such numbers are called **repeating decimals.**

The "block" of repeating digits is called the **repetend.** In the number 8.565656. . . , the repetend is 56.

We can't write all of the decimal places if a decimal doesn't terminate. There are a couple of ways to show that a particular pattern of numbers repeats itself.

1. We can use an **ellipsis,** three dots (. . .), to indicate that a pattern repeats.
2. We can place a bar over all of the digits that repeat, the repetend.

For example, if a nonterminating decimal number has a repeating pattern of the digits *815*, we can represent it as 0.815815815. . . or as $0.\overline{815}$.

Repeating decimals can start with more than just the repeated pattern in the decimal places. Using the same repetend of 815 we might have

1. 4.96815815815 . . . Once the pattern of 815 is clearly established, the ellipsis indicates that just the digits 815 repeat.
2. $4.96\overline{815}$ The bar goes over the repetend only, not the other digits.

Example 9

Write each of these repeating decimals using either the ellipsis or the bar over the repetend, whichever is not shown. *When using the ellipsis, write the repetend at least three times.*

a) 0.93939393 . . . = $0.\overline{93}$ **b)** 8.079079079 . . . = $8.\overline{079}$

c) $12.68\overline{4}$ = 12.6844444 . . . **d)** $1.6\overline{58}$ = 1.658585858 . . .

You Try It 9 **Write each of these repeating decimals using either the ellipsis or the bar over the repetend, whichever is not shown. When using the ellipsis, write the repetend at least three times. Use Example 9 as a guide.**

a) 0.555 . . . = __________ **b)** 7.9131313 . . . = __________

c) $8.0\overline{47}$ = __________ **d)** $0.\overline{905}$ = __________

Pi

Many decimal numbers don't terminate and don't repeat. One such number is represented by the Greek letter, π, pronounced "pie" and spelled **pi**.

The first 12 decimal digits of π are shown at the right. (In this case, the ellipsis lets us know that there are many more decimal digits not shown, but it does not indicate a pattern of any kind.)

$\pi \approx 3.141592653589 \ldots$

We will use this number when calculating the perimeter (circumference) of a circle in Section 5.6. At that time, we'll abbreviate the value of π as 3.14. This value for π is an approximate value shown as $\pi \approx 3.14$.

Decimals on the Number Line

On a number line between the integers are fractions and decimals, both positive and negative. In Section 4.2, we looked at a number line with signed fractions. Let's now look at a number line with signed decimals.

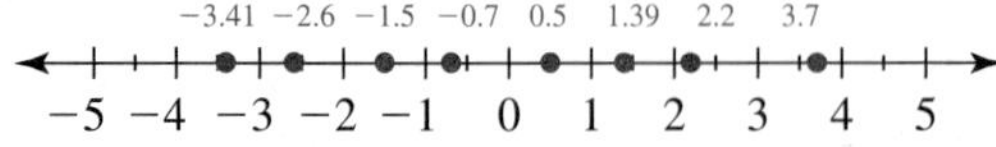

Typically, when locating a decimal number on the number line, we find only an approximate location between two integers. We should recognize which of the integers the number is closest to. We can also use the halfway marks (−3.5, −0.5, 1.5, and 2.5, for example) as guides.

Notice the following in the preceding number line:

- 2.2 is between the integers 2 and 3, closer to 2.
- −2.6 is between −2 and −3, closer to −3.
- −1.5 is between −1 and −2 and directly at the halfway mark.

After the integer, the *tenths* place digit is most helpful in locating a number's position on a number line. (The hundredths place digit does not greatly affect the location of the decimal number.)

When locating a *positive* decimal number such as 4.1 on the number line, we first identify the location of its integer, +4. This positive decimal number, 4.1, is located to the *right* of its integer, $\frac{1}{10}$ of the way toward the next integer, 5.

When locating a *negative* decimal number such as −4.7 on the number line, we first identify the location of its integer, −4. This negative decimal number, −4.7, is located to the *left* of its integer, $\frac{7}{10}$ (more than half) of the way toward the next integer, −5.

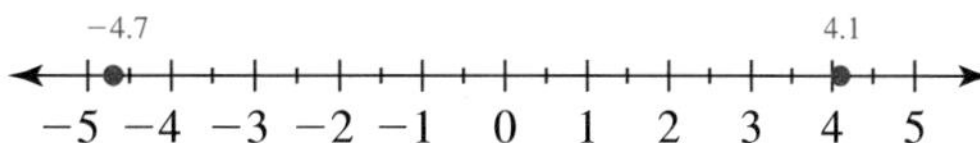

Example 10

Fill in the blanks to complete each sentence. Also locate and label each value on the number line.

a) **2.4** is between the integers ___2___ and ___3___ and is closer to ___2___.

b) **−2.9** is between the integers ___−2___ and ___−3___ and is closer to ___−3___.

c) **−0.3** is between the integers ___0___ and ___−1___ and is closer to ___0___.

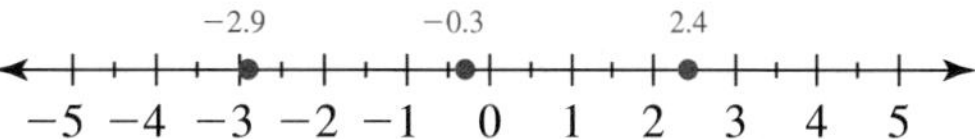

You Try It 10

Fill in the blanks to complete each sentence. Also locate and label each value on the number line provided. Use Example 10 as a guide.

a) **0.8** is between the integers ________ and ________ and is closer to ________.

b) **−1.6** is between the integers ________ and ________ and is closer to ________.

c) **−4.2** is between the integers ________ and ________ and is closer to ________.

d) **3.7** is between the integers ________ and ________ and is closer to ________.

e) **π** is between the integers ________ and ________ and is closer to ________.

f) **$-\pi$** is between the integers ________ and ________ and is closer to ________.

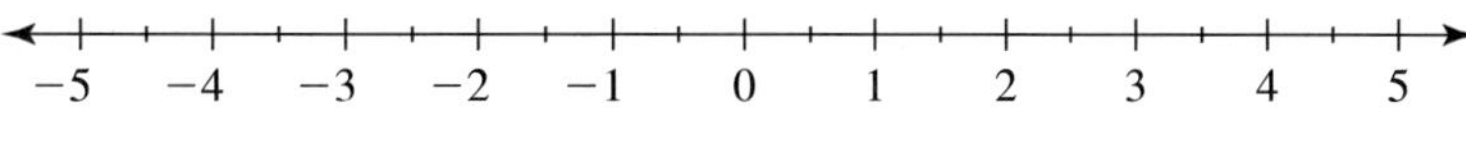

Answers: You Try It and Think About It

You Try It: **1. a)** Tenths; six-tenths **b)** Hundreds; nine hundred **c)** Thousandths; three-thousandths **2. a)** Four-hundredths **b)** Nine-thousandths **c)** Thirty-one thousandths **d)** Twenty-hundredths **3. a)** Six and eight-tenths **b)** Fourteen and nineteen-hundredths **c)** Thirty-five and two-thousandths **4. a)** $\frac{3}{10}$ **b)** $\frac{197}{1{,}000}$ **c)** $\frac{7}{100}$ **d)** $\frac{86}{10{,}000}$ **5. a)** $\frac{98}{10}$ **b)** $\frac{351}{100}$ **c)** $\frac{893}{10}$ **d)** $\frac{7{,}012}{1{,}000}$ **6. a)** 8.7 **b)** 6.19 **c)** 50.4 **d)** 0.06 **e)** 0.098 **f)** 0.005 **7. a)** 6.4300 **b)** 9.1080 **c)** 0.1000 **d)** 5.0000 **8. a)** 6.5 **b)** 9.46 **c)** 0.102 **d)** 0.0008 **9. a)** $0.\overline{5}$ **b)** $7.9\overline{13}$ **c)** 8.0474747... **d)** 0.905905905 . . . **10. a)** 0; 1; 1 **b)** −2; −1; −2 **c)** −5; −4; −4 **d)** 3; 4; 4 **e)** 3; 4; 3 **f)** −4; −3; −3

−4.2 −π −1.6 0.8 π 3.7
−5 −4 −3 −2 −1 0 1 2 3 4 5

Think About It: **1.** One hundred twelve-thousandths is 0.112. One hundred and twelve-thousandths is the mixed number 100.012.

Section 5.1 Exercises

FOR EXTRA HELP
Student Resources on DVD-ROM — Includes ➤ Student's Solutions Manual ➤ Video Lectures ➤ Chapter Test Prep Video | MyMathLab | MathXL

Think Again

Write a sentence or two for each response.

1. What is the difference between one hundred twelve-thousandths and one hundred and twelve-thousandths? *(Refer to Think About It 1)*
One hundred twelve-thousandths is 0.112, and one hundred and twelve-thousandths is a mixed number: 100.012.

2. Is 2.0200220002220000222200000222222. . . a repeating decimal? Explain your answer.
No, the number of 0's and 2's in the pattern is changing. To be a repeating decimal, a repetend must remain consistent.

3. Is 3.14 a terminating decimal? Explain your answer.
Yes, it terminates after only two decimal places. (3.14 should not be confused with π, a decimal that doesn't terminate and doesn't repeat.)

4. On the number line, is -37.25 to the right of -37 or to the left of -37? Explain your answer.
-37.25 is to the left of -37 on the number line. Answers may vary.

Focus Exercises

Fill in the blanks.

5. Two places to the right of the decimal point is called the hundredths place.

6. The three dots at the end of a repeating decimal are called a(n) ellipsis.

7. A decimal number that ends after a few decimal places is called a(n) terminating decimal.

8. A fraction with a denominator of a power of 10 is called a(n) decimal fraction.

True or false.

9. The numbers to the left of the decimal point are called whole numbers.
True

10. The number 1.005 can simplify to 1.5 by eliminating the zeros.
False

11. It's not possible for a repeating decimal to have a repetend with more than three digits.
False

12. The number $\pi = 3.14$.
False

In the given number, name the place value for the digit 3.

13. 12.384
tenths

14. 1,358.47
hundreds

15. 93.806
ones

16. 150.623
thousandths

Name each decimal by writing it in word form.

17. 0.5
five-tenths

18. 0.4
four-tenths

19. 0.29
twenty-nine hundredths

20. 0.62
sixty-two hundredths

21. 0.008
eight-thousandths

22. 0.013
thirteen-thousandths

23. 0.60
sixty-hundredths

24. 0.90
ninety-hundredths

25. 3.07
three and seven-hundredths

26. 2.01
two and one-hundredth

27. 12.008
twelve and eight-thousandths

28. 11.013
eleven and thirteen-thousandths

Write each decimal as a fraction. Do not simplify.

29. 0.2
$\frac{2}{10}$

30. 0.4
$\frac{4}{10}$

31. 0.87
$\frac{87}{100}$

32. 0.14
$\frac{14}{100}$

33. 0.628
$\frac{628}{1,000}$

34. 0.403
$\frac{403}{1,000}$

35. 0.9314
$\frac{9,314}{10,000}$

36. 0.7803
$\frac{7,803}{10,000}$

37. 0.03
$\frac{3}{100}$

38. 0.08
$\frac{8}{100}$

39. 0.07
$\frac{7}{100}$

40. 0.02
$\frac{2}{100}$

41. 0.069
$\frac{69}{1,000}$

42. 0.017
$\frac{17}{1,000}$

43. 0.001
$\frac{1}{1,000}$

44. 0.004
$\frac{4}{1,000}$

45. 9.6
$\frac{96}{10}$

46. 1.3
$\frac{13}{10}$

47. 2.5
$\frac{25}{10}$

48. 16.15
$\frac{1,615}{100}$

49. 8.03
$\frac{803}{100}$

50. 6.05
$\frac{605}{100}$

51. 6.937
$\frac{6,937}{1,000}$

52. 2.802
$\frac{2,802}{1,000}$

Write each fraction as a decimal.

53. $\frac{7}{10}$
0.7

54. $\frac{91}{100}$
0.91

55. $\frac{596}{1,000}$
0.596

56. $\frac{1,762}{10,000}$
0.1762

57. $\frac{4}{10}$
0.4

58. $\frac{6}{100}$
0.06

59. $\frac{43}{10}$
4.3

60. $\frac{139}{100}$
1.39

61. $\frac{2}{100}$
0.02

62. $\frac{31}{1,000}$
0.031

63. $\frac{7}{1,000}$
0.007

64. $\frac{22}{10,000}$
0.0022

65. $\frac{702}{100}$
7.02

66. $\frac{305}{100}$
3.05

67. $\frac{904}{10}$
90.4

68. $\frac{603}{10}$
60.3

Write each of these repeating decimals using the bar over the repetend.

69. 0.33333. . .
$0.\overline{3}$

70. 8.1717171717. . .
$8.\overline{17}$

71. 0.693693693. . .
$0.\overline{693}$

72. 0.411111111. . .
$0.4\overline{1}$

73. 10.85222222. . .
$10.85\overline{2}$

74. 4.9543254325432. . .
$4.9\overline{5432}$

Write each of these repeating decimals using an ellipsis.

75. $0.\overline{6}$
0.66666. . .

76. $2.\overline{94}$
2.94949494. . .

77. $9.\overline{470}$
9.470470470. . .

78. $5.0\overline{7}$
5.07777. . .

79. $0.63\overline{10}$
0.63101010. . .

80. $1.59\overline{07}$
1.59070707. . .

Locate and label each decimal number on the number line.

81. 0.9

82. 1.4

83. 3.5

84. 4.27

85. −0.4

86. −2.5

87. −3.71

88. −4.6

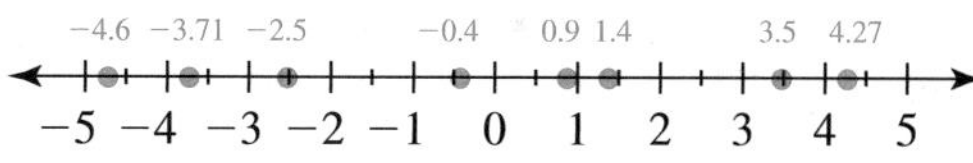

Think Outside the Box

If we can build up a fraction to have a power of 10 in the denominator, we can easily find the decimal equivalent of that fraction. For example, $\frac{3}{5}$ can be multiplied by $\frac{2}{2}$ to get $\frac{6}{10}$, which is 0.6. This indicates that 0.6 is the decimal equivalent of $\frac{3}{5}$. Use this technique to find the decimal equivalent of each of the follow fractions.

89. $\frac{7}{25}$
0.28

90. $\frac{7}{20}$
0.35

91. $\frac{7}{4}$
1.75

92. $\frac{7}{250}$
0.028

Evaluate each product and quotient by rewriting each decimal number as a decimal fraction with a power of 10 in the denominator. Write the final result in decimal form.

93. 0.6×2.8
1.68

94. 0.25×0.4
0.1

95. $0.24 \div 0.8$
0.3

96. $0.0156 \div 0.03$
0.52

SECTION 5.2 Rounding Decimals

OBJECTIVES

In this section, you will learn to

- Round decimals.

You Need to Know

To successfully complete this section, you need to understand

- ☐ Rounding whole numbers (1.1)
- ☐ Decimal place values (5.1)

Introduction

Money is usually represented with two decimal places, such as \$8.36. There are times, however, when a calculated amount has more than two decimal places.

For example, the sales tax on a watch costing \$29.95 might be \$2.321125. The amount beyond \$2.32 is a fraction of a penny. But because nobody can pay that fraction, the tax is rounded to \$2.32.

Similarly, gas stations typically charge an extra $\frac{9}{10}$ of a penny per gallon. So, you might see the price of gas as \$4.169. When this number is multiplied by the number of gallons pumped (say, 9.381), the total price is \$39.109389. When you pay for the gas, though, this number is rounded to the nearest penny: \$39.11. (The gas pump does the rounding automatically.)

In this section, we won't be calculating sales tax and gas prices. Instead, we'll focus on how to round decimal numbers.

Rounding Decimal Numbers to the Nearest Decimal Place

Recall that in Section 1.1, we established the procedure for rounding whole numbers. To round decimals, we use the same procedure with one major difference: In step 4, we don't write the digits after the place digit as zeros; instead, we *eliminate* those ending digits.

Rounding Decimal Numbers to a Given Decimal Place

1. Identify the place digit that is to be rounded.
2. Identify the digit to its immediate right, called the rounding digit.
3. **a)** If the rounding digit is 5 or higher, round *up* (add 1 to the place digit).
   b) If the rounding digit is 4 or lower, round *down* (add 0 to the place digit).
4. If the place digit is after the decimal point, eliminate all digits after the place digit. Rewrite the number showing the appropriate approximation.

Step 4 of this procedure indicates that the rounded decimal number should terminate at the rounded digit's place. For example, if we are rounding to the nearest *hundredth,* the rounded number should be a decimal number that terminates in the *hundredths* place.

Example 1

Round 0.387 to the nearest tenth.

Answer **Step 1:** 0.387: The place digit is in the tenths place: 3.

Step 2: The rounding digit, to the immediate right of the 3, is the number 8.

Step 3: 8 indicates that we must round up (add 1 to 3 to get 4).

Step 4: Everything after the tenths place is eliminated; so, the approximation is 0.4.

Example 2

Round 3.674921 to the nearest hundredth.

Answer **Step 1:** 3.674921: The place digit is in the hundredths place: 7.

Step 2: The rounding digit is the number 4.

Step 3: This indicates that we should round down (add 0 to 7 to get 7).

Step 4: Everything after the hundredths place is eliminated; so, the approximation is 3.67.

Think About It 1 If we round a decimal to the nearest thousandth, in what place will the rounding digit be?

____

You Try It 1 **Round each of these decimal numbers to the place shown. Use Examples 1 and 2 as guides.**

	Number	To the nearest	Approximation
a)	3.832	tenth	________
b)	0.4258	hundredth	________
c)	0.0509	hundredth	________
d)	5.806427	thousandth	________
e)	0.0047186	thousandth	________

You Try It 2 **Recall from Section 5.1 that π (pi) is a decimal number that neither terminates nor repeats. It is used to calculate the area and the circumference of a circle. $\pi \approx 3.141592653589\ldots$. Round this number to the nearest hundredth and finish this sentence:**

The value of π is approximately ________.

Caution If the place digit is rounded to 0, we cannot eliminate it with any other ending zeros.

For example, when we round 1.29604 to the nearest hundredth, the rounding digit, 6, tells us to add 1 to the place digit 9, changing the 29 to 30: $1.29604 \approx 1.30$.

The zero in the hundredths place cannot be eliminated because the rounded approximation must terminate in the hundredths place. This is especially true for rounding dollars and cents: $\$1.29604 \approx \$1.30$.

You Try It 3 **Round each of these numbers to the place shown. Use Examples 1 and 2 as guides. Make sure that the rounding place is represented, even if it is a zero.**

	Number	To the nearest	Approximation
a)	7.038	tenth	________
b)	6.19153	hundredth	________
c)	3.0954	hundredth	________
d)	2.14915	thousandth	________
e)	0.09982	thousandth	________

Example 3

Round each of these dollar amounts to the nearest penny.

a) \$3.12504 b) \$78.06198 c) \$142.29641 d) \$8.99741

Procedure Rounding to the nearest penny is the same as rounding to the nearest hundredth. Remember, the final result must have two decimal places.

Answer a) \$3.13 b) \$78.06 c) \$142.30 d) \$9.00

You Try It 4

Round each of these dollar amounts to the nearest penny. Use Example 3 as a guide.

a) \$1.624908 b) \$0.69502 c) \$251.0523 d) \$56.996021

We also can round to the nearest whole number. The rounding digit is in the tenths place. In rounding to the nearest whole number, no decimal or decimal point is left in the approximation.

Example 4

Round each of these numbers to the nearest whole number.

Procedure In rounding to the nearest one, we look at the tenths place to determine whether we should round up or down.

Round		Answer
a) 0.83	8 causes it to round up (add 1 to the 0 to get 1).	1
b) 2.096	0 causes it to round down (add 0 to the 2 to get 2).	2
c) 13.524	5 causes it to round up (add 1 to the 3 to get 4).	14
d) 751.6	6 causes it to round up (add 1 to the 1 to get 2).	752

You Try It 5

Round each of these numbers to the nearest whole number. Use Example 4 as a guide.

Number	Approximation
a) 1.938	________
b) 4.2381	________
c) 3.0954	________
d) 16.59153	________
e) 23.4599	________

Sometimes, money is rounded to the nearest dollar. What is rounding to the nearest dollar the same as? Explain your answer or show an example that supports your answer.

____

____

Applications

As stated at the beginning of this section, we typically round calculations involving money to the nearest penny, the nearest hundredth. If the sales tax on an item was calculated to be \$1.896125, we would round this up to \$1.90. Even though this is a rounded number, we may state that the sales tax *is* \$1.90 (instead of stating that it is *approximately* \$1.90).

Example 5

Reneé calculated the sales tax on her iPod to be \$21.74925. Round this number to the nearest penny.

Procedure Round the number and restate the first sentence using the rounded number.

Answer \$21.74925 rounds up to \$21.75 (nearest hundredth).

Sentence Reneé calculated the sales tax on her iPod to be \$21.75.

You Try It 6 **April's grade point average (GPA) is 3.1972. Round this to the nearest hundredth.**

Sentence ____

You Try It 7 **Pedro spent \$1,279.51 on office supplies for his business last year. When reporting this amount on his tax form, he rounds it to the nearest dollar. How much should Pedro report on his tax form?**

Sentence ____

You Try It 8 **Tuyn calculated her dinner tip to be \$3.64125. Round this number to the nearest dime (tenth) and write the result in dollars and cents.**

Sentence ____

Answers: You Try It and Think About It

You Try It: **1. a)** 3.8 **b)** 0.43 **c)** 0.05 **d)** 5.806 **e)** 0.005 **2.** 3.14 **3. a)** 7.0 **b)** 6.19 **c)** 3.10 **d)** 2.149 **e)** 0.100 **Note:** In You Try It 3a, 7.0 may not be simplified to 7; in 3c, 3.10 may not be simplified to 3.1; and in 3e, 0.100 may not be simplified to 0.1. **4. a)** \$1.62 **b)** \$0.70 **c)** \$251.05 **d)** \$57.00 **5. a)** 2 **b)** 4 **c)** 3 **d)** 17 **e)** 23 **6.** April's GPA is 3.20. **7.** Pedro should report \$1,280 on his tax form. **8.** Tuyn's dinner tip was \$3.60.

Think About It: **1.** The rounding digit must be in the ten thousandths place. **2.** Rounding to the nearest dollar is the same as rounding to the nearest whole number. Answers may vary.

Section 5.2 Exercises

FOR EXTRA HELP
Student Resources on DVD-ROM
Includes
➤ Student's Solutions Manual
➤ Video Lectures
➤ Chapter Test Prep Video
MyMathLab
MathXL

Think Again

Write a sentence or two for each response.

1. If we round a decimal to the nearest thousandth, in what place will the rounding digit be? *(Refer to Think About It 1)*
   The rounding digit must be in the ten-thousandths place.

2. When filling out tax forms, the Internal Revenue Service (IRS) requires that every money amount be rounded to the nearest dollar. If Billie's postage receipts total \$88.50, to what dollar amount should the number be rounded? Explain your answer.
   \$89. Even though \$88.50 is exactly between \$88.00 and \$89.00, the rounding digit, 5, causes the number to be rounded up to \$89.

Focus Exercises

Each number represents a runner's time in the 100-meter dash. Round each of these numbers to the nearest tenth.

3. 11.592 11.6 **4.** 13.158 13.2 **5.** 10.605 10.6 **6.** 9.9481 9.9

Each number represents a student's GPA. Round each of these numbers to the nearest tenth.

7. 2.036 2.0 **8.** 3.027 3.0 **9.** 2.958 3.0 **10.** 1.962 2.0

Each number represents a student's GPA. Round each of these numbers to the nearest hundredth.

11. 1.2618 1.26 **12.** 3.5947 3.59 **13.** 1.3855 1.39 **14.** 2.4092 2.41

Each number represents a vehicle's miles per gallon. Round each of these numbers to the nearest hundredth.

15. 14.0737 14.07 **16.** 25.0922 25.09

17. 18.04933 18.05 **18.** 31.06509 31.07

Each number represents a newborn's weight in kilograms. Round each of these numbers to the nearest hundredth.

19. 4.1983 4.20 **20.** 3.4971 3.50 **21.** 3.9953 4.00 **22.** 2.9982 3.00

Each number represents a newborn's length in meters. Round each of these numbers to the nearest hundredth.

23. 0.4618 0.46 **24.** 0.4129 0.41 **25.** 0.3873 0.39 **26.** 0.3984 0.40

Each number represents the number of acres a homeowner owns in a rural town. Round each of these numbers to the nearest thousandth.

27. 6.08394 6.084 **28.** 12.0407 12.041 **29.** 1.0096 1.010 **30.** 9.0803 9.080

Each number represents a baseball player's batting average. Round each of these numbers to the nearest thousandth.

31. .21618 0.216 **32.** .31908 0.319 **33.** .3396 0.340 **34.** .4099 0.410

35. .3165 0.317 **36.** .239904 0.240 **37.** .29973 0.300 **38.** .399185 0.399

Each number represents the amount of sales tax applied to a variety of purchases at a home electronics store. Round each dollar amount to the nearest penny.

39. $5.956 $5.96 **40.** $19.0862 $19.09 **41.** $14.0381 $14.04

42. $25.004 $25.00 **43.** $5.99015 $5.99 **44.** $21.9935 $21.99

45. $107.99701 $108.00 **46.** $44.99537 $45.00

Each number represents a 15% tip on a restaurant meal. Round each dollar amount to the nearest dime (tenth), but express the answer in dollars and cents.

47. $6.0085 $6.00 **48.** $15.055 $15.10 **49.** $3.7735 $3.80 **50.** $9.3915 $9.40

Each number represents the average number of SUVs owned per square mile in a variety of suburban areas. Round each of these numbers to the nearest whole number.

51. 31.3895 31 **52.** 48.0892 48 **53.** 82.5714 83

54. 53.61538 54 **55.** 214.9333 215 **56.** 2.5875 3

57. 0.86375 1 **58.** 15.02777 15

Each number represents a business expense that is to be entered on a tax form. Round each of these numbers to the nearest dollar.

59. $31.08 $31 **60.** $142.90 $143 **61.** $20.38 $20 **62.** $40.08 $40

63. $29.60 $30 **64.** $519.73 $520 **65.** $99.56 $100 **66.** $139.74 $140

For each statement, round the number as instructed and restate the first sentence using the rounded number.

67. Yolanda's GPA is 3.4708. Round this number to the nearest tenth.
Yolanda's GPA is 3.5.

68. Carl calculated his car's gas mileage to be 29.539 miles per gallon. Round this number to the nearest tenth.
Carl calculated his car's gas mileage to be 29.5 miles per gallon.

69. Alabama's population density is 88.6953 people per square mile. Round this number to the nearest hundredth.
Alabama's population density is 88.70 people per square mile.

70. Tyra's average monthly electric bill is $127.09533. Round this number to the nearest penny.
Tyra's average monthly electric bill is $127.10.

71. Hank Aaron's batting average in 1969 was .2998171. Round this number to the nearest thousandth.
Hank Aaron's batting average in 1969 was .300.

72. Jackson calculated his monthly insurance payment to be $89.47. Round this number to the nearest dollar.
Jackson calculated his monthly insurance payment to be $89.

Think Outside the Box

Answer each statement as indicated.

73. Round $17.496183 to the nearest dollar.
$17

74. Round $17.496183 to the nearest penny. Then round the result to the nearest dollar.
$17.50, which then rounds to $18

75. Compare the answers to Numbers 71 and 72. If they are different answers, explain why.
Answers may vary.

76. Is it appropriate to round a number that is itself a rounded number? Explain your answer.
Answers may vary.

SECTION 5.3 Adding and Subtracting Decimals

OBJECTIVES

In this section, you will learn to
- Add decimal numbers.
- Subtract decimal numbers.
- Add and subtract signed decimal numbers.

You Need to Know

To successfully complete this section, you need to understand
- ☐ Adding whole numbers (1.2)
- ☐ Subtracting whole numbers (1.2)
- ☐ Solving applications (1.8)
- ☐ Adding signed numbers (2.2)
- ☐ Subtracting signed numbers (2.3)

Introduction

Adding decimals is no different than adding whole numbers. When we are given a list of whole numbers to add, we organize them according to place value.

Example 1

Add 356 + 1,024 + 17 + 8.

Procedure Write the numbers so that the place values are in a line vertically (top to bottom).

Answer We can make the lining up of place values more obvious by including an appropriate number of zeros before each number. Doing so allows all of the numbers to have the same number of digits.

```
 356  →    0356
1024  →    1024
  17  →    0017
+  8  →  + 0008
           1405
```

Of course, we don't have to add whole numbers this way. However, placing zeros in front of whole numbers shows the necessity of adding by lining up the place values.

Adding Decimal Numbers

Lining up the place values is important when adding decimals as well. With decimals, though, we line up the place values *and* the decimal points. Also, we can place more zeros at the *end* of a number instead of at the beginning.

Example 2

Add 4.3 + 1.134 + 7.05 + 0.2861.

Procedure First, write the sum vertically and align the decimal points. Second, identify the number with the highest number of decimal places. The number 0.2861 has four decimal places; so, place enough zeros after the final digit of the other numbers to get four decimal places.

Answer

```
  4.3     — becomes →    4.3000
  1.134       →          1.1340
  7.05        →          7.0500
+ 0.2861      →        + 0.2861
                        12.7701
```

Is it necessary to write all of the ending zeros as shown? No, but it's good practice. It *is* necessary, though, to line up the decimal points; and it's important to write your numbers neatly so that all of the decimal place values line up correctly.

You Try It 1 **Add. Use Example 2 as a guide.**

a) 0.72 + 0.145 **b)** 5.18 + 0.3 + 4.216 **c)** 5.09 + 0.3 + 41.38 + 2.005

Subtracting Decimal Numbers

Subtracting decimals works the same as subtracting whole numbers. Sometimes you can subtract directly, and sometimes you need to regroup.

Also, when subtracting decimals,

1. You need to line up the decimal points one above the other.
2. You may need to place enough zeros after the last decimal place so that the two numbers have the same number of decimal places.

Example 3

Subtract.

a) 85.7 − 62.5 **b)** 3.49 − 0.3 **c)** 2.86 − 1.49

Procedure Place the first number above the second number and line up the decimal points. For part b, place a zero at the end of the second number. For part c, regroup.

Answer

a) 85.7 − 62.5 = 23.2

b) 3.49 − 0.30 = 3.19

c) 2.86 (regrouped: 7 16) − 1.49 = 1.37

You Try It 2 **Subtract. Some of these require regrouping. Use Example 3 as a guide.**

a) 78.9 − 32.8 **b)** 5.3 − 4.8 **c)** 4.358 − 1.3 **d)** 6.1 − 4.43

Think About It 1 Why, when subtracting decimals, is it important to line up the decimal points?

____

____

There are a few ways to write subtraction in words. One way is rather straightforward, *5 minus 3* is written 5 – 3. Likewise, *The difference of 5 and 3* is written 5 − 3.

Another way, however, seems to be written backwards. Although the numbers appear in the reverse order, *Subtract 3 from 5* also means **5 − 3.**

Example 4

Subtract.

a) Subtract 8.4 from 9.7. **b)** Subtract 4.62 from 9.

Procedure Align the numbers vertically. Place the second number above the first. In part b, make 9 into 9.00, align the decimal points, and regroup carefully before subtracting.

Answer

a) 9.7 − 8.4 = 1.3

b) 9.00 − 4.62 → 9.00 (regrouped: 8 10) − 4.62 → 9.00 (regrouped: 8 9 10) − 4.62 = 4.38

You Try It 3 **Evaluate each by first writing it in arithmetic form. Use Example 4 as a guide.**

a) Subtract 1.7 from 9.8. **b)** Subtract 5.28 from 6.2. **c)** Subtract 7.08 from 30.

Adding and Subtracting Signed Decimal Numbers

We use the same guidelines for adding and subtracting signed decimal numbers as we use for other signed numbers.

Adding and Subtracting Signed Decimal Numbers

1. Change subtraction to addition: Add the opposite.
2. If the two numbers have the same sign,
   a) Add their absolute values.
   b) Remember that the resulting sum will have that same sign.
3. If the two numbers have different signs,
   a) Subtract the two numerical values: larger value − smaller value.
   b) Remember that the sign of the sum is the same as the sign of the addend with the larger numerical value.

Keep those guidelines in mind when adding or subtracting signed decimals. Also, whether adding or subtracting decimals, it is best for each number to have the same number of decimal places.

Example 5

Evaluate.

a) $2.9 - 4.2$ b) $1.02 - (-0.38)$ c) $-0.047 + (-0.19)$

Procedure Change all subtraction (if any) to addition. Apply the guidelines to find the sum.

Answer

a) $2.9 - 4.2$ — **The numbers have the same number of decimal places.**
$= 2.9 + (-4.2)$ — **The signs are different; find the difference.** →
$= -1.3$ — **The larger-valued number is negative; so, the result is negative.**

$$\begin{array}{r} 4.2 \\ -\ 2.9 \\ \hline 1.3 \end{array}$$

b) $1.02 - (-0.38)$ — **The numbers have the same number of decimal places. Change subtraction to addition: 1.02 + (+0.38).**
$= 1.02 + (+0.38)$ — **The signs are the same, and they are both positive.**
$= 1.40$, or 1.4

$$\begin{array}{r} 1.02 \\ +\ 0.38 \\ \hline 1.40 \end{array}$$

c) $-0.047 + (-0.19)$ — **Write −0.19 as −0.190 so the numbers have the same number of decimal places.**
$= -0.047 + (-0.190)$ — **The signs are the same, and they are both negative.**
$= -0.237$

$$\begin{array}{r} 0.047 \\ +\ 0.190 \\ \hline 0.237 \end{array}$$

You Try It 4 **Evaluate. Use Example 5 as a guide.**

a) $0.27 + (-1.6)$ b) $-3.8 - 5.2$ c) $0.45 - (-0.29)$ d) $-0.08 + 0.059$

Applications Involving Adding and Subtracting Decimal Numbers

One easy way to demonstrate addition and subtraction of decimals is with money. In each of the following problems,

1. Decide what operation (addition or subtraction) to use.
2. Set up the problem, one number above another. Make sure you line up the decimal points correctly.
3. Place a decimal point and some zeros if needed and then apply the operation.

Example 6

At the grocery store, Ah-Ni buys the following items: milk at $2.89, eggs at $2.55, bacon at $4.78, and a bag of apples at $4. What is Ah-Ni's total bill?

Procedure Decide what operation to use. The word *total* suggests that we need to add the items together. As we set up the problem, we'll be sure to line up the decimal points and to include appropriate zeros.

Answer

Milk	$2.89
Eggs	$2.55
Bacon	$4.78
Apples	+ $4.00
Total	$14.22

Sentence Ah-Ni's total bill is $14.22.

Example 7

Ah-Ni's total grocery bill is $14.22. She pays for it with a $20 bill. How much money will she get back?

Procedure To determine the change that Ah-Ni will get back means that we subtract $14.22 from $20.00.

Answer Make sure you place a decimal point and two zeros after the 10.

```
  $20.00
−  14.22     We must regroup here, but it isn't shown.
  $ 5.78
```

Sentence Ah-Ni will get back 5.78.

You Try It 5 **At Holly's Café, Holly requires the servers to add the tabs by hand. (It gives an old-fashioned touch to the service.) One tab had the items shown at the right.**

What is the total tab for this breakfast?

Omelet	$5.45
Scrambled eggs	4.15
Coffee	1.65
Large juice	1.95
Sales tax	1.02

Sentence ______________________________

You Try It 6 **Tonya bought some school supplies from the college bookstore. The total, with tax, came to $9.15. Tonya paid with a $20 bill. How much change did Tonya get back?**

Show your work:

Sentence ____________________

Recall from Section 1.2 that the perimeter of a geometric figure is the sum of the lengths of the sides of the figure.

Example 8

Find the perimeter of this triangle.

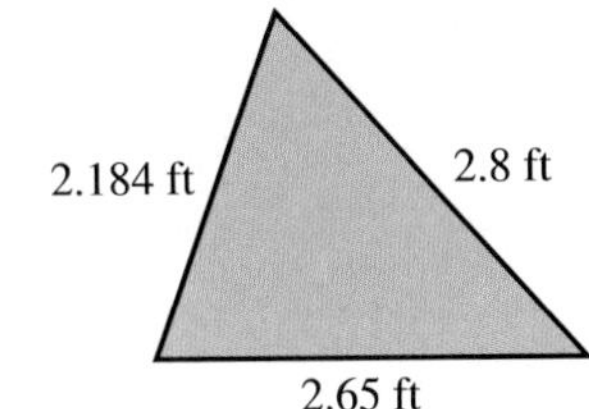

Procedure Add the side measures. (Make sure you line up the decimal points and place zeros as needed.)

Answer

$$\begin{array}{r} 2.184 \\ 2.800 \\ +\ 2.650 \\ \hline 7.634 \end{array}$$

Sentence The perimeter of this triangle is 7.634 feet.

You Try It 7 **Find the perimeter of each geometric figure. Use Example 8 as a guide.**

a) 2 m, 1.36 m, 0.7 m

b)

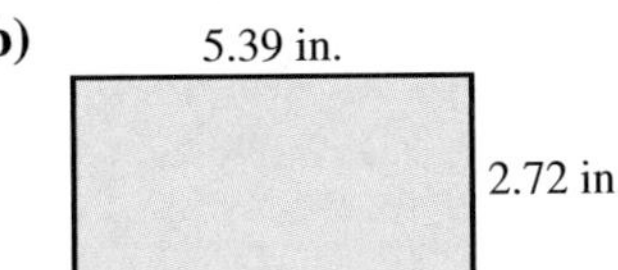

Sentence ____________________

Sentence ____________________

Applications Involving Signed Decimal Numbers

Recall from Section 2.3 a few situations in which signed numbers are common:

Finances: Borrowing money represents a debt, a negative amount. Credits and payments are positive amounts; debits (and purchases) are negative amounts. (The word *debit* is related to *debt.*)

Temperature: Even though an outside temperature of 5° (5 degrees) is cold, −5° is colder by 10°.

Altitude: Below sea level, the altitude is negative. Above sea level, the altitude is positive.

The next few examples and exercises make use of adding and subtracting integers. For each problem, we set up a numerical expression (using addition or subtraction) and then answer the question in a sentence.

Example 9

On her Sears card, Julia has a credit of $65.17 and makes a purchase of $103.45. What is the new balance of her account? Is this new balance a debit or a credit?

Procedure A credit is a positive number (+$65.17), and a purchase is a negative number (−$103.45) added to the account.

Answer $65.17 + (-103.45)$ Rearrange this with the larger-valued number in front.

$= -103.45 + 65.17$ The result is negative. Subtract.

$= -38.28$

$$\begin{array}{r} 103.45 \\ -\ 65.17 \\ \hline 38.28 \end{array}$$

Sentence The new balance is $-\$38.28$; this is a debit.

You Try It 8 **Solve this application. Write a numerical expression and write the answer in the form of a sentence. Use Example 9 as a guide.**

On his MasterCard, Marco has a credit balance of \$43.67. He used his MasterCard to buy groceries worth \$90.25. What is the new balance of Marco's account? Is it a debit or a credit?

Numerical Expression

Sentence ______________________

Example 10

Janine was experimenting with a fast-freezing technique that her company is developing. She started with water at a Celsius temperature of 15.8°. In the experiment, within 1 minute, the temperature fell 24.65°. What was the water temperature after 1 minute?

Procedure Temperature *falling* means *subtracting* the number of degrees it fell from the starting temperature. Apply the techniques used in this section to evaluate.

Answer Numerical expression: $15.8 - 24.65$

YOU FINISH IT: Subtract

Sentence After 1 minute, the temperature was $-8.85°$.

You Try It 9 **Solve this application. Write a numerical expression and write the answer in the form of a sentence. Use Example 10 as a guide.**

At 2 PM, the outside temperature was 3.75°. By 10 PM, the temperature had fallen 5.2°. What was the temperature at 10 PM?

Numerical Expression

Sentence ______________________

Example 11

Find the difference in altitude between a pier that is 2.7 meters above sea level and the ocean floor that is 4.59 meters below sea level.

Procedure To find the difference in altitude, subtract pier − ocean floor.

Answer Numerical expression: $+2.7 - (-4.59)$

YOU FINISH IT: Subtract

Sentence The difference in altitude is 7.29 meters.

You Try It 10 **Solve this application. Write a numerical expression and write the answer in the form of a sentence. Use Example 11 as a guide.**

Find the difference in altitude between the top of a lighthouse 23.6 meters above sea level and an undersea reef 40.56 meters below sea level.

Numerical Expression

Sentence ____________________

Applications and Equations

We can also solve application problems using the formula *The sum of all of the parts equals the whole* and these problem-solving guidelines

1. Read the problem once for general information. Then read it a second time looking for specific information.
2. Identify the parts and the whole and decide what is unknown.
3. Write the unknown in a legend (Let $x = \ldots$).
4. Place the known and unknown values into the formula *The sum of all of the parts equals the whole.*
5. Solve the equation and write a sentence answering the question.

Example 12

American Trophy Shop made engraved awards for the high school marching band's awards ceremony. Dale, the owner, charged a flat rate of $18 per award, including sales tax. If the sales tax on each award is $1.06, what is the actual price of the award?

Procedure In this case, the parts are the price of the award, which is unknown, and the amount of sales tax, $1.06. The whole is the flat charge of $18.

Answer Let $x =$ the price of the award.

$$x + 1.06 = 18.00$$

$$x + 1.06 + (-1.06) = 18.00 + (-1.06) \longrightarrow \begin{array}{r} 18.00 \\ -\ \ 1.06 \\ \hline 16.94 \end{array}$$

$$x = 16.94$$

Check

Award	16.94
+ Tax	+ 1.06
Total	18.00 ✓

Sentence The actual price of the award is $16.94.

You Try It 11 **Senia owns stock in Bilbay Limited. She has instructed her stockbroker to sell her stock when the price reaches $30 per share. Right now the stock's value is $27.58 per share. How much will the share value need to increase before Senia's stockbroker sells the stock?**

Legend

Sentence ____________________

You Try It 12 **Find the length of the third side of the triangle given the information in the diagram.**

Legend

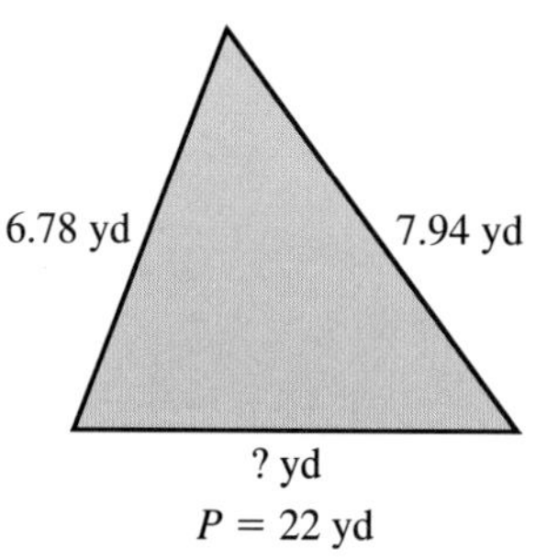

Sentence ______________________________

Answers: You Try It and Think About It

You Try It: **1. a)** 0.865 **b)** 9.696 **c)** 48.775 **2. a)** 46.1 **b)** 0.5 **c)** 3.058 **d)** 1.67 **3. a)** 8.1 **b)** 0.92 **c)** 22.92 **4. a)** −1.33 **b)** −9.0 or −9 **c)** 0.74 **d)** −0.021 **5.** The total tab is $14.22. **6.** Tonya got back $10.85 in change. **7. a)** The perimeter is 4.06 meters. **b)** The perimeter is 16.22 inches. **8.** 43.67 + (−90.25); Marco's new account balance is −$46.58; this is a debit. **9.** 3.75 − 5.2; The temperature at 10 PM was −1.45°. **10.** 23.6 − (−40.56); The difference in altitude is 64.16 meters. **11. Legend:** Let x = the amount of increase in the share value. **Sentence:** The share value will need to increase $2.42 before Senia's stockbroker sells the stock. **12. Legend:** Let x = the length of the third side. **Sentence:** The third side is 7.28 yards long.

Think About It: **1.** If we line up the decimal points, then all of the other places will align as well.

Section 5.3 Exercises

FOR EXTRA HELP
Student Resources on DVD-ROM
Includes
➤Student's Solutions Manual
➤Video Lectures
➤Chapter Test Prep Video
MyMathLab
MathXL

Think Again

Write a sentence or two for each response.

1. Recall from Section 1.3 that multiplication is an abbreviation for repeated addition. Use this idea to evaluate 6 × 2.4.
2.4 + 2.4 + 2.4 + 2.4 + 2.4 + 2.4 = 14.4

2. Recall from Section 1.4 that division is an abbreviation for repeated subtraction. Use this idea to evaluate 7.2 ÷ 1.8.
7.2 − 1.8 = 5.4; 5.4 − 1.8 = 3.6; 3.6 − 1.8 = 1.8; 1.8 − 1.8 = 0. Because we can subtract 1.8 from 7.2 exactly four times, 7.2 ÷ 1.8 = 4.

Focus Exercises

Add.

3. 0.308 + 0.52
0.828

4. 0.73 + 0.4096
1.1396

5. 0.1 + 0.96 + 0.441
1.501

6. 0.07 + 0.6 + 0.559
1.229

Add.

7. Add 0.56 and 0.23.
0.79

8. Add 2.12 and 2.46.
4.58

9. Add 12.89 and 7.52.
20.41

10. Add 10.67 and 5.94.
16.61

11. Add 0.83 and 0.3.
1.13

12. Add 2.76 and 2.6.
5.36

13. Add 11.33 and 8.67.
20

14. Add 0.29 and 0.71.
1

15. Add 2.686 and 1.33.
4.016

16. Add 4.825 and 11.23.
16.055

17. Add 6.31 and 3.696.
10.006

18. Add 14.58 and 5.429.
20.009

19. Add 13.9, 10.018, and 2.86.
26.778

20. Add 0.552, 0.03, and 0.8.
1.382

21. Add 6, 14.3, 1.047, and 1.09.
22.437

22. Add 7.635, 3.24, 22, and 10.3.
43.175

Subtract.

23. $9.78 - 2.7$
7.08

24. $54.52 - 13$
41.52

25. $8.253 - 2.8$
5.453

26. $18.389 - 14.26$
4.129

27. $9.2 - 3.8$
5.4

28. $8.5 - 5.7$
2.8

29. $5.03 - 4.48$
0.55

30. $6.25 - 3.09$
3.16

31. $6.2 - 0.79$
5.41

32. $17.3 - 12.92$
4.38

33. $8 - 5.028$
2.972

34. $4 - 2.408$
1.592

Evaluate each by first writing it in arithmetic form.

35. Subtract 2.2 from 3.8.
1.6

36. Subtract 13.5 from 15.3.
1.8

37. Subtract 0.79 from 6.5.
5.71

38. Subtract 4.03 from 8.1.
4.07

39. Subtract 7.591 from 16.9.
9.309

40. Subtract 9.477 from 13.6.
4.123

41. Subtract 7.53 from 13.
5.47

42. Subtract 7.26 from 28.
20.74

Evaluate.

43. $-3.6 + 6.4$
2.8

44. $-6 + 0.52$
-5.48

45. $-0.53 + (-0.37)$
-0.9

46. $-0.16 + (-0.3)$
-0.46

47. $2.2 + (-0.46)$
1.74

48. $0.81 + (-0.75)$
0.06

49. $5 + (-1.97)$
3.03

50. $5 - 9.86$
-4.86

51. $-2 + (-6.4)$
-8.4

52. $-5.8 + (-3.9)$
-9.7

53. $0.28 - 0.72$
-0.44

54. $1.2 - 7.8$
-6.6

55. $3.2 - (-2.4)$
5.6

56. $0.19 - (-0.48)$
0.67

57. $-9.1 - (-5.2)$
-3.9

58. $-0.41 - (-0.62)$
0.21

59. $-1.7 - (-1.7)$
0

60. $-0.96 - (-0.58)$
-0.38

Combine like terms.

61. $-2.8x + 5.2x$
$2.4x$

62. $-1.2y + 4.83y$
$3.63y$

63. $-1.47w + (-0.68w)$
$-2.15w$

64. $-3.08c + (-1.4c)$
$-4.48c$

65. $5.25h + (-7.03h)$
$-1.78h$

66. $0.91v + (-1.36v)$
$-0.45v$

67. $3.8p - 7.12p$
$-3.32p$

68. $0.75d - 2.1d$
$-1.35d$

69. $2.74w - (-1.6w)$
$4.34w$

70. $3.07x - (-2.9x)$
$5.97x$

71. $-4.68y - (-3.91y)$
$-0.77y$

72. $-6.2w - (-4.55w)$
$-1.65w$

Solve each application by adding or subtracting appropriately. Answer the question with a sentence.

73. Marc needed to purchase items for his desk in his office. Looking through the Staples catalog, he came up with the following items and their prices. What is the total?

Item	Price
Stapler	$27.99
Staples	4.52
Sticky notes	5.96
Paper clips	0.72
Sales tax	2.55

The total is $41.74.

74. At the Summer Olympics, Kari and Jenn received these scores for their synchronized swimming routine. What was their total score for the event?

Judge	Score
Germany	9.41
France	9.55
Spain	9.44
USA	9.78
Russia	9.27

Their total score was 47.45.

75. For their annual Thanksgiving basket donations, members of the Kiwanis Club bought five turkeys. The turkeys weighed 16.89 pounds, 18.18 pounds, 22.06 pounds, 25.23 pounds, and 19.52 pounds. What was the combined weight of the five turkeys?
The combined weight was 101.88 pounds.

76. Carmen purchased doughnuts for her office. The total came to $6.19 including tax. She paid with a $10 bill. How much change did Carmen get back?
Carmen got back $3.81 in change.

77. At the Summer Olympic trials, Sean's time in the 400-meter race was 42.36 seconds and Torii's time was 44.03 seconds. How many seconds behind Sean was Torii?
Torii was 1.67 seconds behind Sean.

78. In March 2006, the price of regular gasoline was \$2.539. Two months later the price was \$3.079. What was the increase in the price per gallon of regular gasoline in those two months?
The increase in price per gallon of regular gasoline was \$0.54.

Find the perimeter of each geometric figure.

79.

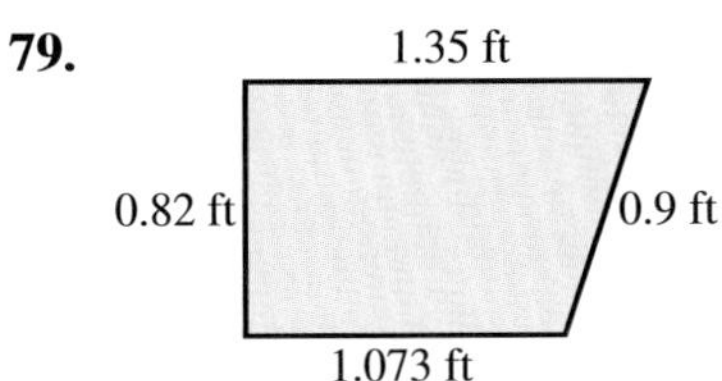

Perimeter = 4.143 feet

80.

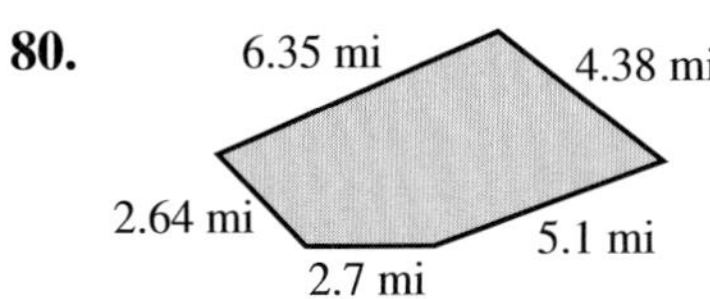

Perimeter = 21.17 miles

For each application, write a numerical expression and then evaluate. Also write a sentence answering the question.

Finances.

81. On her Visa card, Daneice has a debit of \$57.82. She likes to stay ahead by making larger-than-needed payments. Her most recent payment was for \$100. What is Daneice's new account balance? Is this a debit or a credit?
$-57.82 + 100$
Daneice's new account balance is \$42.18.
This is a credit.

82. Arnie had \$38.16 in his checking account when he paid his phone bill and wrote a check for \$52.94. What is Arnie's new account balance? Is this a debit or a credit?
$38.16 - 52.94$
Arnie's new account balance is −\$14.78.
This is a debit.

Temperature.

83. At 2 AM, the outside temperature was −8.4°. By noon, the temperature had risen 15.25°. What was the temperature at noon?
$-8.4 + 15.25$
The temperature at noon was 6.85°.

84. The temperature of a snowball at 10 AM was −3.85° Celsius. Manny put the snowball in the freezer; and by noon, its temperature had fallen 5.9° Celsius. What was the temperature of the snowball at noon?
$-3.85 - 5.9$
The temperature of the snowball at noon was −9.75°C.

Altitude.

85. Find the difference in altitude between a lifeguard tower 4.25 meters above sea level and the underwater beach 1.8 meters below sea level.
$4.25 - (-1.8)$
The difference in altitude is 6.05 meters.

86. Find the difference in altitude between the bottom of a boat 2.65 meters below sea level and a reef 23.9 meters below sea level.
$-2.65 - (-23.9)$
The difference in altitude is 21.25 meters.

For the following applications, set up a legend and an equation. Then solve to answer the question.

87. At the beginning of her diet, Maeve wanted to lose 40 pounds. Because of good eating habits and exercise, she has lost 23.4 pounds so far. How many more pounds does Maeve need to lose to reach her goal?
Let x = the number of pounds Maeve still needs to lose. $23.4 + x = 40$; Maeve needs to lose 16.6 more pounds to reach her goal.

88. Toby swam two laps in a 50-meter race with an overall time of 39.71 seconds. The time on his first lap was 20.84 seconds. What was the time on Toby's second lap?
Let x = the time on Toby's second lap.
$20.84 + x = 39.71$; The time on Toby's second lap was 18.87 seconds.

89. LeRon agreed to pay back his dad for the running shoes he bought last week. The total cost, including tax, was \$107.52. This week he gave his dad his full paycheck of \$79.48. How much does LeRon still owe his dad?
Let x = the amount LeRon still owes his dad.
$79.48 + x = 107.52$; LeRon still owes his dad \$28.04.

90. MapQuest indicated that the distance from Terry's house to his new workplace is 13.24 miles. He also found that the distance from his home to a coffee cart on his route is 5.85 miles. How far is it from the coffee cart to Terry's workplace?
Let x = the distance from the coffee cart to Terry's workplace. $5.85 + x = 13.24$; It is 7.39 miles from the coffee cart to Terry's workplace.

91. Find the length of the third side of the triangle given the information in the diagram.

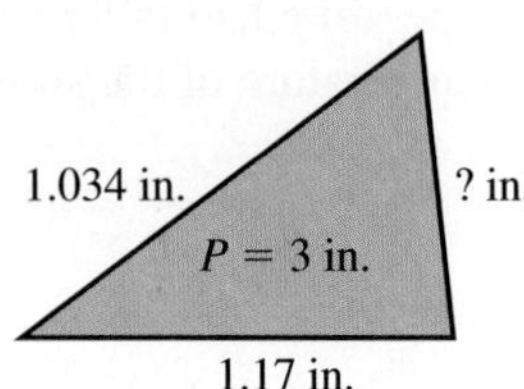

Let x = the length of the third side.
$1.034 + 1.17 + x = 3$; The third side is 0.796 inches.

92. Find the length of the third side of the triangle given the information in the diagram.

1.79 m
? m
$P = 7$ m
2.4 m

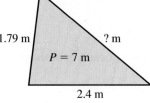

Let x = the length of the third side.
$2.4 + 1.79 + x = 7$; The third side is 2.81 meters.

Think Outside the Box

Evaluate.

93. $3.16 - (0.4 - 5.91)$
8.67

94. $2.8 - [-6.1 - (-4.77)]$
4.13

95. Following is a copy of one incorrect problem from Samantha's decimals quiz. Identify the mistake Samantha made and explain how to correct it.

Subtract: $7.481 - 3.6$

$$\begin{array}{r} 7.481 \\ -\quad 3.6 \\ \hline 7.445 \end{array}$$

(−3)

Answers may vary.

96. Following is a copy of one incorrect problem from Damon's decimals quiz. Identify the mistake Damon made and explain how to correct it.

Subtract: $5.8 - 2.549$

$$\begin{array}{r} 5.8 \\ -\ 2.549 \\ \hline 3.349 \end{array}$$

(−3)

Answers may vary.

SECTION 5.4 Multiplying Decimals

OBJECTIVES

In this section, you will learn to

- Multiply a decimal by a decimal.
- Multiply a decimal by a whole number.
- Multiply a decimal by a power of 10.

You Need to Know

To successfully complete this section, you need to understand

- ☐ Multiplying whole numbers (1.3)
- ☐ Multiplying fractions (4.4)
- ☐ Writing decimals as fractions (5.1)

Introduction

This is a true story, although the names have been changed to protect the innocent.

> Lana went to a small specialty shop to buy a music CD for a friend. While she was there, the power went out and the cash register would not work. Neither the employee nor the manager knew how to calculate sales tax; so, they didn't want to sell the CD to Lana. Lana, the owner of her own retail shop, showed them how to find the sales tax by multiplying the price of the CD, \$15.95, by the sales tax rate, 0.08.

The moral of the story is this: Don't be totally dependent on machines to do all of the work for you. Learn to know what the machine is doing so that you can do it on your own should you ever need to.

Multiplying Decimal Numbers

We know that a decimal can be written as a decimal fraction. For example, 0.3 can be written as $\frac{3}{10}$ and 0.07 can be written as $\frac{7}{100}$.

This concept is helpful when we multiply two decimal numbers. The advantage of writing decimals as fractions is that the numerators are whole numbers and the denominators are powers of 10. We will use this concept to develop a consistent procedure for multiplying decimals.

Consider the product 0.3×0.07. Because each decimal can be written as a fraction, we can multiply using the product rule for fractions.

$$0.3 \times 0.07 = \frac{3}{10} \times \frac{7}{100} = \frac{21}{1{,}000} = 0.021$$

Example 1

Multiply 0.06×0.8 by rewriting each decimal as a fraction.

Answer $0.06 \times 0.8 = \frac{6}{100} \times \frac{8}{10} = \frac{48}{1{,}000} = 0.048$

You Try It 1 **Multiply by rewriting each decimal as a fraction. Multiply the fractions, then convert the answer back into a decimal. Use Example 1 as a guide.**

a) $0.5 \times 0.7 =$ ______________________

b) $0.1 \times 0.8 =$ ______________________

c) $0.04 \times 0.6 =$ ______________________

d) $0.11 \times 0.09 =$ ______________________

Consider 0.8×0.21. Let's think about what we see here.

$$0.8 \times 0.21 = \frac{8}{10} \times \frac{21}{100} = \frac{168}{1{,}000} = 0.168$$

In this product, whether we use fractions or decimals, we are multiplying tenths by hundredths and getting thousandths.

In each factor, if we count the number of zeros in each denominator, we will know the number of denominator zeros in the end result.

Fractions:

Tenths × hundredths = thousandths

$$\frac{8}{10} \times \frac{21}{100} = \frac{168}{1{,}000}$$

One zero, *Two* zeros, *Three* zeros

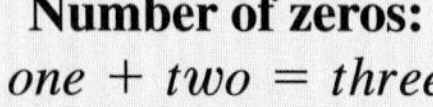

Number of zeros:
one + two = three

In each factor, if we count the number of decimal places in each number, we will know the number of decimal places in the end result.

Decimals:

Tenths × hundredths = thousandths

$$0.8 \times 0.21 = 0.168$$

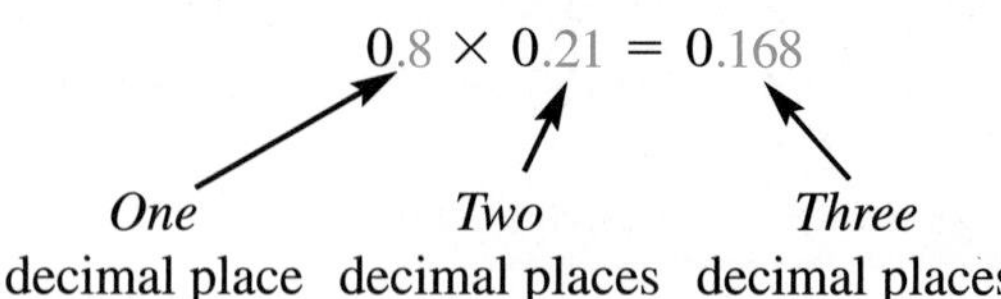

One decimal place, *Two* decimal places, *Three* decimal places

Number of decimal places:
one + two = three

The fractions show us that we are simply multiplying whole numbers 8 × 21. The decimals show us that the answer is a decimal number. The question is, after we multiply 8 × 21 = 168, where do we place the decimal point?

The previous diagrams show that we can count the number of decimal places in the factors and the result will have that total number of decimal places.

Multiplying Two Decimal Numbers

Multiplying decimals is the same as multiplying whole numbers, but we must

1. Temporarily ignore the decimal points and multiply the numbers (factors) as if they were both whole numbers.
2. Count the total number of decimal places in the factors. This total is the number of decimal places in the product (before any simplifying).

Let's practice identifying the number of decimals, as in part 2 of the procedure.

Example 2

Given the following multiplication, decide how many decimal places the product (the result) will have. *Do not multiply.*

a) 0.3 × 5 **b)** 1.2 × 0.4 **c)** 0.16 × 5.3 **d)** 1.25 × 0.12

Procedure Count the total number of decimal places in both numbers. This is the number of decimal places the answer will have.

Answer

a) 0.3 × 5 — 0.3 has one decimal place and 5 has none; so, their product will have a total of one decimal place.

b) 1.2 × 0.4 — 1.2 has one decimal place and 0.4 has one; so, their product will have a total of two decimal places.

c) 0.16 × 5.3 — 0.16 has two decimal places and 5.3 has one; so, their product will have a total of three decimal places.

d) 1.25 × 0.12 — 1.25 has two decimal places and 0.12 has two; so, their product will have a total of four decimal places.

You Try It 2 **Given the following multiplication, decide how many decimal places the product (the result) will have. *Do not multiply*. Use Example 2 as a guide.**

a) 0.45 × 1.2 **b)** 6 × 4.3 **c)** 1.08 × 3.04 **d)** 4.3 × 2.6

Example 3

Multiply. **a)** 13 × 32 **b)** 1.3 × 3.2

Answer

a)

```
   13
 × 32
   26
+ 390
  416
```

b)

```
  1.3
× 3.2
   26
+ 390
 4.16
```

Temporarily ignore the decimal points and multiply as if the decimals were whole numbers. The two decimal places appear in the end result only.

Example 4

Multiply these decimal numbers. (In these, when the whole number is 0, we usually don't multiply it.)

a) 0.3 × 5 **b)** 1.2 × 0.4 **c)** 0.17 × 5.1 **d)** 0.63 × 2.08

Procedure Write the numbers with decimals in place. Temporarily ignore the decimal points and multiply as you would with whole numbers. When you are finished, count up the number of decimals in each product.

Answer

a)

```
  0.3    One decimal place
×   5    No decimal places
  1.5    One decimal place
```

b)

```
  1.2    One decimal place
× 0.4    One decimal place
 0.48    Two decimal places
```

c)

```
  0.17   Two decimal places
× 5.1    One decimal place
    17   ↓
+  850   ↓
 0.867   Three decimal places
```

d)

```
   0.63   Two decimal places
 × 2.08   Two decimal places
    504   ↓
+ 12600   ↓
 1.3104   Four decimal places
```

You Try It 3 **Follow the procedure for multiplying decimals to find the product. Use Examples 3 and 4 as guides.**

a) 0.9 × 7 **b)** 0.8 × 0.4 **c)** 0.17 × 0.6

d) 1.8 × 0.12 **e)** 0.58 × 0.71

Think About It 1 When adding and subtracting decimals, it is important to line up the decimal points, but it is not necessary to do that when multiplying decimals. Why not?

____

____

Caution Two problems in the next example involve working with 0 in one way or another. Be careful!

Example 5

Multiply. **a)** 0.16 × 0.3 **b)** 0.25 × 0.72

Procedure Follow the procedure in Example 4, but notice the following:

Part a will require three decimal places, but there will be only two digits in the product. A third digit will be required before we can place the decimal point in the answer. That third digit is a 0 in front of the answer.

Part b will have enough decimal places, but there will be a few extra zeros at the end of the number that we can eventually eliminate.

Answer

a)

0.16	Two decimal places
× 0.3	One decimal place
0.048	Three decimal places

0.16 × 0.3 = 0.048

b)

0.25	Two decimal places
× 0.72	Two decimal places
50	↓
+ 1750	↓
0.1800	Four decimal places

We can simplify 0.1800 to 0.18.

0.25 × 0.72 = 0.18

Caution We cannot eliminate any ending zeros until after the decimal point is placed in the end result.

You Try It 4 **Follow the procedure for multiplying decimals to find the product. Use Example 5 as a guide.**

a) 0.4 × 0.2 **b)** 0.48 × 0.5 **c)** 0.06 × 0.75

Multiplying Decimal Numbers by Powers of 10

Multiplying a decimal by 10 or by a larger power of 10—such as 100 or 1,000—has the same effect as changing the position of the decimal point in the number. In other words, multiplying a decimal number by a power of 10 moves the decimal point a certain number of places to the right. In particular, multiplying a decimal number

- By 10 moves the decimal point one place to the right.
- By 100 moves the decimal point two places to the right.
- By 1,000 moves the decimal point three places to the right.

To understand why, consider that a decimal number such as 9.413 can be written as a decimal fraction, $\frac{9{,}413}{1{,}000}$. When such a number is multiplied by a power of 10, such as 100, or $\frac{100}{1}$, we can simplify the powers of 10.

$$9.413 \times 100 = \frac{9{,}413}{1{,}000} \times \frac{100}{1} = \frac{9{,}413 \cdot 100}{1{,}000 \cdot 1} = \frac{9{,}413 \cdot 1}{10 \cdot 1} = \frac{9{,}413}{10} = 941.3$$

Divide out a common factor of 100.

In other words, multiplying by 100 has the effect of moving the decimal place two places to the right: **9.413 × 100 = 941.3**.

Here are three examples showing the moving of the decimal point directly.

a) 3.6 × 10 = 36

3.6. = 36

Multiplying by 10, which has only one zero, has the effect of moving the decimal point of 3.6 one place to the right. The product becomes the whole number 36 without the need of a decimal point.

b) $5.24 \times 10 = 52.4$ Again, there is one zero in 10, which has the effect of moving the decimal point of 5.24 one place to the right. This time the decimal point is necessary because 52.4 is still a decimal number.

$5.2.4 = 52.4$

c) $5.9 \times 1{,}000 = 5{,}900$ Multiplying by 1,000 suggests that we need to move the decimal point in 5.9 three places to the right.

$5.900 \times 1{,}000 = 5{,}900$ However, because there aren't enough decimal places in 5.9, we must place some zeros at the end of it so that we can move the decimal point there.

$5.900. = 5{,}900$

Here is the procedure that results from all of this.

Multiplying a Decimal Number by a Power of Ten

1. Count the number of zeros in the power of ten.
2. Move the decimal point that many places to the right.

You Try It 5 **Multiply by moving the decimal point the appropriate number of places to the right.**

a) 8.1×10 **b)** 0.764×100 **c)** 14.6×100

d) $0.0027 \times 1{,}000$ **e)** $0.09 \times 1{,}000$

Multiplying decimals by powers of 10 will prove useful in understanding dividing decimals (Section 5.5) and working with percents (Section 6.4).

Applications

Sales tax is calculated by multiplying the price of an item (usually dollars and cents) by the sales tax rate (a decimal). After the sales tax is calculated, it is rounded to the nearest penny (hundredth) and added to the price of the item.

Example 6 refers to the true story at the beginning of this section.

Example 6

Lana wants to buy a music CD that costs $15.95. If the sales tax rate is 0.08, how much will Lana pay for the CD?

Procedure Calculate the sales tax for the CD and round it to the nearest penny. Then add the sales tax to the cost of the CD to find the amount Lana must pay.

Answer

$$\begin{array}{r} 15.95 \\ \times\ 0.08 \\ \hline 1.2760 \end{array}$$

Round this sales tax to the nearest penny, $1.2760 ≈ $1.28. Add this to the cost.

$$\begin{array}{r} \$15.95 \\ +\ 1.28 \\ \hline \$17.23 \end{array}$$

Sentence Lana will pay a total of $17.23 for the CD.

You Try It 6 **Tim is buying a baseball glove for his son for Little League. The cost of the glove is $34.95, and the tax rate is 0.06. How much in total will Tim pay for the glove?**

Show your work:

Sentence ______________________________

Many more applications involving multiplying decimals appear in Section 5.6.

Answers: You Try It and Think About It

You Try It: **1. a)** $\frac{5}{10} \times \frac{7}{10} = \frac{35}{100} = 0.35$ **b)** $\frac{1}{10} \times \frac{8}{10} = \frac{8}{100} = 0.08$ **c)** $\frac{4}{100} \times \frac{6}{10} = \frac{24}{1{,}000} = 0.024$ **d)** $\frac{11}{100} \times \frac{9}{100} = \frac{99}{10{,}000} = 0.0099$ **2. a)** Three **b)** One **c)** Four **d)** Two **3. a)** 6.3 **b)** 0.32 **c)** 0.102 **d)** 0.216 **e)** 0.4118 **4. a)** 0.08 **b)** 0.240 = 0.24 **c)** 0.0450 = 0.045 **5. a)** 81 **b)** 76.4 **c)** 1,460 **d)** 2.7 **e)** 90 **6.** Tim will pay a total of $37.05 for the glove.

Think About It: **1.** Answers may vary. One possibility: When we multiply decimals, we first treat them like whole numbers. This means that the position of the decimal points don't matter while we are multiplying.

Section 5.4 Exercises

FOR EXTRA HELP
Student Resources on DVD-ROM
Includes
➤Student's Solutions Manual
➤Video Lectures
➤Chapter Test Prep Video
MyMathLab
MathXL

Think Again

Write a sentence or two for each response.

1. When adding and subtracting decimals, it is important to line up the decimal points, but it is not necessary to do that when multiplying decimals. Why not? *(Refer to Think About It 1)*
Answers will vary. One possibility: When multiplying decimals, we ignore the decimal point until the end of the multiplying.

2. Can we multiply a terminating decimal by a repeating decimal? Explain your answer or show an example that supports your answer.
No, not in their decimal forms. We cannot get to the end of the repeating decimal to start the multiplication process.

Focus Exercises

Multiply by rewriting each decimal as a fraction. Multiply the fractions; then write the answer as a decimal.

3. 0.7×6 4.2
4. 0.3×0.8 0.24
5. 1.5×0.4 0.6
6. 1.1×0.9 0.99
7. 1.3×0.05 0.065
8. 0.03×6 0.18

Given the following multiplication, decide how many decimal places the product (the result) will have. Do not multiply.

9. 0.9×3 one
10. 8×0.2 one
11. 0.7×0.6 two
12. 0.1×0.8 two
13. 1.4×0.25 three
14. 3.9×0.86 three
15. 0.05×0.14 four
16. 9.15×1.07 four

Multiply by following the procedure for multiplying decimals.

17. 8×0.4 3.2
18. 5×0.9 4.5
19. 0.3×7 2.1
20. 0.2×2 0.4
21. 0.9×0.7 0.63
22. 0.6×0.8 0.48
23. 0.3×1.5 0.45
24. 3.5×0.4 1.40 or 1.4
25. 9.1×0.6 5.46
26. 0.1×7.2 0.72
27. 1.4×1.1 1.54
28. 5.3×0.1 0.53
29. 0.18×0.5 0.09
30. 0.15×0.9 0.135
31. 0.3×0.2 0.06
32. 0.2×0.14 0.028
33. 0.03×0.5 0.015
34. 0.05×0.4 0.02
35. 0.01×0.9 0.009
36. 0.01×0.7 0.007
37. 0.12×0.04 0.0048
38. 0.15×0.09 0.0135
39. 0.041×0.07 0.00287
40. 0.032×0.06 0.00192
41. 0.05×0.03 0.0015
42. 0.01×0.02 0.0002
43. 9.04×0.05 0.452
44. 2.03×0.05 0.1015
45. 1.25×1.5 1.875
46. 5.25×1.9 9.975
47. 4.25×5.7 24.225
48. 3.25×4.5 14.625
49. 2×0.034 0.068
50. 1.6×1.2 1.92
51. 2.1×4.13 8.673
52. 13.04×0.17 2.2168

Use your knowledge of multiplying signed numbers along with the techniques of multiplying decimals to find each product.

53. -5×0.9 -4.5
54. -6×0.2 -1.2
55. -0.4×0.8 -0.32
56. -0.7×0.6 -0.42
57. $2.5 \times (-0.4)$ -1
58. $4.5 \times (-0.8)$ -3.6

59. $6.5 \times (-0.08)$
-0.52

60. $7.5 \times (-0.04)$
-0.3

61. $-12.5 \times (-1.6)$
20

62. $-6.25 \times (-0.8)$
5

63. $-10.4 \times (-0.05)$
0.52

64. $-2.25 \times (-0.04)$
0.09

Multiply by moving the decimal point the appropriate number of places to the right.

65. 11.9×10
119

66. 6.1×10
61

67. 0.8×10
8

68. 0.1×10
1

69. 9.06×10
90.6

70. 5.07×10
50.7

71. 3.76×100
376

72. 9.41×100
941

73. 0.08×100
8

74. 0.09×100
9

75. 1.206×100
120.6

76. 8.602×100
860.2

77. 4.8×100
480

78. 31.5×100
3,150

79. $0.401 \times 1{,}000$
401

80. $3.159 \times 1{,}000$
3,159

81. $8.05 \times 1{,}000$
8,050

82. $1.56 \times 1{,}000$
1,560

83. $1.3 \times 1{,}000$
1,300

84. $6.2 \times 1{,}000$
6,200

Solve. Round each dollar in your answer to the nearest penny. Answer each question with a complete sentence.

85. In Alabama, the sales tax rate is 0.04. How much sales tax is there on a book that costs $19.50?
There is $0.78 sales tax on the book.

86. In Michigan, the sales tax rate is 0.06. How much sales tax is there on a shirt that costs $22.90?
There is $1.37 sales tax on the shirt.

87. In Minnesota, the sales tax rate is 0.065. How much sales tax is there on a computer printer that costs $104?
There is $6.76 sales tax on the computer printer.

88. In Oklahoma, the sales tax rate is 0.045. How much sales tax is there on a bicycle that costs $252?
There is $11.34 sales tax on the bicycle.

89. At a hardware store, Arash is buying a pack of batteries priced at $24.95. If the sales tax rate is 0.07, what is the total amount for the batteries?
The total amount for the batteries is $26.70.

90. At a music and video store, Linda is purchasing a DVD collection priced at $53.90. If the sales tax rate is 0.065, what is the total amount Linda must pay for the DVD collection?
The total amount Linda must pay for the DVD collection is $57.40.

91. Janelle is buying a used Toyota Camry for $7,600. If the sales tax rate is 0.075, what is the total amount Janelle must pay for the Camry?
The total amount Janelle must pay for the Camry is $8,170.

92. Mark wants to buy a Tony Gwynn autographed baseball card priced at $120. If the sales tax rate is 0.0625, what is the total amount Mark must pay for the card?
The total amount Mark must pay for the card is $127.50.

93. Carlos earns $7.45 per hour. Yesterday he worked 6.5 hours. How much did Carlos earn yesterday?
Carlos earned $48.43 yesterday.

94. Terra earns $8.17 per hour. Yesterday she worked 7.5 hours. How much did Terra earn yesterday?
Terra earned $61.28 yesterday.

95. Oyuki earns $9.60 per hour. Last week she worked 37.25 hours. How much did Oyuki earn last week?
Oyuki earned $357.60 last week.

96. JonRey earns $10.40 per hour. Last week he worked 32.75 hours. How much did JonRey earn last week?
JonRey earned $340.60 last week.

Think Outside the Box

Write a sentence that answers the question.

97. Terrence works for a company that pays each employee a standard hourly wage for the first 40 hours worked each week and 1.25 times the standard wage for each hour worked beyond 40 hours in one week. Terrence's standard wage is $9.60 per hour. If Terrence worked 48 hours last week, how much did he earn for the week?
Terrence earned $480 for the week.

98. Molly is buying a blanket and two pillows for her guest bedroom. The price of the blanket is $22.95, the price of each pillow is $10.95, and the sales tax rate is 0.07. If Molly pays for her whole purchase with two $20 bills and a $10 bill, how much change will she receive after the purchase?
Molly will receive $2.01.

SECTION 5.5 Dividing Decimals

OBJECTIVES

In this section, you will learn to

- Divide decimal numbers by whole numbers.
- Divide decimal numbers by decimals.
- Write fractions as decimal numbers.
- Divide decimal numbers by powers of 10.

You Need to Know

To successfully complete this section, you need to understand

- ☐ Dividing whole numbers (1.4)
- ☐ Repeating decimals (5.1)
- ☐ Terminating decimals (5.1)
- ☐ Writing decimal fractions as decimals (5.1)
- ☐ Multiplying decimals by powers of 10 (5.4)

Introduction

We know that 40 ÷ 4 = 10. What about 36 ÷ 4?

If we think about it, 36 is a little less than 40. So, the quotient of 36 ÷ 4 should be a little less than 10; and it is: 36 ÷ 4 = 9.

Likewise, the quotient of 52 ÷ 4 should be a little *more* than 10 because 52 is a little more than 40.

In fact, 52 ÷ 4 = 13. ⟶

```
   13
4)52
 - 4
   12
 - 12
    0
```

We can check this by multiplying:

```
  13
×  4
  52
```

We also know that 4 ÷ 4 = 1. What can we say about 5.2 ÷ 4? Using the same reasoning, because 5.2 is a little more than 4, the quotient of 5.2 ÷ 4 should be a little bit more than 1; and it is: 5.2 ÷ 4 = 1.3.

The point is this: When we divide a decimal number by a whole number, the quotient will be a decimal number.

```
   1.3
4)5.2
 - 4
   12
 - 12
    0
```

We can check this by multiplying:

```
  1.3
×   4
  5.2
```

Dividing Decimal Numbers by Whole Numbers

Dividing Decimal Numbers by Whole Numbers Using Long Division

1. Place a decimal point in the quotient directly above the decimal point in the dividend.
2. Divide as if each were a whole number.
3. Place zeros at the end of the dividend if needed.

Reminder:

```
        quotient
divisor)dividend
```

Example 1

Divide using long division: 6.52 ÷ 4.

Procedure When the divisor, 4, is a whole number and the dividend (6.52) has a decimal point in it,

1. We place a decimal point in the quotient directly above the decimal point in the dividend.
2. We ignore the decimal point and divide as if each were a whole number.

Answer

$$\begin{array}{r} 1.63 \\ 4\overline{)6.52} \\ -4 \\ \hline 25 \\ -24 \\ \hline 12 \\ -12 \\ \hline 0 \end{array}$$

Notice that the decimal point in the quotient is directly above the decimal point in the dividend.

Check by multiplying:

$$\begin{array}{r} 1.63 \\ \times4 \\ \hline 6.52 \end{array}$$

$6.52 \div 4 = 1.63$

You Try It 1 **Divide using long division. Use Example 1 as a guide.**

a) $6.92 \div 4$ **b)** $10.482 \div 6$ **c)** $50.4 \div 8$

Recall from Section 1.4 that if the divisor doesn't divide into the first digit, we can place a 0 above the first digit and continue to divide into the first two digits and so on.

Example 2

Divide using long division.

a) $0.741 \div 3$ **b)** $\dfrac{0.354}{6}$

Procedure

a)
$$\begin{array}{r} 0.247 \\ 3\overline{)0.741} \\ -6 \\ \hline 14 \\ -12 \\ \hline 21 \\ -21 \\ \hline 0 \end{array}$$

b)
$$\begin{array}{r} 0.059 \\ 6\overline{)0.354} \\ -0 \\ \hline 35 \\ -30 \\ \hline 54 \\ -54 \\ \hline 0 \end{array}$$

(b) We place the decimal point in the quotient along with a 0 in the whole number place. Because 6 won't divide into 3, this quotient also requires a 0 above the 3.

Answer $0.741 \div 3 = 0.247$ $\dfrac{0.354}{6} = 0.059$

Notice that the answer for part a has a 0 *before* the decimal point—in the whole number position—but the answer for part b has one 0 *before* and one 0 *immediately after* the decimal point.

Caution The quotient is commonly written with a whole number even if it is 0. The whole number will be 0 whenever the dividend is less than the divisor.

You Try It 2 **Divide using long division. Use Example 2 as a guide.**

a) $1.099 \div 7$ **b)** $\dfrac{0.056}{8}$ **c)** $1.1415 \div 15$

Sometimes there are not enough decimal places in the dividend to divide fully. When that happens, you need only place as many zeros as you like at the end of the dividend. That way you can continue to divide as long as necessary.

Example 3

Divide using long division: 1.37 ÷ 2.

Procedure Extend the number of decimal places in the dividend by placing as many zeros at the end of the decimal as needed.

Answer

$$\begin{array}{r} 0.685 \\ 2\overline{)1.3700} \\ -1\,2 \\ \hline 17 \\ -16 \\ \hline 10 \\ -10 \\ \hline 0 \end{array}$$

We need to place only one extra zero at the end of the dividend. However, it's okay to place more. (Two extra zeros are shown here.) Usually, we don't know ahead of time how many zeros we will need. Sometimes we need to add more than we originally thought.

Once we get a remainder of zero, we're finished.

1.37 ÷ 2 = 0.685

You Try It 3 **Divide using long division. Use Example 3 as a guide. (In each of these problems, you'll need to place at least one zero at the end of the dividend.)**

a) 3.7 ÷ 5 **b)** $\dfrac{0.5}{4}$ **c)** 0.77 ÷ 8

Think About It 1 When dividing into a decimal, when is it necessary to add zeros to the end of the dividend?

____

____

Reminder:

$$\text{divisor}\overline{)\text{dividend}}^{\text{quotient}}$$

Sometimes we can place as many zeros as we like but we never get a remainder of 0. When this happens, the quotient is a repeating decimal, as you will see in the next example. The repeating pattern will begin to become clear when you get a recurring remainder. Once you see the pattern, you can stop dividing and place a bar over the repeating digits (the repetend), just as you did in Section 5.1.

Example 4

Divide using long division: 2.5 ÷ 3.

Procedure You can probably see that—temporarily ignoring the decimal—3 will not divide evenly into 25. We will need to extend the number of decimal places in the dividend by placing some zeros at the end.

This time, however, we're going to get a repeating decimal. When you recognize that the pattern is repeating, write the quotient with the bar over the repeating part.

Answer

$$\begin{array}{r} 0.833 \\ 3\overline{)2.5000} \\ -2\,4 \\ \hline 10 \\ -9 \\ \hline 10 \\ -9 \\ \hline 1 \end{array}$$

← Place plenty of zeros at the end of 2.5.

← The remainder is 1; bring down the 0.

← The remainder is 1; bring down the 0.

← The remainder of 1 is recurring; let's stop.

$2.5 \div 3 = 0.8\overline{3}$

Sometimes the pattern takes a little longer to develop.

Example 5

Divide using long division: $1.62 \div 11$.

Procedure This time the recurring decimal will take two decimal places. Watch!

Answer

```
     0.147272
 11)1.62000      ← Place plenty of zeros at the end of 1.62.
   -1 1
      52         ← The remainder is 5; bring down the 2.
    - 44
       80        ← The remainder is 8; bring down the 0.
     - 77
        30       ← The remainder is 3; bring down the 0.
      - 22
         80      ← This remainder looks familiar; bring down the 0.
       - 77
          30     ← This remainder also looks familiar. Let's stop.
```

The pattern of 72 repeats in the quotient. $1.62 \div 11 = 0.14\overline{72}$

You Try It 4 **Divide using long division. Use Examples 4 and 5 as guides.**

a) $6.1 \div 9$ **b)** $0.43 \div 6$ **c)** $\dfrac{15.54}{33}$

Dividing When the Divisor Contains a Decimal

Reminder:

$$\text{divisor}\overline{)\text{dividend}}\ \text{with quotient above}$$

$$\text{dividend} \div \text{divisor} = \text{quotient}$$

$$\frac{\text{dividend}}{\text{divisor}} = \text{quotient}$$

To this point, you have learned to divide a decimal number by a whole number, such as $6.52 \div 4$ (as demonstrated in Example 1). What if we need to divide 6.52 by the decimal number 0.4 instead of the whole number 4?

To answer that question, we will use these three ideas to help us develop a procedure for dividing by a decimal number.

1. Any division can be written as a fraction: $6.52 \div 0.4 = \dfrac{6.52}{0.4}$.

2. We can multiply any decimal number by a power of 10, moving the decimal point to the right. $\left\{ \begin{array}{l} 6.52 \times 10 = 65.2 \\ \text{and } 0.4 \times 10 = 4 \end{array} \right.$

3. We can multiply any fraction by a form of 1 without changing the value:

$$\frac{6.52}{0.4} \times \frac{10}{10} = \frac{65.2}{4}, \text{ which is } 65.2 \div 4.$$

Notice that we have changed $6.52 \div 0.4$ into $65.2 \div 4$ by adjusting the decimal point in *each* number, moving it one place to the right.

$$6.52 \div 0.4 \;\rightarrow\; 65.2 \div 4$$

The point is this: If the divisor is a decimal number, we can make it a whole number by moving the decimal point in the dividend *and* the divisor by the same number of places.

In $6.52 \div 0.4$, we can move each decimal point one place to the right in one of three forms:

1. Fractional form

$\frac{6.52}{0.4} \rightarrow \frac{65.2}{4}$

Make the denominator a whole number.

2. Standard division

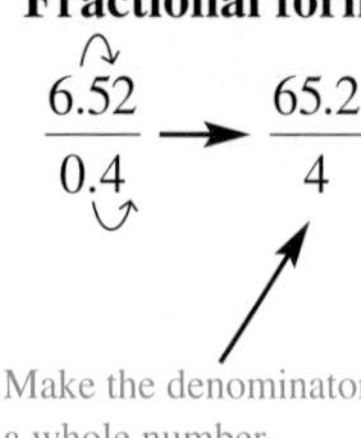

$6.52 \div 0.4$

$65.2 \div 4$

3. Long division form

$0.4\overline{)6.52}$

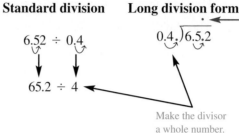

Place the decimal point above the new location right away.

Make the divisor a whole number.

Whichever form you start with, the actual division should be done using long division.

Example 6

Divide.

a) $0.741 \div 0.3$ b) $\frac{0.354}{0.06}$

Procedure First, adjust each number so that the divisor is a whole number.

a) Because the divisor has one decimal place, move each decimal point one place to the right.

$0.741 \div 0.3$

$7.41 \div 3$

b) Because the divisor has two decimal places, move each decimal point two places to the right.

$\frac{0.354}{0.06} \rightarrow \frac{35.4}{6}$

Now set up the long division using these adjusted numbers.

Answer a)

```
   2.47
3)7.41
 - 6
   14
 - 12
    21
  - 21
     0
```

$0.741 \div 0.3 = 2.47$

b)

```
    5.9
6)35.4
 - 30
    54
  - 54
     0
```

$0.354 \div 0.06 = 5.9$

You Try It 5 **Divide. Adjust the decimal points appropriately to make the divisor a whole number. Use Example 6 as a guide.**

a) $0.63 \div 0.9$ b) $\frac{4.85}{0.05}$ c) $0.748 \div 0.0011$

Dividing Decimal Numbers by Powers of 10

As we saw in Section 5.4, multiplying a decimal by a power of 10 has the effect of moving the decimal point to the right. As we'll see next, *dividing* by a power of 10 has the opposite effect, moving the decimal point to the left.

To understand why we would move the decimal point to the left, consider $54.7 \div 100$. We can write this as a fraction, $\frac{54.7}{100}$, and then multiply by a form of 1, $\frac{10}{10}$, so that the numerator becomes a whole number.

$$\frac{54.7}{100} \cdot \frac{10}{10} = \frac{54.7 \times 10}{100 \times 10} = \frac{547}{1{,}000} = 0.547$$

Notice that multiplying the fraction by $\frac{10}{10}$ has two effects: 1. It makes the numerator a whole number by moving the decimal point one place to the right, and 2. It increases the denominator to a higher power of 10.

Here are two examples showing the moving of the decimal point directly.

a) $\frac{3.65}{10} = 0.365$ Dividing by 10, which has only one zero, has the effect of moving the decimal point of 3.65 one place to the left. We can write in 0 as the whole number.

$.3.65 = 0.365$

b) $45.2 \div 1{,}000 = 0.0452$ Dividing by 1,000 suggests that we need to move the decimal point in 45.2 three places to the left.

↓

$00045.2 \div 1{,}000 = 0.0452$ However, because there aren't enough whole numbers in 45.2, we must place some zeros at the beginning of it so that we can move the decimal point there.

$00.045.2 = 0.0452$

Here is the procedure that results from all of this.

Dividing a Decimal by a Power of Ten

1. Count the number of zeros in the power of ten.
2. Move the decimal point that many places to the left.

You Try It 6 **Divide by moving the decimal point the appropriate number of places to the left.**

a) $8.7 \div 10$ **b)** $\frac{34.6}{10}$ **c)** $\frac{76.4}{100}$ **d)** $20.9 \div 1{,}000$

Writing Fractions as Decimal Numbers

Any fraction in which the numerator and the denominator are whole numbers can be written as a decimal number, either a terminating decimal or a repeating decimal.

Start by writing the fraction as division; then divide using long division. For example, $\frac{3}{4} = 3 \div 4$.

Because 4 won't divide into 3 directly, it is necessary to write 3 as 3.000 (possibly with more or with fewer ending zeros) in order to divide.

Example 7

Find the decimal equivalent of each fraction.

a) $\frac{3}{4}$ **b)** $\frac{5}{6}$

Procedure Write the fraction as division and then divide using long division. Write each numerator with a decimal point followed by some ending zeros.

Answer **a)** $\frac{3}{4}$ can be thought of as 3 ÷ 4.

$$\begin{array}{r} 0.75 \\ 4\overline{)3.000} \\ -28 \\ \hline 20 \\ -20 \\ \hline 0 \end{array}$$

$\frac{3}{4} = 0.75$, a terminating decimal

b) $\frac{5}{6}$ can be thought of as 5 ÷ 6.

$$\begin{array}{r} 0.8333 \\ 6\overline{)5.0000} \\ -48 \\ \hline 20 \\ -18 \\ \hline 20 \\ -18 \\ \hline 20 \end{array}$$

$\frac{5}{6} = 0.8\overline{3}$, a repeating decimal

You Try It 7 **Find the decimal equivalent of each fraction. Use Example 7 as a guide.**

a) $\frac{6}{5}$ **b)** $\frac{4}{9}$ **c)** $\frac{23}{40}$

A Strategy for Dividing with Decimal Numbers

Here is the strategy for dividing with decimals.

1. Divide only by a whole number. If the divisor is a decimal number, adjust the dividend and the divisor by moving each decimal point an appropriate number of places to the right.
2. If the quotient has no whole number, write 0 in the quotient.
3. Add as many zeros as needed to the end of any decimal dividend to continue to divide. Sometimes the quotient will be a terminating decimal, and other times it will be a repeating decimal.

Example 8

Follow the preceding strategy guidelines to divide.

a) 0.375 ÷ 0.4 **b)** 29 ÷ 0.03

Procedure **1.** Adjust the dividend and the divisor.

a) Move each decimal point one place to the right.

0.3.75 ÷ 0.4. → 3.75 ÷ 4

b) Move each decimal point two places to the right (giving 29 a decimal point and some zeros).

29.00. ÷ 0.03. → 2,900 ÷ 3

2. Only the quotient in part a requires a zero as the whole number.

3. Add zeros to the end of each dividend for further division.

Answer

a)

$$\begin{array}{r} 0.9375 \\ 4\overline{)3.7500} \\ -3\,6 \\ \hline 15 \\ -12 \\ \hline 30 \\ -28 \\ \hline 20 \\ -20 \\ \hline 0 \end{array}$$

$0.375 \div 0.4 = 0.9375$

b)

$$\begin{array}{r} 0966.66 \\ 3\overline{)2900.00} \\ -27 \\ \hline 20 \\ -18 \\ \hline 20 \\ -18 \\ \hline 20 \\ -18 \\ \hline 20 \end{array}$$

repeats

$29 \div 0.03 = 966.\overline{6}$

(b) Notice that the recurring remainder, 2, began right away. But we can't abbreviate this repeating digit (with a bar over the 6) until *after* the decimal point in the quotient.

You Try It 8 **Divide. Use Example 8 as a guide.**

a) $0.523 \div 8$ **b)** $\dfrac{8.32}{0.3}$ **c)** $0.079 \div 0.05$

d) $7 \div 0.004$ **e)** $4.1062 \div 0.009$ **f)** $\dfrac{3.75}{1.1}$

Following is a chart of the equivalent decimal values of some common fractions.

Decimal Equivalents of Some Fractions

$\frac{1}{2} = 0.5$	$\frac{1}{5} = 0.2$	$\frac{5}{6} = 0.8333\ldots$	$\frac{1}{9} = 0.111\ldots$
$\frac{1}{3} = 0.333\ldots$	$\frac{2}{5} = 0.4$	$\frac{1}{8} = 0.125$	$\frac{2}{9} = 0.222\ldots$
$\frac{2}{3} = 0.666\ldots$	$\frac{3}{5} = 0.6$	$\frac{3}{8} = 0.375$	$\frac{4}{9} = 0.444\ldots$
$\frac{1}{4} = 0.25$	$\frac{4}{5} = 0.8$	$\frac{5}{8} = 0.625$	$\frac{5}{9} = 0.555\ldots$
$\frac{3}{4} = 0.75$	$\frac{1}{6} = 0.1666\ldots$	$\frac{7}{8} = 0.875$	$\frac{7}{9} = 0.777\ldots$

Think About It 2 Based on the previous chart,

a) What is the decimal equivalent of $\frac{8}{9}$? __________

b) What is the decimal equivalent of $\frac{3}{9}$? __________

c) Does the decimal equivalent of $\frac{3}{9}$ appear somewhere else in the chart? Explain your answer.

____

Answers: You Try It and Think About It

You Try It: **1. a)** 1.73 **b)** 1.747 **c)** 6.3 **2. a)** 0.157 **b)** 0.007 **c)** 0.0761 **3. a)** 0.74 **b)** 0.125 **c)** 0.09625 **4. a)** $0.6\overline{7}$ **b)** $0.071\overline{6}$ **c)** $0.4\overline{709}$ **5. a)** 0.7 **b)** 97 **c)** 680 **6. a)** 0.87 **b)** 3.46 **c)** 0.764 **d)** 0.0209 **7. a)** 1.2 **b)** $0.\overline{4}$ **c)** 0.575 **8. a)** 0.065375 **b)** $27.7\overline{3}$ **c)** 1.58 **d)** 1,750 **e)** $456.2\overline{4}$ **f)** $3.\overline{409}$

Think About It: **1.** It is necessary to place extra zeros at the end of the dividend when the divisor does not divide evenly into the dividend. **2. a)** 0.888 . . . **b)** 0.333 . . . **c)** Yes, $\frac{3}{9}$ simplifies to $\frac{1}{3}$, which is already in the chart.

Section 5.5 Exercises

FOR EXTRA HELP
Student Resources on DVD-ROM
Includes
➤ Student's Solutions Manual
➤ Video Lectures
➤ Chapter Test Prep Video
MyMathLab
MathXL

Think Again

1. When dividing into a decimal, when is it necessary to add zeros at the end of the dividend? *(Refer to Think About It 1)*
It is appropriate to add zeros at the end of the dividend when the dividing still has a remainder (other than 0) and the dividend has no more digits.

2. Can a repeating decimal be a divisor? Explain your answer or show an example that supports your answer.
No, not in its decimal form. We cannot move the decimal point to the end of the repeating decimal to start the division process.

3. Can a repeating decimal be a dividend? Explain your answer or show an example that supports your answer.
Yes, as long as the divisor is a terminating decimal. The quotient will be a repeating decimal.

4. The chart at the end of the section shows that when 9 is a denominator, we get repeating decimal values such as $\frac{1}{9} = 0.11111\ldots$ and $\frac{2}{9} = 0.2222\ldots$. Using this reference, what is the decimal equivalent of $\frac{13}{9}$?
1.4444. . . . The fraction must first be written as a mixed number.

Focus Exercises

Divide.

5. $5.4 \div 6$
0.9

6. $3.8 \div 2$
1.9

7. $\dfrac{61.5}{5}$
12.3

8. $74.1 \div 3$
24.7

9. $\dfrac{87.03}{9}$
9.67

10. $51.96 \div 4$
12.99

11. $\dfrac{0.95}{5}$
0.19

12. $0.63 \div 7$
0.09

13. $1.312 \div 8$
0.164

14. $1.737 \div 9$
0.193

15. $2.67 \div 5$
0.534

16. $\dfrac{3.09}{5}$
0.618

17. $1.62 \div 4$
0.405

18. $2.38 \div 4$
0.595

19. $0.532 \div 8$
0.0665

20. $0.676 \div 8$
0.0845

21. $\dfrac{4.7}{3}$
$1.5\overline{6}$

22. $\dfrac{6.5}{3}$
$2.1\overline{6}$

23. $4.8 \div 9$
$0.5\overline{3}$

24. $5.7 \div 9$
$0.6\overline{3}$

25. $3.4 \div 11$
$0.3\overline{09}$

26. $\dfrac{6.2}{11}$
$0.5\overline{63}$

27. $0.49 \div 15$
$0.032\overline{6}$

28. $0.35 \div 6$
$0.058\overline{3}$

29. $\dfrac{2.16}{0.6}$
3.6

30. $3.42 \div 0.9$
3.8

31. $0.468 \div 0.4$
1.17

32. $0.736 \div 0.8$
0.92

33. $0.198 \div 0.05$
3.96

34. $0.267 \div 0.05$
5.34

35. $5.6 \div 0.08$
70

36. $3.6 \div 0.09$
40

37. $\dfrac{48}{0.06}$
800

38. $125 \div 0.05$
2,500

39. $0.24 \div 0.012$
20

40. $0.96 \div 0.032$
30

41. $0.126 \div 0.3$
0.42

42. $\dfrac{0.535}{0.5}$
1.07

43. $0.07128 \div 0.4$
0.1782

44. $0.0579 \div 0.3$
0.193

45. $0.0455 \div 0.09$
$0.50\overline{5}$

46. $0.0316 \div 0.03$
$1.05\overline{3}$

47. $\dfrac{0.073}{0.11}$
$0.6\overline{63}$

48. $0.038 \div 0.11$
$0.3\overline{45}$

49. $0.0015 \div 0.008$
0.1875

50. $0.0053 \div 0.004$
1.325

51. $0.062 \div 0.25$
0.248

52. $\dfrac{0.021}{0.75}$
0.028

Divide by moving the decimal point the appropriate number of places to the left.

53. $25.8 \div 10$
2.58

54. $90.3 \div 10$
9.03

55. $\dfrac{45.7}{10}$
4.57

56. $\dfrac{3.9}{10}$
0.39

57. $74.9 \div 100$
0.749

58. $2.36 \div 100$
0.0236

59. $\dfrac{6.4}{100}$
0.064

60. $\dfrac{0.3}{100}$
0.003

61. $0.45 \div 100$
0.0045

62. $1.03 \div 100$
0.0103

63. $42.1 \div 1{,}000$
0.0421

64. $8.2 \div 1{,}000$
0.0082

Find the decimal equivalent of each fraction.

65. $\frac{5}{2}$ 2.5

66. $\frac{9}{2}$ 4.5

67. $\frac{7}{4}$ 1.75

68. $\frac{13}{4}$ 3.25

69. $\frac{8}{5}$ 1.6

70. $\frac{11}{5}$ 2.2

71. $\frac{9}{8}$ 1.125

72. $\frac{11}{8}$ 1.375

73. $\frac{13}{20}$ 0.65

74. $\frac{29}{20}$ 1.45

75. $\frac{17}{25}$ 0.68

76. $\frac{43}{25}$ 1.72

77. $\frac{7}{6}$ $1.1\overline{6}$

78. $\frac{11}{6}$ $1.8\overline{3}$

79. $\frac{11}{9}$ $1.\overline{2}$

80. $\frac{23}{9}$ $2.\overline{5}$

81. $\frac{9}{11}$ $0.\overline{81}$

82. $\frac{4}{11}$ $0.\overline{36}$

83. $\frac{8}{15}$ $0.5\overline{3}$

84. $\frac{19}{15}$ $1.2\overline{6}$

85. $\frac{4}{37}$ $0.\overline{108}$

86. $\frac{5}{37}$ $0.\overline{135}$

87. $\frac{3}{7}$ $0.\overline{428571}$

88. $\frac{1}{7}$ $0.\overline{142857}$

Use your knowledge of dividing signed numbers along with the techniques of dividing decimals to find each quotient.

89. $-0.24 \div 8$ -0.03

90. $-0.36 \div 4$ -0.09

91. $\frac{-17}{4}$ -4.25

92. $\frac{-27}{5}$ -5.4

93. $2.88 \div (-0.03)$ -96

94. $0.198 \div (-0.06)$ -3.3

95. $\frac{16}{-9}$ $-1.\overline{7}$

96. $\frac{38}{-11}$ $-3.\overline{45}$

97. $-3.62 \div (-0.05)$ 72.4

98. $-15.3 \div (-0.04)$ 382.5

99. $\frac{-41}{-6}$ $6.8\overline{3}$

100. $\frac{-31}{-9}$ $3.\overline{4}$

Think Outside the Box

101. Each of these fractions can be written as a terminating decimal as shown.

$$\frac{3}{4} = 0.75,\quad \frac{1}{8} = 0.125,\quad \frac{9}{10} = 0.9,\quad \frac{7}{16} = 0.4275,\quad \frac{12}{25} = 0.48,$$

$$\frac{13}{40} = 0.325,\quad \frac{27}{50} = 0.54,\quad \frac{87}{100} = 0.87,\quad \text{and}\quad \frac{94}{125} = 0.752.$$

Find the prime factorization of the denominator of each of those fractions. What does each prime factorization have in common?

$\frac{3}{2\cdot 2}, \frac{1}{2\cdot 2\cdot 2}, \frac{9}{2\cdot 5}, \frac{7}{2\cdot 2\cdot 2\cdot 2}, \frac{12}{5\cdot 5}, \frac{13}{2\cdot 2\cdot 2\cdot 5}, \frac{27}{2\cdot 5\cdot 5}, \frac{87}{2\cdot 2\cdot 5\cdot 5}, \frac{94}{5\cdot 5\cdot 5}$

Each denominator has only 2 and 5 as prime factors.

102. Each of these fractions can be written as a repeating decimal as shown.

$$\frac{2}{9} = 0.\overline{2},\quad \frac{3}{14} = 0.2\overline{142857},\quad \frac{13}{22} = 0.5\overline{90},\quad \frac{7}{12} = 0.58\overline{3},$$

$$\frac{10}{33} = 0.\overline{30},\quad \frac{24}{55} = 0.4\overline{36},\quad \frac{49}{66} = 0.7\overline{42},\quad \text{and}\quad \frac{64}{75} = 0.85\overline{3}.$$

Find the prime factorization of the denominator of each of these fractions. What does each prime factorization have in common?

$\frac{2}{3\cdot 3}, \frac{3}{2\cdot 7}, \frac{13}{2\cdot 11}, \frac{7}{2\cdot 2\cdot 3}, \frac{10}{3\cdot 11}, \frac{24}{5\cdot 11}, \frac{49}{2\cdot 3\cdot 11}, \frac{64}{3\cdot 5\cdot 5}$

Each denominator has prime factors other than 2 and 5.

103. Based on your answers to numbers 101 and 102, how can you determine if a fraction will be a terminating decimal or a repeating decimal?

When a fraction's denominator has only 2 or 5 as prime factors, the fraction can be written as a terminating decimal. When a fraction's denominator has a prime factor other than 2 or 5, the fraction can be written as a repeating decimal.

SECTION 5.6 Applications Involving Decimals

OBJECTIVE

In this section, you will learn to

- Solve applications involving multiplication and division of decimals.

You Need to Know

To successfully complete this section, you need to understand

- ☐ Solving equations (1.7 and 3.1)
- ☐ Solving applications (1.8 and 3.5)
- ☐ Multiplying decimals (5.4)
- ☐ Dividing decimals (5.5)

Whenever an answer is in terms of dollars and cents, the result *must* have two decimal places. If in the multiplication process the answer has fewer than two decimal places, add the appropriate number of zeros. If, instead, the answer has more than two decimal places, round it to the nearest penny (hundredth).

Introduction

A variety of situations involve multiplying or dividing decimals. If a situation is related to money or if a formula involves decimals, we can use the techniques learned in the last two sections to answer questions about the situation.

In each of the following applications, you are asked to

- Identify known and unknown values.
- Set up a legend: Let $x = \ldots.$
- Put the values into a formula (equation).
- Solve the equation.
- Write a sentence to answer the question.

Earnings Applications

When a worker is paid the same amount each hour, such as \$7.35 an hour, it is called an hourly wage. To find how much this worker earns in an 8-hour day, we multiply her wage by 8.

Earnings Formula: Number of hours × Hourly wage = Total earnings

Example 1

Nancy is paid \$7.35 an hour. How much does Nancy earn in an 8-hour day?

Procedure Use the earnings formula to calculate her total earnings.

Answer **Legend:** Let x = the amount Nancy earns in an 8-hour day.

Number of hours × Hourly wage = Total earnings

$$8 \times \$7.35 = x$$

YOU FINISH IT: Multiply $8 \times \$7.35$

You should have found that Nancy earns \$58.80 in an 8-hour day.

Example 2

Reza earns \$97.20 for an 8-hour day. How much does he earn per hour?

Procedure Use the earnings formula. Reza's total earnings are \$97.20. We don't know his hourly wage.

Answer Legend: Let h = the amount Reza earns per hour.

\# of hours × Hourly wage = Total earnings

$$8h = \$97.20$$ To isolate the variable, we must divide each side by 8.

$$\frac{8h}{8} = \frac{\$97.20}{8}$$

$$h = \underline{\qquad}$$

YOU FINISH IT: Divide $\$97.20 \div 8$

Sentence Reza earns \$12.15 per hour.

You Try It 1 **For each problem, set up a legend and an equation and solve. Write the answer as a sentence. Use Examples 1 and 2 as guides.**

a) Lori is paid \$10.92 per hour at her job. How much will Lori earn in a day if she works 7.5 hours?

Legend

Equation

Sentence

____

b) Juan earned \$410.40 in a week in which he worked 30 hours. How much did Juan earn per hour?

Legend

Equation

Sentence

____

Rectangle Applications: Area, Length, and Width

As you know, the area of a rectangle can be found using a formula that involves multiplication:

Length × Width = Area

Remember that area is always represented in square units, such as square inches (in.^{2}) or square centimeters (cm^2).

Example 3

A rectangular planter is 2.25 feet wide and 9.5 feet long. What is the area of the planter?

Answer Let A = the area.

$9.5 \times 2.25 = A$

$\_\_\_\_\_\_ = A$

9.5 ft

2.25 ft

Sentence The area of the planter is 21.375 square feet.

YOU FINISH IT: Multiply 9.5×2.25

Example 4

A rectangular yard is 10.5 meters long. Its area is 90.3 square meters. What is the width of the yard?

Answer Let w = the width.

$10.5w = 90.3$

$\frac{10.5w}{10.5} = \frac{90.3}{10.5}$

$w = \_\_\_\_$

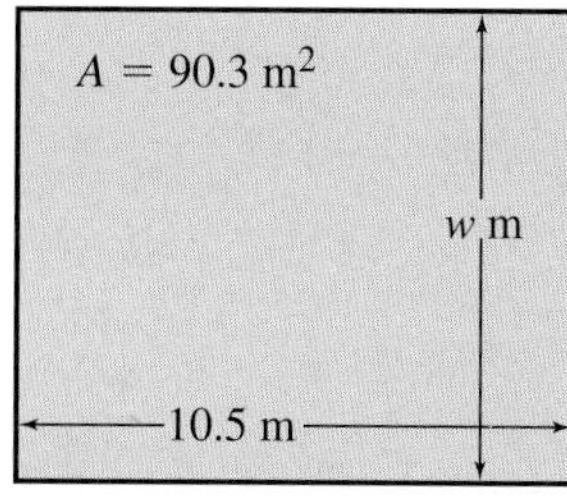

Sentence The width of the yard is 8.6 meters.

YOU FINISH IT: Divide $90.3 \div 10.5$

You Try It 2 **A rectangular bedroom is 18.8 feet long and 9.6 feet wide. What is the area of the bedroom? Set up a legend and an equation and solve.**

Legend

Equation

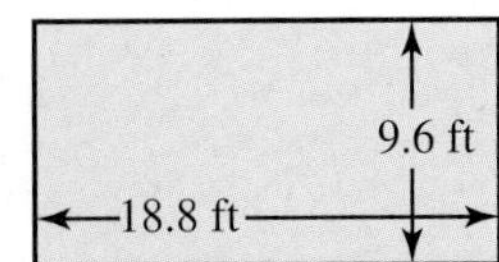

Sentence ______

You Try It 3 **The floor of a rectangular conference room is covered with 40.8 square yards of carpet. The width of the room is 4.8 yards. What is the length of the conference room? Set up a legend and an equation and solve.**

Legend

Equation

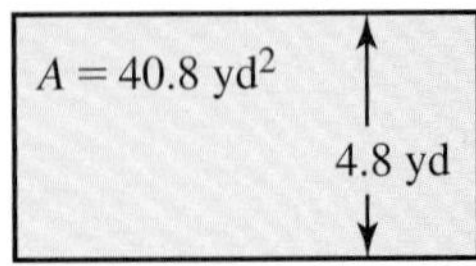

Sentence ______

Circle Applications: Radius and Circumference

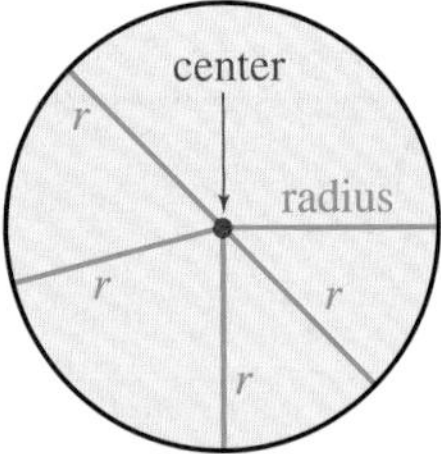

A circle is a round geometric figure with a center point. The center point is the same distance from every point on the circle. This common distance from the center to every point on the circle is called the **radius,** and it is represented in formulas by the letter r.

Recall from Section 5.1 that π is a decimal number that neither terminates nor repeats. Rounded to the nearest hundredth, π is approximately 3.14, written $\pi \approx 3.14$.

The perimeter of a circle is called the **circumference.** It measures the distance around the circle. The formula for the circumference is

$$C = 2 \cdot \pi \cdot r \quad \text{(more commonly read } C = 2\pi r\text{)}$$

$$C \approx 2 \times 3.14 \times r$$

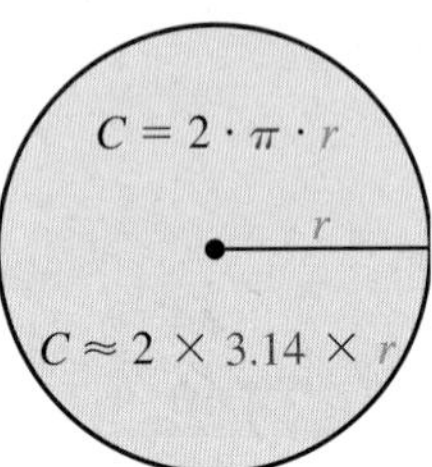

We can multiply $2 \times 3.14 = 6.28$ to get

$$C \approx 6.28 \times r$$

Because this is an approximation, we write the sentence using the word *approximately*.

Example 5

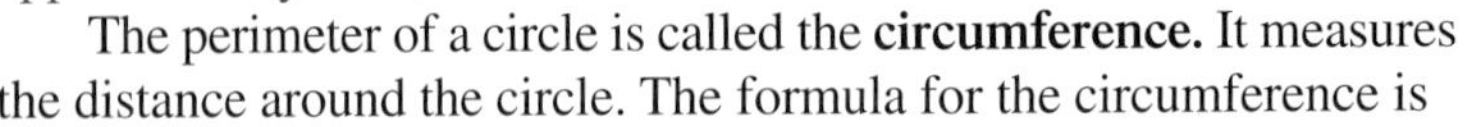

A circular window has a radius of 1.5 meters. What is the circumference of the window?

Answer Let $C =$ the circumference.

$C \approx 6.28 \times 1.5$

$C \approx$ ______

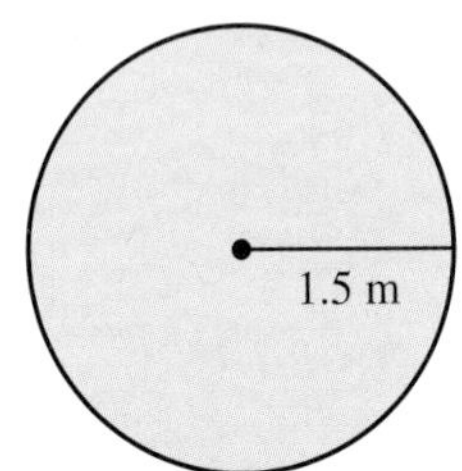

YOU FINISH IT: Multiply 6.28×1.5

Sentence The circumference of the window is approximately 9.42 meters.

Example 6

The cover for a circular spa has a circumference of 37.5 feet. What is the radius of the spa cover? Round the answer to the nearest tenth.

Answer Let r = the radius.

$$37.5 = 6.28r$$

$$\frac{37.5}{6.28} = \frac{6.28r}{6.28}$$

$$5.9713 \approx r$$

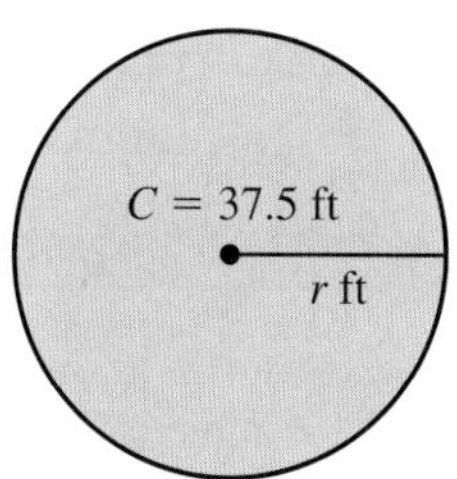

YOU FINISH IT: Divide $37.5 \div 6.28$

Sentence The radius of the spa cover is approximately 6.0 feet.

Think About It 1 Why do the sentence answers to Examples 5 and 6 include the word *approximately*?

____

____

You Try It 4 **A courtyard garden is in the shape of a circle and has a radius of 5 feet. What is the circumference of this garden? Set up a legend and an equation and solve.**

Legend

Equation

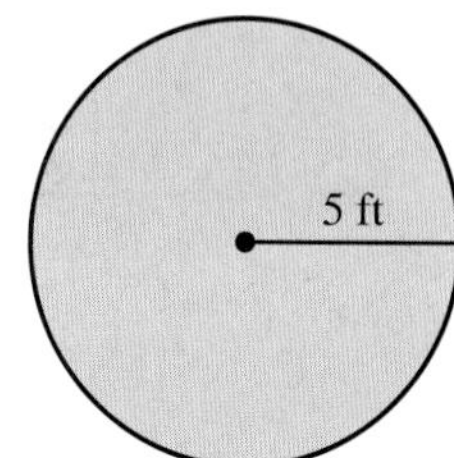

Sentence ______________________________

You Try It 5 **The center jump circle on a youth basketball court has a circumference of 24 feet. What is the length of the radius of this circle? Set up a legend and an equation and solve. Round the answer to the nearest tenth.**

Legend

Equation

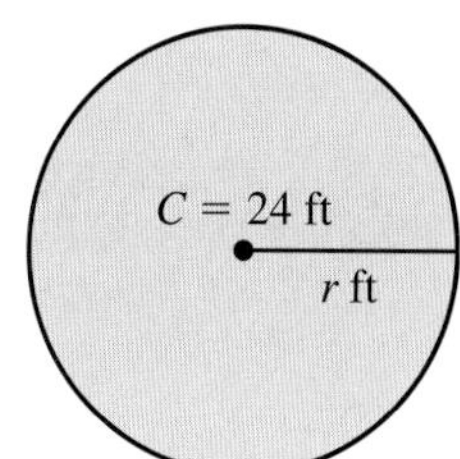

Sentence ______________________________

Miles per Gallon Applications

A car is advertised as getting 24 miles per gallon, on average. This means that if the car has 2 gallons of gas, it should be able to go 24 miles on the first gallon and 24 additional miles on the second gallon. In other words, it should go 2×24 miles = 48 miles.

Of course, different driving conditions, such as freeway driving and city street driving, make a difference in the miles per gallon a car gets, which is why we say *average* miles per gallon.

The gas mileage formula (miles per gallon—mpg) is

Number of gallons × Average miles per gallon = Total number of miles

or

Gallons × mpg = Total miles

Example 7

On a recent trip, Jim started with a full tank of gas and drove 272 total miles. At the end of the first day, he bought 8.5 gallons to refill his tank. What was his car's average miles per gallon on that first day?

Procedure Use the gas mileage formula: gallons × mpg = total miles.

The unknown is the miles per gallon (mpg).

Answer Legend: Let x = the average miles per gallon.

$8.5x = 272$ To isolate the variable, we must divide each side by 8.5.

$$\frac{8.5x}{8.5} = \frac{272}{8.5}$$

$x =$ ______

YOU FINISH IT: Divide $272 \div 8.5$

Sentence Jim's car averaged 32 miles per gallon on that first day.

You Try It 6 **Su-Ki wanted to check out the gas mileage of her new car. To start, she filled up the tank, set the tripometer to 000, then drove as she normally does for a week. At the end of the week, she checked the mileage and filled up the tank. She had driven 194.25 miles, and her car had used 10.5 gallons of gasoline. How many miles per gallon did Su-Ki's car average that week?**

Legend

Equation

Sentence ______________________________

Shopping Applications

Two formulas common to shopping are

1. **Number of items × Price of each item = Total price**

   This is useful if you purchase, for example, 8 ears of corn where each ear has a price of \$0.65.

Number of items	×	Price of each item	=	Total price
8	×	\$0.65	=	\$5.20

   We can use this formula if the unknown value is the price of each item, as in You Try It 7a, which follows.

2. **Number of pounds × Price per pound = Total price**

   We use this formula to find the total price, for example, of a watermelon that costs \$0.38 per pound. If the watermelon weighs 9.4 pounds, what is the total price of the watermelon?

Number of pounds	×	Price per pound	=	Total price
9.4	×	\$0.38	=	\$3.572

   This price must be rounded to the nearest penny. So, the total price is \$3.57.

   We can use this formula if the unknown value is the weight of the item, as in You Try It 7b, which follows.

You Try It 7 **Sara checked her receipt after getting home from shopping. For each situation, set up a legend and an equation and solve.**

a) Sara noticed that 12 boxes of macaroni and cheese cost her a total of \$7.44. What was the price of each box?

Legend

Equation

Sentence

____

b) Sara also noticed that a package of steaks cost her \$7.56. The package showed that the total weight was 2.1 pounds, but the price per pound was not listed. What was the price per pound of the steak?

Legend

Equation

Sentence

____

Distance, Rate, and Time Applications

The distance someone can travel, by car, by bicycle, or on foot, for instance, is based on this formula:

Rate × Time = Distance

In this formula, *rate* is another word for *speed* and is often measured in miles per hour (mph). Time is often measured in hours, and distance is often measured in miles.

The rate is an *average* rate because it's rare to travel the same speed for any length of time. When driving, riding a bike, or jogging, it's common to change speeds for a variety of reasons, such as slowing down to stop at a stop sign.

Someone who rides a bike an average of 12 miles per hour for 3 hours travels a total distance of 36 miles.

Rate $\times$ Time = Distance

12 mph $\times$ 3 hours = 36 miles

We can use this formula if the unknown value is the rate, as in You Try It 8, or the time, as in You Try It 9.

You Try It 8 **Wayne was able to run 11.4 miles in 1.2 hours. What was his average rate?**

Legend

Equation

Sentence ______________________________

You Try It 9 **Kyla races motorcycles. Normally, she is able to average 35 mph on a winding cross-country course. If the course she is racing is 112 miles long, how many hours should it take Kyla to complete the race?**

Legend

Equation

Sentence ______________________________

Answers: You Try It and Think About It

You Try It: **1. a)** Legend: Let x = earnings for the day. Equation: $7.5 \times 10.92 = x$; Sentence: Lori will earn \$81.90 if she works 7.5 hours. **b)** Legend: Let h = Juan's earnings per hour. Equation: $30h = 410.40$; Sentence: Juan earned \$13.68 per hour. **2.** Legend: Let A = the area of the bedroom. Equation: $18.8 \times 9.6 = A$; Sentence: The area of the bedroom is 180.48 square feet. **3.** Legend: Let x = the length of the conference room. Equation: $4.8x = 40.8$; Sentence: The length of the conference room is 8.5 yards. **4.** Legend: Let C = the circumference of the garden. Equation: $C = 6.28 \times 5$ Sentence: The circumference of the garden is approximately 31.4 feet. **5.** Legend: Let r = the length of the radius. Equation: $24 = 6.28r$; Sentence: The radius of the circle is approximately 3.8 feet. **6.** Legend: Let x = the miles per gallon. Equation: $10.5x = 194.25$; Sentence: Su-Ki's car averaged 18.5 miles per gallon that week. **7. a)** Legend: Let p = the price of each box. Equation: $12p = 7.44$; Sentence: The price of each box was \$0.62. **b)** Legend: Let p = the price per pound. Equation: $2.1p = 7.56$; Sentence: The price of the steak was \$3.60 per pound. **8.** Legend: Let r = Wayne's rate. Equation: $1.2r = 11.4$; Sentence: Wayne's average rate was 9.5 miles per hour. **9.** Legend: Let h = the number of hours. Equation: $35h = 112$; Sentence: It should take Kyla 3.2 hours to finish the race.

Think About It: **1.** Because we are using 3.14 for π, each answer is an approximation, not an exact value.

Section 5.6 Exercises

FOR EXTRA HELP

Student Resources on DVD-ROM. Includes ➤Student's Solutions Manual ➤Video Lectures ➤Chapter Test Prep Video | MyMathLab | MathXL

Think Again

1. How many minutes is 1.45 hours? Explain how this is calculated.
87 minutes. We must multiply 1.45 times 60. (Or we multiply 0.45 times 60 and add 60.)

2. What is the decimal equivalent of 21 minutes? Explain how this is calculated.
0.35. We must divide 21 by 60. 21 minutes is a fraction of an hour: $\frac{21}{60}$ of an hour.

Focus Exercises

Solve. Answer each question with a complete sentence.

Earnings

3. Kerri is paid $11.08 per hour at her job, and she works 6.5 hours each day. How much does Kerri earn each day?
Kerri earns $72.02 each day.

4. Mia earns $17.25 per hour working as a truck driver, and she works 8 hours each day. How much does Mia earn each day?
Mia earns $138 each day.

5. Tran is paid $12.30 per hour at his part-time job. Last month he earned $934.80. How many hours did Tran work last month?
Tran worked 76 hours last month.

6. Jorge earns $9.45 per hour at his job. Last week he earned $321.30. How many hours did Jorge work last week?
Jorge worked 34 hours last week.

7. Calvin worked 35 hours last week and earned $355.60. How much does Calvin earn per hour?
Calvin earns $10.16 per hour.

8. Lin worked 7.5 hours yesterday and earned $71.70. How much does Lin earn per hour?
Lin earns $9.56 per hour.

Commission

9. Annie works for an electronics superstore. During her 5-hour shift last Saturday, she earned $91.25 in commission sales. How much commission did Annie earn, on average, each hour?
Annie earned, on average, $18.25 commission each hour.

10. Lee works at an appliance store. He receives a commission on everything he sells. One day he worked 8 hours and earned $134 in commission sales. How much commission did Lee earn, on average, each hour?
Lee earned, on average, $16.75 commission each hour.

11. Hannah works at a flooring store, and she receives a commission on everything she sells. Last week she worked 37.5 hours and earned a total of $780 in commission sales. How much commission did Hannah earn, on average, each hour?
Hannah earned, on average, $20.80 commission each hour.

12. Dimitri is a distributor for Aviator sunglasses. Last week he worked 32 hours and earned a total of $752 in commission sales. How much commission did Dimitri earn, on average, each hour?
Dimitri earned, on average, $23.50 commission each hour.

Area

13. A patio is rectangular. It measures 6.3 yards long and 4.2 yards wide. What is the area of the patio?
The area of the patio is 26.46 square yards.

14. The dance floor in a nightclub is rectangular. It measures 51.8 feet long and 27.5 feet wide. What is the area of the dance floor?
The area of the dance floor is 1,424.5 square feet.

15. A landscaper is designing a yard and wants a small rectangular patch of grass to have an area of 35 square feet. If the width of the patch is 2.8 feet, what is the length?
The length of the patch is 12.5 feet.

16. A rectangular pool house has an area of 182.9 square feet. If the width of the pool house is 11.8 feet, what is the length?
The length of the pool house is 15.5 feet.

Circumference

17. A circular pond has a radius of 35 meters. What is the circumference of the pond? (Round the answer to the nearest whole number.)
The circumference of the pond is approximately 220 meters.

18. A circular tabletop has a radius of 3.2 feet. What is the circumference of the tabletop? (Round the answer to the nearest tenth.)
The circumference of the tabletop is approximately 20.1 feet.

19. A bicycle wheel has a circumference of 219.8 centimeters. What is the radius of the wheel?
The radius of the wheel is approximately 35 centimeters.

20. The circumference of a lighthouse's circular lobby is 78.5 feet. What is the radius of the lobby?
The radius of the lobby is approximately 12.5 feet.

Miles per gallon

21. Dale has a hybrid car that runs on both gasoline and electricity. His car averages 43.5 miles per gallon. How far can Dale drive on 7.8 gallons of gas?
Dale can travel 339.3 miles on 7.8 gallons of gas.

22. Ricky's new car gets 22.6 miles per gallon, and the gas tank holds 12.5 gallons. How many miles can Ricky's car travel on a full tank of gas?
Ricky's car can travel 282.5 miles on a full tank of gas.

23. Beverly drove Dale's hybrid car for one week while Dale was out of town on business. She used it to get around the city. After a week's time, she had driven the car 119.0 miles and it had used 2.8 gallons of gas. How many miles per gallon did the car average for the week?
The car averaged 42.5 miles per gallon for the week.

24. On the first day of his road trip, Milo's car traveled 351 miles before Milo needed to refuel. He refilled the tank with 11.25 gallons of gas. How many miles per gallon did Milo's car average that day?
Milo's car averaged 31.2 miles per gallon that day.

Shopping

25. Ron bought a pack of baseball cards for \$5.28. The pack contained 8 cards. What was the price of each card?
The price of each card was \$0.66.

26. Lynn paid \$32.40 for a crate of 24 organic oranges. What was the price of each orange?
The price of each orange was \$1.35.

27. Laura bought 4.25 yards of linen and paid \$27.20. What was the price of a yard of that fabric?
The price of a yard of that fabric was \$6.40.

28. Thom bought 7 cans of gourmet soup for a total of \$15.33. What was the price of each can of soup?
The price of each can of soup was \$2.19.

Distance, Rate, and Time

29. Jake enjoys cross-country skiing and is able to cover a 9.75-mile course in 0.65 hours. What is Jake's average rate of speed on that course?
Jake's average rate of speed on that course is 15 miles per hour.

30. During rush hour, it takes Linda 1.6 hours to drive 40 miles. What is Linda's average rate of speed?
Linda's average rate of speed is 25 miles per hour.

31. Becca enjoys going horseback riding. Her favorite trail is 9.8 miles long. If her average rate of speed is 5.6 miles per hour, how long does it take Becca to ride the entire trail?
It takes Becca 1.75 hours to ride the entire trail.

32. Kyle rides his bike to the beach. The distance is 27 miles. If he is able to average 15 miles per hour, how long does it take Kyle to reach the beach?
It takes Kyle 1.8 hours to reach the beach.

Miscellaneous

33. Alan works for a department store that asks its employees to work on Thanksgiving. Many choose not to, but some do because the store pays 2.25 times their hourly wage for the day. If Alan normally earns \$8.32 per hour, how much will he earn per hour on Thanksgiving?
Alan will earn \$18.72 per hour on Thanksgiving.

34. Gabriella works in a department store that pays time and a half (regular pay $\times$ 1.5) for anyone who works on Sunday. Gabriella's regular hourly pay is \$7.88. What is Gabriella's hourly pay when she works on Sunday?
Gabriella's hourly pay is \$11.82 when she works on Sunday.

35. Maralee is buying a used car from a local dealer. After her \$2,500 down payment, she still owes \$3,834.72, including interest. She made a deal to pay this remaining balance in 12 equal monthly payments. How much will Maralee pay each month?
Maralee will pay \$319.56 each month.

36. Five friends go out to dinner. They have agreed to split the entire tab (including tax and tip) evenly among themselves. If the total bill is \$117.35, how much does each person pay?
Each person pays \$23.47.

37. Meteor Crater, in Arizona, was formed by a meteorite impact about 49,000 years ago. The rim is circular and has a diameter of nearly 1.2 kilometers. Calculate the circumference of the rim of Meteor Crater.
The circumference of the rim of Meteor Crater is approximately 3.768 kilometers.

38. American Airlines can fly from Los Angeles, California, to Chicago, Illinois, in 3.75 hours. The distance traveled is 1,512 miles. What is the average rate of speed for this flight?
The average rate of speed for this flight is 403.2 miles per hour.

Think Outside the Box

Write a sentence that answers the questions.

39. Terrence works for a company that pays each employee a standard hourly wage for the first 40 hours worked each week and 1.25 times the standard wage for each hour worked beyond 40 hours in one week. Terrence's standard wage is $9.60 per hour. If Terrence earned $456 last week,

a) How many hours beyond 40 hours did he work that week?
Terrence worked 6 hours beyond 40 hours that week.

b) How many total hours did he work that week?
Terrence worked 46 total hours that week.

40. The Hawaiian Ironman competition is a three-sport event in which competitors must swim, bike, and run. The following chart shows the results of the 2005 second-place finisher of the women's competition, Michellie Jones. ***Source:*** **http://ironman.com**

Event	Distance (in miles)	Time (in hours)	Rate
Swimming	2.4	0.9	2.7 mph
Biking	112	4.9	22.9 mph
Running	26.2	3.3	7.9 mph
Total	140.6	9.1	15.5 mph

a) Calculate the total distance and total time.

b) Find Michellie's rate (in miles per hour) for each event and her rate for the total event. (Round each rate to the nearest tenth.)

Chapter 5 Review

Section 5.1 An Introduction to Decimals

Concept	Example
A decimal number is a mixed number with a **decimal point** that separates the **whole number** from the **fractional part.**	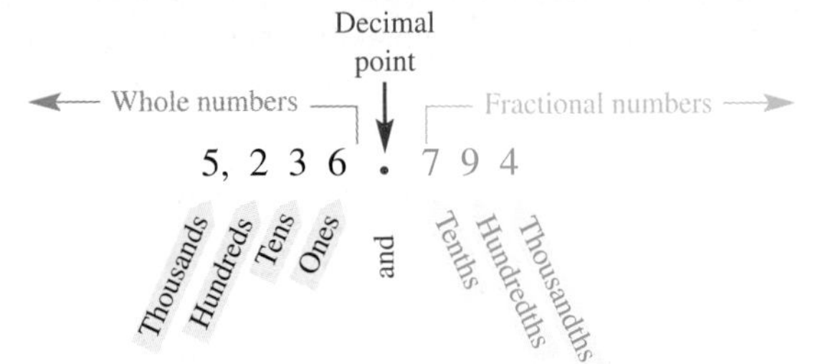
When naming decimals, name the decimal part as you would a whole number, followed by the rightmost place in the decimal.	0.46 is forty-six-**hundredths.**
Any fraction with a denominator that is a power of 10 (such as 10, 100, and 1,000) is called a **decimal fraction.**	$\frac{3}{10}, \frac{52}{100},$ and $\frac{17}{1{,}000}$
The number of decimal places in a decimal number indicates the number of zeros in the power of 10 in the decimal fraction's denominator.	$0.56 = \frac{56}{100}$ and $12.3 = \frac{123}{10}$
The number of zeros in the power of 10 in the denominator of a decimal fraction indicates the number of decimal places the decimal number will have.	$\frac{893}{1{,}000} = 0.893$ and $\frac{71}{10} = 7.1$
We can place zeros at the end of any terminating decimal without changing its value. If a decimal number has *ending* zeros, they may be eliminated, but those zeros *within* the decimal number may not.	$5.6 = 5.6000$ $4.7500 = 4.75$ and $0.1030 = 0.103$
Decimal numbers fit into one of three categories **1.** Terminating decimals **2.** Repeating decimals **3.** Nonterminating, nonrepeating decimals	**1.** 52.9 and 0.625 **2.** 4.383838383838 . . . and $2.1\overline{7}$ **3.** π, which is 3.141592653589 . . .
On a number line between the integers are decimals, both positive and negative.	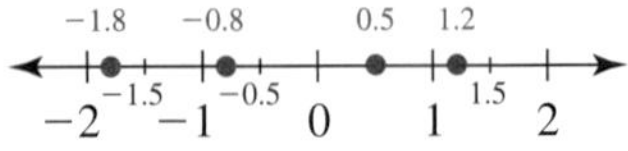

Section 5.2 Rounding Decimals

Concept	Example
To round decimals **1.** Identify the place digit. **2.** Identify the rounding digit **3. a.** If the rounding digit is 5 or higher, round up. **b.** If the rounding digit is 4 or lower, round down. **4.** Eliminate all digits after the place digit.	**a.** Round 28.714 to the nearest tenth. The place digit is 7, and the rounding digit is 1. This number rounds down to **28.7**. **b.** Round 0.3862 to the nearest hundredth. The place digit is 8, and the rounding digit is 6. This number rounds up to **0.39**.

Section 5.3 Adding and Subtracting Decimals

Concept	Example
When adding or subtracting decimals, it is necessary to line up the decimal points in each number. It may also be necessary to place zeros at the end of some of the numbers to make all numbers terminate in the same place.	Add 2.03 + 3.1 + 0.528. $\begin{array}{r} 2.030 \\ 3.100 \\ +\ 0.528 \\ \hline 5.658 \end{array}$ Place zeros at the end of the first two numbers so that each number has three decimal places.
Subtracting decimals works the same as subtracting whole numbers. Sometimes you can subtract directly, and sometimes you need to regroup.	Subtract 1.38 from 4.5. $\begin{array}{r} 4.50 \\ -\ 1.38 \\ \hline 3.12 \end{array}$ Place a zero at the end of the first number so that each number has two decimal places.
Whether adding or subtracting signed decimals, it is best for each decimal to have the same number of decimal places. Then to evaluate, follow the guidelines for adding and subtracting signed numbers.	$-2.7 - (-6.05)$ $= -2.70 + 6.05$ $= +3.35$

Section 5.4 Multiplying Decimals

Concept	Example
To multiply decimals **1.** Temporarily ignore the decimal points and multiply the numbers (factors) as if they were both whole numbers. **2.** Count the total number of decimal places in the factors. This total is the number of decimal places in the product (before any simplifying).	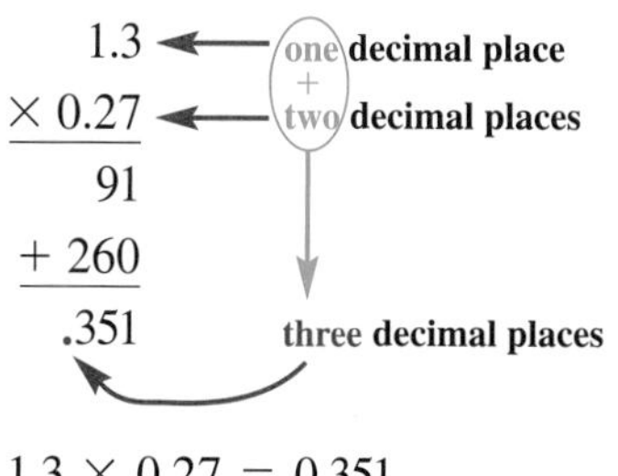 $1.3 \times 0.27 = 0.351$

continued

Section 5.4 Multiplying Decimals

Concept	Example
If there are not enough decimal places in the answer, we need to place the appropriate number of zeros at the front of the answer before placing the decimal point.	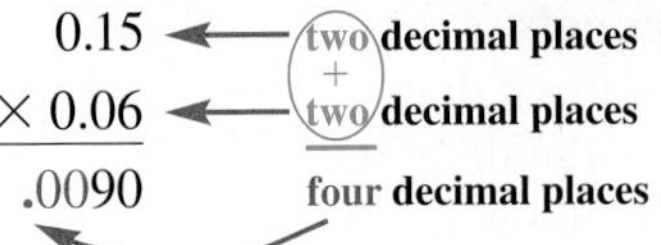Place two extra zeros to get four decimal places.
To multiply a decimal by a power of ten **1.** Count the number of zeros in the power of ten. **2.** Move the decimal point that many places to the right.	3.529 × 100 3.52.9 Move the decimal point two places to the right. = 352.9

Section 5.5 Dividing Decimals

Concept	Example
To divide, the divisor must be a whole number. If the divisor is a decimal number, make it a whole number by moving the decimal points in the dividend *and* the divisor the same number of places.	4.689 ÷ 0.15 ⟶ 4.689 ÷ 0.15 Move each decimal point two places to the right.
To divide a decimal number by a whole number **1.** Place a decimal point in the quotient directly above the decimal point in the dividend. **2.** Divide as if each were a whole number. **3.** Place zeros at the end of the dividend if needed. It's common for the quotient to contain a whole number, even if that number is 0.	0.3126 15)4.68900 − 45 18 − 15 39 − 30 90 − 90 0
To divide a decimal by a power of ten **1.** Count the number of zeros in the power of ten. **2.** Move the decimal point that many places to the left.	6.5 ÷ 100 = 0.065 Move the decimal point in 6.5 two places to the left. ↓ 006.5 ÷ 100 = 0.065 Place some zeros at the beginning of 6.5 before moving the decimal point. 0.06.5 = 0.065

Section 5.6 Applications Involving Decimals

Concept	Example
Earnings formula Number of hours $\times$ Hourly wage $=$ Total earnings	Vy earns \$7.50 per hour. Last week she worked 20 hours. How much did Vy earn last week? $20 \times 7.50 = 150.00$ Vy earned \$150 last week.
Circumference of a circle formula $C = 2 \cdot \pi \cdot r \approx 2 \times 3.14 \times r$ $C \approx 6.28 \times r$	A bicycle wheel has a radius of 12 inches. What is the circumference of the wheel? $C \approx 6.28 \times 12$ $C \approx 75.36$ The circumference of the wheel is approximately 75.36 inches.
Gas mileage formula Gallons $\times$ mpg $=$ Total miles	Last week Jacqui drove her car 324 miles, and it took 13.5 gallons to fill the tank. How many miles per gallon did Jacqui's car average last week? Let $m =$ the miles per gallon. $13.5m = 324$ $\frac{13.5m}{13.5} = \frac{324}{13.5}$ $m = 24$ Jacqui's car averaged 24 miles per gallon last week.
Shopping formulas **1.** Number of items $\times$ Price of each item $=$ Total price **2.** Number of pounds $\times$ Price per pound $=$ Total price	A box of 24 sodas costs \$5.52. What is the price for each can of soda? Let $p =$ the price per can. $24p = 5.52$ $\frac{24p}{24} = \frac{5.52}{24}$ $p = 0.23$ The price of each can of soda is \$0.23.
Distance, rate, and time formula Rate $\times$ Time $=$ Distance	Elda rode her bike 18.9 miles in 1.35 hours. What was Elda's average rate? Let $r =$ Elda's average rate. $r \times 1.35 = 18.9$ $\frac{1.35r}{1.35} = \frac{18.9}{1.35}$ $r = 14$ Elda's average rate was 14 miles per hour.

Chapter 5 Review Exercises

Fill in the blanks.

1. Numbers such as $\frac{43}{100}$ and $\frac{18}{1,000}$ are called ______decimal______ fractions. **(5.1)**

2. When rounding to the nearest hundredth, the ______thousandths______ place determines whether we round up or round down. **(5.2)**

3. To divide by a decimal number, the divisor must first be a written as a(n) ______whole number______. **(5.5)**

4. When dividing decimals, if we never get a remainder of 0, the quotient is a(n) ______repeating decimal______. **(5.5)**

Section 5.1

In the given number, name the value of the digit 7.

5. 1,358.47
hundredths

6. 10.6273
thousandths

Name each decimal by writing it in word form.

7. 0.06
six-hundredths

8. 0.013
thirteen-thousandths

Write each decimal as a fraction. Do not simplify.

9. 0.38
$\frac{38}{100}$

10. 1.2
$\frac{12}{10}$

11. 0.01
$\frac{1}{100}$

12. 2.08
$\frac{208}{100}$

Write each fraction as a decimal.

13. $\frac{72}{100}$
0.72

14. $\frac{3}{100}$
0.03

15. $\frac{57}{10}$
5.7

16. $\frac{8}{1,000}$
0.008

Write each of these repeating decimals using an ellipsis or a bar over the repetend, whichever is not shown.

17. 1.5555...
$1.\overline{5}$

18. 2.61454545...
$2.61\overline{45}$

19. $0.\overline{512}$
0.512512512...

20. $2.3\overline{8}$
2.38888...

Locate and label each decimal number on the number line.

21. 1.7

22. 4.5

23. −0.3

24. −3.1

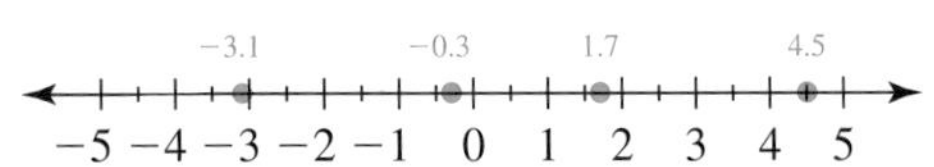

Section 5.2

Round each of these numbers to the indicated place.

25. 24.038 tenths
24.0

26. 1.5239 hundredths
1.52

27. 26.952 tenths
27.0

28. 1.499802 thousandths
1.500

Each number represents the amount of sales tax applied to a variety of purchases at a grocery store. Round each dollar amount to the nearest penny.

29. $1.5853
$1.59

30. $28.79064
$28.79

31. $0.39526
$0.40

32. $11.99702
$12.00

Each number represents a 15% tip on a restaurant meal. Round each dollar amount to the nearest dime (tenth), but express the answer in dollars and cents.

33. $8.9235
$8.90

34. $3.582
$3.60

35. $1.2675
$1.30

36. $24.996
$25.00

Each number represents a business expense that is to be entered on a tax form. Round each of these numbers to the nearest dollar.

37. \$128.06
\$128

38. \$49.53
\$50

39. \$267.91
\$268

40. \$1,599.62
\$1,600

Round the number as instructed and restate the first sentence using the rounded number.

41. Mariko calculated her GPA to be 3.48539. Round this number to the nearest hundredth.
Mariko calculated her GPA to be 3.49.

42. Julia's average daily electricity usage for March was 7.03609 kilowatt-hours. Round this to the nearest thousandth.
Julia's average daily electricity usage for March was 7.036 kilowatt-hours.

Section 5.3

Add.

43. Add 4.7 and 2.38.
7.08

44. Add 0.165 and 1.04.
1.205

45. Add 6.35, 15.2, and 0.081.
21.631

46. Add 3.2, 0.41, 11.907, and 6.
21.517

Subtract.

47. $7.2 - 6.51$
0.69

48. $4 - 1.381$
2.619

49. Subtract 3.5 from 7.9.
4.4

50. Subtract 0.149 from 1.2.
1.051

Evaluate.

51. $-0.32 + (-0.43)$
-0.75

52. $-0.52 + (-3.9)$
-4.42

53. $4 + (-2.8)$
1.2

54. $0.34 + (-0.97)$
-0.63

55. $-1 - (-3.45)$
2.45

56. $3 - (-4.12)$
7.12

Combine like terms.

57. $-0.38x + 4.2x$
$3.82x$

58. $2.1y + (-6.07y)$
$-3.97y$

59. $-5.9w + (-0.24w)$
$-6.14w$

60. $-16.3c - 5.04c$
$-21.34c$

61. $3.72h - (-4.9h)$
$8.62h$

62. $-2.06v - (-4.56v)$
$2.5v$

Solve. Answer the question with a sentence.

63. Find the difference in altitude between a fisherman's pole 3.82 meters above sea level and a school of fish 11.7 meters below sea level.
The difference in altitude is 15.52 meters.

64. The temperature of a frozen chicken was −14.7°C when George placed it in the microwave to thaw. After 2 minutes, the chicken's temperature had risen 8.6°C. What was the temperature of the chicken after 2 minutes?
The temperature of the chicken was −6.1°C.

65. Kelli works in a candy store where customers choose their own quantities and bag the candy themselves. One customer brought four bags of candy to Kelli, and she weighed them one at a time with these results: 1.27 pounds, 0.4 pounds, 0.882 pounds, and 1 pound. What was the total weight of the four bags of candy?
The total weight of the four bags of candy was 3.552 pounds.

66. Zhe-Won, a tax accountant, was working on Samantha's tax return when she came across postage receipts with these totals: \$20.75, \$23.06, \$11.90, \$15.17, and \$8.92. How much did Samantha pay in postage?
Samantha paid \$79.80 in postage.

67. Brothers Malik and Benjou asked their mother to measure their heights. Their mother found that Malik was 1.34 meters tall and that Benjou was 1.61 meters tall. How much taller is Benjou than Malik?
Benjou is 0.27 meters taller than Malik.

68. Jenny purchased a meal from Burger Basket. The total came to \$7.18, including tax; and she paid with a \$20 bill. How much change did Jenny receive?
Jenny received \$12.82 in change.

69. Find the perimeter of this geometric figure.
5.475 feet

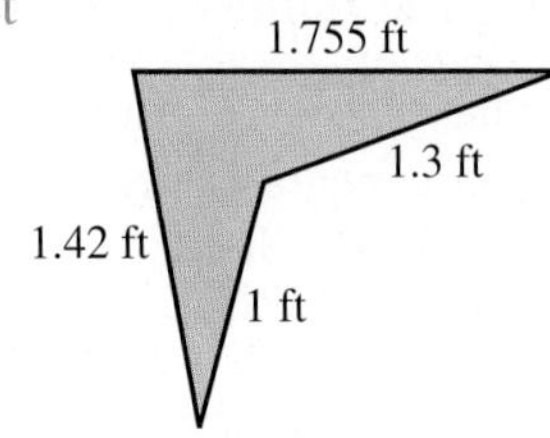

70. Find the length of the third side.
2.31 yards

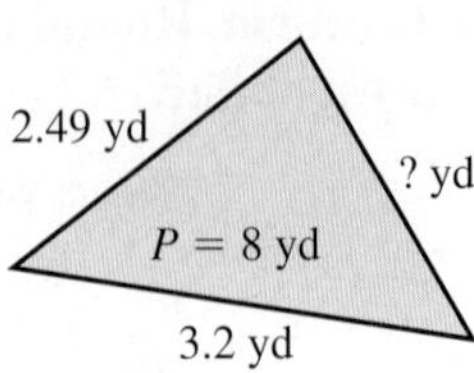

Section 5.4

Follow the procedure for multiplying decimals to find the product.

71. 4×2.1
8.4

72. 0.6×0.73
0.438

73. 1.2×3.5
4.2

74. 2.5×5.6
14

75. 32×0.75
24

76. 1.89×0.06
0.1134

77. 13.4×22.8
305.52

78. 100.6×0.45
45.27

Multiply by moving the decimal point the appropriate number of places to the right.

79. 8.3×10
83

80. 14.12×10
141.2

81. 2.91×100
291

82. 3.7×100
370

83. 0.804×100
80.4

84. $1.691 \times 1{,}000$
1,691

85. $0.089 \times 1{,}000$
89

86. $0.05 \times 1{,}000$
50

Multiply.

87. $-0.3 \times (-0.4)$
0.12

88. $-0.6 \times (-5)$
3

89. $0.8 \times (-0.09)$
-0.072

90. -0.2×0.03
-0.006

Solve. Answer the question with a sentence.

91. Jamali earns \$8.36 per hour. Yesterday she worked 7.25 hours. How much did Jamali earn yesterday?
Jamali earned \$60.61 yesterday.

92. Carlos is buying his girlfriend a necklace that costs \$36.50. If the sales tax rate is 0.06, what is the total amount that Carlos must pay for the necklace?
Carlos must pay a total of \$38.69 for the necklace.

Section 5.5

Divide. Abbreviate any repeating decimals by placing a bar over the repetend.

93. $26.6 \div 4$
6.65

94. $5.7 \div 11$
$0.5\overline{18}$

95. $\dfrac{1.46}{8}$
0.1825

96. $\dfrac{15.7}{6}$
$2.61\overline{6}$

97. $0.153 \div 0.09$
1.7

98. $0.025 \div 0.18$
$0.13\overline{8}$

99. $\dfrac{0.53}{7.5}$
$0.070\overline{6}$

100. $\dfrac{0.0364}{0.65}$
0.056

Find the decimal equivalent of each fraction. Abbreviate any repeating decimals by placing a bar over the repetend.

101. $\dfrac{8}{11}$
$0.\overline{72}$

102. $\dfrac{15}{8}$
1.875

103. $\dfrac{-1}{-12}$
$0.08\overline{3}$

104. $\dfrac{-13}{25}$
-0.52

Section 5.6

Solve. Answer each question with a complete sentence.

105. Wes earns $9.32 per hour as a cashier. How much does Wes earn for a 7-hour shift?
Wes earns $65.24 for a 7-hour shift.

106. Sunil is a server, and he worked 6.5 hours yesterday. Including wages and tips, he earned $109.33. How much did Sunil earn, on average, per hour?
Sunil earned, on average, $16.82 per hour.

107. The rec room in an apartment complex is rectangular. It measures 58.5 feet long and 41.8 feet wide. What is the area of the rec room?
The area of the rec room is 2,445.3 square feet.

108. A snare drum has a radius of 6.5 inches. What is the approximate circumference of the snare drum? (Round the answer to the nearest tenth.)
The circumference of the snare drum is approximately 40.8 inches.

109. Humberto paid $7.68 for 24 cans of soda. What was the price of each can of soda?
The price of each can of soda was $0.32.

110. Ruben's van gets 19.4 miles per gallon, and it can hold 18.5 gallons of gas. How far can Ruben's van travel on a full tank of gas?
Ruben's van can travel 358.9 miles on a full tank of gas.

111. It took Carmen 3.25 hours to walk 11.7 miles. What was Carmen's average walking rate?
Carmen's average walking rate was 3.6 miles per hour.

112. Tom and Margaret are financing the purchase of a new bed that costs $1,257.60, including sales tax. They will pay off the whole amount in 12 months (without interest). How much will Tom and Margaret pay each month?
Tom and Margaret will pay $104.80 each month.

Chapter 5 Test

1. In 9.6531, name the value of the digit 5.
   hundredths

2. Name 0.019 by writing it in word form.
   nineteen-thousandths

Write each decimal number as a decimal fraction.

3. 0.61 $\frac{61}{100}$
4. 0.008 $\frac{8}{1,000}$
5. 13.2 $\frac{132}{10}$

Write each decimal fraction as a decimal number.

6. $\frac{39}{1,000}$ 0.039
7. $\frac{5}{100}$ 0.05
8. $\frac{27}{10}$ 2.7

Locate and label each decimal number on the number line.

9. 1.6
10. −2.8
11. −0.3
12. 3.5

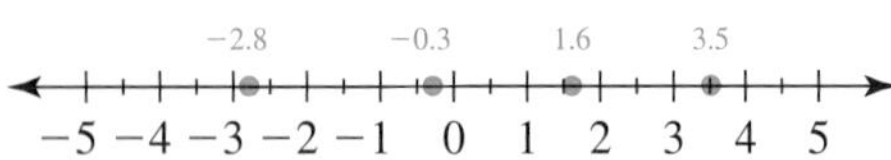

13. Round 49.396 to the nearest whole number.
    49

14. Round 5.9705 to the tenths place.
    6.0

15. Round $12.89503 to the nearest penny.
    $12.90

16. Round $354.053 to the nearest dollar.
    $354

17. Add 1.7 and 3.925.
    5.625

18. Subtract 2.508 from 3.4.
    0.892

Perform the indicated operation. Simplify the answer if possible. Abbreviate any repeating decimals.

19. $-0.9 + (-0.15)$ −1.05
20. $-1.05 - (-3.8)$ 2.75
21. 0.84×2.5 2.1
22. 3.8×100 380

Combine like terms.

23. $-2.9p + 4.05p$ $1.15p$
24. $1.7m - 0.23m$ $1.47m$

Divide.

25. $5.3 \div 6$ $0.88\overline{3}$
26. $\frac{0.364}{0.13}$ 2.8

27. Find the decimal equivalent of $\frac{13}{8}$.
    1.625

28. Find the decimal equivalent of $\frac{-4}{11}$.
    $-0.\overline{36}$

Solve. Answer the question with a sentence.

29. Paul purchased some garden tools for $49.34, including tax, and paid the cashier $60. How much change did Paul get back?
    Paul got back $10.66 in change.

30. Commuting to and from work last week, Silvia drove her car 144 miles on 6.4 gallons of gas. How many miles per gallon did Silvia's car average?
    Silvia's car averaged 22.5 miles per gallon.

31. Rhonda is buying a magazine priced at $4.95. If the sales tax rate is 0.08, what is the total amount Rhonda must pay for the magazine?
    The total amount Rhonda must pay for the magazine is $5.35.

32. Dom earns $10.40 per hour. Last week he worked 33.25 hours. How much did Dom earn last week?
    Dom earned $345.80 last week.

33. Donette is making a rectangular crib quilt that has an area of 19.8 square feet. The quilt is 3.6 feet wide. What is the length of the quilt?
    The length of the quilt is 5.5 feet.

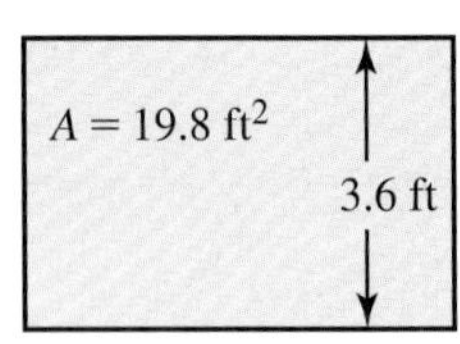

34. The painted circle around a tether ball post has a radius of 5 feet. What is the approximate circumference of this circle?
    The circumference is approximately 31.4 feet.

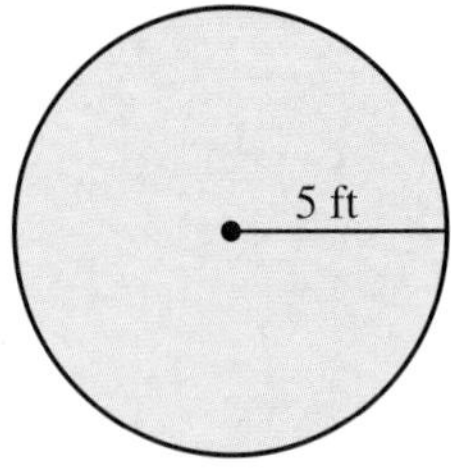

CHAPTER

Ratios, Proportions, and Percents

Introduction

In almost every academic and artistic field, the part and the whole are looked at together.

- In dance, the performer must be aware of and practice the smaller steps (the part) to be able to perform the entire dance (the whole).
- An auto mechanic may look at a car's fuel injection system (the part) to learn how it affects the engine (the whole) working properly.
- Biologists look at cells (the part) to understand the body (the whole).
- Statisticians ask questions of a small number of people (the part) to predict what the population (the whole) believes.
- Writers must understand *parts* of speech to put a sentence, a paragraph, a chapter, and a *whole* story together.

Where do *you* use a part of something to better understand the whole?

In our study of ratios, proportions, and percents, we continually compare a part to its whole.

Preparation Exercises

Section 1.7 Equations
Solve for x.

1. $6x = 114$
   $x = 19$
2. $185 = 5x$
   $x = 37$

Section 4.3 Equivalent Fractions
Simplify completely.

3. $\frac{72}{100}$
   $\frac{18}{25}$
4. $\frac{32}{80}$
   $\frac{2}{5}$

Section 4.3 Equivalent Fractions
Build up each fraction to have a denominator of 100.

5. $\frac{7}{25}$
   $\frac{28}{100}$
6. $\frac{3}{5}$
   $\frac{60}{100}$

Section 5.1 Introduction to Decimals
Write each decimal as a simplified fraction.

7. 0.08
   $\frac{2}{25}$
8. 1.2
   $\frac{6}{5}$

Section 5.1 Introduction to Decimals
Write each fraction as a decimal.

9. $\frac{19}{1{,}000}$
   0.019
10. $\frac{4}{100}$
   0.04

Section 5.2 Rounding Decimals
Round each decimal to the hundredths place.

11. 3.560914
   3.56
12. 8.195347
   8.20

Section 5.4 Multiplying Decimals
Multiply.

13. 1.8×6.5
11.7

14. 0.42×1.03
0.4326

15. 3.7×10
37

16. 0.09×100
9

17. 71.5×100
7,150

18. 0.45×10
4.5

Section 5.5 Dividing Decimals
Divide.

19. $3.43 \div 7$
0.49

20. $8.5 \div 0.04$
212.5

21. $\frac{7}{8}$
0.875

22. $\frac{2}{15}$
$0.1\overline{3}$

23. $\frac{21.3}{100}$
0.213

24. $\frac{6.7}{100}$
0.067

SECTION 6.1 Ratios

OBJECTIVES

In this section, you will learn to
- Define *ratio* and identify the form in which a ratio is written.
- Simplify ratios.
- Use cross multiplication to determine whether two ratios are equivalent.

You Need to know

To successfully complete this section, you need to understand
- ☐ Multiplying whole numbers (1.3)
- ☐ Simplifying fractions (4.3)
- ☐ The Multiplication Property of Equality (4.8)
- ☐ Multiplying decimals (5.4)
- ☐ Dividing decimals (5.5)

Introduction

Consider this: Two women, Sherrie and Jacqui, went into business together selling used books. To start the business, Sherrie invested \$3,000 and Jacqui invested \$2,000, a total investment of \$5,000. When they share profits from the income of their bookstore, Sherrie gets a $\frac{3}{5}$ share and Jacqui gets a $\frac{2}{5}$ share. Does this seem like a fair way to distribute the profits?

If in the first month their profits are \$500, how much of the \$500 should Sherrie and Jacqui receive?

In investment businesses, the greater the amount of the investment, the greater the amount of the *return* when the profits are given out.

We can start to compare Sherrie's and Jacqui's shares by the amount they originally invested: \$3,000 to \$2,000. However, because not all income is based in thousands of dollars, let's look at this in a simplified form: \$3 to \$2.

We can interpret this as for every \$3 that Sherrie invested, Jacqui invested \$2. We can double each amount, and the comparison will be consistent: For every \$6 (twice \$3) Sherrie invested, Jacqui invested \$4 (twice \$2).

This type of comparison of numbers is called a *ratio*. A **ratio** is a comparison between two numbers using division. The symbol used to represent a ratio is the colon.

Ratio Forms

The following table shows two ways to write the ratio between two numbers A and B. Each example compares the amounts of Sherrie's and Jacqui's investments.

1.	Standard ratio form:	*A* : *B*	Sherrie's amount : Jacqui's amount	\$3,000 : \$2,000
2.	Fractional form:	$\frac{A}{B}$	$\frac{\text{Sherrie's amount}}{\text{Jacqui's amount}}$	$\frac{\$3{,}000}{\$2{,}000}$
The ratios are read as →		*A* to *B*	Sherrie's amount to Jacqui's amount	\$3,000 to \$2,000

The two numbers forming a ratio are always specific to a certain unit of measure or count.

Examples of units of measure are miles, gallons, pounds, calories, and acres.

Examples of counts are the number of students, the number of trucks, and the number of candy bars.

In some ratios, the units of measure (or counts) are the same for each number. When the units of measure (or count) are *not* the same, the ratio is called a **rate.** We will discuss rates in Section 6.2, but the discussion in this section applies to *all* ratios.

Sample ratios are

- 760 calories : 20 candy bars $= \dfrac{\text{760 calories}}{\text{20 candy bars}}$
- 350 miles : 14 gallons $= \dfrac{\text{350 miles}}{\text{14 gallons}}$
- 6 dollars : 100 dollars $= \dfrac{\text{6 dollars}}{\text{100 dollars}}$

A variety of ratios can be written based on known information. In our example of Sherrie's and Jacqui's investments, we can create all of these ratios (and others):

Word form	Standard ratio form	Fractional form
Sherrie's amount : Jacqui's amount	\$3,000 : \$2,000	$\dfrac{\$3{,}000}{\$2{,}000}$
Jacqui's amount : Sherrie's amount	\$2,000 : \$3,000	$\dfrac{\$2{,}000}{\$3{,}000}$
Sherrie's amount : Total amount	\$3,000: \$5,000	$\dfrac{\$3{,}000}{\$5{,}000}$
Jacqui's amount : Total amount	\$2,000 : \$5,000	$\dfrac{\$2{,}000}{\$5{,}000}$

Example 1

Joy and Nikia are to receive a total of \$24,000 from their grandfather's will. Of this amount, Joy will receive \$16,000. According to this situation, write the following ratios in standard ratio form and in fractional form.

Procedure We first need to know how much Nikia will receive; so, we subtract:

$$\$24{,}000 - \$16{,}000 = \$8{,}000.$$

Answer

Word form	Standard ratio form	Fractional form
a) Joy's amount : Nikia's amount	\$16,000 : \$8,000	$\dfrac{\$16{,}000}{\$8{,}000}$
b) Nikia's amount : Joy's amount	\$8,000 : \$16,000	$\dfrac{\$8{,}000}{\$16{,}000}$
c) Joy's amount : Total amount	\$16,000 : \$24,000	$\dfrac{\$16{,}000}{\$24{,}000}$
d) Nikia's amount : Total amount	\$8,000 : \$24,000	$\dfrac{\$8{,}000}{\$24{,}000}$

Caution The order in which a ratio is written is important: $A : B \neq B : A$.

You Try It 1 **Sara and Tom are married, and each contributes to the Red Cross through their monthly paychecks. Their total monthly contribution is $160. Each month, Sara contributes $70 to the Red Cross. According to this situation, write the following ratios in standard ratio form and in fractional form. Use Example 1 as a guide.**

	Word form	Standard ratio form	Fractional form
a)	Sara's amount : Tom's amount		
b)	Tom's amount : Sara's amount		
c)	Sara's amount : Total amount		
d)	Tom's amount : Total amount		

Simplifying Ratios

Notice that the simplified ratio is an improper fraction, $\frac{5}{3}$. It is common to leave ratios in *improper fractional form* and not rewrite them as mixed numbers. As a mixed number, it is difficult to build up or to simplify. In a ratio, we must maintain both a numerator and a denominator.

In fractional form, simplifying a ratio is the same as simplifying a fraction: We divide out any common factors. A ratio of 40 to 24, written as $\frac{40}{24}$, can simplify—by a factor of 8—to $\frac{5}{3}$. When the unit of measure for each number is the same, that unit of measure can be divided out as well.

For example, if a rectangular building has a length of 40 feet and a width of 24 feet, the ratio of length : width is 40 feet : 24 feet, written as $\frac{40 \text{ feet}}{24 \text{ feet}}$. We can simplify the numbers by a factor of 8, and we can simplify the units of measure because they are the same. It is as if we have $\frac{\text{feet}}{\text{feet}} = 1$.

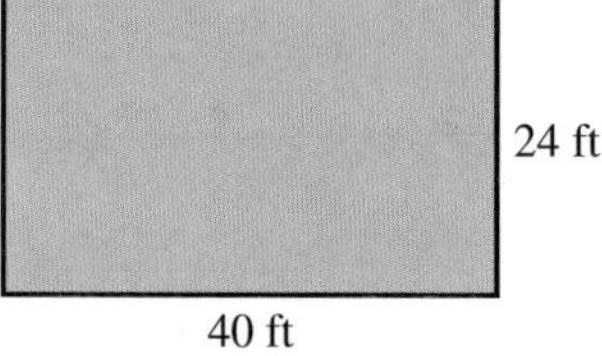

$$\text{In other words, } \frac{40 \text{ feet}}{24 \text{ feet}} = \frac{5 \cdot \cancel{8}\ \cancel{\text{feet}}}{3 \cdot \cancel{8}\ \cancel{\text{feet}}} = \frac{5}{3}.$$

Knowing that this ratio simplifies to $\frac{5}{3}$, an architect can build a model of the building that accurately reflects the building's dimensions.

When the units of measure are not the same, they may not be divided out. For example, in the ratio 40 dollars : 24 feet, the numbers can be simplified but not the units.

$$\frac{40 \text{ dollars}}{24 \text{ feet}} = \frac{5 \cdot \cancel{8} \text{ dollars}}{3 \cdot \cancel{8} \text{ feet}} = \frac{5 \text{ dollars}}{3 \text{ feet}}$$

Sometimes the numbers used in a ratio (the numerator and the denominator) are decimal numbers. When that happens, we need to multiply by 1, using the same power of 10 in the numerator and the denominator, to clear the decimals. Then we can simplify the resulting fraction.

For example, Ryan is an experienced server and knows how to treat his customers so that he generates, on average, \$2.40 per person in tips. Tyler is relatively new on the job, and his average tips are \$1.80 per person. The ratio of Ryan's tips to Tyler's tips can be written as

$$\text{Ryan's tips : Tyler's tips} = \frac{\text{Ryan's tips}}{\text{Tyler's tips}} = \frac{\$2.40}{\$1.80} = \frac{2.4 \text{ dollars}}{1.8 \text{ dollars}} = \frac{2.4}{1.8}$$

To simplify this fraction, we can multiply both the numerator and denominator by a power of 10 to eliminate the decimals. Because each has one decimal place, we should multiply by 10

$$\frac{2.4 \times 10}{1.8 \times 10} = \frac{24}{18}$$

Now that the numerator and the denominator are whole numbers, we can reduce the fraction by a factor of 6.

$$\frac{24}{18} = \frac{4}{3}$$

This means that for every $4 Ryan receives in tips, Tyler receives $3.

Example 2

Write each ratio in fractional form and then simplify it. Make sure you divide out any common units of measure (or counts) as well.

a) 30 inches : 42 inches **b)** 32 trophies : 8 trophies **c)** 5.6 miles : 3.5 liters

Procedure For part c, multiply the fraction by $\frac{10}{10}$ to eliminate the decimals.

Answer

a) $\dfrac{30 \text{ inches}}{42 \text{ inches}} = \dfrac{5 \cdot 6 \text{ inches}}{7 \cdot 6 \text{ inches}} = \dfrac{5}{7}$ We also can rewrite this in standard ratio form, 5 : 7.

b) $\dfrac{32 \text{ trophies}}{8 \text{ trophies}} = \dfrac{4 \cdot 8 \text{ trophies}}{1 \cdot 8 \text{ trophies}} = \dfrac{4}{1}$ This fraction cannot simplify to just 4. As a ratio, it must remain in fractional form or be rewritten in standard form.

c) $\dfrac{5.6 \text{ miles}}{3.5 \text{ liters}} = \dfrac{5.6 \times 10 \text{miles}}{3.5 \times 10 \text{liters}} = \dfrac{56 \text{ miles}}{35 \text{ liters}} = \dfrac{8 \cdot 7 \text{ miles}}{5 \cdot 7 \text{ liters}} = \dfrac{8 \text{ miles}}{5 \text{ liters}}$ In this example, the units of measure are not the same and cannot be divided out.

You Try It 2 **Write each ratio in fractional form and then simplify it. Make sure you divide out any common units of measure (or counts) as well. Use Example 2 as a guide.**

a) 54 inches : 63 inches **b)** 25 boxes : 55 boxes

c) 0.9 grams of fat : 2.7 ounces **d)** 4.8 feet : 0.6 second

Writing Ratios from Phrases

Some words used in phrases commonly indicate a ratio.

1. In the context of fractional form, *per* means "divided by."
2. *For each* means "per 1" or "divided by 1."
3. *For every* means "per many" or "divided by more than 1" (in the denominator).

Typically, when these words are used, the units of measure are different and cannot be divided out.

Example 3

Write each of the following comparisons as a ratio in the form of a fraction. Simplify if possible.

Answer

Phrase	Ratio	Simplified
a) 35 miles per hour **This means "35 miles per 1 hour."** **The units of measure are miles and hours.**	$\dfrac{35 \text{ miles}}{1 \text{ hour}}$	Already simplified
b) 215 calories for each can ***For each* means the same as *per 1*.** **The unit of measure is calories, and the count is the number of cans.**	$\dfrac{215 \text{ calories}}{1 \text{ can}}$	Already simplified
c) 3 boats for every 24 passengers **The counts are the number of boats and passengers.**	$\dfrac{3 \text{ boats}}{24 \text{ passengers}} =$	$\dfrac{1 \text{ boat}}{8 \text{ passengers}}$

You Try It 3 **Write each of the following comparisons as a ratio in the form of a fraction. Simplify if possible. Use Example 3 as a guide.**

	Phrase	Ratio	Simplified
a)	20 inches per second		
b)	$65 for each ticket **Write $65 as "65 dollars."**		
c)	$100 for every 8 people		

Example 4

Write each ratio as a comparison.

a) $\dfrac{\text{5 dollars}}{\text{1 ticket}}$ **b)** $\dfrac{\text{2 doctors}}{\text{15 patients}}$

Answer **a)** $\dfrac{\text{5 dollars}}{\text{1 ticket}}$ $5 per 1 ticket or $5 per ticket or $5 for each ticket

b) $\dfrac{\text{2 doctors}}{\text{15 patients}}$ 2 doctors for every 15 patients

Notice that because the denominator is greater than 1, we use the words *for every* instead of *for each.*

You Try It 4 **Write each ratio as a comparison. Use Example 4 as a guide.**

a) $\dfrac{\text{45 miles}}{\text{1 hour}}$ ____________________

b) $\dfrac{\text{9 children}}{\text{2 adults}}$ ____________________

c) $\dfrac{\text{1 winner}}{\text{20 participants}}$ ____________________

Think About It 1 Where have you heard the word *per* used in everyday English? Write some examples.

____

____

Equivalent Ratios

As you know, when two fractions have the same value, such as $\frac{4}{6} = \frac{2}{3}$, we say that they are equivalent fractions. Likewise, when two ratios have the same value, we say that they are **equivalent ratios.**

If we look at these equivalent fractions, $\frac{4}{6} = \frac{2}{3}$, as two ratios expressed in fractional form (without using any units of measure), we can rewrite them as $4:6 = 2:3$.

Using this standard ratio form, we are able to introduce the notion of *means* and *extremes.*

In the ratio form $4:6 = 2:3$, the *outermost* numbers, 4 and 3, are the **extremes** and the *middle* two numbers, 6 and 2, are the **means.**

Here is a diagram of the means and extremes of equivalent ratios and of equivalent fractions.

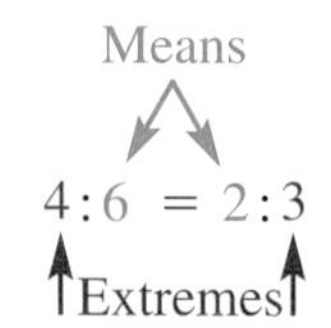

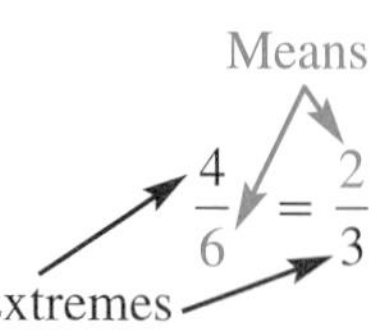

Example 5

Identify the means and the extremes for the pair of equivalent fractions $\frac{10}{15} = \frac{4}{6}$.

Answer The means are 15 and 4. The extremes are 10 and 6.

You Try It 5 **Identify the means and extremes for each pair of equivalent fractions. Use Example 5 as a guide.**

a) $\frac{3}{8} = \frac{20}{45}$ Means: ______ Extremes: ______

b) $\frac{20}{8} = \frac{15}{6}$ Means: ______ Extremes: ______

We can use the means and extremes to test whether two ratios are equivalent.

The Means–Extremes Product Property

If the product of the means equals the product of the extremes, then the two ratios are equivalent.

If the product of the extremes *does not equal* the product of the means, then the two ratios are *not* equivalent.

Example 6

Find the products of the means and extremes for each pair of ratios and determine whether the ratios are equivalent.

a) $\frac{10}{15} \stackrel{?}{=} \frac{4}{6}$

b) $\frac{9}{12} \stackrel{?}{=} \frac{12}{15}$

Answer

a) The product of the means is $15 \times 4 = 60$.

The product of the extremes is $10 \times 6 = 60$.

The ratios $\frac{10}{15}$ and $\frac{4}{6}$ are equivalent.

Note: Both fractions simplify to $\frac{2}{3}$.

b) The product of the means is $12 \times 12 = 144$.

The product of the extremes is $9 \times 15 = 135$.

The ratios $\frac{9}{12}$ and $\frac{12}{15}$ are not equivalent.

Note: $\frac{9}{12}$ simplifies to $\frac{3}{4}$, and $\frac{12}{15}$ simplifies to $\frac{4}{5}$.

You Try It 6 **Find the products of the means and extremes for each pair of fractions, and determine whether the fractions are equivalent. Use Example 6 as a guide.**

Equivalent ratios?

a) $\frac{3}{8} \stackrel{?}{=} \frac{20}{45}$ Product of means ______ Product of extremes ______ ______ Yes/No

b) $\frac{9}{12} \stackrel{?}{=} \frac{15}{20}$ Product of means ______ Product of extremes ______ ______ Yes/No

Cross Multiplication

A direct method for finding the products of the extremes and means is referred to as **cross multiplication.**

Starting with $\frac{4}{3} \stackrel{?}{=} \frac{8}{6}$,

multiply "across" the equal sign.

$$\frac{4}{3} \stackrel{?}{=} \frac{8}{6}$$

to get

$$4 \cdot 6 \stackrel{?}{=} 3 \cdot 8$$

It's true that $24 = 24$ ✓

The product of the extremes = The product of the means

You Try It 7 **Determine whether the ratios are equivalent by cross multiplying.**

a) $\frac{2}{3} \stackrel{?}{=} \frac{8}{12}$ **b)** $\frac{6}{15} \stackrel{?}{=} \frac{8}{25}$ **c)** $\frac{0.4}{1} \stackrel{?}{=} \frac{0.5}{1.25}$

Why Cross Multiplication Works

Cross multiplication is not some magic rule. It is developed from the Multiplication Property of Equations, first introduced in Section 4.8. From the original two ratios, we multiply each side by the product of both denominators.

$$\frac{4}{3} = \frac{8}{6}$$

Multiply each side by $\frac{3 \cdot 6}{1}$: $\frac{4}{3} \cdot \frac{3 \cdot 6}{1} = \frac{3 \cdot 6}{1} \cdot \frac{8}{6}$

Multiply the fractions together: $\frac{3 \cdot 6 \cdot 4}{1 \cdot 3} = \frac{3 \cdot 6 \cdot 8}{1 \cdot 6}$

Divide out the common factors within each fraction. Divide out the 3's on the left side and the 6's on the right side. The result is an equation that is the same as the equation we got when we used cross multiplication.

$$6 \cdot 4 = 3 \cdot 8$$

Answers: You Try It and Think About It

You Try It: **1. a)** \$70 : \$90; $\frac{\$70}{\$90}$ **b)** \$90 : \$70; $\frac{\$90}{\$70}$ **c)** \$70 : \$160; $\frac{\$70}{\$160}$ **d)** \$90 : \$160; $\frac{\$90}{\$160}$ **2. a)** $\frac{6}{7}$ **b)** $\frac{5}{11}$ **c)** $\frac{1 \text{ gram of fat}}{3 \text{ ounces}}$ **d)** $\frac{8 \text{ feet}}{1 \text{ second}}$ **3. a)** $\frac{20 \text{ inches}}{1 \text{ second}}$ **b)** $\frac{65 \text{ dollars}}{1 \text{ ticket}}$ **c)** $\frac{100 \text{ dollars}}{8 \text{ people}} = \frac{25 \text{ dollars}}{2 \text{ people}}$ **4. a)** 45 miles per hour or 45 miles for each hour **b)** 9 children for every 2 adults **c)** 1 winner for every 20 participants **5. a)** The means are 8 and 20; the extremes are 3 and 45. **b)** The means are 8 and 15; the extremes are 20 and 6. **6. a)** The product of the means is $8 \times 20 = 160$; the product of the extremes is $3 \times 45 = 135$. No. **b)** The product of the means is $12 \times 15 = 180$; the product of the extremes is $9 \times 20 = 180$. Yes. **7. a)** Yes **b)** No **c)** Yes

Think About It: **1.** Answers may vary. Two possibilities are: miles per hour and miles per gallon.

Section 6.1 Exercises

FOR EXTRA HELP
Student Resources on DVD-ROM — Includes: Student's Solutions Manual, Video Lectures, Chapter Test Prep Video
MyMathLab
MathXL

Think Again

Write a sentence or two for each response.

1. Where have you heard the word *per* used in everyday English? Write some examples. *(Refer to Think About It 1)*
Answers will vary.

2. This section contains the following caution statement: The order in which a ratio is written is important: $A:B \neq B:A$. Why is this true?
Answers will vary. One possibility: As fractions, the ratio A to B is the reciprocal of the ratio B to A.

3. Is it possible for a ratio to be equivalent to 0? Explain your answer or show an example that supports your answer.
Yes. In a ballet class of 12 students, all girls, the ratio of boys to girls is 0 : 12, or $\frac{0}{12}$.

4. In Section 4.3, we simplified fractions. How can we tell if a fraction has been simplified correctly?
The original fraction can be set equal to the resulting fraction; and if cross multiplication results in a true statement, the fraction was simplified correctly.

Focus Exercises

Fill in the blanks.

5. Feet, meters, ounces, and gallons are examples of __units of measure__.

6. A ratio is a comparison between two numbers using __division__.

7. In the context of fractions, *per* means __divided by__.

8. If two ratios have the same value, we say that they are __equivalent__ ratios.

Write each of the following comparisons as a ratio in the form of a fraction.

	Phrase	Ratio
9.	60 miles per gallon	$\frac{60 \text{ miles}}{1 \text{ gallon}}$
10.	$1.35 for each can	$\frac{\$1.35}{1 \text{ can}}$
11.	$2 for every 3 tickets	$\frac{2 \text{ dollars}}{3 \text{ tickets}}$
12.	20 flights per day	$\frac{20 \text{ flights}}{1 \text{ day}}$
13.	35 miles per hour	$\frac{35 \text{ miles}}{1 \text{ hour}}$
14.	25 children for every 4 adults	$\frac{25 \text{ children}}{4 \text{ adults}}$
15.	3 laps every 10 minutes	$\frac{3 \text{ laps}}{10 \text{ minutes}}$
16.	3 autographs for every 25 cards	$\frac{3 \text{ autographs}}{25 \text{ cards}}$

Write each fraction as a comparison.

17. $\frac{7 \text{ people}}{1 \text{ van}}$
7 people per van

18. $\frac{\$8.50}{1 \text{ hour}}$
$8.50 per hour

19. $\frac{23.4 \text{ miles}}{1 \text{ gallon}}$
23.4 miles per gallon

20. $\frac{8 \text{ rooms}}{1 \text{ house}}$
8 rooms per house

21. $\frac{9 \text{ lemons}}{5 \text{ dollars}}$
9 lemons for every $5

22. $\frac{2 \text{ coffee shops}}{3 \text{ miles}}$
2 coffee shops for every 3 miles

23. $\frac{1 \text{ boy}}{2 \text{ girls}}$
1 boy for every 2 girls

24. $\frac{5 \text{ credits}}{3 \text{ members}}$
5 credits for every 3 members

Write each ratio in fractional form and simplify it completely.

25. 24 feet : 32 feet
$\frac{3}{4}$

26. 20 dollars : 35 dollars
$\frac{4}{7}$

27. 6 hours : 30 hours
$\frac{1}{5}$

28. 9 minutes : 45 minutes
$\frac{1}{5}$

29. 50 miles : 4 hours
$\frac{25 \text{ miles}}{2 \text{ hours}}$

30. 45 calories : 6 ounces
$\frac{15 \text{ calories}}{2 \text{ ounces}}$

31. 120 liters : 30 liters
$\frac{4}{1}$

32. 150 inches : 75 inches
$\frac{2}{1}$

33. 9.6 liters : 0.8 hours
$\frac{12 \text{ liters}}{1 \text{ hour}}$

34. 15.6 meters : 1.2 seconds
$\frac{13 \text{ meters}}{1 \text{ second}}$

35. 12.6 degrees : 0.4 degrees
$\frac{63}{2}$

36. 2.8 years : 0.7 years
$\frac{4}{1}$

Write each described ratio in fractional form and simplify it completely.

37. In 2003, the estimated population of Alabama was 4.5 million and the estimated population of West Virginia was 2.7 million. Write the populations as a ratio, Alabama's population : West Virginia's population. ***Source:*** **www.census.gov**
$\frac{5}{3}$

38. It takes Jupiter about 12 Earth years to orbit the Sun, and it takes Uranus about 84 earth years to orbit the sun. Write the orbits as a ratio, Jupiter's time: Uranus's time. ***Source:*** **www.windows.ucar.edu**

$\frac{1}{7}$

39. In 2004, about 52% of the seniors in Texas and Washington took the SAT exam. In Texas, the average SAT verbal score was about 490; and in Washington, the average verbal score was about 525. Write these SAT verbal scores as a ratio, Texas's scores : Washington's scores. ***Source:*** **www.collegeboard.com**

$\frac{14}{15}$

40. In North Carolina, Henderson County has a land area of about 375 square miles and Perquimans County has a land area of about 250 square miles. Write these land areas as a ratio, Henderson's area : Perquimans' area. ***Source:*** **www.quickfacts.census.gov**

$\frac{3}{2}$

Write the indicated ratios in standard ratio form, in fractional form, and as a simplified fraction.

41. Chris bought school bonds issued by the Austin Unified School District as an investment. He paid $300 for the Series A bond and $150 for the Series B bond.

a) Series A amount : Series B amount

$\$300 : \$150; \frac{\$300}{\$150}; \frac{2}{1}$

b) Series B amount : Series A amount

$\$150 : \$300; \frac{\$150}{\$300}; \frac{1}{2}$

c) Series A amount : Total amount

$\$300 : \$450; \frac{\$300}{\$450}; \frac{2}{3}$

d) Series B amount : Total amount

$\$150 : \$450; \frac{\$150}{\$450}; \frac{1}{3}$

42. Each month, Maria pays an average utility gas and electric bill of $160. The electric portion of the bill averages $120.

a) Electric amount : Gas amount

$\$120 : \$40; \frac{\$120}{\$40}; \frac{3}{1}$

b) Gas amount : Electric amount

$\$40 : \$120; \frac{\$40}{\$120}; \frac{1}{3}$

c) Electric amount : Total amount

$\$120 : \$160; \frac{\$120}{\$160}; \frac{3}{4}$

d) Gas amount : Total amount

$\$40 : \$160; \frac{\$40}{\$160}; \frac{1}{4}$

43. Mae-Li has two children, Cheyenne and Su, in college. Each semester Mae-Li pays a total of $9,600 in tuition costs. She pays $3,600 to Su's college and the rest to Cheyenne's college.

a) Cheyenne's tuition : Su's tuition

$\$6{,}000 : \$3{,}600; \frac{\$6{,}000}{\$3{,}600}; \frac{5}{3}$

b) Su's tuition : Cheyenne's tuition

$\$3{,}600 : \$6{,}000; \frac{\$3{,}600}{\$6{,}000}; \frac{3}{5}$

c) Cheyenne's tuition : Total tuition

$\$6{,}000 : \$9{,}600; \frac{\$6{,}000}{\$9{,}600}; \frac{5}{8}$

d) Su's tuition : Total tuition

$\$3{,}600 : \$9{,}600; \frac{\$3{,}600}{\$9{,}600}; \frac{3}{8}$

44. DeMarco paid $1,050 for a pair of season tickets to the Cucamonga Quakes minor league baseball team for him and his wife, Julia. DeMarco's seats cost him $600, and Julia's were priced at a discount.

a) DeMarco's amount : Julia's amount

$\$600 : \$450; \frac{\$600}{\$450}; \frac{4}{3}$

b) Julia's amount : DeMarco's amount

$\$450 : \$600; \frac{\$450}{\$600}; \frac{3}{4}$

c) DeMarco's amount : Total amount

$\$600 : \$1{,}050; \frac{\$600}{\$1{,}050}; \frac{4}{7}$

d) Julia's amount : Total amount

$\$450 : \$1{,}050; \frac{\$450}{\$1{,}050}; \frac{3}{7}$

Use cross multiplication to determine whether each pair of fractions is equivalent.

45. $\frac{18}{12} \stackrel{?}{=} \frac{3}{2}$ Yes

46. $\frac{15}{6} \stackrel{?}{=} \frac{5}{2}$ Yes

47. $\frac{4}{12} \stackrel{?}{=} \frac{5}{10}$ No

48. $\frac{8}{10} \stackrel{?}{=} \frac{6}{8}$ No

49. $\frac{12}{30} \stackrel{?}{=} \frac{0.8}{1.5}$ No

50. $\frac{15}{20} \stackrel{?}{=} \frac{0.2}{0.3}$ No

51. $\frac{2.5}{1.5} \stackrel{?}{=} \frac{1}{0.6}$ Yes

52. $\frac{0.4}{0.6} \stackrel{?}{=} \frac{1.6}{2.4}$ Yes

Think Outside the Box

In general, we cannot add ratios. To see why, consider the following.

53. Mrs. Nesbitt has a morning dance class with 15 girls and 5 boys and an afternoon dance class with 20 girls and 10 boys. When both classes get together to perform, there are 35 girls and 15 boys.

Use that information to fill in the following tables and to find the requested ratios in simplified form.

Morning Class	Students	Ratio	Simplified Ratio
Girls	15	$\frac{\text{Girls}}{\text{Total}} = \frac{15}{20}$	$\frac{3}{4}$
Boys	5	$\frac{\text{Boys}}{\text{Total}} = \frac{5}{20}$	$\frac{1}{4}$

Afternoon Class	Students	Ratio	Simplified Ratio
Girls	20	$\frac{\text{Girls}}{\text{Total}} = \frac{20}{30}$	$\frac{2}{3}$
Boys	10	$\frac{\text{Boys}}{\text{Total}} = \frac{10}{30}$	$\frac{1}{3}$

Combined Classes	Students	Ratio	Simplified Ratio
Girls	35	$\frac{\text{Girls}}{\text{Total}} = \frac{35}{50}$	$\frac{7}{10}$
Boys	15	$\frac{\text{Boys}}{\text{Total}} = \frac{15}{50}$	$\frac{3}{10}$

54. From the preceding tables, add the simplified ratios as shown and compare them to the combined class ratios. What can you conclude from this?
Answers may vary.

a) $\frac{\text{Morning girls}}{\text{Morning total}} + \frac{\text{Afternoon girls}}{\text{Afternoon total}} \stackrel{?}{=} \frac{\text{Combined girls}}{\text{Combined total}}$

$\frac{3}{4} + \frac{2}{3} \stackrel{?}{=} \frac{7}{10}$; False

b) $\frac{\text{Morning boys}}{\text{Morning total}} + \frac{\text{Afternoon boys}}{\text{Afternoon total}} \stackrel{?}{=} \frac{\text{Combined boys}}{\text{Combined total}}$

$\frac{1}{4} + \frac{1}{3} \stackrel{?}{=} \frac{3}{10}$; False

SECTION 6.2 Rates

OBJECTIVES

In this section, you will learn to
- Define and identify *rates.*
- Simplify rates.
- Define and calculate *unit rates.*
- Compare unit prices.

You Need to Know

To successfully complete this section, you need to understand
- ☐ Dividing whole numbers (1.4)
- ☐ Simplifying fractions (4.3)
- ☐ Dividing decimals (5.5)
- ☐ Writing ratios from phrases (6.1)
- ☐ Simplifying ratios (6.1)

Introduction

Consider this: Gilberto finished the 2004 Baja 1,000-mile off-road race in 41st place. During the race, he was able to average 50 miles per hour. How many hours did it take him to drive all 1,000 miles?

Recall from Section 6.1 that *50 miles per hour* can be written as a ratio, $\frac{50 \text{ miles}}{1 \text{ hour}}$. This ratio has different units of measure in the numerator and the denominator.

A ratio containing different units of measure or counts in the numerator and the denominator is called a **rate.** Some of the ratios we saw in Section 6.1 are rates.

$$\frac{760 \text{ calories}}{20 \text{ candy bars}}, \frac{350 \text{ miles}}{14 \text{ gallons}}, \frac{0.9 \text{ grams of fat}}{2.7 \text{ ounces}}, \text{ and } \frac{4.8 \text{ feet}}{0.6 \text{ seconds}} \text{ are all rates.}$$

For Gilberto, his racing rate of 50 miles in 1 hour suggests that he will go

- 100 miles in 2 hours; as a rate, this is $\frac{100 \text{ miles}}{2 \text{ hours}}$.
- 500 miles in 10 hours; as a rate, this is $\frac{500 \text{ miles}}{10 \text{ hours}}$.
- 1,000 miles in 20 hours; as a rate, this is $\frac{1{,}000 \text{ miles}}{20 \text{ hours}}$.

By the way, the last rate listed is the answer to the question: It took Gilberto 20 hours to drive all 1,000 miles.

Simplifying Rates

Rates are best understood when they are simplified completely. For example, the rate of $\frac{760 \text{ calories}}{20 \text{ candy bars}}$ can be simplified by a factor of 20.

$$\frac{760 \text{ calories } \div 20}{20 \text{ candy bars } \div 20} = \frac{38 \text{ calories}}{1 \text{ candy bar}} = 38 \text{ calories per candy bar}$$

Think About It 1 When a rate is simplified, will the units of measure ever divide out? Why or why not?

____

____

Example 1

Simplify each rate completely and write the result in words.

a) $\frac{63 \text{ granola bars}}{7 \text{ hikers}}$ **b)** $\frac{6 \text{ teachers}}{80 \text{ students}}$ **c)** $\frac{1.5 \text{ miles}}{0.09 \text{ gallons}}$

Procedure For part c, we need to multiply the numerator and the denominator by 100 to eliminate the decimals.

Answer

	Rate	Simplified rate	Word form
a)	$\dfrac{63 \text{ granola bars} \div 7}{7 \text{ hikers} \div 7}$	$= \dfrac{9 \text{ granola bars}}{1 \text{ hiker}}$	9 granola bars per hiker
b)	$\dfrac{6 \text{ teachers} \div 2}{80 \text{ students} \div 2}$	$= \dfrac{3 \text{ teachers}}{40 \text{ students}}$	3 teachers for every 40 students
c)	$\dfrac{1.5 \text{ miles} \times 100}{0.09 \text{ gallons} \times 100}$	$= \dfrac{150 \text{ miles} \div 3}{9 \text{ gallons} \div 3} = \dfrac{50 \text{ miles}}{3 \text{ gallons}}$	50 miles for every 3 gallons

Think About It 2 The answer in Example 1, part c shows multiplying the numerator and the denominator by 100. Why should each be multiplied by 100 and not 10?

___

___

You Try It 1 **Simplify each rate completely and write the result in words. Use Example 1 as a guide.**

	Rate	Simplified rate	Word form
a)	$\dfrac{56 \text{ pounds}}{7 \text{ inches}}$		
b)	$\dfrac{12 \text{ yellowtail}}{15 \text{ sea bass}}$		
c)	$\dfrac{6 \text{ meters}}{1.6 \text{ years}}$		

Unit Rates

In this chapter, we have seen many units of measure, such as ounces, feet, dollars, and gallons. Other types of units include a *single* can of soda in a case of soda, a *single* crayon in a box of crayons, and a *single* apple in a bag of apples. In other words, *unit* often means *one* of something.

A **unit rate** is a rate in which the denominator has been reduced to 1 item. For instance, in Example 1a, the rate of $\dfrac{63 \text{ granola bars}}{7 \text{ hikers}}$ was reduced to a unit rate of $\dfrac{9 \text{ granola bars}}{1 \text{ hiker}}$. In this case, the single unit is 1 hiker and the phrase we can write includes the word *per*, as in 9 granola bars per hiker.

Sometimes, as in the previous example, a completely simplified rate is a unit rate; so, we don't need to do any other work. Another example: A runner who runs at a rate of $\dfrac{16 \text{ miles}}{2 \text{ hours}}$ has a simplified rate of $\dfrac{8 \text{ miles}}{1 \text{ hour}}$, or 8 miles per hour.

Other times, long division is necessary. For example,

$$\frac{55 \text{ miles}}{4 \text{ gallons}}$$

cannot simplify directly, but we can divide as shown to the right.

$$\begin{array}{r} 13.75 \\ 4\overline{)55.00} \\ -4 \\ 15 \\ -12 \\ 30 \\ -28 \\ 20 \\ -20 \\ 0 \end{array}$$

We can then write the unit rate as $\dfrac{13.75 \text{ miles}}{1 \text{ gallon}}$, or 13.75 miles per gallon.

In a rate, the units of measure never divide out; so, we can ignore them for the long division process and include them when we write the unit rate.

Remember, the unit rate always has 1 in the denominator.

Example 2

Simplify each rate to a unit rate and write the unit rate in words.

a) $\frac{80 \text{ people}}{5 \text{ square miles}}$ b) $\frac{\$9.00}{12 \text{ cans}}$ c) $\frac{0.3 \text{ gallons}}{1.2 \text{ miles}}$

Procedure For parts a and b, use long division. For part c, recall that when dividing decimals, we need to adjust the decimal points so that the divisor is a whole number.

Answer

	Rate	Long division	Unit rate	Word form
a)	$\frac{80 \text{ people}}{5 \text{ square miles}}$	$5\overline{)80}$ = 16 →	$\frac{16 \text{ people}}{1 \text{ square mile}}$	16 people per square mile
b)	$\frac{9 \text{ dollars}}{12 \text{ cans}}$	$12\overline{)9.00}$ = 0.75 →	$\frac{0.75 \text{ dollars}}{1 \text{ can}}$	\$0.75 per can
c)	$\frac{0.3 \text{ gallons}}{1.2 \text{ miles}}$	$1.2\overline{)0.3}$ → $12\overline{)3.00}$ = 0.25 →	$\frac{0.25 \text{ gallons}}{1 \text{ mile}}$	0.25 gallons per mile

You Try It 2 **Simplify each rate to a unit rate and write the unit rate in words. Use Example 2 as a guide.**

	Rate	Simplified rate	Word form
a)	$\frac{96 \text{ miles}}{6 \text{ hours}}$		
b)	$\frac{174 \text{ words}}{4 \text{ minutes}}$		
c)	$\frac{\$9.80}{2.8 \text{ feet}}$		

When a rate is included in a problem, it is not always written as a phrase that is immediately obvious. In the following example, the pieces of information—the numerator and the denominator of the rate—are in two different sentences.

Consider this: Melanie was paid \$120 for removing a dead tree. The project took her 7 hours to complete. Write her earnings as an hourly rate.

From the given information, write a fractional rate. Simplify the rate to a unit rate and write it as a phrase.

An hourly rate is how much someone earns per hour. This suggests that the unit of measure, hours, is in the denominator, $\frac{???}{\text{hours}}$. The other unit of measure is dollars; so, from the given information, the rate becomes $\frac{120 \text{ dollars}}{7 \text{ hours}}$.

To write this as a unit rate, we use long division.

$$7\overline{)120.000} = 17.142$$

Because we are dealing with money, we should stop the division after three decimal places and round to the nearest penny: \$17.14. As a unit rate, this can be written as $\frac{17.14 \text{ dollars}}{1 \text{ hour}}$. Then we can answer the question: Melanie earned \$17.14 per hour.

Example 3

On a recent trip, Aden drove 147 miles before refilling the gas tank with 8.4 gallons. Write his car's miles per gallon as a unit rate.

Procedure From the given information, write a fractional rate. Simplify the rate to a unit rate and write it as a phrase.

Per gallon means that the number of gallons is in the denominator.

Answer

Rate	Long division	Unit rate	Word form
$\frac{147 \text{ miles}}{8.4 \text{ gallons}}$	$8.4\overline{)147.0}$ = 17.5 →	$\frac{17.5 \text{ miles}}{1 \text{ gallon}}$	17.5 miles per gallon

You Try It 3 **From the given information, write a fractional rate. Simplify the rate to a unit rate and write it as a phrase. Use Example 3 as a guide.**

a) Last week, Lena worked 15 hours and earned a total of $160, including wages and tips, for delivering pizza. Write her dollars earned per hour as a unit rate.

b) For his morning workout, Zane jogged on a treadmill. The machine showed that he burned 210 calories while jogging for 0.6 hours. Write his calories burned per hour as a unit rate.

Comparing Unit Prices

When shopping for food, it's common to compare prices of items that are similar. The unit rate for the price of an item is called the **unit price.** For example, the unit price for milk may be $2.95 per gallon. We can write this as $\frac{\$2.95}{1 \text{ gallon}}$. If we have a choice between two different brands of the same item, assuming the quality is similar, the one with the lower unit price is a better buy.

It's common to write unit prices as *price per item*. This means that money (dollars or cents) will always be in the numerator.

Consider, for example, two bags of chips, each weighing 28 ounces. If Gorkin's brand costs $2.58 and Byrd's brand costs $2.79, it's clear that Gorkin's chips are a better buy because those chips cost less.

However, if the bags are different in size and price, it's not as easy to know which bag is the better buy. If, for example, at another store Gorkin's bag of chips weighs 28 ounces and costs $2.46 and Byrd's small bag of chips weighs 16 ounces and costs $1.19, it's not immediately clear which is the better buy.

We can determine which is the better buy by finding the unit price for each bag of chips.

Gorkin's: $\frac{2.46 \text{ dollars}}{28 \text{ ounces}}$ $28\overline{)2.460}$ = 0.087 → $\frac{0.087 \text{ dollars}}{1 \text{ ounce}}$, which rounds to \$0.09 per ounce.

Byrd's: $\frac{1.19 \text{ dollars}}{16 \text{ ounces}}$ $16\overline{)1.190}$ = 0.074 → $\frac{0.074 \text{ dollars}}{1 \text{ ounce}}$, which rounds to \$0.07 per ounce.

This time, Byrd's chips are the better buy.

Example 4

A case of 24 cans of Blitz Cola costs \$9.60. A case of 16 cans of Zephyr Cola costs \$5.60. Which is the better buy?

Procedure For each cola, find the unit price of an individual can.

Answer

	Rate	Long division	Unit price	Word form
Blitz:	$\frac{9.60 \text{ dollars}}{24 \text{ cans}}$	$24\overline{)9.60}$ = 0.40 →	$\frac{0.40 \text{ dollars}}{1 \text{ can}}$	\$0.40 per can
Zephyr:	$\frac{5.60 \text{ dollars}}{16 \text{ cans}}$	$16\overline{)5.60}$ = 0.35 →	$\frac{0.35 \text{ dollars}}{1 \text{ can}}$	\$0.35 per can

Sentence Zephyr Cola is the better buy at \$0.35 per can.

You Try It 4

From the given information, write each rate as a unit price, write it as a phrase, and answer the question. Use Example 4 as a guide.

a) The cost of a box of 64 pens is \$28.80. The cost of a box of 48 pens is \$19.20. Which is the better buy?

b) The cost of 15 doughnuts at Lucky Doughnuts is \$5.70. The cost of a box of 18 doughnuts at Do-Nuts to Go is \$7.20. Which is the better buy?

Answers: You Try It and Think About It

You Try It: **1. a)** $\frac{8 \text{ pounds}}{1 \text{ inch}}$; 8 pounds per inch **b)** $\frac{4 \text{ yellowtail}}{5 \text{ sea bass}}$; 4 yellowtail for every 5 sea bass **c)** $\frac{15 \text{ meters}}{4 \text{ years}}$; 15 meters for every 4 years **2. a)** $\frac{16 \text{ miles}}{1 \text{ hour}}$; 16 miles per hour **b)** $\frac{43.5 \text{ words}}{1 \text{ minute}}$; 43.5 words per minute **c)** $\frac{3.5 \text{ dollars}}{1 \text{ foot}}$; \$3.50 per foot **3. a)** $\frac{\$160}{15 \text{ hours}}$; \$10.67 per hour (This number was rounded from 10.666.) **b)** $\frac{210 \text{ calories}}{0.6 \text{ hours}}$; 350 calories per hour **4. a)** Box of 64 pens: $\frac{\$0.45}{1 \text{ pen}}$; \$0.45 per pen; box of 48 pens: $\frac{\$0.40}{1 \text{ pen}}$; \$0.40 per pen; The box of 48 pens is the better buy at \$0.40 per pen. **b)** Lucky Doughnuts: $\frac{0.38 \text{ dollars}}{1 \text{ doughnut}}$; \$0.38 per doughnut; Do-Nuts to Go: $\frac{\$0.40}{1 \text{ doughnut}}$; \$0.40 per doughnut; Lucky Doughnuts is the better buy at \$0.38 per doughnut.

Think About It: **1.** No. In a rate, the units of measure are always different, so they will never divide out. **2.** Because the denominator, 0.09, has two decimal places, we must multiply the fraction by 100 to create whole numbers in both the numerator and the denominator.

Section 6.2 Exercises

FOR EXTRA HELP

Student Resources on DVD-ROM. Includes: Student's Solutions Manual; Video Lectures; Chapter Test Prep Video | MyMathLab | MathXL

Think Again

Write a sentence or two for each response.

1. When a rate is simplified, will the units of measure ever divide out? Why or why not? *(Refer to Think About It 1)*
No. In a rate, the units of measure are always different and will never divide out.

2. Is a rate also a ratio? Explain your answer or show an example that supports your answer.
Yes, it is ratio in which the units of measure are different.

3. When a unit rate is simplified, does it matter whether the value 1 is in the numerator or the denominator? Explain your answer or show an example that supports your answer.
Yes, the value of 1 must always be in the denominator so that we can get *per 1*.

4. Even when a customer knows which is the better buy between two similar products, he or she might not purchase the better buy. What reason(s) might someone have for not purchasing the better buy?
Answers will vary. Some answers may include the quality of the items; the amount in each (buy the smaller quantity so that some of it doesn't go to waste); or the lower price regardless of the amount.

Focus Exercises

True or false.

5. A rate is a special type of ratio.
True

6. In a rate, the units of measure in the numerator and the denominator are always the same.
False

Fill in the blanks.

7. In a fractional unit rate, the __denominator__ is always 1 unit of measure.

8. In a unit price, the numerator's unit of measure is always __money__.

Simplify each rate and write the result in words.

9. $\frac{390 \text{ stitches}}{6 \text{ inches}}$
65 stitches per inch

10. $\frac{210 \text{ dollars}}{9 \text{ months}}$
70 dollars for every 3 months

11. $\frac{63 \text{ gallons}}{12 \text{ weeks}}$
21 gallons for every 4 weeks

12. $\frac{45 \text{ laps}}{75 \text{ minutes}}$
3 laps for every 5 minutes

13. $\frac{12 \text{ supervisors}}{72 \text{ employees}}$
1 supervisor for every 6 employees

14. $\frac{60 \text{ employees}}{24 \text{ printers}}$
5 employees for every 2 printers

15. $\frac{90 \text{ pounds}}{12 \text{ passengers}}$
15 pounds for every 2 passengers

16. $\frac{108 \text{ tourists}}{6 \text{ buses}}$
18 tourists per bus

Simplify each rate to a unit rate and write the unit rate in words.

17. $\frac{256 \text{ voters}}{4 \text{ precincts}}$
64 voters per precinct

18. $\frac{215 \text{ students}}{5 \text{ algebra classes}}$
43 students per algebra class

19. $\frac{10 \text{ yards of fabric}}{8 \text{ doll dresses}}$
1.25 yards of fabric per doll dress

20. $\frac{135 \text{ dollars}}{18 \text{ hours}}$
7.5 dollars per hour

21. $\frac{1.5 \text{ kilowatts}}{0.6 \text{ hours}}$
2.5 kilowatts per hour

22. $\frac{8.5 \text{ pounds}}{3.4 \text{ feet}}$
2.5 pounds per foot

23. $\frac{42.9 \text{ miles}}{2.6 \text{ gallons}}$
16.5 miles per gallon

24. $\frac{\$5.40}{12 \text{ tickets}}$
\$0.45 per ticket

From the given information, write a fractional rate. Simplify the rate to a unit rate and write it as a phrase.

25. Michelle hit a golf ball 276 yards that, unfortunately, landed in a sand trap. The ball took 4 full seconds to travel that far. Write the ball's yards per second as a unit rate.
$\frac{276 \text{ yards}}{4 \text{ second}} = \frac{69 \text{ yards}}{1 \text{ second}}$, 69 yards per second

26. Kevin pays \$396 every 6 months for his auto insurance. Write this payment per month as a unit rate.
$\frac{396 \text{ dollars}}{6 \text{ months}} = \frac{66 \text{ dollars}}{1 \text{ months}}$, \$66 per month

27. It cost Karin \$567 to fly 1,620 miles from Indianapolis, Indiana, to Seattle, Washington. Write the dollars spent per mile as a unit rate.

$\frac{567 \text{ dollars}}{1{,}620 \text{ miles}} = \frac{0.35 \text{ dollars}}{1 \text{ mile}}$, \$0.35 per mile

28. Paolo figured that he has owned 24 cars over the last 42 years. Write the years of ownership per car as a unit rate.

$\frac{42 \text{ years}}{24 \text{ cars}} = \frac{1.75 \text{ years}}{1 \text{ car}}$, 1.75 years per car

29. Marta paid \$9.10 for 3.5 pounds of ground beef. Write the price per pound as a unit rate.

$\frac{9.10 \text{ dollars}}{3.5 \text{ pounds}} = \frac{2.60 \text{ dollars}}{1 \text{ pound}}$, \$2.60 per pound

30. Working a total of 65 hours last month as a server, Taira earned a total of \$728, including wages and tips. Write her dollars earned per hour as a unit rate.

$\frac{728 \text{ dollars}}{65 \text{ hours}} = \frac{11.20 \text{ dollars}}{1 \text{ hour}}$, \$11.20 per hour

31. In 1933, Babe Ruth hit 34 home runs in 459 at bats. Write the number of at bats per home run as a unit rate.

$\frac{459 \text{ at bats}}{34 \text{ home runs}}$, $\frac{13.5 \text{ at bats}}{1 \text{ home run}}$, 13.5 at bats per home run

32. A box of cereal has 6.3 grams of sodium in 18 servings. Write the number of grams of sodium per serving as a unit rate.

$\frac{6.3 \text{ grams}}{18 \text{ servings}}$, $\frac{0.35 \text{ grams}}{1 \text{ serving}}$, 0.35 grams per serving

From the given information, write each rate as a unit price, write it as a phrase, and answer the question.

33. A bag of 18 Goodie candy bars costs \$4.86, and a bag of 25 Yummie candy bars costs \$6.50. Which is the better buy?

Bag of 18 Goodie candy bars: $\frac{0.27 \text{ dollars}}{1 \text{ candy bar}}$, \$0.27 per Goodie candy bar;
Bag of 25 Yummie candy bars: $\frac{0.26 \text{ dollars}}{1 \text{candy bar}}$, \$0.26 per Yummie candy bar;
A bag of Yummie candy bars is the better buy.

34. A tray of 25 Gourmet chocolate chip cookies costs \$3.50, and a tray of 28 Savory chocolate chip cookies costs \$4.20. Which is the better buy?

Tray of 25 Gourmet cookies: $\frac{0.14 \text{ dollars}}{1 \text{ cookie}}$, \$0.14 per Gourmet cookie;
Tray of 28 Savory cookies: $\frac{0.15 \text{ dollars}}{1 \text{ cookie}}$, \$0.15 per Savory cookie;
A tray of Gourmet chocolate chip cookies is the better buy.

35. A 12-pack of bottles of Zapper energy water costs \$13.20, and an 18-pack of bottles of Fizzit energy water costs \$20.70. Which is the better buy?

12-pack of Zapper energy water: $\frac{1.10 \text{ dollars}}{1 \text{ bottle}}$, \$1.10 per Zapper energy water;
18-pack of Fizzit energy water: $\frac{1.15 \text{ dollars}}{1 \text{ bottle}}$, \$1.15 per Fizzit energy water;
A pack of Zapper energy water is the better buy.

36. A box of 8 Lightning energy bars costs \$7.68, and a box of 12 Thunder energy bars costs \$11.04. Which is the better buy?

Box of 8 Lightning energy bars: $\frac{0.96 \text{ dollars}}{1 \text{ bar}}$, \$0.96 per Lightning bar;
Box of 12 Thunder energy bars: $\frac{0.92 \text{ dollars}}{1 \text{ bar}}$, \$0.92 per Thunder bar;
A box of Thunder energy bars is the better buy.

37. A 16.5-ounce bag of Mayse Corn Chips costs \$3.96, and a 12.5-ounce bag of Bluze Corn Chips costs \$2.75. Which is the better buy?

16.5-ounce bag of Mayse Corn Chips: $\frac{0.24 \text{ dollars}}{1 \text{ bag}}$, \$0.24 per ounce;
12.5-ounce bag of Bluze Corn Chips: $\frac{0.22 \text{ dollars}}{1 \text{ bag}}$, \$0.22 per ounce;
A bag of Bluze Corn Chips is the better buy.

38. A 42-ounce jug of orange juice costs \$2.94, and a 64-ounce jug of orange juice costs \$3.84. Which is the better buy?

42-ounce jug of orange juice: $\frac{\$0.07}{1 \text{ ounce}}$, \$0.07 per ounce;
64-ounce jug of orange juice: $\frac{\$0.06}{1 \text{ ounce}}$, \$0.06 per ounce;
The 64-ounce jug of orange juice is the better buy.

Think Outside the Box

39. At Rancho Market, a 16-ounce bag of tortilla chips sells for \$2.88. What should be the price of a 24-ounce bag of tortilla chips so that its unit price is the same as the unit price of the 16-ounce bag?

The price of the 24-ounce bag should be \$4.32.

40. At Rancho Market, an 18-ounce bottle of Fizzy Pop sells for \$0.81. What should be the price of a 24-ounce bottle of Fizzy Pop so that its unit price is 0.5¢ less than the unit price of the 18-ounce bottle?

The price of the 24-ounce bottle should be \$0.96.

SECTION 6.3 Proportions

OBJECTIVES

In this section, you will learn to

- Define and identify *proportions.*
- Set up proportions.
- Solve proportions.
- Work applications involving proportions.
- Solve similar triangles using proportions.

You Need to Know

To successfully complete this section, you need to understand

- ☐ Solving equations (1.7)
- ☐ Simplifying fractions (4.3)
- ☐ Ratios (6.1)
- ☐ Equivalent ratios (6.1)
- ☐ Cross multiplication (6.1)

Introduction

"That's not fair!" cried Alicia. She was complaining to her parents about the number of birthday gifts her sister Juanita received. After all, Alicia received 12 presents on her sixth birthday just last month, but Juanita received 16 gifts on her eighth birthday today.

"Yes, it is fair," countered her mother. "On your eighth birthday, you'll get 16 gifts, too."

"Yes," Alicia argued, "but the month after that, Juanita will receive 20 gifts on her tenth birthday. It's just not fair!"

Are Alicia's parents being fair about the number of gifts given on their daughters' birthdays?

Identifying Proportions

When we show that two ratios are equivalent using an equation, that equation is called a *proportion.* A proportion helps keep things consistent, helps keep things fair.

A **proportion** is a statement that two ratios are equivalent:

$$\frac{a}{b} = \frac{c}{d}$$

A proportion is **true** if the product of the means is equal to the product of the extremes:

$$a \cdot d = b \cdot c$$

In setting up a proportion between two ratios, consistency is the key. This means that

1. The numerators of each fraction must be of the same units of measure (or count).

2. The denominators of each fraction must be of the same units of measure (or count).

Also, using cross multiplication, we must verify that the proportion is true by showing that the product of the means is equal to the product of the extremes.

Let's see if the situation of Alicia's and Juanita's birthday gifts fits a true proportion. In this situation, we are given two sets of information. We know that

1. Alicia is 6 years old and that she received 12 gifts on her birthday.

2. Juanita is 8 years old and that she received 16 gifts on her birthday.

To set up a proportion, we can create a strategy that compares one piece of information, such as the number of gifts a girl receives on her birthday, to the other piece of information, her age.

$$\text{number of gifts} : \text{girl's age} \quad \text{or} \quad \frac{\text{number of gifts}}{\text{girl's age}}$$

To assist us in setting up the proportion correctly, we can use a *proportion table* that outlines the strategy, as shown here.

	First set (Alicia)	Second set (Juanita)
Number of gifts	12	16
Girl's age	6	8

With this proportion table, the ratios become clear and we simply set up the proportion.

$$\frac{12 \text{ gifts}}{6 \text{ years}} = \frac{16 \text{ gifts}}{8 \text{ years}} \quad \text{or} \quad \text{just} \quad \frac{12}{6} = \frac{16}{8}$$

Is this a true proportion? We can use cross multiplication to find out.

$$\frac{12}{6} \stackrel{?}{=} \frac{16}{8}$$ Cross multiply.

$$12 \cdot 8 \stackrel{?}{=} 6 \cdot 16$$

$$96 = 96 \checkmark$$ Yes, the proportion is true.

Example 1

It takes 2 trucks to haul away 5 tons of trash in a day. It takes 6 trucks to haul away 15 tons of trash in a day.

a) Set up a proportion table for the situation.

b) Set up a proportion based on the table.

c) Show that the proportion is true.

Answer **a)** The first set of information is 2 trucks and 5 tons; the second set is 6 trucks and 15 tons.

	First set	Second set
Trucks	2	6
Tons	5	15

b) The proportion, according to the table, is

$$\frac{\text{trucks}}{\text{tons}} = \frac{\text{trucks}}{\text{tons}}$$

$$\frac{2}{5} = \frac{6}{15}$$

c) Use cross multiplication:

$$2 \cdot 15 = 5 \cdot 6$$

$$30 = 30 \checkmark$$

The proportion is true.

You Try It 1 **Mike earns \$288 for 4 days of work. He earns \$432 for 6 days of work.**

a) Set up a proportion table for the situation.

b) Set up the proportion based on the table.

c) Show that the proportion is true.

	First set	Second set

You Try It 2 **Four adults are needed to supervise 18 children on a field trip. Ten adults are needed to supervise 45 children on a field trip.**

a) Set up a proportion table for the situation.

b) Set up the proportion based on the table.

c) Show that the proportion is true.

	First set	Second set

Solving Proportions

In a proportion, what if one of the numerical measurements is not given? If one piece of information is not given, we can solve the proportion as an equation to find the value of the missing measurement.

Before we get into actual situations, though, let's look at solving proportions when one of the values is unknown.

Example 2

Solve each proportion for the unknown value. Check each answer.

a) $\frac{x}{8} = \frac{3}{4}$ **b)** $\frac{5}{4} = \frac{10}{y}$

Procedure Use cross multiplication to put the equation into a more familiar form. Then isolate the variable by dividing by the variable's coefficient.

Answer **a)** $\frac{x}{8} = \frac{3}{4}$

$4 \cdot x = 3 \cdot 8$

$4x = 24$

$\frac{4x}{4} = \frac{24}{4}$

$x = 6$

Check $x = 6$: $\frac{3}{4} \stackrel{?}{=} \frac{6}{8}$; $4 \cdot 6 \stackrel{?}{=} 3 \cdot 8$; $24 = 24$ ✓

b) $\frac{5}{4} = \frac{10}{y}$

$y \cdot 5 = 10 \cdot 4$

$5y = 40$

$\frac{5y}{5} = \frac{40}{5}$

$y = 8$

Check $y = 8$: $\frac{10}{8} \stackrel{?}{=} \frac{5}{4}$; $8 \cdot 5 \stackrel{?}{=} 10 \cdot 4$; $40 = 40$ ✓

You Try It 3 **Solve each proportion for the unknown value. Check each answer. Use Example 2 as a guide.**

a) $\frac{3}{4} = \frac{w}{16}$ **b)** $\frac{5}{3} = \frac{35}{p}$ **c)** $\frac{48}{k} = \frac{8}{5}$

Sometimes the fraction with no variable can be simplified before cross multiplying. Simplifying first is helpful because the numbers are easier to work with.

If we simplify before we cross multiply, we should check our answer in the *original* proportion. (It's possible a mistake was made in simplifying before cross multiplying.) This will guarantee that our answer is correct for the original problem.

Example 3

Solve for x in each proportion.

a) $\frac{21}{12} = \frac{x}{8}$ **b)** $\frac{10}{x} = \frac{25}{30}$

Procedure First, simplify any fraction if possible. Then use cross multiplication to put the equation into a more familiar form. Isolate the variable by dividing by the variable's coefficient.

Answer **a)** $\frac{21}{12} = \frac{x}{8}$ Simplify $\frac{21}{12}$ by a factor of 3.

$\frac{7}{4} = \frac{x}{8}$

$8 \cdot 7 = x \cdot 4$

$56 = 4x$

$\frac{56}{4} = \frac{4x}{4}$

$14 = x$

$x = 14$

Check $x = 14$ in the *original* proportion: $\frac{14}{8} \stackrel{?}{=} \frac{21}{12}$; $8 \cdot 21 \stackrel{?}{=} 14 \cdot 12$; $168 = 168$ ✓

b) $\frac{10}{x} = \frac{25}{30}$ Simplify $\frac{25}{30}$ by a factor of 5.

$\frac{10}{x} = \frac{5}{6}$

$6 \cdot 10 = 5 \cdot x$

$60 = 5x$

$\frac{60}{5} = \frac{5x}{5}$

$12 = x$

$x = 12$

Check $x = 12$ in the *original* proportion: $\frac{25}{30} \stackrel{?}{=} \frac{10}{12}$; $10 \cdot 30 \stackrel{?}{=} 12 \cdot 25$; $300 = 300$ ✓

You Try It 4 **Solve for x in each proportion by simplifying the known fraction. Check the answer. Use Example 3 as a guide.**

a) $\frac{12}{60} = \frac{x}{45}$ **b)** $\frac{21}{x} = \frac{15}{10}$ **c)** $\frac{18}{15} = \frac{30}{x}$

Applying Proportions

Consider the following situation. Yusef is a window washer. He can wash 6 windows in 21 minutes. How many minutes will it take him to wash 10 windows?

We are given two sets of information.

One set is complete: 6 windows in 21 minutes.

The other set is incomplete: 10 windows in how many minutes?

We can answer the question by following this strategy:

- Set up a legend and a proportion table.
- Set up the proportion based on the table.
- Solve the proportion using cross multiplication and other techniques.

Example 4

Yusef can wash 6 windows in 21 minutes. How many minutes will it take him to wash 10 windows?

Procedure Set up a proportion table. From the table, set up a proportion and solve it.

Answer Legend: Let m = the number of minutes it will take Yusef to wash 10 windows.

	First set	Second set
Windows	6	10
Minutes	21	m

Proportion: $\frac{\text{windows}}{\text{minutes}} = \frac{\text{windows}}{\text{minutes}}$

$$\frac{6}{21} = \frac{10}{m}$$ Simplify $\frac{6}{21}$ by a factor of 3

$$\frac{2}{7} = \frac{10}{m}$$

$$2 \cdot m = 7 \cdot 10$$

$$2m = 70$$

$$\frac{2m}{2} = \frac{70}{2}$$

$$m = 35$$

Sentence It will take Yusef 35 minutes to wash 10 windows.

You Try It 5 **At a candy store, the price of 12 inches of licorice is 45¢. What is the price for 20 inches of licorice?**

Legend

	First set	Second set

Sentence

You Try It 6 **D'nae runs 4 miles in 34 minutes. Assuming she can maintain the same rate, how long will it take D'nae to run 6 miles?**

Legend

	First set	Second set

Sentence

You Try It 7 **To create a living fence, Blaga is planting 8 spruce trees every 30 yards. How many spruce trees will Blaga need for a 105-yard fence?**

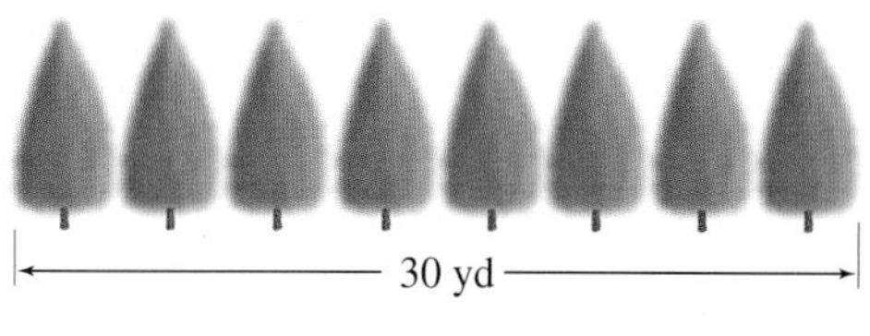

Legend	First set	Second set

Sentence

Similar Triangles

When two triangles have the same shape, they are said to be **similar.** This is true even if one of the triangles is larger than the other.

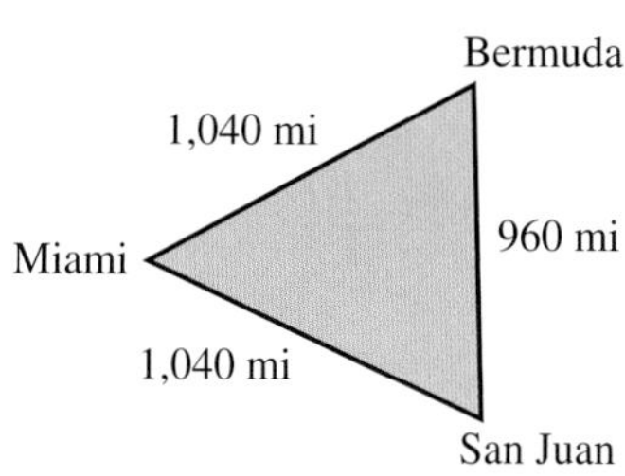

For example, the famous Bermuda Triangle, which extends from Miami, Florida, to San Juan, Puerto Rico, and to the island of Bermuda, can be measured in miles and has the approximate dimensions shown at the right.

However, that same triangle shown on a wall map can be measured in inches with a ruler. These two triangles have the same shape; so, they are similar to each other.

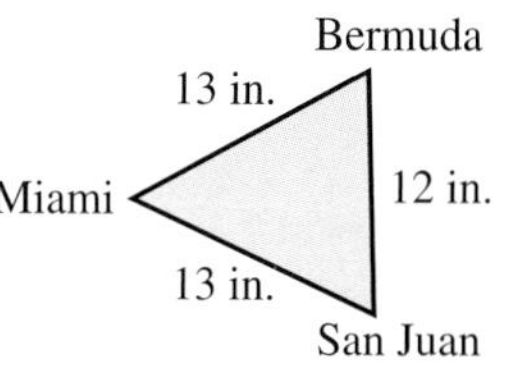

When two triangles are similar to each other, each side of one of the triangles is paired to a side on the other triangle. These two sides are called **corresponding** sides. For example, the side of the large triangle from Miami to Bermuda is corresponding to the side of the small triangle from Miami to Bermuda.

When you take one triangle and make it larger or smaller, the dimensions must be kept proportional to each other so that the new triangle retains its shape and is not distorted. This means that the ratio of one set of corresponding sides is equal to the ratio of another set of corresponding sides.

	First set (Miami to Bermuda)	Second set (Miami to San Juan)	Third set (San Juan to Bermuda)
Large triangle	1,040 mi	1,040 mi	960 mi
Small triangle	13 in.	13 in.	12 in.

$$\frac{1{,}040\text{ mi}}{13\text{ in.}} = \frac{1{,}040\text{ mi}}{13\text{ in.}} = \frac{960\text{ mi}}{12\text{ in.}}$$

Each of those ratios simplifies to $\frac{80\text{ miles}}{1\text{ inch}}$; so, we know they are equivalent to one another.

If we know that two triangles are similar to each other but we don't know all of the side measures of each triangle, we may be able to use proportions to find at least one missing side length. The key is to know each measure of two corresponding sides.

Example 5

Claire knows that the distance between Memphis, Tennessee, and Chicago, Illinois, is 480 miles. She wants to know the distance between Memphis and Buffalo, New York. On a wall map, she measures the distance between Memphis and Chicago and finds it to be 6 inches. On the same map, the distance between Memphis and Buffalo is 10 inches. What is the real distance from Memphis to Buffalo?

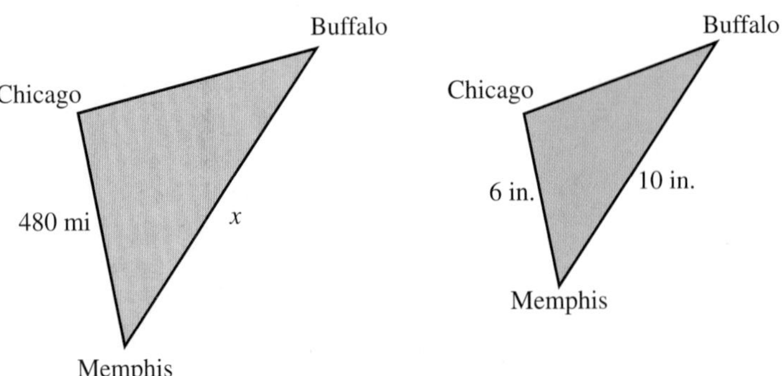

Procedure We know the measures of the two corresponding sides from Memphis to Chicago. We can use those numbers as one set of information. The other set of information will include the known and unknown values from Memphis to Buffalo.

Answer

	First set (Memphis to Chicago)	Second set (Memphis to Buffalo)
Large triangle	480 mi	x mi
Small triangle	6 in.	10 in.

YOU FINISH IT:

$$\frac{480}{6} = \frac{x}{10}$$

Sentence The real distance from Memphis to Buffalo is 800 miles.

You Try It 8 **A 6-foot fence casts a 4-foot shadow and creates a right triangle, as shown. At the same time, a nearby building casts a 22-foot shadow, creating a similar right triangle. What is the height of the building?**

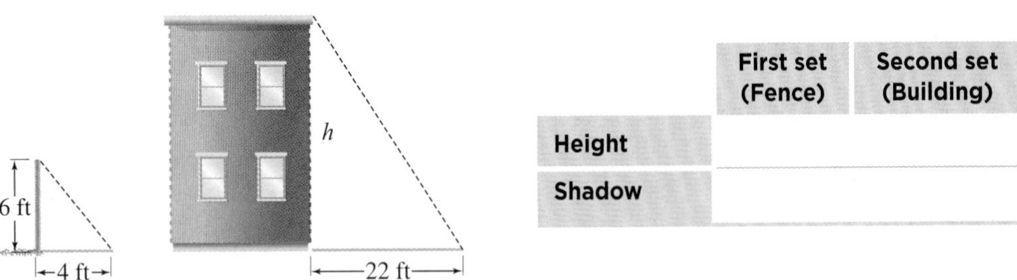

	First set (Fence)	Second set (Building)
Height		
Shadow		

Sentence

Answers: You Try It and Think About It

You Try It: **1. a)**

	First set	Second set
Dollars	288	432
Days	4	6

b) $\frac{288}{4} = \frac{432}{6}$

c) $288 \cdot 6 = 4 \cdot 432$

$1{,}728 = 1{,}728$

2. a)

	First set	Second set
Adults	4	10
Children	18	45

b) $\frac{4}{18} = \frac{10}{45}$

c) $4 \cdot 45 = 18 \cdot 10$

$180 = 180$

3. a) $w = 12$ **b)** $p = 21$ **c)** $k = 30$ **4. a)** $x = 9$ **b)** $x = 14$ **c)** $x = 25$

5. Let p = the price for 20 inches of licorice.

	First set	Second set
Cents	45	p
Inches	12	20

$$\frac{45}{12} = \frac{p}{20}$$

The price for 20 inches of licorice is 75¢.

6. Let m = the number of minutes to run 6 miles.

	First set	Second set
Miles	4	6
Minutes	34	m

$$\frac{4}{34} = \frac{6}{m}$$

It will take D'nae 51 minutes to run 6 miles.

7. Let x = the number of spruce trees needed for 70 yards.

	First set	Second set
Trees	8	x
Yards	30	105

$$\frac{8}{30} = \frac{x}{105}$$

Blaga will need 28 spruce trees for a 105-yard fence.

8. Let h = the height of the building.

	First set (Fence)	Second set (Building)
Height	6	h
Shadow	4	22

$$\frac{6}{4} = \frac{h}{22}$$

The height of the building is 33 feet.

Think About It: No Think About Its in this section.

Section 6.3 Exercises

FOR EXTRA HELP

Student Resources on DVD-ROM — Includes ➤Student's Solutions Manual ➤Video Lectures ➤Chapter Test Prep Video | MyMathLab | MathXL

Think Again

Write a sentence or two for each response.

1. If two ratios are not equivalent, is it still possible to set up a proportion between them? Explain your answer or show an example that supports your answer.
No, proportions are only for equivalent ratios.

2. For this pair of equivalent ratios, $\frac{a}{b} = \frac{c}{d}$, the means are b and c and the extremes are a and d. When we write the same two equivalent ratios as $\frac{c}{d} = \frac{a}{b}$, the means and extremes are switched. Does this affect the outcome of cross multiplication? Show an example that supports your answer.
No. Answers will vary.

Focus Exercises

Solve for x in each proportion. If possible, reduce the known fraction first.

3. $\frac{3}{7} = \frac{x}{14}$ $x = 6$

4. $\frac{2}{9} = \frac{x}{45}$ $x = 10$

5. $\frac{x}{28} = \frac{4}{7}$ $x = 16$

6. $\frac{x}{18} = \frac{5}{3}$ $x = 30$

7. $\frac{11}{5} = \frac{66}{x}$ $x = 30$

8. $\frac{9}{4} = \frac{54}{x}$ $x = 24$

9. $\frac{20}{x} = \frac{5}{4}$ $x = 16$

10. $\frac{36}{x} = \frac{3}{8}$ $x = 96$

11. $\frac{x}{15} = \frac{8}{6}$ $x = 20$

12. $\frac{x}{35} = \frac{12}{15}$ $x = 28$

13. $\frac{8}{x} = \frac{6}{9}$ $x = 12$

14. $\frac{50}{x} = \frac{20}{2}$ $x = 5$

15. $\frac{30}{6} = \frac{x}{5}$
$x = 25$

16. $\frac{42}{14} = \frac{x}{8}$
$x = 24$

17. $\frac{60}{24} = \frac{10}{x}$
$x = 4$

18. $\frac{36}{48} = \frac{15}{x}$
$x = 20$

Solve each application and answer with a complete sentence. (*Hint:* Set up a proportion table, write a legend, and solve the corresponding proportion.)

19. There are 5 seniors for every 3 juniors in the concert choir. If there are 20 seniors in all, how many juniors are in the concert choir?
There are 12 juniors in the concert choir.

20. A group of 48 people can fit in 6 vans. How many vans are needed for a group of 72 people?
9 vans are needed for the group.

21. Karen received 16 votes for every 24 people who voted. If she received a total of 300 votes, how many people voted?
450 people voted.

22. Tom received 22 votes for every 40 people who voted. If a total of 200 people voted, how many votes did Tom receive?
Tom received 110 votes.

23. Cheyenne was able to drive 350 miles on 25 gallons of gasoline. At that rate, how many miles will Cheyenne be able to drive on 15 gallons?
Cheyenne will be able to drive 210 miles.

24. Cheryl was able to drive 360 miles on 16 gallons of gasoline. At that rate, how many gallons of gas will Cheryl use to drive 270 miles?
Cheryl will use 12 gallons of gas.

25. A flooring company sells flooring material for $15 per square yard. How many square yards of flooring material can be purchased for $900?
60 square yards of flooring material can be purchased for $900.

26. A flooring company sells flooring material for $15 per square yard. How much will it cost for 160 square yards of flooring material?
It will cost $2,400 for 160 square yards of flooring material.

27. At a hardware store, the price of 6 feet of chain is $2.80. What length of chain will $14 buy?
$14 will buy 30 feet of chain.

28. At a hardware store, the price of 4 feet of rope is $1.80. What is the price of 10 feet of rope?
The price of 10 feet of rope is $4.50.

29. Delon can ride his bike an average of 8 miles in 28 minutes. At that rate, how many minutes will it take him to ride 20 miles?
It will take him 70 minutes to ride 20 miles.

30. Delon can ride his bike an average of 8 miles in 28 minutes. At that rate, how far can he ride in 42 minutes?
He can ride 12 miles in 42 minutes.

31. Sandy's car can travel 94 miles on 4 gallons of gas. How far can her car travel on 14 gallons of gas?
Her car can travel 329 miles on 14 gallons of gas.

32. Hank's pickup can travel 87 miles on 6 gallons of gas. How many gallons of gas will Hank's pickup need to travel 116 miles?
Hank's pickup will need 8 gallons of gas to travel 116 miles.

33. Tim can run an average of 4 miles in 30 minutes. How far can Tim run in 45 minutes?
Tim can run 6 miles in 45 minutes.

34. Marla works for Kinko's. She needs to make 210 copies of a course pack. Copying the first 42 course packs took 3 hours. How long will it take for the whole job to be completed?
It will take 15 hours for the whole job to be completed.

35. At Charles County Community College, there are 9 math classes for every 15 English classes. If the college offers 24 math classes, how many English classes does it offer?
It offers 40 English classes.

36. At Essex Community College, there are 8 math classes for every 12 English classes. If the college offers 39 English classes, how many math classes does it offer?
It offers 26 math classes.

37. At the gym, Chandra's favorite exercise is a workout on the cross trainer. She has found that she burns 180 calories in a 15-minute workout. How many minutes will Chandra need to work out on the cross trainer to burn 600 calories?
Chandra will need to work out 50 minutes on the cross trainer to burn 600 calories.

38. At the gym, Luana's favorite exercise is a workout on the elliptical machine. She has found that she burns 300 calories in a 25-minute workout. How many calories will Luana burn in 35 minutes on the elliptical machine?
Luana will burn 420 calories in 35 minutes on the elliptical machine.

For each pair of similar triangles, find the missing length and answer with a complete sentence. (*Hint:* Use a proportion table to set up and solve the corresponding proportion.)

39.

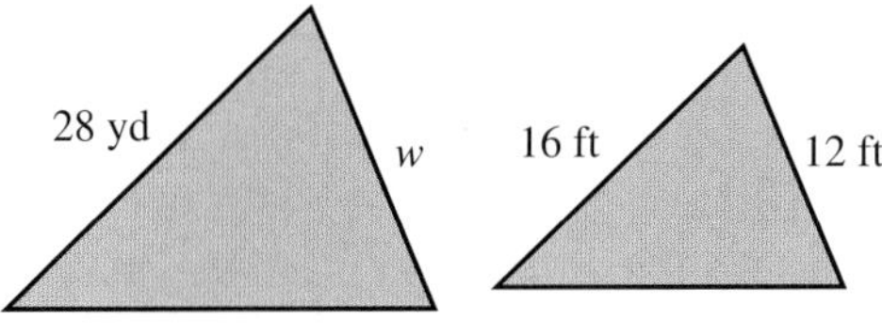

The missing length is 21 yards.

40.

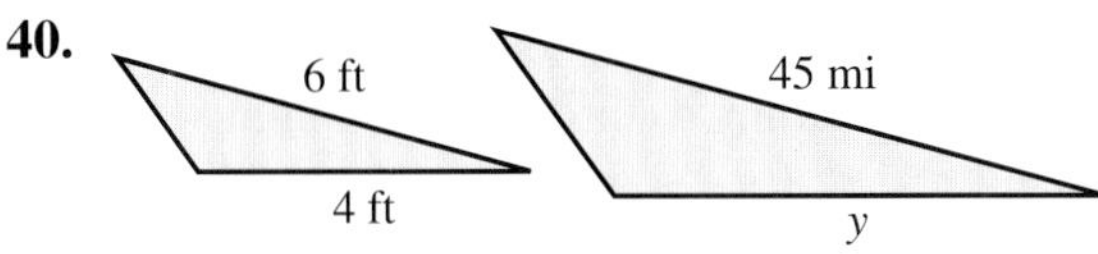

The missing length is 30 miles.

41.

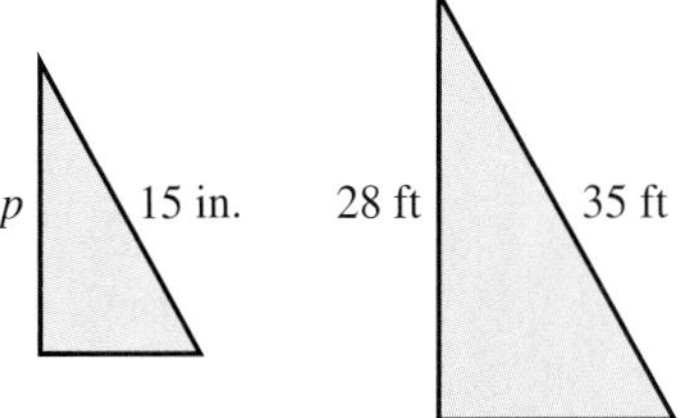

The missing length is 12 inches.

42.

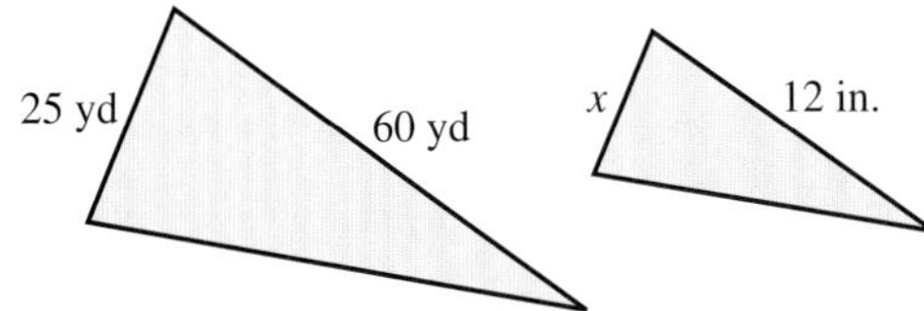

The missing length is 5 inches.

Solve each application and answer with a complete sentence.

43. A 4-foot post casts a 3-foot shadow and creates a right triangle, as shown. At the same time, a nearby tree casts a 15-foot shadow, creating a similar right triangle. What is the height of the tree?

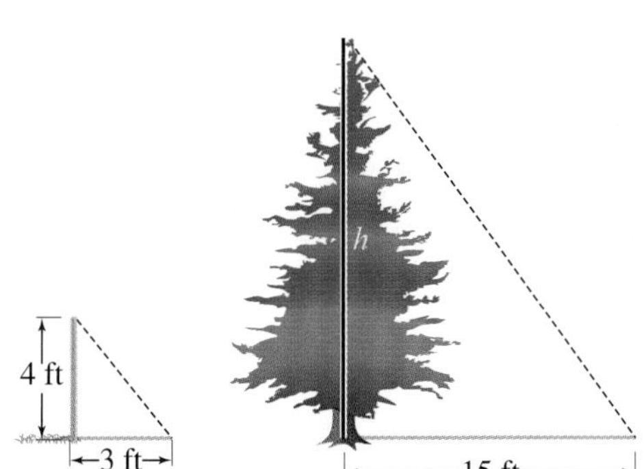

The height of the tree is 20 feet.

44. A 6-foot post casts a 10-foot shadow and creates a right triangle, as shown. At the same time, a nearby clock tower casts a 65-foot shadow, creating a similar right triangle. What is the height of the clock tower?

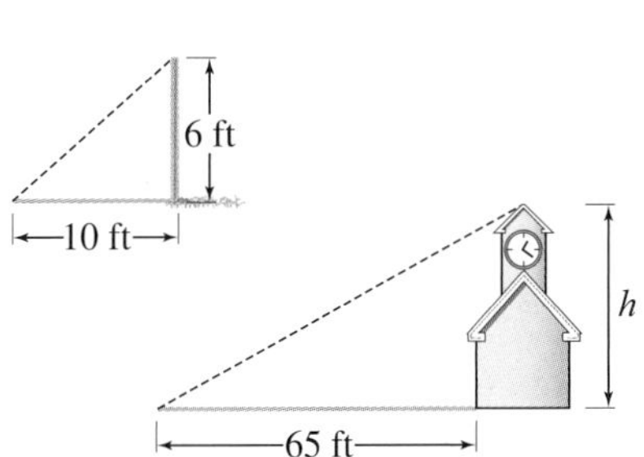

The height of the clock tower is 39 feet.

45. Tony knows that the distance between Dallas, Texas, and Denver, Colorado, is 660 miles. He wants to know the distance between Dallas and Phoenix, Arizona. On a wall map, he measures the distance between Dallas and Denver and finds it to be 12 inches. On the same map, the distance between Dallas and Phoenix is 16 inches. What is the real distance from Dallas to Phoenix?

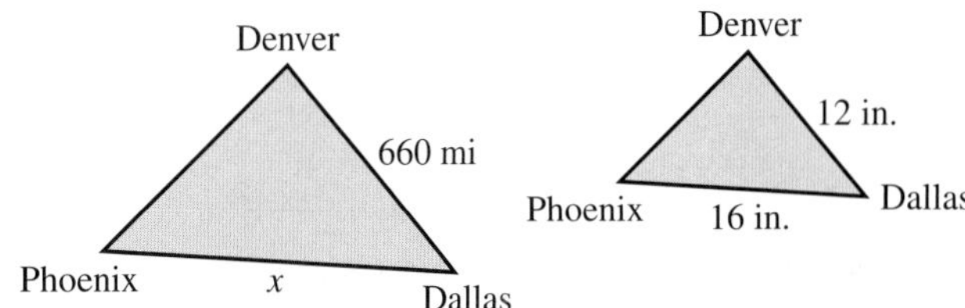

The real distance from Dallas to Phoenix is 880 miles.

46. Calli knows that the distance between Salt Lake City, Utah, and Seattle, Washington, is 700 miles. She wants to know the distance between Salt Lake City and Minneapolis, Minnesota. On a wall map, she measures the distance between Salt Lake City and Seattle and finds it to be 10 inches. On the same map, the distance between Salt Lake City and Minneapolis is 14 inches. What is the real distance from Salt Lake City to Minneapolis?

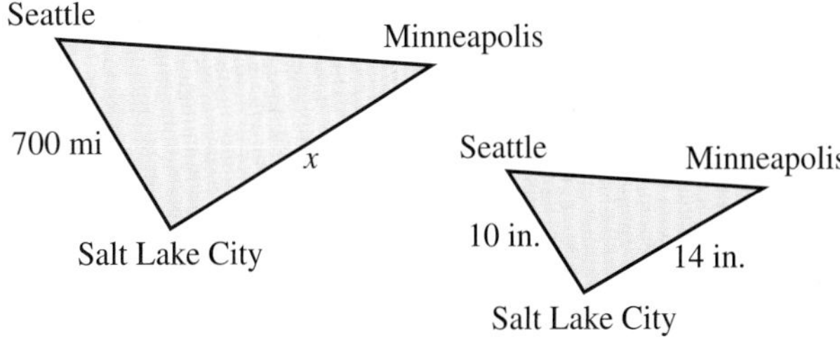

The real distance from Salt Lake City to Minneapolis is 980 miles.

Think Outside the Box

Solve for x in each proportion.

47. $\frac{8}{10} = \frac{x}{x + 3}$
$x = 12$

48. $\frac{6}{4} = \frac{x + 5}{x}$
$x = 10$

49. $\frac{6}{x - 4} = \frac{8}{x + 3}$
$x = 25$

50. $\frac{4x - 3}{7} = \frac{2x + 3}{5}$
$x = 6$

Solve each application and answer with a complete sentence.

51. Suppose Babe Ruth were alive today and still playing professional baseball. Also, suppose that he hit 25 home runs in the first 90 games of the 2008 baseball season. If he kept up that pace, how many home runs would we expect him to hit in a full 162-game season?
We would expect him to hit 45 home runs.

52. Suppose Babe Ruth were alive today and still playing professional baseball. If he hit home runs at a unit rate of 4.5 games per home run, how many home runs would we expect him to hit in a full 162-game season?
We would expect him to hit 36 home runs.

53. Many people don't know that Babe Ruth was an outstanding pitcher as well as a powerful hitter. Suppose that he had a strikeout rate of 5.5 strikeouts thrown for every 9 innings pitched. How many strikeouts would we expect him to throw in 216 innings pitched?
We would expect him to throw 132 strikeouts.

SECTION 6.4 Percents

OBJECTIVES

In this section, you will learn to
- Define *percent.*
- Write a percent as a fraction and a ratio.
- Write a percent as a decimal.
- Write a decimal as a percent.
- Write a fraction as a percent.

You Need to Know

To successfully complete this section, you need to understand
- ☐ Writing decimals as fractions (5.1)
- ☐ Writing fractions as decimals (5.1)
- ☐ Multiplying decimals by powers of 10 (5.4)
- ☐ Dividing decimals by powers of 10 (5.5)
- ☐ Ratios (6.1)

Introduction

You have, no doubt, seen percents. Oftentimes a sales ad will indicate a discount such as "Save 20%." The symbol % means "percent," and the number that is written before the percent symbol is called the **percentage.** For example, in 6%, the percentage is 6.

Recall from Section 6.1 that *per* means "divided by" and indicates a ratio. *Percent*, or *per cent*, means "*per 100*" or "*divided by 100.*" Therefore, a **percent** is a ratio in which the denominator is 100.

As a ratio, 32% means 32 : 100:

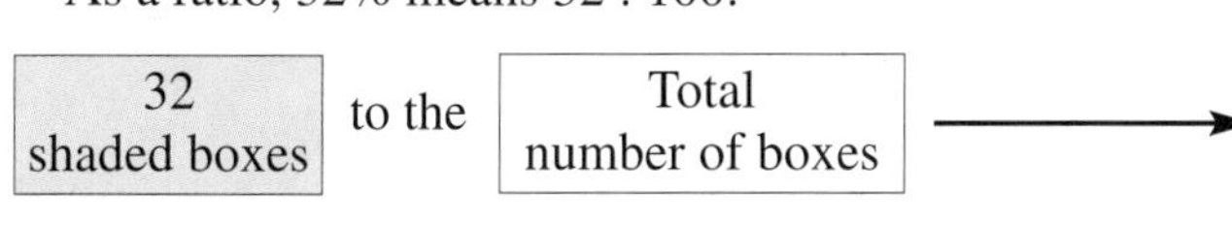

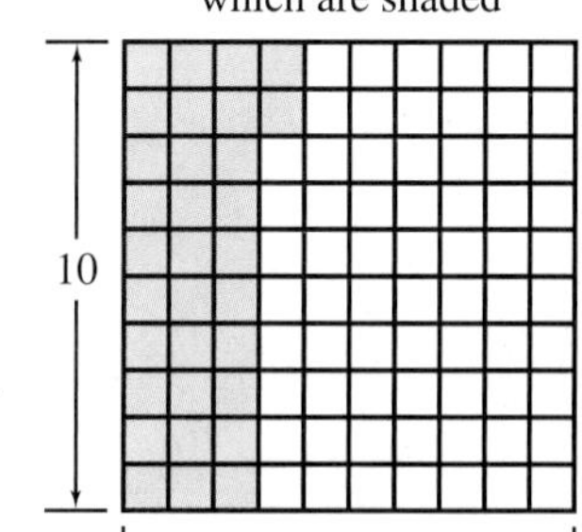

32% of the boxes are shaded.

Writing Percents as Fractions

As a fraction, 32% means $\frac{32}{100}$, or "32 out of 100."

Example 1

Write each percent as a fraction with 100 in the denominator. Do not simplify.

a) 19% **b)** 7.75% **c)** 0.4%

Procedure The percentage is the numerator, and 100 is the denominator.

Answer **a)** $19\% = \frac{19}{100}$ **b)** $7.75\% = \frac{7.75}{100}$ **c)** $0.4\% = \frac{0.4}{100}$

You Try It 1 **Write each percent as a fraction with 100 in the denominator. Do not simplify. Use Example 1 as a guide.**

a) 47% **b)** 8.1% **c)** 3% **d)** 0.9%

Example 2

Write each percent as a fraction. Then simplify the fraction completely.

a) 70% **b)** 1.5% **c)** 100% **d)** 250%

Procedure The percentage is the numerator, and 100 is the denominator.

Answer **a)** $70\% = \frac{70}{100}$ This simplifies by a factor of 10: $\frac{7}{10}$.

b) $1.5\% = \frac{1.5}{100}$ First, multiply by $\frac{10}{10}$ to clear the decimal: $\frac{1.5}{100} \times \frac{10}{10} = \frac{15}{1{,}000}$. That simplifies by a factor of 5: $\frac{3}{200}$.

c) $100\% = \frac{100}{100}$ This simplifies to 1.

d) $250\% = \frac{250}{100}$ This simplifies first by a factor of 10, $\frac{25}{10}$, and then by a factor of 5, $\frac{5}{2}$ or $2\frac{1}{2}$.

You Try It 2 **Write each percent as a fraction. Then simplify the fraction completely. Use Example 2 as a guide.**

a) 60% **b)** 400% **c)** 6.5% **d)** 25%

Example 3

Write this sentence with the percent as a simplified fraction.

60% of the students in Mr. DeGuzman' s class are women.

Procedure Write 60% as a fraction and simplify the fraction. $60\% = \frac{60}{100} = \frac{6}{10} = \frac{3}{5}$

Answer $\frac{3}{5}$ of the students in Mr. DeGuzman' s class are women.

You Try It 3 **Write each sentence with the percent as a simplified fraction. Use Example 3 as a guide.**

a) 35% of the books on Toni's shelf are biographies.

b) 24% of the adults in Cypress County have not registered to vote.

When a fraction has a denominator of 100, the numerator is the percentage and the fraction can be easily written as a percent.

Example 4

Write each fraction as a percent.

a) $\frac{93}{100}$ **b)** $\frac{8}{100}$ **c)** $\frac{1.3}{100}$ **d)** $\frac{184}{100}$

Procedure The numerator is the percentage.

Answer **a)** $\frac{93}{100} = 93\%$ **b)** $\frac{8}{100} = 8\%$ **c)** $\frac{1.3}{100} = 1.3\%$ **d)** $\frac{184}{100} = 184\%$

You Try It 4 **Write each fraction as a percent. Use Example 4 as a guide.**

a) $\frac{9}{100}$ **b)** $\frac{42}{100}$ **c)** $\frac{159}{100}$ **d)** $\frac{1.6}{100}$

Example 5

Write the following sentence with the suggested fraction as a percent.

68 out of 100 fourth-graders prefer ice cream for dessert.

Procedure Write *68 out of 100* first as a fraction and then as a percent: $\frac{68}{100} = 68\%$.

When you are writing the sentence, the words *of all* follow this new percent.

Answer 68% of all fourth-graders prefer ice cream for dessert.

You Try It 5 **Write each sentence with the suggested fraction as a percent. Use Example 5 as a guide.**

a) 13 out of 100 people in the United States live below the poverty level.
Source: **www.census.gov**

b) In the Richmond School District, 62 out of 100 high school students taking an advanced placement exam will pass it.

Writing Percents as Decimals

If the percentage is a whole number, the percent can be written first as a fraction and then as a decimal.

Example 6

Write each percent as a decimal by first writing it as a fraction.

a) 34% **b)** 138% **c)** 300%

Procedure Remember, any fraction with a denominator of 100 can be written as a decimal with two decimal places.

Answer **a)** $34\% = \frac{34}{100} = 0.34$

b) $138\% = \frac{138}{100} = 1.38$

c) $300\% = \frac{300}{100} = 3.00$, or just 3

You Try It 6 **Write each percent as a decimal by first writing it as a fraction. Use Example 6 as a guide.**

a) 52% **b)** 9% **c)** 200% **d)** 168%

Recall from Section 5.5 that dividing by a power of 10 has the effect of moving the decimal point (in the dividend) a number of places to the left. For example, when we divide by 100, we move the decimal point two places to the left.

Example 7

Write each of these percents as a decimal by writing it first as a fraction and then as a decimal.

a) 12.4% **b)** 0.78%

Procedure After writing each as a fraction with a denominator of 100, we can write the number as a decimal by moving the decimal point two places to the left. It may be necessary to place one or more zeros in front of the numerator.

Answer **a)** $12.4\% = \frac{12.4}{100} = .124 = 0.124$

b) $0.78\% = \frac{0.78}{100} = \frac{000.78}{100} = 0.0078$

You Try It 7 **Write each of these percents as a decimal by writing it first as a fraction and then as a decimal. Use Example 7 as a guide.**

a) 7.2% **b)** 28.6% **c)** 0.09% **d)** 5.19%

Is it possible to write percents directly as decimals without having to use fractions?

Yes, it is; and the next example looks closely at doing so. However, it's good to know that we can refer back to fractions when we need to.

Procedure for Writing Percents as Decimals

1. Recognize the percentage and notice whether it contains a decimal point. If it doesn't, write one at its end.	9% → 9.% 83% → 83.% 6.4% → 6.4%
2. If the percentage doesn't have at least two whole numbers before the decimal point, place one or two zeros in front of the percentage.	9% → 09.% 83% → 83.% 6.4% → 06.4%
3. Rewrite the percentage without the % and move the decimal point two places to the left.	09.% → .09 83.% → .83 06.4% → .064

Example 8 follows steps 1 and 3 of that procedure.

Example 8

Write each percent as a decimal.

a) 34% **b)** 138% **c)** 12.4% **d)** 300%

Procedure Follow step 1 of the procedure as necessary and place a decimal point at the end of each whole number percentage. Then follow step 3.

Answer **a)** 34% or 34.% = .34 or 0.34 **b)** 138% = 138.% = 1.38

c) 12.4% = .124 or 0.124 **d)** 300% = 300.% = 3.00, or just 3

12.4% already has two whole number digits in the percentage *and* has its own decimal point.

Example 9 follows steps 2 and 3 of the Procedure for Writing Percents as Decimals.

Example 9

Write each percent as a decimal.

a) 9.4% **b)** 6.38% **c)** 0.58% **d)** 0.07%

Procedure Follow step 2 of the procedure by placing enough zeros in front of the percentage to get two whole numbers. Then follow step 3.

Answer **a)** 9.4% or 09.4% = 0.094 **b)** 6.38% = 06.38% = 0.0638

c) 0.58% = 00.58% = 0.0058 **d)** 0.07% = 00.07% = 0.0007

You Try It 8 **Write each percent as a decimal. Use Examples 8 and 9 as guides.**

a) 82% **b)** 28.5% **c)** 370%

d) 2% **e)** 6.5% **f)** 0.49%

Writing Decimals as Percents

Sometimes we need to convert from a decimal to a percent. Knowing how to do this comes in handy, for instance, when you want to write the results of a study as a percent. For example, if a study finds that 0.126 of all students surveyed are left handed, we can write that as 12.6% of all students surveyed are left handed.

When a decimal number has two decimal places, the number can be written first as a fraction with denominator of 100 and then as a percent. When the number has fewer than two decimal places, we can place one or two ending zeros so that it has two decimal places.

Example 10

Write each number as a percent by first writing it as a fraction.

a) 0.45 **b)** 0.06 **c)** 1.58 **d)** 0.2 **e)** 4

Procedure Remember, any fraction with a denominator of 100 can be written as a decimal with two decimal places. For part d, place one ending zero; and for part e, place a decimal point and two ending zeros.

Answer

a) $0.45 = \frac{45}{100} = 45\%$ **b)** $0.06 = \frac{6}{100} = 6\%$

c) $1.58 = \frac{158}{100} = 158\%$ **d)** $0.2 = 0.20 = \frac{20}{100} = 20\%$

e) $4 = 4.00 = \frac{400}{100} = 400\%$

You Try It 9

Write each number as a percent by first writing it as a fraction. Use Example 10 as a guide.

a) 0.03 **b)** 4.08 **c)** 0.67 **d)** 9 **e)** 0.3

Following is another technique for writing decimals as percents.

First, recall

- From Section 5.4 that multiplying by 100 has the effect of moving the decimal point two places to the right.
- From Example 2 of this section that 100% = 1. In other words, 100% is another form of 1 and we can multiply a number by 100% without changing the value of the number.

Putting those ideas together, we can multiply a number such as 0.45 by 100% without changing the value.

$0.45 \times 100\% = 45\%$. Compare that to the answer in Example 10a.

We multiply 100 by 0.45 to result in 45 (the percentage), and place the percent sign after the percentage.

Writing a Decimal as a Percent

Multiply the decimal number by 100%.

Example 11

Write each number as a percent by multiplying by 100%.

a) 3.28 **b)** 2 **c)** 0.54

d) 0.08 **e)** 0.7 **f)** 0.024

Procedure Multiplying by 100 has the effect of moving the decimal two places to the right.

Answer
a) $3.28 = 3.28 \times 100\% = 328\%$

b) $2 = 2 \times 100\% = 200\%$

c) $0.54 = 0.54 \times 100\% = 54\%$

d) $0.08 = 0.08 \times 100\% = 8\%$

e) $0.7 = 0.7 \times 100\% = 70\%$

f) $0.024 = 0.024 \times 100\% = 2.4\%$

Notice that the percent sign, %, is ignored during the multiplication by 100 and is carried over to the final result.

You Try It 10 **Write each number as a percent. Use Example 11 as a guide.**

a) 0.15 **b)** 0.06 **c)** 0.5

d) 0.041 **e)** 6 **f)** 1.68

Writing Fractions as Percents

We have seen in Example 4 that a fraction with a denominator of 100 can be written immediately as a percent. If the denominator is not 100, we can convert a fraction to a percent by first writing the fraction as a decimal. Once it is in decimal form, we can multiply by 100% to find the percent equivalent.

Example 12

Write each fraction as a percent.

a) $\frac{3}{20}$ **b)** $\frac{3}{8}$

Procedure First, use long division to write each fraction as a decimal. The decimal quotient is equivalent to the fraction.

You show the division:

a) $20\overline{)3.00}$ = 0.15 **b)** $8\overline{)3.000}$ = 0.375

Here is the fraction as a decimal.

$\frac{3}{20} = 0.15$ $\frac{3}{8} = 0.375$

Now multiply the decimal by 100%.

$0.15 \times 100\% = 15\%$ $0.375 \times 100\% = 37.5\%$

Answer **a)** $\frac{3}{20} = 15\%$ **b)** $\frac{3}{8} = 37.5\%$

Example 13

Write $\frac{4}{7}$ as a percent. Round the percentage to the nearest tenth.

Procedure Find the decimal equivalent of $\frac{4}{7}$ using long division. *Rounding the percentage to the nearest tenth* means rounding the quotient to the nearest thousandth. Once the number is in decimal form, multiply by 100%.

You show the division: $7\overline{)4.0000}$ = 0.5714

At four decimal places, we can round to the nearest thousandth. The quotient rounds to 0.571.
Multiply 0.571 by 100% to find the percent equivalent.

$0.571 \times 100\% = 57.1\%$

Because we rounded, this is an *approximate* value and we should use the ≈ sign to indicate that fact.

Answer $\frac{4}{7} \approx 57.1\%$

Think About It 1 In the previous example, when rounding the percentage to the nearest tenth, why must we round the quotient to the nearest thousandth?

____

____

You Try It 11 **Write each fraction as a percent. If necessary, round the percentage to the nearest tenth. Use Examples 12 and 13 as guides.**

a) $\frac{13}{25}$ **b)** $\frac{11}{40}$ **c)** $\frac{5}{6}$ **d)** $\frac{4}{11}$

Following is a chart of the equivalent percent values of some common fractions.

Percent Equivalents of Some Fractions		
$\frac{1}{2} = 50\%$	$\frac{1}{5} = 20\%$	$\frac{5}{6} = 83\frac{1}{3}\%$
$\frac{1}{3} = 33\frac{1}{3}\%$	$\frac{2}{5} = 40\%$	$\frac{1}{8} = 12.5\%$
$\frac{2}{3} = 66\frac{2}{3}\%$	$\frac{3}{5} = 60\%$	$\frac{3}{8} = 37.5\%$
$\frac{1}{4} = 25\%$	$\frac{4}{5} = 80\%$	$\frac{5}{8} = 62.5\%$
$\frac{3}{4} = 75\%$	$\frac{1}{6} = 16\frac{2}{3}\%$	$\frac{7}{8} = 87.5\%$

Some of those values can be written as approximations.

Percent Approximations of Some Fractions			
$\frac{1}{3} \approx 33.3\%$	$\frac{2}{3} \approx 66.7\%$	$\frac{1}{6} \approx 16.7\%$	$\frac{5}{6} \approx 83.3\%$

Example 14

Write the following sentence with the fraction as a percent. Use the preceding percent approximation table.

$\frac{1}{3}$ of all college freshmen take English during their first semester.

Procedure From the table, $\frac{1}{3} \approx 33.3\%$.

Answer 33.3% of all college freshmen take English during their first semester.

Even though 33.3% is an approximation, the word *approximately* is often excluded from this type of sentence.

You Try It 12 **Write each sentence with the fraction as a percent. Use Example 14 as a guide.**

a) In Nebraska, about $\frac{5}{6}$ of the students complete all four years of high school.
Source: **www.manhattan-institute.org**

b) When reading a newspaper, $\frac{3}{8}$ of all men read the local section first.

Answers: You Try It and Think About It

You Try It: **1. a)** $\frac{47}{100}$ **b)** $\frac{8.1}{100}$ **c)** $\frac{3}{100}$ **d)** $\frac{0.9}{100}$ **2. a)** $\frac{3}{5}$ **b)** 4 **c)** $\frac{13}{200}$ **d)** $\frac{1}{4}$ **3. a)** $\frac{7}{20}$ of the books on Toni's shelf are biographies. **b)** $\frac{6}{25}$ of the adults in Cypress County have not registered to vote. **4. a)** 9% **b)** 42% **c)** 159% **d)** 1.6% **5. a)** 13% of all people in the United States live below the poverty level. **b)** 62% of all high school students taking an advanced placement exam will pass it. **6. a)** 0.52 **b)** 0.09 **c)** 2 **d)** 1.68 **7. a)** 0.072 **b)** 0.286 **c)** 0.0009 **d)** 0.0519 **8. a)** .82 or 0.82 **b)** .285 or 0.285 **c)** 3.7 **d)** .02 or 0.02 **e)** .065 or 0.065 **f)** .0049 or 0.0049 **9. a)** 3% **b)** 408% **c)** 67% **d)** 900% **e)** 30% **10. a)** 15% **b)** 6% **c)** 50% **d)** 4.1% **e)** 600% **f)** 168% **11. a)** 52% **b)** 27.5% **c)** ≈ 83.3% **d)** ≈ 36.4% **12. a)** In Nebraska, 83.3% of the students complete all four years of high school. **b)** When reading a newspaper, 37.5% of all men read the local section first.

Think About It: **1.** To have a percentage with one decimal place, the decimal equivalent (the quotient) must have three decimal places.

Section 6.4 Exercises

FOR EXTRA HELP
Student Resources on DVD-ROM
Includes
➤ Student's Solutions Manual
➤ Video Lectures
➤ Chapter Test Prep Video
MyMathLab
MathXL

Think Again

Write a sentence or two for each response.

1. 75%, $\frac{3}{4}$, and 0.75 represent the same value. Which way—as a percent, as a fraction, or as a decimal—do you prefer to see or hear a number reported in a newscast or in the newspaper?
Answers will vary.

2. 75%, $\frac{3}{4}$, and 0.75 represent the same value. Is it necessary to have all three forms of the same number, or is just one or two of these forms enough? Explain your answer or show examples that support your answer.
All three forms are necessary. Answers will vary.

Focus Exercises

Write each percent as a fraction. Then simplify the fraction completely.

3. 45% $\frac{9}{20}$ **4.** 8% $\frac{2}{25}$ **5.** 125% $\frac{5}{4}$ **6.** 7.5% $\frac{3}{40}$

Write each sentence with the percent as a simplified fraction.

7. 35% of all workers in San Francisco use public transportation to commute to and from work.
Source: **www.rideshare.511.org**
$\frac{7}{20}$ of all workers in San Francisco use public transportation to commute to and from work.

8. 52% of a school district's budget is for teacher salaries.
$\frac{13}{25}$ of a school district's budget is for teacher salaries.

9. 32.5% of the book sales at Bookends are romance novels.
$\frac{13}{40}$ of the book sales at Bookends are romance novels.

10. 2% of the cars imported into the United States come from Sweden. ***Source: 2005 World Almanac***
$\frac{1}{50}$ of the cars imported into the United States come from Sweden.

Write each percent as a decimal. You may use any method you like.

11. 16% 0.16 **12.** 19% 0.19 **13.** 4.5% 0.045

14. 8% 0.08 **15.** 2% 0.02 **16.** 5% 0.05

17. 184% 1.84 **18.** 205% 2.05 **19.** 85% 0.85

20. 46% 0.46 **21.** 260% 2.6 **22.** 150% 1.5

23. 4% 0.04 **24.** 9% 0.09 **25.** 3% 0.03

26. 1% 0.01 **27.** 5.2% 0.052 **28.** 6.8% 0.068

29. 17.2% 0.172 **30.** 23.9% 0.239 **31.** 7.75% 0.0775

32. 1.38% 0.0138 **33.** 5% 0.05 **34.** 4% 0.04

Write each decimal as a percent.

35. 0.78
78%

36. 0.35
35%

37. 0.06
6%

38. 0.08
8%

39. 0.9
90%

40. 0.5
50%

41. 2.11
211%

42. 1.64
164%

43. 1.4
140%

44. 2.3
230%

45. 0.128
12.8%

46. 0.875
87.5%

47. 0.062
6.2%

48. 0.047
4.7%

49. 0.001
0.1%

50. 0.003
0.3%

Write each fraction as a percent. If necessary, round each percentage to the nearest tenth.

51. $\frac{2}{5}$
40%

52. $\frac{1}{10}$
10%

53. $\frac{47}{50}$
94%

54. $\frac{11}{20}$
55%

55. $\frac{14}{25}$
56%

56. $\frac{8}{5}$
160%

57. $\frac{5}{4}$
125%

58. $\frac{7}{2}$
350%

59. $\frac{5}{8}$
62.5%

60. $\frac{27}{40}$
67.5%

61. $\frac{9}{16}$
56.3%

62. $\frac{13}{80}$
16.3%

63. $\frac{7}{9}$
77.8%

64. $\frac{5}{11}$
45.5%

65. $\frac{1}{6}$
16.7%

66. $\frac{7}{12}$
58.3%

67. $\frac{4}{15}$
26.7%

68. $\frac{7}{30}$
23.3%

69. $\frac{9}{22}$
40.9%

70. $\frac{5}{18}$
27.8%

Write each sentence with the fraction as a percent. Round the percentage to the nearest tenth.

71. $\frac{1}{40}$ of the U.S. population has a last name of Smith, Johnson, or Williams. ***Source: 2005 World Almanac***
2.5% of the U.S. population has a last name of Smith, Johnson, or Williams.

72. $\frac{3}{10}$ of the medals won by South Korean athletes at the 2004 Olympics were gold medals.
Source: 2005 World Almanac
30% of the medals won by South Korean athletes at the 2004 Olympics were gold medals.

73. By the All-Star break, the Cleveland Cavaliers had won $\frac{5}{9}$ of their games.
By the All-Star break, the Cleveland Cavaliers had won 55.6% of their games.

74. In May 2005, $\frac{3}{16}$ of all commercial flights either departed late or were canceled. ***Source:* www.bts.gov**
In May 2005, 18.8% of all commercial flights either departed late or were canceled.

Think Outside the Box

Write each percent as a decimal. Do not round any of the values.

75. $\frac{1}{2}\%$
0.005

76. $\frac{3}{4}\%$
0.0075

77. $1\frac{5}{8}\%$
0.01625

78. $2\frac{13}{20}\%$
0.0265

SECTION 6.5 Solving Percent Problems Using Proportions

OBJECTIVES

In this section, you will learn to

- Set up a ratio in terms of percentage and base.
- Write a percent equation as a proportion.
- Solve percent proportions.

You Need to Know

To successfully complete this section, you need to understand

- ☐ Solving equations (1.7)
- ☐ Writing ratios as fractions (6.1)
- ☐ Solving proportions (6.3)
- ☐ Writing decimals as percents (6.4)

This section may be presented in addition or as an alternative to Section 6.6.

Introduction

When a ratio is used to compare a part to a whole, we can write the ratio as part : whole or $\frac{\text{part}}{\text{whole}}$.

When the whole is 100 and the part is 80, the ratio of part: whole is 80 : 100 or $\frac{80}{100}$. And as you know, $\frac{80}{100}$ is the same as 80%.

Clearly, more than half the diagram is shaded. →

100 squares, 80 of which are shaded

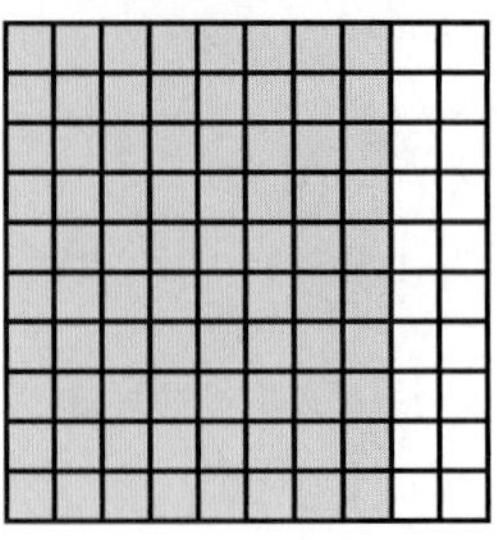

80% of the squares are shaded.

But what if there are 200 squares and only 80 are shaded? In that case, less than half the squares are shaded.

Here 80 of the 200 squares are shaded; but this amounts to only $\frac{80}{200}$, which reduces (by a factor of 2) to $\frac{40}{100}$, only 40%.

So, knowing that only 80 squares are shaded is not enough to determine how much of the whole is shaded. We also need to know how many squares are in the whole.

The whole, then, becomes the basis, or *base*, of our comparison; and this is the denominator of the ratio. In the first diagram, the base is 100 and the ratio $\frac{80}{\text{base}}$ is $\frac{80}{100}$. In the second diagram, the base is 200 and the ratio $\frac{80}{\text{base}}$ is $\frac{80}{200}$.

200 squares, 80 of which are shaded

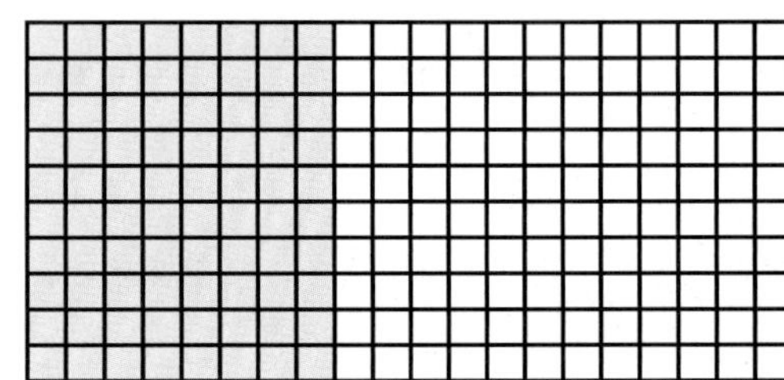

Less than half the squares are shaded.

Recall from Section 6.4 that the number in front of the percent sign (%) is called the *percentage*. As a fraction, 80% becomes $\frac{80}{100}$, or $\frac{\text{percentage}}{100}$.

A ratio that is not a percent can be thought of as $\frac{\text{amount}}{\text{base}}$, where the base is a number other than 100.

Consider the diagram at the left where 4 out of the 5 rectangles are shaded. In this case, the base is 5 and we are interested in the 4 rectangles that are shaded. The shaded-to-total ratio, $\frac{\text{amount}}{\text{base}}$, is $\frac{4}{5}$.

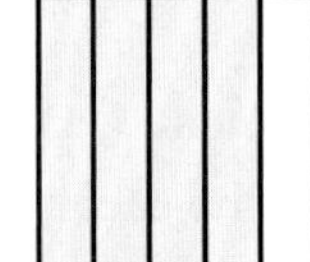

The Percent Proportion

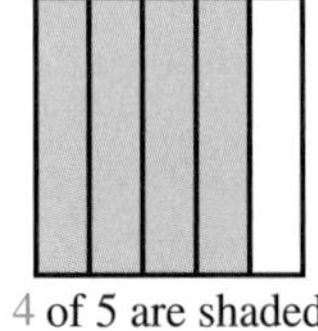

4 of 5 are shaded.

The squares at the left and the right are the same size and have the same amount of shading. The left square is divided into 5 long rectangles, 4 of which are shaded, giving a shaded-to-total ratio of $\frac{4}{5}$. The right square is divided into 100 small squares, 80 of which are shaded, giving a shaded-to-total ratio of $\frac{80}{100}$.

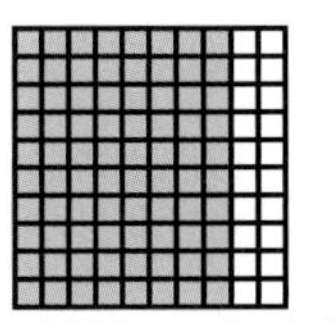

80 of 100 are shaded.

Because the amount of shading in each large square is the same, the shaded-to-total ratios are the same. In other words, the ratios form a proportion.

$$\frac{80}{100} = \frac{4}{5}$$

This proportion can be treated as $\frac{\text{percentage}}{100} = \frac{\text{amount}}{\text{base}}$. In the proportion table of Section 6.3, the table looks like this:

	Percent	Ratio
Percentage/amount	p	a
Base	100	b

As a general formula, we get $\frac{p}{100} = \frac{a}{b}$.

In the formula $\frac{p}{100} = \frac{a}{b}$, there are three variables

p, which represents the percentage
a, which represents the amount
b, which represents the base

For any one problem, two of these variables will be known values and we can set up and solve a proportion to find the third variable.

When the Percentage Is Unknown

The first two examples demonstrate how to use the formula when the percentage is unknown.

Example 1

24 out of the 40 players on the Jefferson High Jaguars varsity football team are seniors. What percent of the players are seniors?

Procedure The question asks "What percent. . . ." So, the unknown variable must be p in the percentage.

	Percent	Players
Seniors (Percentage/amount)	p	24
All players (Base)	100	40

Answer Legend: Let p = the unknown percentage.

$$\frac{p}{100} = \frac{24}{40}$$ $\frac{24}{40}$ simplifies by a factor of 8 to $\frac{3}{5}$.

$$\frac{p}{100} = \frac{3}{5}$$ Cross multiply.

$$p \cdot 5 = 100 \cdot 3$$
$$5p = 300$$
$$\frac{5p}{5} = \frac{300}{5}$$
$$p = 60$$

Check

$p = 60$:

$$\frac{60}{100} \stackrel{?}{=} \frac{24}{40}$$
$$60 \times 40 \stackrel{?}{=} 100 \times 24$$
$$2{,}400 = 2{,}400 \checkmark$$

Sentence 60% of the players are seniors.

You Try It 1 **21 of the 75 singers in the Tulsa Chorale are sopranos. What percent of the singers are sopranos? Use Example 1 as a guide.**

Legend

Sentence

The division in Example 2, $500 \div 6$, can be treated in one of three ways

1. As a fraction, $\frac{500}{6}$, that we write as a mixed number, $83\frac{1}{3}$ (Answer: $83\frac{1}{3}\%$)
2. As a repeating decimal number: $83.333\ldots = 83.\overline{3}$ (Answer: $83.\overline{3}\%$)
3. As an approximate value rounded to the nearest tenth: $83.333\ldots \approx 83.3$ (Answer: Approximately 83.3%, although it is common to leave out the word *approximately* when writing the sentence)

Any of those three are acceptable.

Sometimes the percentage we seek is a bit unusual, as shown in the next example.

Example 2

Of the 24 vehicles sold at Buster Dodge last week, 20 were trucks. What percent of the vehicles sold were trucks?

Procedure Again, p is the unknown value and that is reflected in the legend.

	Percent	Vehicles
Trucks	p	20
All vehicles	100	24

Answer Legend: Let p = the unknown percentage.

$$\frac{p}{100} = \frac{20}{24} \qquad \frac{20}{24} \text{ simplifies by a factor of 4 to } \frac{5}{6}.$$

$$\frac{p}{100} = \frac{5}{6}$$

$$p \cdot 6 = 100 \cdot 5$$

$$\frac{6p}{6} = \frac{500}{6} \longrightarrow 6\overline{)500.000} = 83.333$$

We can round this quotient to the nearest tenth to get 83.3. (This is an approximation.)

$$p \approx 83.3$$

Sentence 83.3% of the vehicles sold were trucks.

You Try It 2 **Of the 176 passengers on a flight from San Francisco to Hawaii, 32 flew first-class. What percent of the passengers flew first-class? Use Example 2 as a guide.**

Legend

Sentence

When the Amount Is Unknown

Example 3 demonstrates how to proceed when we

1. Know the percent of a situation.
2. Know the whole (base) of a certain set.
3. Don't know the part portion (amount).

Example 3

30% of the sixth-grade students at Lincoln Elementary School are eligible for the federal school lunch program. If there is a total of 70 sixth-graders, how many of them are eligible for the federal school lunch program?

Procedure In this situation, we know the percent: 30%. We also know the whole (the base of the second set), 70. The unknown is the *amount* of the second set, the number of sixth-graders who are eligible for the program.

	Percent	Students
Eligible for program	30	a
All sixth graders	100	70

Answer Legend: Let a = the number of sixth-graders eligible for the program.

$$\frac{30}{100} = \frac{a}{70} \quad \text{Cross multiply and solve.}$$
$$30 \cdot 70 = 100 \cdot a$$
$$2100 = 100a$$
$$\frac{2100}{100} = \frac{100a}{100}$$
$$21 = a$$
$$a = 21$$

Sentence 21 sixth-graders are eligible for the program.

In the answer portion of Example 3, can we first simplify $\frac{30}{100}$ by a factor of 10, to $\frac{3}{10}$? Why or why not?

____

____

You Try It 3 **24% of the seniors at Monroe High School are eligible for admission to LSU. If there are 175 seniors at Monroe High, how many are eligible for admission to LSU? Use Example 3 as a guide.**

Legend

Sentence

When the Base Is Unknown

Sometimes the base is unknown.

Example 4

15% of the local Kiwanis Club members are employed by the city. If it is known that 6 members of the club work for the city, how many total members does the club have?

Procedure In this situation, we know the percent: 15%. We also know the part (the amount of the second set). What is unknown is the base of the second set, the total number of members in the Kiwanis Club.

	Percent	Kiwanis members
Work for city	15	6
All members	100	m

Answer Legend: Let m = the total number of members in the Kiwanis Club.

$$\frac{15}{100} = \frac{6}{m} \quad \text{Cross multiply and solve.}$$
$$100 \cdot 6 = 15 \cdot m$$
$$600 = 15m$$
$$\frac{600}{15} = \frac{15m}{15}$$
$$40 = m$$

Sentence The club has a total of 40 members.

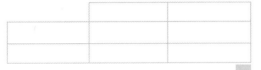

You Try It 4 **72% of the University of Kentucky Wildcats baseball team throw right handed. If 18 members of the team throw right handed, how many players are on the team? Use Example 4 as a guide.**

Legend

Sentence

Answers: You Try It and Think About It

You Try It:

1. Let p = the unknown percentage.

	Percent	Singers
Sopranos	p	21
All singers	100	75

$$\frac{p}{100} = \frac{21}{75}$$

28% of the singers are sopranos.

2. Let p = the unknown percentage.

	Percent	Passengers
First class	p	32
All passengers	100	176

$$\frac{p}{100} = \frac{32}{176}$$

About 18.2% of the passengers flew first-class.

3. Let a = the number of eligible seniors.

	Percent	Seniors
Eligible for Admission	24	a
All seniors	100	175

$$\frac{24}{100} = \frac{a}{175}$$

42 seniors are eligible for admission to LSU.

4. Let b = the total number of players on the team.

	Percent	Players
Throw right-handed	72	18
All players	100	b

$$\frac{72}{100} = \frac{18}{b}$$

25 players are on the University of Kentucky Wildcats baseball team.

Think About It: **1.** Answers may vary. One possibility is: Writing the percent as a fraction allows us to simplify the fraction before we isolate the variable.

Section 6.5 Exercises

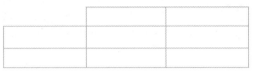

FOR EXTRA HELP: Student Resources on DVD-ROM — Includes ➤Student's Solutions Manual ➤Video Lectures ➤Chapter Test Prep Video | MyMathLab | MathXL

Think Again

Write a sentence or two for each response.

1. By itself, is $\frac{25}{100}$ a proportion? Explain your answer.

No, $\frac{25}{100}$ is a ratio. A proportion is always an equation, one ratio = another (equivalent) ratio.

2. What is the difference between the words *percentage* and *percent*?

Percentage is the number that comes before the percent symbol, %. *Percent* is a ratio in which the denominator is 100. For example, in 38%, the percentage is 38 and the percent is $\frac{38}{100}$.

Focus Exercises

Solve each application and answer with a complete sentence. (*Hint:* Write a legend and set up and solve the proportion.)

3. 72% of the faculty at Lemon Valley College are tenured. If there are 75 faculty members in all, how many are tenured?
54 of the faculty are tenured.

4. 25% of the packs in a box of baseball cards have an autographed card inside. If the box contains 12 packs, how many of them contain an autographed card?
3 of the packs contain an autographed card.

5. 55% of the customers at the Gila Bend Starbucks order a frappuccino. If Starbucks served 160 customers last Saturday, how many ordered a frappuccino?
88 customers ordered a frappuccino.

6. 32% of the female members of a health club attend at least three times a week. If there are 275 female members, how many attend at least three times a week?
88 female members attend at least three times a week.

7. 92% of the Fitchberg Annex Post Office mailboxes have been rented. If the post office has 325 mailboxes, how many of them have been rented?
299 of the mailboxes have been rented.

8. 64% of the voters in Glenn Cove, New York, voted in favor of Proposition A on the ballot. If a total of 225 people voted, how many voted for Proposition A?
144 people voted for Proposition A.

9. 37.5% of the families in a large neighborhood have a backyard pool. If there are 120 families in this neighborhood, how many of them have a backyard pool?
45 families have a backyard pool.

10. 12.5% of the employees in Commandary Publishing Services are managers. If the company has 160 employees, how many are in management?
20 of the employees are in management.

11. 35% of the sports cards that Jesse collects are basketball cards. If Jesse has 42 basketball cards, how many cards does he have in all?
Jesse has 120 cards in all.

12. 15% of the Framingham Post Office mailboxes are extra large. If the post office has 33 extra large mailboxes, how many total mailboxes does it have?
The post office has a total of 220 mailboxes.

13. 56% of the voters in Winooski, Vermont, voted in favor of Measure B on the ballot. If 196 people voted in favor of Measure B, how many total people voted in the town?
A total of 350 people voted in the town.

14. 35% of the customers at the Wayland Coffee Hut consistently order a large-sized drink. Last Sunday, 63 customers ordered a large-sized drink. What was the total number of customers that day?
The total number of customers that day was 180.

15. 15% of the female members of a health club are married. If 39 married women are members at the club, what is the total number of female members?
The total number of female members at the club is 260.

16. 8% of the faculty at Cedar Glen College teach Spanish. If there are 10 Spanish instructors, how many total faculty members are there?
There are a total of 125 faculty members at Cedar Glen College.

17. 52.5% of the employees at a factory have attained senior status, meaning they've worked there at least twenty years. If the factory has 84 senior employees, how many total employees are there?
There are a total of 160 employees at the factory.

18. 62.5% of the families in a neighborhood participate in an annual garage sale. If 45 families participate in the garage sale, how many total families live in this neighborhood?
A total of 72 families live in this neighborhood.

19. Louie made a record of all of the faculty computers on his campus. Of the 125 computers, 45 of them are Macs. What percent of the faculty computers are Macs?
36% of the faculty computers are Macs.

20. Of the 80 chiropractic patients that Dr. James sees regularly, 36 of them get a weekly massage. What percent of his patients receive a weekly massage?
45% of his patients receive a weekly massage.

21. 85 of the 340 Woodland Post Office mailbox renters check their mailbox every Sunday. What percent of the renters check their mailbox every Sunday?
25% of the renters check their mailbox every Sunday.

22. 49 of the 140 sports cards that Jermaine owns are football cards. What percent of his sports card collection is football cards?
35% of his football card collection is football cards.

23. Bala is the chair of the Math and Science Department at Long Beach City College. His department has 21 full-time members on a campus that employs 120 full-time faculty. What percent of the full-time faculty is in Bala's department?
17.5% of the full-time faculty is in Bala's department.

24. In a small town, 260 of the 416 voters voted for Becky Kendall for mayor. What percent of the vote did Ms. Kendall get?
Ms. Kendall got 62.5% of the vote.

25. Of the 81 families in a neighborhood, 36 have children in preschool. What percent of the families have children in preschool?
44.4% of the families have children in preschool.

26. Of the 126 employees at Middlesex Community College, 21 ride the bus to work. What percent of the employees ride the bus to work?
16.7% of the employees ride the bus to work.

Think Outside the Box

27. 20% of the employees at Sterling Motor Parts smoke cigarettes. If Sterling has 165 employees, how many do *not* smoke cigarettes?
132 employees do not smoke cigarettes.

28. Dr. Gutierrez has 80 students in her biology lecture and has an average attendance of 74 students in each class session. On average, what percent of Dr. Gutierrez's students are absent from each class?
On average, 7.5% of Dr. Gutierrez's students are absent from each class.

SECTION 6.6 Solving the Percent Equation

OBJECTIVE

In this section, you will learn to

- Set up and solve a percent equation.

You Need to Know

To successfully complete this section, you need to understand

- ☐ Solving equations involving fractions (4.8)
- ☐ Writing percents as fractions (6.4)
- ☐ Writing percents as decimals (6.4)
- ☐ Writing decimals as percents (6.4)

This section may be presented in addition or as an alternative to Section 6.5.

Introduction

We've seen percent represented as a fraction, as a decimal, and as a ratio. Still, 25% in and of itself doesn't mean very much unless we know "25% of what?" In other words, percent is always *of* something. Furthermore, *percent of* always means *multiply*.

For example, if there are 12 songs on a CD and 25% of them are love songs, there are 25% × 12 = 3 love songs.

When multiplying by a percent, it is common to write the percent as a fraction or a decimal.

As a fraction

$$25\% \times 12 = \frac{25}{100} \times \frac{12}{1} = \frac{1}{4} \times \frac{12}{1} = \frac{12}{4} = 3$$

As a decimal

$$25\% \times 12 = 0.25 \times 12$$

$$\begin{array}{r} 12 \\ \times\ 0.25 \\ \hline 60 \\ +\ 240 \\ \hline 3.00 = 3 \end{array}$$

Example 1

Find each product. Write each percent as a fraction or a decimal.

a) 40% of 45 **b)** 23% of 45

Procedure In part a, 40% would simplify nicely as a fraction: $\frac{40}{100} = \frac{4}{10} = \frac{2}{5}$. In part b, 23% would probably work best as a decimal, 0.23. In either case, we need to multiply.

Answer

a) $\frac{40}{100} \times 45$

$= \frac{2}{5} \times \frac{45}{1}$

$= \frac{2 \cdot 45}{5 \cdot 1}$ ← Reduce by a factor of 5.

$= \frac{2 \cdot 9}{1 \cdot 1}$

$= \frac{18}{1} = 18$

b) 0.23×45

$$\begin{array}{r} 45 \\ \times 0.23 \\ \hline 135 \\ +\ 900 \\ \hline 10.35 \end{array}$$

Sentence 40% of 45 is 18. 23% of 45 is 10.35.

You Try It 1 **Find each product. Write each percent as a fraction or a decimal. Use Example 1 as a guide.**

a) 75% of 20 **b)** 24% of 35 **c)** 125% of 44

The Percent Equation

Refer back to Example 1a. The result is that 40% of 45 is 18.

40%	of	45	is	18
↓			↓	
40%	×	45	=	18
	↓		↓	
percent	×	whole	=	part

This is an example of the *percent equation.*

The Percent Equation

Percent of whole is part

Percent × *Whole* = *Part*

In a typical percent problem, one of the numbers is unknown and we can use the percent equation to find its value. For example, we have seen that 40% of 45 is 18. From this, we can generate three different questions, each with a different unknown value and a slightly different equation. (In this case, we'll use n, for *number*, as the variable.)

Question	Equation	Answer	Sentence
a) When the part is unknown, we ask *40% of 45 is what number?*	$40\% \times 45 = n$	$n = 18$	40% of 45 is 18.
b) When the whole is unknown, we ask *40% of what number is 18?*	$40\% \times n = 18$	$n = 45$	40% of 45 is 18.
c) When the percent is unknown, we ask *What percent of 45 is 18?*	$p \times 45 = 18$	$p = .40$, $p = 40\%$	40% of 45 is 18.

When we don't know all of the information before we ask the question, we must rely on the techniques for solving equations involving multiplication.

When the Part Is Unknown

This is the simplest type of problem to solve because the part, n, is already isolated.

Example 2

Find the missing value. Once the number is found, write a concluding sentence.

a) 15% of 80 is what number? **b)** 48% of 55 is what number?

Procedure As in Example 1, we can write the percent as a fraction or a decimal. Here we'll use a fraction in part a and a decimal in part b.

Answer **a)** Let n = *what number.*

$$15\% \times 80 = n$$

Reduce $\frac{15}{100}$ by a factor of 5 to $\frac{3}{20}$. →

$$\frac{15}{100} \cdot 80 = n$$

$$\frac{3}{20} \cdot \frac{80}{1} = n$$

80 and 20 have a common factor of 20. →

$$\frac{3 \cdot 80}{20 \cdot 1} = n$$

$$\frac{3 \cdot 4}{1 \cdot 1} = n$$

$$12 = n$$

b) Let n = *what number.*

$$48\% \times 55 = n$$

$$0.48 \times 55 = n$$

$$\begin{array}{r} 55 \\ \times\ 0.48 \\ \hline 440 \\ +\ 2200 \\ \hline 26.40 \end{array}$$

$$n = 26.40$$

Sentence 15% of 80 is <u>12</u>. 48% of 55 is <u>26.4</u>.

You Try It 2 **Find the missing value. Once the number is found, write a conclusion. Use Example 2 as a guide. (You may write the percent as a fraction or a decimal.)**

a) 55% of 40 is what number?

Sentence

b) 48% of 75 is what number?

Sentence

When the Whole Is Unknown

When the whole is unknown, the variable, *n*, is on the left side of the equation with the percent, the coefficient of *n*. We'll divide by the coefficient to isolate *n* and solve the equation.

In Example 3, we'll write the percent as a fraction. In Example 4, we'll write the percent as a decimal.

Example 3

Find the missing value. Once the number is found, write a concluding sentence.

15% of what number is 60?

Procedure This time we'll write the percent as a fraction.

Answer Let n = *what number.*

$$15\% \times n = 60 \qquad \text{Write 15\% as } \frac{15}{100}.$$

$$\frac{15}{100} \cdot n = 60 \qquad \text{To isolate } n \text{, multiply each side by the reciprocal of the coefficient.}$$

$$\frac{15}{100} \cdot \frac{100}{15} \cdot n = 60 \cdot \frac{100}{15} \qquad \text{Write 60 as } \frac{60}{1}.$$

$$1 \cdot n = \frac{60}{1} \cdot \frac{100}{15}$$

$$n = \frac{60 \cdot 100}{1 \cdot 15} \qquad \text{60 and 15 have a common factor of 15.}$$

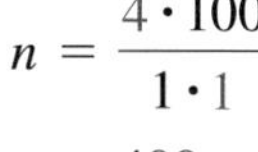

$$n = \frac{4 \cdot 100}{1 \cdot 1}$$

$$n = 400$$

Sentence 15% of <u>400</u> is 60.

Think About It 1 In Example 3, would it make a difference if we wrote the percent as a decimal instead of a fraction? Why or why not?

Example 4

Find the missing value. Once the number is found, write a concluding sentence.

48% of what number is 72?

Procedure Again, *what number* is the whole?

Answer Let n = *what number*.

$48\% \times n = 72$

$0.48n = 72$

$\dfrac{0.48n}{0.48} = \dfrac{72}{0.48}$ (Long division: $48\overline{)72.00}$ → $48\overline{)7200}$)

$n = 150$

$$\begin{array}{r} 150 \\ 48\overline{)7200} \\ -48 \\ \hline 240 \\ -240 \\ \hline 00 \\ -00 \\ \hline 0 \end{array}$$

Sentence 48% of <u>150</u> is 72.

You Try It 3 **Find the missing value. Once the number is found, write a conclusion. Use Examples 3 and 4 as guides. (You may write the percent as a fraction or a decimal.)**

a) 65% of what number is 39? **b)** 40% of what number is 18?

Sentence ____________ **Sentence** ____________

When the Percent Is Unknown

In the percent equation, when the percent is unknown, we let p = the unknown percent and solve the equation for p. However, when the concluding sentence is written, we need to express p as a percent.

Example 5

Find the missing value. Once the number is found, write a concluding sentence.

a) What percent of 140 is 49? **b)** What percent of 120 is 9? **c)** What percent of 12 is 7?

Procedure When the equation is solved, the value of p will be written as a decimal. Make sure you take one more step to write it as a percent.

Answer For each, let p = the *percent*.

a) $p \cdot 140 = 49$ Divide each side by 140.

$\dfrac{140p}{140} = \dfrac{49}{140}$

$p = 49 \div 140$ Use long division.

$p = 0.35$ Since the question is asking *what percent*, write 0.35 as a percent: 0.35 = 35%.

$p = 35\%$

$$\begin{array}{r} 0.35 \\ 140\overline{)49.00} \\ -420 \\ \hline 700 \\ -700 \\ \hline 0 \end{array}$$

Sentence <u>35%</u> of 140 is 49.

b) $p \cdot 120 = 9$ Divide each side by 120.

$\dfrac{120p}{120} = \dfrac{9}{120}$ Use long division.

$p = 0.075$ Write 0.075 as a percent: 0.075 = 7.5%.

$p = 7.5\%$

$$\begin{array}{r} 0.075 \\ 120\overline{)9.0000} \\ -840 \\ \hline 600 \\ -600 \\ \hline 0 \end{array}$$

Sentence <u>7.5%</u> of 120 is 9.

c) $p \cdot 12 = 7$

$$\frac{12p}{12} = \frac{7}{12}$$ Divide each side by 12.

$$p = 0.58\overline{3}$$ Use long division.

$$p = 58.\overline{3}\%$$ Rounding 0.5833 to the nearest thousandth, $0.5833 \approx 58.\overline{3}\%$.

$$\begin{array}{r} 0.5833 \\ 12\overline{)7.0000} \\ -\underline{60} \\ 100 \\ -\underline{96} \\ 40 \\ -\underline{36} \\ 40 \end{array}$$ (repeating . . .)

Sentence $58.\overline{3}\%$ of 12 is 7.

You Try It 4 **Find the missing value. Once the number is found, write a conclusion. For each, let p = percent. Use Example 5 as a guide.**

a) What percent of 85 is 34?

Sentence ____________

b) What percent of 125 is 45?

Sentence ____________

c) What percent of 72 is 45?

Sentence ____________

d) What percent of 81 is 36?

Sentence ____________

Applications Involving the Percent Equation

Typical application situations are not written in English in a way that directly asks: "What percent of the whole is the part?" Often it is up to us to analyze and interpret the wording. For example:

> 24% of the shoppers at a grocery store will use the express lane (with only a few items in their carts). If 75 people are in the store, how many are expected to use the express lane?

In that problem, we are given the percent, 24%, and one other number. That other number is either the whole or the part. Let's think about it: Only some of the customers use the express lane, not all of them. So, the number of shoppers in the express lane is the part and the total number of people in the store is the whole. This is our interpretation:

24% of 75 is what number?

$$24\% \times 75 = n$$

Example 6

Identify the percent, the whole, and the part in each of these situations.

a) In a typical day, 5% of the kindergarten students at Washington Elementary School will be out sick. If 80 kindergarten students are enrolled, how many are expected to be out sick tomorrow?

b) Lucy made cookies for a bake sale and kept 25% of them for her family. If her family got to keep 24 cookies, how many cookies did Lucy make?

c) A local bank's ATM serviced 120 customers one day, and 36 made deposits. What percent of the ATM customers made deposits?

Procedure For each, identify the percent, the whole, and the part.

Answer
a) Percent: 5%; there are 80 students in total, and they won't all be sick. So, whole: 80 and part: unknown.

b) Percent: 25%; if Lucy gave some cookies (24) to her family, she had to have made many more for the bake sale. So, part: 24 and whole: unknown.

c) We don't see the percent; so, it must be the unknown. The other two numbers are the whole (total customers) and the part (customers who made deposits).

So, percent: unknown, whole: 120, and part: 36.

You Try It 5 **Identify the percent, the whole, and the part in each of these situations; but do not solve the application. Use Example 6 as a guide.**

a) 112 voters were asked whether they would support the local school bond on the upcoming ballot. 70 people said yes. What percent of the voters said that they would support the school bond?

Percent: ________ Whole: ________ Part: ________

b) Marco works for a computer store and earns 4% commission on everything he sells. On Tuesday, he sold $2,485 worth of computer equipment. How much commission did he earn?

Percent: ________ Whole: ________ Part: ________

c) 15 students in Ms. Reiner's algebra classes received an A last semester. That is 12% of all of her algebra students. How many algebra students did Ms. Reiner have last semester?

Percent: ________ Whole: ________ Part: ________

d) While on a business trip to Springfield, Illinois, Soyuki purchased gifts for his family. On the flight home, Soyuki became curious about the sales tax rate in Illinois. His receipt indicated that he paid $92.00 for the gifts and $5.75 in sales tax. What is the sales tax rate in Illinois?

Percent: ________ Whole: ________ Part: ________

You Try It 6 **Solve each application by putting the information you found in You Try It 5 into the formula percent × whole = part. Write a legend and a sentence answering the question.**

a) 112 voters were asked whether they would support the local school bond on the upcoming ballot. 70 people said yes. What percent of the voters said that they would support the school bond?

Legend

Sentence

b) Marco works for a computer store and earns 4% commission on everything he sells. On Tuesday, he sold $2,485 worth of computer equipment. How much commission did Marco earn?

Legend

Sentence

c) 15 students in Ms. Reiner's algebra classes received an A last semester. That is 12% of all of her algebra students. How many algebra students did Ms. Reiner have last semester?

Legend

Sentence

d) While on a business trip to Springfield, Illinois, Soyuki purchased gifts for his family. On the flight home, Soyuki became curious about the sales tax rate in Illinois. His receipt indicated that he paid $92.00 for the gifts and $5.75 in sales tax. What is the sales tax rate in Illinois?

Legend

Sentence

Answers: You Try It and Think About It

You Try It: **1. a)** 15 **b)** 8.4 **c)** 55 **2. a)** 55% of 40 is 22. **b)** 48% of 75 is 36. **3. a)** 65% of 60 is 39. **b)** 40% of 45 is 18. **4. a)** 40% of 85 is 34. **b)** 36% of 125 is 45. **c)** 6.25% of 72 is 45. **d)** $44.\overline{4}\%$ of 81 is 36. **5. a)** Percent: unknown; Whole: 112; Part: 70 **b)** Percent: 4%; Whole: $2,485; Part: unknown **c)** Percent: 12%; Whole: unknown; Part: 15 **d)** Percent: unknown; Whole: $92.00; Part: $5.75 **6. a)** Let p = the percent of voters who support the bond. 62.5% of the voters said that they would support the school bond. **b)** Let n = the amount of Marco's commission. Marco earned $99.40 in commission. **c)** Let n = the number of algebra students in Ms. Reiner's classes. Ms. Reiner had a total of 125 algebra students last semester. **d)** Let p = the percent sales tax. The sales tax rate in Illinois is 6.25%.

Think About It: **1.** No, but the equation would be solved differently. Instead of multiplying each side by $\frac{15}{100}$, we would need to divide each side by 0.15.

Section 6.6 Exercises

FOR EXTRA HELP

Student Resources on DVD-ROM — Includes ➤Student's Solutions Manual ➤Video Lectures ➤Chapter Test Prep Video | MyMathLab | MathXL

Think Again

Write a sentence or two for each response.

1. If a men's suit salesperson says that he makes "6¢,on the dollar," what is he referring to and what does it mean?

He is referring to the rate of commission he gets, which is 6% $\left(\frac{6\text{ cents}}{1\text{ dollar}} = \frac{6\text{ cents}}{100\text{ cents}} = \frac{6}{100} = 6\%\right)$.

2. You know that 8% = 0.08. If a state's sales tax rate is 8% and an item costs $3.50, is it appropriate to simply add 8% to the cost it, as in $3.50 + 8%? Explain your answer.

No, percent is always multiplied by a number. In this case, the 8% must be multiplied by the cost of the item, $3.50.

Focus Exercises

Find the product.

3. 20% of 70
14

4. 30% of 120
36

5. 45% of 60
27

6. 85% of 40
34

7. 75% of 96
72

8. 25% of 52
13

9. 64% of 45
28.8

10. 32% of 55
17.6

11. 37.5% of 160
60

12. 12.5% of 240
30

13. 87.5% of 56
49

14. 62.5% of 72
45

For each, find the missing value and write a conclusion.

15. 40% of 125 is what number?
40% of 125 is 50.

16. 60% of 95 is what number?
60% of 95 is 57.

17. 60% of 55 is what number?
60% of 55 is 33.

18. 40% of 35 is what number?
40% of 35 is 14.

19. 44% of 75 is what number?
44% of 75 is 33.

20. 45% of 60 is what number?
45% of 60 is 27.

21. 25% of 62 is what number?
25% of 62 is 15.5

22. 20% of 73 is what number?
20% of 73 is 14.6

23. 75% of what number is 57?
75% of 76 is 57.

24. 35% of what number is 14?
35% of 40 is 14.

25. 40% of what number is 30?
40% of 75 is 30.

26. 45% of what number is 36?
45% of 80 is 36.

27. 30% of what number is 57?
30% of 190 is 57.

28. 60% of what number is 48?
60% of 80 is 48.

29. 35% of what number is 28?
35% of 80 is 28.

30. 42% of what number is 63?
42% of 150 is 63.

31. What percent of 35 is 21?
60% of 35 is 21.

32. What percent of 80 is 56?
70% of 80 is 56.

33. What percent of 120 is 78?
65% of 120 is 78.

34. What percent of 240 is 36?
15% of 240 is 36.

35. What percent of 80 is 34?
42.5% of 80 is 34.

36. What percent of 112 is 98?
87.5% of 112 is 98.

37. What percent of 40 is 23?
57.5% of 40 is 23.

38. What percent of 32 is 20?
62.5% of 32 is 20.

39. What percent of 18 is 5?
$27.\overline{7}\%$ of 18 is 5.

40. What percent of 44 is 25?
$56.\overline{81}\%$ of 44 is 25.

41. What percent of 126 is 77?
$61.\overline{1}\%$ of 126 is 77.

42. What percent of 132 is 54?
$40.\overline{90}\%$ of 132 is 54.

Solve each application using the percent equation. Write a sentence answering the question. (You may want to identify the percent, the whole, and the part before you write the equation.) Use the outline provided with the first problem for the other problems.

43. 20% of the patrons who went roller skating at Skate-A-Rama last Saturday night rented skates. If 29 people rented skates, how many skaters were there that night?

Legend

To help you think it through:

Percent: __________

Whole: __________

Part: __________

Sentence 145 skaters were there that night.

44. Jayne works for a computer store and earns 4% commission on everything she sells. On Wednesday, she made $96 in commissions. What was the dollar value of the computer equipment she sold?
The value of the computer equipment she sold was $2,400.

45. 45% of the runners in a marathon were over 40 years old. If 234 runners were over 40 years old, how many runners were in the marathon?
520 runners were in the marathon.

46. 12% of the cars parked in the college parking lot were in the carpool spaces, reserved for anyone riding in a carpool. If 21 cars were parked in the carpool spaces, how many total cars were in the parking lot?
A total of 175 cars were in the parking lot.

47. 18% of the players in a youth soccer league kick with their left foot. If 27 of the players kick with their left foot, how many players are in the league?
150 players are in the league.

48. 45% of the attendees at Philip and Sharon's wedding were single. If 36 of the attendees were single, how many people were at the wedding?
80 people were at the wedding.

49. 12.5% of the items at a silent auction were gift certificates. If 19 of the items were gift certificates, how many items did the silent auction have?
The silent auction had 152 items.

50. Last Monday, 17.5% of that day's bank customers needed access to their safe-deposit boxes. If 28 people opened their safe-deposit boxes, how many customers did the bank have that day?
The bank had 160 customers that day.

51. In a class of 40 students, only 18 brought their books to class on the first day. What percent of the class brought their books on the first day?
45% of the class brought their books on the first day.

52. Last November, it rained 21 out of 30 days at Big Bear Mountain. What percent of the month of November did it rain at Big Bear Mountain?
It rained 70% of the month of November at Big Bear Mountain.

53. Leonard collects antique clocks. Of the 125 clocks that he owns, 45 of them don't have chimes. What percent of Leonard's clocks don't have chimes?
36% of Leonard's clocks don't have chimes.

54. Of the 75 members of the Riverside Kiwanis Club, 18 have their birthday in June. What percent of the members have their birthday in June?
24% of the members have their birthday in June.

55. Of the 64 e-mail messages Shawna received yesterday, 24 were spam. What percent of Shawna's e-mails were spam?
37.5% of Shawna's e-mails were spam.

56. Of the 72 potential jurors called to Judge Day's courtroom yesterday, 45 got called to the jury box for questioning. What percent of the potential jurors got called to the jury box?
62.5% of the potential jurors got called to the jury box.

57. Of the 360 diners at the King's Grill Buffet, 117 redeemed discount coupons. What percent of the diners used coupons?
32.5% of the diners used coupons.

58. Of the 128 deputies in the Riverside County Sheriff's Department, 16 are women. What percent of the deputies are women?
12.5% of the deputies are women.

Think Outside the Box

59. Of the 120 customers at Donner Party Supplies, 42 ordered a balloon bouquet. What percent of the customers did *not* order a balloon bouquet?
65% of the customers did not order a balloon bouquet.

60. In California, the average price of regular unleaded gasoline in July 2007 was 150% of the average price of regular unleaded gasoline in December 2006, which was $2.30. What was the average price of regular unleaded gasoline in July 2007?
In July 2007, the average price of regular unleaded gasoline was $3.45.

SECTION 6.7 Applications Involving Percents

OBJECTIVES

In this section, you will learn to
- Solve percent applications using arithmetic.

You Need to Know

To successfully complete this section, you need to understand
- ☐ Rounding decimals (5.2)
- ☐ Multiplying decimals (5.4)
- ☐ Dividing decimals (5.5)
- ☐ Writing percents as decimals (6.4)

Introduction

We will not require equations to solve the percent applications in this section. Instead, we will multiply and divide directly *without* using a variable to represent the unknown value.

In each of these applications, the whole is known and we'll do one of the following

1. Multiply the percent by the whole to find the part
2. Divide the part by the whole to find the percent

Let's practice multiplying with percents. In each expression, it's best to rewrite the percent as a decimal.

Example 1

Multiply. For each expression, round the result to the nearest hundredth.

a) 35% × 70 **b)** 9% × 26.65 **c)** 3.2% × 16.95

Answer

a)
$$\begin{array}{r} 70 \\ \times\ 0.35 \\ \hline 350 \\ +\ 2100 \\ \hline 24.50 \end{array}$$
This does not need to be rounded.

b)
$$\begin{array}{r} 26.65 \\ \times\ \ 0.09 \\ \hline 2.3985 \end{array}$$
This rounds up to 2.40.

c)
$$\begin{array}{r} 16.95 \\ \times\ 0.032 \\ \hline 3390 \\ +\ 50850 \\ \hline 0.54240 \end{array}$$
This rounds down to 0.54.

You Try It 1 **Multiply. For each expression, round the result to the nearest hundredth. Use Example 1 as a guide.**

a) 42% × 65 **b)** 4% × 68.75 **c)** 4.8% × 60.55

Using Multiplication to Solve Percent Applications

Sales Tax: We know from Section 5.4 that sales tax is a percentage of the purchase price (whole). The amount of sales tax is calculated by multiplying the sales tax rate (%) by the purchase price. This tax is added to the purchase price at the cash register.

Sales tax amount = Sales tax rate × Price
Total amount to pay = Price + Sales tax amount

Suppose the price of a new computer keyboard is \$60 and the sales tax rate is 8%. The amount of sales tax is found by multiplying 8% by \$60: 0.08 × \$60 = \$4.80. The amount of sales tax, \$4.80, is then added to the price. So, the customer ends up paying \$60.00 + \$4.80 = \$64.80 for the keyboard.

Caution After you calculate the sales tax, round it to the nearest penny. Then add it to the price to get the total amount to be paid.

Example 2

If the sales tax rate is 8%, find the amount of sales tax due on a DVD that has a price of \$23.95. Also, find the total that the customer must pay.

Procedure Write the percent as a decimal and multiply it by the price. Round that amount to the nearest penny. Add the sales tax to the price to find the total amount to be paid.

Answer 8% = 0.08; so, the sales tax = 0.08 × 23.95 = \$1.916.

YOU FINISH IT: Multiply

$$\begin{array}{r} 23.95 \\ \times\ 0.08 \\ \hline \end{array}$$

Round 1.916 to the nearest penny: \$1.92.

Add:	Price	+	Sales tax	=	Total to be paid
	\$23.95	+	\$1.92	=	\$25.87

$$\begin{array}{r} 23.95 \\ +\ 1.92 \\ \hline 25.87 \end{array}$$

Sentence The sales tax is \$1.92, and the customer must pay \$25.87.

You Try It 2 **If the sales tax rate is 7%, find the amount of sales tax due on a carpet that costs \$209.99. Then find the total that the customer must pay. Use Example 2 as a guide.**

Sentence ______________________________

Commission: The amount of commission that a salesperson earns on a single item is always a percentage of that item's price. When a salesperson sells many items throughout the day (or week or month), the amount of commission earned is a percentage of the total sales by that salesperson.

The amount of commission is calculated by multiplying the commission rate (%) by the total sales.

Amount of Commission = Commission rate × Total sales

Let's assume that a furniture salesperson earns a 3% commission on everything he sells. In January, he sold \$136,000 worth of furniture. His commission for that month is found by multiplying 3% by 136,000.

0.03 × \$136,000 = \$4,080

Caution The amount of commission is not added to or subtracted from anything.

Example 3

Connie is a salesperson at an appliance store and earns a 3.5% commission on everything she sells. Last Saturday, she sold appliances totaling \$3,946. How much did she earn in commissions that day?

Procedure Multiply the percent by the total sales and round that amount to the nearest penny.

Answer 3.5% = 0.035; so, the commission is 0.035 × 3,946.

Sentence Connie earned \$138.11 in commissions that day.

YOU FINISH IT: Multiply

$$\begin{array}{r} 3946 \\ \times 0.035 \\ \hline \end{array}$$

You Try It 3 **RaeLynn sells stereo and camera equipment at an electronics store and earns 2.5% on everything she sells. Yesterday, she sold electronic gear totaling $4,087. How much did RaeLynn earn in commissions that day? Use Example 3 as a guide.**

Sentence ______________________________

Discount: The original price of an item is 100% of that price. When a percent discount is given, it can be subtracted directly from 100% before the discounted price is calculated.

New Price = (100% − Disount percent) × Original price

For example, a 25% discount is equal to 100% − 25% = 75% of the original price. To find the new price of the item, we multiply the original price by 75%.

Suppose the original price of a new computer keyboard is $60, but the store is offering a 20% discount. This means that the new price is 100% − 20% = 80% of the price. Multiply 80% by $60: 0.80 × $60 = $48.

The new price of the keyboard is $48. (This is before sales tax is added at the register.)

Example 4

This week at Kasey's Department Store, jackets are discounted 15%. Find the new price of a jacket that originally cost $46.90.

Procedure 100% − 15% = 85%. Multiply the original price, $46.90, by 85% to find the new price. ⟶

Answer

Sentence The new price is $39.87.

YOU FINISH IT: Multiply

$$\begin{array}{r} 46.90 \\ \times 0.85 \\ \hline \end{array}$$

You Try It 4 **Diamond Adventure is offering a 30% discount on all of its tents. Find the new price of a tent that originally cost $169.95. Use Example 4 as a guide.**

Sentence ______________________________

Using Division to Solve Percent Applications

Because percent is a ratio, a fraction, $\frac{\text{percentage}}{\text{base}}$, there is a *division* relationship between the percentage (part) and the base (whole).

So, to find the percent comparison between a part and the whole, we divide as follows:

$$\frac{\textbf{Part}}{\textbf{Whole}} \quad \text{or} \quad \textbf{Part} \div \textbf{Whole} \quad \text{or} \quad \textbf{Whole}\overline{)\textbf{Part}}$$

Example 5

Carlo, a furniture salesperson, sold a total of $8,500 in merchandise last Monday and earned $255 in commissions. What is Carlo's commission rate?

Procedure In this case, we divide $\frac{255}{8{,}500}$, or 255 ÷ 8,500. ⟶

Answer

Sentence Carlo's commission rate is 3%.

YOU FINISH IT: Divide

$$8{,}500\overline{)255}$$

You Try It 5 **Sonya works in a flooring center selling tile and carpet. Last Tuesday, she sold a total of $4,600 worth of goods and earned a commission of $161. What is Sonya's commission rate? Use Example 5 as a guide.**

Sentence ______________________________

Determining Percent Increase and Percent Decrease

Investing in the stock market can be risky. The price of a stock may increase in value, it may decrease in value, or it may not change at all. Often investors are interested in the *percent increase* and the *percent decrease*, each of which is a ratio.

$$\text{Percent increase} = \frac{\text{Amount of increase}}{\text{Original amount}} \qquad \text{Percent decrease} = \frac{\text{Amount of decrease}}{\text{Original amount}}$$

In each ratio, the original amount is the value of an item, such as a stock, before the increase or decrease took place. For example, if we want to know the percent increase (or decrease) from Tuesday to Wednesday, the original amount is Tuesday's amount. Likewise, if we want to know the percent increase from 2004 to 2005, the original amount is the amount from 2004.

If a company's stock rises $3 in one day, is that a lot or a little? It depends on how much the stock was worth the day before.

For example, if Tuesday's ending price for a stock was $30 and the value increases $3 by the end of Wednesday, that is a 10% increase.

$$\text{Percent increase} = \frac{\text{Amount of increase}}{\text{Original value}} = \frac{3}{30} = \frac{1}{10} = 0.10 = 10\%$$

However, if Tuesday's ending price for a stock was $100 and the value increases $3 by the end of Wednesday, that is a 3% increase.

$$\text{Percent increase} = \frac{\text{Amount of increase}}{\text{Original value}} = \frac{3}{100} = 3\%$$

Likewise, if price of a stock at the end of 2004 was $40 and over the course of a year the value *decreases* $2 by the end of 2005, that is a 5% decrease.

$$\text{Percent decrease} = \frac{\text{Amount of decrease}}{\text{Original value}} = \frac{2}{40} = \frac{1}{20} = 0.05 = 5\%$$

When finding a percent increase or decrease, we must know both

1. The amount of increase (or decrease).
2. The original amount to which it is compared.

For example, at the end of 2004, a stock was priced at $24. At the end of 2005, its value had increased to $30. What was the percent increase?

First, we must identify the amount of increase by subtracting the lower amount from the higher amount.

$$\text{Higher} - \text{Lower} = \$30 - \$24 = \$6$$

We then must recognize the original value, the value that came first. Because the $24 value was at the end of 2004 and the $30 value was a year later, we compare the amount of increase ($6) to $24, the value that came first.

$$\text{Percent increase} = \frac{\text{Amount of increase}}{\text{Original value}} = \frac{6}{24} = \frac{1}{4} = 25\%$$

Example 6

At the end of February, a computer store had $18,000 in total sales. At the end of March, it had $20,700 in total sales. What is the percent increase in total sales?

Procedure First, find the amount of increase. Then compare that amount to the original amount, February's sales.

Answer The amount of increase is found by subtracting: $20,700 − $18,000 = $2,700.

This amount is compared to the first total, the total sales in February.

$$\frac{2{,}700}{18{,}000} = \frac{27}{180}$$ This is 27 ÷ 180. ⟶

YOU FINISH IT: Divide

$180\overline{)27}$

Sentence Total sales increased 15% from February to March.

You Try It 6 **At the end of May, a stock's price was $48. At the end of June, the price had increased to $54. What is the percent increase in the price of the stock? Use Example 6 as a guide.**

Sentence ______________________________

Example 7

In 2001, a top-of-the-line laptop computer was priced at $3,600. In 2006, a comparable laptop computer was priced at $2,800. What is the percent decrease in price? (Round the quotient to the nearest thousandth if necessary.)

Procedure First, find the amount of decrease. Then compare that amount to the original price in 2001.

Answer The amount of decrease is found by subtracting:
$3,600 − $2,800 = $800.

$$\frac{800}{3{,}600} = \frac{8}{36} = \frac{2}{9}$$ This is 2 ÷ 9. ⟶

YOU FINISH IT: Divide

$9\overline{)2}$

Sentence The price for a top-of-the-line laptop decreased 22.2% from 2001 to 2006.

You Try It 7 **In 2005, the Cyuga Rotary Club had 33 members. In 2006, the club had only 27 members. What is the percent decrease in membership? (Round the quotient to the nearest thousandth if necessary.) Use Example 7 as a guide.**

Sentence ______________________________

Solving Simple Interest Applications

In the world of banking and finance, money that is deposited into a savings account or that is borrowed as a loan is called the **principal.** Usually, the principal earns **interest,** an amount of money that is automatically added to the principal based on the length of the loan or the savings period.

There are two types of interest: **simple interest,** which is applied to the principal just once, and **compound interest,** which is applied to the principal more than once. In this section, we discuss simple interest only.

Typically, the principal earns interest at a certain yearly interest rate over a period of time. The amount of simple interest earned is based on this formula:

Simple Interest Formula

$$I = P \cdot r \cdot t$$

Where, I is the *amount of interest*.
P is the *principal* (starting amount).
r is the yearly rate (a percent).
t is the amount of time in years.

Example 8

Find the amount of simple interest earned given the principal, the interest rate, and the amount of time the money is left in the account.

a) $P = \$1{,}000$, $r = 8\%$, $t = 2$ years.

b) $P = \$3{,}500$, $r = 10\%$, $t = \frac{1}{2}$ year

c) $P = \$4{,}000$, $r = 6\%$, $t = 8$ months

Procedure Rewrite each percent as a decimal; then multiply appropriately.

Answer

a) $I = P \cdot r \cdot t$ — Substitute the numbers into the formula; $8\% = 0.08$.

$I = 1{,}000 \times (0.08) \times 2$ — Multiply $P \times r$ first: $1{,}000 \times (0.08) = 80$.

$I = 80 \times 2$ — Now multiply $80 \times 2 = 160$.

$I = 160$ dollars

b) $I = P \cdot r \cdot t$ — $10\% = 0.10$

$I = 3{,}500 \times (0.10) \times \frac{1}{2}$ — Multiply $P \times r$ first: $3{,}500 \times (0.10) = 350$.

$I = 350 \times \frac{1}{2}$ — Now multiply $350 \times \frac{1}{2} = 175$.

$I = 175$ dollars

c) $I = P \cdot r \cdot t$ — Rewrite 8 months in terms of years: 8 months $= \frac{8}{12}$ years $= \frac{2}{3}$ years.

$I = 4{,}000 \times (0.06) \times \frac{2}{3}$ — Multiply $P \times r$ frst: $4{,}000 \times (0.06) = 240$.

$I = 240 \times \frac{2}{3}$ — Multiply $240 \times \frac{2}{3} = \frac{240}{1} \times \frac{2}{3} = \frac{80}{1} \times \frac{2}{1} = 160$.

$I = 160$ dollars

You Try It 8 **For each application, find the simple interest based on the given information. Use Example 8 as a guide.**

a) Karen put \$2,000 (the *principal*) in a savings account that gained interest at a rate of 6% (the *rate*). How much interest did Karen's account gain after 3 years (the *time*)?

b) Sondra put \$5,000 in a special account that gained interest at a rate of 8%. How much interest did Sondra's account gain after 9 months? (Hint: Write 9 months as a fraction of one year.)

Answers: You Try It and Think About It

You Try It: **1. a)** 27.3 **b)** 2.75 **c)** 2.91 **2.** 7% × 209.99 = 0.07 × 209.99 = 14.6993; 14.6993 rounds to 14.70. The sales tax is $14.70, and the customer must pay $224.69. **3.** 2.5% × 4,087 = 0.025 × 4,087 = 102.175, which rounds to $102.18. RaeLynn earned $102.18 in commission that day. **4.** 100% − 30% = 70%: 70% × 169.95 = 0.70 × 169.95 = 118.965, which rounds to $118.97. The new price of the tent is $118.97. **5.** 161 ÷ 4,600 = 0.035 = 3.5%; Sonya's commission rate is 3.5%. **6.** First, the amount of increase is $54 − $48 = $6. The percent increase is 6 ÷ 48 = 0.125 = 12.5%. The percent increase in the price of the stock is 12.5%. **7.** First, the amount of decrease is 33 − 27 = 6. The percent decrease is 6 ÷ 33 = 0.1818 . . . , which rounds to 0.182 or 18.2%. The percent decrease in membership is 18.2%. **8. a)** Karen's account gained $360 after 3 years. **b)** Sondra's account gained $300 after 9 months.

Think About It: No Think About Its in this section.

Section 6.7 Exercises

FOR EXTRA HELP: Student Resources on DVD-ROM — Includes ➤ Student's Solutions Manual ➤ Video Lectures ➤ Chapter Test Prep Video | MyMathLab | MathXL

Think Again

In March, the owner of a manufacturing company told his employees that because of financial difficulties, he would have to decrease everyone's pay by 20% for the month of April. However, he promised that in May, he would give everyone a 20% increase. Interestingly, the workers' wages in May will not be the same as their wages in March.

1. Why will the workers' wages be different in May than in March?
The wages will be different in May than in March because the 20% is based on different numbers. In April, the 20% is based on the original March wage. In May, the 20% is based on the new (lower) April wage.

2. Will the workers' wages in May be more or less than they were in March? Explain your answer or show an example that supports your answer. You may use a worker's wage in March as $100 per day, or you may use your own example.
Wages will be lower in May than in March.

Focus Exercises

Solve and write a sentence answering the question.

Sales tax

3. Sue purchased a portable CD player for $75.80 plus tax. If the sales tax rate in her state is 8%, how much did Sue pay at the register?
Sue paid $81.86 at the register.

4. Tanaya bought a jacket that cost $52.10. The sales tax rate in her state is 6%. How much did Tanaya she pay at the cash register?
Tanaya paid $55.23 at the cash register.

5. Maria bought a bouquet of flowers for $48 plus tax. If the sales tax rate in her state is 4.5%, how much did Maria pay at the register?
Maria paid $50.16 at the register.

6. Sadeer rented a car for $98 plus tax. If the sales tax rate on rental cars is 11.5%, how much did Sadeer pay to rent the car?
Sadeer paid $109.27 to rent the car.

Commission

7. Lorena is a salesperson at a mattress store. She receives a 3.5% commission on everything she sells. One day, she sold $3,180 worth of mattresses and bedding supplies. How much commission did Lorena make that day?
Lorena made $111.30 commission that day.

8. Mike is a salesperson at a large home-electronics store. He receives a 2.5% commission on everything he sells. One day, he sold $4,120 worth of stereo equipment. How much commission did Mike make that day?
Mike made $103 commission that day.

9. Gretchen is a salesperson at a furniture store. Yesterday she earned $97.50 on sales of $3,250. What percent is Gretchen's commission?
Gretchen's commission is 3%.

10. Giorgio is a salesperson at a large home appliance store. He earned $129.50 on sales of $3,700. What percent is Giorgio's commission?
Giorgio's commission is 3.5%.

Discounts

11. Tomás found a gas-powered lawn mower with a discount tag of 30%. The original price of the mower was $420. What is the new price of the mower?
The new price of the mower is $294.

12. Ilana saw a sweater that had a discount tag of 35%. The original price of the sweater was $90. What is the new price of the sweater?
The new price of the sweater is $58.50.

13. Eleazar works at a department store As an employee, he receives a 12% discount on any item he purchases in the store. The suit he chose to buy was originally priced at $204.50. What price will Eleazar pay using employee discount?
Eleazar will pay $179.96 using his employee discount.

14. Chloe was shopping for a new MP3 player. She saw an ad for a brand that was 25% off the original price of $168. What was the new price of the MP3 player?
The new price of the MP3 player was $126.

15. Laura purchased $103.50 worth of fabric to make the costumes for a children's musical. At the register, she presented a coupon that gave her a 15% discount on the entire amount. How much was the new price for all of the fabric?
The new price for all of the fabric was $87.98.

16. A car that originally cost $13,600 is discounted 8%. Find the new price of the car.
The new price of the car is $12,512.

Percent

17. Guillermo's company is charged with finding businesses to fill the 120 stores at a shopping mall. So far he has signed leases with 54 businesses. What percent of the stores has Guillermo leased?
Guillermo has leased 45% of the stores.

18. Sarah took a math test and got 51 points out of 60 possible. What is Sarah's grade as a percent?
Sarah's grade is 85%.

19. Mandy served a table of eight guests at her restaurant. The food and drink bill came to $225.00, and she received a tip of $40.50. What percent of the bill did Mandy receive as a tip?
Mandy received 18% of the bill as a tip.

20. Carnel saved $12 on a pair of shoes. The original price was $80. What percent were the shoes discounted?
The shoes were discounted 15%.

21. At the beginning of the week, George had 60 motorcycles in inventory. By the end of the week, he had sold 9 of the motorcycles. What percent of the motorcycles did George sell during the week?
George sold 15% of the motorcycles during the week.

22. Marcus is raising money to go on the marching band's spring tour. He needs to raise $600. So far he has raised $390. What percent of his goal has Marcus attained?
Marcus has attained 65% of his goal.

Percent increase and decrease

23. Last year, Louis was the president of the local Kiwanis Club. During the year, the club grew in membership from 20 to 25. What was the percent increase during Louis's time as club president?
The increase during Louis's time as club president was 25%.

24. The choir budget was $4,000 last year. This year, the budget is $4,800. What is the percent increase in the choir budget?
The increase in the choir budget is 20%.

25. Last year, Daniel and his wife, Sonia, were able to give $500 per month to their synagogue. This year, they have pledged to give $550 a month. What is the percent increase in their monthly gift?
The increase in their monthly gift is 10%.

26. During his junior year in high school, Nate experienced a growth spurt and grew from 64 inches to 72 inches. By what percent did Nate's height increase?
Nate's height increased by 12.5%

27. At the beginning of the month, Alice weighed 225 pounds. After healthy eating and exercise, she weighed 207 pounds at the end of the month. What was the percent decrease of Alice's weight?
The decrease of Alice's weight was 8%.

28. Last year, a Chevrolet Marlin was priced at $24,000. This year its price has dropped to $22,440. What is the percent decrease from last year to this year?
The decrease from last year to this year is 6.5%.

29. The Tomlins' house was up for sale for six months before it finally sold. The original asking price was $360,000, but the house sold for $333,000. What was the percent decrease of the selling price compared to the original asking price?
The decrease of the selling price was 7.5%.

30. For the month of December, sales at Mitch's Sporting Goods were an all-time high of $40,000. As always happens, though, January sales were much lower, at only $28,000. What was the percent decrease in sales from December to January?
The decrease in sales from December to January was 30%.

Simple interest

31. Sally put $800 in a special account that gained 9% interest. How much interest did the account gain after 2 years?
The account gained $144 interest after 2 years.

32. Blanca put $1,200 in a savings account that gained 4.5% interest. How much interest did the account gain after 3 years?
The account gained $162 interest after 3 years.

33. Pradeep put $3,000 in a special account that gained interest at a rate of 7%. How much interest did the account gain after $\frac{1}{2}$ year?
The account gained $105 interest after $\frac{1}{2}$ year.

34. Marilyn put $6,000 in an account that gained interest at a rate of 5.5%. How much interest did the account gain after $\frac{2}{3}$ year?
The account gained $220 after interest $\frac{2}{3}$ year.

35. Lupe put $12,000 in a certificate of deposit that gained interest at a rate of 7.5%. How much interest did the account gain after 4 months?
The account gained $300 interest after 4 months.

36. Conrad put $8,000 in a certificate of deposit that gained interest at a rate of 12%. How much interest did the account gain after 3 months?
The account gained $240 interest after 3 months.

37. Mark put $5,000 in a special account that gained 6% interest. How much interest did the account gain after 8 months?
The account gained $200 interest after 8 months.

38. T'Mara put $4,800 in a special account that gained 5.5% interest. How much interest did the account gain after 9 months?
The account gained $198 interest after 9 months.

Miscellaneous

39. Charlene went to the campus bookstore to purchase a textbook and some school supplies. Here is her shopping list and the price of each item.

Item	Price
Textbook	$103.00
Notebook	4.92
3 pencils	1.83
2 pens	4.26

a) Add up the total price.
The total price is $114.01.

b) Compute the sales tax if the sales tax rate is 8%.
The sales tax is $9.12.

c) Find the total amount that Charlene will pay at the bookstore for these supplies.
The total amount that Charlene will pay is $123.13.

40. Jamal saved $27 to buy a gift for his cousin, Amee. He knew he'd have to pay 6.5% sales tax; so, the gift would need to be priced lower than $27. He found the perfect gift, and it cost $25. With the 6.5% sales tax added on, does Jamal have enough money to buy the gift for Amee? If so, how much will he pay at the register?
Jamal has enough money to buy the gift. He will pay $26.63 at the register.

41. Henry took his family to a nice restaurant to celebrate his daughter's graduation. The meal total came to $134.

a) If the sales tax rate is 7.75%, what is the sales tax on the meal?
The sales tax is $10.39

b) Henry thought the service and the food were good and decided to leave an 18% tip for the server. How much should the tip be? (The tip, by the way, is a percentage of the meal total that should be figured separately from the sales tax.)
The tip should be $24.12.

c) What was the total amount that Henry paid for dinner that night?
The total amount that Henry paid for dinner was $168.51.

42. Bud works for a department store. One fringe benefit the store provides is a 15% discount on all merchandise that employees purchase. Bud found a TV he wants to buy that costs $589.95.

a) What price will Bud pay for the TV?
Bud will pay $501.46 for the TV.

b) If the sales tax rate in his city is 7%, how much will Bud pay in sales tax?
Bud will pay $35.10 in sales tax.

c) How much will Bud pay for the TV, including sales tax?
Bud will pay $536.56 for the TV, including sales tax.

Think Outside the Box

43. On March 31, the price of ten shares of Jozo stock was $420. By the end of April, the total value of the stock had increased 10%. By the end of May, the total price of the stock had decreased 10%. What was the value of the ten shares of Jozo stock at the end of May?
At the end of May, the value of the ten shares of Jozo stock was $415.80.

44. Yu-Lin is purchasing a car. Because of her good credit, she is able to finance the whole amount of the car, including sales tax and fees, for $15,500 at 6% interest. However, over the life of the loan (5 years), the total interest is equal to 16% simple interest.

a) What is the total amount of interest that Yu-Lin will pay on the loan?
The total amount of interest that Yu-Lin will pay is $17,980.

b) What will Yu-Lin's monthly payments be?
Yu-Lin's monthly payment will be $299.67.

Chapter 6 Review

Section 6.1 Ratios

Concept	Example
A **ratio** is a comparison between two numbers using division. The two numbers forming a ratio are always specific to a certain unit of measure or count.	6 dollars : 10 pounds or $\frac{\text{6 dollars}}{\text{10 pounds}}$
To simplify a ratio, divide out any common factors.	$\frac{\text{6 dollars}}{\text{10 pounds}}$ simplifies by a factor of 2 to $\frac{\text{3 dollars}}{\text{5 pounds}}$.
When the unit of measure for each number is the same, that unit of measure can be divided out as well.	$\frac{\text{24 dollars}}{\text{30 dollars}}$ can simplify by a factor of 6 to $\frac{\text{4 dollars}}{\text{5 dollars}}$. The unit of measure, *dollars*, can divide out as well. $\frac{\text{4 dollars}}{\text{5 dollars}} = \frac{4}{5}$, or 4 : 5.
When two ratios have the same value, they are **equivalent ratios.**	$\frac{4}{10}$ and $\frac{2}{5}$ are equivalent ratios.
In the equivalence of two ratios, expressed 4 : 6 = 2 : 3, the *outermost* numbers, 4 and 3, are the extremes. The *middle* two numbers, 6 and 2, are the means.	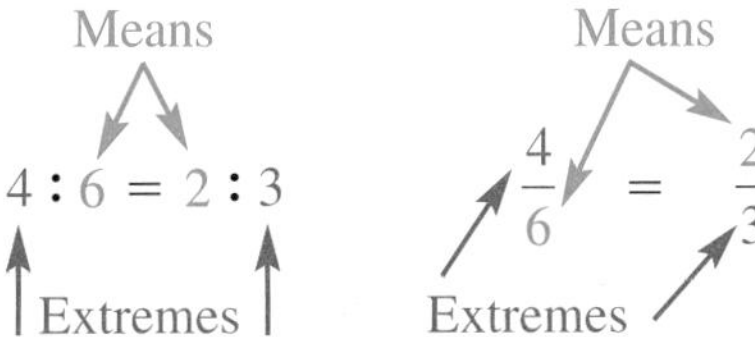
The Means–Extremes Product Property If the product of means equals the product of extremes, then the two ratios are equivalent. This method is called **cross multiplication,** multiplying across the equal sign.	Start with the equality of two ratios. $\frac{2}{3} = \frac{6}{9}$ Multiply "across" the equal sign. $\frac{2}{3} \times \frac{6}{9}$ $2 \cdot 9 = 3 \cdot 6$ $18 = 18$

Section 6.2 Rates

Concept	Example
In a ratio, when the units of measure—or counts—of the numerator and the denominator are different, the ratio is called a **rate.**	$\frac{\text{4.2 miles}}{\text{0.18 gallons}}$
Rates are best understood when they are simplified completely.	$\frac{\text{4.2 miles} \times 100}{\text{0.18 gallons} \times 100} = \frac{\text{420 miles} \div 6}{\text{18 gallons} \div 6} = \frac{\text{70 miles}}{\text{3 gallons}}$ $=$ *70 miles for every 3 gallons*

continued

Section 6.2 Rates

Concept	Example
A **unit rate** is a rate in which the denominator has been reduced to 1 item. Any rate can be reduced to a unit rate by simplifying the fraction or by using long division.	$\frac{105 \text{ miles}}{6 \text{ gallons}}$ cannot simplify directly; but we can divide $6\overline{)105.0}$ = 17.5 and write the unit fraction as $\frac{17.5 \text{ miles}}{1 \text{ gallon}}$ or *17.5 miles per gallon.*
The unit rate for the price of an item is called the **unit price.**	Bag 1: $\frac{3.48 \text{ dollars}}{24 \text{ ounces}}$ $24\overline{)3.480}$ = 0.145; $\frac{0.145 \text{ dollars}}{1 \text{ ounce}} \approx$ *\$0.15 per ounce* Bag 2: $\frac{2.10 \text{ dollars}}{12 \text{ ounces}}$ $12\overline{)2.100}$ = 0.175; $\frac{0.175 \text{ dollars}}{1 \text{ ounce}} \approx$ *\$0.18 per ounce* Bag 1 is the better buy.

Section 6.3 Proportions

Concept	Example
A **proportion** is a statement that two ratios are equivalent. In a proportion, the product of the means is equal to the product of the extremes.	$\frac{a}{b} = \frac{c}{d}$ $a \cdot d = b \cdot c$
In setting up a proportion, the numerators of each fraction must be of the same units of measure (or count) and the denominators of each fraction must be of the same units of measure (or count).	$\frac{2 \text{ inches}}{5 \text{ miles}} = \frac{6 \text{ inches}}{15 \text{ miles}}$
To solve a proportion, use cross multiplication. Then divide by the coefficient of the variable. Before cross multiplying, see if one of the fractions can be simplified.	Solve for x in the proportion. $\frac{20}{x} = \frac{15}{12}$ $\frac{15}{12}$ can be reduced by a factor of 3 to $\frac{5}{4}$. $\frac{20}{x} = \frac{5}{4}$ Cross multiply. $4 \cdot 20 = 5 \cdot x$ $80 = 5x$ $\frac{80}{5} = \frac{5x}{5}$ $16 = x$

Section 6.3 Proportions

Concept	Example
When two triangles have the same shape, they are said to be **similar** to each other. When two triangles are similar to each other, each side of one of the triangles is paired to a side on the other triangle. These two sides are called **corresponding** sides. The ratio of one set of corresponding sides is equal to the ratio of another set of corresponding sides.	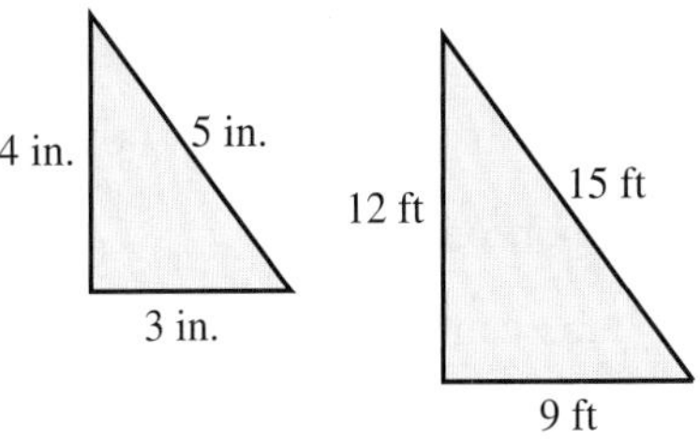 $$\frac{4}{12} = \frac{3}{9} = \frac{5}{15}$$ Each of these ratios simplifies to $\frac{1}{3}$; so, we know they are equivalent to each other.

Section 6.4 Percents

Concept	Example
The symbol % means "percent," and the number that is written before the percent symbol is called the **percentage.**	In 12%, the percentage is 12.
Percent, or *per cent*, means "per 100" or "divided by 100." Therefore, a percent is a *ratio* in which the denominator is 100.	28% means $\frac{28}{100}$, or "28 out of 100."
If the percentage is a whole number, the percent can be written first as a fraction and then as a decimal.	$44\% = \frac{44}{100} = 0.44$
Any percent can be written directly as a decimal by eliminating the % symbol and moving the decimal point two places to the left.	$26.5\% = 0.265$
Procedure for Writing Percents as Decimals **1.** Recognize the percentage and notice whether it contains a decimal point. If it doesn't, write one at its end. **2.** If the percentage doesn't have at least two whole numbers before the decimal point, place one or two zeros in front of the percentage **3.** Write the percentage without the % and move the decimal point two places to the left.	**1.** 5% → 5.% **2.** 5% → 05.% **3.** 5% → 05.% → .05 or 0.05
To write a decimal as a percent, multiply the decimal number by 100%.	$0.62 = 0.62 \times 100\% = 62\%$
To write a fraction as a percent, write the fraction as a decimal (by long division) and then multiply by 100%.	$\frac{3}{4} \rightarrow 4\overline{)3.00}$ (quotient 0.75) $\rightarrow \frac{3}{4} = 0.75 \times 100\% = 75\%$

Section 6.5 Solving Percent Problems Using Proportions

Concept	Example
A ratio that is not a percent can be thought of as $\frac{\textbf{amount}}{\textbf{base}}$, where the base is a number other than 100. The percent proportion is $$\frac{\text{percentage}}{100} = \frac{\text{amount}}{\text{base}},$$ as in the formula $\frac{p}{100} = \frac{a}{b}$. *p* represents the percentage, *a* represents the amount, and *b* represents the base. When two of these variables are known, we can set up and solve a proportion—by first using cross multiplication—to find the third variable.	5 is what percent of 8? $$\frac{p}{100} = \frac{5}{8}$$ $$100 \cdot 5 = p \cdot 8$$ $$500 = 8p$$ $$\frac{500}{8} = \frac{8p}{8}$$ $$62.5 = p$$ 5 is 62.5% of 8.

Section 6.6 Solving the Percent Equation

Concept	Example
Percent is always *of* something, and *percent of* indicates *multiplication.*	If 35 candies are in a box of chocolates and 20% of them are caramels, there are 20% × 35 = 7 caramels.
The percent equation Percent of whole is part ↓ ↓ *Percent* × *Whole* = *Part*	$20\% \times 35 = 7$
When the *part* is unknown, we write the legend: Let $x =$ the part.	$20\% \times 35 = x$ $0.20 \times 35 = x$ $7 = x$
When the *whole* is unknown, we write the legend: Let $x =$ the whole.	$20\% \cdot x = 7$ $0.20 \cdot x = 7$ $x = 35$
When the *percent* is unknown, we write the legend: Let $p =$ the percent.	$p \cdot 35 = 7$ $\frac{35p}{35} = \frac{7}{35}$ $p = 0.20$ $p = 20\%$

Section 6.7 Applications Involving Percents

Concept	Example
Sales tax is a percentage of the purchase price (whole). The amount of sales tax is calculated by multiplying the sales tax rate (%) by the purchase price. This tax is *added* to the purchase price at the cash register. Sales tax amount = Sales tax rate × Price Total to pay = Price + Sales tax amount	If the sales tax rate is 7% and a bookcase costs \$149, how much is the tax on the bookcase? 7% × \$149 = \$10.43. The tax is \$10.43.
The amount of **commission** that a salesperson earns is always a percentage of the total sales (whole). The amount of commission is calculated by multiplying the commission rate (%) by the total sales. Commission = Commission rate × Total sales	Joel earns a commission of 4% on the sales he makes, calculated monthly. In August, he made sales totaling \$14,000. What is Joel's commission for that month? 4% × \$14,000 = \$560. Joel's commission is \$560.
The original price of an item is 100% of that price. When a percent **discount** is given, it can be *subtracted* directly from 100%. This new percent is multiplied by the original price to get the discounted price. Discount = (100% − Discount %) × Price	Tyson is shopping for a pair of sneakers, and he finds a pair that originally cost \$95 but is now discounted 30%. What is the new price of the sneakers? 100% − 30% = 70% 70% × \$95 = \$66.50. The new price is \$66.50.
To find a ratio as a percent, divide the part by the whole.	$\frac{\text{Part}}{\text{Whole}}$ Part ÷ Whole $\text{Whole}\overline{)\text{Part}}$
When finding a percent increase or decrease, we must know the *amount* of increase (or decrease) and the *original value* to which it is compared.	$\text{Percent increase} = \frac{\text{Amount of increase}}{\text{Original value}}$ $\text{Percent decrease} = \frac{\text{Amount of decrease}}{\text{Original value}}$
When we are finding a percent increase or decrease, the comparison value is always the *original* value. When given a original value and a current (or later) value, we can find the *amount* of increase (or decrease) by subtracting the lower amount from the higher amount.	In June, a pond was measured to be 80 inches deep. In September, the pond was measured to be 64 inches deep. What was the percent decrease in the depth of the pond? 80 − 64 = 16 is the *amount* of decrease. The percent decrease is $\frac{\text{Amount of decrease}}{\text{Original value}} = \frac{16}{80} = \frac{2}{10} = 0.2 = 20\%.$

continued

Section 6.7 Applications Involving Percents

Concept	Example
Money that is deposited into a savings account or that is borrowed is called the **principal.** **Simple interest** is a percentage of the principal based on a yearly interest rate. **The formula for simple interest:** $I = P \cdot r \cdot t$ Where, *I* is the *amount of interest.* *P* is the *principal* (starting amount). *r* is the yearly rate (a percent). *t* is the amount of time in years.	If Minh invests \$6,000 in an account earning interest at a rate of 5%, how much interest will the account earn in 10 months? $P = \$6{,}000, \quad r = 5\% = 0.05, \quad t = \frac{10}{12} = \frac{5}{6}$ $I = 6{,}000 \times 0.05 \times \frac{5}{6} = 250$ In 10 months, the account will earn \$250.

Chapter 6 Review Exercises

Vocabulary

Fill in the blanks.

1. In the equation $\frac{25}{60} = \frac{10}{24}$, the extremes are 25 and 24. **(6.1)**
2. We can show that two ratios are equivalent using cross multiplication. **(6.1)**
3. A proportion is an equation showing the equivalence of two ratios. **(6.3)**
4. In the proportion of two similar triangles, each ratio is a comparison of a pair of corresponding sides. **(6.3)**
5. The *of* in *percent of* always means multiply. **(6.6)**
6. In calculating simple interest, the amount invested is called the principal. **(6.7)**

True or false.

7. A rate is a ratio in which the units of measure (or counts) are different. True **(6.2)**
8. Similar triangles have the same shape but can be different sizes. True **(6.3)**
9. The symbol % is called the percentage. False **(6.4)**
10. The sales tax rate is added directly to the price of an item. False **(6.7)**

Section 6.1

Write each of the following comparisons as a ratio in the form of a fraction.

	Phrase	Ratio
11.	12 cookies per box	$\frac{12 \text{ cookies}}{1 \text{ box}}$
12.	45¢ for each doughnut	$\frac{45 \text{ cents}}{1 \text{ doughnut}}$
13.	3 quarts for every 8 children	$\frac{3 \text{ quarts}}{8 \text{ children}}$
14.	$7 for every 10 cards	$\frac{\$7}{10 \text{ cards}}$

Write each fraction as a comparison.

	Phrase	Ratio
15.	$\frac{25 \text{ players}}{1 \text{ game}}$	25 players per game
16.	$\frac{\$85}{1 \text{ week}}$	$85 per week
17.	$\frac{120 \text{ passengers}}{5 \text{ cars}}$	120 passengers for every 5 cars
18.	$\frac{80 \text{ miles}}{3 \text{ gallons}}$	80 miles for every 3 gallons

Write each ratio in fractional form and simplify it completely.

19. 72 dollars : 9 shares $\frac{\$72}{9 \text{ shares}} = \frac{\$8}{1 \text{ share}}$
20. 18 minutes : 45 minutes $\frac{18 \text{ minutes}}{45 \text{ minutes}} = \frac{2}{5}$
21. 2.4 acres : 3 acres $\frac{2.4 \text{ acres}}{3 \text{ acres}} = \frac{4}{5}$
22. 0.6 calories : 1.5 grams $\frac{0.6 \text{ calories}}{1.5 \text{ grams}} = \frac{2 \text{ calories}}{5 \text{ grams}}$

Write each described ratio in fractional form and simplify it completely.

23. Weekly trash collections in Freemont amount to 12 pounds per household; in Clairhaven, 15 pounds per household. Write these collections as a ratio, Freemont amount : Clairhaven amount.
$\frac{12 \text{ pounds}}{15 \text{ pounds}} = \frac{4}{5}$
24. In Montieth, the average yearly rainfall is 20 inches; in Lareno, 32 inches. Write these yearly rainfalls as a ratio, Montieth amount : Lareno amount.
$\frac{20 \text{ inches}}{32 \text{ inches}} = \frac{5}{8}$

25. Each month, Kendall invests \$120 in mutual funds and \$90 in bonds. According to this situation, write the following ratios in standard ratio form; fractional form, and as a simplified ratio.

a) Mutual funds amount : Bonds amount
$\$120 : \$90, \frac{\$120}{\$90}, \frac{4}{3}$

b) Mutual funds amount : Total amount
$\$120 : \$210, \frac{\$120}{\$210}, \frac{4}{7}$

c) Bonds amount : Total amount
$\$90 : \$210, \frac{\$90}{\$210}, \frac{3}{7}$

26. Each month, Rogelio puts money in a savings account for his daughters' college fund. He puts aside \$180 for Maria and \$120 for Gloria. According to this situation, write the following ratios in standard ratio form, in fractional form, and as a simplified ratio.

a) Maria's amount : Gloria's amount
$\$180 : \$120, \frac{\$180}{\$120}, \frac{3}{2}$

b) Maria's amount : Total amount
$\$180 : \$300, \frac{\$180}{\$300}, \frac{3}{5}$

c) Gloria's amount : Total amount
$\$120 : \$300, \frac{\$120}{\$300}, \frac{2}{5}$

Use cross multiplication to determine whether each pair of fractions is equivalent.

27. $\frac{12}{15} \stackrel{?}{=} \frac{4}{5}$
Yes

28. $\frac{25}{10} \stackrel{?}{=} \frac{45}{16}$
No

29. $\frac{12}{8} \stackrel{?}{=} \frac{2.5}{1.5}$
No

30. $\frac{5}{12} \stackrel{?}{=} \frac{1.25}{3}$
Yes

Section 6.2

Simplify each rate and write the result in words.

31. $\frac{520 \text{ dollars}}{8 \text{ days}}$
\$65 per day

32. $\frac{45 \text{ customers}}{6 \text{ hours}}$
15 customers for every 2 hours

33. $\frac{24 \text{ hits}}{72 \text{ times at bat}}$
1 hit for every 3 times at bat

34. $\frac{18 \text{ repetitions}}{30 \text{ minutes}}$
3 repetitions for every 5 minutes

Simplify each rate to a unit rate, and write the unit rate in words.

35. $\frac{48 \text{ golf balls}}{12 \text{ players}}$
4 golf balls per player

36. $\frac{171 \text{ students}}{15 \text{ tutors}}$
11.4 students per tutor

37. $\frac{354 \text{ phone calls}}{30 \text{ days}}$
11.8 phone calls per day

38. $\frac{103.6 \text{ miles}}{5.6 \text{ gallons}}$
18.5 miles per gallon

From the given information, write a fractional rate. Simplify the rate as a unit rate, and write it as a phrase.

39. In 2004, the city of Sunnyvale collected \$1,350 in sales tax for every 6 residents. Write the amount collected as a unit rate per resident.
\$225 per resident

40. In 15 games, Carmello scored a total of 339 points. Write his points scored as a unit rate per game.
22.6 points per game

41. Marcelus trimmed 18 trees and earned \$405. Write his earnings as a unit rate per tree.
\$22.50 per tree

42. Jenice paid \$114 to rent 25 chairs. Write the rental price as a unit rate per chair.
\$4.56 per chair

From the given information, simplify each rate to a unit price, write it as a phrase, and answer the question.

43. A 16-ounce jar of Redd's tomato paste costs \$1.92, and a 24-ounce jar of Green's tomato paste costs \$3.60. Which is the better buy?
16-ounce jar of Redd's tomato paste: \$0.12 per ounce
24-ounce jar of Green's tomato paste: \$0.15 per ounce
The Redd's tomato paste is the better buy.

44. A pack of 16 AlwaysReady batteries costs \$12.96, and a box of 28 LightShine batteries costs \$21.00. Which is the better buy?
Pack of 16 AlwaysReady batteries: \$0.81 per battery
Pack of 28 LightShine batteries: \$0.75 per battery
The LightShine batteries are the better buy.

Section 6.3

Solve for n in each proportion. If possible, first reduce the known fraction.

45. $\frac{9}{6} = \frac{n}{14}$
$n = 21$

46. $\frac{30}{25} = \frac{18}{n}$
$n = 15$

47. $\frac{n}{45} = \frac{16}{10}$
$n = 72$

48. $\frac{21}{n} = \frac{15}{25}$
$n = 35$

For each of the following applications, set up a proportion table. Write a legend and solve the corresponding proportion. Write a sentence answering the question.

49. Kjell's car used 12 gallons of gas to travel 246 miles. How many miles will Kjell's car travel on 16 gallons of gas?
Kjell's car will travel 328 miles.

50. To paint a wall, Lonnie charges \$6 for every 4 square feet of wall space. How much will Lonnie charge to paint 82 square feet of wall space?
Lonnie will charge \$123.

51. At a walk-a-thon, Karrie raised \$13.50 for every 3 laps she walked. If Karrie walked a total of 32 laps, how much money did she raise?
Karrie raised \$144.

52. Tamra's printer can print 6 pages in 15 seconds. How many seconds will it take to print 22 pages?
It will take 55 seconds to print 22 pages.

For each pair of similar triangles, use a proportion table to set up and solve the corresponding proportion. Write a sentence answering the question.

53. Cory knows that the distance between Denver, Colorado, and Los Angeles, California, is 840 miles. Now he wants to know the distance between Denver and El Paso, Texas. On a map, the distance from Los Angeles to Denver is 9 inches and the distance between Denver and El Paso is 6 inches. What is the real distance between Denver and El Paso?

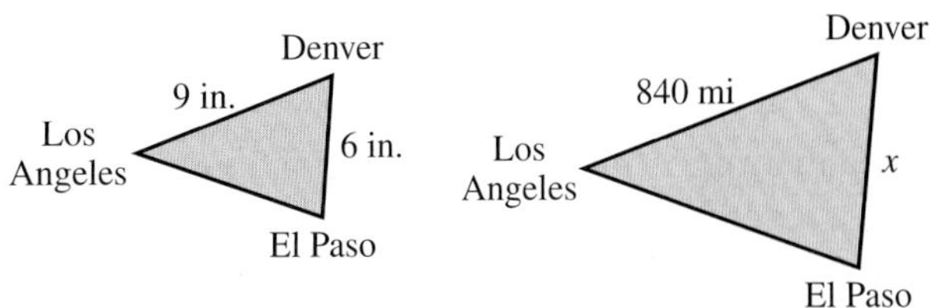

The real distance between Denver and El Paso is 560 miles.

54. A 4-foot post casts a 10-foot shadow. At the same time, a flagpole casts a 35-foot shadow. What is the height of the flagpole?

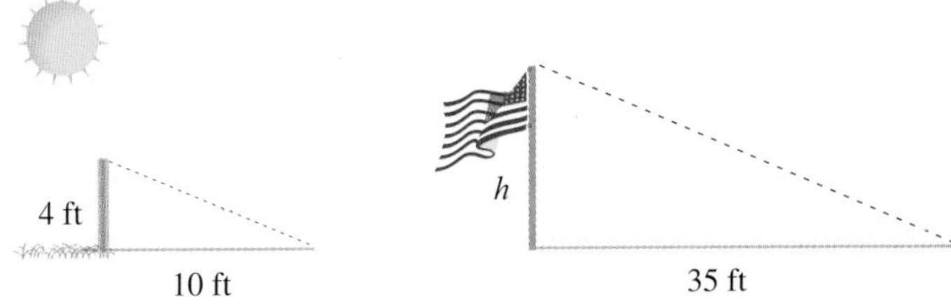

The height of the flagpole is 14 feet.

Section 6.4

Write each sentence with the percent as a simplified fraction

55. 20% of the homes in Westside Estates are ranch style.
$\frac{1}{5}$ of the homes in Westside Estates are ranch style.

56. 15% of all fruit sold in Bryson's Grocery Store is organically grown.
$\frac{3}{20}$ of all fruit sold in Bryson's Grocery Store is organically grown.

57. 8% of all U.S. residents live in Texas.
Source: **www.census.gov**
$\frac{2}{25}$ of all U.S. residents live in Texas.

58. 2.5% of the medals awarded at the 2004 Summer Olympics went to athletes from Ukraine.
Source: **www.sportsillustrated.cnn.com**
$\frac{1}{40}$ of the medals awarded at the 2004 Summer Olympics went to athletes from Ukraine.

Write each percent as a decimal.

59. 47%
0.47

60. 3%
0.03

61. 9.1%
0.091

62. 0.6%
0.006

Write each decimal as a percent.

63. 0.26
26%

64. 0.02
2%

65. 0.175
17.5%

66. 0.7
70%

67. 1.06
106%

68. 0.005
0.5%

Write each sentence with the fraction as a percent. Round the percentage to the nearest tenth.

69. $\frac{3}{20}$ of the Sri Lankan population is Hindu.
Source: The World Almanac, 2005
15% of the Sri Lankan population is Hindu

70. $\frac{16}{25}$ of the shoppers at Lucia's Floristas pay with a credit card.
64% of the shoppers at Lucia's Floristas pay with a credit card.

71. $\frac{7}{8}$ of the students in Ms. Skiba's art class received a passing grade.
87.5% of the students in Ms. Skiba's art class received a passing grade.

72. $\frac{2}{7}$ of all U.S. car sales are small cars.
Source: **www.senate.michigan.gov**
28.6% of all U.S. car sales are small cars.

Section 6.5

For each application, set up a proportion table, write a legend, set up and solve the proportion, and write a sentence answering the question.

73. 15% of the singers in the Lincoln High School choir are tenors. If the choir has 40 members, how many of them are tenors?
6 of the choir members are tenors.

74. 37.5% of the engineers at Hudson Dynamic have been with the company for at least ten years. If the company has a total of 32 engineers, how many have been at Hudson Dynamic at least ten years?
12 of the engineers have been at Hudson Dynamic at least 10 years.

75. 65% of the dinner customers at Chuck's Steakhouse order a salad with their meal. If 91 salads were served last Wednesday night, how many total dinner customers were there?
There were a total of 140 dinner customers.

76. 2.5% of the teachers in the North Unified High School District have a doctorate degree (PhD). If 8 of the teachers have a PhD, how many total teachers are there in the district?
There are a total of 320 teachers in the district.

77. Of the 30 days in April last year, it rained 18 days in the town of Martinville. What percent of the days in April did it rain?
It rained 60% of the days in April.

78. Of the 18 categories for which it was eligible, the 2004 movie *Million Dollar Baby* won 4 Oscars at the 2005 Academy Awards. What percent of the Oscars did the Movie win? ***Source:*** **www.oscars.org**
The movie won 22.2% of the Oscars.

Section 6.6

Find each number.

79. 40% of 95
38

80. 35% of 140
49

81. 8% of 125
10

82. 2.5% of 240
6

For each, find the missing value and write a conclusion.

83. 25% of 76 is what number?
25% of 76 is 19.

84. 60% of 85 is what number?
60% of 85 is 51.

85. 28% of what number is 35?
28% of 125 is 35.

86. 55% of what number is 88?
55% of 160 is 88.

87. What percent of 52 is 39?
75% of 52 is 39.

88. What percent of 120 is 45?
37.5% of 120 is 45.

Solve each application using the percent equation. Write a sentence answering the question. (You may want to identify the percent, the whole, and the part before writing the equation.)

89. A Student Activities report at Cuyama College indicated that only 3% of the student population participate in college clubs. If 120 students are in college clubs, how many students attend Cuyama College?
4,000 students attend Cuyama College.

90. In 2004, of the 160 National League batters who played regularly, 24 of them hit at least 30 home runs. What percent of these players hit at least 30 home runs? ***Source: Sports Illustrated 2005 Almanac***
15% of these players hit at least 30 home runs.

Section 6.7

Solve each application. Write a sentence answering the question. You may want to identify the percent, the whole, and the part before writing the equation.

91. DuJuan is buying a pair of running shoes that costs \$80. The sales tax rate in his state is 7.5%. How much will DuJuan pay at the cash register?
DuJuan will pay \$86 at the cash register.

92. Sandra is a real estate agent. She receives a 2% commission on every house that she sells. One day she sold a house for \$325,000. How much commission did Sandra earn on the sale of that house?
Sandra earned \$6,500 in commission.

93. Shay's Department Store had an after-Christmas clearance sale with decorations discounted 80%. The original price of a box of bulbs was \$24.50. What was the new price of the box of bulbs?
The new price of the box of bulbs was \$4.90.

94. Last season, Torraye scored 25 free throws out of 40 attempts. What was Torraye's free throw success as a percent?
Torraye's free throw success was 62.5%.

95. A television at Ed's TVs and Stuff retailed for \$1,200. Lorraine was able to bargain with the salesperson and save \$300. What was Lorraine's savings as a percent of the retail price?
Lorraine's savings was 25% of the retail price.

96. Connie has a coupon for 25% off any meal at Don Gordo's Mexican Restaurant. She treated her friend Marjorie to dinner one night, and the total food bill came to \$48.

a) With the 25% coupon, how much was the cost of the meal?
The cost of the meal was \$36.

b) If the sales tax rate was 6%, what was the sales tax on the meal?
The sales tax was \$2.16.

c) Connie thought the service and the food were outstanding and decided to leave a 20% tip for the server. She knew to base her tip on the undiscounted meal total of \$48. How much was the tip?
The tip was \$9.60.

d) What was the total amount that Connie paid for dinner that night?
The total amount was \$47.76.

Solve each percent increase and percent decrease application. Write a sentence answering the question.

97. Last year, the Shoreline Chamber of Commerce had 52 members. This year it has 65 members What is the percent increase in the Chamber's membership?
The increase in the Chamber's membership is 25%.

98. Last year at this time, Corrine's home was valued at \$400,000. Now it is worth \$450,000. What is the percent increase in the value of Corrine's home?
The increase in the value of Corrine's home is 12.5%.

99. Gus keeps a large coffee can of water on his porch to check for water evaporation. In one week last August, the water level in the can went from 12 inches to 7.5 inches. What was the percent decrease in the water level for that week?
The decrease in water level that week was 37.5%.

100. At the beginning of the semester, Dr. Ortega had 45 students in his statistics class. By the end of the semester, he had 33 students. What was the percent decrease in the number of students in Dr. Ortega's statistics class?
The decrease in the number of students in Dr. Ortega's statistics class was 26.7%.

101. Rachelle put \$2,400 in an account that gained interest at a rate of 7%. How much interest did the account gain after $\frac{3}{4}$ year?
The account gained \$126 after $\frac{3}{4}$ year.

102. Manuel put \$8,000 in a certificate of deposit that gained interest at a rate of 6.5%. How much interest did the account gain after 3 months?
The account gained \$130 after 3 months.

Chapter 6 Test

Write the described ratio in fractional form and simplify it completely.

1. Kacey gives a weekly allowance to each of her two daughters: \$60 to Alissa and \$36 to Danita. Write their allowances as a ratio, Alissa's amount : Danita's amount, in simplified fractional form.
   $\frac{\$60}{\$36} = \frac{5}{3}$
2. A tractor used 0.8 gallons of diesel to travel 2.8 miles. Write this as a ratio, number of miles : number of gallons, in simplified fractional form.
   $\frac{2.8 \text{ miles}}{0.8 \text{ gallons}} = \frac{7 \text{ miles}}{2 \text{ gallons}}$

Use cross multiplication to determine whether each pair of fractions is equivalent.

3. $\frac{35}{12} \stackrel{?}{=} \frac{25}{8}$
   No
4. $\frac{2.4}{3.2} \stackrel{?}{=} \frac{15}{20}$
   Yes
5. *Simplify the rate* $\frac{63 \text{ seats}}{12 \text{ rows}}$ *and write the result in words.*
   21 seats for every 4 rows
6. A large box of cereal contains 5,200 milligrams of potassium for 25 servings. Write the amount of potassium as a unit rate per serving.
   208 milligrams of potassium per serving
7. *Write each item as a unit price and answer the question.*
   A bag of 15 avocados costs \$8.40, and a box of 12 avocados costs \$7.20. Which is the better buy?
   Bag of 15 avocados: \$0.56 per avocado
   Box of 12 avocados: \$0.60 per avocado
   The bag of 15 avocados is the better buy.
8. *Solve for x*: $\frac{9}{x} = \frac{12}{32}$.
   $x = 24$

Solve each application and answer with a complete sentence.

9. Carl made a long-distance phone call from a pay phone. He was charged \$6.80 to talk for 4 minutes. How much would he have paid to talk for 9 minutes?

Carl would have paid \$15.30.

10. Banjo can finish a 16-pound bag of dog food in 40 days. How many days would it take him to finish a 20-pound bag of food?

It would take him 50 days to finish a 20-pound bag.

11. Given the measures of these two similar triangles, find the value of x.

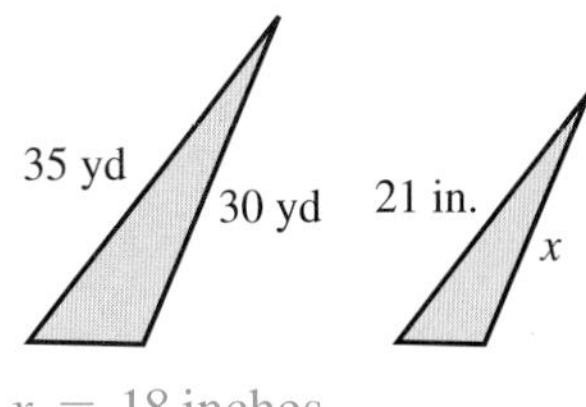

$x = 18$ inches

Write each percent as a decimal.

12. 7%
    0.07
13. 0.5%
    0.005

Write each decimal as a percent.

14. 0.57
    57%
15. 0.2
    20%

Write each percent as a simplified fraction.

16. 16%
    $\frac{4}{25}$
17. 45%
    $\frac{9}{20}$

Write each fraction as a percent. If necessary, round the percentage to the nearest tenth.

18. $\frac{13}{25}$
    52%
19. $\frac{4}{11}$
    36.4%

For each, find the missing value and write a conclusion.

20. 14% of what number is 21?
    14% of 150 is 21.
21. What percent of 80 is 32?
    40% of 80 is 32.

Solve. Write a sentence answering the question.

22. 24% of all students at Riverbend Community College are teenagers. If there are 3,500 students at Riverbend, how many of them are teenagers?
840 of the students at Riverbend are teenagers.

23. Lin-Li bought a textbook that costs $88. The sales tax rate in her state is 7.5%. How much will Lin-Li pay for the textbook at the cash register?
Lin-Li will pay $94.60 for the textbook at the cash register

24. As a member of a bicycle club, Yusef gets a 15% discount on all bicycle gear that he purchases at Cycle-a-Go-Go. The price of a new helmet is $60. What will Yusef's discounted price be?
Yusef's discounted price will be $51.

25. Marcus received a raise from $8 to $10 per hour when he became assistant night manager at Burger Basket. What was the percent increase in his hourly wage?
The increase in his hourly wage was 25%.

26. Lyle put $3,000 in a certificate of deposit that gained interest at a rate of 8%. How much interest did the account gain after 5 months?
The account gained $100 interest after 5 months.

Chapters 1–6 Cumulative Review

Round each number to the indicated place.

1. 3,052 nearest hundred
   3,100

2. 29,602,815 nearest million
   30,000,000

Which property is being demonstrated?

3. $1 \cdot 6 = 6$
   Identity of Multiplication

4. $6 \cdot (10 + 3) = 6 \cdot 10 + 6 \cdot 3$
   Distributive Property of Multiplication over Addition

Based on the bar graph, answer each question with a sentence.

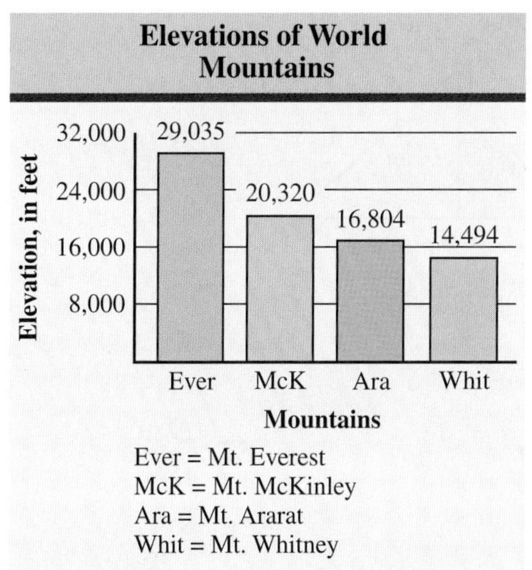

5. How much higher is Mt. Everest (Nepal) than Mt. Ararat (Turkey)?
   Mt. Everest is 12,231 feet higher than Mt. Ararat.

6. How much higher is Mt. McKinley (Alaska) than Mt. Whitney (California)?
   Mt. McKinley is 5,826 feet higher than Mt. Whitney.

Multiply.

7. $400 \cdot 500$
   200,000

Divide.

8. $2,465 \div 17$
   145

9. Expand 8^3 and find its value.
   512

10. Express 10,000,000 as a power of 10.
    10^7

11. Rewrite 700,000 as a product of a whole number and a power of 10.
    $7 \cdot 10^5$

12. Of the numbers 9, 0, 45, 19, 1, and 29, determine which are prime, which are composite, and which are neither.
    Prime: 19, 29
    Composite: 9, 45
    Neither: 0, 1

13. Determine whether 9 is a factor of 2,547. Verify the answer by dividing the number by 9.
    It is a factor.

14. Find the prime factorization of 420. Write the answer two ways: with and without exponents.
    $2 \cdot 2 \cdot 3 \cdot 5 \cdot 7$
    $2^2 \cdot 3 \cdot 5 \cdot 7$

Evaluate each expression.

15. $-8 + (-6)$
    -14

16. $-2 + 15$
    13

17. $-9 - 4$
    -13

18. $5 - 12$
    -7

19. $17 - (-8)$
    25

20. $-2 - (-8)$
    6

21. $(-9) \cdot (-1)$
    9

22. $8 \cdot (-7)$
    -56

23. $-3 \cdot (-2) \cdot (-5) \cdot 4$
    -120

24. $\dfrac{-40}{-8}$
    5

25. $\dfrac{36}{-12}$
    -3

26. $18 - 7 + (-14) - (-6)$
    3

27. Evaluate $x^2 + y$ by replacing x with 3 and y with -10.
    -1

Evaluate each expression using the Order of Operations. Show all steps.

28. $6 - \sqrt{64}$
    -2

29. $9^2 - 10^2$
    -19

30. $|-7 - (-8)|$
    1

Evaluate the numerical value of each formula with the given replacement values.

31. $m = \dfrac{y - k}{x - h}$ $\quad y = 12,\ k = 6,\ x = 3,\ h = 5$

$m = -3$

32. $m = \dfrac{y - k}{x - h}$ $\quad y = 8,\ k = -4,\ x = -2,\ h = 1$

$m = -4$

Simplify by combining like terms, if possible.

33. $-5y^3 - 7y^3$

$-12y^3$

34. $6x - 4x - (-x)$

$3x$

35. $w^3 - 3w - 2w^3 + 6w$

$-w^3 + 3w$

Find the product.

36. $-4(-2x^2)$

$8x^2$

37. $2(-8y)$

$-16y$

Rewrite each expression using the Distributive Property.

38. $6(3w - 5)$

$18w - 30$

39. $-3(-5q + 7)$

$15q - 21$

Solve. Make sure you check each answer.

40. $6y + 4 = 2y - 12$

$y = -4$

41. $3w + 5 - 4w = 7 - 3w - 8$

$w = -3$

For each problem, set up the legend, draw a diagram, identify the formula, and write and solve the equation. Make sure you write the answer as a sentence.

42. Two friends, Becca and Andrea, won a lottery worth \$26,500. From their prearranged agreement, Andrea is to receive \$3,500 more than Becca receives. How much of the money will each woman receive?

Andrea will receive \$15,000, and Becca will receive \$11,500.

43. Find the greatest common factor of 54 and 96.

The GCF is 6.

44. Simplify $\dfrac{90p}{36p^2}$ completely.

$\dfrac{5}{2p}$

Multiply and simplify.

45. $\dfrac{4}{3} \cdot \dfrac{2x}{8}$

$\dfrac{x}{3}$

46. $-\dfrac{3a}{8} \cdot \left(-\dfrac{6b}{2ac}\right) \cdot (-5c)$

$-\dfrac{45b}{8}$

Divide and simplify.

47. $\dfrac{10w}{21} \div \dfrac{5}{3}$

$\dfrac{2w}{7}$

48. $-\dfrac{7x}{10xy} \div \dfrac{49}{5y^2}$

$-\dfrac{y}{14}$

Find the area of each geometric figure.

49.

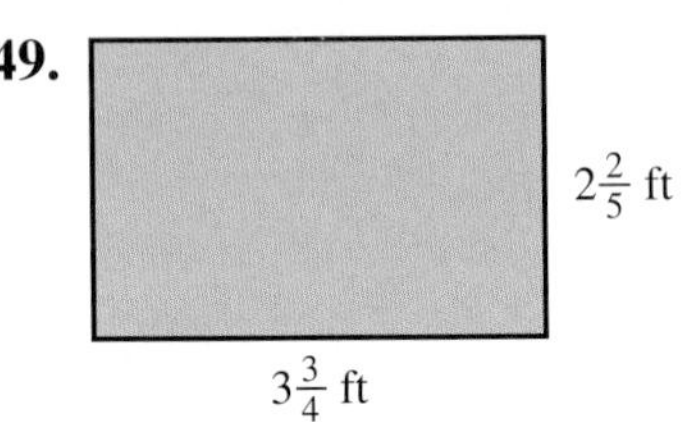

9 square feet

50.

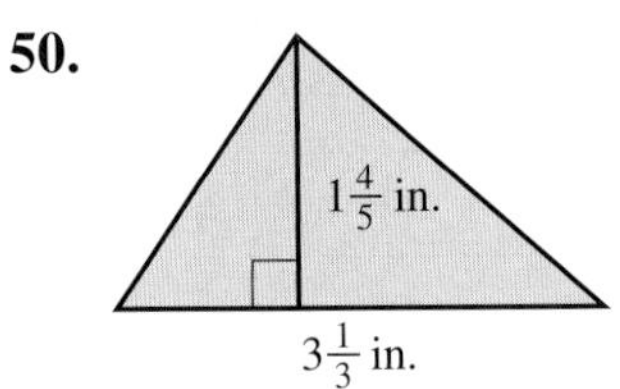

3 square inches

Simplify.

51. $\dfrac{17}{25}p^2 + \dfrac{18}{25}p^2$

$\frac{7}{5}p^2$

52. $\dfrac{19}{15}c - \dfrac{7}{15}c$

$\frac{4}{5}c$

53. $-\dfrac{2}{15}xy - \left(-\dfrac{8}{15}xy\right)$

$\frac{2}{5}xy$

54. $\dfrac{1}{6} + \dfrac{2}{15}$

$\dfrac{3}{10}$

55. $\dfrac{11}{12} - \dfrac{3}{4}$

$\dfrac{1}{6}$

56. $\dfrac{1}{6} - \left(-\dfrac{3}{8}\right)$

$\dfrac{13}{24}$

57. Last weekend, Marco worked $7\frac{11}{12}$ hours on Saturday and $6\frac{3}{4}$ hours on Sunday. How many hours did Marco work that weekend?

Marco worked $14\frac{2}{3}$ hours that weekend.

Solve.

58. $c - \frac{4}{9} = \frac{5}{-6}$
$c = -\frac{7}{18}$

59. $-32 = 40v$
$v = -\frac{4}{5}$

60. $\frac{3}{100}x = 6$
$x = 200$

Answer as indicated.

61. Round \$10.59602 to the nearest penny.
\$10.60

62. Round \$532.19 to the nearest dollar.
\$532

63. Find the perimeter of this geometric figure.

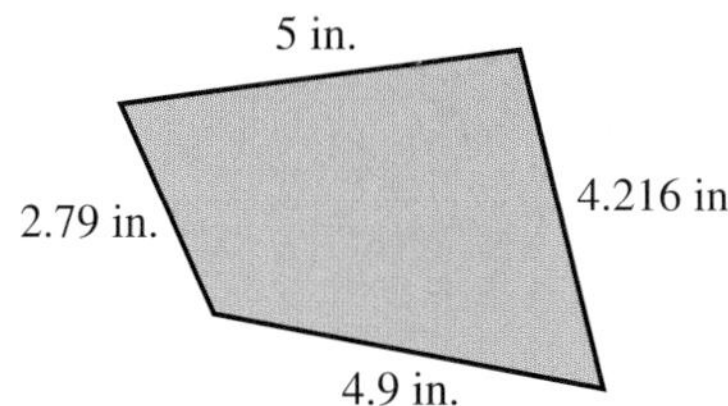

The perimeter is 16.906 inches.

64. Find the length of the third side.

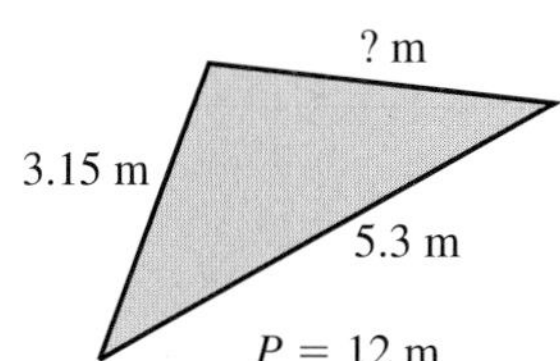

The length of the third side is 3.55 meters.

Evaluate.

65. 0.2×0.36
0.072

66. 0.308×100
30.8

67. $4.92 \div 8$
0.615

68. $2.678 \div 0.13$
20.6

Answer as indicated.

69. Find the decimal equivalent of $\frac{7}{8}$.
0.875

70. Petre owns his own cab. Yesterday in 9.5 hours of driving, he earned \$165.30. How much did Petre earn per hour?
Petre earned \$17.40 per hour.

71. Allison's pool is rectangular, perfect for swimming laps. It measures 24.5 yards long and 8.4 yards wide. What is the area of the pool?

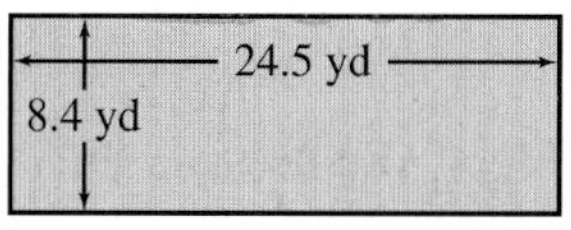

The area of the pool is 205.8 square yards.

72. A round baking dish has a 4.8-inch radius. What is the circumference of the baking dish? Round the answer to the nearest tenth.

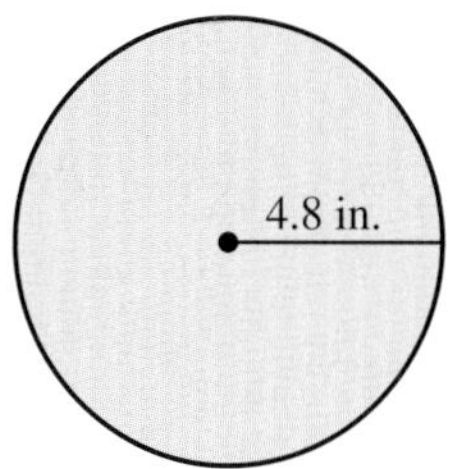

The circumference of the baking dish is about 30.1 inches.

73. Write the ratio 2.7 grams : 0.45 grams in fractional form and simplify it completely.
$\frac{2.7 \text{ grams}}{0.45 \text{ grams}} = \frac{6}{1}$

74. Simplify $\frac{2.85 \text{ calories}}{1.5 \text{ grams}}$ to a unit rate and write the unit rate in words.
$\frac{1.9 \text{ calories}}{1 \text{ gram}}$; 1.9 calories per gram

75. Tomás traveled 195 miles in 4 hours. Write his speed as a unit rate per hour.
48.75 miles per hour

76. A 24-ounce cup of Twinklebuck's coffee costs \$2.28, and a 20-ounce cup of Sunbuck's coffee costs \$1.80. Which has the better value?
24-ounce of Twinklebuck's: \$0.095 per ounce
20-ounce of Sunbuck's: \$0.09 per ounce
Sunbuck's coffee has the better value.

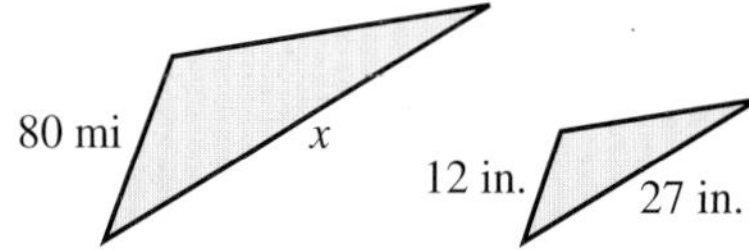

77. Given the two similar triangles, solve for x.
$x = 180$ miles

78. Solve for x in $\frac{24}{9} = \frac{32}{x}$
$x = 12$

79. Write 3.2% as a decimal.
0.032

80. Write 0.4 as a percent.
40%

Solve each problem. Write a sentence answering the question.

81. At the Fireman's Pancake Breakfast, Zia can cook 18 pancakes in 4 minutes. How long will it take Zia to cook 90 pancakes?
It will take Zia 20 minutes to cook 90 pancakes.

82. 45% of what number is 63?
45% of 140 is 63.

83. What percent of 75 is 48?
64% of 75 is 48.

84. Jolanda has a coupon for a 20% discount at Sidewalk's Shoes. She found a pair of shoes that cost $55. What will the discounted price be?
The discounted price will be $44.

85. The Torres family owns three cars, all financed through the local credit union. Last month's car payments totaled $650, but that included the last payment on the Torres's Volvo. This month the family will pay only $390. What is the percent decrease in the Torres's car payment?
The decrease in the Torres's car payment is 40%.

86. Julie put $2,000 in an account that gained interest at a rate of 3.5%. How much interest did the account gain after 9 months?
The account gained $52.50 interest after 9 months.

CHAPTER

Polynomials

Introduction

Polynomials are the backbone of algebra. In this chapter, you'll be introduced to a variety of definitions and vocabulary terms related to polynomials. The best way to learn new vocabulary is to practice using the words. For example, in Chapters 1 and 2, you learned algebra vocabulary words such as *variable*, *constant*, and *coefficient*. Then once they were introduced, you saw them used in the reading. You may even have spoken them aloud when working with other students, your instructor, or a tutor.

Learning about the basic operations of polynomials now, in prealgebra, will ensure that you are well prepared for your future algebra courses.

Preparation Exercises

Section 1.5 Exponents, Square Roots, and the Order of Operations
Expand and evaluate.

1. 5^3 125

2. 3^4 81

Section 1.6 Factors
Find the prime factorization of each number.

3. 40 $2^3 \cdot 5$

4. 54 $2 \cdot 3^3$

Section 2.2 Adding Signed Numbers
Add.

5. $11 + (-6)$ 5

6. $3 + (-9)$ -6

7. $-5 + 13$ 8

8. $-7 + (-12)$ -19

Section 2.3 Subtracting Signed Numbers
Subtract.

9. $-5 - 9$ -14

10. $-2 - (-6)$ 4

Section 2.4 Multiplying and Dividing Signed Numbers
Multiply.

11. $-8 \cdot 6$ -48

12. $(-9) \cdot (-4)$ 36

Section 2.4 Multiplying and Dividing Signed Numbers
Divide.

13. $60 \div (-4)$ -15

14. $\dfrac{-45}{-9}$ 5

Section 2.7 Algebraic Expressions
Simplify.

15. $-6y + 10y$ $4y$

16. $4x - 12x$ $-8x$

Section 4.1 Greatest Common Factors
Find the GCF of each pair of numbers.

17. 20 and 36 4

18. 45 and 60 15

Section 4.3 Equivalent Fractions
Simplify.

19. $\dfrac{16}{40}$ $\frac{2}{5}$

20. $\dfrac{-18}{24}$ $-\frac{3}{4}$

SECTION 7.1 Rules of Exponents

OBJECTIVES

In this section, you will learn to

- Apply the First Power Rule.
- Apply the Product Rule for Exponents.
- Apply the Zero Power Rule.
- Evaluate an expression when an exponent is 0.
- Apply the Power Rule for Exponents.

You Need to Know

To successfully complete this section, you need to understand

☐ Exponents (1.5)
☐ The order of operations (1.5)

Introduction

Anything that has a numerical value can, among other operations, be added, subtracted, and multiplied. This is true for exponents as well. In this section, we will see under what circumstances exponents can be added and separately multiplied. Later in this chapter, in Section 7.5, we will see how exponents can be subtracted, one from another.

The Meaning of the Exponent

Recall from Section 1.5 that an exponent is an abbreviation for repeated multiplication. For example, the expression $2 \cdot 2 \cdot 2 \cdot 2 \cdot 2$ can be abbreviated as 2^5. The number 5 is the *exponent*; and the repeated factor, 2, is the *base*. When written as 2^5, the expression is in **exponential form**. When 2^5 is written out as $2 \times 2 \times 2 \times 2 \times 2$, it is expressed in **expanded form**.

The most important feature about an exponent is its meaning. For example, 2^5 means "five factors of 2" and $2 \cdot 2 \cdot 2 \cdot 2 \cdot 2 = 32$. Because 5 is not a factor of 32, the *value* of 5 is not present in the final result. Instead, the exponent 5 tells us how many factors of 2 to write.

Example 1

For each of the following expressions, state its meaning, write it in expanded form, and evaluate it.

a) 3^4 **b)** $(-5)^3$ **c)** 1^8 **d)** 0^6

Procedure The exponent indicates the number of factors that are to be multiplied. (You can do some of the multiplication off to the side as scratch work.)

Answer

	Meaning	Expanded Form	Value
a) 3^4	means four factors of 3	$= 3 \times 3 \times 3 \times 3$	$= 81$
b) $(-5)^3$	means three factors of -5	$= (-5) \times (-5) \times (-5)$	$= -125$
c) 1^8	means eight factors of 1	$= 1 \cdot 1 \cdot 1 \cdot 1 \cdot 1 \cdot 1 \cdot 1 \cdot 1$	$= 1$
d) 0^6	means six factors of 0	$= 0 \times 0 \times 0 \times 0 \times 0 \times 0$	$= \mathbf{0}$

You Try It 1

For each of the following expressions, state its meaning, write it in expanded form, and evaluate it. Use Example 1 as a guide.

	Meaning	Expanded Form	Value
a) 4^3 means	______	= ______	= ______
b) $(-12)^2$ means	______	= ______	= ______
c) 1^6 means	______	= ______	= ______
d) 0^4 means	______	= ______	= ______

To emphasize the meaning of an exponent, it may be helpful to write the exponent in word form instead of number form. For example, instead of writing 2^5, we could write 2^{five}. Writing the exponential form like this may help you focus on the meaning of the exponent.

Think About It 1 Evaluate each of these expressions; then answer the question.

a) $7 - 2^2$ $= \_\_\_\_\_$ **b)** $4 - 2^2$ $= \_\_\_\_\_$ **c)** $1 - 2^2$ $= \_\_\_\_\_$ **d)** $0 - 2^2$ $= \_\_\_\_\_$ **e)** -2^2 $= \_\_\_\_\_$

Question Is -2^2 the same as or different from $(-2)^2$? Explain your answer.

The First Power Rule

Recall again from Section 1.5 that $b^1 = b$. For example, 7^1 means "one factor of 7." In this case, there are no repeated factors of 7; there is only 7 itself. In other words, $7^1 = 7$.

The First Power Rule

For any base, $x^1 = x$ and $x = x^1$.

This can be seen in the prime factorization of a number such as 45: $45 = 3 \cdot 3 \cdot 5$. Here we can say that 45 has **two** factors of **3** and **one** factor of **5**. For this reason, we also can write the prime factorization as follows:

$45 = 3^2 \cdot 5^1$: **two** factors of **3** and **one** factor of **5**

Example 2

For each expression, if the base has an exponent, rewrite it without the exponent. If it does not have an exponent, rewrite it with one.

a) 8^1 **b)** $(-7)^1$ **c)** y^1 **d)** 4 **e)** w

Answer **a)** $8^1 = 8$ **b)** $(-7)^1 = -7$ **c)** $y^1 = y$

d) $4 = 4^1$ **e)** $w = w^1$

You Try It 2 **For each expression, if the base has an exponent, rewrite it without the exponent. If the base does not have an exponent, rewrite it with one. Use Example 2 as a guide.**

a) $(-2)^1 =$ **b)** $p^1 =$ **c)** $6 =$ **d)** $h =$

The Product Rule for Exponents

Consider $2^3 \cdot 2^2$. We can evaluate this in a couple of different ways.

1. Using the order of operations:

$2^3 \cdot 2^2$ — Apply the exponents first: $2^3 = 8$ and $2^2 = 4$.

$= 8 \cdot 4$

$= 32$

$2^3 \cdot 2^2 = \mathbf{32}$

2. Writing each in expanded form:

$2^3 \cdot 2^2$ — Three factors of 2 times two factors of 2

$= (2 \cdot 2 \cdot 2) \cdot (2 \cdot 2)$ — The Associative Property of Multiplication allows us to drop the parentheses.

$= 2 \cdot 2 \cdot 2 \cdot 2 \cdot 2$ — A total of five factors of 2

$= 2^5$

$2^3 \cdot 2^2 = \mathbf{2^5}$

As we know, $2^5 = 32$; so, each result has the same value.

Think About It 2 We found that $2^3 \cdot 2^2 = 2^5$. How is it that the 3 and 2 can combine to make 5? Explain your answer.

____

____

Think About It 3 Does $2^3 + 2^2 = 2^5$? Explain your answer.

____

____

When each base is the same variable, such as $y^3 \cdot y^2$, we can expand each base and write the product as a single variable with an exponent.

Example 3

Rewrite each product and power in its expanded form. Then combine all of the factors and abbreviate the result using exponential form.

a) $y^3 \cdot y^2$ **b)** $b^3 \cdot b^3$ **c)** $x^7 \cdot x$

Procedure Notice that each factor in each problem has the same base. This is important for recognizing the pattern.

Answer

a) $y^3 \cdot y^2 = (y \cdot y \cdot y) \cdot (y \cdot y) = y^5$ A total of five factors of y

b) $b^3 \cdot b^3 = (b \cdot b \cdot b) \cdot (b \cdot b \cdot b) = b^6$ A total of six factors of b

c) $x^7 \cdot x = (x \cdot x \cdot x \cdot x \cdot x \cdot x \cdot x) \cdot (x) = x^8$ A total of eight factors of x

You Try It 3 **Rewrite each product and power in its expanded form. Then combine all of the factors and abbreviate the result using exponential form. Use Example 3 as a guide.**

a) $x^5 \cdot x^4$ = ______________________ = __________

b) $c^6 \cdot c^2$ = ______________________ = __________

c) $k^1 \cdot k^1$ = ______________________ = __________

d) $m \cdot m^3$ = ______________________ = __________

Because an exponent is an abbreviation for repeated multiplication, when we expand and multiply $y^3 \cdot y^2$, multiplication is the only operation and this product is *more* factors of y.

(three factors of y) times (two *more* factors of y) = a total of five factors of y

This leads to the Product Rule for Exponents.

The Product Rule for Exponents

For any nonzero base,

$$x^a \cdot x^b = x^{a+b}.$$

Example 4

Use the Product Rule for Exponents to write each of these expressions as one base with one exponent.

a) $p^3 \cdot p^4$ **b)** $m^3 \cdot m^3$ **c)** $x^4 \cdot x$

Procedure Notice that each factor in each problem has the same base. This is important for recognizing the pattern.

Answer **a)** $p^3 \cdot p^4 = p^{3+4} = p^7$ (three factors of p and four more factors of p)

b) $m^3 \cdot m^3 = m^{3+3} = m^6$ (three factors of m and three more factors of m)

c) $x^4 \cdot x = x^{4+1} = x^5$ (remember that $x = x^1$)

You Try It 4 **Use the Product Rule for Exponents to write each of these expressions as one base with one exponent. Use Example 4 as a guide.**

a) $x^3 \cdot x^6$ **b)** $w^4 \cdot w^4$ **c)** $k \cdot k^4$ **d)** $p \cdot p$

Caution The Product Rule for Exponents may be used only when the *base* is the same for each factor.

For example, $x^3 \cdot y^5$ can be thought of as three factors of x and five factors of y. But we *don't* have eight factors of any one variable; so, we cannot add the exponents.

The Zero Power Rule

Think About It 4 Recall from Section 1.5 that a whole number exponent of 10 indicates the number of factors of 10 as well as the number of zeros following 1 in the power of 10.

For example,

$10^4 = 10{,}000$ (1 followed by four zeros)

$10^3 = 1{,}000$ (1 followed by three zeros)

$10^2 = 100$ (1 followed by two zeros)

$10^1 = 10$ (1 followed by one zero)

Question Based on that pattern, what is the value of 10^0? Explain your answer.

____

____

Let's explore the value of an expression raised to the 0 power, such as a^0, where a is any nonzero number.

First, recall that 1 is the identity for multiplication. This means that the product of any number, n, and 1 is always that number: $n \cdot 1 = n$.

Second, consider the product $a^4 \cdot a^0$ and the product $a^4 \cdot 1$.

$$a^4 \cdot a^0 = a^{4+0} = a^4$$

$$\text{and} \quad a^4 \cdot 1 = a^4$$

Because both $a^4 \cdot a^0 = a^4$ and $a^4 \cdot 1 = a^4$, a^0 and 1 must be the same. In other words,

$$\mathbf{a^0 = 1}$$

This is the foundation of the Zero Power Rule, which is true when the base is any nonzero constant or variable.

The Zero Power Rule

$$x^0 = 1, x \neq 0$$

In other words, x^0 means "no factors of x."

Example 5

Use the Zero Power Rule to evaluate each expression.

a) b^0 **b)** 9^0 **c)** $(-2)^0$ **d)** m^0

Answer **a)** $b^0 = 1$ **b)** $9^0 = 1$ **c)** $(-2)^0 = 1$ **d)** $m^0 = 1$

You Try It 5 **Use the Zero Power Rule to evaluate each expression. Use Example 5 as a guide.**

a) w^0 **b)** 1^0 **c)** $(-4)^0$ **d)** p^0

To further understand the Zero Power Rule, that $a^0 = 1$, we refer back to $a^4 \cdot a^0 = a^4$ and recognize that a^0 doesn't "add" any more factors to the product. Note that a^0 does *not* make the entire product zero.

Example 6

Use the Product Rule for Exponents to write each of these expressions as one base with one exponent.

a) $b^3 \cdot b^0$ **b)** $y^0 \cdot y^4$ **c)** $x^0 \cdot x$

Procedure Notice that each factor in each problem has the same base. This is important for recognizing the pattern.

Answer

a) $b^3 \cdot b^0 = b^{3+0} = b^3$ This is three factors of b and *no* more factors of b.

b) $y^0 \cdot y^4 = y^{0+4} = y^4$ This is no factors of y and four factors of y.

c) $x^0 \cdot x = x^{0+1} = x^1 = x$ Remember that $x^1 = x$.

You Try It 6 **Use the Product Rule for Exponents to write each of these expressions as one base with one exponent. Use Example 6 as a guide.**

a) $x^5 \cdot x^0$ **b)** $c^0 \cdot c^6$ **c)** $w^4 \cdot w$ **d)** $m^0 \cdot m^0$

The Power Rule for Exponents

As we know, squaring a number or an expression means multiplying the expression times itself. For example, $5^2 = 5 \cdot 5$ and $x^2 = x \cdot x$. In other words, squaring an expression doubles the number of factors in that expression.

When the expression is a base with an exponent, such as x^3, squaring it doubles the number of factors from three to six: $(x^3)^2 = x^3 \cdot x^3 = x^{3+3} = x^6$.

Because an exponent represents the number of factors of the base, doubling the number of factors is the same as doubling the exponent (multiplying the exponent by 2): $(x^3)^2 = x^{3 \cdot 2} = x^6$.

Likewise, cubing an expression—raising the expression to the third power—has the effect of tripling the number of factors. For example, cubing y^5 triples the number of factors from 5 to 15:

$$(y^5)^3 = y^5 \cdot y^5 \cdot y^5 = y^{5+5+5} = y^{15}$$

This can be more easily expressed as $(y^5)^3 = y^{5 \cdot 3} = y^{15}$, which leads to the Power Rule for Exponents.

The Power Rule for Exponents

Also known as a ***Power of a Power***:

$$(x^a)^b = x^{a \cdot b}$$

Example 7

Simplify each expression.

a) $(w^4)^3$ **b)** $(x^2)^5$ **c)** $(p^0)^6$

Procedure For each expression, apply the Power Rule of Exponents by multiplying the exponents in the expression.

Answer **a)** $(w^4)^3 = w^{4 \cdot 3} = w^{12}$ **b)** $(x^2)^5 = x^{2 \cdot 5} = x^{10}$ **c)** $(p^0)^6 = p^{0 \cdot 6} = p^0 = 1$

You Try It 7 **Simplify each expression. Use Example 7 as a guide.**

a) $(c^6)^2$ **b)** $(y^2)^4$ **c)** $(d^5)^1$ **d)** $(q^2)^0$

We know from the Product Rule for Exponents that $x^2 \cdot x^5 = x^{2+5} = x^7$. We also know from the Power Rule for Exponents that $(x^2)^5 = x^{2 \cdot 5} = x^{10}$. What would you say to a classmate who is having difficulty understanding the difference between these two rules?

Answers: You Try It and Think About It

You Try It: **1. a)** **three** factors of $4 = 4 \cdot 4 \cdot 4 = 64$ **b)** **two** factors of $(-12) = (-12) \cdot (-12) = +144$ **c)** **six** factors of $1 = 1 \cdot 1 \cdot 1 \cdot 1 \cdot 1 \cdot 1 = 1$ **d)** **four** factors of $0 = 0 \cdot 0 \cdot 0 \cdot 0 = 0$ **2. a)** -2 **b)** p **c)** 6^1 **d)** h^1 **3. a)** $(x \cdot x \cdot x \cdot x \cdot x) \cdot (x \cdot x \cdot x \cdot x) = x^9$ **b)** $(c \cdot c \cdot c \cdot c \cdot c \cdot c) \cdot (c \cdot c) = c^8$ **c)** $(k) \cdot (k) = k^2$ **d)** $(m) \cdot (m \cdot m \cdot m) = m^4$ **4. a)** x^9 **b)** w^8 **c)** k^5 **d)** p^2 **5. a)** 1 **b)** 1 **c)** 1 **d)** 1 **6. a)** x^5 **b)** c^6 **c)** w^5 **d)** $m^0 = 1$ **7. a)** c^{12} **b)** y^8 **c)** d^5 **d)** $q^0 = 1$

Think About It: **1. a)** 3 **b)** 0 **c)** -3 **d)** -4 **e)** -4; -2^2 is different from $(-2)^2$. Part e) shows us that -2^2 is -4 but $(-2)^2 = (-2)(-2) = +4$. **2.** The only way that 3 and 2 can be combined to make 5 is by addition. **3.** No. $2^3 + 2^2 = 4 + 9 = 13$, but this does not equal 2^5, which is 32. **4.** 10^0 suggests that there is a 1 followed by zero 0's. In other words, 10^0 is just 1. **5.** Answers may vary. One possibility: Use the definition of the exponents to expand each expression. $x^2 \cdot x^5 = (x \cdot x) \cdot (x \cdot x \cdot x \cdot x \cdot x)$, which is 7 factors of x, or x^7, whereas $(x^2)^5 = (x^2) \cdot (x^2) \cdot (x^2) \cdot (x^2) \cdot (x^2)$, which is a total of 10 factors of x, or x^{10}.

FOR EXTRA HELP
Student Resources on DVD-ROM
Includes
➤ Student's Solutions Manual
➤ Video Lectures
➤ Chapter Test Prep Video
MyMathLab
MathXL

Section 7.1 Exercises

Think Again

1. Evaluate each of these expressions; then answer the question. *(Refer to Think About It 1)*

a) $7 - 2^2$ **b)** $4 - 2^2$ **c)** $1 - 2^2$
d) $0 - 2^2$ **e)** -2^2

Is -2^2 the same as or different from $(-2)^2$? Explain your answer.

a) 3 b) 0 c) −3 d) −4 e) −4
-2^2 is −4, but $(-2)^2$ is +4. Because in -2^2 there are no parentheses grouping the negative with the base of 2, the negative is not included in the squaring process and -2^2 is the same as $0 - 2^2$. (This suggests that it doesn't matter how close the negative sign is to the 2—it still is not included in the squaring unless it is grouped within parentheses.)

2. The Zero Power Rule says that x^0 means "no factors of x."

a) How can we represent no factors of 5?
5^0

b) How can we represent no factors of 7?
7^0

3. A factorization of 36 is $2^2 \cdot 3^2$. There are no factors of 5 and no factors of 7 in 36. How can we represent both 5 and 7 in the factorization of 36?
$36 = 2^2 \cdot 3^2 \cdot 5^0 \cdot 7^0$

4. Based on the previous problem, what is the value of 5^0? Explain your answer.
It appears as though $5^0 = 1$. Because $36 = 2^2 \cdot 3^2$, multiplying by 5^0 doesn't change its value, just as multiplying by 1.

Focus Exercises

Evaluate each expression using a rule for exponents.

5. 9^1 9 **6.** 15^1 15 **7.** $\left(\frac{5}{6}\right)^1$ $\frac{5}{6}$ **8.** $\left(\frac{7}{2}\right)^1$ $\frac{7}{2}$

9. y^1 y **10.** b^1 b **11.** $\left(\frac{-1}{9}\right)^1$ $-\frac{1}{9}$ **12.** $\left(\frac{-3}{8}\right)^1$ $-\frac{3}{8}$

13. $(-3)^1$ −3 **14.** $(-1)^1$ −1 **15.** 12^0 1 **16.** 5^0 1

17. x^0 1 **18.** m^0 1 **19.** $(-4)^0$ 1 **20.** $(-6)^0$ 1

21. $\left(\frac{-4}{5}\right)^0$ 1 **22.** $\left(\frac{-2}{3}\right)^0$ 1 **23.** 1^0 1 **24.** 0^1 0

Write each product as one base with one exponent.

25. $x^2 \cdot x^6$ x^8 **26.** $v^5 \cdot v^9$ v^{14} **27.** $y^3 \cdot y^1$ y^4

28. $c^1 \cdot c^4$ c^5 **29.** $v^3 \cdot v^7$ v^{10} **30.** $q^6 \cdot q^4$ q^{10}

31. $m^{13} \cdot m^{12}$ m^{25} **32.** $b^{15} \cdot b^{11}$ b^{26} **33.** $x^5 \cdot x^5$ x^{10}

34. $c^3 \cdot c^3$ c^6 **35.** $p^1 \cdot p^1$ p^2 **36.** $q^2 \cdot q^2$ q^4

37. $h \cdot h^9$ h^{10} **38.** $x \cdot x^6$ x^7 **39.** $r \cdot r$ r^2

40. $m \cdot m$ m^2 **41.** $p^5 \cdot p^0$ p^5 **42.** $k^0 \cdot k^6$ k^6

43. $w^1 \cdot w^0$ w^1 **44.** $y^0 \cdot y^0$ y^0 **45.** $x^0 \cdot y^3$ y^3

46. $n^0 \cdot q^5$ q^5 **47.** $x^2 \cdot y^0$ x^2 **48.** $m^4 \cdot q^0$ m^4

Simplify each expression.

49. $(a^6)^2$ a^{12} **50.** $(y^8)^2$ y^{16} **51.** $(k^7)^3$ k^{21}

52. $(n^4)^3$ n^{12} **53.** $(x^4)^4$ x^{16} **54.** $(m^5)^5$ m^{25}

55. $(b^2)^9$ b^{18} **56.** $(a^3)^8$ a^{24} **57.** $(p^1)^5$ p^5

58. $(w^1)^7$ w^7 **59.** $(x^2)^1$ x^2 **60.** $(q^6)^1$ q^6

61. $(v^0)^5$ 1 **62.** $(c^0)^7$ 1 **63.** $(m^2)^0$ 1 **64.** $(h^6)^0$ 1

Think Outside the Box

Simplify each expression.

65. $q^2 \cdot q^7 \cdot q^5$ q^{14} **66.** $n^3 \cdot n^8 \cdot n^2$ n^{13} **67.** $(x^6)^4 \cdot (y^7)^3$ $x^{24}y^{21}$

68. $(m^5)^8 \cdot (c^9)^4$ $m^{40}c^{36}$ **69.** $(p^3)^4 \cdot (p^5)^2$ p^{22} **70.** $(w^2)^6 \cdot (w^3)^5$ w^{27}

SECTION 7.2 Introduction to Polynomials

OBJECTIVES

In this section, you will learn to
- Use the vocabulary of polynomials.
- Evaluate a polynomial.
- Identify the degree of a term.
- Write a polynomial in descending order.
- Combine like terms.

You Need to Know

To successfully complete this section, you need to understand
- ☐ The order of operations (1.5)
- ☐ Replacing a variable with a number (1.7)
- ☐ The vocabulary of algebra (1.7 and 2.7)
- ☐ Signed numbers and terms (Chapter 2)

Introduction

You've already learned some of the vocabulary of polynomials earlier in this text. For example, in Section 1.7 you learned the following:

Vocabulary	Example
A *variable* is a letter used to represent a number.	x, w, and m
A *constant* is a number.	$5, -8$, and $\frac{3}{4}$
A *coefficient* is a number that is multiplied to a variable.	$\mathbf{5}x$, $\mathbf{-8}w$, and $\mathbf{\frac{3}{4}}m$

In Section 2.7, we learned the following:

Vocabulary	Example
A *term* is the product of a coefficient and a variable, and the variable may have an exponent.	$4y$, $-2x^3$, and $\frac{2}{3}k^2$
A *constant term* has no variable factor in it.	$5, -8$, and $\frac{3}{4}$
Like terms are two or more terms that have the same variable and exponent even when the coefficients are different.	$3y^2$ and $-7y^2$ and $2x$, x, and $4x$
An *algebraic expression* can be a single term, or it can be the sum or difference of two or more terms.	$5y^2 - 2y + 8$

We also learned these two important concepts

1. The sign in front of a term belongs to that term.

For example, there are three terms in the expression $7x^2 - x - 5$.

Expression $\quad 7x^2 \; - x \; - 5$

Terms $\quad +7x^2 \quad -1x \quad -5$

2. We can apply the Commutative Property to the terms in an algebraic expression to reorganize the expression.

For example, $\quad 5y - 2 + 8y^2 = -2 + 8y^2 + 5y \quad$ and $\quad -7 + 4w^2 - 3w = 4w^2 - 3w - 7$

Example 1

Identify the terms in each algebraic expression.

a) $4x^3 - 6xy + 3y^2$ **b)** $-5w - 9$ **c)** $10y^2$

Procedure Remember that the sign in front of the term belongs to that term. Separate the terms using a comma.

Answer

a) $4x^3 - 6xy + 3y^2$ three terms: $+4x^3, -6xy, +3y^2$

b) $-5w - 9$ two terms: $-5w, -9$

c) $10y^2$ one term: $+10y^2$

You Try It 1 **Identify the terms in each algebraic expression. Use Example 1 as a guide.**

Expression	Terms	Expression	Terms
a) $x^2 + 2x - 7$	________	**b)** $2y^3 - y^2 - 5y$	________
c) $-3xy^2 + 4wv$	________	**d)** $-x^3 + 3x^2 + x - 8$	________

As a reminder about reorganizing terms in an algebraic expression, we first recognize the terms, especially noting whether they are positive or negative, and then use the Commutative Property to switch them around.

Example 2

Rewrite each expression so that the constant term appears second.

a) $4 - 2x^3$ **b)** $-5 + 4bc$ **c)** $-6 - 3w^2$

Procedure First, recognize whether the terms are positive or negative. Then rewrite the expression with the variable term first and the constant term second.

Answer

Expression	Terms	With Constant Second
a) $4 - 2x^3$	$+4, -2x^3$	$-2x^3 + 4$
b) $-5 + 4bc$	$-5, +4bc$	$4bc - 5$
c) $-6 - 3w^2$	$-6, -3w^2$	$-3w^2 - 6$

You Try It 2 **Rewrite each expression so that the constant term appears second. Use Example 2 as a guide.**

Expression	Terms	With Constant Second
a) $5 + 2xy$	________	________
b) $-8 + w^3$	________	________
c) $1 - 6y$	________	________
d) $-3 - 7c^2$	________	________

In higher levels of algebra, algebraic expressions take on many more forms, including variables in the denominator of a fraction and variables within a radical such as these:

$$\frac{3x - 1}{p^2 + 5}, \quad 4x + \frac{9}{y^2}, \quad \text{and} \quad \sqrt{1 - 6w}$$

The algebraic expressions presented in this section, called *polynomials*, are similar to the algebraic expressions in Section 2.7.

Polynomials

A **polynomial** is an algebraic expression that is one of the following:

1. A single term
2. The sum of two or more terms

However, a polynomial may not have a variable in the denominator or within a radical.

Some examples of polynomials are shown in the following table.

Polynomial	Explanation
$-3w$	This is a single term.
4	The single term may be just a constant.
$2x^3 + 6x^2 + x + 7$	This is the sum of more than two terms.
$-4y + 5x^2$	It's okay to have different variables in the polynomial.
$\frac{4x^3}{5} + (-9x)$	It's okay to have a number (constant) in the denominator.

Is $4x^2 - 7$ also a polynomial? Yes. Although it is written as the *difference* of two terms, it is still a polynomial. We can express it as a sum by writing it as $4x^2 + (-7)$.

Whether the polynomial is written as $4x^2 - 7$ or as $4x^2 + (-7)$, the terms are still $+4x^2$ and -7.

Some polynomials have special names.

- A **monomial** is a one-term polynomial, such as $4w^3$.
- A **binomial** has two terms (*bi-* means "*two*," as in *bicycle*), such as $4w^3 + 5w$.
- A **trinomial** has three terms (*tri-* means "*three*," as in *triangle*), such as $4w^3 + 5w - 9$.

Example 3

Identify each expression as a *monomial*, a *binomial*, a *trinomial*, or *not a polynomial*.

a) $6x^2 + 4x - 5$ **b)** $-\frac{2}{w} + 3$ **c)** $\frac{3}{4}b^2$ **d)** $-5p - 9$ **e)** -8

Procedure First, determine whether the expression is a polynomial. If it is, count the number of terms to identify its type.

Answer

Expression	Type	
a) $6x^2 + 4x - 5$	Trinomial	Three terms
b) $-\frac{2}{w} + 3$	Not a polynomial	It has a variable in the denominator.
c) $\frac{3}{4}b^2$	Monomial	One term. It's okay to have a constant in the denominator.
d) $-5p - 9$	Binomial	Two terms
e) -8	Monomial	One term

You Try It 3 **Identify each expression as a *monomial*, a *binomial*, a *trinomial*, or *not a polynomial*. Use Example 3 as a guide.**

Expression	Type
a) $4m^5 + 7k$	__________
b) $-\frac{3y}{8}$	__________
c) $-2w^2 + w + 6$	__________
d) $\frac{8x^3 - 1}{x}$	__________
e) $2c^3$	__________

Evaluating Polynomials

Because a variable represents a number, we can assign a value to the variable or variables of a polynomial and evaluate them.

For example, in the polynomial $2x^2 + 3x - 4$, if we choose the value of $x = 3$, the polynomial becomes $2(3)^2 + 3(3) - 4$ and we can use the order of operations to evaluate it.

$2(3)^2 + 3(3) - 4$	Apply the exponent.
$= 2(9) + 3(3) - 4$	Multiply.
$= 18 + 9 - 4$	Add and subtract.
$= 23$	

Example 4

Evaluate each polynomial with the given replacement value.

a) $x^3 - 4x + 1, x = -2$ **b)** $-y^2 + 6y - 3, y = 5$

Procedure Replace the variable with the given value. Make sure you place parentheses around the replacement value.

Answer **a)**

$x^3 - 4x + 1$	Replace each x with (-2).
$= (-2)^3 - 4(-2) + 1$	Apply the exponent: $(-2)^3 = (-2)(-2)(-2) = -8$.
$= -8 - 4(-2) + 1$	Multiply $-4(-2) = +8$.
$= -8 + 8 + 1$	Add.
$= 1$	

b)

$-y^2 + 6y - 3$	Replace each y with (5). Treat $-y^2$ as $-1 \cdot y^2$.
$= -1 \cdot (5)^2 + 6(5) - 3$	Apply the exponent: $(5)^2 = 25$. Notice that the -1 is not included in the squaring.
$= -1(25) + 6(5) - 3$	Multiply $-1(25) = -25$ and $6(5) = 30$.
$= -25 + 30 - 3$	Add.
$= 2$	

Caution When replacing a variable with a number, put parentheses around the number as you place it in the expression, as demonstrated in Example 4. This should be done for both positive and negative replacement values.

You Try It 4 **Evaluate each polynomial with the given replacement value. Use Example 4 as a guide.**

a) $x^2 - 5x, x = 3$ **b)** $2w^3 - 3w + 4, w = 0$ **c)** $-y^2 - 2y + 3, y = -2$

Many situations involve formulas that contain a polynomial. One type of situation deals with an object being dropped or thrown from a certain height.

When a ball or another small object is dropped from a high place, such as the top of a tall building or a cliff, the ball's height above the ground in feet after x seconds is given by the following formula:

$$h = \text{the height of the building (or cliff)} - 16x^2$$

We can evaluate this formula for different values of x seconds. We might want to know how high from the ground the ball is after falling 2 seconds and after falling 5 seconds. In each case, we can replace x in the formula with the number of seconds and find how high the ball is above the ground at that moment.

Example 5

Ignacio dropped a ball off the roof of a 600-foot-high building. The ball's height above the ground in feet after x seconds is given by $h = 600 - 16x^2$.

a) How high above the ground was the ball after falling 2 seconds?

b) How high above the ground was the ball after falling 5 seconds?

Procedure Replace the variable with the given value. Answer the question in sentence form.

Answer **a)**

$h = 600 - 16x^2$	Replace x with 2.
$h = 600 - 16(2)^2$	Apply the exponent: $(2)^2 = 4$.
$h = 600 - 16(4)$	Multiply $16(4) = 64$.
$h = 600 - 64$	Subtract.
$h = 536$	

Sentence The ball was 536 feet above the ground after falling 2 seconds.

b)

$h = 600 - 16x^2$	Replace x with 5.
$h = 600 - 16(5)^2$	Apply the exponent: $(5)^2 = 25$.
$h = 600 - 16(25)$	Multiply $16(25) = 400$.
$h = 600 - 400$	Subtract.
$h = 200$	

Sentence The ball was 200 feet above the ground after falling 5 seconds.

If the ball is thrown downward instead of dropped from the roof (or cliff), the formula changes a little, as shown in the next exercise.

You Try It 5 **Write the answer to each question in sentence form. Use Example 5 as a guide.**

Mai threw a ball downward from the roof of a 900-foot-high building. The ball's height above the ground in feet after x seconds is given by $h = 900 - 16x^2 - 20x$.

a) How high above the ground was the ball after falling 1 second?

b) How high above the ground was the ball after falling 5 seconds?

Sentence ______________________

Sentence ______________________

The Degree of a Term

Every term that is not a constant term has one or more variable factors. For example, $6x^2 = 6 \cdot x \cdot x$ and has two variable factors; $-2x^3y^2 = -2 \cdot x \cdot x \cdot x \cdot y \cdot y$ and has five variable factors.

The number of variable factors in a term is called the **degree** of the term. The degree of the term is related only to the variable factors. The degree of the term is in no way related to the coefficient of the term.

For example, because $6x^2$ contains two variable factors, the degree of $6x^2$ is 2; and because $-2x^3y^2$ contains five variable factors, the degree of $-2x^3y^2$ is 5.

Think About It 1 What is the degree of a constant term? Explain your answer.

____

____

Example 6

For each term, identify the variable factors and indicate the degree.

a) $-3x^5y^2$ **b)** $\frac{1}{2}p^3$ **c)** $8w$ **d)** 6

Procedure Use the exponent to determine the variable factors and count them to indicate the degree of the term.

Answer

Term	Variable Factor(s)	Degree of Term	
a) $-3x^5y^2$	$x \cdot x \cdot x \cdot x \cdot x \cdot y \cdot y$	7	Together, the exponents indicate the total number of variable factors: $5 + 2 = 7$.
b) $\frac{1}{2}p^3$	$p \cdot p \cdot p$	3	The single exponent, 3, indicates the degree of this term.
c) $8w$	w	1	Remember that $x = x^1$; so, the single exponent indicates the degree of this term.
d) 6	None	0	Every constant term has a degree of 0.

You Try It 6 **For each term, identify the variable factors and indicate the degree. Use Example 6 as a guide.**

Term	Variable Factor(s)	Degree of Term
a) $2x^3y^6$	________________	________________
b) -7	________________	________________
c) k^4	________________	________________
d) $-5b^2c$	________________	________________
e) $3x$	________________	________________
f) $-10x^5$	________________	________________

Writing Polynomials in Descending Order

A polynomial that has more than one term can be written more than one way. For example, using the Commutative Property of Addition, $6 + x^2 + 3x$ can be written as $3x + x^2 + 6$, as $x^2 + 6 + 3x$, and so on.

Because there may be many options, we must establish a standard way to write a polynomial. The most typical standard is called *descending order*. A polynomial is in **descending order** when the terms are written according to their *degree*, from highest to lowest.

To write a polynomial in descending order, identify the degree of each term and then rewrite the polynomial with the highest degreed term first, then the next highest, and so on.

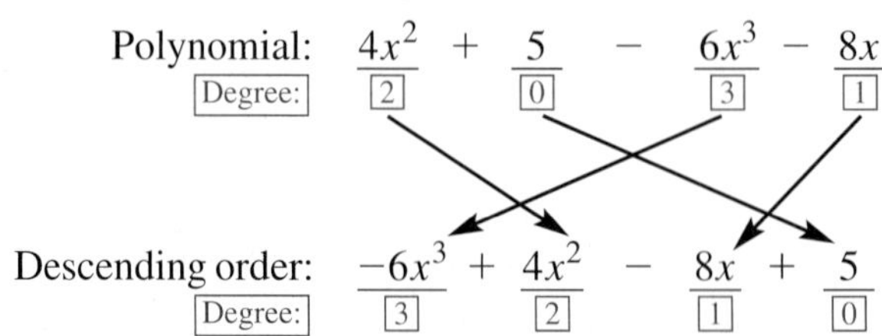

Polynomial: $4x^2 + 5 - 6x^3 - 8x$
Degree: 2 0 3 1
Descending order: $-6x^3 + 4x^2 - 8x + 5$
Degree: 3 2 1 0

Caution Descending order has to do with only the *degree* of the term, not its coefficient.

Example 7

Write each polynomial in descending order.

a) $-7 + 3x^5 - x^2$

b) $5 + 3y^2 - 7y^4 - 10y$

Procedure Recognize the degree of each term and rewrite the polynomial so that the term with the highest degree is first, the next highest degree is second, and so on.

Answer

a) $-7 + 3x^5 - x^2$
$= 3x^5 - x^2 - 7$

The degrees of the terms are 0, 5, and 2. Rearrange the terms so that the degrees are 5, 2, and 0.

b) $5 + 3y^2 - 7y^4 - 10y$
$= -7y^4 + 3y^2 - 10y + 5$

The degrees of the terms are 0, 2, 4, and 1. Rearrange the terms so that the degrees are 4, 2, 1, and 0.

You Try It 7 **Write each polynomial in descending order. Use Example 7 as a guide.**

Polynomial	**Descending Order**
a) $9x + 3x^2$	________
b) $-4x + x^2 - 6$	________
c) $9x^2 - 8 - 6x^4$	________
d) $x - 2x^2 + 5x^3 - 8$	________
e) $-3x^4 + 9 - 2x - x^2$	________

Combining Like Terms

Recall from Section 2.7 that two things are "like" each other when they share the same characteristics. In algebra, two or more terms are considered to be **like terms** when their variable structure is exactly the same. In other words, like terms must have exactly the same variable and exponent even if the coefficients are different. Furthermore, constants, which do not have a variable, are *like* all other constants.

When there is more than one variable in a term, the exponents on each like variable must correspond.

Like terms	**Because . . .**
1. $9w$ and $3w$	they have the same variable with the same exponent: w^1 and w^1.
2. $-4x^3$ and x^3	they have the same variable with the same exponent: x^3 and x^3.
3. 3 and -8	they are both constants.
4. $6x^3y^5$ and $-2x^3y^5$	they have the same two variables. The same exponent is with both x's and the same exponent is with both y's: x^3 and x^3, and y^5 and y^5.

As important as it is to recognize like terms, it is equally important to recognize *unlike* terms.

Unlike terms	**Because . . .**
1. $9w^2$ and $3w$	they have different exponents: w^2 and w^1.
2. $-4x^3$ and y^3	they have different variables: x^3 and y^3.
3. 6 and $-2w$	one is a constant and the other is not.
4. $-x^4y^2$ and $7x^2y^4$	they have the same two variables but the exponents on each like variable are not the same: x^4 and x^2, and y^2 and y^4.

Example 8

In each expression, identify the like terms.

Expression		Like Terms
a) $7x - 6x + 1$	One pair of terms is like.	$7x$ and $-6x$
b) $9 + 5x - 2$	One pair of terms is like.	9 and -2
c) $3a^2 + 2b^3 + 4a^2 + 9b^3$	Two pairs of terms are like.	$3a^2$ and $4a^2$; $2b^3$ and $9b^3$
d) $-15y + 6y - 3y$	All three terms are like.	$-15y$ and $6y$ and $-3y$
e) $4x + 6y - 7$	No terms are like.	(None)
f) $-2x^3 + 6x^2 + 9x$	No terms are like.	(None)

Caution For two terms, a different exponent on the same variable indicates *unlike* terms.

You Try It 8 **In each expression, identify the like terms. Use Example 8 as a guide.**

Expression	Like Terms
a) $5x^2 - x + 3 + 6x$	__________
b) $w^2 - 6w - 8w^2 - 9$	__________
c) $1 - 4y^3 - 4y^2 + 6x^3$	__________
d) $9p^5 + 2p^5 - 6p^5$	__________
e) $-2b^3 + b^2 + 4b - 3b^3 - 5b^2$	__________
f) $5x^2 + 3x - 1 - 4x^2 - 3x - 6$	__________

Recall from Section 2.7 that to combine like terms means to create another like term by adding or subtracting the coefficients.

For example, $8x^3 + 2x^3$ are like terms and create another like term when we add their coefficients: $8 + 2 = 10$. So, $8x^3 + 2x^3 = 10x^3$.

Similarly, $3y - 9y$ can be combined by subtracting: $3 - 9 = -6$. So, $3y - 9y = -6y$.

In a polynomial that has like and unlike terms, it is helpful to rearrange the terms first so that they are in descending order. This puts like terms together and prepares them to be combined. Also, it automatically arranges the final result in descending order.

For example, when we write $3x^2 + 2x - 6 - 5x + 2x^2$ in descending order, both x^2 terms appear before the other terms. (It doesn't matter which x^2 term we write first.) The two x terms follow next, and the single constant term is last.

Descending order: $\underline{3x^2 + 2x^2} + \underline{2x - 5x} - 6$

Combine like terms: $5x^2 \quad -3x \quad -6$

The simplified polynomial: $5x^2 - 3x - 6$

Caution When combining like terms, only the coefficients are combined. The exponents remain the same.

Example 9

Simplify each polynomial by combining like terms wherever possible. Write the simplified polynomial in descending order.

a) $8x^3 + 2x^3 - 5$ **b)** $6y^2 + 3y - 10y^2$

c) $6p^2 - 3k + 2 - 9p^2 + 5k$ **d)** $-3x^2 + 4x - 8 + 2x^2 - 4x - 5$

Procedure Rewrite each polynomial in descending order and combine like terms.

Answer

a) $8x^3 + 2x^3 - 5$ The first two terms are like terms.

$= 10x^3 - 5$

b) $6y^2 + 3y - 10y^2$ The first and third terms are like terms. Put the polynomial in descending order.

$= 6y^2 - 10y^2 + 3y$

$= -4y^2 + 3y$

c) $6p^2 - 3k + 2 - 9p^2 + 5k$ Write the p^2 terms together and the k terms together.

$= 6p^2 - 9p^2 - 3k + 5k + 2$

$= -3p^2 + 2k + 2$

d) $-3x^2 + 4x - 8 + 2x^2 - 4x - 5$ Write the polynomial in descending order.

$= -3x^2 + 2x^2 + 4x - 4x - 8 - 5$

$= -x^2 + 0x - 13$ Notice that the first term of this polynomial could also be written as $-1x^2$.

$= -x^2 - 13$ $0x$ is just 0; so, it is common to omit the term $0x$.

You Try It 9

Simplify each polynomial by combining like terms wherever possible. Write the simplified polynomial in descending order. Use Example 9 as a guide.

a) $5x^6 + 2x^4 - 12x^6$ **b)** $9y - 3y^3 - y^3$

c) $-7c + 2c^2 - c + 8c^2$ **d)** $x^2 + 4x^2 + y^3 - 2y^3$

e) $x^5 - 2x^3 - x - 4x^5 + 7x^3$ **f)** $6h^3 + 2h^2 - 9 - 3h^3 - 2h^2 + 5$

Answers: You Try It and Think About It

You Try It: **1. a)** $x^2, +2x, -7$ **b)** $2y^3, -y^2, -5y$ **c)** $-3xy^2, +4wv$ **d)** $-x^3, +3x^2, +x, -8$ **2. a)** $+5, +2xy; 2xy + 5$ **b)** $-8, +w^3; w^3 - 8$ **c)** $+1, -6y; -6y + 1$ **d)** $-3, -7c^2; -7c^2 - 3$ **3. a)** Binomial **b)** Monomial **c)** Trinomial **d)** Not a polynomial **e)** Monomial **4. a)** -6 **b)** 4 **c)** 3 **5. a)** The ball was 864 feet above the ground after falling 1 second. **b)** The ball was 400 feet above the ground after falling 5 seconds.

6.

	Term	Variable Factor(s)	Degree of Term
a)	$2x^3y^6$	$x \cdot x \cdot x \cdot y \cdot y \cdot y \cdot y \cdot y \cdot y$	9
b)	-7	none	0
c)	k^4	$k \cdot k \cdot k \cdot k$	4
d)	$-5b^2c$	$b \cdot b \cdot c$	3
e)	$3x$	x	1
f)	$-10x^5$	$x \cdot x \cdot x \cdot x \cdot x$	5

7.

	Polynomial	Descending Order
a)	$9x + 3x^2$	$3x^2 + 9x$
b)	$-4x + x^2 - 6$	$x^2 - 4x - 6$
c)	$9x^2 - 8 - 6x^4$	$-6x^4 + 9x^2 - 8$
d)	$x - 2x^2 + 5x^3 - 8$	$5x^3 - 2x^2 + x - 8$
e)	$-3x^4 + 9 - 2x - x^2$	$-3x^4 - x^2 - 2x + 9$

8. a) $-x$ and $6x$ **b)** w^2 and $-8w^2$ **c)** None **d)** $9p^5, 2p^5$, and $-6p^5$ **e)** $-2b^3$ and $-3b^3$; b^2 and $-5b^2$ **f)** $5x^2$ and $-4x^2$; $3x$ and $-3x$; -1 and -6 **9. a)** $-7x^6 + 2x^4$ **b)** $-4y^3 + 9y$ **c)** $10c^2 - 8c$ **d)** $-y^3 + 5x^2$ **e)** $-3x^5 + 5x^3 - x$ **f)** $3h^3 - 4$

Think About It: **1.** The degree of a constant term, such as 5, is 0 because there are zero variable factors.

Section 7.2 Exercises

FOR EXTRA HELP
Student Resources on DVD-ROM
Includes
➤ Student's Solutions Manual
➤ Video Lectures
➤ Chapter Test Prep Video
MyMathLab
MathXL

Think Again

1. What is the degree of a constant term? Explain your answer. *(Refer to Think About It 1)*
   0. The degree of a term is the number of variable factors it has. A constant term has no variable factors; so, its degree must be zero.

2. When combining like terms, as in $7x^3 + 4x^3$, why doesn't the exponent of x increase?
   The exponent of x does not increase because the three factors of x are not being multiplied by the other factor of x.

3. To find the degree of a polynomial such as $x^3 + 3x^2 - 6x + 8$, should we add all of the variable exponents? Explain your answer.
   No, we only add exponents when the variables are being multiplied together.

4. A polynomial is $y^2 + 6$. To evaluate this polynomial for $y = -3$, is it okay to write it as $-3^2 + 6$? Explain your answer.
   No. In the original polynomial, the value y is being squared. When y is -3, we show that it is being squared by including parentheses around -3, as in $(-3)^2 + 6$.

Focus Exercises

Identify the terms in each algebraic expression.

5. $x^2 + 2x$
   x^2 and $2x$

6. $2y^4 + y$
   $2y^4$ and y

7. $-5c^2 - 4c + 1$
   $-5c^2$, $-4c$, and 1

8. $-3p^4 + p^2 - 5$
   $-3p^4$, p^2, and -5

9. $x^4y + \frac{5}{8}xy^4 - \frac{1}{2}$
   x^4y, $\frac{5}{8}xy^4$, and $-\frac{1}{2}$

10. $\frac{3}{4}x^3 - \frac{1}{2}x^2 - \frac{2}{3}x$
    $\frac{3}{4}x^3$, $-\frac{1}{2}x^2$, and $-\frac{2}{3}x$

Identify each expression as a monomial, a binomial, a trinomial, or not a polynomial.

11. $x^2 + 9x - 5$
    trinomial

12. $-8y^3 - 4y^2$
    binomial

13. $2k^3$
    monomial

14. $\frac{-x + 4}{x^2}$
    not a polynomial

15. $\frac{-5d^3}{4}$
    monomial

16. $-x^5 - x^4 + 8x^3$
    trinomial

17. $2m - 9$
    binomial

18. $\frac{4}{5}n^2$
    monomial

19. $\frac{4y^3}{y - 3}$
    not a polynomial

20. $-c^5 - 4c^3 + 10c^2$
    trinomial

21. $-w + 2$
    binomial

22. -7
    monomial

Evaluate each polynomial with the given replacement value.

23. $5q - 7, q = 3$
    8

24. $3k - 15, k = 4$
    -3

25. $4y + 10, y = -5$
    -10

26. $8c + 8, c = -3$
    -16

27. $-5x - 1, x = -9$
    44

28. $-2n - 3, n = -11$
    19

29. $p^2 + 3p, p = 4$
    28

30. $w^2 - 5w, w = 2$
    -6

31. $4v^2 - 3, v = 1$
    1

32. $-6x^2 + 7, x = 1$
    1

33. $-8c^2 - 9, c = 0$
    -9

34. $7y^3 + 4y, y = 0$
    0

35. $x^2 + 6x, x = -3$
    -9

36. $2m^2 + 3m, m = -2$
    2

37. $4h^2 - 5h, h = -3$
    51

38. $3k^2 - 6k, k = -10$
    360

39. $q^2 + 4q + 6, q = 2$
    18

40. $x^2 + 3x - 10, x = 4$
    18

41. $y^2 - 5y + 1, y = -2$
    15

42. $c^2 - 2c + 3, c = -3$
    18

43. $-w^2 + 5w - 9, w = 5$
    -9

44. $-x^2 - 4x + 8, x = -4$
    8

45. $2n^3 - 5n^2, n = -1$
    -7

46. $6m^3 + 4m^2, m = -1$
    -2

For each term, identify the variable factors and indicate the degree.

47. $4x^2$
    $x \cdot x$
    degree 2

48. $-2w^4$
    $w \cdot w \cdot w \cdot w$
    degree 4

49. $\frac{5}{8}y^2$
    $y \cdot y$ degree 2

50. $-c^3$
    $c \cdot c \cdot c$
    degree 3

51. $-6p$
    p degree 1

52. 12
    (none)
    degree 0

53. -10
(none)
degree 0

54. $7k$
k degree 1

55. $10xy^5$
$x \cdot y \cdot y \cdot y \cdot y \cdot y$
degree 6

56. $-9v^3w^2$
$v \cdot v \cdot v \cdot w \cdot w$
degree 5

57. $3ab$
$a \cdot b$
degree 2

58. $-5cd$
$c \cdot d$
degree 2

Write each polynomial in descending order.

59. $4 + 3x$
$3x + 4$

60. $-6 + 2y$
$2y - 6$

61. $-11 - w$
$-w - 11$

62. $13 - 5p$
$-5p + 13$

63. $4b + 7b^3$
$7b^3 + 4b$

64. $-9m + m^2$
$m^2 - 9m$

65. $5x + 4x^2 - 6$
$4x^2 + 5x - 6$

66. $-9 + 2c^2 + 2c$
$2c^2 + 2c - 9$

67. $-w^2 + 6 - w^4$
$-w^4 - w^2 + 6$

68. $w - 12 - w^4$
$-w^4 + w - 12$

69. $-8k^5 + 6k - 2k^3$
$-8k^5 - 2k^3 + 6k$

70. $4x^3 + 15x - 12x^2$
$4x^3 - 12x^2 + 15x$

71. $10a^3 + 9a - 1 - 6a^2$
$10a^3 - 6a^2 + 9a - 1$

72. $-9 + 4n^3 - 2n^2 + 8n$
$4n^3 - 2n^2 + 8n - 9$

73. $1 - y^3 - y - 5y^2$
$-y^3 - 5y^2 - y + 1$

74. $-6 + 3v + 9v^2 - 2v^3$
$-2v^3 + 9v^2 + 3v - 6$

Simplify each polynomial by combining like terms if possible. Write all answers in descending order.

75. $8x^3 + 2x^3$
$10x^3$

76. $6w^2 - 10w^2$
$-4w^2$

77. $-8m + m$
$-7m$

78. $-3c^4 - 2c^4$
$-5x^4$

79. $4p + 3p^2$
$3p^2 + 4p$

80. $-x + 3x^2$
$3x^2 - x$

81. $-2y^2 + 6y^2$
$4y^2$

82. $x^5 - 8x^5$
$-7x^5$

83. $-n^3 - n^3$
$-2n^3$

84. $-k^2 - k^2$
$-2k^2$

85. $-q + q$
0

86. $-x^3 + x^3$
0

87. $7x^4 + 1 + 6x^4$
$13x^4 + 1$

88. $-4h^3 - 3 + 12h^3$
$8h^3 - 3$

89. $3y^2 + 9 - 5y^2$
$-2y^2 + 9$

90. $5p^6 - 4p^3 - 2p^3$
$5p^6 - 6p^3$

91. $7v^2 - v^3 + 6v^2$
$-v^3 + 13v^2$

92. $-6b + 2b^2 - 5b$
$2b^2 - 11b$

93. $4x^2 - 3x^2 + 5y^3 - 6y^3$
$-y^3 + x^2$

94. $3x - 4x^2 + 2x - 8x^2$
$-12x^2 + 5x$

95. $8x + 2x + 9y - 5y$
$10x + 4y$

96. $7x - 6x + 3y - 5y$
$x - 2y$

97. $3x^2 - 4x^2 + 2x - 8x$
$-x^2 - 6x$

98. $5x^2 - x + 3 + 6x$
$5x^2 + 5x + 3$

99. $w^2 - 6w - 8w^2 - 9$
$-7w^2 - 6w - 9$

100. $1 - 4y^3 - 4y^2 + 6x^3$
$2y^3 - 4y^2 + 1$

101. $9p^5 - 10 + 2p^5 - 6p^5 + 3$
$5p^5 - 7$

102. $-4c^3 - c^3 + 3c + 6c^3 - 2c$
$c^3 + c$

103. $5x^2 + 3x - 1 - 4x^2 - 3x - 6$
$x^2 - 7$

104. $-2b^3 + b^2 + 4b - 3b^3 - 5b^2$
$-5b^3 - 4b^2 + 4b$

Terrence dropped a ball from the roof of a 1,200-foot-high building. The ball's height above the ground after x seconds is given by $h = 1{,}200 - 16x^2$. Write a sentence answering the question.

105. How high above the ground was the ball 3 seconds after falling?
The ball was 1,056 feet above the ground.

106. How high above the ground was the ball 6 seconds after falling?
The ball was 624 feet above the ground.

Ronwyn threw a rock upward over the ocean from a 300-foot cliff. The rock's height above the ocean after x seconds is given by $h = 300 + 30x - 16x^2$. Write a sentence answering the question.

107. How high above the ocean was the rock 2 seconds later?
The rock was 296 feet above the ocean.

108. How high above the ocean was the rock 5 seconds later?
The rock was 50 feet above the ocean.

A set of rectangles are constructed in such a way that the area formula for each rectangle is a polynomial: $A = x^2 + 3x + 8$ square inches, where x is measured in inches. Write a sentence answering the question.

109. What is the area of the rectangle when x is 5 inches?
The area of the rectangle is 48 square inches.

110. What is the area of the rectangle when x is 9 inches?
The area of the rectangle is 116 square inches.

Think Outside the Box

A rock is dropped from a 600-foot-high cliff. The rock's height above the ground in feet after x seconds is given by $h = 600 - 16x^2$. Fill in the chart and answer the questions.

	Number of seconds the rock has fallen, x	The rock's height above the ground: $h = 600 - 16x^2$	Distance fallen during one second
	0	600	0 feet
	1	600 − 16 = **584**	600 − 584 = **16 feet**
	2	600 − 64 = **536**	584 − 536 = **48 feet**
111.	3	600 − 144 = **456**	536 − 456 = **80 feet**
112.	4	600 − 256 = **344**	456 − 344 = **112 feet**
113.	5	600 − 400 = **200**	344 − 200 = **144 feet**
114.	6	600 − 576 = **24**	200 − 24 = **176 feet**

115. According to the numbers in the last column, is the distance fallen each second increasing, decreasing, or staying the same? What causes the numbers to occur this way?
The distance fallen each second is increasing. Answers may vary.

SECTION 7.3 Adding and Subtracting Polynomials

OBJECTIVES

In this section, you will learn to

- Multiply a quantity by 1 and −1.
- Add polynomials.
- Subtract polynomials.

You Need to Know

To successfully complete this section, you need to understand

- ☐ Adding signed numbers (2.2)
- ☐ Subtracting signed numbers (2.3)
- ☐ Multiplying signed numbers (2.4)
- ☐ The Distributive Property (2.8)
- ☐ Writing polynomials in descending order (7.2)
- ☐ Combining like terms (7.2)

Introduction

Recall from Section 2.8 the Distributive Property.

The Distributive Property of Multiplication over Addition

$$b(c + d) = b \cdot c + b \cdot d$$

b is called the **multiplier**, and $(c + d)$ is a **quantity**.

For example, we can distribute the multiplier **2** through the quantity $(3x + 4)$:

$$2(3x + 4)$$
$$= 2 \cdot 3x + 2 \cdot 4$$
$$= 6x + 8$$

This distribution can be diagrammed like this:

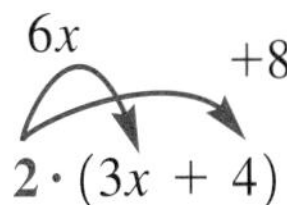

In Section 7.4, we will look at distributing when the multiplier is a term with a variable, such as $2x^3$. In this section, however, our main focus is to add and subtract polynomials; and for those, we need very specific multipliers, 1 and −1.

When the Multiplier Is 1 or −1

Recall that 1 is the multiplicative identity. In other words, $1 \cdot a = a$. For this reason, the simplest distribution is when the multiplier is 1.

$$1(c + d) = 1 \cdot c + 1 \cdot d$$
$$1(c + d) = c + d$$

This suggests that when the multiplier is 1, we may drop the parentheses without needing to distribute.

When the multiplier is −1, the distribution is almost as simple, but we do need to be more careful.

Consider, for example, the product of the multiplier −1 and the quantity $(5x - 2)$. As a product, this looks like $-1 \cdot (5x - 2)$.

$-5x$ $+2$ / $-1 \cdot (5x - 2)$	$-1 \cdot (5x - 2)$	Treat $5x$ as $+5x$ and -2 as -2.
	$= -5x + 2$	$-1 \cdot (+5x) = -5x$; $-1 \cdot (-2) = +2$, or *plus* 2

Notice that multiplying by −1 changes the sign of each term. That is because "−1 times" a term means "the opposite" of that term.

- The opposite of a positive term is a negative term: $-1 \cdot (+5x) = -5x$.
- The opposite of a negative term is a positive term: $-1 \cdot (-2) = +2$.

Example 1

Distribute.

a) $1(5x + 3)$ **b)** $1(2x - 6)$ **c)** $1(-3x^2 - 2x + 4)$

d) $-1(5x + 3)$ **e)** $-1(2x - 6)$ **f)** $-1(-3x^2 - 2x + 4)$

Procedure To distribute 1, simply drop the parentheses. The quantity will be the answer without any changes. To distribute -1, drop the parentheses and take the opposite of each term in the quantity.

Answer

a) $1(5x + 3)$
$= 5x + 3$

b) $1(2x - 6)$
$= 2x - 6$

c) $1(-3x^2 - 2x + 4)$
$= -3x^2 - 2x + 4$

d) $-1(5x + 3)$
$= -5x - 3$

e) $-1(2x - 6)$
$= -2x + 6$

f) $-1(-3x^2 - 2x + 4)$
$= +3x^2 + 2x - 4$
or just $3x^2 + 2x - 4$

You Try It 1 **Distribute. Use Example 1 as a guide.**

a) $1(2x^3 + 3x)$ **b)** $1(w^2 - 6)$ **c)** $-1(3v + 1)$

d) $-1(4y^2 - 4y)$ **e)** $1(6c^3 + c - 4)$ **f)** $-1(-3p^2 + 2p - 5)$

Adding Polynomials

When we add two polynomials, each polynomial has parentheses around it. For example, to add the polynomial $3x^2 + 2x - 6$ and the polynomial $-2x^2 - x + 8$, we might see this:

$$(3x^2 + 2x - 6) + (-2x^2 - x + 8)$$

The first thing we need to do is "clear" the parentheses by distributing. However, since we don't see any multiplier to distribute, it must mean that we are distributing the number 1. Then the expression looks like this:

$$1 \cdot (3x^2 + 2x - 6) + 1 \cdot (-2x^2 - x + 8)$$

Here is the result.

$$3x^2 + 2x - 6 + -2x^2 - x + 8$$

Now we have one large polynomial with six terms. Because some of the terms are like terms, we can simplify this polynomial, as we did in Section 7.2, by writing the polynomial in descending order and combining like terms.

$$\begin{aligned} &3x^2 + 2x - 6 + -2x^2 - x + 8 \\ &= 3x^2 + -2x^2 + 2x - x - 6 + 8 \\ &= x^2 + x + 2 \end{aligned}$$

Example 2

Add. Simplify and write each answer in descending order.

a) $(3x - 5 + x^2) + (2 + 8x - 4x^2)$ **b)** $(2w^2 - 5) + (7w^2 + 3w)$

Procedure Write "1 times" each quantity; then eliminate the parentheses by distributing. Simplify the resulting polynomial and write the answer in descending order.

Answer

a)

$1 \cdot (3x - 5 + x^2) + 1 \cdot (2 + 8x - 4x^2)$ Multiply each quantity by 1 and distribute.
$= 3x - 5 + x^2 + 2 + 8x - 4x^2$ Write this in descending order.
$= x^2 - 4x^2 + 3x + 8x - 5 + 2$ Combine like terms.
$= -3x^2 + 11x - 3$

b)

$1 \cdot (2w^2 - 5) + 1 \cdot (7w^2 + 3w)$ Multiply each quantity by 1 and distribute.
$= 2w^2 - 5 + 7w^2 + 3w$ Write this in descending order.
$= 2w^2 + 7w^2 + 3w - 5$ Combine like terms.
$= 9w^2 + 3w - 5$

You Try It 2 **Add. Simplify and write each answer in descending order. Use Example 2 as a guide.**

a) $(3x - 1) + (x - 6)$ **b)** $(2w^2 + 3) + (w^2 - 8)$

c) $(5c - 2) + (c^2 + 2c)$ **d)** $(x^2 + x - 4) + (-x + 2x^2 - 5)$

Subtracting Polynomials

Subtracting one polynomial from another is similar to adding polynomials in that we usually see the polynomials in parentheses and we need to distribute to remove them. This time, though, as we write a 1 in front of each parenthesis, we need to recognize that we must distribute -1 to the second polynomial.

For example,

$(5x^2 - 4x) - (3x^2 + 6x)$	**Place a "1 times" in front of each quantity.**
$= 1 \cdot (5x^2 - 4x) - 1 \cdot (3x^2 + 6x)$	**Subtracting 1 is the same as adding negative 1.**
$= 1 \cdot (5x^2 - 4x) + (-1) \cdot (3x^2 + 6x)$	**Clear the parentheses.**
$= 5x^2 - 4x + -3x^2 - 6x$	**Write this polynomial in descending order.**
$= 5x^2 - 3x^2 - 4x - 6x$	**Combine like terms.**
$= 2x^2 - 10x$	

Example 3

Subtract. Simplify and write each answer in descending order.

a) $(3x - 5 + x^2) - (2 + 8x - 4x^2)$ **b)** $(2x^2 - 5) - (7x^2 + 3x)$

Procedure Write "1 times" each quantity; then eliminate the parentheses by distributing. Simplify the resulting polynomial and write the answer in descending order.

Answer **a)**

$(3x - 5 + x^2) - (2 + 8x - 4x^2)$	Multiply each quantity by 1.
$= 1 \cdot (3x - 5 + x^2) - 1 \cdot (2 + 8x - 4x^2)$	Change subtraction to "add the opposite."
$= 1 \cdot (3x - 5 + x^2) + (-1) \cdot (2 + 8x - 4x^2)$	Distribute.
$= 3x - 5 + x^2 + -2 - 8x + 4x^2$	Write this polynomial in descending order.
$= x^2 + 4x^2 + 3x - 8x - 5 - 2$	Combine like terms.
$= 5x^2 - 5x - 7$	

b)

$(2x^2 - 5) - (7x^2 + 3x)$	Multiply each quantity by 1.
$= 1 \cdot (2x^2 - 5) - 1 \cdot (7x^2 + 3x)$	Change subtraction to "add the opposite."
$= 1 \cdot (2x^2 - 5) + (-1) \cdot (7x^2 + 3x)$	Distribute.
$= 2x^2 - 5 + -7x^2 - 3x$	Write this polynomial in descending order.
$= 2x^2 - 7x^2 - 3x - 5$	Combine like terms.
$= -5x^2 - 3x - 5$	

You Try It 3 **Subtract. Simplify and write each answer in descending order. Use Example 3 as a guide.**

a) $(4x^3 + 2) - (x^3 - 7)$ **b)** $(9w - 3) - (4 - 2w)$

c) $(y^2 - 4y) - (6y^2 + 2y)$ **d)** $(6x^3 + x - 4) - (-3x^3 + 2x - 5)$

Answers: You Try It and Think About It

You Try It: **1. a)** $2x^3 + 3x$ **b)** $w^2 - 6$ **c)** $-3v - 1$ **d)** $-4y^2 + 4y$ **e)** $6c^3 + c - 4$ **f)** $3p^2 - 2p + 5$ **2. a)** $4x - 7$ **b)** $3w^2 - 5$ **c)** $c^2 + 7c - 2$ **d)** $3x^2 - 9$ **3. a)** $3x^3 + 9$ **b)** $11w - 7$ **c)** $-5y^2 - 6y$ **d)** $9x^3 - x + 1$

Think About It: No Think About Its in this section.

Section 7.3 Exercises

FOR EXTRA HELP: Student Resources on DVD-ROM. Includes ➤Student's Solutions Manual ➤Video Lectures ➤Chapter Test Prep Video. MyMathLab. MathXL

Think Again

1. What is the most important step in subtracting one polynomial from another?
The most important step is to distribute -1 through to the entire second polynomial.

2. As we add or subtract polynomials, why do we write the answer in descending order?
Answers will vary. One possibility: So that everyone will be consistent in the way the resulting polynomial is written.

Focus Exercises

Distribute. Write each answer in descending order.

3. $1(5x + 7x^2)$
$7x^2 + 5x$

4. $1(3 - 8y)$
$-8y + 3$

5. $-1(4c + 6)$
$-4c - 6$

6. $-1(2w + 5w^2)$
$-5w^2 - 2w$

7. $-1(6p - p^2)$
$p^2 - 6p$

8. $-1(-9 - d^3)$
$d^3 + 9$

9. $1(-4x - 5x^2 + 1)$
$-5x^2 - 4x + 1$

10. $1(8 - 4y - 9y^2)$
$-9y^2 - 4y + 8$

11. $-1(-7k^2 - 3k^3 + k)$
$3k^3 + 7k^2 - k$

12. $-1(m^2 + 5 - 4m^4)$
$4m^4 - m^2 - 5$

13. $-1(-2v + 3v^3 - 1)$
$-3v^3 + 2v + 1$

14. $-1(4n^2 - 6n - 8n^3)$
$8n^3 - 4n^2 + 6n$

Add. Combine like terms and write each answer in descending order.

15. $(5x^2 - 3x) + (x - 2x^2)$
$3x^2 - 2x$

16. $(4w^2 - 4w) + (w^2 + 2w)$
$5w^2 - 2w$

17. $(-8n - 2n^2) + (3n^2 + 6n)$
$n^2 - 2n$

18. $(-q^2 + 5q) + (-6q + 4q^2)$
$3q^2 - q$

19. $(-4c - 3) + (c^2 - 6c + 1)$
$c^2 - 10c - 2$

20. $(1 - 9k) + (4k - 3k^2 - 1)$
$-3k^2 - 5k$

21. $(6y^3 + y - 4) + (-3y^3 + 2y - 5)$
$3y^3 + 3y - 9$

22. $(v - 4v^2 + 3) + (4 - v^2 + 6v)$
$-5v^2 + 7v + 7$

Subtract. Combine like terms, and write each answer in descending order.

23. $(2x^2 + 3) - (x^2 - 8)$
$x^2 + 11$

24. $(3h - 1) - (h - 6)$
$2h + 5$

25. $(5a - 6) - (-6 + 5a)$
0

26. $(-9 - 2x^2) - (x^2 - 7)$
$-3x^2 - 2$

27. $(4n - 2) - (n^2 + 2n)$
$-n^2 + 2n - 2$

28. $(6b - 2b^2) - (4b - 9)$
$-2b^2 + 2b + 9$

29. $(-4c^3 - 3c + 8) - (-c^3 + 4c - 1)$
$-3c^3 - 7c + 9$

30. $(p^2 + p - 4) - (-p + 2p^2 - 5)$
$-p^2 + 2p + 1$

31. $(11 - 5q^2) - (-q + 3q^2 - 1)$
$-8q^2 + q + 12$

32. $(6a^3 - 4) - (-2a^3 + a - 6)$
$8a^3 - a + 2$

33. $(y^3 + 4y^2 + 2y) - (4y^2 + 2y - 2)$
$y^3 + 2$

34. $(-x - 3x^2 + 2) - (x^2 - 7x - 1)$
$-4x^2 + 6x + 3$

Perform the indicated operations. Combine like terms and write each answer in descending order.

35. $(6v - 4) + (3v - 2)$
$9v - 6$

36. $(6y - 4) - (3y - 2)$
$3y - 2$

37. $(x + 6) - (x - 5)$
11

38. $(4m - 6) + (3m + 5)$
$7m - 1$

39. $(5w - 8) + (3 - 4w)$
$w - 5$

40. $-6a - (5 - 4a)$
$-2a - 5$

41. $4 - (3h + 4)$
$-3h$

42. $(3 - 2v) + (v + 2)$
$-v + 5$

43. $(6q^3 + q - 4) - (-3q^3 + 2q - 5)$
$9q^3 - q + 1$

44. $(x^2 + x - 4) - (-x + 2x^2 - 5)$
$-x^2 + 2x + 1$

45. $(9 + 2n - 4n^2) + (3n^2 + 11)$
$-n^2 + 2n + 20$

46. $(9y + 6y^3 - 4) - (6 - 2y - 5y^3)$
$11y^3 + 11y - 10$

47. $(5 - 8c) - (3c - 1)$
$-11c + 6$

48. $(2v^2 - 6v + 1) + (3v - 5v^2 - 1)$
$-3v^2 - 3v$

49. $(3p - 8) - (5 + 10p)$
$-7p - 13$

50. $(x^2 + x - 4) - (-x + 2x^2 - 5)$
$-x^2 + 2x + 1$

Think Outside the Box

Simplify each polynomial answer and write it in descending order.

51. Consider the rectangle that has a length given by $(-5x - 7 + 3x^2)$ meters and a width given by $(4 + 8x - x^2)$ meters.

$(-5x - 7 + 3x^2)$ m

$(4 + 8x - x^2)$ m

a) Find the perimeter of the rectangle.
The perimeter is given by $4x^2 + 6x - 6$.

b) Is it possible for x to be 10 meters? Explain your answer.
No. When $x = 10$ meters, the width is negative, which is not possible.

52. Find the length of the third side of this triangle, knowing that the perimeter is given by $(4x^2 - 7x + 9)$ feet.

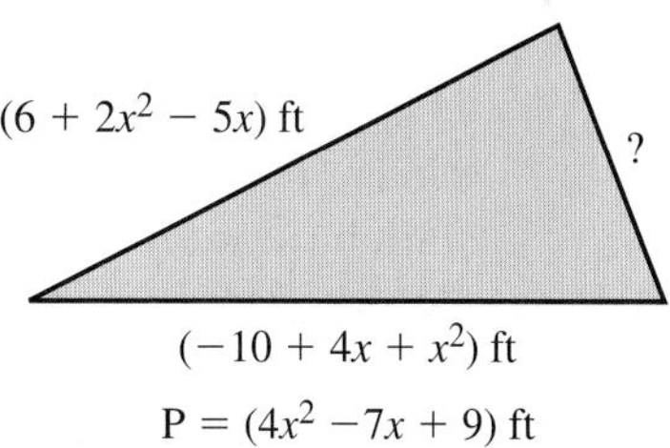

$P = (4x^2 - 7x + 9)$ ft

The third side is $x^2 - 6x + 13$ feet.

SECTION 7.4 Multiplying Polynomials

OBJECTIVES

In this section, you will learn to

- Check that multiplication of polynomials is accurate.
- Multiply two monomials.
- Square a monomial.
- Multiply a monomial and a polynomial.
- Multiply a binomial and a trinomial.
- Multiply two binomials.

You Need to Know

To successfully complete this section, you need to understand

- ☐ Replacing a variable with a number (1.7)
- ☐ Multiplying signed numbers (2.4)
- ☐ The Distributive Property (2.8)
- ☐ Combining like terms (2.7 and 7.2)
- ☐ The Product Rule for Exponents (7.1)

Introduction

To this point, you have been introduced to a variety of definitions related to polynomials. You have also learned to add and subtract polynomials. Now it is time to learn to multiply polynomials.

Here are a few examples of the product of two polynomials:

two monomials: $(-2x^3)(5x^4) = -10x^7$

a monomial and a binomial: $3y^4(6y^2 - 4y) = 18y^6 - 12y^5$

a binomial and a trinomial: $(2c + 1)(5c^2 - c - 4) = 10c^3 + 3c^2 - 9c - 4$

two binomials: $(3w - 5)(2w + 4) = 6w^2 + 2w - 20$

Notice that each resulting product is a simplified polynomial. In this section, we discuss techniques for multiplying polynomials. The key to success in multiplying polynomials is taking a step-by-step approach. Also, it's a good idea to check your work.

Checking the Answer When Multiplying Polynomials

There is a quick technique for checking whether we have multiplied polynomials correctly, and it will be used in each example to verify that the work shown is correct.

Note This method of checking will *not* tell you whether the exponents are correct, but it will tell you whether your coefficients and constant are correct.

To prepare for using this technique, recall from Section 7.2 that we can find the value of a polynomial when we know the value of its variable.

For example, find the value of $6w^2 + 2w - 20$ when $w = 1$.

$$6(1)^2 + 2(1) - 20$$
$$= 6(1) + 2(1) - 20$$
$$= \mathbf{6 + 2 - 20}$$
$$= -12$$

Notice that when $w = 1$, the next-to-last step is just the original polynomial without its variables. In other words, that step is just the coefficients and the constant of the polynomial: **6 + 2 − 20**.

Because $w = 1$, the multiplicative identity, it is as if the variables "melt away" and we can find the sum of the numerical parts of the polynomial. In other words, when the variable is replaced by 1, the variables can be eliminated.

Example 1

Find the value of each polynomial when the variable is 1.

a) $5p^3 + 3p$ **b)** $y^2 - 8y$ **c)** $3x^2 - x + 4$

Procedure To replace the variable with 1 means that the variable(s) can be eliminated. The result is the coefficients and constant of the polynomial. When there is no visible coefficient, the coefficient is 1 (or −1).

Answer

a) $5p^3 + 3p, p = 1$
$5 + 3$
$= 8$

b) $1y^2 - 8y, y = 1$
$1 - 8$
$= -7$

c) $3x^2 - 1x + 4, x = 1$
$3 - 1 + 4$
$= 6$

You Try It 1 **Find the value of each polynomial when the variable is 1. Use Example 1 as a guide.**

a) $2x^3 + 3x$ **b)** $w^2 - 6$ **c)** $6c^3 + c - 4$ **d)** $-3p^2 + 2p - 5$

When a product is complete, we can check to see if the constants and coefficient of the result are correct by replacing the variable with 1 in each polynomial in the product. Again, this check does not indicate whether the exponents are correct.

For example, in the product $(3w - 5)(2w + 4) = 6w^2 + 2w - 20$, we can choose $w = 1$ for each variable. If the product is correct, each side of the equal sign will be the same value.

Original Product $\stackrel{?}{=}$ **Answer**	
$(3w - 5)(2w + 4) \stackrel{?}{=} 6w^2 + 2w - 20$	Replace w with 1, as in Example 1.
$(3 - 5)(2 + 4) \stackrel{?}{=} 6 + 2 - 20$	Evaluate each part separately.
$(-2)(6) \stackrel{?}{=} -12$	
$-12 = -12$ True!	

Because $-12 = -12$, we can be confident that the coefficients and constant of $6w^2 + 2w - 20$ are correct. We also should mentally check our work so that we are confident the exponents of $6w^2 + 2w - 20$ are correct.

This checking process is one that you can do in your head or as scratch work. It is intended only as a quick way to verify that the multiplication of two polynomials is correct.

If we multiply two polynomials and this checking technique shows that the product is not correct, we should search for our mistake or redo the multiplication from the beginning.

Example 2

Replace each variable with 1 in the product to determine whether the multiplication is correct.

a) $3y^4(6y^2 - 4y) = 18y^6 - 12y^5$ **b)** $(4x - 3)(2x + 6) = 8x^2 - 18$

Procedure To replace the variable with 1 means that the variables can be eliminated. Evaluate each side of the equal sign to determine whether the multiplication is correct.

Answer

Original Product $\stackrel{?}{=}$ **Answer**

a)
$$3y^4(6y^2 - 4y) \stackrel{?}{=} 18y^6 - 12y^5$$
$$3\cdot(6 - 4) \stackrel{?}{=} 18 - 12$$
$$3\cdot(2) \stackrel{?}{=} 6$$
$$6 = 6$$

True, this multiplication is correct. Now check mentally to verify that the exponents also are correct.

Original Product $\stackrel{?}{=}$ **Answer**

b)
$$(4x - 3)(2x + 6) \stackrel{?}{=} 8x^2 - 18$$
$$(4 - 3)(2 + 6) \stackrel{?}{=} 8 - 18$$
$$(1)(8) \stackrel{?}{=} -10$$
$$8 \neq -10$$

False, this multiplication is *not* correct. If $8x^2 - 18$ was our answer, we would need to find the error and correct it.

You Try It 2 **Replace each variable with 1 in the product to determine whether the multiplication is correct. Use Example 2 as a guide.**

a) $(-2x^3)(5x^4) = -10x^7$ **b)** $(3v - 4)(2v^2 - 5) = 3v^3 - 8v^2 - 15v + 20$

c) $(y + 6)(y - 5) = y^2 - 30$ **d)** $(2c + 1)(5c^2 - c - 4) = 10c^3 + 3c^2 - 9c - 4$

Multiplying Two Monomials

Finding the product of two monomials relies on three basic principles:

	Example
1. The Associative Property of Multiplication	
If the only operation is multiplication, we can eliminate the parentheses.	$(3x^4)(2x^5)$ $= 3\cdot x^4\cdot 2\cdot x^5$
2. The Commutative Property of Multiplication	
If the only operation is multiplication, we can rearrange the factors so that coefficients are multiplied together and variables are multiplied together.	$= 3\cdot 2\cdot x^4\cdot x^5$ $= 6\cdot x^4\cdot x^5$
3. The Product Rule for Exponents	$= 6\cdot x^{4+5}$
$x^a\cdot x^b = x^{a+b}$	$= 6\cdot x^9$

This process can be abbreviated by multiplying the coefficients and combining the exponents of the variables. In other words, we can write the following:

$$\begin{aligned}&(3x^4)(2x^5)\\&= 3\cdot 2\cdot x^{4+5}\\&= 6x^9\end{aligned}$$

Example 3

Multiply. Check each answer to verify that it is correct.

a) $(2w^2)(6w^4)$ **b)** $(-5x)(x^2)$

Procedure Multiply the coefficients and use the Product Rule for Exponents on the variables. Remember, if a term has no visible coefficient, the coefficient is 1. Likewise, if a variable has no visible exponent, the exponent is 1.

Finally, check each answer by making the variable equal to 1. If true, mentally check the exponents as well.

Answer

a) $(2w^2)(6w^4)$
$= 2\cdot 6\cdot w^{2+4}$
$= 12w^6$

Check. Replace w with 1:
$(2)(6) \stackrel{?}{=} 12$
$12 = 12$ True ✓

b) $(-5x)(x^2)$
$= -5\cdot 1\cdot x^{1+2}$
$= -5x^3$

Check. Replace x with 1:
$(-5)(1) \stackrel{?}{=} -5$
$-5 = -5$ True ✓

You Try It 3 **Multiply. Check each answer to verify that it is correct. Use Example 3 as a guide.**

a) $(2m^3)(8m^4)$ **b)** $(-9p^5)(p^2)$ **c)** $(-4y)(-3y^4)$ **d)** $(w)(-w^5)$

Squaring a Monomial: (Monomial)^2

Squaring a monomial is fairly simple because it is multiplying one monomial by itself.

Example 4

Multiply. Check each answer to verify that it is correct.

a) $(3x^4)^2$ **b)** $(-5y)^2$

Procedure Write each monomial times itself; then multiply as in Example 3.

Answer

a) $(3x^4)^2$
$= (3x^4)\cdot(3x^4)$
$= 3\cdot 3\cdot x^{4+4}$
$= 9x^8$

Check. Replace x with 1:
$(3)^2 \stackrel{?}{=} 9$
$9 = 9$ True ✓

Now mentally check the exponents for accuracy.

b) $(-5y)^2$
$= (-5y)(-5y)$
$= -5\cdot(-5)\cdot y^{1+1}$
$= 25y^2$

Check. Replace y with 1:
$(-5)^2 \stackrel{?}{=} 25$
$25 = 25$ True ✓

Now mentally check the exponents for accuracy.

You Try It 4 **Multiply. Check each answer to verify that it is correct. Use Example 4 as a guide.**

a) $(5c^3)^2$ **b)** $(8w^4)^2$ **c)** $(-9x^5)^2$

Multiplying a Monomial and a Polynomial

In Section 2.8, we practiced the Distributive Property when the multiplier was a constant. If the multiplier is a term, the same principle applies. When we distribute a term to a polynomial, we distribute the multiplier to each term in the quantity.

Keep in mind that *the sign in front of a term belongs to the term.* If a term in the quantity is being subtracted, it is appropriate to think of that term as negative.

Consider, for example, the product of the multiplier $-2x^2$ and the quantity $(4x^3 - 3x)$. As a product, this looks like $-2x^2(4x^3 - 3x)$. The terms in the quantity are $+4x^3$ and $-3x$.

Here is the process step-by-step.

$-2x^2(4x^3 - 3x)$ — Distribute $-2x^2$ to each term in the quantity. Treat $- 3x$ as $-3x$.

$= (-2x^2)(4x^3) + (-2x^2)(-3x)$ — Notice that the original minus sign now appears as part of the $-3x$ term.

$= -2 \cdot 4 \cdot x^{2+3} + (-2) \cdot (-3) \cdot x^{2+1}$

$= -8x^5 + (+6x^3)$ — This step is not necessary, but it emphasizes the multiplication of two negative coefficients: $(-2)(-3) = +6$.

$= -8x^5 + 6x^3$ — Notice that the signs of this binomial are different from the original $(4x^3 - 3x)$ because the multiplier, $-2x^2$, is negative.

We can check this result to see if we multiplied correctly. Replace x with 1.

Original Product $\stackrel{?}{=}$ Answer

$$-2(4 - 3) \stackrel{?}{=} -8 + 6$$
$$-2(1) \stackrel{?}{=} -2$$
$$-2 = -2 \quad \text{True ✓}$$

Now mentally check the exponents for accuracy.

It's possible to distribute in fewer steps, as demonstrated here.

Think of $-2x^2$ being distributed to each term.

$$-8x^5 \qquad +6x^3$$
$$-2x^2(4x^3 - 3x)$$

$-2x^2(4x^3 - 3x)$ — Treat $4x^3$ as $+4x^3$ and $- 3x$ as $-3x$. $(-2x^2)(+4x^3) = -8x^5$; $(-2x^2)(-3x) = +6x^3$, or *plus* $6x^3$.

$= -8x^5 + 6x^3$

The key to successfully multiplying in one step is being aware of the sign of each term being multiplied and of the resulting product of those two terms.

Caution The most common mistake in multiplying a monomial by a polynomial is ending up with an incorrect sign. Make sure you check the signs of each term carefully. If you make a mistake in one of the signs, your check (when the variable is replaced by 1) should come out false.

Example 5

Multiply. Check each answer to verify that it is correct.

a) $(3x^2)(4x^2 - 2x)$ **b)** $(-5y^2)(y^2 + 4y - 1)$

Procedure As we distribute, we'll get a series of monomial products. More steps are shown here than are necessary.

Answer

a) $\mathbf{3x^2}(4x^2 - 2x)$
$= (\mathbf{3x^2}) \cdot (4x^2) + (\mathbf{3x^2}) \cdot (-2x)$
$= 12x^4 - 6x^3$
Check. Replace x with 1.

b) $\mathbf{-5y^2}(y^2 + 4y - 1)$
$= (\mathbf{-5y^2}) \cdot y^2 + (\mathbf{-5y^2}) \cdot (4y) + (\mathbf{-5y^2}) \cdot (-1)$
$= -5y^4 - 20y^3 + 5y^2$
Check. Replace y with 1.

Original Product $\stackrel{?}{=}$ Answer	**Original Product $\stackrel{?}{=}$ Answer**
$(3)(4-2) \stackrel{?}{=} 12-6$	$(-5)(1+4-1) \stackrel{?}{=} -5-20+5$
$(3)(2) \stackrel{?}{=} 6$	$(-5)(4) \stackrel{?}{=} -20$
$6 = 6$ True ✓	$-20 = -20$ True ✓
Now mentally check the exponents for accuracy.	Now mentally check the exponents for accuracy.

You Try It 5 **Multiply. Check each answer to verify that it is correct. Use Example 5 as a guide.**

a) $2c(8c^4 + 3c^2)$ **b)** $-3q^3(q^3 - 5q^2 + 6q)$

c) $-2a^3(-4a^4 + 3a - 1)$ **d)** $4x^2(2x^3 - 3x^2 + x - 1)$

Multiplying a Binomial and a Trinomial

We can think of distributing a term to a polynomial as a repeated action. For example, in the case of multiplying $3y$ by $(4y^2 + 2y + 5)$, we need to *repeatedly multiply by 3y.*

$$\begin{aligned} &\mathbf{3y}(4y^2 + 2y + 5) \\ &= (\mathbf{3y})\cdot(4y^2) + (\mathbf{3y})\cdot(2y) + (\mathbf{3y})\cdot(5) \\ &= 12y^3 + 6y^2 + 15y \end{aligned}$$

Consider for a moment a completely different type of repeated action.

> In the Johnson family, triplets Josh, Brett, and Nate were born on April 30. Every year at that time, each of the boys receives two presents, a special gift from their mom, Laura, and a special gift from their dad, Rob.

To put this scenario in the context of algebra and multiplying polynomials, think of the repeated action of *giving a gift* represented by the arrow, ➤. We might write it like this:

Laura ➤ (Josh & Brett & Nate) This means that Laura gives one gift to Josh, one gift to Brett, and one gift to Nate.

In this case, we are not multiplying, but Laura is *distributing* gifts (repeatedly).

Laura ➤ Josh & Laura ➤ Brett & Laura ➤ Nate

This is truly the distributive property in action. If we were to use letters instead of names, it might look as follows. (Notice the use of a plus sign for &.)

$$\begin{aligned} &\text{L}(\text{J} + \text{B} + \text{N}) \\ &= \text{LJ} + \text{LB} + \text{LN} \end{aligned}$$

Continuing with the analogy, let's look at what happens when we include both of the boys' parents.

(Rob & Laura ➤) (Josh & Brett & Nate)

What this means is that Rob gives a gift to each son and Laura gives a gift to each son. That's six gifts in all. In a sense, Rob *distributes* his gifts and then Laura *distributes* her gifts.

Rob ➤ Josh & Rob ➤ Brett & Rob ➤ Nate & Laura ➤ Josh & Laura ➤ Brett & Laura ➤ Nate

If we were to look at this more algebraically, we might write it as $(\text{R} + \text{L})(\text{J} + \text{B} + \text{N})$.

$$(\text{R} + \text{L})(\text{J} + \text{B} + \text{N})$$

$$= \text{RJ} + \text{RB} + \text{RN} + \text{LJ} + \text{LB} + \text{LN}$$

The point of all of this is that each member of the first polynomial (the givers) distributes to each member of the second polynomial (the receivers). Notice that nowhere do we see that Rob gives a gift to Laura. (It's not *her* birthday.) We also see that since there are two givers and three receivers, there is a total of six gifts.

Let's put that analogy to work in two polynomials where the repeated action is multiplication. Consider multiplying $(2x + 3)$ times $(6x^2 + 4x + 5)$. In this example, the first polynomial is a binomial; it has *two terms* (somewhat like Rob & Laura). The second polynomial is a trinomial; it has *three terms* (somewhat like Josh & Brett & Nate).

When we multiply $(2x + 3)$ times $(6x^2 + 4x + 5)$, we get six individual products and six terms (just like the six gifts). The first term of the binomial, $2x$, will multiply to each term in the trinomial; and the second term of the binomial, $+3$, will multiply to each term in the trinomial.

$$(2x + 3)(6x^2 + 4x + 5)$$

Six products: $2x \cdot 6x^2 + 2x \cdot 4x + 2x \cdot 5 + 3 \cdot 6x^2 + 3 \cdot 4x + 3 \cdot 5$

Six terms: $= 12x^3 + 8x^2 + 10x + 18x^2 + 12x + 15$

Notice that just as Rob doesn't give Laura a gift, we do not multiply the individual terms of the binomial, the $2x$ and the 3, together.

To complete this multiplication, we can put the terms in descending order and combine like terms.

$$= 12x^3 + 8x^2 + 18x^2 + 10x + 12x + 15$$
$$= 12x^3 + 26x^2 + 22x + 15$$

When multiplying two trinomials, how many terms will be in the product before simplifying? Explain your answer or show an example that supports your answer.

____

____

Example 6

Multiply. Write the answer in descending order and combine like terms. Check the answer to verify that it is correct.

$$(3w - 2)(w^2 + 5w - 4)$$

Procedure Multiply each term in the first polynomial by each term in the second polynomial. It's helpful to decide how many terms the product will have after multiplying (and before combining like terms).

Because there is a binomial (2 terms) and a trinomial (3 terms), initially there will be six individual products that become six individual terms.

Answer

$(3w - 2)(w^2 + 5w - 4)$	This will multiply out to become six *products*.
$= (3w) \cdot (w^2) + (3w) \cdot (5w) + (3w) \cdot (-4) +$ $(-2) \cdot (w^2) + (-2) \cdot (5w) + (-2) \cdot (-4)$	From these six products we get six *new terms*.
$= 3w^3 + 15w^2 - 12w - 2w^2 - 10w + 8$	Write the terms in descending order.
$= 3w^3 + 15w^2 - 2w^2 - 12w - 10w + 8$	Combine like terms.
$= 3w^3 + 13w^2 - 22w + 8$	Check. Replace w with 1.

Original Product $\stackrel{?}{=}$ **Answer**

$$(3-2)(1+5-4) \stackrel{?}{=} 3+13-22+8$$
$$(1)(2) \stackrel{?}{=} 2$$
$$2 = 2 \quad \text{True} \checkmark$$

Now mentally check the exponents for accuracy.

We also can multiply $(3w-2)(w^2+5w-4)$ *vertically*, as demonstrated next, according to these guidelines.

1. Distribute $3w$ through to each term in the trinomial and place the three terms on the first line in descending order.
2. Distribute -2 through to each term in the trinomial and place the three terms on the second line in descending order. Align like terms *vertically*, as shown.
3. Combine like terms by adding vertically.

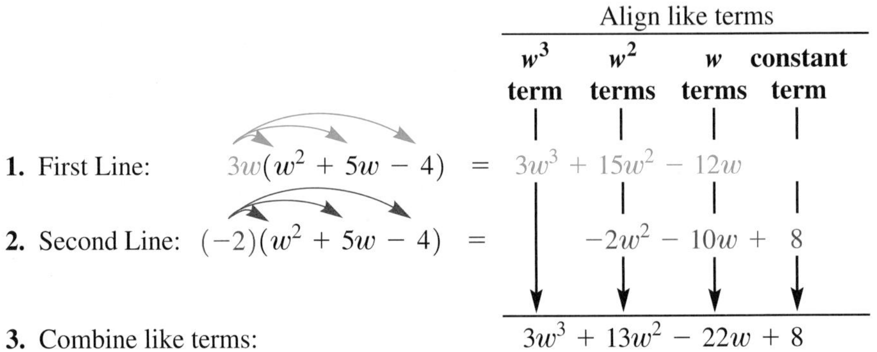

You Try It 6 **Multiply. Write each answer in descending order and combine like terms. Check all answers to verify that they are correct. Use Example 6 as a guide.**

a) $(p+2)(p^2+3p+1)$ **b)** $(y-2)(2y^2-y+5)$

Multiplying Two Binomials

The process for multiplying two binomials is the same as the process for multiplying a binomial and a trinomial. The only difference is that, initially, there will be four terms instead of six.

For example, to find the product of $(4x-3)(5x-2)$, we must multiply each term in the first binomial by each term in the second binomial. Here is a diagram of the distribution process for the product of these two binomials.

$$(4x-3)(5x-2)$$
$$= (4x)\cdot(5x) + (4x)\cdot(-2) + (-3)\cdot(5x) + (-3)\cdot(-2)$$
$$= 20x^2 - 8x - 15x + 6$$
$$= 20x^2 - 23x + 6$$

Example 7

Multiply. Write each answer in descending order and combine like terms. Check all answers to verify that they are correct.

a) $(3x+5)(2x+1)$ **b)** $(x-5)(2x^2-4)$ **c)** $(2x-3)(2x+3)$

Procedure Because each product consists of two binomials (2 terms each), the initial product should have *four* terms. *Think through the previous diagram and multiply directly.*

Answer **a)** $(3x + 5)(2x + 1)$

You do the check: ______

$= (3x)\cdot(2x) + (3x)\cdot(1) + (5)\cdot(2x) + (5)\cdot(1)$

$= 6x^2 + 3x + 10x + 5$ This polynomial is already in descending order; it has one pair of like terms.

$= 6x^2 + 13x + 5$

Answer **b)** $(x - 5)(2x^2 - 4)$

You do the check: ______

$= (x)\cdot(2x^2) + (x)\cdot(-4) + (-5)\cdot(2x^2) + (-5)\cdot(-4)$

$= 2x^3 - 4x - 10x^2 + 20$ Put the polynomial in descending order; there are no like terms.

$= 2x^3 - 10x^2 - 4x + 20$

Answer **c)** $(2x - 3)(2x + 3)$

You do the check: ______

$= (2x)\cdot(2x) + (2x)\cdot(3) + (-3)\cdot(2x) + (-3)\cdot(3)$

$= 4x^2 + 6x - 6x - 9$ The middle two terms are like terms and can be combined.

$= 4x^2 + 0x - 9$ Because the like terms are opposites, they combine to 0; the $0x$ term is not required in the final answer.

$= 4x^2 - 9$

You Try It 7 **Multiply. Write each answer in descending order and combine like terms. Check all answers to verify that they are correct. Use Example 7 as a guide.**

a) $(3x + 4)(x - 2)$ **b)** $(b^2 + 5)(b - 3)$

c) $(2c^2 - 3)(5c^2 - 1)$ **d)** $(9k + 1)(9k - 1)$

Example 8

Multiply. Write each answer in descending order and combine like terms. Check all answers to verify that they are correct.

a) $(x + 9)^2$ **b)** $(2y - 5)^2$

Procedure First, write each expression as the product of two binomials, then multiply.

Answer

a) $(x + 9)^2$

$= (x + 9)(x + 9)$

$= x^2 + 9x + 9x + 81$

$= x^2 + 18x + 81$

Check. Replace x with 1.

Original Product $\stackrel{?}{=}$ Answer

$(1 + 9)^2 \stackrel{?}{=} 1 + 18 + 81$

$(10)^2 \stackrel{?}{=} 100$

$100 \stackrel{?}{=} 100$ True ✓

Now mentally check the exponents for accuracy.

b) $(2y - 5)^2$

$= (2y - 5)(2y - 5)$

$= 4y^2 - 10y - 10y + 25$

$= 4y^2 - 20y + 25$

Check. Replace y with 1.

Original Product $\stackrel{?}{=}$ Answer

$(2 - 5)^2 \stackrel{?}{=} 4 - 20 + 25$

$(-3)^2 \stackrel{?}{=} 9$

$9 \stackrel{?}{=} 9$ True ✓

Now mentally check the exponents for accuracy.

You Try It 8 **Multiply. Write each answer in descending order and combine like terms. Check all answers to verify that they are correct. Use Example 8 as a guide.**

a) $(w - 10)^2$ **b)** $(9y + 1)^2$ **c)** $(3x - 4)^2$

Answers: You Try It and Think About It

You Try It: **1. a)** 5 **b)** −5 **c)** 3 **d)** −6 **2. a)** $-10 = -10$ **True** **b)** $+3 = 0$ **False** **c)** $-28 = -29$ **False** **d)** $0 = 0$ **True** **3. a)** $16m^7$ **b)** $-9p^7$ **c)** $12y^5$ **d)** $-w^6$ **4. a)** $25c^6$ **b)** $64w^8$ **c)** $81x^{10}$ **5. a)** $16c^5 + 6c^3$ **b)** $-3q^6 + 15q^5 - 18q^4$ **c)** $8a^7 - 6a^4 + 2a^3$ **d)** $8x^5 - 12x^4 + 4x^3 - 4x^2$ **6. a)** $p^3 + 5p^2 + 7p + 2$ **b)** $2y^3 - 5y^2 + 7y - 10$ **7. a)** $3x^2 - 2x - 8$ **b)** $b^3 - 3b^2 + 5b - 15$ **c)** $10c^4 - 17c^2 + 3$ **d)** $81k^2 - 1$ **8. a)** $w^2 - 20w + 100$ **b)** $81y^2 + 18y + 1$ **c)** $9x^2 - 24x + 16$

Think About It: **1.** Before simplifying, there will be 9 terms in the product because $3 \times 3 = 9$.

Section 7.4 Exercises

FOR EXTRA HELP
Student Resources on DVD-ROM
Includes
➤ Student's Solutions Manual
➤ Video Lectures
➤ Chapter Test Prep Video
MyMathLab
MathXL

Think Again

1. When multiplying two trinomials, how many terms will be in the product before simplifying? Explain your answer or show an example that supports your answer. (Refer to *Think About It 1*)
   There will be nine terms in the product. Three terms times three terms equals nine terms before simplifying.

2. In some cases, the product of two binomials simplifies to only two terms. For example, $(x + 3)(x - 3) = x^2 - 3x + 3x - 9 = x^2 - 9$. Is it possible that the product of two binomials will ever simplify to just one term? Explain your answer or show an example that supports your answer.
   No. It is not possible that the first term and the last term of a binomial product will be like terms; so, they will not be able to combine to one term.

3. This section provides a technique for checking the results of multiplying polynomials by replacing the variable with 1 in the original problem and the resulting product. This technique, though, does not check for the accuracy of the exponents. Why?
   Answers will vary. One possibility: When replacing the variable, such as x, with 1, it doesn't matter what the exponent is $1^2 = 1$, $1^3 = 1$, $1^4 = 1$, and so on. So, we'd never know if the exponent was correct using this technique.

4. When squaring a binomial, is it okay to square just the first and last terms? Explain your answer or show an example that supports your answer.
   No. When squaring a binomial such as $(x - 4)^2$, there will always be a third (middle) term in addition to the square of each term within the binomial.

Focus Exercises

Multiply. Check each answer to verify that it is correct.

5. $(3y^2)(2y^3)$
   $6y^5$
6. $(6n^5)(4n^4)$
   $24n^9$
7. $(-3b)(b^2)$
   $-3b^3$
8. $(5x^4)(-x)$
   $-5x^5$
9. $(-9k^4)(-2k^2)$
   $18k^6$
10. $(-3m^3)(-5m^5)$
   $15m^8$
11. $(x)(-x^6)$
   $-x^7$
12. $(y^7)(-y)$
   $-y^8$
13. $(4v^4)^2$
   $16v^8$
14. $(7a^5)^2$
   $49a^{10}$
15. $(-p^3)^2$
   p^6
16. $(-d)^2$
   d^2
17. $(3c^2)^2$
   $9c^4$
18. $(4x^7)^2$
   $16x^{14}$
19. $(-2n^6)^2$
   $4n^{12}$
20. $(-8q^8)^2$
   $64q^{16}$

Multiply. Write each answer in descending order. Check all answers to verify that they are correct.

21. $5q(2q^3 - 6)$
   $10q^4 - 30q$
22. $3n(4n^4 - 8)$
   $12n^5 - 24n$
23. $-h^2(h^3 + 4h)$
   $-h^5 - 4h^3$
24. $-c^3(c^3 + 8c)$
   $-c^6 - 8c^4$
25. $-2v(5v^3 - 10v)$
   $-10v^4 + 20v^2$
26. $-5d(5d^5 - 9d)$
   $-25d^6 + 45d^2$
27. $x^3(4x^2 - 5x + 9)$
   $4x^5 - 5x^4 + 9x^3$
28. $m^2(3m^4 - m^3 - 7)$
   $3m^6 - m^5 - 7m^2$
29. $-3y(2y^5 - y^2 + 4y)$
   $-6y^6 + 3y^3 - 12y^2$
30. $-8p^2(2p^2 + 5p - 2)$
   $-16p^4 - 40p^3 + 16p^2$
31. $-a^2(-a^2 - a + 9)$
   $a^4 + a^3 - 9a^2$
32. $-x^3(-x^3 + x - 10)$
   $x^6 - x^4 + 10x^3$

Multiply. Write each answer in descending order and combine like terms. Check all answers to verify that they are correct.

33. $(w + 2)(w^2 + 5w + 3)$
$w^3 + 7w^2 + 13w + 6$

34. $(m - 1)(2m^2 + m - 4)$
$2m^3 - m^2 - 5m + 4$

35. $(4c + 2)(c^2 - 5c - 1)$
$4c^3 - 18c^2 - 14c - 2$

36. $(3x - 2)(5x^2 - 4x + 6)$
$15x^3 - 22x^2 + 26x - 12$

37. $(4x + 3)(x^2 - 3x - 2)$
$4x^3 - 9x^2 - 17x - 6$

38. $(q^2 - 3)(2q + 3 - q^2)$
$-q^4 + 2q^3 + 6q^2 - 6q - 9$

39. $(m + 4)(m + 7)$
$m^2 + 11m + 28$

40. $(v - 8)(v - 5)$
$v^2 - 13v + 40$

41. $(x + 3)(x - 12)$
$x^2 - 9x - 36$

42. $(c - 2)(c + 6)$
$c^2 + 4c - 12$

43. $(4x + 2)(x - 3)$
$4x^2 - 10x - 6$

44. $(9x - 2)(3x + 1)$
$27x^2 + 3x - 2$

45. $(q + 3)(q + 8)$
$q^2 + 11q + 24$

46. $(m - 2)(m - 6)$
$m^2 - 8m + 12$

47. $(m - 6)(m + 6)$
$m^2 - 36$

48. $(y + 5)(y - 5)$
$y^2 - 25$

49. $(3c - 1)(4c - 3)$
$12c^2 - 13c + 3$

50. $(-2x - 3)(5x - 4)$
$-10x^2 - 7x + 12$

51. $(2p + 9)(2p - 9)$
$4p^2 - 81$

52. $(-4m + 1)(-4m - 1)$
$16m^2 - 1$

53. $(4x^2 - 3)(4x^2 + 3)$
$16x^4 - 9$

54. $(5v^3 - 1)(5v^3 + 1)$
$25v^6 - 1$

55. $(x + 7)^2$
$x^2 + 14x + 49$

56. $(w - 9)^2$
$w^2 - 18w + 81$

57. $(y - 1)^2$
$y^2 - 2y + 1$

58. $(2x + 3)^2$
$4x^2 + 12x + 9$

59. $(3v + 5)^2$
$9v^2 + 30v + 25$

60. $(4m - 1)^2$
$16m^2 - 8m + 1$

Think Outside the Box

Multiply. Write each answer in descending order and combine like terms. Check all answers to verify that they are correct.

61. $(5y^2 - 6)^2$
$25y^4 - 60y^2 + 36$

62. $(2p^2 + 3p - 1)^2$
$4p^4 + 12p^3 + 5p^2 - 6p + 1$

Simplify each polynomial answer and write it in descending order.

63. Consider the rectangle that has a length given by $(-5x + 1)$ yards and a width given by $(3x + 8)$ yards.

$(-5x + 1)$ yd

$(3x + 8)$ yd

a) Find the area of the rectangle.
The area is $-15x^2 - 37x + 8$ square yards.

b) Is it possible for x to be -2? Explain your answer.
Yes. When $x = -2$, the length is 11 yards and the width is 2 yards.

64. Consider the square that has a side length given by $(3 - 4x)$ feet.

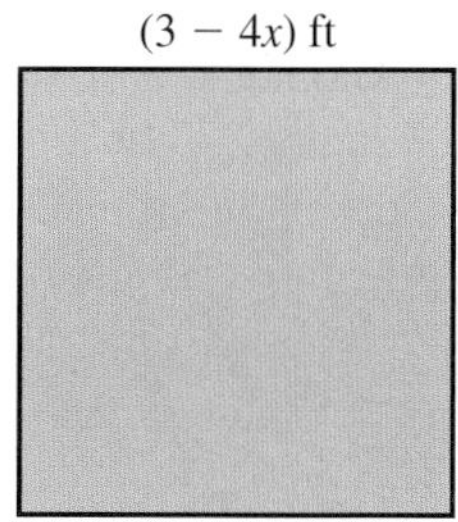

a) Find the area of the square.
The area is $16x^2 - 24x + 9$ square feet.

b) Is it possible for x to be 2? Explain your answer.
No. When $x = 2$, the side length is -5 yards, which is not possible.

SECTION 7.5 Dividing Polynomials

OBJECTIVES

In this section, you will learn to

- Apply the Quotient Rule for Exponents.
- Divide a monomial by a monomial.
- Divide a polynomial by a monomial.

You Need to Know

To successfully complete this section, you need to understand

- ☐ Dividing signed numbers (2.4)
- ☐ Simplifying fractions (4.3)
- ☐ The Product Rule for Exponents (7.1)
- ☐ Terms in a polynomial (7.2)

Introduction

We now turn our attention to the last of the four basic operations, division of polynomials. In this section, you are presented only with division by a monomial. Dividing by a binomial is a completely different process and is introduced in Elementary Algebra.

In preparation for dividing by a monomial, we first look at another rule of exponents called the Quotient Rule for Exponents. You will see that it is similar to the Product Rule for Exponents discussed in Section 7.1.

The Quotient Rule for Exponents

A product is an expression of multiplication; a **quotient** is an expression of *division*. Often, a quotient is expressed as a fraction, and we know that we can simplify fractions by dividing out common factors in the numerator and the denominator.

For example, $\frac{24}{30}$ can reduce by a factor of 6: $\frac{24 \div 6}{30 \div 6} = \frac{4}{5}$. We also can use prime factorization to simplify. Writing the numerator and the denominator in prime factored form allows us to see the common factors that can be divided out.

$$\frac{24}{30} = \frac{2 \cdot 2 \cdot 2 \cdot 3}{2 \cdot 3 \cdot 5} = \frac{\overset{1}{\cancel{2}} \cdot 2 \cdot 2 \cdot \overset{1}{\cancel{3}}}{\underset{1}{\cancel{2}} \cdot \underset{1}{\cancel{3}} \cdot 5} = \frac{1 \cdot 2 \cdot 2 \cdot 1}{1 \cdot 1 \cdot 5} = \frac{4}{5}$$

We can do the same for a quotient such as $\frac{x^5}{x^3}$. If we write it in expanded form, we'll be able to see that many of the factors divide out.

$$\frac{x^5}{x^3} = \frac{x \cdot x \cdot x \cdot x \cdot x}{x \cdot x \cdot x} = \frac{\overset{1}{\cancel{x} \cdot \cancel{x} \cdot \cancel{x}} \cdot x \cdot x}{\underset{1}{\cancel{x} \cdot \cancel{x} \cdot \cancel{x}}} = \frac{x \cdot x}{1} = x^2$$

We can divide out only *common* factors; and because the bases are the same (both x) the numerator and the denominator have plenty of common factors to divide out.

Another way to show the process of dividing out is to group the same number of common factors in the numerator and the denominator. For example, with $\frac{x^5}{x^3}$ we can expand the numerator and the denominator and then group three factors of x in each. When we divide out the three factors of x in the denominator, we are left with 1.

$$\frac{x^5}{x^3} = \frac{x \cdot x \cdot x \cdot x \cdot x}{x \cdot x \cdot x} = \frac{\overset{1}{\cancel{(x \cdot x \cdot x)}} \cdot x \cdot x}{\underset{1}{\cancel{(x \cdot x \cdot x)}}} = \frac{x \cdot x}{1} = x^2$$

Example 1

Write the numerator and the denominator in expanded form. Then simplify the fraction by dividing out any common factors.

a) $\frac{y^8}{y^2}$ **b)** $\frac{w^7}{w^6}$ **c)** $\frac{x^3}{x^3}$ **d)** $\frac{c^5}{c^0}$

Procedure Expand the numerator and the denominator. Using parentheses, group the same number of factors in the numerator and the denominator. Simplify the fraction by dividing out each grouping.

Answer

a) $\frac{y^8}{y^2} = \frac{y \cdot y \cdot y \cdot y \cdot y \cdot y \cdot y \cdot y}{y \cdot y} = \frac{(y \cdot y) \cdot y \cdot y \cdot y \cdot y \cdot y \cdot y}{(y \cdot y)} = \frac{y \cdot y \cdot y \cdot y \cdot y \cdot y}{1} = y^6$

b) $\frac{w^7}{w^6} = \frac{w \cdot w \cdot w \cdot w \cdot w \cdot w \cdot w}{w \cdot w \cdot w \cdot w \cdot w \cdot w} = \frac{(w \cdot w \cdot w \cdot w \cdot w \cdot w) \cdot w}{(w \cdot w \cdot w \cdot w \cdot w \cdot w)} = \frac{w}{1}$ or w^1

c) $\dfrac{x^3}{x^3} = \dfrac{x \cdot x \cdot x}{x \cdot x \cdot x} = \dfrac{(x \cdot x \cdot x)}{(x \cdot x \cdot x)} = \dfrac{1}{1} = 1$ This says that x^3 divided by itself is 1.

d) $\dfrac{c^5}{c^0} = \dfrac{c \cdot c \cdot c \cdot c \cdot c}{1} = c \cdot c \cdot c \cdot c \cdot c = c^5$ Remember, $c^0 = 1$.

Think About It 1 We know that we are not allowed to divide by 0. Why then is it okay to divide by c^0, as in Example 1d?

____

____

You Try It 1 **Write the numerator and the denominator in expanded form. Then simplify the fraction by dividing out any common factors. Use Example 1 as a guide.**

a) $\dfrac{x^5}{x^4} =$ ______________________ = __________

b) $\dfrac{p^4}{p^0} =$ ______________________ = __________

c) $\dfrac{a^6}{a^6} =$ ______________________ = __________

d) $\dfrac{x^7}{x} =$ ______________________ = __________

Let's return to the example of simplifying $\frac{x^5}{x^3}$.

$$\frac{x^5}{x^3} = \frac{x \cdot x \cdot x \cdot x \cdot x}{x \cdot x \cdot x} = \frac{(x \cdot x \cdot x) \cdot x \cdot x}{(x \cdot x \cdot x)} = \frac{x \cdot x}{1} = x^2$$

In this example, the numerator has five factors of x and the denominator has three factors of x. In simplifying the fraction, all three of the denominator factors divide out with three of the numerator factors, leaving two factors of x in the numerator. It's as if we were eliminating three factors of x. This process suggests a subtraction.

five "take away" three is two

$$5 - 3 = 2$$

Of course, in this process, we are *subtracting* the exponents, just as we are *dividing out* the number of common factors. The rule that supports this type of dividing out is called the **Quotient Rule for Exponents.**

The Quotient Rule for Exponents

For any nonzero base, x,

$$\frac{x^a}{x^b} = x^{a-b}.$$

Caution The Quotient Rule for Exponents can be applied only when the base is the same in the numerator and the denominator. When dividing, we subtract the *exponents*.

Example 2

Use the Quotient Rule to simplify each of these quotients.

a) $\dfrac{y^8}{y^2}$ b) $\dfrac{w^7}{w^6}$ c) $\dfrac{x^3}{x^3}$ d) $\dfrac{p^5}{p^0}$

Procedure Make sure the bases are the same. Then subtract the exponents to get the result.

Answer a) $\dfrac{y^8}{y^2} = y^{8-2} = y^6$ b) $\dfrac{w^7}{w^6} = w^{7-6} = w^1$ or just w

c) $\dfrac{x^3}{x^3} = x^{3-3} = x^0 = 1$ d) $\dfrac{p^5}{p^0} = p^{5-0} = p^5$

The Zero Power Rule, Revisited

Refer to Example 2c $\frac{x^3}{x^3}$. The Quotient Rule for Exponents tells us that this is $x^{3-3} = x^0$. However, $\frac{x^3}{x^3}$ is also one value divided by itself, which is equal to 1. This must mean that

$$x^0 = 1, x \neq 0.$$

Think About It 2 In the Zero Power Rule, Revisited box, why does it say $x \neq 0$?

You Try It 2 **Use the Quotient Rule for Exponents to simplify each of these quotient. Use Example 2 as a guide.**

a) $\dfrac{c^6}{c^1}$ b) $\dfrac{p^4}{p^0}$ c) $\dfrac{a^6}{a^6}$

d) $\dfrac{v^9}{v^3}$ e) $\dfrac{m^8}{m^4}$ f) $\dfrac{k^8}{k^7}$

Dividing Two Monomials

Recall from Section 4.4 that we can separate one fraction into the product of two fractions: $\dfrac{a \cdot c}{b \cdot d} = \dfrac{a}{b} \cdot \dfrac{c}{d}$. (This is the reverse of the multiplication of two fractions.)

If a fraction has a monomial in the numerator and the denominator, we can separate the fraction so that the coefficients can be divided separately from the variables. For the variables, we can use the Quotient Rule for Exponents.

Example 3

Simplify each fraction completely.

a) $\dfrac{10y^5}{-2y^2}$ b) $\dfrac{-6p^5}{6p^4}$ c) $\dfrac{30x^3}{5x^3}$

Procedure Separate the fraction into the product of two fractions, separating the coefficients from the variables, and divide separately.

Answer a) $\dfrac{10y^5}{-2y^2} = \dfrac{10}{-2} \cdot \dfrac{y^5}{y^2} = -5 \cdot y^{5-2} = -5y^3$

b) $\dfrac{-6p^5}{6p^4} = \dfrac{-6}{6} \cdot \dfrac{p^5}{p^4} = -1 \cdot p^{5-4} = -1p^1$ or $-p$

c) $\dfrac{30x^3}{5x^3} = \dfrac{30}{5} \cdot \dfrac{x^3}{x^3} = 6 \cdot x^{3-3} = 6x^0 = 6$

You Try It 3 **Simplify each fraction completely. Use Example 3 as a guide.**

a) $\dfrac{12y^2}{6y^2}$ b) $\dfrac{14x^3}{-7x}$ c) $\dfrac{3c^4}{3c}$ d) $\dfrac{-8m^6}{-2m}$

Dividing a Polynomial by a Monomial

Just as we can distribute using multiplication, as in

$$\begin{aligned} &2(x + 5) \\ &= 2 \cdot x + 2 \cdot 5 \\ &= 2x + 10 \end{aligned}$$

We also can distribute using division. For example, consider $(8 + 6) \div 2$ (which is $14 \div 2 = 7$)

$$\begin{aligned} &(8 + 6) \div 2 \\ &= 8 \div 2 + 6 \div 2 \\ &= 4 + 3 \\ &= 7 \end{aligned}$$

This division, and others like it, also can be expressed as a fraction; the distribution is the same.

$$\begin{aligned} &\frac{8 + 6}{2} \\ &= \frac{8}{2} + \frac{6}{2} \\ &= 4 + 3 \\ &= 7 \end{aligned}$$

As you can see, in this type of distribution, the denominator is distributed to every term in the numerator.

This type of distribution also can be called **separating the fraction**.

This is more evident when we see a polynomial such as $15x^3 - 10x^2$ divided by $5x$. Written as a fraction this becomes

$$\frac{15x^3 - 10x^2}{5x}$$

Distribute the denominator to both terms in the numerator: $= \dfrac{15x^3}{5x} - \dfrac{10x^2}{5x}$

Prepare to simplify by separating each fraction: $= \dfrac{15}{5} \cdot \dfrac{x^3}{x} - \dfrac{10}{5} \cdot \dfrac{x^2}{x}$

Simplify: $= 3x^2 - 2x$

Example 4

Divide using distribution.

a) $\dfrac{12y^5 - 8y^2}{4y^2}$ b) $\dfrac{21m^3 + 9m^2 - 6m}{-3m}$

Procedure Distribute the denominator and then simplify each fraction.

Answer a) $\dfrac{12y^5 - 8y^2}{4y^2}$

$= \dfrac{12y^5}{4y^2} - \dfrac{8y^2}{4y^2}$

$= 3y^3 - 2$

b) $\dfrac{21m^3 + 9m^2 - 6m}{-3m}$

$= \dfrac{21m^3}{-3m} + \dfrac{9m^2}{-3m} - \dfrac{6m}{-3m}$

$= -7m^2 - 3m + 2$

Think About It 3 In the answer to Example 4b, the constant term is +2. Why is it positive?

____

____

You Try It 4 **Divide using distribution. Use Example 4 as a guide.**

a) $\dfrac{6x^4 + 12x^2}{6x}$ b) $\dfrac{4p^7 - 2p^6}{-2p^4}$ c) $\dfrac{24m^6 - 4m^4 + 16m^2}{4m^2}$

Answers: You Try It and Think About It

You Try It: **1. a)** $\dfrac{x^5}{x^4} = \dfrac{x \cdot x \cdot x \cdot x \cdot x}{x \cdot x \cdot x \cdot x} = \dfrac{(x \cdot x \cdot x \cdot x) \cdot x}{(x \cdot x \cdot x \cdot x)} = x$ **b)** $\dfrac{p^4}{p^0} = \dfrac{p \cdot p \cdot p \cdot p}{1} = p^4$ **c)** $\dfrac{a^6}{a^6} = \dfrac{a \cdot a \cdot a \cdot a \cdot a \cdot a}{a \cdot a \cdot a \cdot a \cdot a \cdot a} = \dfrac{(a \cdot a \cdot a \cdot a \cdot a \cdot a)}{(a \cdot a \cdot a \cdot a \cdot a \cdot a)} = 1$ **d)** $\dfrac{x^7}{x^1} = \dfrac{x \cdot x \cdot x \cdot x \cdot x \cdot x \cdot x}{x} = \dfrac{(x) \cdot x \cdot x \cdot x \cdot x \cdot x \cdot x}{(x)} = x^6$ **2. a)** c^5 **b)** p^4 **c)** $a^0 = 1$ **d)** v^6 **e)** m^4 **f)** k^1 or just k **3. a)** 2 **b)** $-2x^2$ **c)** c^3 **d)** $4m^5$ **4. a)** $x^3 + 2x$ **b)** $-2p^3 + p^2$ **c)** $6m^4 - m^2 + 4$

Think About It: **1.** It is okay to divide by c^0 because $c^0 = 1$, and it is okay to divide by 1. **2.** If x is 0, then x^3 becomes $0^3 = 0 \cdot 0 \cdot 0 = 0$, and it is not okay to divide by 0. **3.** The constant term is positive because it came from dividing $-6m$ by $-3m$, and a negative divided by a negative is positive.

Section 7.5 Exercises

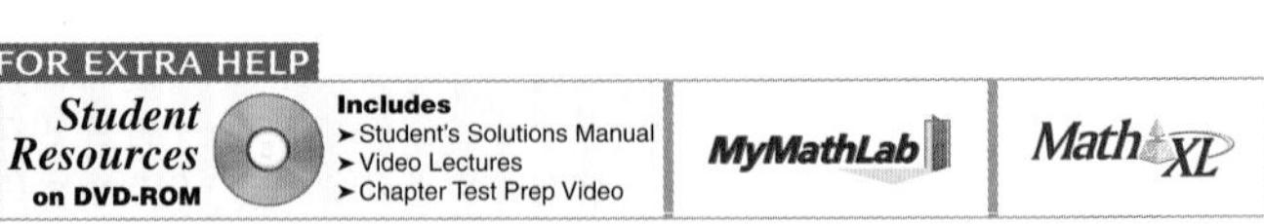

Think Again

1. We know that we are not allowed to divide by 0. Why then is it okay to divide by c^0? *(Refer to Think About It 1)*

   $c^0 = 1$, and it is okay to divide by 1.

2. Use the Quotient Rule to show why $x^0 = 1$. You may use $\frac{x^4}{x^4}$ or your own example in the explanation.

   $\frac{x^4}{x^4}$ can be simplified two ways, each leading to the same result:

   (a) $\frac{x^4}{x^4} = x^{4-4} = x^0$ (b) $\frac{x^4}{x^4} = \frac{x \cdot x \cdot x \cdot x}{x \cdot x \cdot x \cdot x} = 1$

   So, it must be that $x^0 = 1$.

Focus Exercises

Write each quotient as one base with one exponent. You may expand and count the number of factors or use the Quotient Rule for Exponents.

3. $\frac{x^9}{x^5}$ x^4 **4.** $\frac{w^6}{w^3}$ w^3 **5.** $\frac{c^5}{c^1}$ c^4 **6.** $\frac{p^3}{p^0}$ p^3

7. $\frac{y^8}{y}$ y^7 **8.** $\frac{a^7}{a^7}$ $a^0 = 1$ **9.** $\frac{m^{12}}{m^6}$ m^6 **10.** $\frac{x^3}{x^2}$ x^1

11. $\frac{x}{x}$ $x^0 = 1$ **12.** $\frac{c^3}{c^3}$ $c^0 = 1$ **13.** $\frac{y^{10}}{y^2}$ y^8 **14.** $\frac{m^{21}}{m^7}$ m^{14}

15. $\frac{x^4}{x}$ x^3 **16.** $\frac{x^5}{x^4}$ x^1 **17.** $\frac{w^8}{w^0}$ w^8 **18.** $\frac{y^9}{y^7}$ y^2

Simplify each fraction.

19. $\frac{20x^5}{5x^2}$ $4x^3$ **20.** $\frac{16m^7}{2m^3}$ $8m^4$ **21.** $\frac{12y^7}{3y^6}$ $4y$ **22.** $\frac{24q^3}{8q^2}$ $3q$

23. $\frac{-15p^2}{3p^2}$ -5 **24.** $\frac{-30w^4}{6w^4}$ -5 **25.** $\frac{18c^3}{-6c}$ $-3c^2$ **26.** $\frac{28v^5}{-4v}$ $-7v^4$

27. $\frac{-10x^6}{-2x^2}$ $5x^4$ **28.** $\frac{-21d^8}{-7d^4}$ $3d^4$ **29.** $\frac{36y^{12}}{-6y^3}$ $-6y^9$ **30.** $\frac{25w^9}{-5w^3}$ $-5w^6$

Divide using distribution.

31. $\frac{5x^4 + 15x^2}{5x}$ $x^3 + 3x$ **32.** $\frac{6y^6 - 21y^2}{3y^2}$ $2y^4 - 7$

33. $\frac{27w^7 - 9w^3}{-9w^3}$ $-3w^4 + 1$ **34.** $\frac{16k^{12} - 9k^4}{-4k^4}$ $-4k^8 + \frac{9}{4}$

35. $\frac{24m^9 - 3m^6 + 36m^3}{3m^3}$ $8m^6 - m^3 + 12$ **36.** $\frac{25y^7 - 10y^5 - 5y^4}{5y^3}$ $5y^4 - 2y^2 - y$

37. $\frac{6a^4 - 12a^3 + 3a^2}{6a^2}$ $a^2 - 2a + \frac{1}{2}$ **38.** $\frac{-14d^4 - 2d^3 - 10d^2}{-2d}$ $7d^3 + d^2 + 5d$

39. $\frac{3y^4 - 15y^3 + 6y^2}{3y^2}$ $y^2 - 5y + 2$ **40.** $\frac{8v^8 - 16v^4 + 4v^2}{-8v^2}$ $-v^6 + 2v^2 - \frac{1}{2}$

41. $\frac{-14c^3 - 21c^2 + 7c}{-7c}$ $2c^2 + 3c - 1$ **42.** $\frac{6x^6 + 30x^4 - 24x^2}{-6x^2}$ $-x^4 - 5x^2 + 4$

Think Outside the Box

Write each quotient as one base with one exponent, by (a) expanding and counting the number of factors and (b) using the Quotient Rule for Exponents.

43. $\frac{c^4}{c^5}$ (a) $\frac{c^4}{c^5} = \frac{c \cdot c \cdot c \cdot c}{c \cdot c \cdot c \cdot c \cdot c} = \frac{1}{c}$
(b) $\frac{c^4}{c^5} = c^{4-5} = c^{-1}$

44. $\frac{x^5}{x^9}$ (a) $\frac{x^5}{x^9} = \frac{x \cdot x \cdot x \cdot x \cdot x}{x \cdot x \cdot x \cdot x \cdot x \cdot x \cdot x \cdot x \cdot x} = \frac{1}{x^4}$
(b) $\frac{x^5}{x^9} = x^{5-9} = x^{-4}$

Consider negative exponents.

45. Write a rule that includes the results of Exercise 43.
$c^{-1} = \frac{1}{c}$

46. Write a rule that includes the results of Exercise 44.
$x^{-4} = \frac{1}{x^4}$ or $x^{-a} = \frac{1}{x^a}$

SECTION 7.6 Factoring Polynomials: The Greatest Common Factor

OBJECTIVES

In this section, you will learn to
- Write the prime factorization of a term.
- Find the Greatest Common Factor of two terms and three terms.
- Factor out the Greatest Common Factor of a binomial and a trinomial.
- Factor out the Greatest Common Factor when the first term is negative.

You Need to Know

To successfully complete this section, you need to understand
- ☐ Prime factorization (1.6)
- ☐ The Greatest Common Factor (4.1)
- ☐ The Distribution Property (7.4)
- ☐ Dividing by a monomial (7.5)

Introduction

To this point, we have added, subtracted, multiplied, and divided polynomials (by a monomial), the four basic operations. There is, however, one more important aspect to polynomials: *factoring*.

Factoring polynomials is related to multiplication and division of polynomials. Before we begin, though, it's important to review some of the definitions related to factoring and multiplication.

Recall from Section 1.6 that a number is in **factored form** when it is the product of two or more numbers. For Example, $6 \cdot 5$ is a factored form of 30; $2 \cdot 15$ is another factored form of 30.

Recall also that the prime factorization of a number is the product of all of its prime factors. For example, the prime factorization of 30 is $2 \cdot 3 \cdot 5$.

In this section, we will see polynomials put into a factored form.

The Prime Factorization of a Term

A term can be written in prime factorization by

1. Finding the prime factorization of the constant or coefficient.
2. Expanding the variable with an exponent.

For example, $30x$ can be factored into $2 \cdot 3 \cdot 5 \cdot x$ and $30x^3$ can be factored into $2 \cdot 3 \cdot 5 \cdot x \cdot x \cdot x$.

Example 1

Find the prime factorization of each term.

a) 35 **b)** $15x$ **c)** $7y^2$ **d)** $6w^3$

Procedure Write the prime factorization of the constant or coefficients and expand any variable with an exponent.

Answer **a)** $35 = 5 \cdot 7$ **b)** $15x = 3 \cdot 5 \cdot x$

c) $7y^2 = 7 \cdot y \cdot y$ **d)** $6w^3 = 2 \cdot 3 \cdot w \cdot w \cdot w$

You Try It 1 **Find the prime factorization of each term. Use Example 1 as a guide.**

a) 36 **b)** $8y$ **c)** $12x^2$ **d)** $45p^3$

The Greatest Common Factor

Finding the monomial factor that is common to each term is the same as finding the Greatest Common Factor (GCF). In Section 4.1, we discussed a technique for finding the GCF of two numbers using their prime factorization.

For example, to find the GCF of 36 and 54, we can write each number's prime factorization and match as many prime factors as possible.

$$\left.\begin{aligned} 36 &= 2 \cdot 2 \cdot 3 \cdot 3 \\ 54 &= 2 \cdot 3 \cdot 3 \cdot 3 \end{aligned}\right\}$$

36 and 54 have one 2 and two 3's in common; so, the GCF $= 2 \cdot 3 \cdot 3 = 18$.

We also learned that we can find the GCF by using the prime factorization written with exponents. Remember that we need to use the *lesser* of the two exponents on each base.

$$\left.\begin{aligned} 36 &= 2^2 \cdot 3^2 \\ 54 &= 2^1 \cdot 3^3 \end{aligned}\right\}$$

36 and 54 still have one 2 in common and two 3's in common; so, the GCF $= 2^1 \cdot 3^2 = 18$.

Think About It 1 When using exponents to determine the GCF, why do we choose the lesser exponent for each factor? (Hint: Compare the previous two methods.)

____

____

We also can use these methods for finding the GCF of two terms that have variables within them.

Example 2

Find the GCF of each pair of terms.

a) $6x^2$ and $15x$ **b)** $12y^3$ and $18y^2$

Procedure Write the prime factorization of each term and recognize any common factors. This process is shown using the two previous methods.

Answer First, with full prime factorization:

a) $6x^2 = 2 \cdot 3 \cdot x \cdot x$
$15x = 3 \cdot 5 \cdot x$
The GCF $= 3 \cdot x = 3x$.

b) $12y^3 = 2 \cdot 2 \cdot 3 \cdot y \cdot y \cdot y$
$18y^2 = 2 \cdot 3 \cdot 3 \cdot y \cdot y$
The GCF $= 2 \cdot 3 \cdot y \cdot y = 6y^2$.

Second, with exponents:

a) $6x^2 = 2^1 \cdot 3^1 \cdot x^2$
$15x = 3^1 \cdot 5^1 \cdot x^1$
The GCF $= 3^1 \cdot x^1 = 3x$.

b) $12y^3 = 2^2 \cdot 3^1 \cdot y^3$
$18y^2 = 2^1 \cdot 3^2 \cdot y^2$
The GCF $= 2^1 \cdot 3^1 \cdot y^2 = 6y^2$.

When a number or term has no common factors, the terms are *relatively prime* and the GCF is 1.

Example 3

Find the GCF of $8x^2$ and $15y$.

Procedure Write the prime factorization of each term and recognize any common factors.

Answer $8x^2 = 2 \cdot 2 \cdot 2 \cdot x \cdot x$
$15y = 3 \cdot 5 \cdot y$

There are no factors common to these two terms; so, the terms are relatively prime and the GCF = 1.

The greatest common factor can apply to three terms as well. However, any common factor must be common to all three terms.

For example, consider these terms: $6x^4$, $15x^3$, and $10x^2$.

$$6x^4 = 2 \cdot 3 \cdot x \cdot x \cdot x \cdot x$$
$$15x^3 = 3 \cdot 5 \cdot x \cdot x \cdot x$$
$$10x^2 = 2 \cdot 5 \cdot x \cdot x$$

Although 3 is a factor of the first two terms, $6x^4$ and $15x^3$, it is not a factor of the third term, $10x^2$. Likewise, although 5 is a common factor of the last two terms, it is not a factor of the first term.

In the case of $6x^4$, $15x^3$, and $10x^2$, the only common factors are found in the variables and the greatest common factor is the least exponent of the three: x^2.

Example 4

Find the GCF of these three terms: $8w^6$, $14w^4$, and $18w^3$.

Procedure Write the prime factorization of each term and recognize any common factors. The answer shows both processes used together.

Answer First, with full prime factorization:

$$8w^6 = 2 \cdot 2 \cdot 2 \cdot w^6$$ The only common numerical factor is 2.

$$14w^4 = 2 \cdot 7 \cdot w^4$$ The least value of the exponent is 3; so, there is also a factor of w^3 in the GCF.

$$18w^3 = 2 \cdot 3 \cdot 3 \cdot w^3$$ The GCF $= 2 \cdot w^3 = 2w^3$.

You Try It 2 **Find the GCF of each group of terms. Use Examples 2, 3, and 4 as guides.**

a) $12x^2$ and $40x$ **b)** $25m^4$ and $15m^3$ **c)** $5y^3$ and $12x^2$

d) $6c^5$, $15c^4$, and $9c^2$ **e)** $9p^5$, $12p^3$, and $8p$

Finding the GCF of two or more terms requires that we find the GCF of the numbers (the coefficients and the constant) separately from the variables. Of course, it's best if you can "see" the GCF without having to go through all of the work of prime factorization; but it's also nice to know that prime factorization is available.

For example, in the terms $50x^3$ and $30x$, you might be able to see a common factor of **10** (from the 50 and 30) and a common factor of $\boldsymbol{x}$ (the lesser of x^3 and x), thereby making the GCF $\mathbf{10x}$.

Example 5

Identify the GCF of the given terms by identifying it first for the numbers and then for the variables. Make sure you write the GCF, including all common factors.

a) $28x^2$ and $21x$ **b)** $25y^3$ and $12y^4$

c) $15v^4$, $12v^2$, and $33v^5$ **d)** $30c^2$, $60c$, and 20

Procedure Identify (1) the GCF of the numbers and (2) the GCF of the variables.

Answer

a) 1. The GCF of 28 and 21 is **7**.

2. The GCF of x^2 and x is x^1, or $\boldsymbol{x}$.

The GCF is $7x$.

b) 1. 25 and 12 are relatively prime; so, their GCF is **1**.

2. The GCF of y^3 and y^4 is $\boldsymbol{y^3}$.

The GCF is $\mathbf{1y^3}$, or $\boldsymbol{y^3}$.

c) 1. The GCF of 15, 12, and 33 is **3**.

2. The variable GCF is $\boldsymbol{v^2}$.

The GCF is $\mathbf{3v^2}$.

d) 1. The GCF of 30, 60, and 20 is **10**.

2. Not every term has a variable; so, there is no variable factor in the GCF.

The GCF is **10**.

You Try It 3 **Identify the GCF of the given terms by identifying it first for the numbers and then for the variables. Use Example 5 as a guide.**

a) $16x^2$ and $40x$ **b)** $30b^2$ and 45

c) $10y^3$ and $21y^5$ **d)** $9m^2$ and 16

e) $14c^2$, $21c^3$, and $35c$ **f)** $18p^4$, $15p^2$, and $10p^3$

Factoring Out the Greatest Common Factor

To **factor** a polynomial means to write it in a factored form, which is to write the polynomial as a product of its factors. The process of factoring is the opposite, or reverse, of distributing. In fact, we can check the accuracy of our factoring by multiplying the resulting factors.

To better understand the idea of factored form, consider these two examples, one from arithmetic and one from algebra.

Arithmetic	**Algebra**
$2 \cdot 7 = 14$	$3x \cdot (2x + 5) = 6x^2 + 15$
$2 \cdot 7$ is a factored form of 14.	$3x \cdot (2x + 5)$ is a factored form of $6x^2 + 15x$.

In other words, the binomial $6x^2 + 15x$ can be *factored* into $3x \cdot (2x + 5)$.

$$6x^2 + 15x = 3x \cdot (2x + 5)$$

In this case, we have factored out the greatest common factor, $3x$.

Here are the steps to factoring out the GCF.

1. Recognize the GCF. This can be done using any of the techniques presented earlier in this section. Here is the prime factorization method.

$$\text{Prime factorization:} \quad \overbrace{\underline{3} \cdot 2 \cdot \underline{x} \cdot x}^{6x^2} \quad \overbrace{\underline{3} \cdot 5 \cdot \underline{x}}^{15x} \quad \text{GCF} = 3x$$

2. Extract—pull out, factor out—the GCF from each term. The new binomial's terms are made up of the factors that are not a part of the GCF.

$$\overbrace{\underline{3} \cdot 2 \cdot \underline{x} \cdot x}^{6x^2} + \overbrace{\underline{3} \cdot 5 \cdot \underline{x}}^{15x}$$

$$\text{GCF} = 3x \quad \rightarrow \quad 3x(2x + 5)$$

3. Check for accuracy. This can be done by

a) Carefully multiplying the factored form to see if you get the original binomial.

$$3x \cdot (2x + 5) = 6x^2 + 15 \quad \text{True ✓}$$

(with $6x^2$ and $+15x$ shown above as the products)

b) Replacing each variable with 1 in the factored form and in the original.

$$3x \cdot (2x + 5) = 6x^2 + 15x$$
$$3 \cdot (2 + 5) = 6 + 15$$
$$3 \cdot (7) = 21$$
$$21 = 21 \quad \text{True ✓}$$

If the polynomial is a trinomial, we need to find the GCF of all three terms. For example, to factor $10x^3 - 15x^2 + 5x$, we must identify the GCF of $10x^3$, $15x^2$, and $5x$.

The GCF of 10, 15, and 5 is 5; and the least of the variable factors is x^1. So, the GCF is $5x$.

We can write the terms in a factored form with the GCF, $5x$, and then extract $5x$ from each term to get the factored form of the trinomial.

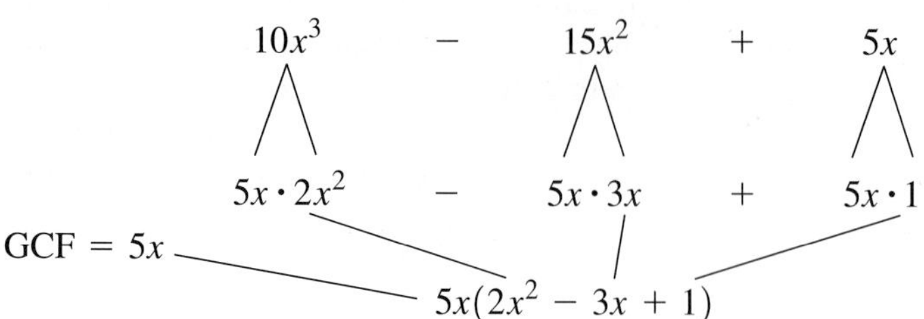

$10x^3$ − $15x^2$ + $5x$
$5x \cdot 2x^2$ − $5x \cdot 3x$ + $5x \cdot 1$
GCF = $5x$
$5x(2x^2 - 3x + 1)$

Again, we can check for accuracy by multiplying the factored form out or by replacing each variable with 1.

Caution When factoring out the GCF from a binomial, the quantity factor is a binomial. When factoring out the GCF from a trinomial, the quantity factor is a trinomial.

In a polynomial, when there is no common factor other than 1, we usually don't factor it and instead say that the polynomial is *prime*.

Example 6

Factor out the GCF from each polynomial. Check all answers for accuracy.

a) $3x + 21$ **b)** $4y^3 - 15y^2$ **c)** $15m^3 + 8$ **d)** $12c^4 - 8c^3 + 4c^2$

Procedure Identify the GCF. Write each term of the polynomial in a factored form, then extract the GCF.

Answer

a) $3x + 21$
$= 3 \cdot x + 3 \cdot 7$
$= 3(x + 7)$

Each term has a factor of 3.

Check: Replace each variable with 1.

$3(1) + 21 \stackrel{?}{=} 3(1 + 7)$
$3 + 21 \stackrel{?}{=} 3(8)$
$24 = 24$ True ✓

b) $4y^3 - 15y^2$
$= y^2 \cdot 4y - y^2 \cdot 15$
$= y^2(4y - 15)$

The numbers 4 and 15 are relatively prime. The least variable is y^2; so, the GCF is $1y^2$, or just y^2.

You do the check:

c) $15m^3 + 8$ is *prime*

The numbers 15 and 8 are relatively prime, and one term has no variable. So, the GCF is 1, and the binomial is prime.

When the polynomial is prime, there is nothing to check.

d) $12c^4 - 8c^3 + 4c^2$
$= 4c^2 \cdot 3c^2 - 4c^2 \cdot 2c + 4c^2 \cdot 1$
$= 4c^2(3c^2 - 2c + 1)$

Each term has a factor of 4. The least variable is c^2; so, the GCF is $4c^2$.

You do the check:

Caution Sometimes the GCF is the same as one of the terms in the polynomial, as in Example 6d. When that happens and we factor out the GCF, we are left with 1 or −1 in that term's place.

If you can recognize the GCF of a polynomial, you can use the following technique to find the polynomial factor.

Division and Factoring

We know that 5 is a factor of 155, but we don't know what the other factor is. We can use division to find it: $\frac{155}{5} = 31$. This tells us that 155 factors into $5 \cdot 31$.

Likewise, if we know that $6x$ is the GCF of $12x^3 + 18x^2 - 6x$ but we don't know what the polynomial factor is, we can use division to find it:

$$\frac{12x^3 + 18x^2 - 6x}{6x}.$$

From our work in Section 7.5, we know that this division can be split up into three fractions.

$$\frac{12x^3}{6x} + \frac{18x^2}{6x} - \frac{6x}{6x} = 2x^2 + 3x - 1.$$ ← This trinomial is the other factor.

This means that $12x^3 + 18x^2 - 6x = 6x(2x^2 + 3x - 1)$.

You Try It 4 **Factor out a common monomial from each of these polynomials. Check the answers for accuracy. If a polynomial cannot be factored, write *prime*. Use Example 6 as a guide.**

a) $25y + 10$ **b)** $6a + 25$ **c)** $21y^3 - 6y^2$ **d)** $14b^3 - 35b^2 - 7b$

Factoring When the First Term Is Negative

When the first term of a polynomial is negative, we have two choices for factoring out the GCF. We can factor using the positive GCF as it is, or we can factor out the negative GCF.

For example, the polynomial $-6x + 21$ has a GCF of 3, and we can factor using the previous techniques.

$$-6x + 21$$
$$= 3 \cdot (-2x) + 3 \cdot 7$$
$$= 3(-2x + 7)$$

We can check our work by multiplying the factored form or by replacing x with 1.

However, because the first term is negative, we can treat the GCF as -3. Then the factoring will look like this:

$$-6x + 21$$ The terms of this binomial are $-6x$ and $+21$.
$$= (-3) \cdot 2x + (-3) \cdot (-7)$$
$$= -3(2x - 7)$$ Note that $(-3)(-7) = +21$.

Let's compare the two factored forms.

GCF $= +3$	GCF $= -3$
$3(-2x + 7)$	$-3(2x - 7)$

Notice that each term in the binomial $(-2x + 7)$ is the opposite of each term in the binomial $(2x - 7)$. This is because factoring out a negative number, just like multiplying by a negative number, changes the sign of each term in the polynomial.

Factoring out a negative, just like dividing by a negative, has the effect of taking the opposite of each term.

The division technique tells us that if we know that -3 is a factor, we can find the other factor by dividing.

$$\frac{-6x + 21}{-3} = \frac{-6x}{-3} + \frac{21}{-3} = +2x + (-7) = 2x - 7$$

This means that the other factor is $2x - 7$; so, again we see that $-6x + 21 = -3(2x - 7)$.

Example 7

Factor out the negative GCF from each polynomial. If the GCF is 1, factor out -1. Check all answers for accuracy.

a) $-5y - 20$ **b)** $-4p^3 + 9$ **c)** $-9c^4 + 6c^3 - 3c^2$

Procedure The first term in each polynomial is negative; so, identify the GCF but factor out the negative of that GCF.

Answer

a) $-5y - 20$ — The terms are $-5y$ and -20. The GCF is 5, but factor out -5.

$= (-5)\cdot y + (-5)\cdot 4$ — Notice that $(-5)\cdot 4 = -20$.

$= -5(y + 4)$ — Check by multiplying or by replacing y with 1.

b) $-4p^3 + 9$ — The terms are $-4p^3$ and $+9$, but these are relatively prime.

$= (-1)\cdot 4p^3 + (-1)\cdot(-9)$ — Because the terms are relatively prime, the GCF is 1. But because the first term is negative, we can factor out -1.

$= -1(4p^3 - 9)$ — Check by multiplying or by replacing y with 1.

c) $-9c^4 + 6c^3 - 3c^2$ — The terms are $-9c^4$, $+6c^3$, and $-3c^2$. Each term has a factor of $3c^2$, but factor out $-3c^2$.

$= (-3c^2)\cdot 3c^2 + (-3c^2)\cdot(-2c) + (-3c^2)\cdot(+1)$

$= -3c^2(3c^2 - 2c + 1)$

You Try It 5 **Factor out the negative GCF from each polynomial. Check all answers for accuracy. Use Example 7 as a guide.**

a) $-10w + 40$ **b)** $-8a^2 - 4a$ **c)** $-2x^3 - 6x^2 + 14x$

Answers: You Try It and Think About It

You Try It: **1. a)** $2\cdot2\cdot3\cdot3$ **b)** $2\cdot2\cdot2\cdot y$ **c)** $2\cdot2\cdot3\cdot x\cdot x$ **d)** $3\cdot3\cdot5\cdot p\cdot p\cdot p$ **2. a)** $4x$ **b)** $5m^3$ **c)** 1 **d)** $3c^2$ **e)** p **3. a)** $8x$ **b)** 15 **c)** y^3 **d)** 1 **e)** $7c$ **f)** p^2 **4. a)** $5(5y + 2)$ **b)** prime **c)** $3y^2(7y - 2)$ **d)** $7b(2b^2 - 5b - 1)$ **5. a)** $-10(w - 4)$ **b)** $-4a(2a + 1)$ **c)** $-2x(x^2 + 3x - 7)$.

Think About It: **1.** The lesser exponent of a prime factor indicating the number of times that prime factors are common to both composite numbers. When finding the GCF, we match up as many common factors as possible. If one number has three factors of 3, 3^3, and another number has only two factors of 3, 3^2, then we can only match up the two factors of 3, the lesser of the two exponents.

Section 7.6 Exercises

FOR EXTRA HELP
Student Resources on DVD-ROM
Includes
➤ Student's Solutions Manual
➤ Video Lectures
➤ Chapter Test Prep Video
MyMathLab
MathXL

Think Again

1. When using exponents to determine the GCF, why do we choose the lesser exponent for each factor? *(Refer to Think About It 1)*
When finding the GCF, we match up as many common factors as possible. If one number has three factors of 3, 3^3, and another number has only two factors of 3, 3^2, we can only match up the two factors of 3, the lesser of the two exponents.

2. In a trinomial, when the lead term is negative and we factor out the negative of the GCF, why do the signs change on the other terms in the trinomial? You may use $-2x^3 - 8x^2 + 6x$ or your own example in the explanation.
Factoring out -1 from a term is like dividing the term by -1; and dividing by -1, just as multiplying by -1, means "the opposite of." So, factoring out -1 has the effect of taking the opposite of every term from which it is factored.

Identify the GCF of each group of terms.

3. $12y^2$ and $18y$
$6y$

4. $21m^3$ and $14m^2$
$7m^2$

5. $36a^3$ and $27a^2$
$9a^2$

6. $15y^4$ and $8y^2$
y^2

7. $25y$ and $14w$
1

8. $18y^3$ and $45y^2$
$9y^2$

9. $9x^5, 21x^4$, and $15x^2$
$3x^2$

10. $35p^6, 20p^4$, and $5p^2$
$5p^2$

11. $24p^7, 14p^5$, and $15p^3$
p^3

12. $24y^3, 16y^2$, and $40y$
$8y$

13. $30a^4, 45a^2$, and 15
15

14. $27a^4, 12a^2$, and 10
1

Factor out the GCF from each polynomial. If the GCF is 1, write *prime*. Check all answers for accuracy.

15. $6x + 9$
$3(2x + 3)$

16. $5y^2 - 15y$
$5y(y - 3)$

17. $16w^3 - 8w^2$
$8w^2(2w - 1)$

18. $15a^3 - 21a$
$3a(5a^2 - 7)$

19. $30w^6 + 20w^3$
$10w^3(3w^3 + 2)$

20. $14x^2 - 7x$
$7x(2x - 1)$

21. $9c^2 + 20c$
$c(9c + 20)$

22. $6w^2 - 15w$
$3w(2w - 5)$

23. $18x^5 - 24x^3$
$6x^3(3x^2 - 4)$

24. $15x^3 - 6x^2$
$3x^2(5x - 2)$

25. $4y^4 - 20y$
$4y(y^3 - 5)$

26. $49c^6 + 14c^3$
$7c^3(7c^3 + 2)$

27. $6x^3 - 15x^2 + 9x$
$3x(2x^2 - 5x + 3)$

28. $8y^6 + 8y^2 - 8y$
$8y(y^5 + y - 1)$

29. $4x^6 + 6x^5 - 2x^3$
$2x^3(2x^3 + 3x^2 - 1)$

30. $10a^4 - 20a^2 - 30a$
$10a(a^3 - 2a - 3)$

31. $18x^2 - 24x - 6$
$6(3x^2 - 4x - 1)$

32. $5y^4 - 10y^2 + 20y$
$5y(y^3 - 2y + 4)$

33. $12x^3 - 9x - 8$
prime

34. $32b^6 + 40b^5 - 16b^4$
$8b^4(4b^2 - 5b - 2)$

35. $12x^5 - 20x^4 + 4x^2$
$4x^2(3x^3 - 5x^2 + 1)$

36. $4y^5 - 8y^2 - 15$
prime

37. $9x^4 - 12x^3 - 16x^2$
$x^2(9x^2 - 12x - 16)$

38. $44b^4 - 33b^3 - 22b$
$11b(4b^3 - 3b^2 - 2)$

Factor out the negative GCF from each polynomial. If the GCF is 1, factor out -1. Check all answers for accuracy.

39. $-3w + 15$
$-3(w - 5)$

40. $-7m + 28$
$-7(m - 4)$

41. $-4y^2 - 12y$
$-4y(y + 3)$

42. $-6x^3 - 18x$
$-6x(x^2 + 3)$

43. $-15y^4 + 5y^2$
$-5y^2(3y^2 - 1)$

44. $-18w^3 + 9w^2$
$-9w^2(2w - 1)$

45. $-2p^2 - 15$
$-1(2p^2 + 15)$

46. $-20y^2 - 9$
$-1(20y^2 + 9)$

47. $-8w^2 - 4w + 2$
$-2(4w^2 + 2w - 1)$

48. $-25x^3 + 30x^2 - 5$
$-5(5x^3 - 6x^2 + 1)$

49. $-m^3 - 2m^2 - 6m$
$-m(m^2 + 2m + 6)$

50. $-12p^6 - 3p^4 - 6p^2$
$-3p^2(4p^4 + p^2 + 2)$

Think Outside the Box

In each expression, the greatest common factor is a binomial. Factor out the GCF and check all answers for accuracy.

51. $5m(2m + 3) + 7(2m + 3)$
$(5m + 7)(2m + 3)$

52. $p^2(3p - 1) + 6(3p - 1)$
$(p^2 + 6)(3p - 1)$

53. $2x^3(4y^2 + 7) - 5(4y^2 + 7)$
$(2x^3 - 5)(4y^2 + 7)$

54. $4x^2(5x - 2) + 6x(5x - 2) - 7(5x - 2)$
$(4x^2 + 6x - 7)(5x - 2)$

Chapter 7 Review

Section 7.1 Rules of Exponents

Concept	Example
The First Power Rule: For any base, $x^1 = x$ and $x = x^1$.	$5^1 = 5$ and $5 = 5^1$
The Product Rule for Exponents: For any *nonzero* base, x, $x^a \cdot x^b = x^{a+b}$.	$x^5 \cdot x^3 = x^{5+3} = x^8$
The Zero Power Rule: For any *nonzero* base, $x^0 = 1$.	$8^0 = 1$
The Power Rule for Exponents: $(x^a)^b = x^{a \cdot b}$	$(x^3)^4 = x^{3 \cdot 4} = x^{12}$

Section 7.2 Introduction to Polynomials

Concept	Example
A **polynomial** is an algebraic expression that is one of the following **1.** A single term **2.** The sum of two or more terms.	Examples of polynomials **1.** $8y^2$ **2.** $-5x^3 + 4x - 9$
A **monomial** is a polynomial with one term. A **binomial** is a polynomial with two terms. A **trinomial** is a polynomial with three terms.	$-2w^4$ is a monomial. $-2w^4 + 6w^2$ is a binomial. $-2w^4 + 6w^2 - 3$ is a trinomial.
To evaluate a polynomial means to replace each variable with a specific number and use the order of operations to evaluate the resulting numerical expression.	Evaluate $y^2 - 3y + 1$ when $y = -4$. $(-4)^2 - 3(-4) + 1$ $= 16 + 12 + 1 = 29$
The **degree** of a term is its number of variable factors. The degree of a constant term is 0.	$-5y^2x^4 = -5 \cdot y \cdot y \cdot x \cdot x \cdot x \cdot x$ $-5y^2x^4$ has six variable factors; so, the degree of this term is 6.

Concept	Example
A polynomial is in **descending order** when the term with the highest degree is written first, the term with the next highest degree is written second, and so on.	In descending order, $3x - 2x^2 + 18$ is equal to $-2x^2 + 3x + 18$.
Two or more terms are like terms when their variable structure is exactly the same.	$2x^2y$ and $-8x^2y$ are like terms.
To combine like terms means to add the coefficients of two or more like terms. The result is another like term.	$4m^3 + (-10m^3) = -6m^3$

Section 7.3 Adding and Subtracting Polynomials

Concept	Example
Multiplying a quantity by 1 does not change the value of the quantity.	$1 \cdot (3v^2 + 5v - 9) = 3v^2 + 5v - 9$
Multiplying a quantity by -1 changes the sign of each term in the quantity.	$-1 \cdot (3v^2 + 5v - 9) = -3v^2 - 5v + 9$
To add polynomials, place $1 \cdot$ (1 times) in front of each polynomial (in parentheses) and distribute. Then write the polynomial in descending order and combine like terms.	$(3x + x^2) + (5 - 2x)$ $= 1 \cdot (3x + x^2) + 1 \cdot (5 - 2x)$ $= 3x + x^2 + 5 - 2x$ $= x^2 + 3x - 2x + 5$ $= x^2 + x + 5$
To subtract polynomials, place $1 \cdot$ (1 times) in front of each polynomial (in parentheses). Distribute 1 to the first polynomial and distribute -1 to the second polynomial. Then write the polynomial in descending order and combine like terms.	$(3x + x^2) - (5 - 2x)$ $= 1 \cdot (3x + x^2) - 1 \cdot (5 - 2x)$ $= 3x + x^2 - 5 + 2x$ $= x^2 + 3x + 2x - 5$ $= x^2 + 5x - 5$

Section 7.4 Multiplying Polynomials

Concept	Example
A technique for checking the accuracy of multiplication of a polynomial is to replace each variable with 1. This reduces the problem to only constants and coefficient. **Caution:** This technique checks the accuracy of only the coefficients and constant; it does *not* check the accuracy of the exponents.	Check $(3x + 5)(4x - 1) = 12x^2 + 4x - 5$. $x = 1$: $(3 + 5)(4 - 1) = 12 + 4 - 5$ $(8) \cdot (3) = 11$ $24 = 11$ False! The multiplication is *incorrect*, and we must try again. Check $(3x + 5)(4x - 1) = 12x^2 + 17x - 5$. $x = 1$: $(3 + 5)(4 - 1) = 12 + 17 - 5$ $(8) \cdot (3) = 24$ $24 = 24$ True! Now check mentally to verify that the exponents are also correct.
To multiply two terms, multiply their coefficients and if the variables are the same, use the Product Rule for Exponents.	$(-2y^3)(6y^4)$ $= -2 \cdot 6 \cdot y^{3+4}$ $= -12y^7$
To square a monomial is to multiply the monomial by itself.	$(-5q^3)^2$ $= (-5q^3)(-5q^3)$ $= (-5) \cdot (-5) \cdot q^{3+3}$ $= 25q^6$
To multiply a monomial and a polynomial with more than one term, we distribute the monomial to each term in the polynomial.	$-4w^3(6w^2 - 5w)$ $= (-4w^3)(6w^2) + (-4w^3)(-5w)$ $= -24w^5 + 20w^4$
To multiply two polynomials, such as two binomials or a binomial and a trinomial, distribute each term of the first polynomial to each term of the second polynomial. Then write the result in descending order and combine like terms. $(x - 5)(2x^2 - 3)$	$(x - 5)(2x^2 - 3)$ $= x \cdot (2x^2) + x \cdot (-3) + (-5) \cdot (2x^2) + (-5) \cdot (-3)$ $= 2x^3 - 3x - 10x^2 + 15$ $= 2x^3 - 10x^2 - 3x + 15$

Section 7.5 Dividing Polynomials

Concept	Example
The Quotient Rule for Exponents: For any nonzero base, x, $\frac{x^a}{x^b} = x^{a-b}$.	$\frac{x^7}{x^4} = x^{7-4} = x^3$
To divide a monomial by a monomial, in a fractional form, divide the coefficients and if the variables are the same, apply the Quotient Rule.	$\frac{12m^4}{-3m} = \frac{12}{-3} \cdot \frac{m^4}{m^1} = -4m^{4-1} = -4m^3$
To divide a polynomial with at least two terms by a monomial in a fractional form, separate the fraction so that each term of the numerator is divided by the original denominator; then simplify each fraction.	$\frac{15w^5 - 20w^4 + 5w^2}{5w^2} = \frac{15w^5}{5w^2} - \frac{20w^4}{5w^2} + \frac{5w^2}{5w^2}$ $= 3w^3 - 4w^2 + 1$

Section 7.6 Factoring Polynomials: The Greatest Common Factor

Concept	Example
The prime factorization of a term is the product of the prime factors of its coefficient and the expanded form of the variable factors.	$18x^4 = 2 \cdot 3 \cdot 3 \cdot x \cdot x \cdot x \cdot x$
The greatest common factor (GCF) of two terms is the product of all of the common prime factors and all of the common variable factors.	$24p^3 = 2 \cdot 2 \cdot 2 \cdot 3 \cdot p \cdot p \cdot p$ $36p^2 = 2 \cdot 2 \cdot 3 \cdot 3 \cdot p \cdot p$ $\text{GCF} = 2 \cdot 2 \cdot 3 \cdot p \cdot p = 12p^2$
If the GCF of all of the terms in a polynomial is not 1, the polynomial can be written as the product of the GCF and a quantity. This is called *factoring out the GCF.*	Because the GCF of $24p^3$ and $36p^2$ is $12p^2$, we can factor $12p^2$ out of $24p^3 + 36p^2$. $24p^3 + 36p^2 = 12p^2(2p + 3)$
When the first term of a polynomial is negative, we can factor out a negative GCF. In doing so, the sign of each term in the polynomial factor is the opposite of the sign in the original polynomial.	The GCF of $6m^2$ and $8m$ is $2m$. Because the first term of $-6m^2 + 8m$ is negative, we can factor out $-2m$. $-6m^2 + 8m = -2m(3m - 4)$

Chapter 7 Review Exercises

Fill in the blanks.

1. $x^2 \cdot x^5 = x^{2+5}$ is an example of the ___Product___ Rule for Exponents. (7.1)

2. The number that multiplies to a variable in a term is called the ___coefficient___. (7.2)

3. A polynomial with only two terms is called a(n) ___binomial___. (7.2)

4. In the polynomial $4y^5 - p^3 + 2$, the degree of the second term is ___3___. (7.2)

5. In the polynomial $4y^5 - p^3 + 2$, the degree of the third term is ___0___. (7.2)

6. $2y^3 + 7y^2 - y - 3$ is an example of a polynomial that is in ___descending order___. (7.2)

7. To multiply a monomial and a trinomial, we must use the ___Distributive___ Property. (7.4)

8. $\frac{x^6}{x^2} = x^{6-2}$ is an example of the ___Quotient___ Rule for Exponents. (7.5)

Section 7.1

Evaluate each expression using a rule of exponents.

9. 5^1
   5

10. $(-6)^1$
    -6

11. 15^0
    1

12. $\left(\frac{5}{6}\right)^0$
    1

Write each product as one base with one exponent.

13. $y^4 \cdot y^6$
    y^{10}

14. $q^3 \cdot q$
    q^4

15. $p^8 \cdot p^0$
    p^8

16. $k \cdot k^4$
    k^5

Simplify each expression.

17. $(c^7)^2$
    c^{14}

18. $(x^3)^3$
    x^9

19. $(w^5)^1$
    w^5

20. $(p^4)^0$
    1

Section 7.2

Identify the terms in each algebraic expression.

21. $3x^4 - 6x^2$
    $3x^4$ and $-6x^2$

22. $-x^4 + 4x^2 - 1$
    $-x^4$, $4x^2$ and -1

Identify each expression as a monomial, a binomial, a trinomial, or not a polynomial.

23. $12 - x^2 + 6x$
    trinomial

24. $q^8 - 3q^4$
    binomial

25. $\dfrac{3y + 9}{y^3}$
    not a polynomial

26. -8
    monomial

Evaluate each polynomial with the given replacement value.

27. $-2k - 9, k = -5$
    1

28. $2m^2 - m, m = 5$
    45

29. $x^2 - 4x - 1, x = -1$
    4

30. $-3y^2 - 5y + 8, y = 0$
    8

At a skydiving camp, Laura jumped out of a plane from a height of 12,500 feet and fell 10 seconds before opening her parachute. Laura's height above the ground after x seconds is given by $h = 12{,}500 - 16x^2$. Write a sentence answering each question.

31. How high above the ground was Laura after falling 5 seconds?
Laura was 12,100 feet above the ground.

32. How high above the ground was Laura after falling 10 seconds?
Laura was 10,900 feet above the ground.

For each term, identify the variable factors and indicate the degree.

33. $-v^4$
$v \cdot v \cdot v \cdot v$ degree 4

34. x^4y^2
$x \cdot x \cdot x \cdot x \cdot y \cdot y$ degree 6

35. $6xy^7$
$x \cdot y \cdot y \cdot y \cdot y \cdot y \cdot y \cdot y$ degree 8

36. -3
none degree 0

Write each polynomial in descending order.

37. $6x + 9x^4 - 3x^6$
$-3x^6 + 9x^4 + 6x$

38. $-6y + 3y^4 + 8 - 12y^3$
$3y^4 - 12y^3 - 6y + 8$

Simplify each polynomial by combining like terms if possible. Write the answer in descending order.

39. $x - 4x^2 - 5x + 6x^2$
$2x^2 - 4x$

40. $-9 + 6p^4 - p^3 + 2p + 2$
$6p^4 - p^3 + 2p - 7$

Section 7.3

Distribute. Write each answer in descending order.

41. $1(2w - 10)$
$2w - 10$

42. $1(-3 - 2x^2 + 6x)$
$-2x^2 + 6x - 3$

43. $-1(-2b^2 - 4b + 1)$
$2b^2 + 4b - 1$

44. $-1(7y^3 + 3y - y^4)$
$y^4 - 7y^3 - 3y$

Add. Combine like terms and write each answer in descending order.

45. $(2v - 6) + (1 - 4v)$
$-2v - 5$

46. $(3 - 9y) + (4y^2 - 2)$
$4y^2 - 9y + 1$

47. $(2x - 5x^2 + 6) + (4 + 3x^2 - x)$
$-2x^2 + x + 10$

48. $(6 - 3q^2 + 2q) + (-4q + 3q^2 - 1)$
$-2q + 5$

Subtract. Combine like terms and write each answer in descending order.

49. $(3x^2 + 6x) - (5x^2 - 4x)$
$-2x^2 + 10x$

50. $(5 + 3x) - (9x^2 + 7)$
$-9x^2 + 3x - 2$

51. $(2k^2 - 6k + 3) - (k^2 + 8k - 7)$
$k^2 - 14k + 10$

52. $(4 + m^2 - 6m) - (m - 3m^2 + 8)$
$4m^2 - 7m - 4$

Section 7.4

Multiply. Check each answer to verify that it is correct.

53. $(2q)(4q^3)$
$8q^4$

54. $(-4x^5)(-2x^6)$
$8x^{11}$

55. $(-4p)^2$
$16p^2$

56. $(w^5)^2$
w^{10}

Multiply. Write each answer in descending order. Check all answers to verify that they are correct.

57. $8x(3x^4 - 5x)$
$24x^5 - 40x^2$

58. $-4w(5w^3 - 10w^5)$
$40w^6 - 20w^4$

59. $p^2(2p^3 - 6p^2 + 4)$
$2p^5 - 6p^4 + 4p^2$

60. $-5y(3y - 6y^2 + 4)$
$30y^3 - 15y^2 - 20y$

Multiply. Write each answer in descending order and combine like terms. Check all answers to verify that they are correct.

61. $(4x + 5)(2x - 3)$
$8x^2 - 2x - 15$

62. $(q - 4)(q - 8)$
$q^2 - 12q + 32$

63. $(m - 9)(m + 9)$
$m^2 - 81$

64. $(5x + 4)(5x - 4)$
$25x^2 - 16$

65. $(p - 4)^2$
$p^2 - 8p + 16$

66. $(2y - 5)^2$
$4y^2 - 20y + 25$

67. $(w + 3)(w^2 - 4w - 2)$
$w^3 - w^2 - 14w - 6$

68. $(3x - 2)(2x^2 - 3x + 1)$
$6x^3 - 13x^2 + 9x - 2$

Section 7.5

Write each quotient as one base with one exponent.

69. $\frac{c^8}{c^5}$
c^3

70. $\frac{k^{10}}{k^5}$
k^5

71. $\frac{x^5}{x}$
x^4

72. $\frac{p^5}{p^5}$
$p^0 = 1$

73. $\frac{v^6}{v^5}$
v^1

74. $\frac{y^9}{y^0}$
y^9

Simplify each fraction.

75. $\frac{27x^6}{9x^2}$
$3x^4$

76. $\frac{-36w^3}{9w^3}$
-4

77. $\frac{-45q^5}{-5q}$
$9q^4$

78. $\frac{16y^6}{-4y^0}$
$-4y^6$

Divide using distribution.

79. $\frac{9y^4 - 24y^2}{3y}$
$3y^3 - 8y$

80. $\frac{6x^6 + 2x^2}{-2x^2}$
$-3x^4 - 1$

81. $\frac{25m^7 - 5m^5 + 15m^3}{5m^3}$
$5m^4 - m^2 + 3$

82. $\frac{16q^6 - 12q^4 + 4q^2}{-4q^2}$
$-4q^4 + 3q^2 - 1$

Section 7.6

Identify the GCF of each group of terms.

83. $24y^3$ and $18y^2$
$6y^2$

84. $28x^4$ and $16x$
$4x$

85. $8x^6$, $12x^2$, and $15x^3$
x^2

86. $18w^3$, $14w^6$, and $8w^4$
$2w^3$

Factor out the GCF from each polynomial. If the GCF is 1, write *prime*. Check all answers for accuracy.

87. $4c - 10$
$2(2c - 5)$

88. $3x^2 - 12x$
$3x(x - 4)$

89. $15w^2 - 8$
prime

90. $45q^3 + 9q^2 - 18q$
$9q(5q^2 + q - 2)$

91. $20y^4 - 15y^3 + 5y^2$
$5y^2(4y^2 - 3y + 1)$

92. $15x^3 - 12x - 2$
prime

Factor out the negative GCF from each polynomial. If the GCF is 1, factor out −1. Check all answers for accuracy.

93. $-8p + 12$
$-4(2p - 3)$

94. $-6m^2 - 3m$
$-3m(2m + 1)$

95. $-10x^4 + 25x^2$
$-5x^2(2x^2 - 5)$

96. $-15y^3 + 9y^2 - 3$
$-3(5y^3 - 3y^2 + 1)$

97. $-c^6 + 6c^4 + 9c^2$
$-c^2(c^4 - 6c^2 - 9)$

98. $-8w^3 - 4w^2 + 4w$
$-4w(2w^2 + w - 1)$

Chapter 7 Test

Evaluate each expression using a rule of exponents.

1. $(-6)^0$
   1
2. 10^1
   10

Write each product as one base with one exponent.

3. $m^3 \cdot m^5$
   m^8
4. $x^4 \cdot x$
   x^5

Simplify each expression.

5. $(w^5)^2$
   w^{10}
6. $(y^6)^0$
   1
7. $(x^1)^4$
   x^4

Evaluate each polynomial with the given replacement value.

8. $-3w^2 + w, w = 4$
   -44
9. $x^2 - 6x + 3, x = -1$
   10

Write a sentence answering the question.

10. From the ground, Diego shot an arrow into the air. The arrow's height in feet above the ground x seconds after being shot is given by $h = 90x - 16x^2$. How high above the ground in feet was the arrow 3 seconds after Diego shot it?
    The arrow was 126 feet above the ground.

For each term, indicate the degree.

11. $-7p^5$
    degree 5
12. $9xy^6$
    degree 7
13. 4
    degree 0

Write the polynomial in descending order.

14. $6m - 12 - 4m^3 + 5m^4$
    $5m^4 - 4m^3 + 6m - 12$

Add or subtract as indicated. Combine like terms and write each answer in descending order.

15. $(3x^2 - 8 + x) + (1 - 4x^2 + 5x)$
    $-x^2 + 6x - 7$
16. $(4c^2 - 3c) - (7c + c^2)$
    $3c^2 - 10c$

Multiply.

17. $(-6m^4)(5m)$
    $-30m^5$
18. $(-7q)^2$
    $49q^2$

Multiply. Write the answer in descending order.

19. $-2y(-4y + 5y^2 - 3)$
    $-10y^3 + 8y^2 + 6y$

Multiply and combine like terms. Write each answer in descending order.

20. $(2w - 6)(w - 4)$
    $2w^2 - 14w + 24$
21. $(x + 4)^2$
    $x^2 + 8x + 16$
22. $(4y - 3)(y^2 - 5y + 1)$
    $4y^3 - 23y^2 + 19y - 3$

Write each quotient as one base with one exponent.

23. $\dfrac{x^9}{x^3}$
    x^6
24. $\dfrac{w^6}{w}$
    w^5

Simplify each fraction.

25. $\dfrac{18p^4}{-6p}$
    $-3p^3$
26. $\dfrac{-35m^4}{-5m^4}$
    7

Divide using distribution.

27. $\dfrac{9x^6 + 3x^2}{-3x^2}$
    $-3x^4 - 1$
28. $\dfrac{12q^5 - 2q^3 + 4q}{2q}$
    $6q^4 - q^2 + 2$

Factor out the GCF from each polynomial. If the GCF is 1, write *prime*.

29. $12x^2 - 20x$
    $4x(3x - 5)$
30. $4y^2 + 9$
    prime
31. $15p^6 + 20p^4 - 5p^2$
    $5p^2(3p^4 + 4p^2 - 1)$

Factor out the negative GCF from each polynomial.

32. $-12y^3 - 18y^2$
    $-6y^2(2y + 3)$
33. $-12x^3 + 6x^2 - 3x$
    $-3x(4x^2 - 2x + 1)$

CHAPTER

Units of Measure and Geometry

Introduction

There are three primary types of learning styles: visual, auditory, and kinesthetic/tactile. Usually, a person learns using all three styles, but one style is more dominant than the other two.

Visual-dominant learners learn through seeing. This may include reading; looking at pictures, diagrams, and graphs; and noticing an instructor's facial expressions.

If you are an auditory-dominant learner, you learn through hearing. You listen to lectures and discussions, you talk about your ideas and listen to others' ideas, and you notice an instructor's voice pitch or level of loudness.

Kinesthetic/tactile-dominant learners learn through touching and using their hands. Examples of learning activities for this group are doing hands-on projects, drawing diagrams and pictures, and dancing or taking part in other physical activities.

Geometry is about shapes and sizes that can be quickly identified and drawn. It is appealing to visual and kinesthetic/tactile learners. Auditory learners master geometry by verbally explaining diagrams and figures to other learners. It's easy to see geometry in the world around us—in architecture, art, and sports. For example, understanding angles is essential to success in games such as billiards and baseball, where a round ball bounces off the side of a pool table or a stadium wall, respectively.

For more information on learning styles, go to your favorite Internet search engine and type in *learning styles.*

Preparation Exercises

Section 4.2 Introduction to Fractions
Write each mixed fraction as an improper fraction.

1. $5\frac{1}{6}$ $\quad\frac{31}{6}$

2. $7\frac{5}{8}$ $\quad\frac{61}{8}$

Section 4.2 Introduction to Fractions
Write each improper fraction as a mixed number.

3. $\dfrac{95}{12}$ $\quad 7\frac{11}{12}$

4. $\dfrac{71}{16}$ $\quad 4\frac{7}{16}$

Section 4.4 Multiplying and Dividing Fractions
Multiply and simplify.

5. $\dfrac{3}{8}\cdot\dfrac{4}{15}$ $\quad\frac{1}{10}$

6. $\dfrac{8}{9}\cdot\dfrac{15}{16}$ $\quad\frac{5}{6}$

Multiply and simplify.

7. $2\frac{1}{4}\cdot 4\frac{2}{3}$ $\quad\frac{21}{2} = 10\frac{1}{2}$

8. $8\frac{1}{3}\cdot 3\frac{3}{5}$ $\quad 30$

Section 4.5 Adding and Subtracting Like Fractions
Add and simplify.

9. $\frac{7}{12} + \frac{11}{12}$
$\frac{3}{2} = 1\frac{1}{2}$

10. $\frac{3}{16} + \frac{9}{16}$
$\frac{3}{4}$

Section 4.7 Adding and Subtracting Unlike Fractions
Add and simplify.

11. $\frac{2}{3} + \frac{5}{6}$
$\frac{3}{2} = 1\frac{1}{2}$

12. $\frac{1}{6} + \frac{3}{8}$
$\frac{13}{24}$

Section 5.1 Introduction to Decimals
Write each decimal as a fraction.

13. 0.47
$\frac{47}{100}$

14. 0.003
$\frac{3}{1{,}000}$

Write each fraction as a decimal.

15. $\frac{63}{100}$
0.63

16. $\frac{2.9}{10}$
0.29

Section 5.4 Multiplying Decimals
Multiply.

17. 1.25×1.6
2

18. 3.14×3.5
10.99

19. 5.8×10
58

20. 0.012×100
1.2

Section 5.5 Dividing Decimals
Divide.

21. $59 \div 4$
14.75

22. $11.7 \div 6$
1.95

23. $5.85 \div 1.5$
3.9

24. $0.86 \div 0.09$
$9.\overline{5}$

SECTION 8.1 U.S. Measures

OBJECTIVES

In this section, you will learn to

- Identify U.S. measures of length, weight, and capacity.
- Convert one U.S. measure to another.
- Add and subtract feet and inches.
- Add and subtract pounds and ounces.

You Need to Know

To successfully complete this section, you need to understand

- ☐ Writing improper fractions as mixed numbers and mixed numbers as improper fractions (4.2)
- ☐ Simplifying fractions (4.3)
- ☐ Multiplying fractions (4.4)

Introduction

Whenever we seek to measure something—find out how long it is or how much it weighs or what its capacity is—we need to use some sort of **unit of measure,** such as feet, pounds, or gallons.

Measures of length, for example, are used to see how long something is. There are both **U.S. measures,** used primarily in the United States, and **metric measures,** used elsewhere in the world. In the fields of science and medicine, however, it is common to use metric measures in the United States as well.

In this section, we concentrate on U.S. measures only. Section 8.2 explores metric measures, and Section 8.3 features a procedure for converting between U.S. and metric measures.

U.S. Measures of Length

Measures of length are called **linear measures.** Here are some U.S. linear measures and their equivalencies.

U.S. Linear Measures	Equivalencies		
inches (in.)	1 ft = 12 in.	and	1 in. = $\frac{1}{12}$ ft
feet (ft)	1 yd = 3 ft	and	1 ft = $\frac{1}{3}$ yd
yards (yd)	1 yd = 36 in.	and	1 in. = $\frac{1}{36}$ yd
miles (mi)	1 mi = 5,280 ft	and	1 ft = $\frac{1}{5{,}280}$ mi

Units of measure are actually multiplied by the number that precedes them.

For example, 2 feet means $2 \cdot (1 \text{ foot})$ and 6 inches means $6 \cdot (1 \text{ inch})$.

Example 1

Write 3 feet in terms of inches.

Procedure From the list of equivalencies we just viewed, 1 foot = 12 inches.

3 feet can be rewritten as $3 \cdot (1 \text{ foot})$.

We can replace 1 foot with 12 inches and multiply.

Answer $3 \text{ feet} = 3 \cdot (1 \text{ foot}) = 3 \cdot (12 \text{ inches}) = 36 \text{ inches}$

12 in. 12 in. 12 in. ← 36 in.

1 ft 1 ft 1 ft ← 3 ft

In this case, we have changed from one unit of measure to another, from feet to inches. We say that we have **converted** from feet to inches.

Example 2

Convert from one measure to the other.

a) 7 yards to feet **b)** 8 inches to feet **c)** $5\frac{1}{4}$ feet to yards

Procedure

1. Write any mixed number as an improper fraction.
2. Write the given unit of measure as a product.
3. Insert the new unit of measure.
4. Multiply and simplify.

Answer

a) $7 \text{ yards} = 7 \cdot (1 \text{ yd}) = 7 \cdot 3 \text{ ft} = 21 \text{ feet}$

b) $8 \text{ inches} = 8 \cdot (1 \text{ in.}) = 8 \cdot \frac{1}{12} \text{ ft} = \frac{8}{12} \text{ ft} = \frac{2}{3} \text{ foot}$

c) $5\frac{1}{4} \text{ feet} = \frac{21}{4} \cdot (1 \text{ ft}) = \frac{21}{4} \cdot \frac{1}{3} \text{ yd} = \frac{21 \cdot 1}{4 \cdot 3} \text{ yd} = \frac{7 \cdot 1}{4 \cdot 1} \text{ yd} = \frac{7}{4} \text{ yd} = 1\frac{3}{4} \text{ yards}$

You Try It 1 **Convert from one measure to the other. Use Examples 1 and 2 as guides.**

a) $4\frac{1}{2}$ feet to inches **b)** $6\frac{1}{3}$ yards to feet **c)** 33 inches to feet

Converting Feet and Inches

Recall that $3\frac{1}{4}$ is a mixed number. It has a whole number, 3, and a fractional part, $\frac{1}{4}$. Also, $3\frac{1}{4}$ means $3 + \frac{1}{4}$.

Units of length can be mixed as well. We can abbreviate 5 ft and 3 in. (which means 5 ft + 3 in.) as 5 ft 3 in. We also can write this in terms of only inches by converting the feet portion to inches.

$$5 \text{ ft} = 5 \cdot (1 \text{ ft}) = 5 \cdot (12 \text{ in.}) = 60 \text{ in.}$$

So, $5 \text{ ft } 3 \text{ in.} = 60 \text{ in.} + 3 \text{ in.} = 63 \text{ inches.}$

This means that a table that is 5 ft 3 in. long is 63 inches long.

12 in. 12 in. 12 in. 12 in. 12 in. 3 in. ← 63 in.

1 ft 1 ft 1 ft 1 ft 1 ft 3 in. ← 5 ft 3 in.

Example 3

Convert 6 feet 2 inches to inches.

Procedure Convert 6 feet to inches and add the extra 2 inches.

Answer

$$6\text{ ft} = 6\cdot(1\text{ ft}) = 6\cdot(12\text{ in.}) = 72\text{ in.}$$

$$72\text{ in.} + 2\text{ in.} = 74\text{ inches}$$

6 ft 2 in. = 74 inches

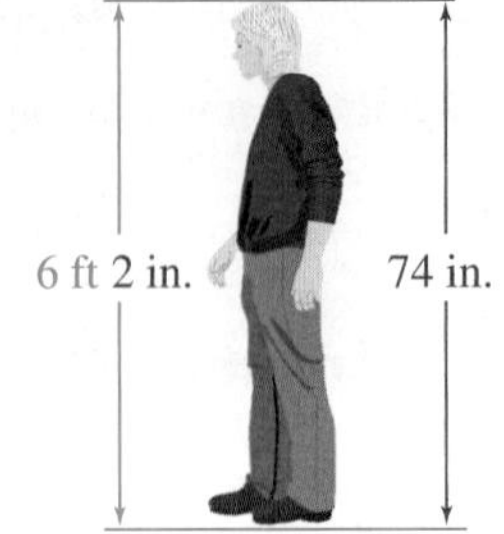

You Try It 2 **Convert each measure to inches. Use Example 3 as a guide.**

a) 2 feet 6 inches **b)** 10 feet 9 inches

You Try It 3 **At 7 ft 5 in., Yao Ming is presently the tallest player in the National Basketball Association. Express this measure in inches.**

Just as 1 foot = 12 inches, 1 inch = $\frac{1}{12}$ foot. This means that we can write 36 inches in terms of feet.

$$36\text{ inches} = 36\cdot\frac{1}{12}\text{ foot} = \frac{36}{12}\text{ foot} = 3\text{ feet}$$

Multiplying the number of inches, 36, by $\frac{1}{12}$ is the same as dividing the number by 12.

$$\begin{array}{r} 3 \\ 12\overline{)36} \\ -\,36 \\ \hline 0 \end{array}$$

← This quotient means that 36 inches = 3 feet.

How does 38 inches compare to 3 feet? 38 inches is 2 inches more than 3 feet: 36 inches + 2 inches = 3 ft 2 in.

This also can be demonstrated by dividing 12 into 38 inches.

$$\begin{array}{r} 3 \\ 12\overline{)38} \\ -\,36 \\ \hline 2 \end{array}$$

3 ← This quotient is the number of feet.

2 ← This remainder is the number of inches. 38 inches = 3 ft 2 in.

Example 4

Convert 53 inches to feet and inches.

Procedure Divide 12 into 53. The quotient is the number of feet, and the remainder is the number of inches.

Answer

$$\begin{array}{r} 4 \\ 12\overline{)53} \\ -\,48 \\ \hline 5 \end{array}$$

53 inches = 4 ft 5 in.

You Try It 4 **Convert each measure to feet and inches. Use Example 4 as a guide.**

a) 70 inches **b)** 109 inches

You Try It 5 **A typical door frame is 80 inches tall. Express this measure in *feet and inches*.**

Adding Linear Measures

Recall from Section 4.5 that we can add like units of measure: 2 feet + 3 feet = 5 feet.

If we have two mixed unit lengths, we can add the feet together and add the inches together. If the sum of the inches is 12 or more, we can convert 12 of those inches into 1 foot. For example,

$$16 \text{ in.} = 12 \text{ in.} + 4 \text{ in.} = 1 \text{ ft } 4 \text{ in.}$$

Example 5

Add. **a)** 6 ft 5 in. + 2 ft 10 in. **b)** 3 ft 8 in. + 1 ft 4 in.

Procedure Set up each sum vertically and line up the feet and inches. Adjust the number of inches if the total includes 12 or more inches.

Answer **a)**

```
    6 ft  5 in.
  + 2 ft 10 in.        Adjust the number of inches.
    8 ft 15 in.   <--  15 inches = 12 in. + 3 in. = 1 ft 3 in.

    8 ft + 1 ft + 3 in.
  = 9 ft 3 in.
```

b)

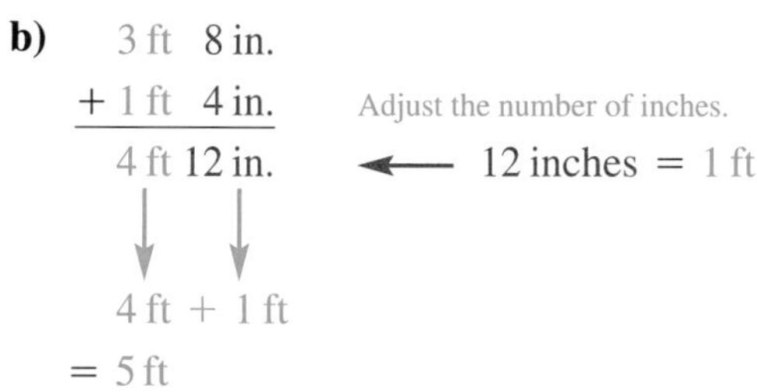

You Try It 6 **Add. Use Example 5 as a guide.**

a) 4 ft 6 in. + 7 ft 3 in. **b)** 2 ft 5 in. + 4 ft 7 in. **c)** 5 ft 8 in. + 3 ft 10 in.

You Try It 7 **Standing on the floor on tiptoes, Ricky can reach 5 feet 9 inches high. How high can Ricky reach if he stands (on tiptoes) on a stool that is 1 foot 6 inches high?**

Subtracting Feet and Inches

Subtracting feet and inches is similar to subtracting whole numbers. Sometimes we can subtract directly, as in

$$\begin{array}{r} 7\text{ ft }9\text{ in.} \\ -\ 3\text{ ft }2\text{ in.} \\ \hline 4\text{ ft }7\text{ in.} \end{array}$$

and sometimes we need to regroup, as in

$$\begin{array}{r} 6\text{ ft }3\text{ in.} \\ -\ 4\text{ ft }7\text{ in.} \\ \hline \end{array}$$

Notice that we can't subtract 7 in. directly from 3 in.

We can, however, regroup by changing 1 foot into 12 inches.

$$\begin{array}{r} \overset{5}{\not{6}}\text{ ft }\overset{15}{\not{3}}\text{ in.} \\ -\ 4\text{ ft }7\text{ in.} \\ \hline 1\text{ ft }8\text{ in.} \end{array}$$

Notice that we add 12 inches to the 3 inches already there.

Example 6

Subtract. **a)** 3 ft 6 in. − 2 ft 1 in. **b)** 9 ft 4 in. − 5 ft 10 in.

Procedure Set up each difference vertically and line up the feet and inches. If necessary, regroup the number of feet and inches if you can't subtract directly.

Answer **a)**

$$\begin{array}{r} 3\text{ ft }6\text{ in.} \\ -\ 2\text{ ft }1\text{ in.} \\ \hline 1\text{ ft }5\text{ in.} \end{array}$$

Subtract directly without regrouping.

b)

$$\begin{array}{r} 9\text{ ft }4\text{ in.} \\ -\ 5\text{ ft }10\text{ in.} \\ \hline \end{array}$$

Regroup the first number. 12 in. + 4 in. = 16 in.

$$\begin{array}{r} 8\text{ ft }16\text{ in.} \\ -\ 5\text{ ft }10\text{ in.} \\ \hline 3\text{ ft }6\text{ in.} \end{array}$$

You Try It 8 **Subtract. Use Example 6 as a guide.**

a) 4 ft 6 in. − 1 ft 3 in. **b)** 25 ft 8 in. − 18 ft 11 in.

The next example uses the equation part = whole − part. When the whole and one part are known, we can subtract to find the other part.

Example 7

Annie cut 2 feet 10 inches from a board that is 8 feet long. How long is the remaining board?

Procedure To start, the length of the board is 8 feet 0 inches. From this, we cut 2 feet 10 inches. This means that we subtract: 8 feet 0 inches − 2 feet 10 inches.

Answer

$$\begin{array}{r} 8\text{ ft }0\text{ in.} \\ -\ 2\text{ ft }10\text{ in.} \\ \hline \end{array}$$

Regroup the first number. 12 in. + 0 in. = 12 in.

$$\begin{array}{r} 7\text{ ft }12\text{ in.} \\ -\ 2\text{ ft }10\text{ in.} \\ \hline 5\text{ ft }2\text{ in.} \end{array}$$

Sentence The remaining board is 5 feet 2 inches long.

You Try It 9 **Taneesha cut a 6-foot-long sub into two pieces. One of the pieces is 3 feet 8 inches. What is the length of the other piece?**

U.S. Measures of Weight

Weight refers to how heavy something is. The basic U.S. units of measure for weight are *ounces*, *pounds*, and *tons*.

U.S. Measures of Weight	Equivalencies
ounces (oz)	1 lb = 16 oz and 1 oz = $\frac{1}{16}$ lb
pounds (lb)	
tons (T)	1 ton = 2,000 lb and 1 lb = $\frac{1}{2{,}000}$ ton

We can convert from one measure of weight to another using multiplication.

Example 8

Convert. **a)** $1\frac{3}{4}$ pounds to ounces **b)** 4,500 pounds to tons

Procedure Multiply using the equivalency. For part a, use 1 lb = 16 oz. For part b, use 1 lb = $\frac{1}{2{,}000}$ ton. In part a, it may be helpful to write $1\frac{3}{4}$ as $\frac{7}{4}$. Also, we can write 16 as $\frac{16}{1}$.

Answer **a)** $1\frac{3}{4}\text{ lb} = 1\frac{3}{4} \times 1\text{ lb} = \frac{7}{4} \times \frac{16}{1}\text{ oz} = \frac{7 \times 16}{4 \times 1}\text{ oz} = \frac{7 \times 4}{1 \times 1}\text{ oz} = 28\text{ oz}$

b) $4{,}500\text{ lb} = 4{,}500 \times 1\text{ lb} = 4{,}500 \times \frac{1}{2{,}000}\text{ ton} = \frac{4{,}500}{2{,}000}\text{ tons} = 2\frac{1}{4}\text{ tons}$

You Try It 10 **Convert one measure to the other. Show all of your work, as in Example 8.**

a) $5\frac{1}{8}$ pounds to ounces **b)** 5,000 pounds to tons **c)** $2\frac{3}{10}$ tons to pounds

Converting Pounds and Ounces

If we have a weight measured in both pounds and ounces, we can convert the weight to only ounces by first converting the pounds to ounces and then adding the ounces already present in the weight.

For example, we can write 5 lb 4 oz in terms of ounces only by converting the pounds portion to ounces.

$5\text{ lb} = 5 \cdot (1\text{ lb}) = 5 \cdot (16\text{ oz}) = 80\text{ oz}$

So, 5 lb 4 oz = 80 oz + 4 oz = 84 ounces.

Pet Best
Dog Food
5 lb 4 oz
(84 oz)

This means that a bag of dog food that weighs 5 lb 4 oz also weighs 84 ounces.

Example 9

Convert 3 pounds 7 ounces to ounces.

Procedure Convert 3 pounds to ounces and add the extra 7 ounces.

Answer $3\text{ lb} = 3 \cdot (1\text{ lb}) = 3 \cdot (16\text{ oz}) = 48\text{ oz}$

$48\text{ oz} + 7\text{ oz} = 55\text{ oz} \longrightarrow 3\text{ lb } 7\text{ oz} = 55\text{ oz}$

You Try It 11 **Convert each measure to ounces. Use Example 9 as a guide.**

a) 2 pounds 6 ounces **b)** 10 pounds 14 ounces

You Try It 12 **Gladys grew a tomato that weighed 4 pounds 15 ounces. How much does this tomato weigh in ounces?**

To write ounces in terms of pounds and ounces, we divide the number of ounces by 16.

For example, to write 100 ounces in terms of pounds and ounces:

$$\begin{array}{r} 6 \\ 16\overline{)100} \\ -\ 96 \\ \hline 4 \end{array}$$

6 ← This quotient is the number of pounds.

4 ← This remainder is the number of ounces.

100 ounces = 6 lb 4 oz

Example 10

Convert 53 ounces to pounds and ounces.

Procedure Divide 16 into 53. The quotient is the number of pounds, and the remainder is the number of ounces.

Answer

$$\begin{array}{r} 3 \\ 16\overline{)53} \\ -\ 48 \\ \hline 5 \end{array}$$

53 ounces = 3 lb 5 oz

You Try It 13 **Convert each measure to pounds and ounces. Use Example 10 as a guide.**

a) 70 ounces **b)** 165 ounces

You Try It 14 **Carla was born prematurely and weighed only 60 ounces at birth. Write her weight in terms of pounds and ounces.**

Adding Weight Measures

Adding pounds and ounces is similar to adding feet and inches. When the sum of the ounces is 16 or more, we can convert 16 of those ounces into 1 pound.

Example 11

Add. **a)** 6 lb 14 oz + 2 lb 6 oz **b)** 3 lb 12 oz + 1 lb 4 oz

Procedure Set up each sum vertically and line up the pounds and ounces. Adjust the number of ounces if the total includes 16 or more ounces.

Answer **a)**

$$\begin{array}{r} 6\text{ lb }14\text{ oz} \\ +\ 2\text{ lb }\ \ 6\text{ oz} \\ \hline 8\text{ lb }20\text{ oz} \end{array}$$

Adjust the number of ounces.

8 lb 20 oz ← 20 ounces = 16 oz + 4 oz = 1 lb 4 oz

8 lb + 1 lb + 4 oz = 9 lb 4 oz

b)

$$\begin{array}{r} 3 \text{ lb } 12 \text{ oz} \\ +\ 1 \text{ lb } \ \ 4 \text{ oz} \\ \hline 4 \text{ lb } 16 \text{ oz} \end{array}$$

Adjust the number of ounces. ← 16 ounces = 1 lb

4 lb + 1 lb = 5 lb

You Try It 15 **Add. Use Example 11 as a guide.**

a) 4 lb 6 oz + 7 lb 3 oz **b)** 2 lb 7 oz + 4 lb 14 oz **c)** 5 lb 7 oz + 3 lb 9 oz

You Try It 16 **One watermelon weighs 23 pounds 11 ounces, and another watermelon weighs 19 pounds 12 ounces. What is the total weight of the two watermelons?**

Subtracting Pounds and Ounces

If regrouping is necessary when subtracting pounds and ounces, we need to write 1 pound as 16 ounces and add it to the ounces already in the weight measure.

Consider 6 lb 3 oz − 4 lb 7 oz ⟶

$$\begin{array}{r} \overset{5}{\cancel{6}} \text{ lb } \ \overset{19}{\cancel{3}} \text{ oz} \\ -\ 4 \text{ lb } \ \ 7 \text{ oz} \\ \hline 1 \text{ lb } 12 \text{ oz} \end{array}$$

← Notice that we add 16 ounces to the 3 ounces already there.

Example 12

Subtract. **a)** 3 lb 6 oz − 2 lb 1 oz **b)** 9 lb 4 oz − 5 lb 10 oz

Procedure Set up each difference vertically and line up the pounds and ounces. If necessary, adjust the number of pounds and ounces if you can't subtract directly.

Answer **a)**

$$\begin{array}{r} 3 \text{ lb } 6 \text{ oz} \\ -\ 2 \text{ lb } 1 \text{ oz} \\ \hline 1 \text{ lb } 5 \text{ oz} \end{array}$$

Subtract directly without regrouping.

b)

$$\begin{array}{r} 9 \text{ lb } \ \ 4 \text{ oz} \\ -\ 5 \text{ lb } 10 \text{ oz} \\ \hline \end{array}$$

← Regroup the first number. ⟶ 16 oz + 4 oz = 20 oz

$$\begin{array}{r} 8 \text{ lb } 20 \text{ oz} \\ -\ 5 \text{ lb } 10 \text{ oz} \\ \hline 3 \text{ lb } 10 \text{ oz} \end{array}$$

You Try It 17 **Subtract. Use Example 12 as a guide.**

a) 4 lb 9 oz − 1 lb 3 oz **b)** 10 lb 2 oz − 4 lb 7 oz **c)** 25 lb − 18 lb 10 oz

Example 13

Mitch is a butcher for a grocery store. He cut 2 pounds 10 ounces of meat from a ham that weighed 5 pounds 7 ounces. What is the weight of the remaining ham?

Procedure To start, the ham weighed 5 pounds 7 ounces. From this, Mitch cut 2 pounds 10 ounces. This means that we subtract: 5 pounds 7 ounces – 2 pounds 10 ounces.

Answer

$$\begin{array}{r} 5 \text{ lb } \ \ 7 \text{ oz} \\ -\ 2 \text{ lb } 10 \text{ oz} \\ \hline \end{array}$$

← Regroup the first number. ⟶ 16 oz + 7 oz = 23 oz

$$\begin{array}{r} 4 \text{ lb } 23 \text{ oz} \\ -\ 2 \text{ lb } 10 \text{ oz} \\ \hline 2 \text{ lb } 13 \text{ oz} \end{array}$$

Sentence The weight of the remaining ham is 2 pounds 13 ounces.

You Try It 18 **Mai's backpack weighs 19 pounds. She took out her history book, which weighs 5 pounds 2 ounces, and left it in her car. How much does her backpack weigh now?**

U.S. Measures of Capacity

Capacity usually refers to a liquid amount, although not always. For example, a pancake recipe includes 1 cup of flour and 1 cup of milk. The flour is dry, and the milk is liquid; but a cook can use the same measuring cup and fill it to the same level for each of the two ingredients.

Capacity refers to the amount a container can hold, usually liquid. A *gallon* of milk, a *quart* of orange juice, a *pint* of ice cream, and a *cup* of sour cream are common grocery items that use measures of capacity.

An ounce is a unit of measure used in both weight and capacity. To distinguish between the two, we use *fluid ounces* for capacity.

1 gallon

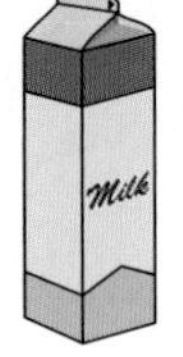

1 quart

1 pint

1 cup

U.S. Measures of Capacity	Equivalencies		
fluid ounces (fl oz)	1 cup = 8 fl oz	and	1 fl oz = $\frac{1}{8}$ cup
cups	1 pt = 2 cups	and	1 cup = $\frac{1}{2}$ pt
pints (pt)	1 qt = 2 pt	and	1 pt = $\frac{1}{2}$ qt
quarts (qt)	1 qt = 4 cups	and	1 cup = $\frac{1}{4}$ qt
gallons (gal)	1 gal = 4 qt	and	1 qt = $\frac{1}{4}$ gal

It is typical to see measures of capacity written in fractional form, such as $1\frac{1}{2}$ cups.

We can convert from one measure of capacity to another using multiplication. Because an *ounce* is both a unit of weight and a unit of capacity, we can refer to *fluid ounces* as just *ounces* only when we know that a given measure is a measure of capacity.

Example 14

Convert by multiplying.

a) $\frac{5}{16}$ cup into ounces **b)** $3\frac{1}{2}$ cups into quarts

Procedure Use the previous equivalencies to convert from one measure to another. For part b, convert from cups to pints, then from pints to quarts.

Answer

a)
$$\frac{5}{16}\text{ cup} = \frac{5}{16}\cdot 1\text{ cup} = \frac{5}{16}\cdot\frac{8}{1}\text{ oz} = \frac{5\cdot 8}{16\cdot 1}\text{ oz} = \frac{5\cdot 1}{2\cdot 1}\text{ oz} = \frac{5}{2}\text{ oz} = 2\tfrac{1}{2}\text{ oz}$$

b)
$$3\tfrac{1}{2}\text{ cups} = \frac{7}{2}\cdot 1\text{ cup} = \frac{7}{2}\cdot\frac{1}{2}\text{ pt} = \frac{7}{4}\text{ pt} = \frac{7}{4}\cdot 1\text{ pt} = \frac{7}{4}\cdot\frac{1}{2}\text{ qt} = \frac{7}{8}\text{ qt}$$

You Try It 19 **Convert by multiplying. Use Example 14 as a guide.**

a) 9 quarts into gallons **b)** $\frac{3}{8}$ quarts into pints **c)** 20 fluid ounces into pints **d)** $1\frac{5}{8}$ quarts into cups

Answers: You Try It and Think About It

You Try It: **1. a)** 54 inches **b)** 19 feet **c)** $2\frac{3}{4}$ feet **2. a)** 30 inches **b)** 129 inches **3.** Yao Ming is 89 inches tall. **4. a)** 5 ft 10 in. **b)** 9 ft 1 in. **5.** A typical bedroom door frame is 6 feet 8 inches tall. **6. a)** 11 ft 9 in. **b)** 7 ft **c)** 9 ft 6 in. **7.** Ricky can reach 7 feet 3 inches high on the stool. **8. a)** 3 ft 3 in. **b)** 6 ft 9 in. **9.** The length of the other piece is 2 feet 4 inches long. **10. a)** 82 oz **b)** $2\frac{1}{2}$ tons **c)** 4,600 lb **11. a)** 38 oz **b)** 174 oz **12.** The tomato weighs 79 oz. **13. a)** 4 lb 6 oz **b)** 10 lb 5 oz **14.** Carla weighed 3 pounds 12 ounces when she was born. **15. a)** 11 lb 9 oz **b)** 7 lb 5 oz **c)** 9 lb **16.** The total weight of the two watermelons is 43 pounds 7 ounces. **17. a)** 3 lb 6 oz **b)** 5 lb 11 oz **c)** 6 lb 6 oz **18.** Mai's backpack now weighs 13 pounds 14 ounces. **19. a)** $2\frac{1}{4}$ gal **b)** $\frac{3}{4}$ pt **c)** $1\frac{1}{4}$ pt **d)** $6\frac{1}{2}$ cups

Think About It: No Think About Its in this section.

Section 8.1 Exercises

FOR EXTRA HELP
Student Resources on DVD-ROM
Includes
➤Student's Solutions Manual
➤Video Lectures
➤Chapter Test Prep Video
MyMathLab
MathXL

Think Again

In tenth-century England, an acre was measured as 1 *furlong* ($\frac{1}{8}$ of a mile) in length and 1 *chain* (22 yards) wide.

1. Using that information, what steps must be taken to calculate the area of 1 acre in square feet?
First, convert 1 mile to feet and then multiply that figure by $\frac{1}{8}$ (to get 660 feet). Then write 22 yards in terms of feet (66 feet).

2. How many square feet is an acre?
An acre is 43,560 square feet.

3. Using that information, what steps must be taken to calculate the area of 1 acre in square yards?
One technique is to convert the 660 feet to yards, and the side (a *chain*) is already in terms of yards.
Another possibility is to divide the 43,560 by 9, which achieves the same result.

4. How many square yards is an acre?
An acre is 4,840 square yards.

Focus Exercises

Convert from one measure to the other.

5. 4 feet to inches
48 inches

6. 8 yards to feet
24 feet

7. $\frac{2}{3}$ feet to inches
8 inches

8. $4\frac{2}{3}$ yards to feet
14 feet

9. 16 feet to yards
$5\frac{1}{3}$ yards

10. $5\frac{1}{4}$ feet to yards
$1\frac{3}{4}$ yards

11. 4 inches to feet
$\frac{1}{3}$ foot

12. 9 inches to feet
$\frac{3}{4}$ foot

Convert each measure to inches.

13. 12 ft
144 in.

14. 2 ft 3 in.
27 in.

15. 6 ft 8 in.
80 in.

16. 15 ft 6 in.
186 in.

Convert each measure to feet and inches.

17. 28 in.
2 ft 4 in.

18. 60 in.
5 ft

19. 93 in.
7 ft 9 in.

20. 106 in.
8 ft 10 in.

21. The rim of a basketball hoop is 10 feet off the floor. Express this measure in inches.
The rim of a basketball hoop is 120 inches off the floor.

22. The record length for a garter snake is 111 inches long. Express this measure in feet and inches.
The record length for a garter snake is 9 feet 3 inches.

Add or subtract as indicated. Express each answer in feet and inches.

23. 4 ft 7 in. + 6 ft 3 in.
10 ft 10 in.

24. 5 ft 2 in. + 9 ft 4 in.
14 ft 6 in.

25. 1 ft 11 in. + 4 ft 1 in.
6 ft

26. 13 ft 8 in. + 12 ft 5 in.
26 ft 1 in.

27. 8 ft 10 in. − 3 ft 2 in.
5 ft 8 in.

28. 12 ft 6 in. − 4 ft 5 in.
8 ft 1 in.

29. 12 ft 2 in. − 8 ft 9 in.
3 ft 5 in.

30. 15 ft − 6 ft 6 in.
8 ft 6 in.

31. Rodney measured 1 foot 9 inches long (tall) at birth. During his first year, he grew 1 foot 5 inches. How tall was Rodney on his first birthday?
Rodney was 3 feet 2 inches tall on his first birthday.

32. As a workshop is being built, one opening measuring 3 feet 4 inches high will include a window on top of an air-conditioning unit. If the air-conditioning unit measures 1 foot 6 inches high, how high will the window portion be?

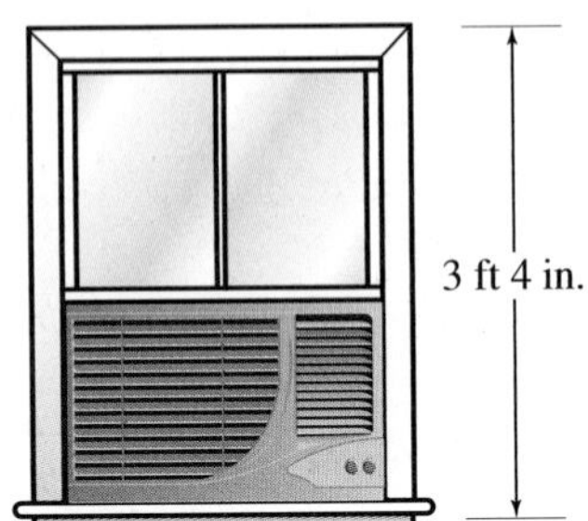

The window portion will be 1 foot 10 inches high.

Convert one measure to the other.

33. 5 pounds to ounces
80 ounces

34. $6\frac{2}{5}$ ounces to pounds
$\frac{2}{5}$ pounds

35. 4 tons to pounds
8,000 pounds

36. 1,500 pounds to tons
$\frac{3}{4}$ ton

Convert each measure to ounces.

37. 9 lb
144 oz

38. 1 lb 9 oz
25 oz

39. 6 lb 8 oz
104 oz

40. 10 lb 6 oz
166 oz

Convert each measure to pounds and ounces.

41. 28 oz
1 lb 12 oz

42. 60 oz
3 lb 12 oz

43. 90 oz
5 lb 10 oz

44. 106 oz
6 lb 10 oz

Answer as indicated.

45. A home run slugger's baseball bat weighs 2 pounds 2 ounces. Express this measure in ounces.
A home run slugger's baseball bat weighs 34 ounces.

46. An African bullfrog weighs 95 ounces. Express this measure in pounds and ounces.
An African bullfrog weighs 5 pounds 15 ounces.

Add or subtract as indicated. Express each answer in pounds and ounces.

47. 4 lb 8 oz + 6 lb 5 oz
10 lb 13 oz

48. 5 lb 12 oz + 9 lb 3 oz
14 lb 15 oz

49. 0 lb 13 oz + 7 lb 3 oz
8 lb

50. 13 lb 10 oz + 12 lb 9 oz
26 lb 3 oz

51. 8 lb 14 oz − 3 lb 2 oz
5 lb 12 oz

52. 12 lb 9 oz − 4 lb 5 oz
8 lb 4 oz

53. 12 lb 2 oz − 8 lb 9 oz
3 lb 9 oz

54. 16 lb − 7 lb 6 oz
8 lb 10 oz

55. Liu is one month old and weighs 9 pounds 12 ounces. His baby carrier weighs 7 pounds 6 ounces. What is the total weight of Liu and his baby carrier?
The total weight of Liu and his baby carrier is 17 pounds 2 ounces.

56. Ginger is a toy poodle. To find her weight, she was put on the vet's scale in her carry crate. Together, Ginger and her crate weigh 10 pounds 8 ounces. By itself, the crate weighs 2 pounds 15 ounces. How much does Ginger weigh by herself?
Ginger weighs 7 pounds 9 ounces by herself.

Convert one measure to the other.

57. 5 quarts to pints
10 pints

58. $3\frac{1}{4}$ cups to fluid ounces
26 fluid ounces

59. $2\frac{3}{8}$ pints to cups
$4\frac{3}{4}$ cups

60. $4\frac{1}{2}$ gallons to quarts
18 quarts

61. $2\frac{3}{4}$ quarts to cups
11 cups

62. $1\frac{1}{2}$ pints to fluid ounces
24 fluid ounces

63. $4\frac{2}{3}$ pints to quarts
$2\frac{1}{3}$ quarts

64. $5\frac{1}{3}$ quarts to gallons
$1\frac{1}{3}$ gallons

Think Outside the Box

65. How many inches are in one mile?
63,360 inches are in one mile.

66. How many yards are in one mile?
1,760 yards are in one mile.

67. How many tons are in 400,000 ounces?
12.5 tons are in 400,000 ounces.

68. How many fluid ounces are in a full pitcher that holds one gallon, three quarts, one pint, and one cup?
248 fluid ounces are in the pitcher.

SECTION 8.2 Metric Measures

OBJECTIVES

In this section, you will learn to

- Identify metric units of measure.
- Identify the prefixes of metric measures.
- Convert between prefixes of metric measures.
- Add and subtract metric measures.

You Need to Know

To successfully complete this section, you need to understand

- ☐ Multiplying fractions (4.4)
- ☐ Decimal fractions (5.1)
- ☐ Adding and subtracting decimals (5.3)
- ☐ Multiplying decimals by powers of 10 (5.4)
- ☐ U.S. system of measurement (8.1)

Introduction

The metric system of measurement is quite different from the U.S. system presented in Section 8.1. In many ways, the metric system is easier to use because it is based on powers of 10, such as 10, 100, and 1,000. This makes converting between measures a simple matter of moving the decimal point to the left or to the right.

The basic metric measurements are as follows:

Length: 1 meter — A **meter** is about $3\frac{3}{8}$ inches longer than a yard. (A yard is 3 feet, about the height of a doorknob from the floor.)

Weight: 1 gram — A **gram** is about the weight of the contents of a packet of artificial sweetener.

Capacity: 1 liter — A **liter** is a little more than 1 quart.

Metric Measures and Prefixes

Here is a portion of a metric ruler. It shows both millimeters (the tiny lines, abbreviated as mm) and centimeters (cm). The centimeters are $\frac{1}{100}$, or 0.01, of a meter (m). The millimeters (mm) are $\frac{1}{1{,}000}$, or 0.001, of a meter.

(Actual size)

The good thing about metric measures is that the relationship between two measures is easy to interpret: It's based on the **prefix** (the beginning part of the word) and the **base unit.**

For example, the prefix *kilo-* means "one thousand." So, a kilometer is 1,000 times as long as a meter, its base unit. Similarly, a kilogram is 1,000 grams and a kiloliter is 1,000 liters.

Here is a table of metric prefixes with different base units of measure. Any one of the base units can be used with any of the prefixes.

Prefix	Meaning	Base Unit	Put Together	Meaning
kilo-	one thousand	gram (g)	kilogram (kg)	1,000 grams
hecto-	one hundred	liter (L)	hectoliter (hL)	100 liters
deka-	ten	meter (m)	dekameter (dam)	10 meters
deci-	one-tenth	liter (L)	deciliter (dL)	$\frac{1}{10}$, or 0.1, of a liter
centi-	one-hundredth	meter (m)	centimeter (cm)	$\frac{1}{100}$, or 0.01, of a meter
milli-	one-thousandth	gram (g)	milligram (mg)	$\frac{1}{1{,}000}$, or 0.001, of a gram

One kilometer is 1,000 meters because of the way units, and now prefixes, multiply.

$$1 \text{ kilometer} = 1{,}000 \cdot (1 \text{ meter}) = 1{,}000 \text{ meters}$$
$$(1 \text{ km} = 1{,}000 \text{ m})$$

This also means that, for example, 6 kilometers $= 6 \cdot 1{,}000 \cdot (1 \text{ meter}) =$ 6,000 meters.

Notice that the base units *meters* (m), *liters* (L), and *grams* (g) have no prefix by themselves.

Example 1

Convert the given unit of measure to its base unit.

a) 7.8 hectometers (hm) **b)** 6.3 dekagrams (dag)

c) 238 centimeters (cm) **d)** 45 milliliters (mL)

Procedure Write the given unit of measure as a product based on the prefix, then multiply.

Answer **a)** 7.8 hectometers $= 7.8 \cdot (100 \text{ meters}) = 780 \text{ m}$

b) 6.3 dekagrams $= 6.3 \cdot (10 \text{ grams}) = 63 \text{ g}$

c) 238 centimeters $= 238 \cdot (0.01 \text{ meters}) = 2.38 \text{ m}$

d) 45 milliliters $= 45 \cdot (0.001 \text{ liters}) = 0.045 \text{ L}$

You Try It 1 **Convert the given unit of measure to its base unit. Use Example 1 as a guide.**

a) 5.21 kilometers **b)** 34 deciliters **c)** 514 milligrams

The abbreviation for liters is the capital letter L. It is the only metric measure abbreviated with a capital letter. Why do you think that is so?

The different metric measures are connected to each other in this way: If you were to make a plastic box with a length, width, and height of 1 decimeter (about 4 inches), it would hold exactly 1 liter of water; and that amount of water would weigh exactly 1 kilogram.

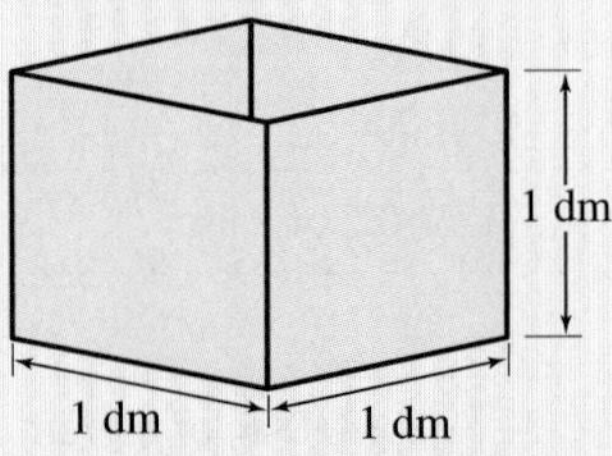

There is a slight connection between U.S. measures: 1 fluid ounce of water (capacity) weighs about 1 ounce (weight).

A sentence that may help in remembering the order of metric prefixes is "King Henry Died Monday Drinking Chocolate Milk." This stands for kilo-, hecto-, deka-, meter, deci-, centi-, milli-.

Direct Decimal Conversions of Metric Measures

Another good thing about metric measures is that they are based on the decimal system. This means that a conversion from, say, hectometers to decimeters is simply a matter of moving the decimal point to the left or to the right. It's all based on the prefix, which represents a power of 10 in one form or another.

For example, we already know that *kilo-* means "1,000 times larger" and that 6 kilometers = 6,000 meters. If we think of 6 as 6.000, converting kilometers to meters has the effect of moving the decimal point *three places to the right.*

6.000 kilometers = 6,000 meters

Similarly, we know that *centi-* means 0.01 meters; and from Example 1c, we know that 238 centimeters = 2.38 meters. So, if we think of the decimal point at the end of 238., converting from centimeters to meters has the effect of moving the decimal point *two places to the left.*

238. centimeters = 2.38 meters

The number of places the decimal point must be moved is based on the prefix. Whether to move left or right, though, might be hard to remember. Here is a diagram that will assist us in knowing which way to move the decimal point.

kilo-	hecto-	deka-	*meter*	deci-	centi-	milli-
km	hm	dam	*m*	dm	cm	mm

To use this diagram, locate the starting measure (the "converting from" measure) and count left or right to the "converting to" measure, as demonstrated in the next example.

Example 2

Convert each measure according to the diagram.

a) 52 centimeters to dekameters **b)** 3.6 hectograms to centigrams

Procedure Locate the first measure on the chart and count (right or left) the number of places it takes to get to the "converting to" measure.

Answer **a)** Start at centimeters (cm) and move three places to the left to get to the dekameters (dam) position.

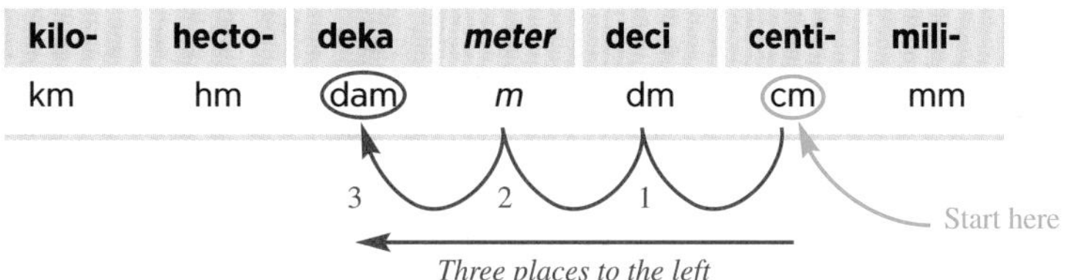

This means that we move the decimal point three places to the left.

052. centimeters = 0.052 dekameters

Moving the decimal point three places to the left requires more digits than we have. So, we need to place a 0 in front of the number.

b) Start at hectograms (hg) and move four places to the right to get to the centigrams (cg) position.

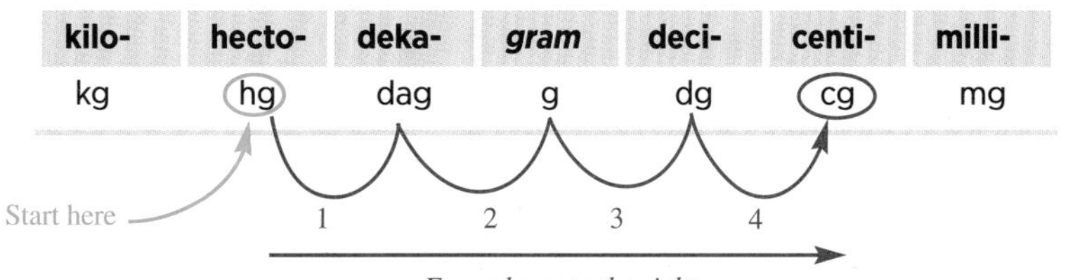

This means that we move the decimal point four places to the right.

3.6000 hectograms = 36,000. centigrams

Moving the decimal point four places to the right requires more digits than we have. So, we need to place 0's at the end of the number.

You Try It 2 **Convert each measure to its base unit of measure: meters, grams, or liters. You may use the conversion chart as demonstrated in Example 2.**

a) 8 deciliters **b)** 18.5 dekagrams **c)** 0.59 hectometers

You Try It 3 **Convert the given unit of measure to the measure shown. You may use the conversion chart as demonstrated in Example 2.**

a) 5.21 kilograms to dekagrams **b)** 28 deciliters to milliliters **c)** 7.95 hectometers to kilometers

You Try It 4 **A chewable tablet contains 500 mg of vitamin C. Express this number in grams.**

You Try It 5 **A soda bottle contains 2.3 L of soda. Express this number in milliliters.**

Adding and Subtracting Metric Measures

Adding and subtracting metric measures is simply a matter of adding and subtracting decimals. We must be sure, though, that the measures are expressed in the same units.

For example, to add 3.56 meters + 2.8 meters, we write them vertically and line up the decimal points.

$$\begin{array}{r} 3.56\text{ m} \\ +\,2.80\text{ m} \\ \hline 6.36\text{ m} \end{array}$$

← Place a 0 on the end to give each number the same number of decimal places.

However, to add 3.56 meters + 2.8 *decimeters*, we must write one in terms of the other; then we may add. Let's write 2.8 dm in terms of meters. (Use the direct method and move the decimal point.)

2.8 dm = 0.28 m We can now add 3.56 m + 0.28 m.

$$\begin{array}{r} 3.56\text{ m} \\ +\,0.28\text{ m} \\ \hline 3.84\text{ m} \end{array}$$

Example 3

Convert each measure to the one that is underlined and then add or subtract as indicated.

a) 6.4 kilograms + 850 grams **b)** 5.1 deciliters − 325 milliliters

Procedure First, convert from one measure to another. Then add or subtract.

Answer **a)** First, convert 850 grams to kilograms. 850 g = 0.850 kg

Now add.

$$\begin{array}{r} 6.400\text{ kg} \\ +\,0.850\text{ kg} \\ \hline 7.250\text{ kg} \end{array}$$

b) First, convert 5.1 deciliters to milliliters. 5.1 dL = 510 mL

Now subtract.

$$\begin{array}{r} 510\text{ mL} \\ -\,325\text{ mL} \\ \hline 185\text{ mL} \end{array}$$

You Try It 6 **Convert each measure to the one that is underlined and then add or subtract as indicated. Use Example 3 as a guide.**

a) 48 millimeters + 5.6 centimeters **b)** 10.1 kilograms + 540 grams

c) 1.65 liters − 820 milliliters

In the next two examples, we'll use the following equations:

Part + Part = Whole — When two parts are known, we can add to find the whole.

and

Part = Whole − Part — When the whole and one part are known, we can subtract to find the other part.

Example 4

When Riley was born, she weighed 34 hectograms. During her first week of life, she gained 650 grams. How many kilograms did Riley weigh after the first week?

Procedure The last sentence indicates that we want everything in terms of kilograms; so, first we must convert.

1. 34 hectograms to kilograms: 34 hg = 3.4 kg
2. 650 grams to kilograms: 650 g = 0.650 kg = 0.65 kg

Next, we'll add 3.4 kg + 0.65 kg.

Answer

$$\begin{array}{r} 3.40 \text{ kg} \\ +\,0.65 \text{ kg} \\ \hline 4.05 \text{ kg} \end{array}$$

Sentence Riley weighed 4.05 kilograms after the first week.

Example 5

Wendy was mixing punch for her son's eighth birthday party. The punch will be a 2-liter mixture of orange juice and grapefruit soda. The first thing she poured into the punch bowl was a 725-milliliter bottle of grapefruit soda. How many milliliters of orange juice must Wendy add?

Procedure The last sentence indicates that we want everything in terms of milliliters. So, we must first convert 2 liters to milliliters: 2 L = 2,000 mL.

Next, we'll subtract 2,000 mL − 725 mL.

Answer

$$\begin{array}{r} 2{,}000 \text{ mL} \\ -\,725 \text{ mL} \\ \hline 1{,}275 \text{ mL} \end{array}$$

Sentence Wendy must add 1,275 milliliters of orange juice.

You Try It 7 **A pharmacist is making a single pain pill by combining 45 centigrams of ibuprofen with 325 milligrams of ibuprofen. How many grams of ibuprofen will the pill contain?**

You Try It 8 **Harvey the snail is in a race to climb a 0.3-meter pole. After 1 minute, he has climbed 1.2 decimeters. How many centimeters does Harvey have left to climb?**

Answers: You Try It and Think About It

You Try It: **1. a)** 5,210 m **b)** 3.4 L **c)** 0.514 g **2. a)** 0.8 L **b)** 185 g **c)** 59 m **3. a)** 521 dag **b)** 2,800 mL **c)** 0.795 km **4.** A chewable tablet contains 0.5 gram of vitamin C. **5.** The soda bottle contains 2,300 milliliters of soda. **6. a)** 10.4 cm **b)** 10.64 kg **c)** 830 mL **7.** The pill will contain 0.775 gram of ibuprofen. **8.** Harvey has 18 centimeters left to climb.

Think About It: **1.** Liters is abbreviated with a capital *L* so that it is not confused with the numeral 1.

Section 8.2 Exercises

FOR EXTRA HELP
Student Resources on DVD-ROM
Includes
➤Student's Solutions Manual
➤Video Lectures
➤Chapter Test Prep Video
MyMathLab
MathXL

Think Again

1. The abbreviation for liters is the capital letter *L*. It is the only metric measure abbreviated with a capital letter. Why do you think that is so? *(Refer to Think About It 1)*

Answers will vary. One possibility: We use a capital *L* so that we don't confuse it with the number 1.

2. In this section, the connection between the three types of metric measures was mentioned: A plastic box with a length, width, and height of 1 decimeter would hold exactly 1 liter of water; and that amount of water would weigh exactly 1 kilogram. If a plastic box had a length, width, and height of 2 decimeters, how much water (in liters) would it hold and how much would that water weigh? Explain your answer.

That box would be 8 times the size of the original box; so, it would hold 8 liters of water and weigh 8 kilograms.

Focus Exercises

Convert the given unit of measure to its base unit—meters, grams, or liters.

3. 3.08 km
3,080 m

4. 28 dL
2.8 L

5. 0.45 kL
450 L

6. 6.8 dg
0.68 g

7. 870 mg
0.87 g

8. 2.4 hm
240 m

9. 91 mL
0.091 L

10. 0.89 hL
89 L

11. 365 cm
3.65 m

12. 3.1 dag
31 g

13. 49 cg
0.49 g

14. 0.87 dam
8.7 m

15. 6 cL
0.06 L

16. 27 kL
27,000 L

17. 4 dm
0.4 m

18. 40 kg
40,000 g

Convert the given unit of measure to the measure shown.

19. 3.6 kilometers to dekameters
360 dekameters

20. 2 decigrams to milligrams
200 milligrams

21. 312 milligrams to decigrams
3.12 decigrams

22. 1.2 hectometers to kilometers
0.12 kilometer

23. 0.9 dekaliters to kiloliters
0.009 kiloliter

24. 14 centiliters to milliliters
140 milliliters

25. 2.5 meters to decimeters
25 decimeters

26. 45 meters to kilometers
0.045 kilometer

27. 1,810 milliliters to deciliters
18.1 deciliters

28. 52 centigrams to milligrams
520 milligrams

29. 0.83 kilometers to hectometers
8.3 hectometers

30. 3.6 hectometers to centimeters
36,000 centimeters

31. 37 centigrams to decigrams
3.7 decigrams

32. 0.82 kiloliters to deciliters
8,200 deciliters

33. 6 milliliters to centiliters
0.6 centiliter

34. 16 centigrams to decigrams
1.6 decigrams

Answer each problem with a sentence containing the requested measurement.

35. A child weighs 15.6 kilograms. Express this number in grams.
A child weighs 15,600 grams.

36. Randi's thermos can hold 355 milliliters. Express this number in liters.
Randi's thermos can hold 0.355 liter.

37. Veronica's rosebush is 83.5 centimeters high. Express this number in meters.
Veronica's rosebush is 0.835 meter high.

38. A baseball weighs 155 grams. Express this number in kilograms.
A baseball weighs 0.155 kilogram.

39. A dwarf pygmy goby (the smallest known freshwater fish) has a diameter of 0.099 decimeters. Express this number in millimeters.
A dwarf pygmy goby has a diameter of 9.9 millimeters.

40. A pen holds 8.3 milliliters of ink. Express this number in deciliters.
A pen holds 0.083 deciliter of ink.

Convert each measure to the one that is underlined and then add.

41. 1.42 dekagrams + 591 <u>centigrams</u>
2,011 centigrams

42. 745 milliliters + 0.862 <u>liters</u>
1.607 liter

43. 3.07 <u>dekaliters</u> + 15 deciliters
3.22 dekaliters

44. 28.7 <u>decimeters</u> + 64 millimeters
29.34 decimeters

Convert each measure to the one that is underlined and then subtract.

45. 41 <u>centigrams</u> − 253 milligrams
15.7 centigrams

46. 125 <u>centimeters</u> − 0.72 meter
53 centimeters

47. 2 hectoliters − 684 <u>deciliters</u>
1,316 deciliters

48. 1,020 milligrams − 0.045 <u>dekagrams</u>
0.057 dekagram

Add or subtract as indicated. Write a sentence answering the question.

49. At a hospital, a fluid drip bag contains 1.5 liters of saline solution. A nurse adds 12.5 centiliters of pain medicine to the bag. How many milliliters of fluid are now in the bag?
1,625 milliliters of fluid are now in the bag.

50. For the first three days of a winter storm, the town of Bedrock received 0.95 decimeter of rain. On the storm's fourth day, the town received 18 millimeters. How many centimeters of rain did Bedrock receive in those four days?
Bedrock received 11.3 centimeters of rain in those four days.

51. Jermaine put 750 milliliters of a fuel additive into his gas tank. He then pumped in 1.265 dekaliters of gas. How many total liters did Jermaine add to his gas tank?
Jermaine added a total of 13.4 liters to his gas tank.

52. A single serving of cereal contains 0.35 gram of sodium. With milk, the total amount of sodium is 3.97 decigrams. How many milligrams of sodium are in the serving of milk?
47 milligrams of sodium are in the serving of milk.

53. On her tenth birthday, Tamayra was 1.15 meters tall. On her eleventh birthday, she was 1,237 millimeters tall. How many centimeters did Tamayra grow that year?
Tamayra grew 8.7 centimeters that year.

54. A small apple weighs 1.42 hectograms, and a large apple weighs 0.22 kilogram. How many grams heavier is the large apple than the small apple?
The large apple is 78 grams heavier than the small apple.

Think Outside the Box

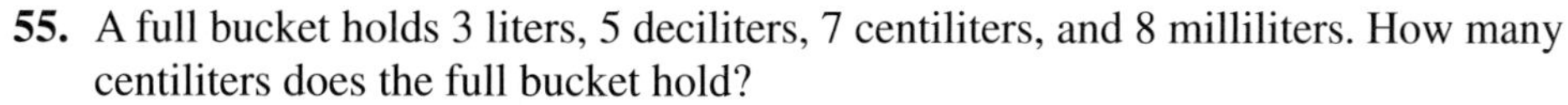

55. A full bucket holds 3 liters, 5 deciliters, 7 centiliters, and 8 milliliters. How many centiliters does the full bucket hold?
The full bucket holds 357.8 centiliters.

56. A rectangle has a length of 3.6 centimeters and a width of 0.17 decimeter.

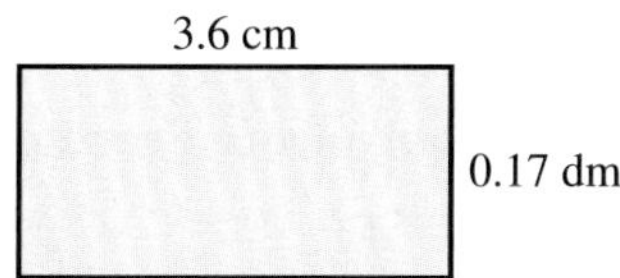

a) Express the perimeter of this rectangle in meters.
The perimeter is 0.106 meter.

b) Express the area of this rectangle in square millimeters.
The area is 612 square millimeters.

SECTION 8.3 Converting U.S. and Metric Measures

OBJECTIVES

In this section, you will learn to

- Create and use equivalence fractions to convert between measures.
- Convert between U.S. and metric measures.

You Need to Know

To successfully complete this section, you need to understand

- ☐ Writing improper fractions as mixed numbers and mixed numbers as improper fractions (4.2)
- ☐ Simplifying fractions (4.3)
- ☐ U.S. measures (8.1)
- ☐ Metric measures (8.2)

Introduction

One technique for converting measures relies on multiplying by 1. As you know, any fraction in which the numerator and the denominator have the same value is equivalent to 1.

Because 1 foot = 12 inches, the fraction $\frac{1\text{ ft}}{12\text{ in.}}$ and its reciprocal $\frac{12\text{ in.}}{1\text{ ft}}$ are both equal to 1. Each of these is called an **equivalence fraction** because it expresses the value 1. Equivalence fractions are used to convert from one measure to another, as demonstrated throughout this section.

Converting Measures Using Equivalence Fractions

Here are a variety of equivalence fractions based on the measures we used in the past two sections. This is just a sample of the many possible equivalence fractions, with both U.S. and metric measures.

Notice that an equivalence fraction can be expressed two ways. One fraction is the reciprocal of the other.

Equivalency	**Equivalence Fractions**
1 ft = 12 in.	$\frac{1\text{ ft}}{12\text{ in.}} = 1$ and $\frac{12\text{ in.}}{1\text{ ft}} = 1$
1 gal = 4 qt	$\frac{1\text{ gal}}{4\text{ qt}} = 1$ and $\frac{4\text{ qt}}{1\text{ gal}} = 1$
1 lb = 16 oz	$\frac{1\text{ lb}}{16\text{ oz}} = 1$ and $\frac{16\text{ oz}}{1\text{ lb}} = 1$
1 km = 1,000 m	$\frac{1\text{ km}}{1{,}000\text{ m}} = 1$ and $\frac{1{,}000\text{ m}}{1\text{ km}} = 1$
100 cL = 1 L	$\frac{100\text{ cL}}{1\text{ L}} = 1$ and $\frac{1\text{ L}}{100\text{ cL}} = 1$
1,000 mg = 1 g	$\frac{1{,}000\text{ mg}}{1\text{ g}} = 1$ and $\frac{1\text{ g}}{1{,}000\text{ mg}} = 1$

Example 1

Given the following equivalencies, create an equivalence fraction equal to 1. Then write its reciprocal equal to 1.

a) 1 pennyweight = 24 grains **b)** 8 furlongs = 1 mile

Procedure Make a fraction from each equivalency and set it equal to 1. Then do the same for its reciprocal.

Answer **a)** $\frac{1\text{ pennyweight}}{24\text{ grains}} = 1$ and $\frac{24\text{ grains}}{1\text{ pennyweight}} = 1$

b) $\frac{8\text{ furlongs}}{1\text{ mile}} = 1$ and $\frac{1\text{ mile}}{8\text{ furlongs}} = 1$

You Try It 1 **Given the following equivalencies, create an equivalence fraction equal to 1. Then write its reciprocal equal to 1. Use Example 1 as a guide.**

a) 1 day = 24 hours **b)** 1 dekagram = 10 grams **c)** 2,000 pounds = 1 ton

We can multiply anything by 1 and not change its value. For example, if we need to write 2 feet in terms of inches, we can write 2 feet as a fraction, $\frac{2 \text{ feet}}{1}$, and then multiply $\frac{2 \text{ feet}}{1}$ by a special form of 1, namely $\frac{12 \text{ inches}}{1 \text{ foot}}$.

Recall from Section 8.1 that the unit of measure is multiplied by the number that precedes it. So, we are able to divide out the unit of measure as we would any other common factor, as shown here.

$$\frac{2 \text{ feet}}{1} \times \frac{12 \text{ inches}}{1 \text{ foot}} = \frac{2 \cdot \cancel{\text{feet}} \times 12 \cdot \text{inches}}{1 \times 1 \cdot \cancel{\text{foot}}} = \frac{24 \cdot \text{inches}}{1} = 24 \text{ inches}$$

Notice that *feet* is in the numerator and *foot* is in the denominator; so, they can be divided out.

Caution We must choose the correct form of 1 to get the results we seek. Not just any fraction of 1 will do. For example, if we're trying to convert from feet to inches, we must work with the equivalency 1 foot = 12 inches.

Our choice of fraction should always be $\frac{\text{New unit of measure}}{\text{Original unit of measure}}$.

Example 2

Use an equivalence fraction to convert from one measure to another. Write $11\frac{1}{4}$ feet in terms of yards.

Procedure First, write $11\frac{1}{4}$ feet as an improper fraction: $11\frac{1}{4} \text{ feet} = \frac{45 \text{ ft}}{4}$.

Notice that the units are placed in the numerator.

Second, decide which equivalency to make into 1 so that we can convert properly. Third, divide out the common units, multiply, and simplify.

Answer Multiply $\frac{45 \text{ ft}}{4}$ by $\frac{1 \text{ yd}}{3 \text{ ft}}$. ← New unit / ← Original unit

$$\frac{45 \text{ ft}}{4} \cdot \frac{1 \text{ yd}}{3 \text{ ft}} = \frac{45 \text{ ft} \cdot 1 \text{ yd}}{4 \cdot 3 \text{ ft}} = \frac{45 \cdot 1 \text{ yd}}{4 \cdot 3} = \frac{15 \cdot 1 \text{ yd}}{4 \cdot 1} = \frac{15}{4} \text{ yd} = 3\frac{3}{4} \text{ yards}$$

Divide out feet. Divide out a factor of 3.

Example 3

Use an equivalence fraction to convert from one metric measure to another. Write 3.5 liters in terms of deciliters.

Procedure First, write 3.5 liters as a fraction.

Second, notice that the equivalency is between liters and deciliter: 10 deciliters = 1 liter.

Use $\frac{10 \text{ dL}}{1 \text{ L}}$. ← New unit / ← Original unit

Third, divide out the common units, multiply, and simplify.

Answer

$$\frac{3.5 \text{ L}}{1} \times \frac{10 \text{ dL}}{1 \text{ L}} = \frac{3.5 \text{ L} \times 10 \text{ dL}}{1 \times 1 \text{ L}} = \frac{35 \text{ dL}}{1} = 35 \text{ deciliters}$$

Divide out liters.

You Try It 2 **Use an equivalence fraction to convert from one measure to another. Use Examples 2 and 3 as guides. (You may want to refer back to the equivalencies presented in Sections 8.1 and 8.2.)**

a) Convert 24 fluid ounces to cups. **b)** Convert 350 meters to kilometers.

Converting between U.S. and Metric Linear Measures

We also can convert between U.S. measures and metric measures. Usually, though, these conversions are approximations and are not exact. Instead of the equal sign, we use $\approx$ to represent "approximately equal to."

Here is a list of some common U.S. and metric "equivalents." Notice in the first one that 2 inches is *approximately equal to* 5 centimeters. It's appropriate to say that 2 inches is *about* 5 centimeters.

	Approximation	**As a fraction ≈ 1**
Measures of length/distance	2 inches ≈ 5 centimeters	$\frac{2\text{ in.}}{5\text{ cm}} \approx \frac{5\text{ cm}}{2\text{ in.}} \approx 1$
	10 yards ≈ 9 meters	$\frac{10\text{ yd}}{9\text{ m}} \approx \frac{9\text{ m}}{10\text{ yd}} \approx 1$
	5 miles ≈ 8 kilometers	$\frac{5\text{ mi}}{8\text{ km}} \approx \frac{8\text{ km}}{5\text{ mi}} \approx 1$

Example 4

Convert 9 miles into kilometers.

Procedure First, write 9 miles as a fraction over 1.

Second, use an approximate equivalency to make 1; we want to eliminate *miles*.

Third, multiply the fractions. Write the answer as a decimal approximation.

Answer

$9\text{ mi} = \frac{9\text{ mi}}{1}$ We want to eliminate miles; so, use $\frac{8\text{ km}}{5\text{ mi}}$. (8 km ← New unit; 5 mi ← Original unit)

$\approx \frac{9\text{ mi}}{1} \cdot \frac{8\text{ km}}{5\text{ mi}}$

$\approx \frac{9\text{ mi} \cdot 8\text{ km}}{1 \cdot 5\text{ mi}}$ Divide out the miles.

$\approx \frac{9 \cdot 8\text{ km}}{1 \cdot 5}$ This is $\frac{72\text{ km}}{5}$, but it is also $\frac{72}{5}$ km.

$\approx \frac{72}{5}\text{ km}$ Divide: $5\overline{)72.0}$ = 14.4

$\approx 14.4\text{ km}$ 9 miles ≈ 14.4 kilometers

The fraction $\frac{72}{5}$ can be written as a mixed number, $14\frac{2}{5}$, or as a decimal, 14.4. In the previous example, we wrote the answer as a decimal. Would it have been better to write the answer as a mixed number, $14\frac{2}{5}$ km? Why or why not?

____

____

Example 5

Convert 6 kilometers to miles.

Procedure First, write 6 km as a fraction over 1.

Second, use an approximate equivalency to make 1; we want to eliminate kilometers. Third, multiply the fractions and simplify. Write the answer as a mixed number.

Answer

$$6\text{ km} = \frac{6\text{ km}}{1}$$ Use $\frac{5\text{ mi}}{8\text{ km}}$ (5 mi ← New unit; 8 km ← Original unit).

$$\approx \frac{6\text{ km}}{1} \cdot \frac{5\text{ mi}}{8\text{ km}}$$ Multiply the fractions.

$$\approx \frac{6\text{ km} \cdot 5\text{ mi}}{1 \cdot 8\text{ km}}$$ Divide out the kilometers.

$$\approx \frac{6 \cdot 5\text{ mi}}{1 \cdot 8}$$ This is $\frac{30\text{ mi}}{8} = \frac{15\text{ mi}}{4} = \frac{15}{4}\text{ mi}$.

$$\approx \frac{15}{4}\text{ mi}$$ As a mixed number, $\frac{15}{4} = 3\frac{3}{4} = 3.75$.

6 kilometers $\approx 3\frac{3}{4}$ miles

It is common to write metric values in decimal form and U.S. measures in fractional or mixed number form, but this is not a requirement.

You Try It 3 **Convert each measure to its alternate U.S. or metric form. Use Examples 4 and 5 as guides. Write a metric measure answer as a decimal approximation and a U.S. measure answer as a mixed number.**

a) Convert 8 inches to centimeters.

b) Convert 15 meters to yards.

c) Convert $17\frac{1}{2}$ miles to kilometers.

Example 6

A computer screen is 17 inches wide. Approximately how many centimeters wide is the screen?

Procedure This problem is asking us to convert inches to centimeters. Write the first sentence with the converted measurement.

Answer

$$17\text{ in.} = \frac{17\text{ in.}}{1}$$ Use $\frac{5\text{ cm}}{2\text{ in.}}$ (5 cm ← New unit; 2 in. ← Original unit).

$$\approx \frac{17\text{ in.}}{1} \cdot \frac{5\text{ cm}}{2\text{ in.}}$$

$$\approx \frac{17\text{ in.} \cdot 5\text{ cm}}{1 \cdot 2\text{ in.}}$$ Divide out the inches.

$$\approx \frac{17 \cdot 5\text{ cm}}{1 \cdot 2}$$ This is $\frac{85\text{ cm}}{2} = \frac{85}{2}\text{ cm}$.

$$\approx \frac{85}{2}\text{ cm}$$ Divide: $2\overline{)85.0}$ = 42.5

$$\approx 42.5\text{ cm}$$

Sentence The computer screen is approximately 42.5 cm wide.

You Try It 4 **Soo-Kim jogs 4 kilometers every morning. Approximately how many miles does Soo-Kim jog?**

You Try It 5 **The longest goldfish in captivity is 37 centimeters long. Express this number in inches.** ***Source: Guinness World Records, 2004***

Think About It 2 How tall are you?

a) Write your height in feet and inches.

b) Convert your height to inches only.

c) Write your height in centimeters.

d) Write your height in meters.

Measures of Capacity and Measures of Weight

Here is a list of some common U.S. and metric equivalents for capacity and weight.

	Approximation	As a fraction ≈ 1
Measures of capacity	20 quarts ≈ 19 liters	$\frac{20\text{ qt}}{19\text{ L}} \approx \frac{19\text{ L}}{20\text{ qt}} \approx 1$
	5 gallons ≈ 19 liters	$\frac{5\text{ gal}}{19\text{ L}} \approx \frac{19\text{ L}}{5\text{ gal}} \approx 1$
Measures of weight	1 ounce ≈ 28 grams	$\frac{1\text{ oz}}{28\text{ g}} \approx \frac{28\text{ g}}{1\text{ oz}} \approx 1$
	11 pounds ≈ 5 kilograms	$\frac{11\text{ lb}}{5\text{ kg}} \approx \frac{5\text{ kg}}{11\text{ lb}} \approx 1$

Example 7

Convert 4 kilograms to pounds.

Procedure First, write 4 kilograms as a fraction over 1.
Second, use an equivalence fraction to make 1.
Third, multiply the fractions and simplify. Write the answer as a mixed number.

Answer

$4\text{ kg} = \frac{4\text{ kg}}{1}$ Use $\frac{11\text{ lb}}{5\text{ kg}} \leftarrow \frac{\text{New unit}}{\text{Original unit}}$.

$\approx \frac{4\text{ kg}}{1} \cdot \frac{11\text{ lb}}{5\text{ kg}}$ Multiply the fractions.

$\approx \frac{4\text{ kg} \cdot 11\text{ lb}}{1 \cdot 5\text{ kg}}$ Divide out the kilograms.

$\approx \frac{4 \cdot 11\text{ lb}}{1 \cdot 5}$

$\approx \frac{44}{5}\text{ lb}$ Write the fraction as a mixed number.

$\approx 8\frac{4}{5}\text{ lb}$

4 kilograms ≈ $8\frac{4}{5}$ pounds

You Try It 6 **Convert each measure to its alternate U.S. or metric form. Use Example 7 as a guide. Write improper fractions as mixed numbers for U.S. measures and as decimals for metric measures.**

a) Convert 5 ounces to grams.

b) Convert 15 pounds to kilograms.

c) Convert 50 quarts to liters.

You Try It 7 **A prehistoric bird called the teratorn is believed to have weighed about 80 kilograms. Express this number in pounds.** ***Source: Guinness World Records,* 2004**

You Try It 8 **The largest gelatin dessert ever made contained 35,000 liters of watermelon flavored Jell-O mix. Express this number in gallons. (Round to the nearest whole number.)** ***Source: Guinness World Records,* 2004**

Answers: You Try It and Think About It

You Try It: **1. a)** $\frac{1\text{ day}}{24\text{ hours}} = 1$; $\frac{24\text{ hours}}{1\text{ day}} = 1$ **b)** $\frac{1\text{ dekagram}}{10\text{ grams}} = 1$; $\frac{10\text{ grams}}{1\text{ dekagram}} = 1$ **c)** $\frac{2{,}000\text{ pounds}}{1\text{ ton}} = 1$; $\frac{1\text{ ton}}{2{,}000\text{ pounds}} = 1$ **2. a)** 24 oz = 3 cups **b)** 350 m = 0.35 km **3. a)** 8 in. ≈ 20 cm **b)** 15 m ≈ $16\frac{2}{3}$ yd **c)** $17\frac{1}{2}$ mi ≈ 28 km **4.** Soo-Kim jogs about $2\frac{1}{2}$ miles. **5.** The longest goldfish in captivity is about $14\frac{4}{5}$ inches long. **6. a)** 5 oz ≈ 140 g **b)** 15 lb ≈ 6.82 kg **c)** 50 qt ≈ 47.5 L **7.** The teratorn weighed about 176 pounds. **8.** The dessert contained about 9,211 gallons.

Think About It: **1.** No. To convert quickly between metric measures, such as from kilometers to meters, it is better to write all metric measures in decimal form. **2.** Answers may vary.

Section 8.3 Exercises

FOR EXTRA HELP
Student Resources on DVD-ROM
Includes
➤ Student's Solutions Manual
➤ Video Lectures
➤ Chapter Test Prep Video
MyMathLab
MathXL

Think Again

1. The fraction $\frac{72}{5}$ can be written as a mixed number, $14\frac{2}{5}$, or as a decimal, 14.4. Is it better to write $\frac{72}{5}$ kilometers as $14\frac{2}{5}$ km or as 14.4 km? Explain your answer. *(Refer to Think About It 1)*
Answers will vary. One possibility: It's okay to write it as $14\frac{2}{5}$; but because the unit of measure is metric, it is more common to write the number (or approximation) in decimal form. So, 14.4 is preferred.

2. Fahrenheit and Celsius are two temperature scales, one of which is metric. Which is the metric temperature scale? Explain your answer or show an example that supports your answer.
(This may require a little research.) Celsius is the metric temperature scale. It includes 0° for the freezing point of water (or melting point of ice) and 100° for the boiling point of water.

Focus Exercises

Use an equivalence fraction to write each measure in the requested unit of measure. (You may want to refer back to the equivalencies in Sections 8.1 and 8.2.)

3. Convert 42 feet to yards.
14 yards

4. Convert 60 inches to feet.
5 feet

5. Convert 28 quarts to gallons.
7 gallons

6. Convert 3,000 pounds to tons.
$1\frac{1}{2}$ tons

7. Convert 36 fluid ounces to cups.
$4\frac{1}{2}$ cups

8. Convert 40 ounces to pounds.
$2\frac{1}{2}$ pounds

9. Convert 95 centimeters to meters.
0.95 meter

10. Convert 86 decimeters to meters.
8.6 meters

11. Convert 60 milligrams to grams.
0.06 gram

12. Convert 2,850 grams to kilograms.
2.85 kilograms

13. Convert 45 deciliters to liters.
4.5 liters

14. Convert 65.4 liters to hectoliters.
0.654 hectoliter

Convert each measure to its alternate U.S. or metric form. Write improper fractions as mixed numbers for U.S. measures and as decimals for metric measures.

15. Convert 12 inches to centimeters.
30 centimeters

16. Convert 36 meters to yards.
40 yards

17. Convert 24 kilometers to miles.
15 miles

18. Convert 25 centimeters to inches.
10 inches

19. Convert $13\frac{1}{3}$ yards to meters.
12 meters

20. Convert $22\frac{1}{2}$ miles to kilometers.
36 kilometers

21. Convert $12\frac{1}{2}$ gallons to liters.
47.5 liters

22. Convert $3\frac{3}{4}$ ounces to grams.
105 grams

23. Convert $8\frac{1}{4}$ pounds to kilograms.
3.75 kilograms

24. Convert 85.5 liters to gallons.
$22\frac{1}{2}$ gallons

25. Convert 413 grams to ounces.
$14\frac{3}{4}$ ounces

26. Convert 45 kilograms to pounds.
99 pounds

Answer each problem with a sentence containing the requested unit of measure. Write improper fractions as mixed numbers for U.S. measures and as decimals for metric measures.

27. A foot race is 10 kilometers. Express this number in miles.
A foot race is $6\frac{1}{4}$ miles.

28. An auto race is 120 miles. Express this number in kilometers.
An auto race is 192 kilometers.

29. A window is 48 inches wide. Express this number in centimeters.
A window is 120 centimeters wide.

30. A rectangular field is 40 meters long. Express this number in yards.
A rectangular field is $44\frac{4}{9}$ yards long.

31. Sammy weighs 27.5 kilograms. Express this number in pounds.
Sammy weighs $60\frac{1}{2}$ pounds.

32. A pint of potato salad contains $1\frac{1}{2}$ ounces of fat. Express this number in grams.
A pint of potato salad contains 42 grams of fat.

33. Papi's pickup truck can hold 30 gallons of gas. Express this number in liters.
Papi's pickup truck can hold 114 liters of gas.

34. Tina's fish tank holds 7.6 liters of water. Express this number in quarts.
Tina's fish tank holds 8 quarts of water.

Think Outside the Box

35. Myra is 9 months old and weighs 15 pounds 13 ounces. Express her weight in kilograms.
Myra weighs 7.188 kilograms.

36. A rectangle has a length of 5 feet 8 inches and a width of 4 feet 2 inches.

a) Express the perimeter of this rectangle in meters.
The perimeter is 5.9 meters.

b) Express the area of this rectangle in square meters. (*Hint:* First convert each side measure to meters.)
The area is 2.125 square meters.

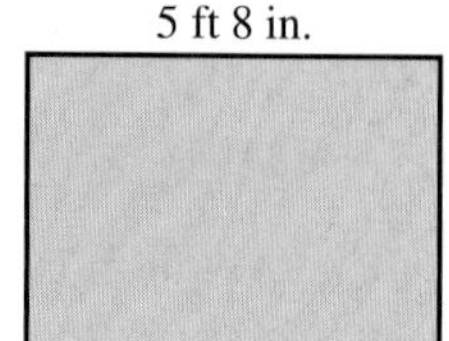

SECTION 8.4 Lines and Angles

OBJECTIVES

In this section, you will learn to

- Identify the parts of a line.
- Name geometric figures.
- Recognize the different types of angles.
- Find angle measures.

You Need to Know

To successfully complete this section, you need to understand

- ☐ Adding and subtracting whole numbers (1.2)
- ☐ Adding and subtracting decimals (5.3)
- ☐ Adding and subtracting measures (8.1 and 8.2)

Introduction

Geometry is the study of the shapes and sizes of things such as these:

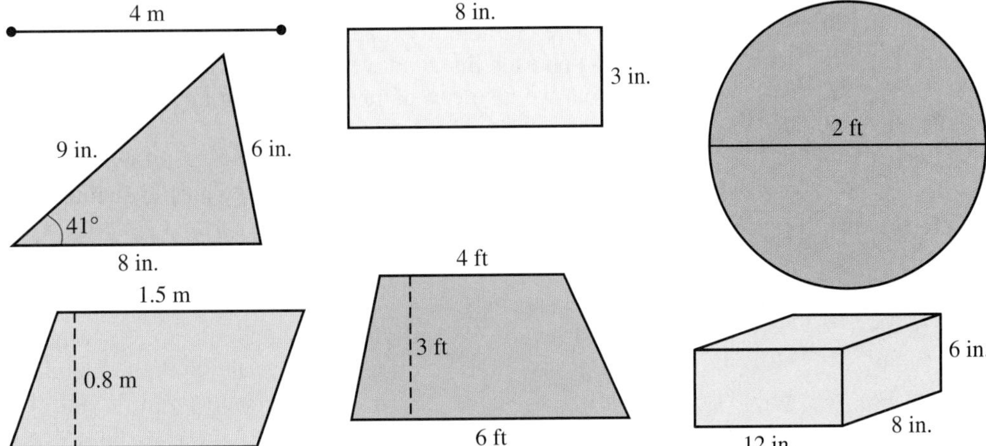

We'll study each of those shapes in this chapter.

Most of the work in the geometry portion of this chapter deals with shapes in a *plane*. In geometry, a **plane** is any flat surface that extends forever in each direction.

Although a wall, a floor, and a tabletop have definite lengths and widths, we might think of those things as part of a plane, as if each went on endlessly.

The Bonneville Salt Flats in Utah might give you a better idea of a geometric plane. It is over 30,000 acres in size, it is flat, and its surface contains (among other minerals) potassium and sodium chloride (table salt).

According to the definition of a plane, is the Bonneville Salt Flats a real plane? Why or why not?

____

____

Lines and Points

Some highways in America are straight for miles and miles. A line is usually painted on the side of the highway, and this line seems to go on forever.

In geometry, we consider a **line** to be straight, and we say that it goes on forever. To indicate this, we put arrows on each end. We also say that the line has *infinite* length. To the right are four examples of lines.

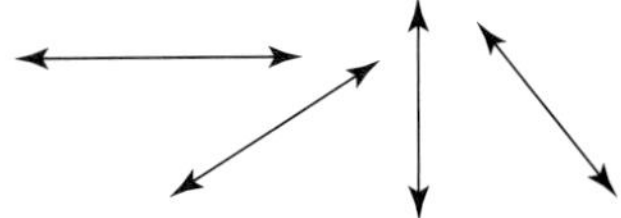

When two lines cross each other, they meet in a single **point** of intersection. We say that the lines *intersect* at the point.

A point is very small (it really has no size at all), but we often represent a point with a dot • and label it with a capital letter.

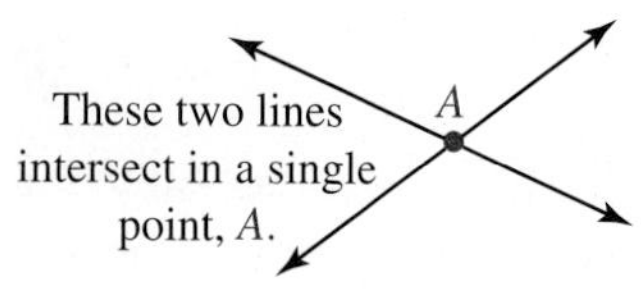

These two lines intersect in a single point, A.
A

There are actually many points (an *infinite* number of points) on each line, and we might label a few of them. Notice that we also can name a line using any two points on that line, such as this line.

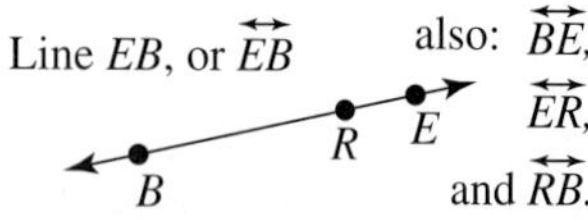

Two lines in the same plane that won't *ever* intersect each other are called **parallel lines.**

$\overleftrightarrow{MP}$ and $\overleftrightarrow{QR}$ are parallel to each other.

However, $\overleftrightarrow{XY}$ and $\overleftrightarrow{BC}$ are not parallel to each other.

Because lines go on forever, we can think of extending each line to see if they'll ever intersect. Even though the lines don't cross on this page, if our page were the size of the Bonneville Salt Flats, we'd see that the lines eventually intersect.

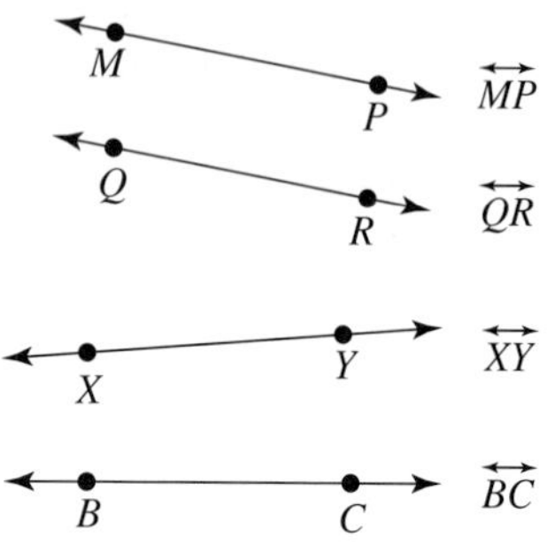

Line Segments

The portion of a line that is drawn between two points is called a **line segment.** A line segment doesn't go on forever in each direction. It has a definite, or *finite*, length. A line segment has definite start and stop points called **endpoints.** We use the two endpoints to name the line segment, and either endpoint can be written first.

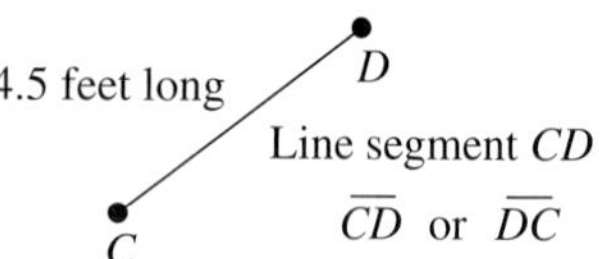

Because line segments don't extend forever, many line segments don't intersect. However, even though they don't intersect, line segments such as $\overline{AB}$ and $\overline{JK}$ are not considered parallel.

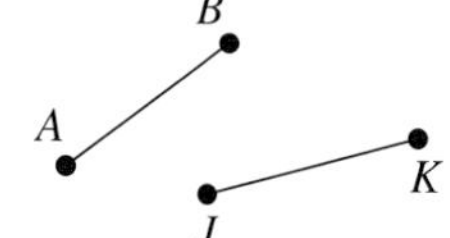

Two line segments are parallel if and only if the lines that contain them are parallel.

So, because *lines* $\overleftrightarrow{AB}$ and $\overleftrightarrow{JK}$ are not parallel, line *segments* $\overline{AB}$ and $\overline{JK}$ also are not parallel.

Rays

A piece of a line that starts at one of its points and goes on forever in only one directions is called a **ray.** The starting point of the ray is called the endpoint. The ray will have at least one other point labeled on it indicating the direction of the ray.

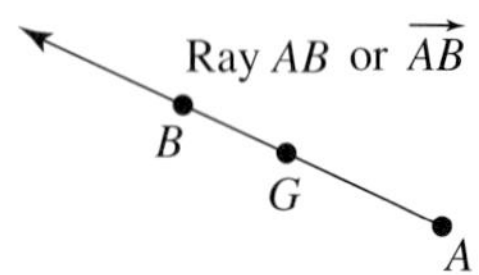

Notice that in naming the ray, *the endpoint is always written first.*

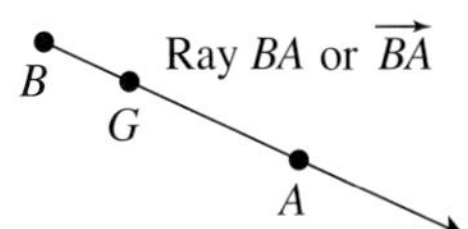

You Try It 1 **Use a straightedge (a ruler) to draw the requested figure.**

a) Draw $\overleftrightarrow{PR}$

b) Draw $\overline{WX}$

c) Draw $\overrightarrow{KH}$

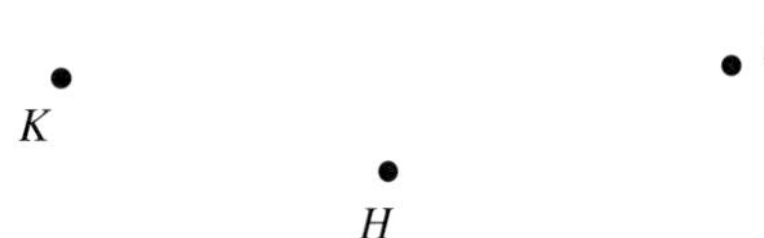

Angles

When two rays share the same endpoint, they form an **angle.** The common endpoint is called the **vertex** of the angle. The two rays that form the angle are called the sides of the angle.

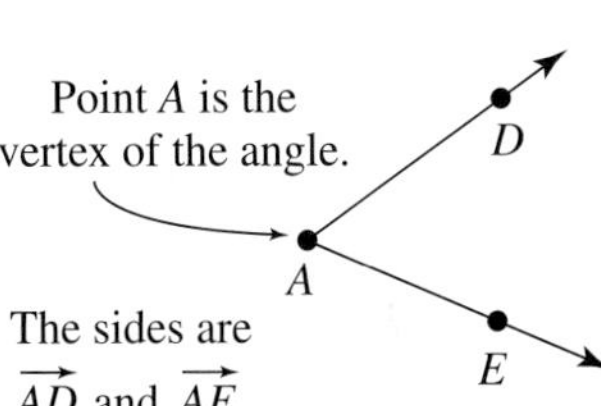

An angle can also be formed when two line segments share the same endpoint. Here the vertex is G and the sides are line segments $\overline{GT}$ and $\overline{GM}$.

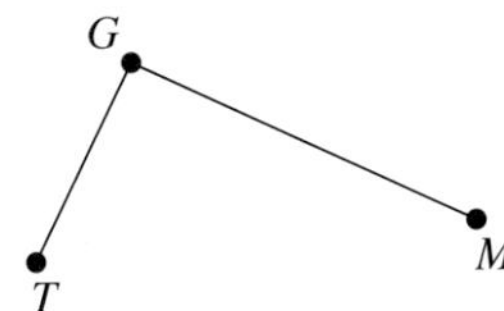

The angle at the right can be named in a variety of ways using the angle symbol, $\angle$. We can name it $\angle G$, $\angle 1$, $\angle TGM$, or $\angle MGT$. They all name the same angle according to the diagram. (The number *1* within the angle is only a label; it has nothing to do with the size of the angle.)

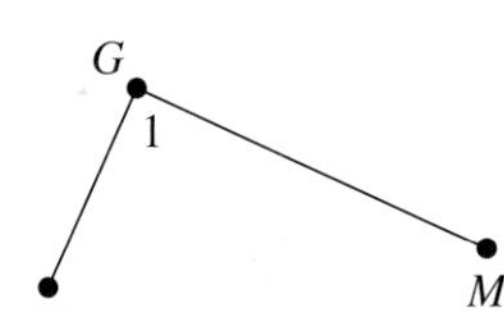

Notice that the last two names, $\angle TGM$ and $\angle MGT$, use the three endpoints of the line segments that define the angle and that the vertex, **G**, is in the middle.

The diagram at the right shows what is meant by the *interior* of an angle; it is the shaded portion. Point T is in the interior of the angle.

Just as the rays continue indefinitely, the interior of the angle continues indefinitely and never crosses to the other side of the rays.

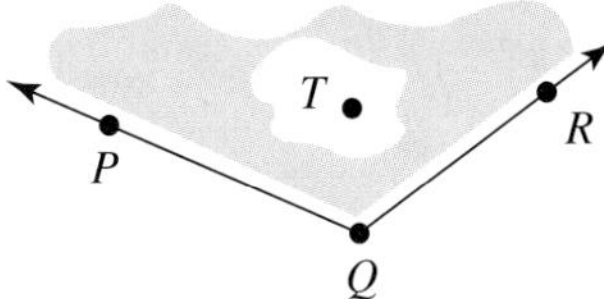

You Try It 2 **Use a straightedge to draw the angles named. (Keep in mind that the vertex is always in the middle of the name.) You may use line segments for the sides of each angle. Also, *lightly* shade the interior of each angle.**

a) Draw $\angle PDH$ **b)** Draw $\angle PHD$.

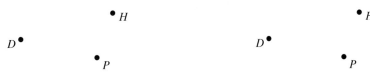

When two lines cross, they form four angles, each with the same vertex, as in the diagram at the right. It would not be appropriate to name any of these angles $\angle G$ because it would be unclear which of the four angles is being named.

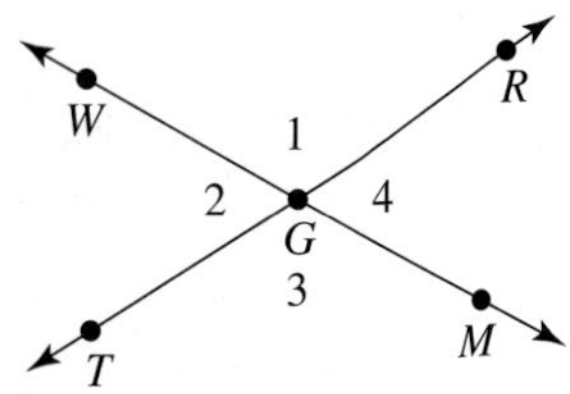

Example 1

Identify two other names for each given angle.

Answer **a)** $\angle TGW$ can also be named $\underline{\angle 2 \text{ or } \angle WGT}$.

b) $\angle TGM$ can also be named $\underline{\angle 3 \text{ or } \angle MGT}$.

c) $\angle 1$ can also be named $\underline{\angle RGW \text{ or } \angle WGR}$.

d) $\angle 4$ can also be named $\underline{\angle RGM \text{ or } \angle MGR}$.

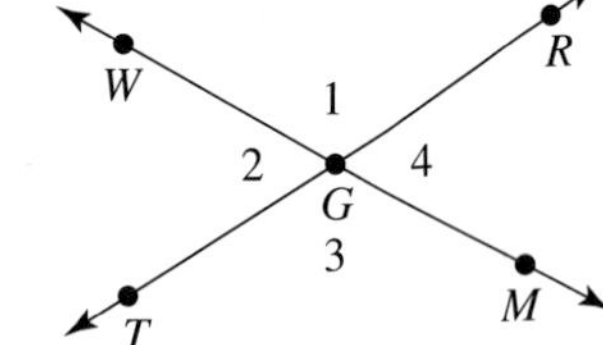

You Try It 3 **Identify two other names for each given angle. Use Example 1 as a guide.**

a) $\angle CDE$ can also be named ______________.

b) $\angle CDA$ can also be named ______________.

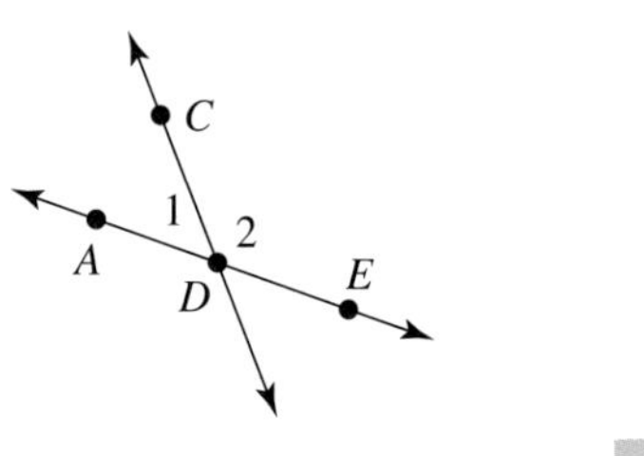

Measures of Angles

When a ballet dancer performs a pirouette, he or she is making a complete turn, a complete rotation, so that when the turn is finished, the dancer is facing the same direction as when he or she started.

Circles have 360 degrees—instead of 100 degrees or 400 degrees—because 360 has many whole number factors, including 12. This means that it can be easily divided into 12 equal parts, the number of months in a year and the number of hours on a standard clock face.

A complete turn is also seen as a complete circle, and it is measured in *degrees*.

360°

One full revolution is 360°, read "360 degrees."

This small raised circle is the symbol for *degrees*, the unit used to measure angles.

Half of a turn is 180°, and one-fourth of a turn (or a *quarter turn*) is 90°.

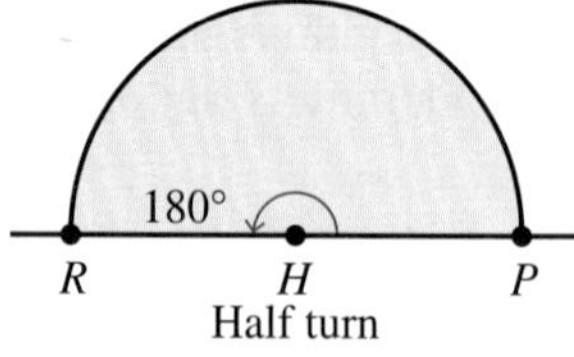

Half turn

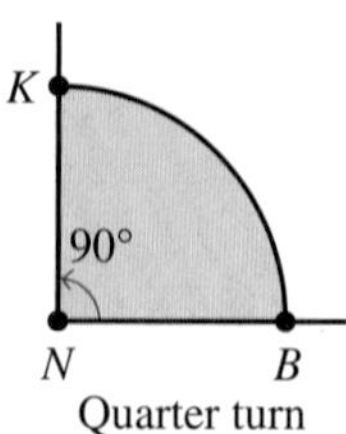

Quarter turn

Any point on a line can be the vertex of a straight angle. Here are two examples.

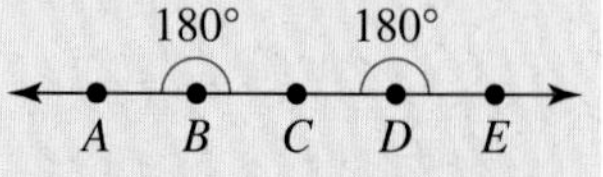

$\angle ABC$ is a straight angle. B is the vertex.
$\angle CDE$ is a straight angle. D is the vertex.

The half turn creates a 180° angle in which the two sides, $\overrightarrow{HR}$ and $\overrightarrow{HP}$, for a straight line; so, it is called a straight angle. Every line forms a straight angle.

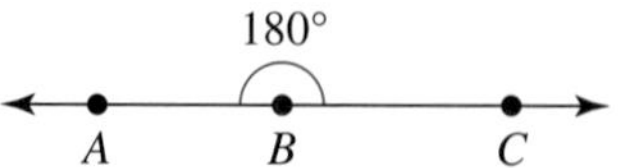

$\angle ABC$ is a straight angle. B is the vertex. It forms $\overleftrightarrow{AC}$.

Think About It 2 What does a 0° angle look like? Does it have another name? Draw one, if you can.

____

____

A quarter turn—like making a right turn at an intersection—creates a 90° angle and is called a **right angle.** We use a little square in the angle to show that it is a right angle. (The corner of a page of this textbook fits perfectly into a right angle.)

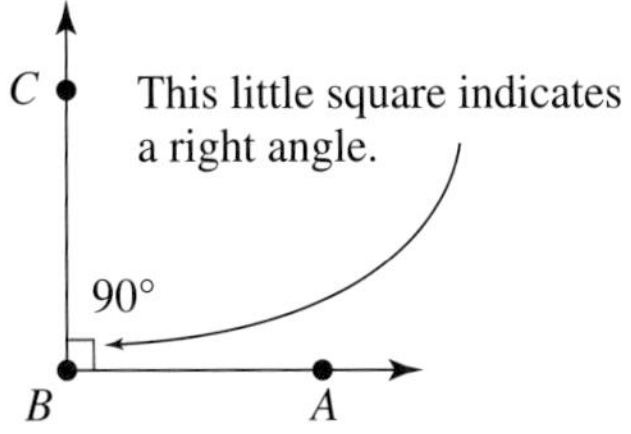

An **acute angle** has a measure greater than 0° and less than 90°.

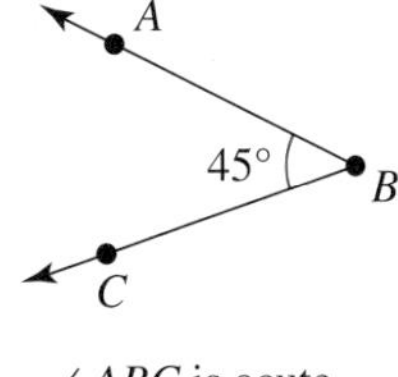

∠*ABC* is acute.

An **obtuse angle** has a measure greater than 90° and less than 180°.

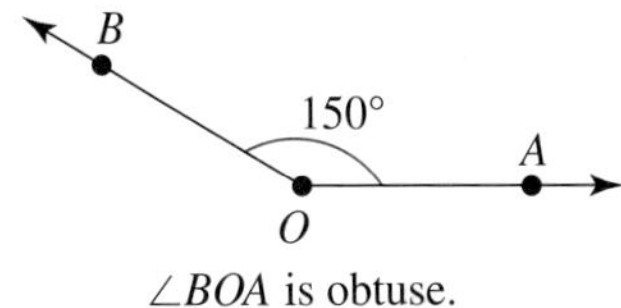

∠*BOA* is obtuse.

Example 2

Use a straightedge to complete the requested angle. Let the point *P* be the vertex and be mindful of the interior of the angle.

a) A right angle **b)** An obtuse angle **c)** An acute angle

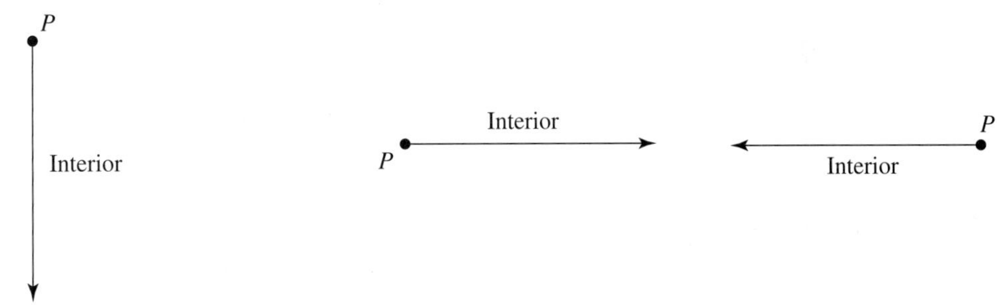

Procedure Use the definitions of *right, obtuse,* and *acute* to assist you in drawing each angle.

The dashed lines in parts b and c that follow are to show you where a right angle would be for the obtuse and acute angles. The obtuse angle is larger than 90°, and the acute angle is smaller than 90°.

Answer There is only one possible answer for part a but many possible answers for parts b and c.

a) A right angle is equal to 90°.

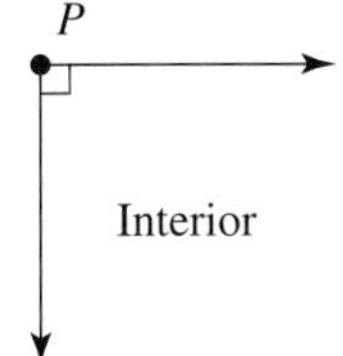

b) An obtuse angle is greater than 90° but less than 180°.

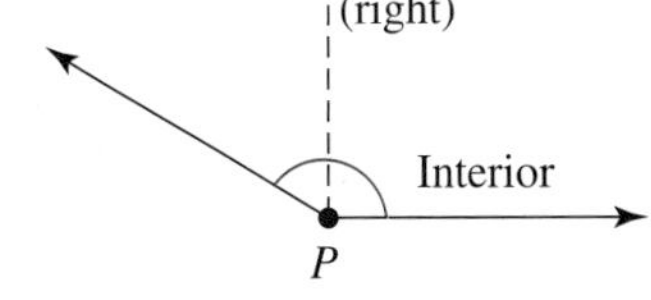

c) An acute angle is less than 90°.

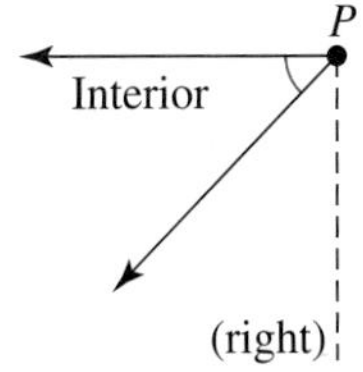

You Try It 4 **Use a straightedge to complete the requested angle. Let the point P be the vertex and be mindful of the interior of the angle. Use Example 2 as a guide.**

a) An acute angle **b)** A right angle **c)** An obtuse angle

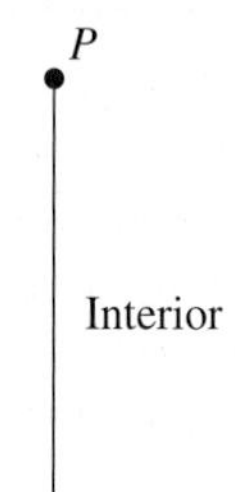

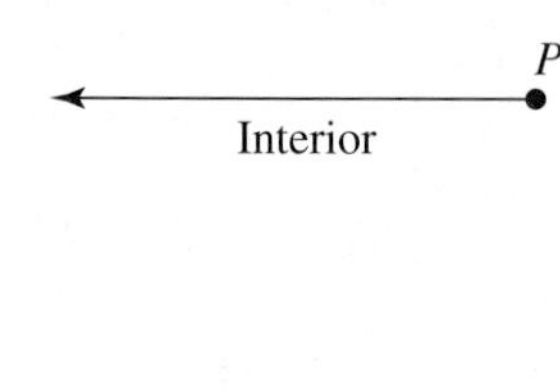

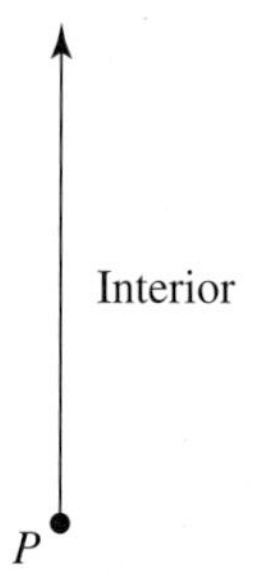

Adjacent Angles

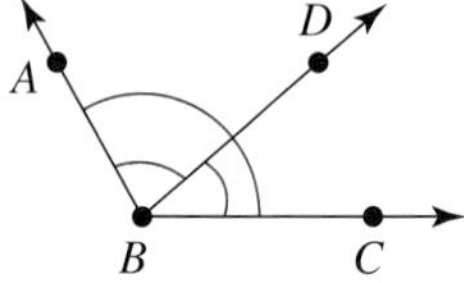

Adjacent angles are any two angles that share a common side, forming an even larger angle. All three angles have the same vertex, and the shared side must be in the interior of the larger angle. In the drawing at the right, there are three angles.

$\angle ABD$ and $\angle DBC$ are adjacent (they share the common side $\overrightarrow{BD}$; together they form the larger angle $\angle ABC$.

In fact, the formula *The sum of the parts equals the whole* applies to adding adjacent angles.

The sum of the measures of the two smaller adjacent angles is equal to the measure of the larger angle.

$$\angle ABD + \angle DBC = \angle ABC$$
$$80° + 40° = 120°$$

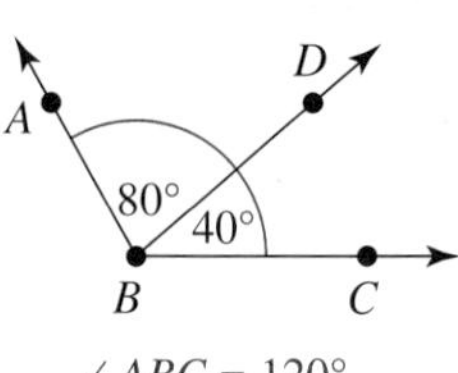

$\angle ABC = 120°$

Likewise, the measure of one of the smaller angles is the difference between the measure of the larger angle and the measure of the other smaller angle.

$$\angle DBC = \angle ABC - \angle ABD$$
$$40° = 120° - 80°$$

Example 3

Given the measures of two angles, find the measure of the third angle.

a)

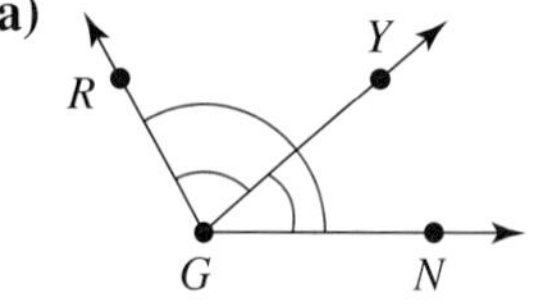

$\angle RGY = 84°$ and $\angle NGY = 39°$

b)

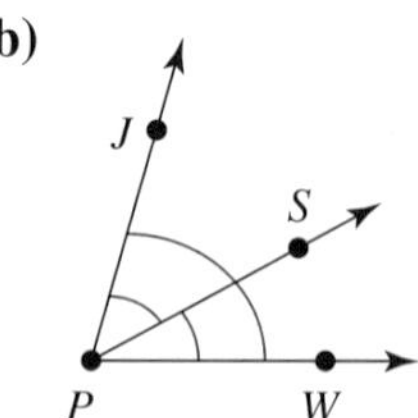

$\angle JPS = 46°$ and $\angle JPW = 74°$

Procedure Label each diagram with the angle measures given and decide whether you need to

1. Add to find the measure of the larger angle.

2. Subtract to find the measure of one of the smaller angles.

Answer a)

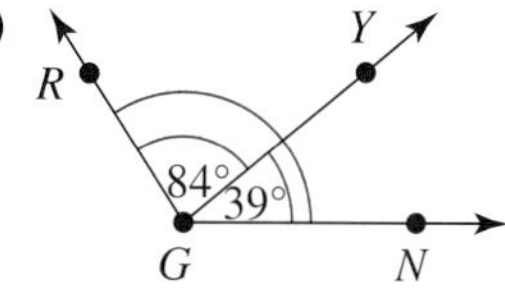

In this diagram, we know the measures of the two smaller angles and need to find the measure of the larger angle, $\angle RGN$. So, we add.

$$\angle RGN = 84° + 39° = 123°$$

b)

In this diagram we know the measures of the larger angle and one of the smaller angles, $\angle JPS$. So, we subtract to find $\angle SPW$.

$$\angle SPW = 74° - 46° = 28°$$

You Try It 5 **Given the measures of two angles, find the measure of the third angle. Use Example 3 as a guide. (These angles might not be drawn exactly to scale.)**

a)

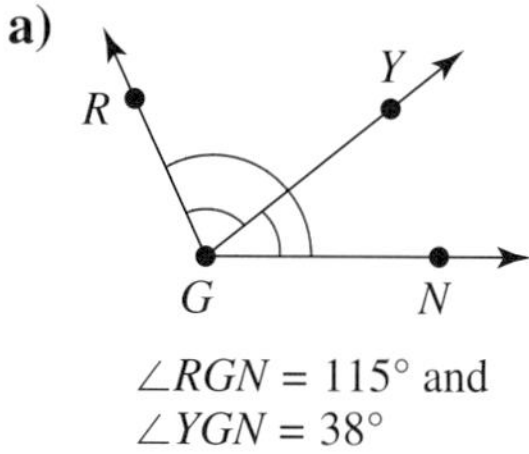

$\angle RGN = 115°$ and
$\angle YGN = 38°$

b)

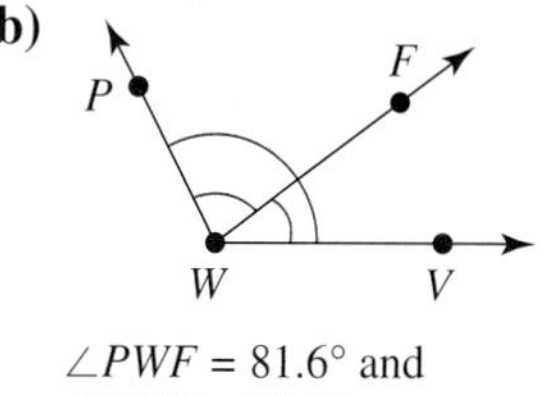

$\angle PWF = 81.6°$ and
$\angle FWV = 35.9°$

Supplementary and Complementary Angles

Supplementary angles are *any* two angles with measures that add to 180°. Supplementary angles can be adjacent, but they don't have to be.

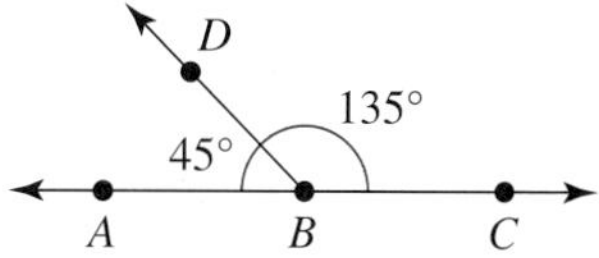

$\angle ABD$ and $\angle DBC$ are supplementary.
$\angle DBC$ *is the supplement* of $\angle ABD$.
$\angle ABD + \angle DBC = 180°$

$\angle ABD$ and $\angle XYZ$ are supplementary.
$\angle XYZ$ *is the supplement* of $\angle ABD$.
$\angle ABD + \angle XYZ = 180°$

If we know that $\angle PQR$ and $\angle STU$ are supplementary angles and if we know the measure of one of the angles (say, $\angle PQR = 70°$), its supplement, $\angle STU$, can be found by subtracting 70° from 180°:

$$\angle STU = 180° - 70° = 110°$$

Example 4

The following pairs of angles are supplementary. Given the measure of one of the angles, find the measure of the other angle.

a) $\angle DEF = 43°$

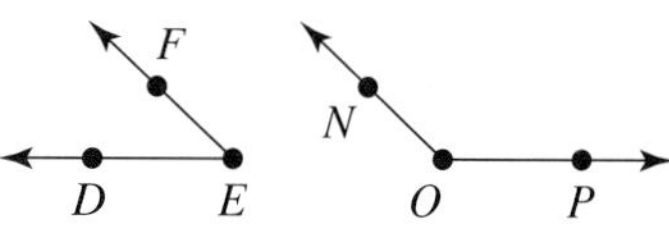

b) $\angle JKL = 112.7°$

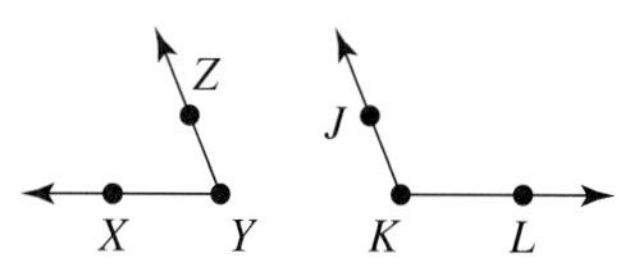

Procedure Label the known angle measure and subtract it from 180°.

Answer a) $\angle NOP = 180° - 43°$

$$\begin{array}{r} 180 \\ -\ 43 \\ \hline 137 \end{array}$$

$\angle NOP = 137°$

b) $\angle XYZ = 180° - 112.7°$

$$\begin{array}{r} 180.0 \\ -\ 112.7 \\ \hline 67.3 \end{array}$$

$\angle XYZ = 67.3°$

You Try It 6 **The following pairs of angles are supplementary. Given the measure of one of the angles, find the measure of the other angle. Use Example 4 as a guide.**

a) $\angle EFG = 146°$

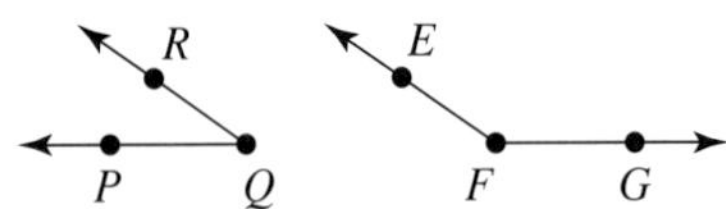

b) $\angle UVW = 78.8°$

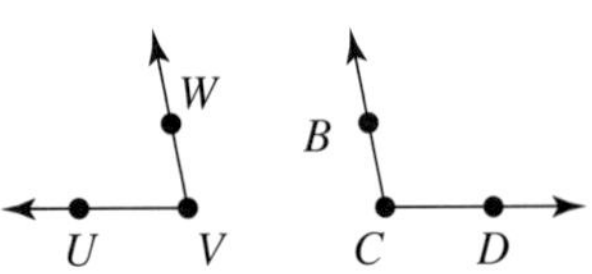

Complementary angles are *any* two angles with measures that add to 90°. Complementary angles can be adjacent, but they don't have to be.

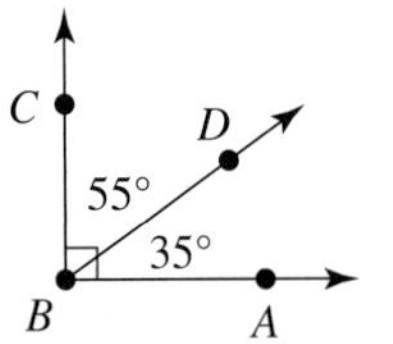

$\angle ABD$ and $\angle DBC$ are complementary.
$\angle DBC$ is the complement of $\angle ABD$.
$\angle ABD + \angle DBC = 90°$

$\angle CBD$ and $\angle XYZ$ are complementary.
$\angle XYZ$ is the complement of $\angle CBD$.
$\angle CBD + \angle XYZ = 90°$

If we know that $\angle PQR$ and $\angle STU$ are complementary angles and if we know the measure of one of these angles (say, $\angle PQR = 70°$), its complement, $\angle STU$, can be found by subtracting 70° from 90°.

$$\angle STU = 90° - 70° = 20°$$

Example 5

The following pairs of angles are complementary. Given the measure of one of the angles, find the measure of the other angle.

a) $\angle DEF = 43°$

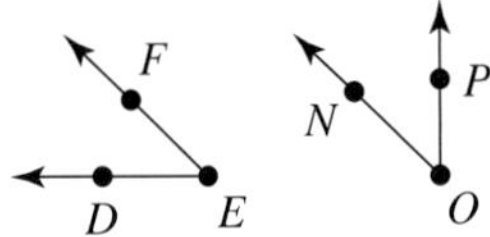

b) $\angle XYZ = 22.7°$

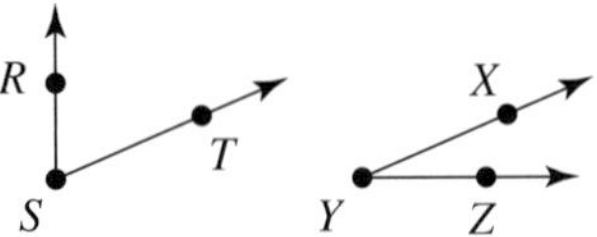

Procedure Label the known angle measure and subtract it from 90°.

Answer a) $\angle NOP = 90° - 43°$

$$\begin{array}{r} 90 \\ -\ 43 \\ \hline 47 \end{array}$$

$\angle NOP = 47°$

b) $\angle RST = 90° - 22.7°$

$$\begin{array}{r} 90.0 \\ -\ 22.7 \\ \hline 67.3 \end{array}$$

$\angle RST = 67.3°$

You Try It 7 **The following pairs of angles are complementary. Given the measure of one of the angles, find the measure of the other angle. Use Example 5 as a guide.**

a) $\angle TUV = 48°$

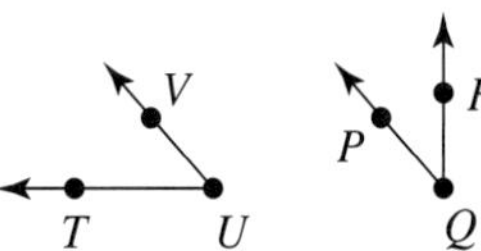

b) $\angle DCB = 59.2°$

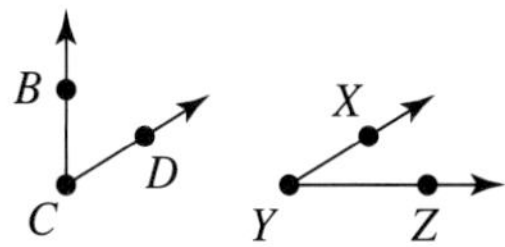

Vertical Angles

When two lines intersect, several pairs of angles are formed, all sharing the same vertex. Some pairs are adjacent and supplementary (forming a straight line). The other pairs are **vertical angles,** angles that are not adjacent to each other.

Pairs of adjacent, supplementary angles:

$\angle 1$ and $\angle 2$ $\angle 2$ and $\angle 3$

$\angle 3$ and $\angle 4$ $\angle 4$ and $\angle 1$

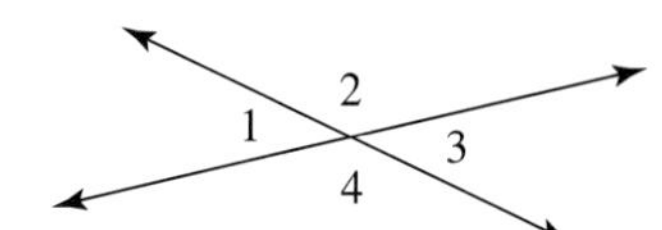

Pairs of vertical angles:

$\angle 1$ and $\angle 3$ $\angle 2$ and $\angle 4$

Think About It 3 In the diagram, are $\angle 1$ and $\angle 3$ vertical angles? Why or why not?

____

____

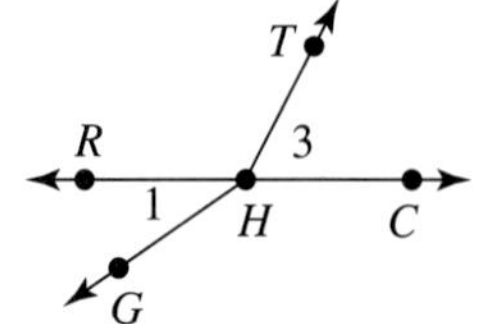

It can easily be shown that *vertical angles have the same measure*. For example, consider the following diagram:

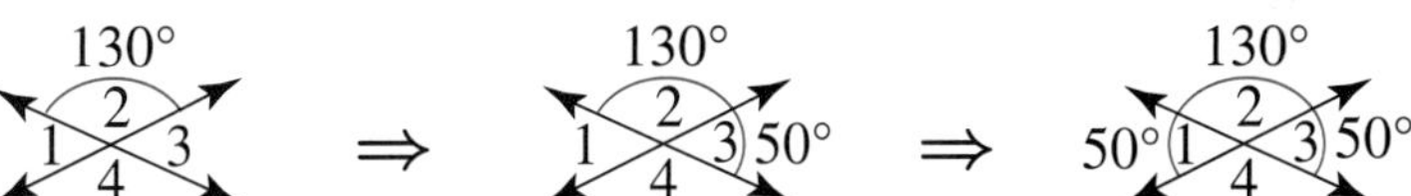

Given: $\angle 2 = 130°$

$\angle 3$ is supplementary to $\angle 2$; so, $\angle 3$ is $180° - 130° = 50°$.

Likewise, $\angle 1$ is supplementary to $\angle 2$; so, $\angle 1$ is also 50°.

Notice that vertical angles $\angle 1$ and $\angle 3$ have the same measure; in other words, $\angle 1 = \angle 3$.

It follows that $\angle 4$ is also 130°. So, vertical angles $\angle 4$ and $\angle 2$ have the same measure: $\angle 4 = \angle 2$.

Knowing the measure of one of the angles, it is possible to determine the measures of all four angles.

Example 6

Given $\angle 1 = 27°$, find the measures of the other three angles.

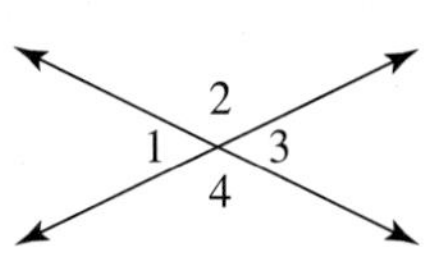

Procedure Vertical angles have the same measure, and the angle vertical with $\angle 1$ is $\angle 3$. Also, $\angle 2$ and $\angle 4$ are supplementary to $\angle 1$. We can find their value by subtracting $\angle 1$ from 180°.

Answer $\angle 3 = \angle 1$; so, $\angle 3 = 27°$, too.

$\angle 2 = 180° - 27° = 153°$; therefore, $\angle 4 = 153°$, too.

You Try It 8 **Given the measure of one angle, find the measures of the other three angles. Use Example 6 as a guide.**

a) $\angle 3 = 58°$

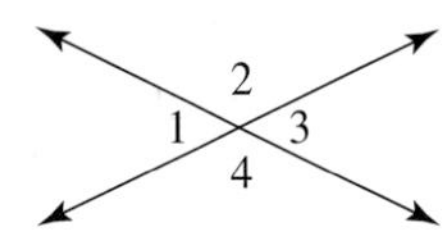

b) $\angle 3 = 142.5°$

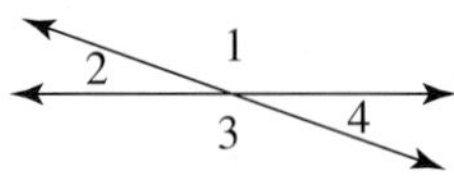

This next exercise leads into the last topic in this section.

You Try It 9 **Given that $\angle 1 = 90°$, find the measures of the other three angles. Use Example 6 as a guide.**

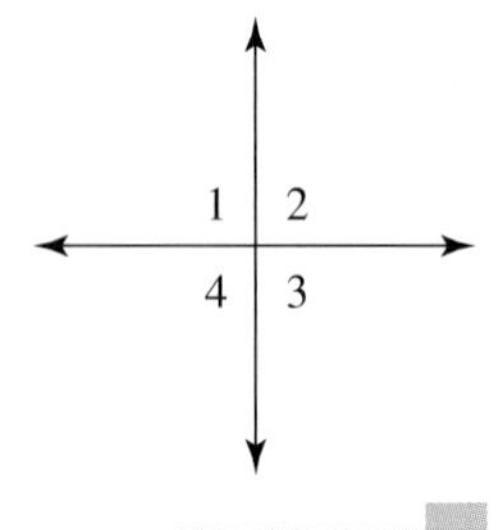

Perpendicular Lines and Line Segments

Two lines are **perpendicular** if they intersect to form four right angles. We use the symbol $\perp$ to indicate that two lines (or line segments) are perpendicular to each other.

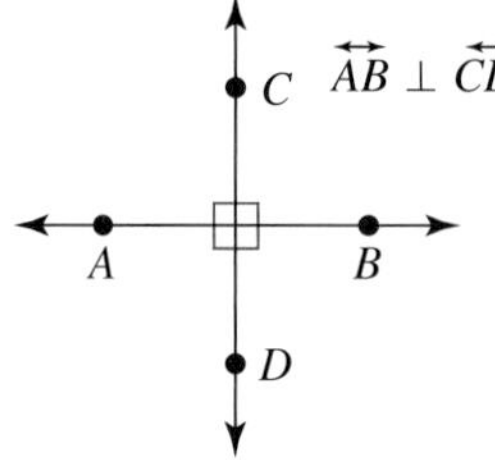

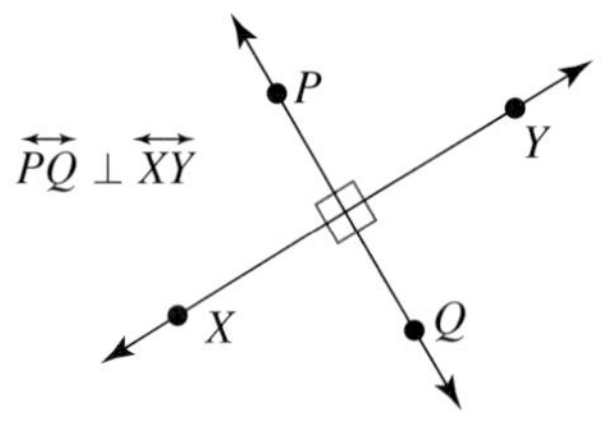

Here, the vertical angles have the same measure as the adjacent, supplementary angles.

Two line *segments* that are perpendicular may form *one*, *two*, or *four* right angles.

One right angle

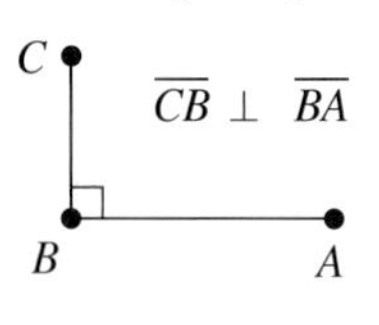

Two right angles

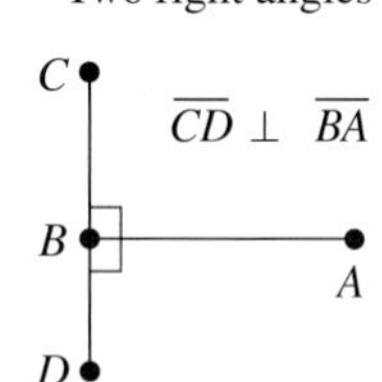

Four right angles

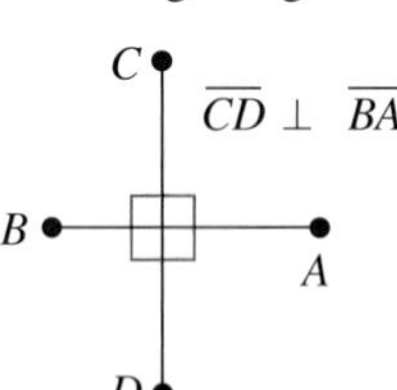

Answers: You Try It and Think About It

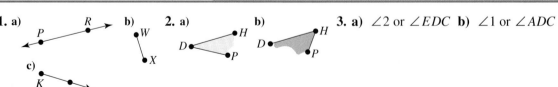

You Try It: **1. a)** P R **b)** W X **c)** K H **2. a)** D H P **b)** D H P **3. a)** $\angle 2$ or $\angle EDC$ **b)** $\angle 1$ or $\angle ADC$

4. There are many correct answers for parts a and c; part b has only one answer.

a) An acute angle: b) A right angle: c) An obtuse angle:

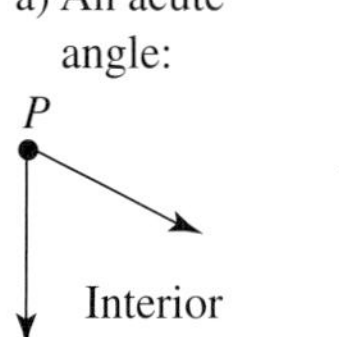

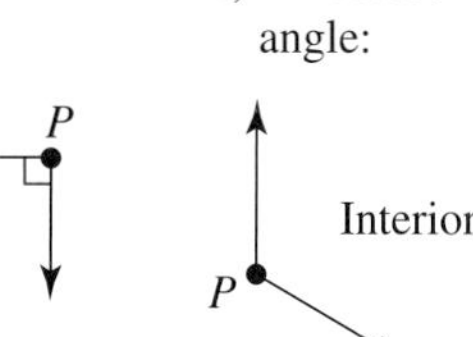

5. a) $\angle RGY = 77°$ **b)** $\angle PWV = 117.5°$
6. a) $\angle PQR = 34°$ **b)** $\angle BCD = 101.2°$
7. a) $\angle PQR = 42°$ **b)** $\angle XYZ = 30.8°$
8. a) $\angle 1 = 58°$; $\angle 2$ and $\angle 4 = 122°$
b) $\angle 1 = 142.5°$; $\angle 2$ and $\angle 4 = 37.5°$
9. Every angle measures 90°.

Think About It: **1.** No. A real plane extends indefinitely, and the Bonneville Salt Flats has mountains surrounding it. **2.** A 0° angle is also known as a ray. **3.** No. Vertical angles are formed by two lines, and $\angle GHT$ is not a straight angle, not a line.

Section 8.4 Exercises

FOR EXTRA HELP
Student Resources on DVD-ROM
Includes
➤ Student's Solutions Manual
➤ Video Lectures
➤ Chapter Test Prep Video
MyMathLab
MathXL

Think Again

1. According to the definition of a plane, is the Bonneville Salt Flats a real plane? Why or why not? *(Refer to Think About It 1)*
   No. A real plane has no outside boundaries, and the Bonneville Salt Flats has mountains on at least one side. (Also, a student might mention that because the Earth is round and a plane is flat, the Flats cannot be a real plane.)
2. What does a 0° angle look like? Does it have another name? *(Refer to Think About It 2)*
   (Drawing not shown here.) A 0° angle is also called a *ray*.
3. The metric degree measure is called a *grad*. There are 400 grads in a full circle. How many grads are in a right angle?
   100 grads are in a right angle.
4. Using grads instead of degrees, what is the sum of the three angles in a triangle? Explain your answer or show an example that supports your answer.
   The sum of the angles in a triangle is 200 grads, just like two right angles.

Focus Exercises

Answer as indicated.

5. Write four names for this angle.

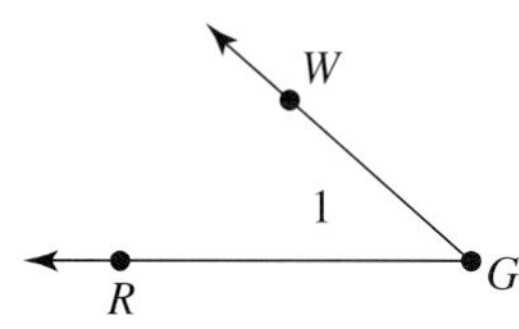

$\angle 1$, $\angle G$, $\angle RGW$, and $\angle WGR$

6. What rays form this angle?

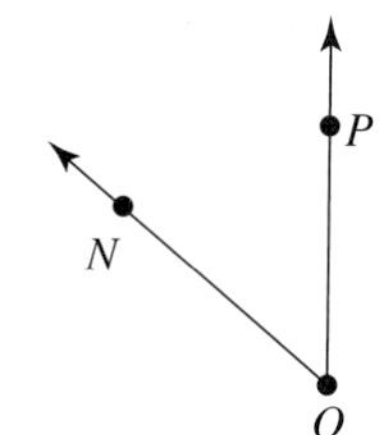

$\overrightarrow{ON}$ and $\overrightarrow{OP}$

7. Name two lines in this diagram.

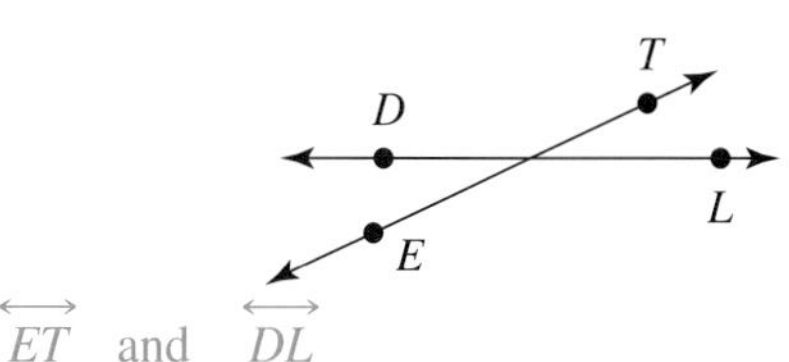

$\overleftrightarrow{ET}$ and $\overleftrightarrow{DL}$

8. Name two acute and two obtuse angles in this diagram.

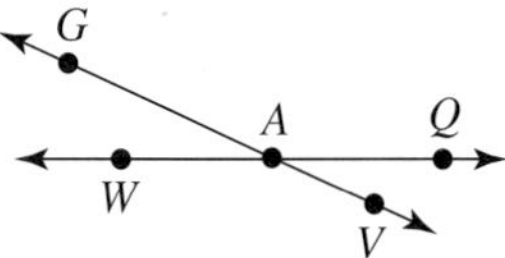

Acute: $\angle WAG$ and $\angle VAQ$
Obtuse: $\angle GAQ$ and $\angle WAV$

Given the measure of two of the angles, find the measure of the third angle.

9. $\angle FRB = 18°$ and $\angle KRF = 35°$

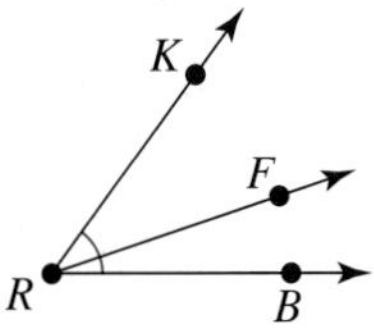

$\angle KRB = 53°$

10. $\angle WTR = 60.8°$ and $\angle VTW = 19.6°$

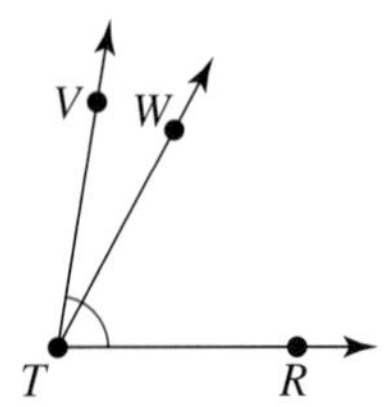

$\angle VTR = 80.4°$

11. $\angle SAX = 153°$ and $\angle PAX = 59°$

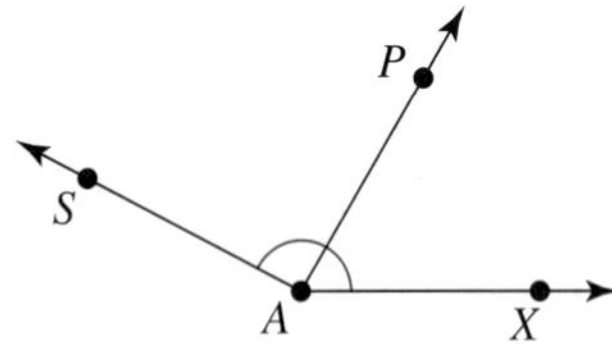

$\angle SAP = 94°$

12. $\angle QME = 160.4°$ and $\angle QMD = 62.8°$

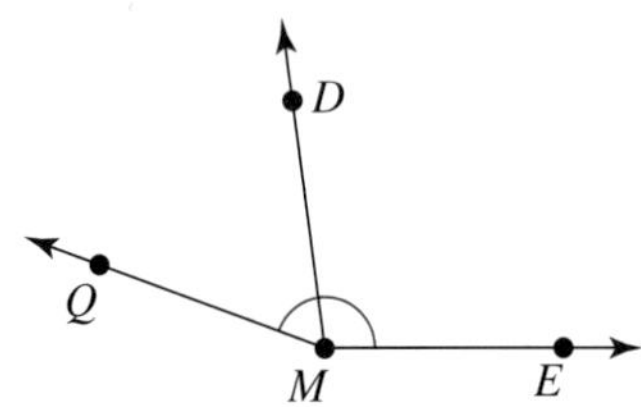

$\angle DME = 97.6°$

Each pair of angles is supplementary. Given the measure of one of the angles, find the measure of the other angle.

13. $\angle DSZ = 77°$

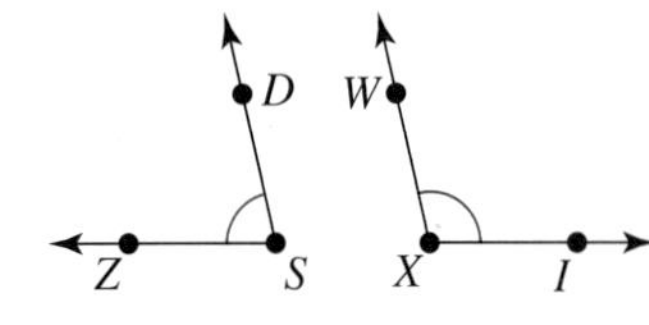

$\angle WXI = 103°$

14. $\angle AQP = 72.1°$

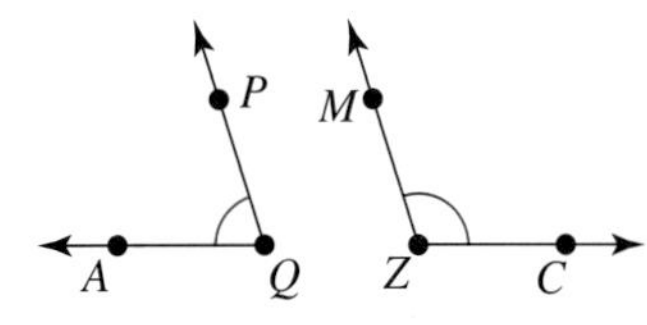

$\angle MZC = 107.9°$

15. $\angle DKH = 132°$

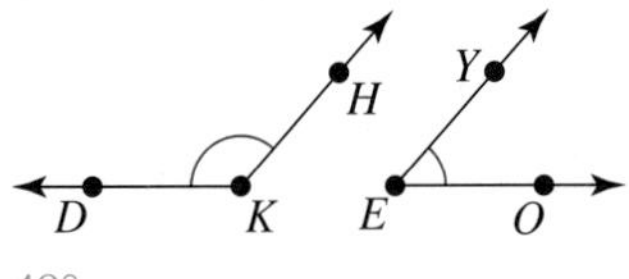

$\angle YEO = 48°$

16. $\angle WFZ = 118.4°$

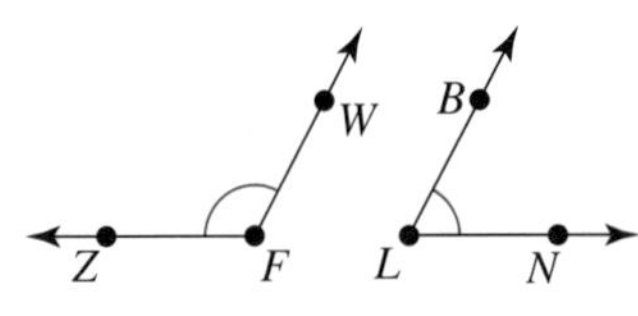

$\angle BLN = 61.6°$

Each pair of angles is complementary. Given the measure of one of the angles, find the measure of the other angle.

17. $\angle RVT = 27°$

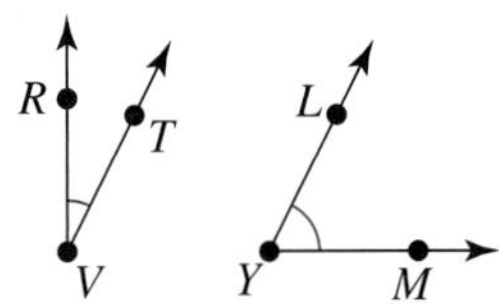

$\angle LYM = 63°$

18. $\angle RFX = 29.5°$

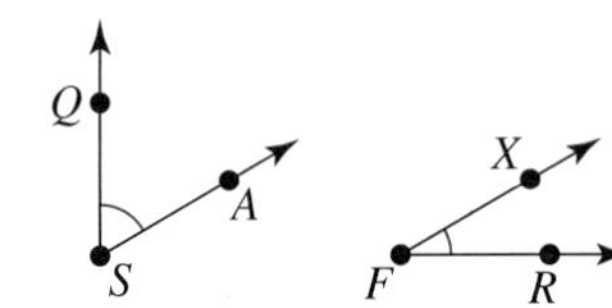

$\angle QSA = 60.5°$

19. $\angle BGY = 72°$

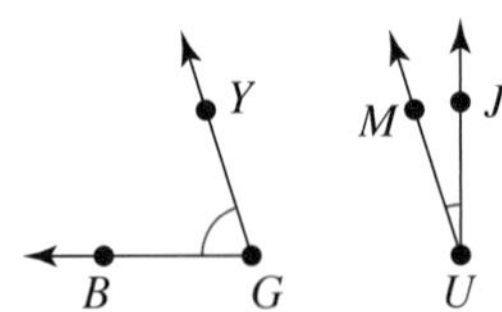

$\angle MUJ = 18°$

20. $\angle LOP = 49.2°$

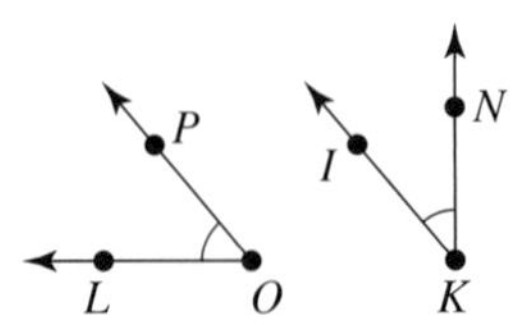

$\angle IKN = 40.8°$

Given the measure of one angle, find the measure of the other three angles.

21. $\angle 3 = 32°$

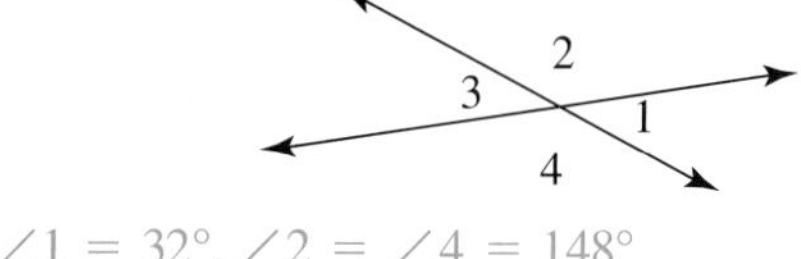

$\angle 1 = 32°, \angle 2 = \angle 4 = 148°$

22. $\angle 1 = 47.9°$

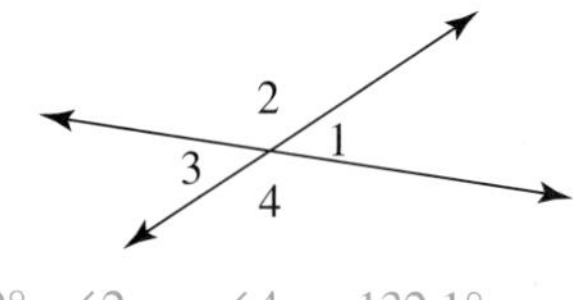

$\angle 3 = 47.9°, \angle 2 = \angle 4 = 132.1°$

23. $\angle 4 = 111°$

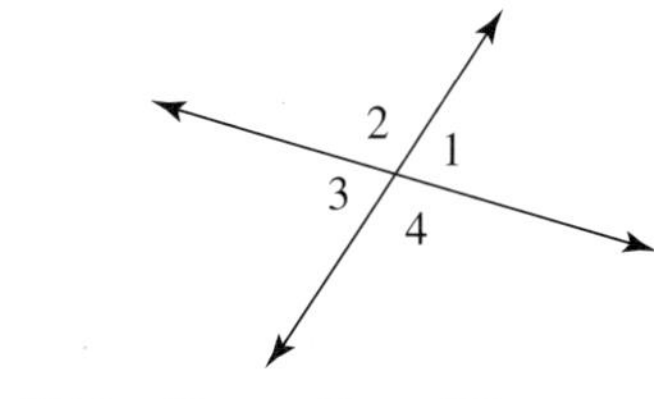

$\angle 2 = 111°, \angle 1 = \angle 3 = 69°$

24. $\angle 2 = 127.8°$

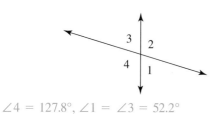

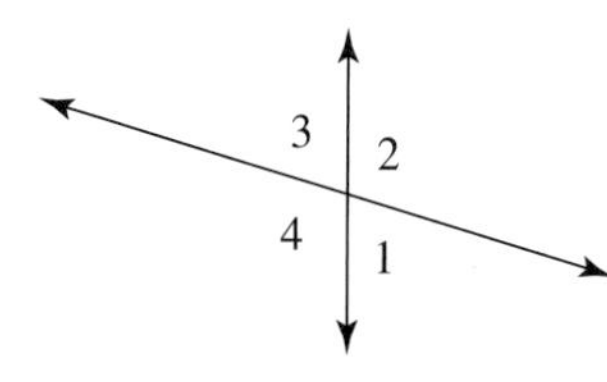

$\angle 4 = 127.8°, \angle 1 = \angle 3 = 52.2°$

Fill in the blanks. (You are encouraged to draw a picture to help you understand the statement better.)

25. Two angles that are supplementary and adjacent form a(n) __straight__ angle.

26. Two angles that are complementary and adjacent form a(n) __right__ angle.

27. When two lines intersect, the nonadjacent angles are called __vertical__ angles.

28. Two perpendicular lines form four __right__ angles.

True or false.

29. Complementary angles must be adjacent to each other. __False__

30. Vertical angles are never adjacent to each other. __True__

31. Acute angles always measure less than 90°. __True__

32. Obtuse angles may measure more than 180°. __False__

33. Two line segments that don't intersect must be parallel. __False__

34. Straight angles always measure 180°. __True__

35. If two nonzero angles are complementary to each other, both must be acute angles. __True__

36. If two lines meet to form a right angle, the lines must be perpendicular. __True__

Think Outside the Box

37. If two angles have the same measure and are supplementary to each other, what is the measure of each angle? The measure of each angle is 90°.

38. If two angles have the same measure and are complementary to each other, what is the measure of each angle? The measure of each angle is 45°.

39. If point W is in the interior of $\angle ABC$, is it possible to place another point, Z, not in the interior of $\angle ABC$ so that $\overline{WZ}$ does not intersect either ray of $\angle ABC$?

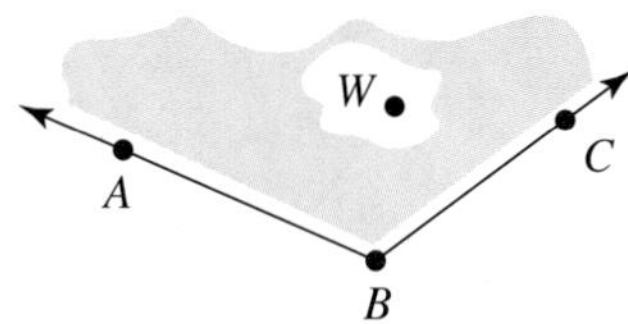

No, a line segment drawn from a point in the interior of an angle to the exterior of the angle must cross one of the sides of the angles.

40. If three lines intersect in one point, as shown, how many angles are formed? List as many as you can find.

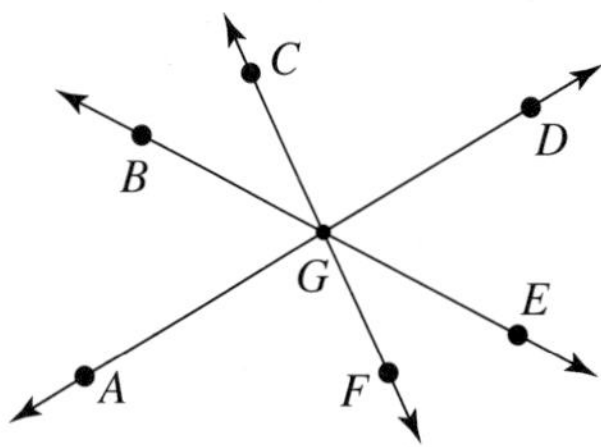

There are a total of 18 angles. Here is a partial list: $\angle AGB$, $\angle AGC$, $\angle AGD$, $\angle BGC$, $\angle BGD$, and $\angle BGE$.

SECTION 8.5 Geometric Shapes

OBJECTIVES

In this section, you will learn to

- Identify different types of triangles.
- Find the measures of the angles in a triangle.
- Identify some types of quadrilaterals (parallelogram, rectangle, square, and trapezoid).
- Find the measures of the angles in quadrilaterals.

You Need to Know

To successfully complete this section, you need to understand

- ☐ Solving equations (1.7 and Ch. 3)
- ☐ Complementary angles (8.4)
- ☐ Supplementary angles (8.4)
- ☐ Right angles (8.4)
- ☐ Parallel line segments (8.4)

Introduction

Many geometric shapes are made from connected line segments that form a *polygon*. Each line segment is called a *side*, and every pair of connected sides forms an angle.

In general, a **polygon** is a closed plane figure composed of at least three straight line segments. In a polygon, two line segments meet only at an endpoint, called a **vertex,** and never cross each other. Each vertex is the endpoint of only two sides.

Some examples of polygons follow. The type of polygon is based on the number of sides it has. Typically, each vertex is labeled with a capital letter.

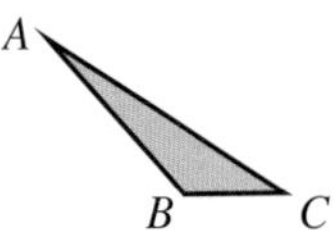

Triangle: *Three* sides

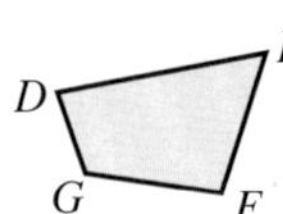

Quadrilateral: *Four* sides

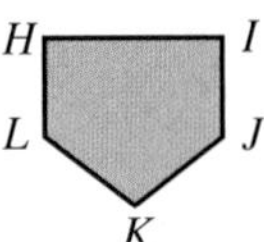

Pentagon: *Five* sides

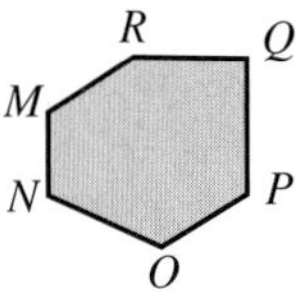

Hexagon: *Six* sides

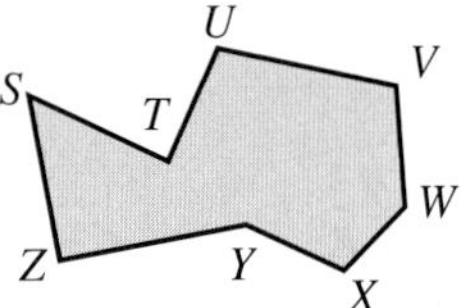

Octagon: *Eight* sides

A polygon must be *closed*: Each endpoint of a side must share an endpoint with one other side.

This is *not* a polygon because it is not closed.

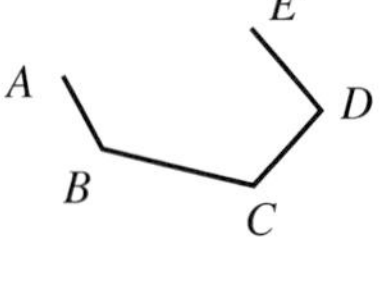

The line segments of a polygon must not cross each other.

This is *not* a polygon because two sides cross each other.

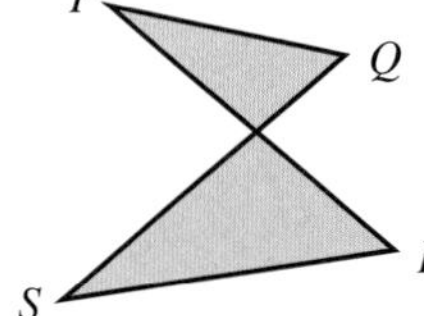

Also, in a polygon, each side must be a line segment and not be curved.

This is *not* a polygon because one of the sides is not a line segment.

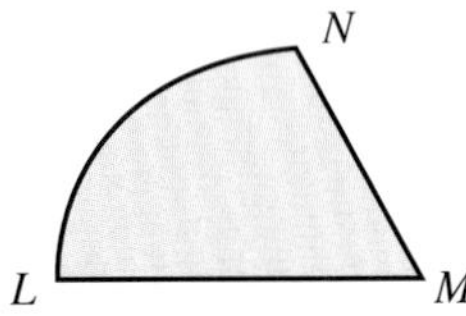

A **triangle** is a polygon that has three sides and three angles. The letters we use to label each vertex also name the triangle: ΔABC.

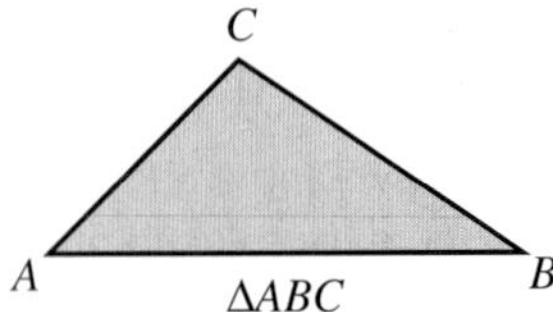

ΔABC

The Sum of the Angles in a Triangle

The sum of the angles in a triangle is 180°, no matter the shape of the triangle.

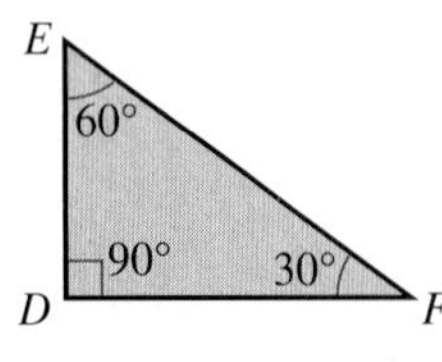

$30° + 60° + 90° = 180°$

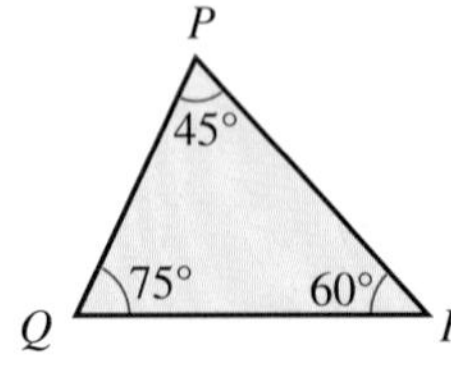

$45° + 60° + 75° = 180°$

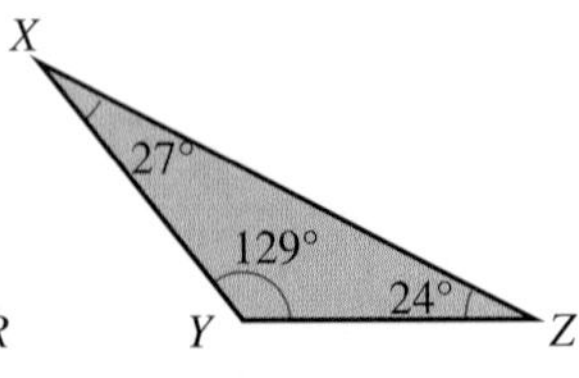

$27° + 129° + 24° = 180°$

In general, we can say that, in ΔABC, $\angle A + \angle B + \angle C = 180°$.

When the measures of two angles of a triangle are known, the measure of the third angle can be found, as shown in Example 1.

Example 1

In ΔABC, $\angle A = 52°$ and $\angle B = 67°$. Find the measure of $\angle C$.

Procedure One technique is to set up an equation with the measure of $\angle C$ being the unknown value.

Answer Legend: Let n = the measure of $\angle C$. Then use the formula for the sum of the angles in a triangle.

$$\begin{aligned} 52 + 67 + n &= 180 \\ n + 119 &= 180 \\ n + 119 - 119 &= 180 - 119 \\ n &= 61 \end{aligned}$$

The measure of $\angle C$ is 61°.

Notice that it's appropriate to write the answer as 61 *degrees* (61°), not just 61.

You Try It 1 **Given the measures of two angles, find the measure of the third angle. Use Example 1 as a guide.**

a) $\angle J = 60°$
$\angle B = 95°$

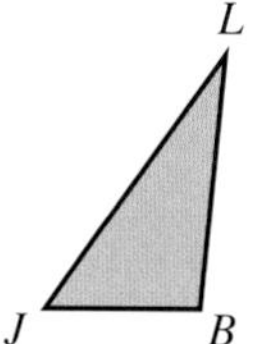

b) $\angle R = 18.9°$
$\angle M = 23.4°$

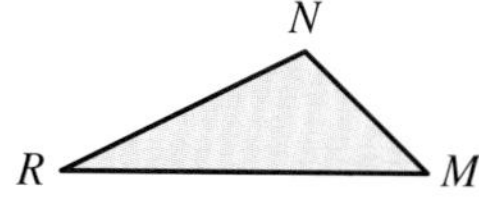

Think About It 1 Is it possible for a triangle to contain two obtuse angles? Why or why not?

Can you draw a triangle with two obtuse angles?

____

____

Is it possible for a triangle to contain two right angles? Why or why not?

Can you draw a triangle with two right angles?

____

____

Types of Triangles

We can describe triangles based on (1) their side measures or (2) their angle measures.

Triangles named by their side measures include the following:

Equilateral triangle: A triangle in which all three sides are of equal length and all three angles are of equal measure.

Each angle in an equilateral triangle measures 60°.

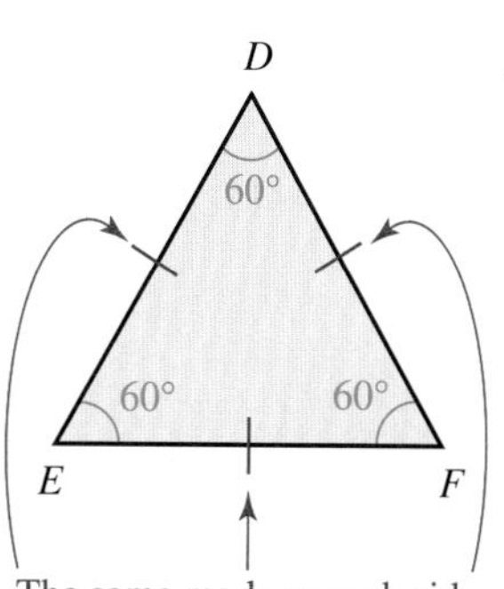

The same mark on each side indicates that each side has the same length.

Isosceles triangle: A triangle in which two (or more) sides are of equal length. The sides with equal length are called **legs,** and the third side is called the **base.**

The angles opposite the legs are called **base angles** and are of equal measure.

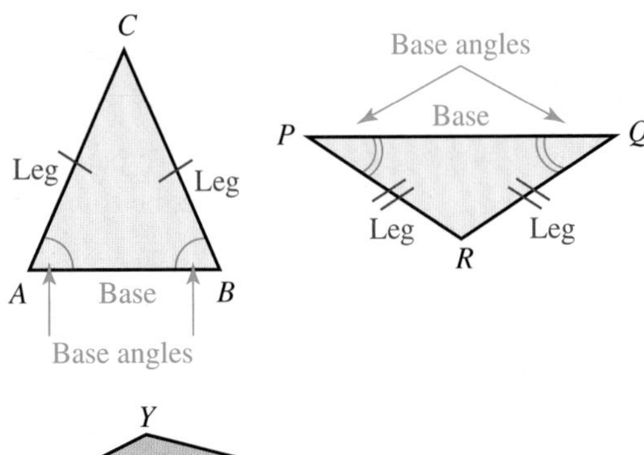

Scalene triangle: A triangle in which no two sides have the same length and no two angles are of equal measure.

Y

X Z

You Try It 2 **For each triangle, based on the side measures given, decide whether it is equilateral, isosceles, or scalene.**

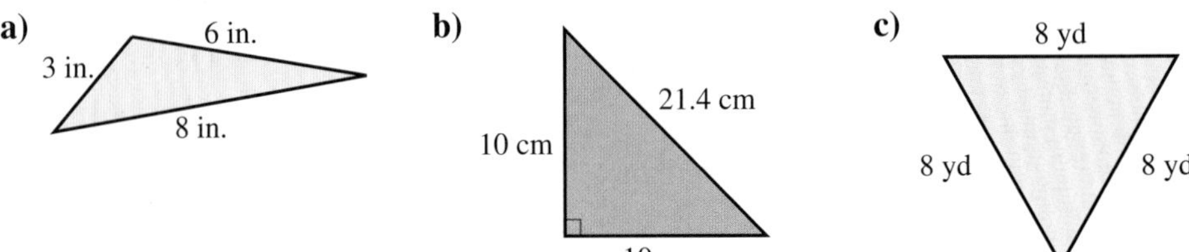

______________ ______________ ______________

Think About It 3 Is an equilateral triangle also isosceles? Why or why not?

____

____

Right Triangles

A type of triangle named by its angle measure is the *right triangle.* A triangle in which one angle is a right angle (90°) is called a **right triangle.**

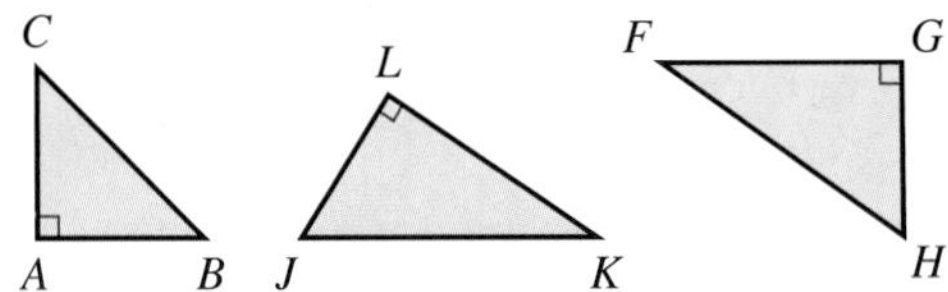

In a right triangle, the other two angles are acute. In fact:

The two acute angles in a right triangle are *complementary* to each other.

For example, in ΔABC, if $\angle C$ is the right angle, we could write the angle equation as follows:

$$\angle A + \angle B + 90 = 180$$ Subtract 90 from each side.

$$\angle A + \angle B + 90 - 90 = 180 - 90$$ Simplify each side.

$$\angle A + \angle B = 90$$ Since the sum of $\angle A + \angle B$ is 90°, it must be that $\angle A$ and $\angle B$ are complementary to each other.

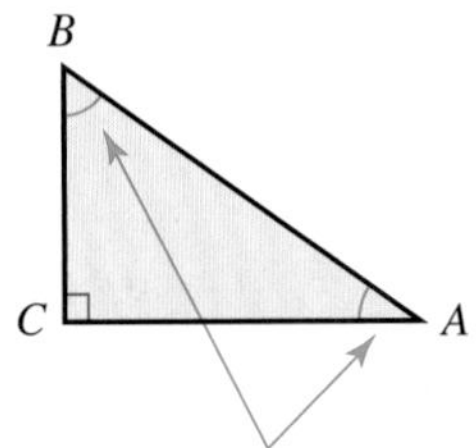

These two acute angles are complementary.

Example 2

In ΔABC, $\angle A$ is a right angle. If $\angle B = 67°$, find the measure of $\angle C$.

Procedure Legend: Let n = the measure of $\angle C$.

We can use all three angles, knowing that $\angle A = 90°$ and $\angle B = 67°$.

$$90 + 67 + n = 180$$
$$n + 157 = 180$$
$$n + 157 - 157 = 180 - 157$$
$$n = 23$$

Or we can use only the two acute angles, knowing that they add to 90°.

$$67 + n = 90$$
$$n + 67 - 67 = 90 - 67$$
$$n = 23$$

Answer Either way we get the same result: the measure of $\angle C$ is 23°.

You Try It 3 **Given the measure of one acute angle, find the measure of the other acute angle. Use Example 2 as a guide.**

a) $\angle R = 51°$

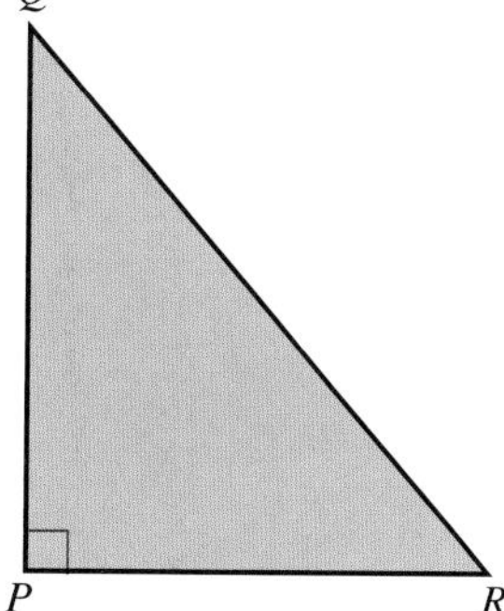

b) $\angle U = 41.8°$

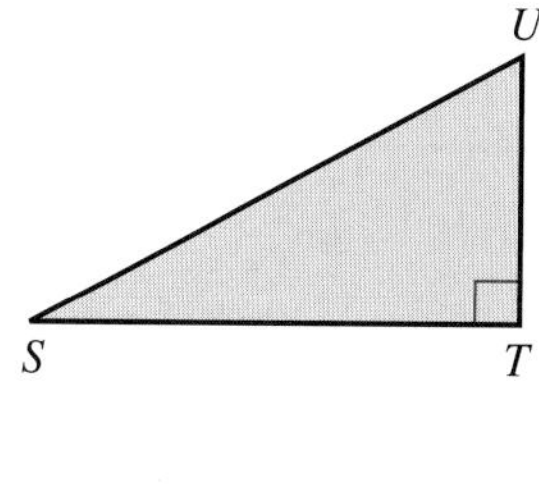

Quadrilaterals

A **quadrilateral** is a four-sided polygon. It also has four angles, and the sum of those angles is 360°. There are many special types of quadrilaterals.

A **parallelogram** has two pairs of parallel sides.

The arrows pointing in the same direction indicate that the sides are parallel to each other.

The parallel sides are also the same length, as indicated by the number of marks on each side.

When a parallelogram is not a rectangle, it has a pair of acute angles and a pair of obtuse angles.

The acute angles are opposite each other and have equal measure. The same is true for the obtuse angles.

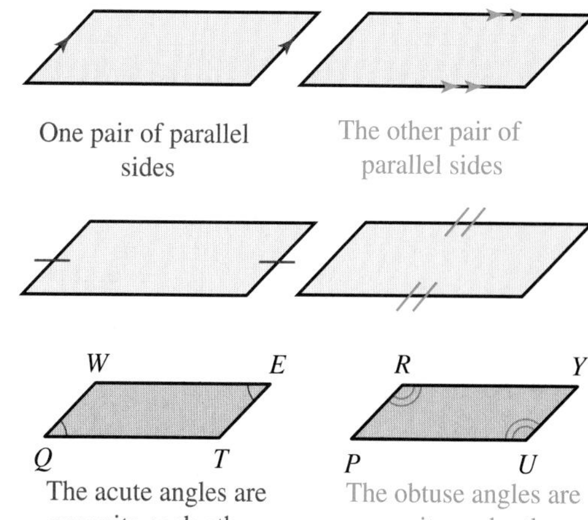

Also, an acute angle and an obtuse angle are supplementary.

When you know one angle in a parallelogram, you can find the others.

The acute angle is supplementary to the obtuse angle.

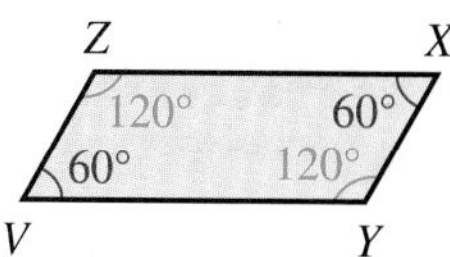

Example 3

In this parallelogram, $\overline{CD}$ is 10 centimeters long, $\overline{BC}$ is 6 centimeters long, and $\angle A = 45°$. Complete this diagram by finding the measures of the other three angles and the lengths of the other two sides.

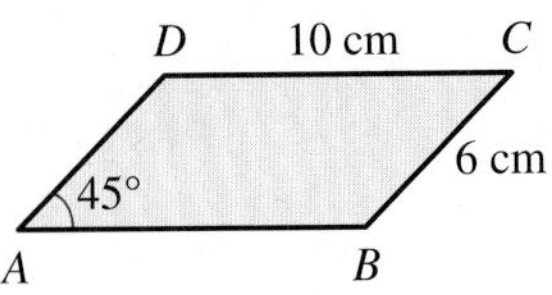

Procedure Opposite sides of a parallelogram have the same length, and opposite angles have the same measure. $\angle A$ (acute) and $\angle B$ (obtuse) are supplementary.

Answer $\angle A$ and $\angle C$ are both acute; so, they have the same measure. Because $\angle A = 45°$, $\angle C = 45°$, too.

We can find the measure of $\angle B$ by subtracting $\angle A$ from 180°.

$$\begin{array}{r} 180 \\ -\ 45 \\ \hline 135 \end{array}$$

So, $\angle B = 135°$, which means that $\angle D = 135°$ as well.

Also, $\overline{AB}$ is 10 centimeters long, and $\overline{AD}$ is 6 centimeters long.

A completed diagram looks like this:

D 10 cm C
6 cm 135° 45° 6 cm
45° 135°
A 10 cm B

You Try It 4 **For each parallelogram, find the measures of the other three angles and the lengths of the other two sides.**

a) $\angle D = 127°$,
$\overline{AD} = 9$ in.
$\overline{DC} = 15$ in.

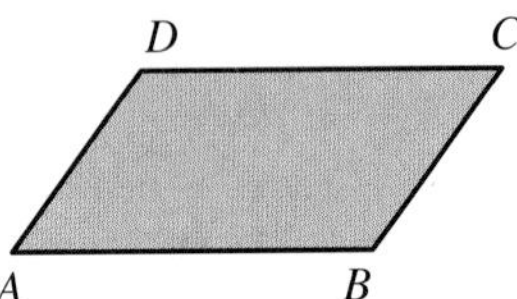

b) $\angle S = 52.4°$,
$\overline{RS} = 3.2$ mm
$\overline{RQ} = 5.9$ mm

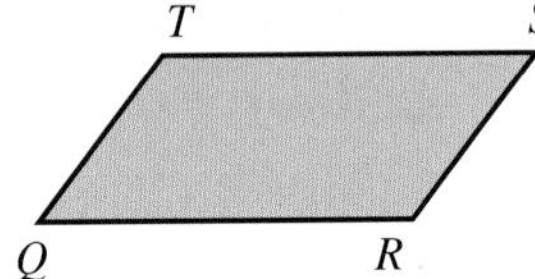

c) $\angle W = 90°$,
$\overline{WZ} = 14$ ft
$\overline{XW} = 33$ ft

A **rectangle** is a parallelogram with four right angles.

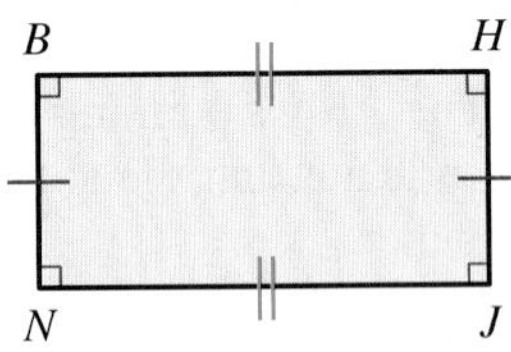

A **square** is a rectangle with four sides of equal length.

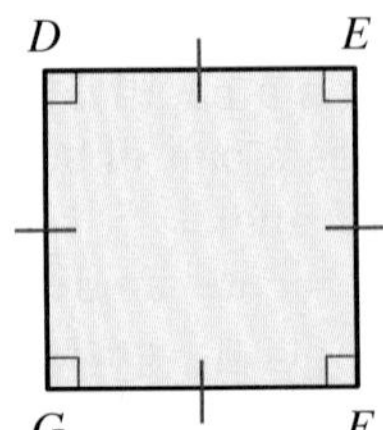

Think About It 4 Is a square also a parallelogram? Why or why not?

____

____

A **trapezoid** is *not* a parallelogram; it is a quadrilateral with only one pair of parallel sides. The parallel sides are called *bases* and have different lengths.

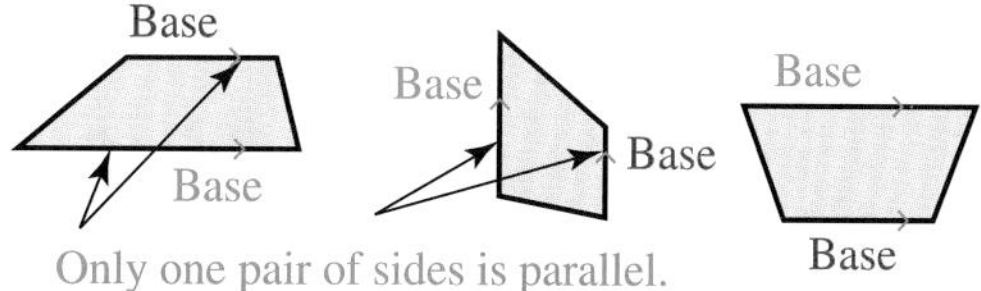

Example 4

For each quadrilateral, decide whether it is a parallelogram, a rectangle, a square, a trapezoid, or none of these.

a)

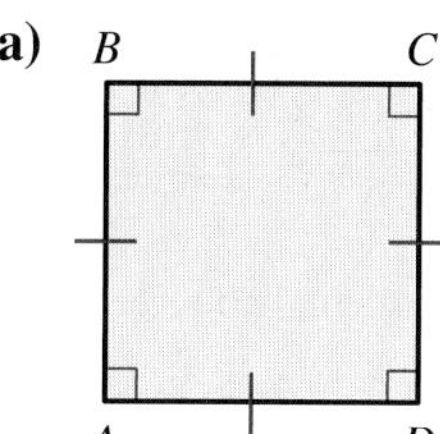

b)

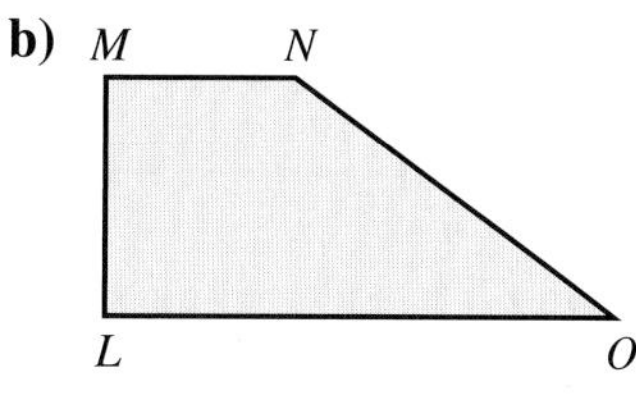

c)

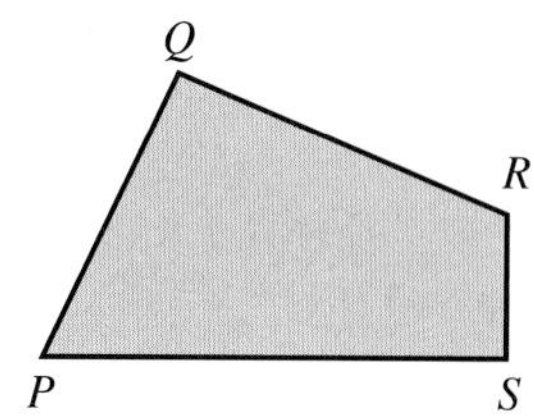

Answer

a) square, rectangle, and parallelogram

b) trapezoid (it has only one pair of parallel sides)

c) none of these (it has no parallel sides)

You Try It 5

For each quadrilateral, decide whether it is a parallelogram, a rectangle, a square, a trapezoid, or none of these.

a)

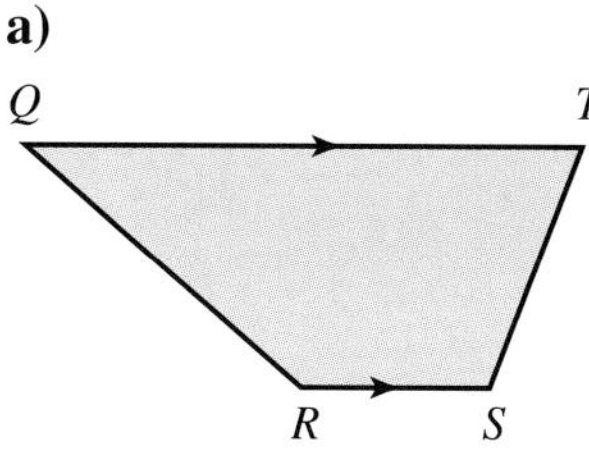

____

b)

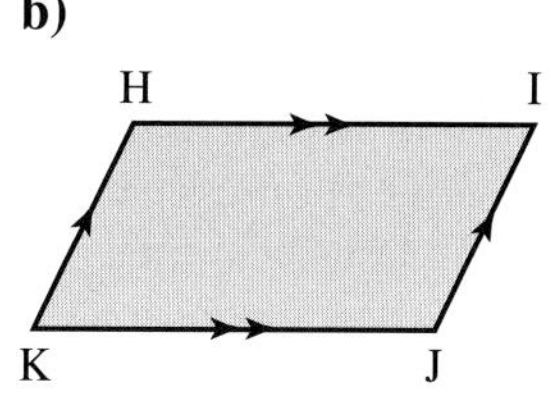

____

c)

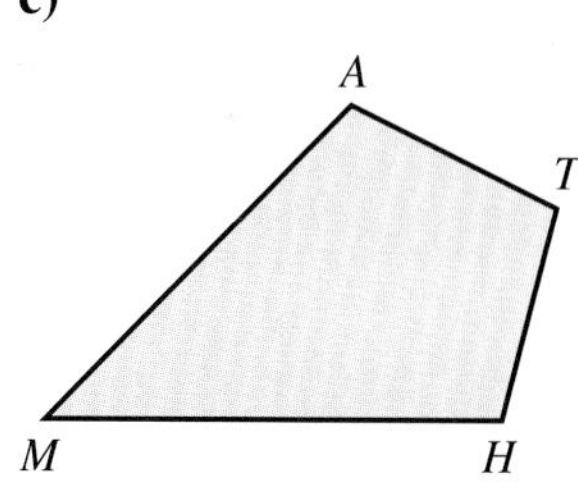

____

d)

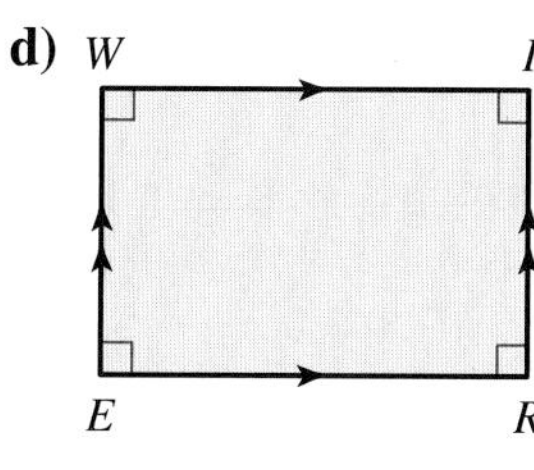

____

e)

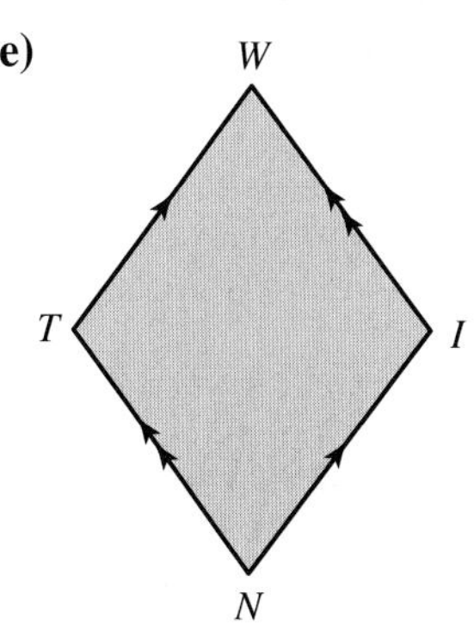

____

Regular Polygons

A **regular polygon** is a polygon in which all of the sides are the same length and all of the angles have the same measure.

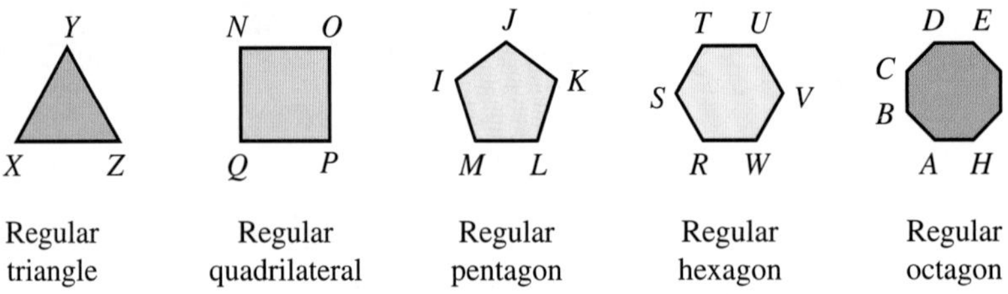

Regular triangle | Regular quadrilateral | Regular pentagon | Regular hexagon | Regular octagon

You Try It 6 **What is the special name given to**

a) A regular triangle? ______________ **b)** A regular quadrilateral? ______________

You Try It 7 **Complete each regular polygon by filling in the side measures and the angle measures.**

a) $\angle M = 108°$, $\overline{JK} = 9$ m **b)** $\angle S = 120°$, $\overline{RS} = 1.2$ yd

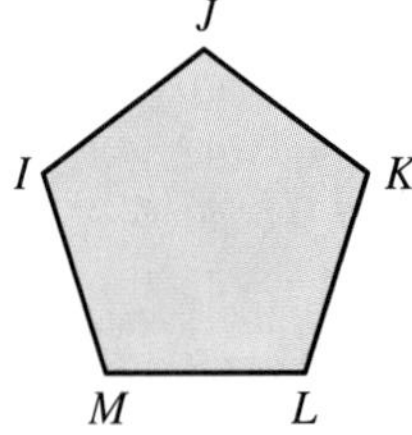

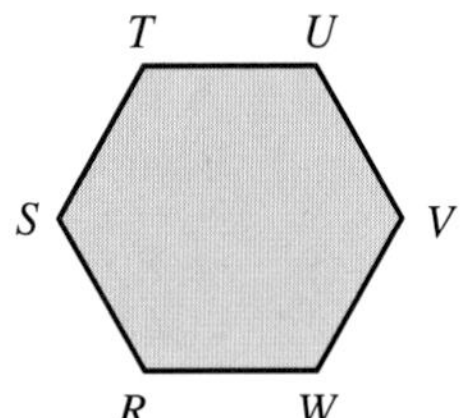

Circles

Recall from Section 5.6 that a circle is a round geometric figure with a center point. The center point is the same distance from every point on the circle.

The center names the circle. The circle at the right is named *circle C*, or ⊙ *C*. The center point, *C*, is not *on* the circle; it is a point that is in the *interior* of the circle.

Recall also that a line segment from the center point to any point on the circle (for example, $\overline{CA}$ at the right) is called a *radius*. The word *radius* also refers to the length of a radius.

A
radius
C
center

There are many radii (pronounced *ray′–dee–eye*, the plural of radius) in the circle, and each one has the center as one of its endpoints. Here are four radii: $\overline{CG}$, $\overline{CA}$, $\overline{CM}$, and $\overline{CX}$.

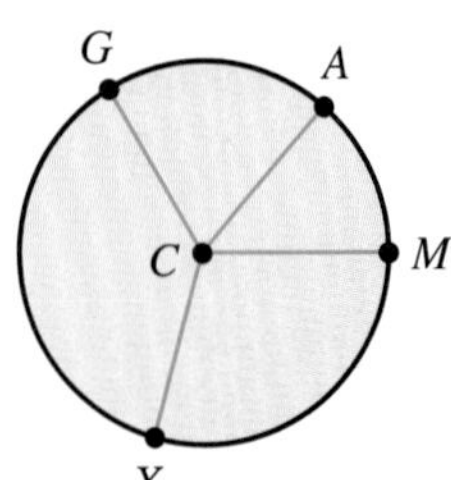

Two radii that form a straight angle (180°) are called a *diameter*. A **diameter** is a line segment that contains the center of the circle and that has endpoints on the circle. Here we see diameter $\overline{RT}$. As with the radius, *diameter* refers to the segment and its length.

The length of a diameter, *d*, is always twice the length of a radius, *r*: $d = 2 \cdot r$.

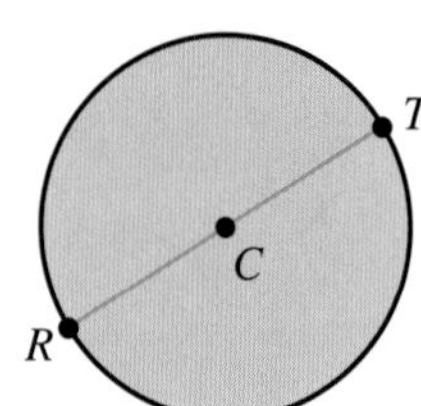

Example 5

Given the radius of the circle, find the diameter.

a)

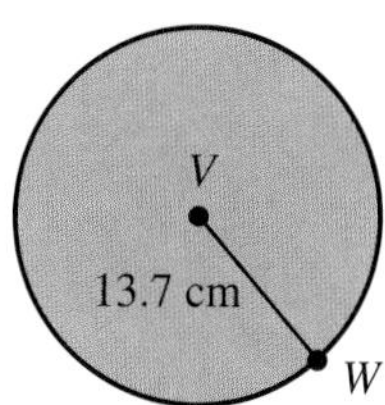

b)

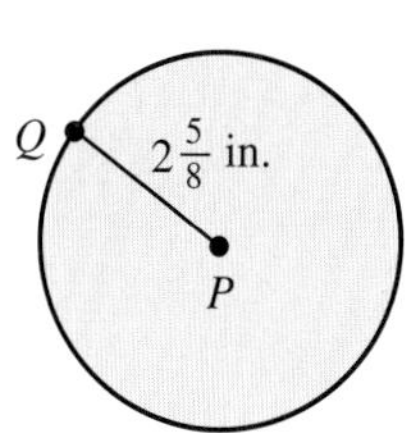

Procedure The diameter is twice the radius; so, use $d = 2 \cdot r$.

For part b, we can write $2\frac{5}{8}$ as $\frac{21}{8}$.

Answer a) $d = 2 \cdot 13.7\text{ cm}$

$d = 27.4\text{ cm}$

b) $d = 2 \cdot \frac{21}{8}\text{ in.}$

$d = \frac{2}{1} \cdot \frac{21}{8}\text{ in.}$

$d = \frac{21}{4}\text{ in.}$

$d = 5\frac{1}{4}\text{ in.}$

You Try It 8 **Given the radius of the circle, find the diameter. Use Example 5 as a guide.**

a)

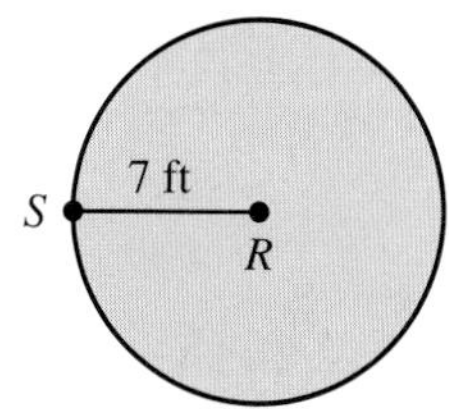

b)

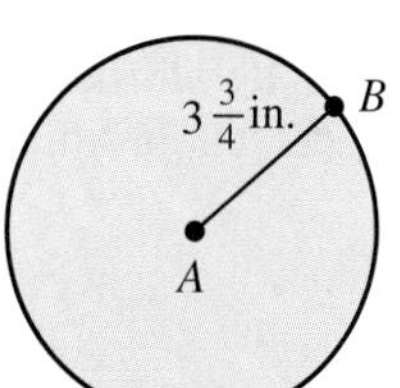

c)

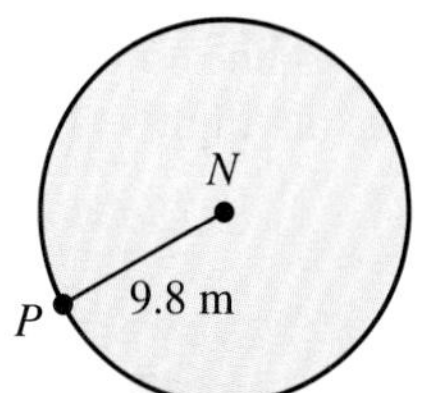

Likewise, the length of a radius, r, is always half the length of the diameter, d.

$$r = \frac{d}{2} \quad \text{or} \quad r = \frac{1}{2} \cdot d$$

Example 6

Given the diameter of the circle, find the radius.

a)

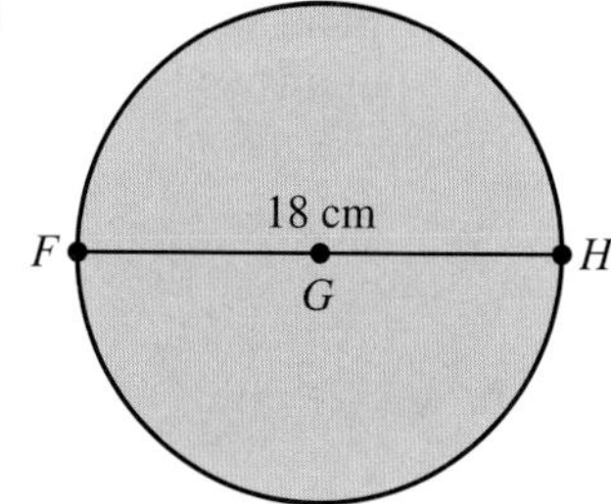

b)

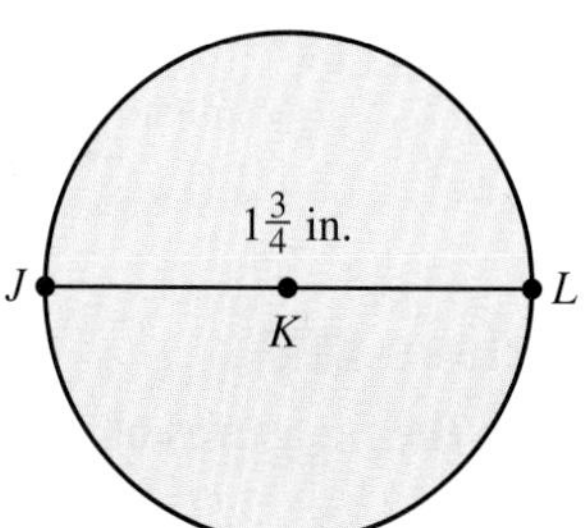

Procedure The radius is half the diameter; so, use either $r = \frac{d}{2}$ or $r = \frac{1}{2} \cdot d$.

For part b, we can write $1\frac{3}{4}$ as $\frac{7}{4}$.

Answer a) $r = \frac{18}{2}\text{ cm} = 9\text{ cm}$

b) $r = \frac{1}{2} \cdot \frac{7}{4}\text{ in.} = \frac{7}{8}\text{ in.}$

You Try It 9 **Given the diameter of the circle, find the radius. Use Example 6 as a guide.**

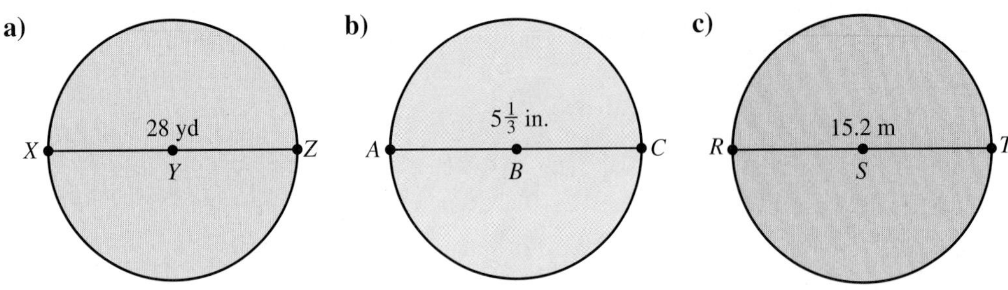

Answers: You Try It and Think About It

You Try It: **1. a)** The measure of $\angle L$ is 25°. **b)** The measure of $\angle N$ is 137.7°. **2. a)** scalene **b)** isosceles **c)** equilateral **3. a)** The measure of $\angle Q$ is 39°. **b)** The measure of $\angle S$ is 48.2°. **4. a)** $\angle B = 127°$, $\angle A = 53°$, and $\angle C = 53°$; $\overline{BC} = 9$ in. and $\overline{AB} = 15$ in. **b)** $\angle Q = 52.4°$, $\angle R = 127.6°$, and $\angle T = 127.6°$; $\overline{TQ} = 3.2$ mm and $\overline{TS} = 5.9$ mm **c)** All angles measure 90°; $\overline{XY} = 14$ ft and $\overline{YZ} = 33$ ft **5. a)** trapezoid **b)** parallelogram **c)** none of these **d)** rectangle and parallelogram **e)** parallelogram **6. a)** equilateral triangle **b)** square **7. a)** Each angle measures 108°; each side measures 9 meters. **b)** Each angle measures 120°; each side measures 1.2 yards. **8. a)** $d = 14$ ft **b)** $d = 7\frac{1}{2}$ in. **c)** $d = 19.6\,m$ **9. a)** $r = 14$ yd **b)** $r = 2\frac{2}{3}$ in. **c)** $r = 7.6$ m

Think About It: **1.** No. Two obtuse angles (each more than 90°) add to more than 180°, too many for a triangle. **2.** No. Two right angles (each equal to 90°) add to 180°, with no room for a third angle. **3.** Yes, an isosceles triangle has two *or more* sides of equal length. An equilateral triangle has three equal sides of equal length. **4.** Yes. A parallelogram is defined as having two pairs of parallel sides, and a square fits that description.

Section 8.5 Exercises

FOR EXTRA HELP
Student Resources on DVD-ROM
Includes
➤ Student's Solutions Manual
➤ Video Lectures
➤ Chapter Test Prep Video
MyMathLab
MathXL

Think Again

1. Is an equilateral triangle also isosceles? Why or why not? *(Refer to Think About It 3)*
   Yes, any two of the angles in an equilateral triangle can be chosen to be the base angles of an isosceles triangle.

2. Can a trapezoid also be a rectangle? Explain your answer or show an example that supports your answer.
   No. A trapezoid may have only one pair of parallel sides, whereas a rectangle must have two pairs of parallel sides.

3. Is it possible for a right triangle also to be isosceles? Explain your answer or show an example that supports your answer.
   Yes, the two sides that form the right angle can be the same length.

4. Is it possible for a right triangle also to be equilateral? Explain your answer or show an example that supports your answer.
   No, all three angles of an equilateral triangle measure 60°.

Focus Exercises

For $\triangle ABC$, given the measures of $\angle A$ and $\angle B$, find the measure of $\angle C$.

5. $\angle A = 30°$ and $\angle B = 70°$
   $\angle C = 80°$

6. $\angle A = 20°$ and $\angle B = 110°$
   $\angle C = 50°$

7. $\angle A = 65°$ and $\angle B = 45°$
   $\angle C = 70°$

8. $\angle A = 125°$ and $\angle B = 15°$
   $\angle C = 40°$

9. $\angle A = 76°$ and $\angle B = 92°$
   $\angle C = 12°$

10. $\angle A = 61°$ and $\angle B = 48°$
    $\angle C = 71°$

11. $\angle A = 106°$ and $\angle B = 28°$
    $\angle C = 46°$

12. $\angle A = 73°$ and $\angle B = 99°$
$\angle C = 8°$

13. $\angle A = 62.9°$ and $\angle B = 41.7°$
$\angle C = 75.4°$

14. $\angle A = 79.1°$ and $\angle B = 27.5°$
$\angle C = 73.4°$

15. $\angle A = 80.3°$ and $\angle B = 29.5°$
$\angle C = 70.2°$

16. $\angle A = 85.8°$ and $\angle B = 65.4°$
$\angle C = 28.8°$

17. $\angle A = 13.06°$ and $\angle B = 81.79°$
$\angle C = 85.15°$

18. $\angle A = 119.47°$ and $\angle B = 43.06°$
$\angle C = 17.47°$

19. $\angle A = 91.24°$ and $\angle B = 33.7°$
$\angle C = 55.06°$

20. $\angle A = 15.65°$ and $\angle B = 103.5°$
$\angle C = 60.85°$

For each triangle, based on the side lengths given, decide whether it is equilateral, isosceles, or scalene.

21.

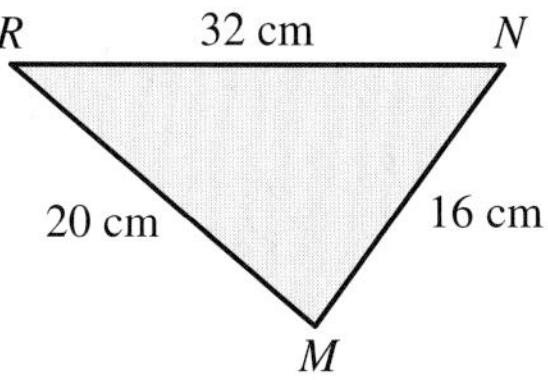

scalene

22.

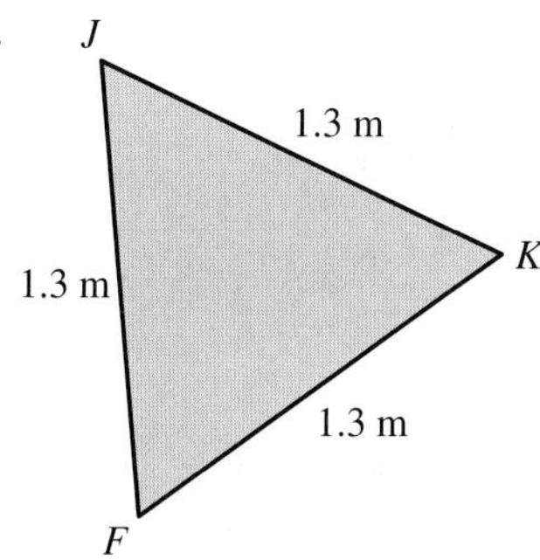

equilateral and isosceles

23.

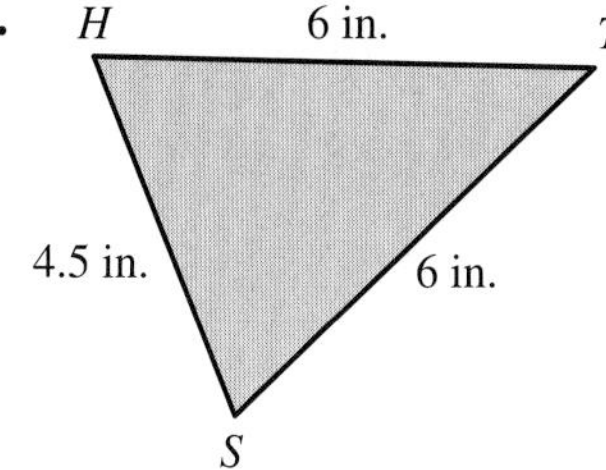

isosceles

24.

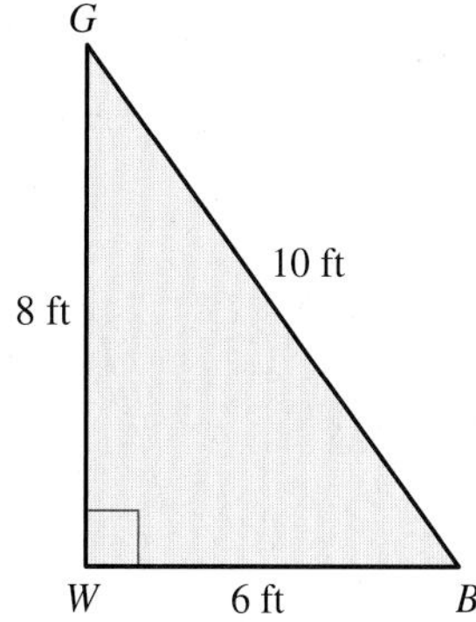

scalene

25.

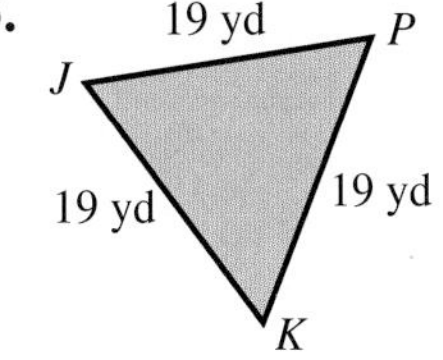

equilateral and isosceles

26.

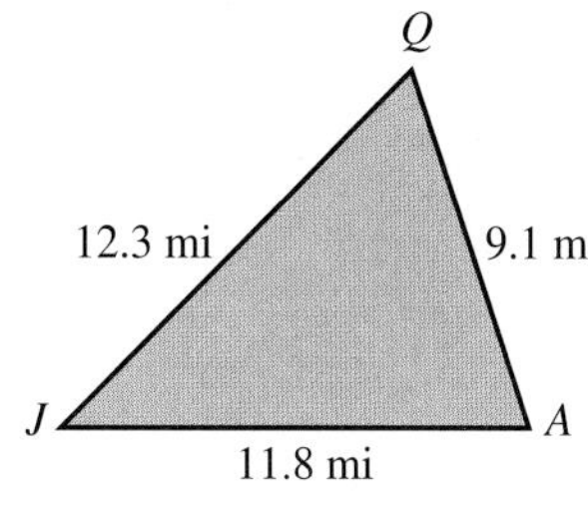

scalene

27.

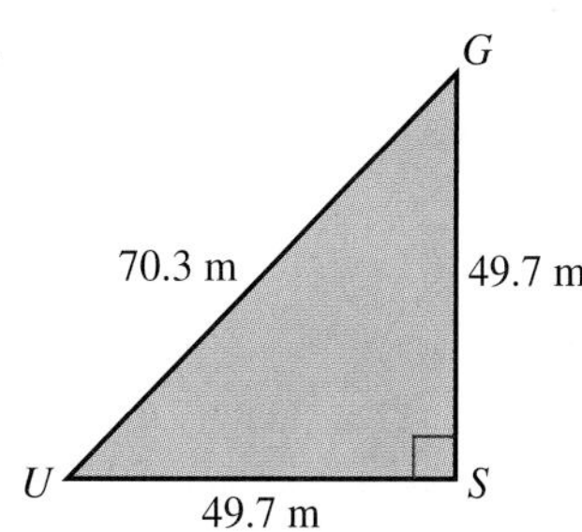

isosceles

28.

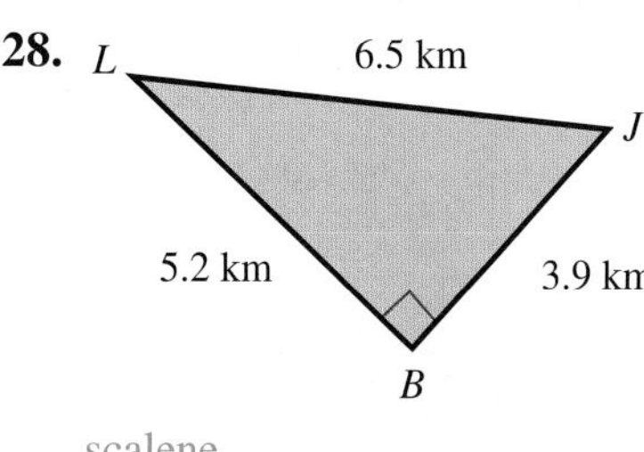

scalene

Given the measure of one acute angle, find the measure of the other acute angle.

29. $\angle A = 36°$

$\angle C = 54°$

30. $\angle G = 21°$

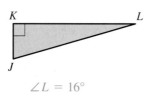

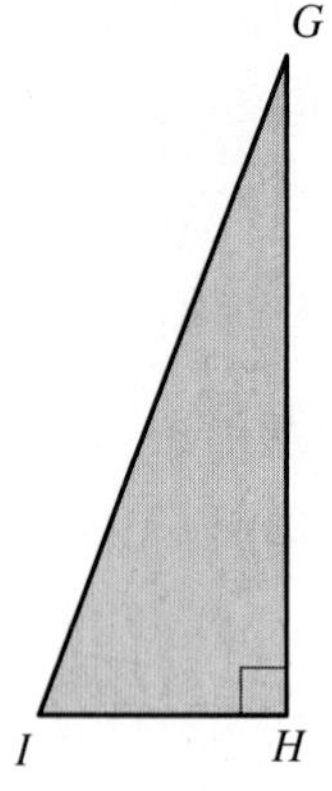

$\angle I = 69°$

31. $\angle D = 73°$

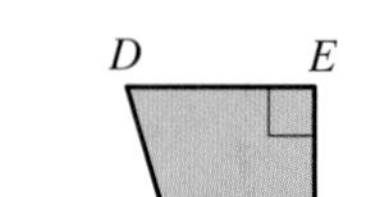
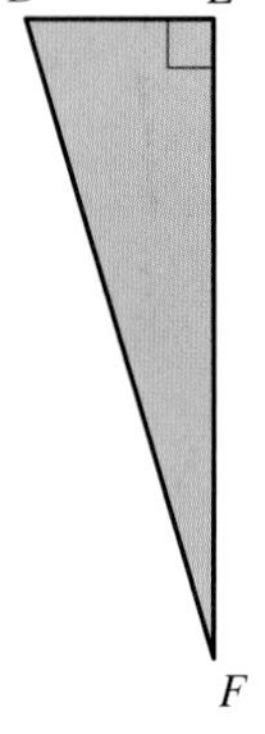

$\angle F = 17°$

32. $\angle J = 74°$

K L J

$\angle L = 16°$

33. $\angle M = 28.3°$

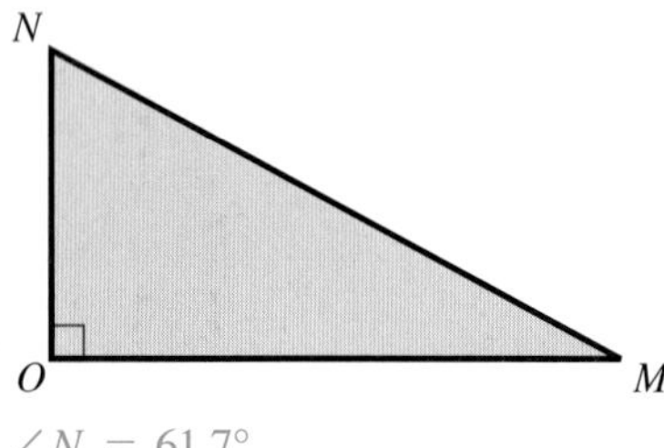

$\angle N = 61.7°$

34. $\angle R = 41.6°$

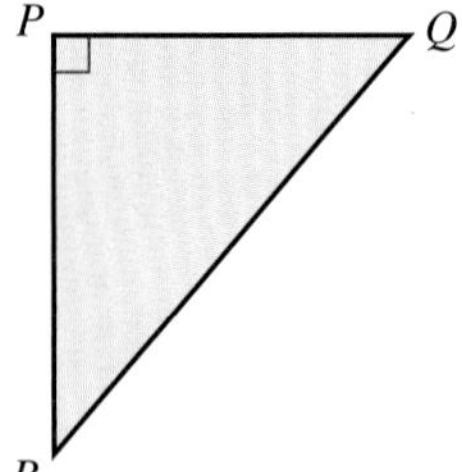

$\angle Q = 48.4°$

35. $\angle U = 52.5°$

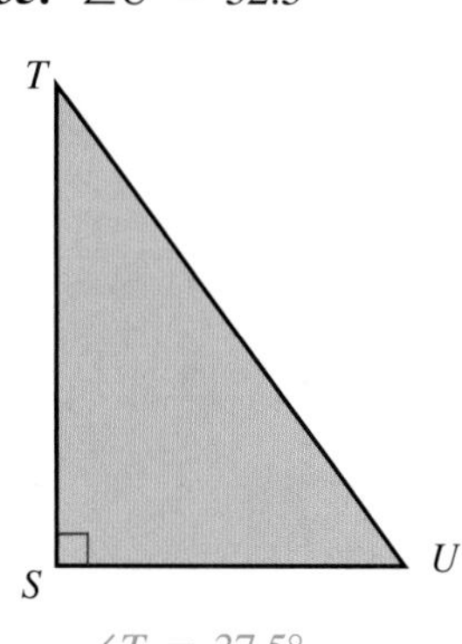

$\angle T = 37.5°$

36. $\angle V = 44.5°$

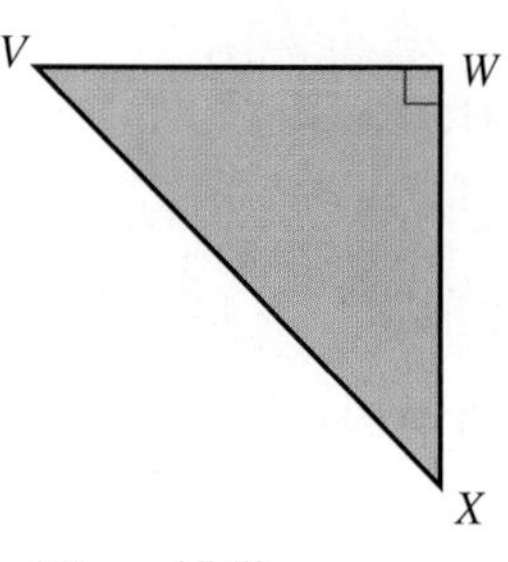

$\angle X = 45.5°$

Find the measures of the other three angles and the lengths of the other two sides.

37. $\angle N = 35°$
$\overline{NQ} = 3$ in.
$\overline{NO} = 7$ in.

Q P N O

$\angle P = 35°$, $\angle Q = \angle O = 145°$,
$\overline{OP} = 3$ in., and $\overline{QP} = 7$ in.

38. $\angle X = 132°$
$\overline{WZ} = 5$ mi
$\overline{XW} = 8$ mi

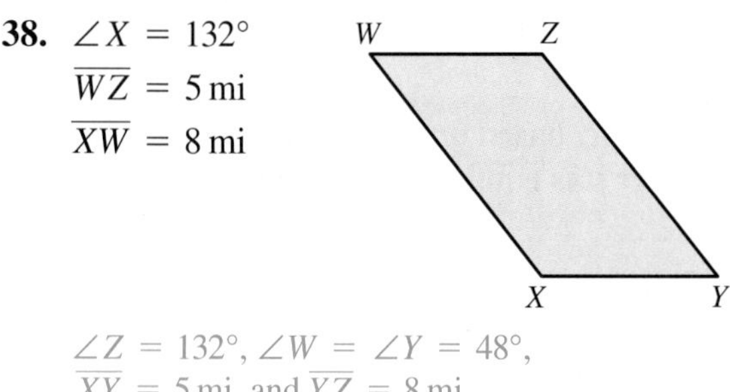

$\angle Z = 132°$, $\angle W = \angle Y = 48°$,
$\overline{XY} = 5$ mi, and $\overline{YZ} = 8$ mi

39. $\angle H = 127.4°$
$\overline{HG} = 3.9$ m
$\overline{IH} = 6.4$ m

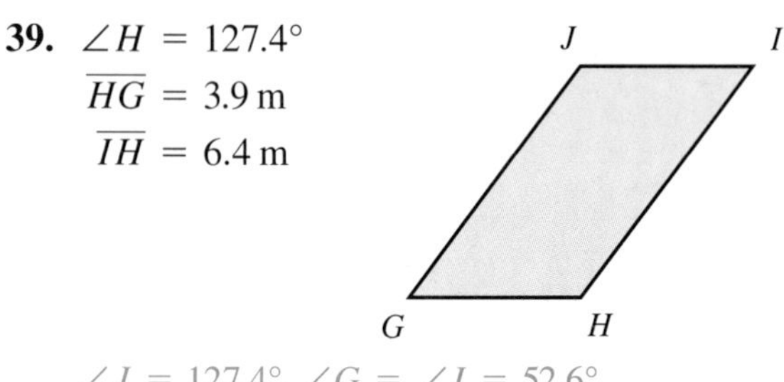

$\angle J = 127.4°$, $\angle G = \angle I = 52.6°$,
$\overline{IJ} = 3.9$ m, and $\overline{JG} = 6.4$ m

40. $\angle C = 29.1°$
$\overline{BC} = 15.6$ cm
$\overline{AB} = 11.7$ cm

A D B C

$\angle A = 29.1°$, $\angle B = \angle D = 150.9°$,
$\overline{AD} = 15.6$ cm, and $\overline{DC} = 11.7$ cm

True or false.

41. Some trapezoids are also parallelograms. False

42. Every square is also a parallelogram. True

43. In a trapezoid, it's possible for the two parallel sides (the bases) to be the same length. False

44. When one angle in a parallelogram is a right angle, the parallelogram is a rectangle. True

45. In a polygon, the number of sides and the number of angles is always the same. True

46. In an equilateral triangle, all three sides are equal in length, but the angles can be of different measures. False

47. A triangle can contain three acute angles. True

48. A regular octagon has eight sides, but an octagon that is not regular can have more than eight sides. False

Given the radius, *r*, of a circle, find the diameter; or given the diameter, *d*, of a circle, find the radius.

49. $r = 6$ in.
$d = 12$ in.

50. $r = 3.8$ m
$d = 7.6$ m

51. $d = 20$ in.
$r = 10$ in.

52. $d = 27$ in.
$r = 13\frac{1}{2}$ in.

53. $r = 6.5$ cm
$d = 13$ cm

54. $r = 5\frac{3}{4}$ in.
$d = 11\frac{1}{2}$ in.

55. $d = 7\frac{1}{3}$ ft
$r = 3\frac{2}{3}$ ft

56. $d = 26.5$ in.
$r = 13.25$ in.

Think Outside the Box

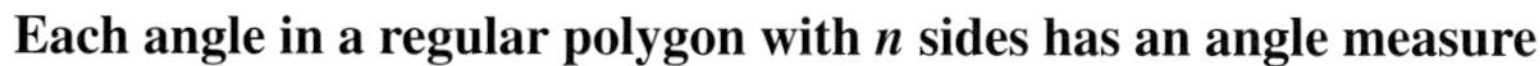

Each angle in a regular polygon with n sides has an angle measure according to this formula: $\frac{(n-2)\cdot 180}{n}$ degrees. Use this formula to find the measure of each angle of the following regular polygons. Then use the value found to determine the total angle measure of the associated regular polygon.

57. A regular triangle
Each angle measures 60°.
The total angle measure is 180°.

58. A regular quadrilateral
Each angle measures 90°.
The total angle measure is 360°.

59. A regular pentagon
Each angle measures 108°.
The total angle measure is 540°.

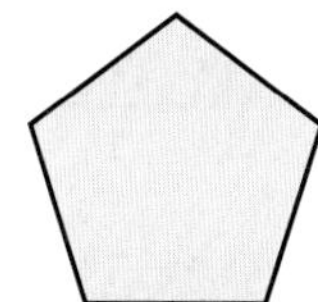

60. A regular hexagon
Each angle measures 120°.
The total angle measure is 720°.

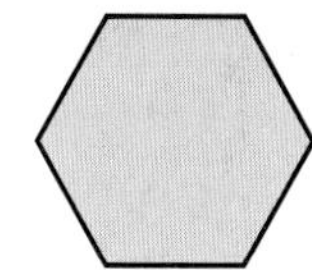

SECTION 8.6 Perimeter

OBJECTIVES

In this section, you will learn to

- Find the perimeter of a polygon.
- Apply the perimeter formulas for a rectangle and for regular polygons.
- Determine the circumference of a circle.
- Find the perimeter of unusual shapes.

You Need to Know

To successfully complete this section, you need to understand

- ☐ Solving equations (1.7 and Ch. 3)
- ☐ Multiplying fractions (4.4)
- ☐ Adding and subtracting fractions (4.5, 4.7)
- ☐ Adding and subtracting decimals (5.3)
- ☐ Multiplying decimals (5.4)
- ☐ Geometric shapes (8.5)
- ☐ Regular polygons (8.5)

Introduction

Recall from Section 1.2 that we add the lengths of the sides of a figure to find its perimeter.

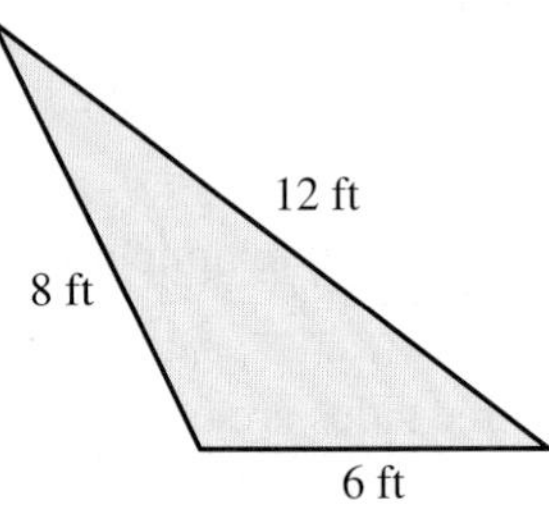

Perimeter: $P = 8\text{ ft} + 12\text{ ft} + 6\text{ ft}$

$P = 26\text{ ft}$

$P = 26\text{ ft}$

Recall from Section 2.6 that the formula for the perimeter of a rectangle is $P = 2 \cdot L + 2 \cdot W$.

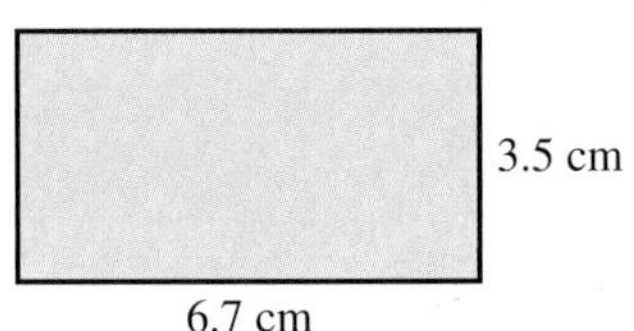

Perimeter: $P = 2 \times 3.5\text{ cm} + 2 \times 6.7\text{ cm}$

$P = 7.0\text{ cm} + 13.4\text{ cm}$

$P = 20.4\text{ cm}$

You Try It 1 **Find the perimeter of each polygon.**

a)

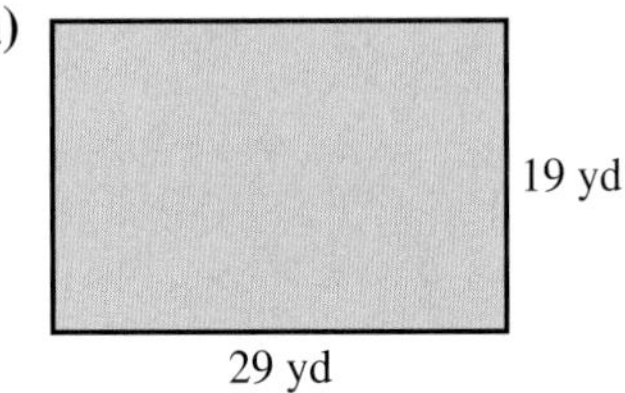

b)

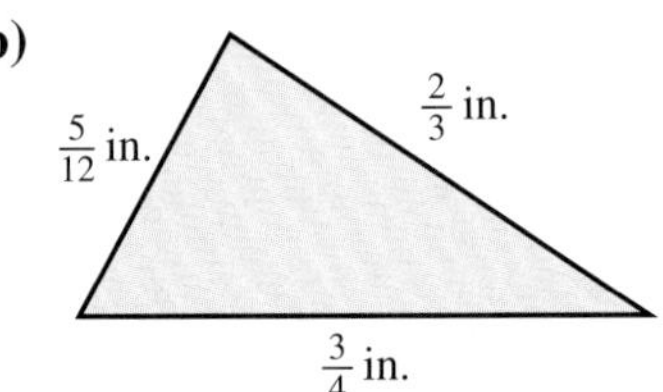

Perimeter of a Regular Polygon

Recall from Section 8.5 that all of the sides of a regular polygon are the same length. If a polygon is regular, then knowing the length of one side means that you know the length of all of the sides, as demonstrated in this equilateral triangle.

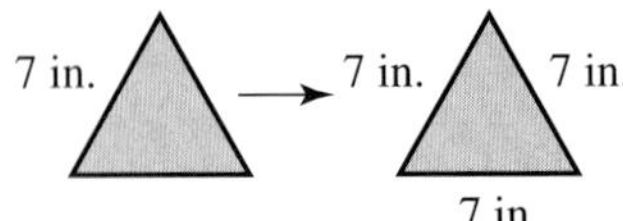

$P = 7\text{ in.} + 7\text{ in.} + 7\text{ in.} = 3 \cdot 7\text{ in.} = 21\text{ in.}$

Some perimeter formulas for regular polygons are as follows:

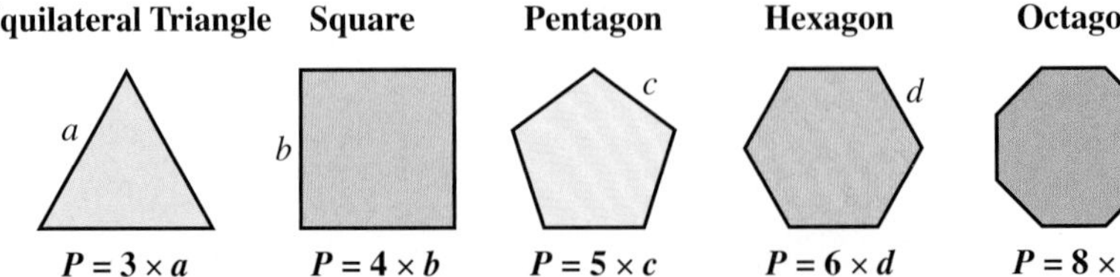

Example 1

Draw a regular hexagon and label one side 5 feet long. What is the perimeter of this hexagon?

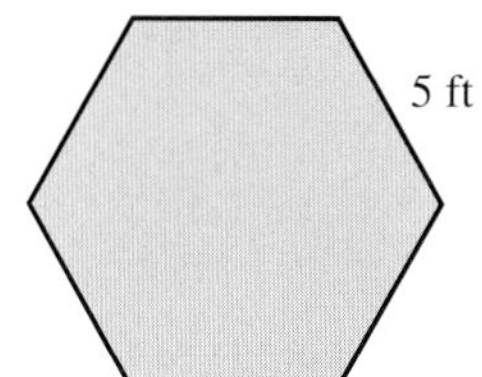

Procedure No one expects you to draw a perfect regular hexagon; just do your best. Try to copy the previous hexagon. Then label one side 5 ft.

Answer Using the formula for the perimeter of a regular hexagon (six sides), $P = 6 \cdot 5\text{ ft} = 30\text{ ft}$.

You Try It 2 **Draw a regular polygon and label one side as requested. What is the perimeter of this polygon? Use Example 1 as a guide.**

a) An equilateral triangle; one side is 14.8 cm long.

b) A regular pentagon; one side is $3\frac{1}{2}$ in. long.

Example 2

Draw a regular octagon. Given that the perimeter of this octagon is 116 centimeters, find the length of each side.

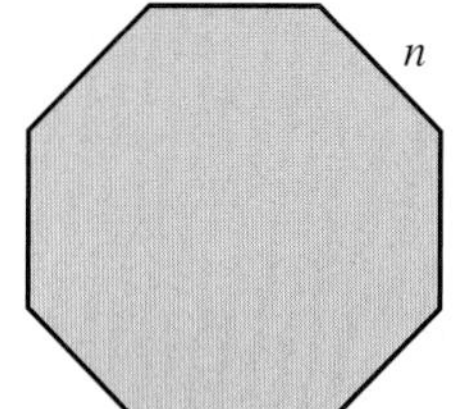

Procedure Do your best to draw a regular octagon. If the whole perimeter is 116 cm and the side value is unknown (n), we can use the formula to set up an equation to find the length of each side.

Answer Legend: Let n = one side of the regular octagon.

$$8 \cdot n = 116 \quad \text{Divide each side by 8.}$$
$$n \cdot 8 \div 8 = 116 \div 8$$
$$n = 14.5$$

$$\begin{array}{r} 14.5 \\ 8\overline{)116.0} \\ -8 \\ \hline 36 \\ -32 \\ \hline 4\,0 \\ -4\,0 \\ \hline 0 \end{array}$$

Each side of the octagon is 14.5 centimeters long.

You Try It 3 **Draw a regular polygon. Given the perimeter of the polygon, find the length of each side. Use Example 2 as a guide.**

a) A regular hexagon; the perimeter is 84 ft.

b) An equilateral triangle; the perimeter is 73.5 cm.

Circumference and Pi

In finding the perimeter of a polygon, we determine the sum of all of its sides. Because a circle has only one "side" and it isn't straight, the perimeter of the circle—the **circumference**—must be evaluated differently, as we'll soon see.

Imagine a pebble dropped into a pond. A circular ripple is created that grows increasingly larger. As this circle grows, both the diameter and the circumference get larger *proportionally*.

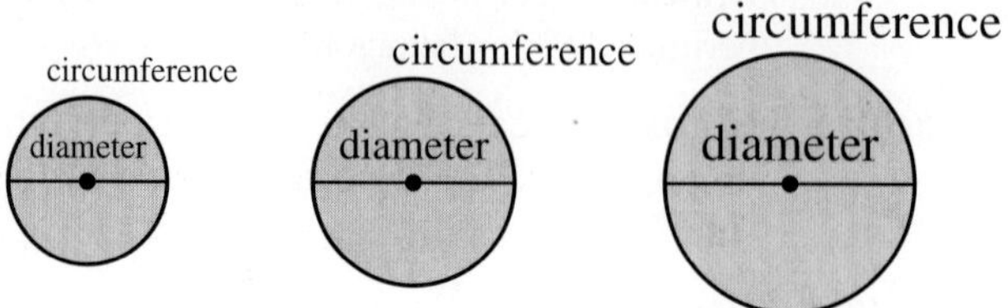

This means that there is a ratio, circumference : diameter or $\frac{\text{circumference}}{\text{diameter}}$, that is the same number no matter how large the circle. This ratio is labeled with the Greek letter pi (π).

If the circumference and diameter of a circle could be precisely measured, we'd find that

$$\pi = \frac{\text{Circumference}}{\text{Diameter}} = 3.14159265358979\ldots$$

Recall from Sections 5.1 and 5.6 that π is a decimal number that never terminates and never repeats; so, we cannot write out an exact decimal value of π.

There are two accepted close approximations to π. One is the fraction $\frac{22}{7} = 3.\overline{142857}$; the other is the rounded decimal 3.14. In other words,

$$\pi \approx \frac{22}{7} \text{ and } \pi \approx 3.14$$

Think About It 1 Notice that $\frac{22}{7}$ is a repeating decimal, $3.\overline{142857}$. Does this mean that π is also a repeating decimal? Explain your answer.

____

____

The formula for the circumference of a circle is circumference $= \pi \cdot$ diameter,

abbreviated as $C = \pi \cdot d$

or as $C = \pi \cdot 2 \cdot r$ Because $d = 2 \cdot r$.

or as $C = 2 \cdot \pi \cdot r$

To represent an exact value of the circumference, we use the symbol π. For example, if the diameter of a circle is 7 inches, the exact value of the circumference is written $C = 7 \cdot \pi$ inches, or 7π inches.

However, we can't really measure anything that has the symbol π in it. When we want to measure the circumference, we need to use one of the approximate values for π.

$C \approx 7 \cdot \frac{22}{7}$ inches or $C \approx 7 \times 3.14$ inches

$C \approx 22$ inches $C \approx 21.98$ inches

Think About It 2 Notice that the values we found for the circumference, 22 and 21.98, are very close to each other. Why aren't they exactly the same?

____

____

In deciding which approximate value to use for π, it's common to use

- $\frac{22}{7}$ when the radius or diameter is in a fractional form or is a whole number easily divisible by 7.
- 3.14 when the measure is a decimal or a whole number not divisible by 7.

Furthermore, when using the decimal approximation, it's common to round the resulting value to an easily measurable amount, usually the tenths or the hundredths place.

Example 3

For each circle, find the circumference. Use an approximate value for π.

a)

3.5 m, Q, R, S

b)

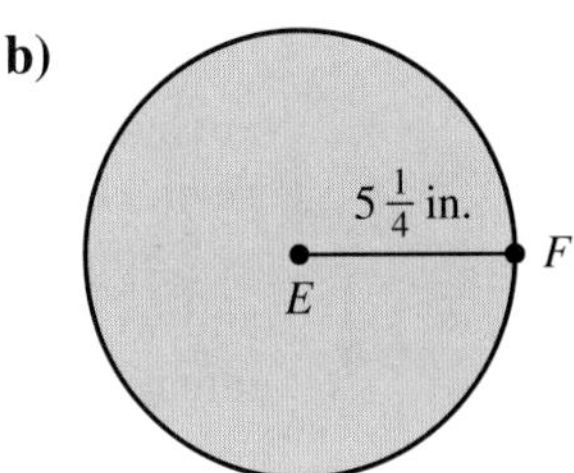

Procedure For part a, we'll use $C = \pi \cdot d$; for part b, we'll use $C = \pi \cdot 2 \cdot r$.
For part a, use $\pi \approx 3.14$ and round the answer to the tenths place.
For part b, use $\pi \approx \frac{22}{7}$. Also for part b, we can write $5\frac{1}{4}$ as $\frac{21}{4}$.

Answer

a) $\overline{QS}$ is a diameter.

$C \approx 3.14 \times 3.5\text{ m}$

$C \approx 10.99\text{ m}$

Rounded: $C \approx 11.0\text{ m}$

YOU FINISH IT: Multiply

$$\begin{array}{r} 3.14 \\ \times\ \ 3.5 \\ \hline \end{array}$$

b) $\overline{EF}$ is a radius.

$C \approx \frac{22}{7} \cdot 2 \cdot \frac{21}{4}$ in.

$C \approx \frac{22 \cdot 2 \cdot 21}{7 \cdot 1 \cdot 4}$ in. 7 and 21 have a common factor of 7; 2 and 4 have a common factor of 2.

$C \approx \frac{22 \cdot 1 \cdot 3}{1 \cdot 1 \cdot 2}$ in.

$C \approx \frac{66}{2}$ in. $\frac{66}{2} = 33$

$C \approx 33$ in.

You Try It 4 **For each circle, find an approximate value of the circumference. Use Example 3 as a guide. Round each decimal answer, if necessary, to the nearest tenth.**

a)

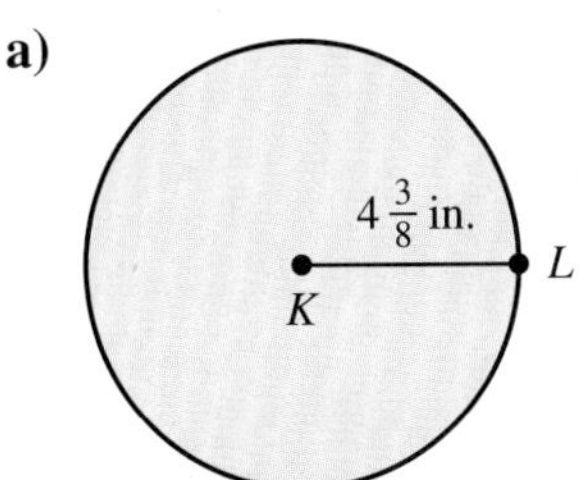

b)

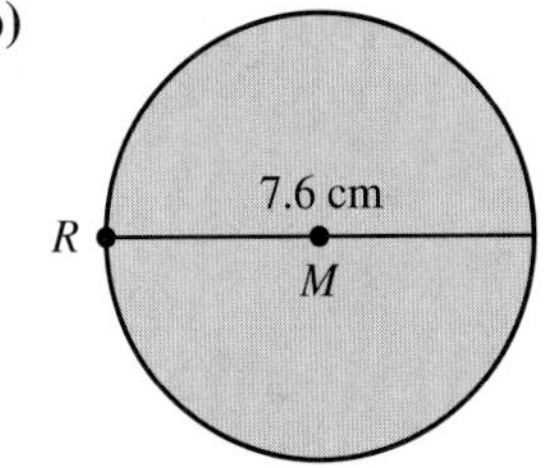

Perimeters of Unusual Shapes

Consider this situation: Arturo and Terry are laying wood flooring in Janet's living room. After laying the flooring, they will place baseboards along each wall. According to the floor plan's dimensions, how many feet of baseboard are needed to complete the room? Assume that all angles are right angles.

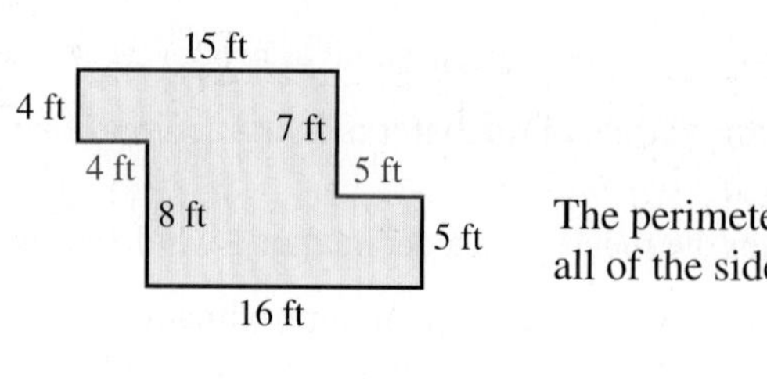

The perimeter is found by adding all of the side measures.

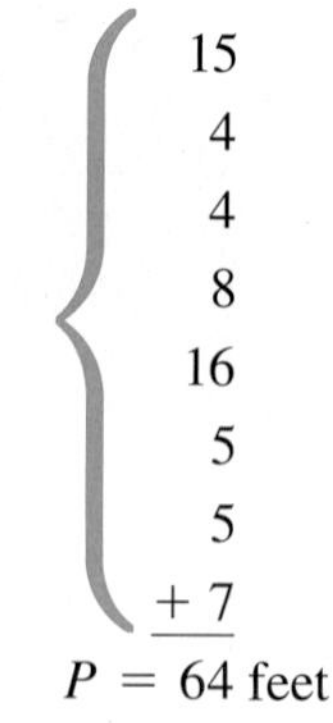

$$\begin{array}{r} 15 \\ 4 \\ 4 \\ 8 \\ 16 \\ 5 \\ 5 \\ +\ 7 \\ \hline P = 64 \text{ feet} \end{array}$$

Arturo and Terry will need 64 feet of baseboard to complete the room.

As you see from this example, not all geometric shapes are standard shapes; but that doesn't stop us from finding the perimeter. We just need to know all of the dimensions.

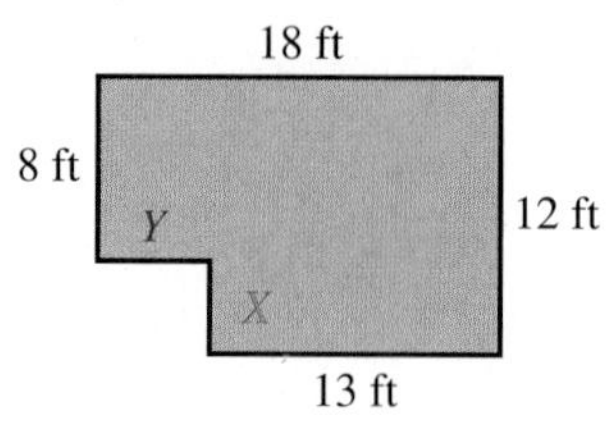

Sometimes, though, not all of the dimensions are shown and we must figure out any missing side measures based on other information in the diagram.

For example, in this diagram, we know the total length to be 18 feet and the total width to be 12 feet. Using other measures in the diagram, fill in the missing side measures, one marked with an *X* and the other marked with a *Y*.

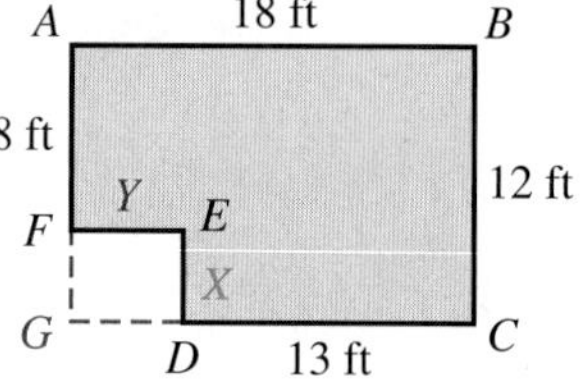

We can label this floor plan as a polygon at each vertex. We can even show a point *G* that is not a vertex by creating a rectangle out of the floor plan, rectangle *ABCG*. In fact, a smaller rectangle is formed by *DEFG*.

In the larger rectangle, $\overline{GC}$ is the same measure as $\overline{AB}$; and in the smaller rectangle, $\overline{FG}$ is the same measure as $\overline{ED}$. Explain why.

____

____

The two lower measures, $\overline{FE}$ and $\overline{DC}$, add to the total measure of $\overline{AB}$, 18 ft.

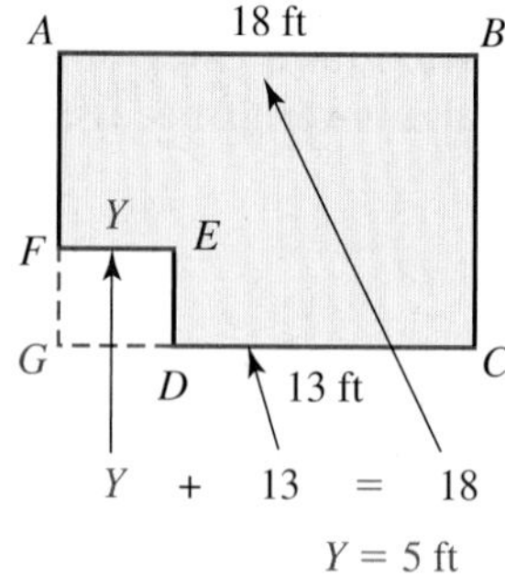

The two left measures, $\overline{AF}$ and $\overline{ED}$, add to the total measure of $\overline{BC}$, 12 ft.

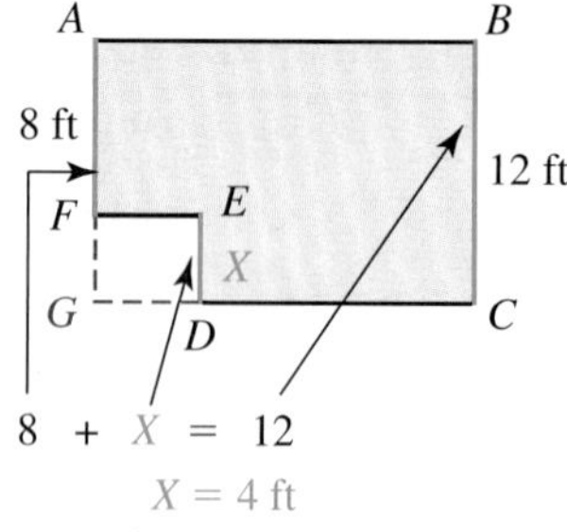

We can now put those measures on the diagram and add to find the perimeter.

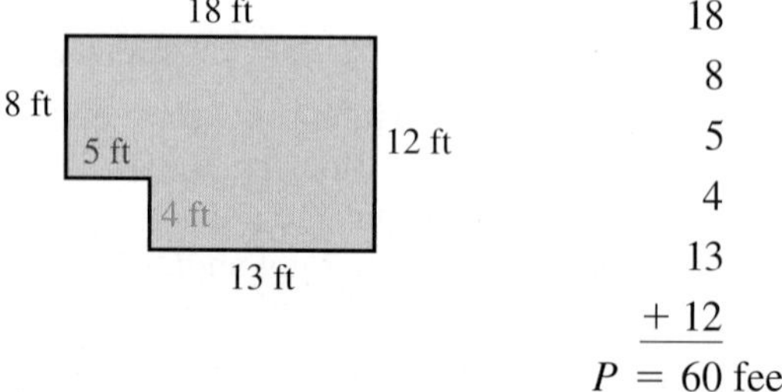

$$\begin{array}{r} 18 \\ 8 \\ 5 \\ 4 \\ 13 \\ +\ 12 \\ \hline P = 60 \text{ feet} \end{array}$$

Example 4

Find the measures of the marked sides (one with an X, the other with a Y). Then find the perimeter of the polygon. Assume that all angles are right angles.

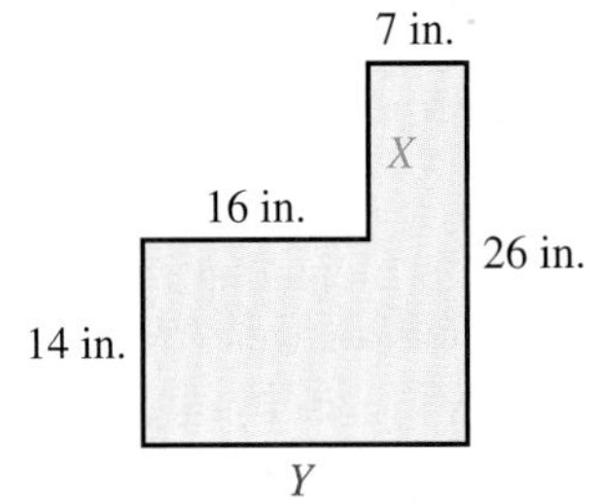

Procedure Notice that the side marked with a Y is across the *whole* width of the figure. The side marked with an X is one of the *parts*, not the whole. (You may want to label each vertex in the polygon, including a point that completes the rectangle.)

Answer Since X is a *part*, we'll use this equation:

$$14 + X = 26$$
$$X + 14 - 14 = 26 - 14$$
$$X = 12$$

Since Y is a *whole*, we'll use this equation:

$$16 + 7 = Y$$
$$23 = Y$$

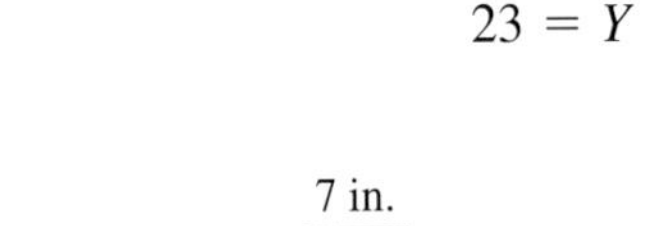

Putting these values into the polygon, we get:

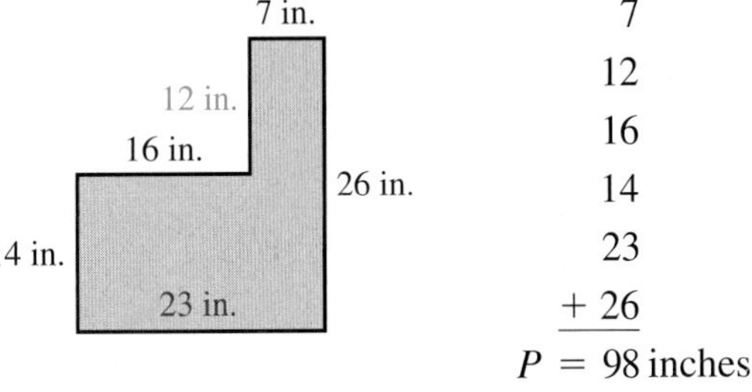

$$\begin{array}{r} 7 \\ 12 \\ 16 \\ 14 \\ 23 \\ +\ 26 \\ \hline P = 98 \text{ inches} \end{array}$$

You Try It 5 **Find the measures of the marked sides (with an X and a Y) and then find the perimeter of the polygon. Use Example 4 as a guide. Assume that all angles are right angles.**

a)

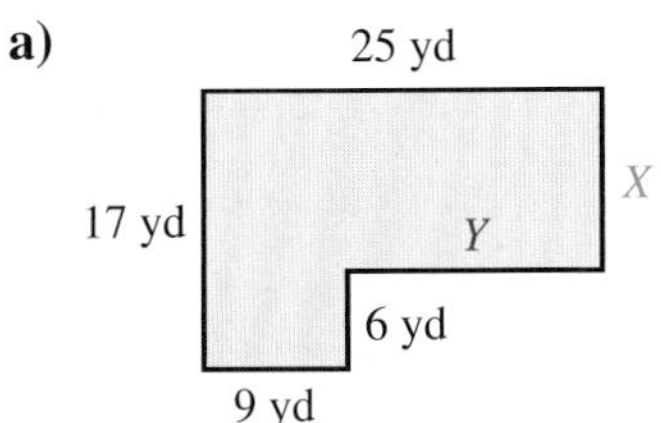

b)

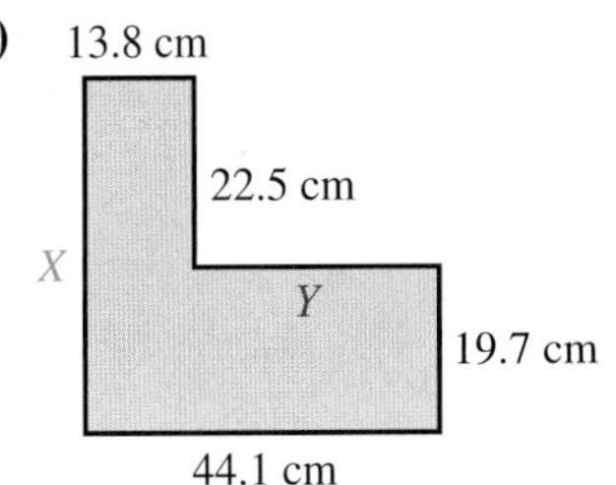

Answers: You Try It and Think About It

You Try It: **1. a)** $P = 96$ yd **b)** $P = 1\frac{5}{6}$ in.

2. a)

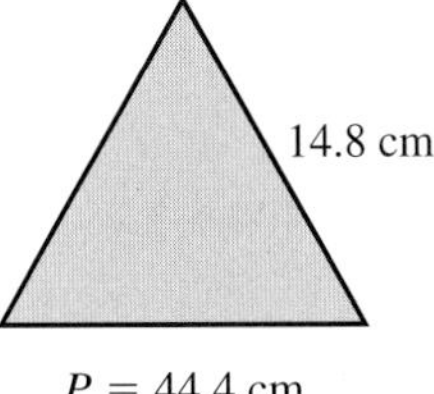

$P = 44.4$ cm

b)

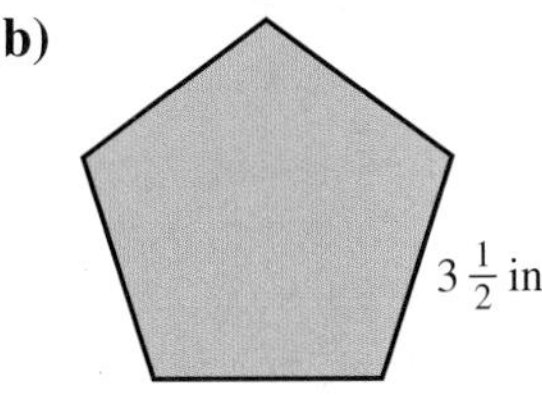

$P = 17\frac{1}{2}$ in.

3. a)

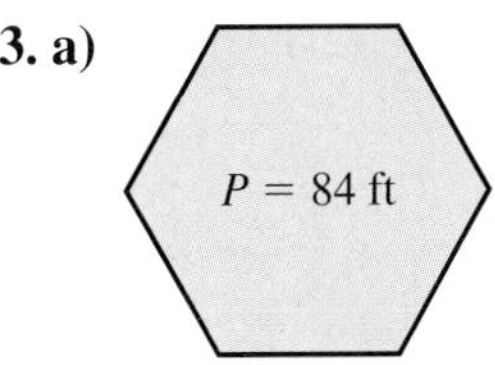

Side = 14 ft

b)

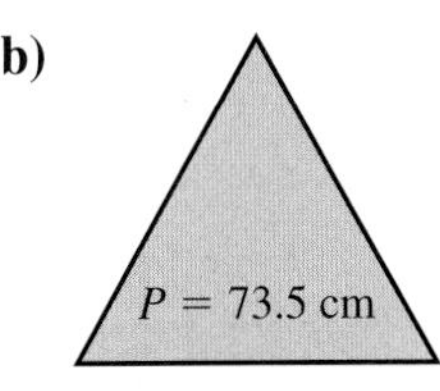

Side = 24.5 cm

4. a) $C \approx 27\frac{1}{2}$ in. or $C \approx 27.5$ in. **b)** $C \approx 23.9$ cm **5. a)** $X = 11$ yd, $Y = 16$ yd, $P = 84$ yd **b)** $X = 42.2$ cm, $Y = 30.3$ cm, $P = 172.6$ cm

Think About It: **1.** No. π is approximately $\frac{22}{7}$, but it is not equal to $\frac{22}{7}$. π is a decimal that never terminates and never repeats. **2.** They aren't exactly the same because the approximations for π, $\frac{22}{7}$ and 3.14, aren't exactly the same either. **3.** Every rectangle is a parallelogram, and the opposite sides of a parallelogram are the same length. This is true for opposite side $\overline{GC}$ and $\overline{AB}$ as well as for $\overline{FG}$ and $\overline{ED}$.

Section 8.6 Exercises

Think Again

1. $\pi \approx \frac{22}{7}$ and $\frac{22}{7}$ is a repeating decimal, $3.\overline{142857}$. Does this mean that π is also a repeating decimal? Explain your answer. *(Refer to Think About It 1)*
No, $\frac{22}{7}$ is only an approximation for π. π is not a repeating decimal.

2. In a circle, is it possible for both the diameter and the circumference to be whole numbers? Explain your answer or show an example that supports your answer.
No. The ratio of the circumference to the diameter is always π, a nonterminating, nonrepeating decimal number. If the diameter is a whole number, the circumference will be a decimal number and vice versa.

Focus Exercises

For each application, find the perimeter of the polygon described. Answer the question with a sentence.

3. In her design class, Karin created this unusual shape for a lamp stand. What is the perimeter of Karin's lamp stand?

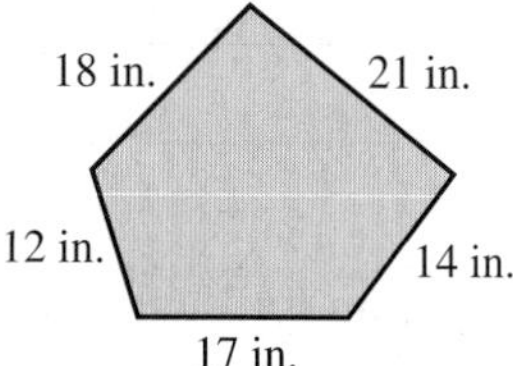

The perimeter of Karin's lamp stand is 82 inches.

4. Hoi Mai's backyard is in the shape of a trapezoid (at the right). What is the perimeter of Hoi Mai's backyard?

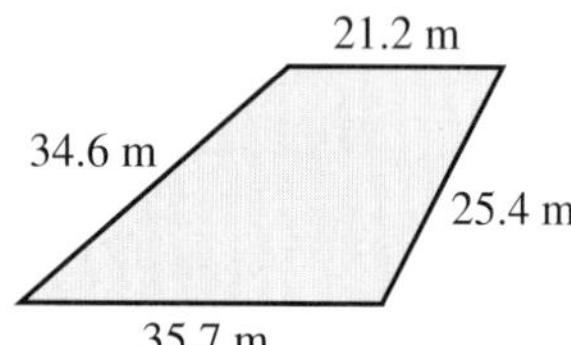

The perimeter of Hoi Mai's backyard is 116.9 meters.

5. The dimensions of home plate for professional baseball are shown at the right. What is the perimeter of home plate?

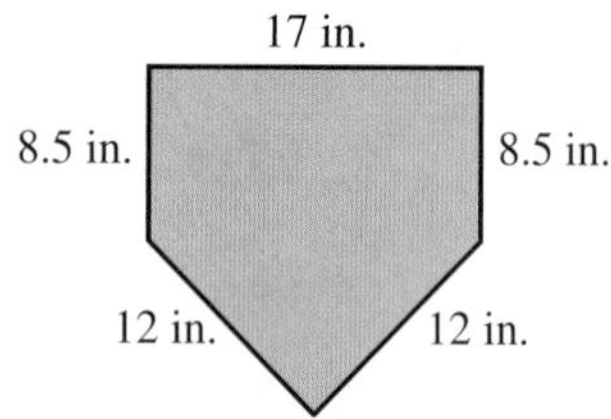

The perimeter of home plate is 58 inches.

6. A room in a modern home has the unusual shape of a hexagon, as shown at the right. What is the perimeter of this room?

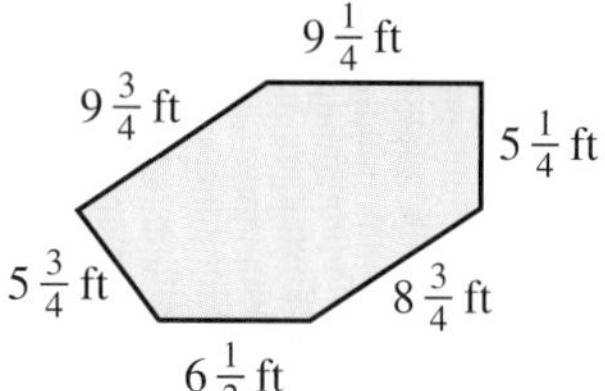

The perimeter of this room is $45\frac{1}{4}$ feet.

7. The Pentagon, the Washington, D.C., building that houses the U.S. Department of Defense, is in the shape of a regular pentagon. The outer wall of each side is 921 feet long. What is the outer perimeter of the Pentagon?

The outer perimeter of the Pentagon is 4,605 feet.

8. A stop sign is a regular octagon that has a side length of 4.75 decimeters. What is the perimeter of the stop sign?

The perimeter of the stop sign is 38 dm.

9. A long-neck 5-string banjo has a resonator (the round part) with a diameter of 12 inches. What is the circumference of the resonator?

The circumference of the resonator is about 37.7 inches.

10. The largest Ferris wheel in Europe (and maybe the world) is the London Eye, on the River Thames in London, England. This Ferris wheel has a diameter of 135 meters. What is the circumference of the wheel?
Source: **www.londoneye.com**

The circumference of the wheel is about 423.9 meters.

11. A hula hoop has a diameter of 2 feet. What is the circumference of this hula hoop?

The circumference of this hula hoop is about 6.3 feet.

12. A child's bike has a wheel with a diameter of 20 inches. What is the circumference of this wheel?

The circumference of this wheel is about 62.8 inches.

13. A crop circle of barley has a diameter of 400 feet. What is the circumference of this crop circle?

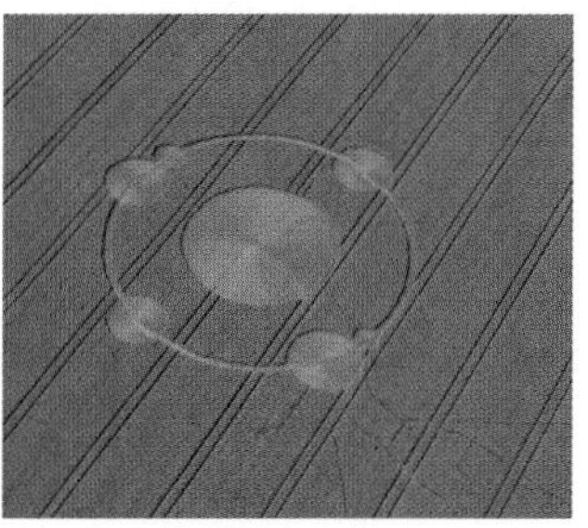

The circumference of this crop circle is about 1,256 feet.

14. Stonehenge, in southern England, is an ancient collection of huge stones that form a circular monument. The outer circle of Stonehenge has a diameter of 330 ft. What is the circumference of this circle?
Source: **http://witcombe.sbc.edu**

The circumference of this circle is about 1036.2 feet.

Find the perimeter of each triangle. (Write all improper fraction answers as mixed numbers.)

15.

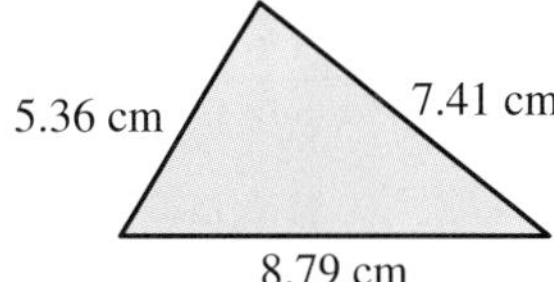

$P = 21.56$ cm

16.

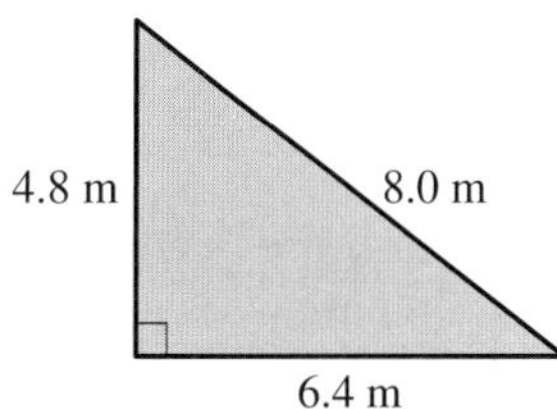

$P = 19.2$ m

17.

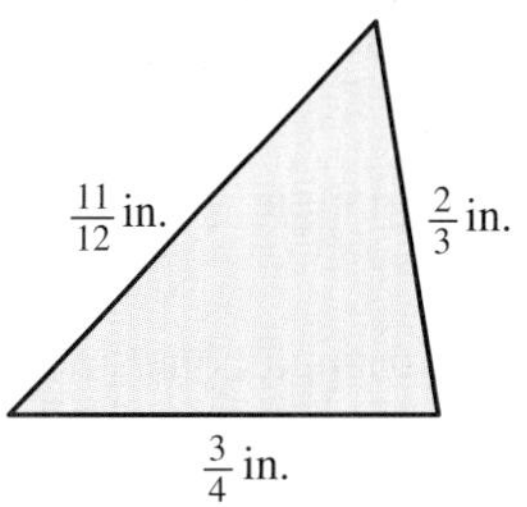

$P = 2\frac{1}{3}$ in.

18.

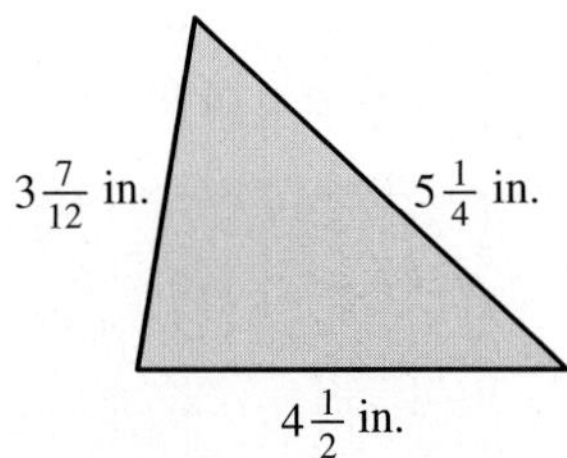

$P = 13\frac{1}{3}$ in.

Find the perimeter of each rectangle.

19.

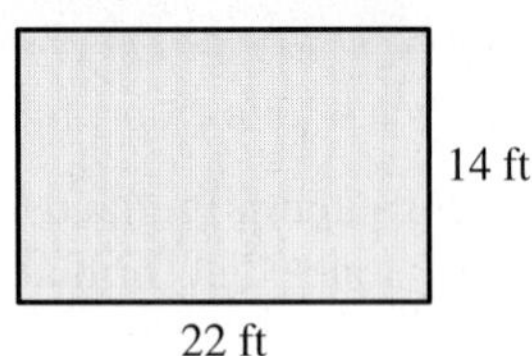

$P = 72$ ft

20.

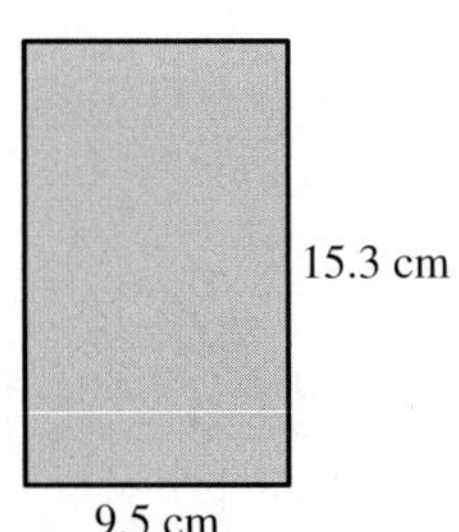

$P = 49.6$ cm

21.

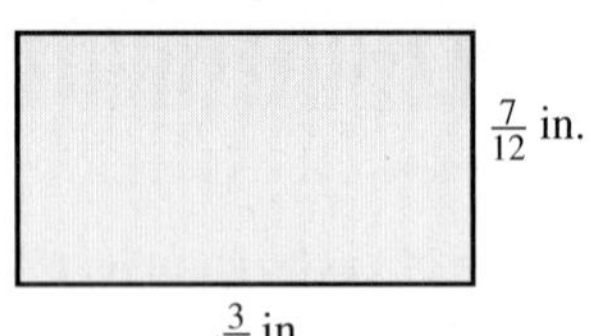

$P = 2\frac{2}{3}$ in.

22.

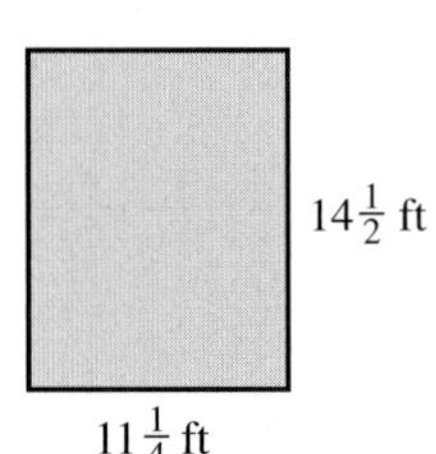

$P = 51\frac{1}{2}$ ft

For each polygon, find the perimeter described. (Write all improper fraction answers as mixed numbers.)

23. A square; one side is 13 ft long.
$P = 52$ ft

24. A square; one side is 28.5 m long.
$P = 114$ m

25. An equilateral triangle; one side is 31.4 m long.
$P = 94.2$ m

26. An equilateral triangle; one side is $3\frac{5}{12}$ in. long.
$P = 10\frac{1}{4}$ in.

27. A regular pentagon; one side is 148 yd long.
$P = 740$ yd

28. A regular octagon; one side is 4.65 cm long.
$P = 37.2$ cm

29. A regular hexagon; one side is $2\frac{5}{12}$ in. long.
$P = 14\frac{1}{2}$ in.

30. A regular octagon; one side is $6\frac{3}{4}$ in. long.
$P = 54$ in.

Given the perimeter of the regular polygon, find the length of each side. (Write all improper fraction answers as mixed numbers.)

31. An equilateral triangle; $P = 51$ in.
side $= 17$ in.

32. An equilateral triangle; $P = 17.25$ m
side $= 5.75$ m

33. A square; $P = 71.6$ cm
side $= 17.9$ cm

34. A square; $P = 55$ ft
side $= 13\frac{3}{4}$ ft

35. A regular hexagon; $P = 105$ yd
side $= 17\frac{1}{2}$ yd

36. A regular pentagon; $P = 16.7$ m
side $= 3.34$ m

37. A regular octagon; $P = 10\frac{2}{3}$ ft
side $= 1\frac{1}{3}$ ft

38. A regular octagon; $P = 25\frac{1}{3}$ in.
side $= 3\frac{1}{6}$ in.

For each circle with the given radius, approximate the circumference using $\pi \approx \frac{22}{7}$.

39. $r = 14$ in.
$C \approx 88$ in.

40. $r = 42$ in.
$C \approx 264$ in.

41. $r = 2.8$ cm
$C \approx 17.6$ cm

42. $r = 6.3$ cm
$C \approx 39.6$ cm

43. $r = \frac{7}{8}$ in.
$C \approx 5\frac{1}{2}$ in.

44. $r = 5\frac{1}{4}$ in.
$C \approx 33$ in.

45. $r = 3\frac{1}{2}$ ft
$C \approx 22$ ft

46. $r = 4\frac{2}{3}$ ft
$C \approx 29\frac{1}{3}$ ft

For each circle with the given radius, approximate the circumference using $\pi \approx 3.14$. Round the result to the tenths place.

47. $r = 5$ yd
$C \approx 31.4$ yd

48. $r = 6$ yd
$C \approx 37.7$ yd

49. $r = 4.5$ m
$C \approx 28.3$ m

50. $r = 3.2$ m
$C \approx 20.1$ m

For each circle with the given diameter, approximate the circumference using $\pi \approx 3.14$. Round the result to the tenths place.

51. $d = 8$ ft
$C \approx 25.1$ ft

52. $d = 15$ ft
$C \approx 47.1$ ft

53. $d = 6.5$ cm
$C \approx 20.4$ cm

54. $d = 10.5$ cm
$C \approx 33.0$ cm

Find the measures of the marked sides (with an X and a Y). Then find the perimeter of the polygon. Assume that all angles are right angles.

55.

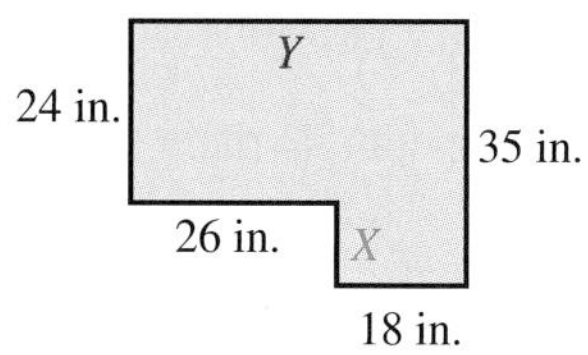

$X = 11$ in., $Y = 44$ in., $P = 158$ in.

56.

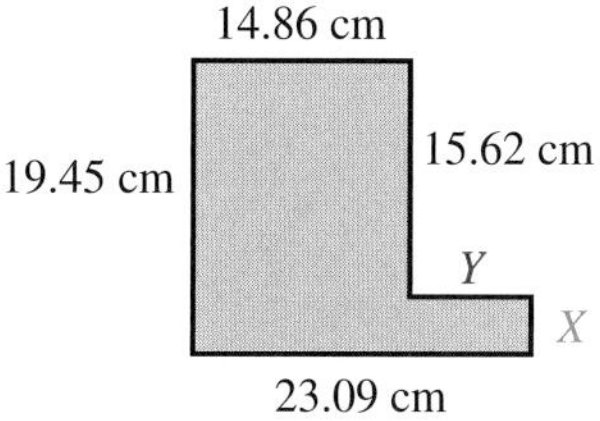

$X = 3.83$ cm, $Y = 8.23$ cm, $P = 85.08$ cm

57.

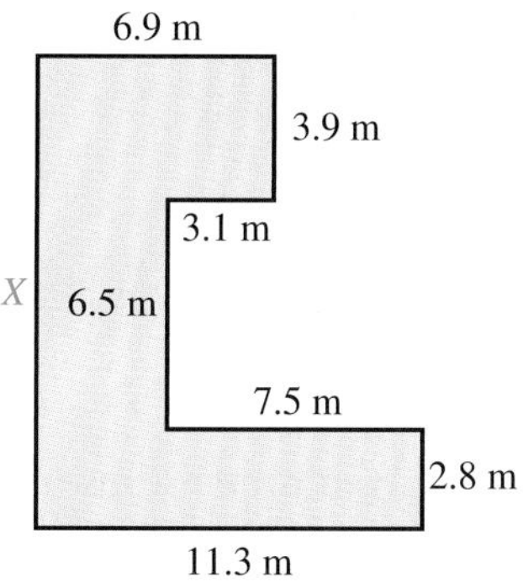

$X = 13.2$ m, $P = 55.2$ m

58.

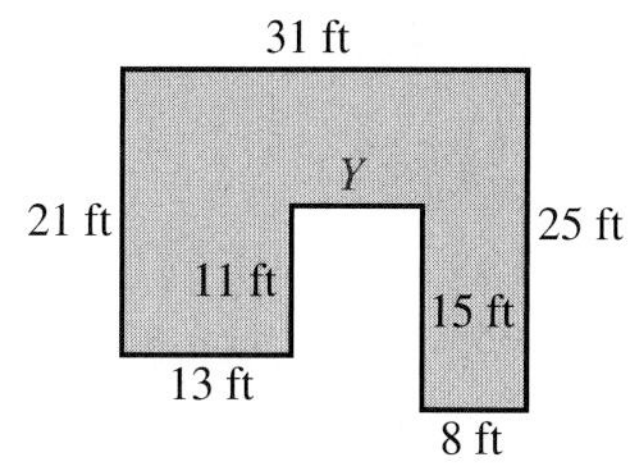

$Y = 10$ ft, $P = 134$ ft

59.

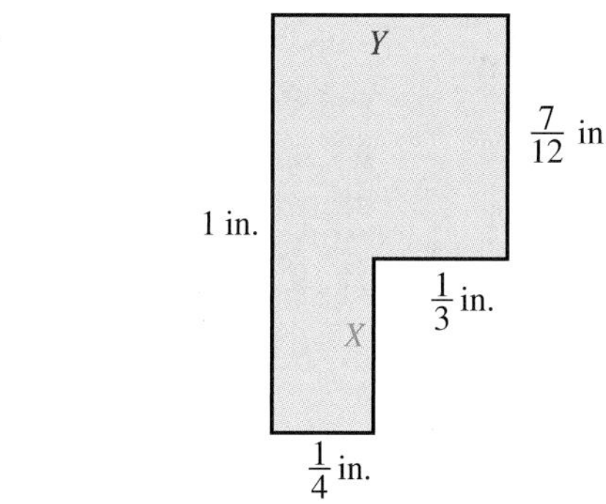

$X = \frac{5}{12}$ in., $Y = \frac{7}{12}$ in., $P = 3\frac{1}{6}$ in.

60.

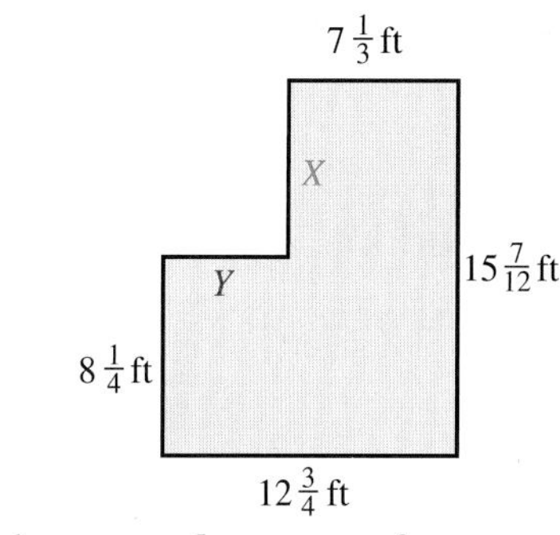

$X = 7\frac{1}{3}$ ft, $Y = 5\frac{5}{12}$ ft, $P = 56\frac{2}{3}$ ft

Think Outside the Box

A semicircle is half a circle. Use that fact to respond to the following applications. Write each answer as a sentence.

61. Marta's garden is in the shape of a semicircle with a diameter of 20 feet. Approximate the perimeter of Marta's garden.

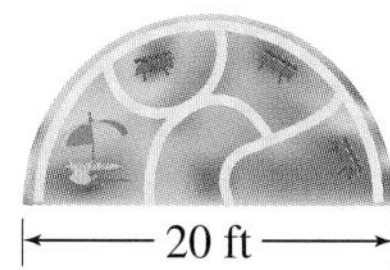

The perimeter of Marta's garden is approximately 51.4 feet.

62. An ice-skating rink has the shape of two semicircles attached to a square. Each side of the square measures 30 yards. Approximate the perimeter of the skating rink.

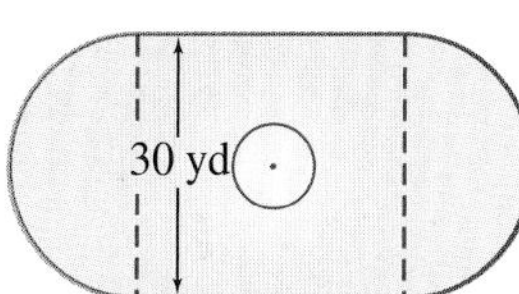

The perimeter of the skating rink is approximately 154.2 yards.

SECTION 8.7 Area

OBJECTIVES

In this section, you will learn to
- Find the area of a rectangle and a square.
- Find the area of a parallelogram.
- Find the area of a triangle.
- Find the area of a trapezoid.
- Find the area of a circle.
- Find the area of unusual shapes.

You Need to Know

To successfully complete this section, you need to understand
- ☐ Multiplying fractions (4.4)
- ☐ Multiplying mixed numbers (4.4)
- ☐ Multiplying decimals (5.4)
- ☐ Perpendicular lines (8.4)
- ☐ Geometric shapes (8.5)

Introduction

Remember: A plane is a flat surface that extends indefinitely in all directions.

The **area** of a geometic figure is the amount of a region in a plane that the figure covers, measured in square units.

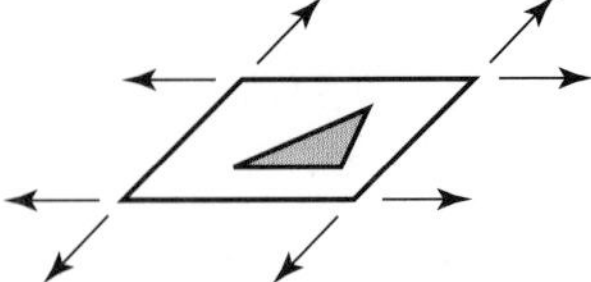

Area of a Rectangle

Recall from Section 1.3 that the area of a rectangle is the product of its length and width: $A = L \cdot W$.

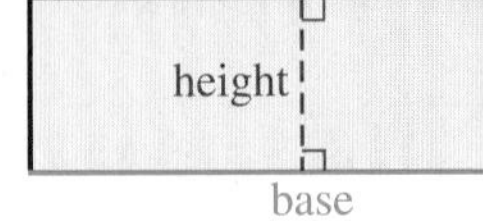

Sometimes we choose to call the dimensions of a rectangle the *base* and *height*, and **Area = base × height.**

The height can be measured from anywhere within the rectangle. The height is always *perpendicular* (forms a right angle) to the base, the bottom side.

Example 1

Draw the rectangle with the given dimensions and find its area.

a) base = 12 in., height = 5 in.

b) base = 6.5 m, height = 4 m

Procedure Multiply base times height. Recall that area is always expressed in square units.

Answer

a)

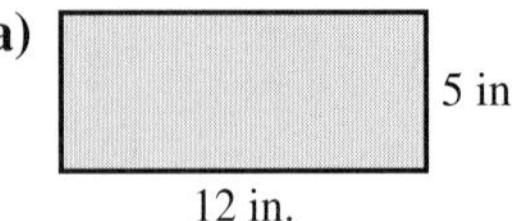

Area = 12 in. · 5 in.
= 60 in.^{2}

b)

4 m
6.5 m

Area = 6.5 m · 4 m
= 26 m^2

Multiply: $\begin{array}{r} 6.5 \\ \times\ 4 \\ \hline 26.0 \end{array}$

You Try It 1

Sketch a rough drawing of a rectangle with the given dimensions and find its area. Use Example 1 as a guide.

a) base = 15 yd
height = 9 yd

b) base = 11.25 cm
height = 9.2 cm

Area of a Square

A square is a special type of rectangle in which all four sides have the same measure. The area formula for a square is the same as for a rectangle, but here the formula can be abbreviated as $A = s \cdot s = s^2$, where s = the length of each side.

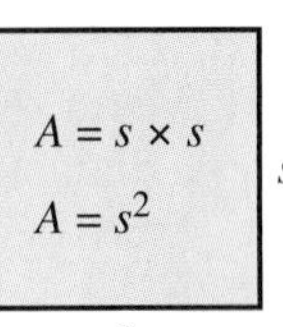

You Try It 2

Find the area of a square with the given side measure.

a) side = 7 ft

b) side = $\frac{5}{8}$ in.

Area of a Parallelogram

The formula for the area of a parallelogram is, just like that for a rectangle, the product of the base and height. After all, a rectangle is just a special type of parallelogram.

Area of a parallelogram: **Area = Base × Height**

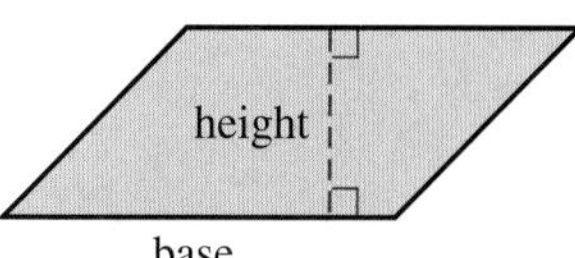

However, the height of a parallelogram is perpendicular to both the top side and the bottom side (the base).

Caution The slanted side measure of a parallelogram is not used in calculating the area.

To demonstrate why a parallelogram has the same area formula as that of a rectangle, consider this:

We can cut off this triangle and put it on the other side, forming a rectangle.

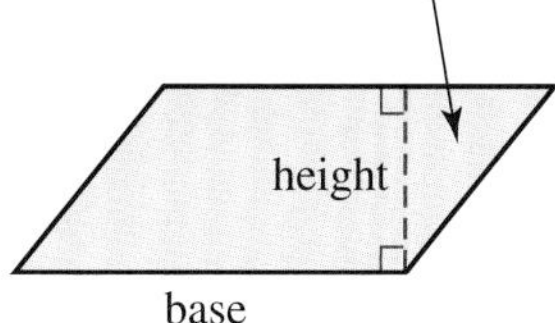

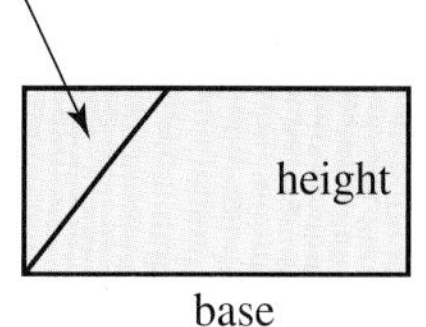

The slanted sides match up perfectly because they were originally parallel to each other and of the same length. In the rectangle that is formed, the base and height do not change, so

Area of parallelogram = Base × Height

With some parallelograms, the height needs to be shown outside the parallelogram. When that happens, imagine extending the base until the height can be drawn perpendicular to it.

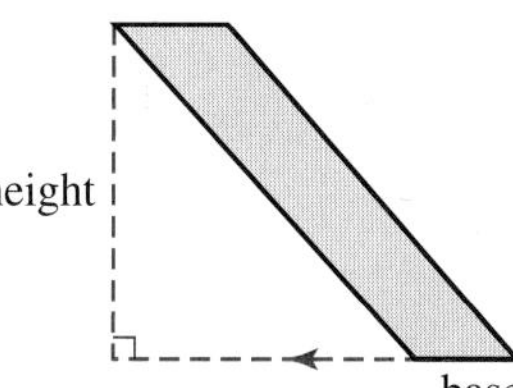

Example 2

Find the area of each parallelogram.

a)

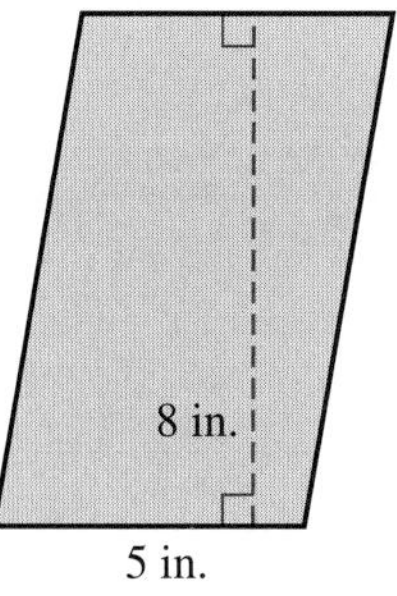

b)

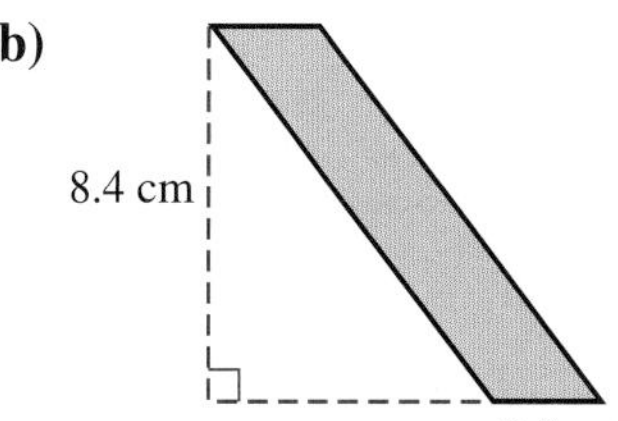

Answer

a) Area = 5 in. · 8 in.
= 40 in.^{2}

b) Area = 2.5 cm · 8.4 cm
= 21 cm^2

Multiply:

```
   2.5
 × 8.4
   100
+ 2000
 21.00
```

You Try It 3 **Find the area of each parallelogram. Use Example 2 as a guide.**

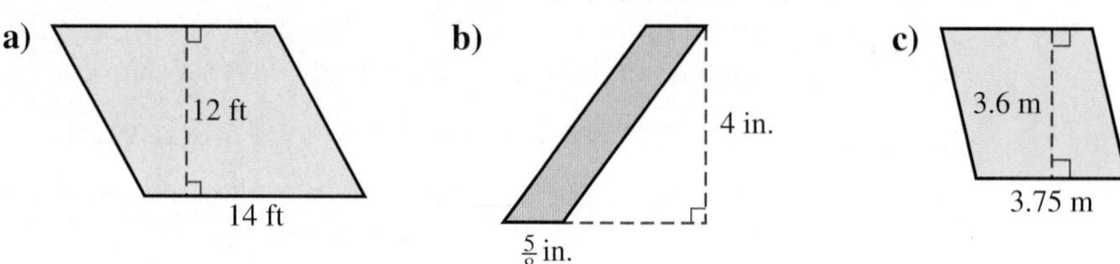

Area of a Triangle

To find the formula for the area of a triangle, two copies of the same triangle can be put together to form a parallelogram.

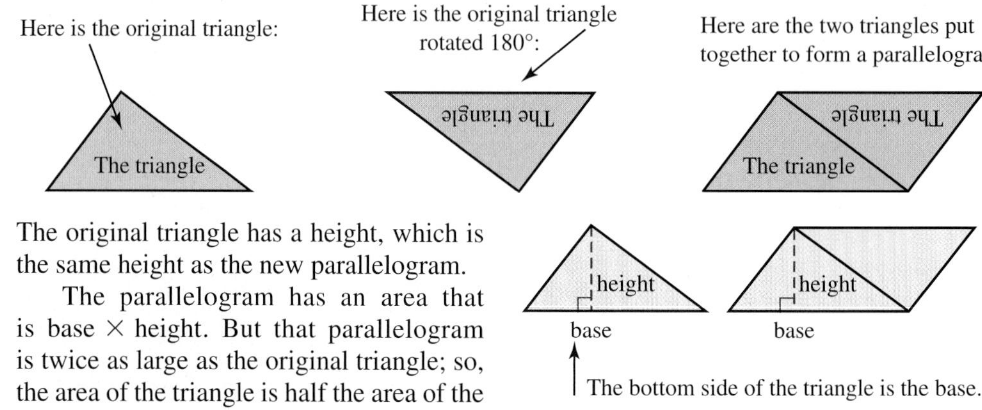

The original triangle has a height, which is the same height as the new parallelogram.

The parallelogram has an area that is base × height. But that parallelogram is twice as large as the original triangle; so, the area of the triangle is half the area of the parallelogram.

Here are two different forms of the formula for the area of a triangle, where $\boldsymbol{b}$ = base and $\boldsymbol{h}$ = height:

1. $A = \frac{1}{2} \cdot b \cdot h$ 2. $A = \frac{b \cdot h}{2}$

Generally, we choose a form that seems easy to work with based on the numbers given for the base and height.

Example 3

Find the area of each triangle.

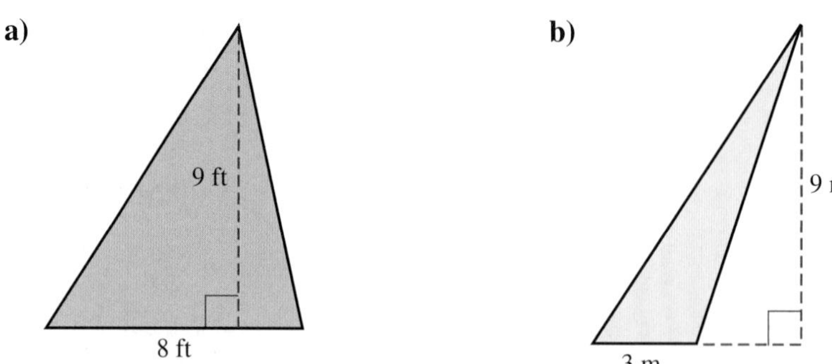

Procedure For each triangle, we need to choose one of the forms of the formula for the area of a triangle.

Answer **a)** Let's use form 1: Area $= \frac{1}{2} \cdot 8 \cdot 9$ First, multiply $\frac{1}{2} \cdot 8 = 4$.

This is a good form to use because the base is an even number.

$A = 4 \cdot 9$ Now complete the multiplication.

$A = 36 \text{ ft}^2$ The area is stated in square feet.

b) Let's use form 2: $A = \frac{3 \cdot 9}{2}$ First, multiply the numerator.

This is a good form to use because neither number is even.

$A = \frac{27}{2}$ Simplify this fraction to a mixed number.

$A = 13\frac{1}{2} \text{ m}^2$ or 13.5 m^2 The area is stated in square meters.

You Try It 4 **Find the area of each triangle. Use Example 3 as a guide.**

a)

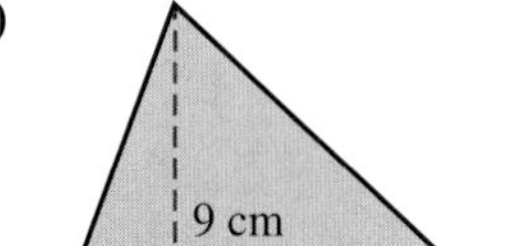

b)

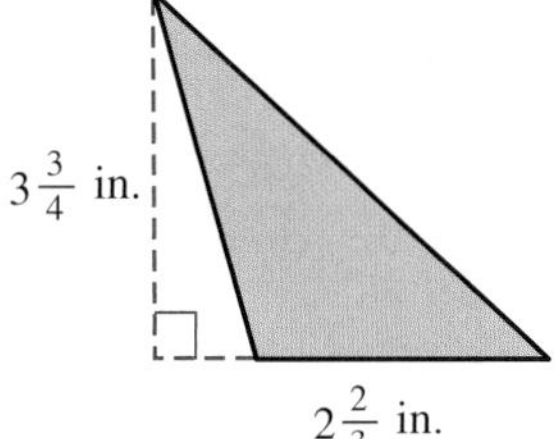

c)

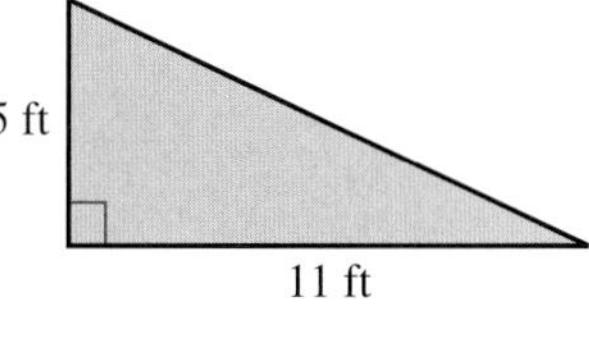

Area of a Right Triangle

The height of a geometric shape is always perpendicular to the base.

In a right triangle, the two sides that form the right angle are perpendicular to each other. This means that one of the sides can be thought of as the base and that the other side can be thought of as the height.

Here is the same triangle rotated in different directions. Notice that we can label either of the perpendicular sides as the base (sometimes it's the 4-inch side; sometimes it's the 6-inch side) and the other side as the height.

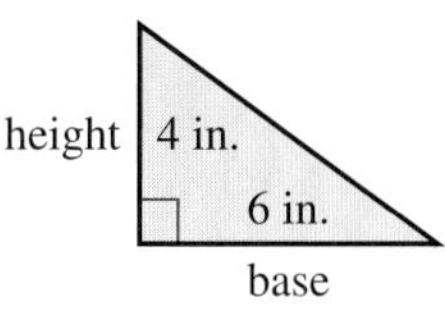

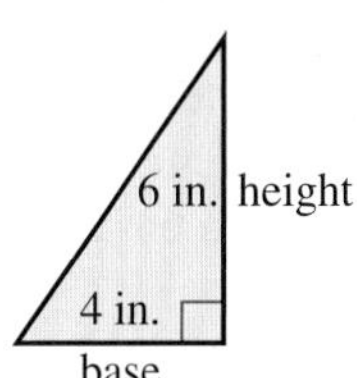

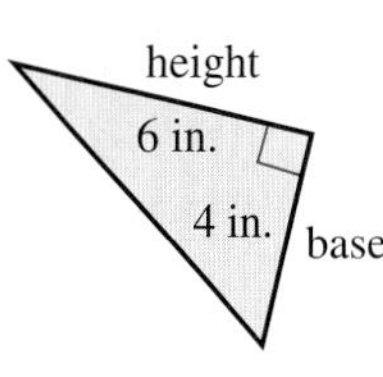

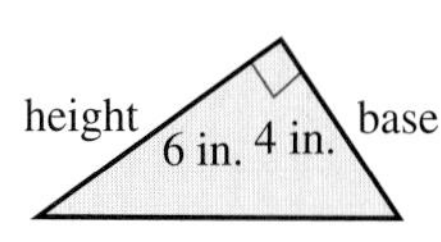

Because they are the same triangle, they have the same area:

$$A = \frac{1}{2} \cdot \text{base} \cdot \text{height}$$

$$A = \frac{1}{2} \cdot 6 \text{ in.} \cdot 4 \text{ in.}$$

$$A = 3 \cdot 4 \text{ in.}^2$$

$$A = 12 \text{ in.}^2$$

You Try It 5 **Find the area of each right triangle.**

a)

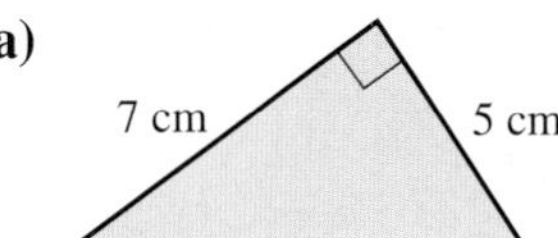

b)

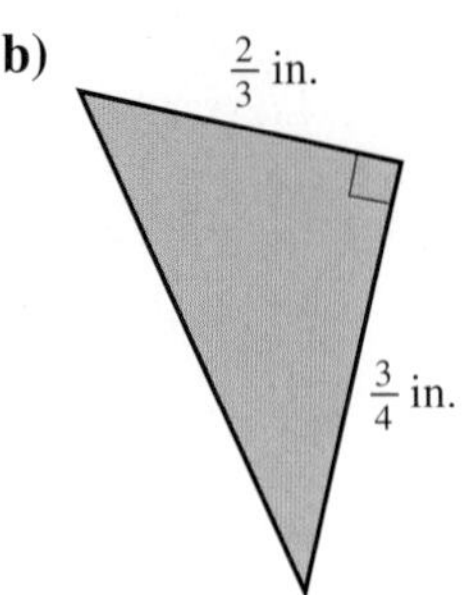

Area of a Trapezoid

Recall that a trapezoid has two bases—a top base, labeled a, and a bottom base, labeled b. A trapezoid also has a height, labeled h.

To find the formula for the area of a trapezoid, we see that two copies of the same trapezoid can be put together to form a parallelogram.

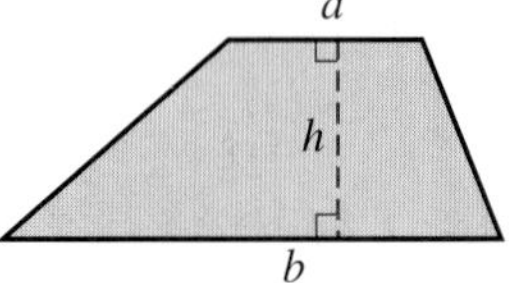

Here is the original trapezoid:

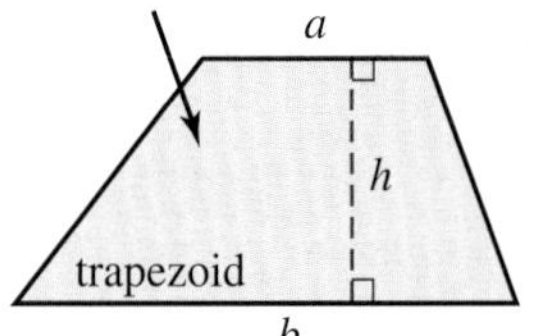

Here is the original trapezoid rotated 180°:

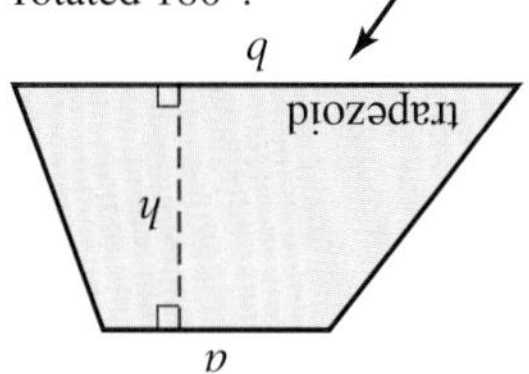

Here are the two trapezoids put together to form a parallelogram:

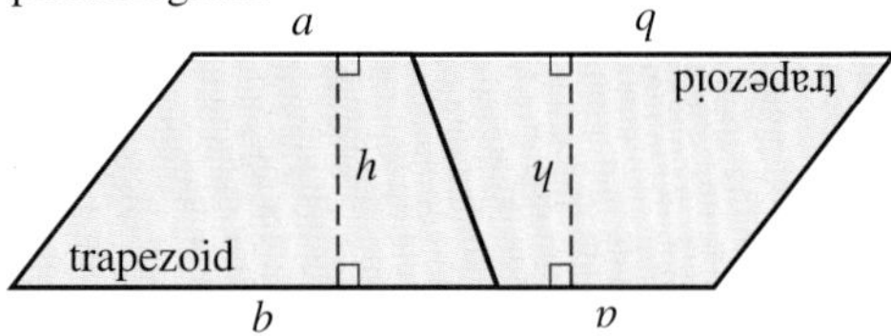

The height of the original trapezoid is the same height as the new parallelogram.

Also, the base of the new parallelogram measures $a + b$.

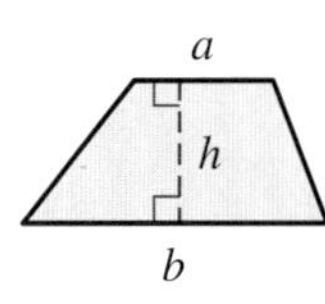

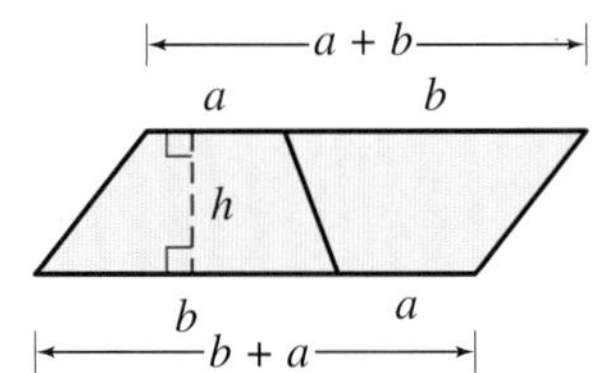

So, the area of this new parallelogram is base · height: $A = (a + b) \cdot h$ or $A = h \cdot (a + b)$.

However, because the parallelogram is twice as large as the original trapezoid, the area of the trapezoid is half the area of the new parallelogram.

Here are two different forms of the same formula, area of a trapezoid, where $\boldsymbol{a}$ = the top base, $\boldsymbol{b}$ = the bottom base, and $\boldsymbol{h}$ = the height.

1. $A = \frac{1}{2} \cdot h \cdot (a + b)$ 2. $A = \frac{a + b}{2} \cdot h$

Example 4

Find the area of the given trapezoid.

a)

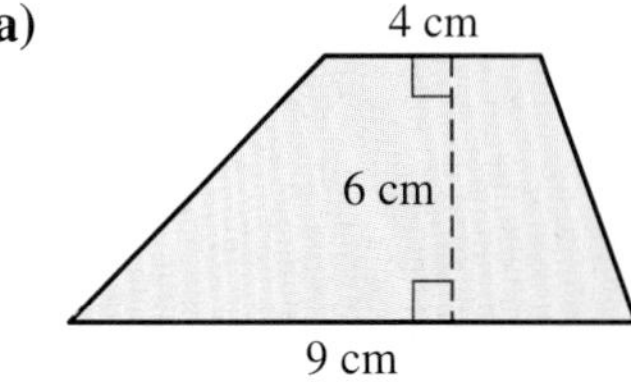

b)

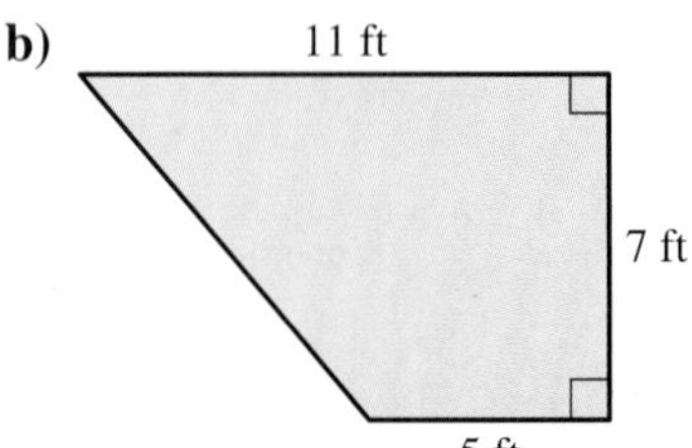

Procedure For each trapezoid, we need to choose one of the forms of the formula for the area of a trapezoid.

In the trapezoid in part b, the right side is perpendicular to both bases (the top *and* the bottom); so, that side (which measures 7 ft) is the height.

Answer **a)** Let's use form 1: Area $= \frac{1}{2} \cdot 6 \cdot (4 + 9)$ — First, apply the order of operations and add within the parentheses.

This is a good form to use because h is an even number.

$A = \frac{1}{2} \cdot 6 \cdot 13$ — Next, multiply $\frac{1}{2} \cdot 6 = 3$.

$A = 3 \cdot 13$

$A = 39\ \text{cm}^2$ — The area is stated in square centimeters.

b) Let's use form 2: $A = \frac{11 + 5}{2} \cdot 7$ — First, add the numerator.

This is a good form to use because h is not an even number.

$A = \frac{16}{2} \cdot 7$ — Simplify the fraction.

$A = 8 \cdot 7$

$A = 56\ \text{ft}^2$ — The area is stated in square feet.

You Try It 6 **Find the area of each trapezoid. Use Example 4 as a guide.**

a)

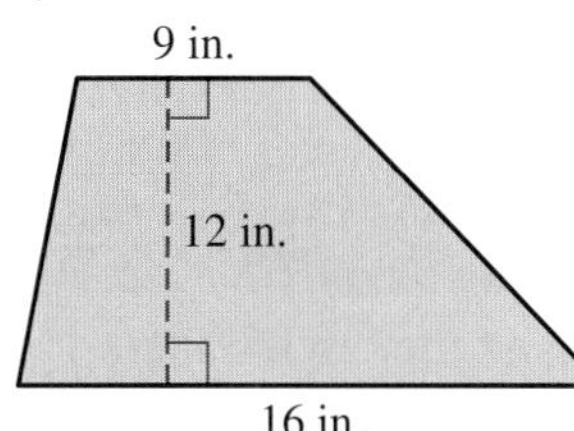

b)

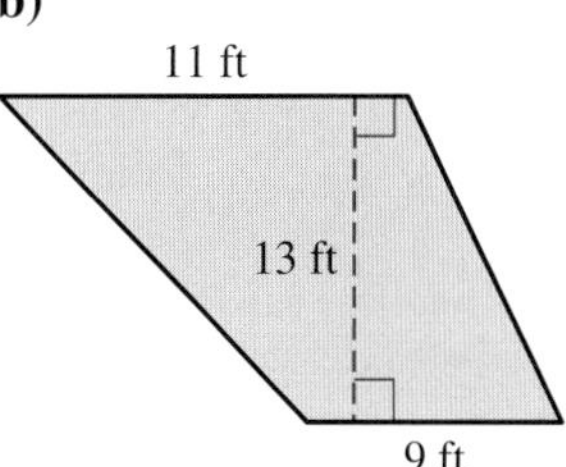

c)

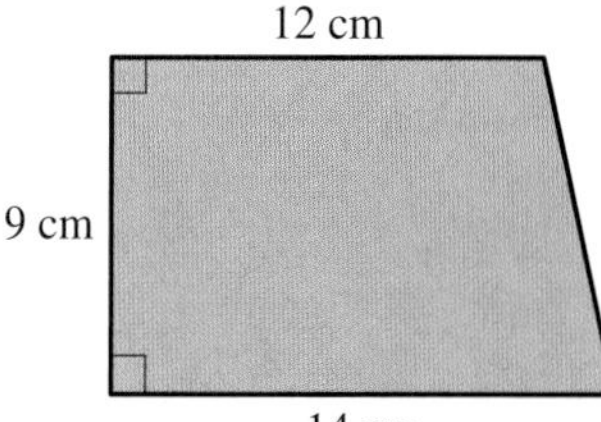

Area of a Circle

It can be shown, using higher-level mathematics, that the area of a circle is $A = \pi \cdot r \cdot r$, or $A = \pi \cdot r^2$. Notice that the formula for the area requires the value of the radius, not the diameter. So, if you're given the diameter, you must find the radius before using the area formula.

Also, as with rectangles and triangles, the area of a circle is stated in square units because when we square the radius, we also are squaring the units of the radius.

For example, if the radius is 2 meters, then

$$A = \pi \cdot (2\ \text{m})^2 = \pi \cdot 2\ \text{m} \cdot 2\ \text{m} = \pi \cdot 4\ \text{m}^2$$

Example 5

For each circle, approximate the area. Use $\pi \approx 3.14$ and round the result to the nearest tenth.

a)

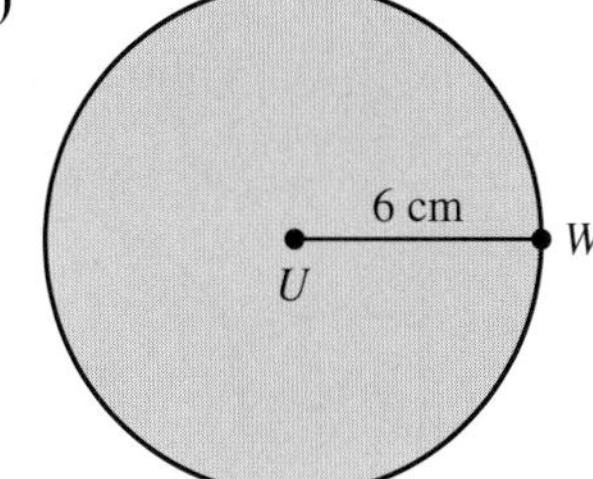

b)

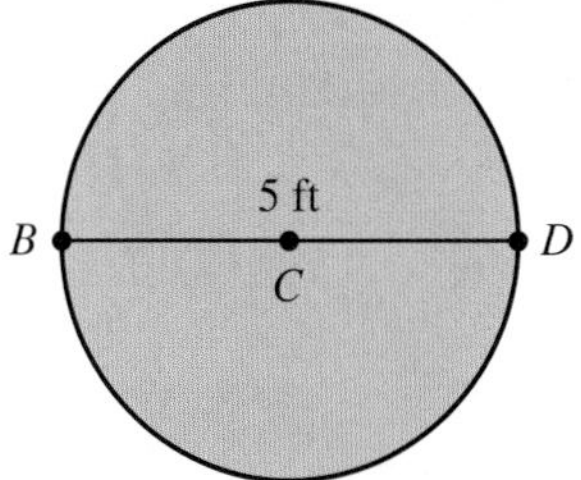

Procedure For part b, we'll need to find the value of the radius.

Answer

a) $\overline{UW}$ is a radius.

$A = \pi \cdot 6^2\ \text{cm}^2$

$A \approx 3.14 \cdot 36\ \text{cm}^2$

$A \approx 113.04\ \text{cm}^2$

Rounded: $A \approx 113.0\ \text{cm}^2$

b) $\overline{BD}$ is a *diameter*. radius $= \frac{5}{2}$ ft $= 2.5$ ft

$A = \pi \cdot (2.5)^2\ \text{ft}^2$

$A \approx 3.14 \cdot 6.25\ \text{ft}^2$

$A \approx 19.625\ \text{ft}^2$

$A \approx 19.6\ \text{ft}^2$

Multiply: 2.5 × 2.5 = 125 + 500 = 6.25

You Try It 7 **For each circle, approximate the area using $\pi \approx 3.14$. Round the result to the tenths place. Use Example 5 as a guide.**

a)

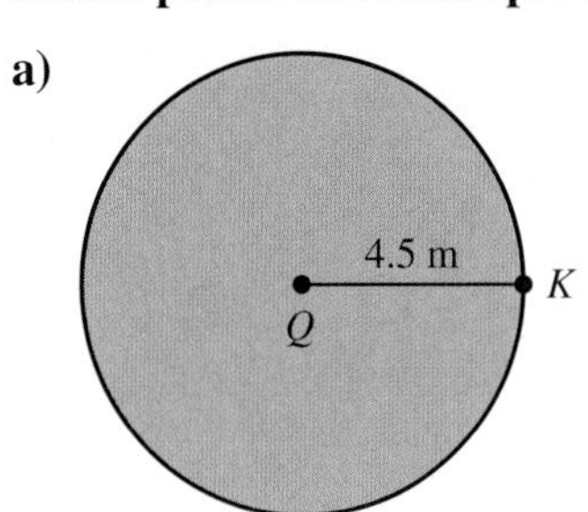

b)

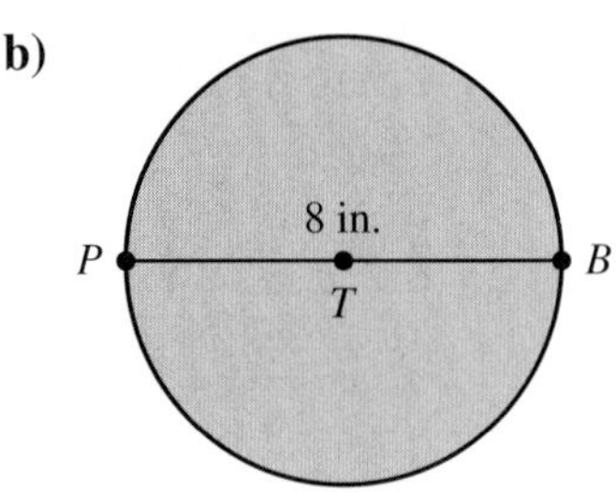

Subtracting Areas

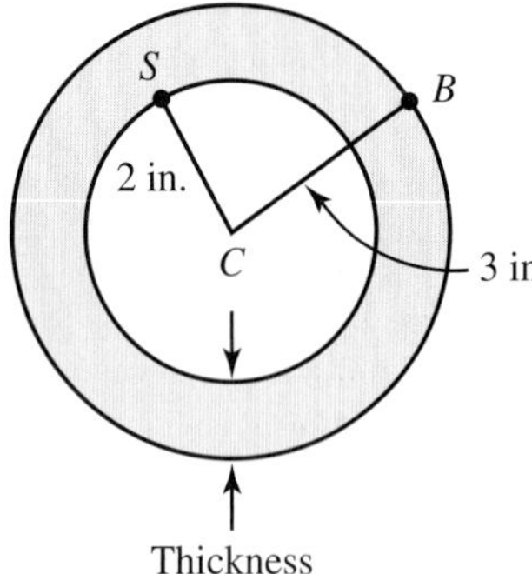

Sometimes the area of interest isn't from a single geometric figure. For example, consider a ring. A ring is, typically, the region between two circles with the same center but different radii.

In the diagram at the right, the radius $\overline{CS}$ (2 in.) is shorter than the radius $\overline{CB}$ (3 in.) and the circles share the same center.

The region between the two circles, the thickness of the rings, is shaded. Its area (A_{ring}) can be found by subtracting the inner circle's area (A_{inner}) from the outer circle's area (A_{outer}).

Area of shaded ring = Area of outer circle − Area of inner circle

$$A_{\text{ring}} = A_{\text{outer}} - A_{\text{inner}}$$

$$A_{\text{ring}} \approx \pi \cdot (3)^2 - \pi \cdot (2)^2$$

$$A_{\text{ring}} \approx 28.26 - 12.56$$

$$A_{\text{ring}} \approx 15.7 \text{ square inches, or } 15.7\ \text{in.}^2$$

Outer area: $\pi \cdot (3)^2 \approx 3.14 \cdot 9 = 28.26$

Inner area: $\pi \cdot (2)^2 \approx 3.14 \cdot 4 = 12.56$

Subtract: 28.26 − 12.56 = 15.70

Example 6

Find the area of the shaded region formed by a circle inside a square. Use $\pi \approx 3.14$ and round the result to the nearest tenth.

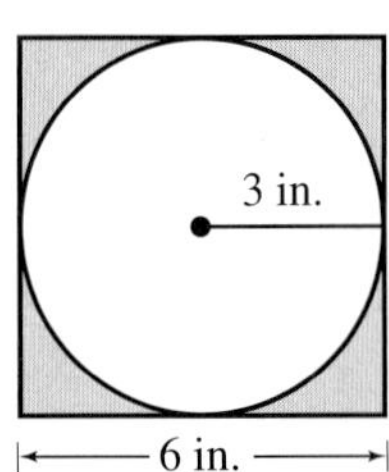

Procedure Find the area of the circle and the square and subtract the inner area (circle) from the outer area (square).

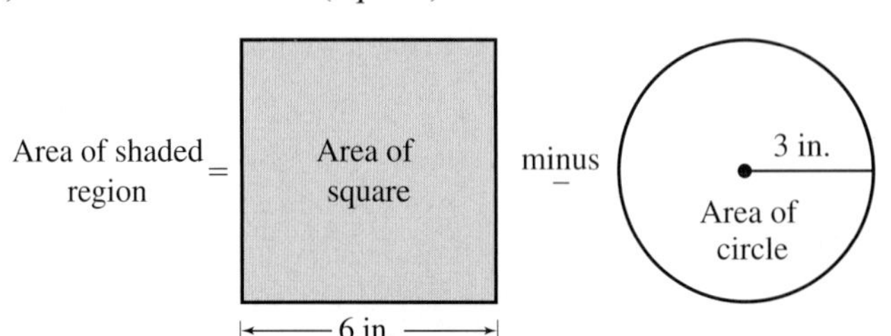

Answer Area of square $= 6^2 = 36$ Area of circle $= \pi \cdot (3)^2 \approx 3.14 \cdot 9 = 28.26$
Subtract: Area $\approx 36 - 28.26 = 7.74$; rounded, Area ≈ 7.7 in.^{2}

Caution Do not round until the last step of the process.

You Try It 8 **For each figure, approximate the area of the shaded region. Use $\pi \approx 3.14$ and round the result to the tenths place. Use Example 6 as a guide.**

a)

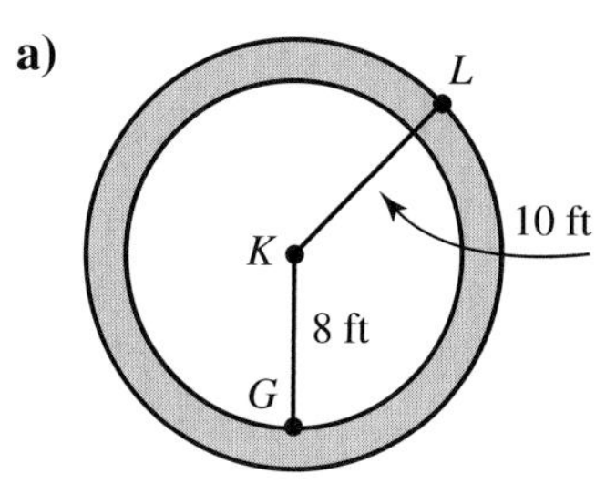

b)

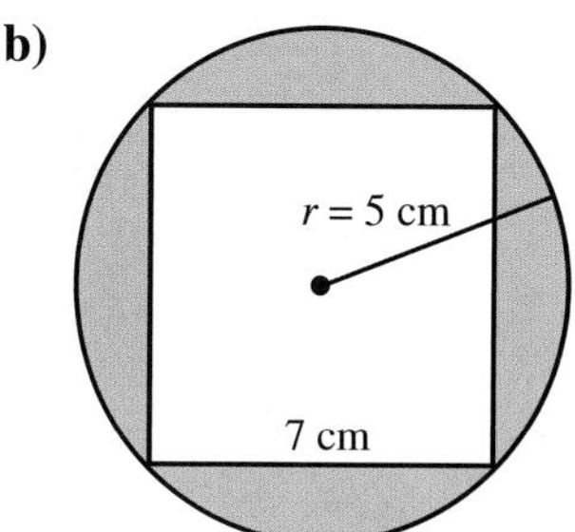

Areas of Unusual Shapes

Consider this situation: Arturo and Terry are laying wood flooring in Janet's living room. They must measure the room (in square feet) to know how much flooring they'll need. According to the diagram and dimensions shown, how many square feet of flooring is needed to cover the whole room?

Here is the original living room floor plan:

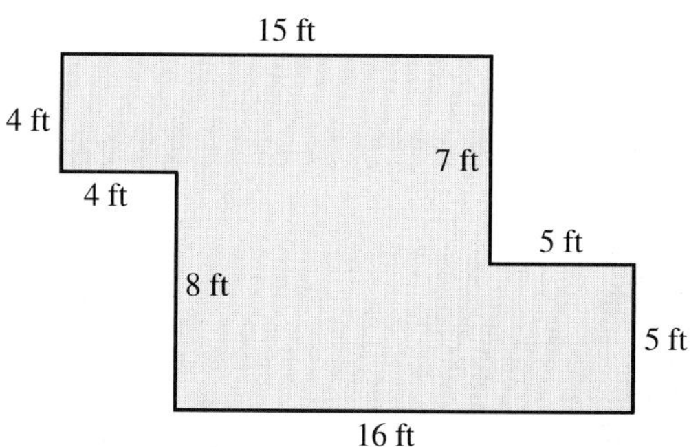

We can break up the room into three rectangles, each having its own dimensions and area.

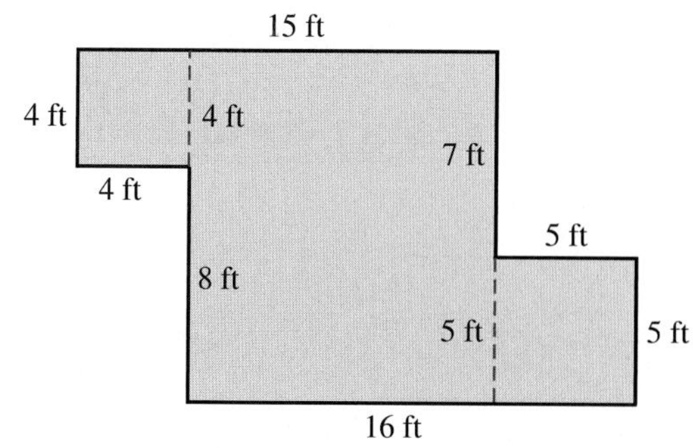

Caution We need to be careful when dividing the room into rectangles. New dimensions will appear that weren't originally shown, such as the 11 ft across the top of the large rectangle and the 12 ft height of the large rectangle.

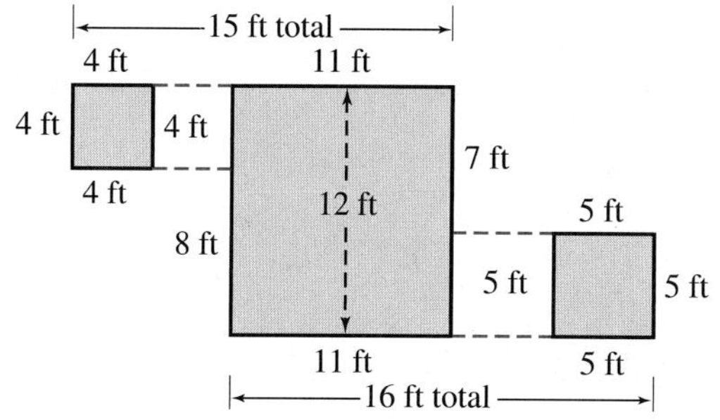

	Areas of the rectangles	Total Area:
4 ft by 4 ft rectangle:	$A = 4 \times 4 = 16\text{ ft}^2$	16
12 ft by 11 ft rectangle:	$A = 12 \times 11 = 132\text{ ft}^2$	132
5 ft by 5 ft rectangle:	$A = 5 \times 5 = 25\text{ ft}^2$	+ 25
		173

Arturo and Terry will need 173 square feet of flooring to cover the whole room.

Example 7

Find the area of the polygon with the given dimensions

20 in. 15 in. 12 in.

Procedure Break up the polygon into a rectangle with a triangle on top.

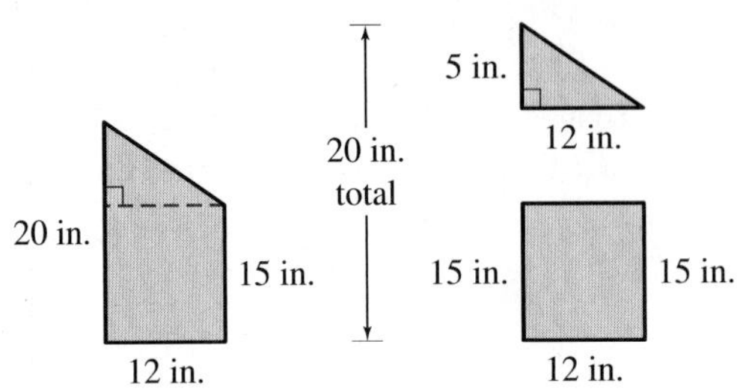

Area of triangle:	Area of rectangle:	Total Area
$A = \frac{1}{2} \times 12 \times 5 = 30 \text{ in.}^2$	$A = 12 \times 15 = 180 \text{ in.}^2$	30 + 180 210

Answer The area of the polygon is 210 square inches.

You Try It 9 **Find the area of each polygon. Use Example 7 as a guide.**

a)

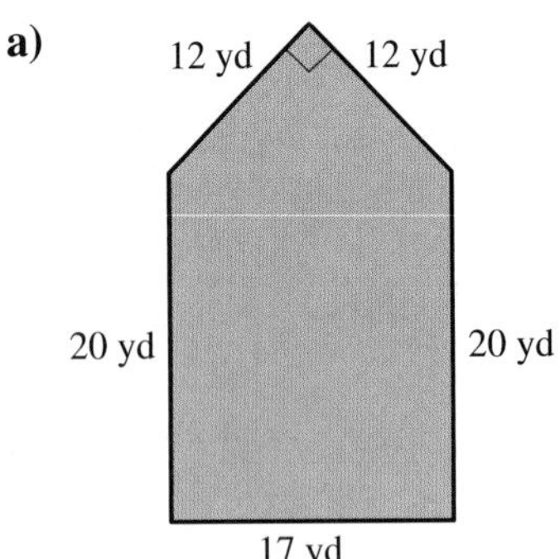

b)

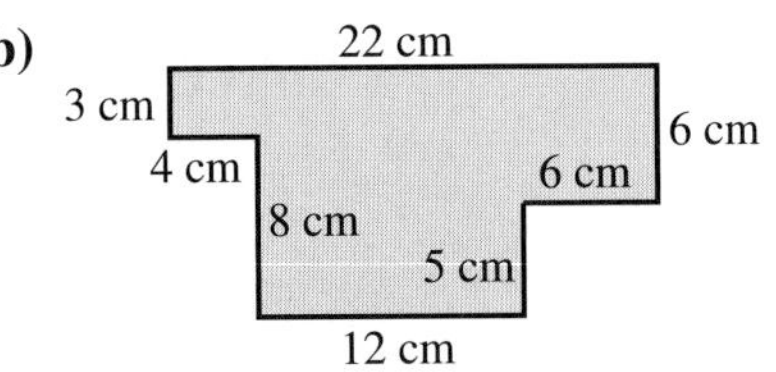

Answers: You Try It and Think About It

You Try It:

1. a) 9 yd, 15 yd, $A = 135 \text{ yd}^2$

b)

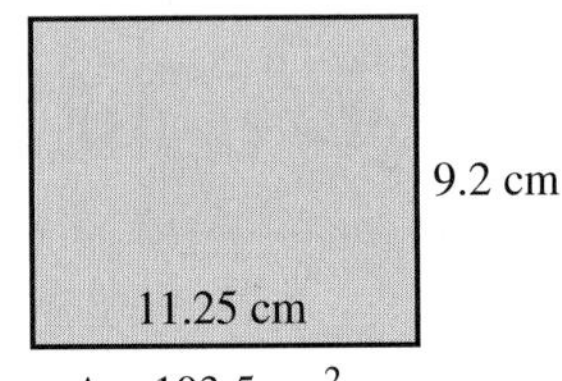

$A = 103.5 \text{ cm}^2$

2. a) $A = 49 \text{ ft}^2$ **b)** $A = \frac{25}{64} \text{ in.}^2$ **3. a)** $A = 168 \text{ ft}^2$ **b)** $A = 2\frac{1}{2} \text{ in.}^2$ **c)** $A = 13.5 \text{ m}^2$ **4. a)** $A = 54 \text{ cm}^2$ **b)** $A = 5 \text{ in.}^2$ **c)** $A = 27.5 \text{ ft}^2$ **5. a)** $A = 17.5 \text{ cm}^2$ **b)** $A = \frac{1}{4} \text{ in.}^2$ **6. a)** $A = 150 \text{ in.}^2$ **b)** $A = 130 \text{ ft}^2$ **c)** $A = 117 \text{ cm}^2$ **7. a)** $A \approx 63.6 \text{ m}^2$ **b)** $A \approx 50.2 \text{ in.}^2$ **8. a)** $A \approx 113.0 \text{ ft}^2$ **b)** $A \approx 29.5 \text{ cm}^2$ **9. a)** $A = 412 \text{ yd}^2$ **b)** $A = 180 \text{ cm}^2$

Think About It: No Think About Its in this section.

FOR EXTRA HELP
Student Resources on DVD-ROM
Includes
➤ Student's Solutions Manual
➤ Video Lectures
➤ Chapter Test Prep Video
MyMathLab
MathXL

Section 8.7 Exercises

Think About It

1. The area formula for a parallelogram, $A = b \cdot h$, is the same formula as the area of a rectangle. Why?
Answers will vary.

2. In this section, we were given two forms of the area formula for a trapezoid:

$$A = \frac{1}{2} \cdot h \cdot (a + b) \qquad A = \frac{a + b}{2} \cdot h$$

What steps would you take to show that the first formula can be written as the second formula?
Answers will vary.

Focus Exercises

Find the area of each rectangle with the given dimensions.

3. $b = 12$ yd
$h = 8$ yd
$A = 96\text{ yd}^2$

4. $h = 9$ yd
$b = 23$ yd
$A = 207\text{ yd}^2$

5. $b = \frac{7}{8}$ in.
$h = \frac{1}{4}$ in.
$A = \frac{7}{32}\text{ in.}^2$

6. $b = \frac{3}{4}$ in.
$h = \frac{2}{3}$ in.
$A = \frac{1}{2}\text{ in.}^2$

7. $h = 5.2$ cm
$b = 8.5$ cm
$A = 44.2\text{ cm}^2$

8. $b = 10.25$ m
$h = 5.2$ m
$A = 53.3\text{ m}^2$

9. $h = 1\frac{1}{3}$ ft
$b = 3\frac{3}{8}$ ft
$A = 4\frac{1}{2}\text{ ft}^2$

10. $b = 5\frac{1}{4}$ ft
$h = 2\frac{2}{3}$ ft
$A = 14\text{ ft}^2$

Find the area of each square with the given side measure.

11. side $= 13$ ft
$A = 169\text{ ft}^2$

12. side $= 4.5$ cm
$A = 20.25\text{ cm}^2$

13. side $= \frac{7}{12}$ in.
$A = \frac{49}{144}\text{ in.}^2$

14. side $= 2\frac{1}{4}$ ft
$A = 5\frac{1}{16}\text{ ft}^2$

Find the area of each parallelogram.

15.
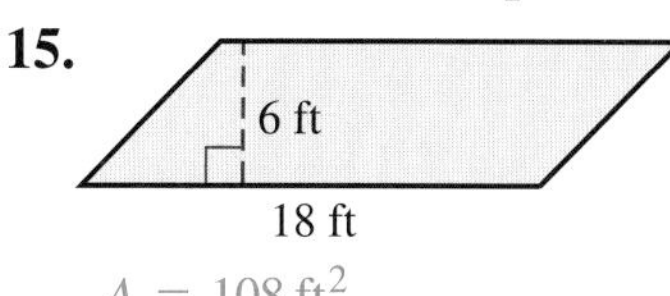

$A = 108\text{ ft}^2$

16.
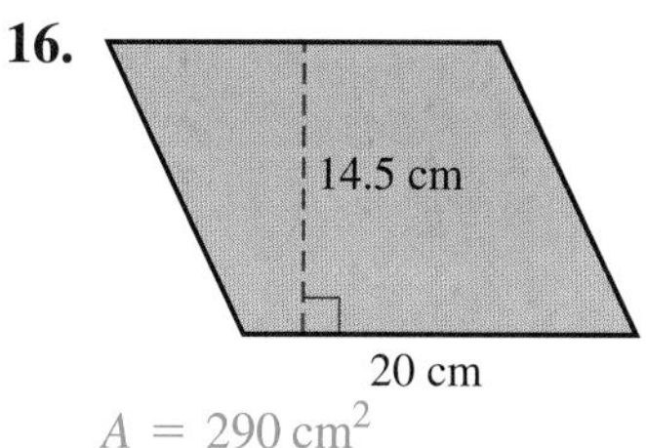

$A = 290\text{ cm}^2$

17.
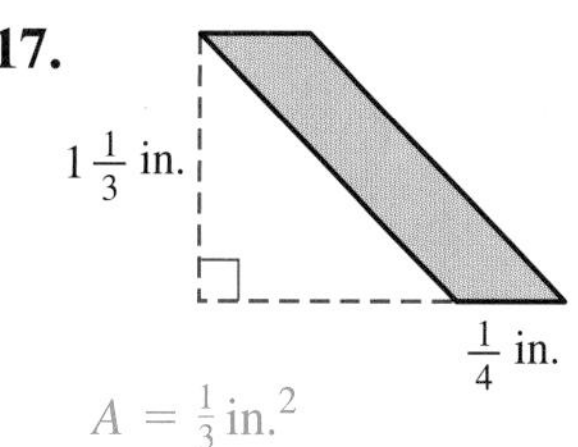

$A = \frac{1}{3}\text{ in.}^2$

18.
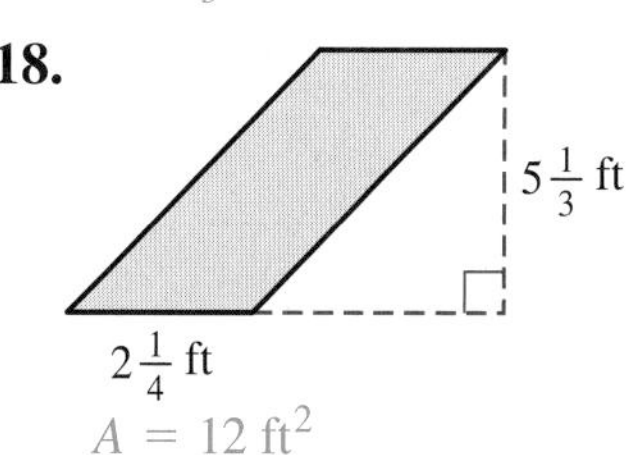

$A = 12\text{ ft}^2$

Find the area of each triangle.

19.
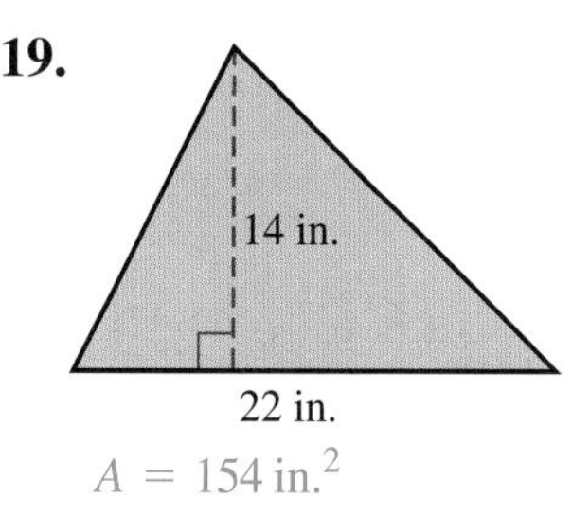

$A = 154\text{ in.}^2$

20.
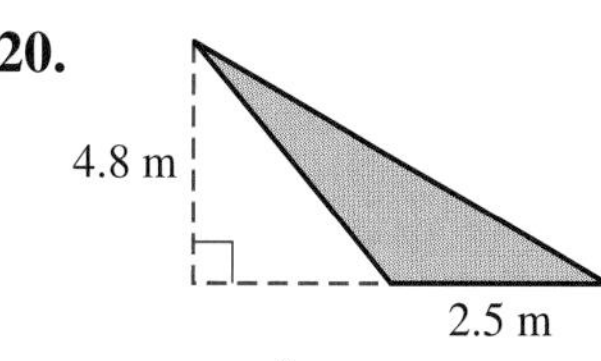

$A = 6\text{ m}^2$

21.
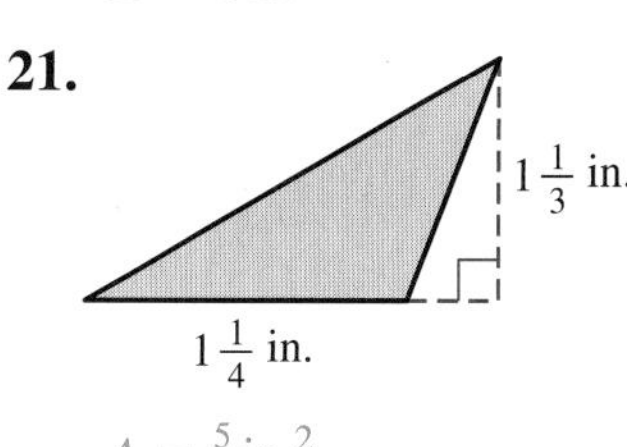

$A = \frac{5}{6}\text{ in.}^2$

22.

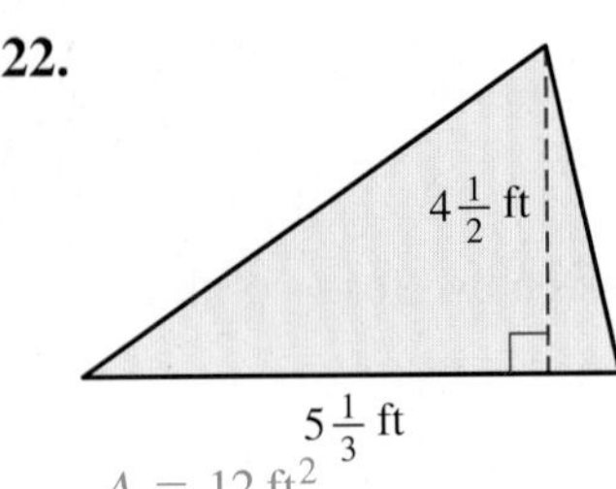

$A = 12\text{ ft}^2$

Find the area of each right triangle.

23.

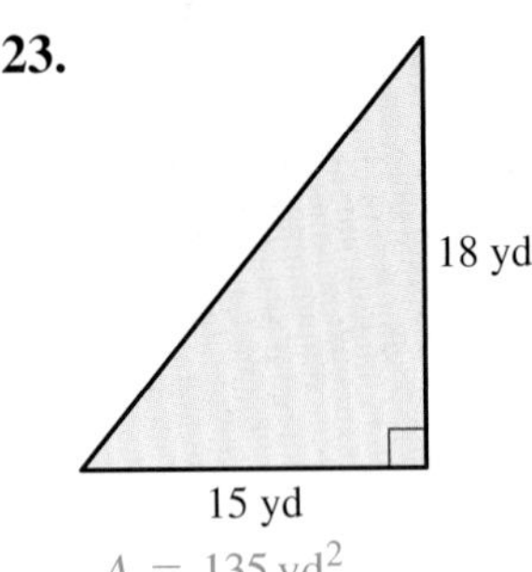

$A = 135\text{ yd}^2$

24.

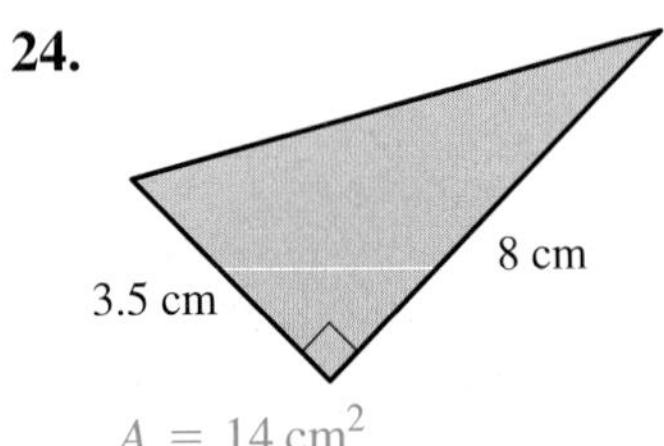

$A = 14\text{ cm}^2$

25.

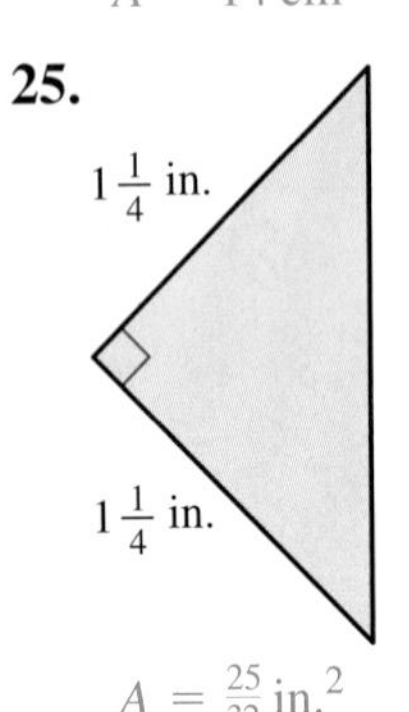

$A = \frac{25}{32}\text{ in.}^2$

26.

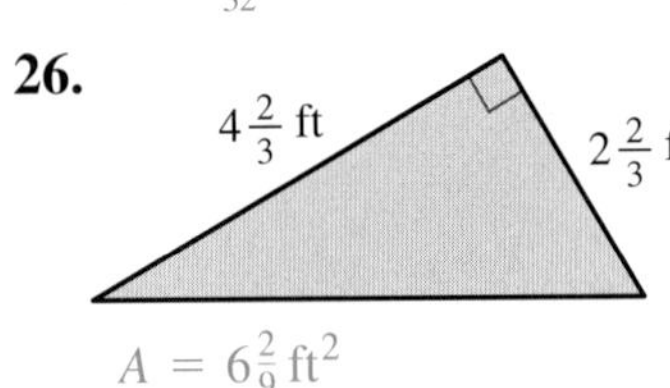

$A = 6\frac{2}{9}\text{ ft}^2$

Find the area of each trapezoid.

27.

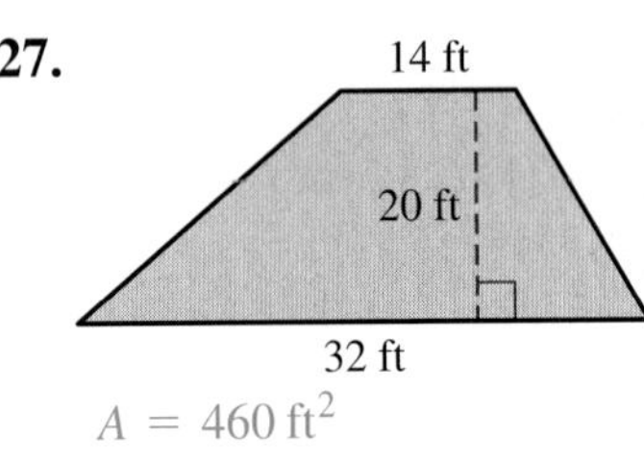

$A = 460\text{ ft}^2$

28.

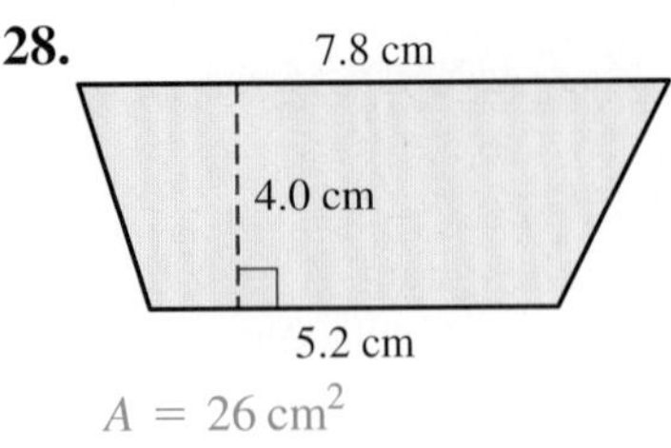

$A = 26\text{ cm}^2$

29.

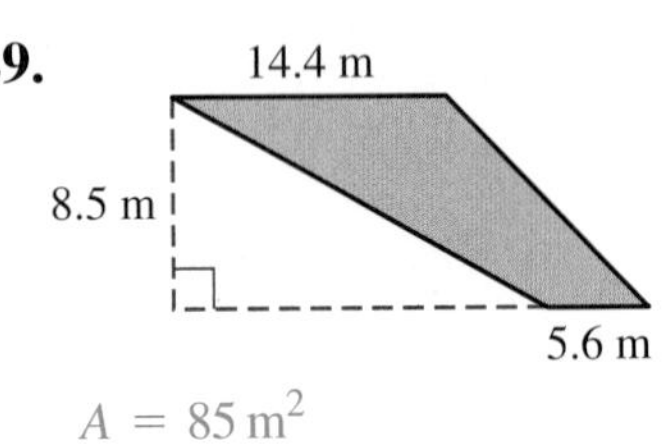

$A = 85\text{ m}^2$

30.

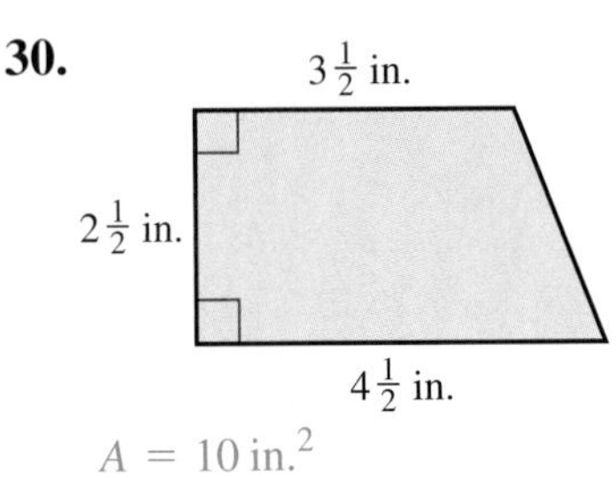

$A = 10\text{ in.}^2$

Approximate the area of each circle with the given radius. Use $\pi \approx 3.14$ and round the result to the tenths place.

31. $r = 3$ in.
$A \approx 28.3\text{ in.}^2$

32. $r = 5$ in.
$A \approx 78.5\text{ in.}^2$

33. $r = 1.2$ m
$A \approx 4.5\text{ m}^2$

34. $r = 2.5$ m
$A \approx 19.6\text{ m}^2$

Approximate the area of each circle with the given diameter. Use $\pi \approx 3.14$ and round the result to the tenths place.

35. $d = 8$ yd
$A \approx 50.2\text{ yd}^2$

36. $d = 12$ yd
$A \approx 113.0\text{ yd}^2$

37. $d = 2.2$ cm
$A \approx 3.8\text{ cm}^2$

38. $d = 3$ cm
$A \approx 7.1\text{ cm}^2$

Find the area of each polygon.

39.

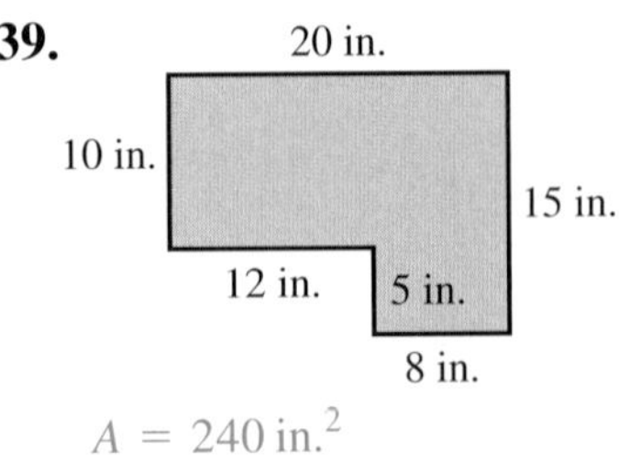

$A = 240\text{ in.}^2$

40.

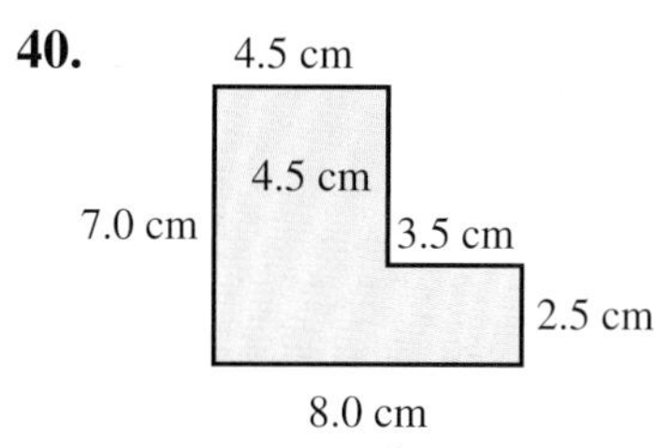

$A = 40.25 \text{ cm}^2$

41.

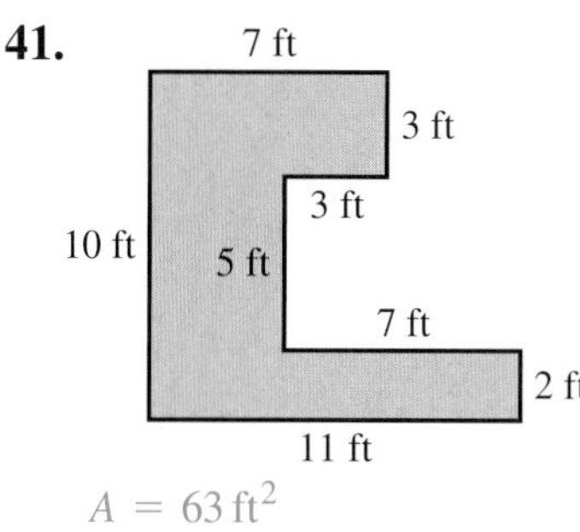

$A = 63 \text{ ft}^2$

42.

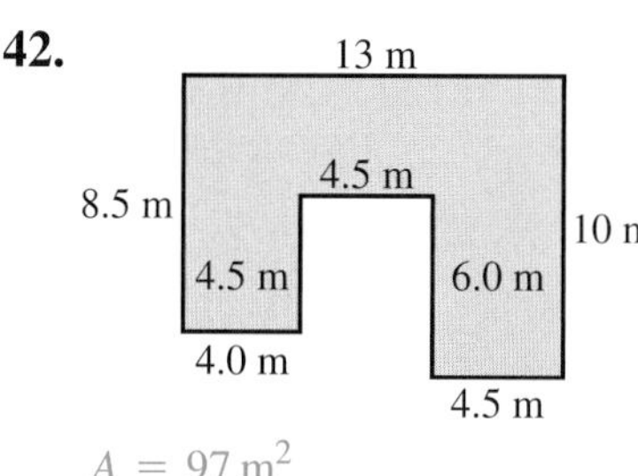

$A = 97 \text{ m}^2$

Solve the following applications.

43. In her design class, Rebekah created a parallelogram shape for the side of a magazine rack. What is the area of this side?

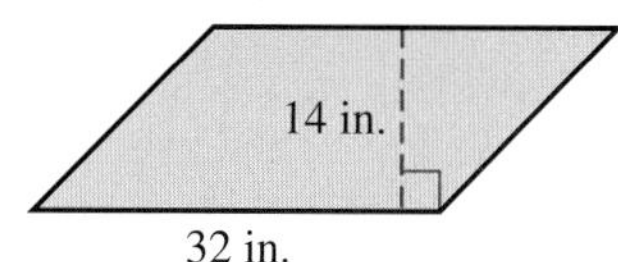

The area of this side is 448 square inches.

44. Kevin's backyard is in the shape of a trapezoid. What is the area of Kevin's backyard?

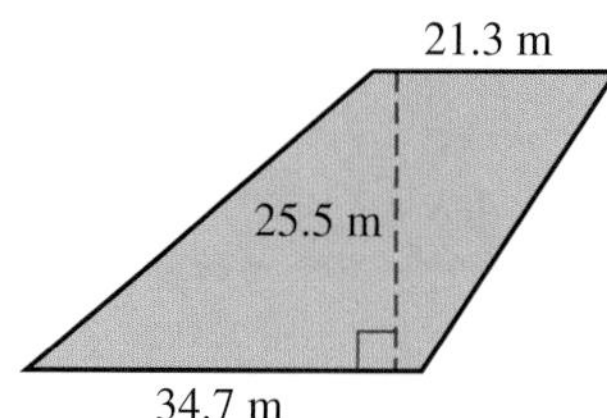

The area of Kevin's backyard is 714 square meters.

45. A bedroom is in the shape of a rectangle. What is the area of this room?

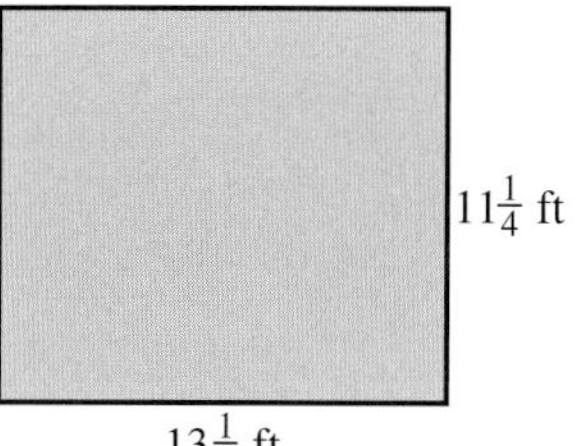

The area of this room is 150 square feet.

46. The dimensions of home plate for professional baseball are shown here. What is the area of home plate?

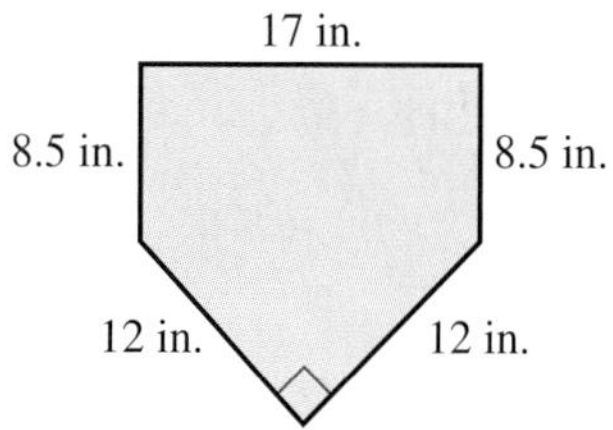

The area of home plate is 216.5 square inches.

Solve the following applications. Use $\pi \approx 3.14$ and round each result to the nearest whole number.

47. A hot tub has a circular cover with a diameter of 5 feet. What is the area of the hot tub cover?

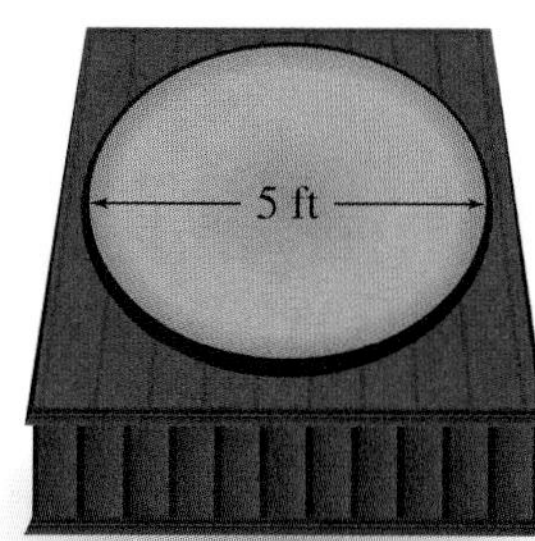

The area of the hot tub cover is about 20 square feet.

48. Nate's drum set contains a drum that is 14 inches in diameter. What is the area of the top of the drum?

The area of the top of the drum is about 154 square inches.

49. A crop circle of wheat has a diameter of 80 meters. What is the area of this crop circle?

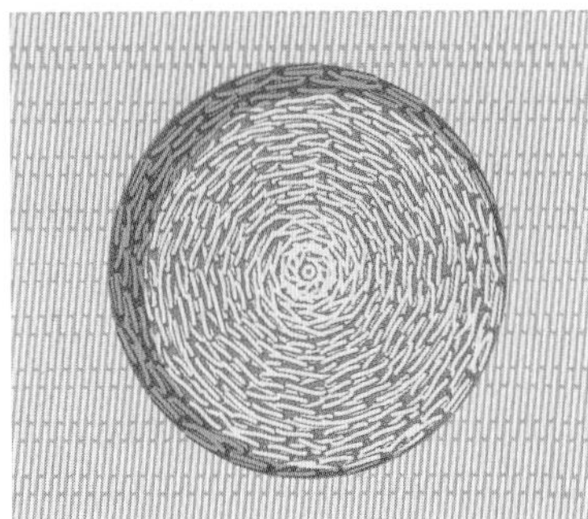

The area of this crop circle is about 5,024 square meters.

50. A circular planetarium window has a diameter of 90 centimeters. What is the area of this window?

The area of this window is about 6,359 square centimeters.

Answer as indicated. Round each result to the nearest tenth.

51. Find the area of the ring.

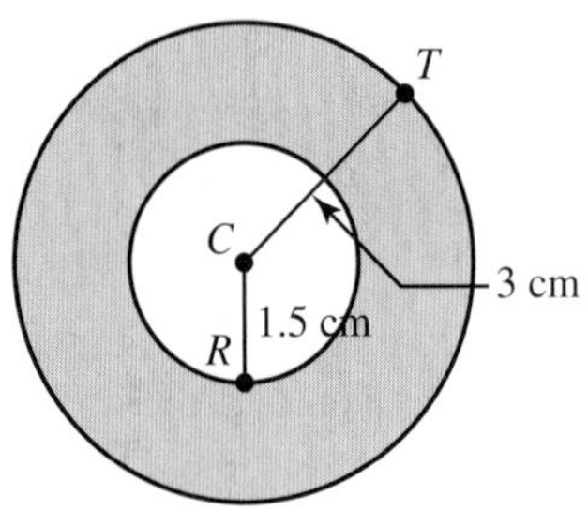

$A \approx 21.2\ \text{cm}^2$

52. Find the area of the shaded region.

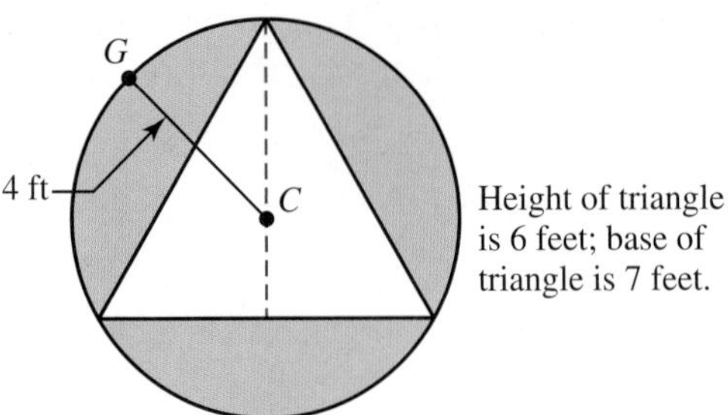

$A \approx 29.2\ \text{ft}^2$

53. A farmer created a small "crop circle" that is a square inside of a circle. The square contains the crop of wheat, and the circular part is flat. The circle has a radius of 25 meters, and the square has a side length of 35 meters. What is the area of the flat region within this crop circle?

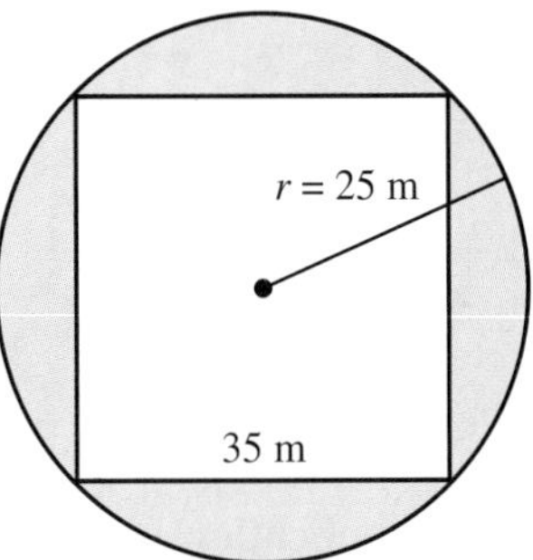

The area of the flat region is about 737.5 square meters.

54. LaTanya is building a model of the solar system. To represent the rings of Saturn, she is using a single piece of Styrofoam in the shape of a flat ring. The inner radius of the ring is 3 inches, and its outer radius is 5 inches. What is the area of the top of the ring?

The area of the top of the ring is about 50.2 square inches.

Think Outside the Box

A semicircle is half a circle. Use that fact to solve the following applications. Write each answer as a sentence.

55. Marta's garden is in the shape of a semicircle with a diameter of 20 feet. Approximate the area of Marta's garden.

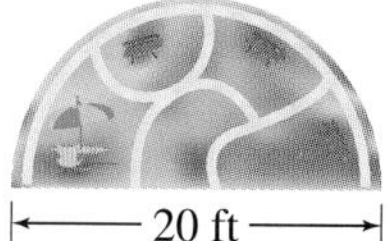

The area of Marta's garden is approximately 157 square feet.

56. An ice-skating rink has the shape of two semicircles attached to a square. Each side of the square measures 30 yards. Approximate the area of the skating rink.

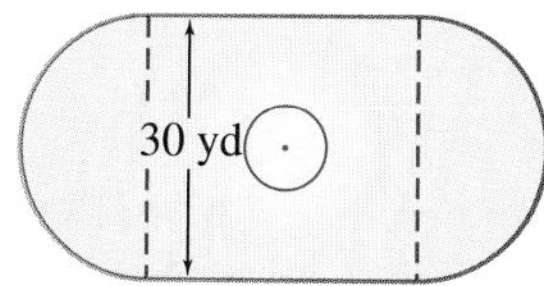

The area of the skating rink is approximately 1,606.5 square yards.

57. Armando's property is in the shape of a triangle atop a rectangle and has the dimensions shown in the diagram. Find the area of Armando's property.

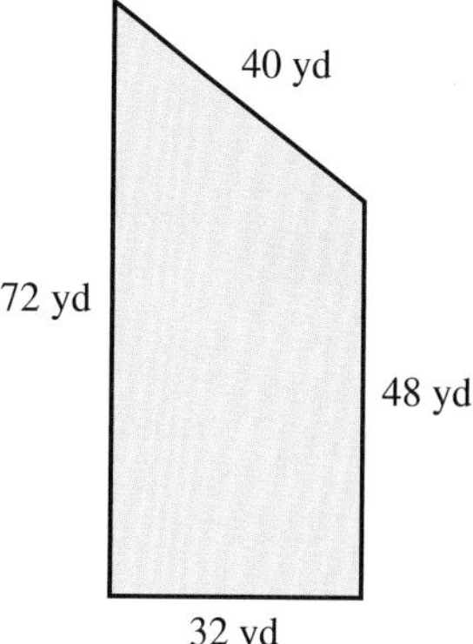

The area of Armando's property is 1,920 square yards.

SECTION 8.8 Volume

OBJECTIVES

In this section, you will learn to
- Recognize different dimensions.
- Find the volume of a rectangular solid.
- Apply the volume formulas for a cylinder, cone, pyramid, and sphere.

You Need to Know

To successfully complete this section, you need to understand
- ☐ Multiplying fractions (4.4)
- ☐ Multiplying decimals (5.4)
- ☐ Parts of a circle (8.5)
- ☐ Area (8.7)

Introduction

In geometry, **dimension** is the visible, linear measure of an object, including its length (or breadth), its width (or height), and its depth. A variety of geometric figures demonstrate dimension.

A *point* is infinitely small and has no length, width, or depth. Points have no dimension; they cannot be measured.

A *line segment* has length but no height or depth. A line segment has only one measurement, one dimension. Its measure is in simple units such as centimeters.

A *rectangle* has two measurements, *two* dimensions: length and height. Its area is measured in square units such as square inches, or in.^{2}

Parallelograms, triangles, and *circles* are also two-dimensional; and their area is measured in square units, or units^2.

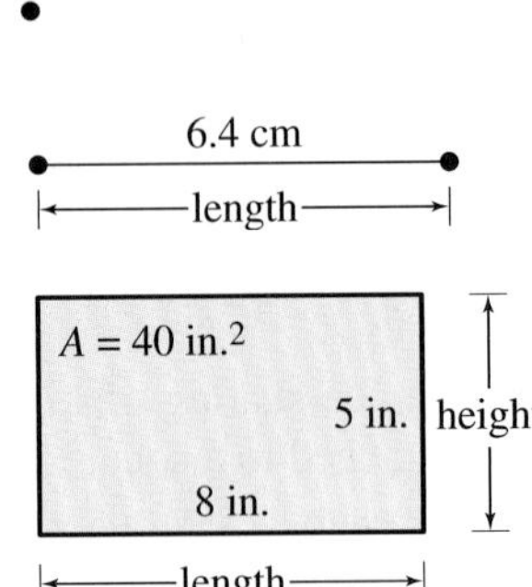

Volume of a Rectangular Solid

A **solid** is a *three-dimensional* object. The box shown here is a solid because it has three distinct measures—three distinct dimensions—length, height, and depth.

Its **volume**—a measure indicating the amount a solid (such as a box) can hold—is the product of those three dimensions.

$$\text{Volume} = \text{Length} \cdot \text{Depth} \cdot \text{Height}$$
$$V = 6 \text{ in.} \cdot 5 \text{ in.} \cdot 4 \text{ in.}$$
$$V = 6 \cdot 5 \cdot 4 \cdot \text{in.} \cdot \text{in.} \cdot \text{in.}$$
$$V = 120 \text{ in.}^3$$

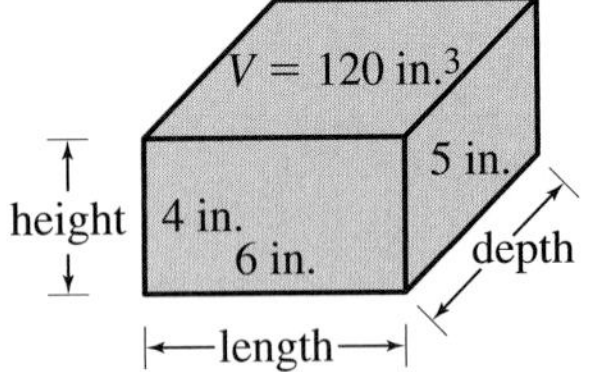

This is read, "The volume is 120 cubic inches."

For three-dimensional figures, we commonly refer to the flat bottom as the **base.** The base is always two-dimensional. In the case of a box, the base is the rectangle formed by the length and the depth. In some solid figures, the base is the flat surface on the top.

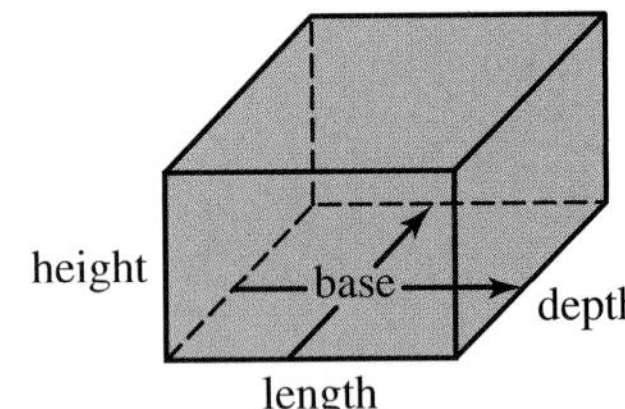

Many of the volume formulas presented in this section require us to find the area of the base. In a volume formula, the **Base** (with a capital B) means the area of the two-dimensional base of the solid figure.

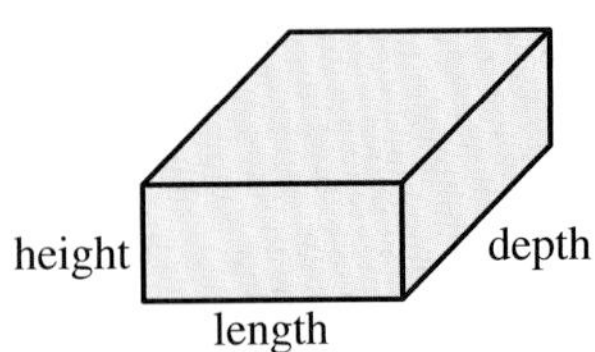

Base = Length × Depth

A box in which all three measures are the same is called a **cube.** The cube at the right has a volume of 1 cubic cm. We'll use this cube as a building block to help us visually understand the volume of a box.

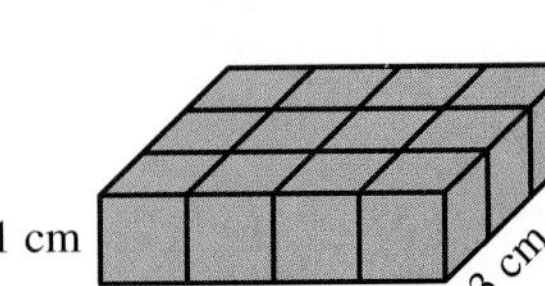

We can put many of these cubes together to form one layer. This layer has a volume of 12 cubic cm.

$$V = 4 \text{ cm} \cdot 3 \text{ cm} \cdot 1 \text{ cm} = 12 \text{ cm}^3$$

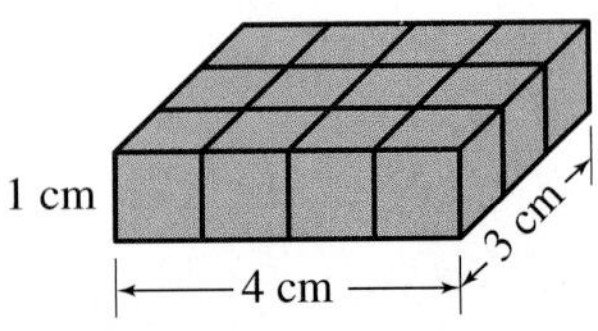

We can stack more cubes on top of this layer to begin to form another layer.

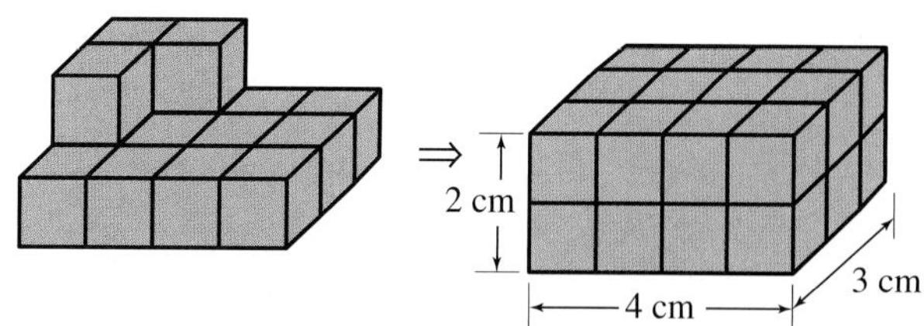

This second layer creates a rectangular solid that has a volume of 24 cubic cm.

$$V = 4\text{ cm} \cdot 3\text{ cm} \cdot 2\text{ cm} = 24\text{ cm}^3$$

We can think of the volume of a box as follows:

Volume = (Length · Depth) · Height

Volume = The area of the base (**Base**) · The height (***h***)

$V = \textbf{\textit{Base}} \cdot \boldsymbol{h}$

Example 1

Find the volume of this box:

6 in. base 15 in. 8 in.

Procedure Apply the formula for the volume of a box. Remember, the volume will be in cubic inches.

Base = 8 in. × 15 in. = 120 sq in.

height = 6 in.

Answer $V = 120 \times 6$ cubic inches

$V = 720\text{ in.}^3$

The volume of the box is 720 cubic inches.

You Try It 1 **Given the dimensions of each box, find the volume. Use Example 1 as a guide. (First, find the Base.)**

a)

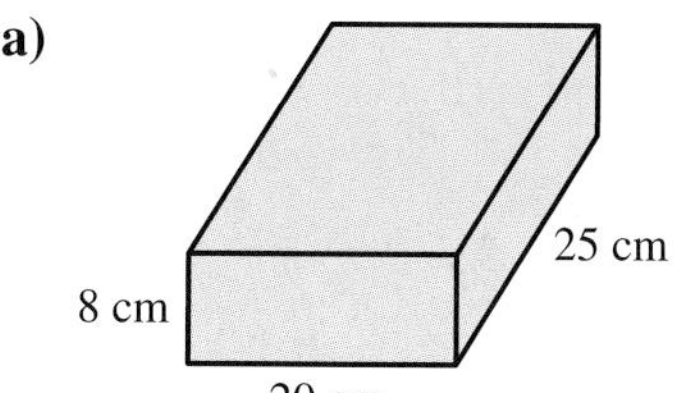

b)

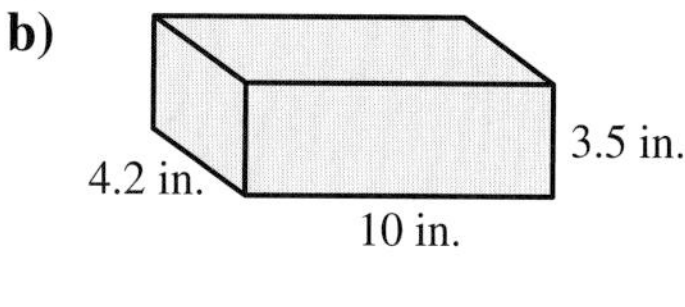

Volumes of Other Solids

Some common solids are the cylinder, cone, pyramid, and sphere. Each has its own volume formula.

The formulas for the cylinder, cone, and pyramid require finding the area of the base.

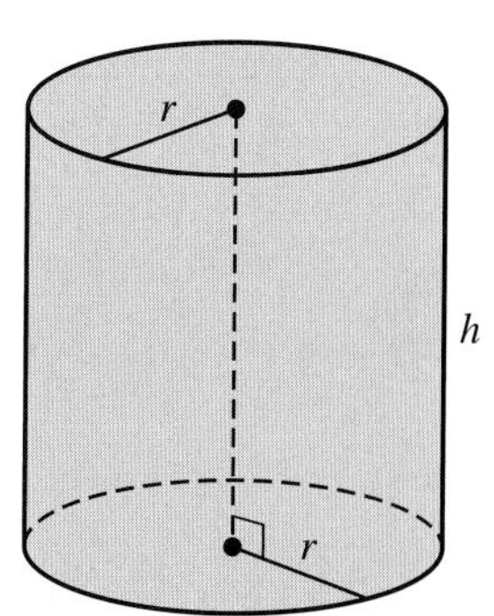

A **cylinder** is the rounded version of a box, such as a can of soup.

Like a box, its volume is Base · height.

$V = \text{Base} \cdot h$

Because the base is a circle, the base area is $\pi \cdot r^2$.

This makes the volume formula $V = \pi \cdot r^2 \cdot h$.

Example 2

Find the volume of this cylinder. Use $\pi \approx 3.14$ and round to the nearest tenth.

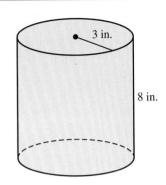

Procedure Apply the volume of a cylinder formula. The radius is 3 inches, and the height is 8 inches; so, the volume will be in cubic inches.

Answer

$V = \pi \times 3^2 \times 8 \text{ in.}^3$

$V \approx 3.14 \times 9 \times 8 \text{ in.}^3$

$V \approx 226.08 \text{ in.}^3 \approx 226.1 \text{ in.}^3$

Remember the order of operations and apply the exponent to 3 first.

The volume of the cylinder is about 226.1 cubic inches.

3 in.

8 in.

You Try It 2 **Find the volume of each cylinder. Use $\pi \approx 3.14$ and round the result to the nearest tenth. Use Example 2 as a guide.**

a)

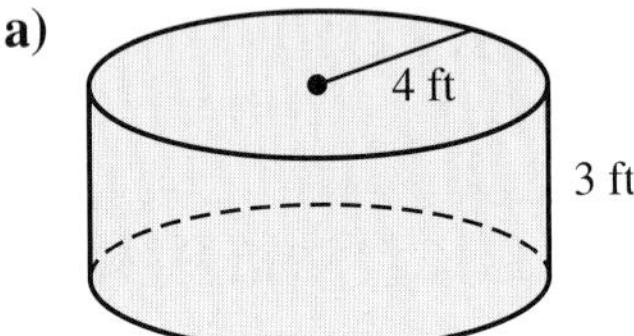

b)

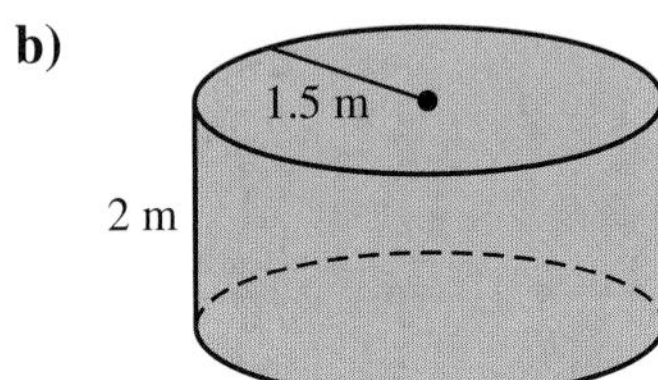

Here are two ways to view a **cone**: like the orange cones used on the highway in a construction zone or like a snow-cone cup.

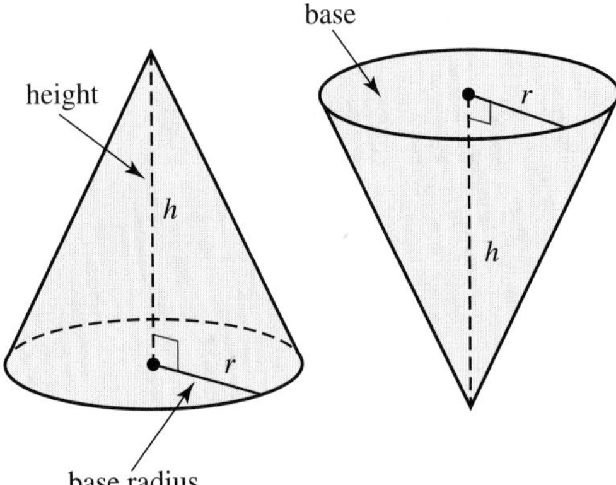

Either way, the base is a circle and has base radius r.

If a cylinder and a cone have the same radius and the same height, we might imagine the cone fitting inside the cylinder.

The volume of a cone is $\frac{1}{3}$ the volume of such a cylinder.

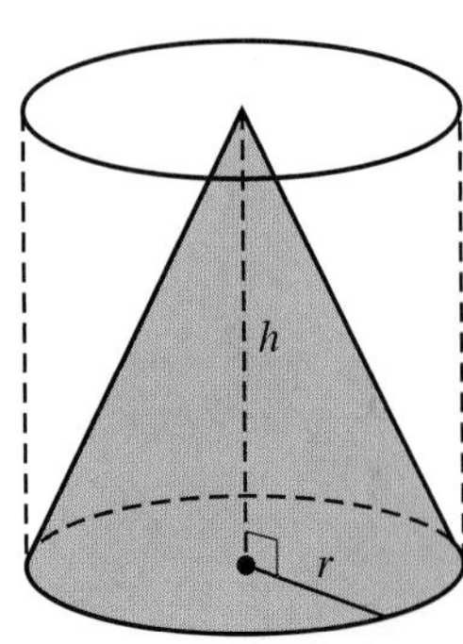

$$V = \frac{1}{3} \cdot \text{Base} \cdot h \longrightarrow V = \frac{1}{3} \cdot \pi \cdot r^2 \cdot h$$

Example 3

Find the volume of this cone. Use $\pi \approx 3.14$.

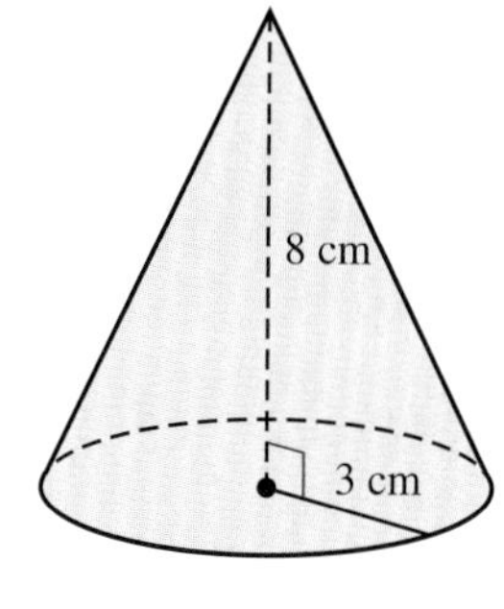

Procedure Apply the formula for the volume of a cone. The base radius is 3 centimeters, and the height is 8 centimeters; so, the volume will be in cubic centimeters.

Answer
$$V = \frac{1}{3} \times \pi \times 3^2 \times 8\,\text{cm}^3$$
$$V \approx \frac{1}{3} \times 3.14 \times 9 \times 8\,\text{cm}^3$$
$$V \approx 3.14 \times 3 \times 8\,\text{cm}^3$$
$$V \approx 75.36\,\text{cm}^3 \approx 75.4\,\text{cm}^3$$

$\frac{1}{3} \times 9 = 3$

The volume of the cone is about 75.4 cubic centimeters.

You Try It 3 **Find the volume of each cone. Use $\pi \approx 3.14$ and round the result to the nearest tenth. Use Example 3 as a guide.**

a)

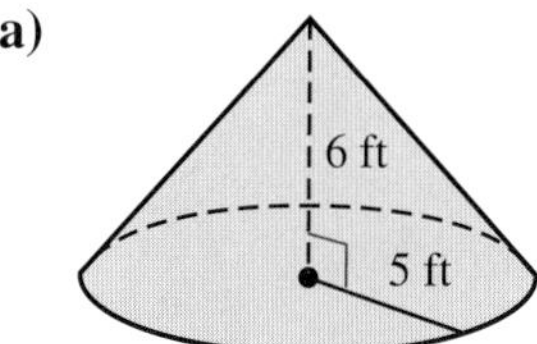

b)

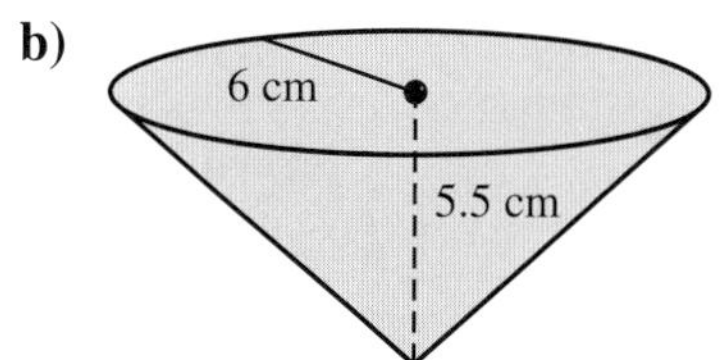

A **pyramid** is like a cone with a rectangular base.

If a pyramid and a box have the same base and same height, we might imagine the pyramid fitting inside the box.

The volume of a pyramid is $\frac{1}{3}$ the volume of such a box.

$$V = \frac{1}{3} \cdot \text{Base} \cdot h \longrightarrow V = \frac{1}{3} \cdot l \cdot d \cdot h$$

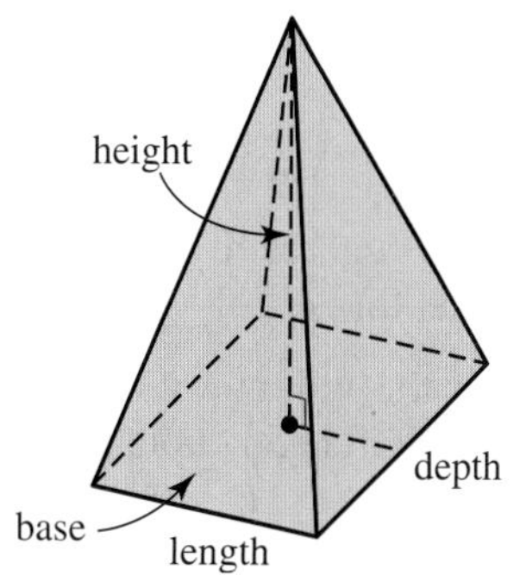

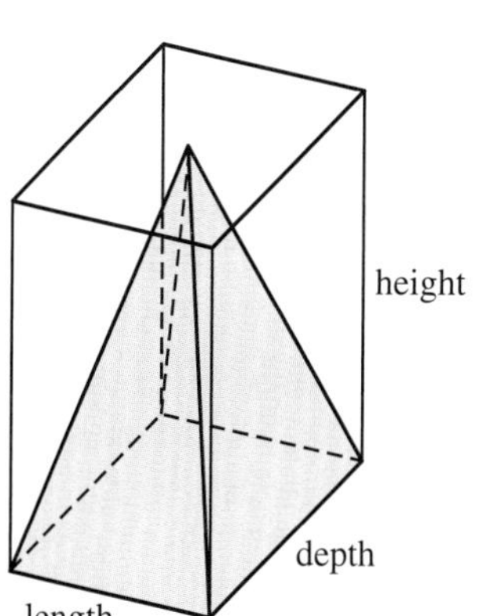

Example 4

Find the volume of this pyramid.

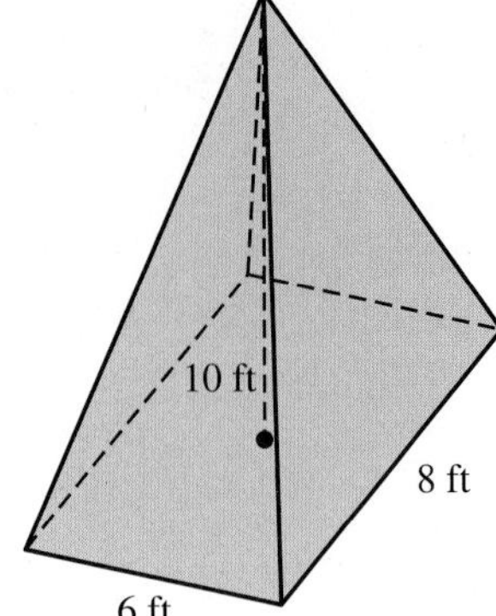

Procedure Apply the formula for the volume of a pyramid. The base length is 6 feet, its base depth is 8 feet, and the height is 10 feet; so, the volume will be in cubic feet.

Answer $V = \frac{1}{3} \cdot 6 \cdot 8 \cdot 10 \text{ ft}^3$

$V = 2 \cdot 8 \cdot 10 \text{ ft}^3$ $\qquad \frac{1}{3} \cdot 6 = 2$

$V = 160 \text{ ft}^3$

The volume of the pyramid is 160 cubic feet.

You Try It 4 **Find the volume of each pyramid. Use Example 4 as a guide.**

a)

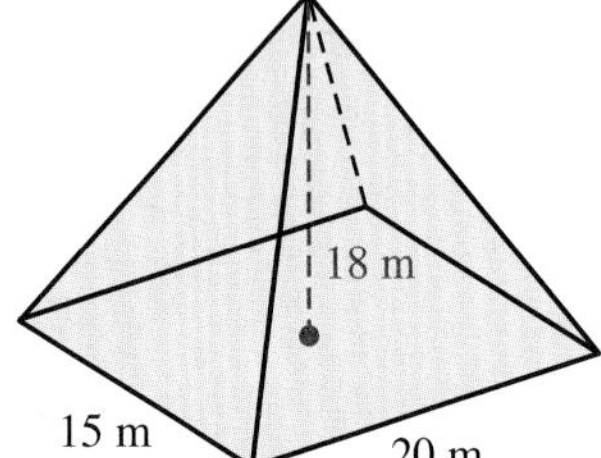

b)

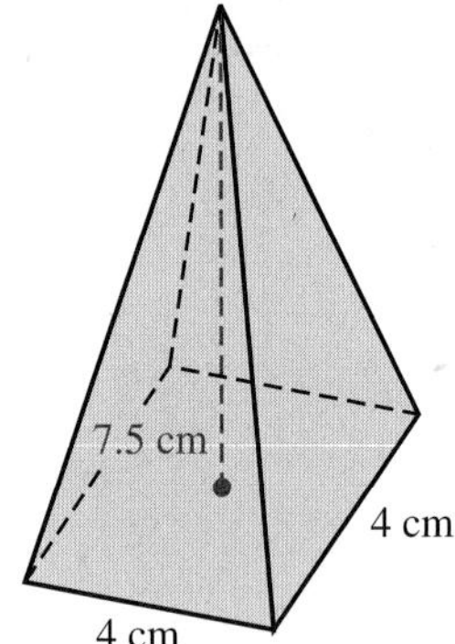

A **sphere** is the three-dimensional version of a circle, such as a ball.

Like a circle, it has many radii, all the same length.

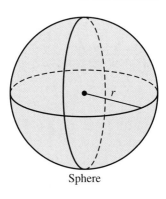

Volume of a sphere: $V = \frac{4}{3} \cdot \pi \cdot r^3$

We typically use $\pi \approx 3.14$ when finding the volume of a sphere. We can also use a decimal approximation for $\frac{4}{3}$: $\frac{4}{3} \approx 1.333$.

Sphere

Example 5

Find the volume of this sphere. Use $\pi \approx 3.14$ and round the result to the nearest tenth.

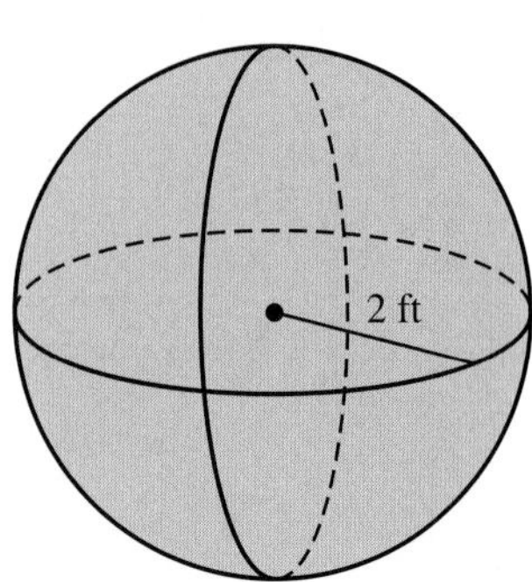

Procedure Use the formula for the volume of a sphere. The radius is 2 feet; so, the volume will be in cubic feet.

Answer $V = \frac{4}{3} \times \pi \times 2^3 \text{ ft}^3$

$V \approx 1.333 \times 3.14 \times 8 \text{ ft}^3$

$V \approx 33.48496 \text{ cubic feet} \approx 33.5 \text{ ft}^3$

Remember the order of operations and apply the exponent to 2 first.

The volume of the sphere is about 33.5 cubic feet.

You Try It 5 **Find the volume of each sphere. Use $\pi \approx 3.14$ and round the result to the nearest tenth. Use Example 5 as a guide.**

a)

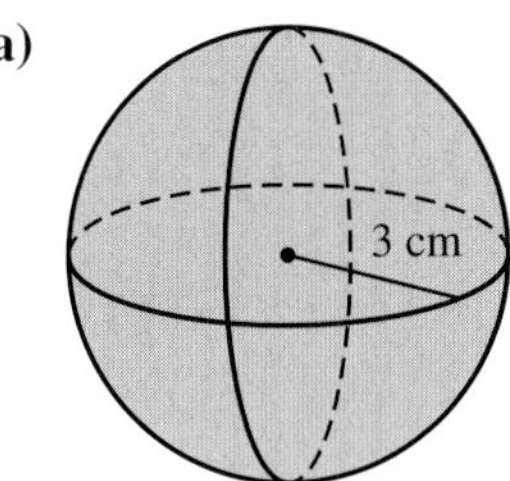

b)

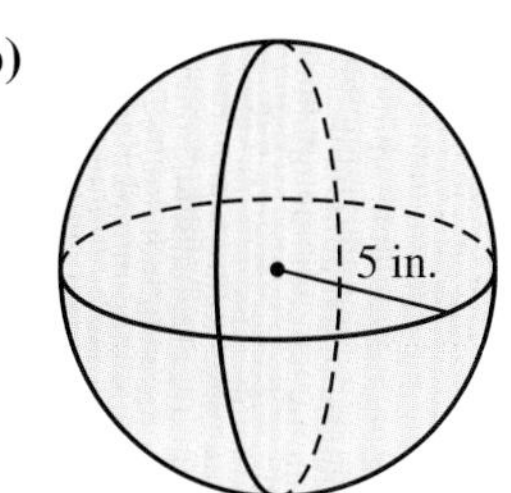

Answers: You Try It and Think About It

You Try It: Each decimal answer has been rounded to the nearest tenth. **1. a)** $V = 4{,}000\text{ cm}^3$ **b)** $V = 147\text{ in.}^3$ **2. a)** $V \approx 150.7\text{ ft}^3$ **b)** $V \approx 14.1\text{ m}^3$ **3. a)** $V \approx 157\text{ ft}^3$ **b)** $V \approx 207.2\text{ cm}^3$ **4. a)** $V = 1{,}800\text{ m}^3$ **b)** $V = 40\text{ cm}^3$ **5. a)** $V \approx 113.0\text{ cm}^3$ **b)** $V \approx 523.3\text{ in.}^3$

Think About It: No Think About Its in this section.

Section 8.8 Exercises

FOR EXTRA HELP
Student Resources on DVD-ROM
Includes
➤ Student's Solutions Manual
➤ Video Lectures
➤ Chapter Test Prep Video
MyMathLab
MathXL

Think Again

1. A cylinder is like a rounded box. How are the volume formulas for a cylinder and a box similar?
The volume formulas for the cylinder and the box both are Area of Base × height.

2. A cone is like a rounded pyramid. How are the volume formulas for a cone and a pyramid similar?
The volume formulas for the pyramid and the cone both are $\frac{1}{3}$ × Area of Base × height.

Focus Exercises

Given the dimensions of each box, find the volume.

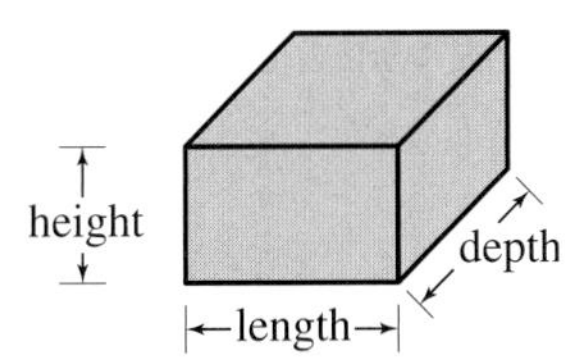

3. $l = 8\text{ in.}, d = 6\text{ in.}, h = 5\text{ in.}$
$V = 240\text{ in.}^3$

4. $l = 11\text{ m}, d = 9\text{ m}, h = 4\text{ m}$
$V = 396\text{ m}^3$

5. $l = 20\text{ ft}, d = 12\text{ ft}, h = 10\text{ ft}$
$V = 2{,}400\text{ ft}^3$

6. $l = 15\text{ cm}, d = 12\text{ cm}, h = 8\text{ cm}$
$V = 1{,}440\text{ cm}^3$

7. $l = 10.5\text{ cm}, d = 9.6\text{ cm}, h = 20\text{ cm}$
$V = 2{,}016\text{ cm}^3$

8. $l = 5.2\text{ m}, d = 6.5\text{ m}, h = 4.0\text{ m}$
$V = 135.2\text{ m}^3$

9. $l = 4\text{ in.}, d = 2\frac{1}{2}\text{ in.}, h = 1\frac{3}{4}\text{ in.}$
$V = 17\frac{1}{2}\text{ in.}^3$

10. $l = 6\frac{1}{4}\text{ ft}, d = 7\frac{1}{2}\text{ ft}, h = 10\frac{2}{3}\text{ ft}$
$V = 500\text{ ft}^3$

Given the dimensions of each cylinder, find the volume. Use $\pi \approx 3.14$ and round the result to the nearest tenth.

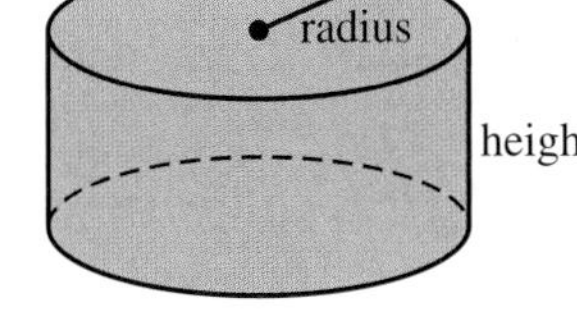

11. $r = 2\text{ ft}, h = 4\text{ ft}$
$V \approx 50.2\text{ ft}^3$

12. $r = 3\text{ in.}, h = 6\text{ in.}$
$V \approx 169.6\text{ in.}^3$

13. $r = 1.2\text{ m}, h = 4.5\text{ m}$
$V \approx 20.3\text{ m}^3$

14. $r = 1.6\text{ cm}, h = 5.0\text{ cm}$
$V \approx 40.2\text{ cm}^3$

Given the dimensions of each cone, find the volume. Use $\pi \approx 3.14$ and round the result to the nearest tenth.

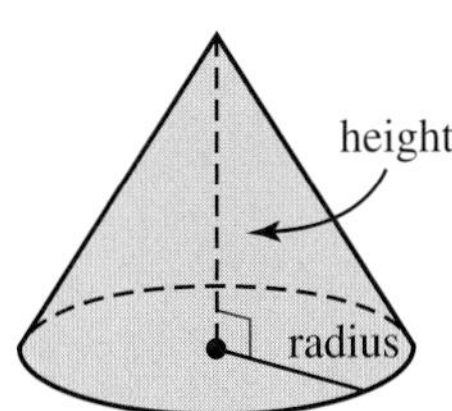

15. $r = 3\text{ ft}, h = 5\text{ ft}$
$V \approx 47.1\text{ ft}^3$

16. $r = 5\text{ in.}, h = 6\text{ in.}$
$V \approx 157.0\text{ in.}^3$

17. $r = 2\text{ m}, h = 3.3\text{ m}$
$V \approx 13.8\text{ m}^3$

18. $r = 1.5\text{ cm}, h = 4.0\text{ cm}$
$V \approx 9.4\text{ cm}^3$

Given the dimensions of each pyramid, find the volume.

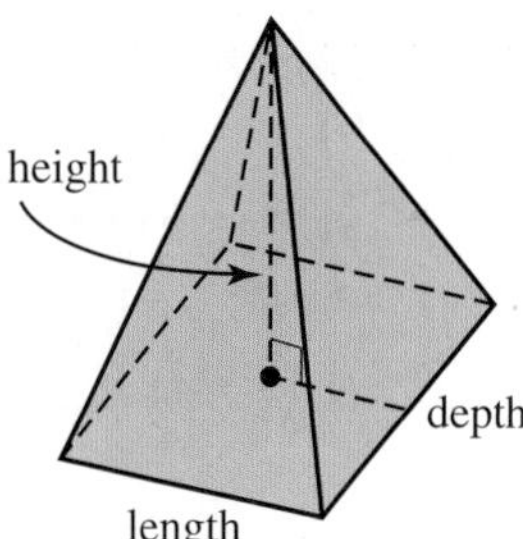

19. $l = 8$ in., $d = 6$ in., $h = 5$ in.
$V = 80$ in.^{3}

20. $l = 12$ ft, $d = 9$ ft, $h = 7$ ft
$V = 252$ ft^3

21. $l = 1.5$ m, $d = 2$ m, $h = 7$ m
$V = 7$ m^3

22. $l = 1.2$ cm, $d = 5$ cm, $h = 4$ cm
$V = 8$ cm^3

Find the volume of each sphere. Use $\pi \approx 3.14$ and round the result to the nearest tenth.

23. $r = 1$ ft
$V \approx 4.2$ ft^3

24. $r = 3$ in.
$V \approx 113.0$ in.^{3}

25. $r = 1.2$ m
$V \approx 7.2$ m^3

26. $r = 2.5$ cm
$V \approx 65.4$ cm^3

Solve the following applications. Round each result to the nearest whole number.

27. A storage rental business rents a space that is 8 feet long, 6 feet wide, and 12 feet high. What is the volume of this space?

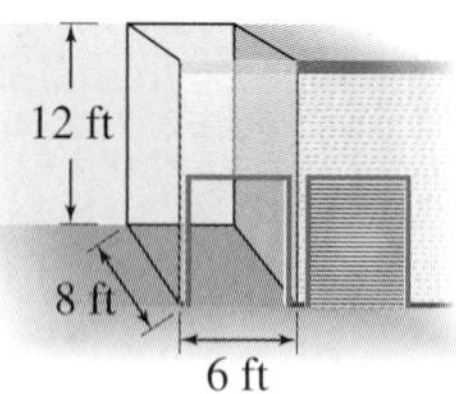

The volume of this space is about 576 cubic feet.

28. An NBA basketball has a radius of 4.7 inches. What is its volume?

Its volume is about 435 cubic inches.

29. A soup can has a base radius of 4 centimeters and a height of 12 centimeters. What is the volume of the can?

The volume of the can is about 603 cubic centimeters.

30. An ice cream sugar cone is in the shape of—that's right—a cone! A cone has a base radius of 3.5 centimeters and a height of 10 centimeters. What is the volume of the cone?

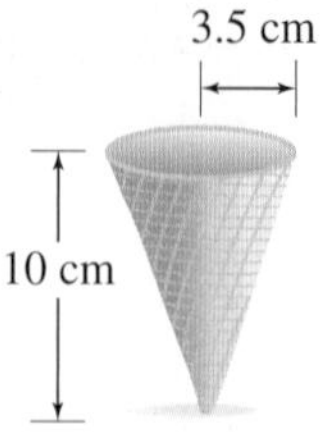

The volume of the cone is about 128 cubic centimeters.

31. Rebecca is building a model of the great pyramids of Egypt. One of her pyramids has a square base with a side length of 6 inches and a height of 3.9 inches. What is the volume of Rebecca's pyramid?

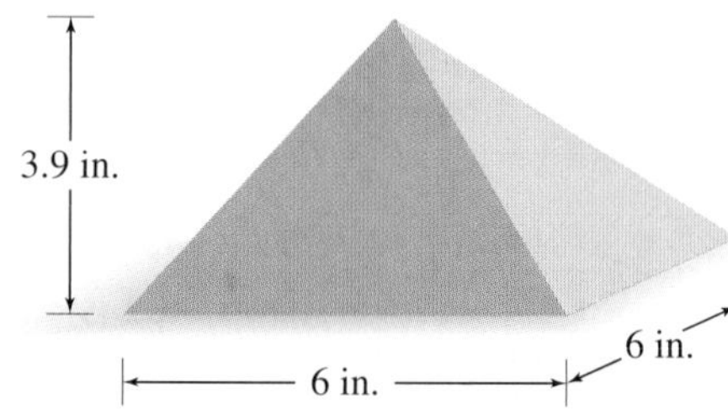

The volume of Rebecca's pyramid is about 47 cubic inches.

32. Gregor's aboveground pool is in the shape of a cylinder with a height of 6 feet and a diameter of 24 feet. If the pool is filled to 5 feet deep, what is the volume of water in Gregor's pool?

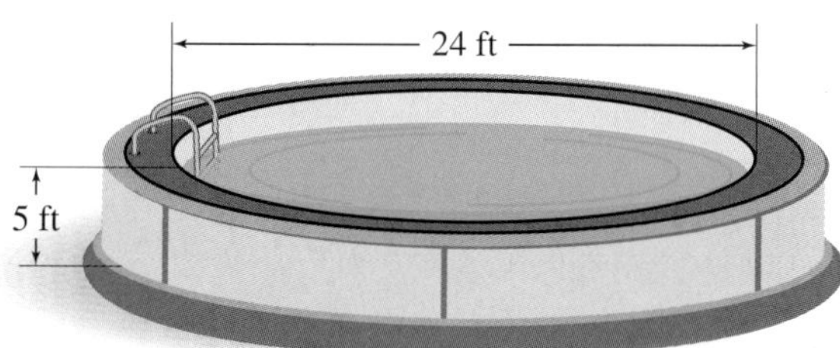

The volume of water in Gregor's pool is about 2,261 cubic feet.

Think Outside the Box

A hemisphere is half a sphere. Use that fact to solve the following applications. Write the answer as a sentence.

33. A bowl is in the shape of a hemisphere with a diameter of 8 inches. Find the volume of the bowl.

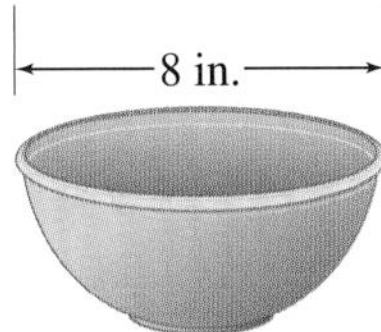

The volume of the bowl is approximately 134.0 cubic inches.

34. Jackson has a silo on his farm that is a cone on top of a cylinder. The cylinder is 9 meters tall and has a base radius of 2 meters. The whole silo is 11 meters tall. Find the volume of the whole silo.

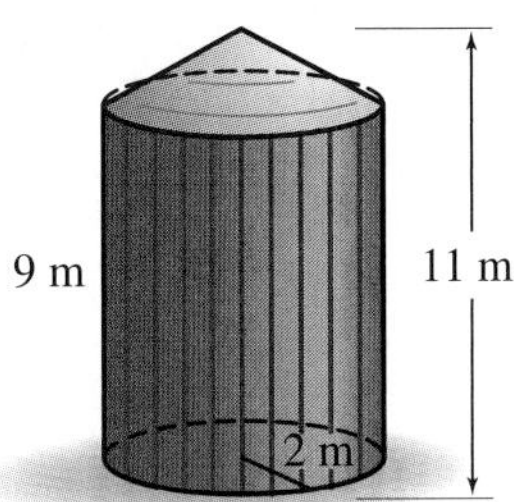

The volume of the whole silo is approximately 121.4 cubic meters.

Chapter 8 Review

Section 8.1 U.S. Measures

Concept	Example
To add feet and inches, set up the addition vertically and line up the feet and inches. When the sum is found, adjust the number of inches if the total includes 12 or more inches.	6 ft 5 in. + 2 ft 10 in. = 8 ft 15 in. 8 ft + 1 ft + 3 in. = 9 ft 3 in.
To subtract feet and inches, regroup (if necessary) by changing 1 foot into 12 inches.	~~4~~ 3 ft ~~2~~ 14 in. − 2 ft 5 in. = 1 ft 9 in. Notice that we add 12 inches to the 2 inches already there.

Section 8.2 Metric Measures

Concept	Example
Converting from one metric measure to another is a matter of moving the decimal point to the left or right.	Convert 83 centimeters to dekameters. 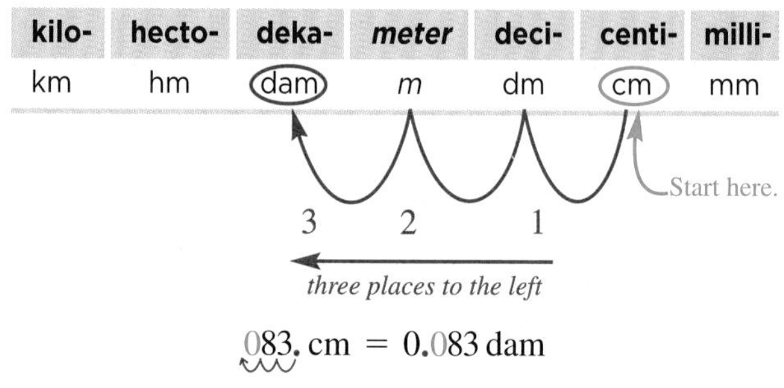 083. cm = 0.083 dam
Adding and subtracting metric measures is a matter of adding and subtracting decimals. First, write one in terms of the other.	Add 7.12 meters + 1.4 decimeters 7.12 m + 0.14 m = 7.26 m ← 1.4 dm = 0.14 m

Section 8.3 Converting U.S. and Metric Measures

Concept	Example
In an **equivalence fraction,** the numerator and the denominator are of equal value but of a different unit measure. Equivalence fractions always have the value 1. We can convert from one measure to another by multiplying by an equivalence fraction.	Convert 4 kilograms to miles using $\frac{5\text{ mi}}{8\text{ km}}$. $\frac{4\text{ km}}{1} \times \frac{5\text{ mi}}{8\text{ km}} = \frac{4\cancel{\text{km}} \times 5\text{ mi}}{1 \times 8\cancel{\text{km}}} = \frac{20\text{ mi}}{8} = 2\frac{1}{2}\text{ miles}$

Section 8.4 Lines and Angles

Concept	Example
A **plane** is any flat surface that extends forever in each direction.	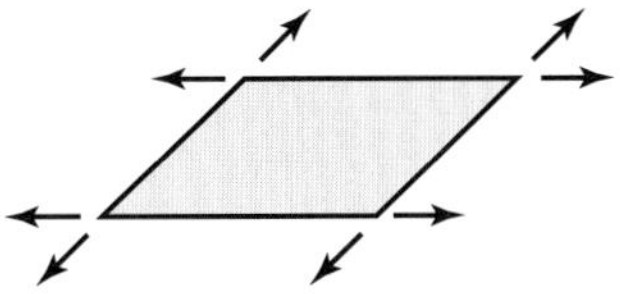
A **line** is straight, and it goes on forever.	
When two distinct lines intersect each other, they meet in a single **point.** A point is infinitely small.	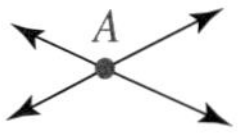 A
Two lines (in the same plane) that never intersect each other are called **parallel lines.**	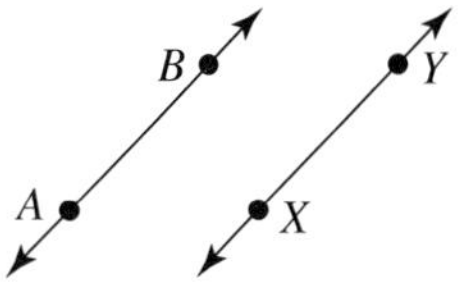 B, A, Y, X
The portion of a line that is drawn between two points (called *endpoints*) is a **line segment.**	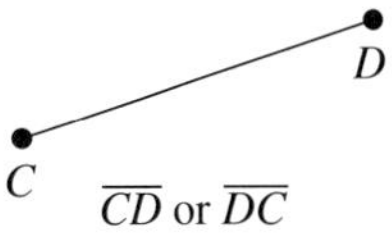 D, C $\overline{CD}$ or $\overline{DC}$
A **ray** is the portion of a line that starts at a point on the line (the endpoint) and goes on forever in only one direction.	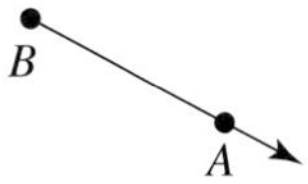 B, A $\overrightarrow{BA}$
When two rays (or line segments) share the same endpoint, they form an **angle.** The common endpoint is the **vertex** of the angle. The two rays (line segments) that form the angle are the **sides** of the angle.	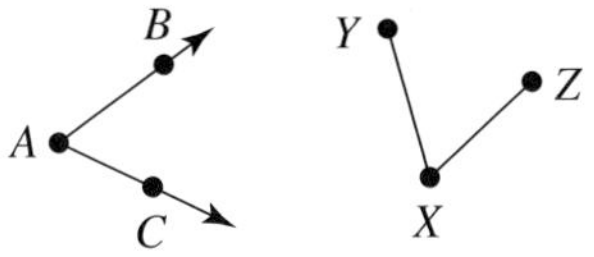 B, A, C, Y, Z, X $\angle BAC$ $\angle YXZ$
The **interior** of an angle is the portion between the rays that form the angle. Just as the rays continue indefinitely, the interior of the angle continues indefinitely.	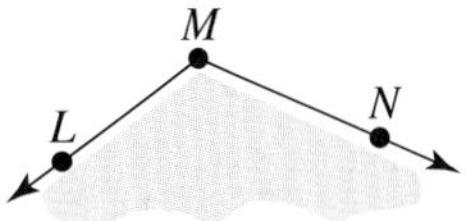 M, L, N The interior of $\angle LMN$
If the rays forming an angle go in opposite directions so as to form a line, the angle is called a **straight angle.** A straight angle has a measure of 180°.	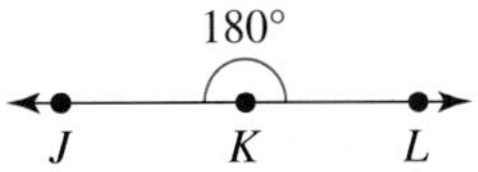 180° J K L $\angle JKL$ is a straight angle.
A **right angle** measures 90°.	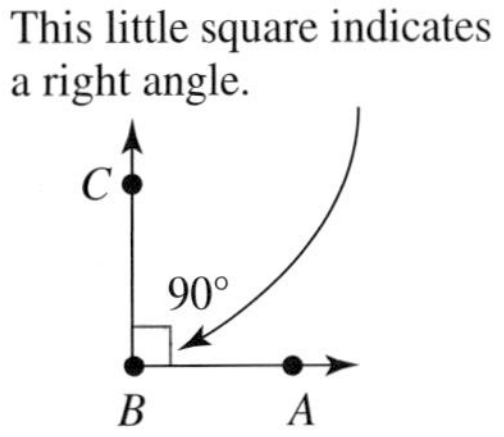 This little square indicates a right angle. C, 90°, B, A

continued

Section 8.4 Lines and Angles

Concept | **Example**

An **acute** angle has a measure greater than 0° and less than 90°.

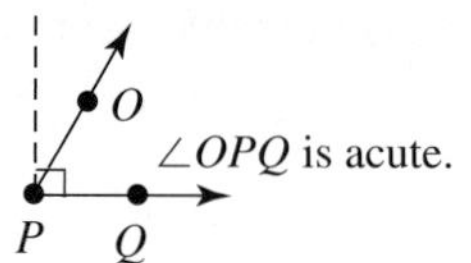

An **obtuse** angle has a measure greater than 90° and less than 180°.

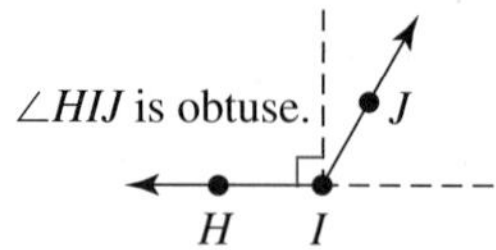

Adjacent angles are any two angles that share a common side, forming an even larger angle. All three angles have the same vertex, and the shared side must be in the **interior** of the larger angle.

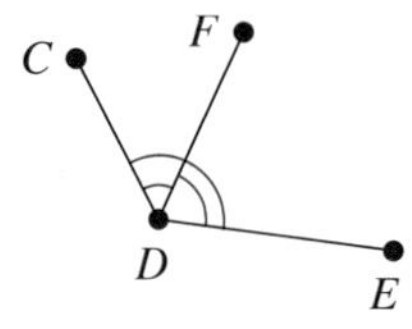

$\angle CDF$ and $\angle FDE$ are adjacent (they share the common side $\overline{DF}$), and together they form the larger angle $\angle CDE$.

Supplementary angles are any two angles with measures that add to 180°. Supplementary angles can be adjacent, but they don't have to be.

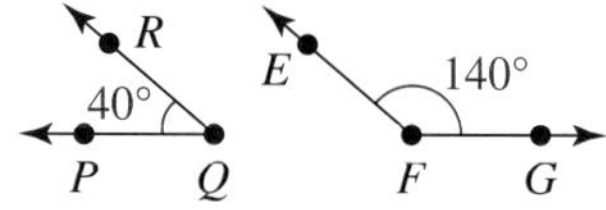

Complementary angles are *any* two angles with measures that add to 90°. Complementary angles can be adjacent, but they don't have to be.

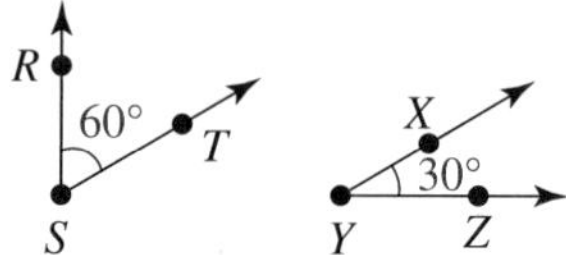

When two lines intersect, several pairs of angles are formed, all sharing the same vertex. Some pairs are adjacent and supplementary (forming a straight line). The other pairs are **vertical angles,** angles that are not adjacent to each other.

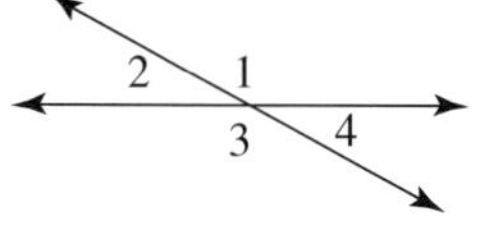

$\angle 1$ and $\angle 3$ are vertical angles; $\angle 2$ and $\angle 4$ are vertical angles.

Two lines are **perpendicular** if they intersect to form four right angles. The symbol $\perp$ indicates that two lines (or line segments) are perpendicular to each other.

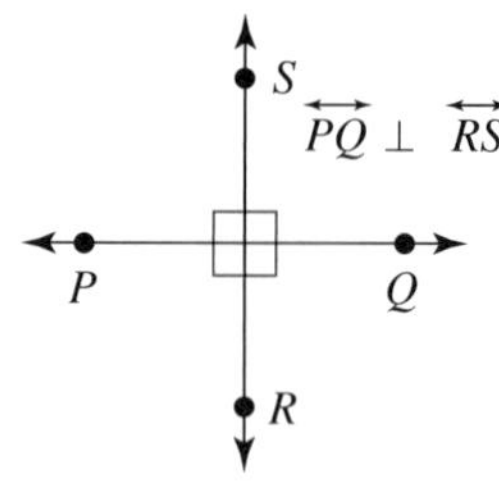

Section 8.5 Geometric Shapes

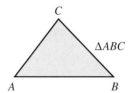

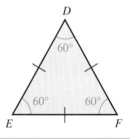

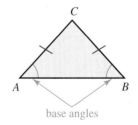

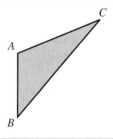

Concept	Example
A **polygon** is a closed plane figure composed of at least three straight line segments. In a polygon, two line segments meet only at an endpoint, called a **vertex,** and never cross each other. Each vertex is the endpoint of only two sides.	
A **triangle** is a polygon that has three sides and three angles. The sum of the angles in a triangle is 180° no matter the shape of the triangle.	C; ΔABC; A; B
In an **equilateral triangle,** all three sides are of equal length and all three angles are of equal measure. Each angle in an equilateral triangle has a measure of 60°.	D; 60°; 60°; 60°; E; F
An **isosceles triangle** has two (or more) sides that are of equal measure. The angles opposite those sides, called **base angles,** also are of equal measure.	C; A; B; base angles
A **scalene triangle** is a triangle in which no two sides have the same length.	C; A; B
A triangle in which one angle is a right angle (90°) is called a **right triangle.** The other two angles are both acute and complementary.	B; C; A; These two acute angles are complementary.
A **quadrilateral** is a four-sided polygon. It has four angles, and the sum of those angles is 360°.	

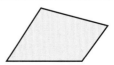

continued

Section 8.5 Geometric Shapes

Concept	Example
A **parallelogram** has two pairs of parallel sides. Each side in a pair has the same length. The angles opposite each other have the same measure. When a parallelogram is not a rectangle, it has a pair of acute angles and a pair of obtuse angles. Also, in a parallelogram, an acute angle and an obtuse angle are supplementary.	The acute angle is supplementary to the obtuse angle.
A **rectangle** is a parallelogram with four right angles.	
A **square** is a rectangle with four sides of equal length.	
A **trapezoid** is a quadrilateral with only one pair of parallel sides. The parallel sides are called **bases** and have different lengths. A trapezoid is not a parallelogram.	
A **regular polygon** is a polygon in which all of the sides have the same length and all of the angles have the same measure.	
A **circle** is a round geometric figure with a center point. The center point is the same distance from every point on the circle. A line segment from the center point to any point on the circle is a **radius.** The word *radius* also refers to the length of a radius.	
Two radii that form a straight angle $(180°)$ are called a **diameter.** A diameter always intersects the center of the circle. The length of a diameter, d, is twice the length of a radius, r: $d = 2 \cdot r$. The length of a radius is half the length of the diameter: $r = \frac{d}{2}$ or $r = \frac{1}{2} \cdot d$.	

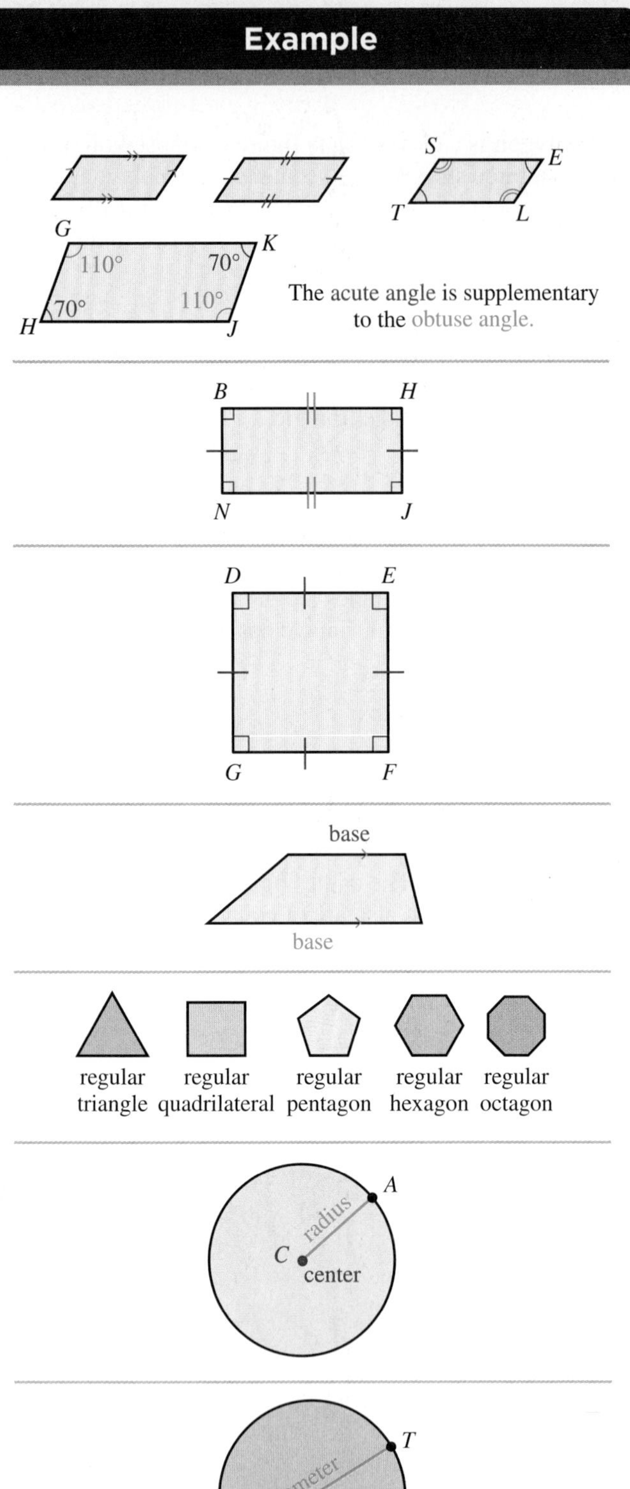

Section 8.6 Perimeter

Concept	Example
The **perimeter** of a polygon is the total length (sum) of its side measures.	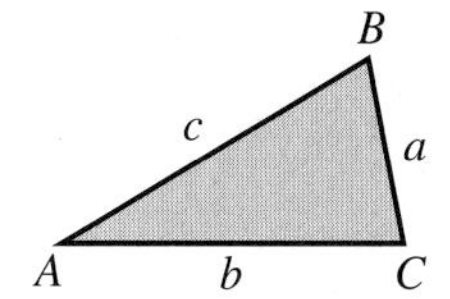 The perimeter of ΔABC is $P = a + b + c$.
The perimeter formula for a rectangle, $P = L + W + L + W$, can be abbreviated as $P = 2 \cdot L + 2 \cdot W$.	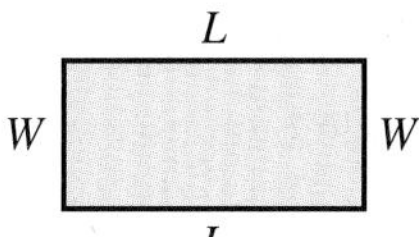
In a regular polygon, the perimeter is the number of sides times the length of one side.	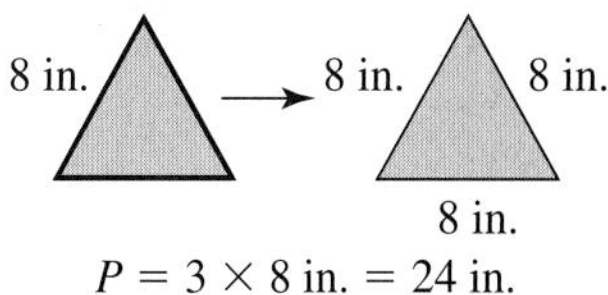 $P = 3 \times 8$ in. $= 24$ in.
The perimeter of the circle, the **circumference,** relies on π. The circumference formula is $C = 2 \cdot \pi \cdot r$ or $C = \pi \cdot d$. We commonly use two approximations for π: $\pi \approx \frac{22}{7}$ and $\pi \approx 3.14$	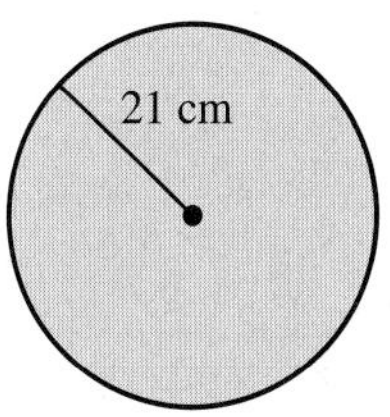 Find the circumference of the circle with radius 21 centimeters. The exact value is $C = 2 \cdot \pi \cdot 7 = 14\pi$. As an approximation, use $\pi \approx \frac{22}{7}$. $C \approx 2 \cdot \frac{22}{7} \cdot \frac{21}{1}$ cm ≈ 132 centimeters

Section 8.7 Area

Concept	Example
The area of a rectangle is the product of its length and width: $A = L \times W$. The dimensions of a rectangle can also be labeled base and height, in which case the area is base $\times$ height. In a rectangle, the height can be measured from anywhere within the rectangle. The height is always *perpendicular* to the **base.**	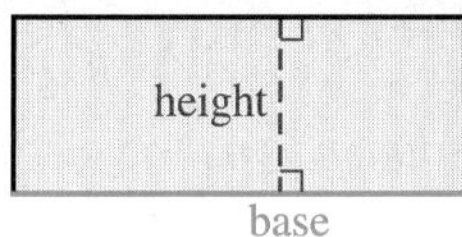

continued

Section 8.7 Area

Concept	Example
The area of a square with side length s is $$A = s \cdot s = s^2.$$	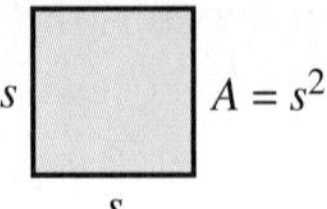
The area of a parallelogram is $$\text{Area} = \text{Base} \times \text{Height}.$$ The height of a parallelogram is perpendicular to both the top side and the bottom side (the base).	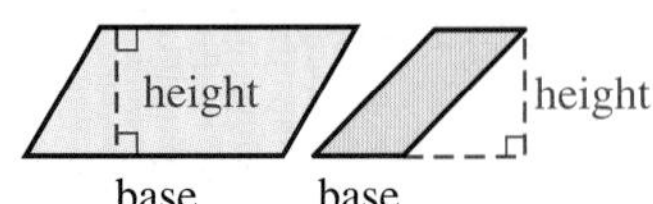
The area of a triangle, where ***b*** = base and ***h*** = height, can be written as follows: $$A = \frac{1}{2} \cdot b \cdot h \text{ or } A = \frac{b \cdot h}{2}$$	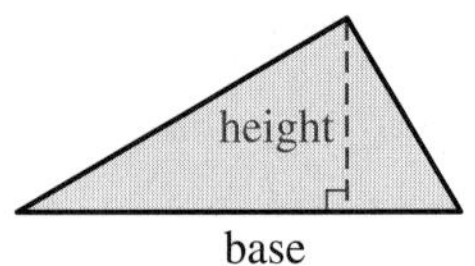
In a *right* triangle, the two perpendicular sides are the base and the height no matter how the triangle might be turned.	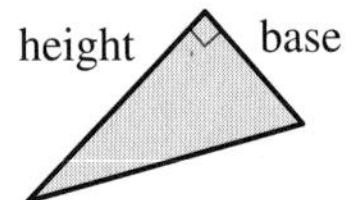
Two different forms of the formula for the area of a trapezoid, where ***a*** = the top base, ***b*** = the bottom base, and ***h*** = height are **1.** $A = \frac{1}{2} \cdot h \cdot (a + b)$ **2.** $A = \frac{a + b}{2} \cdot h$	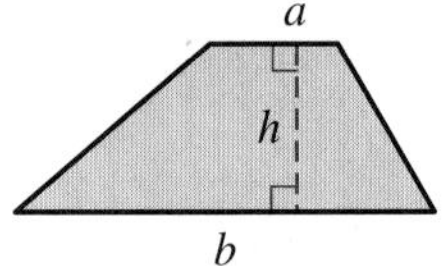
The formula for the area of a circle is $A = \pi \cdot r^2$.	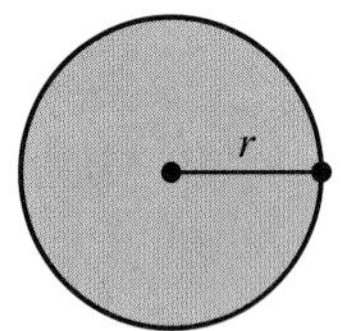

Section 8.8 Volume

Concept	Example
A **solid** is a *three-dimensional* object having three distinct measures—three distinct dimensions—length, depth, and height. The **volume** of a solid indicates the amount the solid can hold. The volume of a rectangular solid is the product of its three dimensions: $$\text{Volume} = \text{Length} \cdot \text{Depth} \cdot \text{Height}$$ $$V = l \cdot d \cdot h \text{ or } V = \text{Base} \cdot h$$	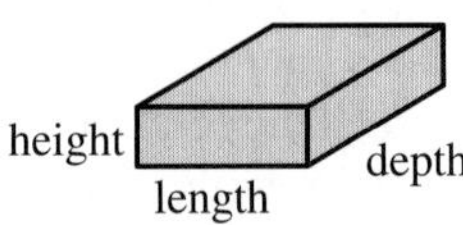 Base = Length × Depth

Section 8.8 Volume

Concept	Example
A **cylinder** is the rounded version of a box. Like a rectangular solid, its volume is Base · height. $V = \text{Base} \cdot h$ Because the base is a circle, the base area is $\pi \cdot r^2$. This makes the volume formula $V = \pi \cdot r^2 \cdot h$.	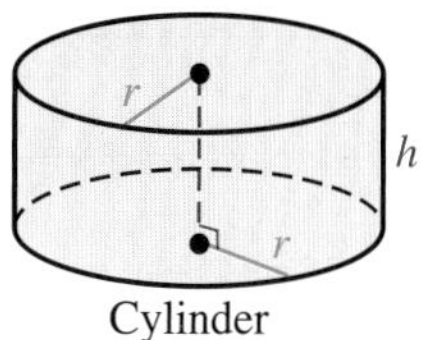 Cylinder
A **cone** has a base that is a circle with radius r. The volume of a cone is $\frac{1}{3}$ the volume of a cylinder. $V = \frac{1}{3} \cdot \text{Base} \cdot h$ Because the base is a circle, the area is $\pi \cdot r^2$. This makes the volume formula $V = \frac{1}{3} \cdot \pi \cdot r^2 \cdot h$.	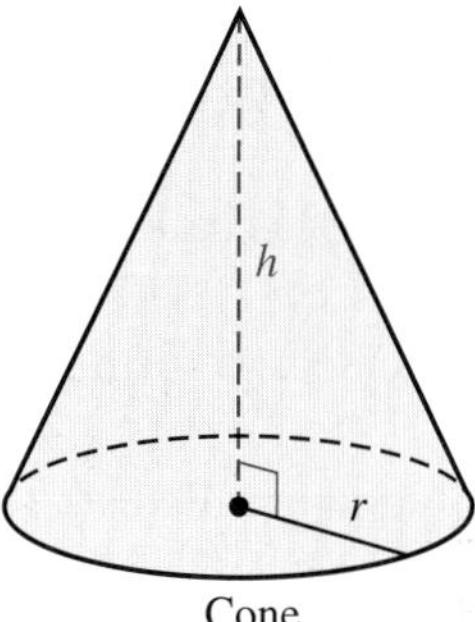 Cone
A **pyramid** is like a cone with a rectangular base. The volume of a pyramid is $\frac{1}{3}$ the volume of a rectangular solid. $V = \frac{1}{3} \cdot \text{Base} \cdot h$ Because the base is a rectangle, the base area is $l \cdot d$. This makes the volume formula $V = \frac{1}{3} \cdot l \cdot d \cdot h$.	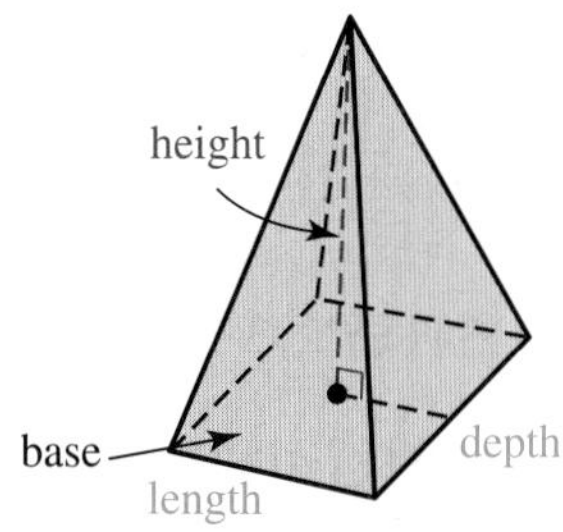 Pyramid
A **sphere**, such as a ball, is the three-dimensional version of a circle. Like a circle, it has many radii, all the same length. The volume of a sphere is $V = \frac{4}{3} \cdot \pi \cdot r^3$	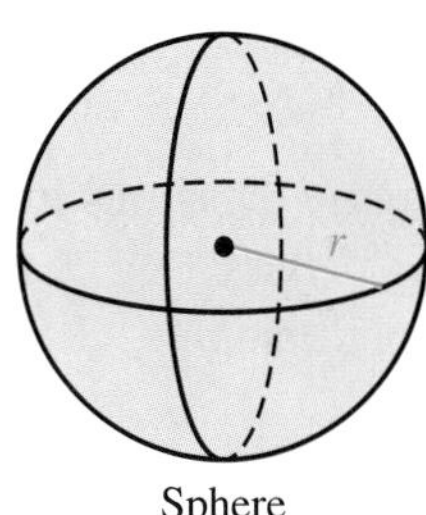 Sphere

Chapter 8 Review Exercises

Fill in the blanks.

1. Pounds, feet, gallons, and meters are examples of units of measure. (8.1)

2. The metric prefix that means "one thousand" is kilo-. (8.2)

3. The metric prefix that means "one-thousandth" is milli-. (8.2)

4. A flat surface that extends forever in each direction is called a(n) plane. (8.4)

5. The point where two rays meet to form an angle is called the vertex. (8.4)

6. A line measuring 180° is called a(n) straight angle. (8.4)

7. Two lines that intersect to form four right angles are perpendicular to each other. (8.4)

8. A four-sided polygon is called a(n) quadrilateral. (8.5)

True or false.

9. A parallelogram contains only one pair of parallel sides. False (8.4)

10. A straight angle measures 180°. True (8.4)

11. Two angles that are complementary to each other must be adjacent to each other. False (8.4)

12. A triangle can have two right angles. False (8.4)

13. In a trapezoid, the parallel sides are never the same length. True (8.5)

14. A square is both a rectangle and a parallelogram. True (8.5)

15. The sum of the angle measures in every triangle is 180°. True (8.5)

16. An octagon has eight angles. True (8.5)

Section 8.1

Convert each measure to inches.

17. 18 ft
216 in.

18. 9 ft 11 in.
119 in.

Convert each measure to feet and inches.

19. 50 in.
4 ft 2 in.

20. 130 in.
10 ft 10 in.

Answer as indicated.

21. A tennis racket is 2 feet 3 inches long. Express this measure in inches.
A tennis racket is 27 inches long.

22. At 86 inches tall, Margo Dydek is the tallest woman professional basketball player in the WNBA. Express this measure in feet and inches.
Source: **www.tallwomen.org**
At 7 feet 2 inches tall, Margo Dydek is the tallest woman professional basketball player in the WNBA.

Add or subtract as indicated. Express each answer in feet and inches.

23. 8 ft 4 in. + 3 ft 8 in.
12 ft

24. 7 ft 4 in. − 5 ft 10 in.
1 ft 6 in.

25. In a one-story house, the distance from the floor to the ceiling is 8 feet 3 inches and the distance from the ceiling to the top of the roof is 4 feet 11 inches. How tall is the house?
The house is 13 feet 2 inches tall.

26. An alligator measures 17 feet 2 inches from its nose to the tip of its tail. Its tail alone is 8 feet 5 inches long. How long is the rest of its body?
The rest of its body is 8 feet 9 inches long.

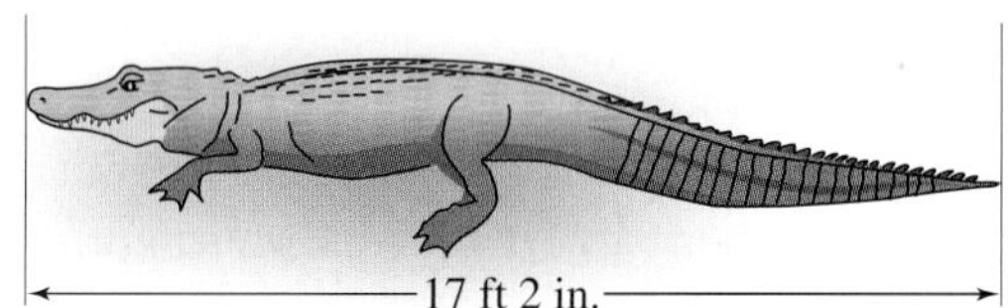

Convert one measure to the other.

27. Convert $15\frac{2}{3}$ yards to feet
47 feet

28. Convert 51 feet to yards.
17 yards

29. Convert 5 quarts to pints.
10 pints

30. Convert 12 ounces to pounds.
$\frac{3}{4}$ pound

31. Convert 10 cups to pints.
5 pints

32. Convert $\frac{7}{16}$ quarts to cups.
$1\frac{3}{4}$ cups

33. Convert 3 pints to fluid ounces.
48 fluid ounces

34. Convert 24 pints to gallons.
3 gallons

35. Convert $2\frac{1}{4}$ cups to fluid ounces.
18 fluid ounces

36. Convert $12\frac{4}{5}$ ounces to pounds.
$\frac{4}{5}$ pound

37. Convert $\frac{3}{8}$ pounds to ounces.
6 ounces

38. Convert $1\frac{45}{100}$ tons to pounds.
2,900 pounds

Convert each measure to ounces.

39. 15 lb
240 oz

40. 5 lb 2 oz
82 oz

Convert each measure to pounds and ounces.

41. 40 oz
2 lb 8 oz

42. 142 oz
8 lb 14 oz

Answer as indicated.

43. A roasting chicken weighs 3 pounds 13 ounces. Express this measure in ounces.
A roasting chicken weighs 61 ounces.

44. A double box of cereal weighs 76 ounces. Express this measure in pounds and ounces.
A double box of cereal weighs 4 pounds 12 ounces.

Add or subtract as indicated. Express each answer in pounds and ounces.

45. 8 pounds 9 ounces + 3 pounds 7 ounces
12 pounds

46. 7 pounds 3 ounces − 5 pounds 7 ounces
1 pound 12 ounces

47. Jena's math book weighs 4 pounds 11 ounces, and her chemistry book weighs 6 pounds 9 ounces. What is the total weight of these two books?
The total weight of these two books is 11 pounds 4 ounces.

48. A large block of cheese weighs 17 pounds 3 ounces. Ralph cuts off a portion that weighs 3 pounds 12 ounces. How much does the remaining block of cheese weigh?
The remaining block of cheese weighs 13 pounds 7 ounces.

Section 8.2

Convert the given unit of measure into its base unit—meters, grams, or liters.

49. 3.08 dag
30.8 g

50. 6.8 kg
6,800 g

51. 870 cm
8.7 m

52. 2.4 dl
0.24 L

Convert the given unit of measure to the measure shown.

53. 2 centimeters to millimeters
20 mm

54. 0.9 decigrams to centigrams
9 cg

55. 312 milliliters to deciliters
3.12 dL

56. 1.2 hectoliters to kiloliters
0.12 kL

57. 3.6 hectometers to kilometers
0.36 km

58. 0.82 kilometers to dekameters
82 dam

Write a sentence with the requested measurement.

59. Jakhil ran 0.4 kilometers in one minute. Express this number in meters.
Jakhil ran 400 meters in one minute.

60. A hummingbird weighs 1.6 grams. Express this number in milligrams.
A hummingbird weighs 1,600 milligrams.

Add or subtract as indicated. Convert each measure to the one that is underlined.

61. 14.2 grams + 591 centigrams
2,011 centigrams

62. 52 deciliters − 395 milliliters
48.05 deciliters

Solve each application. Write a sentence answering the question.

63. In preparation for a medical test, Martin had to drink 2.25 liters of a sugary liquid and 850 milliliters of water. How many total liters did Martin drink for his test?
Martin drank 3.1 liters for his test.

64. Foofi, a toy poodle, weighed 2.95 kilograms on her first birthday. On her second birthday, she weighed 3.43 kilograms. How many grams did Foofi gain during the year?
Foofi gained 480 grams during the year.

Section 8.3

Use an equivalence fraction to write each measure in the terms requested.

65. Convert 14 cups to quarts.
$3\frac{1}{2}$ quarts

66. Convert $2\frac{1}{4}$ pounds to ounces.
36 ounces

67. Convert 54 inches to feet.
$4\frac{1}{2}$ feet

68. Convert 27 kilograms to grams.
27,000 grams

69. Convert 95 hectometers to meters.
9,500 meters

70. Convert 120 deciliters to liters.
12 liters

Convert each measure to its alternate U.S. or metric form. Write the answer as a decimal approximation. Write improper fractions as mixed numbers for U.S. measures and as decimals for metric measures.

71. Convert 22 inches to centimeters.
55 centimeters

72. Convert 60 centimeters to inches.
24 inches

73. Convert $9\frac{1}{2}$ liters to quarts.
10 quarts

74. Convert $7\frac{1}{2}$ gallons to liters.
28.5 liters

75. Convert 35 grams to ounces.
$1\frac{1}{4}$ ounces

76. Convert $16\frac{2}{3}$ yards to meters.
15 meters

Convert each measure to its alternate U.S. or metric form. Write improper fractions as mixed numbers for U.S. measures and as decimals for metric measures.

77. A sack of potatoes weighs 20 pounds. Express this number in kilograms.
The sack of potatoes weighs about 9.1 kilograms.

78. Jorge is 6 feet tall. (a) How many inches tall is Jorge? (b) Approximately how many centimeters tall is Jorge?
(a) Jorge is 72 inches tall. (b) Jorge is about 180 centimeters tall.

Section 8.4

Given the measure of two of the angles, find the measure of the third angle.

79. $\angle FRB = 18.6°$ and $\angle KRF = 29.8°$

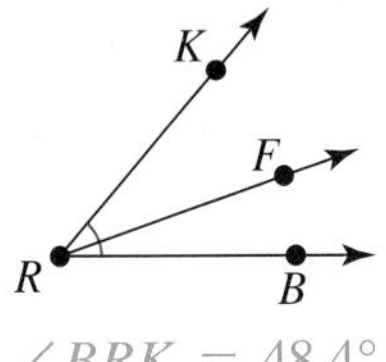

$\angle BRK = 48.4°$

80. $\angle QME = 165.3°$ and $\angle QMD = 67.8°$

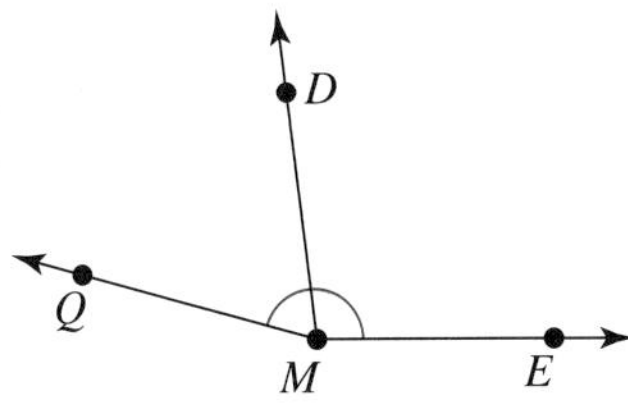

$\angle EMD = 97.5°$

81. $\angle WFZ$ and $\angle BLN$ are supplementary. Given $\angle WFZ = 125.2°$, find $\angle BLN$.

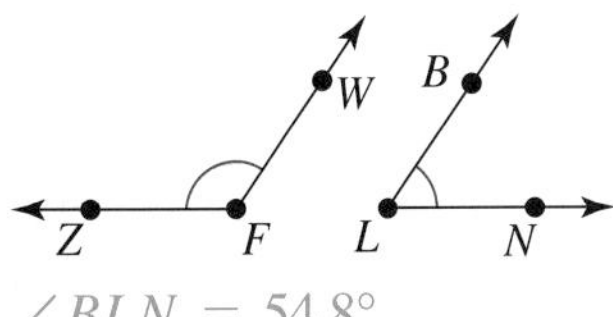

$\angle BLN = 54.8°$

82. $\angle QSA$ and $\angle RFX$ are complementary. Given $\angle QSA = 62.5°$, find $\angle RFX$.

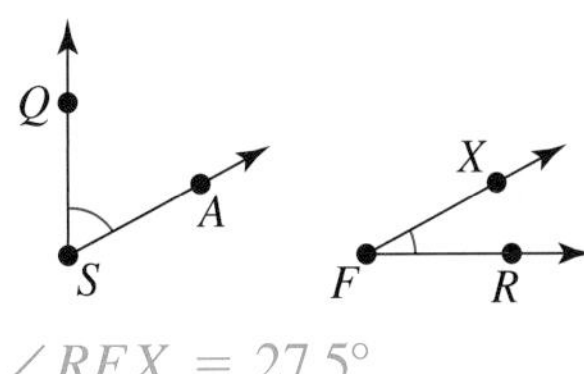

$\angle RFX = 27.5°$

Answer as indicated.

83. Given $\angle 3 = 32°$, find the measures of the other three angles.

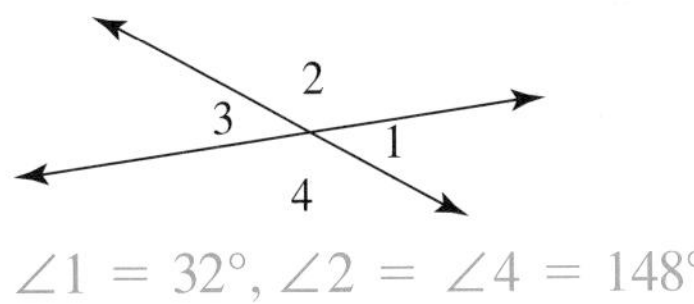

$\angle 1 = 32°, \angle 2 = \angle 4 = 148°$

84. In $\triangle ABC$, given $\angle A = 25°$ and $\angle B = 103°$, find the measure of $\angle C$.
$\angle C = 52°$

Section 8.5

Based on the side measures given, decide whether each triangle is equilateral, isosceles, or scalene.

85.

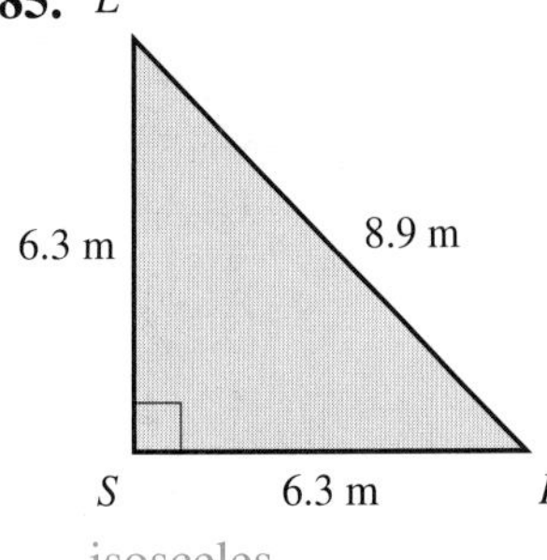

isosceles

86.

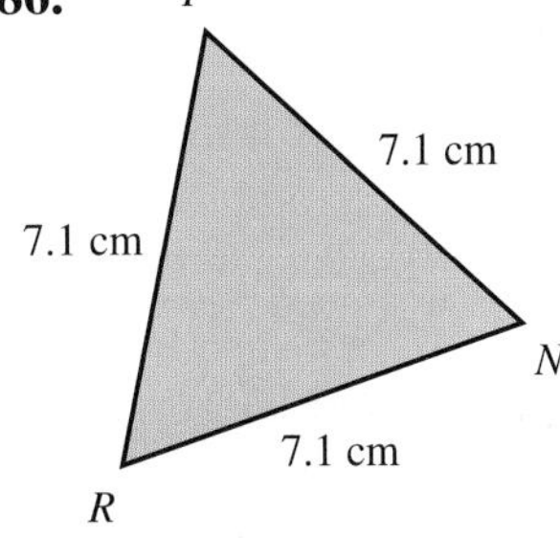

equilateral and isosceles

In each right triangle, given the measure of one acute angle, find the measure of the other acute angle.

87. $\angle I = 59.4°$

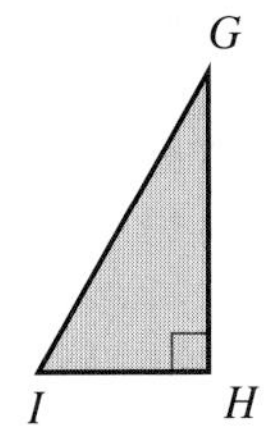

$\angle G = 30.6°$

88. $\angle L = 23.8°$

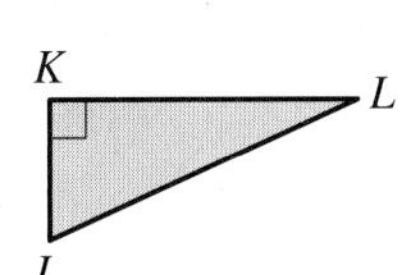

$\angle J = 66.2°$

In the parallelogram *NOPQ*, find the measures of the other three angles and the measures of the other two sides.

89. $\angle N = 40°$; $\overline{NQ} = 5\text{ cm}$; $\overline{NO} = 9\text{ cm}$

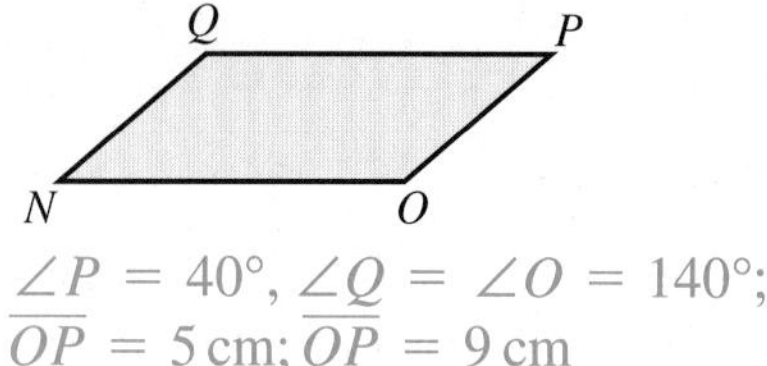

$\angle P = 40°$, $\angle Q = \angle O = 140°$; $\overline{OP} = 5\text{ cm}$; $\overline{QP} = 9\text{ cm}$

Given the radius, *r*, of a circle, find the diameter; or given the diameter, *d*, of a circle, find the radius.

90. $r = 9\text{ ft}$
$d = 18\text{ ft}$

91. $d = 7\frac{1}{3}\text{ in.}$
$r = 3\frac{2}{3}\text{ in.}$

92. $d = 2.3\text{ m}$
$r = 1.15\text{ m}$

Section 8.6

Find the perimeter of the polygon described.

93. Leon's pool is in the shape of a trapezoid. What is the perimeter of Leon's pool?

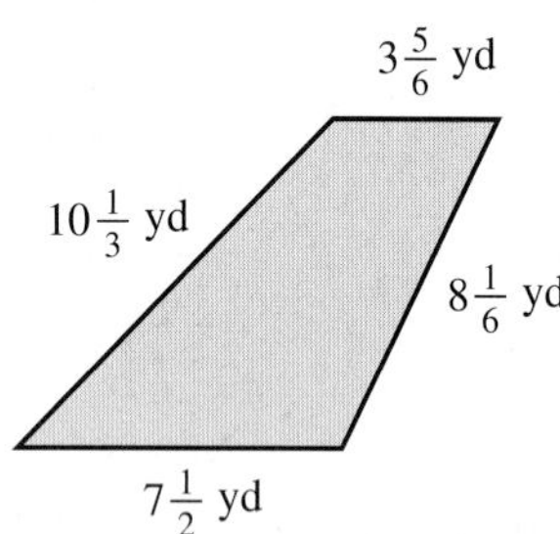

The perimeter of Leon's pool is $29\frac{5}{6}$ yards.

94. The ramp support for model wooden car races is in the shape of a triangle. What is the perimeter of this support structure?

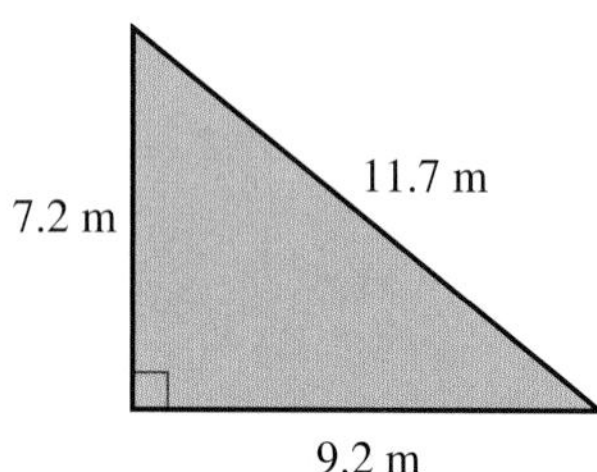

The perimeter of this support structure is 28.1 meters.

95. Tai's garage is in the shape shown below. First, find the values of *X* and *Y* and then calculate its perimeter.

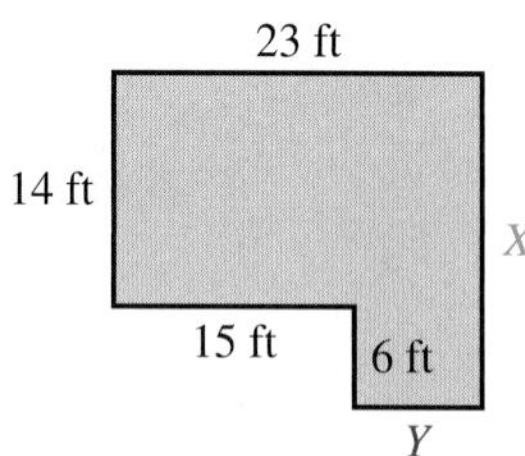

$X = 20\text{ ft}$, $Y = 8\text{ ft}$; the perimeter is 86 feet.

96. The dance floor at the Jitterbug Coffeehouse is in the shape of a rectangle. What is perimeter of the dance floor?

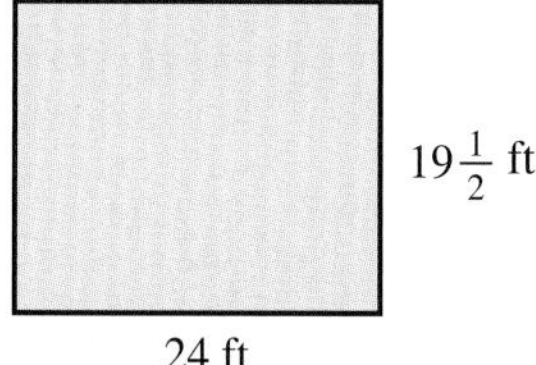

The perimeter of the dance floor is 87 feet.

Find the perimeter of the described figure. (Write all improper fraction answers as mixed numbers.)

97. An equilateral triangle; one side is $2\frac{1}{3}$ ft.
$P = 7\text{ ft.}$

98. A regular octagon; one side is 3.75 yd.
$P = 30\text{ yd.}$

Given the perimeter of the regular polygon, find the length of each side. (Write all improper fraction answers as mixed numbers.)

99. An equilateral triangle; $P = 42\text{ cm}$
Each side $= 14\text{ cm}$

100. A regular hexagon; $P = 210\text{ in.}$
Each side $= 35\text{ in.}$

Solve each application. Write a sentence answering the question.

101. Marcus wants to fence in a circular garden that has a radius of 21 feet. How much fencing does Marcus need to fit around the garden? Use $\pi \approx \frac{22}{7}$.
Marcus needs about 132 feet of fencing to fit around the garden.

102. In her living room, Janet has a circular window with a diameter of 30 inches. For better insulation, she wants to place weather stripping around the window. How many inches of weather stripping is needed to go around the window? Use $\pi \approx 3.14$ and round the result to the nearest tenth.
About 94.2 inches of weather stripping is needed to go around the window.

Section 8.7

Find the area of each rectangle.

103. $b = 5$ yd
$h = 4$ yd
$A = 20\text{ yd}^2$

104. $h = 2.5$ m
$b = 1.2$ m
$A = 3\text{ m}^2$

Find the area of each square.

105. side $= \frac{3}{8}$ in.
$A = \frac{9}{64}\text{ in.}^2$

106. side $= 3\frac{1}{2}$ ft
$A = 12\frac{1}{4}\text{ ft}^2$

Find the area of each polygon.

107.

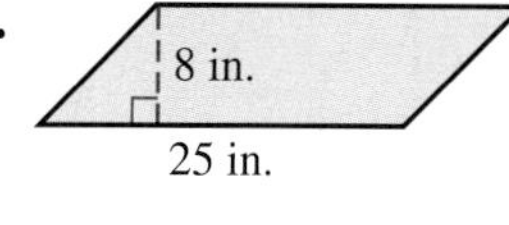

$A = 200\text{ in.}^2$

108.

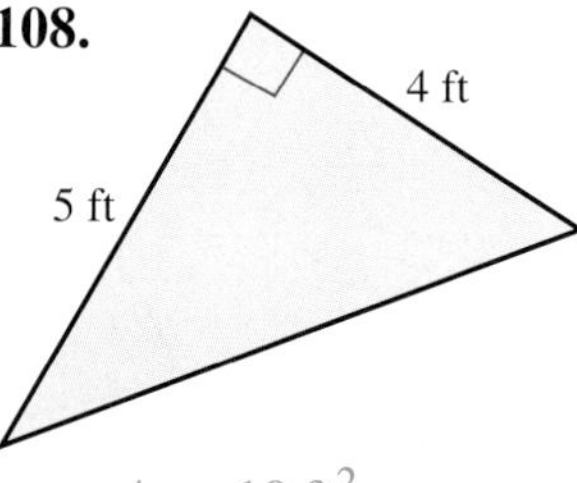

$A = 10\text{ ft}^2$

109.

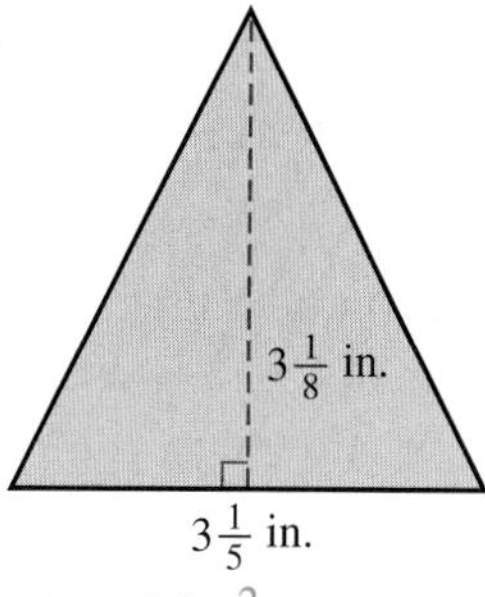

$A = 5\text{ in.}^2$

110.

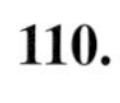
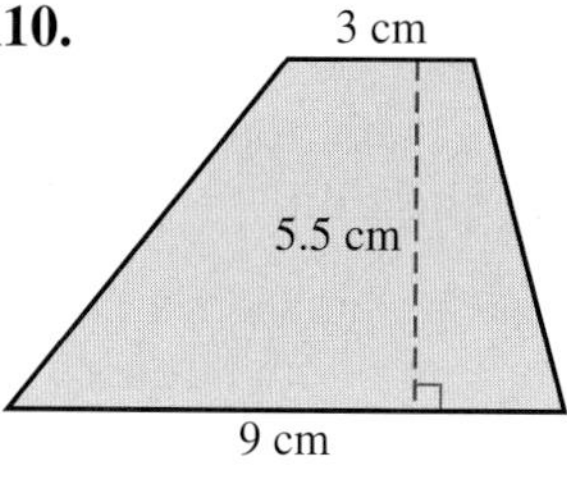

$A = 33\text{ cm}^2$

Approximate the area of each circle with the given radius or diameter. Use $\pi \approx 3.14$ and round the result to the tenths place.

111. $r = 2$ ft
$A \approx 12.6\text{ ft}^2$

112. $d = 3.2$ cm
$A \approx 8.0\text{ cm}^2$

Answer as indicated.

113. A round pond has the basic shape of a circle with a diameter of 60 feet. What is the area of the surface of this pond?
The area of the surface of this pond is about 2,826 feet.

114. Erica's living room is in the shape shown here. What is the area? (All measures are in feet.)

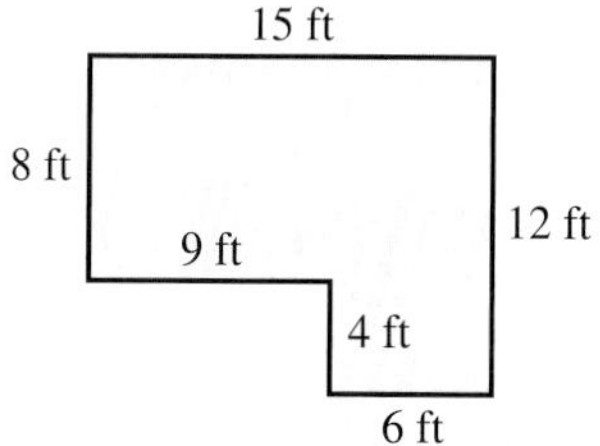

The area of Erica's living room is 144 ft^2.

115. This diagram shows a circle within a square. Find the area of the shaded part.

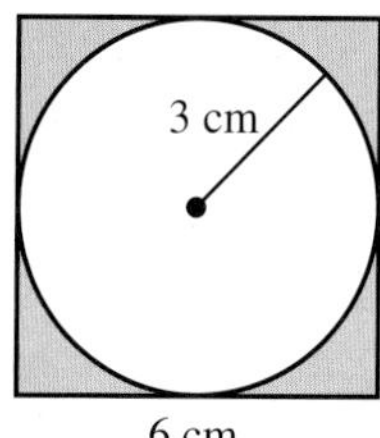

$A \approx 7.74\text{ cm}^2$

Section 8.8

Find the volume of the box.

116. $l = 9$ in., $d = 12$ in., $h = 5$ in.

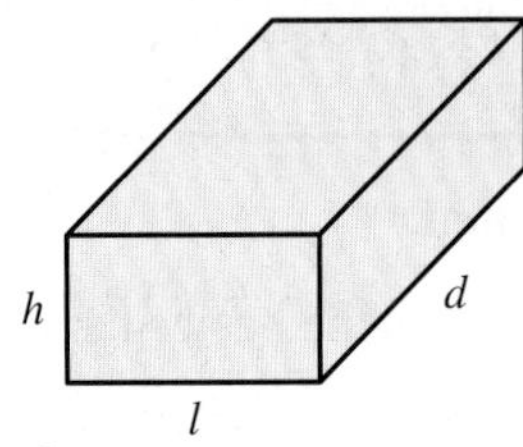

$V = 540$ in.^{3}

Find the volume of the sphere. Use $\pi \approx 3.14$ and round the result to the nearest tenth.

117. $r = 2$ ft

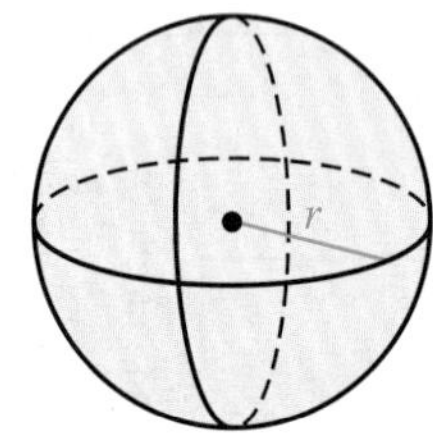

$V \approx 33.5$ ft^3

Find the volume of the cylinder. Use $\pi \approx 3.14$ and round the result to the nearest tenth.

118. $r = 5$ in., $h = 4$ in.

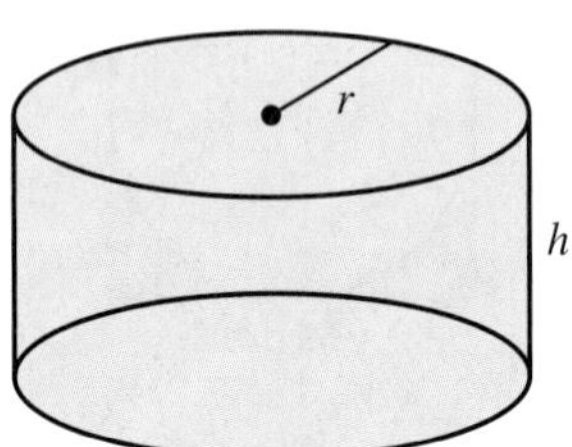

$V \approx 314$ in.^{3}

Find the volume of the cone. Use $\pi \approx 3.14$ and round the result to the nearest tenth.

119. $r = 1.5$ ft, $h = 3$ ft

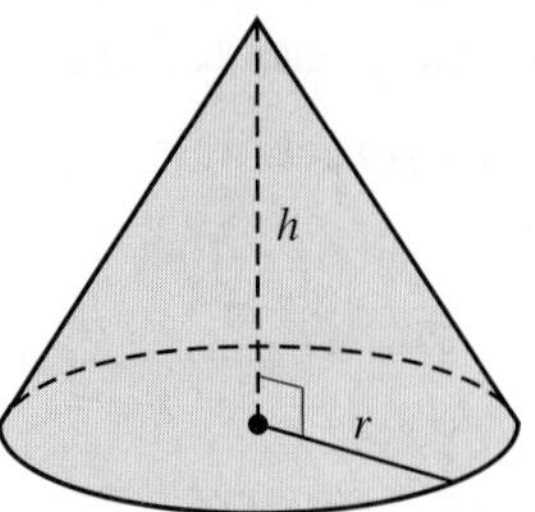

$V \approx 7.1$ ft^3

Find the volume of the pyramid.

120. $l = 6$ m, $d = 8$ m, $h = 10$ m

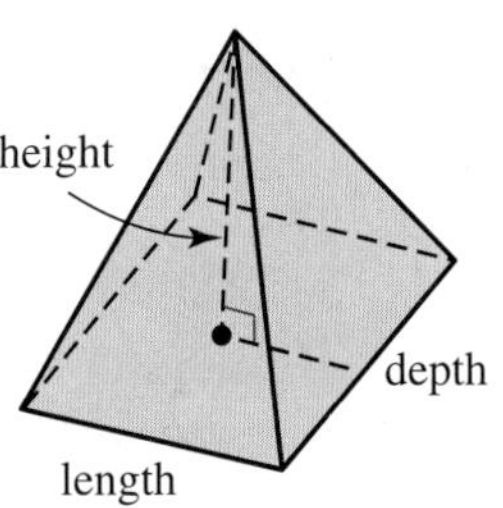

$V = 160$ m^3

Solve each application. Write a sentence answering the question.

121. An amusement park has a display that is a pyramid with a square base 20 yards by 20 yards. The pyramid is 15 yards high. What is the volume of this pyramid?
The volume of this pyramid is 2,000 cubic yards.

122. The storage area in a U-Drive-It truck is 8 feet wide, 7 feet high, and 12 feet long. What is the volume of this storage area?
The volume of this storage area is 672 cubic feet.

Chapter 8 Test

True or false.

1. Two line segments that meet to form a right angle are perpendicular to each other. True
2. Two angles that are vertical angles always have the same measure. True
3. In a parallelogram, all four sides are parallel to each other. False
4. In a right triangle, one of the angles may be obtuse. False

Add or subtract as indicated. Express each answer in feet and inches.

5. 9 ft 6 in. + 8 ft 11 in.
   18 ft 5 in.
6. 20 ft − 6 ft 8 in.
   13 ft 4 in.

Convert one measure to the other.

7. Convert 6 ft 9 in. to inches.
   81 inches
8. Convert 142 in. to feet and inches.
   11 feet 10 inches
9. Convert 3 pounds to ounces.
   48 ounces
10. Convert $11\frac{1}{4}$ feet to yards.
    $3\frac{3}{4}$ yards

Convert the given unit of measure into its base unit—meters, grams, or liters.

11. 28 cL
    0.28 L
12. 0.45 hm
    45 m
13. 91 mm
    0.091 m
14. 0.89 kg
    890 g

Add or subtract as indicated. Convert each measure to the one that is underlined.

15. 28.7 <u>millimeters</u> + 0.64 decimeters
    92.7 mm
16. 900 dekagrams − 3.75 <u>kilograms</u>
    5.25 kg

Convert one measure to the other.

You may use these equivalencies:

5 mi ≈ 8 km, 11 lb ≈ 5 kg, 5 gal ≈ 19 L

17. Convert 20 miles to kilometers.
    32 kilometers
18. Convert $6\frac{1}{4}$ kilograms to pounds.
    $13\frac{3}{4}$ pounds
19. Elisa put 9.5 liters of gas into her motorcycle. Convert this number to gallons.
    Elisa put 2.5 gallons of gas into her motorcycle.

Given the measure of two of the angles, find the measure of the third angle.

20. $\angle VTR = 77.6°$ and $\angle VTW = 15.7°$

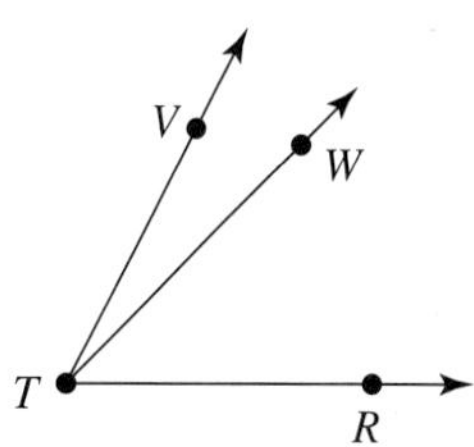

$\angle RTW = 61.9°$

21. $\angle PAS = 92.1°$ and $\angle PAX = 52.4°$

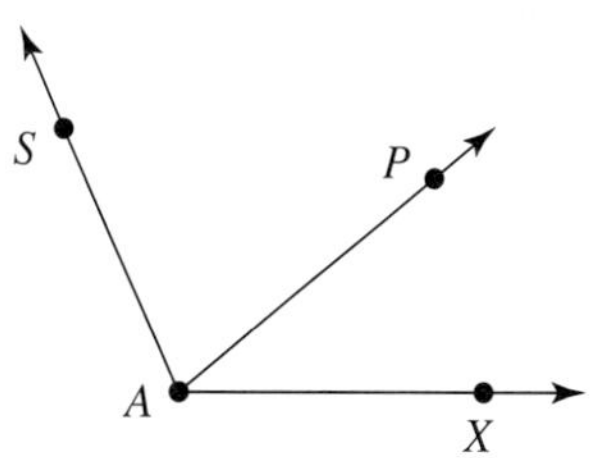

$\angle XAS = 144.5°$

Find the measure of the angle.

22. $\angle DSZ$ and $\angle WXI$ are supplementary. Given $\angle DSZ = 68.3°$, find the measure of $\angle WXI$.

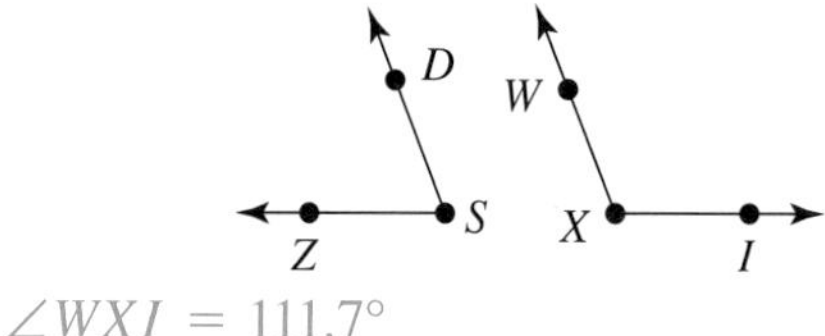

$\angle WXI = 111.7°$

23. $\angle RVT$ and $\angle LYM$ are complementary. Given $\angle RVT = 31.1°$, find the measure of $\angle LYM$.

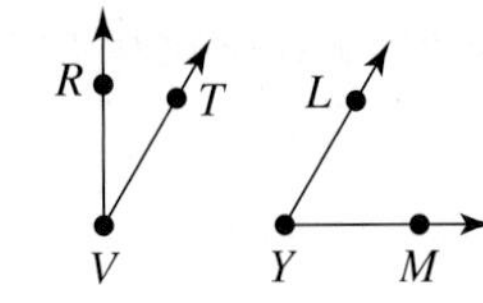

$\angle LYM = 58.9°$

Solve.

24. In $\angle ABC$, $\angle A = 47.2°$ and $\angle B = 26.9°$. Find the measure of $\angle C$.

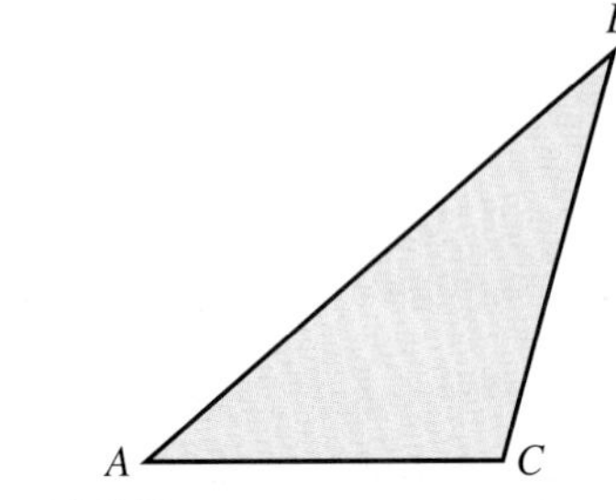

$\angle C = 105.9°$

25. Siobhán painted a picture on a canvass in the shape of a parallelogram, as shown. What is the perimeter of the canvass?

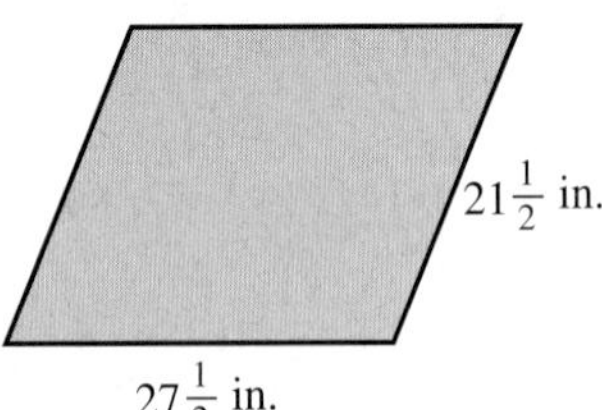

The perimeter of the canvass is 98 inches.

26. Find the perimeter of a square with side measure 7.5 cm.
$P = 30$ cm

27. Find the area of a square with side measure $2\frac{1}{3}$ ft.
$A = 5\frac{4}{9}\text{ ft}^2$

28. A manhole cover is in the shape of a circle and has a radius of $17\frac{1}{2}$ inches. Approximate the circumference of the manhole cover. Use $\pi \approx \frac{22}{7}$.
$C \approx 110$ in.

29. A window is in the shape of a regular octagon and has a perimeter of $10\frac{2}{3}$ ft. What is the length of each side of the window?
Each side $= 1\frac{1}{3}$ ft

30. A piece of tile is in the shape of the triangle shown here. What is the area of this piece of tile?

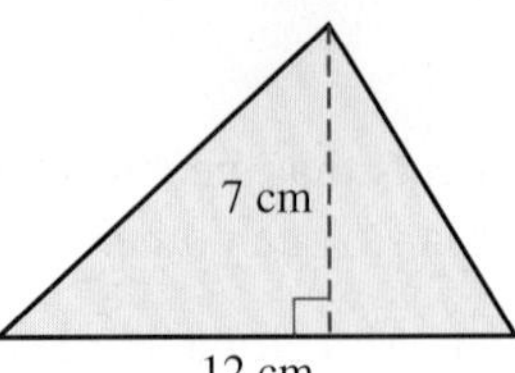

$A = 42\text{ cm}^2$

31. The kitchen at the Kountry Folks Restaurant is in the shape shown here. What is the area of this kitchen?

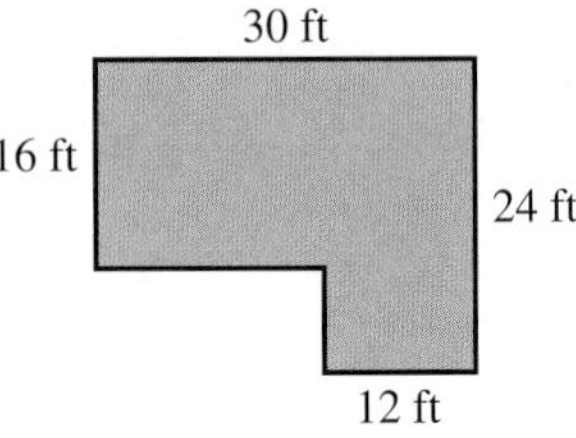

The area of this kitchen is 576 square feet.

32. A can of tomato sauce is in the shape of a cylinder. Its radius is 4 cm, and its height is 7.5 cm. Find the volume of the can. Use $\pi \approx 3.14$.

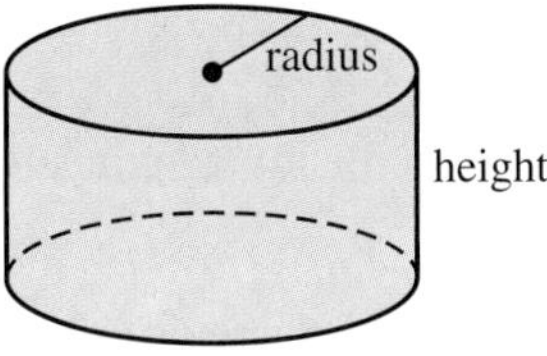

$V \approx 376.8\text{ cm}^3$

33. A souvenir pyramid is made of solid brass. The length of the pyramid is 3 inches. Its depth is $2\frac{2}{3}$ in., and its height is $3\frac{3}{4}$ in. Find the volume of this pyramid.

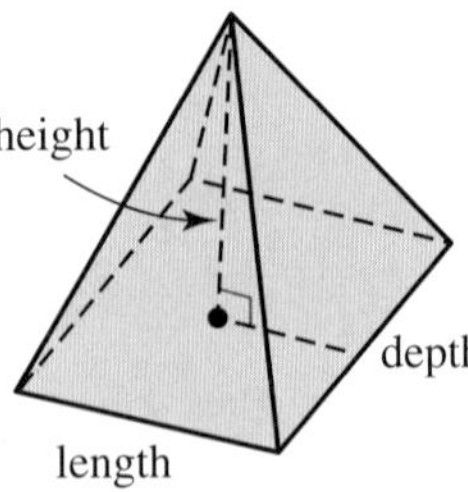

$V = 10\text{ in.}^3$

Chapters 1–8 Cumulative Review

Answer as indicated

1. Which property is being demonstrated?
   a. $8 \cdot (5 \cdot 6) = (8 \cdot 5) \cdot 6$
   Associative Property of Multiplication
   b. $6 \cdot 1 = 6$
   Identity for Multiplication

2. Expand 5^4 and find its value.
   625

3. A youth-sized basketball court is 63 feet long and 35 feet wide. What is the area of the court?

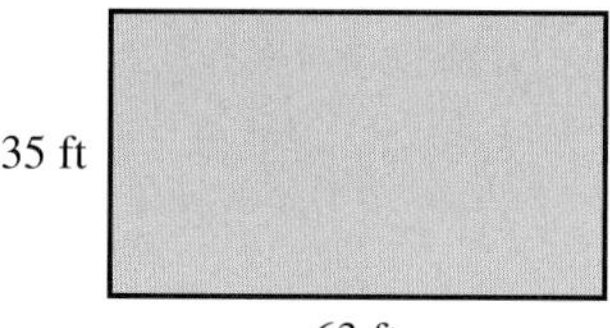

The area of the court is 2,205 square feet.

Find the prime factorization of the following numbers. Write the answer two ways: with and without exponents.

4. 280
   $2 \cdot 2 \cdot 2 \cdot 5 \cdot 7$
   $2^3 \cdot 5 \cdot 7$

5. 396
   $2 \cdot 2 \cdot 3 \cdot 3 \cdot 11$
   $2^2 \cdot 3^2 \cdot 11$

6. Dr. Sternburg has been given a budget of $17,580 for the six biology classes in his department. If the budget is divided evenly, how much will each class receive?
   Each class will receive $2,930.

Evaluate the following according to the replacement value given.

7. $5 \cdot x - x \div y$; replace x with 6 and y with 2.
   27

8. $\dfrac{c^2 - 4}{c + 2}$; replace c with 4.
   2

Evaluate each expression.

9. $10 - 12 + (-7) - (-5)$
   -4

10. $-3 \cdot (-2) \cdot (-5) \cdot 4$
    -120

Evaluate each expression using the order of operations. Show all steps.

11. $\sqrt{5^2 - 4^2} - 10$
    -7

12. $|-9 + 3| + |-6|$
    12

Simplify by combining like terms if possible.

13. $-8y - (-7y) + 2y$ y

Solve. Make sure you check your answer.

14. $-5y + 10 = 3y - 14$ $y = 3$

Set up the legend, draw a diagram, identify the formula, and write and solve the equation. Make sure you write the answer as a sentence.

15. On Monday, Star Coffee sold a total of 360 cups of coffee in two sizes: small and large. If the coffee shop sold 80 more large cups than small cups, how many cups of each size did it sell that day?
    It sold 140 small cups and 220 large cups.

Evaluate. Simplify completely. Write improper fractions as mixed numbers.

16. $7\frac{1}{2} \times 1\frac{1}{9}$
    $8\frac{1}{3}$

17. $8\frac{3}{4} \div 1\frac{7}{8}$
    $4\frac{2}{3}$

18. $4\frac{4}{9} + 3\frac{5}{6}$
    $8\frac{5}{18}$

19. $6\frac{1}{12} - 4\frac{5}{12}$
    $1\frac{2}{3}$

20. two-thirds of $8\frac{1}{4}$
    $5\frac{1}{2}$

21. $-\dfrac{7}{6} \div \left(-\dfrac{14}{9}\right)$
    $\frac{3}{4}$

22. $\dfrac{7}{12} - \left(-\dfrac{1}{6}\right)$
    $\frac{3}{4}$

23. $-\dfrac{41}{72} + \dfrac{11}{72}$
    $-\frac{5}{12}$

Evaluate. Simplify completely.

24. $\dfrac{-2 \cdot 6}{-15 + (-3)}$ $\frac{2}{3}$

Find the area of this triangle.

25.

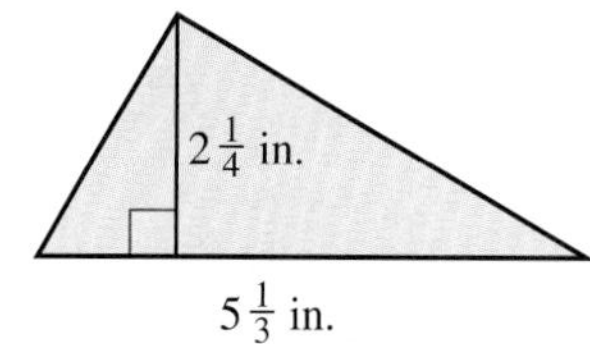

The area of this triangle is 6 square inches.

Solve each equation.

26. $-15 = -18y$
$y = \frac{5}{6}$

27. $\frac{5x}{14} = \frac{10}{21}$
$x = 1\frac{1}{3}$

Solve each application.

28. On Saturday at the Kansas State Fair, 5 members of the local Elks Club agreed to staff their booth for a total of $11\frac{3}{4}$ hours. If this time was broken up into 5 equal shifts, how long was each shift?
Each shift was $2\frac{7}{20}$ hours.

29. NaShana has a piece of scrap lumber that is $52\frac{1}{8}$ inches long. She wants to use that piece as a cross brace for the gate she is building for her fence. If the cross brace is to be $49\frac{3}{4}$ inches long, how much does she need to cut off the end of the scrap lumber?
She needs to cut $2\frac{3}{8}$ inches off the end.

Answer as indicated.

30. Write $\frac{218}{100}$ as a decimal.
2.18

31. Write 0.09 as a fraction.
$\frac{9}{100}$

32. Round $28.096125 to the nearest penny.
$28.10

33. Round $89.53 to the nearest dollar.
$90

34. Gloria's grade point average is 2.948065. Round this to the nearest tenth.
2.9

35. Add 1.076, 3.9, and 8.85.
13.826

36. DuJuan paid (including tax) $23.18 for a book, and he gave the cashier $30. How much change did he get back?
He got back $6.82 in change.

37. Perform the indicated operation.
a. 0.083×100
8.3
b. $9.84 \div 1.2$
8.2

38. Find the decimal equivalent of $\frac{5}{16}$.
0.3125

39. Last night, Tara worked 6.5 hours at her job as a server and earned $72.15 in tips. How much did Tara earn in tips per hour?
Tara earned $11.10 in tips per hour.

40. Karin made a small custom canvas on which to create her next painting. The canvas measures 18.5 centimeters long and 10.2 centimeters wide. What is the area of this canvas?

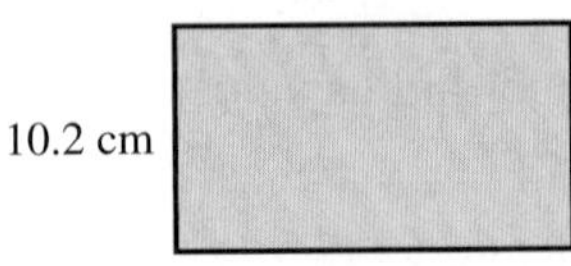

The area of this canvas is 188.7 square centimeters.

41. The cover on Todd's hot tub is circular and has a radius of 1.4 meters. What is the area of the top of this cover? *(Round the answer to the nearest hundredth.)*

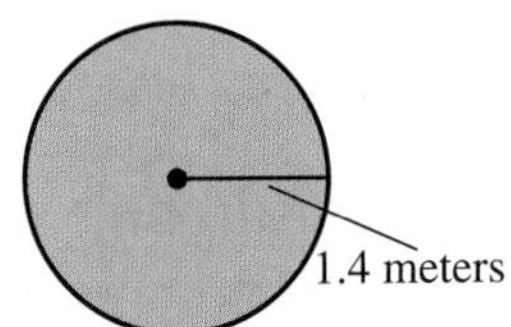

The area of the top of the cover is approximately 6.15 square meters.

42. Write the ratio 0.21 liters : 3.5 liters in fractional form and simplify it completely.
$\frac{3}{50}$

43. Simplify $\frac{12.88 \text{ dollars}}{1.4 \text{ hours}}$ to a unit rate and write the unit rate in words.
9.2 dollars per hour

44. A 16-pound bag of Nutrichow dog food costs $25.60, and a 12-pound bag of Kibblefest dog food costs $18.60.
a. What is the unit price of the Nutrichow?
$1.60 per pound
b. What is the unit price of the Kibblefest?
$1.55 per pound
c. Which is the better value?
Kibblefest is the better value.

45. Solve for x in this proportion: $\frac{16}{20} = \frac{28}{x}$.
$x = 35$

46. Write each percent as a decimal.
  a. 9% 0.09
  b. 1.5% 0.015

47. Write each decimal as a percent.
  a. 0.6 60%
  b. 0.04 4%

Solve each problem. Write a sentence answering the question.

48. Yesterday, Tomás invested \$3,000 in an account that earns simple interest at a rate of 6% per year. How much interest will he earn
  a. in one year? \$180
  b. in 8 months? \$120

49. Employees at Mike's Bikes can put together 10 bikes in 6 hours. How many hours will it take them to put together 35 bikes?
It will take them 21 hours to put together 35 bikes.

50. 60% of what number is 51?
60% of 85 is 51.

51. As an employee at Soy House, Mika was earning \$8 per hour. When she was promoted to assistant manager, her wage went up to \$10 per hour. What is the percent increase of this hourly wage?
The percent increase of this hourly wage is 25%.

Evaluate each expression using a rule of exponents.

52. -5^1
-5

53. 4^0
1

Write each product as one base with one exponent.

54. $p^4 \cdot p$
p^5

55. $x^3 \cdot x^3$
x^6

Evaluate.

56. $2x^2 + x - 5$ when $x = -2$
1

Write all polynomial answers in descending order and combine like terms.

Simplify.

57. Simplify $3x - x^2 - 7x + 4x^2$.
$3x^2 - 4x$

Subtract.

58. $(5q + 8 - 6q^2) - (3 - 4q^2 + 6q)$
$-2q^2 - q + 5$

Multiply.

59. $(-5n^2)(3n^3)$
$-15n^5$

60. $-4q(2 - 3q^2 + 5q)$
$12q^3 - 20q^2 - 8q$

61. $(5x + 1)(4x - 2)$
$20x^2 - 6x - 2$

62. $(y - 3)^2$
$y^2 - 6y + 9$

Simplify.

63. $\dfrac{m^7}{m^2}$
m^5

64. $\dfrac{k^{12}}{k^6}$
k^6

65. $\dfrac{-18p^4}{-3p}$
$6p^3$

66. $\dfrac{25x^6}{-5x^0}$
$-5x^6$

Divide using distribution.

67. $\dfrac{20q^6 + 12q^4 - 4q^2}{4q^2}$
$5q^4 + 3q^2 - 1$

Factor out the GCF from each polynomial.

68. $18y^3 - 6y^2$
$6y^2(3y - 1)$

69. $30m^3 + 10m^2 - 15m$
$5m(6m^2 + 2m - 3)$

Convert one measure to the other.

70. Convert $2\frac{3}{4}$ yards to inches.
99 inches

71. Convert $3\frac{5}{8}$ pounds to ounces.
58 ounces

72. Convert 3.6 milliliters to deciliters.
0.036 deciliters

73. Convert 3.96 kilometers to meters.
3,960 meters

74. A turkey weighs $19\frac{1}{4}$ pounds. Express this number in kilograms.
$19\frac{1}{4}$ pounds $\approx 8\frac{3}{4}$ kilograms

75. The distance between Birmingham, Alabama, and Miami, Florida, is 1,208 kilometers. Express this number in miles.
1,208 kilometers $\approx$ 755 miles

76. Derrick is 5 foot 8 inches tall. Express this number in centimeters.
5 foot 8 inches $\approx$ 170 centimeters

Answer as indicated.

77. $\angle CDE = 124.3°$ and $\angle FDE = 82.6°$. Find $\angle CDF$.

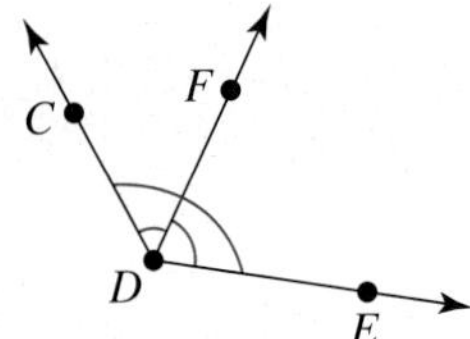

$\angle CDF = 41.7°$

78. A regular hexagon has a perimeter of $8\frac{1}{4}$ in. What is the length of each side of this hexagon?

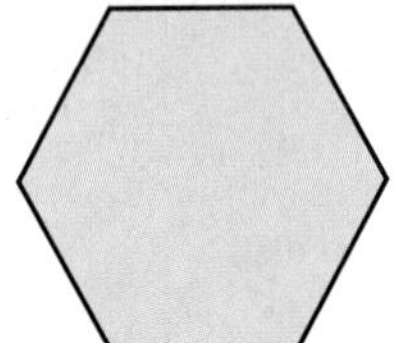

Each side is $1\frac{3}{8}$ inches.

Find the area of each polygon.

79.

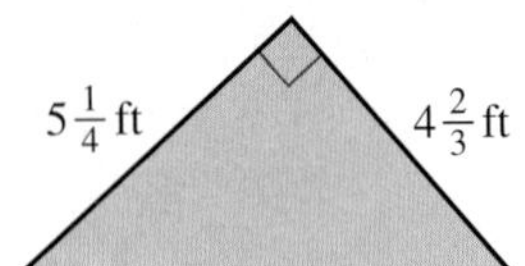

$12\frac{1}{4}$ square feet

80.

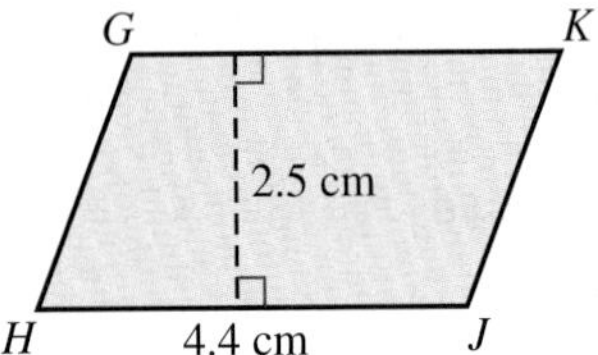

11 square centimeters

A compact disc (CD) has a total diameter of 12 centimeters. In the center is a small hole and a clear plastic piece that have a total radius of 2 centimeters. (Round each answer to the nearest tenth.)

81. What is the area of the silver part of the CD?

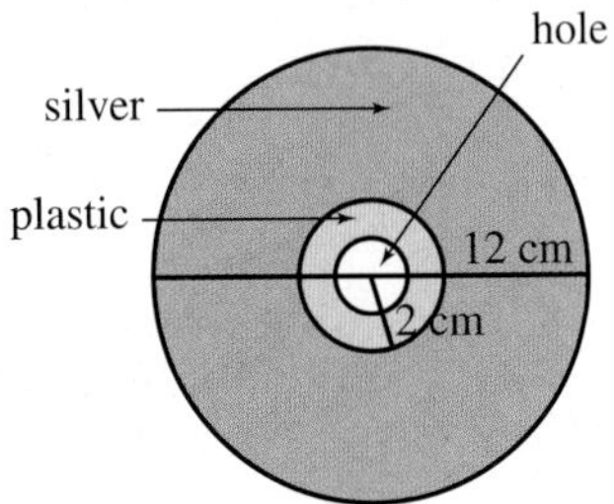

The area is approximately 100.5 sq cm.

82. What is the outer circumference of the CD?
The outer circumference is approximately 37.7 cm.

Solve each application. Write a sentence answering the question. (Round each answer to the nearest tenth.)

83. A coffee can in the shape of a cylinder has a height of 5 inches and a base radius of 2 inches. What is the volume of this can?

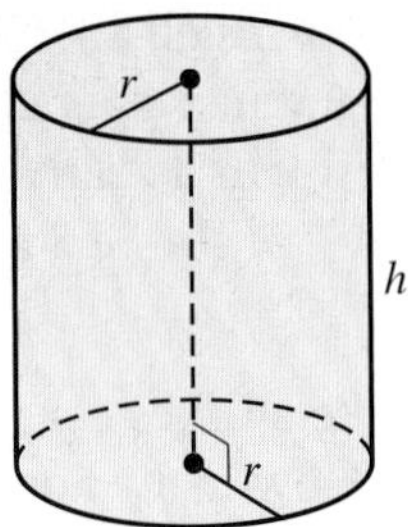

The volume of this can is approximately 62.8 cubic inches.

84. A beach ball has a radius of 12 inches. What is the volume of this beach ball?

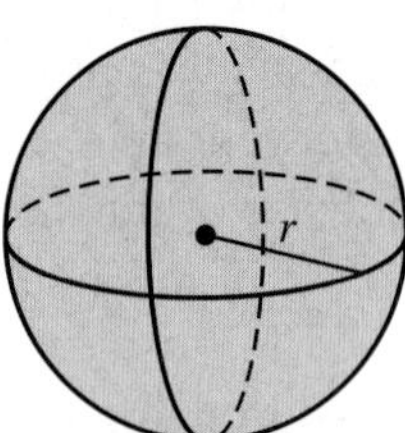

The volume of this beach ball is approximately 7,234.6 cubic inches.

CHAPTER 9

Graphing Lines, Statistics, and Probability

Introduction

Graphing points on a grid or a chart helps us visualize numerical information. In the first part of this chapter (Sections 9.1 and 9.2), we use the foundations of geometry from Section 8.4 to plot points and draw lines in a plane.

In the second part of this chapter, we explore one of the most real-life branches of mathematics, *statistics*. Statistics has been given a bad name because some people use statistics to twist facts and/or alter the conclusions drawn from a given set of data. In fact, there is a joke that 86.2% of all statistics are made up on the spot! If that is the case, you might ask yourself, "Why study statistics at all?" It's important to study statistics so that you can be wise about the statistics you hear and read. In this course, you'll learn about data (pieces of numerical information) and how to organize and interpret them.

The last part of the chapter covers the topic of probability—the study of how likely events are to occur. One foolish man remarked, "I figure the chances of winning the lottery is 50%; either I win or I lose." Do you think he's right? Well, he's right in saying that he'll either win or lose, but just because something has only two possible outcomes doesn't mean that each outcome is equally likely to happen.

No, the real chance of winning the grand prize in the Powerball Mega Millions lottery with a $1 wager is 1 in over 175 million—and that's a statistic that's *not* made up!

Preparation Exercises

Section 2.2 Adding Signed Numbers
Add.

1. $-18 + 9$
-9

2. $-12 + (-8)$
-20

Section 2.3 Subtracting Signed Numbers
Subtract.

3. $-2 - 9$
-11

4. $-7 - (-10)$
3

Section 2.5 The Order of Operations for Signed Numbers
Evaluate.

5. $8 + 20 \div 4$
13

6. $30 \div 3 \cdot 2$
20

Section 2.6 Formulas
Evaluate the formula when $x = -2$.

7. $A = 5x + 8$
$A = -2$

8. $h = 3x - 1$
$h = -7$

Section 3.2 Solving Equations Involving Two Operations
Solve. Check your answer.

9. $2x - 1 = 7$
$x = 4$

10. $1 = 5x - 3$
$x = \frac{4}{5}$

Section 4.4 Multiplying and Dividing Fractions
Multiply and simplify.

11. $\frac{3}{10} \cdot \frac{5}{9}$
$\frac{1}{6}$

12. $\frac{15}{-8} \cdot \frac{-4}{-25}$
$-\frac{3}{10}$

Section 4.7 Adding and Subtracting Unlike Fractions
Add.

13. $\frac{3}{5} + 2$
$2\frac{3}{5}$ or $\frac{13}{5}$

14. $3 + \frac{3}{4}$
$3\frac{3}{4}$ or $\frac{15}{4}$

Section 5.2 Rounding Decimals
Round each decimal to the hundredths place.

15. 0.85307
0.85

16. 0.09841
0.10

Section 5.3 Adding and Subtracting Decimals
Add.

17. 1.408 + 0.63
2.038

18. 0.49 + 2.357
2.847

Subtract.

19. 2.3 − 1.84
0.46

20. 6 − 3.195
2.805

Section 5.4 Multiplying Decimals
Multiply.

21. 0.7 × 0.3
0.21

22. 0.25 × 0.16
0.04

Section 5.5 Dividing Decimals
Divide.

23. 7.11 ÷ 0.9
7.9

24. 3.83 ÷ 0.04
95.75

Section 6.4 Percents
Write each decimal as a percent.

25. 0.47
47%

26. 0.03
3%

27. 0.279
27.9%

28. 0.005
0.5%

Write each fraction as a percent.

29. $\frac{43}{100}$
43%

30. $\frac{9}{100}$
9%

31. $\frac{3}{4}$
75%

32. $\frac{1}{6}$
≈16.7%

SECTION 9.1 The Rectangular Coordinate System

OBJECTIVES

In this section, you will learn to
- Verify a solution for an equation with two variables.
- Create and interpret the values in an ordered pair.
- Plot points in the *x-y* plane.
- Recognize points in a quadrant and on an axis.
- Plot points that are on the same line and draw the line that passes through them.

You Need to Know

To successfully complete this section, you need to understand
- ☐ Number lines (2.1)
- ☐ Adding and subtracting signed numbers (2.2 and 2.3)
- ☐ Solving equations (3.1)
- ☐ Checking equations (3.1)

Introduction

While watching the local news recently, Sergio began to realize that the temperature in Sunny Meadows seemed to be 4° C warmer than the temperature in Green Valley. Curious about that, he did some online research and found it to be true.

Sergio wrote down the temperatures from six days throughout the year and made this chart of his findings. He also wrote in four other temperatures and searched for days that matched what he wrote.

Temperature in Sunny Meadows	Temperature in Green Valley
27	23
7	3
5.5	1.5
4	0
1	−3
−3	−7
	16
	−2.5
0	
−5.5	

You Try It 1 Knowing that the temperature in Sunny Meadows is 4° warmer than the temperature in Green Valley, fill in the rest of Sergio's chart.

As Sergio discovered, the relationship between the two temperatures is consistent. Because the difference in the temperatures is always 4°, we can represent this relationship using algebra.

For example, we could let x = the temperature in Sunny Meadows and y = the temperature in Green Valley. This allows us to write an equation that represents their relationship:

$$x - y = 4$$

As you can see, $x - y = 4$ is an equation with two variables. From our experience, equations are to be solved, but how do we solve an equation that contains two variables?

Equations with Two Variables

In Chapter 3, you learned how to solve equations with one variable, such as $x - 1 = 4$.

Without too much work, you can find that the solution is 5. This is the one and only value of x that makes the equation true.

$$x - 1 = 4$$
$$5 - 1 \stackrel{?}{=} 4$$
$$4 = 4 \text{ True!}$$

However, an equation with two variables can have many solutions. For example, consider an equation that represents that the *difference* of x and y is 4: $x - y = 4$.

This equation has many solutions because there are many combinations of x- and y-values whose difference is 4. Each pair of numbers will have one x-value and one y-value, as shown in the table at the right.

Other pairs of numbers are solutions to this equation, as demonstrated in Example 1.

x	y	$x - y = 4$
7	3	$7 - 3 = 4$ True!
5.5	1.5	$5.5 - 1.5 = 4$ True!
4	0	$4 - 0 = 4$ True!
3	−1	$3 - (-1) = 4$ True!
−1	−5	$-1 - (-5) = 4$ True!

Example 1

Show that the given pair of numbers is a solution for $x - y = 4$.

a) $x = 7.25$ and $y = 3.25$ **b)** $x = 2$ and $y = -2$

Procedure Replace the variables in the equation with the given x- and y-values. Evaluate each equation to show that the pair is a solution.

Answer

a) $x = 7.25$ and $y = 3.25$

$$7.25 - 3.25 \stackrel{?}{=} 4$$
$$4 = 4 \text{ True!}$$

b) $x = 2$ and $y = -2$

$$2 - (-2) \stackrel{?}{=} 4$$
$$2 + 2 \stackrel{?}{=} 4$$
$$4 = 4 \text{ True!}$$

You Try It 2 **Show that the given pair of numbers is a solution for $x - y = 4$. Use Example 1 as a guide.**

a) $x = 5.4$ and $y = 1.4$ **b)** $x = -1$ and $y = -5$

Ordered Pairs

One solution for the equation $x - y = 4$ is $x = 5$ and $y = 1$. It is more common to represent this solution as an *ordered pair* (5, 1).

An **ordered pair** is a pair of numbers—grouped in parentheses and separated by a comma—in which the order in which they are written in is important. For an ordered pair of x- and y-values, the x-value is always written first and the y-value is always written second.

Example 2

Write each pair of x- and y-values as an ordered pair.

a) $x = 7.2, y = 3.1$ **b)** $y = 8, x = -2$

c) $y = -6, x = 0$ **d)** $x = \frac{3}{5}, y = -\frac{1}{4}$

Procedure In every ordered pair, the x-value is always written first and the y-value is always written second. (Notice that the x-value is not always written first in each of the preceding pairs.)

Answer a) $(7.2, 3.1)$ **b)** $(-2, 8)$ **c)** $(0, -6)$ **d)** $\left(\frac{3}{5}, -\frac{1}{4}\right)$

You Try It 3 **Write each pair of x- and y-values as an ordered pair. Use Example 2 as a guide.**

a) $x = -1, y = 4$ **b)** $y = 0, x = 5.2$

c) $x = -2\frac{5}{8}, y = 1\frac{3}{8}$ **d)** $y = -10, x = -8$

Example 3

Identify the x- and y-values in each ordered pair.

a) $(-4, -5)$ **b)** $(3, 0)$ **c)** $\left(\frac{1}{8}, \frac{2}{3}\right)$

Procedure In every ordered pair, the x-value is always written first and the y-value is always written second.

Answer **a)** $x = -4, y = -5$ **b)** $x = 3, y = 0$ **c)** $x = \frac{1}{8}, y = \frac{2}{3}$

You Try It 4 **Identify the x- and y-values in each ordered pair. Use Example 3 as a guide.**

a) $(-5.1, -3)$ **b)** $(4, 0)$ **c)** $\left(-6, -\frac{7}{10}\right)$

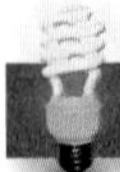

Think About It 1 What is the best way to remember which is the x-value and which is the y-value in an ordered pair?

An ordered pair (x, y) is a set of two replacement values, one for x and the other for y. It's possible that an ordered pair is not a solution to a particular equation.

To determine whether an order pair is a solution to an equation, we first identify the x- and y-values and then place those values in the equation. If we get a true statement, the ordered pair is a solution. If we do not get a true statement, the ordered pair is not a solution.

Example 4

Determine whether the given ordered pair is a solution to the equation $2x - y = 4$.

a) $(3, 1)$ **b)** $\left(\frac{1}{2}, -3\right)$

Procedure First, identify the x-value and the y-value in each ordered pair. Then place those values in the equation.

Answer

a) $x = 3$ and $y = 1$

$$2(3) - 1 \stackrel{?}{=} 4$$
$$6 - 1 \stackrel{?}{=} 4$$
$$5 = 4 \text{ False!}$$

No, $(3, 1)$ is *not* a solution.

b) $x = \frac{1}{2}$ and $y = -3$

$$2\left(\frac{1}{2}\right) - (-3) \stackrel{?}{=} 4$$
$$1 + 3 \stackrel{?}{=} 4$$
$$4 = 4 \text{ True!}$$

Yes, $\left(\frac{1}{2}, -3\right)$ is a solution.

You Try It 5 **Determine whether the given ordered pair is a solution to the equation $3x + 2y = 6$. Use Example 4 as a guide.**

a) $(4, -3)$ **b)** $(2, -6)$ **c)** $\left(\frac{2}{3}, 2\right)$

The *x*-*y* Plane and Plotting Points

In Section 2.1, we were introduced to number lines that represent both positive and negative integers. Recall that 0 is the **origin**. On a vertical number line, positive numbers are above 0 and negative numbers are below 0. On a horizontal number line, positive numbers are to the right of 0 and negative numbers are to the left of 0.

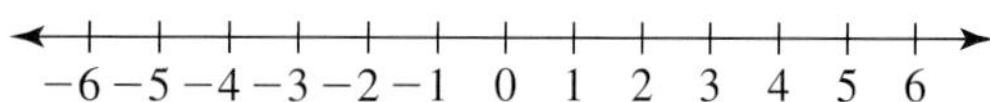

Because we now consider equations with two variables, we will combine a horizontal number line, the ***x*-axis** representing *x*-values, with a vertical number line, the ***y*-axis** representing *y*-values. (By the way, the plural of axis is *axes*, pronounced "ax-eze.")

The horizontal number line:

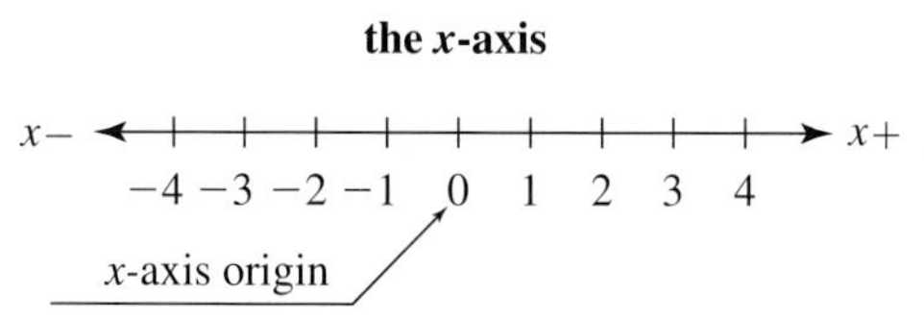

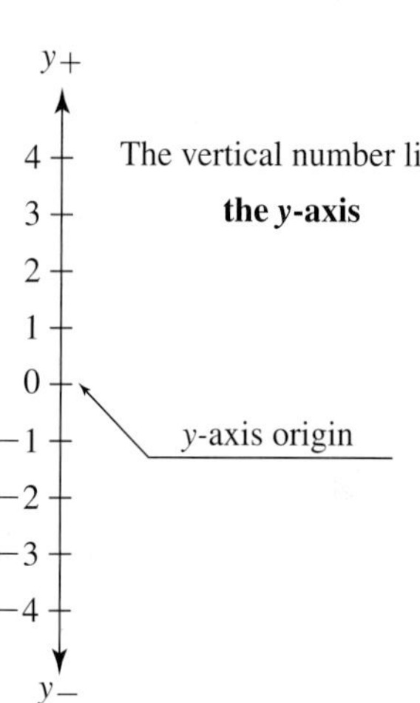

We can join these two number lines at the origins to create a grid called the ***x*-*y* plane**. Recall from Section 8.4 that a plane is any flat surface that extends forever in each direction, and this is true of the *x*-*y* plane.

Notice that each axis is labeled with *x* or *y* and notice where each axis is positive (+) and negative (−). This grid is called the **rectangular coordinate system.**

On the grid, there are infinitely many places where the vertical and horizontal lines intersect (cross each other). Each place of intersection is called a **point**, and it represents exactly one ordered pair, such as (6, 4).

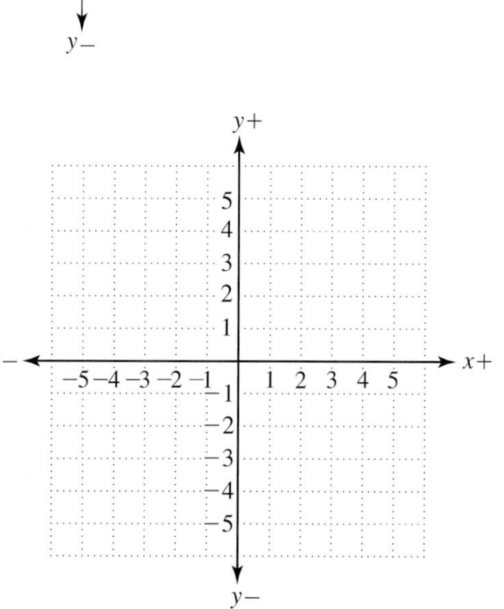

The grid lines in this graph show only integer values, but it's also possible to have points with fractional or decimal values. For example, (5.5, 1.5) is one of the solutions to the equation $x - y = 4$, and it can be represented by a point in the *x*-*y* plane.

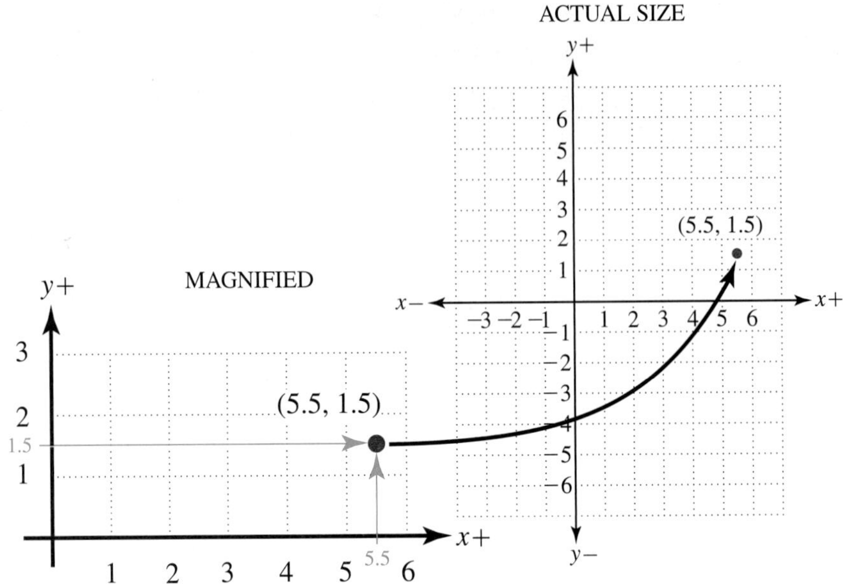

When we speak of an ordered pair as a *solution* to an equation, such as (6, 4), the numbers are called values: The *x-value* is 6, and the *y-value* is 4.

However, when we consider the ordered pair (6, 4) as a *point* in the *x*-*y* plane, the numbers are called **coordinates:** The *x-coordinate* is 6, and the *y-coordinate* is 4.

One special point is the **origin** of the *x*-*y* plane. It is where the two axes meet, and the ordered pair it represents is (0, 0). In other words, the origin has an *x*-coordinate of 0 and a *y*-coordinate of 0.

Notice that the point (6, 4) forms a rectangle with the *x*-axis, the *y*-axis, and the origin (0, 0). This is why the grid is called the rectangular coordinate system.

(6, 4)
The origin (0, 0)

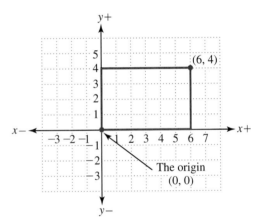

Placing and labeling points in the *x*-*y* plane is called **plotting points.** We have seen the point (6, 4) placed in the plane, but how did it get there?

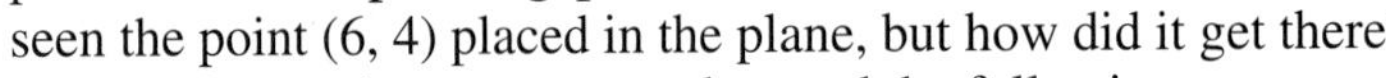

To plot a point, we must understand the following:

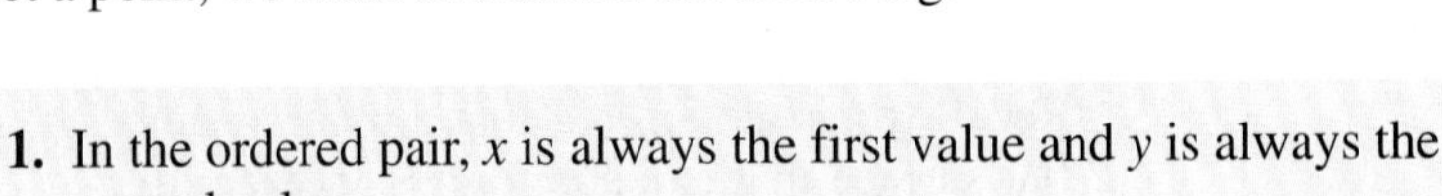

1. In the ordered pair, *x* is always the first value and *y* is always the second value.
2. The *x*-axis is the horizontal axis, and the *y*-axis is the vertical axis.
3. On the *x*-axis,
   a) The positive values are to the right of the origin.
   b) The negative values are to the left of the origin.
4. On the *y*-axis,
   a) The positive values are above the origin.
   b) The negative values are below the origin.

Guidelines:

1. Start at the origin (the beginning) and move left or right to the *x*-value on the *x*-axis.
2. From there, move up or down according to the *y*-value.
3. Place a round dot and label the point with its ordered pair.

Example: Plot the point (6, 4).

1. From the origin, move 6 to the right.
2. From there, move up 4.
3. Draw a dot at the final location. Also write the ordered pair (6, 4) near the point.

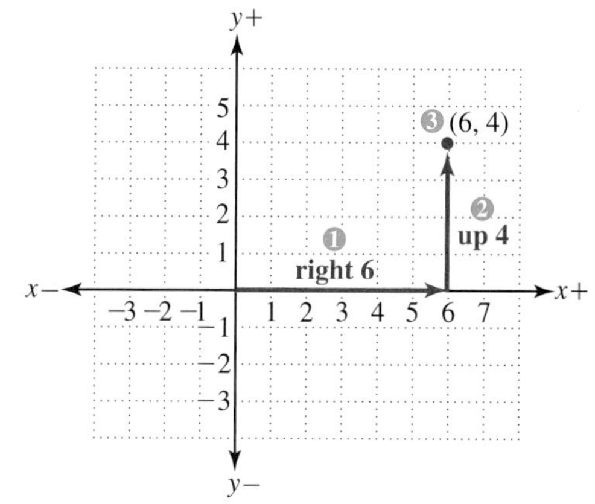

Example: Plot the point $(-5, -3)$.

1. From the origin, move 5 to the left.
2. From there, move down 3.
3. Draw a dot at the final location. Also write the ordered pair $(-5, -3)$ near the point.

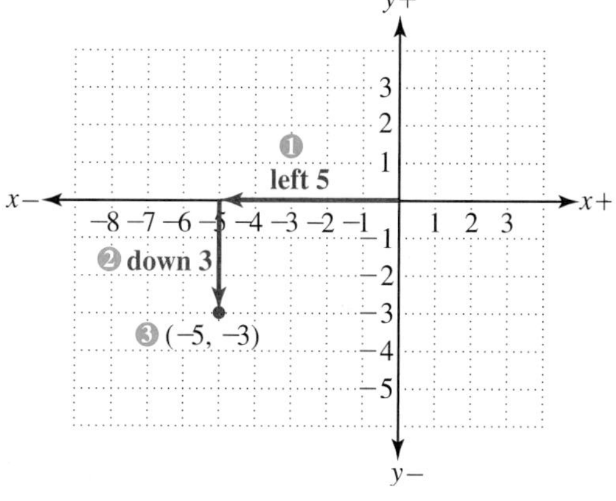

Example 5

Plot each of these points in the *x*-*y* plane. Label each one with its ordered pair as well.

a) $(-5, 6)$ **b)** (4, 3) **c)** $(7, -2)$ **d)** $(-1, -5)$

Procedure In each ordered pair, recognize the *x*-value and the *y*-value and verify that these points have been placed properly in the grid.

Answer

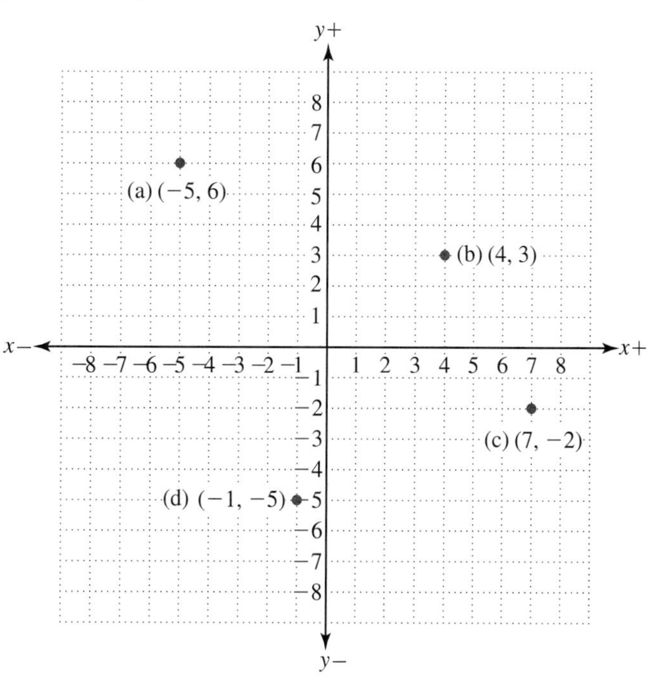

Caution It is easy to get the x-coordinate confused with the y-coordinate. One way to avoid this confusion is to place a little (x, y) over each ordered pair, such as $\overset{(x,\,y)}{(2, 3)}$.

You Try It 6 **Plot each of these points in the x-y plane provided. Make sure you label each one with its ordered pair. Use Example 5 as a guide.**

a) $(3, 2)$ b) $(-1, 5)$

c) $(6, -3)$ d) $(-7, -4)$

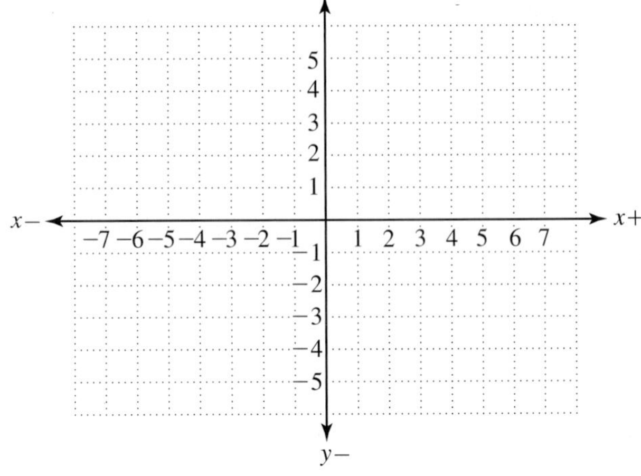

You Try It 7 **Given the graph, identify the ordered pair of each point shown. (Make sure you place parentheses around each ordered pair.)**

a) A b) B

c) C d) D

e) E f) F

g) G h) H

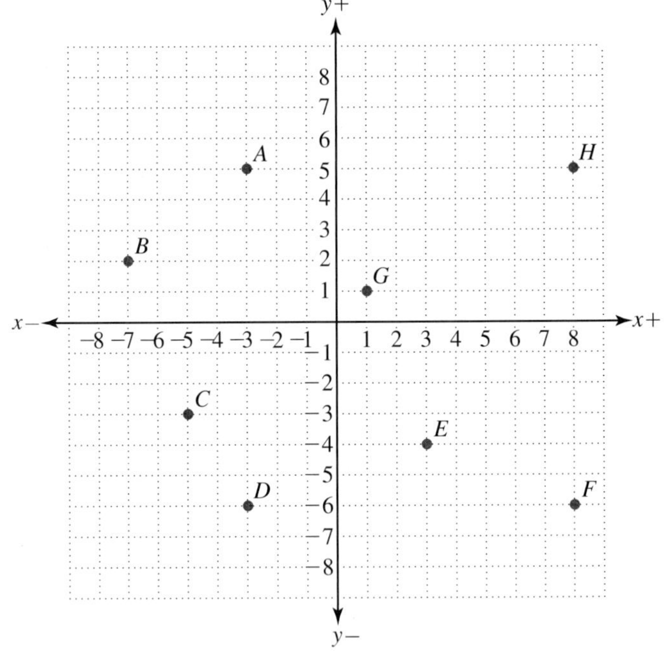

Special Features of the x-y Plane

The x-axis and the y-axis divide the x-y plane into four distinct regions called **quadrants.** We label the quadrants using Roman numerals I, II, III, and IV.

In the following diagram, notice that the first quadrant, Quadrant I, begins in the *upper right* part and the other quadrants follow in a counterclockwise direction (the opposite direction of what a clock's hands move).

Also notice that the first quadrant, Quadrant I, is the part of the grid where both x and y are positive. Similarly, points in Quadrant II represent negative x-coordinates but positive y-coordinates.

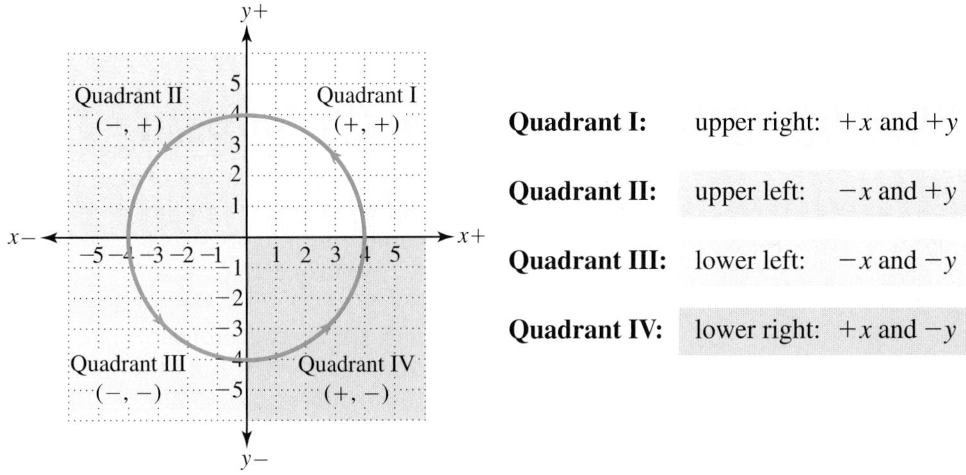

Example 6

In which quadrant would you find these points?

a) $(2, -3)$ **b)** $(-5, 4)$ **c)** $(3, 5)$ **d)** $(-7, -6)$

Procedure Look at the positive and negative nature of the x- and y-coordinates to determine the quadrant. You may draw a simple x-y plane to help determine where each point lies.

Answer

a) $(2, -3)$ The x-coordinate is positive, and the y-coordinate is negative. This suggests that the point is in the *lower-right* quadrant, **Quadrant IV.**

b) $(-5, 4)$ A negative x-coordinate means *left,* and a positive y-coordinate means *up.* So, the point is in the *upper left* quadrant, **Quadrant II.**

c) $(3, 5)$ x: right; y: up; the point is in the *upper-right* quadrant, **Quadrant I.**

d) $(-7, -6)$ x: left; y: down; *lower-left* quadrant, **Quadrant III.**

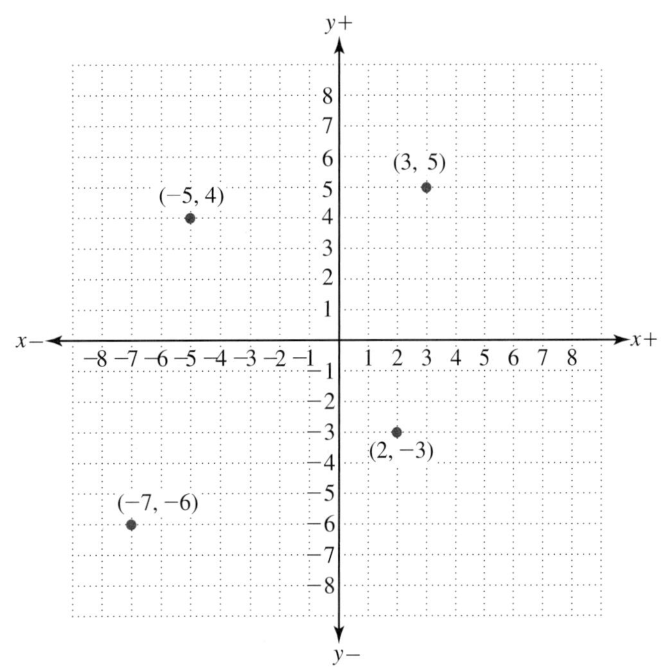

You Try It 8 **Determine in which quadrant each point lies. Use Example 6 as a guide.**

a) $(-5, 6)$ **b)** $(2, -9)$ **c)** $(-8, -4)$ **d)** $(10, 2)$

Other Points in the *x*-*y* Plane

Some points in the *x*-*y* plane are not in any of the quadrants. Instead, they are on one of the axes. We may plot points on the *x*-axis and separately on the *y*-axis. We may even plot the point (0, 0), the origin.

Consider the point (5, 0): The *x*-coordinate is positive; so, the point is to the *right*. But because the *y*-coordinate is neither positive nor negative, we can't consider it *up* or *down*. In other words, it is not in Quadrant I and it is not in Quadrant IV. The point (5, 0) is *on the positive x-axis.*

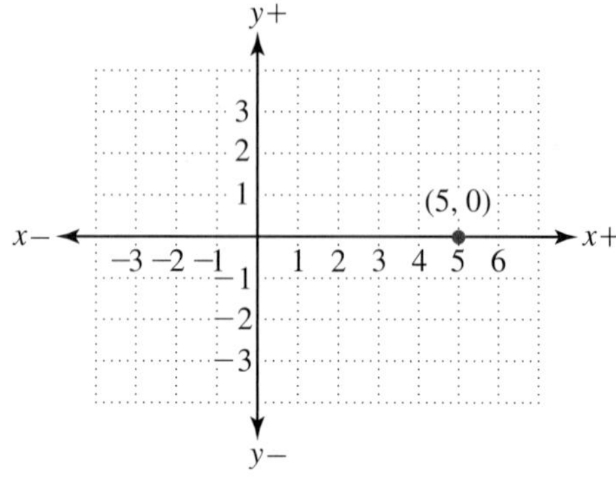

Example 7

Describe where each point is located in the *x*-*y* plane.

a) $(-3, 0)$ **b)** $(0, 6)$ **c)** $(0, -4)$ **d)** $(7, 0)$ **e)** $(0, 0)$

Procedure Because at least one of the coordinates in each point is 0, none of these is in one of the quadrants. Each point is on one (or both) of the axes.

Answer

a) $(-3, 0)$ This point is on the **negative *x*-axis.**

b) $(0, 6)$ This point is on the **positive *y*-axis.**

c) $(0, -4)$ This point is on the **negative *y*-axis.**

d) $(7, 0)$ This point is on the **positive *x*-axis.**

e) $(0, 0)$ This point is at the **origin;** it is on both the *x*- and *y*-axes.

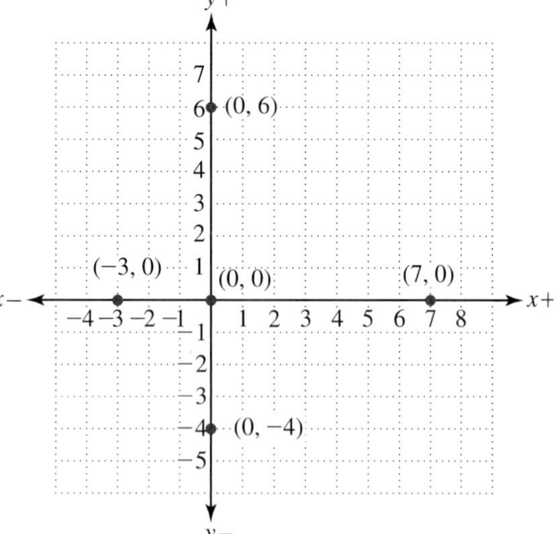

You Try It 9 **Describe where each point is located in the *x*-*y* plane. Use Example 7 as a guide.**

a) $(-5, 0)$ **b)** $(0, 9)$

c) $(0, -8)$ **d)** $(10, 0)$

You Try It 10 **Plot each of these points in the x-y plane. Make sure you label each one as well.**

a) $(0, 5)$ **b)** $(-4, 0)$ **c)** $(0, 0)$ **d)** $(5, 0)$ **e)** $(0, -3)$

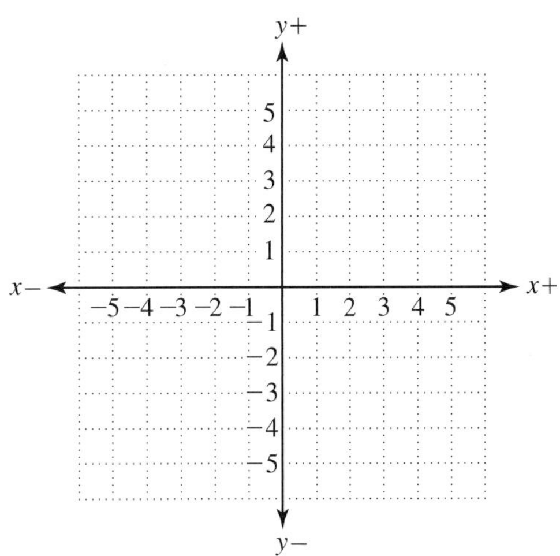

You Try It 11 **Given the graph at the right, identify the ordered pair of each point.**

a) A **b)** B

c) C **d)** D

e) E **f)** F

g) G **h)** H

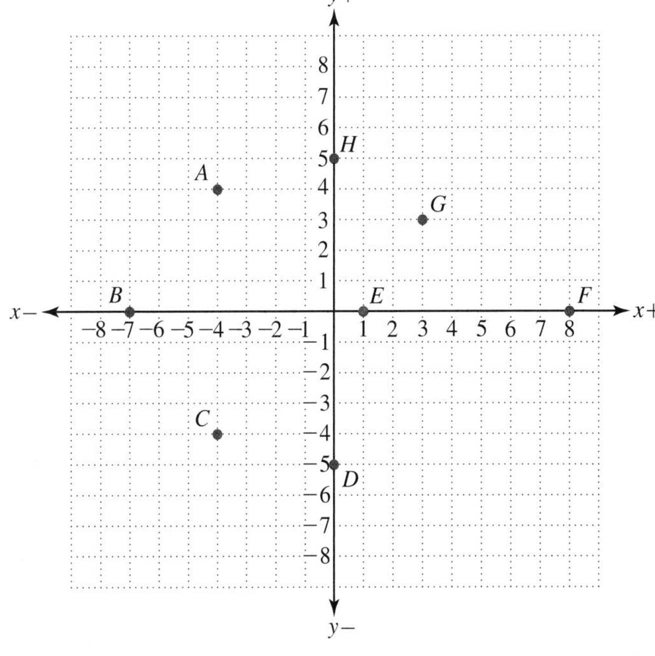

Points on a Line

Sometimes the points plotted in the x-y plane lie on a straight line. Recall from Section 8.4 that a line in a plane is straight and goes on forever. We indicate this by placing arrows on each end of the line.

When we know that some points are on a line, we can plot the points and draw the line that passes through them. For example, throughout this section, we have seen a number of solutions to the equation $x - y = 4$, such as $(7, 3)$, $(5.5, 1.5)$, $(4, 0)$, and $(1, -3)$.

Because these ordered pairs are all solutions of the same equation, they can be represented as points in the *x-y* plane and can be connected to form a straight line. We place arrowheads at the end of each line to indicate that the line continues indefinitely beyond the confines of the *x-y* plane shown. (Use something with a straight edge, such as a ruler, to draw the line.)

The reason we draw the line is to show all of the possible solutions to an equation or a situation, such as the comparison of temperatures between Sunny Meadows and Green Valley.

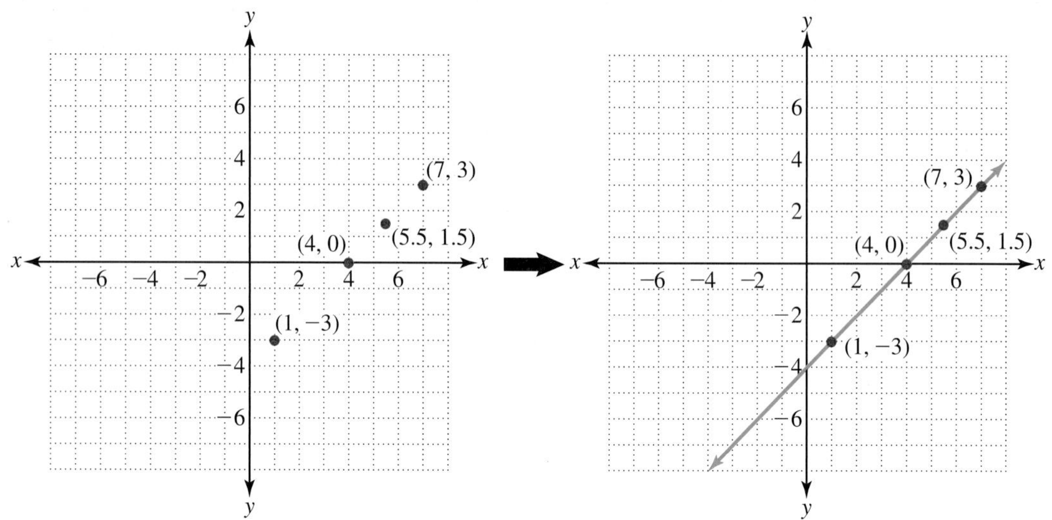

These points represent some of the temperature comparisions between Sunny Meadows and Green Valley.

This line represents all of the possible temperatures of Sunny Meadows and Green Valley.

The fewest number of points we need to draw a line is two. In other words,

> Any two points determine one straight line.

However, to help us draw the line more accurately, we usually plot at least three points.

In Section 9.2, you will learn a technique for finding points on a line. For now, you are given three points known to be on a line, asked to plot the points, and then asked to draw a line through the three points.

You Try It 12 **Three solutions to the line equation $y = \frac{1}{2}x + 3$ are $(-4, 1)$, $(0, 3)$, and $(2, 4)$. Plot these points in the *x-y* plane and draw the line that passes through them.**

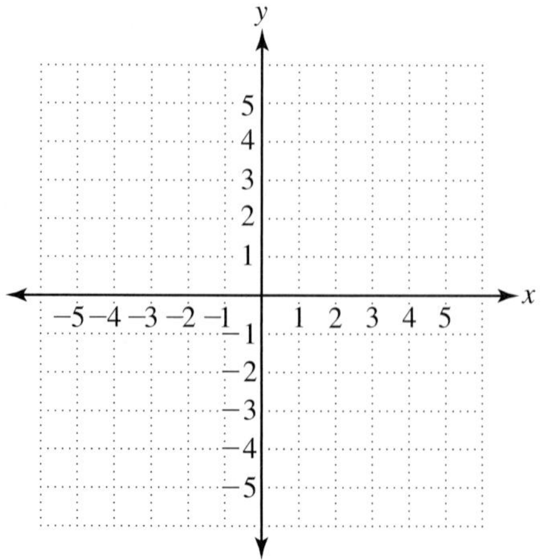

You Try It 13 **Three solutions to the line equation $3x + y = -2$ are $(-2, 4)$, $(0, -2)$, and $(1, -5)$. Plot these points in the x-y plane and draw the line that passes through them.**

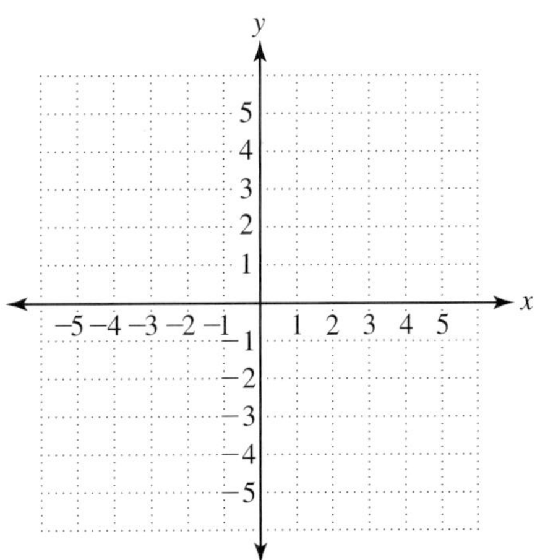

Answers: You Try It and Think About It

You Try It: **1.**

Temperature in Sunny Meadows	Temperature in Green Valley
20	16
1.5	−2.5
0	**−4**
−5.5	**−9.5**

2. a) $5.4 - 1.4 \stackrel{?}{=} 4$
$4 = 4$ True! **b)** $-1 - (-5) \stackrel{?}{=} 4$
$-1 + 5 \stackrel{?}{=} 4$
$4 = 4$ True!
3. a) $(-1, 4)$ **b)** $(5.2, 0)$ **c)** $(-2\frac{5}{8}, 1\frac{3}{8})$ **d)** $(-8, -10)$
4. a) $x = -5.1, y = -3$ **b)** $x = 4, y = 0$
c) $x = -6, y = -\frac{7}{10}$ **5. a)** $6 = 6$ True! Yes, $(4, -3)$ is a solution. **b)** $-6 = 6$, False! No, $(2, -6)$ is not a solution.
c) $6 = 6$ True! Yes, $(\frac{2}{3}, 2)$ is a solution.
7. a) $(-3, 5)$ **b)** $(-7, 2)$ **c)** $(-5, -3)$ **d)** $(-3, -6)$
e) $(3, -4)$ **f)** $(8, -6)$ **g)** $(1, 1)$ **h)** $(8, 5)$
8. a) Quardrant II **b)** Quardrant IV **c)** Quardrant III
d) Quardrant I **9. a)** $(-5, 0)$ is on the negative x-axis.
b) $(0, 9)$ is on the positive y-axis. **c)** $(0, -8)$ is on the negative y-axis. **d)** $(10, 0)$ is on the positive x-axis.

6.

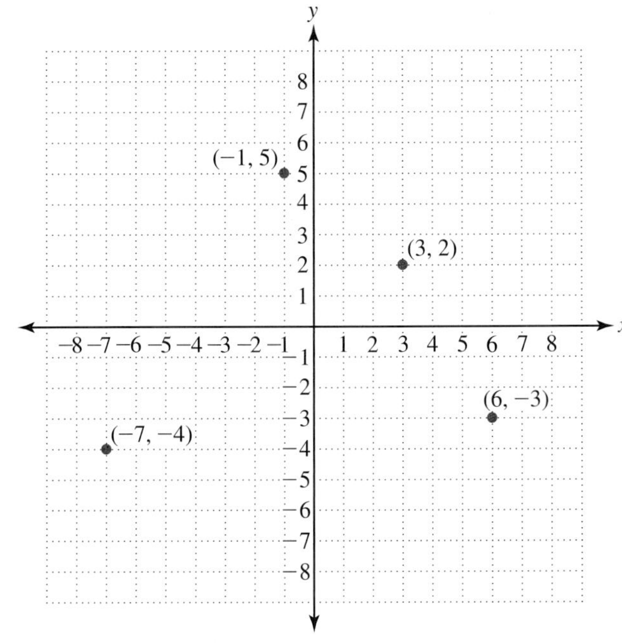

10.

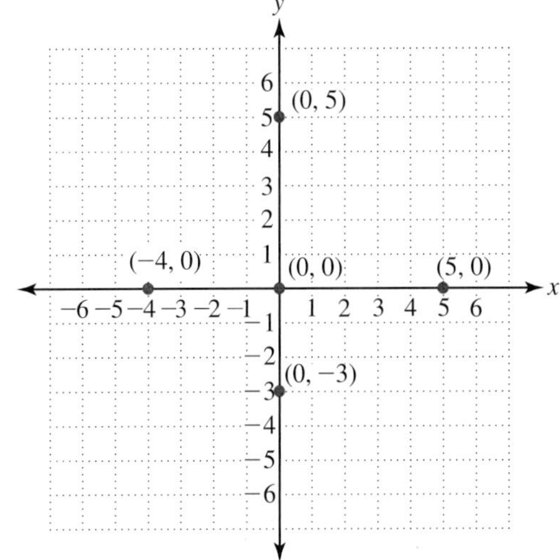

11. a) $(-4, 4)$ **b)** $(-7, 0)$ **c)** $(-4, -4)$ **d)** $(0, -5)$ **e)** $(1, 0)$ **f)** $(8, 0)$ **g)** $(3, 3)$ **h)** $(0, 5)$

12.

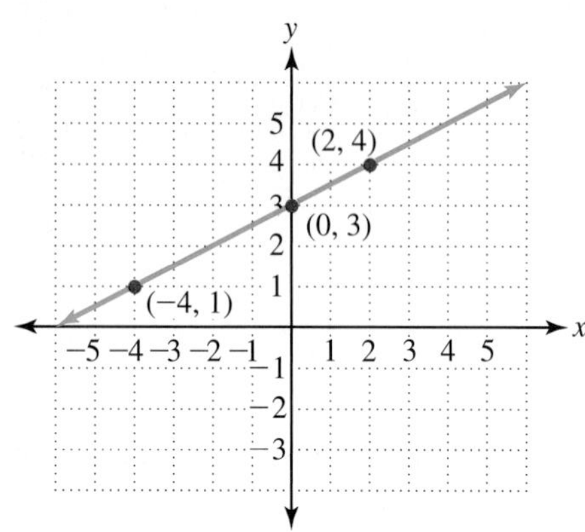

13.

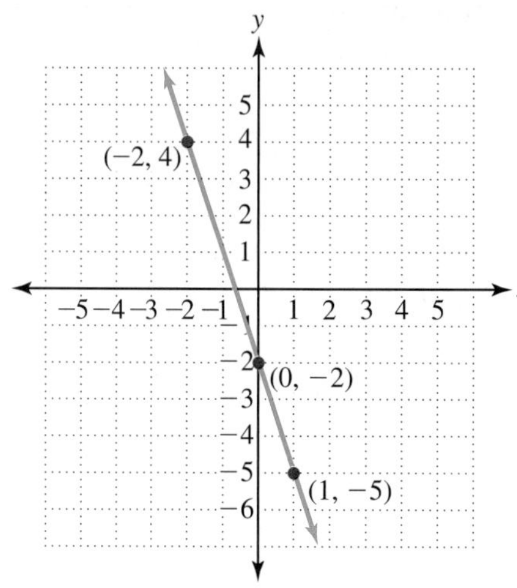

Think About It: **1.** Answers may vary. One possibility: The numbers in an orderd pair are listed alphabetically; so, the x-value is written first and the y-value is written second.

Section 9.1 Exercises

FOR EXTRA HELP
Student Resources on DVD-ROM
Includes
➤ Student's Solutions Manual
➤ Video Lectures
➤ Chapter Test Prep Video
MyMathLab
MathXL

Think Again

1. What would you tell a classmate is the best way to remember which is the x-value and which is the y-value in an ordered pair? (*Refer to Think About It 1*)
   Answers will vary.

2. In which quadrant(s) in the x-y plane would you find all points for which the x-coordinate is the same as the y-coordinate? Explain your answer or show an example that supports your answer.
   The x-coordinate and the y-coordinate are the same in Quadrant I, as in (3, 3) and (5, 5), and in Quadrant III, as in $(-2, -2)$ and $(-4, -4)$. (They also are the same at the origin.)

3. In which quadrant(s) in the x-y plane would you find all points for which the x-coordinate is the opposite of the y-coordinate? Explain your answer or show an example that supports your answer.
   The x-coordinate and the y-coordinate are opposites in Quadrant II, as in $(-4, 4)$ and $(-6, 6)$, and in Quadrant IV, as in $(1, -1)$ and $(7, -7)$.

4. What types of lines pass through only two quardrants? Explain your answer or show an example that supports your answer.
   Horizontal lines and vertical lines pass through only two quardrants (the exceptions being $x = 0$ and $y = 0$). Also, all lines that pass through the origin pass through only two quardrants. All other lines pass through three quardrants.

Focus Exercises

Determine whether the given ordered pair is a solution to the equation $2x - 5y = 10$.

5. (10, 2)
   $10 = 10$ True! Yes, $(10, 2)$ is a solution.

6. $(0, -2)$
   $10 = 10$ True! Yes, $(0, -2)$ is a solution.

7. $(-5, 0)$
   $-10 = 10$ False! No, $(-5, 0)$ is not a solution.

8. (5, 4)
   $-10 = 10$ False! No, $(5, 4)$ is not a solution.

Determine whether the given ordered pair is a solution to the equation $y = \frac{3}{2}x - 4$.

9. (6, 0)
   $0 = 5$ False! No, $(6, 0)$ is not a solution.

10. $(-4, -2)$
    $-2 = -10$ False! No, $(-4, -2)$ is not a solution.

11. $(2, -1)$
    $-1 = -1$ True! Yes, $(2, -1)$ is a solution.

12. $(0, -4)$
    $-4 = -4$ True! Yes, $(0, -4)$ is a solution.

Plot each point in the *x-y* plane. Make sure you label each one with its ordered pair.

13. (5, 4)

14. $(-3, 1)$

15. $(2, -5)$

16. $(-2, -4)$

17. $(-7, 0)$

18. (5, 0)

19. $(0, -2)$

20. (0, 6)

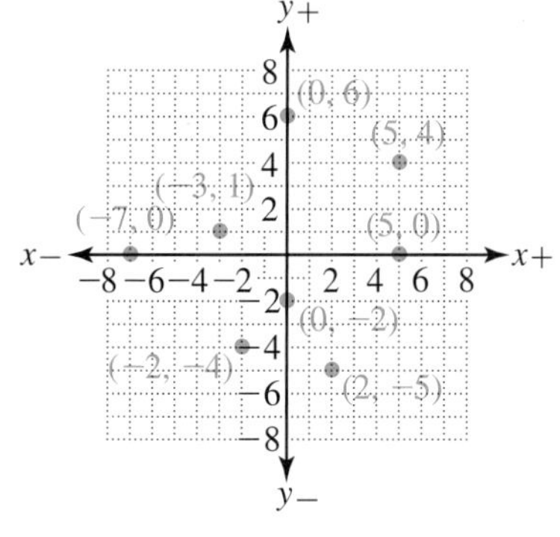

Plot each point in the *x-y* plane. Make sure you label each one with its ordered pair.

21. $(-6, 2)$

22. (1, 4)

23. $(7, -1)$

24. $(-3, -5)$

25. (4, 0)

26. $(-3, 0)$

27. (0, 1)

28. $(0, -6)$

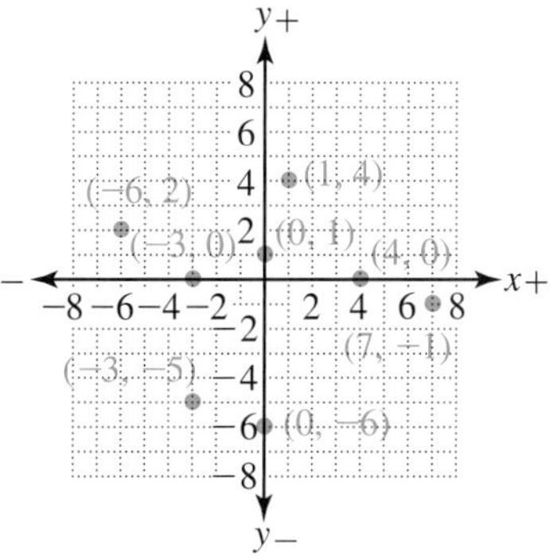

Given the graph below, identify the ordered pair of each point shown.

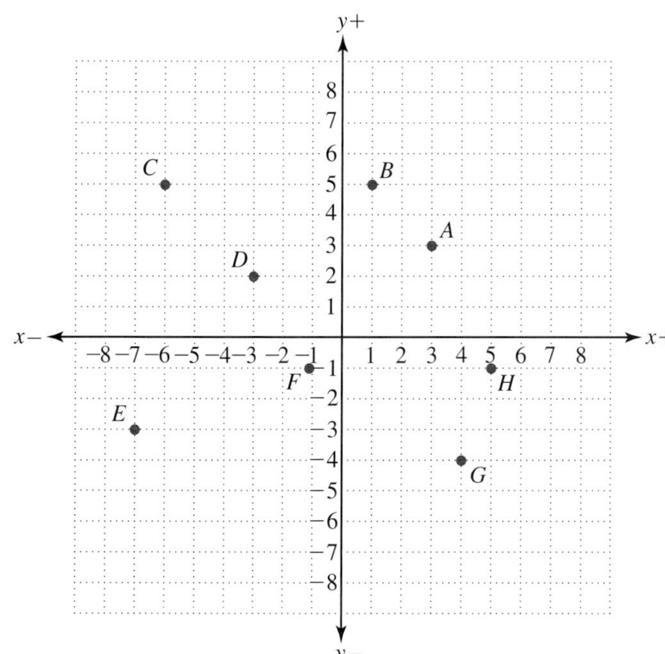

29. *A*
(3, 3)

30. *B*
(1, 5)

31. *C*
$(-6, 5)$

32. *D*
$(-3, 2)$

33. *E*
$(-7, -3)$

34. *F*
$(-1, -1)$

35. *G*
$(4, -4)$

36. *H*
$(5, -1)$

Determine in which quardrant each point lies.

37. $(8, -2)$
Quadrant IV

38. $(-2, -2)$
Quadrant III

39. $(-3, 5)$
Quadrant II

40. (1, 4)
Quadrant I

41. $(-10, -15)$
Quadrant III

42. $(-8, 12)$
Quadrant II

43. $(1, -20)$
Quadrant IV

44. (26, 34)
Quadrant I

45. $(1, -7)$
Quadrant IV

46. $(-4, -9)$
Quadrant III

47. (2, 11)
Quadrant I

48. $(-5, 14)$
Quadrant II

Describe where each point is located in the *x-y* plane.

49. (0, 6)
On the positive *y*-axis

50. $(-5, 0)$
On the negative *x*-axis

51. (3, 0)
On the positive *x*-axis

52. $(0, -1)$
On the negative *y*-axis

53. $(0, -7)$
On the negative *y*-axis

54. (8, 0)
On the positive *x*-axis

55. (0, 0)
At the origin

56. (0, 3)
On the positive *y*-axis

Given the graph below, identify the ordered pair of each point shown.

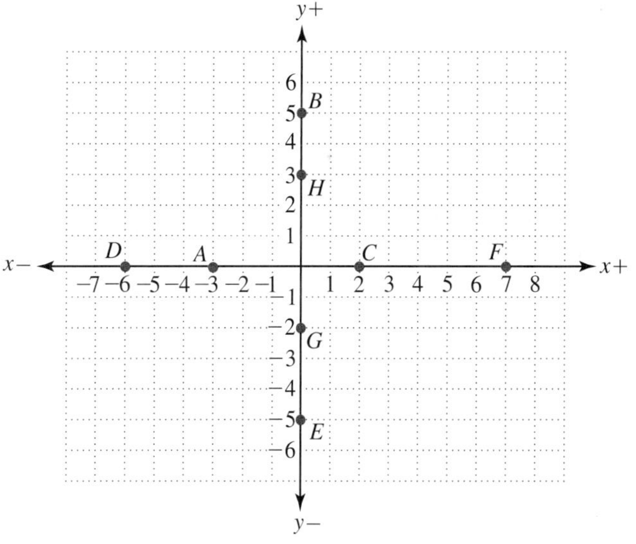

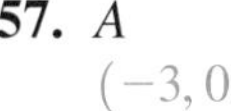

57. *A*
$(-3, 0)$

58. *B*
(0, 5)

59. *C*
(2, 0)

60. *D*
$(-6, 0)$

61. *E*
$(0, -5)$

62. *F*
(7, 0)

63. *G*
$(0, -2)$

64. *H*
(0, 3)

Plot the three points of the given equation and draw the line that passes through them.

65. The line equation $y = 2x - 5$ passes through the points $(2, -1)$, $(0, -5)$, and $(5, 5)$.

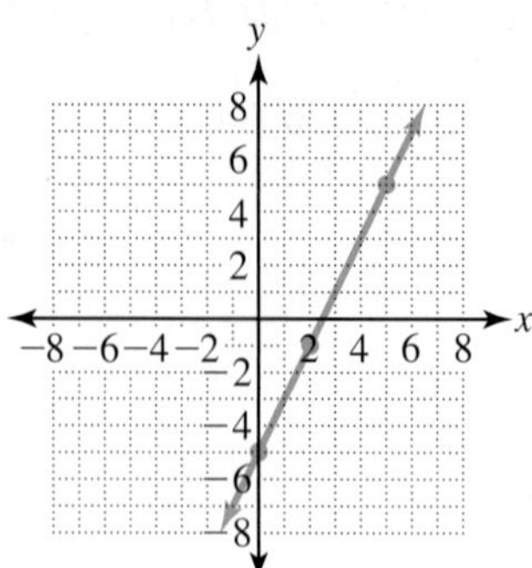

66. The line equation $y = -2x + 3$ passes through the points $(-2, 7)$, $(0, 3)$, and $(2, -1)$.

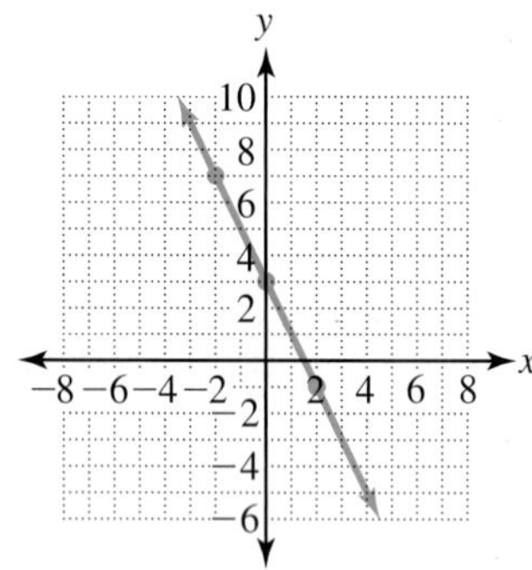

67. The line equation $y = \frac{2}{3}x - 1$ passes through the points $(-3, -3)$, $(0, -1)$, and $(3, 1)$.

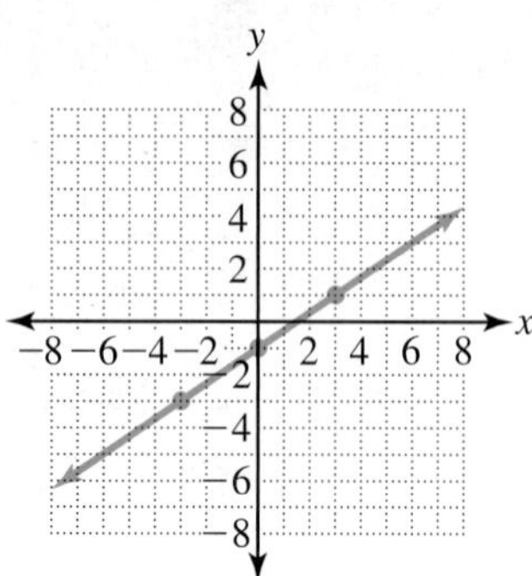

68. The line equation $y = -\frac{3}{4}x + 2$ passes through the points $(-4, 5)$, $(0, 2)$, and $(8, -4)$.

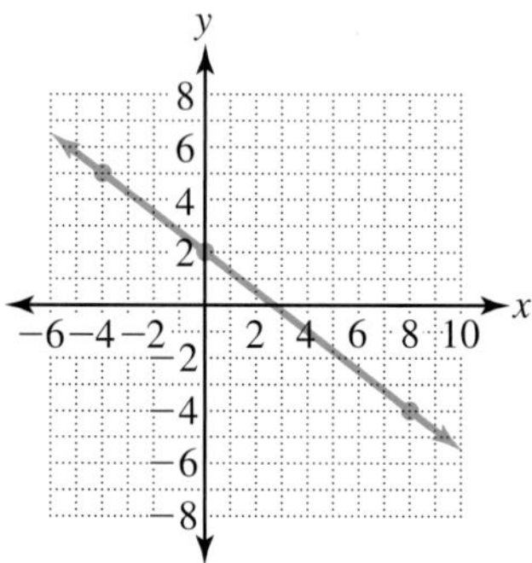

Think Outside the Box

Determine whether the given ordered pair is a solution to the equation $y = x^2 - 3x - 4$.

69. $(3, 4)$
No

70. $(5, 6)$
Yes

71. $(-2, 6)$
Yes

72. $(-4, 0)$
No

SECTION 9.2 Graphing Lines

OBJECTIVES

In this section, you will learn to

- Find ordered pair solutions of an equation of the form $ax + by = c$.
- Graph the line of an equation of the form $ax + by = c$.
- Find ordered pair solutions of an equation of the form $y = mx + b$.
- Graph the line of an equation of the form $y = mx + b$.
- Graph a line using a different scale.
- Graph a horizontal line.
- Graph a vertical line.

You Need to Know

To successfully complete this section, you need to understand

- ☐ The order of operations (1.5 and 2.5)
- ☐ Placing values into formulas (2.6)
- ☐ Simplifying fractions (4.3)
- ☐ Multiplying fractions (4.4)
- ☐ Adding fractions and whole numbers (4.7)
- ☐ Plotting points in the x-y plane (9.1)
- ☐ Drawing a line in the x-y plane (9.1)

Introduction

A **linear** (lih´-nee-er) **equation** of two variables, x and y, is an equation in which the ordered pair solutions are points on a line in the x-y plane. Because the ordered pairs are both solutions and points, there will be times in this text when they are referred to as *points*.

In Section 9.1, we saw a few linear equations, including $x - y = 4$, $y = \frac{1}{2}x + 3$, and $3x + y = -2$. Each of these equations was graphed as a line in the x-y plane. And even though the fewest number of points needed to draw a line is two, we usually plot at least three points to help us draw the line more accurately.

Linear Equation of the Form $ax + by = c$

In Section 9.1, you were introduced to the equation $x - y = 4$. This is a member of a family of linear equations of the form $ax + by = c$, where a, b, and c are numbers and x and y are variables.

> Linear equations with two variables, x and y, can be written in this form:
>
> $$ax + by = c$$
>
> Here $\boldsymbol{a}$, $\boldsymbol{b}$, and $\boldsymbol{c}$ are numbers and $\boldsymbol{x}$ and $\boldsymbol{y}$ are variables.

We can graph any line that has this form by finding three ordered pair solutions, and we often have a choice as to which three points are to be plotted. However, not all points on a line are easy to graph. For example, a point such as (21, 17) would not normally fit on our x-y plane.

We can find the coordinates of a point by doing one of the following

1. Choosing a value of x, placing it in the equation, and then solving for y
2. Choosing a value of y, placing it in the equation, and then solving for x

Often, one point can be found by choosing $x = 0$ and another by choosing $y = 0$. Then by choosing one other value of x or y, we will have three points to plot and can draw the line associated with the linear equation.

Example 1

Graph the line of the equation $3x + y = -2$ by finding three ordered pair solutions.

Procedure For this example, let's choose $x = 0$, $y = 0$, and $x = 1$. We'll set up a table to keep our work organized. (Be careful to place the x- and y-coordinates correctly in the ordered pair.)

Answer

Choose a value	$3x + y = -2$	(x, y)
$x = 0$	$3(0) + y = -2$ $0 + y = -2$ $y = -2$	$(0, -2)$
$y = 0$	$3x + 0 = -2$ $3x = -2$ $\frac{3x}{3} = -\frac{2}{3}$ $x = -\frac{2}{3}$	$\left(-\frac{2}{3}, 0\right)$
$x = 1$	$3(1) + y = -2$ $3 + y = -2$ $3 + (-3) + y = -2 + (-3)$ $y = -5$	$(1, -5)$

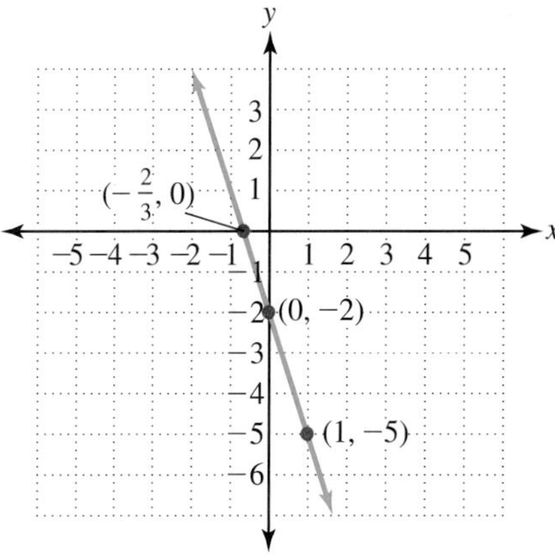

Caution If the three points we plot are not on the same line, we made a mistake somewhere and must correct it before we can graph the line.

You Try It 1 **Graph the line of the equation $2x - y = 4$ by finding three sets of ordered pair solutions. Use these values: $x = 0, y = 0,$ and $y = 2$. Use Example 1 as a guide.**

Choose a value	$2x - y = 4$	(x, y)
$x = 0$		
$y = 0$		
$y = 2$		

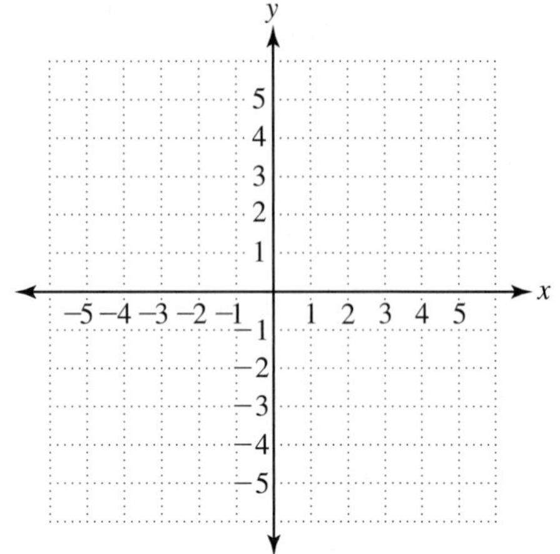

You Try It 2 **Graph the line of the equation $x + 2y = 6$ by finding three sets of ordered pair solutions. Choose the three values to use. Use Example 1 as a guide.**

Choose a value	$x + 2y = 6$	(x, y)

y
x
5 4 3 2 1
−5 −4 −3 −2 −1
1 2 3 4 5
−1 −2 −3 −4 −5

Linear Equations of the Form $y = mx + b$

Another form of a linear equation is $y = mx + b$. Written in this form, we say that *y is in terms of x*. Because y is already isolated on the left side, we can choose three values of x to find the three points.

Linear equations written with y in terms of x have the form

$$y = mx + b.$$

Here m and b are numbers and x and y are variables.

Example 2

Graph the line of the equation $y = 2x + 1$ by finding three ordered pair solutions.

Procedure For this example, let's choose $x = 0, 2,$ and -2. We'll set up a table to keep our work organized.

Answer

x	y = 2x + 1	(x, y)
0	$y = 2(0) + 1$ $y = 0 + 1$ $y = 1$	(0, 1)
2	$y = 2(2) + 1$ $y = 4 + 1$ $y = 5$	(2, 5)
−2	$y = 2(-2) + 1$ $y = -4 + 1$ $y = -3$	(−2, −3)

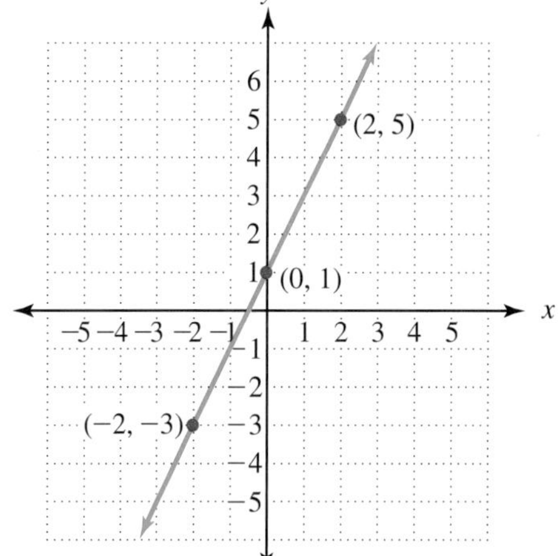

Sometimes we must be careful with the values of x that we choose.

For example, in the equation $y = 2x + 1$, choosing $x = 10$ gives $y = 21$. The ordered pair (10, 21) is a solution and such a point can be plotted, but it would not fit on a typical x-y plane.

Instead, choose relatively small values of x, somewhat close to 0. In fact, a good value to choose for x is 0 itself.

You Try It 3 **Graph the line of each given equation by finding three sets of ordered pairs solutions. Use Example 2 as a guide.**

a) $y = -2x + 3$ **b)** $y = 3x - 1$

a)

x	y = −2x + 3	(x, y)

b)

x	y = 3x − 1	(x, y)

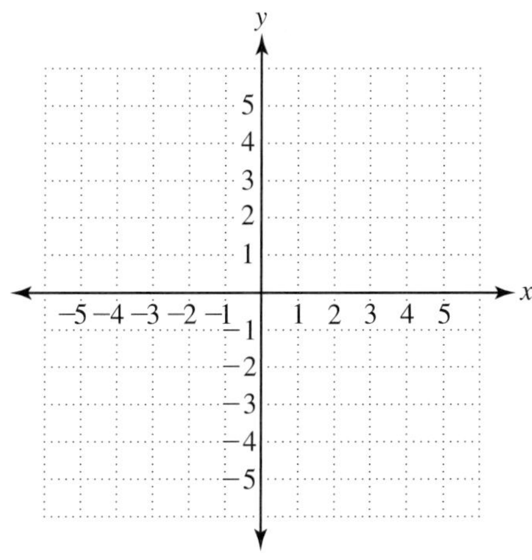

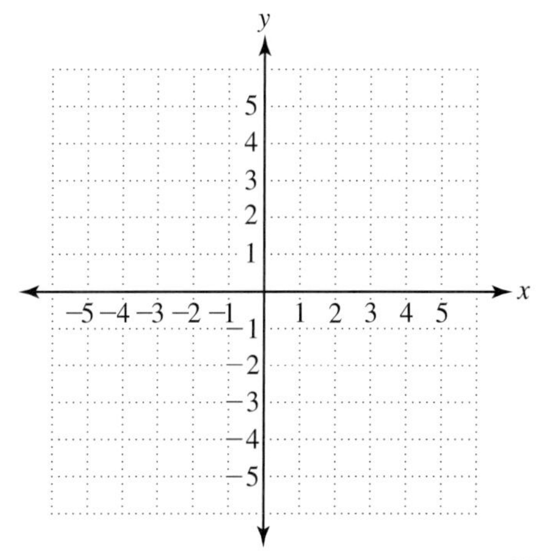

Consider the equation $y = \frac{2}{3}x + 5$. What values of x would you choose to find *four* points on the line? Explain your answer.

____

____

Graphing Lines with a Large Scale

Sometimes a graph will have x- or y-coordinates that do not fit easily in our typical x-y plane. Consider, for example, the equation $y = 10x + 20$. If we choose values of x such as -1, 0, and 1, as shown at the right, all of the points found will be above or below our typical graph.

We can still graph the line, but we must create an x-y plane that has a larger scale on the y-axis. To create a larger scale, we make each grid line 5 or 10 (or more) times the normal y-value. For this example, we'll see it done with two different scales.

x	$y = 10x + 20$	(x, y)
-1	$y = 10(-1) + 20$ $y = -10 + 20$ $y = 10$	$(-1, 10)$
0	$y = 10(0) + 20$ $y = 0 + 20$ $y = 20$	$(0, 20)$
1	$y = 10(1) + 20$ $y = 10 + 20$ $y = 30$	$(1, 30)$

5 times the normal y-value scale

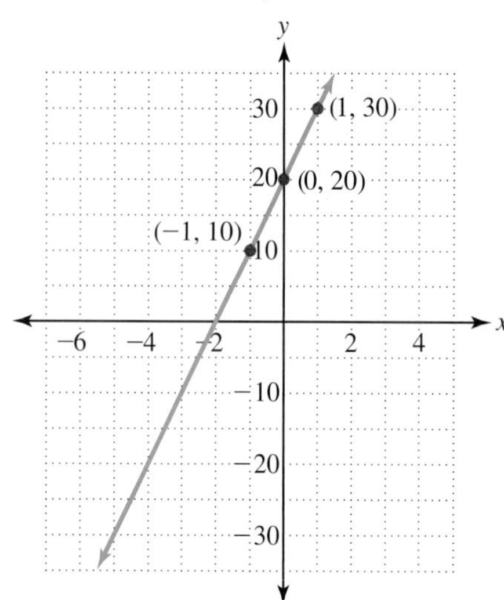

10 times the normal y-value scale

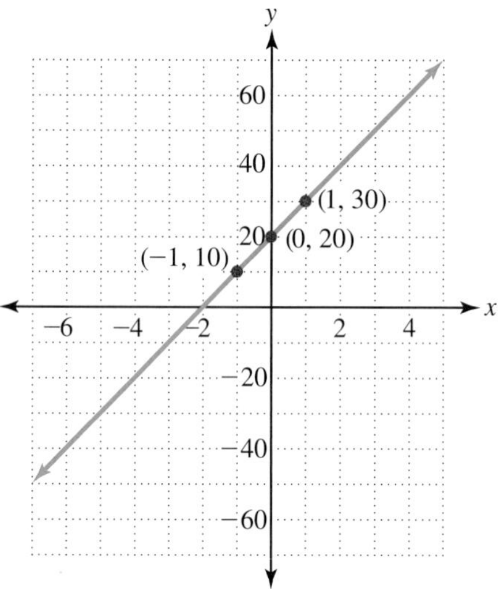

These two lines represent the same line, $y = 10x + 20$, even though they appear to have a different slant to them. The different slant is due to the different scales being used.

You Try It 4 **Graph the line of the given equation by finding three ordered pairs solutions. Use the values of x given in the table.**

a) $y = 10x - 5$

b) $y = -40x + 20$

a)

x	$y = 10x - 5$	(x, y)
-2		
0		
2		
3		

b)

x	$y = -40x + 20$	(x, y)
-1		
0		
1		
2		

5 times the normal y-value scale

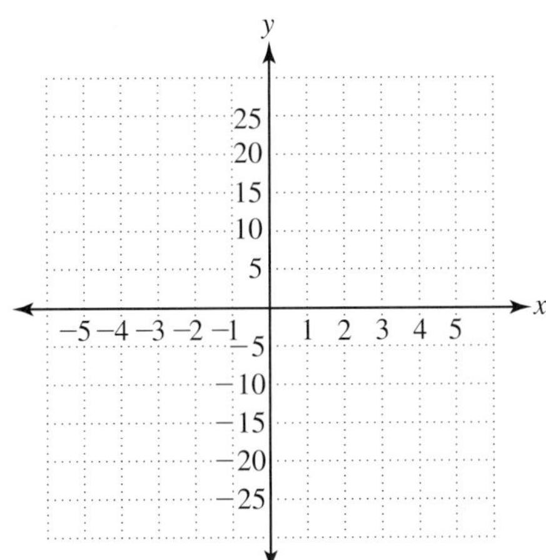

10 times the normal y-value scale

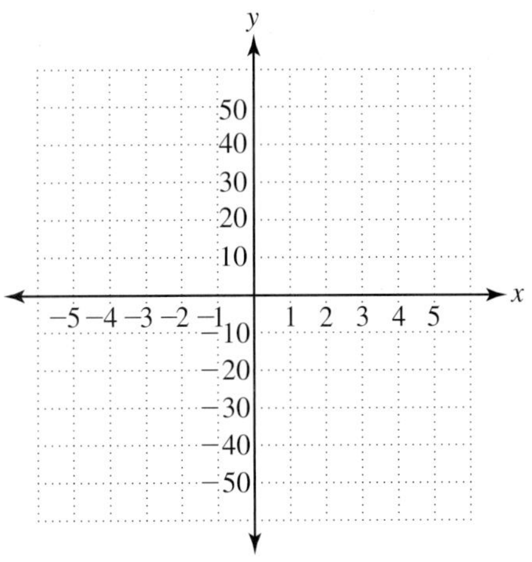

Graphing Horizontal and Vertical Lines

An equation that, at first, may seem a bit unusual is one in which no x appears, such as $y = -3$. In truth, the equation does have an x; we just don't see it because the coefficient of x is 0. The equation is really

$$0x + y = -3.$$

As we see in the next example, no matter what value is chosen for x, the y-value is always -3. This is what the equation actually says: $y = -3$, which means "y is -3."

Example 3

Graph the line of equation $y = -3$.

Procedure First, think of the equation as $0x + y = -3$. Then choose three values of x, find the corresponding values of y, plot the points, and draw the line that passes through them.

Answer

x	$0x + y = -3$	(x, y)
8	$0(8) + y = -3$ $y = -3$	$(8, -3)$
0	$0(0) + y = -3$ $y = -3$	$(0, -3)$
−6	$0(-6) + y = -3$ $y = -3$	$(-6, -3)$

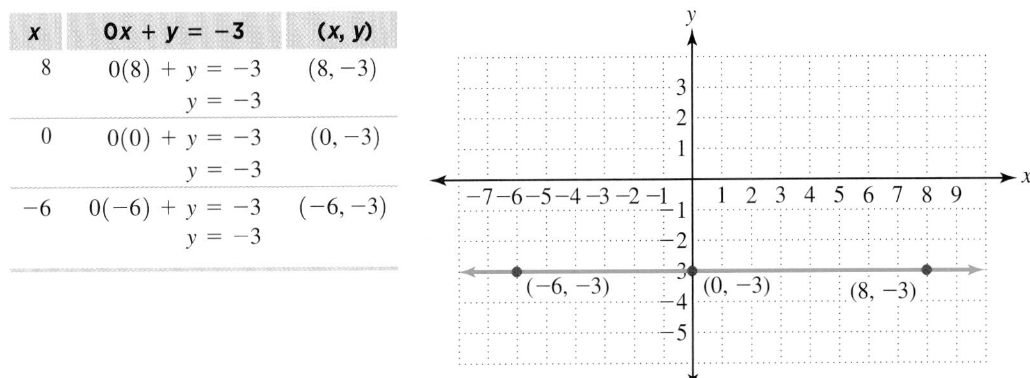

The graph of $y = -3$ is a horizontal line, and each y-coordinate is -3. This means that in identifying three points on the line, we can choose any x-coordinate as long as we write -3 for the y-coordinate.

Let's draw the graph of $y = 4$. This graph will be another horizontal line, and we won't have to create a table of values. Instead, we can assign any three points to it as long as the y-coordinate is 4.

Example 4

Graph the line of equation $y = 4$.

Procedure This equation says that y is always 4 no matter the value of x. The graph is a horizontal line. We can choose a variety of x-values, but the y-value is always 4:

$$(-2, 4), (0, 4), \text{ and } (6, 4)$$

Answer

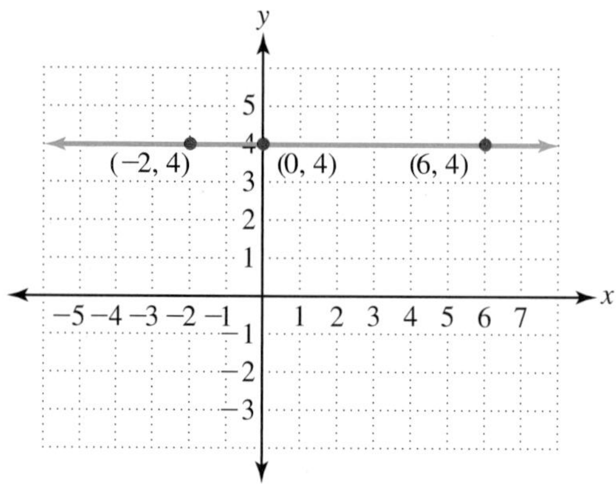

How would the graph be different if the equation had an x-term but no y-term, such as $x = 5$ or $x = -2$? First, each of them could be written containing a y-term with coefficient 0:

$$x + 0y = 5 \text{ and } x + 0y = -2$$

Second, although similar in form to the horizontal lines, they would not be horizontal. Instead, they would be vertical lines, straight up and down. In each case, the x-value would never change.

As in Example 4, let's graph these vertical lines. There is no need for a table, but we should identify three points on the line.

Example 5

Graph the line of each equation.

a) $x = 5$ **b)** $x = -2$

Procedure Each is a vertical line.

Answer

a) The value of x is always 5. For example, these three points are on the line: (5, 2), (5, 0), and (5, −4).

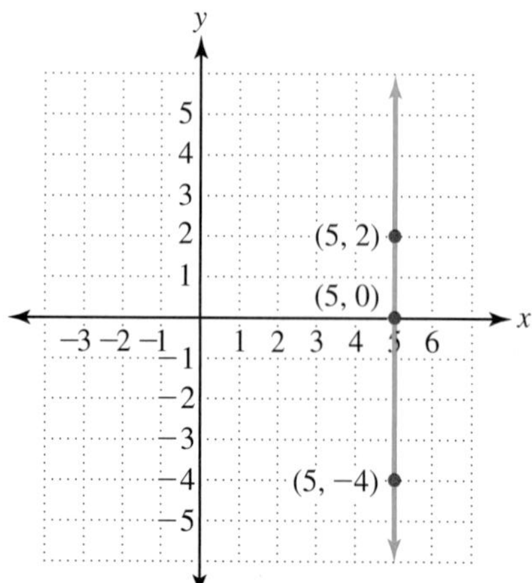

b) The value of x is always −2. For example, these three points are on the line: (−2, 4), (−2, 0), and (−2, −5).

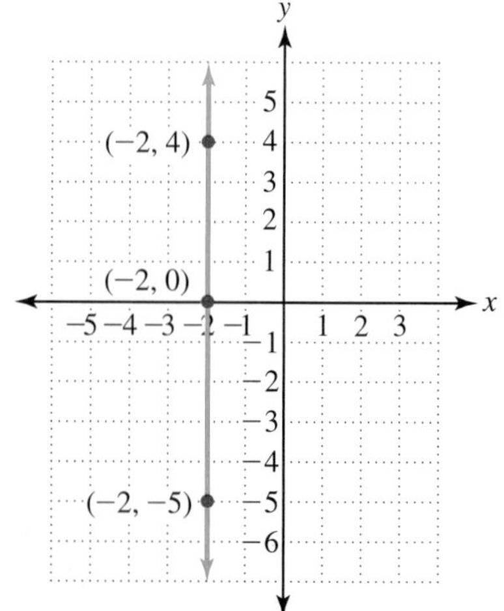

You Try It 5 **Graph each of these lines. Use Examples 3, 4, and 5 as guides.**

a) $y = 2$

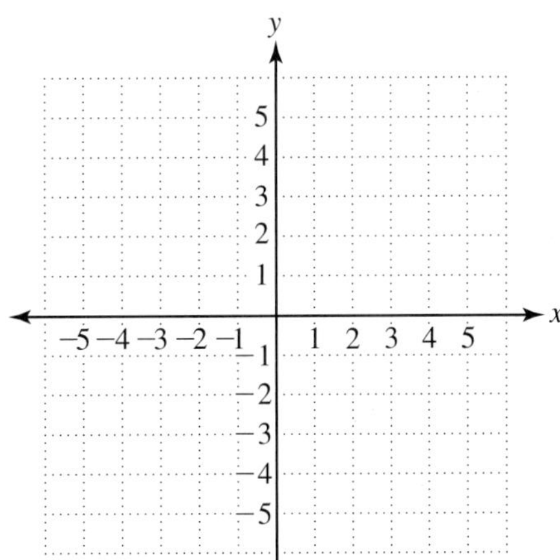

b) $x = -4$

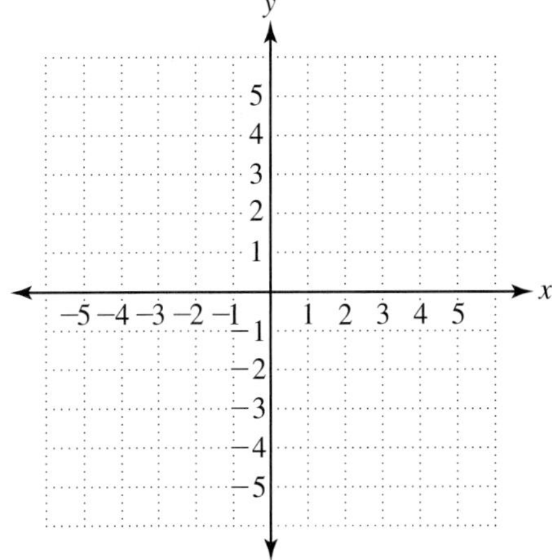

Answers: You Try It and Think About It

You Try It: *Some of the points you find may be different from the ones displayed here, but they should be on the same line.*

1.

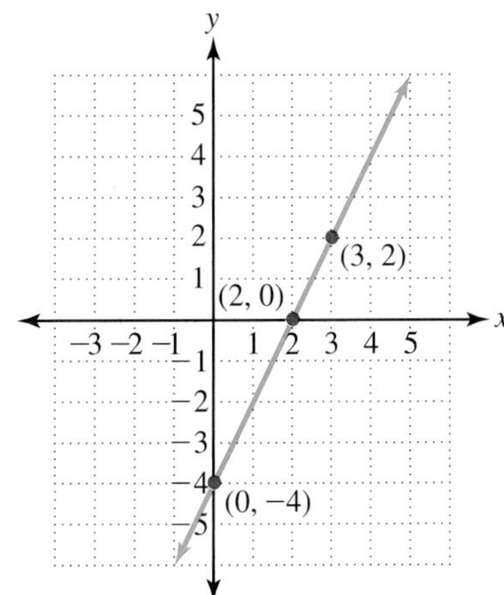

2.

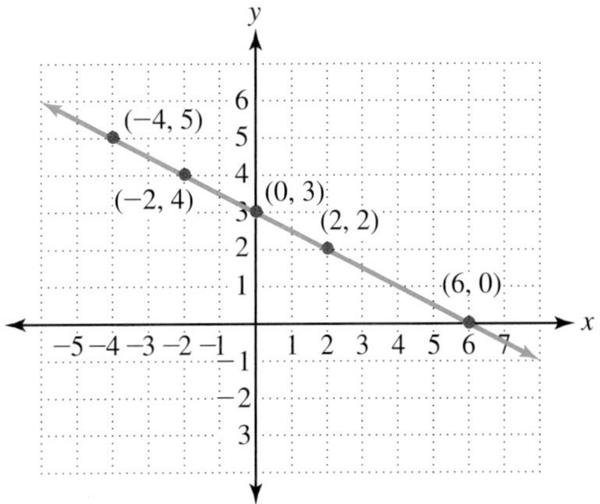

3. a.

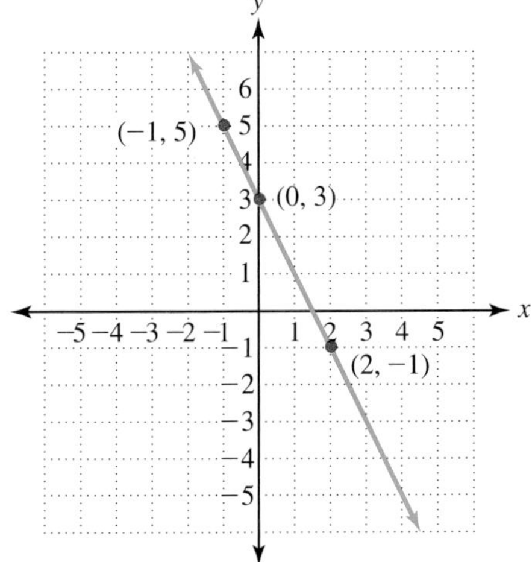

b.

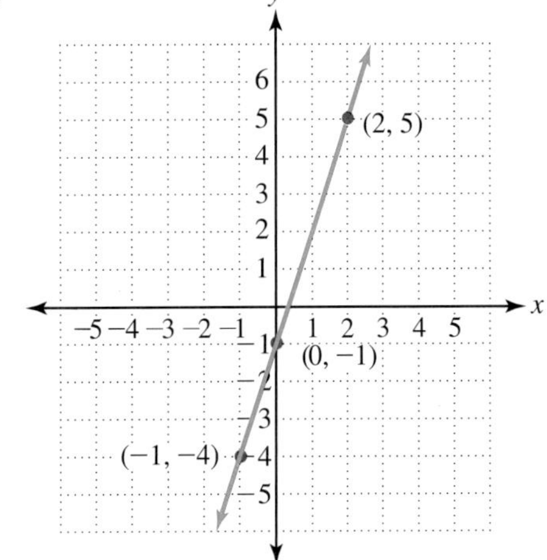

4. a. 5 times the normal y-value scale

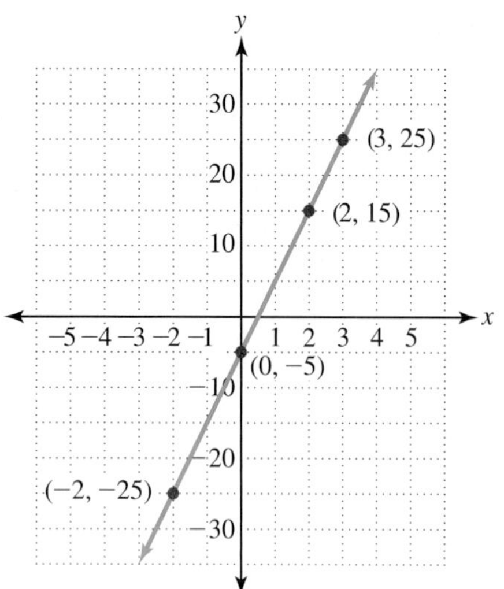

b. 10 times the normal y-value scale

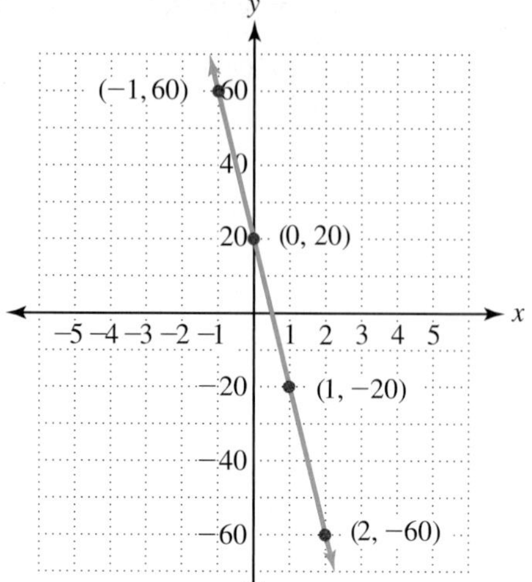

5. a.

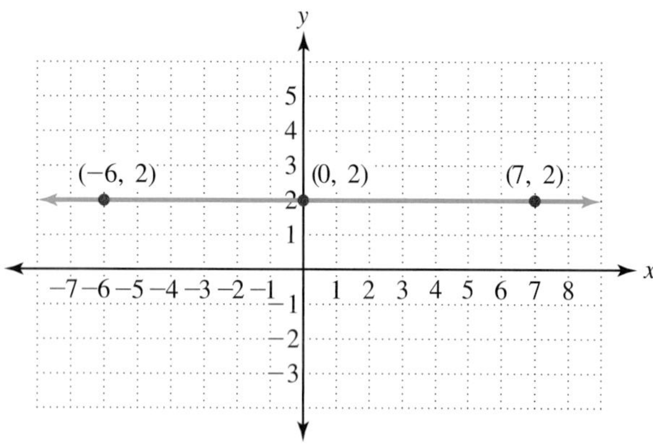

b.

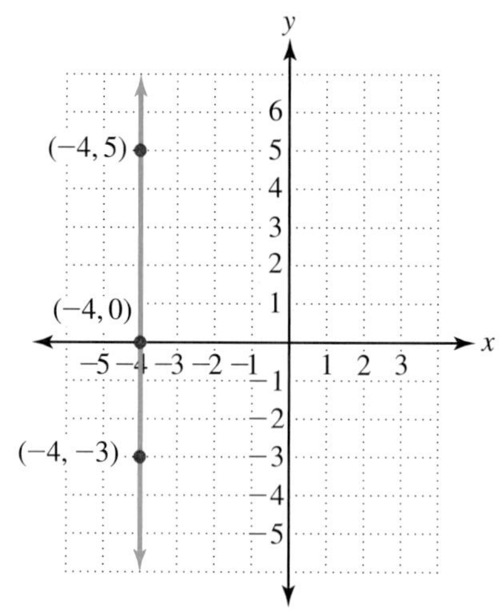

Think About It: **1.** Answers may vary. One possibility: After choosing 0 for x, it is best to choose positive or negative values of x that are multiples of 3, such as 3, -3, 6, and -6, to divide out with the denominator of 3.

Section 9.2 Exercises

FOR EXTRA HELP
Student Resources on DVD-ROM
Includes
➤ Student's Solutions Manual
➤ Video Lectures
➤ Chapter Test Prep Video
MyMathLab
MathXL

Think Again

1. Consider the equation $y = \frac{2}{3}x + 5$. What values of x would you choose to find *four* points on the line? Explain your answer. (*Refer to Think About It 1*)
   Answers will vary. Some good choices are any multiples of 3, such as -6, -3, 0, 3, and 6. Multiples of 3 are good choices because the denominator of the fraction can be completely divided out and each y-value will be an integer.

2. What is another name for the graph of $y = 0$?
   The x-axis

3. Three points in the x-y plane are not on the same line. How many lines can be drawn in the plane that pass through at least two of the points? Explain your answer or show an example that supports your answer. You may want to draw a graph to assist you in your explanation.
   Three lines can be drawn. If the points are represented by A, B, and C, one line can be drawn through A and B, a second line through A and C, and a third line through B and C.

4. Why is it good to find at least three points on a line before graphing the line?
   Finding a third point is a way of checking for accuracy. A line can pass through the two points; but if a third point is found and is not on that line, at least one of the three points is incorrect.

Focus Exercises

Graph the line with the given equation by finding three points on the line.

5. $x + y = 6$

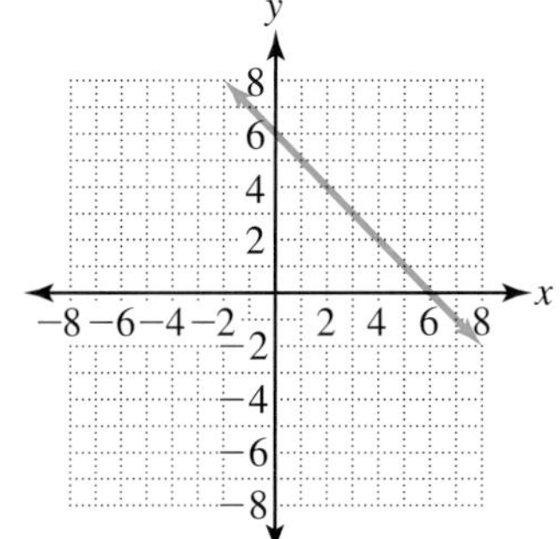

6. $x + y = -1$

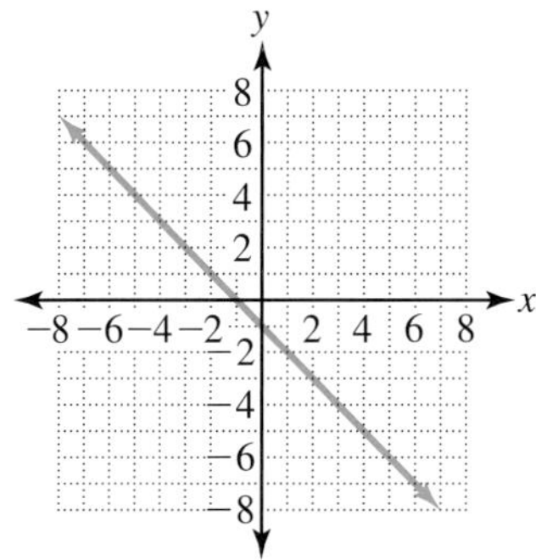

7. $x - y = -3$

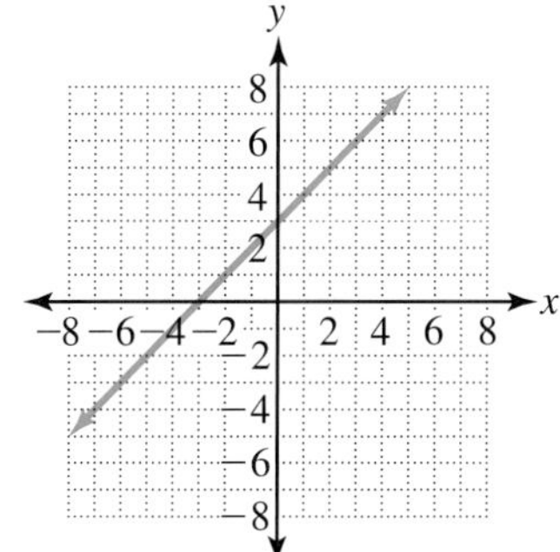

8. $x - y = 2$

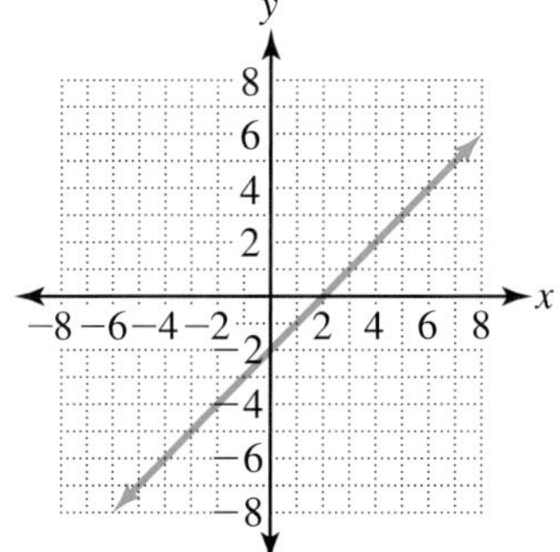

9. $x + y = 0$

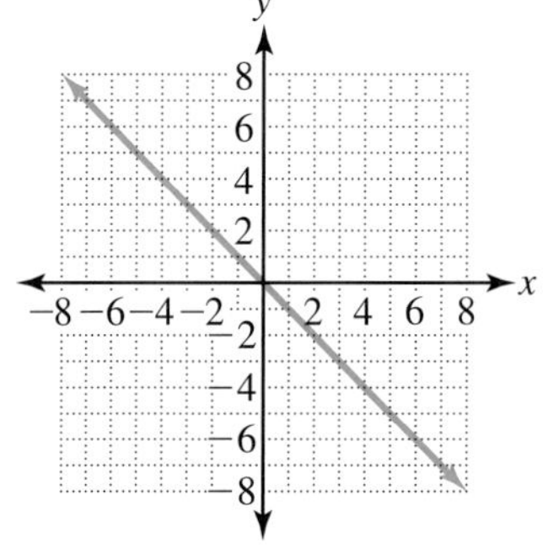

10. $x - y = 0$

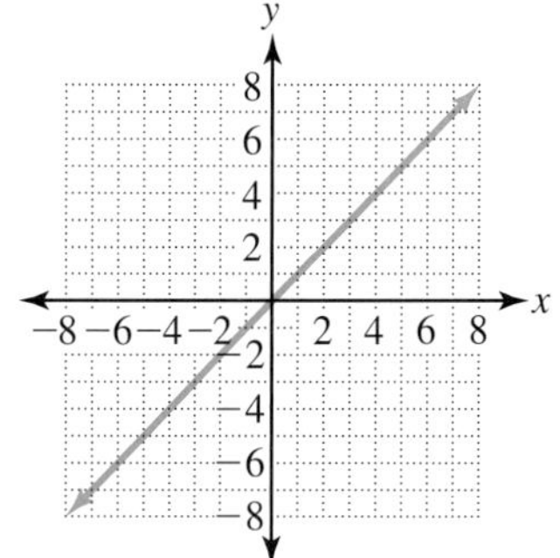

11. $x + 2y = 4$

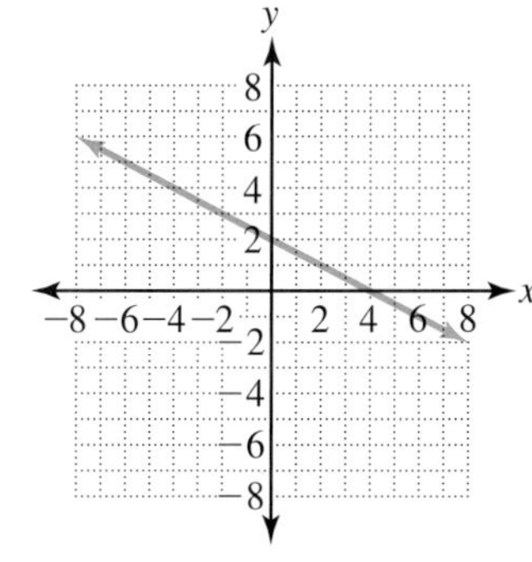

12. $x + 3y = 6$

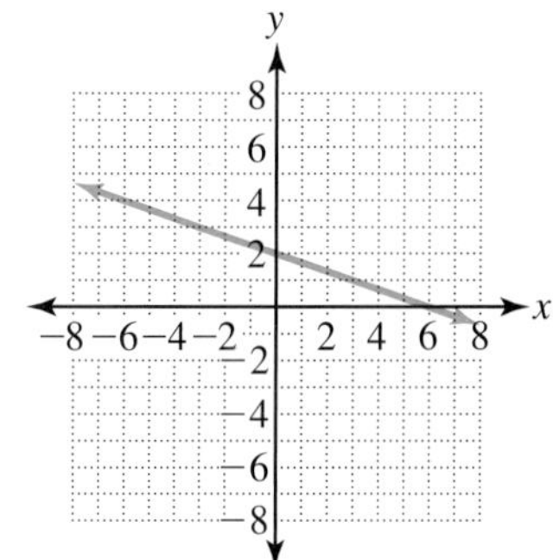

13. $x - 2y = -2$

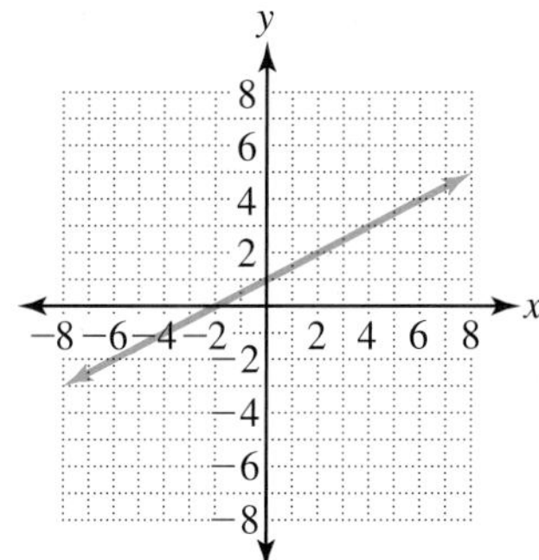

14. $x - 3y = -3$

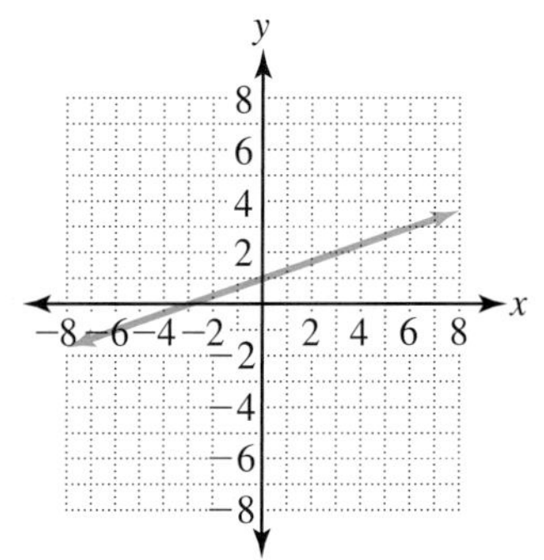

15. $2x + y = -2$

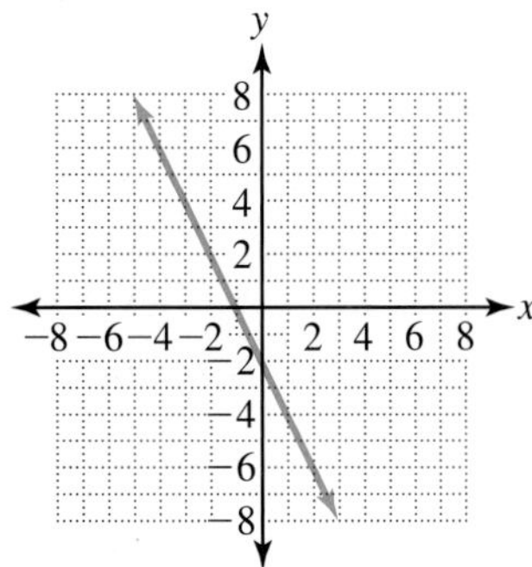

16. $3x + y = -3$

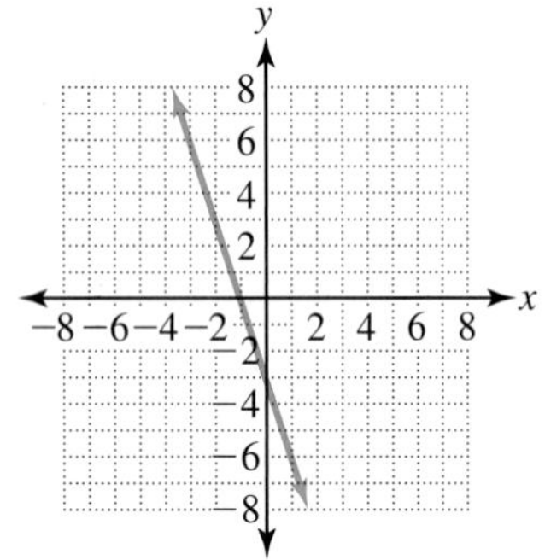

17. $4x + y = 4$

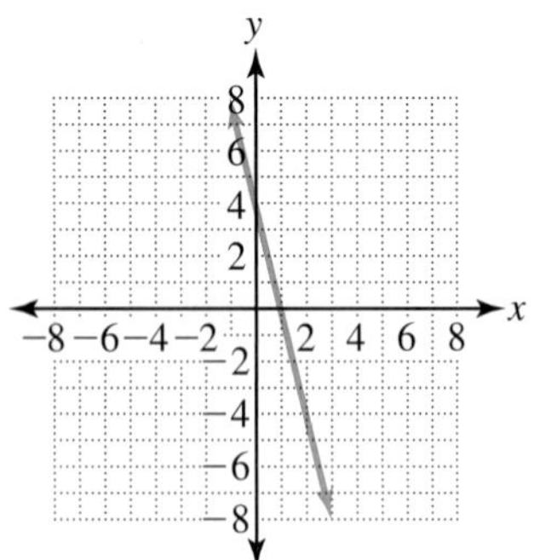

18. $2x + y = 2$

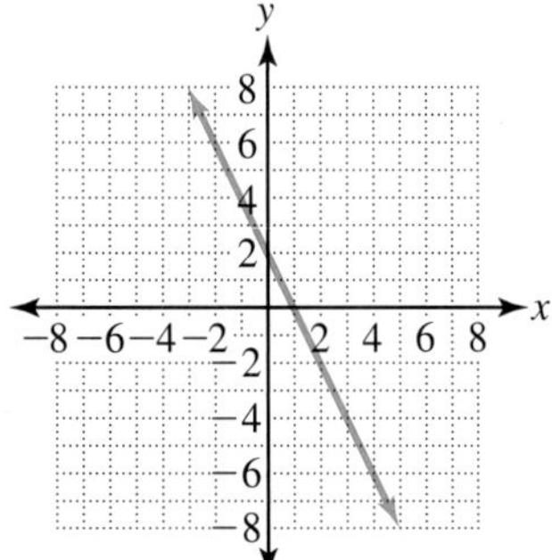

19. $4x - y = 0$

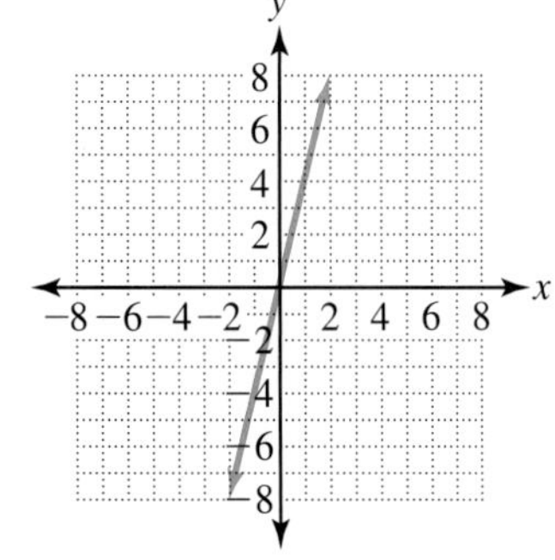

20. $5x - y = 0$

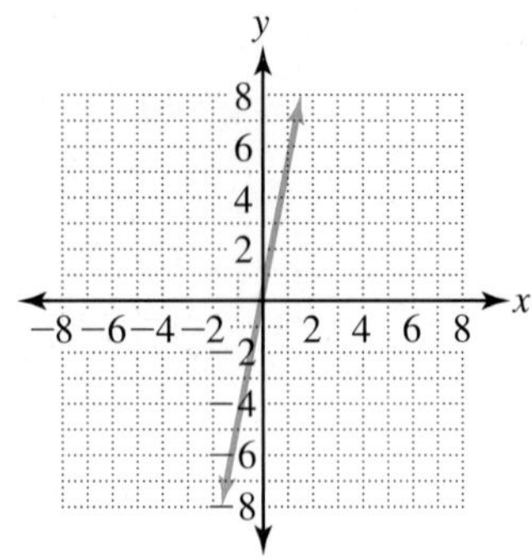

Graph the line with the given equation by finding three points on the line.

21. $y = x + 3$

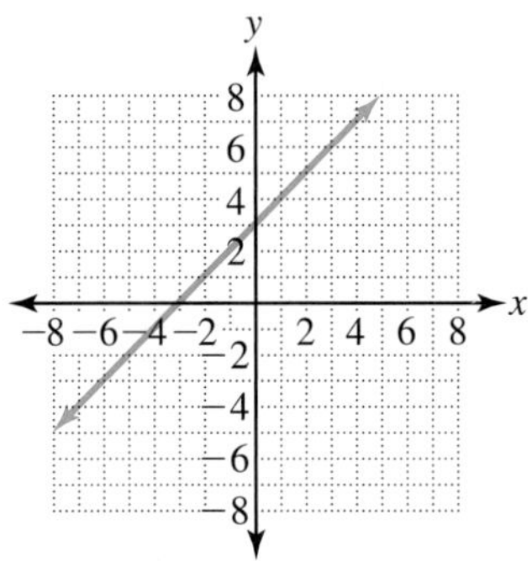

22. $y = x + 5$

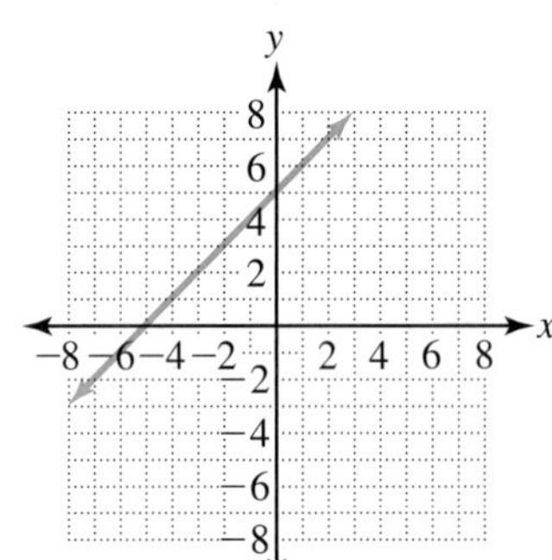

23. $y = x - 2$

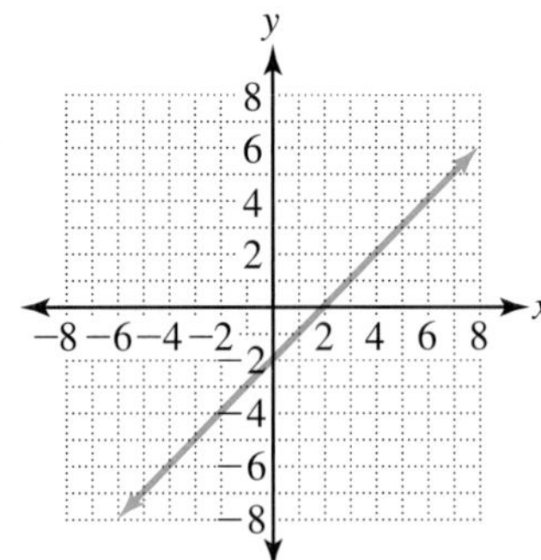

24. $y = x - 6$

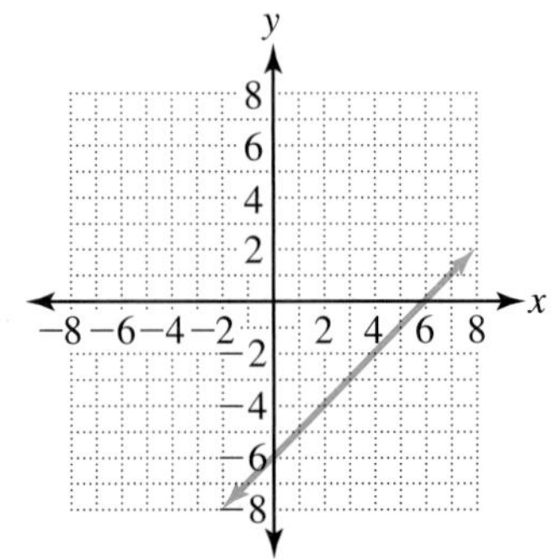

25. $y = -x + 4$

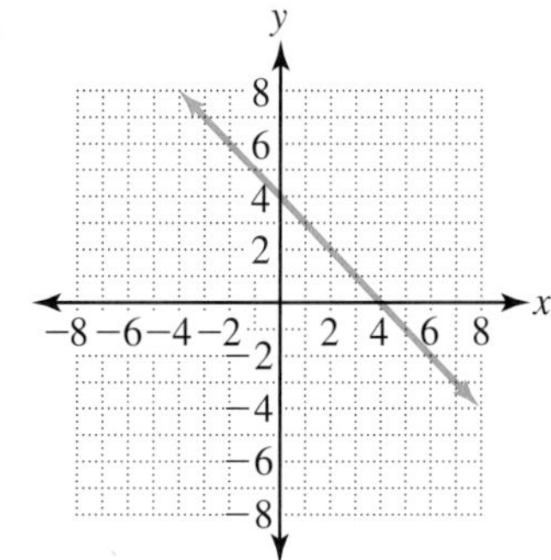

26. $y = -x + 1$

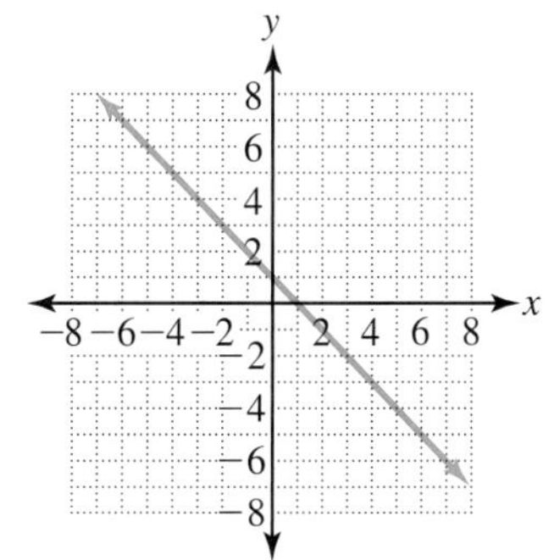

27. $y = -x - 3$

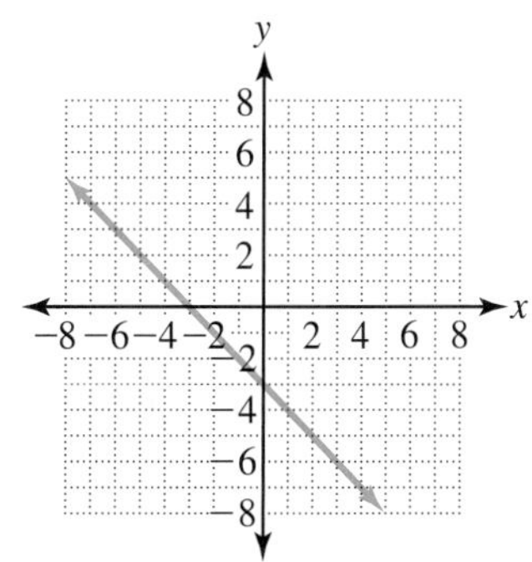

28. $y = -x$

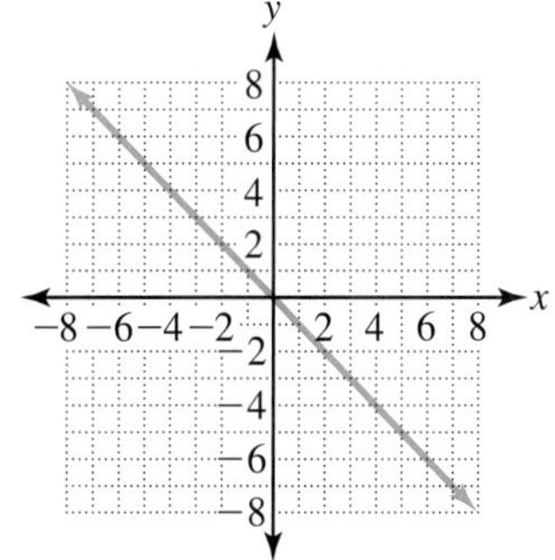

29. $y = 2x - 3$

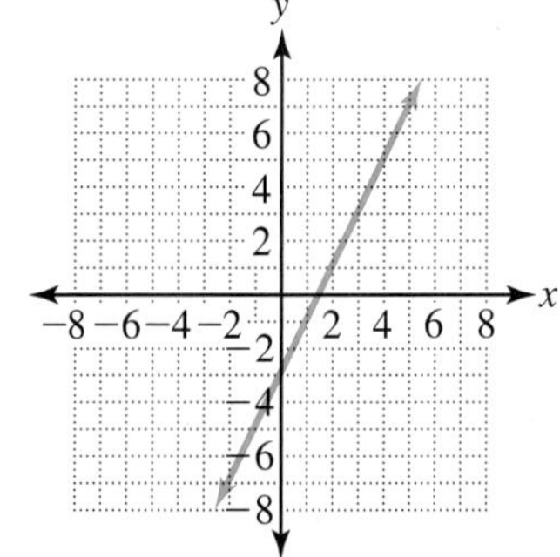

30. $y = 3x - 1$

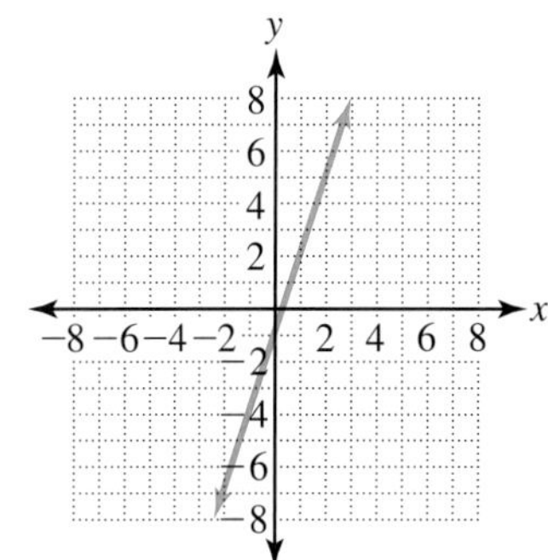

31. $y = -3x + 4$

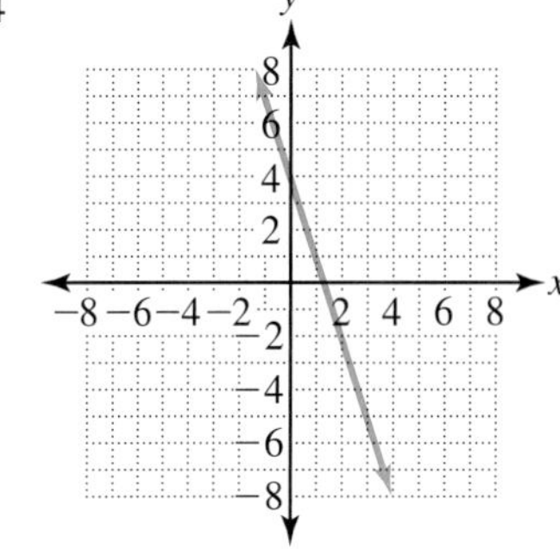

32. $y = -2x + 3$

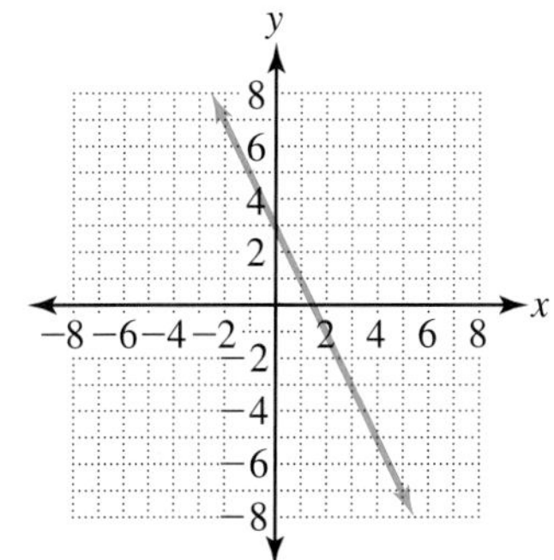

33. $y = 4x$

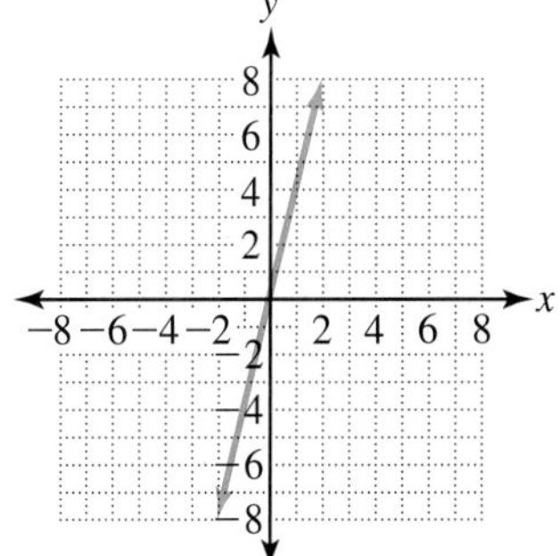

34. $y = 5x$

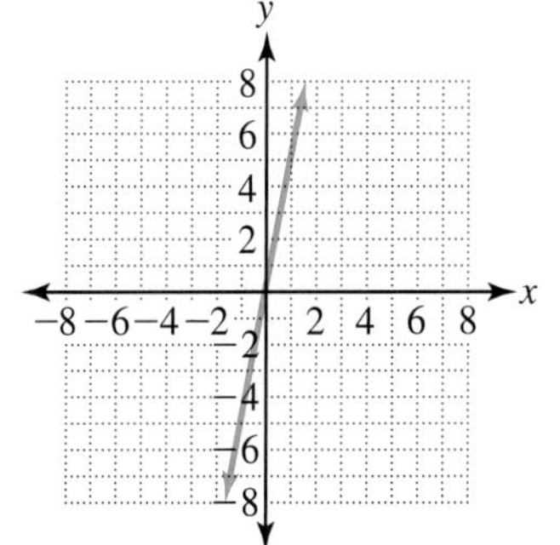

35. $y = -2x$

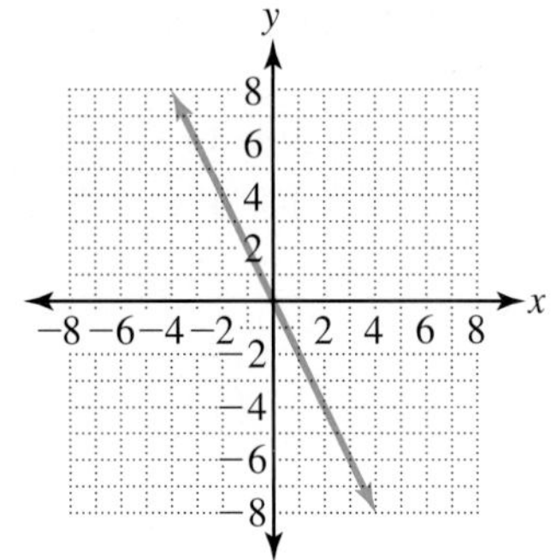

36. $y = -3x$

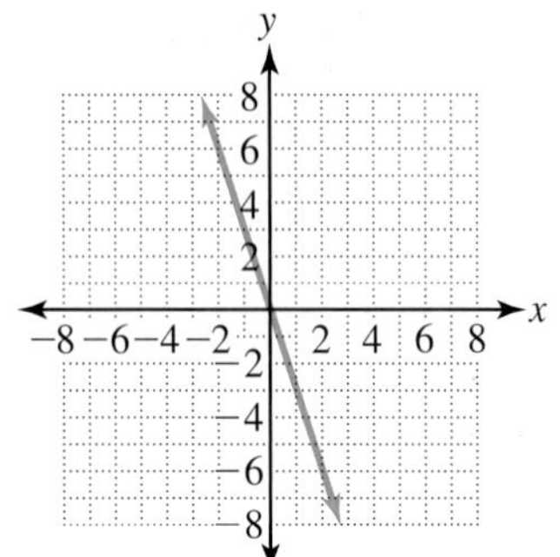

37. $y = \frac{1}{2}x - 5$

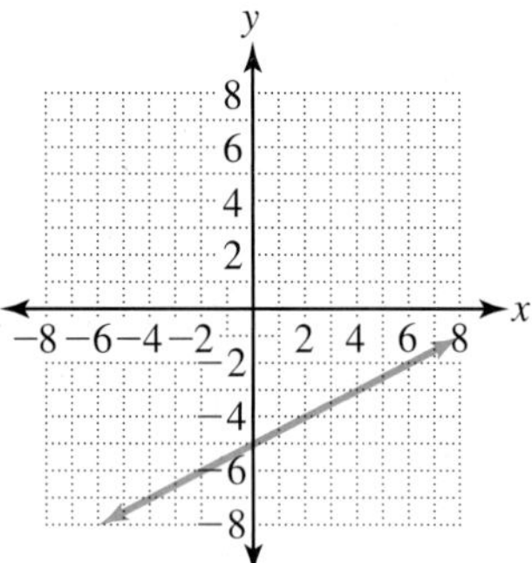

38. $y = \frac{1}{3}x - 4$

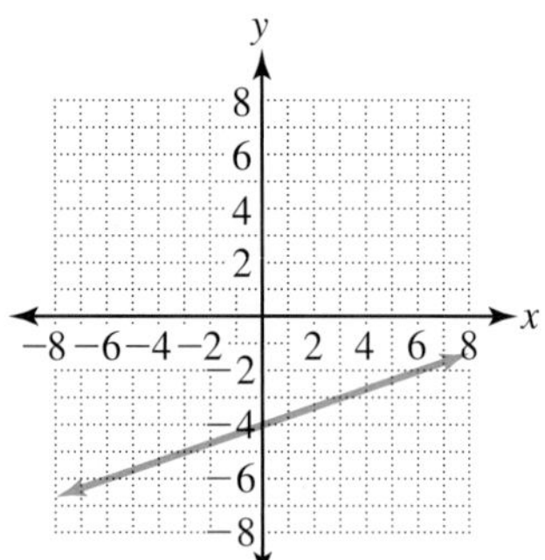

39. $y = -\frac{1}{2}x + 2$

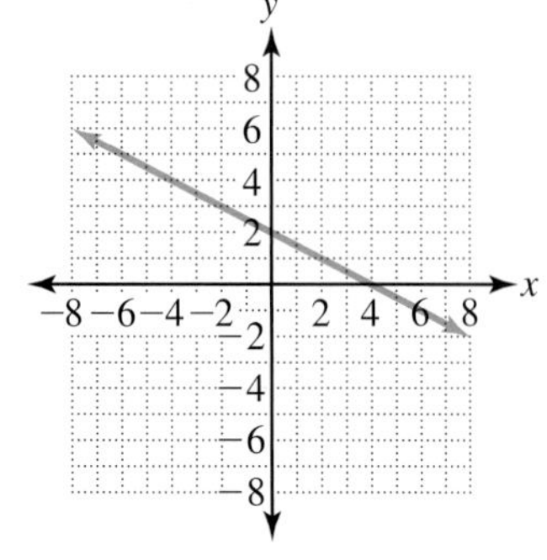

40. $y = -\frac{1}{3}x + 5$

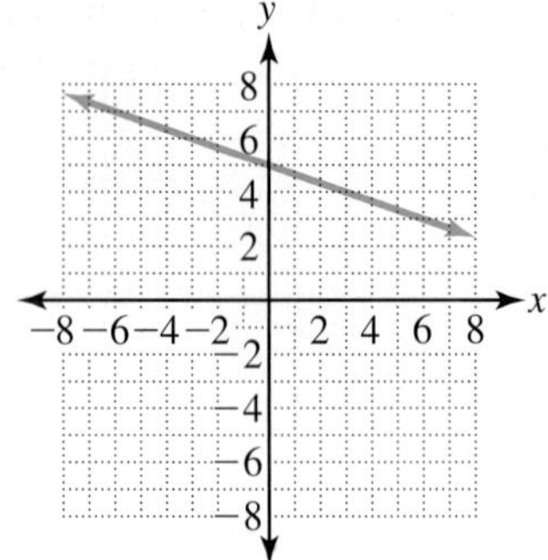

41. $y = \frac{3}{2}x - 4$

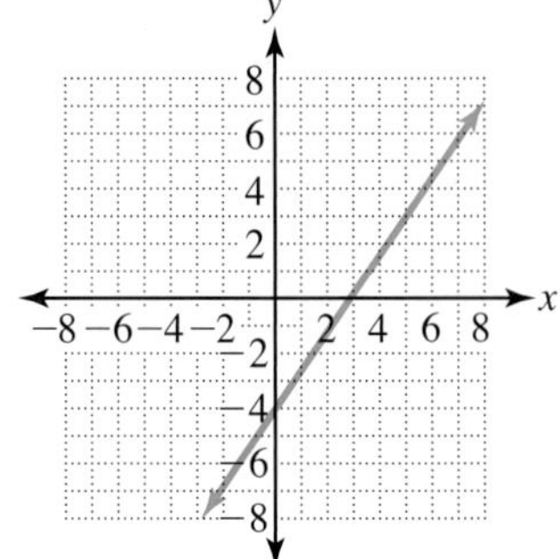

42. $y = \frac{2}{3}x + 3$

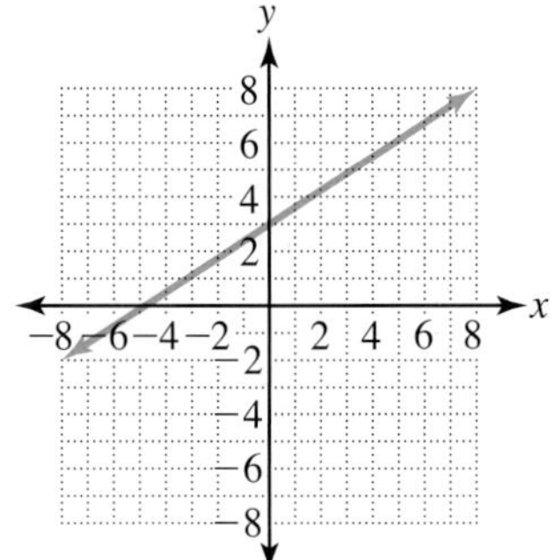

43. $y = \frac{1}{4}x - 3$

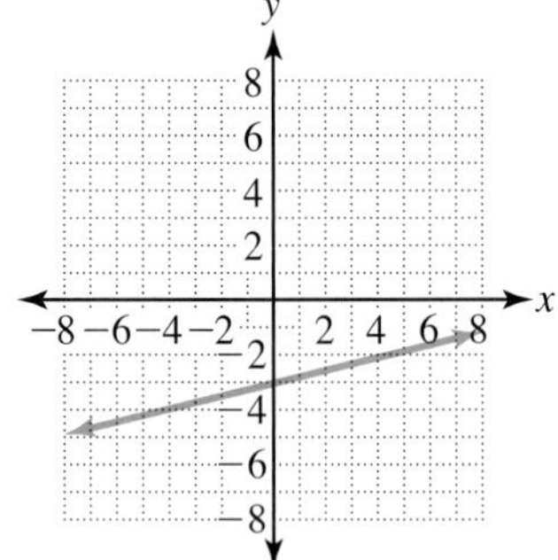

44. $y = -\frac{3}{4}x + 1$

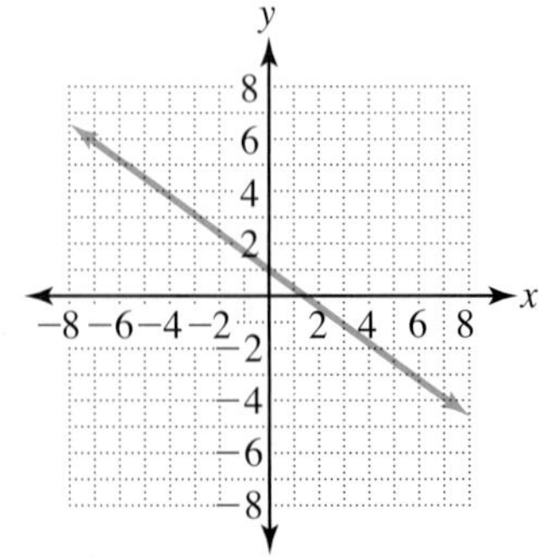

Graph the line with the given equation by finding three points on the line. For each, a suggested larger scale is given.

45. $y = 10x - 15$

Use a scale that is 5 times the normal scale.

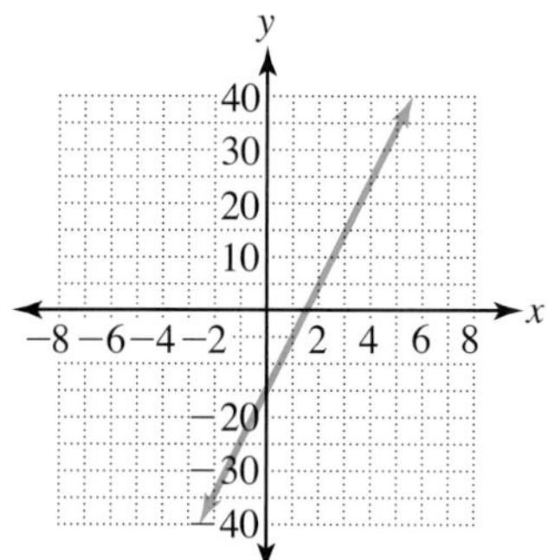

46. $y = 30x - 20$

Use a scale that is 10 times the normal scale.

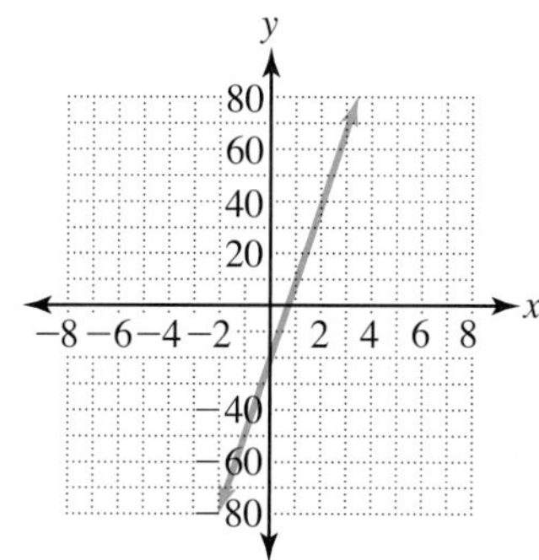

47. $y = -40x - 20$

Use a scale that is 20 times the normal scale.

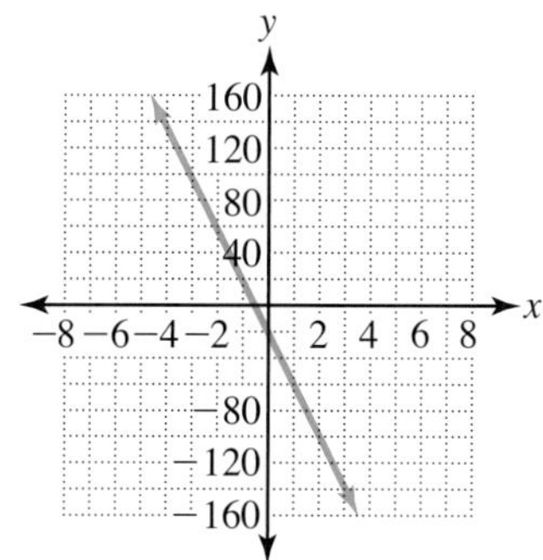

48. $y = -150x + 50$

Use a scale that is 50 times the normal scale.

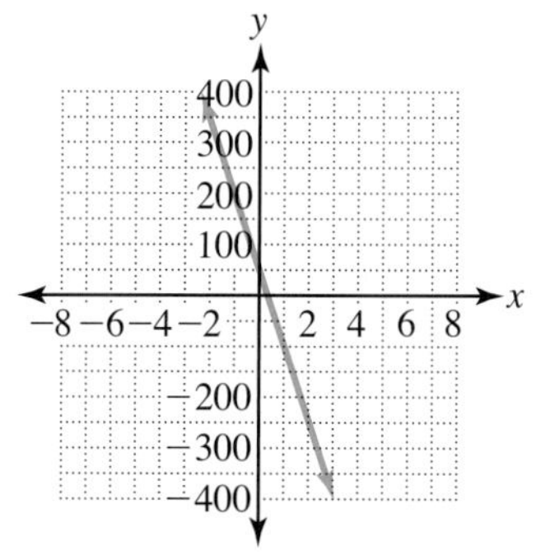

Graph the line with the given equation by identifying three points on the line.

49. $y = 3$

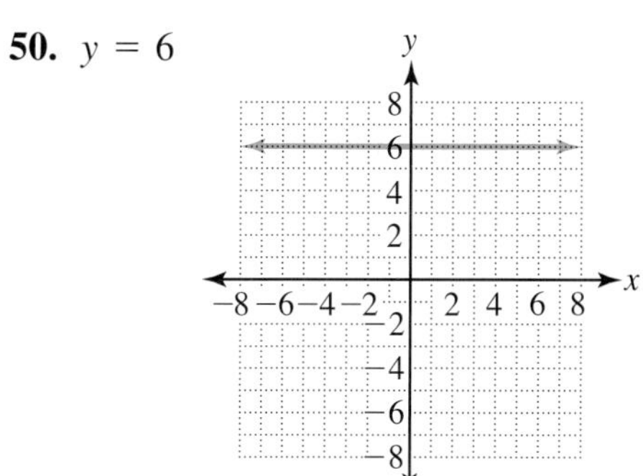

50. $y = 6$

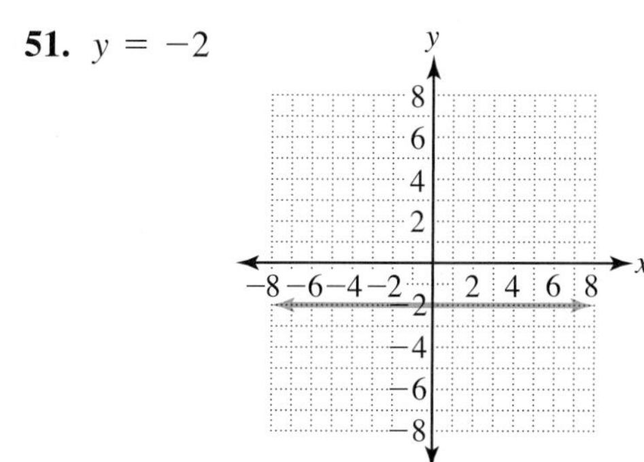

51. $y = -2$

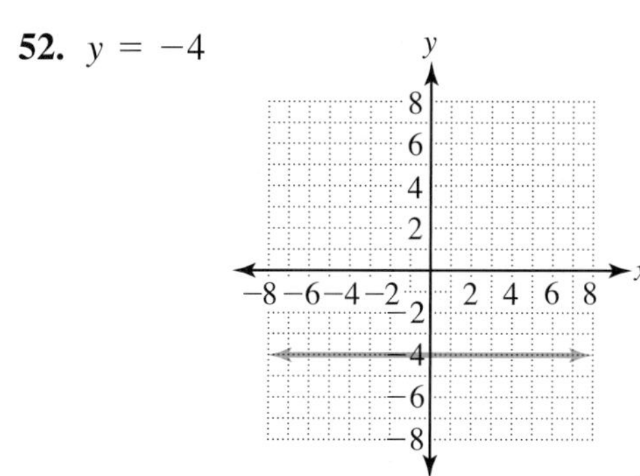

52. $y = -4$

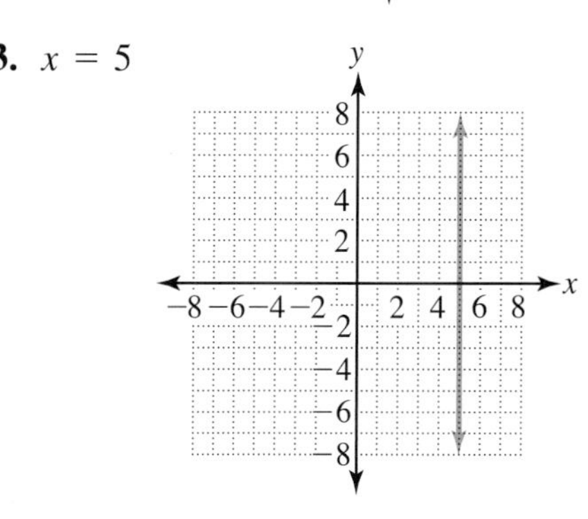

53. $x = 5$

54. $x = 3$

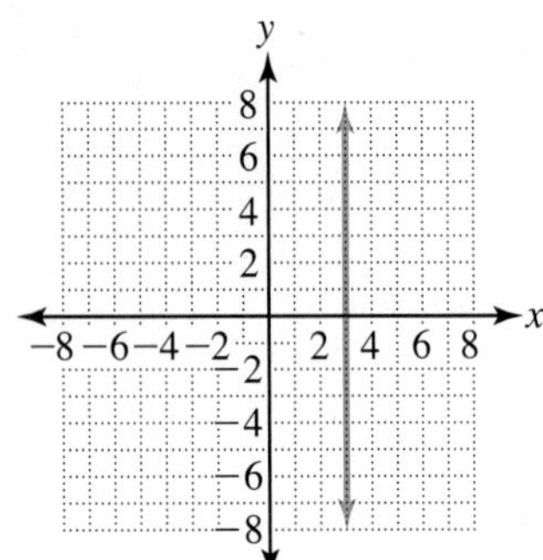

55. $x = -1$

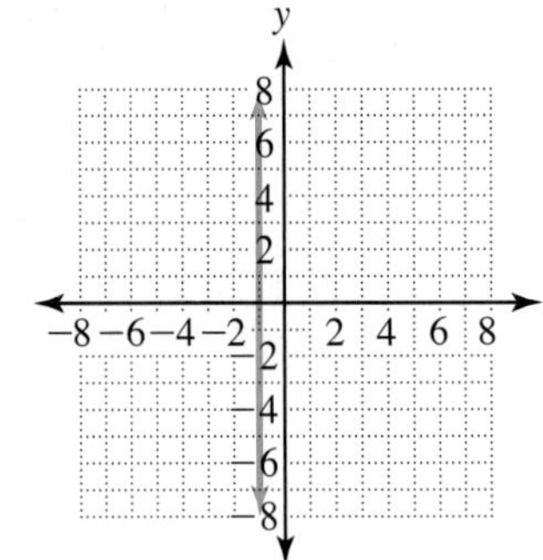

56. $x = -7$

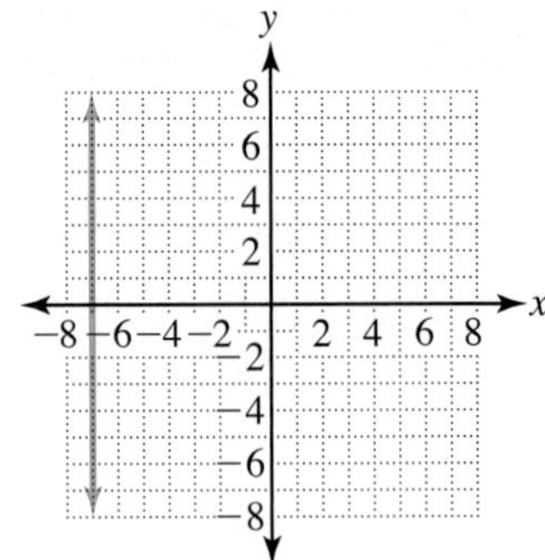

Think Outside the Box

The graph of these equations are not lines. Instead, the points are connected by smooth curves. The graphs are called *parabolas*. For each equation, find the ordered pairs and plot them in the *x*-*y* plane. Then connect the dots as smoothly as possible to form a parabola.

57. $y = x^2 - 4x + 3$

x	$y = x^2 - 4x + 3$	(x, y)
−1		
0		
1		
2		
3		
4		
5		

58. $y = x^2 + 2x - 3$

x	$y = x^2 + 2x - 3$	(x, y)
−4		
−3		
−2		
−1		
0		
1		
2		

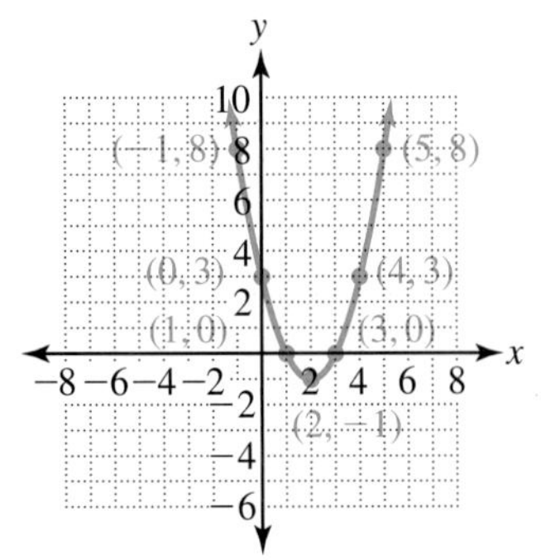

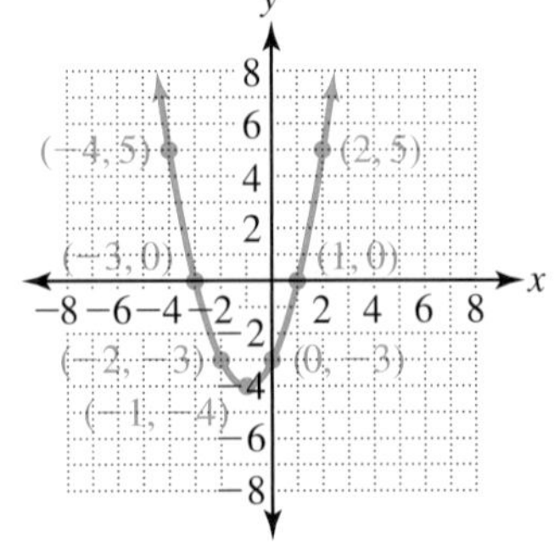

SECTION 9.3 Graphing Data: Line Graphs and Bar Graphs

OBJECTIVES

In this section, you will learn to

- Interpret a data table.
- Draw and interpret the data in a line graph.
- Calculate percent increase and decrease from data.
- Draw and interpret the data in a bar graph.

You Need to Know

To successfully complete this section, you need to understand

- ☐ Simplifying fractions (4.3)
- ☐ Multiplying fractions (4.4)
- ☐ Multiplying decimals (5.4)
- ☐ Writing decimals as percents (6.4)
- ☐ Percent increase and decrease (6.7)
- ☐ Plotting points in the *x-y* plane (9.1)
- ☐ Graphing lines (9.2)

Introduction

The lines we graphed in Section 9.2 were straight and consistent in their direction. Each of those lines was based on a formula, an equation, that compared x- and y-values, such as $y = 2x + 1$. From this equation, we built a table of values and the table of values gave us an ordered pair, (x, y).

$y = 2x + 1$

x	y	(x, y)
−2	−3	(−2, −3)
0	1	(0, 1)
2	5	(2, 5)

In this section, there are some similarities to graphing the lines of Section 9.2, such as having horizontal and vertical axes and plotting points on a grid. There are also many differences, as you'll soon see. In this section, we still have a table of values, but the values are not typically x- and y-values.

For example, at a bank, we might have some information that compares the time of day to the length of time a customer must wait in line. Instead of x and y, the values might be *Time of day (Hour)* and *Wait time, in seconds.*

Hour	Wait time, in seconds
9:00	30
10:00	150
11:00	120

To graph this information, we use a grid, like the x-y plane, but the axes are labeled differently and the graph is usually not a straight and consistent line.

Drawing and Reading Line Graphs

Consider this: Eli is the manager at Redlands Bank. He did a study of the length of time a customer must wait in line at the bank at certain times of the day. Eli collected the following information, called **data,** and organized it in a **data table,** shown at the right.

Hour	Wait time, in seconds
9:00	30
10:00	160
11:00	120
12:00	240
1:00	80
2:00	60
3:00	105
4:00	180

Eli then made a graph of the data by plotting the points on a grid and connecting the points one by one as shown.

This type of graph is called a **line graph.** A line graph is typically used to compare the progression of data over a period of time, which is how we'll use a line graph in this text.

The line graph gives us the same information as the data in the table, but it allows us to see changes more easily from one point in time to the next. In the case of waiting time at Redlands Bank, we can see that a customer entering the bank at 12:00 will have to wait longer than a customer entering the bank at 1:00.

Based on the data, Eli can make decisions about how many teller windows should be open at any one time or the best time for a teller to take a break.

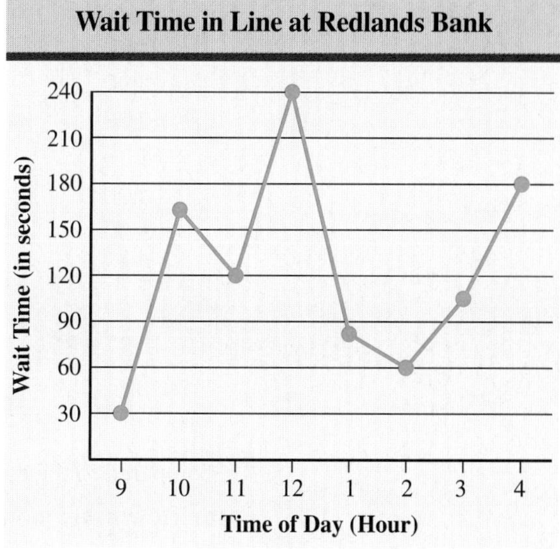

Example 1

Beatriz normally comes to Redlands Bank at about 11:00. According to the graph or the table of wait times for Redlands Bank customers, answer each of the following questions.

a) How many more or fewer seconds will Beatriz have to wait in line if she enters the bank at 4:00 instead?

b) How many more or fewer seconds will Beatriz have to wait in line if she enters the bank at 9:00 instead?

c) At what time would Beatriz need to wait twice as long in line?

d) At what time would Beatriz need to wait only half as long in line?

Procedure For parts a and b, to find more or fewer seconds requires that we subtract. For parts c and d, we must multiply to find the number of seconds asked for and then identify the corresponding time. For each, Beatriz's wait time at 11:00 is 120 seconds.

Answer **a)** The wait time at 4:00 is 180 seconds (a longer wait): $180 - 120 = 60$.

Beatriz will wait 60 more seconds if she enters the bank at 4:00.

b) The wait time at 9:00 is 30 seconds (a shorter wait): $120 - 30 = 90$.

Beatriz will wait 90 fewer seconds if she enters the bank at 9:00.

c) Twice (2 times) 120 seconds is 240 seconds: $2 \times 120 = 240$, and that occurs at 12:00.

Beatriz must wait twice as long if she enters the bank at 12:00.

d) Half of 120 seconds is 60 seconds: $\frac{1}{2} \times 120 = 60$, and that occurs at 2:00.

Beatriz will need to wait only half as long if she enters the bank at 2:00.

You Try It 1 **The Book Tree, a local seller of children's books, also sells books online. Each hour, Jackie prints out the new online orders and prepares the books for shipping. The table represents the number of online book orders received, on average, each hour. Draw the associated line graph and answer each question based on this data.**

a) Draw the line graph for this data table.

Hour	Number of online book orders
10:00	75
11:00	21
12:00	38
1:00	63
2:00	46
3:00	25
4:00	25
5:00	7

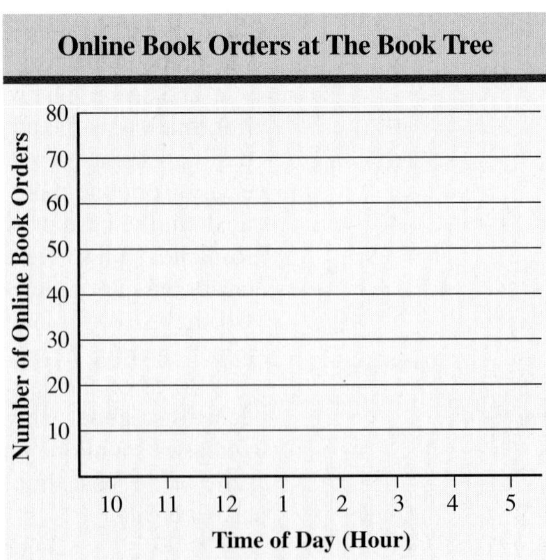

Compared to the number of book orders received at 11:00,

b) How many more book orders were received at 2:00?

c) At what time did Jackie receive three times as many book orders?

d) At what time did Jackie receive one-third as many book orders?

A line graph may have other time measures along the horizontal axis, such as days, months, or years. Also, sometimes the data is hard to read on a full vertical scale that starts at 0. When this happens, it is common to show a break in the vertical axis and to magnify its scale.

For example, consider the minimum wage progression in the state of Washington from 1999 to 2006. If we graph the data using a vertical scale starting at 0, we might get something like this:

Year	Minimum wage
1999	$5.70
2000	$6.50
2001	$6.72
2002	$6.90
2003	$7.01
2004	$7.16
2005	$7.35
2006	$7.63

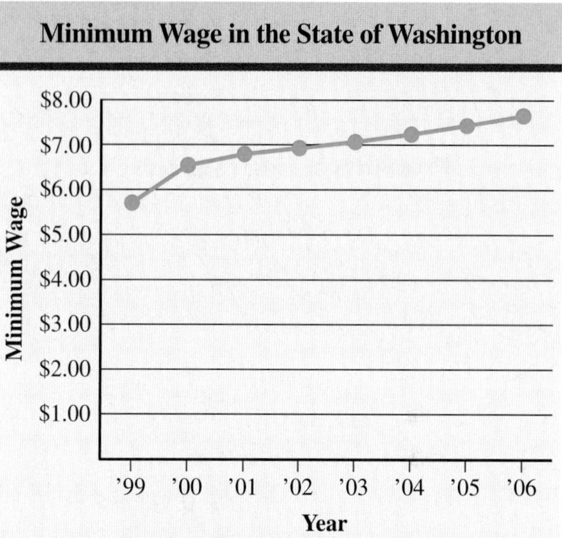

Breaking the vertical axis so that it doesn't start at 0 allows us to better see the increases and decreases between the data.

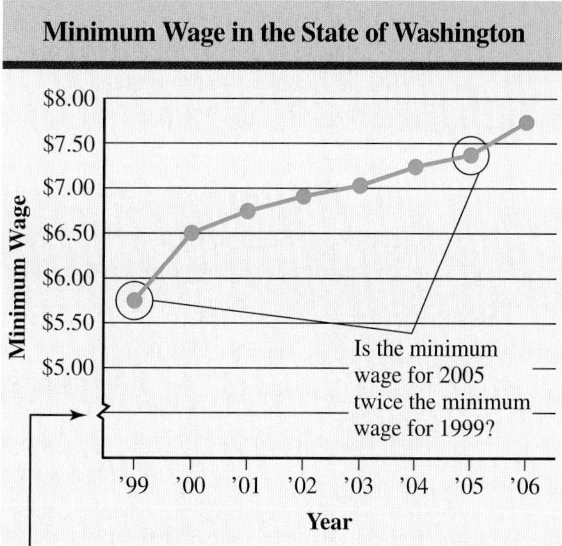

Caution When we break the vertical axis, we must be careful when comparing data. For example, in looking at the previous graph, it *appears* as though the minimum wage for 2005 ($7.35) is twice the minimum wage for 1999 ($5.70) when, in fact, it isn't.

You Try It 2 **The table indicates the daily attendance for one week at Parkway Theaters. Draw the associated line graph and answer each question based on this data.**

a) Draw the line graph for this data table.

Day of week	Attendance
Monday	1,910
Tuesday	1,820
Wednesday	2,350
Thursday	2,105
Friday	2,730
Saturday	2,960
Sunday	2,245

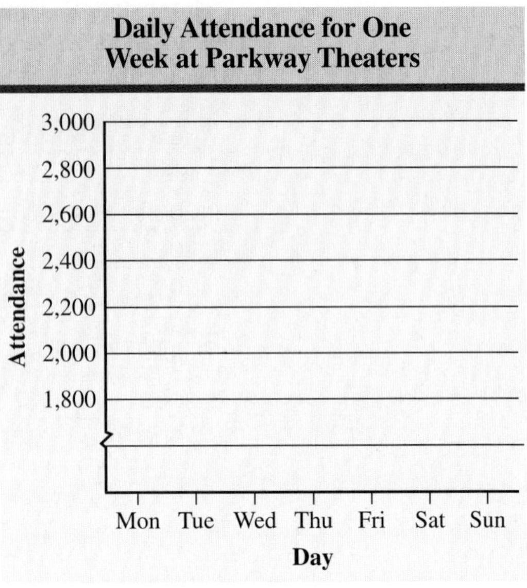

b) What was the difference between the highest and lowest attendance?

c) Which day's attendance was 1.5 times Tuesday's attendance?

d) Which day showed the biggest increase in attendance from the previous day? What was the difference in attendance?

e) Which day showed the biggest decrease in attendance from the previous day? What was the difference in attendance?

Calculating Percent Increase and Decrease from Data

Recall from Section 6.7 that the percent increase (or decrease) is a ratio of the amount of increase (or decrease) compared to the original amount.

$$\text{Percent increase} = \frac{\text{Amount of increase}}{\text{Original amount}}$$

$$\text{Percent decrease} = \frac{\text{Amount of decrease}}{\text{Original amount}}$$

In each ratio, the original amount is the value of an item before the increase or decrease took place. The original amount is not necessarily the value of the item when it is brand new.

Example 2

The following data table shows the estimated yearly decline in value (depreciation) of a $20,000 car.

Age of car, in years	Estimated value of car
0	$20,000
1	$15,000
2	$13,200
3	$11,600
4	$10,000
5	$8,500
6	$7,200

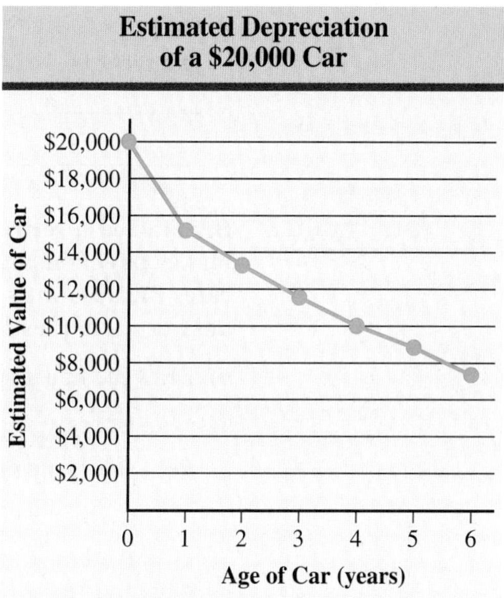

a) What was the percent decrease in value during the first year?

b) What was the percent decrease in value between the third and fourth years?

c) What was the percent decrease in value during the first six years?

Procedure For each question, we must consider the starting year's value and the ending year's value. We must subtract to find the amount of decrease and then compare that to the starting year's value.

Answer **a)** *During the first year* means that the starting value is the value at Year 0 and the ending value is the value at Year 1.

$$\text{Amount of decrease} = \text{Year 0 value} - \text{Year 1 value} = \$20{,}000 - \$15{,}000 = \$5{,}000$$

$$\text{Percent of decrease} = \frac{\text{Amount of decrease}}{\text{Original amount}} = \frac{\$5{,}000}{\$20{,}000} = 0.25 = 25\%$$

The value of the car decreased 25% during the first year.

b) *Between the third and fourth year* means that the "original value" is the Year 3 value.

$$\text{Amount of decrease} = \text{Year 3 value} - \text{Year 4 value} = \$11{,}600 - \$10{,}000 = \$1{,}600$$

$$\text{Percent of decrease} = \frac{\text{Amount of decrease}}{\text{Original amount}} = \frac{\$1{,}600}{\$11{,}600} \approx 0.138 = 13.8\%$$

The value of the car decreased 13.8% between the third and fourth years.

c) *During the first six years* means that the starting value is the value at Year 0 and the ending value is the value at Year 6.

$$\text{Amount of decrease} = \text{Year 0 value} - \text{Year 6 value} = \$20{,}000 - \$7{,}200 = \$12{,}800$$

$$\text{Percent of decrease} = \frac{\text{Amount of decrease}}{\text{Original amount}} = \frac{\$12{,}800}{\$20{,}000} = 0.64 = 64\%$$

The value of the car decreased 64% during the first six years.

In Example 2b, we found that between the third and fourth years, the amount of decrease was $1,600 and the percent decrease in value was 13.8%. From the data table, we can find that between the second and third years, the amount of decrease is also $1,600. Is the percent decrease in value between the second and third years also 13.8%? Explain your answer.

You Try It 3 **Bik started a retirement account by making an initial investment of $800. She also makes payments of at least $40 to the account each month. The following table indicates the account's value for the first five years. Draw the associated line graph and answer each question based on the data.**

a) Draw the line graph for this data table.

Years of investment	Value of investment
0	$800
1	$1,360
2	$2,000
3	$2,750
4	$3,600
5	$4,600

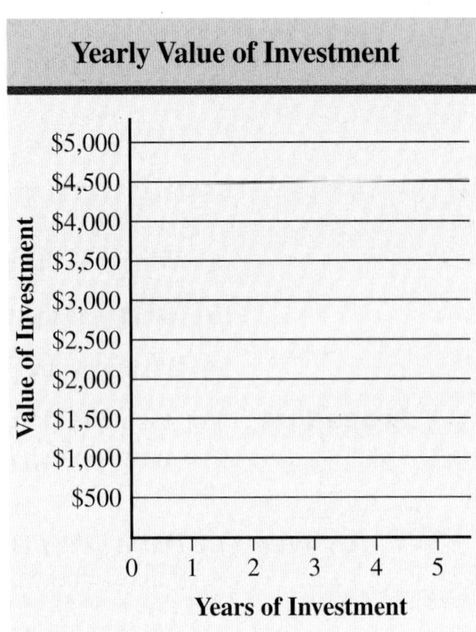

b) What was the percent increase in value during the first year?

c) What was the percent increase in value during the first two years?

d) What was the percent increase in value between the second and third years?

e) What was the percent increase in value between the fourth and fifth years?

Drawing Bar Graphs

Not all collected data is related to a progression of time, as is the data usually associated with line graphs. Sometimes data is collected and organized into preestablished categories. The number of items in a given category is called the **frequency** of the category. (*Frequency* means how frequently—how many times—something occurs.)

For example, Sarah, a newspaper reporter, is interested in the number of drive-up lunch customers each local fast-food restaurant attracts. She gathers data by asking the manager of each restaurant for the Tuesday count from noon to 2 P.M. After collecting the data, she puts the numbers in a table, the categories (fast-food restaurants) on the left and the frequency (the number of cars reported) on the right.

Categories → | ← Frequency

Restaurant	Number of Drive-up Customers
Burger King	40
Carl's Jr.	32
Del Taco	28
In-N-Out Burger	36
Jack-in-the-Box	52
KFC	52
McDonald's	64
Taco Bell	36
Wendy's	60

A table is one way to organize information so that it can be easily read. Another way to organize data values is to put them into a graph, providing a visual representation.

Sarah decides to put the data she collected into a **bar graph.** In a bar graph, the categories in the study are written across the bottom or along the left side. A rectangle, or bar, is drawn for each category, indicating the number of data values for that category. The length of the bar indicates the frequency of each category.

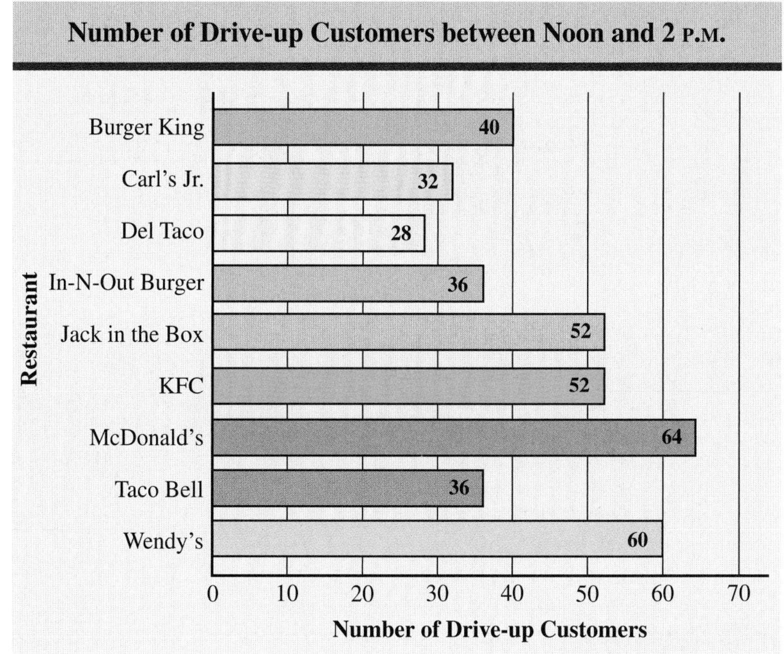

It's also common to see bar graphs with vertical rectangles, as shown here. One way is not any better than the other; each gives you the same information.

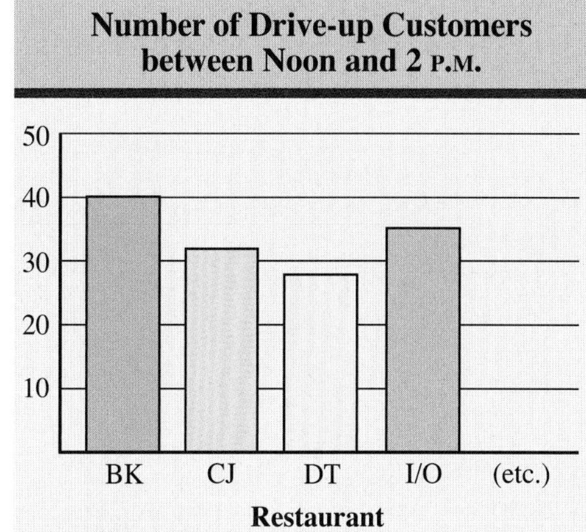

Notice these characteristics of the bar graph

- The bars are all the same width, and they don't touch each other. The length of each bar is determined by the data for each restaurant—the number of drive-up customers.
- The outline of the graph has numbers on the bottom, a simple standard measure of 10. Two of the bars meet up with the standard measure lines exactly—those of Burger King (40) and Wendy's (60).
- The bars for Jack-in-the-Box and KFC extend a little beyond the 50 line, and the bar for Del Taco falls a little bit short of the 30 line.

- Both In-N-Out Burger and Taco Bell had a total of 36 customers, which is a little more than halfway between 30 and 40. If we were to draw a line at 35, we'd see that the bar for In-N-Out Burger (and Taco Bell) would extend a little beyond the 35 line.

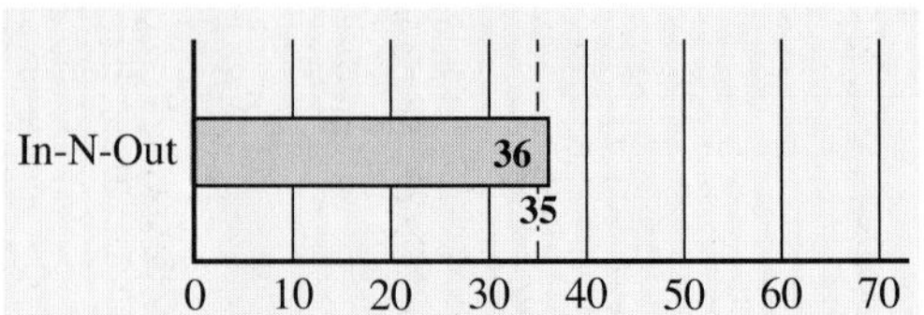

Keep those characteristics in mind when drawing your own bar graph. If possible, use a straightedge and graph paper so that your work is neat and organized.

You Try It 4 A local movie center has seven theaters, each of which plays older movies. Last Wednesday the manager asked Mark to count the number of people in each theater at 5 PM. The following table shows the data collected.

Draw a bar graph based on the given data. (Keep in mind the characteristics of a bar graph when you draw it.) A grid is provided to help you draw the bar for each category.

Movie	Number of People
The Mummy	51
The Matrix	45
Ghostbusters	39
City Slickers	42
The Rugrats	54
Airplane II	30
Ghost	39

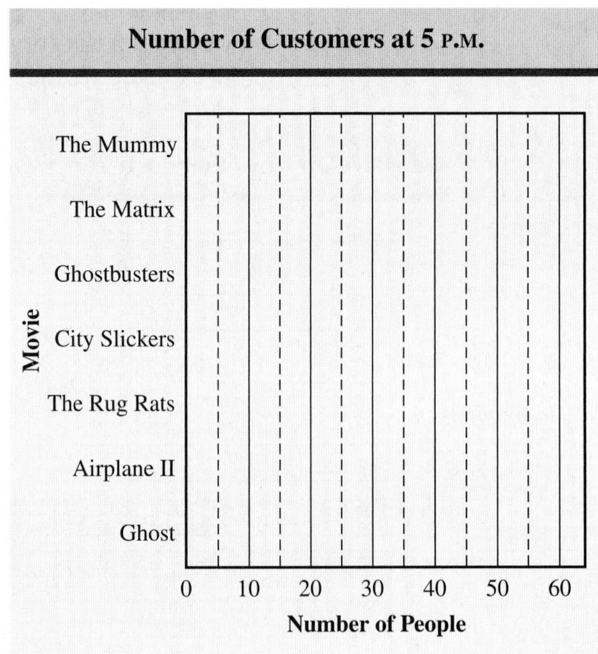

Reading Bar Graphs

If we are shown a bar graph of data but we don't know the individual amounts, we can estimate using the numbers on the bottom and their lines.

For example, here is the bar graph for the number of new cars sold during the month of May at each dealership in the Myriad Mile of Cars. Together, the dealerships sold a total of 180 new cars.

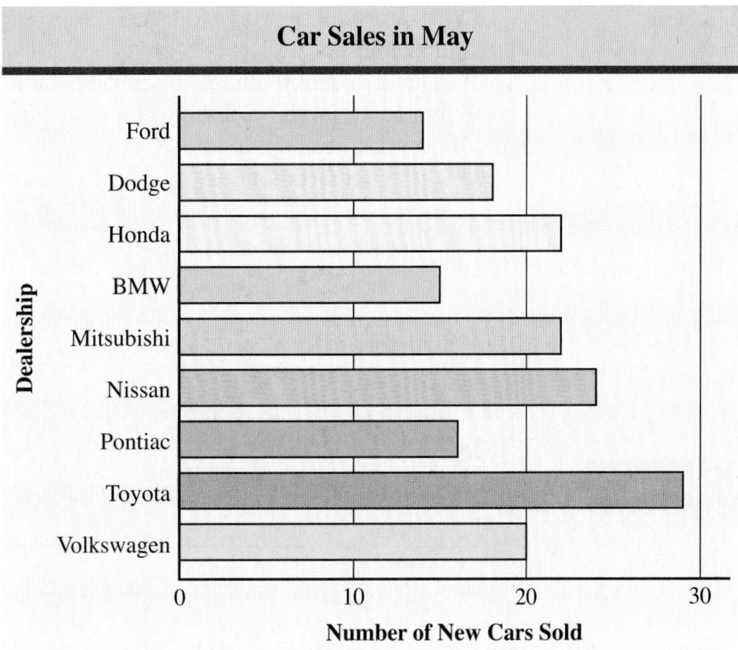

One data value is obvious: 20 Volkswagens were sold. All of the other values will take some educated visual guessing, called **interpolation,** to determine.

To assist us in this interpolation, we can draw our own lines halfway between the standard measures. In other words, we can draw lines representing 15 and 25 new cars sold. We can then compare each bar to the lines that we have drawn to see whether a bar is close (or very close) to one of the lines.

Also, a bar may look as though it is in the middle of, for example, 15 and 20. However, the very middle of 15 and 20 is $17\frac{1}{2}$ and car dealers can't sell half of a car. So the number is 17 or 18 depending on whether it appears closer to 15 or to 20.

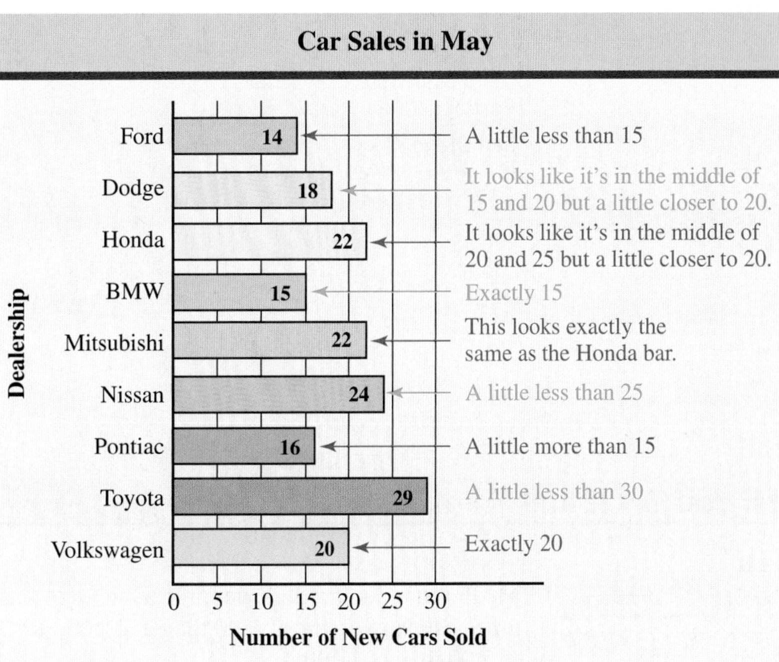

After our interpolation, we can put the values we find into a table.

We can also double-check to see if the table is accurate by adding up the column of the number of cars sold. If accurate, it will add to 180.

Do these numbers total 180? ________
Yes or no

Dealership	Number of New Cars Sold
Ford	14
Dodge	18
Honda	22
BMW	15
Mitsubishi	22
Nissan	24
Pontiac	16
Toyata	29
Volkswagen	20

Think About It 2 What should you do if the numbers in the table do *not* total 180?

____

____

You Try It 5 Following is a bar graph of the annual rainfall in selected cities throughout the United States. (Each value has been rounded to the nearest whole number.) ***Source: The World Almanac, 2005.***

Based on the bar graph, complete the data table.

City	Number of Inches of Rain
Albuquerque, NM	
Austin, TX	
Bismarck, ND	
Buffalo, NY	
Denver, CO	
Detroit, MI	
Phoenix, AZ	
Portland, OR	
San Diego, CA	

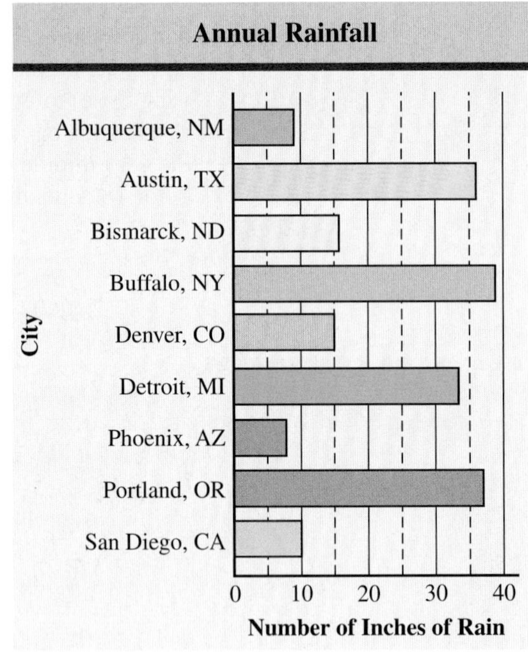

Answers: You Try It and Think About It

You Try It:

1. a)

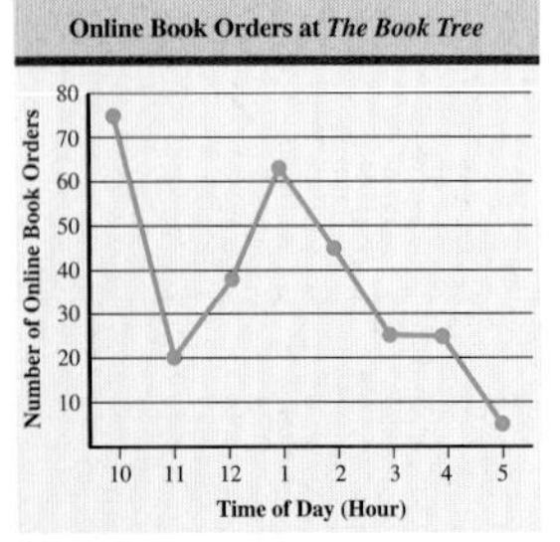

b) 25 more book orders were received at 2:00 than at 11:00. **c)** Jackie received three times as many orders at 1:00 than at 11:00. **d)** Jackie received one-third as many orders at 5:00 than at 11:00.

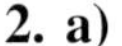

2. a)

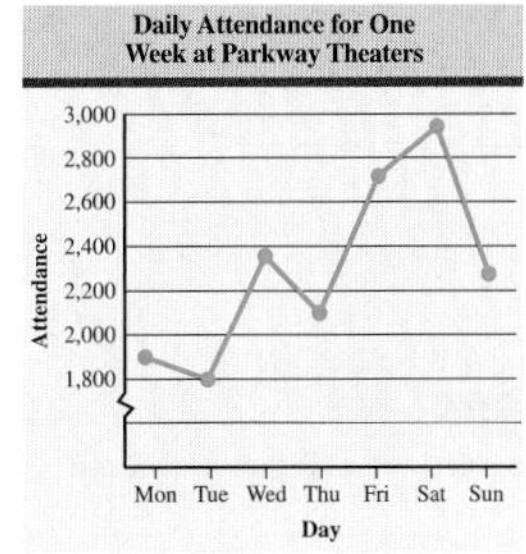

b) The difference between the highest and lowest attendance was 1,140 people. **c)** Friday's attendance was 1.5 times Tuesday's attendance. **d)** The biggest increase in attendance was from Thursday to Friday. The difference was 625 people. **e)** The biggest decrease in attendance was from Saturday to Sunday. The difference was 715 people.

3. a)

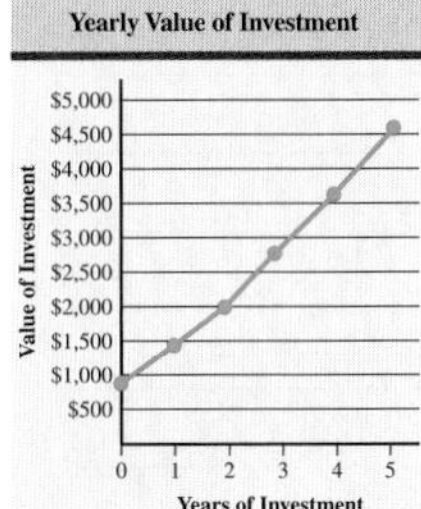

b) The value of the account increased 70% during the first year. **c)** The value of the account increased 150% during the first two years. **d)** The value of the account increased 37.5% between the second and third years. **e)** The value of the account increased 27.8% between the fourth and fifth years.

4.

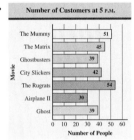

Number of Customers at 5 P.M.

Movie
The Mummy 51
The Matrix 45
Ghostbusters 39
City Slickers 42
The Rugrats 54
Airplane II 30
Ghost 39
0 10 20 30 40 50 60
Number of People

5.

City	Number of Inches of Rain
Albuquerque, NM	9
Austin, TX	36
Bismarck, ND	16
Buffalo, NY	39
Denver, CO	15
Detroit, MI	34
Phoenix, AZ	8
Portland, OR	37
San Diego, CA	10

Think About It: **1.** No, the percent decrease between the second and third years is based on a different original year (the second year). **2.** Answers may vary. One possibility: If the numbers do not total 180, a mistake has been made and it must be found and corrected.

Section 9.3 Exercises

FOR EXTRA HELP
Student Resources on DVD-ROM
Includes
➤Student's Solutions Manual
➤Video Lectures
➤Chapter Test Prep Video
MyMathLab
MathXL

Think Again

1. Consider the data in Example 2 in this section. Would it be possible to graph this data as a bar graph instead? Explain your answer or show an example that supports your answer.
   Yes, this data could be graphed as a bar graph. The categories would be the age of the cars, and the height of each bar would be the value of the car during that year.

2. Consider the data about the number of drive-in customers at fast-food restaurants presented in this section. Would it be possible to graph this data as a line graph instead? Explain your answer or show an example that supports your answer.
   No. A line graph is used to compare the progression of data over a period of time. Although the restaurant data include the number of customers from 12:00–2:00, the data do not include information from 2:00–4:00, 4:00–6:00, and so on.

Focus Exercises

The following data table shows the average high temperature in degrees Celsius for Salt Lake City, Utah. Draw the line graph associated with this data table.
***Source:* www.weather.com**

Month	Temperature, in Celsius
January	4
February	7
March	12
April	16
May	21
June	28
July	32
August	31
September	26
October	18
November	10
December	4

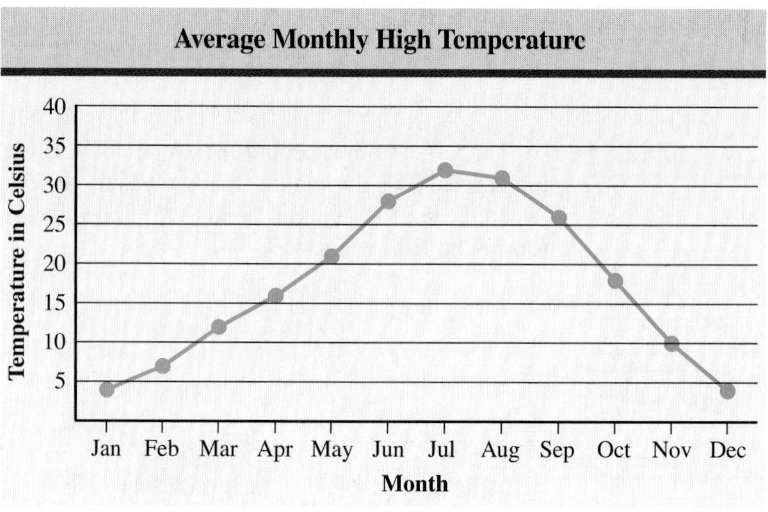

3. Between which two months did the greatest increase in temperature occur?
The greatest increase in temperature occurred between May and June.

4. Between which two months did the greatest decrease in temperature occur?
The greatest decrease in temperature occurred between September and October and between October and November.

5. In what month was the average temperature twice the average temperature for April?
In July, the average temperature was twice the average temperature for April.

6. In what month was the average temperature three times the average temperature for February?
In May, the average temperature was three times the average temperature for February.

7. In what month was the average temperature three-fourths the average temperature for June?
In May, the average temperature was three-fourths the average temperature for June.

8. In what month was the average temperature $1\frac{1}{2}$ times the average temperature for March?
In October, the average temperature was $1\frac{1}{2}$ times the average temperature for March.

9. What was the percent increase in the average temperature from March to April?
The percent increase in the average temperature from March to April was $33\frac{1}{3}\%$.

10. What was the percent decrease in the average temperature from November to December?
The percent decrease in the average temperature from November to December was 60%.

The following data table shows the minimum wage progression in Connecticut from 2000 to 2007. Draw the line graph associated with this data table.

Year	Minimum wage
2000	$6.15
2001	$6.40
2002	$6.70
2003	$6.90
2004	$7.10
2005	$7.10
2006	$7.40
2007	$7.65

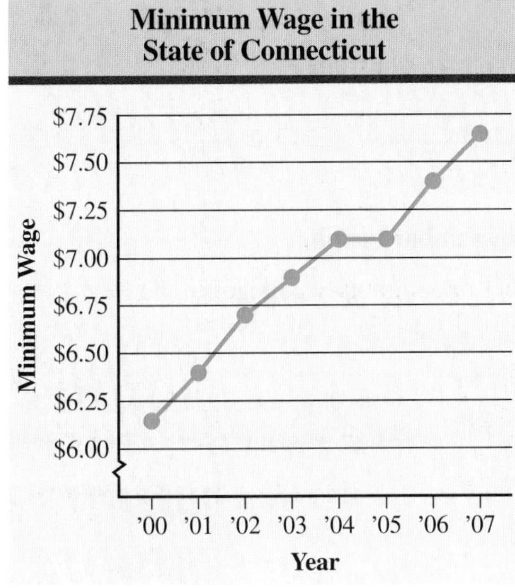

11. In which year(s) did the greatest amount of increase in the minmum wage occur?
The greatest amount of increase in the minimum wage occurred in 2002 and 2006.

12. In which year(s) did the least amount of increase in the minimum wage occur?
The least amount of increase in the minimum wage occurred in 2005.

13. What was the percent increase in the minimum wage from 2001 to 2002?
The percent increase in the minimum wage from 2001 to 2002 was approximately 4.7%.

14. What was the percent increase in the minimum wage from 2000 to 2007?
The percent increase in the minimum wage from 2000 to 2007 was approximately 24.4%.

When John turned 40 in 2000, his metabolism changed and he began to gain weight. Finally, in 2004, he began working out on a regular basis and changed to more healthy eating habits. The following data table shows John's weight each January 1 from 2000 to 2007. Draw the line graph associated with this data table.

Year	Weight in pounds
2000	200
2001	215
2002	225
2003	240
2004	270
2005	250
2006	235
2007	200

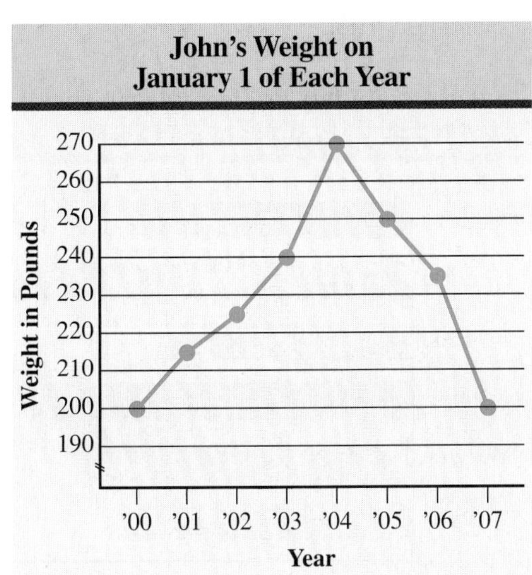

15. During which year did John gain the most weight?
John gained the most weight during 2003.

16. During which year did John lose the most weight?
John lost the most weight during 2006.

17. What was the percent increase in John's weight from 2003 to 2004?
The percent increase in John's weight from 2003 to 2004 was 12.5%.

18. What was the percent decrease in John's weight from 2004 to 2005?
The percent decrease in John's weight from 2004 to 2005 was approximately 7.4%.

19. What was the percent decrease in John's weight from 2004 to 2007?
The percent decrease in John's weight from 2004 to 2007 was approximately 25.9%.

20. What was the percent increase in John's weight from 2000 to 2004?
The percent increase in John's weight from 2000 to 2004 was 35%.

Given each table of values, draw the related bar graph.

21. The following table shows the number of community colleges in the Rocky Mountain states.

Mountain State	Number of Community Colleges
Arizona	20
Colorado	15
Idaho	3
Montana	14
Nevada	4
New Mexico	19
Utah	4
Wyoming	8

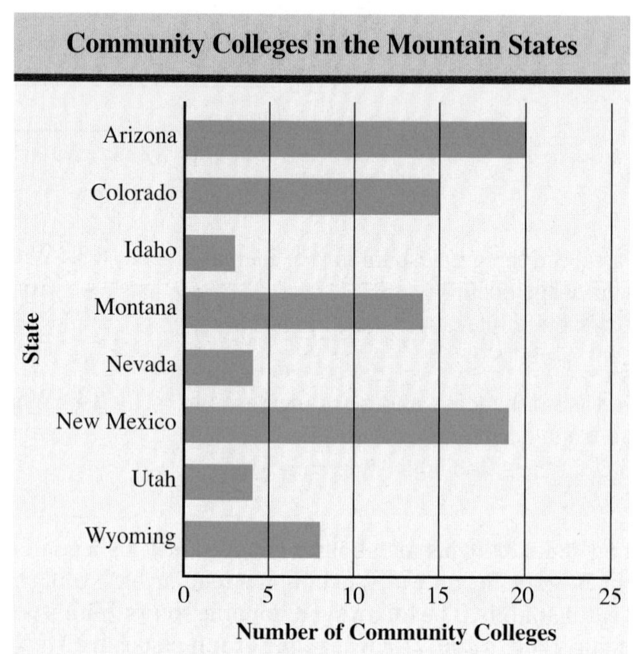

22. The following table shows the number of student computers at six different elementary schools in the Hartin Unified School District.

Elementary School	Number of Computers
Amber Lane	23
Garfield	34
Hartin	20
Kemper	31
Morrison	27
Wells	25

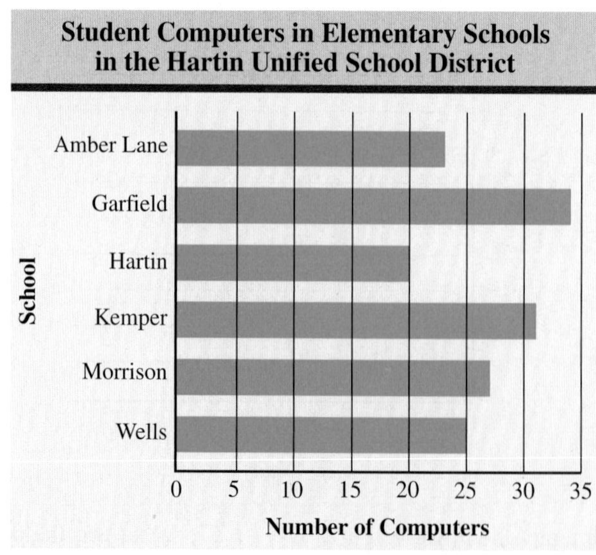

23. The following table shows the number of athletes representing some South American countries in the 2004 Summer Olympics.

Country	Number of Athletes
Bolivia	7
Chile	22
Colombia	57
Ecuador	17
Paraguay	32
Uruguay	16
Venezuela	51

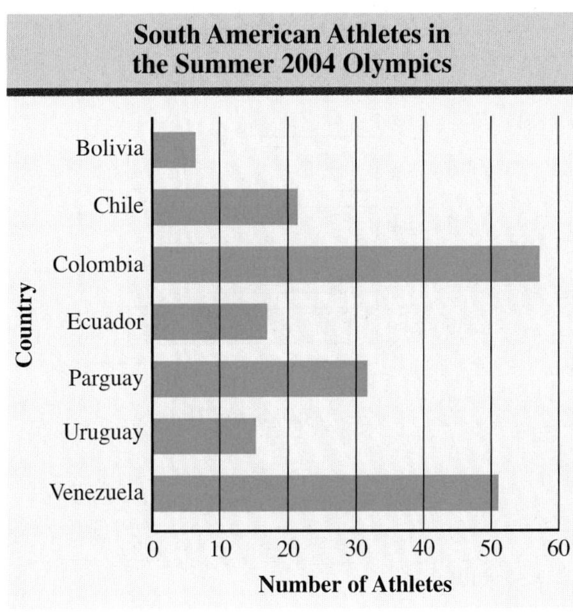

24. The following table shows by geographic region the number of votes in the U.S. Electoral College.

U.S. Region	Number of Electoral Votes
New England	34
Mid-Atlantic	83
South	132
Midwest	124
Southwest	56
West	109

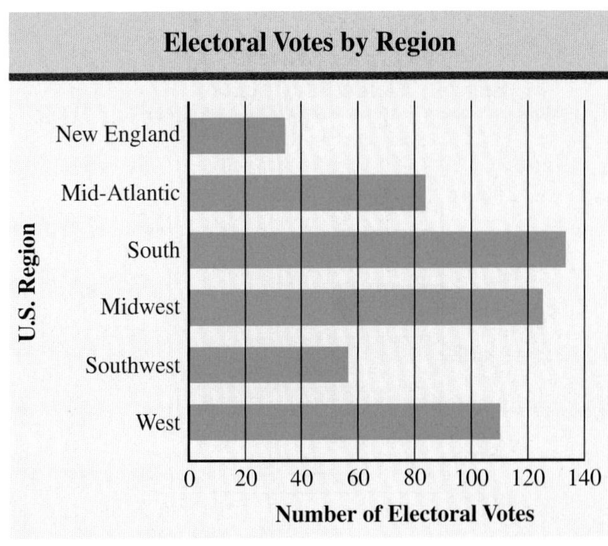

Given each bar graph, write the related table of values. Compare the table values to the total number indicated.

25. The combined age of all five of the charter jets in the Glory Aviators fleet is 33 years. The graph shows the age of each jet.

Jet	Age, in Years
Freedom	8
Glory	13
Patriot	6
Spirit	2
Victory	4

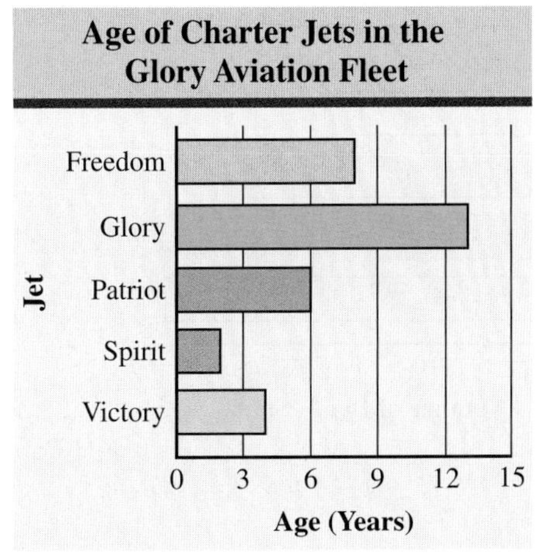

26. At Lincoln Elementary School, a total of 81 children had perfect attendance for the entire school year. Here is a bar graph of the data for each grade level.

Grade	Number of Children
First	3
Second	11
Third	10
Fourth	17
Fifth	24
Sixth	16

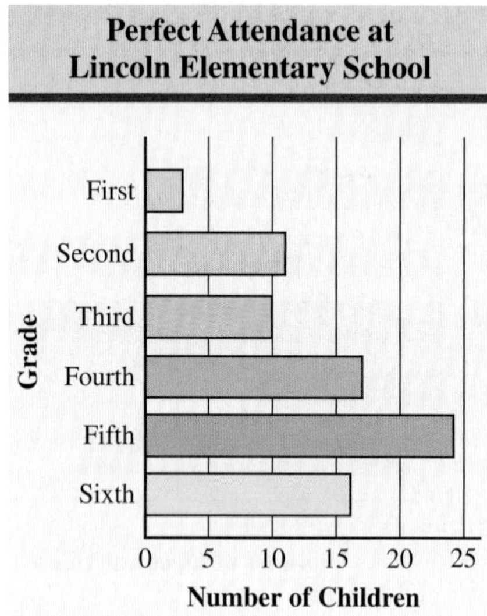

27. KenTen Stables, near the border of Kentucky and Tennessee, boards a variety of horses. Here is a bar graph of the 85 horses currently boarded at KenTen Stables.

Breed	Number of Horses
Andalusian	14
Clydesdale	8
Friesian	22
Paint	16
Palomino	25

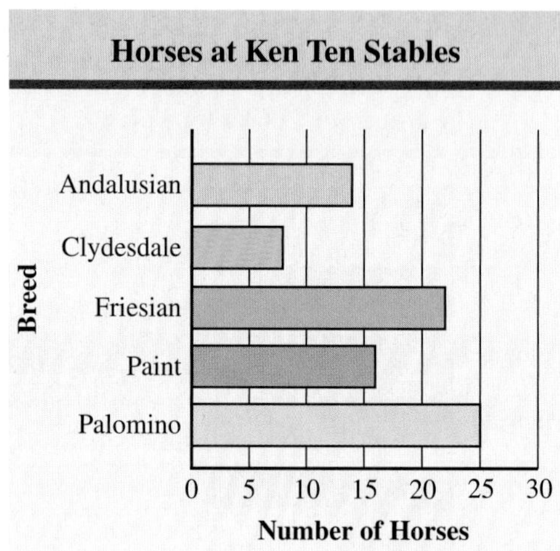

28. Last week Bargain Book Barn sold a total of 145 books by these best-selling authors. The bar graph indicates, by author, the number of books sold.

Author	Number of Books Sold
Albom	20
Brown	32
Crichton	19
Grisham	26
King	15
Steele	33

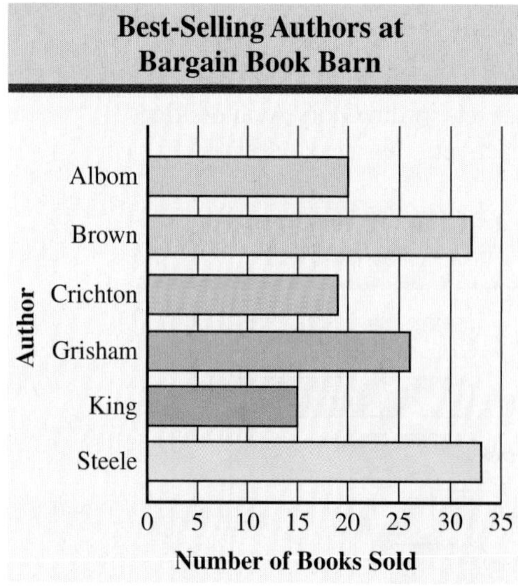

Think Outside the Box

Following are two tables showing the total population (rounded to the nearest thousand), at census, for Alaska and Wyoming from 1960 to 2000. The associated line graphs are shown on the same grid. (Alaska is represented by the triangles, and Wyoming is represented by the squares.)

Source: 2005 World Almanac

▲ *Alaska*

Census Year	Total Population	Total change in Population
1960	226,000	
1970	303,000	77,000
1980	402,000	99,000
1990	550,000	148,000
2000	627,000	77,000

■ *Wyoming*

Census Year	Total Population	Total change in Population
1960	330,000	
1970	332,000	2,000
1980	470,000	138,000
1990	454,000	−16,000
2000	494,000	40,000

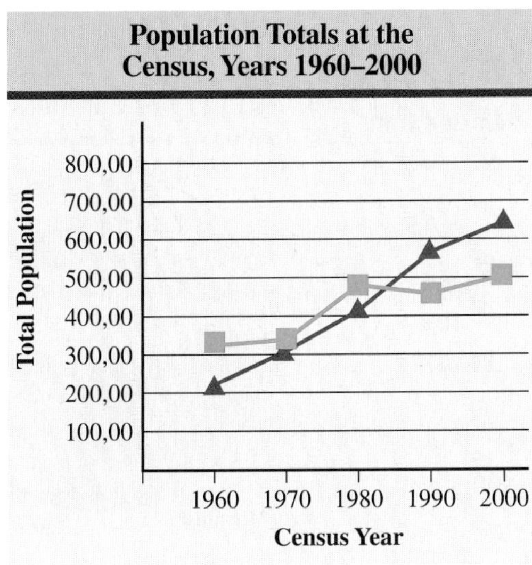

29. Fill in the population totals for each state from 1970 to 2000.

Year	Alaska	Wyoming
1970	303,000	332,000
1980	402,000	470,000
1990	550,000	454,000
2000	627,000	494,000

30. Which state had the highest increase in population from 1970 to 1980?
Wyoming had the highest increase in population at 138,000 (compared to Alaska's 99,000).

31. Which state had the highest percent increase in population from 1970 to 1980?
Wyoming had a 41.6% increase in population. (Alaska had a 32.7% increase.)

32. According to the line graphs, in which year does it appear that Alaska and Wyoming had the same population? Estimate the population that year.
1984 (or 1985); the population was about 465,000 in each state that year.

SECTION 9.4 Graphing Data: Circle Graphs and Histograms

OBJECTIVES

In this section, you will learn to

- Read and draw circle graphs.
- Read frequency distribution tables.
- Read and draw histograms.

You Need to Know

To successfully complete this section, you need to understand

- ☐ Adding whole numbers (1.2)
- ☐ Dividing whole numbers (1.4)
- ☐ Writing decimals as percents (6.4)
- ☐ Reading and drawing bar graphs (9.3)

Introduction

In this section, you are introduced to two more types of graphs for data. The first, *circle graphs*, involves the same types of categories as bar graphs; but instead of using the frequency of each category, we calculate the percentage that each category has compared to the whole distribution. The second, *histograms*, involves categories that are numbers.

Reading Circle Graphs

Another popular graph is the **circle graph**, sometimes called a *pie chart*. The categories are displayed in a circle, and it is typical to show each category as a percent rather than the actual frequency. Each region in the circle graph, called a **sector**, is defined by two radii.

For example, in a small city, an auto mall offered the following makes of automobiles: Buick, Chrysler, Ford, Mitsubishi, and Toyota. Brian, the city's director of economic development, compiled the following new vehicle sales statistics for 2005 and created a circle graph.

Dealership	Percent of Total New Vehicles Sold
Buick	25%
Chrysler	20%
Ford	5%
Mitsubishi	10%
Toyota	40%
Total	**100%**

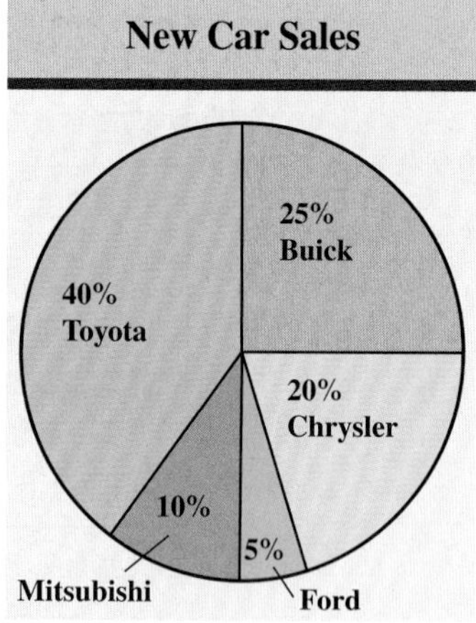

Notice that the total percent of all categories is 100%. Also, the larger the percent, the larger the part of the circle a sector covers. Furthermore, the radii of a 50% sector form a straight angle and the radii of a 25% sector form a right angle.

Because the categories are measured in percents, it's appropriate to use 50% (half the circle) and 25% (one-fourth of the circle) as standards by which to compare other percentages in the circle.

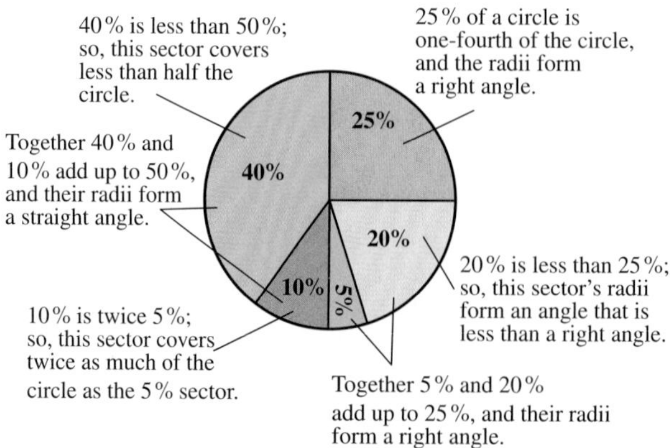

Note Where we place the radius that divides the 40% and 10% sectors is just an estimate.

If all of the dealerships combined sold 2,400 new cars, how many new cars did each dealership sell?

We can answer that question using the table of percents. Recall from Section 6.6 that percent always needs to be *of* something. For each item in the circle graph, the percent means "of the total number of cars."

Also recall that *percent of* means "percent times," so, when the Ford dealership sold 5%, that number is 5% of 2,400 new cars: 5% × 2,400 = 120.

Dealership	Percent of Total New Vehicles Sold	Total New Vehicles Sold (× 2,400)		Write the percent as a decimal and multiply by 2,400.
Buick	25%	600	←	0.25 × 2,400 = 600
Chrysler	20%	480	←	0.20 × 2,400 = 480
Ford	5%	120	←	0.05 × 2,400 = 120
Mitsubishi	10%	240	←	0.10 × 2,400 = 240
Toyota	40%	960	←	0.40 × 2,400 = 960
Total	**100%**	**2,400**	←	Make sure this column adds to 2,400.

You Try It 1 **Following is a circle graph with the grade distribution for Dr. Garcia's algebra classes last spring semester. Dr. Garcia had a total of 150 algebra students that semester.**

Complete the table by listing the percent for each grade as well as the number of students receiving that grade.

Grades in Dr. Garcia's Algebra Classes

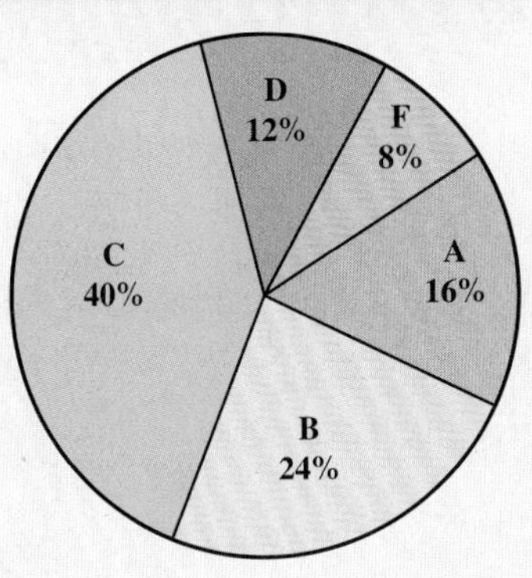

Grade	Percent of Students Receiving the Grade	Total Number of Students Receiving the Grade
A		
B		
C		
D		
F		
Total		

Think About It 1 Can the values in this table be made into a circle graph? Why or why not?

____

____

____

____

Restaurant Breakfast Customers

Beverage	Percent That Have This Beverage with Breakfast
Coffee	58%
Juice	34%
Tea	17%
Milk	10%
Other	16%

Drawing Circle Graphs

Sometimes our task is to make a circle graph from a table of data. Here are some guidelines to help us make our circle graph as accurate as possible.

1. We can use the 25% and 50% standards as guidelines. First, we can lightly draw in a horizontal and a vertical diameter as shown.

   We can use these standards to estimate where a 23% sector (a little less than 25%) and a 42% sector (between 25% and 50%) might fit.

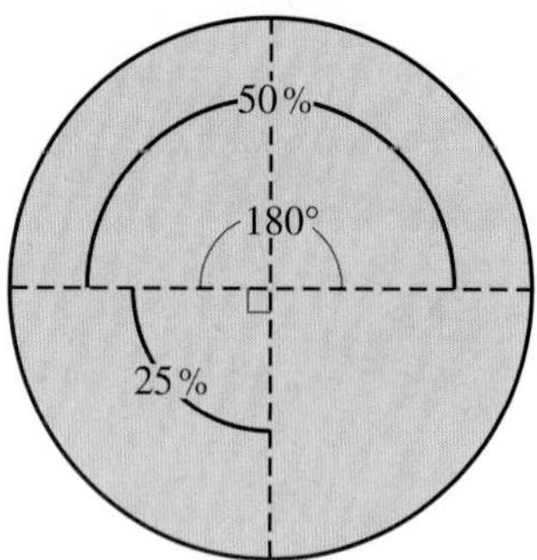

2. It is also helpful to know that 12.5% is halfway between 0 and 25% and forms a 45° angle.

   So, a 12% sector is about halfway between 0 and 25%.

   Furthermore, 37.5% is halfway between 25 and 50%; so, a 42% sector (more than halfway between 25 and 50%) will be closer to 50% than to 25%.

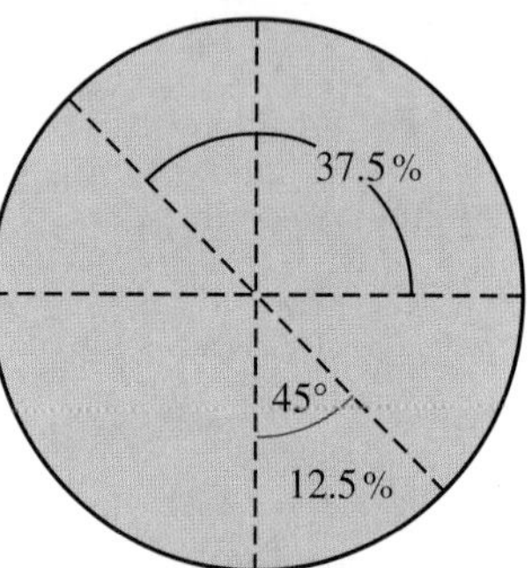

3. If a circle graph has, for example, sectors measuring 12%, 23%, and 42%, we might place them in the circle first (and erase the diameters that we drew in lightly).

   These larger sectors do not need to be adjacent to each other. The other smaller sectors can fit in next to and around these larger sectors.

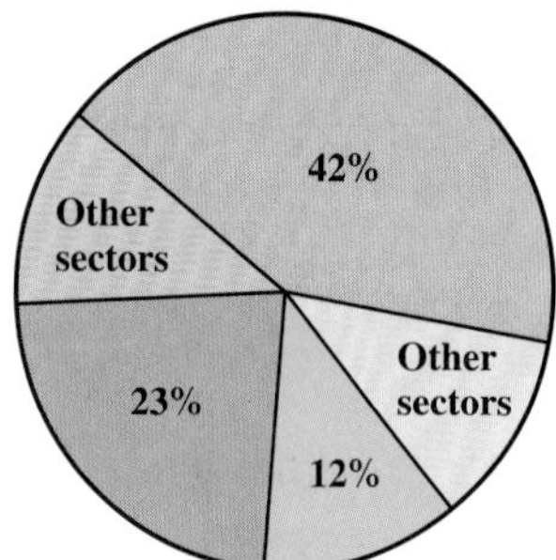

4. Another helpful guideline is to look for two (or more) percents that add up to 25% or 50%.

   For example, 17% and 8% add to 25%; so, those two sectors can fit into one of the 25% sectors formed by the two lightly drawn diameters.

   Also, 39% and 11% add to 50% and can, together, fit into one-half of the circle. The remaining sectors will fit in where they can.

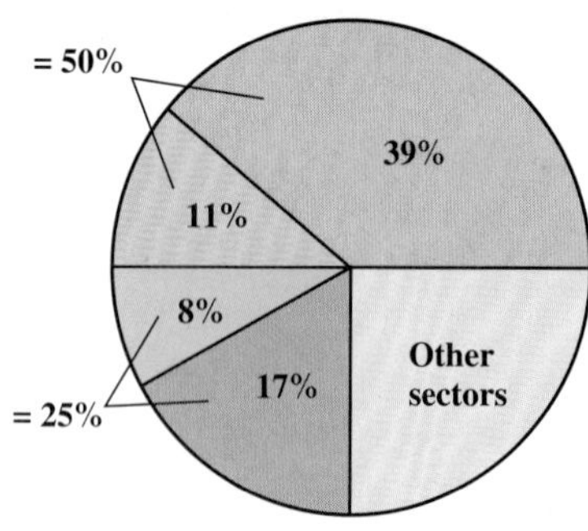

You Try It 2 **Here is a table of second graders' favorite colors and the percentage of those students who favor each color. Fit the percent data into a circle graph.**

Color	Percent of Students Who Favor Each
Black	8%
Blue	35%
Green	9%
Orange	15%
Red	16%
Yellow	12%
Other	5%
Total	**100%**

Second-Graders' Favorite Color

If we are given a table of data values from a study, then we can calculate the percent in each category before drawing its related circle graph. For example, Miko is the manager at a record store. She took the sales data from one weekend and compiled the following information, with the number of CDs sold in parentheses:

> Classical (8); Country (18); Gospel (24); Jazz/Blues (40); Rap/Hip Hop (34); Rock/Soft Rock (60); Other (16). (The "other" category includes foreign language and children's CDs.)

Type of Music	Number of CDs Sold
Classical	8
Country	18
Gospel	24
Jazz/Blues	40
Rap/Hip-Hop	34
Rock/Soft Rock	60
Other	16
Total	**200**

From the information above we can find the percentage for each category by dividing each count (the number of CDs sold) by the total of 200 CDs sold.

For example, for the Gospel category, the percentage would be $24 \div 200$. This can be shown as a fraction, $\frac{24}{200}$, which simplifies by a factor of 2 to $\frac{12}{100}$, which is 12%. We could also use long division to find the percentage: $200\overline{)24.000}$. Whichever procedure you choose, make sure you write the answer as a percent. Don't leave it as a decimal or fraction.

Type of Music	Number of CDs Sold	Percent of Total Sales
Classical	8	4%
Country	18	9%
Gospel	24	12%
Jazz/Blues	40	20%
Rap/Hip-Hop	34	17%
Rock/Soft Rock	60	30%
Other	16	8%
Total	**200**	**100%**

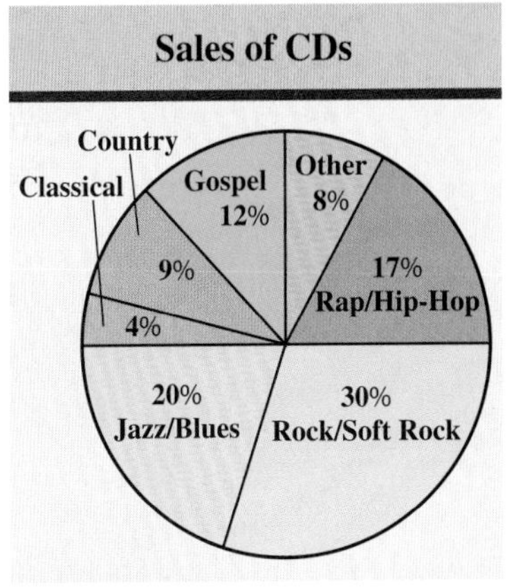

Note: 8% and 17% add to 25%; 20% and 30% add to 50%.

You Try It 3 **Following is a table showing the number of people in Minnesota (in 1,000s) that speak a language other than English. For example, the table shows that about 16,000 people in Minnesota speak an African language.** *Source:* **www.mla.org**

Complete the table by listing the percent for each language group. Then place the percent data appropriately to complete the circle graph.

Language Classification	Number of People (in 1,000s)	Percent of Total
Spanish	136	
Asian Languages	100	
Germanic	56	
Russian/Slavic	24	
African Languages	16	
French	24	
Other	44	
Total	**400**	

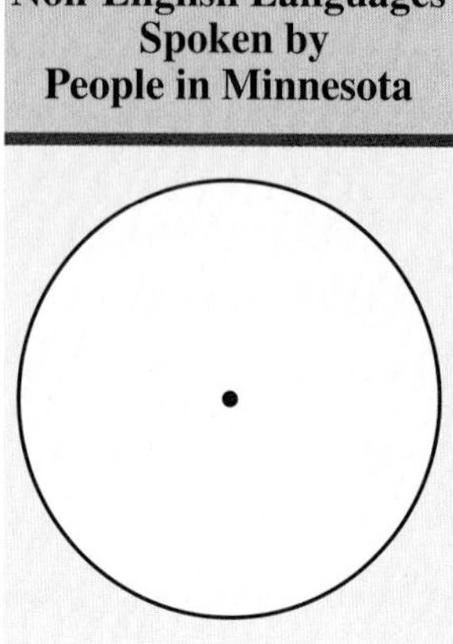

Non-English Languages
Spoken by
People in Minnesota

Frequency Distribution Tables

In each of the examples and exercises pertaining to bar graphs and circle graphs, the categories were *things* such as *restaurants, movies, dealerships,* and *cities.*

When the categories are numbers, we organize the data differently and use a different type of table, called a *frequency distribution table*, and a different type of graph, called a *histogram.*

In the following example, each category is a range of numbers called a **class interval.** Each class interval has a lowest value and a highest value. The class interval also includes every whole number between those two values.

Also, the difference between the highest and lowest values in each class interval is the same for each interval.

Consider this: Nate took a survey of people attending the last game of the Houston Astros 2006 baseball season. He asked 50 Astros fans how many games they had attended that season. For two fans, the answer was only one game; others had attended more than 20 games; and one man attended 67 games!

Nate organized the data in numerical order and put each group into a class interval. He decided to make the class intervals—representing the number of games a fan had attended—from 1 to 10, from 11 to 20, and so on. The last class interval was from 61 to 70.

He then counted the number of fans that fit into each class interval—called the **class frequency**—and created the following **frequency distribution table.** (Recall that *frequency* means how many times something occurs.)

(Class Interval) Games Attended	(Class Frequency) Number of Fans
1-10	12
11-20	10
21-30	15
31-40	5
41-50	3
51-60	4
61-70	1
Total	**50**

Two fans attended only *one* game, but others attended 3 games, some attended 4 games, and so on. A total of 12 fans fit into this first class interval: 1–10 games attended.

From the information provided, we know that one man attended 67 games. Looking at the table only, though, we can't tell how many games he attended, just that he attended anywhere from 61 to 70 games.

Reading Histograms

Nate then displayed these data in a graph called a **histogram.** Notice that the categories (across the bottom) are the class intervals and the height of each bar shows the corresponding class frequency. (As with a bar graph, the bars in a histogram are of the same width.)

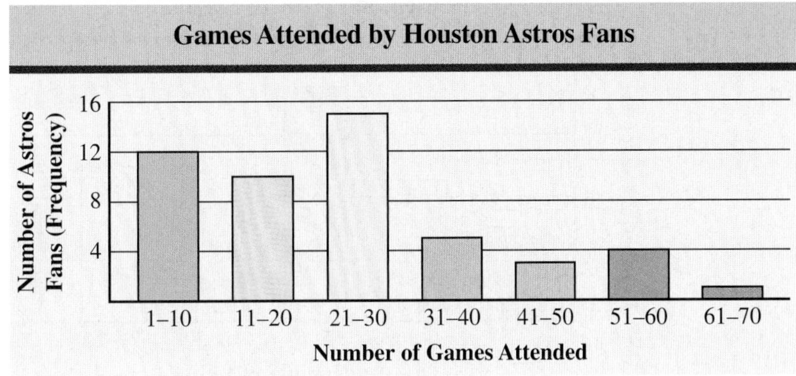

From the frequency distribution table or from the histogram, we can combine class intervals to answer questions about the distribution. For example,

a) How many fans attended more than 40 games?

"More than 40" includes the last three classes:

	3	in the 41–50 class
	4	in the 51–60 class
	+ 1	in the 61–70 class
8 fans attended more than 40 games. ←	8	Total

b) How many fans attended from 11 to 30 games?

"From 11 to 30" includes the second and third classes:

	10	in the 11–20 class
	+ 15	in the 21–30 class
25 fans attended from 11 to 30 games. ←	25	Total

You Try It 4 **On July 31, Tammy checked the accounts of 50 Corona Video customers and looked at how many DVDs they rented that month. Following is a frequency distribution table of the data she collected.**

a) Draw a histogram based on the given data. (Use the outline provided and remember that the width of each bar should be the same.)

DVDs Rented	Number of Customers
1-5	14
6-10	10
11-15	12
16-20	5
21-25	6
26-30	3
Total	**50**

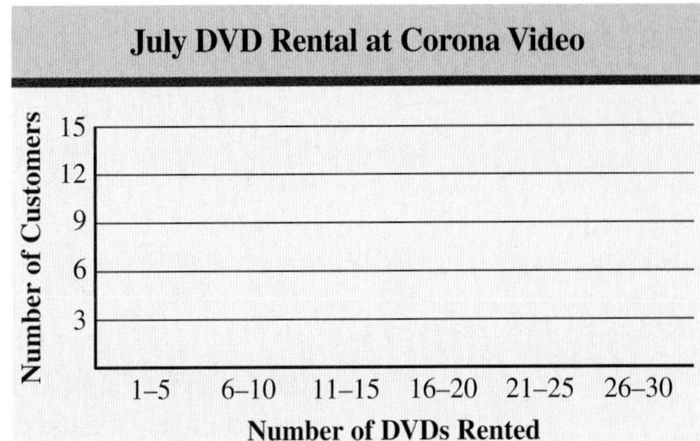

b) How many customers rented from 16 to 25 DVDs?

c) How many customers rented more than 10 DVDs?

d) How many customers rented fewer than 16 DVDs?

You Try It 5 **On September 30, Tanya asked 60 Starbucks customers how many times they had visited Starbucks during the month. Following is a histogram of the data she collected.**

a) Fill in the frequency distribution table with the data shown in the histogram.

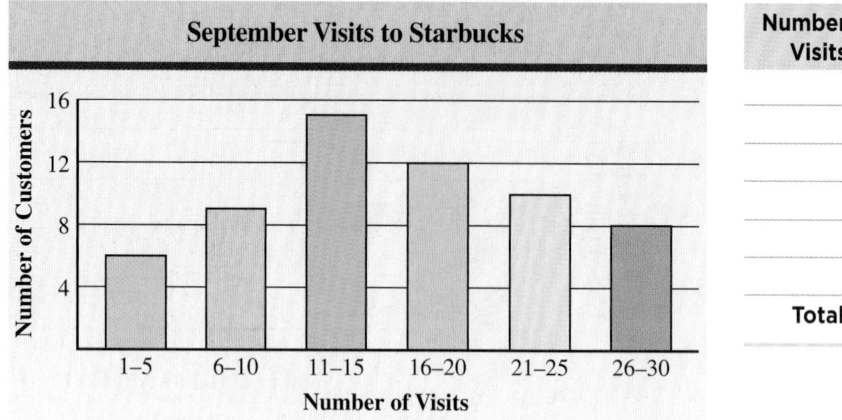

Number of Visits	Number of Customers
Total	**60**

b) How many customers visited more than 20 times?

c) How many customers visited fewer than 11 times?

d) How many customers visited from 11 to 25 times?

Drawing Histograms

If you were conducting a survey, you might ask a question that has a numerical answer, such as, "How many pairs of shoes do you own?"

The people answering your question will have answers such as 3, 10, 6, 1, and 15. In other words, the numbers you get would not be in numerical order. Your job would be to write down the numbers as you hear them and then determine into which class interval each number (data value) fits.

Let's say that these are the data you collected for the number of pairs of shoes people own. We can organize these data into a frequency distribution table by making a tally mark each time a number fits into a class interval.

3	10	6	1	15	4
7	6	12	20	13	5
23	4	19	8	2	6
14	20	25	16	9	7
4	12	9	2	8	14

Number of Pairs of Shoes	Tally	Number of People
1-5	~~||||~~ |||	8
6-10	~~||||~~ ~~||||~~	10
11-15	~~||||~~ |	6
16-20	||||	4
21-25	||	2
Total		**30**

At the end, total your frequency column to make sure that it adds—in this case—to 30, the total number of data values. You can now create a histogram from this table.

You Try It 6 **Ms. Chung wanted to create a histogram of her algebra students' Chapter 5 test results. (She has 40 students in this class.) Following is the data from her grade book. (Use the previous discussion as a guide.)**

a) Fill in the frequency distribution table and draw its related histogram.

83	90	76	81	65	87	96
77	96	82	70	53	75	85
93	74	59	88	82	89	79
74	80	75	66	79	70	83
84	92	79	62	58	62	71
71	83	90	68	72		

Scores on Test	Tally	Number of Students
50–59		
60–69		
70–79		
80–89		
90–99		
Total		

Ms. Chung's grading scale is **A** (90–100), **B** (80–89), **C** (70–79), **D** (60–69), and **F** (below 60).

b) How many students got an A on the test?

c) How many students got an A or a B on the test?

d) How many students got below a C on the test?

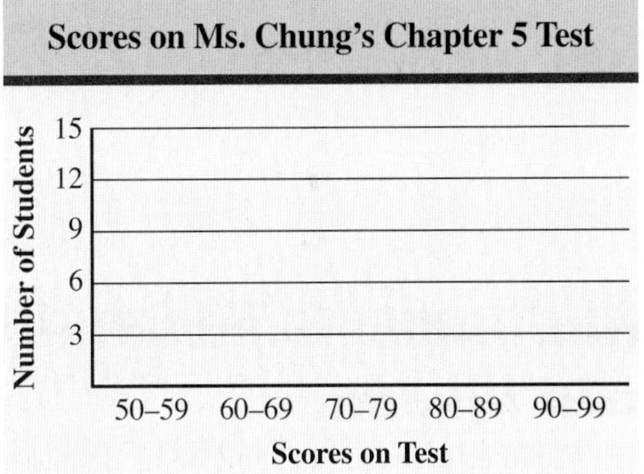

Answers: You Try It and Think About It

You Try It:

1.

Grade	Percent of Students Receiving the Grade	Total Number of Students Receiving the Grade
A	16%	24
B	24%	36
C	40%	60
D	12%	18
F	8%	12
Total	**100%**	**150**

2.

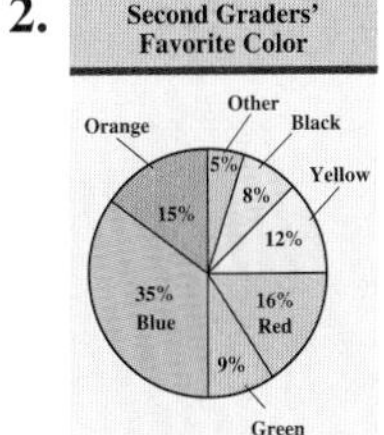

Note: 9% and 16% add to 25% and fit into one-fourth of the circle.

Similarly, 15% and 35% add to 50% and fit into one-half of the circle.

The placement of the sectors may be different than what is shown here.

3.

Language Classification	Number of People (in 1,000s)	Percent of Total
Spanish	136	34%
Asian Languages	100	25%
Germanic	56	14%
Russian/Slavic	24	6%
African Languages	16	4%
French	24	6%
Other	44	11%
Total	**400**	**100%**

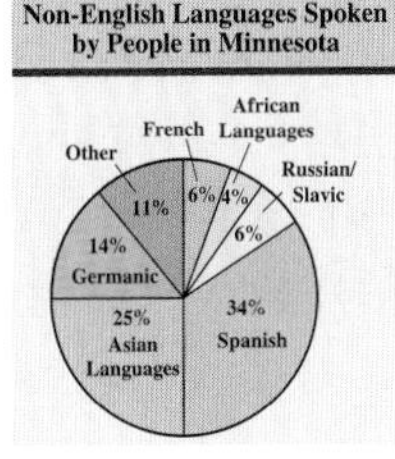

4. a)

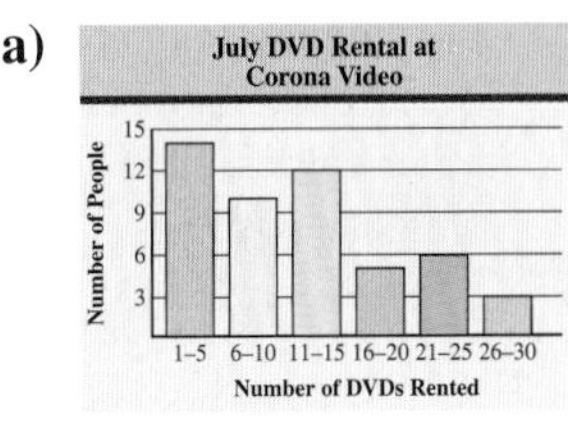

b) 11 **c)** 26 **d)** 36

5. a)

Number of Visits	Number of Customers
1-5	6
6-10	9
11-15	15
16-20	12
21-25	10
26-30	8
Total	**60**

b) 18 **c)** 15 **d)** 37

6. a)

Scores on Test	Tally	Number of Students
50-59	///	3
60-69	HHT	5
70-79	HHT HHT ////	14
80-89	HHT HHT //	12
90-99	HHT /	6
Total		**40**

b) 6 **c)** 18 **d)** 8

Scores on Ms. Chung's Chapter 5 Test

Number of Students: 3, 6, 9, 12, 15

50–59 60–69 70–79 80–89 90–99

Scores on Test

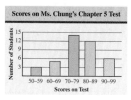

Think About It: **1.** No. The percents add to more than 100%.

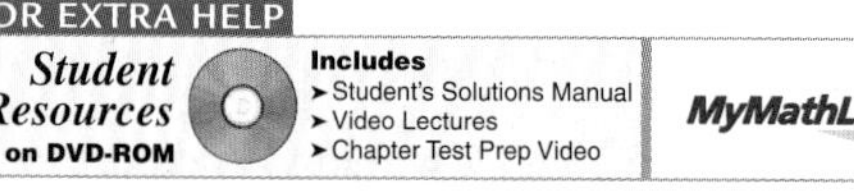

Section 9.4 Exercises

Think Again

1. Can the values in this table be made into a circle graph? Why or why not? (*Refer to Think About It 1*)

Beverage	Percent That Have This Beverage with Breakfast
Coffee	58%
Juice	34%
Tea	17%
Milk	10%
Other	16%

No, because the percent values add to more than 100%.

2. How does a histogram differ from a bar graph?
A histogram has numerical categories that are often grouped into numerical intervals. Also, in a histogram, the height of each rectangle is based on the frequency of occurrence. In a bar graph, the height can be based on other values, such as dollars, heights of buildings, and weights of dogs.

3. Zig is the instructor at Ride Rite, a motorcycle riding class. He asked his students what motorcycle brand they owned, and the results are shown in this table. Assuming that the data is real, how is it possible?

Motorcycle Brand	Percent That Own This Brand
Harley-Davidson	24%
Honda	18%
Kawasaki	22%
Suzuki	26%
Yamaha	20%
Other	10%
Total	

The second column adds to 120%, but this means that some of the riders own more than one brand of motorcycle.

4. In a circle graph, write the steps you would take to draw a sector for 12%. You may want to draw a circle graph to assist you in your explanation.
Half a circle is 50%. Half of that (one quarter of a circle) is 25%. Half of that (one-eighth of a circle) is 12.5%. So, a 12% sector is a little bit less than one-eighth of a circle.

Focus Exercises

Complete each table by listing the percent for each category as well as the number of items in that category.

5. The marital status of 120 women who signed up for Rancho College's Sociology 20 class, Marriage and Family, is shown in the following circle graph.

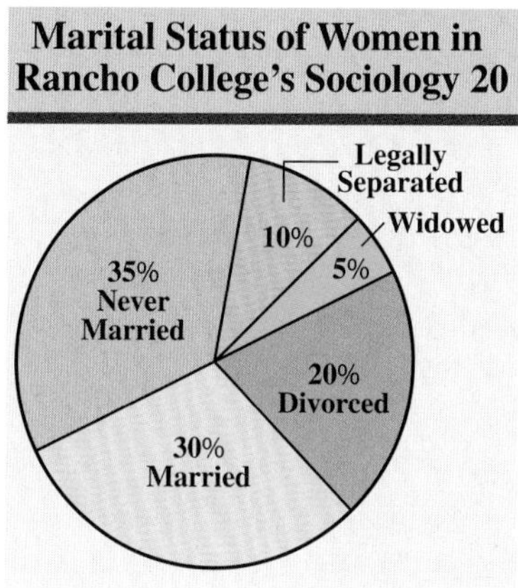

Marital Status	Percent of Women	Total Number of Women
Married	30%	36
Divorced	20%	24
Widowed	5%	6
Legally Separated	10%	12
Never Married	35%	42
Total	**100%**	**120**

6. Following is a circle graph indicating the education level of the 300 parent members of the PTA at Goodhew Middle School.

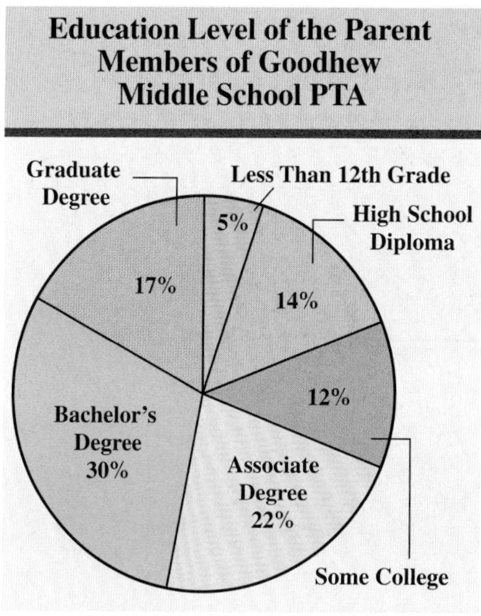

Level of Education	Percent of Parents	Total Number of Parents
Less Than 12th Grade	5%	15
High School Diploma	14%	42
Some College	12%	36
Associate Degree	22%	66
Bachelor's Degree	30%	90
Graduate Degree	17%	51
Total	**100%**	**300**

7. Following is a circle graph indicating the type of doctors employed at Vista General Hospital. A total of 250 doctors work at Vista General.

Doctors at Vista General Hospital

Anesthesiologists 6%
Surgeons 8%
16% Gynecologists
18% Internists
32% General Practitioners
Other 20%

Type of Doctor	Percent of Doctors	Total Number of Doctors
Anesthesiologists	6%	15
General Practitioners	32%	80
Gynecologists	16%	40
Internists	18%	45
Surgeons	8%	20
Others	20%	50
Total	**100%**	**250**

8. Following is a circle graph indicating the brand of golf balls sold at Putt 'N Cup Golf Supplies last week. Putt 'N Cup sold a total of 150 boxes of golf balls.

Brands of Golf Balls Sold at Putt 'N Cup

34% Titleist
16% Callaway
18% Dunlop
8% Maxfli
Top–Flite 24%

Golf Ball Brand	Percent of Golf Balls	Total Number of Golf Balls
Callaway	16%	24
Dunlop	18%	27
Maxfli	8%	12
Titleist	34%	51
Top-Flite	24%	36
Total	**100%**	**150**

Complete each table by listing the percent of the whole for each category. Then place the percent data appropriately in the circle graph.

9. At a liberal arts college, the 80 students living in one dorm had the majors shown in the table.

Major	Total Number of Students	Percent of Students
Arts/Humanities	40	50%
English/Languages	16	20%
Life Science	4	5%
Social Science	8	10%
Other	12	15%
Total	**80**	**100%**

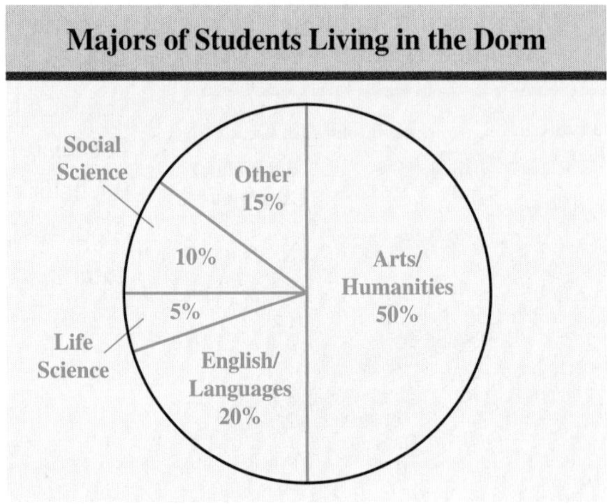

10. At Pick and Strum Guitars, the following acoustic guitar brands were sold last month.

Brand	Total Number of Guitars	Percent of Guitars
Fender	6	5%
Gibson	24	20%
Martin	30	25%
Ovation	18	15%
Taylor	42	35%
Total	**120**	**100%**

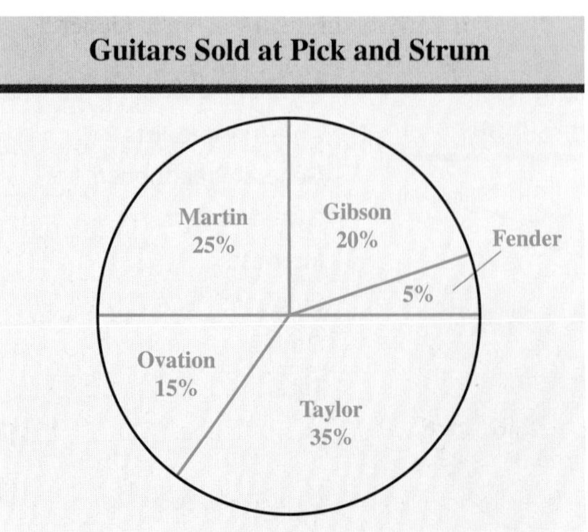

11. A total of 200 children signed up to play Little League in Norco. On the application form, each player was asked to indicate the primary position for which he or she wanted to try out: pitcher, catcher, infielder, or outfielder. Some of the children left that answer blank. A table of their responses follows.

Position	Total Number of Players	Percent of Players
Picher	56	28%
Catcher	10	5%
Infielder	74	37%
Outfielder	44	22%
Left blank	16	8%
Total	**200**	**100%**

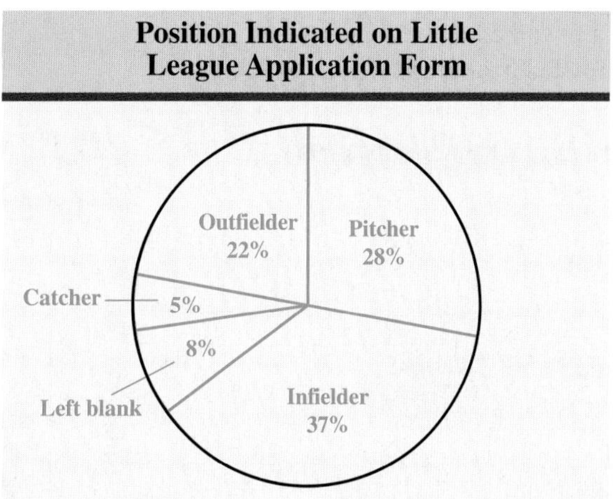

12. At the annual Kiwanis International Convention, one session—with 150 delegates—focused on worldwide child hunger. Following is a table of the number of delegates from each of the six continental regions with Kiwanis Clubs.

Continental Region	Total Number of Delegates	Percent of Delegates
Oceania	12	8%
Africa	45	30%
Asia	21	14%
Europe	18	12%
North America	30	20%
South America	24	16%
Total	**150**	**100%**

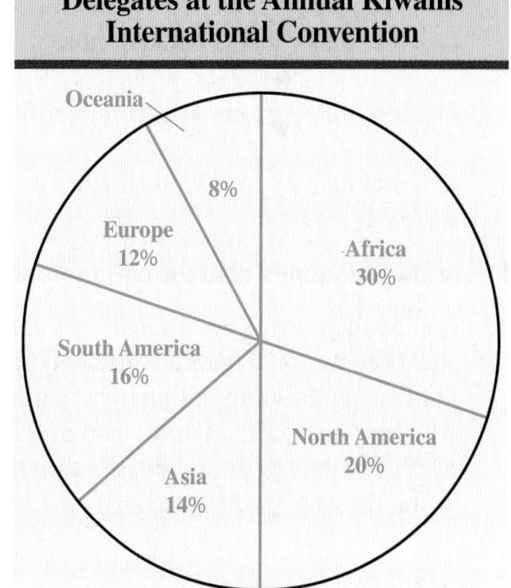

Draw a histogram based on the given data.

13. Sixty night-light bulbs were tested to determine their lifetime. Following is a frequency distribution table of the data collected. **a)** Draw the related histogram.

Number of Hours	Number of Lightbulbs
170–179	2
180–189	8
190–199	15
200–209	19
210–219	13
220–229	3
Total	**60**

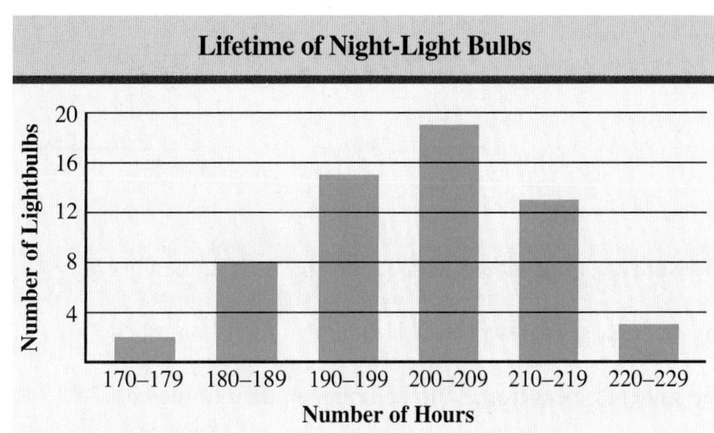

b) How many lightbulbs lasted longer than 189 hours?
50

c) How many lightbulbs lasted fewer than 190 hours?
10

d) How many lightbulbs lasted from 190 to 209 hours?
34

e) How many lightbulbs lasted at least 200 hours?
35

14. At the annual Relay for Life cancer walk, 40 teams walked for 24 hours straight. (At least one person from each team was on the track throughout the entire 24 hours.) Following is a frequency distribution of the number of miles each team walked.

a) Draw the related histogram.

Number of Miles	Number of Teams
35-39	4
40-44	3
45-49	12
50-54	9
55-59	5
60-64	7
Total	**40**

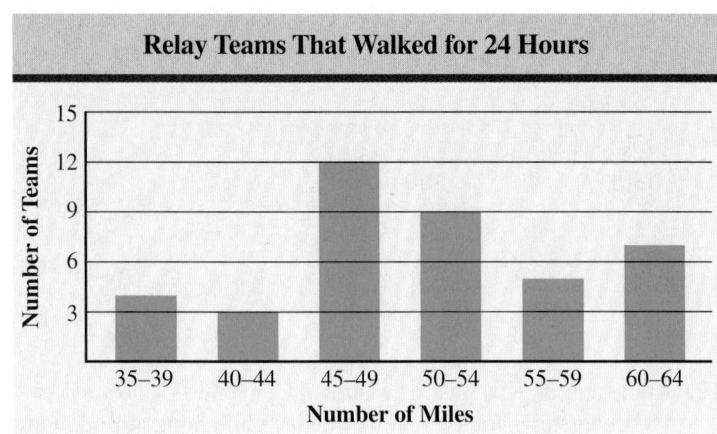

b) How many teams walked more than 49 miles?
21

c) How many teams walked fewer than 50 miles?
19

d) How many teams walked from 40 to 59 miles?
29

Fill in the frequency distribution table with the data shown in the histogram.

15. Ms. Mojica's algebra students (90 students in all) were required to do online homework. Algebra software kept track of the number of hours each student spent doing online homework. Following is a histogram of the data Ms. Mojica collected at the end of the semester.

a) Fill in the frequency distribution table based on the data in the histogram.

Number of Hours	Number of Students
1-5	9
6-10	20
11-15	18
16-20	26
21-25	12
26-30	5
Total	**90**

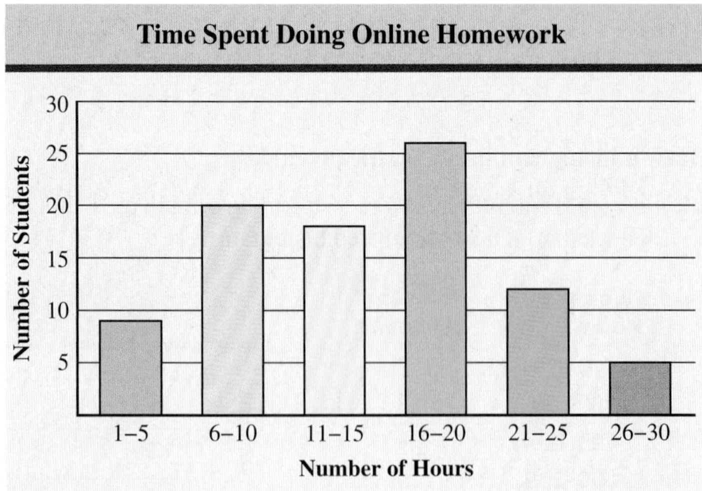

b) How many students spent more than 20 hours on their homework?
17

c) How many students spent fewer than 11 hours on their homework?
29

d) How many students spent from 11 to 25 hours on their homework?
56

16. Leslie's junior bowling league has 40 children in it. Following is a histogram showing the bowlers' best scores of the season.

a) Fill in the frequency distribution table based on the data in the histogram.

Best Score	Number of Bowlers
100–109	7
110–119	10
120–129	8
130–139	13
140–149	0
150–159	2
Total	**40**

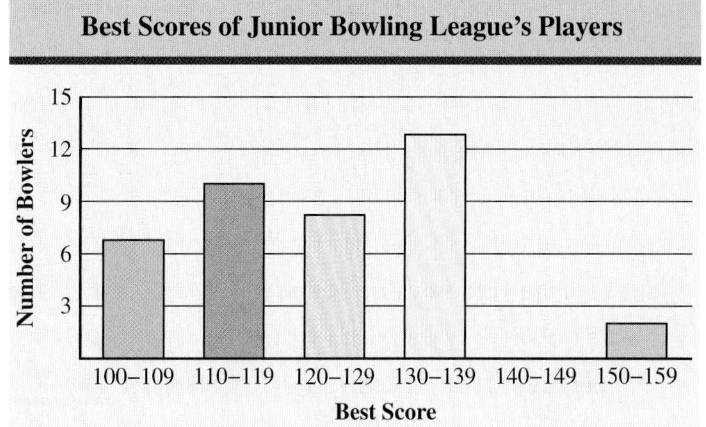

b) How many bowlers' best score was more than 129?
15

c) How many bowlers' best score was less than 120?
17

d) How many bowlers' best score was from 130 to 149?
13

Fill in the frequency distribution table and draw its related histogram.

17. Jane is a manager at 24-Hour Fitness. She wanted to know how many minutes clients were spending at the gym each time they visited. Following are the data collected from 50 different clients.

52	47	61	80	50	65	73	39
45	56	72	51	62	43	35	67
64	52	49	32	71	58	52	89
70	65	58	42	60	54	33	79
38	46	62	62	51	86	83	48
69	49	75	77	34	59	50	62
40	61						

a) Fill in the frequency distribution table and draw the related histogram.

Number of Minutes	Tally	Number of Clients
30–39	~~////~~ /	6
40–49	~~////~~ ////	9
50–59	~~////~~ ~~////~~ //	12
60–69	~~////~~ ~~////~~ //	12
70–79	~~////~~ //	7
80–89	////	4
Total		**50**

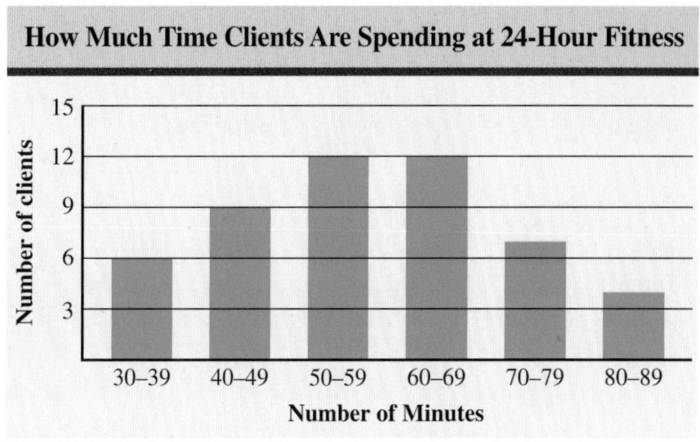

b) How many clients spend less than 50 minutes?
15

c) How many clients spend from 40 to 69 minutes?
33

d) How many clients spend at least one hour?
23

18. Gail, Aldersgate Church's secretary, was asked to compile the 2008 Sunday service and special events attendance for 40 families. (Special events include services on Ash Wednesday, Christmas Eve, and other special days.) Following are the data she collected.

38	15	20	5	42	33	29	53
46	21	18	7	35	3	19	42
40	56	25	30	10	12	41	30
29	49	47	26	15	8	27	25
46	32	49	13	41	33	46	28

a) Fill in the frequency distribution table and draw the related histogram.

Number of Services	Tally	Number of Families
0–9	////	4
10–19	~~////~~ //	7
20–29	~~////~~ ////	9
30–39	~~////~~ //	7
40–49	~~////~~ ~~////~~ /	11
50–59	//	2
Total		**40**

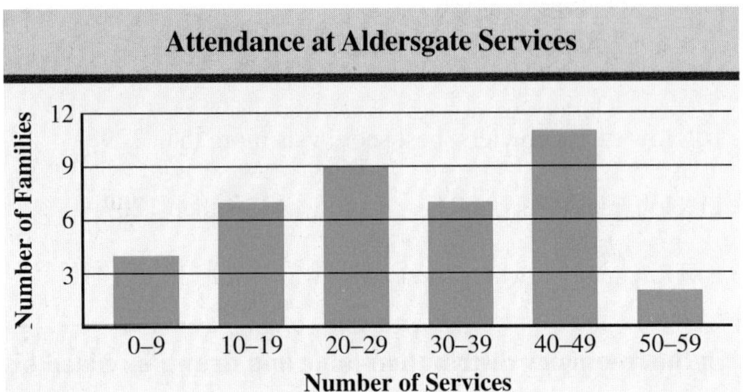

b) How many families attended fewer than 20 services?
11

c) How many families attended from 40 to 59 services?
13

d) How many families attended at least 30 services?
20

Think Outside the Box

19. BJ, a waitress at Polly's Pies, kept track of the number of cups of coffee her customers drank at breakfast one morning. The results of the 40 customers' coffee consumption are shown in the bar graph.

a) Complete the table of values, including the number of cups of coffee for each category and the percent of the total 40 customers.

b) Draw a circle graph based on the percentages in the table.

Number of Cups	Number of Customers	Percent of Customers
None	6	15%
One	10	25%
Two	15	37.5%
Three	5	12.5%
More Than Three	4	10%
Total	**40**	**100%**

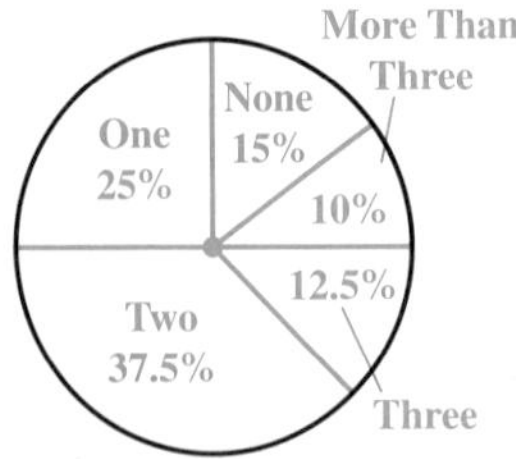

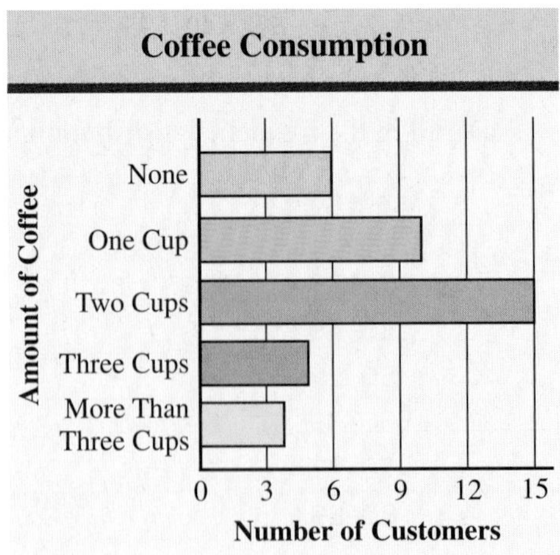

SECTION 9.5 Mean, Median, and Mode

OBJECTIVES

In this section, you will learn to

- Use a formula to find the mean of a set of numerical data values.
- Find the median of a set of numerical data values.
- Find the mode of a set of numerical data values.

You Need to Know

To successfully complete this section, you need to understand

- ☐ Adding whole numbers (1.2)
- ☐ Rounding decimals (5.2)
- ☐ Adding decimals (5.3)
- ☐ Dividing decimals (5.5)

Introduction

In the previous section, we looked at a set of numerical data and graphed it as a histogram. The histogram provided a visual description of the data. In this section, we look at three numerical descriptions of the data.

As an example, a few years ago a newspaper article in the *Cucamonga Community College Chronicle* described the school's typical student as follows

- 25 years old
- having a 3.152 grade point average
- seeking an AA degree
- living 5.2 miles from the college
- female

Some young men just out of high school thought, "That doesn't sound anything like me! Am I not a typical student?"

The truth is, they are rather typical; they just don't fit this one description of being typical. In fact, there might be just one or two students at the whole college who fit that description. So how did the *Chronicle* determine those characteristics of the college's typical student?

The *Chronicle* used three types of statistical measures to determine the typical student: the *mean,* the *mode,* and the *median*

1. The **mean** is the balancing point for all of the data. There is a formula for finding the mean.
2. When the data are written in numerical order, the **median** is the middle number of all of the data. For the most part, the median is found by observation.
3. The **mode** is the data value that occurs most often. This is found by observation.

Estimate the mean, the *balance point*, for 75, 89, and 94

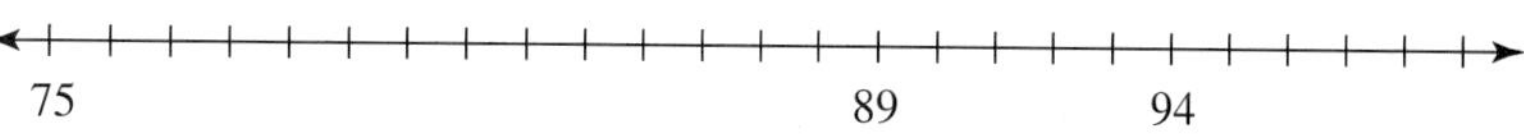

Although the median is the only one of the three that requires it, writing the data in numerical order can be helpful for finding the mean and mode as well.

We'll explore mean, median, and mode one at a time.

The Mean

The **mean** is a numerical average that is found by addition and division.

$$\text{Mean} = \frac{\text{Sum of all of the data values}}{\text{Total number of data values collected}}$$

You use this formula to figure out the average score on your math tests. Let's say that on three 100-point tests, your scores were 89, 75, and 94. Your average score, your *mean* score is as follows:

$$\text{Mean} = \frac{89 + 75 + 94}{3}$$

← The sum of the data values: your three scores

← The total number of data values: your three tests

$\text{Mean} = \frac{258}{3} = 86$ (the balance point of the three numbers) Your average score is 86.

For 75, 89, and 94, the mean is 86. Why do you think the mean is called the *balance point*?

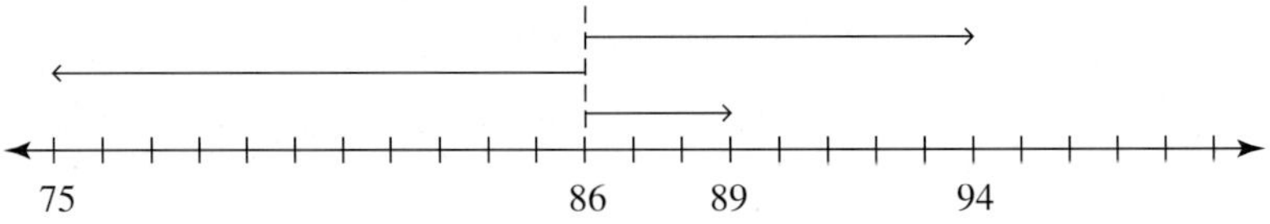

In this text, we round the mean to have one more decimal place than the data values have. If the data values are whole numbers, round to the nearest tenth. If the data values have one decimal place, round to the nearest hundredth. It's common, though, to round money to the nearest penny.

Example 1

Calculate the mean of each of these quiz test scores. If necessary, round to one more decimal place than the data values have.

a) Four 25-point quizzes: 17, 24, 16, 21 **b)** Six 50-point tests: 41, 37, 33, 39, 42, 37

Procedure Add the quiz/test scores and divide by the number of quizzes/tests. Use long division and decimals if necessary. Because the data values are whole numbers, round the mean to the nearest tenth.

Answer **a)** Mean $= \dfrac{17 + 24 + 16 + 21}{4} = \dfrac{78}{4} = 19.5$ Rounding is not necessary.

b) Mean $= \dfrac{41 + 37 + 33 + 39 + 42 + 37}{6} = \dfrac{229}{6} = 38.166 \approx 38.2$

You Try It 1 **Calculate the mean of each of these quiz or test scores. If necessary, round to one more decimal place than the data values have. Use Example 1 as a guide.**

a) Three 20-point quizzes: 17, 20, 15 **b)** Five 10-point quizzes: 9, 6, 8, 10, 6

The mean can be found for any set of numerical data values in the very same way: Add all of the data values together and divide that sum by the number of data values.

Example 2

Calculate the mean of each set of data values. If necessary, round to one more decimal place than the data values have.

a) Eight retired men were asked to recall the number of cars they have owned over their lifetime. Here are their answers: 10, 17, 8, 12, 15, 23, 11, 16.

b) Elton keeps track of the distance (in miles) from his office to his various clients. Here are the distances for six clients: 9.3, 5.2, 7.6, 11.4, 1.9, 8.8.

Answer **a)** Mean $= \dfrac{10 + 17 + 8 + 12 + 15 + 23 + 11 + 16}{8} = \dfrac{112}{8} = 14 \text{ or } 14.0$

b) Mean $= \dfrac{9.3 + 5.2 + 7.6 + 11.4 + 1.9 + 8.8}{6} = \dfrac{44.2}{6} = 7.3666 \approx 7.37$

You Try It 2 **Calculate the mean of each set of data values. If necessary, round to one more decimal place than the data values have. Use Example 2 as a guide.**

a) One week of high temperatures in Phoenix: 103°, 105°, 99°, 100°, 106°, 110°, 101°

b) Five runners' times (in seconds) in the 100-meter dash: 11.1, 10.9, 11.2, 10.6, 10.8

c) Tips that four servers earned at Denny's: $41.20, $38.50, $35.90, $43.70

Knowing how to calculate the mean is helpful in finding a grade point average (GPA). Each grade in college has a point value according to this table:

Grade	A	B	C	D	F
Point Value	4	3	2	1	0

Question What is the mean between an A, a B, and a C?

Answer It depends on whether the grades are weighted the same or weighted differently.

Consider these two situations:

1. Trevor is taking a history class and has had to write three term papers so far this semester. His grades on those three papers are C, B, and A. The three grades average to a B.

$$\text{Mean} = \frac{C + A + B}{3} = \frac{2 + 4 + 3}{3} = \frac{9}{3} = 3 = B$$

2. Stacy took three courses last semester: Spanish (5 units), sociology (3 units), and yoga (1 unit). Here is a table of her grades.

Course	Spanish	Sociology	Yoga
Grade	C	B	A

At first, it might look as though Stacy has a B average from these three courses. However, because each class has a different number of units, the grades are weighted differently.

For example, an A in Stacy's Spanish class has 5 times more grade point value than an A in her yoga class. In other words, we must multiply each grade by the unit value of the class.

Course	Grade	Point Value	Units	Grade Points (Point Value x Units)
Spanish	C	2	5	10
Sociology	B	3	3	9
Yoga	A	4	1	4
		Total:	9	23

Here is how we figure Stacy's grade point average for last semester:

$$\text{Grade point average} = \frac{\text{Total grade points}}{\text{Total units}} = \frac{23}{9} = 2.5555$$

It's common to round the grade point average to two decimal places: GPA = 2.56.

A GPA of 2.56 is below a B average (which is 3.00). That's because the 5-unit C grade is much stronger than the 1-unit A grade.

You Try It 3 **Calculate the grade point average for this student's spring grade distribution. Round the answer to two decimal places. Use the previous example as a guide.**

Course	Grade	Point Value	Units	Grade Points (Point Value × Units)
Biology	B		4	
Algebra	A		5	
Guidance	A		1	
History	C		3	
Art	D		2	
		Total:		

Grade Point Average = ______________

The Median

After a list of data has been put in numerical order, the **median** is found to be the middle number of the ordered list. This means that there will be an equal number of data values to its right and to its left.

For example, here are the data collected from 15 art students who were asked their age:

21, 17, 33, 24, 67, 18, 45, 19, 27, 30, 19, 32, 22, 41, 28

When put in numerical order, the list becomes

17, 18, 19, 19, 21, 22, 24, 27, 28, 30, 32, 33, 41, 45, 67

With an odd number of data values (15), once the **middle** number is found, there is an even number of values (14) remaining to divide equally between the left and the right of the middle number. In this case, there are seven values on each side.

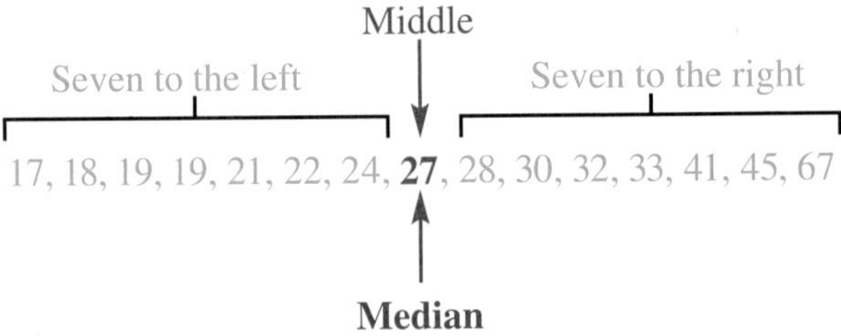

Example 3

Put each list in numerical order and identify the median.

a) Nine quiz scores: 8, 5, 9, 10, 4, 5, 2, 9, 6

b) Ages of 13 students: 18, 26, 39, 22, 22, 19, 23, 21, 18, 42, 21, 25, 28

Procedure Put the data in numerical order and find the middle number.

Answer **a)** Quiz scores: 2, 4, 5, 5, **6**, 8, 9, 9, 10

Four to the left | Four to the right

Median = 6

b) Ages of students: 18, 18, 19, 21, 21, 22, **22**, 23, 25, 26, 28, 39, 42

Six to the left | Six to the right

Median = 22

Caution Notice in part b that the median is not the only 22 in the list. Another 22 is on the left side of the median, but that doesn't matter when finding the middle number.

You Try It 4 **Put each list in numerical order and identify the median. Use Example 3 as a guide.**

a) Pairs of shoes that 11 people own: 10, 9, 1, 4, 18, 12, 7, 5, 4, 15, 2

b) Number of children in 15 families: 2, 3, 2, 0, 4, 1, 1, 7, 0, 5, 1, 2, 8, 3, 2

An Even Number of Data Values

When the list contains an even number of data values, the median splits the two middlemost numbers. (Keep in mind that the list of data values must be put in numerical order first.)

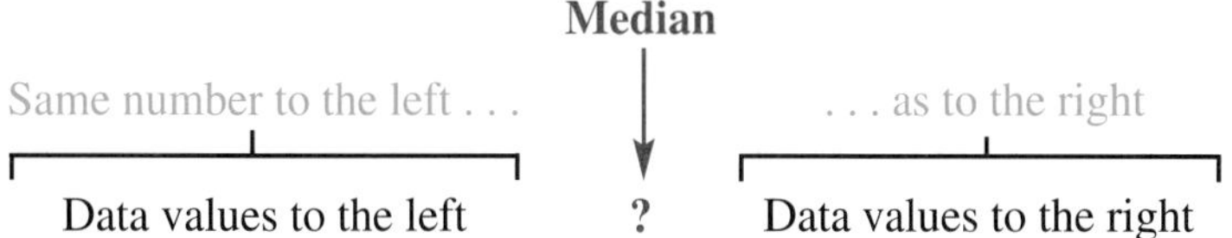

Consider this list of ten numbers representing the ages of cousins at a family gathering (already placed in numerical order):

1, 3, 4, 6, 7, 9, 10, 10, 11, 15

These ten data values can be evenly split with the first five data values on the left and the second five data values on the right. In this case, the median lands between 7 and 9.

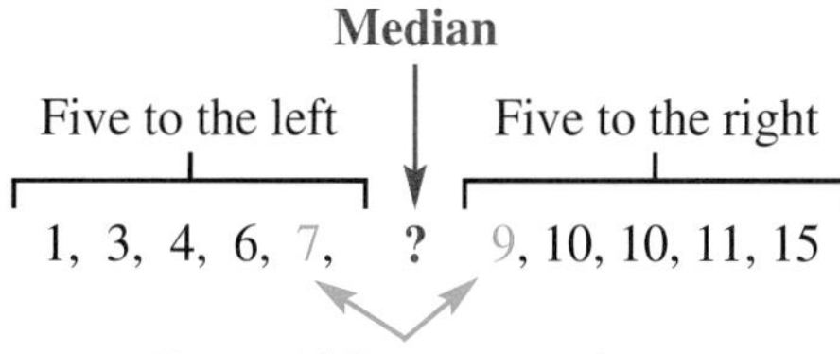

The median is the number in the middle of 7 and 9. It is the *mean* of just those *two* middlemost numbers.

$$\text{Median} = \frac{7 + 9}{2} = 8$$

Caution When finding the *mean* of the two middlemost numbers, we're *not* determining the mean of the whole list—just those two numbers—so, we divide by 2.

Sometimes the median is a decimal even though all of the data values are whole numbers. Consider this list of data values representing the number of hours that eight members of the Kiwanis Club worked in preparation for their silent auction: 15, 19, 8, 6, 14, 3, 21, 9.

In numerical order, the list is 3, 6, 8, 9, 14, 15, 19, 21. The two middlemost data values are 9 and 14; so,

$$\text{Median} = \frac{9 + 14}{2} = \frac{23}{2} = 11.5$$

Also, it's possible that the two middlemost data values are the same number, as in this list:

2, 6, 9, 11, 11, 14, 19, 23

Here the two middlemost numbers are 11; so, the median is 11.

Example 4

Put each list in numerical order and identify the median.

a) 10 test scores: 77, 81, 63, 92, 82, 88, 75, 95, 60, 71

b) 16 quiz scores: 8, 6, 9, 10, 8, 5, 2, 9, 7, 6, 10, 9, 3, 4, 9, 6

Procedure Put the data in numerical order, underline the two middlemost data values, and find the mean of just those two numbers.

Answer a) Test scores: 60, 63, 71, 75, <u>77</u>, <u>81</u>, 82, 88, 92, 95 Median $= \frac{77 + 81}{2} = \frac{158}{2} = 79$

b) Quiz scores: 2, 3, 4, 5, 6, 6, 6, <u>7</u>, <u>8</u>, 8, 9, 9, 9, 9, 10, 10 Median $= \frac{7 + 8}{2} = \frac{15}{2} = 7.5$

You Try It 5 **Put each list in numerical order and identify the median. Use Example 4 as a guide.**

a) High temperatures in San Diego for the first two weeks in July: 86°, 85°, 82°, 75°, 72°, 77°, 86°, 81°, 75°, 76°, 75°, 86°, 83°, 77°

b) Number of college classes that eight students took: 14, 9, 23, 17, 6, 11, 5, 20

The Mode

How is it that the typical student at a college can be female? This is where the mode comes in. Simply put, the **mode** is the value (not necessarily numerical) that occurs most often.

If 3,289 women and 2,956 men are enrolled at Cucamonga College, there are more women than men and the mode is *female*. In other words, according to the mode, the typical student is a woman.

In a numerical list of data values, we can search out the mode as the number that occurs most often. It's best to put the list in numerical order so that the mode can be easily identified. Sometimes there is more than one mode, as shown in the next example.

Example 5

Put each list in numerical order and identify the mode.

a) 16 quiz scores: 8, 6, 9, 10, 8, 5, 2, 9, 7, 6, 10, 9, 3, 4, 9, 6

b) Ages of 14 students: 18, 26, 39, 22, 18, 19, 23, 21, 18, 42, 21, 25, 28, 21

c) 15 test scores: 95, 46, 77, 81, 63, 92, 80, 88, 75, 93, 60, 71, 82, 90, 79

Procedure Put the data in numerical order and underline the value(s) that occur most often.

Answer a) Quiz scores: 2, 3, 4, 5, 6, 6, 6, 7, 8, 8, <u>9</u>, <u>9</u>, <u>9</u>, <u>9</u>, 10, 10 Mode = 9

b) Ages of students: <u>18</u>, <u>18</u>, <u>18</u>, 19, <u>21</u>, <u>21</u>, <u>21</u>, 22, 23, 25, 26, 28, 39, 42

This list has *two modes*, 18 and 21.

c) Test scores: 46, 60, 63, 71, 75, 77, 79, 80, 81, 82, 88, 90, 92, 93, 95

Every number on this list occurs only once; so, the list has no mode.

You Try It 6 **Put each list in numerical order and identify the mode by underlining it. If there is no mode, state so. Use Example 5 as a guide.**

a) Number of daily student absences from Dr. Chavez's weekend class: 0, 2, 1, 5, 6, 2, 4, 1, 5, 3, 4, 6, 5, 0, 2, 5

b) High temperatures in Baltimore for the last two weeks in March: 72, 70, 67, 59, 57, 62, 71, 66, 60, 61, 63, 74, 68, 65

c) Number of handbags that 18 women own: 7, 9, 3, 1, 4, 9, 9, 12, 4, 7, 5, 9, 4, 15, 2, 4, 9, 4

Think About It 3 Explain in your own words the differences between mean, median, and mode and give an example of each.

You Try It 7 **The following list is the number of quarter-mile laps that 12 physical education students completed in 30 minutes. Put this list in numerical order and identify the mean, median, and mode. (If there is no mode, state so.)**

12, 5, 9, 4, 9, 20, 14, 18, 6, 9, 15, 16

Numerical Order:

a) Mean:

b) Median:

c) Mode:

Answers: You Try It and Think About It

You Try It: **1. a)** ≈17.3 **b)** 7.8 **2. a)** 103.4° **b)** 10.92 seconds **c)** $39.83

3.

Course	Grade	Point Value	Units	Grade Points (Point Value × Units)
Biology	B	3	4	12
Algebra	A	4	5	20
Guidance	A	4	1	4
History	C	2	3	6
Art	D	1	2	2
		Total:	**15**	**44**

$$\textbf{Grade Point Average} = \frac{44}{15} \approx 2.93$$

4. a) 1, 2, 4, 4, 5, 7, 9, 10, 12, 15, 18; Median = 7 **b)** 0, 0, 1, 1, 1, 2, 2, 2, 2, 3, 3, 4, 5, 7, 8; Median = 2 **5. a)** 79° **b)** 12.5 **6. a)** 5 **b)** No mode **c)** 4 and 9 **7.** Numerical Order: 4, 5, 6, 9, 9, 9, 12, 14, 15, 16, 18, 20 **a)** ≈11.4 **b)** 10.5 **c)** 9

Think About It: **1.** Anwers may vary. The mean is 86. **2.** Answers may vary. **3.** Answers may vary. One possibility: The mean is the average of all of the values, the median is the value that is in the middle of the ordered data, and the mode is the data value that occurs most often.

FOR EXTRA HELP
Student Resources on DVD-ROM
Includes
➤ Student's Solutions Manual
➤ Video Lectures
➤ Chapter Test Prep Video
MyMathLab
MathXL

Section 9.5 Exercises

Think Again

1. Explain in your own words the differences between mean, median, and mode and give an example of each. (*Refer to Think About It 3*)
   Answers will vary.

2. What steps are required to find the median of a set of data values?
   a) Organize the data in numerical order and count the number of data values.
   b) If there is an odd number of data values, find the middle data value. This is the median.
   c) If there is an even number of data values, find the two middle data values and find the mean of those two values only. This is the median.

3. If there is an odd number of data values, will the median always be one of the data values? Explain your answer or show an example that supports your answer.
   The median will always be one of the data values. It is the middle data value in the ordered set of values.

4. If there is an even number of data values, will the median ever be one of the data values? Explain your answer or show an example that supports your answer.
   The median will be one of the data values only if the middle two values are the same number.

Focus Exercises

Calculate the mean of each of these quiz or test scores. If necessary, round to one more decimal place than the data values have.

5. Four 100-point tests: 92, 86, 70, 84
   83

6. Five 25-point tests: 23, 20, 18, 22, 17
   20

7. Six 50-point tests: 40, 47, 48, 42, 39, 42
   43

8. Eight 10-point quizzes: 6, 9, 10, 8, 7, 10, 5, 9
   8

9. Five 100-point tests: 72, 81, 86, 75, 83
   79.4

10. Four 20-point quizzes: 18, 12, 15, 20
    16.3

11. Three 200-point tests: 157, 172, 174
    167.7

12. Seven 10-point quizzes: 10, 8, 9, 6, 6, 4, 9
    7.4

Calculate the mean of each set of data values. If necessary, round to one more decimal place than the data values have.

13. The number of miles that eight runners ran in two hours: 15, 19, 23, 17, 17, 20, 19, 22
    19 miles

14. The number of students receiving A's in seven algebra classes: 5, 8, 12, 4, 9, 11, 6
    7.9 students

15. The number of cars sold in a six-month period: 33, 28, 26, 35, 25, 31
    29.7 cars

16. The number of European cities that nine tourists visited: 8, 3, 6, 12, 5, 7, 7, 9, 4
    6.8 cities

17. The number of kilometers that five runners ran in two hours: 26.3, 21.8, 22.6, 19.5, 20.1
    22.06 kilometers

18. The price of a seedless watermelon at five grocery stores: $3.80, $4.50, $5.10, $4.20, $4.90
    $4.50

19. The height, in meters, of four children: 1.26, 1.32, 1.41, 1.08
    1.268 meters

20. The number of miles that seven students live from school: 5.7, 6.3, 10.1, 4.2, 0.8, 1.9, 2.4
    4.49 miles

Calculate the grade point average for each student's fall grade distribution. Round the answer to two decimal places.

21.

Course	Grade	Point Value	Units	Grade Points (Point Value × Units)
Anatomy	C	2	5	10
Guitar	A	4	1	4
English	B	3	4	12
Sociology	A	4	3	12
		Total:	**13**	**38**

Grade Point Average ≈ 2.92

22.

Course	Grade	Point Value	Units	Grade Points (Point Value × Units)
Physics	B	3	4	12
Calculus	A	4	5	20
Political Science	C	2	3	6
Music Theory	B	3	3	9
Art Appreciation	D	1	2	2
		Total:	**17**	**49**

Grade Point Average ≈ 2.88

Identify the median and the mode.

23. The 18-hole score for 13 golfers:
82, 71, 75, 73, 80, 69, 71, 78, 76, 71, 79, 81, 77
Median is 76.
Mode is 71.

24. The number of sponsors for 11 people in a walk-a-thon: 23, 18, 25, 36, 24, 17, 41, 32, 19, 38, 14
Median is 24.
There is no mode.

25. The number of pets of 15 children:
3, 5, 0, 2, 6, 1, 2, 4, 4, 3, 1, 3, 6, 0, 3
Median is 3.
Mode is 3.

26. The number of cousins of 17 children:
8, 4, 2, 9, 0, 4, 4, 11, 9, 15, 9, 0, 1, 4, 8, 10, 9
Median is 8.
Modes are 4 and 9.

27. The number of rooms in 12 houses:
9, 15, 13, 8, 6, 4, 6, 12, 14, 6, 10, 11
Median is 9.5.
Mode is 6.

28. The number of strikeouts for eight pitchers in a 40-game season: 20, 32, 40, 26, 52, 49, 29, 41
Median is 36.
There is no mode.

29. The number of hits for 16 batters in a 40-game season: 53, 42, 38, 32, 45, 54, 38, 42, 38, 56, 48, 31, 39, 38, 45, 35
Median is 40.5.
Mode is 38.

30. The number of minutes that 14 students took to finish a final exam:
63, 85, 92, 65, 63, 78, 92, 100, 63, 98, 92, 88, 70, 62
Median is 81.5.
Modes are 63 and 92.

Identify the mean, the median, and the mode.

31. The number of push-ups that 9 sixth-graders completed:
31, 15, 19, 7, 2, 6, 10, 14, 20
Mean ≈13.8.
Median is 14.
There is no mode.

32. The number of Reuben sandwiches that a deli sold each day for 15 days:
8, 3, 6, 7, 2, 5, 4, 2, 6, 7, 4, 6, 5, 6, 4
Mean is 5.
Median is 5.
Mode is 6.

33. The number of hours of flight for 12 novice pilots:
11, 13, 5, 9, 5, 7, 5, 13, 18, 16, 5, 19
Mean is 10.5.
Median is 10.
Mode is 5.

34. The number of innings the starting pitcher lasted in 20 consecutive baseball games:
6, 4, 8, 7, 6, 9, 2, 3, 5, 1, 10, 6, 8, 7, 7, 4, 3, 6, 2, 4
Mean is 5.4.
Median is 6.
Mode is 6.

Think Outside the Box

35. In her algebra class, Myla's instructor always gives 100-point tests. The mean of Myla's first three algebra test scores is 89, and her fourth test score is 81. What is Myla's test average after the fourth test?
After the fourth test, Myla's test average is 87.

36. In his geometry class, Corren's instructor always gives 50-point tests. The mean of Corren's first four geometry test scores is 47, and the mean of Corren's first five geometry test scores is 45. How many points did Corren score on the fifth test?
Corren scored 37 points on the fifth test.

37. Miguel's 12 homework scores for his chemistry class are listed in numerical order:
1, 5, 6, 6, 7, 7, 8, 9, 9, 9, 10, 10

a) Calculate the mean, the median, and the mode of this set of numbers.
Mean = 7.25; Median = 7.5; Mode = 9

b) If the chemistry instructor throws out the highest and lowest homework scores,
i. will the mean increase, decrease, or stay the same?
Increase
ii. will the median increase, decrease, or stay the same?
Stay the same
iii. will the mode increase, decrease, or stay the same?
Stay the same

c) Throw out the highest and lowest homework scores and find the mean, median, and mode of this distribution.
Mean = 7.6; Median = 7.5; Mode = 9

d) Are any of your answers in part b inaccurate? If so, which one(s) and why?
Answers may vary.

SECTION 9.6 Probability

OBJECTIVES

In this section, you will learn to

- Use the vocabulary of probability.
- Use a simple formula to find probabilities.
- Use a bar graph to find probabilities.

You Need to Know

To successfully complete this section, you need to understand

- ☐ Simplifying fractions (4.3)
- ☐ Writing a fraction as a percent (6.4)
- ☐ How to read a bar graph (9.3)

Introduction

Has this ever happened to you?

Tim, a college freshman, wants to take an English class, but the class was already full when it came time for him to register for classes. He decides to go to the class the first day to see if the instructor will add him to the class.

A total of four students want to take the class, but the instructor says that she will add only one student. She takes their add cards and, placing them upside down, chooses a card at random.

Question What is the probability that Tim's card will be chosen?

Answer Tim has a 1 out of 4 chance—a 25% chance—of being selected: $\frac{1}{4} = 25\%$.

The Vocabulary of Probability

Situations such as the one Tim experienced cause us to think about **probability**—a numerical way to express the likelihood that some particular thing, or *event,* will happen.

In Tim's probability situation, let's look at some of the vocabulary related to probability.

- An **experiment** is the act of doing something to create a result.

  In our example, the experiment is the instructor choosing a card at random.

- The **possible outcomes** of an experiment are all of the different results that can occur, although usually, only one outcome can occur for each experiment.

  In our example, the possible outcomes are the four different cards the instructor has to choose from, but only one card will be selected.

- An **event** is the outcome that we're most interested in occurring.

  In our example, the event is that Tim's card is selected. We're not interested in anyone else's card. Choosing Tim's card is considered a success.

- The **probability** of an event is a ratio, abbreviated as P(event), and is calculated as follows:

$$P(\text{event}) = \frac{\text{Number of successes in the event}}{\text{Total number of possible outcomes}}$$

  Probability is commonly written as a fraction or a percent. In our example, there's only one way our event can occur—because Tim has only 1 card in the drawing—and a total of 4 outcomes are possible.

$$P(\text{Tim's card}) = \frac{1}{4} = 25\%$$

  An important aspect of probability is the idea that the selection is random.

- A **random selection** is the act of choosing something so that each possible outcome has an *equal chance* of being selected and is *equally likely* to be selected.

  In our example, the winning card is selected randomly when there is no way to distinguish between one card or another in the selection process. Tim's card has just as much chance of being selected as Rey's card or Amy's card or Dionne's card.

Two Common Probabilities

Here are two experiments that have simple outcomes. In each experiment, to ensure that a possible outcome has an equal chance of being selected, we assume that the experiment is being conducted fairly.

1. Flip a coin and let it land on the floor. There are only two possible outcomes, *heads* or *tails*. An event could be *landing heads up* or *landing tails up*. Each possible outcome is equally likely if the coin is fair and evenly balanced.
2. Roll a single die (singular of *dice*). There are six possible outcomes: 1, 2, 3, 4, 5, and 6. An event could be one of the following
   - Rolling a 4 (one success, the number 4)
   - Rolling an even number (three successes, the numbers 2, 4, and 6)
   - Rolling a number higher than 4 (two successes, the numbers 5 and 6)

   Each possible outcome is equally likely if the die is evenly balanced.

Example 1

A quarter is flipped in the air. What is the probability that it will land tails up?

Procedure Assume the coin is fair and

1. Identify the experiment.
2. Identify the total number of possible outcomes.
3. Identify the event.
4. Identify the number of successes described in the event.

Answer The experiment is flipping a quarter, a total of 2 outcomes are possible, the event is the quarter landing tails up, and this event has only **1** success. So, the probability ratio is

$$\mathbf{P}(\text{tails}) = \frac{\text{Number of successes (tails)}}{\text{Total possible outcomes}} = \frac{1}{2}$$

There's a 1 out of 2 chance of the coin landing tails up.

Example 2

Find the probability of each event. Simplify the fraction if possible.

a) A die is rolled. What is the probability of getting a 3?

b) A die is rolled. What is the probability of getting a number greater than 4?

Procedure The experiment is to roll a single die. A total of **6** outcomes are possible.

Assume the die is fair and

1. Identify the event.
2. Identify the number of successes described in the event.

Answer a) The event is to roll a 3. This event has only 1 success. The probability ratio is

$$\mathbf{P}(\text{roll a } 3) = \frac{\text{Number of successes (rolling a 3)}}{\text{Total possible outcomes}} = \frac{1}{6}$$

There's a 1 out of 6 chance of rolling a 3.

b) The event is to roll a number higher than 4. This event has 2 successes: rolling a 5 or a 6. The probability ratio is

$$\mathbf{P}(\text{more than } 4) = \frac{\text{Number of successes (rolling a 5 or 6)}}{\text{Total possible outcomes}} = \frac{2}{6} = \frac{1}{3}$$

There's a 1 out of 3 chance of rolling a number higher than 4.

You Try It 1 **Find the probability of each event. Assume that the coin and die are fair. Use Examples 1 and 2 as guides.**

a) A nickel is flipped in the air. What is the probability that it will land heads up?

b) A die is rolled. What is the probability of getting a 5?

c) A die is rolled. What is the probability of getting an odd number?

Example 3

A bag contains 20 marbles. Four are red, 5 are yellow, 3 are green, and 8 are blue. The experiment is to pick one marble out of the bag.

What is the probability that the chosen marble is

a) red? b) green?

Procedure Assume that each marble has an equally likely chance of being selected.

Answer a) A total of 20 outcomes are possible. The event, *selecting a red marble,* has 4 possible successes because there are 4 red marbles. The probability ratio is

$$\mathbf{P}(\text{red}) = \frac{\text{Number of red marbles}}{\text{Total number of marbles}} = \frac{4}{20} = \frac{1}{5}$$

There's a 1 out of 5 chance of selecting a red marble.

b) A total of 20 outcomes are possible. The event, *selecting a green marble,* has 3 possible successes because there are 3 green marbles. The probability ratio is

$$\mathbf{P}(\text{green}) = \frac{\text{Number of green marbles}}{\text{Total number of marbles}} = \frac{3}{20}$$

There's a 3 out of 20 chance of selecting a green marble.

You Try It 2 **A bag has 12 marbles. Two are red, 1 is yellow, 3 are green, and 6 are blue. The experiment is to randomly select one marble from the bag.**

What is the probability that the chosen marble is

a) red? b) yellow? c) green? d) blue? Use Example 3 as a guide.

Bar Graphs and Probability

Consider the bag of marbles in Example 3. The bag has 4 red, 5 yellow, 3 green, and 8 blue marbles; and there is a total of 20 marbles.

We could create a bar graph representing the bag of marbles.

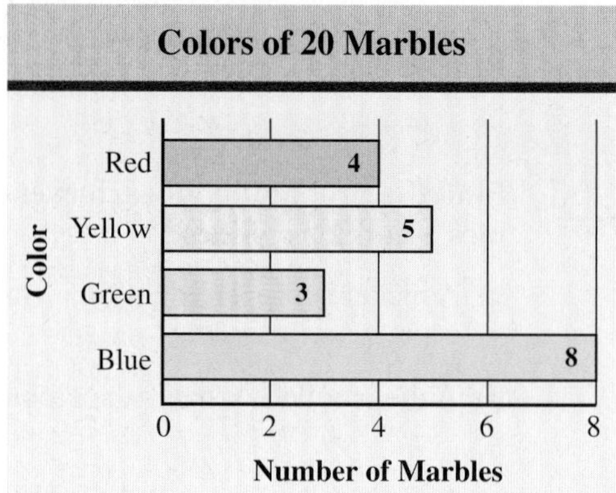

We could ask the same questions as in Example 3 but use the bar graph to help us find the answers. It is important to know that there are 20 marbles in all, as this is the denominator for each probability ratio.

One marble from the bag is randomly chosen.

a) What is the probability that the marble will be red? $\mathbf{P}(\text{red}) = \frac{4}{20} = \frac{1}{5}$

b) What is the probability that the marble will be blue? $\mathbf{P}(\text{blue}) = \frac{8}{20} = \frac{2}{5}$

We can use bar graphs to answer questions of probability that involve a random selection of one kind or another.

Example 4

Here is a bar graph representing the number of lunch-hour employees at each restaurant in a shopping mall food court. Use it to answer the probability questions that follow.

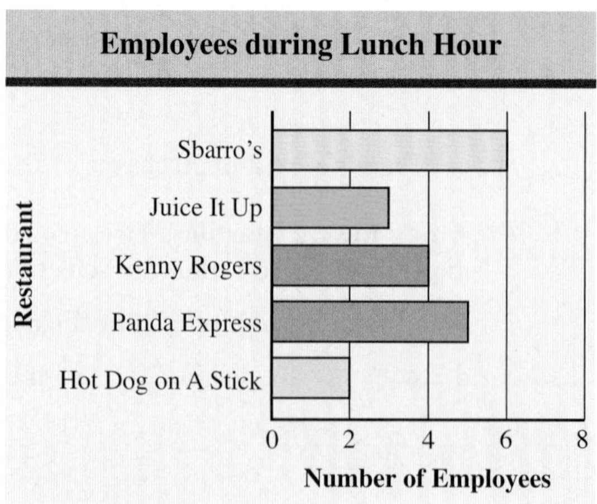

Question If the mall chooses a food court employee at random to win a mall gift certificate, what is the probability that the employee works for

a) Panda Express? **b)** Sbarro's?

Procedure Determine how many employees each bar represents and determine the total number of employees in the diagram. (This is work you can do in the original bar graph.)

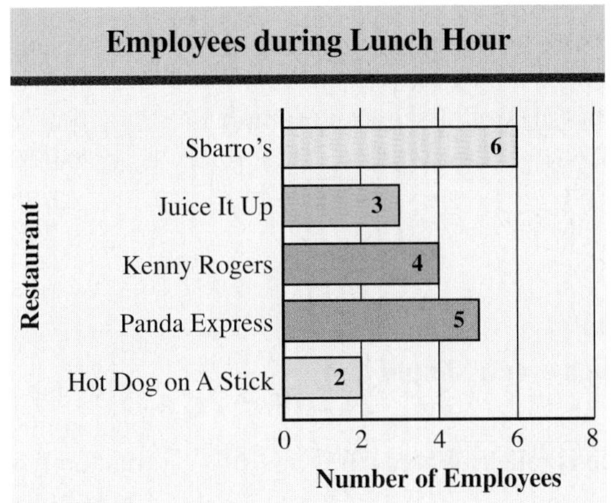

Total:
6
3
4
5
+2
20

Answer **a)** $\mathbf{P}(\text{Panda Express}) = \dfrac{\text{Number of Panda express employees}}{\text{Total number of employees}} = \dfrac{5}{20} = \dfrac{1}{4}$

b) $\mathbf{P}(\text{Sbarro's}) = \dfrac{\text{Number of Sbarro's employees}}{\text{Total number of employees}} = \dfrac{6}{20} = \dfrac{3}{10}$

You Try It 3 **The bar graph shows the number of singers by section in Aldersgate's chancel choir. One member is randomly selected to be the guest director for the July 4th patriotic anthem.**

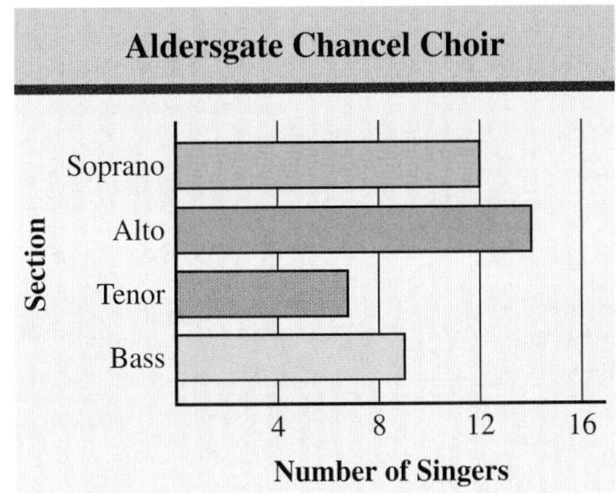

What is the probability that the chosen singer is

a) a soprano?

b) a tenor?

c) an alto?

Use Example 4 as a guide.

Answers: You Try It and Think About It

You Try It: **1. a)** $\mathbf{P}(\text{heads}) = \frac{1}{2}$ **b)** $\mathbf{P}(\text{roll is a 5}) = \frac{1}{6}$ **c)** $\mathbf{P}(\text{roll is odd}) = \frac{3}{6} = \frac{1}{2}$ **2. a)** $\mathbf{P}(\text{red}) = \frac{2}{12} = \frac{1}{6}$ **b)** $\mathbf{P}(\text{yellow}) = \frac{1}{12}$ **c)** $\mathbf{P}(\text{green}) = \frac{3}{12} = \frac{1}{4}$ **d)** $\mathbf{P}(\text{blue}) = \frac{6}{12} = \frac{1}{2}$ **3. a)** $\mathbf{P}(\text{soprano}) = \frac{12}{42} = \frac{2}{7}$ **b)** $\mathbf{P}(\text{tenor}) = \frac{7}{42} = \frac{1}{6}$ **c)** $\mathbf{P}(\text{alto}) = \frac{14}{42} = \frac{1}{3}$

Think About It: No Think About Its in this section.

Section 9.6 Exercises

Think Again

1. What is the probability of an event that is guaranteed to happen (such as rolling a number less than 7 on a single die)? Explain your answer or show an example that supports your answer.
100% or 1. In the example of the die, P(getting a number less than 7) $= \frac{6}{6} = 1$.

2. What is the probability of an event that cannot happen (such as rolling a number less than 1 on a single die)? Explain your answer or show an example that supports your answer.
0% or 0. In the example of the die, P(getting a number less than 1) $= \frac{0}{6} = 0$.

Focus Exercises

Find the probability of each event. Simplify all fractions completely.

Experiment: A six-sided die is rolled. What is the probability that it comes up

3. 2?
$\frac{1}{6}$

4. less than 5?
$\frac{2}{3}$

5. greater than 1?
$\frac{5}{6}$

6. less than 7?
1

7. an odd number that is greater than 2?
$\frac{1}{3}$

8. an even number that is less than 3?
$\frac{1}{6}$

Experiment: The spinner is spun; and, when the needle comes to rest, it is pointing toward one of the numbers. What is the probability that it will point toward

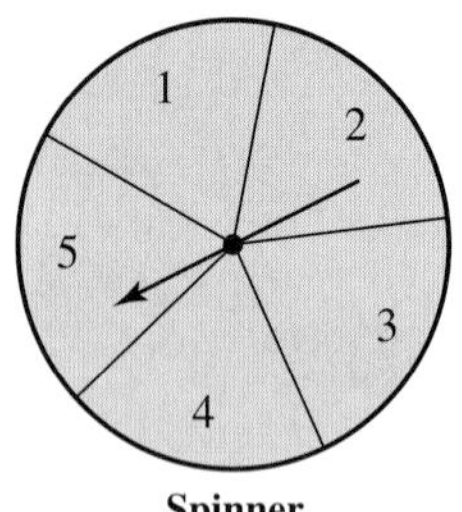

Spinner

In this diagram, the spinning needle is pointing toward 5.

Assume that each number has an equal chance of being pointed to.

9. 3?
$\frac{1}{5}$

10. an even number?
$\frac{2}{5}$

11. an odd number?
$\frac{3}{5}$

12. a number greater than 4?
$\frac{1}{5}$

13. a number less than 4?
$\frac{3}{5}$

14. a number greater than 0?
1

A bag has 20 candies. Eight are cherry, 5 are lemon, 4 are sour apple, and 3 are watermelon. The experiment is to randomly select one candy out of the bag. What is the probability that the chosen candy is

15. cherry?
$\frac{2}{5}$

16. lemon?
$\frac{1}{4}$

17. sour apple?
$\frac{1}{5}$

18. watermelon?
$\frac{3}{20}$

The bar graph below shows the number of players by position on two competing softball teams. One player is randomly selected to lead everyone in singing "Take Me Out to the Ball Game."

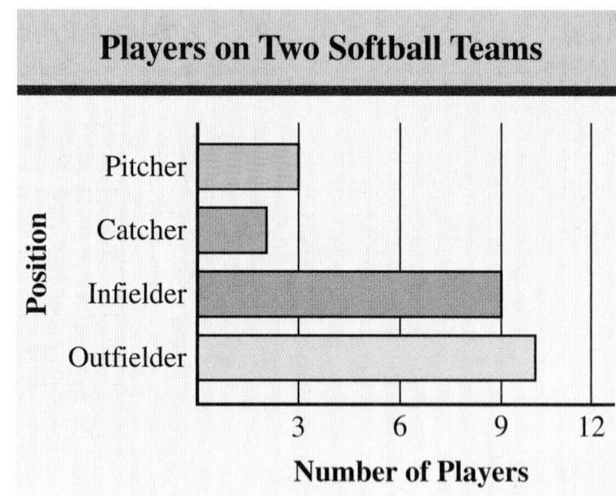

What is the probability that the chosen player's position is

19. pitcher?
$\frac{1}{8}$

20. catcher?
$\frac{1}{12}$

21. infielder?
$\frac{3}{8}$

22. outfielder?
$\frac{5}{12}$

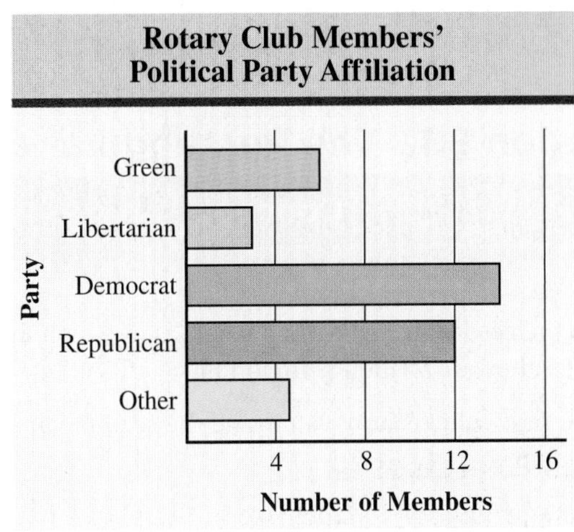

Every year the Eastvale Rotary Club holds a special dinner to install its new board members.

For this event, a member is chosen at random to give a short speech about the American flag and its meaning to him or her.

The bar graph at the right shows the number of members by political party affiliation.

What is the probability that the chosen member's political party is

23. Republican ? $\frac{3}{10}$

24. Green? $\frac{3}{20}$

25. Libertarian? $\frac{3}{40}$

26. Democrat? $\frac{7}{20}$

Think Outside the Box

Following is a table of the sums of two dice, a blue one (B) and a red one (R), when rolled at the same time. Use this table to answer the following questions.

B↓ \ R→	1	2	3	4	5	6
1	2	3	4	5	6	7
2	3	4	5	6	7	8
3	4	5	6	7	8	9
4	5	6	7	8	9	10
5	6	7	8	9	10	11
6	7	8	9	10	11	12

What is the probability that when two dice are rolled, the sum of the dice is

27. 2? $\frac{1}{36}$

28. 4? $\frac{1}{12}$

29. 7? $\frac{1}{6}$

30. 9? $\frac{1}{9}$

31. less than 5? $\frac{1}{6}$

32. more than 10? $\frac{1}{12}$

33. 2 or 12? $\frac{1}{18}$

34. 7 or 11? $\frac{2}{9}$

Chapter 9 Review

Section 9.1 The Rectangular Coordinate System

Concept	Example
An **ordered pair** is a pair of numbers in which the order they are written is important. The first value is an x-value, and the second value is a y-value.	$(5, -8)$ means $x = 5$ and $y = -8$.
An ordered pair may or may not be a solution to an equation.	Determine whether $(-2, 6)$ is a solution to $2x + y = 2$. $x = -2$ and $y = 6$: $2(-2) + 6 \stackrel{?}{=} 2$ $-4 + 6 \stackrel{?}{=} 2$ $2 = 2$ True! ✓ Yes, $(-2, 6)$ is a solution.
The x-y plane, also called the **rectangular coordinate system,** is a grid based on two number lines—the x-axis and the y-axis. Where the two axes meet is called the origin. The origin is represented by the ordered pair (0, 0). The axes separate the x-y plane into four regions called **quadrants.**	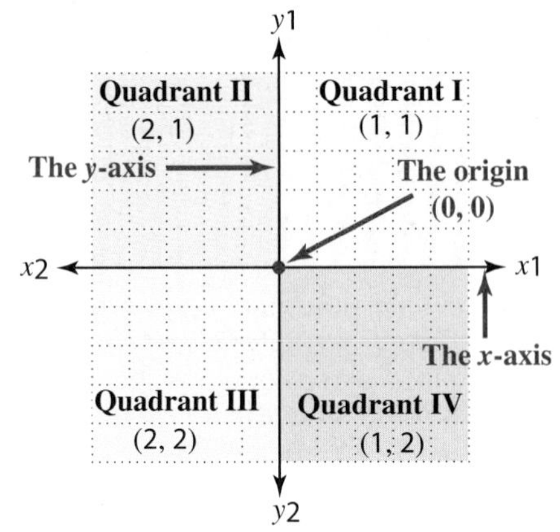
We use the x-y plane to plot points representing ordered pairs. As a point, the x- and y-values are called **coordinates.**	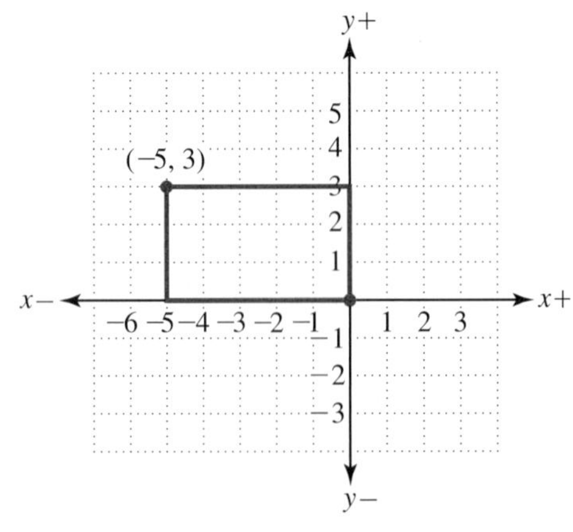

Section 9.1 The Rectangular Coordinate System

Concept	Example
Points on the x-axis always have a y-coordinate of 0, and points on the y-axis always have an x-coordinate of 0. A point on one of the axes is not in any of the quadrants.	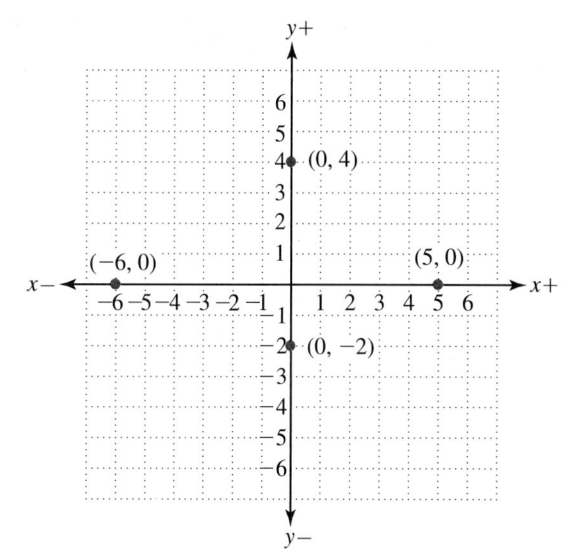
When we know that two or more points are on the same line, we can draw the line that passes through them.	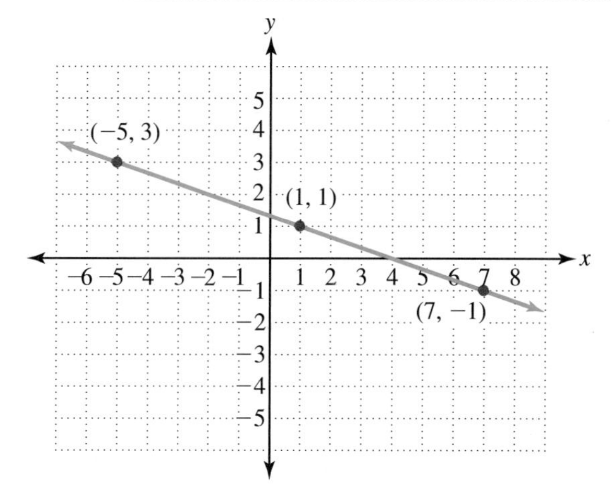

Section 9.2 Graphing Lines

Concept	Example
A **linear equation** of two variables, x and y, is an equation in which the ordered pair solutions are points on a line in the x-y plane.	$x - y = 4$, $y = \frac{1}{2}x + 3$, and $y = -2$ are examples of linear equations.

continued

Section 9.2 Graphing Lines

Concept	Example

Concept

A linear equation can be put in this form:

$$ax + by = c$$

We can graph any line that has this form by finding three ordered pair solutions.

Example

$$x + 2y = 5$$

Choose a value	$x + 2y = 5$	(x, y)
$x = -1$	$(-1) + 2y = 5$ $y = 3$	$(-1, 3)$
$y = 0$	$x + 2(0) = 5$ $x = 5$	$(5, 0)$
$x = 3$	$(3) + 2y = 5$ $y = 1$	$(3, 1)$

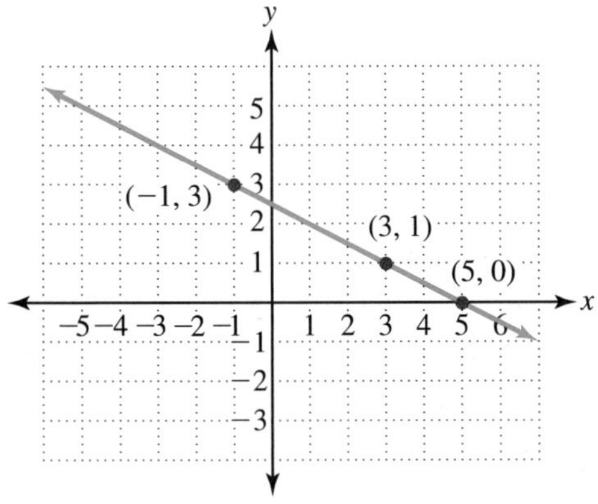

Concept

Another form of a linear equation is $y = mx + b$. Written in this form, we say that *y is in terms of x*.

Here m and b are real numbers and x and y are variables.

To find ordered pair solutions to an equation, choose three x-values, place them in the equation, and solve for each corresponding y-value.

Example

Find three sets of ordered pairs for $y = 2x + 3$.

x	$y = 2x + 3$	(x, y)
-1	$y = 2(-1) + 3$ $y = 1$	$(-1, 1)$
0	$y = 2(0) + 3$ $y = 3$	$(0, 3)$
2	$y = 2(2) + 3$ $y = 7$	$(2, 7)$

Section 9.2 Graphing Lines

Concept	Example
If a linear equation has coordinates that do not fit easily in a typical x-y plane, we must create an x-y-plane that has a larger scale on the y-axis.	Draw the graph of $y = 5x + 30$. 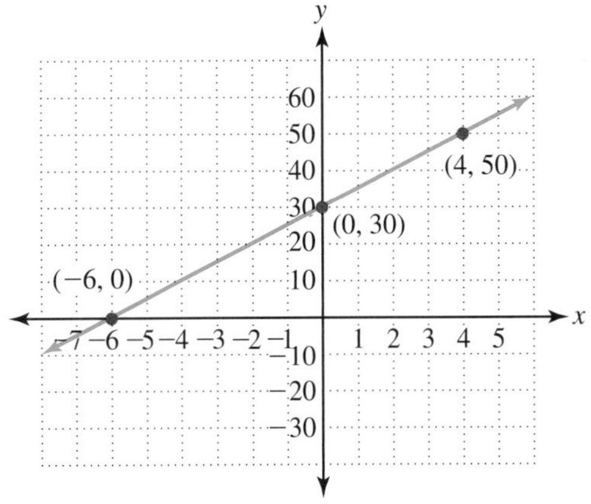
The graph of a linear equation with just one variable is a horizontal or vertical line. When the equation is of the form $y = b$, the graph is a horizontal line. When the equation is of the form $x = c$, the graph is a vertical line.	Draw the graph of $y = 5$ and draw the graph of $x = 3$ on the same set of axes. 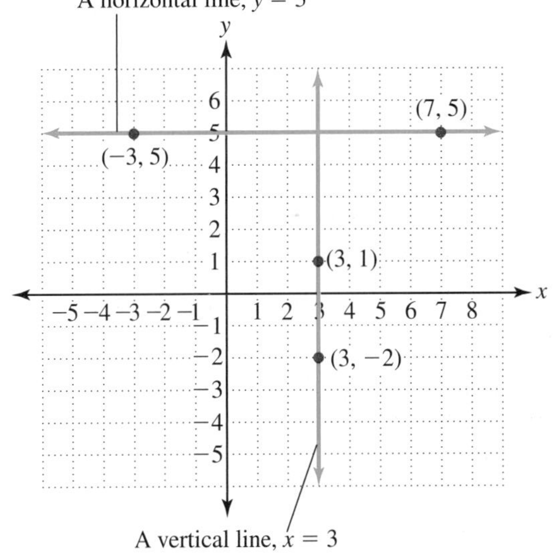

Section 9.3 Graphing Data: Line Graphs and Bar Graphs

Concept	Example
Data is collected information, often numerical, and is usually organized in a **data table.**	The owner of Giani's Deli kept track one week of the number of sandwiches made during the lunch hours:

Day	Number of Sandwiches
Monday	83
Tuesday	95
Wednesday	106
Thursday	90
Friday	98

A **line graph** is typically used to compare the progression of data over a period of time.

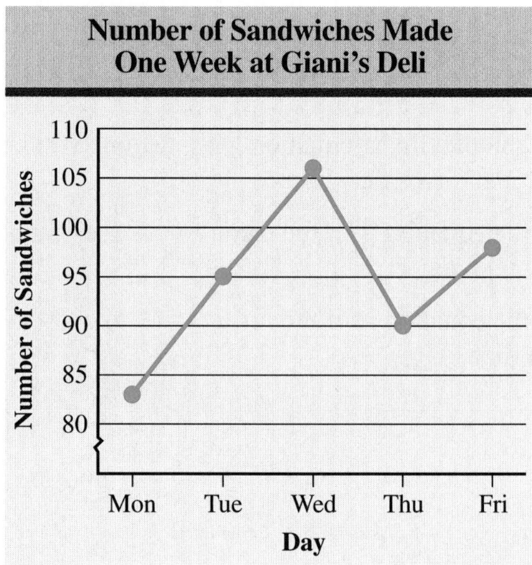

We can calculate the percent increase or percent decrease between any two time periods in the data table.

$$\text{Percent increase} = \frac{\text{Amount of increase}}{\text{Original amount}}$$

$$\text{Percent decrease} = \frac{\text{Amount of decrease}}{\text{Original amount}}$$

The percent increase in sandwiches made from Tuesday to Wednesday is calculated as follows:

Amount of increase: $106 - 95 = 11$

Original amount (from Tuesday) is 95.

$$\text{Percent increase} = \frac{11}{95} \approx 0.116 = 11.6\%$$

There was an 11.6% increase in the number of sandwiches made from Tuesday to Wednesday.

Section 9.3 Graphing Data: Line Graphs and Bar Graphs

Concept

When a study is conducted, items of information—called **data**—are collected, organized in a table, and drawn as a graph. A **bar graph** uses bars to represent the number of responses to a certain category.

Example

50 members of the Iowa Association of Birders went bird-watching on the first day of spring. Each member recorded the first bird he or she saw that day. Kami collected the results, organized them in a table, and drew a bar graph showing the data.

First Bird Spotted	Number of Bird-Watchers
Blue Jay	6
Nuthatch	10
Swallow	13
Thrush	7
Wren	14
Total	**50**

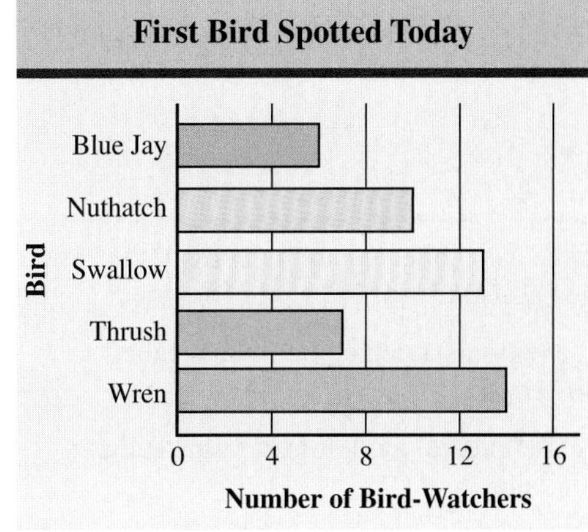

Section 9.4 Graphing Data: Circle Graphs and Histograms

Concept

In a **circle graph,** the categories are displayed in a circle and we show each category as a percent of the whole instead of the actual number of data values.

Example

First Bird Spotted	Number of Bird Watchers	Percent
Blue Jay	6	12%
Nuthatch	10	20%
Swallow	13	26%
Thrush	7	14%
Wren	14	28%
Total	**50**	**100%**

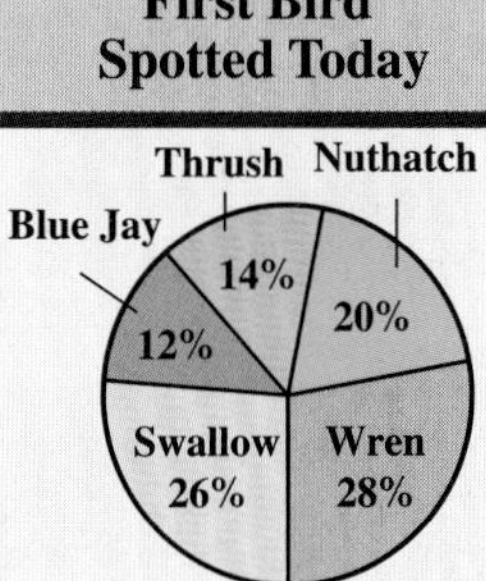

When the data collected are numerical, the bar graph is called a **histogram.**

Each category is a range of numbers called a **class interval.**

Each class interval has a lowest value and a highest value.

The class interval also includes every whole number between the lowest and highest values.

The difference between the highest and lowest values in each class interval is the same for each interval.

The number of data values that fit into each class interval is called the **class frequency.**

A table of class frequencies is called a **frequency distribution table.**

The categories (across the bottom) are the class intervals, and the height of each bar indicates the corresponding class frequency.

Number of Birds Identified	Frequency
0–4	2
5–9	8
10–14	13
15–19	10
20–24	6
25–29	11
Total	**50**

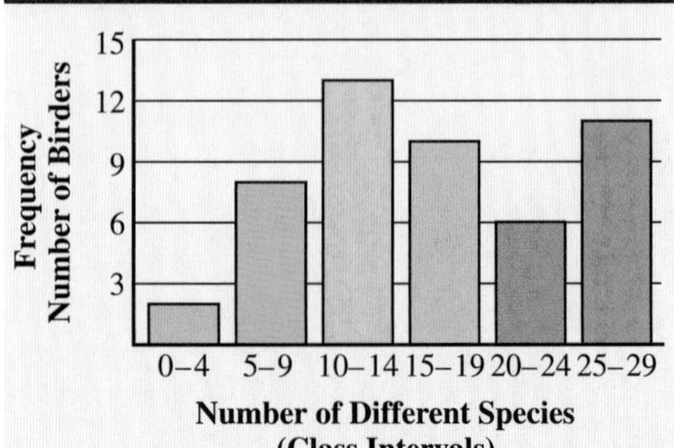

Section 9.5 Mean, Median, and Mode

Concept	Example
The **mean** is the balancing point for all of the data. It is a numerical average that is found by addition and division $$\text{Mean} = \frac{\text{Sum of all of the data values}}{\text{Total number of data values}}$$	**Note: The example presented here runs throughout the review for Section 9.5.** 12 members of the Ohio Bird-watchers Club went to Crane Creel in Ohio on the 3rd of May. Each member recorded the *number* of different warbler species that he or she identified. Stan collected the number of different species and put them in numerical order. The number of different species the 12 members identified: 3, 5, 6, 8, 8, 8, 11, 12, 12, 14, 15, 18 $$\text{Mean} = \frac{3 + 5 + 6 + 8 + 8 + 8 + 11 + 12 + 12 + 14 + 15 + 18}{12}$$ $$= \frac{120}{2} = 10$$
When the data values are written in numerical order, the **median** is the middle number. After the median is found, there will be an equal number of data values to its right and to its left. With an odd number of data values, the median is the middle number. With an even number of data values, the median splits the two middlemost numbers. The median is the mean of the two middlemost data values.	With 12 data values collected, the median will be the mean of the two middlemost data values. 3, 5, 6, 8, 8, 8, ↓ 11, 12, 12, 14, 15, 18 $$\text{Median} = \frac{8 + 11}{2} = \frac{19}{2} = \mathbf{9.5}$$
The **mode** is the data value (not necessarily numerical) that occurs most often.	The data value that occurs most is 8; so, mode = 8. 3, 5, 6, <u>8</u>, <u>8</u>, <u>8</u>, 11, 12, 12, 14, 15, 18

Section 9.6 Probability

Concept	Example
Probability is a numerical way to express the likelihood that some particular thing, or event, will happen.	**Note: The example presented here runs throughout the review for Section 9.6.** Elsie is raffling a fruit basket and sells 10 raffle tickets. Carlo bought 1 raffle ticket, Dan bought 2, Ravi bought 3, and Eli bought 4. What is the probability that Ravi will win the raffle?
An **experiment** is the act of doing something to create a result.	The experiment is Elsie choosing a raffle ticket at random.
The **possible outcomes** of an experiment are all of the different results that can occur. Usually, only one outcome can occur for each experiment.	The possible outcomes are the 10 different raffle tickets Elsie has to choose from. Only one will actually be selected.
An **event** is the outcome that we're most interested in occurring.	The event is that one of Ravi's tickets will be selected.
Each possible outcome within the event is called a **success.** Within an event, there may be just one success or many successes.	Ravi has three tickets; so, he has three chances to win, or three successes.
The probability of an event is a ratio, abbreviated as **P**(event), and is calculated as follows: $\mathbf{P}(\text{event}) = \dfrac{\text{Number of successes in the event}}{\text{Total number of possible outcomes}}$ Probability is commonly written as a fraction or a percent.	$\mathbf{P}(\text{Ravi wins}) = \dfrac{\text{His 3 successes}}{\text{10 total tickets in the drawing}} = \dfrac{3}{10}$ There is a 3 out of 10 chance that Ravi will win the raffle.
A **random** selection is the act of choosing something such that each possible outcome has an equal chance of being selected and is equally likely to be selected.	The winning ticket is selected randomly when there is no way to distinguish between one ticket and another in the selection process and when each ticket has an equally likely chance of being selected.

Chapter 9 Review Exercises

Fill in the blank with the word or words that correctly complete the sentence.

1. In the rectangular coordinate system, the main horizontal axis is the *x*-axis. **(9.1)**

2. The *x*-coordinate and the *y*-coordinate form a(n) ordered pair. **(9.1)**

3. In a data table, how often a category occurs is called the frequency. **(9.3)**

4. The type of graph that displays a category's percent of occurrence is a(n) circle graph. **(9.4)**

5. In a list of data, the item that occurs most often is the mode. **(9.5)**

6. When a list of data is in numerical order, the middle number is the median. **(9.5)**

True or false.

7. In the rectangular coordinate system, we can plot points only if each coordinate is an integer. False **(9.1)**

8. In the rectangular coordinate system, a point on the *y*-axis must have an *x*-coordinate that is 0. True **(9.1)**

9. The graph of $x = -3$ is a horizontal line. False **(9.2)**

10. To find the mean of a set of data, the data does not need to be in numerical order. True **(9.5)**

Section 9.1

Determine whether each ordered pair is a solution to the equation $y = -\frac{3}{5}x + 3$.

11. $(10, 3)$
$3 = -3$ False! No, $(10, 3)$ is not a solution.

12. $(0, -3)$
$-3 = 3$ False! No, $(0, -3)$ is not a solution.

13. $(5, 0)$
$0 = 0$ True! Yes, $(5, 0)$ is a solution.

14. $(-5, 6)$
$6 = 6$ True! Yes, $(-5, 6)$ is a solution.

Given the graph at the right, identify the ordered pair of each point shown.

15. *A*
$(3, 5)$

16. *B*
$(0, 1)$

17. *C*
$(-5, 3)$

18. *D*
$(-3, 0)$

19. *E*
$(-8, -4)$

20. *F*
$(0, -5)$

21. *G*
$(3, -3)$

22. *H*
$(6, 0)$

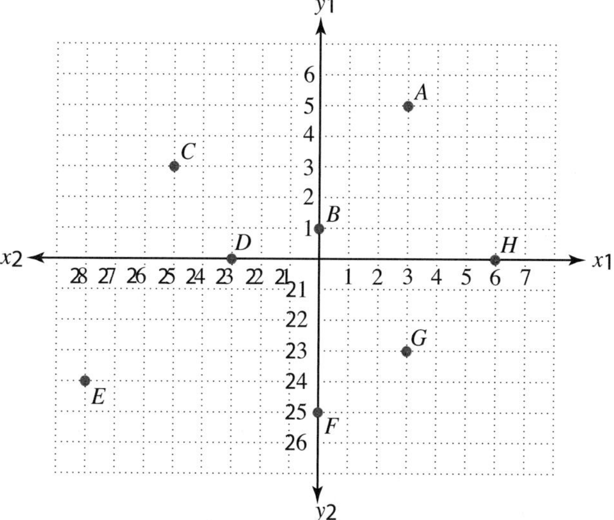

Determine in which quadrant each point lies.

23. $(-9, 15)$
Quadrant II

24. $(7, 10)$
Quadrant I

25. $(-6, -12)$
Quadrant III

26. $(11, -20)$
Quadrant IV

Describe where each point is located in the *x*-*y* plane.

27. $(-10, 0)$
On the negative x-axis

28. $(15, 0)$
On the positive x-axis

29. $(0, 18)$
On the positive y-axis

30. $(0, -12)$
On the negative y-axis

Section 9.2

Graph each line with the given equation by finding three points.

31. $x + y = 3$

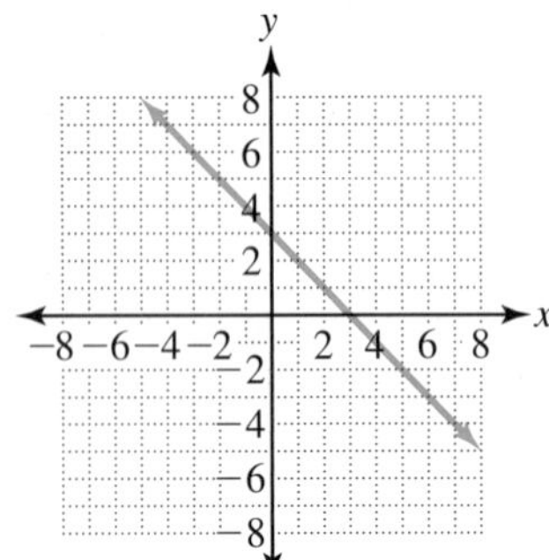

32. $x - y = -4$

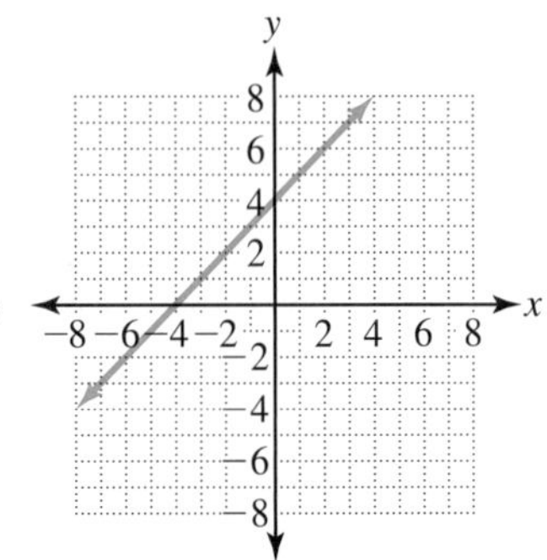

37. $y = 3x - 2$

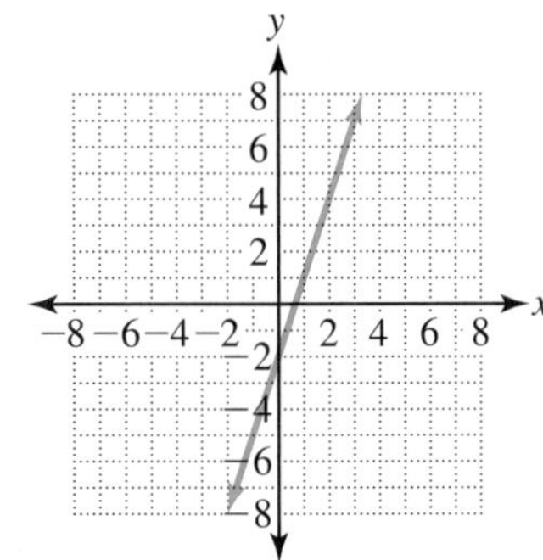

38. $y = 4x$

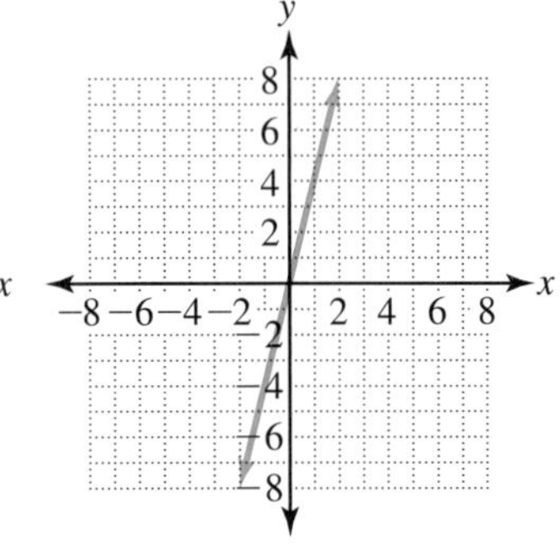

33. $2x - 4y = -8$

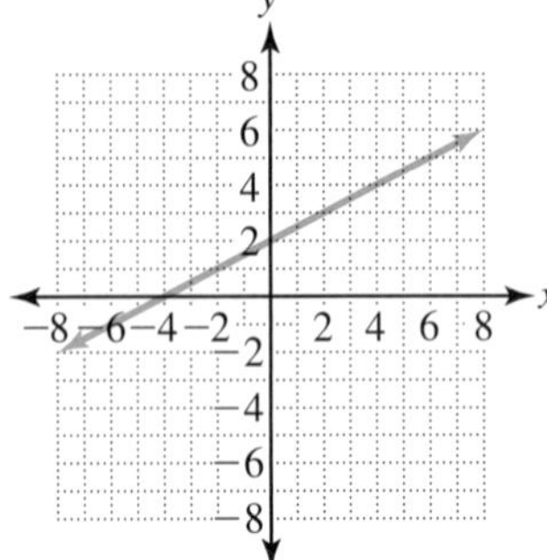

34. $3x + 2y = 6$

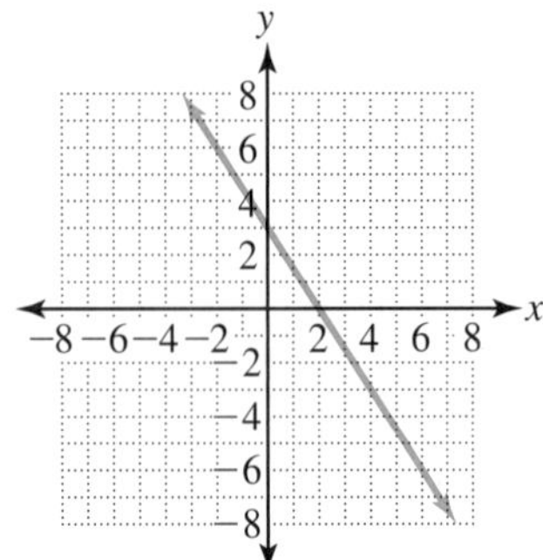

39. $y = 6$

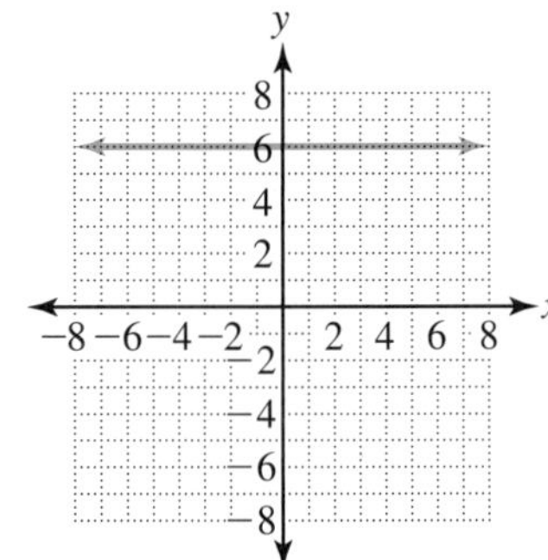

40. $y = -3$

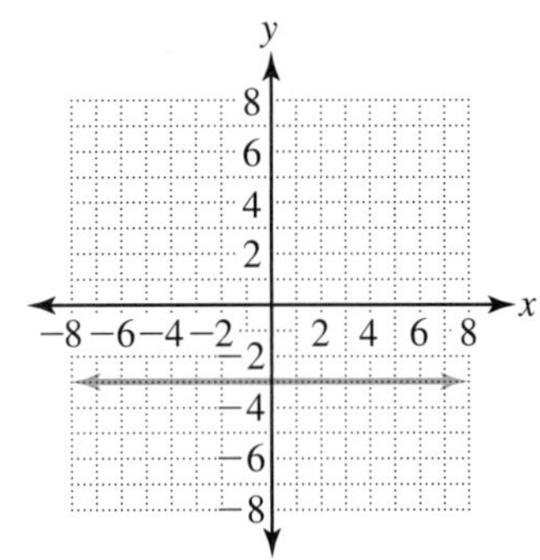

35. $y = -x - 4$

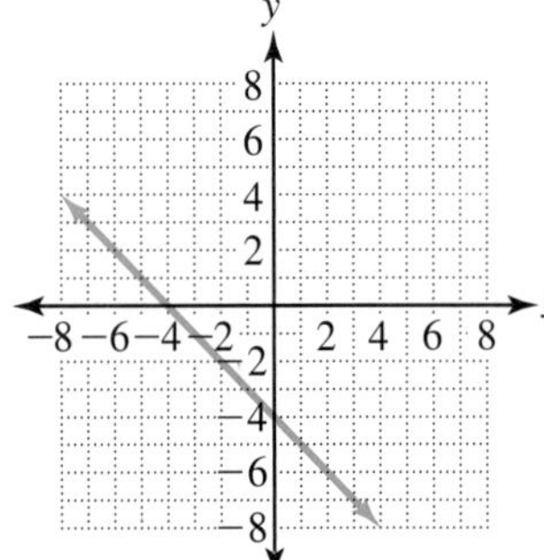

36. $y = -2x + 1$

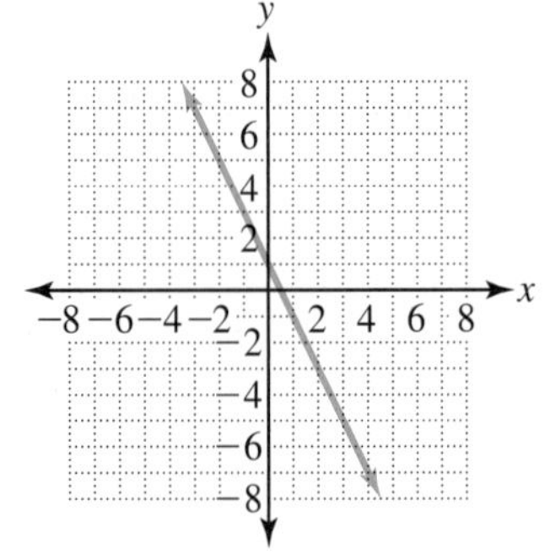

41. $x = -4$

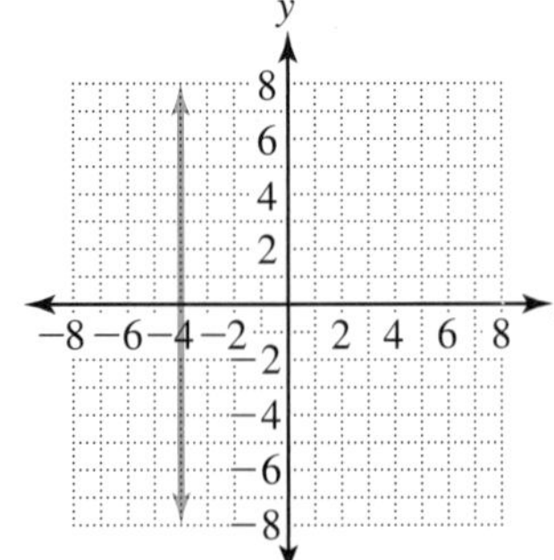

42. $x = 2$

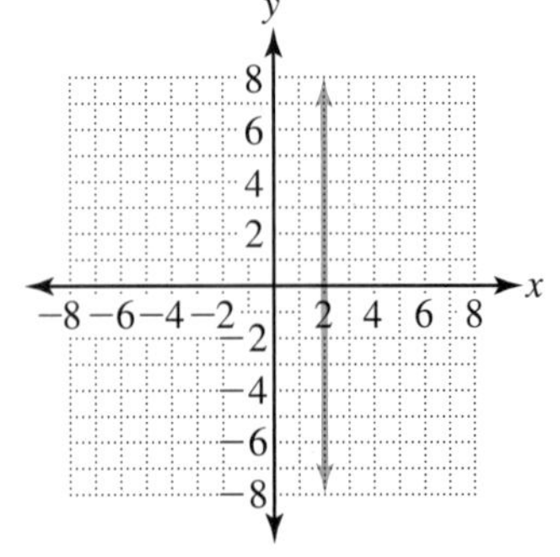

Section 9.3

The following data table shows the monthly average gasoline prices (regular unleaded) for California in 2006. Draw the line graph associated with this data table.

Month	Average Price of Gasoline
January	$2.36
February	$2.63
March	$2.57
April	$2.90
May	$3.42
June	$3.30
July	$3.31
August	$3.27
September	$3.02
October	$2.63
November	$2.50
December	$2.46

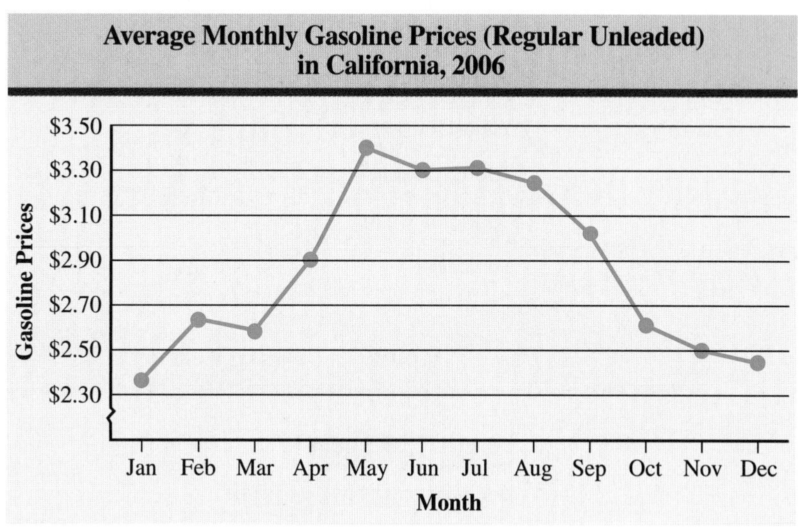

43. What was the largest price increase from one month to the next, and when did it occur?
The largest price increase was $0.52, and it occurred from April to May.

44. What was the largest price decrease from one month to the next, and when did it occur?
The largest price decrease was $0.39, and it occurred from September to October.

45. What was the percent increase in price from March to May?
The increase from March to May was approximately 33.1%.

46. What was the percent decrease in price from July to December?
The decrease from July to December was approximately 25.7%.

Given each table of values, draw the related bar graphs.

47. The New York Yankees have played in the World Series 39 times, winning 26 of them throughout the years. Following is a table of other teams that have a high number of World Series appearances (through the 2007 season).
***Source:* ESPN.com**

Team	Number of World Series Appearances
Athletics	14
Cardinals	17
Cubs	10
Dodgers	18
Giants	17
Red Sox	11

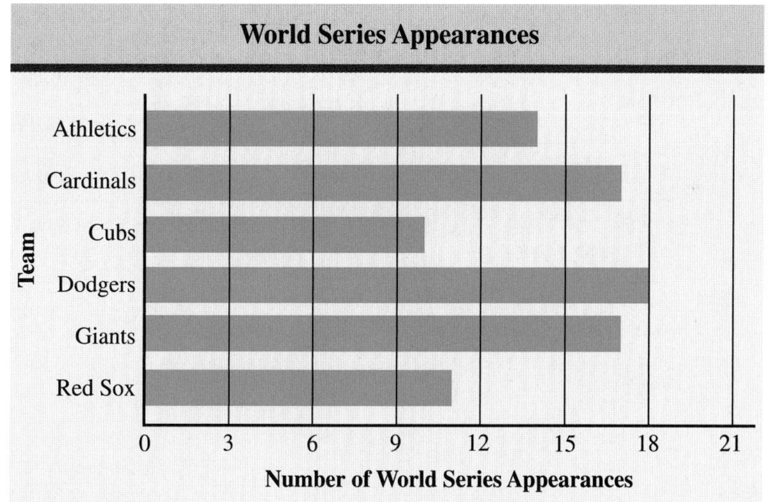

48. Following is a table of the number of teams in the various bowling leagues offered at Pinto Lanes.

Bowling League	Number of Teams
12 and under	10
13 through 17	14
Young Singles	20
Young Couples	16
Young at Heart	18
Seniors	24

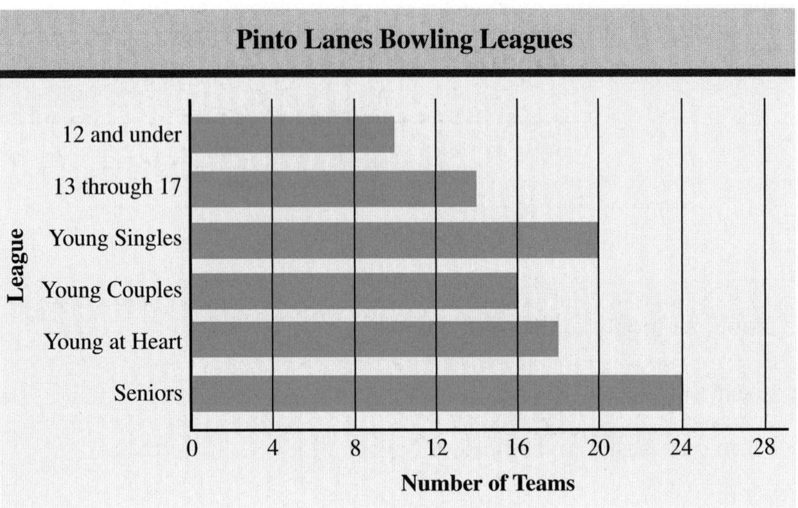

Given each bar graph, write the related table of values.

49. Following is a table of the top six U.S. food companies, which total $96 billion in revenue, based on 2003 revenues (in billions of dollars).
Source: The World Almanac, 2005

Company	Revenue (in $ billions)
ConAgra Foods	22
General Mills	11
H. J. Heinz	9
Kellogg	9
PepsiCo	27
Sara Lee	18
Total	**96**

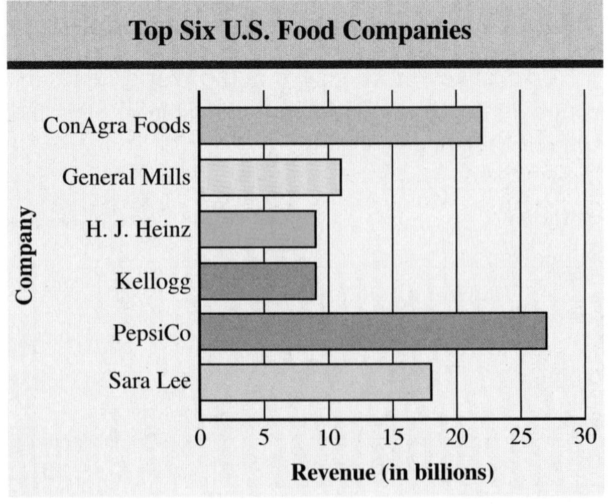

50. There have been 111 runnings of the Boston Marathon, from 1897 to 2007. Of the 111 men's events, a runner outside the United States has won the race 68 times. Following is a bar graph of the top four foreign countries with winning runners (men's category) as well as all other countries grouped together.
Source: **bostonmarathon.org**

Country	Number of Winners
Canada	16
Great Britain	4
Japan	7
Kenya	16
Other	25
Total	**68**

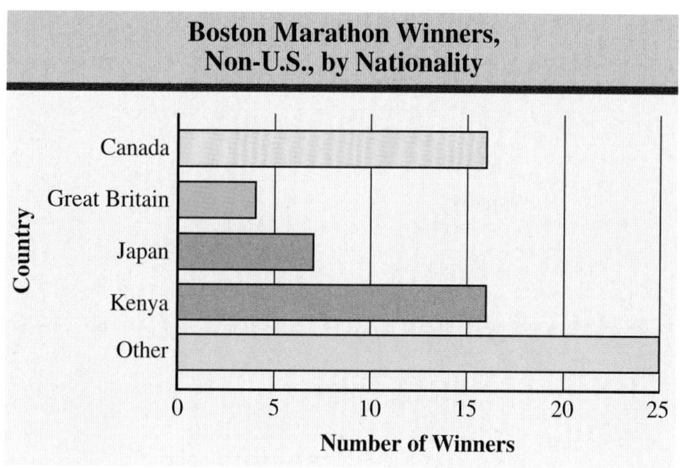

Section 9.4

Complete each table.

51. A company's advertising budget is $200,000. The circle graph shows the percent of that budget devoted to different advertising opportunities.

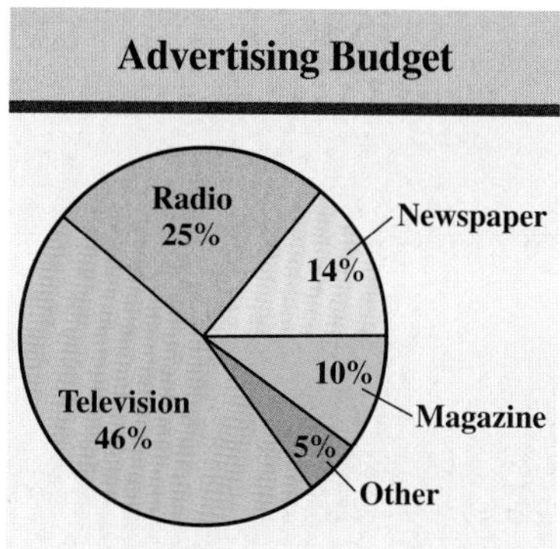

Advertising Budget Item	Percentage of Budget	Total Amount of Budget
Magazine	10%	$20,000
Newspaper	14%	$28,000
Radio	25%	$50,000
Television	46%	$92,000
Other	5%	$10,000
Total	**100%**	**$200,000**

52. A construction company is putting up 500 homes in a new development, offering a variety of home styles such as a two-story, three-bedroom house. The circle graph shows the percent of the development for each home style.

Styles of New Homes

3 BR, 2 Story 35%; 4 BR, 1 Story 19%; 4 BR, 2 Story 12%; 2 BR, 1 Story 8%; 3 BR, 1 Story 26%

Style of New Home	Percent of Development	Total Number of Homes
2BR, 1 Story	8%	40
3BR, 1 Story	26%	130
3BR, 2 Story	35%	175
4BR, 1 Story	19%	95
4BR, 2 Story	12%	60
Total	**100%**	**500**

Complete each table by listing the percent of the whole for each category. Then place the percent data appropriately in the circle graph.

53. Following is a table of the ethnic diversity of the 120 players in the six-team Connerville High School Baseball League.

Ethnicity	Total Number of Players	Percent of Players
African American	42	35%
Asian	18	15%
Hispanic	30	25%
White	24	20%
Other	6	5%
Total	**120**	**100%**

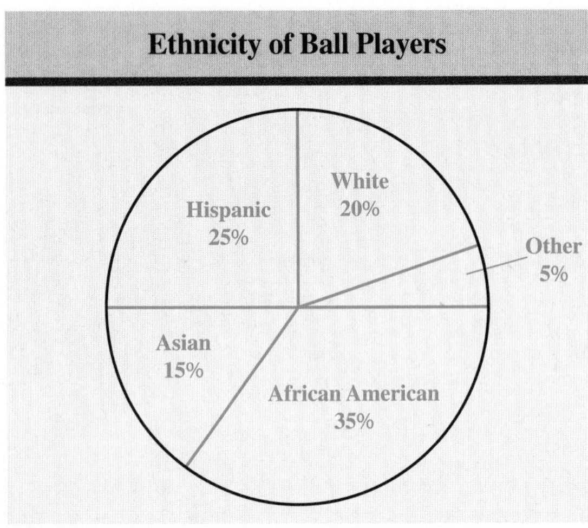

54. Following is a table of the religious affiliations of the attendees at an ecumenical (all-inclusive) Thanksgiving service.

Religion	Total Number of Attendees	Percent of Attendees
Buddhist	8	4%
Christian	68	34%
Hindu	24	12%
Jewish	50	25%
Muslim	34	17%
Other	16	8%
Total	**200**	**100%**

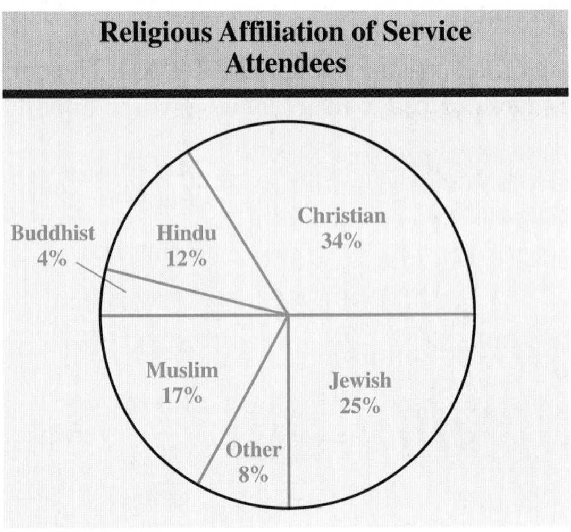

Fill in each frequency distribution table and/or draw the related histogram as appropriate.

55. The distance that 75 employees drive to work is recorded in the following frequency distribution table.

Number of Miles	Number of Employees
1-5	12
6-10	15
11-15	19
16-20	14
21-25	9
26-30	6
Total	**75**

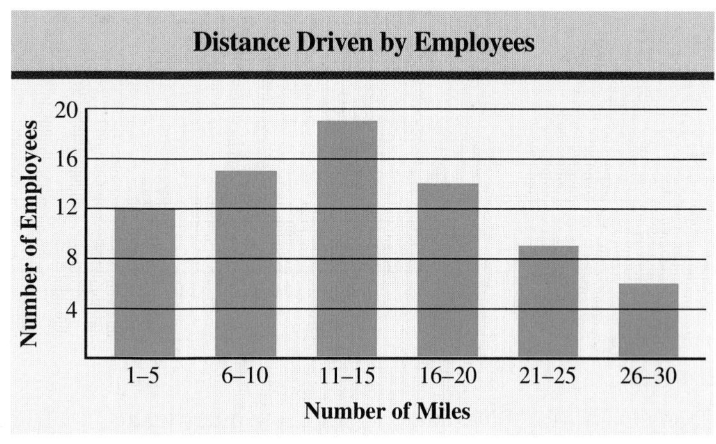

a) How many employees drive fewer than 11 miles?
27

b) How many employees drive from 6 to 20 miles?
48

c) How many employees drive at least 16 miles?
29

56. Following is the number of years of reign (in numerical order) of the 62 rulers of England and Great Britain before Queen Elizabeth II. (Note: Those who are credited with reigning 0 years actually reigned less than 1 year.)
Source: The World Almanac, 2005

10	6	5	35	13	6	3	7
19	9	2	10	9	5	13	63
2	3	24	17	39	44	6	9
6	17	0	56	22	22	12	25
5	4	21	35	0	24	13	1
28	37	13	20	2	5	33	15
25	0	35	50	24	1	59	
16	19	19	22	38	25	10	

Number of Years	Tally	Number of Rulers
0-9	~~IIII~~ ~~IIII~~ ~~IIII~~ ~~IIII~~ III	23
10-19	~~IIII~~ ~~IIII~~ ~~IIII~~	15
20-29	~~IIII~~ ~~IIII~~ II	12
30-39	~~IIII~~ II	7
40-49	I	1
50-59	III	3
60-69	I	1
Total		**62**

a) How many rulers reigned more than 39 years?
5

b) How many rulers reigned from 20 to 39 years?
19

c) How many rulers reigned fewer than 20 years?
38

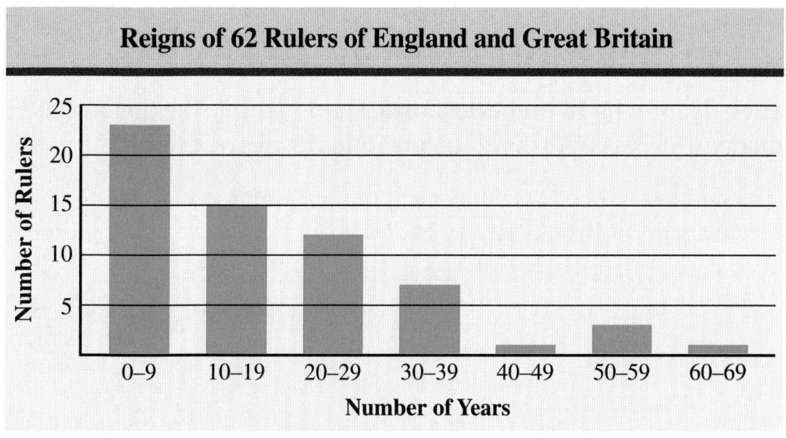

57. Following is a histogram of the number of wins for the 30 teams in Major League Baseball in the 2004 season. *Source:* **http://mlb.mlb.com**

Number of Wins	Number of Teams
50–57	1
58–65	2
66–73	8
74–81	3
82–89	6
90–97	7
98–105	3
Total	**30**

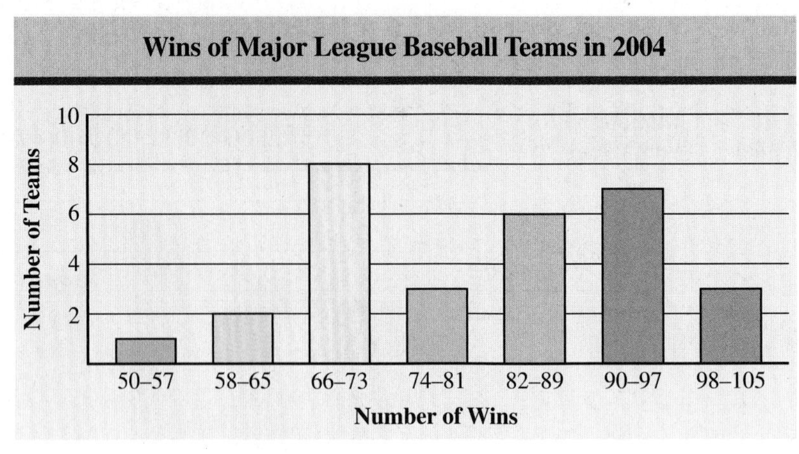

a) How many teams had more than 89 wins?
10

b) How many teams had from 66 to 89 wins?
17

c) How many teams had fewer than 74 wins?
11

58. A chapter of Alcoholics Anonymous (AA) was started 3 years ago. Following is a histogram of the number of consecutive months that the 45 attendees have remained sober. (Yea!)

Number of Months	Number of Attendees
1–6	4
7–12	3
13–18	12
19–24	8
25–30	11
31–36	7
Total	**45**

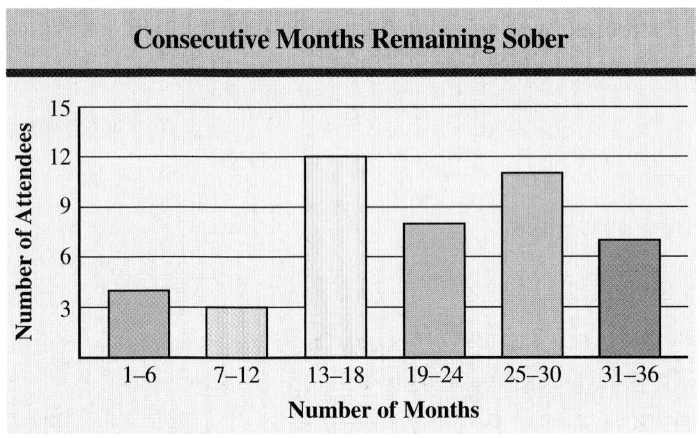

a) How many attendees were sober more than 30 months?
7

b) How many attendees were sober from 19 to 30 months?
19

c) How many attendees were sober fewer than 13 months?
7

Section 9.5

Put each list of data in numerical order and identify the mean, the median, and the mode. If necessary, round the mean to the nearest tenth.

59. The number of homes sold by each of the 11 real estate agents at Brooks Realty in 2008:
9, 3, 7, 6, 4, 6, 8, 10, 6, 7, 12
Mean $\approx$ 7.1; median: 7; mode: 6

60. The number of minutes it takes Jennifer to get from home to work on nine days in February:
17, 23, 29, 15, 20, 28, 16, 21, 22
Mean $\approx$ 21.2; median: 21; mode: None

61. The number of DVDs rented each month by the Park family in 2008.
6, 2, 3, 0, 4, 3, 5, 6, 7, 4, 5, 6
Mean ≈ 4.3; median: 4.5; mode: 6

62. The number of flights 14 business people flew during the month of May:
9, 17, 6, 22, 14, 7, 9, 18, 17, 9, 12, 17, 20, 19
Mean: 14; median: 15.5; modes: 9 and 17

Calculate the grade point average for each students s pring grade distribution. Round the answer to two decimal places.

63.

Course	Grade	Point Value	Units	Grade Points (Point Value × Units)
French	B	3	5	15
Computer Lab	A	4	1	4
CIS	A	4	3	12
Psychology	C	2	3	6
		Total:	12	37

Grade Point Average ≈ 3.08

64.

Course	Grade	Point Value	Units	Grade Points (Point Value × Units)
Chemistry	C	2	4	8
Trigonometry	B	3	4	12
History	D	1	3	3
Physical Education	A	4	1	4
Music Appreciation	B	3	2	6
		Total:	14	33

Grade Point Average ≈ 2.36

Section 9.6

Find the probability of each event. Simplify all fractions completely.

Experiment: A six-sided die is rolled. What is the probability that it comes up

65. an even number?
$\frac{1}{2}$

66. less than 3?
$\frac{1}{3}$

67. greater than 2?
$\frac{2}{3}$

68. less than 1?
0

69. an odd number greater than 4?
$\frac{1}{6}$

70. an even number less than 5?
$\frac{1}{3}$

Experiment: The spinner has 8 numbers on it. When the needle is spun and comes to rest, it is pointing toward one of the numbers. What is the probability that it will point toward

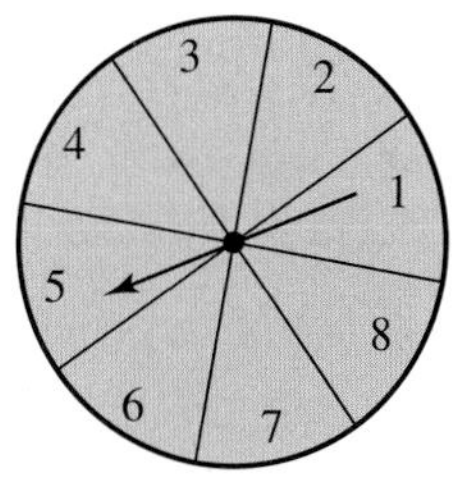

In this diagram, the spinning needle is pointing toward 5.

Assume that each number has an equal chance of being pointed to.

71. an odd number?
$\frac{1}{2}$

72. a number less than 4?
$\frac{3}{8}$

73. an even number less than 7?
$\frac{3}{8}$

74. a number greater than 6?
$\frac{1}{4}$

75. an odd number less than 4?
$\frac{1}{4}$

76. a number greater than 0?
1

For Exercises 77–80. A bag has 30 raffle tickets. Nine are red, 8 are yellow, 2 are green, 5 are pink, and 6 are blue. The experiment is to randomly select one ticket out of the bag. What is the probability that the chosen ticket is

77. pink?
$\frac{1}{6}$

78. blue?
$\frac{1}{5}$

79. red?
$\frac{3}{10}$

80. yellow?
$\frac{4}{15}$

For Exercises 81–84. At a St. Louis Cardinals baseball game one Friday night, a prize is to be awarded to the ugliest car in the parking lot. Here is a bar graph of the make of the 50 finalists.

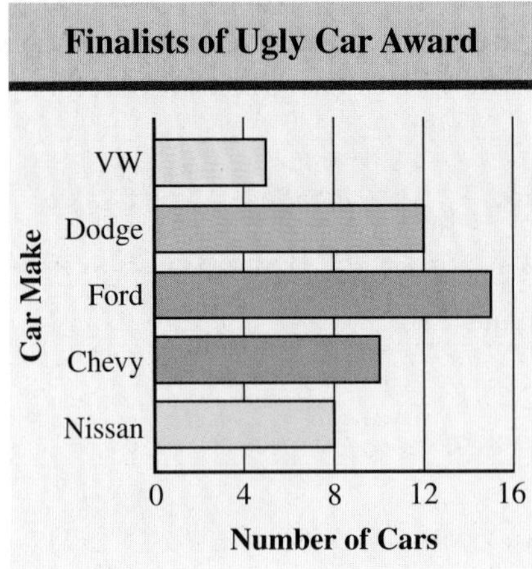

Assuming the final car was randomly selected, what is the probability that the chosen car is a

81. Chevy?
$\frac{1}{5}$

82. Ford?
$\frac{3}{10}$

83. VW?
$\frac{1}{10}$

84. Dodge?
$\frac{6}{25}$

Chapter 9 Test

1. Determine whether $(-3, 5)$ is a solution to the equation $4x + 2y = 2$. Show all of your work.
$4(-3) + 2(5) = 2$
$-12 + 10 = 2$
$-2 = 2$ False
No, $(-3, 5)$ is not a solution.

2. Determine whether $(-3, 5)$ is a solution to the equation $y = \frac{2}{3}x + 7$. Show all of your work.
$5 = \frac{2}{3}(-3) + 7$
$5 = -2 + 7$
$5 = 5$ True
Yes, $(-3, 5)$ is a solution.

Determine in which quadrant each point lies.

3. $(3, -9)$
Quadrant IV

4. $(-6, 10)$
Quadrant II

Given the graph below, identify the ordered pair of each point shown.

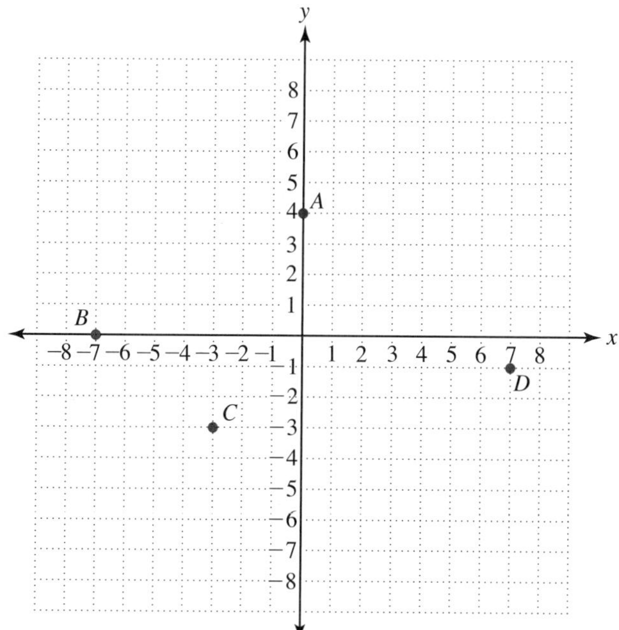

5. A
$(0, 4)$

6. B
$(-7, 0)$

7. C
$(-3, -3)$

8. D
$(7, -1)$

Graph the line with the given equation.

9. 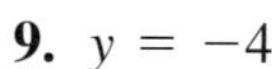 $y = -4$

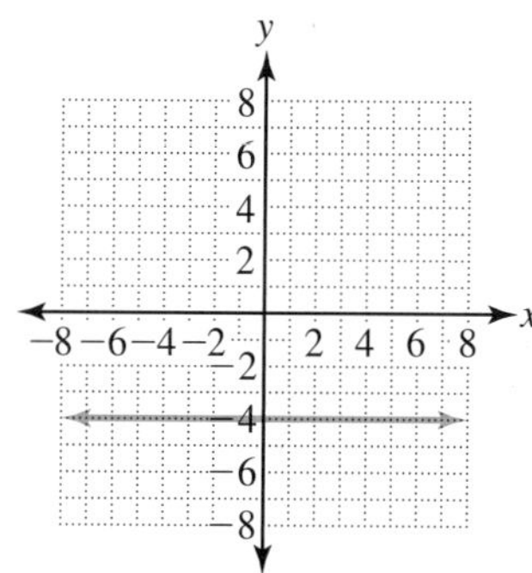

10. $x = 3$

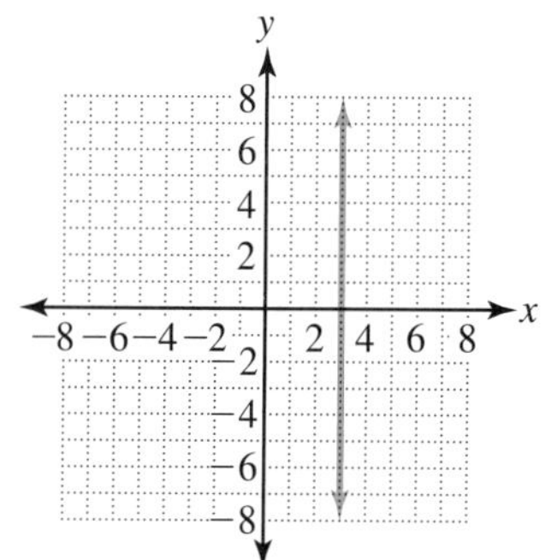

Graph the line with the given equation. Use the chart to find three points first.

11.

Choose a value	3x + y + = −3	(x, y)

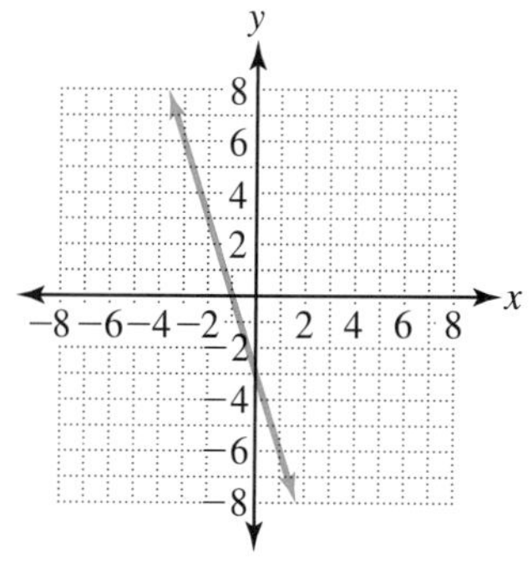

12.

x	$y = -2x + 4$	(x, y)

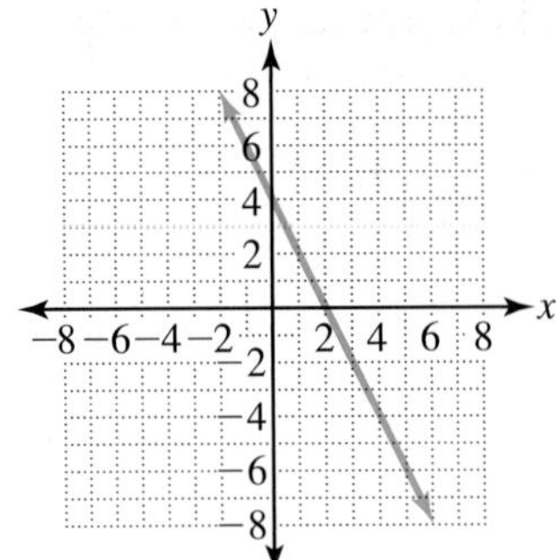

13. It's not widely known that California is a large farming state. The table shows the amount of some crops (in millions of bushels) produced by California farms in 2003. Draw the bar graph related to this table. ***Source: World Almanac, 2005***

Crop	Bushels (in millions)
Barley	4
Corn	27
Hay	9
Oats	3
Wheat	34

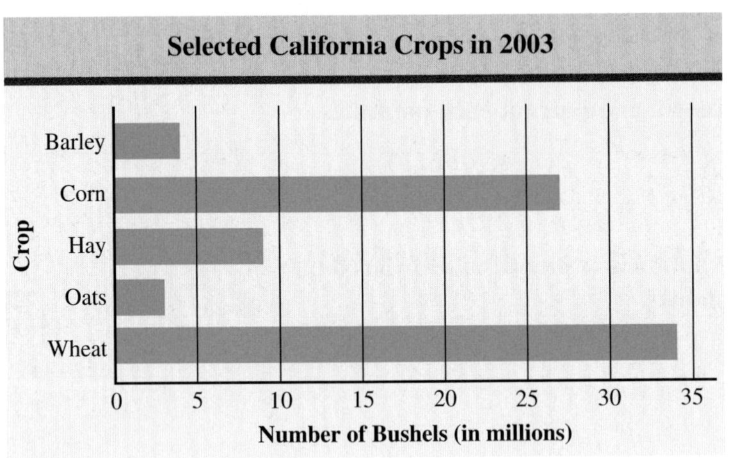

14. The following data table shows the average monthly rainfall in centimeters in Honolulu, Hawaii.

a) Draw the line graph associated with this data table.

Month	Average Monthly Rainfall, in cm
January	6.9
February	6.0
March	4.8
April	2.8
May	2.0
June	1.1
July	1.3
August	1.2
September	1.9
October	5.5
November	5.8
December	7.2

Average Monthly Rainfall in Honolulu Hawaii

Rainfall (cm): 8.0, 7.0, 6.0, 5.0, 4.0, 3.0, 2.0, 1.0

Month: Jan, Feb, Mar, Apr, May, Jun, Jul, Aug, Sep, Oct, Nov, Dec

b) In what month is the average rainfall one-fourth the average rainfall for March?
August has an average rainfall that is one-fourth the average rainfall for March.

c) In what month is the average rainfall $1\frac{1}{2}$ times the average rainfall for March?
December has an average rainfall that is $1\frac{1}{2}$ times the average rainfall for March.

d) What was the percent decrease in rainfall from February to March?
The percent decrease in rainfall from February to March was 20%.

15. Following is a bar graph of the average annual snowfall of six U.S. cities. (These six cities total 85 annual inches of snow.) Write its related table of values.
Source: **www.weathertoday.net**

City	Average Annual Snowfall
Albuquerque, NM	11
Baltimore, MD	21
Evansville, IN	14
Nashville, TN	10
Roanoke, VA	23
Wichita Falls, TX	6
Total	**85**

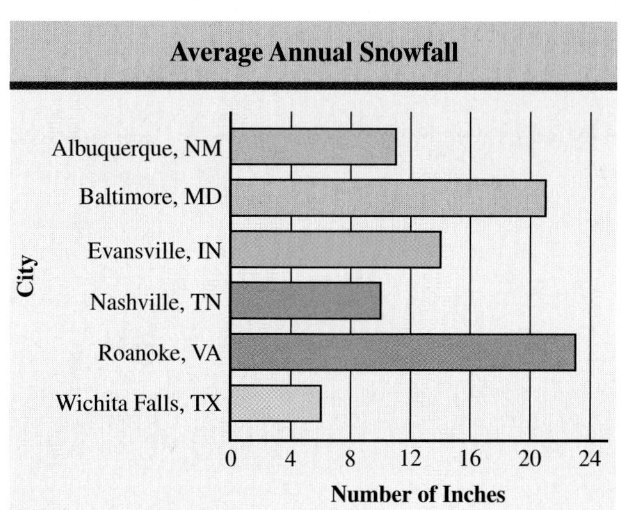

16. A city's economic development plan calls for 150 acres to be set aside for new businesses. Following is a circle graph of the acreage (as a percent) allotted to various types of businesses. Complete the table.

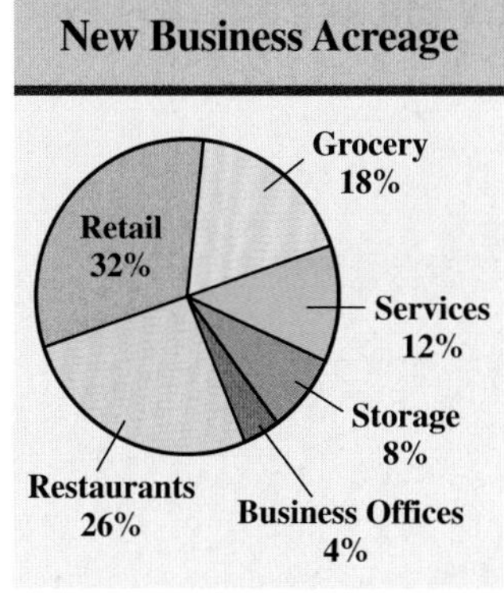

Type of Business	Percent of Acreage	Total Number of Acres
Business Offices	4%	6
Grocery	18%	27
Restaurants	26%	39
Retail	32%	48
Services	12%	18
Storage	8%	12
Total	**100%**	**150**

17. Following is the number of wins of the 29 National Basketball Association teams during the 2003–2004 regular season.

a) Fill in the frequency distribution table and draw its related histogram.

b) How many teams won more than 43 games?
10

c) How many teams won from 44 to 55 games?
6

d) How many teams won fewer than 38 games?
12

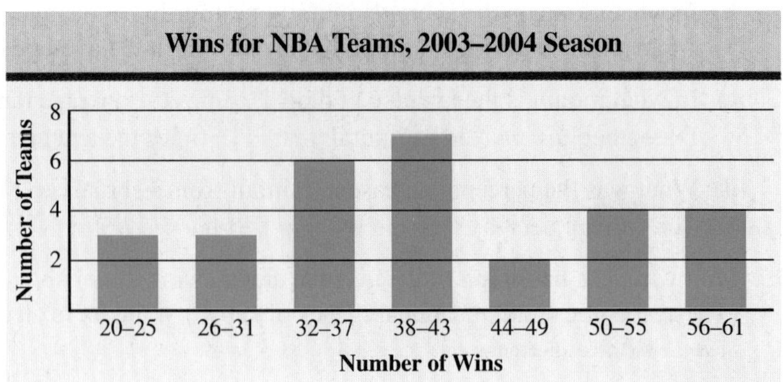

61	47	58	56
54	42	57	55
41	39	52	41
41	36	50	37
35	33	45	37
33	25	43	29
28	21	42	28
23			

Number of Wins	Tally	Number of Teams
20-25	///	3
26-31	///	3
32-37	~~////~~ /	6
38-43	~~////~~ //	7
44-49	//	2
50-55	////	4
56-61	////	4
Total		**29**

18. Following is a table of the annual giving of the 300 families at the community church. Complete the table by listing the percent of the whole for each category. Then place the percent data appropriately around the circle graph.

Annual Giving	Total Number of Families	Percent of Families
Less than $200	12	4%
$200-$999	36	12%
$1,000-$4,999	75	25%
$5,000-$9,999	96	32%
$10,000-$19,999	54	18%
$20,000 and above	27	9%
Total	**300**	**100%**

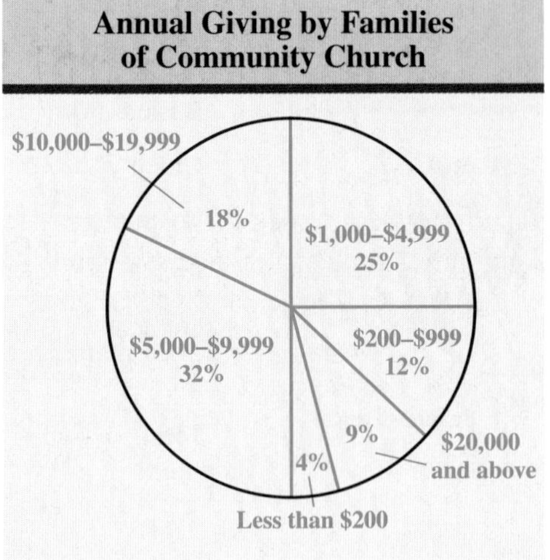

19. Following is a histogram of the grade distribution for all of the scores of Ms. Garcia's final algebra exam.

a) Fill in the frequency distribution table based on the data in the histogram.

Final Exam Score	Number of Students
44-51	4
52-59	7
60-67	11
68-75	14
76-83	19
84-91	16
92-99	9
Total	**80**

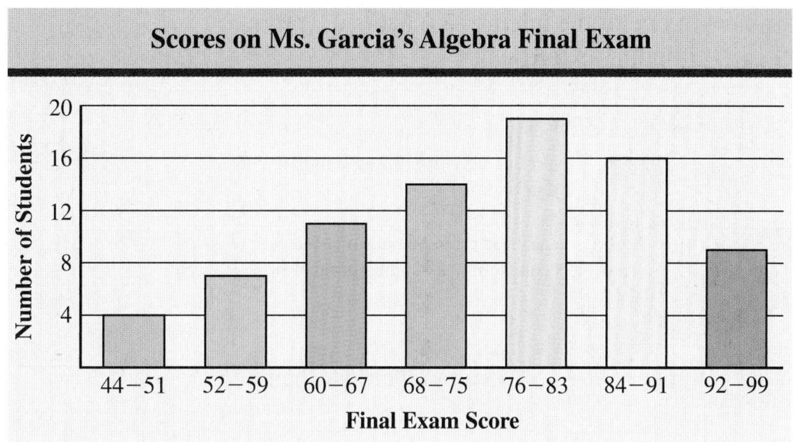

b) How many students scored more than 75 points?
44

c) How many students scored from 60 to 75 points?
25

d) How many students scored fewer than 68 points?
22

Put each list of data in numerical order and identify the mean, the median, and the mode. If necessary, round the mean to the nearest tenth.

20. Mr. Singh teaches a sociology class with 75 students. Following are the number of absences he had for the 13 class meetings in March.
5, 8, 2, 6, 8, 10, 4, 1, 11, 8, 3, 8, 6
Mean ≈ 6.2; median: 6; mode: 8

21. Marci put a grandfather clock up for sale on eBay. Following are the number of hits her auction site generated during each of the ten days the clock was for sale.
6, 16, 18, 7, 4, 9, 10, 16, 15, 16
Mean: 11.7; median: 12.5; mode: 16

Calculate the grade point average for this student's fall and spring grade distribution. Round the answer to two decimal places.

22.

Course	Grade	Point Value	Units	Grade Points (Point Value × Units)
Microbiology	C	2	5	10
Algebra	A	4	4	16
CIS	B	3	2	6
Karate	C	2	1	2
Geometry	C	2	3	6
Spanish	B	3	5	15
Guidance	A	4	2	8
Speech	B	3	3	9
		Total:	25	72

Grade Point Average ≈ 2.88

Find the probability of each event. Simplify all fractions completely.

Experiment: The spinner has nine numbers on it. The needle is spun; and when the needle comes to rest, it is pointing toward one of the numbers. What is the probability that it will point toward

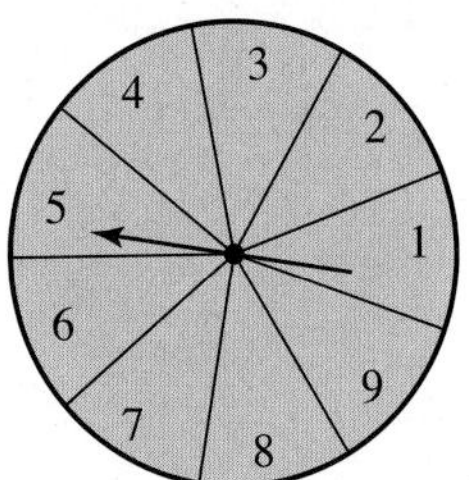

In this diagram, the spinning needle is pointing toward 5.

Assume that each number has an equal chance of being pointed to.

23. an even number?
$\frac{4}{9}$

24. a number less than 6?
$\frac{5}{9}$

25. an even number less than 7?
$\frac{1}{3}$

26. a number greater than 6?
$\frac{1}{3}$

27. an odd number less than 4?
$\frac{2}{9}$

28. a number divisible by 3?
$\frac{1}{3}$

In his wallet, Carl has three \$20 bills, two \$10 bills, four \$5 bills, and six \$1 bills. If Carl's son, Jason, sneaks in during the night and randomly takes one bill from Carl's wallet, what is the probability that Jason grabbed

29. a \$5 bill?
$\frac{4}{15}$

30. a \$20 bill?
$\frac{1}{5}$

31. a \$1 bill?
$\frac{2}{5}$

32. a \$10 bill?
$\frac{2}{15}$

In the Smithson household, the family has 40 DVDs. The number of each type of DVD is shown in the bar graph. If a DVD is chosen randomly, what is the probability that it is

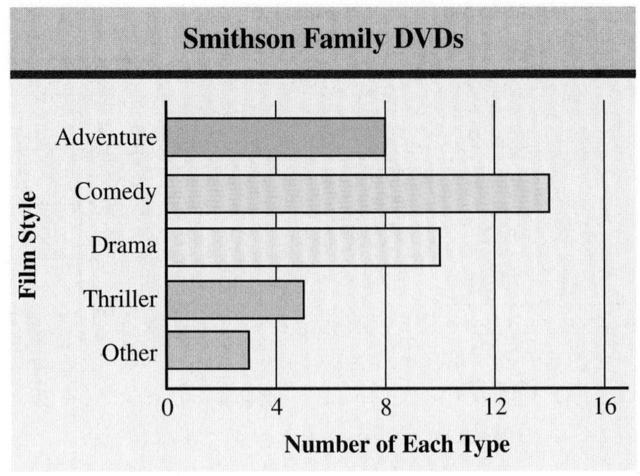

33. a comedy?
$\frac{7}{20}$

34. a drama?
$\frac{1}{4}$

35. an adventure?
$\frac{1}{5}$

36. a thriller?
$\frac{1}{8}$

Chapters 1–9 Cumulative Review

Answer as indicated.

1. Which property is being demonstrated?
   a) $3 \cdot (5 + 6) = 3 \cdot 5 + 3 \cdot 6$
   Distributive Property of Multiplication Over Addition
   b) $7 \cdot 4 = 4 \cdot 7$
   Commutative Property of Multiplication

2. Evaluate $2^3 + 4 \cdot 3$
   20

3. The estimated 2005 population of Texas was 22,859,968. Round this number to the nearest
   a) thousand
   22,860,000
   b) hundred thousand
   22,900,000

4. Find the prime factorization of 126. Write the answer two ways: with and without exponents.
   $2 \cdot 3 \cdot 3 \cdot 7$ and $2 \cdot 3^2 \cdot 7$

5. Jimmy is hauling a total of 2,660 pounds of cargo in his truck. The cargo is 95 boxes of caramel candies. How much does each box weigh?
   Each box weighs 28 pounds.

The bar graph below shows the number of strikeouts Hall of Famer Nolan Ryan pitched in his last five seasons, from 1989–1993.

Source: **www.baseballhalloffame.org**

Answer each question with a sentence.

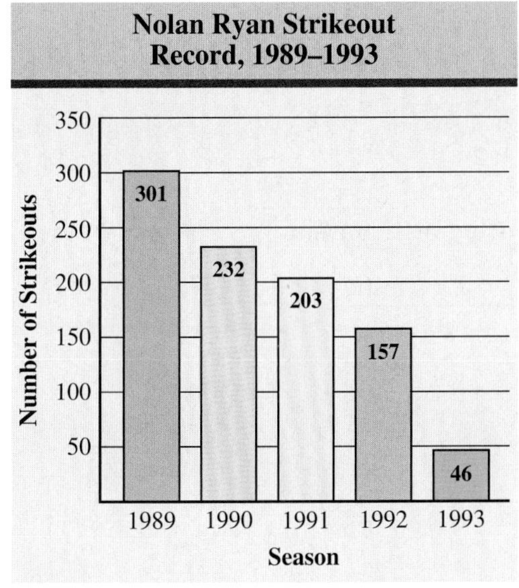

6. How many more strikeouts did Nolan Ryan have in 1989 than in 1990?
   Nolan Ryan had 69 more strikeouts in 1989 than in 1990.

7. How many total strikeouts did Nolan Ryan have from 1989 to 1993?
   Nolan Ryan had 939 total strikeouts from 1989 to 1993.

Answer as indicated.

8. The formula to convert from Fahrenheit degrees to Celsius degrees is $C = 5 \cdot \frac{F - 32}{9}$. What is the Celsius temperature when the temperature is 113° F?
   The Celsius temperature is 45° C.

Evaluate each expression using the order of operations. Show all steps.

9. $-36 \div (-8 \div 2)$
   9

10. $20 - [-5 \cdot (-6)]$
    -10

11. $|3| - |-8 + 4|$
    $|3| - |-4|$
    $3 - 4$
    -1

12. $\sqrt{20 - x^2} - x^2$; replace x with 2.
    $\sqrt{20 - 2^2} - 2^2$
    $\sqrt{16} - 4$
    $4 - 4$
    0

Simplify by combining like terms.

13. $-6x - (-9x) + 3x$
    $6x$

Solve. Make sure you check each answer.

14. $7y + 36 = y - 12$
    $y = -8$

Set up the legend, draw a diagram, identify the formula, and write and solve the equation. Make sure you write the answer as a sentence.

15. The floor of a rectangular room has a perimeter of 64 feet. Find the dimensions of the floor if the length is 6 feet more than the width.
    The width is 13 feet, and the length is 19 feet.

Evaluate. Simplify completely. Write improper fractions as mixed numbers.

16. $\frac{18}{25} \times \frac{15}{6}$
    $\frac{9}{5}$ or $1\frac{4}{5}$

17. $-\frac{9}{16} \div \frac{21}{8}$
    $-\frac{3}{14}$

18. $\frac{7}{20} + \frac{7}{30}$
    $\frac{7}{12}$

19. $-\frac{5}{6} + \frac{1}{18}$
    $-\frac{7}{9}$

20. Solve for w: $\frac{3w}{8} = -\frac{9}{4}$

$w = -6$

Solve each application.

21. Every Saturday morning, Tu walks a course that is $2\frac{3}{8}$ miles. That same evening he walks a course that is $1\frac{5}{6}$ miles. How many total miles does Tu walk each Saturday?

Tu walks a total of $4\frac{5}{24}$ miles each Saturday.

22. Marcela rode her bike 21.7 miles to the beach, and it took her 1.4 hours to get there. What was Marcela's average rate of speed on this ride?

Marcela's average rate of speed was 15.5 miles per hour.

Answer as indicated.

23. Find the decimal equivalent of $\frac{7}{8}$.

0.875

24. Simplify $\frac{18.2 \text{ calories}}{3.5 \text{ grams}}$ to a unit rate and write the unit rate in words.

5.2 calories per gram

25. Last night Dorie treated herself to dinner out. Her bill included steak, \$13.85; dessert, \$3.95, and a beverage, \$1.95.

a) Find Dorie's meal total.

The meal total was \$19.75.

b) Calculate 6% sales tax on her meal total. *(Round this value to the nearest penny.)*

The sales tax was \$1.19.

c) Calculate a 15% tip on her meal total. *(Round this value to the nearest dollar.)*

The tip was \$2.96.

d) Dorie paid for the meal and tip with \$30. How much change did Dorie receive?

Dorie received \$6.10 in change.

Solve each problem. Write a sentence answering the question.

26. This morning Lauren invested \$2,400 in an account that earns simple interest at a rate of 5% per year. How much interest will she earn

a) in one year?

She'll earn \$120.

b) in 9 months?

She'll earn \$90.

27. At a high-rise hotel, Chitra's crew can clean and prepare 27 rooms in 6 hours. How many rooms can they clean and prepare in 8 hours?

They can clean and prepare 36 rooms.

28. What percent of 65 is 39?

39 is 60% of 65.

29. In June, the average price of gasoline was \$3.20 per gallon. Six months later in December, the average price was \$2.40. What was the percent decrease in the average price of gasoline in the six-month period?

The percent decrease in the average price of gasoline was 25%.

Write all polynomial answers in descending order and combine like terms.

Add.

30. $(-4y + 3y^2 - 8) + (y^2 + 6 - y)$

$4y^2 - 5y - 2$

Simplify.

31. $\frac{m^{10}}{m^2}$

m^8

32. $\frac{-16v^4}{4v}$

$-4v^3$

Multiply.

33. $(-p)(-6p^4 + 3p)$

$6p^5 - 3p^2$

34. $(2x - 3)(x^2 + 5x - 2)$

$2x^3 + 7x^2 - 19x + 6$

35. $(5w - 3)(5w + 3)$

$25w^2 - 9$

Divide using distribution.

36. $\frac{3q^5 + 15q^3 - 6q}{-3q}$

$-q^4 - 5q^2 + 2$

Factor out the GCF.

37. $15y^6 - 25y^3$

$5y^3(3y^3 - 5)$

Answer as indicated.

38. A floor-to-ceiling window is 7 feet 8 inches high. Express this measure in inches.
The height is 92 inches.

39. Dionne's horse is 4 feet 8 inches tall. When Dionne sits on her horse, her body extends 2 feet 9 inches above the saddle. When Dionne sits on her horse, how far is the top of her head from the ground?
The top of her head is 7 feet 5 inches above the ground.

40. When Arend was 15 years old, his best long jump was 13 feet 7 inches. When he was 18 years old, his best long jump was 20 feet 3 inches. How much in feet and inches did his long jump improve in those three years?
His long jump improved 6 feet 8 inches.

Convert one measure to the other.

41. Convert $6\frac{1}{2}$ quarts to pints.
13 pints

42. Convert 20 ounces to pounds.
1.25 pounds

43. Convert 3.6 centimeters to millimeters.
36 millimeters

44. Convert 85.6 deciliters to liters.
8.56 liters

45. Convert 47.2 grams to kilograms.
0.0472 kilograms

46. Convert $1\frac{3}{4}$ ounces to grams.
49 grams

47. A jug holds $2\frac{1}{2}$ gallons of water. Express this number in liters.
9.5 liters

48. The distance between Cleveland, Ohio, and Columbus, Ohio, is 140 miles. Express this number in kilometers.
224 kilometers

49. Martine's beagle is 55 centimeters long. Express this number in inches.
22 inches

Answer as indicated.

50. $\angle WFZ$ and $\angle BLN$ are supplementary. Given $\angle BLN = 71.3°$, find $\angle WFZ$.

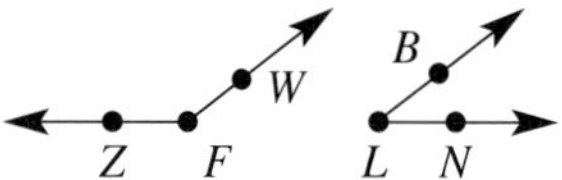

$\angle WFZ = 108.7°$

51. $\triangle ABC$ is a right triangle. One acute angle measures 31.9°. Find the measure of the other acute angle.
The other acute angle is 58.1°.

52. One side of a regular octagon is $5\frac{1}{4}$ in. What is the perimeter of this octagon?

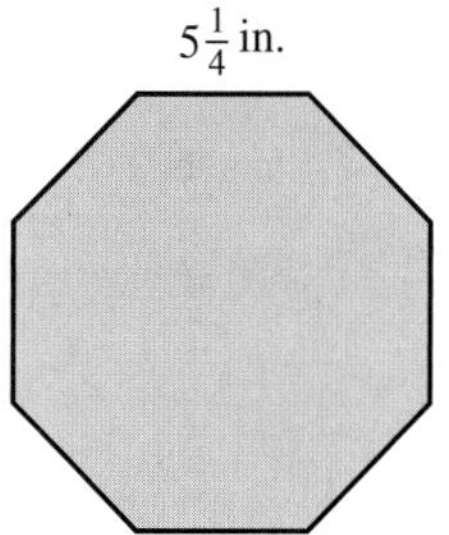

The perimeter is 42 inches.

53. A regular pentagon has a perimeter of 86.3 cm. What is the length of each side of this pentagon?

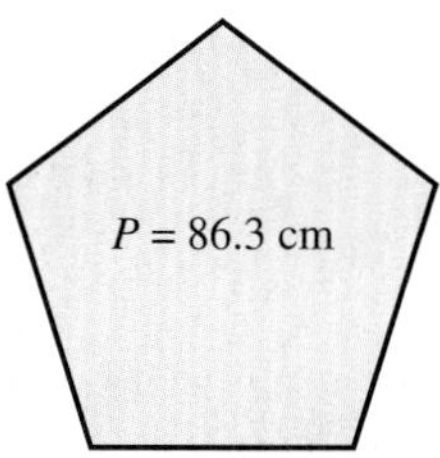

Each side is 17.26 cm.

Find the area of each polygon.

54.

7.5 cm

8.4 cm

The area is 31.5 square cm.

55.

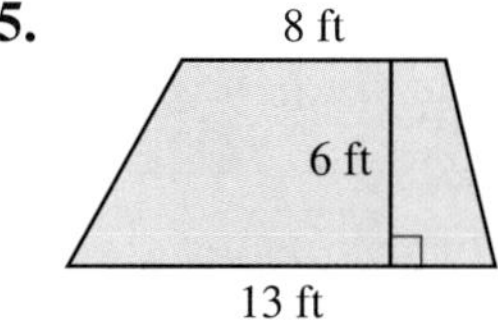

The area is 63 square feet.

56. At a fair, the floor of a carousel is made from two circles. The inside circle, where the machinery is kept, has a radius of 10 feet. The outer circle has a radius of 20 feet. What is the area of the floor of this merry-go-round? Use $\pi \approx 3.14$ and round the result to the nearest tenth.

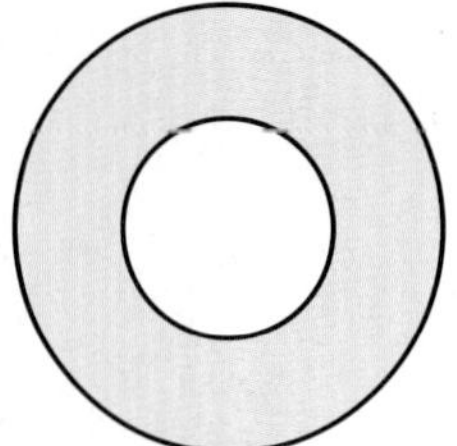

The area is 942 square feet.

Find the circumference.

57. A Ferris wheel has a radius of 25 feet. What is the circumference of this Ferris wheel? Use $\pi \approx 3.14$ and round the result to the nearest tenth.

The circumference is 157 feet.

Solve each application. Write a sentence answering the question.

58. A pile of gravel is in the shape of a cone. Its height is 10 feet, and its base radius is 6 feet. What is the volume of this cone?

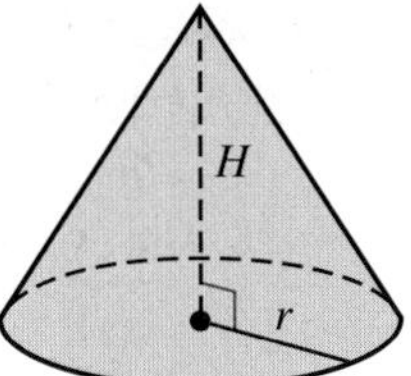

The volume is 120π (≈ 376.8) cubic feet.

59. A baking pan is 28 cm wide, 20 cm deep, and 5 cm high. What is the volume of this baking pan?

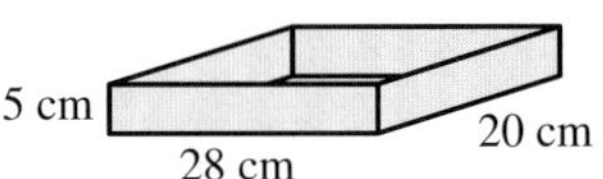

The volume is 2,800 cubic cm.

Graph the line with the given equation by first finding three points.

60. $y = 3x - 6$

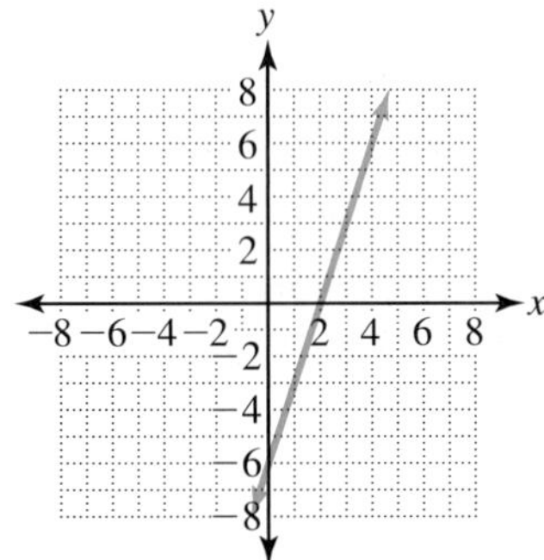

61. $y = -\frac{1}{3}x + 2$

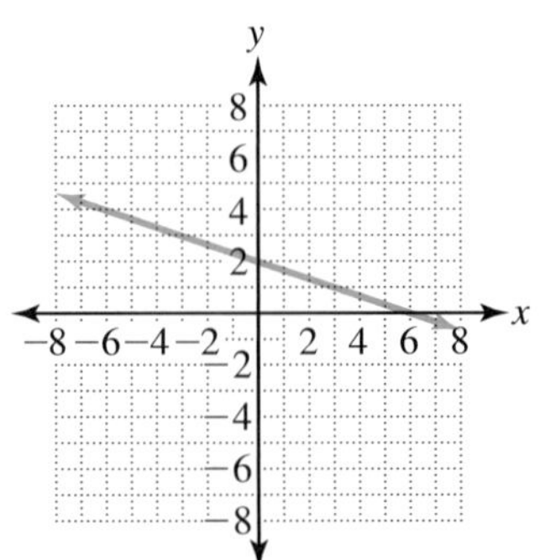

Answer as indicated.

62. Following is a histogram of the number of years U.S. senators have served in the senate as of 2006.

a) Fill in the frequency distribution table based on the data in the histogram.

Year Served	Number of Senators
1–6	22
7–12	29
13–18	16
19–24	14
25–30	11
31–36	3
37–42	2
43–48	3
Total	**100**

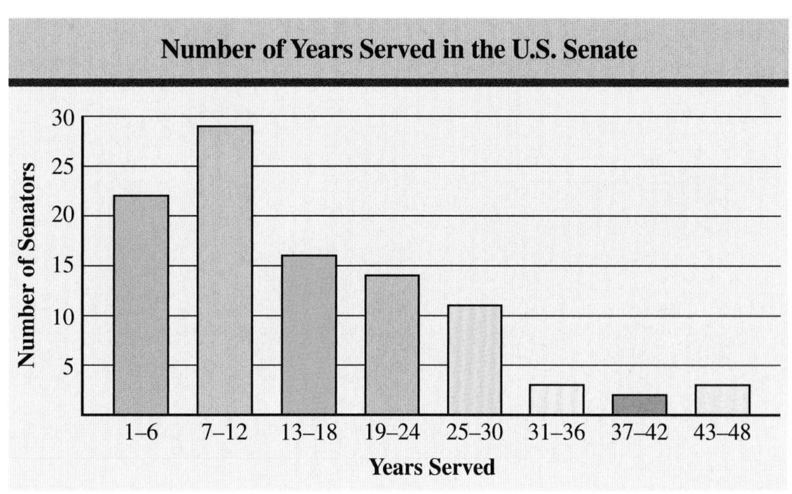

b) How many senators have served more than 24 years?
19

c) How many senators have served from 7 to 18 years?
45

Put the list of data in numerical order and identify the mean, the median, and the mode. If necessary, round the mean to the nearest tenth.

63. At a racetrack, 12 owners were asked how many racehorses they owned. Here are the data:
15, 6, 3, 7, 9, 6, 12, 5, 3, 8, 6, 10
Mean is 7.5.
Median is 6.5.
Mode is 6.

Answer as indicated.

64. Experiment: At a church picnic, it was time for the children to swing at the piñata. A month was to be chosen randomly, and any child whose birthday was in that month got to line up first. What is the probability that the first letter of the month chosen was

a) a *J*? $\frac{1}{4}$

b) an *S*? $\frac{1}{12}$

c) an *M*? $\frac{1}{6}$

Appendix

A.1 Scientific Notation: Large Numbers

OBJECTIVES

In this section, you will learn to

- Write a large number in scientific notation.
- Expand a large number from scientific notation.

You Need to Know

To successfully complete this section, you need to understand

- ☐ Rounding whole numbers (1.1)
- ☐ Exponents and powers of 10 (1.5)
- ☐ Decimals (5.1)

Introduction

Scientific notation means what it says: It is the notation used in many areas of science. It is used so that scientists and mathematicians can work relatively easily with very large and very small numbers and their related computations. Here is a sample of some very large numbers and what they refer to:

The Earth is about 93,000,000 (93 million) miles from the Sun.

It takes about 588,000,000,000,000,000,000,000 (588 billion trillion) atoms of hydrogen to make 1 gram of hydrogen.

Light travels at a rate of about 300,000,000 (300 million) meters per second. (A meter is about 39.6 inches.)

These numbers are much too large when we are trying to do calculations, but we can use scientific notation to make them appear less intimidating. Our work here is to understand scientific notation as it applies to large numbers. Later, we work with scientific notation and very small decimal numbers in A.3; and in A.4, we multiply numbers in scientific notation.

Working with Powers of 10

Recall from Section 1.5 that when a power of 10 is expanded, the exponent of 10 indicates the number of zeros that follow the 1.

For example, 10^6 is a 1 followed by six zeros: 1,000,000. We say that 1,000,000 is the 6th power of 10.

You Try It 1 **Expand each power of ten.**

a) 10^1 **b)** 10^4 **c)** 10^7 **d)** 10^9

Recall, also, that when a number ends in one or more zeros, it can be written as a product of a whole number and a power of 10.

For example, 400,000 has five zeros following the 4. So, the *fifth* power of 10, or 10^5, is a factor: $400{,}000 = 4 \cdot 10^5$.

Example 1

Write each expression as a product of a whole number and a power of 10. Use an exponent to represent the power of 10.

a) 900 **b)** 100,000,000 **c)** 5,000

Procedure Be sure to count the number of zeros.

Answer

a) 900
$= 9 \times 100$
$= 9 \times 10^2$

b) 100,000,000
$= 1 \times 100{,}000{,}000$
$= 1 \times 10^8$

c) 5,000
$= 5 \times 1{,}000$
$= 5 \times 10^3$

You Try It 2 **Write each number as a product of a whole number and a power of 10. Use Example 1 as a guide.**

a) 60,000 **b)** 20,000,000 **c)** 700,000,000,000 **d)** 40

Defining Scientific Notation

When a number is written as the product of a number—called a **coefficient**—and a power of 10, we say that the number is in **scientific notation**.

So, the *scientific notation form* of 300 is 3×10^2.

The coefficient The power of 10

> **Scientific Notation** is a product of two factors:
>
> a **numerical coefficient** and a **power of 10**.
>
> In **scientific notation**,
> the coefficient is a number between 1.00000 and 9.99999.
> (In other words, it has a *whole* number 1 through 9.)

Here is a table of some numbers that are proper coefficients and some numbers that are not.

Proper coefficients	Not proper coefficients (with reasons)	
3.11	23.11	The whole number has two digits; it can have only one.
6	0.6	The number is less than 1; the whole number can't be 0.
1.308		
9.96		

Example 2

Identify which of the following numbers could be a proper coefficient in scientific notation. If the number cannot be a proper coefficient, state why.

Answer

	Number	Could it be a coefficient?
a)	9.75	Yes, the whole number, 9, is only one digit.
b)	15.4	No, the whole number, 15, has more than one digit.
c)	4	Yes, the whole number doesn't need to be followed by a decimal.
d)	0.56	No, the whole number can't be zero.

You Try It 3 **Identify which of the following numbers could be a proper coefficient in scientific notation. If a number cannot be a proper coefficient, state why. Use Example 2 as a guide.**

a) 3 **b)** 2.09 **c)** 31.4 **d)** 0.91 **e)** 10

Writing Large Numbers in Scientific Notation

To write a large number in scientific notation we need to ask the following questions:

1. Is the coefficient a single-digit whole number 1 through 9?
2. How many spaces must we move the decimal point to create a proper coefficient?

A number such as 3,000,000 is easy to write in scientific notation because the coefficient is the whole number 3 with no decimals. In this case, we need only count the number of zeros. There are six.

$$3{,}000{,}000 = 3 \times 10^6$$

However, a number such as 3,580,000 is a little more challenging because we can't just count the number of zeros. Instead, we need to recognize that

a) The proper coefficient is 3.58.

b) To get the coefficient, the decimal point (at the end of the whole number) needs to move to between the 3 and 5. It must move 6 places to the left.

$$3580000. = 3.58 \times 10^6$$

654321

Example 3

Write each number in scientific notation.

a) 960,000 **b)** 745,000,000

Procedure Decide on the proper coefficient and count the number of places the decimal point must move.

Answer **a)** The proper coefficient is 9.6; the decimal point must move *five* places.

$$960{,}000 = 9.6 \times 10^5$$

b) The proper coefficient is 7.45; the decimal point must move *eight* places.

$$745{,}000{,}000 = 7.45 \times 10^8$$

You Try It 4 **Write each number in scientific notation. Use Example 3 as a guide.**

a) 28,000 **b)** 413,000 **c)** 9,070,000

d) 62,150,000,000 **e)** 580 **f)** 73

Applications

In Section 1.1, we rounded large numbers. We can now represent those same large numbers in scientific notation. What follows are some statistical data involving large numbers. Your job is to round them and write a sentence that shows the approximation in scientific notation.

Example 4

In July 2006, the U.S. population was believed to be 299,125,000. Round this number to the nearest million and write a sentence indicating the approximation in scientific notation. ***Source:*** **www.census.gov**

Procedure First, round the number to the nearest million. Write this approximation in scientific notation. Then write a sentence containing the approximation. Use the word *approximately* in the sentence.

Answer Rounded to the nearest million: 299,000,000
In scientific notation: 2.99×10^8

Sentence In July 2006, the U.S. population was approximately 2.99×10^8.

You Try It 5 **Write the underlined sentence with the requested approximation in scientific notation. Use Example 4 as a guide.**

In 2006, the population of Hawaii was 1,275,194. Round this number to the nearest hundred thousand and write a sentence indicating the approximation in scientific notation. ***Source:*** **http://www.quickfacts.census.gov**

Sentence ______________________

You Try It 6 **Write the underlined sentence with the requested approximation in scientific notation. Use Example 4 as a guide.**

The speed of light is 187,370 miles per second. Round this number to the nearest ten thousand and write a sentence indicating the approximation in scientific notation.

Sentence ______________________________

Expanding from Scientific Notation

Sometimes when a number is written in scientific notation, it's easier to work with but understanding how large the number actually is can be difficult. Of course, if you work with any system long enough, interpreting it becomes second nature.

Since we are new to scientific notation, though, it's good to be able to write numbers in their more familiar form. In other words, whereas scientific notation abbreviates very large and very small numbers, expanding them to their natural form "*un*-abbreviates" them.

Example 5

Expand 4.6×10^4 to its natural form.

Procedure First, place a number of zeros at the end of the coefficient so that the decimal point can be moved easily. The exponent indicates the number of places to the right to move the decimal point.

Answer

4.6×10^4	This is going to be a *large* number; so, place some zeros *after* the coefficient.
4.60000×10^4	Move the decimal four places to create a large number.
46000.0	Eliminate the unnecessary zeros after the decimal point and insert a comma.
46,000	Verify that this is correct by reversing the process: $46{,}000 = 4.6 \times 10^4$.

You Try It 7 **Expand each number to its natural form. Use Example 5 as a guide.**

a) 6.1×10^3 **b)** 4.33×10^5 **c)** 2.084×10^8

Answers: You Try It and Think About It

You Try It: **1. a)** 10 **b)** 10,000 **c)** 10,000,000 **d)** 1,000,000,000 **2. a)** 6×10^4 **b)** 2×10^7 **c)** 7×10^{11} **d)** 4×10^1 **3. a)** Yes. (The whole number is a single digit.) **b)** Yes. (The whole number is a single digit.) **c)** No. (The whole number has more than one digit.) **d)** No. (The whole number can't be 0.) **e)** No. (The whole number has more than one digit.) **4. a)** 2.8×10^4 **b)** 4.13×10^5 **c)** 9.07×10^6 **d)** 6.215×10^{10} **e)** 5.8×10^2 **f)** 7.3×10^1 **5.** In 2006, the population of Hawaii was approximately 1.3×10^6. **6.** The speed of light is approximately 1.9×10^5 miles per second. **7. a)** 6,100 **b)** 433,000 **c)** 208,400,000

Think About It: No Think About Its in this section.

A.1 Exercises

FOR EXTRA HELP
Student Resources on DVD-ROM — Includes ➤Student's Solutions Manual ➤Video Lectures ➤Chapter Test Prep Video
MyMathLab
MathXL

Write each expression as a product of a whole number and a power of 10. Use exponents to represent the power of 10.

1. 700,000,000
7×10^8

2. 4,000,000
4×10^6

3. 900
9×10^2

4. 200,000
2×10^5

5. 10,000,000,000
1×10^{10}

6. 50
5×10^1

7. 80,000
8×10^4

8. 6,000
6×10^3

Write each number in scientific notation.

9. 330,000
3.3×10^5

10. 78,000
7.8×10^4

11. 5,090,000
5.09×10^6

12. 4,020
4.02×10^3

13. 130
1.3×10^2

14. 74
7.4×10^1

15. 901,000
9.01×10^5

16. 3,002,000
3.002×10^6

17. 83,000,000,000
8.3×10^{10}

18. 6,400,000,000
6.4×10^9

19. 10,900,000
1.09×10^7

20. 403,000,000
4.03×10^8

Expand each number to its natural form.

21. 5.6×10^7
56,000,000

22. 2.9×10^9
2,900,000,000

23. 7.32×10^3
7,320

24. 9.51×10^4
95,100

25. 2.03×10^5
203,000

26. 4.01×10^6
4,010,000

27. 1.89×10^{10}
18,900,000,000

28. 3.201×10^8
320,100,000

29. 5.3×10^2
530

30. 6.12×10^2
612

31. 2.7×10^1
27

32. 1.0×10^1
10

Write the underlined phrase with the requested approximation in scientific notation.

33. In 2002, gas consumption statistics for all international flights showed that the total number of gallons of gas consumed was **4,990,797,640**. Round this number to the nearest ten million and write a sentence indicating the approximation in scientific notation. ***Source:*** **www.bts.gov**
The total number of gallons of gas consumed was approximately 4.99×10^9.

34. In fall 2004, **923,500** students were enrolled in grades K–12 in Tennessee. Round this number to the nearest ten thousand and write a sentence indicating the approximation in scientific notation.
Source: **http://nces.ed.gov**
Approximately 9.2×10^5 students were enrolled in grades K–12 in Tennessee.

35. In December, 2006, the U.S. national debt was **$8,656,279,633,102**. Round this number to the nearest ten billion and write a sentence indicating the approximation in scientific notation. ***Source:*** **www.publicdebt.treas.gov**
The U.S. national debt was approximately 8.66×10^{12}.

36. It is estimated that the world's rain forests are being destroyed at a rate of **77,893,900** acres per year. Round this yearly acreage to the nearest hundred thousand and write a sentence indicating the approximation in scientific notation. ***Source:*** **www.rain-tree.com**
The world's rain forests are being destroyed at a rate of approximately 7.79×10^7 acres per year.

A.2 Negative Exponents

OBJECTIVES

In this section, you will learn to

- Apply negative exponents to constants and variables.
- Use the Product Rule for Exponents with negative exponents.
- Use the Quotient Rule for Exponents with negative exponents.
- Use the Power Rule for Exponents with negative exponents.

You Need to Know

To successfully complete this section, you need to understand

- ☐ Exponents (1.5)
- ☐ Adding signed numbers (2.2)
- ☐ Subtracting signed numbers (2.3)
- ☐ Reciprocals (4.2)
- ☐ The Product Rule for Exponents (7.1)
- ☐ The Power Rule for Exponents (7.1)
- ☐ The Quotient Rule of Exponents (7.5)

Introduction

Recall from Section 4.2 that the reciprocal of a fraction, $\frac{a}{b}$, is the fraction, $\frac{b}{a}$, as long as a and b are not 0. That is, the reciprocal of a fraction results from interchanging (inverting) the numerator and the denominator.

$$\text{So, the reciprocal of } \frac{3}{5} \text{ is } \frac{5}{3}.$$

When a variable is in the fraction, the reciprocal can be represented the same way. For example, the reciprocal of $\frac{4}{x}$ is $\frac{x}{4}$.

The reciprocal of a whole number can be found by writing the whole number as a fraction with a denominator of 1, as in $7 = \frac{7}{1}$.

$$\text{The reciprocal of } \frac{7}{1} \text{ is } \frac{1}{7}.$$

Likewise, we can write x as a fraction, $\frac{x}{1}$, before writing its reciprocal: The reciprocal of x is $\frac{1}{x}$.

Example 1

Find the reciprocal of each.

a) $\frac{2}{m}$ **b)** $\frac{1}{w}$ **c)** $5p$ **d)** y^2

Procedure For parts c and d, write a denominator of 1, then invert the fraction.

Answer **a)** The reciprocal of $\frac{2}{m}$ is $\mathbf{\frac{m}{2}}$.

b) The reciprocal of $\frac{1}{w}$ is $\mathbf{\frac{w}{1}}$.

c) The reciprocal of $5p = \frac{5p}{1}$ is $\mathbf{\frac{1}{5p}}$.

d) The reciprocal of $y^2 = \frac{y^2}{1}$ is $\mathbf{\frac{1}{y^2}}$.

You Try It 1 **Find the reciprocal of each. Use Example 1 as a guide.**

a) The reciprocal of $\frac{1}{y}$ is ________. **b)** The reciprocal of $\frac{x}{8}$ is ________.

c) The reciprocal of $7k$ is ________. **d)** The reciprocal of m^3 is ________.

Understanding Negative Exponents

In this section, we're going to look at numbers and variables with negative exponents, such as x^{-2}, y^{-5}, and 2^{-3}. Recall from Section 7.1 that the most important feature about an exponent is its meaning. This also is true for the negative in an exponent.

In an exponent, the negative sign means *reciprocal*. We can develop this idea in several ways. Two are presented here. Showing you two methods is not intended to confuse you. Instead, it is intended to cement the meaning of the negative exponent, to give it greater validity. From these, we will develop the Rule of Negative Exponents.

First, let's start with something rather familiar: powers of 3.

We know that	$3^0 = 1$	
and that	$3^1 = 3$	
and that multiplying by another 3	$3 \cdot 3 = 9 = 3^2$	increases the exponent by 1
multiplying by another 3	$3 \cdot 3 \cdot 3 = 27 = 3^3$	increases the exponent by 1
multiplying by another 3	$3 \cdot 3 \cdot 3 \cdot 3 = 81 = 3^4$	increases the exponent by 1
and so on		

We'll now work in the reverse direction starting with $3^4 = 81$. Instead of multiplying by 3, we'll multiply by $\frac{1}{3}$; and instead of increasing the exponent by 1, the exponent will decrease by 1.

Start with	$81 = 3^4$	
multiply by $\frac{1}{3}$	$27 = 3^3$	and decrease the exponent by 1
multiply by $\frac{1}{3}$	$9 = 3^2$	and decrease the exponent by 1
multiply by $\frac{1}{3}$	$3 = 3^1$	and decrease the exponent by 1
multiply by $\frac{1}{3}$	$1 = 3^0$	and decrease the exponent by 1
multiply by $\frac{1}{3}$	$\frac{1}{3} = 3^{-1}$	and decrease the exponent by 1
multiply by $\frac{1}{3}$	$\frac{1}{9} = 3^{-2}$	and decrease the exponent by 1
multiply by $\frac{1}{3}$	$\frac{1}{27} = 3^{-3}$	and decrease the exponent by 1
multiply by $\frac{1}{3}$	$\frac{1}{81} = 3^{-4}$	and decrease the exponent by 1
and so on		

You Try It 2 **Duplicate the previous process using a base of 2. Follow the outline given.**

Start with	$32 = 2^5$			
multiply by $\frac{1}{2}$	16	=	2^4	and decrease the exponent by 1
a) multiply by $\frac{1}{2}$	______	=	______	and decrease the exponent by 1
b) multiply by $\frac{1}{2}$	______	=	______	and decrease the exponent by 1
c) multiply by $\frac{1}{2}$	______	=	______	and decrease the exponent by 1
d) multiply by $\frac{1}{2}$	______	=	______	and decrease the exponent by 1
e) multiply by $\frac{1}{2}$	______	=	______	and decrease the exponent by 1

f) multiply by $\frac{1}{2}$ ______ = ______ and decrease the exponent by 1

g) multiply by $\frac{1}{2}$ ______ = ______ and decrease the exponent by 1

Let's look graphically at some of the powers of 2 you just generated.

2^{-3} 2^{-1} 2^{-2} 2^0 2^1 2^2

0 $\frac{1}{8}$ $\frac{1}{4}$ $\frac{1}{2}$ 1 2 4

Notice that all of the powers of 2 are to the right of 0. That means they are all positive.

Also notice that all of the values between 0 and 1, $\frac{1}{8}$, $\frac{1}{4}$, and $\frac{1}{2}$, are reciprocals of powers of 2.

We know that $2^3 = 2 \cdot 2 \cdot 2 = 8$. Is it possible that $2^{-3} = -8$? Or is it possible that $2^{-3} = -\frac{1}{8}$? Explain.

____

____

The Rule of Negative Exponents

Recall from Section 7.5 the Quotient Rule for Exponents: $\frac{x^a}{x^b} = x^{a-b}$. This rule is easy to apply to something like $\frac{x^5}{x^3} = x^{5-3} = x^2$. It's also easy to understand when we expand the numerator and the denominator and divide out common factors.

$$\frac{x^5}{x^3} = \frac{x \cdot x \cdot x \cdot x \cdot x}{x \cdot x \cdot x} = \frac{\cancel{x \cdot x \cdot x} \cdot x \cdot x}{\cancel{x \cdot x \cdot x}} = \frac{x \cdot x}{1} = x^2$$

The numerator had five factors of x; but we divided out three of them, leaving the numerator with only two factors of x.

We can use this same process when the denominator has more factors than the numerator. For example, the fraction $\frac{x^3}{x^5}$ can be simplified by expanding the numerator and the denominator and dividing out any common factors.

$$\frac{x^3}{x^5} = \frac{x \cdot x \cdot x}{x \cdot x \cdot x \cdot x \cdot x} = \frac{\cancel{x \cdot x \cdot x}}{\cancel{x \cdot x \cdot x} \cdot x \cdot x} = \frac{1}{x \cdot x} = \frac{1}{x^2}$$

This time the result is two factors of x in the denominator.

Instead of expanding, by applying the Quotient Rule, we find that $\frac{x^3}{x^5} = x^{3-5} = x^{-2}$, which appears to be a different result compared to that found when dividing out the common factors.

However, the results are actually two different forms of the same answer. The first method shows that $\frac{x^3}{x^5} = \frac{1}{x^2}$, and the second method shows that $\frac{x^3}{x^5} = x^{-2}$. This means that $x^{-2} = \frac{1}{x^2}$.

This leads us to the Rule of Negative Exponents.

The Rule of Negative Exponents

For any real number x, $x^{-n} = \frac{1}{x^n}$, $x \neq 0$.

In other words, the negative in the exponent of x^{-n} means the *reciprocal* of x^n.

Example 2

Rewrite each expression so that it contains a positive exponent. Evaluate if possible.

a) 4^{-3} **b)** 5^{-2} **c)** y^{-6} **d)** x^{-1}

Procedure Apply the Rule of Negative Exponents. The negative in an exponent means "reciprocal."

Answer

a) $4^{-3} = \frac{1}{4^3} = \frac{1}{64}$ **b)** $5^{-2} = \frac{1}{5^2} = \frac{1}{25}$

c) $y^{-6} = \frac{1}{y^6}$ **d)** $x^{-1} = \frac{1}{x^1} = \frac{1}{x}$

You Try It 3 Rewrite each expression so that it contains a positive exponent. Evaluate if possible. Use Example 2 as a guide.

a) 2^{-4} **b)** 3^{-3} **c)** 7^{-2} **d)** h^{-4} **e)** q^{-5} **f)** w^{-1}

Using Negative Exponents with Other Rules of Exponents

Recall from Chapter 7 these three rules of exponents:

The Product Rule	The Power Rule	The Quotient Rule
For any base, $x^a \cdot x^b = x^{a+b}$	$(x^a)^b = x^{a \cdot b}$	For any nonzero base, x, $\frac{x^a}{x^b} = x^{a-b}$

We can apply each of these rules to bases with negative exponents too. In applying any of the rules, if the result gives a base with a negative exponent, it is common practice to write the result with positive exponents only.

Let's apply these rules one at a time.

Caution With negative numbers of any kind, we need to be careful. That is especially true when working with the rules of exponents. For the sake of accuracy, it is recommended that you do all of the steps and show all of your work.

Example 3

Simplify each expression.

a) $x^7 \cdot x^{-3}$ **b)** $w^{-2} \cdot w^{-5}$

Procedure Carefully apply the Product Rule for Exponents. If the result is a base with a negative exponent, apply the Rule of Negative Exponents and write the answer with a positive exponent.

Answer **a)** $x^7 \cdot x^{-3} = x^{7+(-3)} = x^4$ **b)** $w^{-2} \cdot w^{-5} = w^{-2+(-5)} = w^{-7} = \frac{1}{w^7}$

You Try It 4 Simplify each expression. Make sure you write the result with positive exponents only. Use Example 3 as a guide.

a) $v^{-4} \cdot v^5$ **b)** $y^2 \cdot y^{-8}$ **c)** $w^6 \cdot w^{-6}$

Example 4

Simplify each expression.

a) $\dfrac{c^3}{c^{-4}}$ b) $\dfrac{x^{-3}}{x^{-1}}$

Procedure Carefully apply the Quotient Rule for Exponents. If the result is a base with a negative exponent, apply the Rule of Negative Exponents and write the answer with a positive exponent.

Answer a) $\dfrac{c^3}{c^{-4}} = c^{3-(-4)} = c^{3+4} = c^7$ b) $\dfrac{x^{-3}}{x^{-1}} = x^{-3-(-1)} = x^{-3+1} = x^{-2} = \dfrac{1}{x^2}$

You Try It 5 Simplify each expression. Make sure you write the result with positive exponents only. Use Example 4 as a guide.

a) $\dfrac{m^{-4}}{m^{-7}}$ b) $\dfrac{c^{-1}}{c^6}$ c) $\dfrac{p^{-3}}{p^{-3}}$

Example 5

Simplify each expression.

a) $(y^4)^{-2}$ b) $(x^{-5})^{-1}$

Procedure Carefully apply the Power Rule for Exponents. If the result is a base with a negative exponent, apply the Rule of Negative Exponents and write the answer with a positive exponent.

Answer a) $(y^4)^{-2} = y^{4\cdot(-2)} = y^{-8} = \dfrac{1}{y^8}$ b) $(x^{-5})^{-1} = x^{-5\cdot(-1)} = x^{+5}$ or x^5

You Try It 6 Simplify each expression. Make sure you write the result with positive exponents only. Use Example 5 as a guide.

a) $(c^{-3})^4$ b) $(q^{-2})^{-3}$ c) $(w^4)^{-1}$

Answers: You Try It and Think About It

You Try It: **1. a)** y **b)** $\frac{8}{x}$ **c)** $\frac{1}{7k}$ **d)** $\frac{1}{m^3}$ **2. a)** $8 = 2^3$ **b)** $4 = 2^2$ **c)** $2 = 2^1$ **d)** $1 = 2^0$ **e)** $\frac{1}{2} = 2^{-1}$ **f)** $\frac{1}{4} = 2^{-2}$ **g)** $\frac{1}{8} = 2^{-3}$ **3. a)** $\frac{1}{2^4} = \frac{1}{16}$ **b)** $\frac{1}{3^3} = \frac{1}{27}$ **c)** $\frac{1}{7^2} = \frac{1}{49}$ **d)** $\frac{1}{h^4}$ **e)** $\frac{1}{q^5}$ **f)** $\frac{1}{w^1} = \frac{1}{w}$ **4. a)** v^1 or v **b)** $y^{-6} = \frac{1}{y^6}$ **c)** $w^0 = 1$ **5. a)** m^3 **b)** $c^{-7} = \frac{1}{c^7}$ **c)** $p^0 = 1$ **6. a)** $c^{-12} = \frac{1}{c^{12}}$ **b)** q^{+6} or q^6 **c)** $w^{-4} = \frac{1}{w^4}$

Think About It: **1.** No. Answers may vary. One possibility: 2^{-3} is to the right of 0; so, that means that it is positive and cannot be negative.

A.2 Exercises

FOR EXTRA HELP
Student Resources on DVD-ROM
Includes
➤ Student's Solutions Manual
➤ Video Lectures
➤ Chapter Test Prep Video
MyMathLab
MathXL

Apply the Rule of Negative Exponents and evaluate.

1. 11^{-2} $\frac{1}{121}$ **2.** 9^{-2} $\frac{1}{81}$ **3.** 10^{-3} $\frac{1}{1{,}000}$ **4.** 10^{-4} $\frac{1}{10{,}000}$

5. 2^{-4} $\frac{1}{16}$ **6.** 2^{-5} $\frac{1}{32}$ **7.** 3^{-4} $\frac{1}{81}$ **8.** 5^{-3} $\frac{1}{125}$

9. 6^{-1} $\frac{1}{6}$ **10.** 8^{-1} $\frac{1}{8}$ **11.** 15^{-1} $\frac{1}{15}$ **12.** 20^{-1} $\frac{1}{20}$

Apply the Rule of Negative Exponents and write each expression with a positive exponent.

13. w^{-2} $\frac{1}{w^2}$ **14.** b^{-2} $\frac{1}{b^2}$ **15.** q^{-3} $\frac{1}{q^3}$ **16.** p^{-4} $\frac{1}{p^4}$

17. m^{-6} $\frac{1}{m^6}$ **18.** x^{-5} $\frac{1}{x^5}$ **19.** y^{-8} $\frac{1}{y^8}$ **20.** c^{-10} $\frac{1}{c^{10}}$

21. k^{-1} $\frac{1}{k}$ **22.** d^{-1} $\frac{1}{d}$ **23.** a^{-1} $\frac{1}{a}$ **24.** n^{-1} $\frac{1}{n}$

Apply the Product Rule and simplify each expression. Write the results with positive exponents only.

25. $b^6 \cdot b^{-2}$ b^4
26. $m^5 \cdot m^{-3}$ m^2
27. $c^{-1} \cdot c^2$ c^1
28. $w^{-5} \cdot w^6$ w^1
29. $y^{-6} \cdot y^3$ $\frac{1}{y^3}$
30. $x^{-8} \cdot x^4$ $\frac{1}{x^4}$
31. $p^{-7} \cdot p^6$ $\frac{1}{p}$
32. $x^{-4} \cdot x^3$ $\frac{1}{x}$
33. $k^6 \cdot k^{-8}$ $\frac{1}{k^2}$
34. $n^2 \cdot n^{-9}$ $\frac{1}{n^7}$
35. $h^{-8} \cdot h^{-5}$ $\frac{1}{h^{13}}$
36. $a^{-3} \cdot a^{-6}$ $\frac{1}{a^9}$
37. $q^4 \cdot q^{-4}$ $q^0 = 1$
38. $d^2 \cdot d^{-2}$ $d^0 = 1$
39. $g^{-7} \cdot g^7$ $g^0 = 1$
40. $b^{-3} \cdot b^3$ $b^0 = 1$

Apply the Quotient Rule and simplify each expression. Write the results with positive exponents only.

41. $\frac{x^3}{x^{-2}}$ x^5
42. $\frac{c^2}{c^{-4}}$ c^6
43. $\frac{y^1}{y^{-5}}$ y^6
44. $\frac{a^1}{a^{-6}}$ a^7
45. $\frac{k^{-4}}{k^6}$ $\frac{1}{k^{10}}$
46. $\frac{q^{-3}}{q^3}$ $\frac{1}{q^6}$
47. $\frac{d^{-1}}{d^8}$ $\frac{1}{d^9}$
48. $\frac{n^{-1}}{n^1}$ $\frac{1}{n^2}$
49. $\frac{m^{-5}}{m^{-9}}$ m^4
50. $\frac{w^{-3}}{w^{-6}}$ w^3
51. $\frac{v^{-1}}{v^{-4}}$ v^3
52. $\frac{p^{-1}}{p^{-7}}$ p^6
53. $\frac{h^{-7}}{h^{-4}}$ $\frac{1}{h^3}$
54. $\frac{x^{-6}}{x^{-2}}$ $\frac{1}{x^4}$
55. $\frac{y^{-3}}{y^{-1}}$ $\frac{1}{y^2}$
56. $\frac{c^{-2}}{c^{-1}}$ $\frac{1}{c}$
57. $\frac{a^{-5}}{a^{-5}}$ $a^0 = 1$
58. $\frac{p^{-3}}{p^{-3}}$ $p^0 = 1$
59. $\frac{x^5}{x^5}$ $x^0 = 1$
60. $\frac{y^2}{y^2}$ $y^0 = 1$

Apply the Power Rule and simplify each expression. Write the results with positive exponents only.

61. $(x^{-1})^6$ $\frac{1}{x^6}$
62. $(v^{-1})^7$ $\frac{1}{v^7}$
63. $(p^{-8})^2$ $\frac{1}{p^{16}}$
64. $(c^{-6})^3$ $\frac{1}{c^{18}}$
65. $(q^{-4})^{-1}$ q^4
66. $(x^{-2})^{-1}$ x^2
67. $(m^{-5})^{-2}$ m^{10}
68. $(d^{-3})^{-3}$ d^9
69. $(w^9)^{-1}$ $\frac{1}{w^9}$
70. $(n^1)^{-1}$ $\frac{1}{n}$
71. $(x^4)^{-4}$ $\frac{1}{x^{16}}$
72. $(y^6)^{-2}$ $\frac{1}{y^{12}}$
73. $(p^{-1})^0$ $p^0 = 1$
74. $(m^{-3})^0$ $m^0 = 1$
75. $(y^0)^{-5}$ $y^0 = 1$
76. $(q^0)^{-3}$ $q^0 = 1$
77. $(x^{-4})^5$ $\frac{1}{x^{20}}$
78. $(y^{-3})^{-8}$ y^{24}
79. $(a^{-6})^{-5}$ a^{30}
80. $(n^7)^{-4}$ $\frac{1}{n^{28}}$

A.3 Scientific Notation: Small Numbers

OBJECTIVES

In this section, you will learn to
- Write a small number in scientific notation.
- Expand a small number from scientific notation.

You Need to Know

To successfully complete this section, you need to understand
- ☐ Decimals (5.1)
- ☐ Scientific notation (A.1)
- ☐ Negative exponents (A.3)

Introduction

In A.1, we worked with scientific notation and large numbers. We also can use scientific notation to express very small numbers such as these:

Grass grows at a rate of 0.00000002 (2 hundred-millionths) meters per second.

The mass of an atom of hydrogen is 0.00000000000000000000000000167. (This number is so small it doesn't even have a name.)

The only difference in scientific notation for small numbers compared to large numbers is that the exponent is negative.

Working with Base 10 and Negative Exponents

The Rule of Negative Exponents says that $x^{-n} = \frac{1}{x^n}$; so,

$$10^{-1} = \frac{1}{10^1} = \frac{1}{10} = 0.1 \quad \text{(one decimal place)}$$

$$10^{-2} = \frac{1}{10^2} = \frac{1}{100} = 0.01 \quad \text{(two decimal places)}$$

$$10^{-3} = \frac{1}{10^3} = \frac{1}{1{,}000} = 0.001 \quad \text{(three decimal places)}$$

and $$10^{-4} = \frac{1}{10^4} = \frac{1}{10{,}000} = 0.0001 \quad \text{(four decimal places)}$$

From these examples, we can see that in base 10, a negative exponent indicates a small (sometimes very small) number. Also, the absolute value of the negative exponent indicates the number of decimal places for a base of 10.

You Try It 1 **Write each power of ten as a decimal number.**

a) 10^{-2} **b)** 10^{-4} **c)** 10^{-6} **d)** 10^{-8}

Recall from A.1 that a large number such as 50,000 can be written as the product of a whole number and a power of 10: $50{,}000 = 5 \times 10^4$.

We can represent very small numbers (decimals) in the same way. The difference is that for small numbers, the powers of 10 have negative exponents.

For example,

$0.005 = 5 \times 0.001$ (which has *three* decimal places)

$= 5 \times 10^{-3}$ (notice the exponent of negative *three*)

Example 1

Write each expression as a product of a whole number and a power of 10. Use an exponent to represent the power of 10.

a) 0.04 **b)** 0.00009

Procedure Make sure you count the number of decimal places. In each of these, the number of decimal places is one more than the number of zeros after the decimal point.

Answer In part a, there are two decimal places; in part b, there are five decimal places.

a) $0.04 = 4 \times 0.01 = 4 \times 10^{-2}$ **b)** $0.00009 = 9 \times 0.00001 = 9 \times 10^{-5}$

You Try It 2 **Write each expression as a product of a whole number and a power of 10. Use Example 1 as a guide.**

a) 0.006 (six-thousandths) **b)** 0.00002 (two-hundred thousandths)

c) 0.000007 (seven-millionths) **d)** 0.00000003 (three-hundred millionths)

Writing Small Numbers in Scientific Notation

Recall from A.1 that we used scientific notation to abbreviate a large number as the product of a coefficient and a power of 10. Also recall that the coefficient must have a single digit whole number from 1 to 9.

Writing small numbers (positive numbers less than 1) in scientific notation is similar to writing large numbers. We still have the three things to consider:

1. The coefficient must have a single-digit whole number, 1 through 9.
2. We must count the number of spaces needed to move a decimal point.
3. A small number has a power of 10 with a negative exponent.

A number such as 0.000003 is easy to write in scientific notation because the coefficient is the whole number 3 with no decimals.

$$0.000003 = 3 \times 10^{-6}$$

However, a number such as 0.00000358 is a little more challenging because we can't just count the number of decimal places. Instead, we need to recognize that

1. The coefficient is 3.58.
2. We must count the number of spaces needed to move the decimal point between the 3 and 5. It must move 6 places to the right.

   0.00000358
   1 2 3 4 5 6

3. Because 0.00000358 is a small number, it will have a negative power of 10.

So, in scientific notation, $0.00000358 = 3.58 \times 10^{-6}$.

Example 2

Write each number in scientific notation.

a) 0.00096 **b)** 0.00000000745

Procedure Decide on the coefficient and count the number of places the decimal point must be moved to get a proper coefficient. Because both of these numbers are small, the exponents will be negative.

Answer **a)** The coefficient is 9.6; the decimal must move *four* places: $0.00096 = 9.6 \times 10^{-4}$

b) The coefficient is 7.45; the decimal must move *nine* places: $0.00000000745 = 7.45 \times 10^{-9}$

You Try It 3 **Write each number in scientific notation. Use Example 2 as a guide.**

a) 0.0028 **b)** 0.00000006215 **c)** 0.92

Example 3

Rewrite this sentence using scientific notation.

The chance of winning the Powerball Mega Millions Lottery grand prize with a $1 wager is 0.00000000569.

Procedure First, write the number in scientific notation. Then write a sentence for the approximation.

Answer In scientific notation, the number is 5.69×10^{-9}.

Sentence The chance of winning the Powerball Mega Millions Lottery grand prize with a $1 wager is 5.69×10^{-9}.

You Try It 4 **Rewrite the sentence using scientific notation. Use Example 3 as a guide.**

Grass grows at a rate of 0.00000002 meters per second.

Sentence ____________________

You Try It 5 **Rewrite the sentence using scientific notation. Use Example 3 as a guide.**

The mass of one trillion atoms of hydrogen is 0.00000000000000167.

Sentence ____________________

Expanding from Scientific Notation

If a number is written in scientific notation, we can expand it to see its value represented without exponents. Expanding a number from scientific notation is like *un*-abbreviating it.

Example 4

Expand 8.07×10^{-5} to its natural form.

Procedure As we expand this number, it will be very small and the digits in the coefficient will follow a number of zeros. For this reason, we should place a number of zeros at the beginning of the coefficient so that the decimal point has somewhere to go as it is being moved.

Answer You can check your answer by thinking about how it would be if you were to abbreviate it back into scientific notation.

8.07×10^{-5}	Because of the negative exponent, this is a small number. Place some zeros *before* the coefficient.
$000000008.07 \times 10^{-5}$	Move the decimal five places to create a small number.
0000.0000807	Eliminate the unnecessary zeros (before the decimal point).
0.0000807	Verify that this is correct by reversing the process: $0.0000807 = 8.07 \times 10^{-5}$.

You Try It 6 **Expand each number to its natural form. Use Example 3 as a guide.**

a) 9.2×10^{-2} **b)** 3.06×10^{-4} **c)** 4.138×10^{-7}

Answers: You Try It and Think About It

You Try It: **1. a)** 0.01 **b)** 0.0001 **c)** 0.000001 **d)** 0.00000001 **2. a)** 6×10^{-3} **b)** 2×10^{-5} **c)** 7×10^{-6} **d)** 3×10^{-8} **3. a)** 2.8×10^{-3} **b)** 6.215×10^{-8} **c)** 9.2×10^{-1} **4.** Grass grows at a rate of 2×10^{-8} meters per second. **5.** The mass of 1 trillion atoms of hydrogen is 1.67×10^{-15}. **6. a)** 0.092 **b)** 0.000306 **c)** 0.0000004138

Think About It: No Think About Its in this section.

A.3 Exercises

FOR EXTRA HELP
Student Resources on DVD-ROM
Includes
➤ Student's Solutions Manual
➤ Video Lectures
➤ Chapter Test Prep Video
MyMathLab
MathXL

Write each expression as a product of a whole number and a power of 10. Use an exponent to represent the power of 10.

1. 0.03
3×10^{-2}

2. 0.2
2×10^{-1}

3. 0.0005
5×10^{-4}

4. 0.006
6×10^{-3}

5. 0.000004
4×10^{-6}

6. 0.00008
8×10^{-5}

7. 0.0000007
7×10^{-7}

8. 0.00000009
9×10^{-8}

Write each number in scientific notation.

9. 0.28
2.8×10^{-1}

10. 0.042
4.2×10^{-2}

11. 0.0053
5.3×10^{-3}

12. 0.00096
9.6×10^{-4}

13. 0.00913
9.13×10^{-3}

14. 0.708
7.08×10^{-1}

15. 0.000407
4.07×10^{-4}

16. 0.0000813
8.13×10^{-5}

17. 0.002914
2.914×10^{-3}

18. 0.06152
6.152×10^{-2}

19. 0.0002009
2.009×10^{-4}

20. 0.00003006
3.006×10^{-5}

Expand each number to its natural form.

21. 2.3×10^{-2}
0.023

22. 5.6×10^{-3}
0.0056

23. 1.8×10^{-1}
0.18

24. 9.2×10^{-1}
0.92

25. 6.32×10^{-4}
0.000632

26. 2.19×10^{-3}
0.00219

27. 8.86×10^{-5}
0.0000886

28. 1.33×10^{-4}
0.000133

29. 4.01×10^{-6}
0.00000401

30. 3.06×10^{-5}
0.0000306

31. 4.915×10^{-4}
0.0004915

32. 8.023×10^{-6}
0.000008023

Rewrite each sentence using scientific notation.

33. The average hummingbird weighs 0.0032 kilograms.
The average hummingbird weighs 3.2×10^{-3} kilograms.

34. Light travels 1 mile in 0.00000534 seconds.
Light travels 1 mile in 5.34×10^{-6} seconds.

A.4 Scientific Notation: Multiplication

OBJECTIVES

In this section, you will learn to

- Multiply numbers in scientific notation.

You Need to Know

To successfully complete this section, you need to understand

- ☐ Multiplying decimals (5.4)
- ☐ The Product Rule for Exponents (7.1)
- ☐ Scientific notation: large numbers (A.1)
- ☐ Negative exponents (A.2)
- ☐ Scientific notation: small numbers (A.3)

Introduction

How far is it to the center of our galaxy, the Milky Way? To get there, a spaceship would have to travel 1 light-year, then another light-year, and then another 27,998 light-years.

A light-year is the distance light can travel in one year. Light travels over 187,000 miles per *second*, which is more than 5.9×10^{12} miles per year. That's a lot of miles! And that's just one light-year. To get to the center of our galaxy, a spaceship would have to travel 28,000 light-years.

The actual distance from Earth to the center of the Milky Way will never be known precisely, but we can still calculate it using scientific notation. You'll find the answer to the question above in Example 3, where you will learn to multiply numbers written in scientific notation.

Before we learn to multiply in scientific notation, we must prepare for the different results that we may encounter. For example, we may multiply two numbers in scientific notation and end up with a coefficient that is not in proper form. If that happens, we must adjust the coefficient to get it into proper form.

Adjusting the Coefficient

If a number is written in scientific notation but the coefficient is not in *proper form,* we need to adjust the coefficient to put it into proper form.

For example, the coefficient in 12×10^5 is not in proper form. (The coefficient is too large.) If we expand it, we can better see what it will be when it is written in proper form.

12×10^5 is 12 followed by five zeros $=$ 1,200,000.
This, however, can be written as 1.2×10^6.

In other words, $12 \times 10^5 = 1.2 \times 10^6$ and the coefficient is now in proper form.

We call this—going from an improper form to a proper one—***adjusting the coefficient.*** Whether the coefficient is too large or too small, we can make this adjustment by

1. Writing the coefficient in scientific notation.
2. Combining it with the power of 10 already present in the number.

For example, because $12 = 1.2 \times 10^1$, we can adjust 12×10^5 to be

$(1.2 \times 10^1) \times 10^5$ We can then use the Associative Property to write it as
$= 1.2 \times (10^1 \times 10^5)$ We then use the Product Rule for Exponents to get
$= 1.2 \times 10^6$

Example 1

Adjust each problem so that the coefficient is in proper form.

a) 56×10^4 **b)** 30.7×10^{-8} **c)** 0.043×10^6

Procedure In parts a and b, the coefficient is too large. In part c, the coefficient is too small.

Answer

a) 56×10^4
$= (5.6 \times 10^1) \times 10^4$
$= 5.6 \times (10^1 \times 10^4)$
$= 5.6 \times 10^5$

b) 30.7×10^{-8}
$= (3.07 \times 10^1) \times 10^{-8}$
$= 3.07 \times (10^1 \times 10^{-8})$
$= 3.07 \times 10^{-7}$

c) 0.043×10^6
$= (4.3 \times 10^{-2}) \times 10^6$
$= 4.3 \times (10^{-2} \times 10^6)$
$= 4.3 \times 10^4$

You Try It 1 **Adjust each problem so that the coefficient is in proper form. Use Example 1 as a guide.**

a) 61×10^3 **b)** 49.2×10^{-5} **c)** 0.35×10^9 **d)** 0.019×10^{-3}

Multiplying with Scientific Notation

Recall from Section 1.3 that multiplying $300 \times 4{,}000$ is the same as multiplying $3 \times 4 = 12$ and placing the total number of zeros after the 12.

$300 \times 4{,}000 = 12$ followed by a total of five zeros: 1,200,000.

If each of those numbers was written in scientific notation first, the product would look like this:

$300 \times 4{,}000$	
$= (3 \times 10^2) \times (4 \times 10^3)$	Written in scientific notation
$= (3 \times 4) \times (10^2 \times 10^3)$	Using Associative and Commutative Properties
$= 12 \times 10^5$	Applying the Product Rule: $10^2 \times 10^3 = 10^{2+3} = 10^5$
$= (1.2 \times 10^1) \times 10^5$	Adjusting the coefficient: $12 = 1.2 \times 10^1$
$= 1.2 \times (10^1 \times 10^5)$	
$= 1.2 \times 10^6$	If we wanted to do so, we could expand this to be 1,200,000 (the same answer that we found earlier).

If two numbers are already written in scientific notation, the process of *multiplying* them together requires that we

1. Multiply the coefficients together.
2. Multiply the powers of 10 by adding the exponents (even if one is positive and the other is negative).
3. Adjust the coefficient, if necessary.

Example 2

Perform the indicated operation. Write the answer in scientific notation.

a) $(1.2 \times 10^6) \times (2.7 \times 10^{-8})$ **b)** $3.5 \times (6.4 \times 10^7)$

Procedure The multiplications are not shown here. You may do the multiplying of the coefficients off to the side or on a calculator (if allowed). In part b, the first factor, 3.5, is not in scientific notation; so, it will multiply to only the second number's coefficient, 6.4. *Make sure you adjust the coefficient if necessary.*

Answer

a)
$$(1.2 \times 10^6) \times (2.7 \times 10^{-8})$$
$$= (1.2 \times 2.7) \times (10^{6+(-8)})$$
$$= 3.24 \times 10^{-2}$$
We don't need to adjust this coefficient.

b)
$$3.5 \times (6.4 \times 10^7)$$
$$= (3.5 \times 6.4) \times 10^7$$
$$= 22.4 \times 10^7$$
$$= (2.24 \times 10^1) \times 10^7$$
$$= 2.24 \times 10^8$$

You Try It 2

Perform the indicated operation. Write the answer in proper scientific notation. Use Example 2 as a guide.

a) $(1.5 \times 10^9) \times (4.4 \times 10^5)$ **b)** $(2.5 \times 10^{-3}) \times (6.8 \times 10^6)$

c) $(2.5 \times 10^3) \times 9.8$ **d)** $(7.6 \times 10^{-10}) \times (4.0 \times 10^6)$

Applications

Following are some applications that involve multiplying very large numbers. We do so using scientific notation. Most of the numbers you see here have already been rounded and written in scientific notation. If a number is not written in scientific notation, you should do so before multiplying.

The beginning of this appendix mentioned that a light-year is 5.9×10^{12} miles and that the center of the Milky Way Galaxy is about 28,000 light-years from Earth. In Example 3, we calculate the distance from Earth to the center of the Milky Way using scientific notation.

Example 3

To calculate the distance from Earth to the center of our galaxy, the Milky Way, we multiply 28,000 light-years times the distance of a light-year, 5.9×10^{12} miles, that is, $(2.8 \times 10^4) \times (5.9 \times 10^{12})$ miles. Calculate this distance using scientific notation.

Procedure First, multiply the coefficients. Then use the Product Rule for Powers of 10. Adjust the coefficient if necessary and write a sentence for the answer using scientific notation.

Answer

$(2.8 \times 10^4) \times (5.9 \times 10^{12})$
$= (2.8 \times 5.9) \times (10^4 \times 10^{12})$
$= 16.52 \times 10^{4+12}$
$= 16.52 \times 10^{16}$ Adjust the coefficient.
$= (1.652 \times 10^1) \times 10^{16}$
$= 1.652 \times 10^{17}$

YOU FINISH IT: Multiply 2.8×5.9

Sentence The center of the Milky Way is 1.652×10^{17} miles from Earth.

You Try It 3 **Write a sentence containing scientific notation for the answer.**

There were 8.25×10^4 students enrolled in Iowa community colleges in the fall of 2005. Each student averaged 9.6 credit hours that semester. Multiply those numbers to determine the total number of enrolled credit hours for Iowa students in the fall of 2005. ***Source:*** **www.state.ia.us/educate.html**

Sentence ______________________________

You Try It 4 **Write a sentence containing scientific notation for the answer.**

The fastest lizard can travel 5×10^{-3} miles per second. Multiply this by 3.6×10^3 to see how fast the lizard can travel in miles per hour. ***Source:*** **www.sandiegozoo.org**

Sentence ______________________________

Answers: You Try It and Think About It

You Try It: **1. a)** 6.1×10^4 **b)** 4.92×10^{-4} **c)** 3.5×10^8 **d)** 1.9×10^{-5} **2. a)** 6.6×10^{14} **b)** 1.7×10^4 **c)** 2.45×10^4 **d)** 3.04×10^{-3} **3.** In the fall of 2005, Iowa students enrolled a total of 7.92×10^5 credit hours. **4.** The fastest lizard can travel 1.8×10 miles per hour.

Think About It: No Think About Its in this section.

A.4 Exercises

FOR EXTRA HELP

Student Resources on DVD-ROM — Includes ➤Student's Solutions Manual ➤Video Lectures ➤Chapter Test Prep Video | MyMathLab | MathXL

Adjust each problem so that the coefficient is in proper form.

1. 85×10^6
8.5×10^7

2. 93×10^3
9.3×10^4

3. 20×10^{-4}
2.0×10^{-3}

4. 90×10^{-7}
9.0×10^{-6}

5. 45.3×10^5
4.53×10^6

6. 60.2×10^2
6.02×10^3

7. 30.9×10^{-2}
3.09×10^{-1}

8. 70.1×10^{-1}
7.01×10^0 or 7.01

Perform the indicated operation. Write each answer in proper scientific notation.

9. $(1.5 \times 10^6) \times 2.4$
3.6×10^6

10. $(1.8 \times 10^5) \times 3.5$
6.3×10^5

11. $6.5 \times (2.02 \times 10^{-3})$
1.313×10^{-2}

12. $4.6 \times (8.5 \times 10^{-2})$
3.91×10^{-1}

13. $(2.3 \times 10^4) \times (3.0 \times 10^5)$
6.9×10^9

14. $(4.6 \times 10^3) \times (2.0 \times 10^7)$
9.2×10^{10}

15. $(3.5 \times 10^8) \times (2.4 \times 10^{-2})$
8.4×10^6

16. $(7.5 \times 10^{10}) \times (1.2 \times 10^{-6})$
9.0×10^4

17. $(6.4 \times 10^{-9}) \times (2.5 \times 10^1)$
1.6×10^{-7}

18. $(8.2 \times 10^{-4}) \times (3.5 \times 10^2)$
2.87×10^{-1}

19. $(3.1 \times 10^{-7}) \times (9.2 \times 10^{-4})$
2.852×10^{-10}

20. $(5.02 \times 10^{-9}) \times (6.8 \times 10^{-2})$
3.4136×10^{-10}

Write a sentence containing scientific notation for each answer.

21. The average annual amount of money spent on fuel for all commercial fights in the United States from 1979 to 2005 was 8.71×10^9. Multiply that by 27 (2.7×10^1) to find the total amount of money spent on fuel during those 27 years. ***Source:*** **www.bts.gov**
The total amount of money spent on fuel during those 27 years was 2.3517×10^{11}.

22. The U.S. Postal Service (USPS) delivers mail 305 (3.05×10^2) days each year. The USPS delivers 8.6×10^7 pounds of mail every day. Multiply those numbers to find the total number of pounds of mail delivered each year. ***Source:*** **http://www.usps.com**
The total number of pounds of mail delivered each year is 2.623×10^{10}.

23. The giant tortoise can travel 6.4×10^{-5} miles per second. Multiply that by 3,600 to see how far it can travel in one hour. ***Source: Guinness World Records***
The giant tortoise can travel 2.304×10^{-1} miles per hour.

24. In December 2006, the U.S. population was 3.0×10^8. If each U.S. citizen contributed 2.885×10^4, the amount would equal the U.S. national debt. Multiply those two numbers to find the amount of the U.S. national debt. ***Source:*** **www.publicdebt.treas.gov**
The amount of the U.S. national debt in December 2006 was 8.655×10^{12}.

25. In Hawaii, there are about 4.04×10^5 households. Each household consumes, on average, 16.6 cans of Spam per year. Multiply those numbers to find the total number of cans of Spam consumed in Hawaii each year. ***Source:*** **AP, 2006**
The total number of cans of Spam consumed in Hawaii each year is 6.7064×10^6.

A.5 Factoring Trinomials of the Form $x^2 + bx + c$

OBJECTIVES

In this section, you will learn to

- Play the Factor Game.
- Factor a trinomial using the Factor Game.

You Need to Know

To successfully complete this section, you need to understand

- ☐ Factor pairs (1.6)
- ☐ Adding signed numbers (2.2)
- ☐ Multiplying signed numbers (2.4)
- ☐ Multiplying binomials (7.4)

Introduction

In Section 7.4, we learned that the product of two binomials will always be, at first, four terms. Often in such a product, two of the terms are like terms that can be combined so that the end result is a trinomial.

For example, when we multiply $(x + 5)(x - 3)$, at first, we get $x^2 - 3x + 5x - 15$. The middle two terms are like terms that combine to $2x$. The end result is then $x^2 + 2x - 15$.

$$(x + 5)(x - 3) = x^2 + 2x - 15$$

This suggests that $(x + 5)(x - 3)$ is a factored form of $x^2 + 2x - 15$. In other words,

$$x^2 + 2x - 15 = (x + 5)(x - 3)$$

Notice that both factors are binomials. Each factor has a single variable as its first term, and each binomial has a constant as its second term. This pattern is consistent with all of the factorable trinomials presented in this section. (In later algebra courses, trinomials with different characteristics appear and the factoring techniques are slightly different.)

In this text, we focus only on trinomials of the form $x^2 + bx + c$, where b and c are integers.

A number of techniques can be used to factor trinomials, but only one technique is presented here. Before we learn to factor trinomials, we must learn how to play a number game, called the Factor Game, that is directly related to factoring trinomials.

The Factor Game

The Factor Game

Given two numbers, a **product number** and a **sum number**, find a factor pair (two factors) of the product number that will add to the sum number.

The correct factor pair is called the **solution.**

Special Notes:

1. It's possible that there is no solution; in other words, there is no factor pair that works.
2. If there is a solution, there will be only *one* solution.

To identify a possible factor pair solution, we can list the factor pairs of the product number and then identify which pair will add to get the sum number.

You can understand this game best through examples. Look over each example carefully and follow it through to get a full understanding.

Example 1

Find the solution of the Factor Game with the given product and sum numbers.

a) Product # = 12 and sum # = 7

b) Product # = 36 and sum # = 15

Procedure Write out the factor pairs of the product # and see if any of them add to the sum #. When writing factor pairs, it is best to write the smaller factors on the same side, all on the left side or all on the right side.

Answer **a)** Factor pairs of 12:

12		**Sum**	
1	12	13	Too large.
2	6	8	Close but not quite.
3	4	7	This is it!

The solution is: 3 and 4. Check: $3 \cdot 4 = 12$ ✓ and $3 + 4 = 7$ ✓

b) Factor pairs of 36:

36		**Sum**	
1	36	37	
2	18	20	
3	12	15	This is it!
4	9	13	Once you discover the correct factor pair, you do not need to continue searching.
6	6	12	

The solution is: 3 and 12. Check: $3 \cdot 12 = 36$ ✓ and $3 + 12 = 15$ ✓

It's possible that the factors in the solution are the same number. Also, as already mentioned, it's possible that there is no solution.

Example 2

Find the solution to the Factor Game with the given product and sum numbers.

a) Product # = 16 and sum # = 8

b) Product # = 30 and sum # = 12

Procedure Write out the factor pairs of the product # and see if any of them add to the sum #.

Answer **a)** Factor pairs of 16:

16		**Sum**	
1	16	17	
2	8	10	
4	4	8	This is it!

The solution is: 4 and 4. Check: $4 \cdot 4 = 16$ ✓ and $4 + 4 = 8$ ✓

b) Factor pairs of 30:

30		**Sum**	
1	30	31	
2	15	17	
3	10	13	
5	6	11	None of these add to 12.

There is no solution.

You Try It 1 **Find the solution to the Factor Game with the given product and sum numbers. Use Examples 1 and 2 as guides.**

a) product # = 24 and sum # = 10

b) product # = 36 and sum # = 12

c) product # = 15 and sum # = 7

In Examples 1 and 2, both the product number and the sum number are positive. It's possible, though, to have one or both of them be negative. Let's look at how the Factor Game works when the sum number is negative.

The Factor Game and a Negative Sum Number

You know that when the signs of two numbers are the same, their product is positive.

$$(+2)\cdot(+3) = +6 \quad \text{and} \quad (-2)\cdot(-3) = +6$$

It is also true that when the *product* of two factors is positive, the *factors* must have the same sign, either both positive or both negative.

$$+6 = (+2)\cdot(+3) \quad \text{and} \quad +6 = (-2)\cdot(-3)$$
$$+6 = (+1)\cdot(+6) \quad \text{and} \quad +6 = (-1)\cdot(-6)$$

In the Factor Game, when the product number is positive, the two numbers in the factor pair must have the same sign, either both positive or both negative. The sign of the sum number determines the signs of the factor pair.

For example, if the product number is +12 and the sum number is +7, the solution is +3 and +4, as we saw in Example 1.

If the product number is +12 but the sum number is −7, the solution is −3 and −4.

Example 3

Find the solution of the Factor Game with the given product and sum numbers.

a) Product # = +20 and sum # = −12 **b)** Product # = +30 and sum # = −11

Procedure When the product is positive and the sum is negative, both factors in the factor pair must be negative.

Answer **a)** Factor pairs of +20:

+20		**Sum**	
−1	−20	−21	
−2	−10	−12	This is it!
−4	−5		

The solution is: −2 and −10. Check: $-2\cdot(-10) = +20$ ✓ and $-2+(-10) = -12$ ✓

b) Factor pairs of +30:

+30		**Sum**	
−1	−30	−31	
−2	−15	−17	
−3	−10	−13	
−5	−6	−11	This is it!

The solution is: −5 and −6. Check: $-5\cdot(-6) = +30$ ✓ and $-5 + (-6) = -11$ ✓

You Try It 2 **Find the solution of the Factor Game with the given product and sum numbers. Use Example 3 as a guide.**

a) product # = +25 and sum # = −10

b) product # = +15 and sum # = −16

c) product # = +40 and sum # = −14

The Factor Game and a Negative Product Number

You know that when the signs of two numbers are different, their product is negative.

$$(-2)\cdot(+3) = -6 \quad \text{and} \quad (+2)\cdot(-3) = -6$$

It is also true that when the product of two factors is *negative*, the factors must have *different* signs; one must be positive, and the other must be negative.

$$-6 = (-2)\cdot(+3) \quad \text{and} \quad -6 = (+2)\cdot(-3)$$
$$-6 = (-1)\cdot(+6) \quad \text{and} \quad -6 = (+1)\cdot(-6)$$

Furthermore, the sum of two numbers with different signs can be either positive or negative. The number with the largest numerical value determines the sign of the sum.

For example,

- The sum of **+9** and **−5** is +4. +9 has the larger numerical value; and because it is positive, the result is positive.
- Likewise, the sum of **+3** and **−10** is −7. −10 has the larger numerical value; and because it is negative, the result is negative.

In the Factor Game, when the product number is negative, the two numbers in the factor pair must have different signs. The sign of the sum number indicates which sign the larger factor will have.

So, when trying to find two factors of, say, −12 that *add* to equal +4, we need to recognize that the signs of the factors will be *different* and that we need to look for factors of 12 that have a *difference* of 4. Those factors are 6 and 2.

However, the sum of +6 and +2 is +8; so, we need to consider the sum of +6 and −2 to get a sum of +4. In this way, the factors +6 and −2 have *different* signs and their product is negative: −12.

Example 4

Find the solution of the Factor Game with the given product and sum numbers.

a) Product # = −12 and sum # = +1

b) Product # = −30 and sum # = −13

Procedure When the product is negative, one of the factors must be positive and the other negative. The factor with the larger numerical value will have the same sign as the sum number.

Answer **a)** The sum # is positive; so, the factor with the larger numerical value is positive and the factor with the smaller numerical value is negative.

Factor pairs of −12:

−12		**Sum**	
−1	+12	+11	
−2	+6	+4	
−3	+4	+1	This is it!

The solution is: −3 and +4. Check: $-3 \cdot (+4) = -12$ ✓ and $-3 + (+4) = +1$ ✓

b) The sum # is negative; so, the factor with the larger numerical value is negative and the factor with the smaller numerical value is positive.

Factor pairs of −30:

−30		**Sum**	
+1	−30	−29	
+2	−15	−13	This is it!
+3	−10	−7	
+5	−6	−1	

The solution is: +2 and −15. Check: $+2 \cdot (-15) = -30$ ✓ and $+2 + (-15) = -13$ ✓

$-3 + (-10) = -13$. So, why isn't the solution in Example 4b −3 and −10?

____

____

You Try It 3 **Find the solution of the Factor Game with the given product and sum numbers. Use Example 4 as a guide.**

a) product # = −28 and sum # = +3

b) product # = −36 and sum # = −5

c) product # = −20 and sum # = +12

Trinomials and the Factor Game

Besides the Factor Game being a great mental exercise—one that helps you learn more about multiplying and adding signed numbers—its main purpose is to assist you in factoring trinomials. Here is the connection.

> **Trinomials and the Factor Game**
>
> In $x^2 + bx + c$, the product number $= c$
> and the sum number $= b$.

Example 5

Given the trinomial, find the solution to its associated Factor Game.

a) $y^2 - 6y + 8$ **b)** $x^2 + x - 12$ **c)** $w^2 + 5w + 14$

Procedure Identify the values of b and c and write the product number and the sum number.

Answer **a)** $y^2 - 6y + 8$ $b = -6$ and $c = 8$ product # $= +8$ and sum # $= -6$

The product # is positive, and the sum # is negative. So, both factors are negative.

Factor pairs of $+8$:

+8		Sum	
−1	−8	−9	
−2	−4	−6	This is it!

The solution is -2 and -4.

b) $x^2 + x - 12$ $b = +1$ and $c = -12$ product # $= -12$ and sum # $= +1$

The product # is negative, and the sum # is positive. So, the factor with the larger numerical value is positive and the other factor is negative.

Factor pairs of -12:

−12		Sum	
−1	+12	+11	
−2	+6	+4	
−3	+4	+1	This is it!

The solution is -3 and $+4$.

c) $w^2 + 5w + 14$ $b = +5$ and $c = +14$ product # $= +14$ and sum # $= +5$

The product # is positive, and the sum # is positive. So, both factors are positive.

Factor pairs of $+14$:

+14		Sum
+1	+14	+15
+2	+7	+9

Neither factor pair adds to the value of b, $+5$.

There is no solution.

You Try It 4 **Given the trinomial, find the solution to its associated Factor Game. Use Example 5 as a guide.**

a) $x^2 + 9x + 18$ **b)** $p^2 - 10p + 21$ **c)** $m^2 - 3m - 40$ **d)** $y^2 - y + 20$

Factoring Trinomials

The great thing about factoring trinomials using the Factor Game is that it is foolproof. If you play the Factor Game correctly, you'll be able to factor any trinomial of the form $x^2 + bx + c$ that isn't prime. The Factor Game will also help you identify trinomials that *are* prime.

For example, factor the trinomial $x^2 + 8x + 15$ using the Factor Game.

First, prepare the binomial factors by setting up the parentheses and the first terms (the x's): $(x \quad)(x \quad)$

Second, identify the product and sum numbers and play the Factor Game. The product number is c, and the sum number is b: product # $= +15$, sum # $= +8$ → $\underline{+5 \text{ and } +3}$

Third, the constant terms are the factors of the solution to the Factor Game: $\mathbf{(x + 5)(x + 3)}$

Fourth, check this result by multiplying the binomials together:
$= x^2 + 3x + 5x + 15$
$= x^2 + 8x + 15$ ✓ It checks!

Example 6

Factor each trinomial using the Factor Game. If the trinomial is not factorable, write *prime*.

a) $x^2 - 7x + 10$ **b)** $m^2 - m - 20$ **c)** $y^2 + 5y + 30$

Procedure **First**, prepare the binomial factors by setting up the parentheses and the variables.
Second, identify the product and sum numbers and play the Factor Game.
Third, place the factor pair solution as the constant terms in the binomials.
Fourth, check this result by multiplying the binomials together.

Answer **a)** $x^2 - 7x + 10$ product # $= +10$, sum # $= -7$ → $\underline{-5 \text{ and } -2}$ *You show the work:*

$= (x \quad)(x \quad)$
$= \mathbf{(x - 5)(x - 2)}$

Check $(x - 5)(x - 2)$
$= x^2 - 2x - 5x + 10$
$= x^2 - 7x + 10$ ✓ It checks!

b) We don't see a coefficient of the middle term; so, it must be -1.

$m^2 - m - 20$ product # $= -20$, sum # $= -1$ → $\underline{+4 \text{ and } -5}$ *You show the work:*

$= (m \quad)(m \quad)$
$= \mathbf{(m + 4)(m - 5)}$

Check $(m + 4)(m - 5)$
$= m^2 - 5m + 4m - 20$
$= m^2 - m - 20$ ✓ It checks!

c) $y^2 + 5y + 30$ product # $= +30$, sum # $= +5$ → No solution! *You show the work:*

$= (y \quad)(y \quad)$
prime

Because the Factor Game has no solution (no factor pair of 30 will add to get 5), the trinomial is prime and cannot be factored.

Caution The challenge with factoring trinomials is that it is easy to make a mistake in one of the constants—in either the sign of the constant (positive or negative) or its numerical value. That is why the check is so important.

Think About It 2

In Example 6a, $x^2 - 7x + 10$ factored as $(x - 5)(x - 2)$. Would it be okay to write the factoring as $(x - 2)(x - 5)$ instead? Explain your answer.

You Try It 5 Factor each trinomial. If a trinomial is not factorable, write *prime*. Use Example 6 as a guide.

a) $x^2 + 12x + 20$

b) $y^2 + 11y - 12$

c) $q^2 - 13q + 40$

d) $x^2 - 3x + 16$

e) $n^2 + 6n - 40$

f) $w^2 - 5w - 24$

Answers: You Try It and Think About It

You Try It: **1. a)** 6 and 4 **b)** 6 and 6 **c)** No solution **2. a)** -5 and -5 **b)** -1 and -15 **c)** -4 and -10 **3. a)** -4 and $+7$ **b)** $+4$ and -9 **c)** No solution **4. a)** Product # = $+18$; sum # = $+9$. The solution is $+3$ and $+6$. **b)** Product # = $+21$; sum # = -10. The solution is -3 and -7. **c)** Product # = -40; sum # = -3. The solution is $+5$ and -8. **d)** Product # = $+20$; sum # = -1. There is no solution. **5. a)** $(x + 10)(x + 2)$ **b)** $(y + 12)(y - 1)$ **c)** $(q - 5)(q - 8)$ **d)** prime **e)** $(n + 10)(n - 4)$ **f)** $(w + 3)(w - 8)$

Think About It: **1.** -3 and -10 is not the solution because the product of these factors is $+30$ and the product number in Example 4b is -30. **2.** Yes, the Commutative Property allows us to write the two factors in any order.

A.5 Exercises

FOR EXTRA HELP
Student Resources on DVD-ROM
Includes
➤ Student's Solutions Manual
➤ Video Lectures
➤ Chapter Test Prep Video
MyMathLab
MathXL

Find the solution of the Factor Game with the given product and sum numbers.

1. product # = +10 and sum # = +7
5 and 2

2. product # = +8 and sum # = +9
8 and 1

3. product # = +30 and sum # = +13
3 and 10

4. product # = +50 and sum # = +15
5 and 10

5. product # = +30 and sum # = +7
No solution

6. product # = +24 and sum # = +4
No solution

7. product # = +16 and sum # = −10
−8 and −2

8. product # = +36 and sum # = −13
−4 and −9

9. product # = +18 and sum # = −19
−1 and −18

10. product # = +24 and sum # = −25
−24 and −1

11. product # = +48 and sum # = −16
−12 and −4

12. product # = +42 and sum # = −23
−2 and −21

13. product # = −20 and sum # = +8
+10 and −2

14. product # = −28 and sum # = +12
+14 and −2

15. product # = −30 and sum # = +1
+6 and −5

16. product # = −42 and sum # = +1
+7 and −6

17. product # = −20 and sum # = −9
No solution

18. product # = −25 and sum # = −10
No solution

19. product # = −60 and sum # = −4
−10 and +6

20. product # = −24 and sum # = −5
−8 and +3

21. product # = −36 and sum # = 0
−6 and +6

23. product # = −49 and sum # = 0
−7 and +7

23. product # = +16 and sum # = 0
No solution

24. product # = +25 and sum # = 0
No solution

Factor each trinomial using the Factor Game. If the trinomial is not factorable, write *prime*.

25. $p^2 + 13p + 30$
$(p + 3)(p + 10)$

26. $x^2 + 9x + 20$
$(x + 5)(x + 4)$

27. $v^2 + 7v + 12$
$(v + 3)(v + 4)$

28. $x^2 + 15x + 36$
$(x + 3)(x + 12)$

29. $v^2 - 11v + 30$
$(v - 6)(v - 5)$

30. $w^2 - 9w + 18$
$(w - 6)(w - 3)$

31. $x^2 - 8x + 16$
$(x - 4)(x - 4)$

32. $q^2 - 13q + 36$
$(q - 9)(q - 4)$

33. $v^2 - 12v + 36$
$(v - 6)(v - 6)$

34. $h^2 - 10h + 25$
$(h - 5)(h - 5)$

35. $y^2 - 12y + 30$
prime

36. $q^2 - 8q + 24$
prime

37. $d^2 - 10d + 9$
$(d - 9)(d - 1)$

38. $m^2 - 7m + 6$
$(m - 6)(m - 1)$

39. $r^2 - 23r + 60$
$(r - 20)(r - 3)$

40. $w^2 - 18w + 32$
$(w - 16)(w - 2)$

41. $x^2 - 20x + 36$
$(x - 2)(x - 18)$

42. $n^2 - 19n + 48$
$(n - 3)(n - 16)$

43. $y^2 + 4y - 12$
$(y + 6)(y - 2)$

44. $k^2 + 3k - 10$
$(k + 5)(k - 2)$

45. $q^2 + 3q - 18$
$(q + 6)(q - 3)$

46. $h^2 + 7h - 30$
$(h + 10)(h - 3)$

47. $x^2 + 3x - 24$
prime

48. $r^2 + 6r - 18$
prime

49. $n^2 + 4n - 32$
$(n + 8)(n - 4)$

50. $x^2 + 10x - 24$
$(x + 12)(x - 2)$

51. $p^2 - 9p - 36$
$(p - 12)(p + 3)$

52. $x^2 - 4x - 45$
$(x - 9)(x + 5)$

53. $w^2 - w - 30$
$(w - 6)(w + 5)$

54. $y^2 - y - 56$
$(y - 8)(y + 7)$

55. $v^2 - 12v - 32$
prime

56. $y^2 - 13y - 40$
prime

57. $m^2 - 3m - 70$
$(m - 10)(m + 7)$

58. $q^2 - 13q - 30$
$(q - 15)(q + 2)$

59. $y^2 - 7y - 60$
$(y - 12)(y + 5)$

60. $x^2 - 11x - 60$
$(x - 15)(x + 4)$

Answers

Basic Skills Pretest

1. 927 **2.** 422 **3.** 7,000 **4.** 2,241 **5.** 925 **6.** 1,930,498 **7.** 40 **8.** 36 **9.** 54 **10.** 560 **11.** 342 **12.** 2,745 **13.** 7 **14.** 4 **15.** 9 **16.** 14 **17.** 48 **18.** 27 **19.** 12 r 6 **20.** 85 r 2 **21.** 81 **22.** 48 r 6 **23.** 940 **24.** 860 **25.** 800 **26.** 4,000 **27.** 40,000 **28.** 540,000 **29.** 280,000 **30.** 100,000 **31.** Tamari earned \$1,625 last month. **32.** The rental price for each table was \$18. **33.** Gilberto will pay \$5,100 rent for one year. **34.** The price of the couch is \$246 more than the price of the recliner. **35.** B **36.** C **37.** E **38.** A **39.** D **40.** G **41.** H **42.** F

Chapter 1

Section 1.1 **1.** Answers will vary. **3.** The parentheses indicate a single value, and that one value should be known before the whole expression is evaluated.

Period	Place	Value
5. ones	ten	70 (seventy)
7. thousands	hundred	700,000 (seven hundred thousand)

9. 400 + 80 + 6 **11.** 200,000 + 3,000 + 50 + 8 **13.** four hundred ninety-eight **15.** five hundred seven thousand, ninety-three **17.** 518 **19.** 280,034 **21.** 1,000,426 **23.** $8 \div 4 = 2$ **25.** $2 + 5 = 7$ **27.** difference **29.** product **31.** 8 **33.** 6 **35.** 4 **37.** 7 **39.** 1 **41.** 64 **43.** Associative Property of Addition **45.** Identity of Addition **47.** Commutative Property of Addition **49.** 70 **51.** 680 **53.** 1,940 **55.** 200 **57.** 600 **59.** 1,700 **61.** 14,400 **63.** 7,000 **65.** 3,000 **67.** 36,000 **69.** 159,000 **71.** 290,000 **73.** 30,000 **75.** 100,000 **77.** 150,000 **79.** 900,000 **81.** 600,000 **83.** 1,000,000 **85.** 800,000 **87.** 5,000,000 **89.** 4,000,000 **91.** 4,000,000 **93.** 9,000,000 **95.** 30,000,000 **97.** The Rialto Unified School District's budget revenues were approximately \$193,000,000. **99.** The total number of gallons of gas consumed was approximately 5,000,000,000. **101.** Approximately 6,200,000 students were enrolled in grades K–12 in California. **103.** There will be 41 different rounded numbers.

Section 1.2 **1.** If we add the length and width together, we can double that value to find the perimeter. **3.** 59 **5.** 557 **7.** 887 **9.** 6,875 **11.** 62,925 **13.** 757,597 **15.** 104,799 **17.** 1,000,000 **19.** 104 **21.** 817 **23.** 17,011 **25.** 143,900 **27.** 252 **29.** 102 **31.** 3,312 **33.** 25 **35.** 5 **37.** 54 **39.** 57 **41.** 93 **43.** 73 **45.** 293 **47.** 5,425 **49.** 24,835 **51.** 6,731 **53.** 4,293 **55.** 133,343 **57.** 193,422 **59.** Dionne contributed \$2,133 to the two charities. **61.** Debbie's monthly payment was \$79 less in 2008 than in 2007. **63.** The four agents have 80 combined years of experience. **65.** The total cost of the car is \$17,933. **67. a)** The total attendance was 105,210 for the three-game series. **b)** There were 12,777 more fans in attendance on Saturday than on Sunday. **69. a)** The perimeter is 595 meters. **b)** The perimeter is 190 inches. **c)** The perimeter is 82 feet. **71. a)** The sum is 43,639. Rounded to the nearest thousand, it is 44,000. **b)** The rounded numbers are 1,000, 12,000, 0, and 30,000. The sum of the rounded numbers is 43,000. **c)** Answers may vary.

Section 1.3 **1.** Yes. The multiples of 3 are all of the products involving 3 and some other whole number, and 0 is a whole number. So, $0 \cdot 3 = 0$ is a multiple of 3. **3.** No. The perimeter is 16 inches, but the area is 16 *square* inches. Those are not the same. **5.** 20 **7.** 28 **9.** 36 **11.** 54 **13.** 27 **15.** 35 **17.** 64 **19.** 49 **21.** 240 **23.** 420 **25.** 300 **27.** 2,400 **29.** 6,300 **31.** 2,700 **33.** 36,000 **35.** 480,000 **37.** 342 **39.** 1,849 **41.** 9,810 **43.** 88,825 **45.** 17,136 **47.** 32,768 **49.** Commutative Property of Multiplication **51.** Associative Property of Multiplication **53.** 168 cars can fit in the parking lot. **55.** Ignacio's truck will go 364 miles before it runs out of gas. **57.** The jet will travel 7,872 miles in 16 hours. **59.** Diane earned \$594 that first week. **61.** The total income for PTA's field trip program is \$5,508. **63.** The area of the soccer field is 1,680 square yards. **65.** The approximate area of Wyoming is 97,200 square miles. **67.** There are 92,160 cans of soup stored in the warehouse.

Section 1.4 **1.** The results are both identities. 0 is the identity for addition, and 1 is the identity for multiplication. **3.** No. 0 divided by itself, $0 \div 0$, is undefined. **5.** 5 **7.** 0 **9.** 4 **11.** 6 **13.** 9 **15.** 7 **17.** 10 **19.** 4 **21.** 15 **23.** 29 **25.** 37 **27.** 13 **29.** 12 r 7 **31.** 33 r 1 **33.** 118 r 7 **35.** 304 r 2 **37.** 6,009 r 2 **39.** 3,004 **41.** 13,050 **43.** 23 **45.** 73 **47.** 1,002 r 6 **49.** 42 **51.** 84 **53.** 2,003 **55.** 1,025 **57.** There are 14 campers in each group. **59.** Each person paid \$23 to attend the concert. **61.** 23 boxcars are needed to ship the motorcycles. **63.** She will need 57 boxes for all of the booklets. **65.** 1,111 **67.** 4,321

Section 1.5 **1.** Yes, we can apply addition before multiplication when the addition is within a pair of grouping symbols. **3.** $\sqrt{20}$ is between 4 and 5 because $\sqrt{20}$ is between $\sqrt{16}$ and $\sqrt{25}$. **5.** $6 \cdot 6 \cdot 6 = 216$ **7.** $15 \cdot 15 = 225$ **9.** 12 **11.** $5 \cdot 5 \cdot 5 \cdot 5 = 625$ **13.** $10 \cdot 10 \cdot 10 = 1{,}000$ **15.** 10^5 **17.** 10^4 **19.** $3 \cdot 10^2$ **21.** $48 \cdot 10^3$ **23.** $95 \cdot 10^5$ **25.** 7 **27.** 6 **29.** 9 **31.** 12 **33.** 7 **35.** 26 **37.** 10 **39.** 45 **41.** 72 **43.** 8 **45.** 45 **47.** 51 **49.** 93 **51.** 72 **53.** 8 **55.** 18 **57.** 10 **59.** 36 **61.** 45 **63.** 31 **65.** 11 **67.** 12 **69.** 8 **71.** 26 **73.** 32 **75.** 90 **77.** 25 **79.** 8 **81. a)**

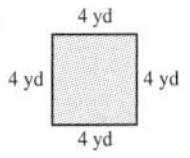

b) The perimeter is 16 yards. **c)** The area is 16 square yards.

83. a)

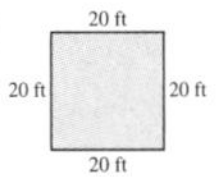

b) The perimeter is 80 feet. **c)** The area is 400 square feet. **85. a)** The length of each side is 6 feet.

b) The perimeter is 24 feet. **87. a)** The length of each side is 3 feet. **b)** The perimeter is 12 feet. **89.** Odd. The product of any two or more odd numbers is another odd number. **91.** Answers may vary. $4 \times 3 - 2 + 1$ **93.** Answers may vary. $4 \times (3 + 2) + 1$

Section 1.6 **1.** Not necessarily. The number might be prime, but it might have larger prime factors. For example, $169 = 13 \times 13$; so, 169 is not prime. **3.** In that case, the prime factorization of 7 could be $7 \cdot 1$, $7 \cdot 1 \cdot 1 \cdot 1$, and so on. In other words, we can have as many factors of 1 as we want. (This is why 1 is not a prime number.) **5.** 4, 8, 12, 16, 20, 24, 28, 32 **7.** 7, 14, 21, 28, 35, 42, 49, 56 **9.**

32	
1	32
2	16
4	8

11.

28	
1	28
2	14
4	7

13. 2 **15.** None **17.** 2 **19.** 5 **21.** 3 and 5 **23.** 2 and 3 **25.** 3 **27.** 2, 3, and 5 **29.** No **31.** Yes **33.** Prime: 7, 23, 41 Composite: 9, 8, 40, 15, 33, 32, 12, 51, 50 Neither: 0 **35.** $2 \cdot 3 \cdot 3$ or $2 \cdot 3^2$ **37.** $3 \cdot 3 \cdot 7$ or $3^2 \cdot 7$ **39.** $3 \cdot 5 \cdot 7$ **41.** $2 \cdot 2 \cdot 2 \cdot 2 \cdot 31$ or $2^4 \cdot 31$ **43.** $2 \cdot 2 \cdot 2 \cdot 2 \cdot 3 \cdot 3 \cdot 5$ or $2^4 \cdot 3^2 \cdot 5$ **45.** $3 \cdot 3 \cdot 3 \cdot 5 \cdot 7$ or $3^3 \cdot 5 \cdot 7$ **47.** $2 \cdot 2 \cdot 5$ or $2^2 \cdot 5$ **49.** $2 \cdot 2 \cdot 13$ or $2^2 \cdot 13$ **51.** $2 \cdot 2 \cdot 19$ or $2^2 \cdot 19$ **53.** $2 \cdot 7 \cdot 7$ or $2 \cdot 7^2$ **55.** $2 \cdot 2 \cdot 31$ or $2^2 \cdot 31$ **57.** $2 \cdot 2 \cdot 2 \cdot 5 \cdot 5$ or $2^3 \cdot 5^2$ **59.** 6 and 4 **61.** 4 and 3 **63.** Composite: $91 = 7 \times 13$ **65.** Prime

Section 1.7 **1.** Something that is *constant* is never changing. **3.** When we apply the Subtraction Property of Equality, we always attain the number 0. **5.** $45 = 55$ No **7.** $80 = 80$ Yes **9.** $n = 13$ **11.** $p = 35$ **13.** $q = 198$ **15.** $v = 123$ **17.** $x = 6{,}016$ **19.** $x = 119$ **21.** $y = 1{,}101$ **23.** $c = 19$ **25.** $n = 14$ **27.** $a = 13$ **29.** $w = 140$ **31.** $y = 1{,}480$ **33.** $w = 178$ **35.** $m = 36$ **37.** $x = 57$ **39.** $x = 14$ **41.** $x = 32$ **43.** $b = 47$ **45.** Yes, 7 is the solution. **47.** $w = 11$

Section 1.8 **1.** The legend helps define what number the variable represents. **3.** Answers will vary. **5.** Let $m =$ how much more money Nate needs. $73 + m = 162$ Nate needs \$89 more dollars. **7.** Let $m =$ how much money Adam will have left over. $380 + 140 + 115 + m = 900$ Adam will have \$265 left to spend on vacation. **9.** Let $x =$ how much more money the city must collect. $19{,}629 + x = 21{,}000$ The city must collect \$1,371 more. **11.** Let $c =$ the number of copies left to produce. $1{,}758 + 1{,}365 + 1{,}259 + c = 5{,}000$ There are 618 copies left to produce. **13.** Let $a =$ the amount Ajay pays each month. $12 \cdot a = 252$ Ajay pays \$21 each month for dues. **15. a)** Let $m =$ the number of minutes to prepare 1 box for shipping. $18 \cdot m = 72$ Carnell takes, on average, 4 minutes to prepare 1 box. **b)** Let $j =$ the number of jars of jam in all of the boxes. $24 \cdot 18 = j$ There are 432 jars of jam in the 18 boxes. **17. a)** Let $x =$ the total weight of the cargo. $25 \cdot 42 = x$ Mansour's plane will be carrying 1,050 pounds. **b)** Let $x =$ the average weight of each box. $24 \cdot x = 864$ The average weight of each box is 36 pounds. **19.** Let $f =$ the number of flyers printed on each machine. $5 \cdot f = 3{,}200$ Each machine will print, on average, 640 flyers. **21.** Let $x =$ the measure of the shortest side. $42 + 31 + x = 90$ The shortest side measures 17 feet. **23.** Let $x =$ the length of the top side. $7 + 10 + 10 + 7 + x = 48$ The length of the top side is 14 inches. **25.** Let $a =$ the area of the dance floor. $30 \cdot 24 = a$ The area of the dance floor is 720 square feet. **27.** Let $x =$ the length of the rectangle. $40 \cdot x = 4{,}840$ The length of the rectangle is 121 yards. **29.** Let $c =$ the number of candy canes. $7 \cdot c = 945$ He can make 135 candy canes. **31.** The perimeter of the rectangle is 68 inches.

Chapter 1 Review **1.** number **2.** numeral **3.** digits **4.** operations **5.** evaluate **6.** identity **7.** approximation **8.** perimeter **9.** addition **10.** factors **11.** multiple **12.** Area **13.** quotient **14.** remainder **15.** composite **16.** variable **17.** constant **18.** solution **19.** coefficient **20.** legend **21.** $700 + 20 + 4$ **22.** $6{,}000 + 800 + 7$ **23.** four hundred eight **24.** nine thousand, fifty-one **25.** two hundred six thousand, five **26.** five million, four hundred seventy thousand **27.** 107 **28.** 2,005 **29.** 508,041 **30.** 1,000,652 **31.** Identity of Multiplication **32.** Associative Property of Addition **33.** Commutative Property of Addition **34.** Identity of Addition **35.** 640 **36.** 300 **37.** 1,450 **38.** 3,000 **39.** 600 **40.** 30,300 **41.** 126,500 **42.** 4,900 **43.** 31,000 **44.** 54,000 **45.** 250,000 **46.** 1,000 **47.** The total number of full-time airline employees was approximately 510,000. **48.** In 2004, the U.S. population was approximately 294,500,000. **49.** 530 **50.** 550 **51.** 2,032 **52.** 10,012 **53.** 110 **54.** 2,865 **55.** Brian burned 682 calories on the exercise bike. **56.** Kaira received a total of 256 points on the three tests. **57.** 564 cm **58.** 98 feet **59.** 141 **60.** 74 **61.** 647 **62.** 15,584 **63.** 178 **64.** 903,795 **65.** There were, on average, 10,895 more fans in attendance at New Orleans home games than Oakland home games. **66.** Nevada is 13,845 square miles larger than Michigan. **67.** Mt. Shasta received 48 more inches of snow in March than in November. **68.** Mt. Shasta received a total of 249 inches of snow. **69.** 35 **70.** 28 **71.** 40 **72.** 54 **73.** 56 **74.** 81 **75.** 210 **76.** 2,400 **77.** 7,200 **78.** 1,600 **79.** 63,000 **80.** 10,000. **81.** 376 **82.** 1,204 **83.** 2,088 **84.** 58,752 **85.** Associative Property of Multiplication **86.** Zero Product Property **87.** Distributive Property of Multiplication over Addition **88.** Commutative Property of Multiplication **89.** Marley drove 874 miles to and from work in March. **90.** The area of Colin's basement floor is 646 square feet. **91.** 9 **92.** 7 **93.** 0 **94.** 8 **95.** 8 **96.** 7 **97.** 8 **98.** 9 **99.** 6 r 6 **100.** 6 r 1 **101.** 23 **102.** 42 **103.** 171 r 1 **104.** 314 r 6 **105.** 26 **106.** 64 **107.** 140 **108.** 73 r 23 **109.** 72 r 2 **110.** 108 r 45 **111.** Each member's contribution was \$265. **112.** 13 classrooms are needed. **113.** $1 \cdot 1 \cdot 1 \cdot 1 \cdot 1 \cdot 1 \cdot 1$, 1 **114.** $2 \cdot 2 \cdot 2 \cdot 2$, 16 **115.** $3 \cdot 3 \cdot 3 \cdot 3 \cdot 3$, 243 **116.** $4 \cdot 4 \cdot 4$, 64 **117.** 16 **118.** $17 \cdot 17$, 289 **119.** $20 \cdot 20 \cdot 20$, 8,000 **120.** $10 \cdot 10 \cdot 10 \cdot 10 \cdot 10 \cdot 10 \cdot 10$, 10,000,000 **121.** 10^3 **122.** 10^7 **123.** 10^5 **124.** 10^1 **125.** $7 \cdot 10^1$ **126.** $84 \cdot 10^2$ **127.** $3 \cdot 10^5$ **128.** $12 \cdot 10^5$ **129.** 6 **130.** 2 **131.** 3 **132.** 10 **133.** 7 **134.** 16 **135.** 6 **136.** 21 **137.** 9 **138.** 30 **139.** 30 **140.** 9 **141.** 16 **142.** 16 **143.** 19 **144.** 5 **145.** 3, 6, 9, 12, 15 **146.** 6, 12, 18, 24, 30 **147.** 11, 22, 33, 44, 55 **148.** 12, 24, 36, 48, 60

149.

1	18
2	9
3	6

150.

1	36
2	18
3	12
4	9
6	6

151.

1	45
3	15
5	9

152.

1	60
2	30
3	20
4	15
5	12

153. Prime: 17, 29, 11; 15, 81, 45; 0 **154.** 2, 61, 43, 31; 70, 62, 57; 1 **155.** 3 **156.** 2 and 5 **157.** 5 **158.** 2 and 3 **159.** Yes **160.** Yes **161.** No **162.** Yes **163.** $2 \cdot 2 \cdot 3 \cdot 3 \cdot 3, 2^2 \cdot 3^3$ **164.** $2 \cdot 3 \cdot 3 \cdot 3 \cdot 3, 2 \cdot 3^4$ **165.** $2 \cdot 3 \cdot 5 \cdot 7$ **166.** $5 \cdot 43$ **167.** 12 is not the solution. **168.** 12 is the solution. **169.** 12 is the solution. **170.** 12 is not the solution. **171.** $c = 8$ **172.** $x = 33$ **173.** $m = 4{,}390$ **174.** $p = 14$ **175.** $y = 98$ **176.** $w = 36$ **177.** Carlotta's sales need to be $3,160 on Monday. **178.** Rhani will pay her parents $125 each month. **179.** On average, Antonio earned $19 in tips from each table. **180.** The length of the carpet is 27 yards.

Chapter 1 Test **1.** 700 **2.** 10,000 **3.** 580,000 **4.** 8,640 **5.** 4,505 **6.** 558 **7.** 28 **8.** Identity of Multiplication **9.** Distributive Property of Multiplication over Addition **10.** Associative Property of Addition **11.** Commutative Property of Addition **12.** $5 \cdot 5 \cdot 5$, 125 **13.** $20 \cdot 20$, 400 **14.** $74 \cdot 10^4$ **15.** $9 \cdot 10^2$ **16.** 4 **17.** 9 **18.** 27 **19.** 17 **20.** 50 **21.** 41, 19, 2; 77, 38; 1 **22.** 3 and 5 **23.** 2 and 3 **24.** None **25.** 2 **26.** $2 \cdot 2 \cdot 3 \cdot 7, 2^2 \cdot 3 \cdot 7$ **27.** $2 \cdot 2 \cdot 2 \cdot 2 \cdot 5, 2^4 \cdot 5$ **28.** $2 \cdot 2 \cdot 3 \cdot 3 \cdot 3 \cdot 5, 2^2 \cdot 3^3 \cdot 5$ **29.** $x = 15$ **30.** $y = 49$ **31.** $w = 19$ **32.** Alfre needs to collect 126 more signatures. **33.** The length of the sandbox is 25 yards. **34.** Jerry spent $168 on his team's tickets. **35.** The average amount of money that each member raised was $215.

Chapter 2

Preparation Exercises **1.** 111 **2.** 3,050 **3.** 15 **4.** 229 **5.** 676 **6.** 945 **7.** 39 **8.** 264 **9.** 8 **10.** 25 **11.** 7 **12.** 9 **13.** 8 **14.** 8 **15.** 39 **16.** 13

Section 2.1 **1.** -0 is the same as 0. It is in the same position as 0 on the number line. **3.** No, the absolute value of a positive number is not the opposite of the number. **5.** 10 **7.** 2 **9.** 14 **11.** 22 **13.** 28 **15.** 3 **17.** 25 **19.** 10 **21.** $x - 15$ **23.** $20 + x$ **25.** $x \div 18$ **27.** x^2 **29.** $x \cdot 9$ or $9x$ **31.** $|x|$ **33.** -8 **35.** 18 **37.** $<$ **39.** $<$ **41.** $<$ **43.** $<$ **45.** 15 **47.** 21 **49.** 0 **51.** 65 **53.** 6, to the right **55.** 12, to the left **57.** 0, it has no direction **59.** 10, to the right **61.** No. Replacing x with 2 would make the expression $4 \div 0$, and that is undefined. **63.** $8x(x - 9)$

Section 2.2 **1.** Because the result is 0, neither sign should be given to the result. **3.** Yes, but only if a and b are opposites, such as 2 and -2. **5.** 2 **7.** 8 **9.** -3 **11.** -8 **13.** -13 **15.** -9 **17.** 12 **19.** 19 **21.** 5 **23.** 5 **25.** -3 **27.** -5 **29.** 0 **31.** 0 **33.** 0 **35.** 0 **37.** 1 **39.** 7 **41.** -1 **43.** -5 **45.** 7 **47.** -5 **49.** -2 **51.** 0 **53.** 24 **55.** -18 **57.** 9 **59.** 4 **61.** 0 **63.** -31 **65.** -1 **67.** 16 **69.** 7 **71.** -40 **73.** -7 **75.** -86 **77.** -23 **79.** 0 **81.** 5 **83.** -4 **85.** 3 **87.** -5 **89.** -8 **91.** 0 **93.** -5 **95.** 5 **97.** 0 **99.** -6 **101.** The perimeter is 80 inches.

Section 2.3 **1.** Yes, when the first negative number is to the right of the second negative number. For example, $-2 - (-8) = -2 + 8 = +6$ **3.** 0. Examples may vary. One example is $-3 - (-3)$ **5.** 6 **7.** -6 **9.** -8 **11.** -8 **13.** 4 **15.** -7 **17.** 7 **19.** 8 **21.** 10 **23.** 2 **25.** 2 **27.** -10 **29.** -90 **31.** -54 **33.** -17 **35.** 0 **37.** 70 **39.** 91 **41.** 4 **43.** -1 **45.** -17 **47.** -4 **49.** 4 **51.** -35 **53.** 25 **55.** 2 **57.** -8 **59.** -8 **61.** -14 **63.** 3 **65.** 15 **67.** 0 **69.** 4 **71.** -6 **73.** 13 **75.** -8 **77.** $-375 + 2{,}825 = 2{,}450$ Mike's new account balance was $2,450. **79.** $58 - 72 + 33 - 39 - 29 + 115 = 66$; Joni's account balance on Saturday was $66. **81.** $-16 + 7 = -9$ The temperature was $-9°$ at 10 AM. **83.** $-2 - 16 = -18$; The new temperature was $-18°$ F. **85.** $5{,}280 - 1{,}416 = 3{,}864$; The difference in altitude between the two mountains is 3,864 feet. **87.** $4{,}638 - (-784) = 4{,}638 + (+784) = 5{,}422$; The difference in altitude between the mountain and the ocean floor is 5,422 feet. **89.** -15; answers may vary. **91. a)** 7 **b)** 7 units **c)** -7 **d)** 7 **e)** Answers may vary.

Section 2.4 **1.** It will be negative because there is an odd number of negative factors. **3.** The dividend and the divisor must have the same sign, both positive or both negative. **5.** 20 **7.** -63 **9.** 60 **11.** 16 **13.** 0 **15.** -28 **17.** -100 **19.** -36 **21.** 24 **23.** -24 **25.** -9 **27.** -60 **29.** -170 **31.** 112 **33.** 120 **35.** 60 **37.** -96 **39.** -54 **41.** 120 **43.** 0 **45.** 60 **47.** -5 **49.** 7 **51.** -3 **53.** 2 **55.** Undefined **57.** -20 **59.** 0 **61.** -7 **63.** 9 **65.** -9 **67.** -6 **69.** -9 **71.** 2 **73.** -5 **75.** 14 **77.** -1 **79.** -20.

Section 2.5 **1.** $\sqrt{0} = 0$ because $0 \cdot 0 = 0$ **3.** When one number is positive and the other number is negative. **5.** 8 **7.** 9 **9.** 5 **11.** 7 **13.** 1 **15.** -4 **17.** -64 **19.** $-1{,}000$ **21.** 36 **23.** 81 **25.** -8 **27.** -32 **29.** 7 **31.** -20 **33.** 18 **35.** 18 **37.** 7 **39.** 8 **41.** -26 **43.** 16 **45.** 16 **47.** 12 **49.** 23 **51.** -4 **53.** -44 **55.** -5 **57.** -50 **59.** 4 **61.** 7 **63.** -13 **65.** -2 **67.** 1 **69.** 4 **71.** 14 **73.** 11 **75.** -4 **77.** 12 **79.** -2 **81.** -2 **83.** -6 **85.** 7 **87.** -2 **89.** 3 **91.** -4 **93.** 16 **95.** -1.

Section 2.6 **1.** They are both variations of the same formula. The order of operations groups the quantity in the numerator $(F - 32)$, 5 multiplies this quantity, and it is divided by 9. **3.** 41° F **5.** 10° C **7.** $z = 3$ **9.** $A = 29$ **11.** $a = 8$ **13.** $A = 20$ **15.** $P = 42$ **17.** $r = 12$ **19.** $A = 18$ **21.** $W = 8$ **23.** $C = 4$ **25.** $P = 30$ **27.** $I = 12$ **29.** $m = -2$ **31.** $m = -4$ **33.** $R = 8$ **35.** Mai's average rate of speed was 67 miles per hour. **37.** Bertie's average rate of speed was 59 miles per hour. **39.** It should take Hank 9 hours to complete the trip. **41.** Padam's average rate of speed will need to be 64 miles per hour. **43. a)** 108° F **b)** 60° C

Section 2.7 **1.** $9x^3$ and $9x^2$ aren't like terms because the exponents on the variables are different. **3.** Yes. If x is negative, then $5x$ is negative. **5.** 4, m, 2 **7.** -3, y, 1 **9.** 9, None, None **11.** 1, y, 3 **13.** -1, d, 5 **15.** -1, n, 1 **17. a)** $3x^4$ **b)** $-5x^3$ **c)** 1 **d)** -9 **e)** 4 **f)** 1 **19. a)** $-8y$ **b)** y^2 **c)** -3 **d)** 1 **e)** 1 **f)** 4 **21.** $7x$ **23.** $13x^2$ **25.** $-a$ **27.** Not like terms **29.** $2y^2$ **31.** $-3m$ **33.** 0 **35.** $8p^2$ **37.** $-6k$ **39.** $-v^2$ **41.** Not like terms **43.** 0 **45.** $10y$ **47.** $-5x$ **49.** $6c$ **51.** $-7x$ **53.** $-8y^2$ **55.** $-28p$ **57.** $11x + 3$ **59.** $-x - 2$ **61.** $4a - 5$ **63.** $5w^2 + 5w$ **65.** $-3x - 1$ **67.** $-4b - 4c$ **69.** $16x + 54$ yards represents the perimeter. **71.** If $x = 1$, the length is 17 yards and the width is 18 yards. This means that the width is longer than the length.

Section 2.8 **1.** Yes. Answers will vary. **3.** Answers will vary. **5.** $20x$ **7.** $4y$ **9.** $-28c^2$ **11.** $-48a^4$ **13.** $8w + 10$ **15.** $20y + 32$ **17.** $7x^2 + 14$ **19.** $2x - 16$ **21.** $14p - 21$ **23.** $32x^2 - 8$ **25.** $-8c + 6$ **27.** $-15x^2 - 5$ **29.** $-10y - 35$ **31.** $-36x + 12$ **33.** $-4 + 8b^2$ **35.** $12x - 30$ **37.** $-4y - 4$ **39.** $-x - 3$ **41.** $48w^2 + 8$ **43.** $-m - 8$ **45.** $-7p + 7$ **47.** $9w + 12$ **49.** $c + 6$ **51.** $w^2 - 12$ **53.** $-h + 10$ **55.** $-y - 4$ **57.** $5k + 45$ **59.** $2p^2 - 2$ **61.** $-4m + 32$ **63.** $-7y - 14$ **65.** $-10q^2 - 20$ **67.** $-2n + 10$ **69.** $8v^2 - 8$ **71.** $11x + 33$ **73.** x^5 **75.** $64x^2$ **77.** $15x^2 + 12x$ **79.** $16x^3 + 12x^2$ **81.** Yes, as long as the value of x is negative.

Chapter 2 Review Exercises **1.** algebraic expression **2.** replacement value **3.** negative **4.** integers **5.** negative **6.** coefficient **7.** False **8.** False **9.** True **10.** True **11.** 24 **12.** 28 **13.** 30 **14.** 33 **15.** $8 - x$ **16.** $x + 14$ **17.** $>$ **18.** $<$ **19.** $<$ **20.** $>$ **21.** 9 **22.** 11 **23.** 0 **24.** 21 **25.** -6 **26.** -4 **27.** -13 **28.** -5 **29.** 0 **30.** -20 **31.** 16 **32.** -31 **33.** 71 **34.** -52 **35.** 74 **36.** -27 **37.** -16 **38.** -3 **39.** -2 **40.** -4 **41.** -7 **42.** -12 **43.** 9 **44.** 4 **45.** 30 **46.** -13 **47.** -8 **48.** 55 **49.** 0 **50.** 91 **51.** 2 **52.** -16 **53.** $97 - (-64) = 97 + (+64) = 161$; The difference in altitude between the seaside cliff and the ocean floor is 161 feet. **54.** $-13 + 8 = -5$; The temperature at 6 AM was $-5°$. **55.** $10 - 17 = -7$; Adele's new checkbook balance is $-\$7$. **56.** $-26 - 15 = -41$; Adele's checkbook balance is now $-\$41$. **57.** 6 **58.** 99 **59.** -77 **60.** 40 **61.** 4 **62.** 0 **63.** 30 **64.** 40 **65.** -48 **66.** -28 **67.** -11 **68.** 6 **69.** 6 **70.** -5 **71.** 3 **72.** -2 **73.** -12 **74.** 81 **75.** -2 **76.** -8 **77.** -8 **78.** 5 **79.** 30 **80.** -6 **81.** 7 **82.** 17 **83.** 4 **84.** 2 **85.** 212° F **86.** 59° F **87.** 50° C **88.** 15° C **89.** $A = 84$ **90.** $W = 16$ **91.** $A = 25$ **92.** $z = 2$ **93.** $a = 5$ **94.** $C = 6$ **95.** Tracey's average rate of speed was 13 miles per hour. **96.** Timara can fly 870 miles. **97.** It will take him 9 hours to get there. **98.** Peetey's average rate of speed was 29 cm per minute. **99.** $2w$ **100.** $-8w^2$ **101.** -1 **102.** 4 **103.** 1 **104.** 3 **105.** $12x$ **106.** $-6y$ **107.** $-4w^2$ **108.** $6b^2$ **109.** x **110.** 0 **111.** Not like terms **112.** $-14x$ **113.** $-4x$ **114.** $-2y - 1$ **115.** $-7b - 5c$ **116.** $-5x + 4$ **117.** $-12x^2$ **118.** $81y$ **119.** $42p$ **120.** $-15a^3$ **121.** $9y + 18$ **122.** $-12y + 18$ **123.** $15h - 5$ **124.** $-2y - 2$ **125.** $24y - 21$ **126.** $-8p + 28$

Chapter Test 2 **1.** $15 - x$ **2.** $\sqrt{x}$ **3.** $<$ **4.** $>$ **5.** -8 **6.** 14 **7.** -2 **8.** -100 **9.** -10 **10.** -6 **11.** 15 **12.** -36 **13.** -16 **14.** -5 **15.** -7 **16.** Undefined **17.** 5 **18.** -7 **19.** -1 **20.** 16 **21.** -27 **22.** -2 **23.** 120 **24.** $-162 - (-458) = -162 + (+458) = 296$; The difference in altitude between the undersea hilltop and the ocean floor is 296 feet. **25.** $95 - (-8) = 95 + (+8) = 103$; The difference between the highest and lowest temperatures was 103°. **26.** -4 **27.** 1 **28.** 203° F **29.** 35° C **30.** $A = 25$ **31.** It will take him 7 hours to get there. **32.** $7p^2$ **33.** $-5a$ **34.** $-10x - 5y$ **35.** $32k^4$ **36.** $-10y$ **37.** $6w - 15$ **38.** $-14c + 21$.

Chapters 1 and 2 Cumulative Review **1.** 70,000 (seventy thousand) **2.** Five hundred thousand, twenty-six **3.** 500 **4.** 8,000 **5.** 8,000 **6.** 210,000 **7.** Commutative Property of Multiplication **8.** Identity of Addition **9.** Distributive Property of Multiplication over Addition **10.** Associative Property of Addition **11.** 3,036 **12.** 2,001 **13.** 875 **14.** 9,429 **15.** 112 cm **16.** 262 in. **17.** 450 square yards **18.** Gore received 254,921 more votes than Bush. **19.** 25 more people prefer Jack in the Box than McDonald's. **20.** 127 total people were surveyed. **21.** 56,000 **22.** 4,620 **23.** 47 r 3 **24.** 52 **25.** $9 \cdot 9 \cdot 9 = 729$ **26.** $5 \cdot 5 \cdot 5 \cdot 5 = 625$ **27.** 10^4 **28.** 10^9 **29.** $6 \cdot 10^4$ **30.** $52 \cdot 10^5$ **31.** 8 **32.** 11 **33.** 7 **34.** 147 **35.** 8 **36.** 34 **37.** 5, 31, 43; 18, 55; 1 **38.** 2 and 3 and 5 **39.** 2 and 3 **40.** None **41.** 3 and 5 **42.** 3 **43.** Yes **44.** Yes **45.** $2 \cdot 2 \cdot 2 \cdot 3 \cdot 7$; $2^3 \cdot 3 \cdot 7$ **46.** $2 \cdot 2 \cdot 3 \cdot 5 \cdot 5$; $2^2 \cdot 3 \cdot 5^2$ **47.** $x = 27$ **48.** $y = 35$ **49.** A total of 510 pages were photocopied. **50.** Each teacher received 12 markers. **51.** Ben and Adrian need to raise \$113. **52.** Each girl received 35 Red Vines. **53.** 3 **54.** 6 **55.** $x + 9$ **56.** $6 - x$ **57.** $24 \div x$ **58.** $8 \cdot x$ **59.** $\sqrt{x}$ **60.** $-x$ **61.** $<$ **62.** $>$ **63.** $<$ **64.** $<$ **65.** 7 **66.** 8 **67.** 0 **68.** 5 **69.** 0 **70.** -2 **71.** -16 **72.** -3 **73.** 15 **74.** 26 **75.** -94 **76.** -16 **77.** -35 **78.** -35 **79.** 36 **80.** -28 **81.** -50 **82.** 8 **83.** -7 **84.** -6 **85.** 6 **86.** -7 **87.** -120 **88.** -42 **89.** -4 **90.** 4 **91.** $6 - 10 = -4$; The temperature at 2 AM was $-4°$. **92.** $-123 + 75 = -48$; The new balance is $-\$48$. **93.** -5 **94.** 81 **95.** -19 **96.** -8 **97.** -9 **98.** 6 **99.** -4 **100.** 5 **101.** 140° F **102.** 45° C **103.** $a = 6$ **104.** $c = 4$ **105.** The jet's average speed was 435 miles per hour. **106.** Jasper will go 285 miles. **107.** $10x^3$ **108.** $-y^2$ **109.** $-9y$ **110.** Not like terms **111.** 0 **112.** $-4w$ **113.** $a - 3$ **114.** $-4x^4 + 4x$ **115.** $21y$ **116.** $-6x^2$ **117.** $-18y^2$ **118.** $40y^3$ **119.** $-6x + 4$ **120.** $-4x - 12$ **121.** $-18a - 30$ **122.** $-45w + 10$

Chapter 3

Preparation Exercises **1.** $p = 13$ **2.** $y = 15$ **3.** $x = 67$ **4.** $n = 15$ **5.** -2 **6.** -15 **7.** -16 **8.** 12 **9.** -72 **10.** 50 **11.** -5 **12.** 4 **13.** $5x$ **14.** $-16x$ **15.** $20x - 28$ **16.** $-15x + 10$

Section 3.1 **1.** We can never divide by 0; so, we need to make sure that the Division Property of Equations reflects that. **3.** To clear a coefficient, we need to attain a coefficient of 1. That can be achieved only by dividing by the exact coefficient, not its opposite. **5.** $x = 18$ **7.** $w = -3$ **9.** $y = 1$ **11.** $x = 3$ **13.** $p = -6$ **15.** $m = -10$ **17.** $x = 6$ **19.** $c = 3$ **21.** $p = -8$

23. $x = -14$ **25.** $w = -4$ **27.** $m = 0$ **29.** $m = -90$ **31.** $x = -210$ **33.** $h = -340$ **35.** $y = 99$ **37.** $k = 48$ **39.** $m = -170$ **41.** $n = 7$ **43.** $y = -6$ **45.** $k = 12$ **47.** $p = 2$ **49.** $v = -8$ **51.** $y = 3$ **53.** $y = 8$ **55.** $h = -3$ **57.** $c = 8$ **59.** $w = -9$ **61.** $x = 6$ **63.** $k = -22$ **65.** $b = -4$ **67.** $m = 4$ **69.** $x = 0$ **71.** $y = -22$ **73.** $d = 65$ **75.** $q = -24$ **77.** $x = 15$ **79.** $w = -23$ **81.** $h = -33$ **83.** $x = 25$ **85.** $v = 205$ **87.** $x = -174$ **89.** The length of the sides are 58 feet, 87 feet, and 116 feet.

Section 3.2 **1.** The like terms need to be combined on the left side. **3.** $x = 4$ **5.** $x = 0$ **7.** $w = 8$ **9.** $x = 9$ **11.** $v = -3$ **13.** $y = 2$ **15.** $p = 20$ **17.** $x = 0$ **19.** $y = 3$ **21.** $x = -3$ **23.** $k = 3$ **25.** $w = -6$ **27.** $m = -12$ **29.** $p = -17$ **31.** $k = -5$ **33.** $y = -30$ **35.** $x = 22$ **37.** $m = 9$ **39.** The width of the rectangle is 31 centimeters.

Section 3.3 **1.** Answers will vary. **3.** $x = 2$ **5.** $y = 1$ **7.** $p = -3$ **9.** $x = -9$ **11.** $y = -1$ **13.** $x = 2$ **15.** $w = 1$ **17.** $h = -3$ **19.** $m = 3$ **21.** $y = 4$ **23.** $x = 1$ **25.** $y = -10$ **27.** $x = -5$ **29.** $c = 0$ **31.** $y = 2$ **33.** $x = -3$ **35.** $k = 6$ **37.** $r = -2$ **39.** $x = 3$ **41.** $c = -1$ **43.** $x = 0$ **45.** $w = -4$ **47.** $y = -10$ **49.** $x = -6$ **51.** The length of the rectangle is 27 feet.

Section 3.4 **1.** Answers will vary. One possibility: The sum of twice a number and 10; 10 more than twice a number. Also, twice a number increased by 10. **3.** $x + 9$ **5.** $x - 5$ **7.** $2x$ **9.** $18 - x$ **11.** $-9 + x$ **13.** $3 - x$ **15.** $8x$ **17.** $2x + 4$ **19.** $6x + 7$ **21.** $3x + 1$ **23.** $2x - 8$ **25.** The number is −3. **27.** The number is −14. **29.** The number is 5. **31.** The number is −6. **33.** The number is 6. **35.** The number is 3. **37.** The number is −7. **39.** The number is 7. **41.** The number is −1. **43.** The number is −5. **45.** The number is 5. **47.** The number is −3. **49.** The number is 8. **51.** The number is −5. **53.** The number is 6. **55.** The number is −5. **57.** The number is −1. **59.** The number is 11. **61.** The number is −7. **63.** The number is 23.

Section 3.5 **1.** The legend helps define the number that the variable represents. **3.** Answers will vary. One possibility: The sentence should include many of the same words in the question and the answer along with the unit of measure. **5.** The measure of the third angle is 35°. **7.** The width of the rectangle is 31 inches. **9.** The width of the garden should be 13 feet. **11.** Adrienne's score on the first test was 73. **13.** It will take Margo 3 hours to get to the beach. **15.** The length of the third side is 15 inches. **17.** The length of the floor will be 14 feet. **19.** The length of the canvas will be 48 inches. **21.** Hannah drove 485 miles. **23.** Roberto earns $1,838 each month. **25.** The length of the rectangle is 41 inches.

Section 3.6 **1.** Yes. Answers will vary. **3.** Mr. Daniels received 68 votes, and Mrs. Jenkins received 52 votes. **5.** The shorter piece is 6 feet, and the longer piece is 18 feet. **7.** Eli is to paid $850, and Chandra was paid $1,350. **9.** The smallest angle is 28°, and the middle angle is 52°. **11.** Carol's score on the second test was 81, and her score on the first test was 89. **13.** The length of each shorter side is 9 inches. **15.** The width of the playground is 39 feet, and the length is 51 feet. **17.** The smaller one weighs 13 pounds, and the larger watermelon weighs 21 pounds. **19.** Shufen drove 439 miles, and her dad, Zhe, drove 1,033 miles. **21.** Marisa received 331 votes, and Shandrell received 977 votes. **23.** The sides of the triangle are 22 inches, 27 inches, and 44 inches. **25.** The angles of the triangle are 27°, 72°, and 81°.

Chapter 3 Review Exercises **1.** True **2.** False **3.** legend **4.** formula **5.** No, it is not a solution. **6.** Yes, it is a solution. **7.** $w = 8$ **8.** $d = 19$ **9.** $n = -13$ **10.** $h = -4$ **11.** $x = -17$ **12.** $y = -15$ **13.** $m = 23$ **14.** $c = 20$ **15.** $x = 10$ **16.** $w = 1$ **17.** $m = -3$ **18.** $y = -4$ **19.** $k = 3$ **20.** $x = -14$ **21.** $y = -1$ **22.** $w = 4$ **23.** $n = -11$ **24.** $x = -5$ **25.** $y = 29$ **26.** $m = 8$ **27.** $y = -9$ **28.** $x = -5$ **29.** $p = 4$ **30.** $y = 12$ **31.** $x = 9$ **32.** $w = -8$ **33.** $p = -3$ **34.** $k = -2$ **35.** $x = 5$ **36.** $x = 3$ **37.** $y = 1$ **38.** $c = -6$ **39.** $k = -3$ **40.** $y = 2$ **41.** $p = 8$ **42.** $h = 11$ **43.** $y = 5$ **44.** $x = -4$ **45.** $y = 1$ **46.** $p = 4$ **47.** $x = 11$ **48.** $x = -2$ **49.** $y = -7$ **50.** $x = 12$ **51.** $y = -27$ **52.** $x = 5$ **53.** $x = -38$ **54.** $p = 4$ **55.** $4x$ **56.** $x - 12$ **57.** $23 - x$ **58.** $9 + x$ **59.** $x - 8$ **60.** $5 + 2x$ **61.** $2x - 6$ **62.** $6x + 15$ **63.** The number is −10. **64.** The number is 8. **65.** The number is 33. **66.** The number is −9. **67.** The number is −3. **68.** The number is −11. **69.** The number is 9. **70.** The number is 0. **71.** The number is 12. **72.** The number is −26. **73.** It will take him 7 hours to get there. **74.** $467 is available for Berenda's second semester. **75.** The measure of the middle angle is 71°. **76.** The length of the canvas is 23 inches. **77.** Kristen rode 29 miles, and Karl rode 34 miles. **78.** The length of the third side is 18 yards, and the length of the second side is 11 yards. **79.** The smallest angle is 20°, and the middle angle is 60°. **80.** The width of the rectangle is 5, cm and the length is 13 cm.

Chapter 3 Test **1.** $y = 7$ **2.** $x = -4$ **3.** $k = 1$ **4.** $x = 8$ **5.** $x = 7$ **6.** $x = -14$ **7.** $w = -18$ **8.** $x = 3$ **9.** $x = 2$ **10.** $y = -2$ **11.** $x + 14 = -11$. The number is −25. **12.** $2x - 9 = 13$. The number is 11. **13.** Let $x =$ the money after the second day. $185 + x = 450$; The Elsons will pay Josh $265 after the second day. **14.** Let $x =$ the number of points on second test. $89 + x = 185$; Timon scored 96 points on the second test. **15.** Let $x =$ the length of the shorter piece. $2x =$ the length of the longer piece. $x + 2x = 240$; The shorter piece is 80 cm. The longer piece is 160 cm. **16.** Let $x =$ the length of the third side. $x + 2 =$ the length of the second side. $18 + (x + 2) + x = 42$; The length of the second side is 13 inches. The length of the third side is 11 inches. **17.** Let $x =$ the number of miles on the second day. $x + 180 =$ the number of miles on the first day. $(x + 180) + x = 1{,}050$; Greta drove 435 miles on the second day and 615 miles on the first day. **18.** Let $x =$ the money Marta earns. $x - 1{,}500 =$ the money Tunde earns. $x + (x - 1{,}500) = 47{,}500$; Marta earns $24,500. Tunde earns $23,000. **19.** Let $x =$ the measure of the second angle. $2x - 12 =$ the measure of the first angle. $(2x - 12) + x = 90$; The measure of the second angle is 34°. The measure of the first angle is 56°. **20.** Let x = the length of the rectangle. $x - 24$ = the width of the rectangle. $128 = 2x + 2(x - 24)$; The length of the rectangle is 44 feet. The width of the rectangle is 20 feet.

Chapter 4

Preparation Exercises **1.** 15 **2.** 1 **3.** 12 **4.** 24 **5.** 24 **6.** 36 **7.** 60 **8.** 18 **9.** 18 **10.** 14 **11.** 19 **12.** 22 **13.** 20 **14.** −12 **15.** −3 **16.** $3^2 \cdot 5$ or $3 \cdot 3 \cdot 5$ **17.** $2^2 \cdot 3 \cdot 5$ or $2 \cdot 2 \cdot 3 \cdot 5$ **18.** $2^3 \cdot 3^2$ or $2 \cdot 2 \cdot 2 \cdot 3 \cdot 3$ **19.** $2^4 \cdot 5$ or $2 \cdot 2 \cdot 2 \cdot 2 \cdot 5$ **20.** −11 **21.** −10 **22.** −5 **23.** 4 **24.** 24 **25.** −24 **26.** −6 **27.** 1 **28.** $x = 6$ **29.** $y = 12$ **30.** $w = 18$

Section 4.1 **1.** Yes. Two prime numbers have only 1 as a common factor, making them relatively prime. **3.** No. In writing out the GCF, we'd definitely see $2 \cdot 3$ (because 6 is a factor) and we'd definitely see $3 \cdot 3$ (because 9 is a factor); but this does not guarantee that we would see $2 \cdot 3 \cdot 3 \cdot 3$ (which is 54). For example, 18 and 36 have 6 and 9 as factors, but 54 is not a factor of either one. **5.** 5 **7.** 3 **9.** 15 **11.** relatively prime **13.** 15 **15.** 8 **17.** 12 **19.** 24 **21.** 2 **23.** 5 **25.** 6 **27.** 12 **29.** 10 **31.** relatively prime **33.** 18 **35.** 15 **37.** 8 **39.** 16 **41.** 15 **43.** 24 **45.** relatively prime **47.** relatively prime **49.** 30 **51.** 50 **53.** 90 **55.** 18 **57.** 35 **59.** 8 **61.** 4 **63.** 15 **65.** 24

Section 4.2 **1.** We may never have 0 in the denominator, and that is true for the fraction $\frac{0}{0}$. **3.** *An improper fraction.* In a proper fraction, the denominator is greater than the numerator. In its reciprocal, the numerator is greater than the denominator, making it an improper fraction. **5.** Improper fraction **7.** Proper fraction **9.** Mixed number **11.** Improper fraction **13.** $\frac{13}{9}$ **15.** $-\frac{19}{8}$ **17.** $\frac{10}{3}$ **19.** $\frac{32}{7}$ **21.** $-\frac{21}{4}$ **23.** $\frac{61}{10}$ **25.** $\frac{22}{3}$ **27.** $-\frac{39}{4}$ **29.** $\frac{95}{8}$ **31.** $\frac{123}{8}$ **33.** $2\frac{1}{5}$ **35.** $-3\frac{1}{4}$ **37.** $3\frac{3}{7}$ **39.** $3\frac{1}{3}$ **41.** $-2\frac{3}{8}$ **43.** $6\frac{2}{5}$ **45.** $7\frac{2}{5}$ **47.** $11\frac{5}{7}$ **49.** $\frac{8}{45}$ **51.** $-\frac{9}{25}$ **53.** $\frac{8}{15}$ **55.** $-\frac{35}{48}$ **57.** $1\frac{1}{6}$ **59.** $5\frac{5}{8}$ **61.** $\frac{5}{3}$, product is 1 **63.** $\frac{6}{7}$, product is 1 **65.** $-\frac{10}{7}$, product is 1 **67.** 6, product is 1 **69.** $\frac{1}{8}$, product is 1 **71.** No reciprocal **73.** $5\frac{5}{8}$ **75.** The reciprocal is $\frac{4}{11}$; the product is 1.

Section 4.3 **1.** No. We may divide out only common factors, and 2 is not a factor of the numerator or the denominator because it is not being multiplied. **3.** $\frac{3}{5}$ **5.** $\frac{4}{9}$ **7.** $\frac{3}{4}$ **9.** $-\frac{3}{8}$ **11.** $-\frac{1}{4}$ **13.** $\frac{2}{5}$ **15.** $-\frac{2}{3}$ **17.** $\frac{4}{9}$ **19.** $\frac{2}{5}$ **21.** $-\frac{2}{9}$ **23.** $\frac{1}{5}$ **25.** $\frac{1}{7}$ **27.** $-\frac{1}{5}$ **29.** $\frac{11}{4}$ **31.** $\frac{1}{3}$ **33.** $\frac{2}{3}$ **35.** $\frac{5}{18}$ **37.** $\frac{1}{3}$ **39.** $\frac{6}{5}$ **41.** $\frac{5}{16}$ **43.** $\frac{2}{5}$ **45.** $\frac{1}{4}$ **47.** $\frac{11x}{5}$ **49.** $\frac{2a}{7c}$ **51.** $\frac{1}{4x}$ **53.** $\frac{w}{4v}$ **55.** $\frac{7}{10} \times \frac{3}{3} = \frac{21}{30}$ **57.** $-\frac{4}{9} \times \frac{2}{2} = \frac{-8}{18}$ **59.** $\frac{9}{7} \times \frac{6}{6} = \frac{54}{42}$ **61.** $\frac{4}{15} \times \frac{3}{3} = \frac{12}{45}$ **63.** $-\frac{3}{20} \times \frac{5}{5} = \frac{-15}{100}$ **65.** $\frac{3p}{8} \times \frac{10}{10} = \frac{30p}{80}$ **67.** $-\frac{7w}{12} \times \frac{5}{5} = \frac{-35w}{60}$ **69.** $\frac{8m}{25} \times \frac{4}{4} = \frac{32m}{100}$ **71.** $\frac{5}{8} \times \frac{4x}{4x} = \frac{20x}{32x}$ **73.** $\frac{k}{6mp} \times \frac{7mp}{7mp} = \frac{7kmp}{42m^2p^2}$

Section 4.4 **1.** We can simplify the fraction by expanding the numerator and the denominator and dividing out common factors. **a)** m^4 **b)** x^6 **c)** 1 **d)** $\frac{1}{y^5}$ **e)** $\frac{1}{w^3}$ **3.** The result will be less than the original number. Multiplying by 1 keeps the value the same, and multiplying by a number less than (a proper fraction) gives a result that is less than the original number. **5.** $\frac{4}{45}$ **7.** $-\frac{25}{42}$ **9.** $\frac{5a}{6b}$ **11.** $\frac{5}{28}$ **13.** $-\frac{4}{33}$ **15.** $\frac{21}{50}$ **17.** $\frac{2}{5}$ **19.** −12 **21.** $\frac{1}{6}$ **23.** $\frac{q}{3}$ **25.** $\frac{6}{35}$ **27.** $-\frac{4}{5x}$ **29.** $\frac{21m}{10}$ **31.** $\frac{2}{3}$ **33.** $\frac{2}{3}$ **35.** $\frac{3}{2}$ or $1\frac{1}{2}$ **37.** $\frac{4}{5}$ **39.** $\frac{4p}{5}$ **41.** $-\frac{16}{9}$ or $-1\frac{7}{9}$ **43.** $-\frac{3}{4}$ **45.** $\frac{3}{20b}$ **47.** 10 **49.** 6 **51.** 1 **53.** $4\frac{1}{2}$ **55.** 45 **57.** $9\frac{3}{5}$ **59.** $1\frac{1}{15}$ **61.** 4 **63.** $\frac{12}{25}$ **65.** $1\frac{1}{2}$ **67.** $1\frac{3}{7}$ **69.** $\frac{5}{14}$ **71.** $\frac{1}{4} \cdot \frac{8}{15} = \frac{2}{15}$ **73.** $\frac{1}{5} \cdot \frac{15}{16} = \frac{3}{16}$ **75.** $\frac{2}{3} \cdot \frac{9}{16} = \frac{3}{8}$ **77.** $\frac{1}{3} \cdot \frac{18}{1} = 6$ **79.** $\frac{3}{8} \cdot \frac{48}{1} = 18$ **81.** $\frac{1}{10} \cdot \frac{1}{100} = \frac{1}{1{,}000}$ **83.** $\frac{2}{3} \cdot 2\frac{5}{8} = 1\frac{3}{4}$ **85.** $\frac{1}{2} \cdot 3\frac{1}{5} = 1\frac{3}{5}$ **87.** $\frac{3}{7} \cdot 4\frac{1}{5} = 1\frac{4}{5}$ **89.** $\frac{1}{6}$ of a cup will wash $\frac{1}{4}$ of a load. **91.** Pedro must get 36 correct to pass the class. **93.** The middle of the photo is $2\frac{3}{16}$ inches from the side and $1\frac{7}{8}$ inches from the bottom. **95.** $8\frac{1}{3}$ square yards **97.** $\frac{3}{5}$ square miles **99.** $\frac{2}{5}$ **101.** $\frac{1}{4}$

Section 4.5 **1.** Yes, like terms have the same variable with the same exponent regardless of the coefficient. **3.** 1 **5.** $\frac{4}{3}$ or $1\frac{1}{3}$ **7.** $\frac{4}{3}$ or $1\frac{1}{3}$ **9.** $\frac{4}{7}$ **11.** $\frac{10}{7}$ or $1\frac{3}{7}$ **13.** $\frac{7}{5}$ or $1\frac{2}{5}$ **15.** $\frac{1}{2}$ **17.** 1 **19.** $\frac{2}{3}$ **21.** $\frac{2}{5}$ **23.** $\frac{4}{5}k$ **25.** $\frac{1}{2}y$ **27.** v **29.** $\frac{3}{2}n$ **31.** $\frac{2}{5}hk$ **33.** $\frac{3}{5}xy$ **35.** $\frac{4}{3}q^2$ **37.** $\frac{7}{6}x^2$ **39.** $\frac{1}{4}$ **41.** 0 **43.** $\frac{1}{3}$ **45.** $\frac{2}{5}$ **47.** $\frac{4}{9}h$ **49.** $\frac{1}{5}x$ **51.** $\frac{3}{4}ab$ **53.** $\frac{3}{5}d^2$ **55.** $\frac{2}{5}m^2$ **57.** $-\frac{1}{2}$ **59.** −1 **61.** $\frac{1}{3}$ **63.** $-\frac{1}{2}$ **65.** $\frac{2}{3}$ **67.** $-\frac{3}{4}b$ **69.** $-\frac{1}{4}v$ **71.** $-\frac{2}{5}ab$ **73.** $-\frac{1}{5}y^2$ **75.** $9\frac{4}{5}$ **77.** $12\frac{7}{9}$ **79.** 12 **81.** 14 **83.** $37\frac{1}{2}$ **85.** $5\frac{1}{2}$ **87.** $2\frac{5}{9}$ **89.** $9\frac{4}{7}$ **91.** $1\frac{2}{5}$ **93.** $4\frac{5}{9}$ **95.** $7\frac{11}{15}$ **97.** $4\frac{1}{3}$ **99.** Marius is adding $1\frac{1}{2}$ inches to the top of the cabinets. **101.** Carlita was $1\frac{3}{4}$ inches taller than Juanita. **103.** $\frac{10}{3}$ or $1\frac{3}{10}$ **105.** $\frac{4}{3}$ or $1\frac{1}{3}$

Section 4.6 **1.** There are many common denominators for $\frac{5}{6}$ and $\frac{4}{15}$, including 60, 90, 120, 600, and 900. There is no *greatest* common denominator. **3.** The LCD is the product of the two denominators. **5.** LCD is 12, $\frac{3}{12}$ and $\frac{2}{12}$ **7.** LCD is 28, $\frac{21}{28}$ and $\frac{18}{28}$ **9.** LCD is 30, $\frac{25}{30}$ and $\frac{27}{30}$ **11.** LCD is 60, $\frac{10}{60}$ and $\frac{39}{60}$ **13.** LCD is 40, $\frac{15}{40}$ and $\frac{4}{40}$ **15.** LCD is 40, $\frac{35}{40}$ and $\frac{22}{40}$ **17.** LCD is 60, $\frac{45}{60}$ and $\frac{28}{60}$ **19.** LCD is 40, $\frac{35}{40}$ and $\frac{12}{40}$ **21.** LCD is 12, $\frac{9}{12}$ and $\frac{7}{12}$ **23.** LCD is 30, $\frac{26}{30}$ and $\frac{23}{30}$ **25.** LCD is 42, $\frac{27}{42}$ and $\frac{17}{42}$ **27.** LCD is 20, $\frac{12}{20}$ and $\frac{5}{20}$ **29.** LCD is 36, $\frac{27}{36}$ and $\frac{20}{36}$ **31.** LCD is 100, $\frac{25}{100}$ and $\frac{84}{100}$ **33.** LCD is 30, $\frac{25}{30}$ and $\frac{12}{30}$ **35.** LCD is 60, $\frac{37}{60}$ and $\frac{44}{60}$ **37.** LCD is 48, $\frac{15}{48}$ and $\frac{26}{48}$ **39.** LCD is 72, $\frac{34}{72}$ and $\frac{21}{72}$ **41.** LCD = 12, $\frac{8}{12}$, $\frac{9}{12}$, and $\frac{2}{12}$ **43.** LCD = 48, $\frac{8}{48}$, $\frac{44}{48}$, and $\frac{39}{48}$

Section 4.7 **1.** Answers will vary. **3.** Yes. When simplified, each fraction will have a denominator of 8, thereby making the denominators common denominators. **5.** $\frac{9}{10}$ **7.** $\frac{8}{9}$ **9.** $\frac{7}{24}$ **11.** $\frac{8}{33}$ **13.** $1\frac{1}{6}$ **15.** $1\frac{1}{4}$ **17.** $\frac{1}{3}$ **19.** $\frac{2}{9}$ **21.** $\frac{19}{28}$ **23.** $1\frac{11}{20}$ **25.** $\frac{7}{18}$ **27.** $\frac{17}{36}$ **29.** $1\frac{1}{15}$

31. $\frac{8}{15}$ **33.** $\frac{11}{18}c$ **35.** $\frac{31}{36}y$ **37.** $\frac{8}{15}q$ **39.** $\frac{19}{45}x$ **41.** $\frac{17}{30}ab$ **43.** $\frac{21}{50}p^2$ **45.** $\frac{1}{8}$ **47.** $\frac{4}{9}$ **49.** $-\frac{33}{20}m$ **51.** $-\frac{19}{20}$ **53.** $\frac{13}{24}$ **55.** $-\frac{1}{10}x$ **57.** $25\frac{5}{8}$ **59.** $25\frac{7}{18}$ **61.** $10\frac{5}{12}$ **63.** $21\frac{4}{9}$ **65.** $18\frac{7}{24}$ **67.** $66\frac{5}{12}$ **69.** $2\frac{9}{20}$ **71.** $1\frac{13}{36}$ **73.** $2\frac{2}{5}$ **75.** $5\frac{5}{9}$ **77.** $1\frac{3}{4}$ **79.** $11\frac{11}{12}$ **81.** There are $1\frac{3}{8}$ pounds of meat in this recipe. **83.** Rohin burned $\frac{11}{12}$ of a cord during the winter. **85.** The total weight of the two watermelons is $33\frac{5}{8}$ pounds. **87.** Carin must pour $\frac{7}{12}$ cup of milk from a new carton. **89.** The door should be $35\frac{3}{16}$ inches wide. **91.** Sanjeer took $11\frac{3}{8}$ pounds of aluminum cans to be recycled. **93.** $1\frac{7}{16}$ inches should be added to the length of the driver. **95.** $\frac{1}{12}$ **97.** $-5\frac{1}{15}$

Section 4.8 **1.** We must first rewrite the mixed number as an improper fraction so that we can identify the reciprocal of the coefficient. **3.** $y = 1$ **5.** $k = \frac{1}{3}$ **7.** $m = -\frac{2}{5}$ **9.** $x = -\frac{2}{3}$ **11.** $p = -\frac{4}{3} = -1\frac{1}{3}$ **13.** $x = -\frac{11}{6} = -1\frac{5}{6}$ **15.** $y = \frac{4}{3} = 1\frac{1}{3}$ **17.** $h = -\frac{3}{2} = -1\frac{1}{2}$ **19.** $c = \frac{8}{5} = 1\frac{3}{5}$ **21.** $m = \frac{2}{3}$ **23.** $n = \frac{3}{2} = 1\frac{1}{2}$ **25.** $p = \frac{3}{5}$ **27.** $x = \frac{3}{8}$ **29.** $w = \frac{3}{4}$ **31.** $v = \frac{3}{4}$ **33.** $y = -35$ **35.** $x = 12$ **37.** $k = 4$ **39.** $x = 10$ **41.** $w = -24$ **43.** $v = \frac{1}{3}$ **45.** $n = \frac{2}{3}$ **47.** $x = \frac{1}{14}$ **49.** $y = -\frac{1}{14}$ **51.** $x = 6$ **53.** $v = 10$ **55.** $m = \frac{5}{2} = 2\frac{1}{2}$ **57.** $x = \frac{5}{8}$ **59.** $y = \frac{2}{3}$ **61.** $n = 7$ **63.** Marnay gives 10 lessons each day. **65.** Each dose will be $\frac{1}{8}$ cup of syrup. **67.** 9 bows can be made from one spool. **69.** Yuan will need 20 afternoon sessions. **71.** Tim takes $6\frac{1}{4}$ hours to clean all 25 classrooms. **73.** $x = -12$ **75.** $w = -12$

Chapter 4 Review Exercises **1.** 8 **2.** 14 **3.** 18 **4.** Relatively prime **5.** Mixed number **6.** Improper fraction **7.** Proper fraction **8.** Improper fraction **9.** $\frac{31}{9}$ **10.** $-\frac{31}{6}$ **11.** $4\frac{2}{3}$ **12.** $-7\frac{7}{9}$ **13.** Reciprocal is $\frac{4}{9}$; Product is 1 **14.** Reciprocal is 8; Product is 1 **15.** Reciprocal is $-\frac{1}{4}$; Product is 1 **16.** No reciprocal **17.** $\frac{3}{4}$ **18.** $-\frac{2}{9}$ **19.** $\frac{2}{5}$ **20.** $\frac{5x}{4}$ **21.** $\frac{3}{5} \times \frac{4}{4} = \frac{12}{20}$ **22.** $-\frac{1}{10} \times \frac{6}{6} = \frac{-6}{60}$ **23.** $-\frac{5}{12} \times \frac{4}{4} = \frac{-20}{48}$ **24.** $\frac{4}{9} \times \frac{7}{7} = \frac{28}{63}$ **25.** $\frac{2}{3}$ **26.** $\frac{3}{14}$ **27.** $-\frac{5}{6}$ **28.** 5 **29.** 1 **30.** 6 **31.** $-3\frac{1}{2}$ **32.** $\frac{10}{3v}$ **33.** 9 **34.** $7\frac{3}{5}$ **35.** $\frac{9}{16}$ **36.** $3\frac{1}{3}$ **37.** $\frac{2}{3} \cdot \frac{15}{14} = \frac{5}{7}$ **38.** $\frac{3}{8} \cdot \frac{4}{15} = \frac{1}{10}$ **39.** $\frac{3}{4} \cdot 2\frac{2}{9} = \frac{3}{4} \cdot \frac{20}{9} = \frac{5}{3} = 1\frac{2}{3}$ **40.** $\frac{5}{6} \cdot 4\frac{1}{5} = \frac{5}{6} \cdot \frac{21}{5} = \frac{7}{2} = 3\frac{1}{2}$ **41.** There are 28 women in Ms. Grecu's algebra class. **42.** There are 8 pounds of meat in 12 patties. **43.** 9 square feet **44.** 3 square inches **45.** $\frac{3}{4}$ **46.** $\frac{4}{5}$ **47.** $\frac{8}{5}k$ **48.** $2ab$ **49.** $\frac{2}{3}$ **50.** $\frac{1}{2}$ **51.** $\frac{3}{5}q$ **52.** $\frac{1}{6}d^2$ **53.** $-\frac{1}{3}$ **54.** $-n$ **55.** $11\frac{1}{2}$ **56.** $\frac{2}{3}$ **57.** A total of $4\frac{1}{5}$ inches of rain fell in Garberville that day. **58.** Marika gained $16\frac{1}{4}$ pounds during her first year. **59.** LCD is 18; $\frac{15}{18}$ and $\frac{8}{18}$ **60.** LCD is 42; $\frac{15}{42}$ and $\frac{20}{42}$ **61.** LCD is 54; $\frac{33}{54}$ and $\frac{29}{54}$ **62.** LCD is 60; $\frac{9}{60}$ and $\frac{28}{60}$ **63.** $\frac{2}{3}$ **64.** $\frac{11}{12}$ **65.** $\frac{44}{45}x$ **66.** $\frac{43}{36}h^2$ **67.** $\frac{1}{6}$ **68.** $\frac{1}{6}$ **69.** $\frac{3}{5}ab$ **70.** $\frac{7}{30}q$ **71.** $-1\frac{13}{20}$ **72.** $-\frac{1}{6}m$ **73.** $-\frac{19}{20}d^2$ **74.** $-\frac{1}{10}$ **75.** $3\frac{1}{4}$ **76.** $8\frac{3}{10}$ **77.** Mary burned $\frac{7}{10}$ of a cord of wood during the winter. **78.** $2\frac{5}{6}$ yards of fabric are left on the bolt. **79.** $y = \frac{2}{3}$ **80.** $m = -\frac{2}{5}$ **81.** $x = \frac{1}{3}$ **82.** $p = \frac{1}{12}$ **83.** $w = -1\frac{1}{2}$ **84.** $m = \frac{7}{10}$ **85.** $x = 1\frac{1}{5}$ **86.** $w = 1\frac{1}{3}$ **87.** Keith must work 24 Saturdays. **88.** Each row in the dispay is $9\frac{1}{3}$ feet long.

Chapter 4 Test **1.** 8 **2.** 9 **3.** 14 **4.** $\frac{37}{5}$ **5.** $7\frac{3}{7}$ **6.** $\frac{7}{10}$ **7.** $\frac{4}{5}$ **8.** $\frac{3x}{2y}$ **9.** $\frac{3}{4}$ **10.** $\frac{d}{2}$ **11.** $7\frac{1}{2}$ **12.** $1\frac{1}{2}$ **13.** 21 square yards **14.** The original picture is 8 inches high. **15.** $-\frac{2}{3}$ **16.** $2\frac{2}{5}$ **17.** LCD is 45; $\frac{12}{45}$ and $\frac{25}{45}$ **18.** LCD is 60; $\frac{35}{60}$ and $\frac{33}{60}$ **19.** $\frac{2}{3}$ **20.** $\frac{3}{4}y$ **21.** $4\frac{2}{3}$ **22.** $1\frac{2}{5}$ **23.** $\frac{13}{15}$ **24.** $\frac{7}{36}$ **25.** $10\frac{1}{6}$ **26.** $3\frac{7}{24}$ **27.** $\frac{1}{2}$ **28.** $-\frac{3}{10}p$ **29.** Timina walked $6\frac{4}{15}$ total miles that day. **30.** It rained $1\frac{11}{20}$ more inches that day. **31.** $y - \frac{1}{6}$ **32.** $p = -1\frac{1}{3}$ **33.** $y = 21$ **34.** $x = 7$ **35.** Let $x =$ the weight of each chocolate bar. $18 \cdot x = 22\frac{1}{2}$; Each chocolate bar weighs $1\frac{1}{4}$ pounds.

Chapters 1–4 Cumulative Review **1. a)** 1,970,000 **b)** 2,000,000 **2. a)** Identity for Addition **b)** Commutative Property of Multiplication **3.** Ms. Bohler received $8,921 more from business groups than from police/fire groups. **4.** Ms. Bohler received $62,164 in total contributions. **5.** 405 square inches **6.** $3 \cdot 3 \cdot 3 \cdot 3 = 81$ **7.** Each employee will receive $152. **8.** Prime 2, 43; Composite 9, 42, 115; Neither 1 **9.** 2 and 5 **10.** 2 and 3 **11.** 9 is not a factor. **12.** $2 \cdot 2 \cdot 3 \cdot 3 \cdot 3$; $2^2 \cdot 3^3$ **13.** 2 **14.** -3 **15.** -210 **16.** 7 **17.** -8 **18.** -5 **19.** 10 **20.** 14° F **21.** −15° C **22.** $-6y$ **23.** $-4p^3 + 4p$ **24.** $36k$ **25.** $-16m + 30$ **26.** $y = 9$ **27.** $y = -5$ **28.** $7 + 2x = -15$; $x = -11$ **29.** There were 215 copies from the newer machine and 135 copies from the older machine. **30.** The second side is 21 inches, and the third side is 14 inches. **31.** 14 **32.** $\frac{k}{3h}$ **33.** $4\frac{3}{4}$ **34.** $\frac{53}{8}$ **35.** LCD is 90; $\frac{21}{90}$ and $\frac{16}{90}$ **36.** $\frac{2}{3}$ **37.** $\frac{4}{9}$ **38.** $1\frac{7}{8}$ **39.** $\frac{3}{4}m$ **40.** $\frac{3}{7}$ **41.** $6\frac{1}{6}$ **42.** $\frac{2}{3}$ **43.** $-\frac{1}{4}r^2$ **44.** $2\frac{1}{3}$ **45.** $x = -\frac{2}{3}$ **46.** $m = \frac{4}{9}$ **47.** Each team member must run $1\frac{3}{4}$ miles. **48.** Kahlil surpassed his previous mark by $1\frac{7}{16}$ inches.

Chapter 5

Preparation Exercises **1.** 71,000 **2.** 7,000 **3.** 617 **4.** 709 **5.** 577 **6.** 1,847 **7.** 1,016 **8.** 1,560 **9.** 47 **10.** 35 **11.** $x = 183$ **12.** $x = 19$ **13.** $\frac{7}{20}$ **14.** $\frac{11}{15}$ **15.** $3\frac{1}{3}$ **16.** $\frac{63}{100}$ **17.** $\frac{3}{10}$ **18.** 2 **19.** $\frac{1}{3}$ **20.** $\frac{13}{20}$ **21.** $\frac{3}{8}$ **22.** $\frac{57}{100}$

Section 5.1 **1.** One hundred twelve-thousandths is 0.112, and one hundred and twelve-thousandths is a mixed number: 100.012. **3.** Yes, it terminates after only two decimal places. (3.14 should not be confused with π, a decimal that doesn't terminate and doesn't repeat.) **5.** hundredths **7.** terminating **9.** True **11.** False **13.** tenths **15.** ones **17.** five-tenths **19.** twenty-nine hundredths **21.** eight-thousandths **23.** sixty-hundredths **25.** three and seven-hundredths **27.** twelve and eight-thousandths **29.** $\frac{2}{10}$ **31.** $\frac{87}{100}$ **33.** $\frac{628}{1,000}$ **35.** $\frac{9,314}{10,000}$ **37.** $\frac{3}{100}$ **39.** $\frac{7}{100}$ **41.** $\frac{69}{1,000}$ **43.** $\frac{1}{1,000}$ **45.** $\frac{96}{10}$ **47.** $\frac{25}{10}$ **49.** $\frac{803}{100}$ **51.** $\frac{6,937}{1,000}$ **53.** 0.7 **55.** 0.596 **57.** 0.4 **59.** 4.3 **61.** 0.02 **63.** 0.007 **65.** 7.02 **67.** 90.4 **69.** $0.\overline{3}$ **71.** $0.\overline{693}$ **73.** $10.85\overline{2}$ **75.** 0.66666. . . **77.** 9.470470470. . . **79.** 0.63101010. . . **81–87.** −3.71 −0.4 0.9 3.5 (−5 −4 −3 −2 −1 0 1 2 3 4 5) **89.** 0.28 **91.** 1.75 **93.** 1.68 **95.** 0.3

Section 5.2 **1.** The rounding digit must be in the ten thousandths place. **3.** 11.6 **5.** 10.6 **7.** 2.0 **9.** 3.0 **11.** 1.26 **13.** 1.39 **15.** 14.07 **17.** 18.05 **19.** 4.20 **21.** 4.00 **23.** 0.46 **25.** 0.39 **27.** 6.084 **29.** 1.010 **31.** 0.216 **33.** 0.340 **35.** 0.317 **37.** 0.300 **39.** $5.96 **41.** $14.04 **43.** $5.99 **45.** $108.00 **47.** $6.00 **49.** $3.80 **51.** 31 **53.** 83 **55.** 215 **57.** 1 **59.** $31 **61.** $20 **63.** $30 **65.** $100 **67.** Yolanda's GPA is 3.5. **69.** Alabama's population density is 88.70 people per square mile. **71.** Hank Aaron's batting average in 1969 was .300. **73.** $17 **75.** Answers may vary.

Section 5.3 **1.** $2.4 + 2.4 + 2.4 + 2.4 + 2.4 + 2.4 = 14.4$ **3.** 0.828 **5.** 1.501 **7.** 0.79 **9.** 20.41 **11.** 1.13 **13.** 20 **15.** 4.016 **17.** 10.006 **19.** 26.778 **21.** 22.437 **23.** 7.08 **25.** 5.453 **27.** 5.4 **29.** 0.55 **31.** 5.41 **33.** 2.972 **35.** 1.6 **37.** 5.71 **39.** 9.309 **41.** 5.47 **43.** 2.8 **45.** −0.9 **47.** 1.74 **49.** 3.03 **51.** −8.4 **53.** −0.44 **55.** 5.6 **57.** −3.9 **59.** 0 **61.** $2.4x$ **63.** $-2.15w$ **65.** $-1.78h$ **67.** $-3.32p$ **69.** $4.34w$ **71.** $-0.77y$ **73.** The total is $41.74. **75.** The combined weight was 101.88 pounds. **77.** Torii was 1.67 seconds behind Sean. **79.** Perimeter = 4.143 feet **81.** $-57.82 + 100$; Daneice's new account balance is $42.18.; This is a credit. **83.** $-8.4 + 15.25$; The temperature at noon was 6.85°. **85.** $4.25 - (-1.8)$; The difference in altitude is 6.05 meters. **87.** Let x = the number of pounds Maeve still needs to lose. $23.4 + x = 40$; Maeve needs to lose 16.6 more pounds to reach her goal. **89.** Let x = the amount LeRon still owes his dad. $79.48 + x = 107.52$; LeRon still owes his dad $28.04. **91.** Let x = the length of the third side. $1.034 + 1.17 + x = 3$; The third side is 0.796 inches. **93.** 8.67 **95.** Answers may vary.

Section 5.4 **1.** Answers will vary. One possibility: When multiplying decimals, we ignore the decimal point until the end of the multiplying. **3.** 4.2 **5.** 0.6 **7.** 0.065 **9.** one **11.** two **13.** three **15.** four **17.** 3.2 **19.** 2.1 **21.** 0.63 **23.** 0.45 **25.** 5.46 **27.** 1.54 **29.** 0.09 **31.** 0.06 **33.** 0.015 **35.** 0.009 **37.** 0.0048 **39.** 0.00287 **41.** 0.0015 **43.** 0.452 **45.** 1.875 **47.** 24.225 **49.** 0.068 **51.** 8.673 **53.** −4.5 **55.** −0.32 **57.** −1 **59.** −0.52 **61.** 20 **63.** 0.52 **65.** 119 **67.** 8 **69.** 90.6 **71.** 376 **73.** 8 **75.** 120.6 **77.** 480 **79.** 401 **81.** 8,050 **83.** 1,300 **85.** There is $0.78 sales tax on the book. **87.** There is $6.76 sales tax on the computer printer. **89.** The total amount for the batteries is $26.70. **91.** The total amount Janelle must pay for the Camry is $8,170. **93.** Carlos earned $48.43 yesterday. **95.** Oyuki earned $357.60 last week. **97.** Terrence earned $480 for the week.

Section 5.5 **1.** It is appropriate to add zeros at the end of the dividend when the dividing still has a remainder (other than 0) and the dividend has no more digits. **3.** Yes, as long as the divisor is a terminating decimal. The quotient will be a repeating decimal. **5.** 0.9 **7.** 12.3 **9.** 9.67 **11.** 0.19 **13.** 0.164 **15.** 0.534 **17.** 0.405 **19.** 0.0665 **21.** $1.5\overline{6}$ **23.** $0.5\overline{3}$ **25.** $0.3\overline{09}$ **27.** $0.03\overline{26}$ **29.** 3.6 **31.** 1.17 **33.** 3.96 **35.** 70 **37.** 800 **39.** 20 **41.** 0.42 **43.** 0.1782 **45.** $0.50\overline{5}$ **47.** $0.6\overline{63}$ **49.** 0.1875 **51.** 0.248 **53.** 2.58 **55.** 4.57 **57.** 0.749 **59.** 0.064 **61.** 0.0045 **63.** 0.0421 **65.** 2.5 **67.** 1.75 **69.** 1.6 **71.** 1.125 **73.** 0.65 **75.** 0.68 **77.** $1.1\overline{6}$ **79.** $1.\overline{2}$ **81.** $0.\overline{81}$ **83.** $0.5\overline{3}$ **85.** $0.\overline{108}$ **87.** $0.\overline{428571}$ **89.** −0.03 **91.** −4.25 **93.** −96 **95.** $-1.\overline{7}$ **97.** 72.4 **99.** $6.8\overline{3}$ **101.** $\frac{3}{2\cdot 2}, \frac{1}{2\cdot 2\cdot 2}, \frac{9}{2\cdot 5}, \frac{7}{2\cdot 2\cdot 2\cdot 2}, \frac{12}{5\cdot 5}, \frac{13}{2\cdot 2\cdot 2\cdot 5}, \frac{27}{2\cdot 5\cdot 5}, \frac{87}{2\cdot 2\cdot 5\cdot 5}, \frac{94}{5\cdot 5\cdot 5}$; Each denominator has only 2 and 5 as prime factors. **103.** When a fraction's denominator has only 2 or 5 as prime factors, the fraction can be written as a terminating decimal. When a fraction's denominator has a prime factor other than 2 or 5, the fraction can be written as a repeating decimal.

Section 5.6 **1.** 87 minutes. We must multiply 1.45 times 60. (Or we multiply 0.45 times 60 and add 60.) **3.** Kerri earns $72.02 each day. **5.** Tran worked 76 hours last month. **7.** Calvin earns $10.16 per hour. **9.** Annie earned, on average, $18.25 commission each hour. **11.** Hannah earned, on average, $20.80 commission each hour. **13.** The area of the patio is 26.46 square yards. **15.** The length of the patch is 12.5 feet. **17.** The circumference of the pond is approximately 220 meters. **19.** The radius of the wheel is approximately 35 centimeters. **21.** Dale can travel 339.3 miles on 7.8 gallons of gas. **23.** The car averaged 42.5 miles per gallon for the week? **25.** The price of each card was $0.66. **27.** The price of a yard of that fabric was $6.40. **29.** Jake's average rate of speed on that course is 15 miles per hour. **31.** It takes Becca 1.75 hours to ride the entire trail. **33.** Alan will earn $18.72 per hour on Thanksgiving. **35.** Maralee will pay $319.56 each month. **37.** The circumference of the rim of Meteor Crater is approximately 3.768 kilometers. **39. a)** Terrence worked 6 hours beyond 40 hours that week **b)** Terrence worked 46 total hours that week.

Chapter 5 Review Exercises **1.** decimal **2.** thousandths **3.** whole number **4.** repeating decimal **5.** hundredths **6.** thousandths **7.** six-hundredths **8.** thirteen-thousandths **9.** $\frac{38}{100}$ **10.** $\frac{12}{10}$ **11.** $\frac{1}{100}$ **12.** $\frac{208}{100}$ **13.** 0.72 **14.** 0.03 **15.** 5.7 **16.** 0.008 **17.** $1.\overline{5}$ **18.** $2.6\overline{145}$ **19.** 0.512512512 . . . **20.** 2.38888 . . . **21–24.** −3.1 −0.3 1.7 4.5 (−5 −4 −3 −2 −1 0 1 2 3 4 5) **25.** 24.0 **26.** 1.52 **27.** 27.0 **28.** 1.500 **29.** $1.59 **30.** $28.79 **31.** $0.40 **32.** $12.00 **33.** $8.90 **34.** $3.60 **35.** $1.30 **36.** $25.00 **37.** $128 **38.** $50 **39.** $268 **40.** $1,600 **41.** Mariko calculated her GPA to be 3.49. **42.** Julia's average daily electricity usage for March was 7.036 kilowatt-hours. **43.** 7.08 **44.** 1.205 **45.** 21.631 **46.** 21.517 **47.** 0.69 **48.** 2.619 **49.** 4.4 **50.** 1.051 **51.** −0.75

52. −4.42 **53.** 1.2 **54.** −0.63 **55.** 2.45 **56.** 7.12 **57.** $3.82x$ **58.** $-3.97y$ **59.** $-6.14w$ **60.** $-21.34c$ **61.** $8.62h$ **62.** $2.5v$ **63.** The difference in altitude is 15.52 meters. **64.** The temperature of the chicken was −6.1°C. **65.** The total weight of the four bags of candy was 3.552 pounds. **66.** Samantha paid $79.80 in postage. **67.** Benjou is 0.27 meters taller than Malik. **68.** Jenny received $12.82 in change. **69.** 5.475 feet **70.** 2.31 yards **71.** 8.4 **72.** 0.438 **73.** 4.2 **74.** 14 **75.** 24 **76.** 0.1134 **77.** 305.52 **78.** 45.27 **79.** 83 **80.** 141.2 **81.** 291 **82.** 370 **83.** 80.4 **84.** 1,691 **85.** 89 **86.** 50 **87.** 0.12 **88.** 3 **89.** −0.072 **90.** −0.006 **91.** Jamali earned $60.61 yesterday. **92.** Carlos must pay a total of $38.69 for the necklace. **93.** 6.65 **94.** $0.5\overline{18}$ **95.** 0.1825 **96.** $2.61\overline{6}$ **97.** 1.7 **98.** $0.13\overline{8}$ **99.** $0.070\overline{6}$ **100.** 0.056 **101.** $0.\overline{72}$ **102.** 1.875 **103.** $0.08\overline{3}$ **104.** −0.52 **105.** Wes earns $65.24 for a 7-hour shift. **106.** Sunil earned, on average, $16.82 per hour. **107.** The area of the rec room is 2,445.3 square feet. **108.** The circumference of the snare drum is approximately 40.8 inches. **109.** The price of each can of soda was $0.32. **110.** Ruben's van can travel 358.9 miles on a full tank of gas. **111.** Carmen's average walking rate was 3.6 miles per hour. **112.** Tom and Margaret will pay $104.80 each month.

Chapter 5 Test **1.** hundredths **2.** nineteen-thousandths **3.** $\frac{61}{100}$ **4.** $\frac{8}{1,000}$ **5.** $\frac{132}{10}$ **6.** 0.039 **7.** 0.05 **8.** 2.7 **9–12.**
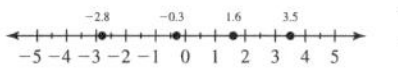

13. 49 **14.** 6.0 **15.** $12.90 **16.** $354 **17.** 5.625 **18.** 0.892 **19.** −1.05 **20.** 2.75 **21.** 2.1 **22.** 380 **23.** $1.15p$ **24.** $1.47m$ **25.** $0.88\overline{3}$ **26.** 2.8 **27.** 1.625 **28.** $-0.\overline{36}$ **29.** Paul got back $10.66 in change. **30.** Silvia's car averaged 22.5 miles per gallon. **31.** The total amount Rhonda must pay for the magazine is $5.35. **32.** Dom earned $345.80 last week. **33.** The length of the quilt is 5.5 feet. **34.** The circumference is approximately 31.4 feet.

Chapter 6

Preparation Exercises **1.** $x = 19$ **2.** $x = 37$ **3.** $\frac{18}{25}$ **4.** $\frac{2}{5}$ **5.** $\frac{28}{100}$ **6.** $\frac{60}{100}$ **7.** $\frac{2}{25}$ **8.** $\frac{6}{5}$ **9.** 0.019 **10.** 0.04 **11.** 3.56 **12.** 8.20 **13.** 11.7 **14.** 0.4326 **15.** 37 **16.** 9 **17.** 7,150 **18.** 4.5 **19.** 0.49 **20.** 212.5 **21.** 0.875 **22.** $0.1\overline{3}$ **23.** 0.213 **24.** 0.067

Section 6.1 **1.** Answers will vary. **3.** Yes. In a ballet class of 12 students, all girls, the ratio of boys to girls is 0 : 12, or $\frac{0}{12}$ **5.** units of measure **7.** divided by **9.** $\frac{60\text{ miles}}{1\text{ gallon}}$ **11.** $\frac{2\text{ dollars}}{3\text{tickets}}$ **13.** $\frac{35\text{ miles}}{1\text{ hour}}$ **15.** $\frac{3\text{ laps}}{10\text{ minutes}}$ **17.** 7 people per van **19.** 23.4 miles per gallon **21.** 9 lemons for every $5 **23.** 1 boy for every 2 girls **25.** $\frac{3}{4}$ **27.** $\frac{1}{5}$ **29.** $\frac{25\text{ miles}}{2\text{ hours}}$ **31.** $\frac{4}{1}$ **33.** $\frac{12\text{ liters}}{1\text{ hour}}$ **35.** $\frac{63}{2}$ **37.** $\frac{5}{3}$ **39.** $\frac{14}{15}$ **41. a.** \$300 : \$150; $\frac{\$300}{\$150}$; $\frac{2}{1}$ **b.** \$150 : \$300; $\frac{\$150}{\$300}$; $\frac{1}{2}$ **c.** \$300 : \$450; $\frac{\$300}{\$450}$; $\frac{2}{3}$ **d.** \$150 : \$450; $\frac{\$150}{\$450}$; $\frac{1}{3}$ **43. a.** \$6,000 : \$3,600; $\frac{\$6,000}{\$3,600}$; $\frac{5}{3}$ **b.** \$3,600 : \$6,000; $\frac{\$3,600}{\$6,000}$; $\frac{3}{5}$ **c.** \$6,000 : \$9,600; $\frac{\$6,000}{\$9,600}$; $\frac{5}{8}$ **d.** \$3,600 : \$9,600; $\frac{\$3,600}{\$9,600}$; $\frac{3}{8}$ **45.** Yes **47.** No **49.** No **51.** Yes

53.

Morning Class	Students	Ratio	Simplified Ratio
Girls	15	$\frac{\text{Girls}}{\text{Total}} = \frac{15}{20}$	$\frac{3}{4}$
Boys	5	$\frac{\text{Boys}}{\text{Total}} = \frac{5}{20}$	$\frac{1}{4}$

Afternoon Class	Students	Ratio	Simplified Ratio
Girls	20	$\frac{\text{Girls}}{\text{Total}} = \frac{20}{30}$	$\frac{2}{3}$
Boys	10	$\frac{\text{Boys}}{\text{Total}} = \frac{10}{30}$	$\frac{1}{3}$

Combined Classes	Students	Ratio	Simplified Ratio
Girls	35	$\frac{\text{Girls}}{\text{Total}} = \frac{35}{50}$	$\frac{7}{10}$
Boys	15	$\frac{\text{Boys}}{\text{Total}} = \frac{15}{50}$	$\frac{3}{10}$

Section 6.2 **1.** No. In a rate, the units of measure are always different and will never divide out. **3.** Yes, the value of 1 must always be in the denominator so that we can get *per 1*. **5.** True **7.** denominator **9.** 65 stitches per inch **11.** 21 gallons for every 4 weeks **13.** 1 supervisor for every 6 employees **15.** 15 pounds for every 2 passengers **17.** 64 voters per precinct **19.** 1.25 yards of fabric per doll dress **21.** 2.5 kilowatts per hour **23.** 16.5 miles per gallon **25.** $\frac{276\text{ yards}}{4\text{ seconds}} = \frac{69\text{ yards}}{1\text{ second}}$, 69 yards per second **27.** $\frac{567\text{ dollars}}{1,620\text{ miles}} = \frac{0.35\text{ dollars}}{1\text{ mile}}$, \$0.35 per mile **29.** $\frac{9.10\text{ dollars}}{3.5\text{ pounds}} = \frac{2.60\text{ dollars}}{1\text{ pound}}$, \$2.60 per pound **31.** $\frac{459\text{ at bats}}{34\text{ home runs}}$, $\frac{13.5\text{ at bats}}{1\text{ home run}}$, 13.5 at bats per home run **33.** Bag of 18 Goodie candy bars: $\frac{0.27\text{ dollars}}{1\text{ candy bar}}$, \$0.27 per Goodie candy bar; Bag of 25 Yummie candy bars: $\frac{0.26\text{ dollars}}{1\text{ candy bar}}$, \$0.26 per Yummie candy bar; A bag of Yummie candy bars is the better buy. **35.** 12-pack of Zapper energy water: $\frac{1.10\text{ dollars}}{1\text{ bottle}}$, \$1.10 per Zapper energy water; 18-pack of Fizzit energy water: $\frac{1.15\text{ dollars}}{1\text{ bottle}}$, \$1.15 per Fizzit energy water; A pack of Zapper energy water is the better buy. **37.** 16.5-ounce bag of Mayse Corn Chips: $\frac{0.24\text{ dollars}}{1\text{ bag}}$, \$0.24 per ounce; 12.5-ounce bag of Bluze Corn Chips: $\frac{0.22\text{ dollars}}{1\text{ bag}}$, \$0.22 per ounce; A bag of Bluze Corn Chips is the better buy. **39.** The price of the 24-ounce bag should be $4.32.

Section 6.3 **1.** No, proportions are only for equivalent ratios. **3.** $x = 6$ **5.** $x = 16$ **7.** $x = 30$ **9.** $x = 16$ **11.** $x = 20$ **13.** $x = 12$ **15.** $x = 25$ **17.** $x = 4$ **19.** There are 12 juniors in the concert choir. **21.** 450 people voted. **23.** Cheyenne will be able to drive 210 miles. **25.** 60 square yards of flooring material can be purchased for $900. **27.** $14 will buy 30 feet of chain. **29.** It will take him 70 minutes to ride 20 miles. **31.** Her car can travel 329 miles on 14 gallons of gas. **33.** Tim can run 6 miles in 45 minutes. **35.** It offers 40 English classes. **37.** Chandra will need to work out 50 minutes on the cross trainer to burn 600 calories. **39.** The missing length is 21 yards. **41.** The missing length is 12 inches. **43.** The height of the tree is 20 feet. **45.** The real distance from Dallas to Phoenix is 880 miles. **47.** $x = 12$ **49.** $x = 25$ **51.** We would expect him to hit 45 home runs. **53.** We would expect him to throw 132 strikeouts.

Section 6.4 **1.** Answers will vary. **3.** $\frac{9}{20}$ **5.** $\frac{5}{4}$ **7.** $\frac{7}{20}$ of all workers in San Francisco use public transportation to commute to and from work. **9.** $\frac{13}{40}$ of the book sales at Bookends are romance novels. **11.** 0.16 **13.** 0.045 **15.** 0.02 **17.** 1.84 **19.** 0.85 **21.** 2.6 **23.** 0.04 **25.** 0.03 **27.** 0.052 **29.** 0.172 **31.** 0.0775 **33.** 0.05 **35.** 78% **37.** 6% **39.** 90% **41.** 211% **43.** 140% **45.** 12.8% **47.** 6.2% **49.** 0.1% **51.** 40% **53.** 94% **55.** 56% **57.** 125% **59.** 62.5% **61.** 56.3% **63.** 77.8% **65.** 16.7% **67.** 26.7% **69.** 40.9% **71.** 2.5% of the U.S. population has a last name of Smith, Johnson, or Williams. **73.** By the All-Star break, the Cleveland Cavaliers had won 55.6% of their games. **75.** 0.005 **77.** 0.01625

Section 6.5 **1.** No, $\frac{25}{100}$ is a ratio. A proportion is always an equation, one ratio = another (equivalent) ratio. **3.** 54 of the faculty are tenured. **5.** 88 customers ordered a frappuccino. **7.** 299 of the mailboxes have been rented. **9.** 45 families have a backyard pool. **11.** Jesse has 120 cards in all. **13.** A total of 350 people voted in the town. **15.** The total number of female members at the club is 260. **17.** There are a total of 160 employees at the factory. **19.** 36% of the faculty computers are Macs. **21.** 25% of the renters check their mailbox every Sunday. **23.** 17.5% of the full-time faculty is in Bala's department. **25.** 44.4% of the families have children in preschool. **27.** 132 employees do not smoke cigarettes.

Section 6.6 **1.** He is referring to the rate of commission he gets, which is 6% $\left(\frac{6\text{ cents}}{1\text{ dollar}} = \frac{6\text{ cents}}{100\text{ cents}} = \frac{6}{100} = 6\%\right)$. **3.** 14 **5.** 27 **7.** 72 **9.** 28.8 **11.** 60 **13.** 49 **15.** 40% of 125 is 50. **17.** 60% of 55 is 33. **19.** 44% of 75 is 33. **21.** 25% of 62 is 15.5 **23.** 75% of 76 is 57. **25.** 40% of 75 is 30. **27.** 30% of 190 is 57. **29.** 35% of 80 is 28. **31.** 60% of 35 is 21. **33.** 65% of 120 is 78. **35.** 42.5% of 80 is 34. **37.** 57.5% of 40 is 23. **39.** $27.\overline{7}\%$ of 18 is 5. **41.** $61.\overline{1}\%$ of 126 is 77. **43.** 145 skaters were there that night. **45.** 520 runners were in the marathon. **47.** 150 players are in the league. **49.** The silent auction had 152 items. **51.** 45% of the class brought their books on the first day. **53.** 36% of Leonard's clocks don't have chimes. **55.** 37.5% of Shawna's e-mails were spam. **57.** 32.5% of the diners used coupons. **59.** 65% of the customers did not order a balloon bouquet.

Section 6.7 **1.** The wages will be different in May than in March because the 20% is based on different numbers. In April, the 20% is based on the original March wage. In May, the 20% is based on the new (lower) April wage. **3.** Sue paid $81.86 at the register. **5.** Maria paid $50.16 at the register. **7.** Lorena made $111.30 commission that day. **9.** Gretchen's commission is 3%. **11.** The new price of the mower is $294. **13.** Eleazar will pay $179.96 using his employee discount. **15.** The new price for all of the fabric was $87.98. **17.** Guillermo has leased 45% of the stores. **19.** Mandy received 18% of the bill as a tip. **21.** George sold 15% of the motorcycles during the week. **23.** The increase during Louis's time as club president was 25%. **25.** The increase in their monthly gift is 10%. **27.** The decrease of Alice's weight was 8%. **29.** The decrease of the selling price was 7.5%. **31.** The account gained $144 interest after 2 years. **33.** The account gained $105 interest after $\frac{1}{2}$ year. **35.** The account gained $300 interest after 4 months. **37.** The account gained $200 interest after 8 months. **39. a)** The total price is $114.01. **b)** The sales tax is $9.12. **c)** The total amount that Charlene will pay is $123.13. **41. a)** The sales tax is $10.39. **b)** The tip should be $24.12. **c)** The total amount that Henry paid for dinner was $168.51. **43.** At the end of May, the value of the ten shares of Jozo stock was $415.80.

Chapter 6 Review Exercises **1.** 25, 24 **2.** cross multiplication **3.** ratios **4.** corresponding **5.** multiply **6.** principal **7.** True **8.** True **9.** False **10.** False **11.** $\frac{12\text{ cookies}}{1\text{ box}}$ **12.** $\frac{45\text{ cents}}{1\text{ doughnut}}$ **13.** $\frac{3\text{ quarts}}{8\text{ children}}$ **14.** $\frac{\$7}{10\text{ cards}}$ **15.** 25 players per game **16.** $85 per week **17.** 120 passengers for every 5 cars **18.** 80 miles for every 3 gallons **19.** $\frac{\$72}{9\text{ shares}} = \frac{\$8}{1\text{ share}}$ **20.** $\frac{18\text{ minutes}}{45\text{ minutes}} = \frac{2}{5}$ **21.** $\frac{2.4\text{ acres}}{3\text{ acres}} = \frac{4}{5}$ **22.** $\frac{0.6\text{ calories}}{1.5\text{ grams}} = \frac{2\text{ calories}}{5\text{ grams}}$ **23.** $\frac{12\text{ pounds}}{15\text{ pounds}} = \frac{4}{5}$ **24.** $\frac{20\text{ inches}}{32\text{ inches}} = \frac{5}{8}$ **25. a.** $120 : $90, $\frac{\$120}{\$90}, \frac{4}{3}$ **b.** $120 : $210, $\frac{\$120}{\$210}, \frac{4}{7}$ **c.** $90 : $210, $\frac{\$90}{\$210}, \frac{3}{7}$ **26. a.** $180 : $120, $\frac{\$180}{\$120}, \frac{3}{2}$ **b.** $180 : $300, $\frac{\$180}{\$300}, \frac{3}{5}$ **c.** $120 : $300, $\frac{\$120}{\$300}, \frac{2}{5}$ **27.** Yes **28.** No **29.** No **30.** Yes **31.** $65 per day **32.** 15 customers for every 2 hours **33.** 1 hit for every 3 times at bat **34.** 3 repetitions for every 5 minutes **35.** 4 golf balls per player **36.** 11.4 students per tutor **37.** 11.8 phone calls per day **38.** 18.5 miles per gallon **39.** $225 per resident **40.** 22.6 points per game **41.** $22.50 per tree **42.** $4.56 per chair **43.** 16-ounce jar of Redd's tomato paste: $0.12 per ounce; 24-ounce jar of Green's tomato paste: $0.15 per ounce; The Redd's tomato paste is the better buy. **44.** Pack of 16 AlwaysReady batteries: $0.81 per battery; Pack of 28 LightShine batteries: $0.75 per battery; The LightShine batteries are the better buy. **45.** $n = 21$ **46.** $n = 15$ **47.** $n = 72$ **48.** $n = 35$ **49.** Kjell's car will travel 328 miles. **50.** Lonnie will charge $123.

51. Karrie raised \$144. **52.** It will take 55 seconds to print 22 pages. **53.** The real distance between Denver and El Paso is 560 miles. **54.** The height of the flagpole is 14 feet. **55.** $\frac{1}{5}$ of the homes in Westside Estates are ranch style. **56.** $\frac{3}{20}$ of all fruit sold in Bryson's Grocery Store is organically grown. **57.** $\frac{2}{25}$ of all U.S. residents live in Texas. **58.** $\frac{1}{40}$ of the medals awarded at the 2004 Summer Olympics went to athletes from Ukraine. **59.** 0.47 **60.** 0.03 **61.** 0.091 **62.** 0.006 **63.** 26% **64.** 2% **65.** 17.5% **66.** 70% **67.** 106% **68.** 0.5% **69.** 15% of the Sri Lankan population is Hindu **70.** 64% of the shoppers at Lucia's Floristas pay with a credit card. **71.** 87.5% of the students in Ms. Skiba's art class received a passing grade. **72.** 28.6% of all U.S. car sales are small cars. **73.** 6 of the choir members are tenors. **74.** 12 of the engineers have been at Hudson Dynamic at least 10 years. **75.** There were a total of 140 dinner customers. **76.** There are a total of 320 teachers in the district. **77.** It rained 60% of the days in April. **78.** The movie won 22.2% of the Oscars. **79.** 38 **80.** 49 **81.** 10 **82.** 6 **83.** 25% of 76 is 19. **84.** 60% of 85 is 51. **85.** 28% of 125 is 35. **86.** 55% of 160 is 88. **87.** 75% of 52 is 39. **88.** 37.5% of 120 is 45. **89.** 4,000 students attend Cuyama College. **90.** 15% of these players hit at least 30 home runs. **91.** DuJuan will pay \$86 at the cash register. **92.** Sandra earned \$6,500 in commission. **93.** The new price of the box of bulbs was \$4.90. **94.** Torraye's free throw success was 62.5%. **95.** Lorraine's savings was 25% of the retail price. **96. a.** The cost of the meal was \$36. **b.** The sales tax was \$2.16. **c.** The tip was \$9.60. **d.** The total amount was \$47.76. **97.** The increase in the Chamber's membership is 25%. **98.** The increase in the value of Corrine's home is 12.5%. **99.** The decrease in water level that week was 37.5%. **100.** The decrease in the number of students in Dr. Ortega's statistics class was 26.7%. **101.** The account gained \$126 after $\frac{3}{4}$ year. **102.** The account gained \$130 after 3 months.

Chapter Test 6 **1.** $\frac{\$60}{\$36} = \frac{5}{3}$ **2.** $\frac{2.8 \text{ miles}}{0.8 \text{ gallons}} = \frac{7 \text{ miles}}{2 \text{ gallons}}$ **3.** No **4.** Yes **5.** 21 seats for every 4 rows **6.** 208 milligrams of potassium per serving **7.** Bag of 15 avocados: \$0.56 per avocado; Box of 12 avocados: \$0.60 per avocado; The bag of 15 avocados is the better buy. **8.** $x = 24$ **9.** Carl would have paid \$15.30. **10.** It would take him 50 days to finish a 20-pound bag. **11.** $x = 18$ inches **12.** 0.07 **13.** 0.005 **14.** 57% **15.** 20% **16.** $\frac{4}{25}$ **17.** $\frac{9}{20}$ **18.** 52% **19.** 36.4% **20.** 14% of 150 is 21. **21.** 40% of 80 is 32. **22.** 840 of the students at Riverbend are teenagers. **23.** Lin-Li will pay \$94.60 for the textbook at the cash register. **24.** Yusef's discounted price will be \$51. **25.** The increase in his hourly wage was 25%. **26.** The account gained \$100 interest after 5 months.

Chapters 1–6 Cumulative Review **1.** 3,100 **2.** 30,000,000 **3.** Identity of Multiplication **4.** Distributive Property of Multiplication over Addition **5.** Mt. Everest is 12,231 feet higher than Mt. Ararat. **6.** Mt. McKinley is 5,826 feet higher than Mt. Whitney. **7.** 200,000 **8.** 145 **9.** 512 **10.** 10^7 **11.** $7 \cdot 10^5$ **12.** Prime: 19, 29; Composite: 9, 14; Neither: 0, 1 **13.** It is a factor. **14.** $2 \cdot 2 \cdot 3 \cdot 5 \cdot 7$ **15.** -14 **16.** 13 **17.** -13 **18.** -7 **19.** 25 **20.** 6 **21.** 9 **22.** -56 **23.** -120 **24.** 5 **25.** -3 **26.** 3 **27.** -1 **28.** -2 **29.** -19 **30.** 1 **31.** $m = -3$ **32.** $m = -4$ **33.** $-12y^3$ **34.** $3x$ **35.** $-w^3 + 3w$ **36.** $8x^2$ **37.** $-16y$ **38.** $18w - 30$ **39.** $15q - 21$ **40.** $y = -4$ **41.** $w = -3$ **42.** Andrea will receive \$15,000, and Becca will receive \$11,500. **43.** The GCF is 6. **44.** $\frac{5}{2p}$ **45.** $\frac{x}{3}$ **46.** $-\frac{45b}{8}$ **47.** $\frac{2w}{7}$ **48.** $-\frac{y}{14}$ **49.** 9 square feet **50.** 3 square inches **51.** $\frac{7}{5}p^2$ **52.** $\frac{4}{5}c$ **53.** $\frac{2}{5}xy$ **54.** $\frac{3}{10}$ **55.** $\frac{1}{6}$ **56.** $\frac{13}{24}$ **57.** Marco worked $14\frac{2}{3}$ hours that weekend. **58.** $c = -\frac{7}{18}$ **59.** $v = -\frac{4}{5}$ **60.** $x = 200$ **61.** \$10.60 **62.** \$532 **63.** The perimeter is 16.906 inches. **64.** The length of the third side is 3.55 meters. **65.** 0.072 **66.** 30.8 **67.** 0.615 **68.** 20.6 **69.** 0.875 **70.** Petre earned \$17.40 per hour. **71.** The area of the pool is 205.8 square yards. **72.** The circumference of the baking dish is about 30.1 inches. **73.** $\frac{2.7 \text{ grams}}{0.45 \text{ grams}} = \frac{6}{1}$ **74.** $\frac{1.9 \text{ calories}}{1 \text{ gram}}$; 1.9 calories per gram **75.** 48.75 miles per hour **76.** 24-ounce of Twinklebuck's: \$0.095 per ounce; 20-ounce of Sunbuck's: \$0.09 per ounce; Sunbuck's coffee has the better value. **77.** $x = 180$ miles **78.** $x = 12$ **79.** 0.032 **80.** 40% **81.** It will take Zia 20 minutes to cook 90 pancakes. **82.** 45% of 140 is 63. **83.** 64% of 75 is 48. **84.** The discounted price will be \$44. **85.** The decrease in the Torres's car payment is 40%. **86.** The account gained \$52.50 interest after 9 months.

Chapter 7

Preparation Exercises **1.** 125 **2.** 81 **3.** $2^3 \cdot 5$ **4.** $2 \cdot 3^3$ **5.** 5 **6.** -6 **7.** 8 **8.** -19 **9.** -14 **10.** 4 **11.** -48 **12.** 36 **13.** -15 **14.** 5 **15.** $4y$ **16.** $-8x$ **17.** 4 **18.** 15 **19.** $\frac{2}{5}$ **20.** $-\frac{3}{4}$

Section 7.1 **1.** -2^2 is -4, but $(-2)^2$ is $+4$. Because in -2^2 there are no parentheses grouping the negative with the base of 2, the negative is not included in the squaring process and -2^2 is the same as $0 - 2^2$. (This suggests that it doesn't matter how close the negative sign is to the 2—it still is not included in the squaring unless it is grouped within parentheses.) **3.** $36 = 2^2 \cdot 3^2 \cdot 5^0 \cdot 7^0$ **5.** 9 **7.** $\frac{5}{6}$ **9.** y **11.** $-\frac{1}{9}$ **13.** -3 **15.** 1 **17.** 1 **19.** 1 **21.** 1 **23.** 1 **25.** x^8 **27.** y^4 **29.** v^{10} **31.** m^{25} **33.** x^{10} **35.** p^2 **37.** h^{10} **39.** r^2 **41.** p^5 **43.** w^1 **45.** y^3 **47.** x^2 **49.** a^{12} **51.** k^{21} **53.** x^{16} **55.** b^{18} **57.** p^5 **59.** w^2 **61.** 1 **63.** 1 **65.** q^{14} **67.** $x^{24}y^{21}$ **69.** p^{22}

Section 7.2 **1.** 0. The degree of a term is the number of variable factors it has. A constant term has no variable factors; so, its degree must be zero. **3.** No, we only add exponents when the variables are being multiplied together. **5.** x^2 and $2x$

7. $-5c^2, -4c$, and 1 **9.** $x^4y, \frac{5}{8}xy^4$, and $-\frac{1}{2}$ **11.** trinomial **13.** monomial **15.** monomial **17.** binomial **19.** not a polynomial **21.** binomial **23.** 8 **25.** -10 **27.** 44 **29.** 28 **31.** 1 **33.** -9 **35.** -9 **37.** 51 **39.** 18 **41.** 15 **43.** -9 **45.** -7 **47.** $x \cdot x$ degree 2 **49.** $y \cdot y$ degree 2 **51.** p degree 1 **53.** (none) degree 0 **55.** $x \cdot y \cdot y \cdot y \cdot y \cdot y$ degree 6 **57.** $a \cdot b$ degree 2 **59.** $3x + 4$ **61.** $-w - 11$ **63.** $7b^3 + 4b$ **65.** $4x^2 + 5x - 6$ **67.** $-w^4 - w^2 + 6$ **69.** $-8k^5 - 2k^3 + 6k$ **71.** $10a^3 - 6a^2 + 9a - 1$ **73.** $-y^3 - 5y^2 - y + 1$ **75.** $10x^3$ **77.** $-7m$ **79.** $3p^2 + 4p$ **81.** $4y^2$ **83.** $-2n^3$ **85.** 0 **87.** $13x^4 + 1$ **89.** $-2y^2 + 9$ **91.** $-v^3 + 13v^2$ **93.** $-y^3 + x^2$ **95.** $10x + 4y$ **97.** $-x^2 - 6x$ **99.** $-7w^2 - 6w - 9$ **101.** $5p^5 - 7$ **103.** $x^2 - 7$ **105.** The ball was 1,056 feet above the ground. **107.** The rock was 296 feet above the ocean. **109.** The area of the rectangle is 48 square inches. **111.** $600 - 144 =$ **456** $536 - 456 =$ **80 feet**; **113.** $600 - 400 =$ **200** $344 - 200 = 144$ feet; **115.** The distance fallen each second is increasing. Answers may vary.

Section 7.3 **1.** The most important step is to distribute -1 through to the entire second polynomial **3.** $7x^2 + 5x$ **5.** $-4c - 6$ **7.** $p^2 - 6p$ **9.** $-5x^2 - 4x + 1$ **11.** $3k^3 + 7k^2 - k$ **13.** $-3v^3 + 2v + 1$ **15.** $3x^2 - 2x$ **17.** $n^2 - 2n$ **19.** $c^2 - 10c - 2$ **21.** $3y^3 + 3y - 9$ **23.** $x^2 + 11$ **25.** 0 **27.** $-n^2 + 2n - 2$ **29.** $-3c^3 - 7c + 9$ **31.** $-8q^2 + q + 12$ **33.** $y^3 + 2$ **35.** $9v - 6$ **37.** 11 **39.** $w - 5$ **41.** $-3n$ **43.** $9q^3 - q + 1$ **45.** $-n^2 + 2n + 20$ **47.** $-11c + 6$ **49.** $-7p - 13$ **51. a)** The perimeter is given by $4x^2 + 6x - 6$ **b)** No. When $x = 10$ meters, the width is negative, which is not possible.

Section 7.4 **1.** There will be nine terms in the product. Three terms times three terms equals nine terms before simplifying. **3.** Answers will vary. One possibility: When replacing the variable, such as x, with 1, it doesn't matter what the exponent is $1^2 = 1, 1^3 = 1, 1^4 = 1$, and so on. So, we'd never know if the exponent was correct using this technique. **5.** $6y^5$ **7.** $-3b^3$ **9.** $18k^6$ **11.** $-x^7$ **13.** $16v^8$ **15.** p^6 **17.** $9c^4$ **19.** $4n^{12}$ **21.** $10q^4 - 30q$ **23.** $-h^5 - 4h^3$ **25.** $-10v^4 + 20v^2$ **27.** $4x^5 - 5x^4 + 9x^3$ **29.** $-6y^6 + 3y^3 - 12y^2$ **31.** $a^4 + a^3 - 9a^2$ **33.** $w^3 + 7w^2 + 13w + 6$ **35.** $4c^3 - 18c^2 - 14c - 2$ **37.** $4x^3 - 9x^2 - 17x - 6$ **39.** $m^2 + 11m + 28$ **41.** $x^2 - 9x - 36$ **43.** $4x^2 - 10x - 6$ **45.** $q^2 + 11q + 24$ **47.** $m^2 - 36$ **49.** $12c^2 - 13c + 3$ **51.** $4p^2 - 81$ **53.** $16x^4 - 9$ **55.** $x^2 + 14x + 49$ **57.** $y^2 - 2y + 1$ **59.** $9v^2 + 30v + 25$ **61.** $25y^4 - 60y^2 + 36$ **63. a)** The area is $-15x^2 - 37x + 8$ square yard **b)** Yes. When $x = -2$, the length is 11 yards and the width is 2 yards.

Section 7.5 **1.** $c^0 = 1$, and it is okay to divide by 1. **3.** x^4 **5.** c^4 **7.** y^7 **9.** m^6 **11.** $x^0 = 1$ **13.** y^8 **15.** x^3 **17.** w^8 **19.** $4x^3$ **21.** $4y$ **23.** -5 **25.** $-3c^2$ **27.** $5x^4$ **29.** $-6y^9$ **31.** $x^3 + 3x$ **33.** $-3w^4 + 1$ **35.** $8m^6 - m^3 + 12$ **37.** $a^2 - 2a + \frac{1}{2}$ **39.** $y^2 - 5y + 2$ **41.** $2c^2 + 3c - 1$ **43. a)** $\frac{c^4}{c^5} = \frac{c \cdot c \cdot c \cdot c}{c \cdot c \cdot c \cdot c \cdot c} = \frac{1}{c}$ **b)** $\frac{c^4}{c^5} = c^{4-5} = c^{-1}$ **45.** $c^{-1} = \frac{1}{c}$

Section 7.6 **1.** When finding the GCF, we match up as many common factors as possible. If one number has three factors of 3, 3^2, we can only match up the two factors of 3, the lesser of the two exponents. **3.** $6y$ **5.** $9a^2$ **7.** 1 **9.** $3x^2$ **11.** p^3 **13.** 15 **15.** $3(2x + 3)$ **17.** $8w^2(2w - 1)$ **19.** $10w^3(3w^3 + 2)$ **21.** $c(9c + 20)$ **23.** $6x^3(3x^2 - 4)$ **25.** $4y(y^3 - 5)$ **27.** $3x(2x^2 - 5x + 3)$ **29.** $2x^3(2x^3 + 3x^2 - 1)$ **31.** $6(3x^2 - 4x - 1)$ **33.** prime **35.** $4x^2(3x^3 - 5x^2 + 1)$ **37.** $x^2(9x^2 - 12x - 16)$ **39.** $-3(w - 5)$ **41.** $-4y(y + 3)$ **43.** $-5y^2(3y^2 - 1)$ **45.** $-1(2p^2 + 15)$ **47.** $-2(4w^2 + 2w - 1)$ **49.** $-m(m^2 + 2m + 6)$ **51.** $(5m + 7)(2m + 3)$ **53.** $(2x^3 - 5)(4y^2 + 7)$

Chapter 7 Review Exercises **1.** Product **2.** coefficient **3.** binomial **4.** 3 **5.** 0 **6.** descending order **7.** Distributive **8.** Quotient **9.** 5 **10.** -6 **11.** 1 **12.** 1 **13.** y^{10} **14.** q^4 **15.** p^8 **16.** k^5 **17.** c^{14} **18.** x^9 **19.** w^5 **20.** 1 **21.** $3x^4$ and $-6x^2$ **22.** $-x^4, 4x^2$ and -1 **23.** trinomial **24.** binomial **25.** not a polynomial **26.** monomial **27.** 1 **28.** 45 **29.** 4 **30.** 8 **31.** Laura was 12,100 feet above the ground. **32.** Laura was 10,900 feet above the ground. **33.** $v \cdot v \cdot v \cdot v$ degree 4 **34.** $x \cdot x \cdot x \cdot x \cdot y \cdot y$ degree 6 **35.** $x \cdot y \cdot y \cdot y \cdot y \cdot y \cdot y \cdot y$ degree 8 **36.** none degree 0 **37.** $-3x^6 + 9x^4 + 6x$ **38.** $3y^4 - 12y^3 - 6y + 8$ **39.** $2x^2 - 4x$ **40.** $6p^4 - p^3 + 2p - 7$ **41.** $2w - 10$ **42.** $-2x^2 + 6x - 3$ **43.** $2b^2 + 4b - 1$ **44.** $y^4 - 7y^3 - 3y$ **45.** $-2v - 5$ **46.** $4y^2 - 9y + 1$ **47.** $-2x^2 + x + 10$ **48.** $-2q + 5$ **49.** $-2x^2 + 10x$ **50.** $-9x^2 + 3x - 2$ **51.** $k^2 - 14k + 10$ **52.** $4m^2 - 7m - 4$ **53.** $8q^4$ **54.** $8x^{11}$ **55.** $16p^2$ **56.** w^{10} **57.** $24x^5 - 40x^2$ **58.** $40w^6 - 20w^4$ **59.** $2p^5 - 6p^4 + 4p^2$ **60.** $30y^3 - 15y^2 - 20y$ **61.** $8x^2 - 2x - 15$ **62.** $q^2 - 12q + 32$ **63.** $m^2 - 81$ **64.** $25x^2 - 16$ **65.** $p^2 - 8p + 16$ **66.** $4y^2 - 20y + 25$ **67.** $w^3 - w^2 - 14w - 6$ **68.** $6x^3 - 13x^2 + 9x - 2$ **69.** c^3 **70.** k^5 **71.** x^4 **72.** $p^0 = 1$ **73.** v^1 **74.** y^9 **75.** $3x^4$ **76.** -4 **77.** $9q^4$ **78.** $-4y^6$ **79.** $3y^3 - 8y$ **80.** $-3x^4 - 1$ **81.** $5m^4 - m^2 + 3$ **82.** $-4q^4 + 3q^2 - 1$ **83.** $6y^2$ **84.** $4x$ **85.** x^2 **86.** $2w^3$ **87.** $2(2c - 5)$ **88.** $3x(x - 4)$ **89.** prime **90.** $9q(5q^2 + q - 2)$ **91.** $5y^2(4y^2 - 3y + 1)$ **92.** $-5x^2(2x^2 - 5)$ **93.** $-4(2p - 3)$ **94.** $-3m(2m + 1)$ **95.** $-5x^2(2x^2 - 5)$ **96.** $-3(5y^3 - 3y^2 + 1)$ **97.** $-c^2(c^4 - 6c^2 - 9)$ **98.** $-4w(2w^2 + w - 1)$

Chapter 7 Test **1.** 1 **2.** 10 **3.** m^8 **4.** x^5 **5.** w^{10} **6.** 1 **7.** x^4 **8.** -44 **9.** 10 **10.** The arrow was 126 feet above the ground. **11.** degree 5 **12.** degree 7 **13.** degree 0 **14.** $5m^4 - 4m^3 + 6m - 12$ **15.** $-x^2 + 6x - 7$ **16.** $3c^2 - 10c$ **17.** $-30m^5$ **18.** $49q^2$ **19.** $-10y^3 + 8y^2 + 6y$ **20.** $2w^2 - 14w + 24$ **21.** $x^2 + 8x + 16$ **22.** $4y^3 - 23y^2 + 19y - 3$ **23.** x^6 **24.** w^5 **25.** $-3p^3$ **26.** 7 **27.** $-3x^4 - 1$ **28.** $6q^4 - q^2 + 2$ **29.** $4x(3x - 5)$ **30.** prime **31.** $5p^2(3p^4 + 4p^2 - 1)$ **32.** $-6y^2(2y + 3)$ **33.** $-3x(4x^2 - 2x + 1)$

Chapter 8

Preparation Exercises **1.** $\frac{31}{6}$ **2.** $\frac{61}{8}$ **3.** $7\frac{11}{12}$ **4.** $4\frac{7}{16}$ **5.** $\frac{1}{10}$ **6.** $\frac{5}{6}$ **7.** $\frac{21}{2} = 10\frac{1}{2}$ **8.** 30 **9.** $\frac{3}{2} = 1\frac{1}{2}$ **10.** $\frac{3}{4}$ **11.** $\frac{3}{2} = 1\frac{1}{2}$ **12.** $\frac{13}{24}$ **13.** $\frac{47}{100}$ **14.** $\frac{3}{1,000}$ **15.** 0.63 **16.** 0.29 **17.** 2 **18.** 10.99 **19.** 58 **20.** 1.2 **21.** 14.75 **22.** 1.95 **23.** 3.9 **24.** $9.\overline{5}$

Section 8.1 **1.** First, convert 1 mile to feet and then multiply that figure by $\frac{1}{8}$ (to get 660 feet). Then write 22 yards in terms of feet (66 feet). **3.** One technique is to convert the 660 feet to yards, and the side (a *chain*) is already in terms of yards. Another possibility is to divide the 43,560 by 9, which achieves the same result. **5.** 48 inches **7.** 8 inches **9.** $5\frac{1}{3}$ yards **11.** $\frac{1}{3}$ foot **13.** 144 in. **15.** 80 in. **17.** 2 ft 4 in. **19.** 7 ft 9 in. **21.** The rim of a basketball hoop is 120 inches off the floor. **23.** 10 ft 10 in. **25.** 6 ft **27.** 5 ft 8 in. **29.** 3 ft 5 in. **31.** Rodney was 3 feet 2 inches tall on his first birthday. **33.** 80 ounces **35.** 8,000 pounds **37.** 144 oz **39.** 104 oz **41.** 1 lb 12 oz **43.** 5 lb 10 oz **45.** A home run slugger's baseball bat weighs 34 ounces. **47.** 10 lb 13 oz **49.** 8 lb **51.** 5 lb 12 oz **53.** 3 lb 9 oz **55.** The total weight of Liu and his baby carrier is 17 pounds 2 ounces. **57.** 10 pints **59.** $4\frac{3}{4}$ cups **61.** 11 cups **63.** $2\frac{1}{3}$ quarts **65.** 63,360 inches are in one mile. **67.** 12.5 tons are in 400,000 ounces.

Section 8.2 **1.** Answers will vary. One possibility: We use a capital *L* so that we don't confuse it with the number 1. **3.** 3,080 m **5.** 450 L **7.** 0.87 g **9.** 0.091 L **11.** 3.65 m **13.** 0.49 g **15.** 0.06 L **17.** 0.4 m **19.** 360 dekameters **21.** 3.12 decigrams **23.** 0.009 kiloliter **25.** 25 decimeters **27.** 18.1 deciliters **29.** 8.3 hectometers **31.** 3.7 decigrams **33.** 0.6 centiliter **35.** A child weighs 15,600 grams. **37.** Veronica's rosebush is 0.835 meter high. **39.** A dwarf pygmy goby has a diameter of 9.9 millimeters. **41.** 2,011 centigrams **43.** 3.22 dekaliters **45.** 15.7 centigrams **47.** 1,316 deciliters **49.** 1,625 milliliters of fluid are now in the bag. **51.** Jermaine added a total of 13.4 liters to his gas tank. **53.** Tamayra grew 8.7 centimeters that year. **55.** The full bucket holds 357.8 centiliters.

Section 8.3 **1.** Answers will vary. One possibility: It's okay to write it as $14\frac{2}{5}$; but because the unit of measure is metric, it is more common to write the number (or approximation) in decimal form. So, 14.4 is preferred. **3.** 14 yards **5.** 7 gallons **7.** $4\frac{1}{2}$ cups **9.** 0.95 meter **11.** 0.06 gram **13.** 4.5 liters **15.** 30 centimeters **17.** 15 miles **19.** 12 meters **21.** 47.5 liters **23.** 3.75 kilograms **25.** $14\frac{3}{4}$ ounces **27.** A foot race is $6\frac{1}{4}$ miles. **29.** A window is 120 centimeters wide. **31.** Sammy weighs $60\frac{1}{2}$ pounds. **33.** Papi's pickup truck can hold 114 liters of gas. **35.** Myra weighs 7.188 kilograms.

Section 8.4 **1.** No. A real plane has no outside boundaries, and the Bonneville Salt Flats has mountains on at least one side. (Also, a student might mention that because the earth is round and a plane is flat, the "Flats" cannot be a real plane.) **3.** 100 grads are in a right angle. **5.** $\angle 1$, $\angle G$, $\angle RGW$, and $\angle WGR$ **7.** $\overleftrightarrow{ET}$ and $\overleftrightarrow{DL}$ **9.** $\angle KRB = 53°$ **11.** $\angle SAP = 94°$ **13.** $\angle WXI = 103°$ **15.** $\angle YEO = 48°$ **17.** $\angle LYM = 63°$ **19.** $\angle MUJ = 18°$ **21.** $\angle 1 = 32°$, $\angle 2 = \angle 4 = 148°$ **23.** $\angle 2 = 111°$, $\angle 1 = \angle 3 = 69°$ **25.** straight **27.** vertical **29.** False **31.** True **33.** False **35.** True **37.** The measure of each angle is 90°. **39.** No, a line segment drawn from a point in the interior of an angle to the exterior of the angle must cross one of the sides of the angles.

Section 8.5 **1.** Yes, any two of the angles in an equilateral triangle can be chosen to be the base angles of an isosceles triangle. **3.** Yes, the two sides that form the right angle can be the same length. **5.** $\angle C = 80°$ **7.** $\angle C = 70°$ **9.** $\angle C = 12°$ **11.** $\angle C = 46°$ **13.** $\angle C = 75.4°$ **15.** $\angle C = 70.2°$ **17.** $\angle C = 85.15°$ **19.** $\angle C = 55.06°$ **21.** scalene **23.** isosceles **25.** equilateral and isosceles **27.** isosceles **29.** $\angle C = 54°$ **31.** $\angle F = 17°$ **33.** $\angle N = 61.7°$ **35.** $\angle T = 37.5°$ **37.** $\angle P = 35°$, $\angle Q = \angle O = 145°$, $\overline{OP} = 3$ in., and $\overline{QP} = 7$ in. **39.** $\angle J = 127.4°$, $\angle G = \angle I = 52.6°$, $\overline{IJ} = 3.9$ m, and $\overline{JG} = 6.4$ m **41.** False **43.** False **45.** True **47.** True **49.** $d = 12$ in. **51.** $r = 10$ in. **53.** $d = 13$ cm **55.** $r = 3\frac{2}{3}$ ft **57.** Each angle measures 60°. The total angle measure is 180°. **59.** Each angle measures 108°. The total angle measure is 540°.

Section 8.6 **1.** No, $\frac{22}{7}$ is only an approximation for π. π is not a repeating decimal. **3.** The perimeter of Karin's lamp stand is 82 inches. **5.** The perimeter of home plate is 58 inches. **7.** The outer perimeter of the Pentagon is 4,605 feet. **9.** The circumference of the resonator is about 37.7 inches. **11.** The circumference of this hula hoop is about 6.3 feet. **13.** The circumference of this crop circle is about 1,256 feet. **15.** $P = 21.56$ cm **17.** $P = 2\frac{1}{3}$ in. **19.** $P = 72$ ft **21.** $P = 2\frac{2}{3}$ in. **23.** $P = 52$ ft **25.** $P = 94.2$ m **27.** $P = 740$ yd **29.** $P = 14\frac{1}{2}$ in. **31.** side $= 17$ in. **33.** side $= 17.9$ cm **35.** side $= 17\frac{1}{2}$ yd **37.** side $= 1\frac{1}{3}$ ft **39.** $C \approx 88$ in. **41.** $C \approx 17.6$ cm **43.** $C \approx 5\frac{1}{2}$ in. **45.** $C \approx 22$ ft **47.** $C \approx 31.4$ yd **49.** $C \approx 28.3$ m **51.** $C \approx 25.1$ ft **53.** $C \approx 20.4$ cm **55.** $X = 11$ in., $Y = 44$ in., $P = 158$ in. **57.** $X = 13.2$ m, $P = 55.2$ m **59.** $X = \frac{5}{12}$ in., $Y = \frac{7}{12}$ in., $P = 3\frac{1}{6}$ in. **61.** The perimeter of Marta's garden is approximately 51.4 feet.

Section 8.7 **1.** Answers will vary. **3.** $A = 96\text{ yd}^2$ **5.** $A = \frac{7}{32}\text{ in.}^2$ **7.** $A = 44.2\text{ cm}^2$ **9.** $A = 4\frac{1}{2}\text{ ft}^2$ **11.** $A = 169\text{ ft}^2$ **13.** $A = \frac{49}{144}\text{ in.}^2$ **15.** $A = 108\text{ ft}^2$ **17.** $A = \frac{1}{3}\text{ in.}^2$ **19.** $A = 154\text{ in.}^2$ **21.** $A = \frac{5}{6}\text{ in.}^2$ **23.** $A = 135\text{ yd}^2$ **25.** $A = \frac{25}{32}\text{ in.}^2$ **27.** $A = 460\text{ ft}^2$ **29.** $A = 85\text{ m}^2$ **31.** $A \approx 28.3\text{ in.}^2$ **33.** $A \approx 4.5\text{ m}^2$ **35.** $A \approx 50.2\text{ yd}^2$ **37.** $A \approx 3.8\text{ cm}^2$ **39.** $A = 240\text{ in.}^2$ **41.** $A = 63\text{ ft}^2$ **43.** The area of this side is 448 square inches. **45.** The area of this room is 150 square feet. **47.** The area of the hot tub cover is about 20 square feet. **49.** The area of this crop circle is about 5,024 square meters. **51.** $A \approx 21.2\text{ cm}^2$

53. The area of the flat region is about 737.5 square meters. **55.** The area of Marta's garden is approximately 314 square feet. **57.** The area of Armando's property is 1,920 square yards.

Section 8.8 **1.** The volume formulas for the cylinder and the box is Area of Base × height. **3.** $V = 240 \text{ in.}^3$ **5.** $V = 2{,}400 \text{ ft}^3$ **7.** $V = 2{,}016 \text{ cm}^3$ **9.** $V = 17\frac{1}{2} \text{ in.}^3$ **11.** $V \approx 50.2 \text{ ft}^3$ **13.** $V \approx 20.3 \text{ m}^3$ **15.** $V \approx 47.1 \text{ ft}^3$ **17.** $V \approx 13.8 \text{ m}^3$ **19.** $V = 80 \text{ in.}^3$ **21.** $V = 7 \text{ m}^3$ **23.** $V \approx 4.2 \text{ ft}^3$ **25.** $V \approx 7.2 \text{ m}^3$ **27.** The volume of this space is about 576 cubic feet. **29.** The volume of the can is about 603 cubic centimeters. **31.** The volume of Rebecca's pyramid is about 47 cubic inches. **33.** The volume of the bowl is approximately 134.0 cubic inches.

Chapter 8 Review Exercises **1.** units of measure **2.** kilo- **3.** milli- **4.** plane **5.** vertex **6.** straight **7.** perpendicular **8.** quadrilateral **9.** False **10.** True **11.** False **12.** False **13.** True **14.** True **15.** True **16.** True **17.** 216 in. **18.** 119 in. **19.** 4 ft 2 in. **20.** 10 ft 10 in. **21.** A tennis racket is 27 inches long. **22.** At 7 feet 2 inches tall, Margo Dydek is the tallest woman professional basketball player in the WNBA. **23.** 12 ft **24.** 1 ft 6 in. **25.** The house is 13 feet 2 inches tall. **26.** The rest of its body is 8 feet 9 inches long. **27.** 47 feet **28.** 17 yards **29.** 10 pints **30.** $\frac{3}{4}$ pound **31.** 5 pints **32.** $1\frac{3}{4}$ cups **33.** 48 fluid ounces **34.** 3 gallons **35.** 18 fluid ounces **36.** $\frac{4}{5}$ pound **37.** 6 ounces **38.** 2,900 pounds **39.** 240 oz **40.** 82 oz **41.** 2 lb 8 oz **42.** 8 lb 14 oz **43.** A roasting chicken weighs 61 ounces. **44.** A double box of cereal weighs 4 pounds 12 ounces. **45.** 12 pounds **46.** 1 pound 12 ounces **47.** The total weight of these two books is 11 pounds 4 ounces. **48.** The remaining block of cheese weighs 13 pounds 7 ounces. **49.** 30.8 g **50.** 6,800 g **51.** 8.7 m **52.** 0.24 L **53.** 20 mm **54.** 9 cg **55.** 3.12 dL **56.** 0.12 kL **57.** 0.36 km **58.** 82 dam **59.** Jakhil ran 400 meters in one minute. **60.** A hummingbird weighs 1,600 milligrams. **61.** 2,011 centigrams **62.** 48.05 deciliters **63.** Martin drank 3.1 liters for his test. **64.** Foofi gained 480 grams during the year. **65.** $3\frac{1}{2}$ quarts **66.** 36 ounces **67.** $4\frac{1}{2}$ feet **68.** 27,000 grams **69.** 9,500 meters **70.** 12 liters **71.** 55 centimeters **72.** 24 inches **73.** 10 quarts **74.** 28.5 liters **75.** $1\frac{1}{4}$ ounces **76.** 15 meters **77.** The sack of potatoes weighs about 9.1 kilograms. **78. a.** Jorge is 72 inches tall. **b.** Jorge is about 180 centimeters tall. **79.** $\angle BRK = 48.4°$ **80.** $\angle EMD = 97.5°$ **81.** $\angle BLN = 54.8°$ **82.** $\angle RFX = 27.5°$ **83.** $\angle 1 = 32°, \angle 2 = \angle 4 = 148°$ **84.** $\angle C = 52°$ **85.** isosceles **86.** equilateral and isosceles **87.** $\angle G = 30.6°$ **88.** $\angle J = 66.2°$ **89.** $\angle P = 40°, \angle Q = \angle O = 140°; \overline{OP} = 5 \text{ cm}; \overline{QP} = 9 \text{ cm}$ **90.** $d = 18$ ft **91.** $r = 3\frac{2}{3}$ in. **92.** $r = 1.15$ m **93.** The perimeter of Leon's pool is $29\frac{5}{6}$ yards. **94.** The perimeter of this support structure is 28.1 meters. **95.** $X = 20$ ft, $Y = 8$ ft; the perimeter is 86 feet. **96.** The perimeter of the dance floor is 87 feet. **97.** $P = 7$ ft. **98.** $P = 30$ yd. **99.** Each side $= 14$ cm **100.** Each side $= 35$ in. **101.** Marcus needs about 132 feet of fencing to fit around the garden. **102.** About 94.2 inches of weather stripping is needed to go around the window. **103.** $A = 20 \text{ yd}^2$ **104.** $A = 3 \text{ m}^2$ **105.** $A = \frac{9}{64} \text{ in.}^2$ **106.** $A = 12\frac{1}{4} \text{ ft}^2$ **107.** $A = 200 \text{ in.}^2$ **108.** $A = 10 \text{ ft}^2$ **109.** $A = 5 \text{ in.}^2$ **110.** $A = 33 \text{ cm}^2$ **111.** $A \approx 12.6 \text{ ft}^2$ **112.** $A \approx 8.0 \text{ cm}^2$ **113.** The area of the surface of this pond is 2,826 feet. **114.** The area of Erica's living room is 144 ft^2. **115.** $A \approx 7.74 \text{ cm}^2$ **116.** $V = 540 \text{ in.}^3$ **117.** $V \approx 33.5 \text{ ft}^3$ **118.** $V \approx 314 \text{ in.}^3$ **119.** $V \approx 7.1 \text{ ft}^3$ **120.** $V = 160 \text{ m}^3$ **121.** The volume of this pyramid is 2,000 cubic yards. **122.** The volume of this storage area is 672 cubic feet.

Chapter 8 Test **1.** True **2.** True **3.** False **4.** False **5.** 18 ft 5 in. **6.** 13 ft 4 in. **7.** 81 inches **8.** 11 feet 10 inches **9.** 48 ounces **10.** $3\frac{3}{4}$ yards **11.** 0.28 L **12.** 45 m **13.** 0.091 m **14.** 890 g **15.** 92.7 mm **16.** 5.25 kg **17.** 32 kilometers **18.** $13\frac{3}{4}$ pounds **19.** Elisa put 2.5 gallons of gas into her motorcycle. **20.** $\angle RTW = 61.9°$ **21.** $\angle XAS = 144.5°$ **22.** $\angle WXI = 111.7°$ **23.** $\angle LYM = 58.9°$ **24.** $\angle C = 105.9°$ **25.** The perimeter of the canvas is 98 inches. **26.** $P = 30$ cm **27.** $A = 5\frac{4}{9} \text{ ft}^2$ **28.** $C \approx 110$ in. **29.** Each side $= 1\frac{1}{3}$ ft **30.** $A = 42 \text{ cm}^2$ **31.** The area of this kitchen is 576 square feet. **32.** $V \approx 376.8 \text{ cm}^3$ **33.** $V = 10 \text{ in.}^3$

Chapters 1–8 Cumulative Review **1. a.** Associative Property of Multiplication **b.** Identity for Multiplication **2.** 625 **3.** The area of the court is 2,205 square feet. **4.** $2 \cdot 2 \cdot 2 \cdot 5 \cdot 7$ $2^3 \cdot 5 \cdot 7$ **5.** $2 \cdot 2 \cdot 3 \cdot 3 \cdot 11$ $2^2 \cdot 3^2 \cdot 11$ **6.** Each class will receive \$2,930. **7.** 27 **8.** 2 **9.** −4 **10.** −120 **11.** −7 **12.** 12 **13.** y **14.** $y = 3$ **15.** It sold 140 small cups and 220 large cups. **16.** $8\frac{1}{3}$ **17.** $4\frac{2}{3}$ **18.** $8\frac{5}{18}$ **19.** $1\frac{2}{3}$ **20.** $5\frac{1}{2}$ **21.** $\frac{3}{4}$ **22.** $\frac{3}{4}$ **23.** $-\frac{5}{12}$ **24.** $\frac{2}{3}$ **25.** The area of this triangle is 6 square inches. **26.** $y = \frac{5}{6}$ **27.** $x = 1\frac{1}{3}$ **28.** Each shift was $2\frac{7}{20}$ hours. **29.** She needs to cut $2\frac{3}{8}$ inches off the end **30.** 2.18 **31.** $\frac{9}{100}$ **32.** \$28.10 **33.** \$90 **34.** 2.9 **35.** 13.826 **36.** He got back \$6.82 in change. **37. a.** 8.3 **b.** 8.2 **38.** 0.3125 **39.** Tara earned \$11.10 in tips per hour. **40.** The area of this canvas is 188.7 square centimeters. **41.** The area of the top of the cover is approximately 6.15 square meters. **42.** $\frac{3}{50}$ **43.** 9.2 dollars per hour **44. a.** \$1.60 per pound **b.** \$1.55 per pound **c.** Kibblefest is the better value. **45.** $x = 35$ **46. a.** 0.09 **b.** 0.015 **47. a.** 60% **b.** 4% **48. a.** \$180 **b.** \$120 **49.** It will take them 21 hours to put together 35 bikes. **50.** 60% of 85 is 51. **51.** The percent increase of this hourly wage is 25%. **52.** −5 **53.** 1 **54.** p^5 **55.** x^6 **56.** 1 **57.** $3x^2 - 4x$ **58.** $-2q^2 - q + 5$ **59.** $-15n^5$ **60.** $12q^3 - 20q^2 - 8q$ **61.** $20x^2 - 6x - 2$ **62.** $y^2 - 6y + 9$ **63.** m^5 **64.** k^6 **65.** $6p^3$ **66.** $-5x^6$ **67.** $5q^4 + 3q^2 - 1$ **68.** $6y^2(3y - 1)$ **69.** $5m(6m^2 + 2m - 3)$ **70.** 99 inches **71.** 58 ounces **72.** 0.036 deciliters **73.** 3,960 meters **74.** $19\frac{1}{4}$ pounds $\approx 8\frac{3}{4}$ kilograms **75.** 1,208 kilometers ≈ 755 miles **76.** 5 foot 8 inches ≈ 170 centimeters **77.** $\angle CDF = 41.7°$ **78.** Each side is $1\frac{3}{8}$ inches. **79.** $12\frac{1}{4}$ square feet **80.** 11 square centimeters **81.** The area is approximately 100.5 sq cm. **82.** The outer circumference is approximately 37.7 cm. **83.** The volume of this can is approximately 62.8 cubic inches. **84.** The volume of this beach ball is approximately 7,234.6 cubic inches.

Chapter 9

Preparation Exercises **1.** −7 **2.** −20 **3.** −11 **4.** 3 **5.** 13 **6.** 20 **7.** $A = -2$ **8.** $h = -7$ **9.** $x = 4$ **10.** $x = \frac{4}{5}$ **11.** $\frac{1}{6}$ **12.** $-\frac{3}{10}$ **13.** $2\frac{3}{5}$ or $\frac{13}{5}$ **14.** $3\frac{3}{4}$ or $\frac{15}{4}$ **15.** 0.85 **16.** 0.10 **17.** 2.038 **18.** 2.847 **19.** 0.46 **20.** 2.805 **21.** 0.21 **22.** 0.04 **23.** 7.9 **24.** 95.75 **25.** 47% **26.** 3% **27.** 27.9% **28.** 0.5% **29.** 43% **30.** 9% **31.** 75% **32.** ≈16.7%

Section 9.1 **1.** Answers will vary. **3.** The x-coordinate and the y-coordinate are opposites in Quadrant II, as in $(-4, 4)$,and $(-6, 6)$, and in Quadrant IV, as in $(1, -1)$ and $(7, -7)$. **5.** $10 = 10$ True! Yes, $(10, 2)$ is a solution. **7.** $-10 = 10$ False! No, $(-5, 0)$ is not a solution. **9.** $0 = 5$ False! No, $(6, 0)$ is not a solution. **11.** $-1 = -1$ True! Yes, $(2, -1)$ is a solution. **13.–19.** **21.–27.**

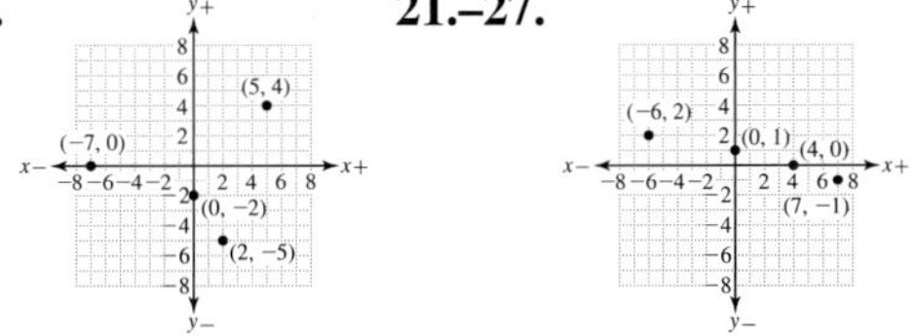

29. $(3, 3)$ **31.** $(-6, 5)$ **33.** $(-7, -3)$ **35.** $(4, -4)$

37. Quadrant IV **39.** Quadrant II **41.** Quadrant III **43.** Quadrant IV **45.** Quadrant IV **47.** Quadrant I **49.** On the positive y-axis **51.** On the positive x-axis **53.** On the negative y-axis **55.** At the origin **57.** $(-3, 0)$ **59.** $(2, 0)$ **61.** $(0, -5)$ **63.** $(0, -2)$

65. **67.**

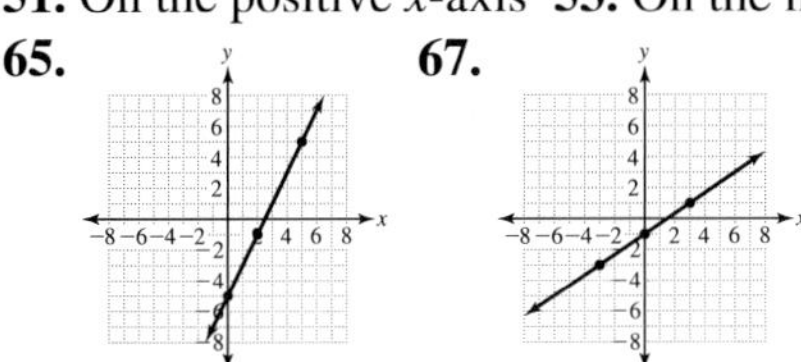

69. No **71.** Yes

Section 9.2 **1.** Answers will vary. Some good choices are any multiples of 3, such as −6, −3, 0, 3, and 6. Multiples of 3 are good choices because the denominator of the fraction can be completely divided out and each y-value will be an integer. **3.** Three lines can be drawn. If the points are represented by A, B, and C, one line can be drawn through A and B, a second line through A and C, and a third line through B and C.

5. **7.** **9.** **11.** **13.** **15.**

17. **19.** **21.** **23.** **25.** **27.**

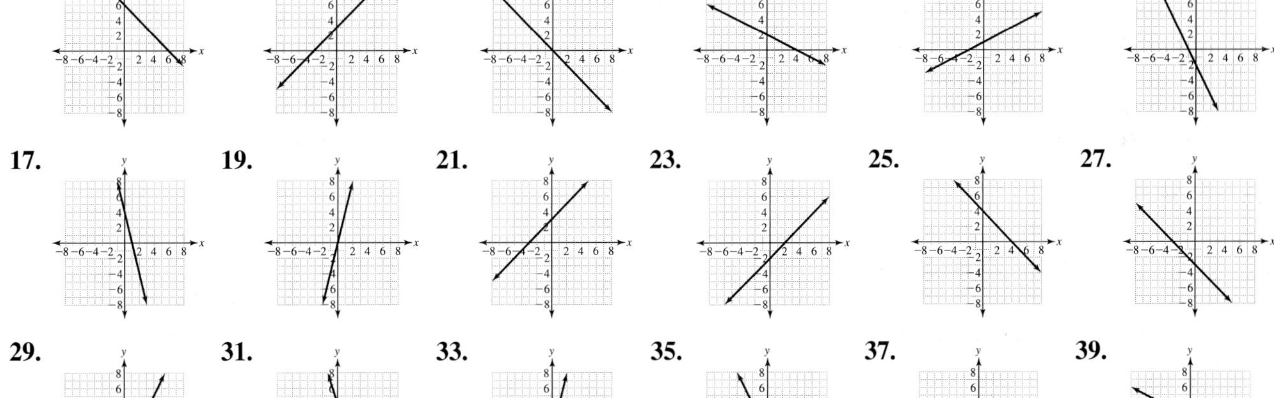

29. **31.** **33.** **35.** **37.** **39.**

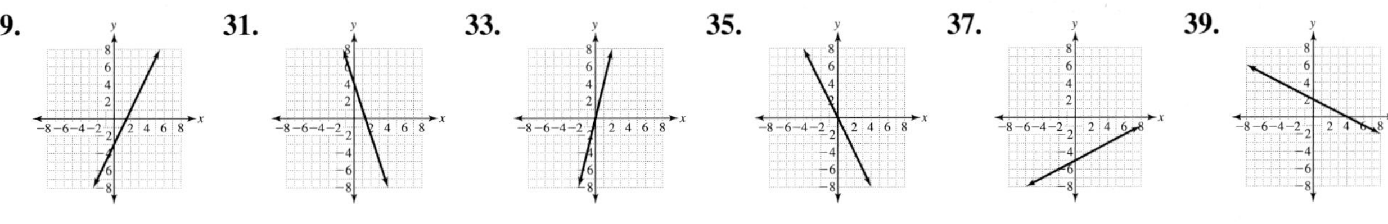

41. **43.** **45.** **47.** **49.** **51.**

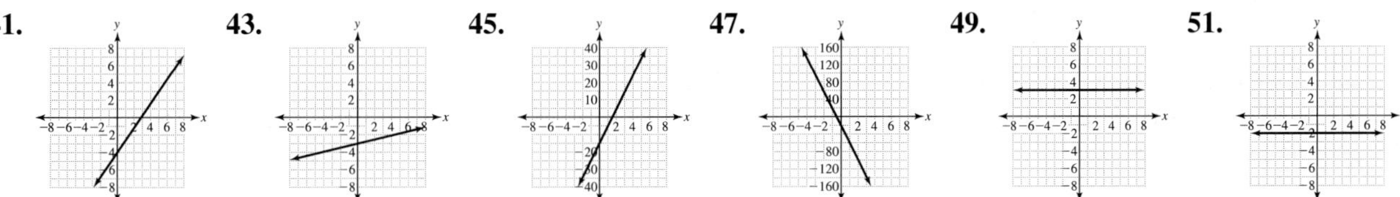

53. **55.** **57.**

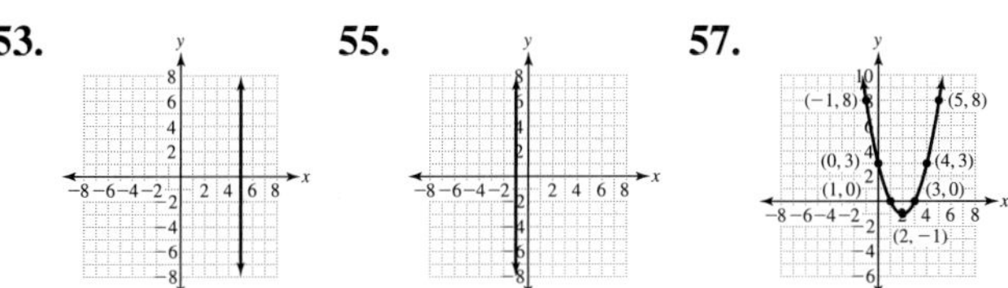

Section 9.3 **1.** Yes, this data could be graphed as a bar graph. The categories would be the age of the cars, and the height of each bar would be the value of the car during that year.

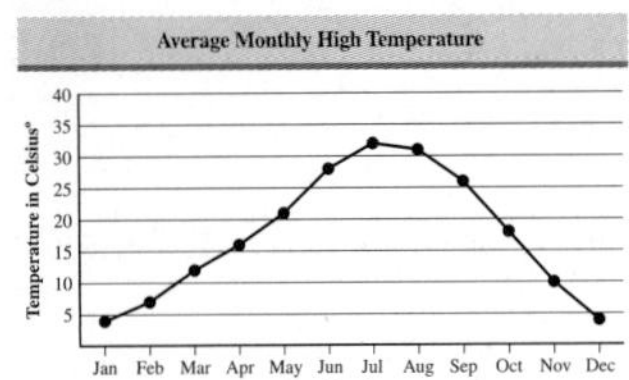

3. The greatest increase in temperature occurred between May and June. **5.** In July, the average temperature was twice the average temperature for April. **7.** In May, the average temperature was three-fourths the average temperature for June. **9.** The percent increase in the average temperature from March to April was $33\frac{1}{3}\%$.

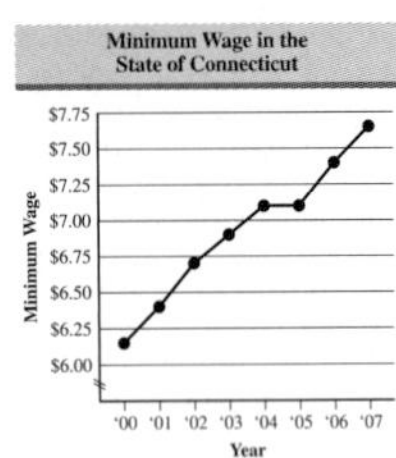

11. The greatest amount of increase in the minimum wage occurred in 2002 and 2006. **13.** The percent increase in the minimum wage from 2001 to 2002 was approximately 4.7%.

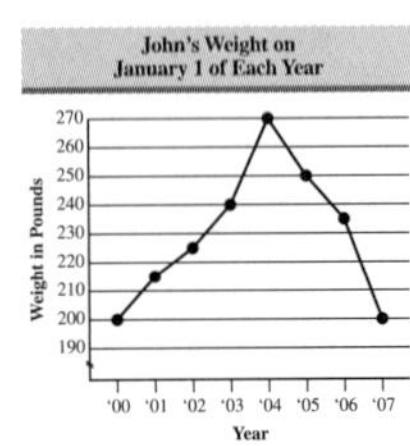

15. John gained the most weight during 2003. **17.** The percent increase in John's weight from 2003 to 2004 was 12.5%. **19.** The percent decrease in John's weight from 2004 to 2007 was approximately 25.9%.

21.

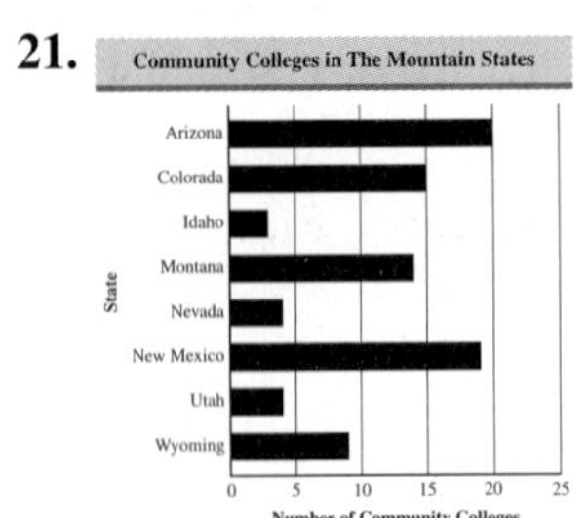

23.

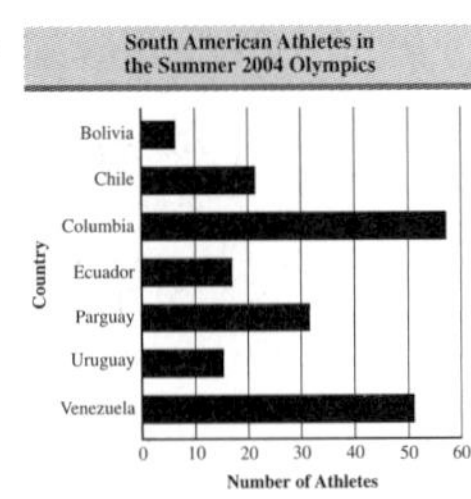

25.

Jet	Age, in Years
Freedom	8
Glory	13
Patriot	6
Spirit	2
Victory	4

27.

Breed	Number of Horses
Andalusian	14
Clydesdale	8
Friesian	22
Paint	16
Palomino	25

29.

▲ *Alaska*

Census Year	Total Population	Total change in Population
1960	226,000	
1970	303,000	77,000
1980	402,000	99,000
1990	550,000	148,000
2000	627,000	77,000

■ *Wyoming*

Census Year	Total Population	Total change in Population
1960	330,000	
1970	332,000	2,000
1980	470,000	138,000
1990	454,000	−16,000
2000	494,000	40,000

31. Wyoming had a 41.6% increase in population. (Alaska had a 32.7% increase.)

Section 9.41. No, because the percent values add to more than 100%. **3.** The second column adds to 120%, but this means that some of the riders own more than one brand of motorcycle.

5.

Marital Status	Percent of Women	Total Number of Women
Married	30%	36
Divorced	20%	24
Widowed	5%	6
Legally Separated	10%	12
Never Married	35%	42
Total	**100%**	**120**

7.

Type of Doctor	Percent of Doctors	Total Number of Doctors
Anesthesiologists	6%	15
General Practitioners	32%	80
Gynecologists	16%	40
Internists	18%	45
Surgeons	8%	20
Others	20%	50
Total	**100%**	**250**

9.

Major	Total Number of Students	Percent of Students
Arts/Humanities	40	50%
English/Languages	16	20%
Life Science	4	5%
Social Science	8	10%
Other	12	15%
Total	**80**	**100%**

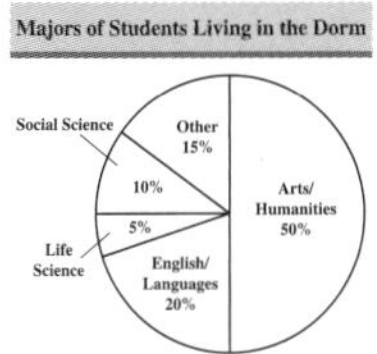

11.

Position	Total Numbers of Players	Percents of Players
Pitcher	56	28%
Catcher	10	5%
Infielder	74	37%
Outfielder	44	22%
Left blank	16	8%
Total	**200**	**100%**

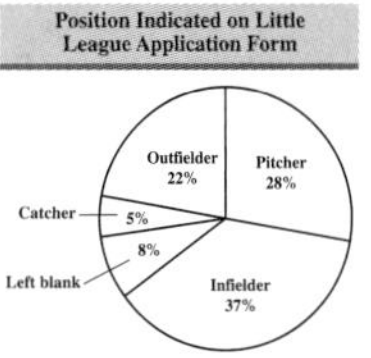

13. a.

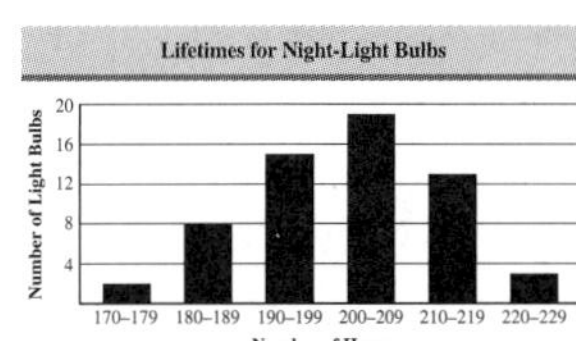

b. 50 **c.** 10 **d.** 34 **e.** 35

15. a.

Number of Hours	Number of Students
1–5	9
6–10	20
11–15	18
16–20	26
21–25	12
26–30	5
Total	**90**

b. 17 **c.** 29 **d.** 56

17. a.

Number of Minutes	Tally	Number of Clients
30–39	~~IIII~~ I	6
40–49	~~IIII~~ IIII	9
50–59	~~IIII~~ ~~IIII~~ II	12
60–69	~~IIII~~ ~~IIII~~ II	12
70–79	~~IIII~~ II	7
80–89	IIII	4
Total		**50**

b. 15 **c.** 33 **d.** 23

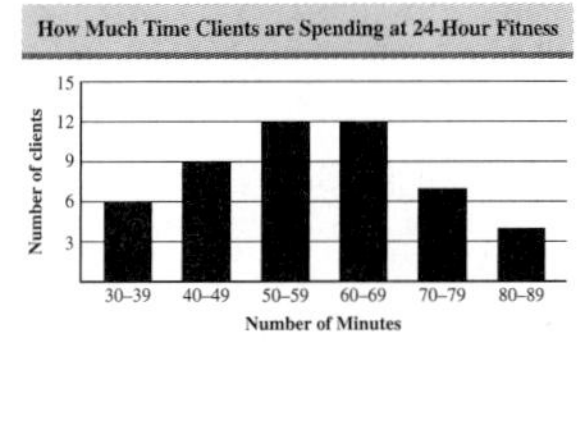

19. a.

Number of Cups	Number of Customers	Percent of Customers
None	6	15%
One	10	25%
Two	15	37.5%
Three	5	12.5%
More Than Three	4	10%
Total	**40**	**100%**

b.

More Than Three
None 15%
One 25%
10%
12.5%
Two 37.5%
Three

Section 9.5 **1.** Answers will vary. **3.** The median will always be one of the data values. It is the middle data value in the ordered set of values. **5.** 83 **7.** 43 **9.** 79.4 **11.** 167.7 **13.** 19 miles **15.** 29.7 cars **17.** 22.06 kilometers **19.** 1.268 meters

21.

Course	Grade	Point Value	Units	Grade Points (Point Value × Units)
Anatomy	C	2	5	10
Guitar	A	4	1	4
English	B	3	4	12
Sociology	A	4	3	12
		Total:	**13**	**38**

Grade Point Average ≈ 2.92

23. Median is 76; Mode is 71.

25. Median is 3; Mode is 3. **27.** Median is 9.5; Mode is 6. **29.** Median is 40.5; Mode is 38. **31.** Mean ≈13.8; Median is 14; There is no mode. **33.** Mean is 10.5; Median is 10; Mode is 5. **35.** After the fourth test, Myla's test average is 87. **37. a.** Mean = 7.25; Median = 7.5; Mode = 9 **b. i.** Increase **ii.** Stay the same **iii.** Stay the same **c.** Mean = 7.6; Median = 7.5; Mode = 9 **d.** Answers may vary

Section 9.6 **1.** 100% or 1. In the example of the die, P(getting a number less than 7) $= \frac{6}{6} = 1$. **3.** $\frac{1}{6}$ **5.** $\frac{5}{6}$ **7.** $\frac{1}{3}$ **9.** $\frac{1}{5}$ **11.** $\frac{3}{5}$ **13.** $\frac{3}{5}$ **15.** $\frac{2}{5}$ **17.** $\frac{1}{5}$ **19.** $\frac{1}{8}$ **21.** $2\frac{3}{8}$ **23.** $\frac{3}{10}$ **25.** $\frac{3}{40}$ **27.** $\frac{1}{36}$ **29.** $\frac{1}{6}$ **31.** $\frac{1}{6}$ **33.** $\frac{1}{18}$

Chapter 9 Review Exercises **1.** x-axis **2.** ordered pair **3.** frequency **4.** circle graph **5.** mode **6.** median **7.** False **8.** True **9.** False **10.** True **11.** $3 = -3$ False! No, $(10, 3)$ is not a solution. **12.** $-3 = 3$ False! No, $(0, -3)$ is not a solution. **13.** $0 = 0$ True! Yes, $(5, 0)$ is a solution. **14.** $6 = 6$ True! Yes, $(-5, 6)$ is a solution. **15.** $(3, 5)$ **16.** $(0, 1)$ **17.** $(-5, 3)$ **18.** $(-3, 0)$ **19.** $(-8, -4)$ **20.** $(0, -5)$ **21.** $(3, -3)$ **22.** $(6, 0)$ **23.** Quadrant II **24.** Quadrant I **25.** Quadrant III **26.** Quadrant IV **27.** On the negative x-axis **28.** On the positive x-axis **29.** On the positive y-axis **30.** On the negative y-axis

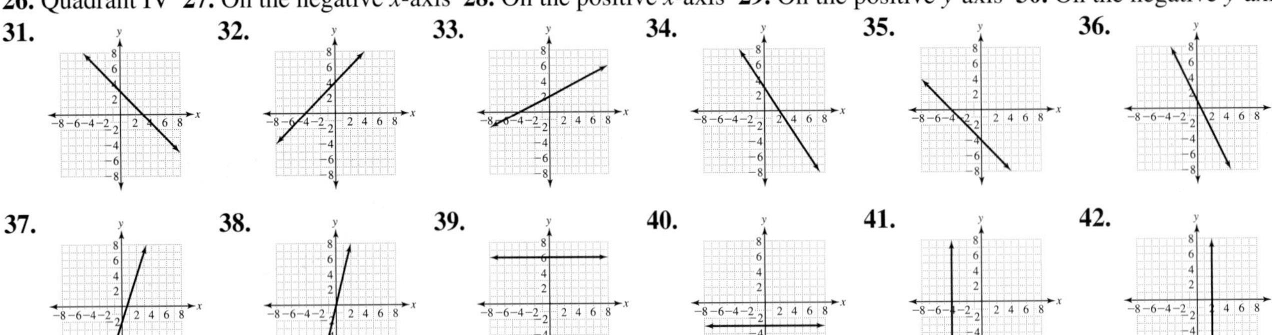

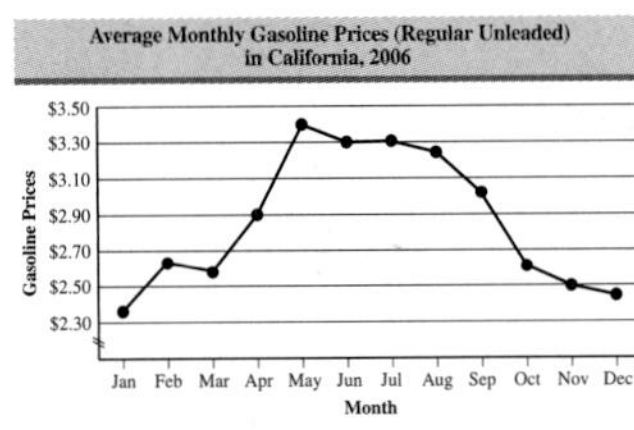

43. The largest price increase was \$0.52, and it occurred from April to May **44.** The largest price decrease was \$0.39, and it occurred from September to October. **45.** The increase from March to May was approximately 33.1%. **46.** The decrease from July to December was approximately 25.7%.

47. **48.**

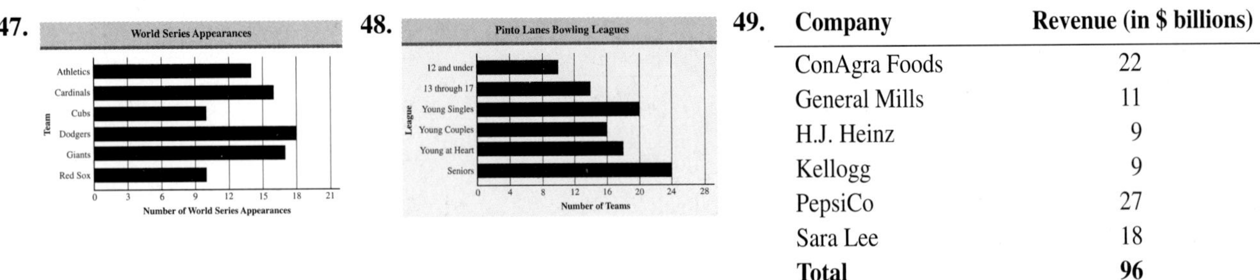

49.

Company	Revenue (in \$ billions)
ConAgra Foods	22
General Mills	11
H.J. Heinz	9
Kellogg	9
PepsiCo	27
Sara Lee	18
Total	**96**

50.

Country	Number of Winners
Canada	16
Great Britain	4
Japan	7
Kenya	16
Other	25
Total	**68**

51.

Advertising Budget Item	Percentage of Budget	Total Amount of Budget
Magazine	10%	\$20,000
Newspaper	14%	\$28,000
Radio	25%	\$50,000
Television	46%	\$92,000
Other	5%	\$10,000
Total	**100%**	**\$200,000**

52.

Style of New Home	Percent of Development	Total Number of Homes
2BR, 1 Story	8%	40
3BR, 1 Story	26%	130
3BR, 2 Story	35%	175
4BR, 1 Story	19%	95
4BR, 2 Story	12%	60
Total	**100%**	**500**

53.

Ethnicity	Total Number of Players	Percent of Players
African American	42	35%
Asian	18	15%
Hispanic	30	25%
White	24	20%
Other	6	5%
Total	**120**	**100%**

Ethnicity of Ball Players
Hispanic 25%
White 20%
Other 5%
Asian 15%
African-American 35%

54.

Religion	Total Number of Attendees	Percent of Attendees
Buddhist	8	4%
Christian	68	34%
Hindu	24	12%
Jewish	50	25%
Muslim	34	17%
Other	16	8%
Total	**200**	**100%**

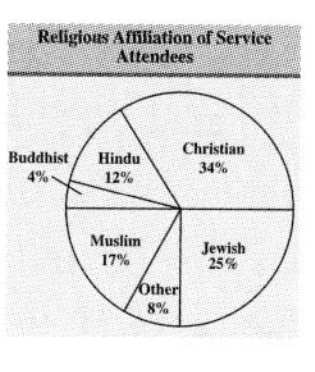

55.

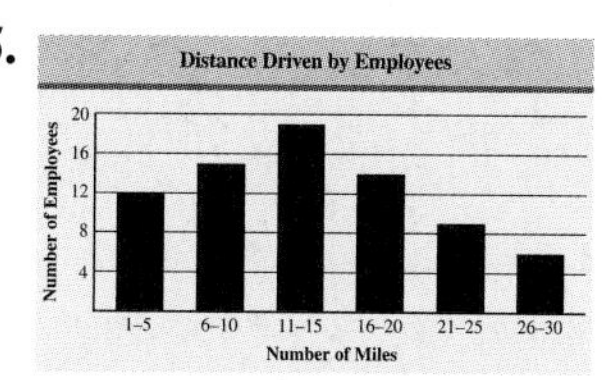

a) 27 **b)** 48 **c)** 29

56.

Number of Years	Tally	Number of Rulers
0–9	/𝍸/𝍸/𝍸/𝍸///	23
10–19	/𝍸/𝍸/𝍸	15
20–29	/𝍸/𝍸//	12
30–39	/𝍸//	7
40–49	/	1
50–59	///	3
60–69	/	1
Total		**62**

a) 5 **b)** 19 **c)** 38

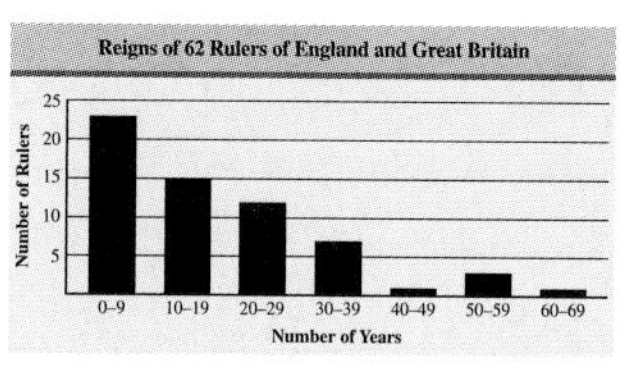

57.

Number of Wins	Number of Teams
50–57	1
58–65	2
66–73	8
74–81	3
82–89	6
90–97	7
98–105	3
Total	**30**

a) 10 **b)** 17 **c)** 11

58.

Number of Months	Number of Attendees
1–6	4
7–12	3
13–18	12
19–24	8
25–30	11
31–36	7
Total	**45**

a) 7 **b)** 19 **c)** 7

59. Mean ≈ 7.1; median: 7; mode: 6 **60.** Mean ≈ 21.2; median: 21; mode: None **61.** Mean ≈ 4.3; median: 4.5; mode: 6 **62.** Mean: 14; median: 15.5; modes: 9 and 17

63.

Course	Grade	Point Value	Units	Grade Points (Point Value × Units)
French	B	3	5	15
Computer Lab	A	4	1	4
CIS	A	4	3	12
Psychology	C	2	3	6
		Total:	**12**	**37**

Grade Point Average ≈ 3.08

64.

Course	Grade	Point Value	Units	Grade Points (Point Value × Units)
Chemistry	C	2	4	8
Trigonometry	B	3	4	12
History	D	1	3	3
Physical Education	A	4	1	4
Music Appreciation	B	3	2	6
		Total:	**14**	**33**

Grade Point Average ≈ 2.36

65. $\frac{1}{2}$ **66.** $\frac{1}{3}$ **67.** $\frac{2}{3}$ **68.** 0 **69.** $\frac{1}{6}$ **70.** $\frac{1}{3}$ **71.** $\frac{1}{2}$ **72.** $\frac{3}{8}$ **73.** $\frac{3}{8}$ **74.** $\frac{1}{4}$ **75.** $\frac{1}{4}$ **76.** 1 **77.** $\frac{1}{6}$ **78.** $\frac{1}{5}$ **79.** $\frac{3}{10}$ **80.** $\frac{4}{15}$ **81.** $\frac{1}{5}$ **82.** $\frac{3}{10}$ **83.** $\frac{1}{10}$ **84.** $\frac{6}{25}$

Chapter 9 Test **1.** $4(-3) + 2(5) = 2$; $-12 + 10 = 2$; $-2 = 2$; False; No, $(-3, 5)$ is not a solution. **2.** $5 = \frac{2}{3}(-3) + 7$; $5 = -2 + 7$; $5 = 5$ True; Yes, $(-3, 5)$ is a solution. **3.** Quadrant IV **4.** Quadrant II **5.** $(0, 4)$ **6.** $(-7, 0)$

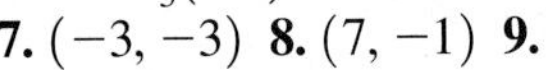

7. $(-3, -3)$ **8.** $(7, -1)$ **9.**

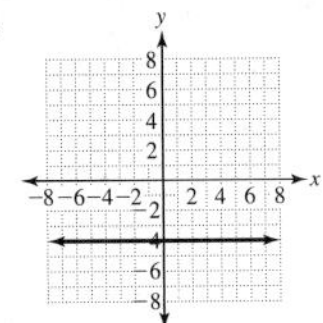

10.

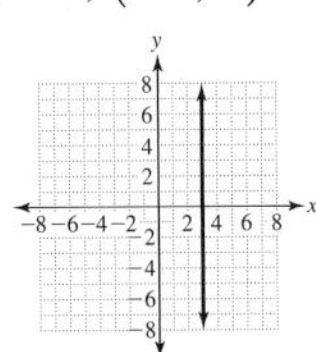

11. **12.**

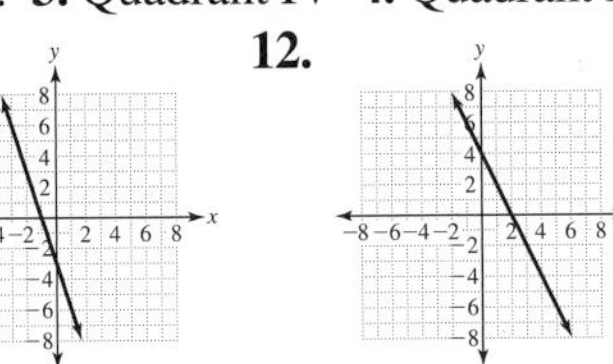

13.

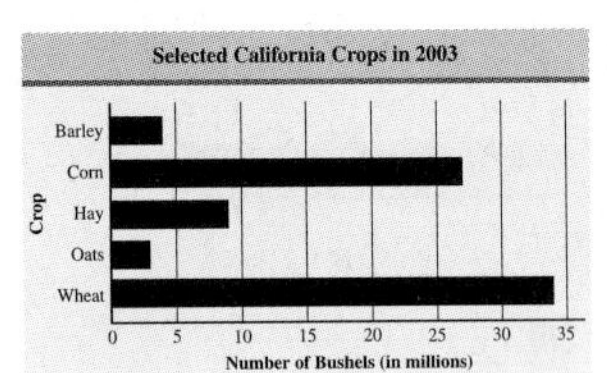

14. a)

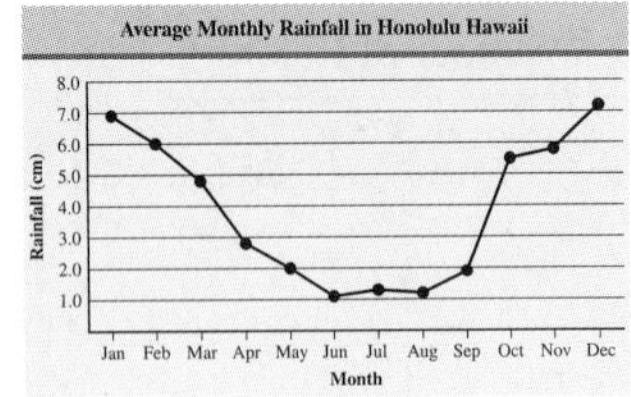

b) August has an average rainfall that is one-fourth the average rainfall for March.
c) December has an average rainfall that is $1\frac{1}{2}$ times the average rainfall for March.
d) The percent decrease in rainfall from February to March was 20%.

15.

City	Average Annual Snowfall
Albuquerque, NM	11
Baltimore, MD	21
Evansville, IN	14
Nashville, TN	10
Roanoke, VA	23
Wichita Falls, TX	6

16.

Type of Business	Percent of Acreage	Total Number of Acres
Business Offices	4%	6
Grocery	18%	27
Restaurants	26%	39
Retail	32%	48
Services	12%	18
Storage	8%	12
Total	100%	150

17. a)

Number of Wins	Tally	Number of Teams
20–5	///	3
26–31	///	3
32–37	~~////~~ /	6
38–43	~~////~~ //	7
44–49	//	2
50–55	////	4
56–61	////	4
Total		29

b) 10 **c)** 6 **d)** 12

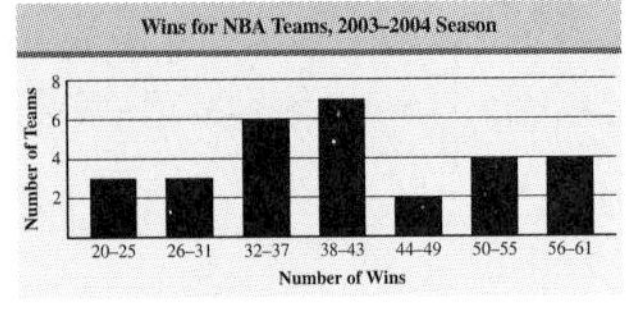

18.

Annual Giving	Total Number of Families	Precent of Families
Less than \$200	12	4%
\$200–\$999	36	12%
\$1,000–\$4,999	75	25%
\$5,000–\$9,999	96	32%
\$10,000–\$19,999	54	18%
\$20,000 and above	27	9%
Total	300	100%

19. a)

Final exam Score	Number of Students
44–51	4
52–59	7
60–67	11
68–75	14
76–83	19
84–91	16
92–99	9
Total	80

b) 44 **c)** 25 **d)** 22

20. Mean $\approx$ 6.2; median: 6; Mode: 8 **21.** Mean: 11.7; median: 12.5; Mode: 16

22.

Course	Grade	Point Value	Units	Grade Points (Point Value × Units)
Microbiology	C	2	5	10
Algebra	A	4	4	16
CIS	B	3	2	6
Karate	C	2	1	2
Geometry	C	2	3	6
Spanish	B	3	5	15
Guidance	A	4	2	8
Speech	B	3	3	9
		Total:	25	72

Grade Point Average = 2.88

23. $\frac{4}{9}$ **24.** $\frac{5}{9}$ **25.** $\frac{1}{3}$ **26.** $\frac{1}{3}$ **27.** $\frac{2}{9}$ **28.** $\frac{1}{3}$ **29.** $\frac{4}{15}$ **30.** $\frac{1}{5}$ **31.** $\frac{2}{5}$

32. $\frac{2}{15}$ **33.** $\frac{7}{20}$ **34.** $\frac{1}{4}$ **35.** $\frac{1}{5}$ **36.** $\frac{1}{8}$

Chapters 1–9 Cumulative Review **1. a)** Distributive Property of Multiplication Over Addition **b)** Commutative Property of Multiplication **2.** 20 **3. a)** 22,860,000 **b)** 22,900,000 **4.** $2 \cdot 3 \cdot 3 \cdot 7$ and $2 \cdot 3^2 \cdot 7$ **5.** Each box weighs 28 pounds. **6.** Nolan Ryan had 69 more strikeouts in 1989 than in 1990. **7.** Nolan Ryan had 939 total strikeouts from 1989 to 1993. **8.** The Celsius temperature is 45° C. **9.** 9 **10.** -10 **11.** $|3| - |-4|;\ 3 - 4;\ -1$ **12.** $\sqrt{20 - 2^2} - 2^2;\ \sqrt{16} - 4;\ 4 - 4;\ 0$ **13.** $6x$ **14.** $y = -8$ **15.** The width is 13 feet, and the length is 19 feet. **16.** $\frac{9}{5}$ or $1\frac{4}{5}$ **17.** $-\frac{3}{14}$ **18.** $\frac{7}{12}$ **19.** $-\frac{7}{9}$ **20.** $w = -6$ **21.** Tu walks a total of $4\frac{5}{24}$ miles each Saturday. **22.** Marcela's average rate of speed was 15.5 miles per hour. **23.** 0.875 **24.** 5.2 calories per gram **25. a)** The meal total was $19.75. **b)** The sales tax was $1.19. **c)** The tip was $2.96. **d)** Dorie received $6.10 in change. **26. a)** She'll earn $120. **b)** She'll earn $90. **27.** They can clean and prepare 36 rooms. **28.** 39 is 60% of 65. **29.** The percent decrease in the average price of gasoline was 25%. **30.** $4y^2 - 5y - 2$ **31.** m^8 **32.** $-4v^3$ **33.** $6p^5 - 3p^2$ **34.** $2x^3 + 7x^2 - 19x + 6$ **35.** $25w^2 - 9$ **36.** $-q^4 - 5q^2 + 2$ **37.** $5y^3(3y^3 - 5)$ **38.** The height is 92 inches. **39.** The top of her head is 7 feet 5 inches above the ground. **40.** His long jump improved 6 feet 8 inches. **41.** 13 pints **42.** 1.25 pounds **43.** 36 millimeters **44.** 8.56 liters **45.** 0.0472 kilograms **46.** 49 grams **47.** 9.5 liters **48.** 224 kilometers **49.** 22 inches **50.** $\angle WFZ = 108.7°$ **51.** The other acute angle is 58.1°. **52.** The perimeter is 42 inches. **53.** Each side is 17.26 cm. **54.** The area is 31.5 square cm. **55.** The area is 63 square feet. **56.** The area is 942 square feet. **57.** The circumference is 157 feet. **58.** The volume is 120π cubic feet. **59.** The volume is 2,800 cubic cm.

60.

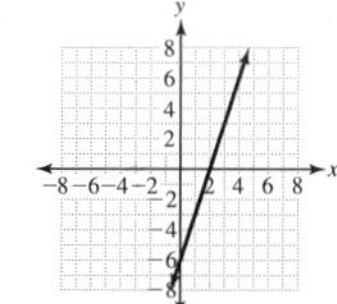

61.

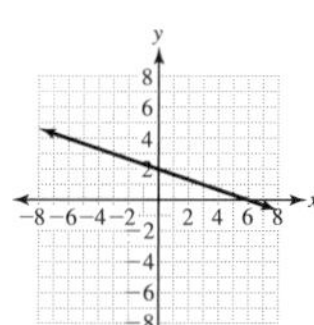

62. a)

Year Served	Number of Senators
1–6	22
7–12	29
13–18	16
19–24	14
25–30	11
31–36	3
37–42	2
43–48	3
Total	100

b) 19 **c)** 45 **63.** Mean is 7.5; Median is 6.5; Mode is 6. **64. a)** $\frac{1}{4}$ **b)** $\frac{1}{12}$ **c)** $\frac{1}{6}$

Appendix

A.1 **1.** 7×10^8 **3.** 9×10^2 **5.** 1×10^{10} **7.** 8×10^4 **9.** 3.3×10^5 **11.** 5.09×10^6 **13.** 1.3×10^2 **15.** 9.01×10^5 **17.** 8.3×10^{10} **19.** 1.09×10^7 **21.** 56,000,000 **23.** 7,320 **25.** 203,000 **27.** 18,900,000,000 **29.** 530 **31.** 27 **33.** The total number of gallons of gas consumed was approximately 4.99×10^9. **35.** The U.S. national debt was approximately 8.66×10^{12}.

A.2 **1.** $\frac{1}{121}$ **3.** $\frac{1}{1,000}$ **5.** $\frac{1}{16}$ **7.** $\frac{1}{81}$ **9.** $\frac{1}{6}$ **11.** $\frac{1}{15}$ **13.** $\frac{1}{w^2}$ **15.** $\frac{1}{q^3}$ **17.** $\frac{1}{m^6}$ **19.** $\frac{1}{y^8}$ **21.** $\frac{1}{k}$ **23.** $\frac{1}{a}$ **25.** b^4 **27.** c^1 **29.** $\frac{1}{y^3}$ **31.** $\frac{1}{p}$ **33.** $\frac{1}{k^2}$ **35.** $\frac{1}{h^{13}}$ **37.** $q^0 = 1$ **39.** $g^0 = 1$ **41.** x^5 **43.** y^6 **45.** $\frac{1}{k^{10}}$ **47.** $\frac{1}{d^9}$ **49.** m^4 **51.** v^3 **53.** $\frac{1}{h^3}$ **55.** $\frac{1}{y^2}$ **57.** $a^0 = 1$ **59.** $x^0 = 1$ **61.** $\frac{1}{x^6}$ **63.** $\frac{1}{p^{16}}$ **65.** q^4 **67.** m^{10} **69.** $\frac{1}{w^9}$ **71.** $\frac{1}{x^{16}}$ **73.** $p^0 = 1$ **75.** $y^0 = 1$ **77.** $\frac{1}{x^{20}}$ **79.** a^{30}

A.3 **1.** 3×10^{-2} **3.** 5×10^{-4} **5.** 4×10^{-6} **7.** 7×10^{-7} **9.** 2.8×10^{-1} **11.** 5.3×10^{-3} **13.** 9.13×10^{-3} **15.** 4.07×10^{-4} **17.** 2.914×10^{-3} **19.** 2.009×10^{-4} **21.** 0.023 **23.** 0.18 **25.** 0.000632 **27.** 0.0000886 **29.** 0.00000401 **31.** 0.0004915 **33.** The average hummingbird weighs 3.2×10^{-3} kilograms.

A.4 **1.** 8.5×10^7 **3.** 2.0×10^{-3} **5.** 4.53×10^6 **7.** 3.09×10^{-1} **9.** 3.6×10^6 **11.** 1.313×10^{-2} **13.** 6.9×10^9 **15.** 8.4×10^6 **17.** 1.6×10^{-7} **19.** 2.852×10^{-10} **21.** The total amount of money spent on fuel during those 27 years was 2.3517×10^{11}. **23.** The giant tortoise can travel 2.304×10^{-1} miles per hour. **25.** The total number of cans of Spam consumed in Hawaii each year is 6.7064×10^6.

A.5 **1.** 5 and 2 **3.** 3 and 10 **5.** No solution **7.** -8 and -2 **9.** -1 and -18 **11.** -12 and -4 **13.** $+10$ and -2 **15.** $+6$ and -5 **17.** No solution **19.** -10 and $+6$ **21.** -6 and $+6$ **23.** No solution **25.** $(p + 3)(p + 10)$ **27.** $(v + 3)(v + 4)$ **29.** $(v - 6)(v - 5)$ **31.** $(x - 4)(x - 4)$ **33.** $(v - 6)(v - 6)$ **35.** prime **37.** $(d - 9)(d - 1)$ **39.** $(r - 20)(r - 3)$ **41.** $(x - 2)(x - 18)$ **43.** $(y + 6)(y - 2)$ **45.** $(q + 6)(q - 3)$ **47.** prime **49.** $(n + 8)(n - 4)$ **51.** $(p - 12)(p + 3)$ **53.** $(w - 6)(w + 5)$ **55.** prime **57.** $(m - 10)(m + 7)$ **59.** $(y - 12)(y + 5)$

Glossary

Absolute value The absolute value of a number is the distance the number is from zero on a number line. The absolute value of a nonzero number is always positive.
Acute angle An acute angle has a measure greater than 0° and less than 90° .
Addition Property of Equality The Addition Property of Equality allows the same value to be added to each side of an equation without the solution of the equation being affected.
Additive identity For addition, the identity is zero: $a + 0 = a$ and $0 + a = a$. (See *identity*.)
Adjacent angles Adjacent angles are any two angles that share a common side, forming an even larger angle. All three angles have the same vertex, and the shared side is in the interior of the larger angle.
Algebraic expression An algebraic expression is a combination of numbers and letters connected by operations and grouping symbols.
Angle When two rays share the same endpoint, they form an angle.
Area Area is the amount of surface in an enclosed region and is always measured in square units.
Associative Property of Addition The Associative Property of Addition allows the grouping of the numbers to be changed without the resulting sum being affected: $a + (b + c) = (a + b) + c$.
Associative Property of Multiplication The Associative Property of Multiplication allows the grouping of the numbers to be changed without the resulting product being affected: $a \cdot (b \cdot c) = (a \cdot b) \cdot c$.

Bar graph A bar graph is a visual way to quickly compare numbers of things in a particular category. In a bar graph, a rectangle (or bar) is drawn for each category to indicate the size or the number of data values for that category.
Base The base of an exponent is the number that is raised to a power.
Base angles The two angles with equal measures in an isosceles triangle are known as base angles.
Base-ten numbering system A base-ten numbering system uses ten digits: 0, 1, 2, 3, 4, 5, 6, 7, 8, and 9.
Binomial A binomial is a polynomial that has two terms.
Building up Multiplying a fraction by a fractional form of 1, such as $\frac{2}{2}$ and $\frac{5}{5}$, is called building up a fraction.

Capacity Capacity refers to the amount a container can hold, usually liquid.
Circle A circle is a round geometric figure with a center point. The center point is the same distance from every point on the circle.
Circle graph In a circle graph, the categories are displayed as sectors in a circle and each category is shown as a percent of the whole.
Circumference The circumference is the perimeter of a circle.
Class frequency The number of data values that fits into each class interval is called the class frequency.
Class interval In a histogram, each category is a range of numbers called a class interval.
Coefficient When a variable, n, is multiplied by a number, that number is called a coefficient.
Commission The amount of commission that a salesperson earns on a single item is a percentage of that item's price. The amount of commission is calculated by multiplying the commission rate (%) by the total amount sold.
Common denominators Two or more fractions have common denominators when their denominators are the same.
Common factors The common factors of two numbers are all of the numbers that are factors of both numbers.
Commutative property of addition The commutative property of addition states that when two numbers are added, it doesn't matter which number is written first, the resulting sum will be the same: $a + b = b + a$.
Commutative property of multiplication The commutative property of multiplication states that when two numbers are multiplied, it doesn't matter which number is written first, the resulting product will be the same: $a \cdot b = b \cdot a$.
Complementary angles Complementary angles are any two angles with measures that add to 90°.
Complex fraction A complex fraction is a fraction in which the numerator or the denominator (or both) contains a fraction.
Composite number A composite number is a natural number that has more than two distinct factors.
Compound interest Compound interest is interest that is applied to the principal on a loan or savings account more than once.
Cone A cone is a solid object with a circular base and sides that taper to a common point.
Constant A constant is a known value, a number.
Coordinates Coordinates are a pair of x- and y-values that show the exact position of a point in the x-y plane.
Counting numbers Numbers used for counting are called counting numbers: 1, 2, 3, 4, 5, 6, 7, 8, and so on; also called natural numbers.
Cross multiplication In a proportion, the direct method for finding the products of the extremes and means is referred to as cross multiplication.
Cube A cube is a solid object with a square base and parallel sides. A cube has six square sides, such as a die.
Cylinder A cylinder is a solid object with a circular base and parallel sides.

Data When a study is conducted, pieces of information called data are collected.
Data table A table used to organize data is called a data table.

Decimal fraction Any fraction with a denominator that is a power of 10 is called a decimal fraction.
Decimal number Any number that contains a decimal point is called a decimal number.
Decimal point A decimal point is the separator between the whole number and the fractional part of a decimal number.
Degree of term The number of variable factors in a term is called the degree of the term.
Denominator The denominator is the bottom part of a fraction. The denominator represents the number of equal-sized parts contained in the whole.
Diameter In a circle, two radii that form a straight angle (180°) are called a diameter.
Difference A difference is the written subtraction and the result of subtracting.
Digits The ten symbols 0, 1, 2, 3, 4, 5, 6, 7, 8, and 9 are digits.
Dimension Dimension is the visible linear measure of an object. A line segment has one dimension (length), a rectangle has two dimensions (length and width), and a cube has three dimensions (length, width, and depth).
Distributive property of multiplication over addition A multiplier, a, can be distributed to a sum $(b + c)$ so that it multiplies both numbers in the sum: $a \cdot (b + c) = a \cdot b + a \cdot c$.
Dividend In division, the dividend is the number to be divided. In a fraction, the numerator also is called the dividend.
Divisible When the divisor divides evenly into the dividend (there is no remainder), the dividend is divisible by the divisor.
Division Division is the inverse operation of multiplication.
Division bar A division bar, also called a fraction bar, is a line that separates the dividend (top) and the divisor (bottom). (See *fraction bar.*)
Division method The division method is a technique for finding the prime factors of a composite number, the greatest common factor of two numbers, and the least common denominator of two fractions.
Division Property of Equality The Division Property of Equality allows each side of an equation to be divided by the same nonzero number without the solution of the equation being affected.
Divisor In division, the divisor is the number by which another number is divided.
Double negative A double negative is the use of two negative signs next to each other. A double negative can be rewritten as a positive number. For example, $-(-5)$ can be written as $+5$ or 5.

Endpoints A line segment has definite start and stop points called endpoints.
Equation An equation is a mathematical sentence in which one expression equals another expression.
Equilateral triangle An equilateral triangle has three sides of equal length, and all three angles are of equal measure. Each angle in an equilateral triangle measures 60°.
Equivalence fraction In an equivalence fraction, the numerator and the denominator are of equal value but of different units of measure, such as $\frac{12 \text{ inches}}{1 \text{ foot}}$ or $\frac{1 \text{ foot}}{12 \text{ inches}}$. Equivalence fractions always have the value 1.
Equivalent When two things have equal value, they are equivalent.
Equivalent ratios When two ratios have the same value, they are equivalent ratios.
Evaluate To evaluate means to "find the value of."
Event In probability, an event is the outcome that one is most interested in occurring.
Experiment An experiment is the act of doing something to create a result.
Exponent An exponent is used to abbreviate repeated multiplication (repeated factors). An exponent indicates the number of factors of the base.
Expression An expression is the written form of an operation.
Extremes In a proportion in which the ratios are written in standard form ($a:b = c:d$), the outermost numbers (a and d) are called the extremes.

Factor Each number or quantity in a product is called a factor. To factor a number or a polynomial means to write it in a factored form.
Factor pair Any whole number can be written as the product of two whole numbers, called a factor pair.
Factor tree A factor tree is used to break up a composite number into its prime factors.
Factored form A number is in factored form when it is written as a product of its factors.
Fraction A fraction is typically the comparison of a part of something to its whole, where all of the parts are the same size.
Fraction bar A fraction bar is a line that separates the numerator and the denominator.
Frequency The number of items in a given category is called the frequency of the category. Frequency means how frequently—how many times—something occurs.
Frequency distribution table A frequency distribution table is a table of class intervals and their corresponding class frequencies.

Gram A gram is the standard unit of weight in the metric system. A gram is about the weight of the contents of a packet of artificial sweetener.
Greater than On a horizontal number line, *greater than* means "to the right of" and is represented by the symbol $>$.
Greatest common factor (GCF) The greatest common factor (GCF) of two numbers is the highest, or greatest, of their common factors.
Grouping symbols Parentheses and brackets are considered grouping symbols. The expression they group is called a quantity.

Histogram When each category is a range of numbers, a bar graph is called a histogram.

Identity An identity is a number that, when applied, won't change the value of another number or quantity.

Improper fraction An improper fraction has a numerator that is equal to or greater than the denominator.
Integers The whole numbers and their opposites are called integers.
Interest Interest, or earned interest, is an amount of money that is automatically added to the principal based on the length of the loan or savings period.
Interior of an angle The interior of an angle is the portion of the plane that lies between the rays that form the angle. The interior of an angle continues indefinitely.
Interpolation To make an educated visual guess is known as interpolation.
Intersect When two lines cross each other, they meet in a single point of intersection. The lines *intersect* at the point.
Invert To invert a fraction is to write the reciprocal of the fraction.
Isolating the variable In an equation, isolating the variable means getting the variable by itself on one side of the equation and all constants on the other side of the equation.
Isosceles triangle An isosceles triangle has two or more sides of equal measure.

Least common denominator The least common denominator (LCD) of two or more fractions is the least common multiple of the denominators.
Legend The legend describes the unknown value in an application problem.
Less than On a horizontal number line, *less than* means "to the left of" and is represented by the symbol $<$.
Like fractions Two or more fractions are like fractions when they have the same denominator.
Like terms Two or more terms are like terms when they have exactly the same variable and exponent.
Line A line is straight and goes on indefinitely.
Line graph A line graph is used to compare the progression of data over a period of time.
Line segment The portion of a line that is drawn between two points is called a line segment. It has a definite length.
Linear equation A linear equation of two variables, x and y, is an equation in which the ordered pair solutions are points on a line in the x-y plane.
Linear measures Measures of length are called linear measures.
Liter A liter is the standard unit of capacity in the metric system. A liter is a little more than 1 quart.
Long division Long division is used when more than one step is needed to divide.
Long Division Algorithm The Long Division Algorithm is a process for finding the quotient of two numbers.
Lowest terms A fraction is in lowest terms when the numerator and the denominator are relatively prime.

Mean The mean is the average of a set of data values. It is calculated by dividing the sum of all of the data values by the number of data values.
Means In a proportion in which the ratios are written in standard form ($a:b = c:d$), the middle two numbers (b and c) are called the means.
Means-Extremes Product Property For two ratios, the Means-Extremes Product Property states that if the product of the means equals the product of the extremes, the ratios are equivalent.
Median When data values are written in numerical order, the median is the middle number.
Meter A meter is the standard unit of length in the metric system. A meter is about $3\frac{3}{8}$ inches longer than a yard.
Metric system The metric system is a set of measures of length, of weight, and of capacity that can be easily converted to larger or smaller amounts using powers of 10.
Mixed number A mixed number is the sum of a whole number and a fraction.
Mode The mode is the data value that occurs most often in a data set.
Monomial A monomial is a one-term polynomial.
Multiples The multiples of any whole number, b, are all of the products involving b and another whole number.
Multiplication For whole numbers, multiplication is an abbreviation for repeated addition.
Multiplication Property of Equality The Multiplication Property of Equality allows each side of an equation to be multiplied by the same nonzero number without the solution of the equation being affected.
Multiplication Property of Zero The Multiplication Property of Zero says that the product of 0 and any number is always 0.
Multiplicative identity For multiplication, the identity is one: $a \cdot 1 = a$ and $1 \cdot a = a$. (See *identity*.)
Multiplicative inverse The multiplicative inverse of a number is the reciprocal of the number. The product of a nonzero number and its multiplicative inverse is always 1.

Natural numbers Natural numbers are the numbers used for counting: 1, 2, 3, 4, 5, 6, 7, 8, and so on.
Negative numbers Numbers less than zero are called negative numbers. On a number line, negative numbers are to the left of 0.
Nonterminating decimal A decimal number that does not have a final decimal place (the decimal part goes on and on and never ends) is called a nonterminating decimal.
Number A number describes how many items there are.
Number line A number line is a line on which there are marked points to show the relative positions of numbers.
Numeral A numeral is a symbol that represents a number.
Numerator The numerator is the top part of a fraction. The numerator represents the number of equal-sized parts being considered.
Numerical value The numerical value of a number is the same as its absolute value.

Obtuse angle An obtuse angle has a measure greater that 90° and less than 180°.
Operations There are four basic operations in mathematics: addition, subtraction, multiplication, and division.
Opposites Two signed numbers, one positive and one negative, with the same numerical value are called opposites. They are on opposite sides of 0 and are the same distance from 0 on the number line.

Order of operations A system that tells which operation to evaluate first in an expression.
Ordered pair An ordered pair is a pair of numbers grouped in parentheses and separated by a comma in which the order they are written is important.
Origin Origin means beginning. On a number line, the origin is 0; in the x-y plane, the origin is the point (0, 0).

Parallel lines Two lines in the same plane that never intersect each other are called parallel lines.
Parallelogram A parallelogram is a quadrilateral with two pairs of parallel sides.
Percent A percent is a ratio in which the denominator is 100.
Percentage In a percent, the number that is written before the percent symbol (%) is called the percentage. In the fractional form of a percent, the percentage is the numerator and 100 is the denominator.
Perfect square As a number, a perfect square is a whole number that is the product of a whole number and itself.
Perimeter The perimeter of a geometric figure is the total measurement around the figure.
Perpendicular Two lines are perpendicular if they intersect to form four right angles.
Place value Every digit in a whole number has a place value based on the position it holds in the number.
Place digit When rounding a number, the digit occupying the place digit being rounded to is called the place digit.
Plane A plane is any flat surface that extends forever in each direction.
Point When two lines cross each other, they meet in a single point of intersection. The lines intersect at the point.
Polygon A polygon is a closed plane figure composed of at least three straight line segments.
Polynomial A polynomial is an algebraic expression that is a single term or the sum of two or more terms.
Positive numbers Numbers greater than zero are called positive numbers. On a number line, positive numbers are to the right of 0.
Possible outcomes The possible outcomes of an experiment are all of the different results that can occur.
Power The word *power* is sometimes used in place of the word *exponent*. The power of a number is the result of applying an exponent to the number (base) and evaluating. For example, the third power of 2 is written 2^3 and, when applied, is 8; so, 8 is the third power of 2.
Powers of 10 Numbers such as 10, 100, and 1,000 are powers of 10. A power of 10 is any number that starts with 1 and is followed by at least one zero.
Prime factorization The prime factorization of a composite number is the product of the prime factors (including repetitions) of the number.
Prime number A whole number is a prime number when it has exactly two distinct whole number factors, 1 and itself.
Principal Money that is deposited into a savings account or that is borrowed as a loan is called the principal.
Principal square root The positive square root of a number is called the principal square root. For example, 3 and -3 are square roots of 9, but 3 is the principal square root.
Probability Probability is a numerical way to express the likelihood that some particular thing or event will happen.
Product A product is the written multiplication and the result of multiplying.
Proper fraction A proper fraction has a numerator that is less than the denominator.
Proportion A proportion is an equation showing that two ratios are equivalent.
Pyramid A pyramid is a solid object with a rectangular base and sides that taper up to a common point.

Quadrants The x-axis and the y-axis divide the x-y plane into four distinct regions called quadrants.
Quadrilateral A quadrilateral is a four-sided polygon that has four angles, and the sum of the angles is 360°.
Quantity A quantity is an expression, often denoted by a pair of parentheses, that is considered to be one value.
Quotient A quotient is the written division and the result of dividing.

Radical A radical, $\sqrt{\ }$, is the symbol that represents a square root.
Radicand A radicand is the number within a radical.
Radius A radius of a circle is any line segment from the center of the circle to any point on the circle. The word *radius* also is used to refer to the length of a radius. The plural of *radius* is *radii*.
Random selection A random selection is the act of choosing something so that each possible outcome has an equal chance of being selected and is equally likely to be selected.
Rate A rate is a ratio containing different units of measure, or counts, in the numerator and the denominator.
Ratio A ratio is a comparison between two numbers, using division.
Ray A ray is a piece of a line that starts at one of its points and goes on indefinitely in only one direction.
Reciprocal The reciprocal of a fraction results from interchanging the numerator and the denominator.
Rectangle A rectangle is a parallelogram with four right angles.
Rectangular coordinate system A grid based on two perpendicular number lines, the x-axis and the y-axis is called the rectangular coordinate system. It also is known as the x-y plane.
Regular polygon A regular polygon is a polygon in which all of the sides are the same length and all of the angles have the same measure.
Relatively prime Two natural numbers that have no common prime factor are said to be relatively prime.
Remainder A remainder is the amount left over after the last digit in the dividend is covered by the quotient.

Repeating decimals A nonterminating decimal in which a pattern, or block, of one or more digits is repeated indefinitely is called a repeating decimal.
Repetend In a repeating decimal, the digit or block of digits that repeats indefinitely is called the repetend.
Replacement value When the numerical value of a variable is given, that number is called a replacement value.
Right angle A right angle is an angle that measures 90°.
Right triangle A right triangle has one angle that is a right angle.
Rounding digit When rounding a number to a certain place, the digit to the immediate right of the place digit is called the rounding digit. When the rounding digit is 0, 1, 2, 3, or 4, the number is rounded down. When the rounding digit is 5, 6, 7, 8, or 9, the number is rounded up.

Sales tax Sales tax is a percentage of the purchase price. The amount of sales tax is calculated by multiplying the sales tax rate (%) by the purchase price.
Scalene triangle A triangle in which no two sides have the same length and no two angles have equal measure is called a scalene triangle.
Sector Within a circle, the region between two radii is called a sector.
Short division Short division is used when only one step is needed to divide.
Signed numbers Together, positive and negative numbers are called signed numbers.
Simple interest Simple interest is applied just once and is a percentage of the principal based on a yearly interest rate.
Simplify To simplify a fraction means to divide out any factors greater than 1 that are common to both the numerator and the denominator.
Solid A solid is a three-dimensional object.
Solution A solution is the value of a variable that makes an equation true.
Solve To solve an equation means to find the number that makes the equation true.
Sphere A sphere is the three-dimensional version of a circle. As with a circle, a sphere has a center point that is the same distance from every point on the sphere.
Square A square is a rectangle with four sides of equal length.
Square root The square root of a perfect square is a number that, when multiplied by itself, will result in that perfect square. Geometrically, the square root of a perfect square is the length of one side of the square.
Straight angle When the rays forming an angle go in opposite directions to form a line, the angle is a straight angle. A straight angle has a measure of 180°.
Substitute To substitute a value into a formula means to replace a variable with the value.
Subtraction Property of Equality The Subtraction Property of Equality allows the same number to be subtracted from each side of an equation without the solution of the equation being affected.
Sum A sum is the written addition and the result of adding.
Supplementary angles Supplementary angles are any two angles with measures that add to 180°.

Term A term can be a constant and can be the product of a number (coefficient) and a variable.
Terminating decimal Any decimal number that has a final decimal place is called a terminating decimal.
Trapezoid A trapezoid is a quadrilateral with only one pair of parallel sides.
Triangle A triangle is a polygon that has three sides and three angles. The sum of the angles in a triangle is 180° no matter the size or shape of the triangle.
Trinomial A trinomial is a polynomial that has three terms.

Unit of measure A unit of measure is a standard of basic amount, such as miles, ounces, and liters, by which something is measured.
Unit price The unit rate for the price of an item is called unit price.
Unit rate A unit rate is a rate in which the denominator has been simplified to one item.
Unknown value An unknown value is a number that is not yet known, typically represented by a variable.
Unlike fractions Two fractions with different denominators are unlike fractions.
U.S. measures Measurements based on feet, pounds, gallons, and Fahrenheit are referred to as U.S. measures.

Variable A variable is a letter that represents a number.
Vector A number can be represented on a number line by a vector. A vector has a linear length indicating its numerical value and an arrow indicating its direction.
Vertex A vertex is the common endpoint that two rays share when they form an angle.
Vertical angles When two lines intersect and form four angles, each pair of nonadjacent angles is called vertical angles.
Volume Volume indicates the amount of space a solid object occupies or can hold.

Whole numbers The set of whole numbers is 0, 1, 2, 3, 4, 5, 6, 7, 8, and so on.

x-axis The x-axis is a horizontal number line that represents x-values in the x-y plane.
x-y plane A grid based on two number lines, the x-axis and the y-axis is called the x-y plane. It also is known as the rectangular coordinate system.

y-axis The y-axis is a vertical number line that represents y-values in the x-y plane.

Photograph Credits: p. 1: Shutterstock; p. 111: Shutterstock; p. 195: Shutterstock; p. 241: Shutterstock; p. 333: Shutterstock; p. 397: Photodisc/Getty Royalty Free; p. 477: Bob Prior; p. 535: Bob Prior; p. 561: Shutterstock; p. 592: U.S. Department of Defense; p. 593: Shutterstock; p. 593: iStockphoto; p. 593: Corbis/Bettmann; p. 608: NASA; p. 639: Bob Prior; p. 746: Shutterstock; p. 746: Shutterstock

Index of Applications

Agriculture
Area of land, 348, 406, 609
Crop circle, 593, 608
Gardening, 224, 593, 609
Plot of land, 224, 279, 305
Silo, 617

Automotive/Transportation
Airplane travel, 164, 386, 414
Aviators, 683
Car ownership, 414
Car travel, 164, 225, 232, 386, 422
Driving distances, 359, 422
Jumbo jet, 40
Miles driven, 93, 107
Miles per gallon, 40, 348, 386
Miles per hour, 162, 164
Motorcycles, 694
New car, 28, 457, 458
Parking lot, 448
Purchasing a used car, 109, 367, 386
SUVs, 348
Train travel, 225

Banking
Adjustable rate mortgage, 28
ATM, 140
Certificate of deposit, 458
Checking account, 140, 359
Credit cards, 359
Interest earned, 458
Safe-deposit box, 448
Simple interest, 457

Business
Billing department, 93
Boarding horses, 684
Bookstore, 684
Business expenses, 348
Coffee shop, 439
Computer store, 448
Construction firm, 232
Courier service, 93, 164
Department store, 386
Electronics store, 348
Embroidery business, 93
Flooring company, 422
Furniture business, 109
Hardware store, 367, 422
Inventory, 40, 457
Music store, 696
Party rental, 94
Printing company, 92, 93, 315, 422
Shopping mall, 457
Sporting goods store, 457
Storage rental, 616

Business Equipment
Computers, 40
Office furniture, 28
Office supplies, 358

Construction
Air-conditioning, 546
Carpenter's union, 93
Ceramic tile, 224
Construction, 290, 395
Furniture construction, 225
House construction, 225
Steel bridge, 93
Tile layers, 231, 290
Window, 94

Consumers
Auto insurance, 92, 413
Clothing prices, 359
Cost of a gift, 458
Cost of a meal, 386, 458
Cost of food, 414
Discounts, 449, 456, 457
Gasoline prices, 359, 449
Insurance payment, 349
Shopping, 386
Utility bill, 28, 349, 406

Domestic
Church attendance, 700
Family picnic, 305
Firewood, 305
Laundry, 2790
Marital status, 694
Neighborhood families, 439, 440
Neighborhood garage sale, 439
Recipes, 305
Thanksgiving gathering, 40
Walking the dog, 270
Wedding, 448
Will, 232

Economics
Choir budget, 457
Saving money, 92
Stocks, 458

Education
Attendance, 684
College enrollment, 28
College majors, 696
College tuition, 406
Community college, 422, 682
Elementary school, 682
Faculty members, 439, 440
Final exam, 279
GPA, 348, 709
Homework, 698, 710
Lecture, 440
Math test, 457
PTA, 40, 231
SAT, 406
School bonds, 406
School budget, 16
Test scores, 105, 224, 231, 232, 710
Textbooks, 448, 458
Tuition, 231
Visiting a college, 107

Employment/Employee Benefits
City manager, 92
Commission, 385, 456
Earnings, 225, 385
Employee discount, 457, 458
High school counselor, 315
Hourly wages, 367, 387, 414
Hours worked, 305
Janitor, 315
Manager, 439
Pension, 225
Restaurant tips, 348, 457
Restaurant waiter, 109
Senior employees, 439

Entertainment/Sports
Amusement park, 305, 314
Baseball, 28, 406, 422, 710
Basketball, 545, 616
Batting average, 348
Bicycling, 224, 386, 422
Board games, 716
Bowling, 699
College basketball, 40
Dance class, 407
Ferris wheel, 593
Film documentary, 232
Football, 106
Golf, 306, 413, 695
Guitar lessons, 314
Hiking, 224
Home plate, 94
Home runs, 414
Horseback riding, 386
Ice-skating, 595, 609
Little league, 696
Lottery tickets, 40
Marathon, 448
NBA careers, 28
Parks and recreation, 314
Running, 422
Skating, 448
Skiing, 315, 386
Soccer, 40, 448
Softball, 716
Sporting event, 387
Summer camp, 50
Summer Olympics, 358, 683
Swimming, 359
Vacation, 92

Environment
Altitude, 140, 359
Elevation, 28
Rain forest, 16
Rain, 448
Recycling, 305
Snow, 106

Food Service Industry

Beef, 414
Beverages, 694
Candy, 94, 716
Catering, 94
Coffee consumption, 700
Doughnuts, 358
Jars of jam, 93
Soup company, 40, 616
Sugar cones, 616
Watermelons, 232

General Interest

Antique clocks, 448
Bermuda Triangle, 94
Bullfrog, 546
Charity, 28, 231, 358, 448, 457, 697
Dog care, 315, 546
Emailing, 449
Garter snake, 545
Gift wrapping, 314
Kiwanis Club, 94
Rotary Club, 107, 717
Silent auction, 448

Geometry

Angle measurements, 231, 232, 422
Area, 40, 106, 107, 109, 385
Circumference, 385, 386, 593
Dimensions of a rectangle, 205, 224, 225, 232, 280, 305
Length, 231, 232
Measure of a triangle, 224, 232
Parallelogram, 607
Perimeter, 232, 306, 592
Trapezoid, 607
Volume, 616, 617

Government

Federal taxes, 92
Jurors, 449
Post office, 439
Sales tax rate, 367, 456
Sheriff's department, 449
Voting, 422, 439, 440, 683

Health Care

Cancer walk, 698
Chiropractic office, 40, 439
Cough medicine, 314
Dieting, 359, 457
Doctors, 695
Exercising, 105, 422
Growth spurt, 457
Health club, 439, 699
Height of a child, 290, 305, 546
Metabolism, 681
Nutritionist, 93
Smoking, 440
Sodium intake, 414
Weight of a newborn, 348, 546

Hobbies

Artist, 225
Baseball cards, 367, 439
Craft fair, 39, 92
Football cards, 439
Sewing, 305

Real Estate

Apartment lease, 40
Housing prices, 457
Realtors, 28
Rent, 93

Science/Technology

Galilean moons of Jupiter, 29
Lightbulbs, 697
Meteor, 386
Orbits, 406
Solar system, 608
Temperature, 140, 164, 359, 559

Statistics

AIDS cases, 16
Gas consumption, 16
Population, 16, 349, 405, 685
Students enrolled in grades K–12, 16

Index

A

Absolute value, 118, 179
Absolute value bars, 155, 179, 182
Acute angles, 576, 577, 620
Addition, 5
  applications involving, 20, 85–86
  Associative Property of, 8, 9, 96
  by carrying numbers over, 18
  Commutative Property of, 7, 96, 126, 490
  of decimal numbers, 350, 389
  formula for, 87, 103
  in geometry: perimeter, 20–21
  identity for, 10
  of like fractions, 281, 320–321
  of linear measures, 539
  of metric measures, 550, 618
  of mixed numbers
    with like denominators, 284
    with unlike denominators, 300
  of more than two whole numbers, 19
  of negative fractions, 299
  of numbers
    when signs are different, 128
    when signs are the same, 127
  of polynomials, 498, 527
  of pounds and ounces, 542
  repeated. *See* Multiplication
  of signed decimal numbers, 352
  of signed numbers, 180
    on number line, 124
    using Commutative Property, 126
  of unlike fractions, using LCD for, 298
  of whole numbers, 17, 97
  writing subtraction as, 134
Addition Property of Equations, 197, 233, 307
Additive identity, 10, 96
Adjacent angles, 566, 620
Algebra, 2
  backbone of, 477
  symbols used in, 112
  translating
    between English and, 114
    from English to, 210, 211
  uses for, 111
  vocabulary of, 112–113
Algebraic expressions, 166, 179, 485
  applying Commutative Property to terms in, 485
  combining like terms, 183–184
  definition of, 112
  examples of, 113
  multiplication of, using Distributive Property, 174, 184
Algorithm, 45
Altitude, 137
Amount
  over base, 462
  solving for unknown, 436
Angles, 619
  acute, 576, 577, 620
  adjacent, 566, 620
  complementary, 568, 620
  in equilateral triangles, 575
  interior of, 563
  in isosceles triangles, 576
  measures of, 564
  naming, 563
  obtuse, 577, 620
  in right triangles, 576
  in scalene triangles, 576
  straight, 564, 619, 622
  sum of, in triangles, 574–575
  supplementary, 567, 620
  vertical, 569, 620
Applications
  adding and subtracting decimal numbers, 353
  adding or subtracting fractions and mixed numbers, 287
  addition, 20, 85–86
  averages, 89–90
  distance, rate, and time, 383–384
  division, 48
  earnings, 378
  and equations, 356
  geometry, 90, 276
  with integers, 136–137
  involving more than two parts, 86–87
  involving percent equation, 445
  involving percents, 450–455
  key steps for solving problems with, 84
  miles per gallon, 382
  multiplication, 37, 87–88
  multiplication of fractions, 275
  multiplying and dividing fractions, 310
  multiplying very large numbers, 765–766
  rectangle, 379
  rounding calculations involving money, 347
  for rounding large numbers, 13
  sales tax calculations, 365
  shopping, 383
  signed decimal numbers, 354
  solving
    involving one unknown value, 217–221, 234
    involving two unknown values, 226–230, 235–236
    simple interest, 454–455
    with two unknowns, 229–230
    using equations, 102–103
  subtraction, 25, 85–86
  using division to solve percent, 452
  using multiplication to solve percent, 450
  using scientific notation, 751
  when answer is improper fraction, 302
Approximations, 10
Area, 38, 98, 623–624
  of circle, 601
  of geometric figures, 596
  of parallelogram, 597
  of rectangle, 379, 596
  of right triangle, 599
  of square, 596
  subtracting, 602
  of trapezoid, 600
  of triangle, 598
  of unusual shapes, 603
Arithmetic, 2
  multiplication sign in, 30
Associative Property, 8, 96, 146, 284
Associative Property of Addition, 8, 9, 96
Associative Property of Multiplication, 8, 9, 96, 144–145, 173
Auditory-dominant learners, 535
Averages, applications involving, 89–90
Axis, 643

B

Balanced equation, 78
Bar graphs, 723
  characteristics of, 675–676
  drawing, 674–675
  and probability, 714
  reading, 26, 676–678
Base, 51, 99
  amount over, 462
  of exponent, negative number as, 182
  of isosceles triangle, 576
  of rectangle, 623
  solving for unknown, 437
  of three-dimensional figures, 610
  in volume formula, 610
Base angles, of isosceles triangle, 576
Base-ten numbering system, 2
  and negative exponents, 760
Base unit, 547
Binomials, 487
  caution regarding factoring out GCF from, 522
  multiplying
    and trinomials, 506–507
    two, 508
Building up, of fraction, 265, 319

C

Calculators, 60
  tips for negative numbers used on, 149
Calculus, 2
Capacity
  definition of, 544
  U.S. and metric equivalents for, 558
  U.S. measures of, and equivalencies, 544
Carrying, 17
  numbers over, adding by, 18
Centimeters, 547
Circle graphs, 686, 724
  drawing, 688
  reading, 686–687
Circle(s), 623
  applications, 380
  area of, 601, 624
  center point of, 580
  circumference and pi, 587–588
  definition of, 622
  degrees in, 564
  formula for circumference of, 391
  measurement of, 610
Circumference, 380, 391, 623
  and pi, 587–588
Class frequency, 690, 724

Class interval, 690, 724
Clearing the constant, 197, 200, 233
caution regarding, 198
Coefficient, 80, 102, 165, 183, 485
clearing, 198, 200, 203, 233
fractional, 308, 321
in scientific notation, 750
adjusting, 764
Colons, 398
Commission, calculating, 451, 463
Common denominators, 291–296, 322, 323
Common factors, 242, 316
Commutative Property, 7, 96, 146, 284
applying, to terms in algebraic expression, 485
Commutative Property of Addition, 7, 96, 490
adding signed numbers using, 126
Commutative Property of Multiplication, 7, 96, 173
Complementary angles, 568, 620
Composite numbers, 66, 101
relatively prime, 244
Compound interest, 454
Cones, 612, 625
Constant, 77, 102, 165, 485
clearing, 197, 200, 233
caution regarding, 198
Constant term, 485
degree of, 526
Coordinates, 644, 718
of point, finding, 655
x- and *y*-, avoiding confusion with, 646
Corresponding sides, in triangles, 419, 461
Counting numbers, 2, 95
Counts, examples of, 398
Cross multiplication, 404, 459
to solve a proportion, 460
Cubes, 610
Cylinders, 611, 625

D

Dash
first three meanings of, 133, 181
fourth meaning of, 144, 181
Data, 669
calculating percent increase and decrease from, 672
graphing
circle graphs and histograms, 724
line graphs and bar graphs, 722–723
values, even number of, 705
Data table, 669, 724
Decimal fractions, 336, 388
as decimals, writing, 337–338
writing mixed numbers as, 336–337
Decimal numbers, 334. *See also* Decimal(s)
adding, 350, 389
applications involving, 353
categories of, 388
definition of, 388
dividing, by powers of 10, 373
multiplying, 361–362
by powers of 10, 364–365
rounding, to nearest decimal place, 344
strategy for dividing with, 374
subtracting, 351
applications involving, 353
writing fractions as, 373
zeros at end of, 339
Decimal place
rounding decimal numbers to nearest, 344
use of last (rightmost), 335
and zeros in denominator, 336
Decimal point, 334, 548
when converting from one metric measure to another, 618
Decimal(s). *See also* Decimal numbers
adding and subtracting, 389
applications involving, 391
converting metric measures to, 548
dividing when divisor contains, 371–372
monetary use of, 334
multiplying, 389–390
naming, 335, 388
on number line, 340–341
as percents, writing, 428–429, 461
rounding, 389
terminating and repeating, 339–340
uses for, 333
writing decimal fractions as, 337–338
writing percents as, 427, 441, 461
procedure for, 428
Degree, 564
of constant term, 526
of term, 489
definition of, 526
writing terms according to their, 490
Denominators, 249, 316, 317
mixed numbers with like
adding, 284
subtracting, 285
mixed numbers with unlike
adding, 300
subtracting, 301
in unit rates, 409
Descending order, 527
caution regarding, 490
writing polynomials in, 490
Diagrams, for solving application problems, 217, 219, 235
Diameter, 580, 601, 622
Difference, 6, 95
Digit(s), 2
of 9, rounding place, 12
rounding, 11
Dimension, 610
Direction, 119, 179
Discounts, calculating, 452, 463
Distance, rate, and time formula, 160, 183, 383–384, 391
Distributive Property
multiplying algebraic expressions using, 174
with negative multiplier, 176
Distributive Property of Multiplication over Addition, 34–35, 98, 173, 174, 184, 497
Distributive Property of Multiplication over Subtraction, 175, 184
Dividend, 41, 99
Divisibility, 41, 64
tests. *See* Divisibility tests
Divisibility tests
for 2, 67, 101
for 3, 68, 101
for 5, 67, 101
for 9, 68, 101
for 10, 68, 101
Division, 5, 41–42
as alternative to using factor tree, 73
applications involving, 48
of decimal numbers
by powers of 10, 373
by whole numbers using long division, 368, 390
definition of, 41
and factoring, 523
finding Greatest Common Factor (GCF) using, 245–246
finding least common denominator (LCD) using, 294–295, 322
of fractions, 319–320
fractions as, 250
of mixed numbers, 273
of polynomial by monomial, 515
of polynomials, 529
quotient as expression of, 512
rewriting, as multiplication, 272
of signed numbers, 147–148, 181
of two monomials, 514
using, to solve percent applications, 452
when divisor contains decimal, 371–372
of whole numbers, 99
Division bar, 154, 182, 317
Division Property of Equality, 81, 102
Division Property of Equations, 198–199, 233, 308
Division sign, 5
Divisor, 41, 99
containing decimal, dividing, 371–372
determining prime to use as, 101
inability to use zero as, 43–44
as two-digit number, 47
Double negative, 133–134

E

Earnings formula, 391
Ellipsis, 340
Endpoints, 619
in angles, 563
definition of, 562
in line segments, 562
in polygons, 574
in rays, 562
Equal sign, 77
with question mark over it, 196
with slash through it, 199
Equations, 102. *See also* Linear equations
Addition Property of, 197, 233, 307
and applications, 356
involving fractions, 323–324
balancing, 78
definition of, 77, 233
developing, for problem solving, 217, 220, 235
Division Property of, 198–199, 233, 308
with fractional solutions, 308
guidelines to solving simple, 200
involving more than two operations, solving, 206–208, 234
involving multiplication, solving, 80

involving one operation, solving, 196–200, 233
involving two operations, solving, 203, 233
Multiplication Property of, 309, 323
ordered pairs as solution to, 642, 720
solving applications using, 102–103
translating
containing two operations, 213
from English to algebra, 211
with two variables, 641
Equilateral triangles, 575, 621
Equivalence fractions
converting measures using, 554
definition of, 554
value of, 618
Equivalent, 77
Equivalent fractions, 259, 318–319
Equivalent ratios, 402, 459
Estimates, 10
Evaluating, 95
with decimals, 333
expressions, 6
polynomials, 488
quantities with parentheses, 7
Event, 711, 726
probability of, 711
Expanded form, 478
Experiment(s), 711, 726
with simple outcomes, 712
Exponential form, 478
writing, in words, 478
Exponents, 51, 99, 112. *See also* Power
meaning of, 478
negative. *See* Negative exponents
and negative numbers, 152, 182
Power Rule for, 482–483, 526, 757
Product Rule for, 479, 480, 526, 757
Quotient Rule for, 512, 513, 756, 757
Rule of Negative, 756
Expression(s), 5, 95
algebraic, 166. *See also* Algebraic expressions
containing more than one variable, 113
containing two operations, translating, 212
containing two sets of unrelated grouping symbols, 153
evaluating, 6–7
simplifying, before solving, 207
translating from English to algebra, 210
with two pairs of like terms, 169–170
Extremes, 402, 459

F

Factored form, 31, 97, 518, 521
Factor Game, 768
and negative product number, 770–771
and negative sum number, 770
trinomials and, 772
Factoring. *See also* Factor Game
division and, 523
out the Greatest Common Factor, 521–522
caution regarding, 522
polynomials, 518, 529
trinomials, 768, 772–773
when first term is negative, 523
Factor pairs, 65, 100
Factor pair table, 65
Factors, 6, 31, 65, 97
definition of, 64
product having more than two nonzero, 181
Factor trees, 71, 101
division as alternative to using, 73
Feet and inches, 537, 618
adding, 539
converting, 537
subtracting, 540
First Power Rule, 479, 526
Fluid ounces, 544
Formula(s)
for adding adjacent angles, 566
for addition, 87, 103
for area
of circle, 601
of parallelogram, 597
of rectangle, 38, 379, 596
of right triangle, 599
of square, 596
of trapezoid, 600
of triangle, 598
circumference, 391, 623
common to shopping, 383
distance, rate, and time, 160, 183, 383–384, 391
earnings, 378, 391
gas mileage, 391
identifying, in problem solving, 217, 220, 235
involving multiplication and division of fractions, 324
for multiplication, 87, 103
perimeter, for regular polygons, 586
shopping, 391
simple interest, 455, 464
substituting a value into, 158
temperature conversion, 183
using math, 158
for volume, 611, 612, 613, 614
Base, in, 610
Fraction bar, 249, 316
Fractions. *See also* Like fractions
applications involving multiplication of, 275
applications involving multiplying and dividing, 310
building blocks of, 242. *See also* Common factors
building up, 265
coefficients as, 308, 321
containing negative signs, 263
containing variables, 264
combining, 283
decimal. *See* Decimal fractions
as decimal numbers, writing, 373
definition of, 249, 316
dividing, 319–320
as division, 250
equations and applications involving, 323–324
equivalent, 259
and lowest terms, 259
measures of capacity, 544
multiplying, 269, 319–320
negative
adding and subtracting, 299
combining, 282
on number line, 249–250
as percents, writing, 430, 461
principles of, 254
Product Rule for, 254, 318
proper, 251, 317
reciprocals of, 255, 318
simplifying. *See* Simplifying fractions
types of, 251
undefined, 251
unlike. *See* Unlike fractions
uses of, 241
writing percents as, 425, 441
writing probability as, 711
Frequency distribution tables, 690, 724. *See also* Tables

G

Geometric shapes, 574. *See also specific figures*
Geometry, 2
addition in: perimeter, 20–21
applications involving, 90, 276
dimension in, 610
learning styles and, 535
multiplication in: area, 38
planes in, 561
Grams, 547
Graphing
data
circle graphs and histograms, 724
line graphs and bar graphs, 722–723
lines, 720–721
horizontal and vertical, 659
with a large scale, 658
points, 639
Graphs. *See also* Histograms
bar, 26, 674–675, 675–676, 676–678, 714, 723
circle, 686–687, 688, 724
line, 669, 724
of linear equations with one variable, 721
Greater than symbol, 117, 179
Greatest Common Factor (GCF), 316, 518
factoring out, 521–522, 529
finding, using prime factorization, 243
methods of finding, 316
using division method to find, 245–246
Grouping symbols, 6, 115
absolute value bars, 155, 179, 182
division bar, 154, 182, 317
forming quantities, 100
fraction bars, 249, 316
unrelated, expressions containing two sets of, 153

H

Histograms, 690, 724. *See also* Graphs
drawing, 692
reading, 691
Hourly wage, 378

I

Identity, 10, 96
Improper fractions

definition of, 251, 317
simplifying, 265
writing
as mixed numbers, 253, 317
mixed numbers as, 252
Integers, 117, 179
applications with, 136–137
Interior, of angle, 563, 619
Interpolation, 677
Inverse, multiplicative, 256
Inverse operations, 21, 99. *See also* Addition; Subtraction
Invert and multiply process, 272
Isolating the variable, 78, 233
Addition Property of Equations, 197
Division Property of Equations, 198–199
in simple equations, 200
solving equations by, 196
using Division Property of Equality, 81
using Subtraction Property of Equality, 79
Isosceles triangles, 576, 621

K

Kilograms, 547
Kiloliters, 547
Kilometers, 547
Kinesthetic/tactile-dominant learners, 535

L

Learning styles, 535
Least common denominator (LCD), 322
definition of, 291
finding, by listing multiples, 291
methods of finding, 322
using division method to find, 294–295
using prime factorization to find, 292–293
using, to add and subtract unlike fractions, 298
Legend, 84, 102
for one unknown value, determining, 218
for two unknown values, determining, 226
using, for problem solving, 217, 234, 235
Legs, of triangle, 576
Less than symbol, 117, 179
Like fractions
adding and subtracting, 320–321
combining, 281
Like terms, 166, 283, 485, 527
combining, 167–168, 183–184, 234, 491, 527
caution regarding, 492
definition of, 183
expressions with two pairs of, 169–170
with fractional coefficients, 321
Linear equations, 720. *See also* Equations
of the form $ax + b = c$, 655
of the form $y = mx + b$, 656
graphs of, with one variable, 721
with two variables, 655
Linear measures
adding, 539
converting, 537
between metric measures and, 556
U.S., and their equivalencies, 536
Line graphs
definition of, 669
drawing and reading, 669
uses for, 724
Lines
definition of, 561, 619
graphing, 720–721
horizontal and vertical, 659
with a large scale, 658
parallel, 562, 619
and points, 561–562
points on, 649–650
Line segments, 619
definition of, 562
measurement of, 610
parallel, 562
perpendicular lines and, 570
in polygons, 574
Liters, 547
Long division, 42
dividing decimal numbers by whole numbers using, 368, 390
Long Division Algorithm, 45, 99
Lowest terms, 259, 318

M

Mathematics, 2
terminology, basic, 5–6
Mean(s), 725
definition of, 701
finding, 701
Means-Extremes Product Property, 402–403, 459
Median, 701, 725
finding, 704
Meters, 547
Metric measures, 536, 547, 618. *See also* U.S. measures; Units of measure
adding and subtracting, 550
converting, 618
direct decimal conversions of, 548
and prefixes, 547
Millimeters, 547
Minus sign, 5, 133
Mixed numbers
definition of, 251, 317
with like denominators, adding, 284
multiplying and dividing, 273, 320
regrouping within, 286
subtracting, using regrouping, 287
with unlike denominators
adding, 300
subtracting, 301
writing
as decimal fractions, 336–337
as improper fractions, 252, 317
improper fractions as, 253
Mode, 701, 725
identifying, 706
Monomial(s), 487
definition of, 526
dividing
in fractional form, 529
polynomial by, 515
two, 514
multiplying
polynomial and, 505
two, 503–504
squaring, 504, 528
Multiples, 31, 64, 97
of 10, multiplying by, 33
finding least common denominator (LCD) by listing, 291
Multiplication, 5, 30, 112
by 10 and by 100, 32–33, 98
of algebraic expressions using Distributive Property, 174, 184
applications involving, 37, 87–88
Associative Property of, 8, 9, 96, 144–145, 173
of binomial and trinomial, 506–507
Commutative Property of, 7
converting measures of weight using, 541
cross, 404, 459
of decimal numbers, 361–362, 389–390
by powers of 10, 364–365, 390
formula for, 87, 103
of fractions, 269, 319–320
in geometry: area, 38
identity for, 10
of mixed numbers, 273
of monomial and polynomial, 505
by multiples of 10, 33
of polynomials, 528
checking answer for, 502
rewriting division as, 272
with scientific notation, 765
of signed numbers, 181
and simplifying fractions, 269
simplifying fractions before, 270
of single-digit numbers, 35
by two-digit number, 98
solving equations involving, 80
of terms, 173
of two binomials, 508
two-digit by two-digit, 36, 98
of two monomials, 503–504
of two or more signed numbers, 147
of two signed numbers, 145
use of term *of* in, 274
using, to solve percent applications, 450
of whole numbers, 97
Multiplication Property of 0, 32, 97
Multiplication Property of Equality, 404
Multiplication Property of Equations, 309, 323
Multiplication sign, 5, 30
Multiplicative identity, 10, 96, 254, 256, 497
Multiplicative inverse, 256
Multiplier, 497
of one or minus one, 497

N

Natural numbers, 2, 95, 318
Negative exponents
Rule of, 756
understanding, 754–755
using, with other rules of exponents, 757
working with base 10 and, 760
Negative numbers, 116, 179
debits as, 136
exponents and, 152
product of positive numbers and, 142
product of two, 144–145
reciprocals of, 255
temperature represented by, 137
uses for, 111
Nonterminating decimals, 339–340

Number line, 115–116, 179, 180
- adding signed numbers on, 124
- decimals on, 340–341
- fractions on, 249–250, 317
- representing whole numbers on, 2

Numbers, 2
- absolute value of, 118
- adding
  - when signs are different, 128
  - when signs are the same, 127
- definition of, 119
- divisors as two-digit, 47
- in expanded form, writing, 4
- in factored form, 31
- multiples of, 31
- numerical value and direction of nonzero, 119, 122, 179
- prime and composite, 66
- properties of, 173
- recognizing relatively prime, 244
- relatively prime, 316
- using commas to separate, 3
- using x to represent, 114

Numerals, 2
Numerators, 249, 316, 317
- zeros as, 251

Numerical coefficient, 165, 183
Numerical value, 119, 179

O

Obtuse angles, 577, 620
Operation(s), 95
- inverse, 21
- in mathematics, basic, 5–6
- more than two, solving equations involving, 206–208, 234
- one, solving equations involving, 196–200, 233
- rank of, 151
- results of, 5–6
- translating equations containing two, 213
- translating expressions containing two, 212
- two, solving equations involving, 203, 233
- written form of, 5, 95

Opposites, 116, 179
Ordered pairs, 641
- definition of, 718
- as solutions to equations, 642

Order of operations, 56–57, 100, 151
- key to applying, 152
- rules for, 57
- for signed numbers, 182

Origin, 116, 179, 718
- of x-y plane, 644

Ounces, 541, 544

P

Parabolas, 668
Parallel lines, 562, 619
Parallelograms, 577, 578, 622
- area of, 597, 624
- measurement of, 610

Parentheses, 6, 95
- for adding two polynomials, 498
- clearing, by distributing, 498
- evaluating quantities with, 7
- ordered pairs grouped in, 641
- separating double negatives with, 133
- as separators, 115
- using, when replacing variable with number, 488

Part(s)
- guidelines for recognizing whole, part, and number of, 88
- over whole, 249
- unknown, in percent equation, 442
- and whole
  - looked at together, 397
  - as same unit of measure, 324

Percentage, 425, 434, 461
- solving for unknown, 435

Percent equation
- applications involving, 445
- example of, 442
- solving, 441–445, 462
- when part is unknown, 442
- when percent is unknown, 444
- when whole is unknown, 443

Percent proportion, 434–435
- when amount is unknown, 436
- when base is unknown, 437
- when percentage is unknown, 435

Percent(s)
- applications, 450–455
  - using division to solve, 452
  - using multiplication to solve, 450
- as decimals, writing, 427, 461
  - procedure for, 428
- definition of, 425
- finding ratio as, 463
- as fractions, writing, 425
- increase or decrease, determining, 453, 672
- multiplying by, 441
- proportion, 434–435
- symbol, 425, 434, 461
- unknown, in percent equation, 444
- writing decimals as, 428–429, 461
- writing fractions as, 430, 461
- writing probability as, 711

Perfect squares, 54
Perimeter, 20–21, 97, 623
- of circle. *See* Circumference
- of regular polygon, 586, 623
- of unusual shapes, 589–590

Periods, 3
Perpendicular lines, 620
- and line segments, 570
- symbol for, 570

P(event) (probability of an event), 711, 726
Pi, 380, 623
- approximate value of, 340, 588
- circumference and, 587–588

Place value, 3, 95
- and powers of 10, 338–339

Place-value chart, 3
Planes, 561, 596
- definition of, 619
- x-y, 643

Plotting points, 644–645
- x-y plane and, 643–645

Plus sign, 5
Points, 619
- definition of, 562, 610
- finding coordinates of, 655
- graphing, 639
- guidelines for plotting, 643–645
- of intersection, 643
- on lines, 649–650
- lines and, 561–562
- other, in x-y plane, 648
- on x-axis, 719
- x-y plane and plotting, 643–645
- on y-axis, 719

Polygons, 574, 621. *See also* Regular polygons
- definition of, 574
- four-sided, 577, 621

Polynomial(s), 477
- adding, 498, 527
- definition of, 486, 526
- in descending order, 527
- dividing, 529
  - by monomial, 515
- evaluating, 488
- examples of, 487
- factoring, 518, 529
- multiplying, 528
  - checking answer when, 502
  - two, 528
- multiplying monomial and, 505
  - caution regarding, 505
- with special names, 487
- subtracting, 499, 527
- vocabulary of, 485
- writing, in descending order, 490

Positive numbers, 116, 179
- absolute value of, 118
- product of negative numbers and, 142

Possible outcomes, 711, 726
Pounds and ounces, 541
- adding, 542
- converting, 541
- subtracting, 543

Power, 51, 99. *See also* Exponents
Power of 1, 100
Power of a Power, 483
Power Rule
- for Exponents, 482–483, 526, 757
- First, 479, 526
- Zero, 481–482, 526

Powers of 10, 52–53, 100
- dividing decimal numbers by, 373, 390
- multiplying decimal numbers by, 364–365, 390
- place value and, 338–339
- in scientific notation, 750
- working with, 749

Prefixes, 547, 548
Prime divisors, 101
Prime factorization, 70–71, 101, 479
- finding Greatest Common Factor (GCF) using, 243
- finding least common denominator (LCD) using, 292–293, 322
- of a term, 518

Prime numbers, 66, 101
- finding lowest possible, 73
- recognizing relatively, 244

Principal, 454, 464
Principal square root, 182
Probability/ies, 639, 726
- bar graphs and, 714
- two common, 712
- vocabulary of, 711

Problems. *See also* Problem solving
application. *See also* Applications
guidelines for mastering, 217–221, 235, 236
key steps for solving, 84
how to read division, 41
Problem solving, 210–213, 234. *See also* Problems
applications involving two unknown values, 226–230
equations for, 356
guidelines for mastering an application problem, 217–221, 235
using legend for, 234. *See also* Legend
Product, 6, 95
finding, of two monomials, 503–504
having more than two nonzero factors, 181
of more than two signed numbers, 146–147
of positive number and negative number, 142
of two negative numbers, 144–145
Product number
in Factor Game, 768
Factor Game and negative, 770–771
Product Rule for Exponents, 479, 480, 526, 757
caution regarding, 481
Product Rule for Fractions, 254, 318
extended, 269, 319
Proper fractions, definition of, 251, 317
Proportion(s), 460–461
applying, 418
consistency and, 415
definition of, 415
identifying, 415–416
percent, 434–435
solving, 416–417
solving percent problems using, 462
true, 415
Pyramids, 613, 625

Q

Quadrants, 646–647, 718
Quadrilaterals, 577, 579, 621
Quantity, 7, 95, 497
double, 153
formed by grouping symbols, 100
multiplying
by minus one, 527
by one, 527
Quotient, 6, 41, 95
expressing, 6
as expression of division, 512
negative and positive, 181
as prime number, 101
Quotient Rule for Exponents, 512, 529, 756, 757
caution regarding, 513
definition of, 513
Quotient Rule of Fractions, 272, 319

R

Radical, 55, 100, 112
placing negative outside, 182
Radicand, 55
Radius, 380, 391, 580, 601, 622
Random selection, 711, 726
Rate(s), 399, 459
definition of, 408
simplifying, 408
unit, 409
Ratio(s), 459
definition of, 398
equivalent, 402
expressed in fractional form, 402
forms, 398–399
as percent, finding, 463
sample, 399
simplifying, 400
writing, from phrases, 401
Rays, 562, 619
Reciprocal, 754
of fraction, 255, 318. *See also* Multiplicative inverse
of negative number, 255
Rectangle
area of, 38, 379, 596, 623
definition of, 578, 622
measurement of, 610
perimeter formula for, 623
Rectangular coordinate system, 643, 718–719. *See also x-y* plane
Regrouping
within mixed numbers, 286
subtracting by, 22–23
subtracting mixed numbers using, 287
in subtraction, 17, 540
Regular polygons, 580
definition of, 622
perimeter of, 586, 623
Relatively prime numbers, 244, 316
Remainder, 41, 44, 99
Repeating decimals, 340
Replacement values, 113, 158, 179, 183
Right angles, 619
Right triangles, 576, 621
area of, 599, 624
Rounding, 10–11, 96
calculations involving money, 347
decimal numbers to nearest decimal place, 344
decimals, 389
digit, 11
large numbers, applications for, 13
rules for, 11
when place digit is 9, 12
Rule of Negative Exponents, 756

S

Sales tax, 365
calculating, 450, 463
Scalene triangles, 576, 621
Scientific calculators, 60
negative key on, 149
Scientific notation, 749
adjusting coefficient in, 764
defining, 750
expanding from, 752, 762
multiplying with, 765
writing large numbers in, 750–751
writing small numbers in, 761
Sector, 686
Separators, 115
Short division, 42
Sides, of angle, 619
Signed decimal numbers
adding and subtracting, 352
applications involving, 354
Signed numbers
adding, 180
on number line, 124
using Commutative Property, 126
comparing, 117
decimal, adding and subtracting, 352
definition of, 116
dividing, 147–148
guidelines for adding and subtracting, 135
multiplying and dividing, 181
multiplying two, 145
or more, 147
order of operations for, 182
product of more than two, 146–147
subtracting, 181
Similar triangles, 419, 461
Simple interest, 454, 464
applications, solving, 454–455
formula, 455
Simplifying fractions, 260, 318, 259
improper, 265
before multiplying, 270
multiplying and, 269
Solid(s)
definition of, 610, 624
volume of rectangular, 610–611
volumes of other, 611
Solution(s). *See also* Problem solving
to equations, 77, 196, 233
in Factor Game, 768
writing, in sentence form, 217, 221, 235
Spheres, 614, 625
Square Root Property, 55
Square roots, 55, 100
negative, 153
Square root symbol, 55–56
Squares, 578
area of, 596, 624
definition of, 622
Statistics, 2, 639
Straight angles, 564, 619, 622
Substitution
in formulas, 158
of values for variables, 113
Subtraction, 5
as addition, writing, 134
applications involving, 25, 85–86
of area, 602
of decimal numbers, 351, 389
of feet and inches, 540
finding difference in altitudes using, 137
of like fractions, 281, 320–321
of metric measures, 550, 618
of mixed numbers
with like denominators, 285
with unlike denominators, 301
using regrouping, 287
of negative fractions, 299
of polynomials, 499, 527
of pounds and ounces, 543
by regrouping, 22–23
of signed decimal numbers, 352
of signed numbers, 181
of unlike fractions, using LCD for, 298

of whole numbers, 21, 97
Subtraction Property of Equality, 79, 102
Subtraction sign, 133
Success, 711, 726
Sum, 6, 95
Sum number
in Factor Game, 768
Factor Game and negative, 770
Supplementary angles, 567, 620
Symbols
angle, 563
approximately equal to, 556
division sign, 5
equal sign, 77
greater than, 117, 179
less than, 117, 179
minus sign, 5, 133
multiplication sign, 5, 30
percent, 425, 434, 461
for perpendicular lines, 570, 620
plus sign, 5
for remainder, 44
representing numbers, 2
square root, 55–56, 100
used in algebra, 112
used to represent ratios, 398

T

Tables, 675. *See also* Frequency distribution tables
Temperature, 137
conversion formula, 183
Terminating decimals, 339
Term(s), 165, 183, 485
in algebraic expression, applying Commutative Property to, 485
constant, 485, 526
degree of, 489
like, 166. *See also* Like terms
multiplying, 173
two, 528
negative first, factoring for, 523
prime factorization of, 518
solving linear equations containing more than one variable, 206
written according to degree, 490
Tons, 541
Trapezoids, 579
area of, 600, 624
definition of, 622
Triangle(s), 621
area of, 598, 624
corresponding sides in, 419, 461
definition of, 574
equilateral, 575, 621
isosceles, 576, 621
measurement of, 610
right, 576, 621
scalene, 576, 621
similar, 419, 461
sum of angles in, 574–575
types of, 575–576
Trinomials, 487
caution regarding factoring out GCF from, 522
definition of, 526
and Factor Game, 772
factoring, 768, 772–773
multiplying binomials and, 506–507

U

Undefined fraction, 251
Unit prices, 460
comparing, 411–412
definition of, 411
Unit rates, 409, 460
Units of measure, 536. *See also* Metric measures; U.S. measures
combining like, 281
converting, 537
using equivalence fractions, 554
examples of, 398
in proportions, 460
in ratios, 398
Unknown value(s), 77. *See also* Value(s)
applications involving solution of, 217–221, 226–230, 234, 235–236
determining legend for, 218, 226
Unlike fractions
using least common denominator (LCD) to add and subtract, 298
with variables, combining, 300
U.S. measures, 536, 618. *See also* Metric measures; Units of measure
of capacity, 544
converting, 618
between metric measures and, 556
of length, 536
of weight, 541

V

Value(s), 644. *See also* Unknown value(s)
even number of data, 705
grouping of different, 6
interpreting, first unknown, in a comparison, 227
numerical, 179
one unknown, determining legend for, 218
rounding, 10–11
solving applications involving one unknown, 217–221
two unknown, determining legend for, 226
unknown, 77
Variable(s), 77, 102, 113, 485
caution regarding replacement of, 488
combining unlike fractions with, 300
equations with two, 641
fractions containing, 264
combining, 283
graph of linear equation with one, 721
isolating, 78
linear equation with two, 655
term, solving linear equations containing more than one, 206
x as common, 114
Vectors, 122, 180
Vertex
of adjacent angles, 566
of angle, 563, 619
in polygons, 574, 621
Vertical angles, 569, 620
Visual-dominant learners, 535
Volume, 624–625
definition of, 610
of other solids, 611–614
of rectangular solid, 610–611

W

Weight
adding measures of, 542
converting, 541
U.S. and metric equivalents for, 558
U.S. measures of, and equivalencies, 541
Whole
guidelines for recognizing, part, and number of parts, 88
part and, looked at together, 397
and part as same unit of measure, 324
part over, 249
unknown, in percent equation, 443
Whole numbers, 2, 95
adding, 17
more than two, 19
and subtracting, 97
dividing, 99
dividing decimal numbers by, 368, 390
multiplying, 97
representing, on number line, 2
rounding, 96
subtracting, 21
writing, in words, 4
Words
indicating a part, 89
indicating subtraction, 25
key, meaning division, 48
pertaining to addition, 20
writing exponential form in, 478
writing whole numbers in, 4

X

x-axis, 643, 646
points on, 719
x-*y* plane, 643, 718. *See also* Rectangular coordinate system
origin of, 644
other points in, 648
and plotting points, 643–645
quadrants of, 718
special features of, 646–647

Y

y-axis, 643, 646
points on, 719

Z

Zero, 2, 99, 179. *See also* Zero Power Rule
in decimal fractions, 337
effect on decimals of adding, 388
at end of decimal number, 339
inability to divide by, 43–44
Multiplication Property of, 32, 97
negative numbers, positive numbers, and, 119
on number line, 116
as numerator, 251
as a vector, 122
Zero Power Rule, 481–482, 526
foundation of, 482
revisited, 514